A Brief, Hands-On Lab Manual with Frequent Opportunities to Practice

Visual Anatomy & Physiology Lab Manual uses stunning images from Ric Martini's *Visual Anatomy & Physiology* textbook and active-learning exercises to get students practicing in the lab.

B The two ventricles are thick-walled chambers that pump blood into the great arteries.

1. Identify the two inferior heart chambers:
   Left ventricle ________
   Right ventricle ________

2. Within the right ventricle, identify the following regions:

   The inferior portion receives blood from the right atrium. Its walls are covered by an irregular network of muscular elevations called the **trabeculae carneae.**

   Superiorly, the right ventricle narrows into a cone-shaped chamber, the **conus arteriosus,** which leads to the pulmonary trunk. The wall of the conus arteriosus is smooth and lacks trabeculae carneae.

3. Inside the left ventricle, the **aortic vestibule** is the smooth-walled, superior region that leads to the aorta. It is similar to the conus arteriosus in the right ventricle. Describe another structural similarity between the two ventricles.
   ▶ ________

4. Place the thumb and index finger of one hand on either side of the wall that separates the two ventricles. This structure is the **interventricular septum.** Notice that this wall is much thicker than the interatrial septum.

5. On the surface of the heart, what two sulci form the anterior and posterior margins of the interventricular septum?
   ▶ ________

Multiple opportunities for students to label, draw, and analyze.

IN THE CLINIC

Atrial Septal Defect

Inside the right atrium, along the interatrial septum, there is an oval depression called the fossa ovalis. This depression marks the site of the **foramen ovale,** an opening that connects the atria in the fetal heart. The foramen ovale has a valve that allows blood to travel from the right atrium to the left atrium but not in the reverse direction. This specialization in the fetal circulation allows most of the oxygen-rich blood from the placenta to bypass the lungs and pulmonary circulation and pass directly to other vital organs via the systemic circulation. At birth, the foramen ovale closes when the valve fuses with the interatrial septum. Incomplete closure of the foramen ovale, called an **atrial septal defect,** allows oxygen-rich blood in the left atrium to mix with oxygen-poor blood in the right atrium. This malformation can be repaired surgically to prevent the two blood supplies from blending.

MAKING CONNECTIONS

Speculate on the function of the trabeculae carneae.

▶ ________

Want more practice? Go to: **MasteringA&P** > Study Area > Menu > Lab Tools > PAL >
- Anatomical Models > Cardiovascular System > Heart
- Human Cadaver > Cardiovascular System > Heart

NEW! Clear directions to find study tools in MasteringA&P.™

Prepare, Practice, and Put It All

PRE-LAB QUIZ

Before you begin, read all the activities in Exercise 2 and the required reading in your textbook that is assigned by your instructor.

1. When carrying a microscope, you should hold it securely with both hands. One hand should be on the __________, and the other hand should be under the __________.
   a. arm . . . base
   b. head . . . stage
   c. arm . . . stage
   d. stage . . . base
2. On a compound microscope, where is the light source located?
   a. on the oculars
   b. on the base
   c. on the stage
   d. on the cord
3. True or False: The space between the objective lens and the microscope stage is called the working distance. __________
4. True or False: When an image is approximately in focus, you can use the coarse adjustment knob to bring it to exact focus. __________
5. The illuminated area that you view with a microscope is called the __________.
6. The nosepiece on a microscope is a revolving structure that holds the __________.
7. The __________ concentrates the light before it travels through the tissue on the slide.
8. During this laboratory exercise, you will use a prepared slide with the letter *e* to demonstrate
   a. depth of field.
   b. the relationship between total magnification and field diameter.
   c. the working distance.
   d. inversion of image.
9. The thickness of the tissue layer that is in focus is called
   a. resolving power.
   b. image inversion.
   c. depth of field.
   d. field diameter.
10. During this laboratory exercise, you will use a clear millimeter ruler to
   a. estimate the diameter of the field of view.
   b. measure the working distance.
   c. estimate the resolving power of the microscope.
   d. estimate the size of a structure on a tissue section.

NEW! Pre-Lab Quizzes open each exercise by asking students questions that will help ensure they come to lab prepared. These quizzes are also assignable in MasteringA&P.™

MAKING CONNECTIONS

During this activity, you observed that the lumen of a blood vessel, a passageway for blood, is lined by a simple squamous epithelium, which is a very thin cell layer. You also observed that the lumen of the esophagus, a passageway for food to the stomach, is lined by a stratified squamous epithelium, which is much thicker. Why do you think these two structures have epithelia that are so strikingly different?

▶ ______________________________

____

____

____

____

____

____

Making Connections
give students an opportunity to pause, internalize information, and apply their understanding.

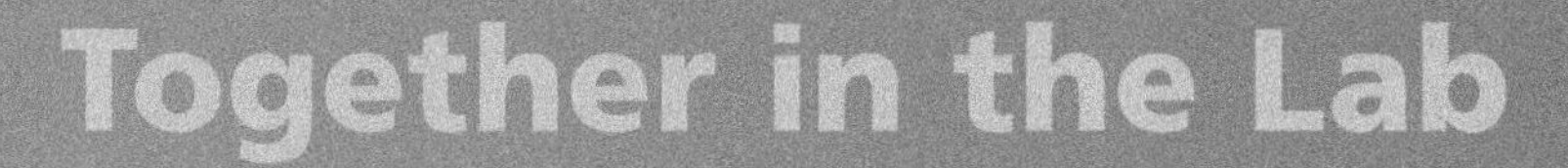

In the Clinic boxes throughout the lab manual help students connect what they learn in lab to the real world.

IN THE CLINIC

Malignant Melanoma

Malignant melanoma is a type of skin cancer that causes unregulated reproduction of melanocytes. Because these cells produce the pigment melanin, melanomas are often easily spotted because of their dark coloration. Although these malignancies may grow slowly at first, they become aggressive and, if not caught early, have a high mortality rate. Normally, melanin protects us from genetic damage caused by ultraviolet radiation, which can lead to melanoma and other skin cancers. People who have skin with a high level of melanin have more protection against this damage. The risk of melanoma is about 20 times higher for people of European descent than for people of African descent. Because the occurrence of melanoma in people of African descent is relatively rare, its diagnosis is often delayed. As a result, when the disease is discovered, it is usually at a more advanced stage, and the chances of survival are reduced. Early detection is the key, and that requires frequent self-examination of your skin, including areas that are not exposed to sunlight.

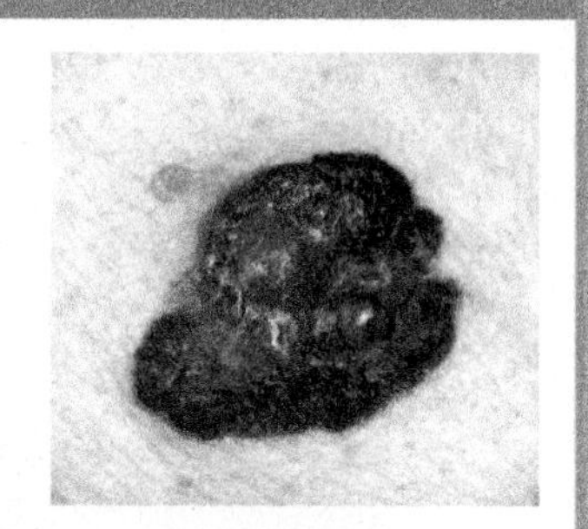

BEFORE YOU MOVE ON . . .

« LOOKING BACK

The appendicular skeleton comprises the bones of the upper and lower limbs (the appendages). In this laboratory exercise, you learned that each upper limb includes a clavicle and scapula (1/2 pectoral girdle) and the bones of the arm, forearm, wrist, and hand. Each lower limb includes a coxal (hip) bone (1/2 pelvic girdle) and the bones of the thigh, leg, and foot. You also observed that the organization of the bones in the upper limb is comparable to those in the lower limb. For example, the arm and thigh each contain one large long bone; the forearm and leg each contain two smaller long bones that are roughly parallel.

Despite these similarities, there are important structural and functional differences between the upper and lower limbs.

Consider these questions: ►

1. **Why are the bones of the lower limb larger than those of the upper limb?**

   ____

2. **What is the fundamental difference in function between the foot and the hand?**

   ____

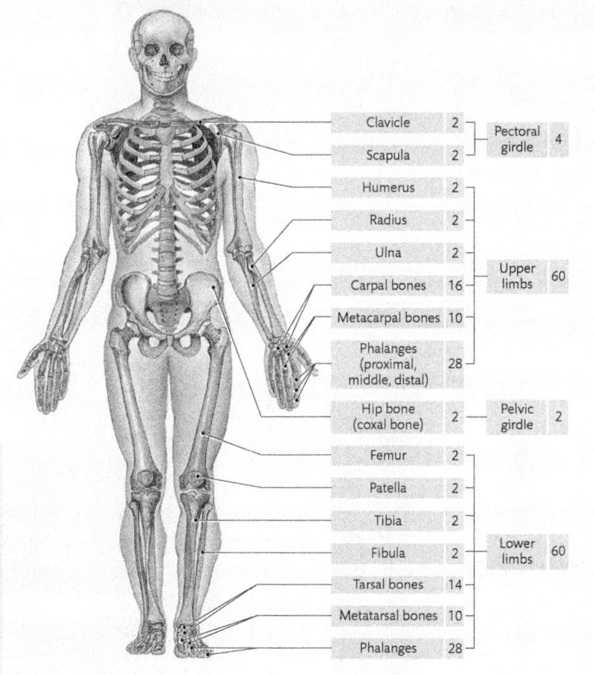

LOOKING FORWARD »

Be aware that the skeletal system not only includes all the bones of the body, but also the cartilage, tendons, and ligaments associated with the articulations (joints). Tendons and ligaments are both composed of dense regular connective tissue. At a joint, a tendon connects a muscle to a bone; a ligament connects one bone to another bone. In the next laboratory exercise (Laboratory Exercise 9), you will study articulations. Think of articulations as the functional junctions between bones. They bind various parts of the skeletal system together, are locations on the body where movement occurs, allow bone growth and development, and permit parts of the skeleton to change shape.

NEW! Before You Move On feature wraps up each exercise by asking students to think critically about the lab they just completed, and then connect that information to next lab.

Continuous Learning Before, During, and After Lab

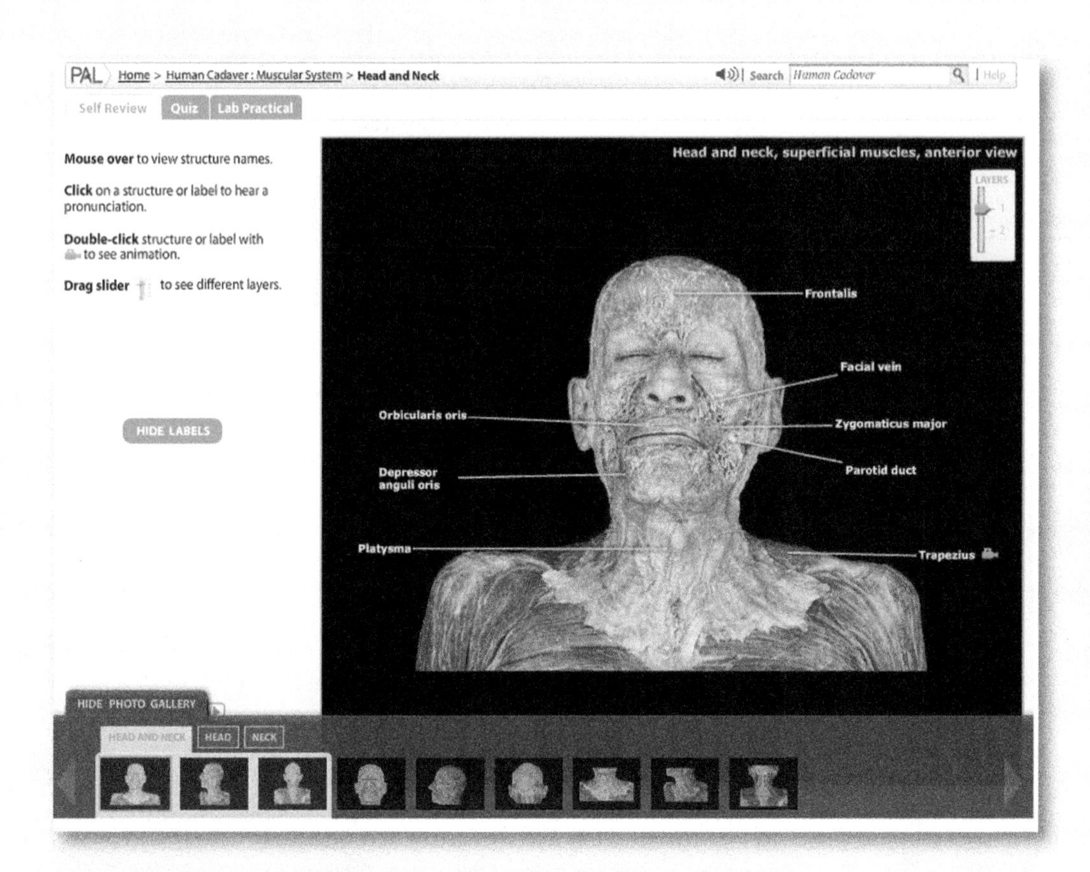

Practice Anatomy Lab (PAL™ 3.0) is a virtual anatomy study and practice tool that gives students 24/7 access to the most widely used lab specimens, including the human cadaver, anatomical models, histology, cat, and fetal pig.

PAL 3.0 is easy to use and includes built-in audio pronunciations, rotatable bones, and simulated fill-in-the-blank lab practical exams.

PhysioEx 9.1 is an easy-to-use lab simulation program that allows students to conduct experiments that are difficult in a wet lab environment because of time, cost, or safety concerns.

Students are able to repeat labs as often as they like, can perform experiments without animals, and are asked to stop frequently and predict within the labs.

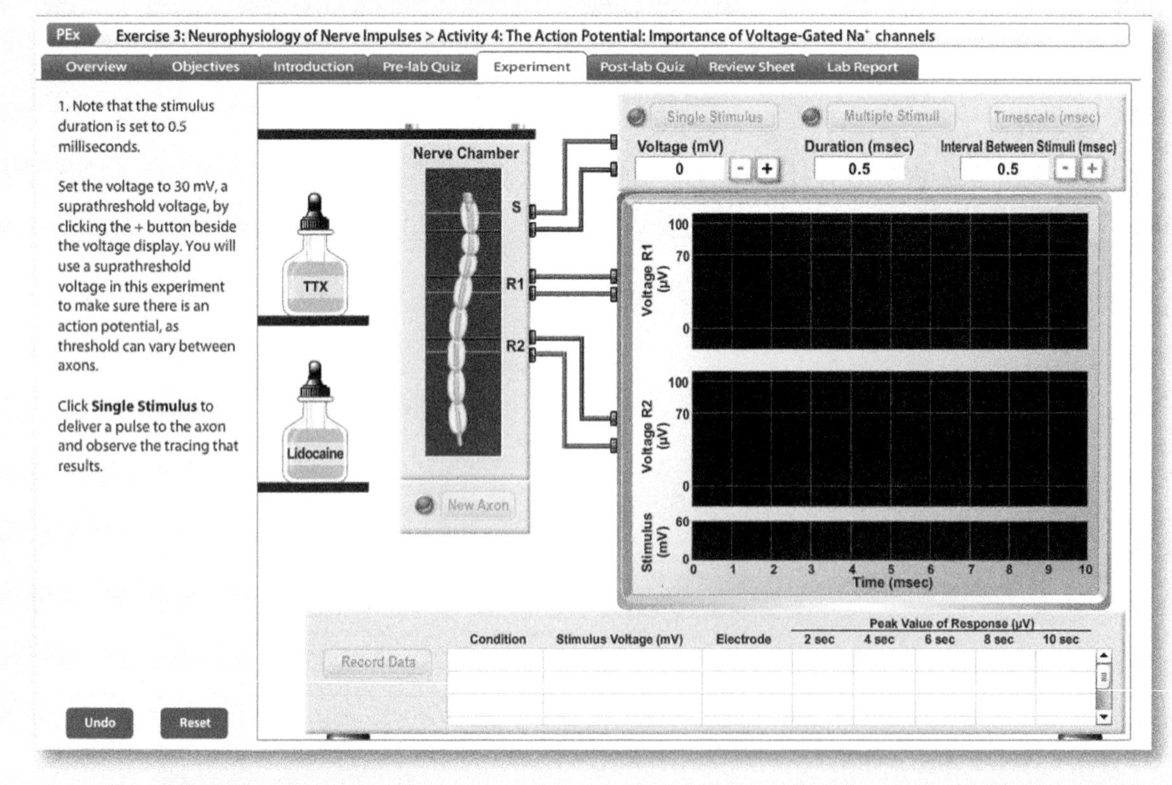

with MasteringA&P™

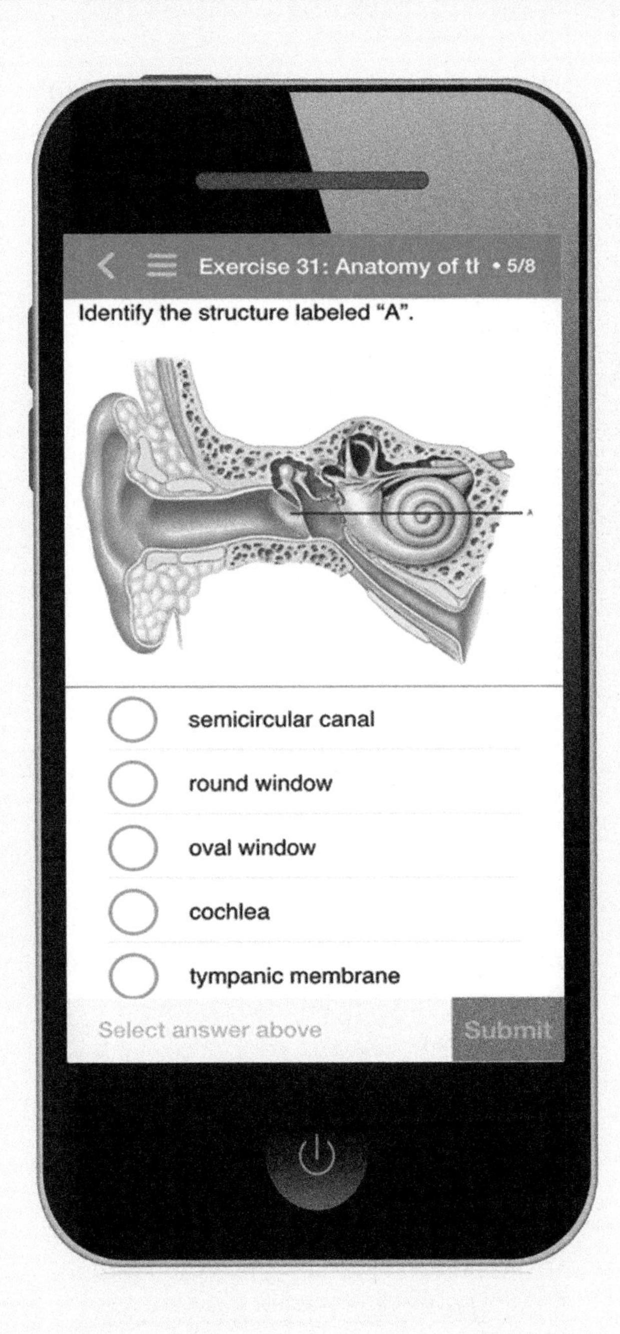

Dynamic Study Modules enable students to study more effectively on their own. With the Dynamic Study Modules mobile app, students can quickly access and learn the concepts they need to be more successful on quizzes and exams.

NEW! Instructors can now select which questions to assign to students.

Bone and Dissection Videos help students identify bones and learn how to do organ dissections.

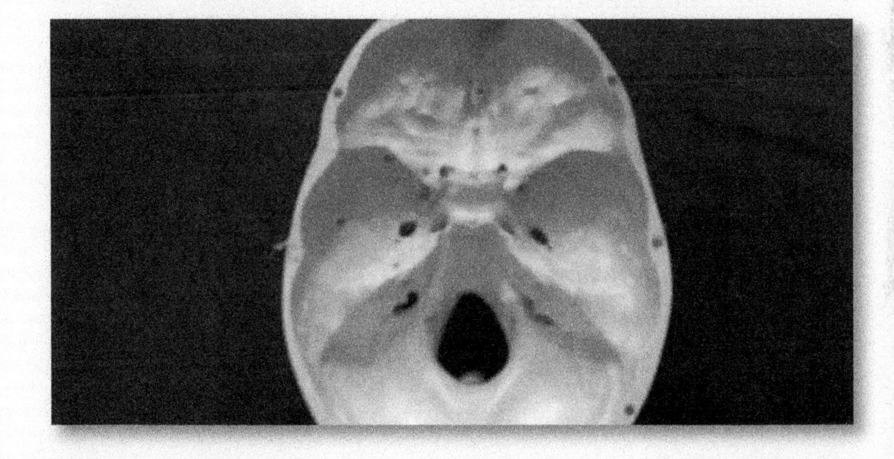

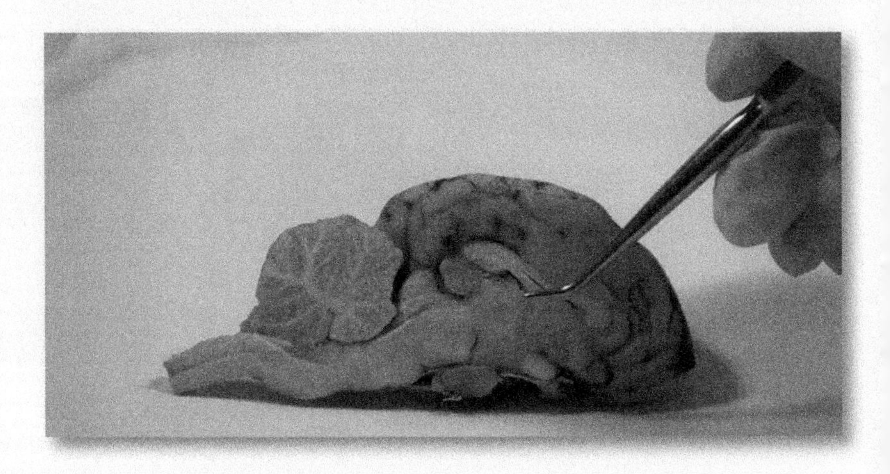

MasteringA&P™

Name ____________________
Lab Section ____________________
Date ____________________

EXERCISE 21 REVIEW SHEET
Gross Anatomy of the Heart

1. The apex of the heart is formed by the
   a. right atrium.
   b. left atrium.
   c. right ventricle.
   d. left ventricle.
2. Which heart groove travels between the atria and the ventricles?
   a. anterior interventricular sulcus
   b. posterior interventricular sulcus
   c. coronary sulcus
   d. both (a) and (b)
   e. (a), (b), and (c)
3. The epicardium and the ____________________ are the same structure.
4. The ____________________ artery forms an anastomosis with the right coronary artery.
5. The adult heart structure that marks the location of an opening between the two atria in the fetal heart is called the ____________________.

QUESTIONS 6–10: Answer the following questions by selecting the correct labeled structure. Answers may be used once or not at all.

6. This structure pumps deoxygenated blood into the pulmonary trunk. ____________
7. The pulmonary veins deliver oxygenated blood to this structure. ____________
8. This structure delivers deoxygenated blood to the right atrium. ____________
9. This structure pumps oxygenated blood into the aorta. ____________
10. This structure and its branches deliver deoxygenated blood to the lungs. ____________

a, b, c, d, e, f, g

Assignable Review Sheets, based on the Review Sheets that appear at the end of each lab exercise, are available in a gradable format in MasteringA&P so that instructors can easily assign them for homework.

Additional assignable MasteringA&P activities include:

- Bone & Dissection Video Coaching Activities
- A&P Flix™ for Anatomy Topics
- PAL™ Assessments
- PhysioEx™ Assessments
- Pre-Lab and Post-Lab Quizzes
- And More!

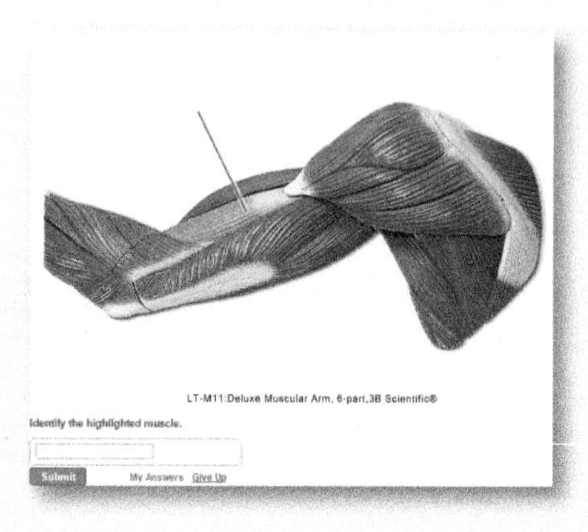

A&P Flix: Trapezius

Watch the animation, then answer the questions.

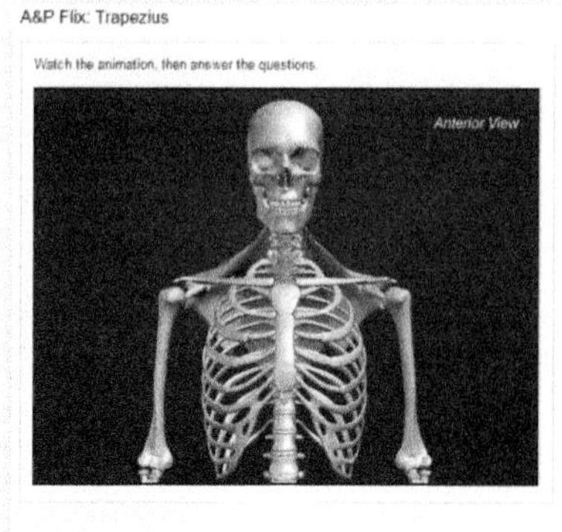

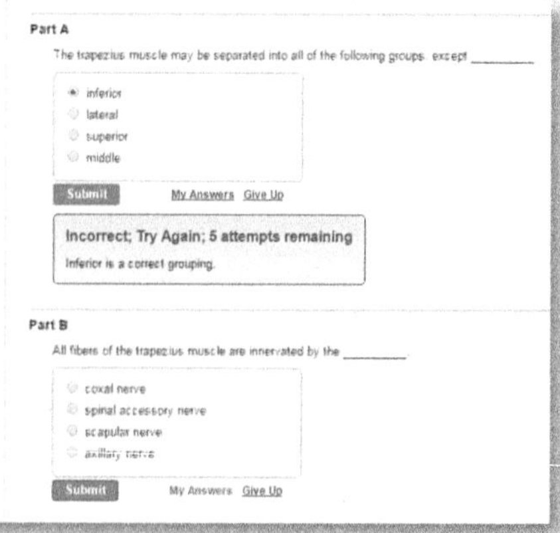

Access the Complete Lab Manual On or Offline with eText 2.0

NEW! The **Second Edition** is available in Pearson's fully-accessible eText 2.0 platform.*

NEW! The eText 2.0 mobile app offers offline access and can be downloaded for most iOS and Android phones and tablets from the iTunes or Google Play stores.

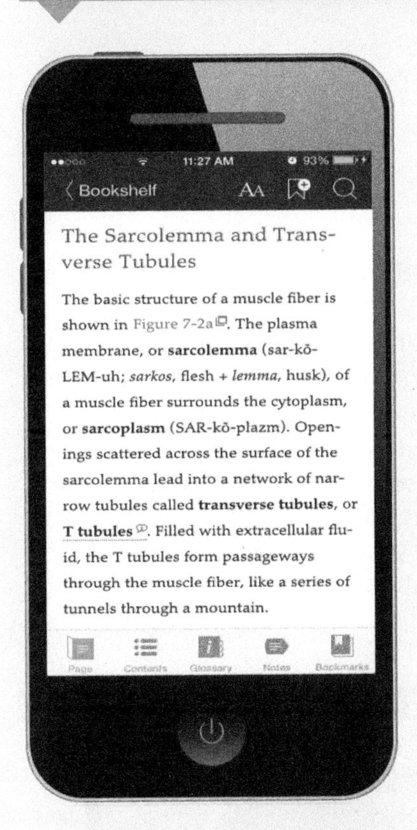

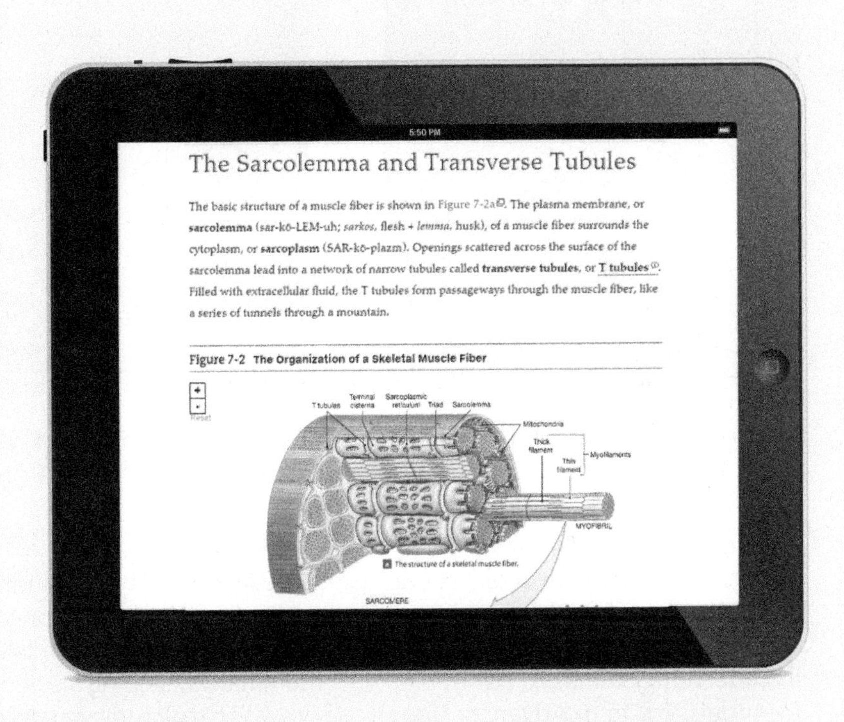

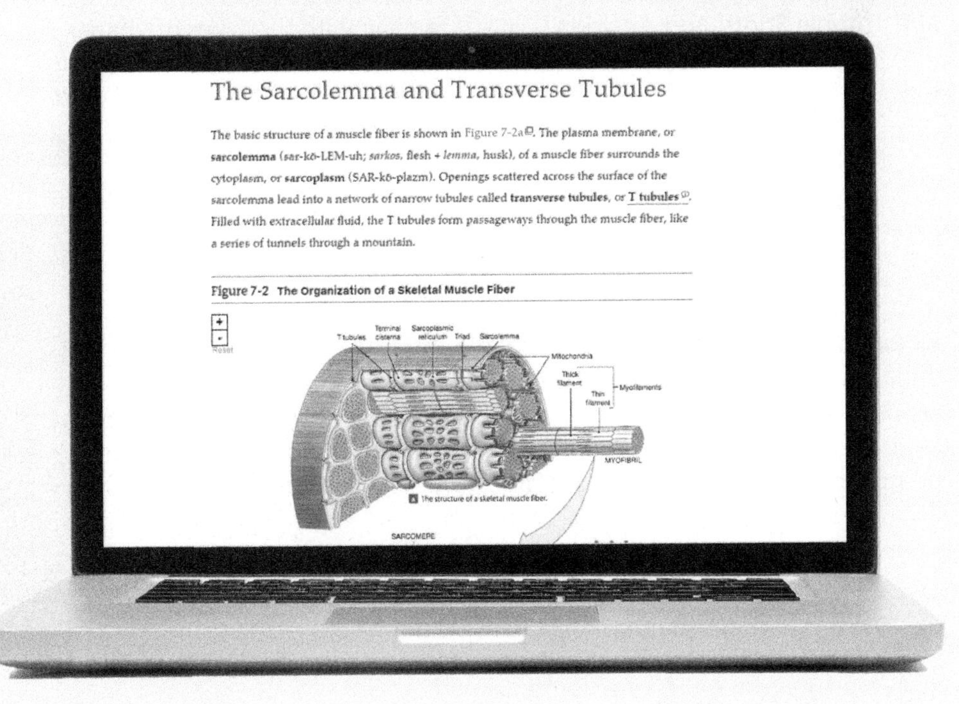

Powerful interactive and customization functions include instructor and student note-taking, highlighting, bookmarking, search, and links to glossary terms.

*The eText 2.0 edition will be live for Fall 2017 classes.

Instructor and Student Support

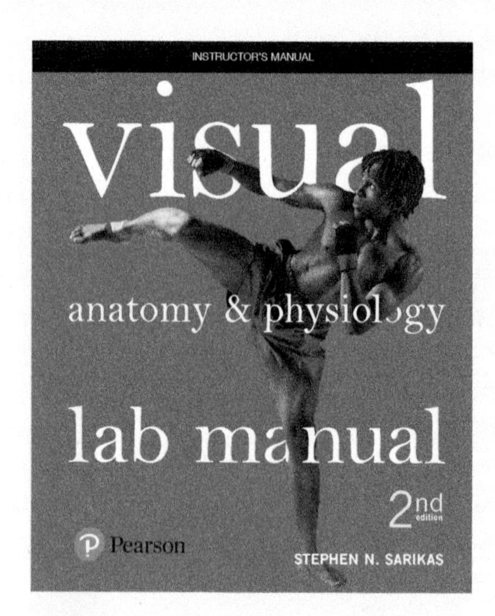

Instructor's Manual by Lori K. Garrett

978-0-13-454796-1 / 0-13-454796-9

This resource includes a wealth of materials to help instructors set up and run successful lab activities. Sections for every lab exercise include Time Estimates, List of Materials, To Do in Advance, Tips and Trouble Spots, and Answers.

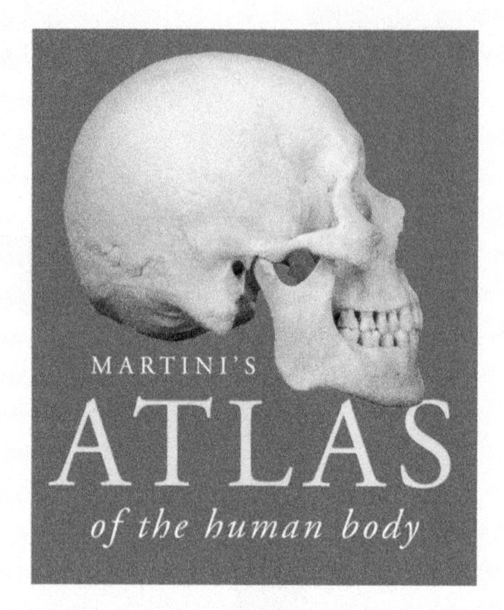

Martini's Atlas of the Human Body by Frederic H. Martini

978-0-321-94072-8 / 0-321-94072-5

The Atlas offers an abundant collection of anatomy photographs, radiology scans, and embryology summaries, helping students visualize structures and become familiar with the types of images they might encounter in a clinical setting. Free when packaged with the textbook.

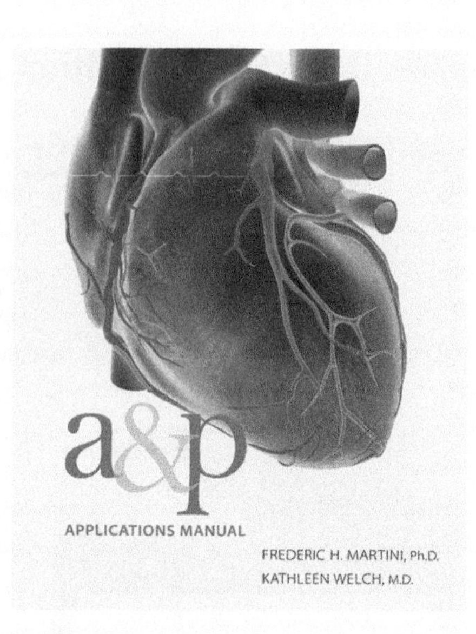

A&P Applications Manual by Frederic H. Martini and Kathleen Welch

978-0-321-94973-8 / 0-321-94973-0

This manual contains extensive discussions on clinical topics and disorders to help students apply the concepts of anatomy and physiology to daily life and their future health professions. Free when packaged with the textbook.

Create a Lab Manual for Your Course

Your lab manual should match your course, not the other way around. With Pearson Collections, you can quickly and easily create a lab manual with only the content you use. Visit https://www.pearsonhighered.com/collections/ for more information.

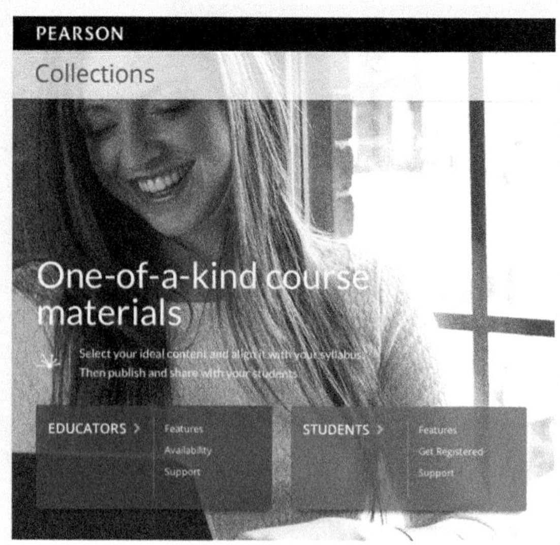

visual

anatomy & physiology

lab manual

2nd edition

STEPHEN N. SARIKAS, Ph.D.
Lasell College, Newton, Massachusetts

Using the visual approach and modified art from

Visual Anatomy & Physiology

by

Frederic H. Martini, Ph.D.
University of Hawaii at Manoa

William C. Ober, M.D.
Washington and Lee University

Judi L. Nath, Ph.D.
Lourdes University, Sylvania, Ohio

Edwin F. Bartholomew, M.S.

Kevin Petti, Ph.D.
San Diego Miramar College

with

Claire E. Ober, R.N.
Illustrator

Kathleen Welch, M.D.
Clinical Consultant

Ralph T. Hutchings
Biomedical Photographer

Editor-in-Chief: Serina Beauparlant
Portfolio Manager: Cheryl Cechvala
Content Producer: Lauren Bakker
Managing Producer: Nancy Tabor
Courseware Director, Content Development: Barbara Yien
Development Editor: Kari Hopperstead
Courseware Editorial Assistant: Kimberly Twardochleb
Rich Media Content Producer: Patrice Fabel
Full-Service Vendor: Cenveo Publisher Services
Production Management: Mary Tindle
Design Manager: Mark Ong
Interior and Cover Designer: John Walker Design
Photographer and Photo Researcher: Kristin Piljay
Rights & Permissions Project Manager: Kathy Zander
Rights & Permissions Management: Cenveo Publisher Services
Manufacturing Buyer: Stacey J. Weinberger
Executive Marketing Manager: Allison Rona

Cover Photo Credit: Nanettegrebe/Fotolia

Library of Congress Cataloging-in-Publication Data

Names: Sarikas, Stephen N. | Martini, Frederic. Visual anatomy & physiology.
Title: Visual anatomy & physiology lab manual. Cat version / Stephen N. Sarikas, Ph.D., Lasell College, Newton, Massachusetts ; using the visual approach and modified art from Visual anatomy & physiology / by Frederic H. Martini, Ph.D., University of Hawaii at Manoa [and four others] ; with Claire E. Ober, R.N., illustrator, Kathleen Welch, M.D., clinical consultant, Ralph T. Hutchings, biomedical photographer.
Other titles: Visual anatomy and physiology lab manual. Cat version | Anatomy & physiology lab manual. Cat version
Description: 2 [2nd edition]. | New York, NY : Pearson, [2018] | Includes bibliographical references and index.
Identifiers: LCCN 2016043183 | ISBN 9780134403854
Subjects: LCSH: Cats—Physiology—Laboratory manuals. | Cats--Anatomy—Laboratory manuals. | Cats—Dissection—Laboratory manuals. | Visual learning.
Classification: LCC SF767.C29 S27 2018 | DDC 636.8/0891—dc23 LC record available at https://lccn.loc.gov/2016043183

ISBN 10: 0-13-440385-1; ISBN 13: 978-0-13-440385-4 (Cat Version)
ISBN 10: 0-13-454797-7; ISBN 13: 978-0-13-454797-8 (Instructor's Review Copy)

www.pearsonhighered.com

13 2022

Preface

Visual Anatomy & Physiology Lab Manual, 2e brings all the strengths of the revolutionary *Visual Anatomy & Physiology* textbook to the lab. This lab manual combines a visual approach with a modular organization to maximize learning.

The goal of *Visual Anatomy & Physiology Lab Manual* is to create a better lab experience by presenting anatomy and physiology in an innovative way. Hands-on activities in the lab manual combine with assignable content in MasteringA&P™ to offer students frequent practice that reinforces important concepts. Whether you are using *Visual Anatomy & Physiology* or a different textbook for your lecture, *Visual Anatomy & Physiology Lab Manual* provides students with the powerful tools they need to succeed in the lab.

KEY FEATURES

The following are the distinctive features of this lab manual:

- **The visual approach** breaks out of the text-heavy model of other two-semester A&P lab manuals. Instead of long columns of narrative that refer to visuals, this lab manual features visuals with brief integrated text built around them. Students can't read without seeing the corresponding visual, and they can't look at a visual without reading.
- **The modular organization** presents each lab exercise in a series of two-page lab activity spreads. The top-left page of each module begins with the lab activity number and title, the bottom-right page ends with a set of self-check Making Connections questions, and the guided lab activity fills the rest of the two-page spread. Students can see everything for a lab activity at a glance without the page flipping that often contributes to students losing their way and then getting confused.
- Four new **Biopac activities** allow instructors to assign labs using the most popular Biopac programs and equipment. New labs include:
  - Electromyography: Recording an electromyogram (EMG) to study motor unit recruitment (Activities 13.2 and 13.3 in Exercise 13).
  - Electroencephalography: Recording an electroencephalogram (EEG) to study the brain's electrical activity (Activity 15.6 in Exercise 15).
  - Electrocardiography: Recording an electrocardiogram (ECG) to study the heart's electrical activity during the cardiac cycle (Activities 23.5 and 23.6 in Exercise 23).
  - Measuring respiratory volumes (Activity 26.5 in Exercise 26).
- A new **10-question Pre-Lab Quiz** for each exercise provides an opportunity for students to check their content knowledge and answer questions about lab procedure, thereby encouraging them to read the lab and review the lecture content before coming to lab. Each quiz can be assigned in MasteringA&P as pre-lab homework to help ensure that students spend less lab time reviewing lecture content and more time on the activities in that exercise.
- Frequent **pencil-to-paper tasks** within each lab activity are marked with a black arrow (▶) to indicate where students need to write answers, fill in tables, record data, label, or draw.
- **Making Connections questions** wrap up each two-page lab activity and give students the opportunity to pause, internalize information, and then apply their understanding.
- **In the Clinic** boxes provide clinical context for the material students are learning.
- **Word Origins** boxes simplify learning by connecting the terminology used in anatomy and physiology to word roots.
- **Review Sheets** at the end of each lab exercise offer a series of questions that assess students on all the activities in the lab exercise. Assignable versions of the Review Sheet questions (without the coloring questions) are also available in MasteringA&P.
- New **Before You Move On** sections at the end of each exercise help summarize key takeaway points from the exercise and point out how those concepts will inform the next topic, prompting students to appreciate how topics relate to one another.
- **Learning Outcomes** on each exercise-opening page indicate to students what they should be able to do by the end of the exercise. Learning Outcomes are coordinated by number to the lab activities, thus allowing students to check their understanding by both Learning Outcomes and lab activity topics. Additionally, the assessments in MasteringA&P are similarly organized, allowing instructors to assign, assess, and demonstrate teaching results by Learning Outcomes.
- New **Mastering A&P prompts** within each exercise direct students to online animations, tutorials, and video that will help further their understanding of difficult concepts.

MasteringA&P is an online homework, tutorial, and assessment environment designed to improve results by helping students quickly master concepts. Students benefit from self-paced tutorials that feature immediate wrong-answer feedback and hints that emulate the office-hour experience to help keep students on track. With a wide range of interactive, engaging, and assignable activities, students are encouraged to actively learn and retain tough course concepts.

Please turn to the front pages for a visual walkthrough of ***Visual Anatomy & Physiology Lab Manual***, 2e and MasteringA&P.

About the Author

STEPHEN N. SARIKAS received his Ph.D. in Anatomy from Boston University School of Medicine. He is a professor of biology at Lasell College in Newton, Massachusetts, where he has taught courses in anatomy and physiology, general biology, human reproduction, environmental science, and history of science and has conducted seminars on Charles Darwin, AIDS in America, and medical ethics. He is also a lecturer of occupational therapy at Boston School of Occupational Therapy, Tufts University, where he teaches a graduate-level anatomy course. In 2008, he was selected to be Lasell College's fifth Joan Weiler Arnow '49 Professor, a three-year endowed professorship that recognizes a scholar-teacher for his or her commitment to teaching and personal interest in students.

Dr. Sarikas's past research interests and publications have included studies on the histochemistry of egg capsules in two salamander species (*Ambystoma* sp.); the development, maturation, and distribution of small-granule APUD cells in the mammalian lung; and membrane-intermediate filament interactions in transitional epithelium during the contraction-expansion cycle of the mammalian urinary bladder. His current research interests are in the area of HIV/AIDS awareness among college students. Dr. Sarikas is the author of the first and second editions of *Laboratory Investigations in Anatomy & Physiology* and the first edition of *Visual Anatomy & Physiology Lab Manual.*

Dr. Sarikas is a member of the Human Anatomy and Physiology Society, the American Association of Anatomists, and the American Association for the Advancement of Science. He lives in Chelsea, Massachusetts, where he serves as chairperson of the Chelsea Conservation Commission. He and his wife, Marlena, enjoy gardening, running, entertaining friends, and watching the Red Sox beat the Yankees. They regularly travel to New York City, where their son, Anthony, lives and performs comedy, and to Montreal, where they have many close friends.

Acknowledgments

I am very privileged to have been given the opportunity to write a lab manual that takes a distinctively different and unique approach to anatomy and physiology lab instruction. The inspiration for this manual comes from the innovative and groundbreaking work of the Martini author and illustrator team who developed the two-page modular format and the visual approach found in *Visual Anatomy & Physiology*. This pedagogical approach stimulates learning and promotes success and self-confidence in the anatomy and physiology course. Therefore, I gratefully acknowledge the author/illustrator team that conceptualized and developed *Visual Anatomy & Physiology*—Frederic H. Martini, William C. Ober, Judi L. Nath, Edwin F. Bartholomew, Kevin Petti, and Claire E. Ober—and thank each of them for giving me the tools I needed to create this lab manual.

I would also like to thank Robert Tallitsch, Augustana College, Rock Island, Illinois, who provided several of the stunning light micrographs in this manual. The talented dissector/photographer team of Shawn Miller and Mark Nielsen, University of Utah, contributed the highest-quality cat and fetal pig dissection photographs in any anatomy and physiology lab manual. I also thank Kristin Piljay, photo researcher, who located several hard-to-find photos.

The publication of this laboratory manual was a team effort and represents the contributions of many dedicated and creative individuals. I am especially thankful and fortunate to have had the opportunity to collaborate with a team of talented people from Pearson who supported my work throughout every phase of its development. Special recognition goes to Cheryl Cechvala, Senior Acquisitions Editor, who directed all aspects of this project with professionalism, skill, and competence. I am sincerely grateful for Cheryl's support and confidence in my work, and I thank her for believing in and advocating for my ideas.

I am indebted to Lauren Bakker, Content Producer, who skillfully managed to keep this project, with all its complexities, on the right course and on schedule from start to finish. Lauren's tireless efforts, keen insights, and attention to detail made my job a lot easier. It was a pleasure working with her.

I would like to thank Kari Hopperstead, Development Editor, for critically reviewing the manuscript and providing invaluable suggestions for improving the presentation of the two-page lab activities. I also express my gratitude to Kimberly Twardochleb, Portfolio Management Assistant, who adeptly managed all the behind-the-scenes tasks that often went unnoticed but yet were vitally important to the success of this project.

A special thank-you goes out to Mark Ong, Design Manager, for his expertise in page design and for his patience in teaching me how to use InDesign. Appreciation is also due to Barbara Yien, Courseware Director, Content Development, who provided expert management in the development of the manual's content, and to John Walker for his special skills and insights in text design. I am also thankful for the significant contributions of Stacey J. Weinberger, Senior Procurement Specialist, who handled the physical manufacturing of the manual.

Allison Rona, Executive Marketing Manager, planned and directed the presentation of this lab manual to my anatomy and physiology colleagues at colleges and universities across the United States. I also express my gratitude to all the Pearson sales representatives for skillfully representing this lab manual to instructors.

I express my sincere gratitude to Kathleen M. Lafferty of Roaring Mountain Editorial Services for her careful and attentive review of the manuscript. A special thank-you goes to Mary Tindle, Senior Project Manager at Cenveo Publisher Services, for her attentiveness and commitment to make each page perfect during the production of this lab manual.

A special note of thanks goes to Lori K. Garrett, Parkland College, author of *Get Ready for A & P* and an exceptional writer, who wrote the *Instructor's Manual* that accompanies this lab manual.

I am most fortunate to have in my life three special individuals who have given me priceless gifts of their unconditional love and emotional support. My stepson, Anthony Atamanuik, who performs improvisational comedy in New York City, guided me through some of the more difficult periods with his comic relief. Anthony often engaged me in lengthy and highly entertaining discussions on particular topics or sections that I was writing. His unique perspective helped me articulate my ideas more clearly, and perhaps I provided him with material for new comedy skits! Anthony's wife, Flossie Arend, has given me invaluable emotional support. Her sharp insights and refreshing ideas on life have helped me be a better person. I might add that Flossie is a proficient knitter of anatomical structures; her gift to me, Ned the Neuron, hangs proudly on the wall of my office. Marlena Yannetti, my wife and life partner, offered tender guidance, sound advice, and loving support. Her calming influence guided me through the many weekends and late nights at the computer and the recurring periods of escalating pressure and stress when weekly deadlines approached. Marlena's unending encouragement and support are just two of the countless reasons I love her so much.

Many of my colleagues from across the United States and Canada donated their time, expertise, and thoughtfulness to review this lab manual. Their valuable comments and suggestions guided me through the writing process. I am profoundly indebted to the following instructors for their support:

INSTRUCTOR REVIEWS OF THE SECOND EDITION

Erin Bailey, *Kent State University*
Jill Bigos, *Elms College*
James Gunipero, *Central Carolina Community College*
Greta Herin, *Eastern Mennonite University*
Karen Dunbar Kareiva, *Ivy Tech Community College*
Jared LeMaster, *Cuyahoga Community College East*
Shawn Macauley, *Muskegon Community College*
Annie McKinnon, *Howard College Big Spring*
John Mecham, *Meredith College*
April Murphy, *Columbus Technical College*
Hamid Nawaz, *Eastern Gateway Community College*
Eileen Roark, *Manchester Community College*
Amy Ryan, *Clinton Community College*
Kris Schwab, *Indiana State University*

INSTRUCTOR REVIEWS OF THE FIRST EDITION

Ticiano Alegre, *North Lake College*
Ken Anyanechi, *Southern University at Shreveport*
Dena Berg, *Tarrant County College–Northwest*
Chris Brandon, *Georgia Gwinnett College*
Carol Britson, *University of Mississippi*
David Brooks, *East Texas Baptist University*
Nishi Bryska, *University of North Carolina at Charlotte*
Stephen Burnett, *Clayton State University*
Robert Byrer, *University of West Florida*
Yavuz Cakir, *Benedict College*
Ronald Canterbury, *University of Cincinnati*
Maria Carles, *Northern Essex Community College*
Joann Chang, *Arizona Western College*
Roger Choate, *Oklahoma City Community College*
Debra Claypool, *Mid Michigan Community College*
Richard Coppings, *Jackson State Community College*
Elaine Cox, *Bossier Parish Community College*
Cassy Cozine, *University of Saint Mary*
Kenneth Crane, *Texarkana College*
Donna Crapanzano, *Stony Brook University*
David Doe, *Westfield State University*
Rick Doolin, *Daytona State College*
Miranda Dunbar, *Southern Connecticut State University*
Bernadette Dunphy, *Monmouth University*
Abdeslem El Idrissi, *College of Staten Island–CUNY*
Kurt Elliott, *Northwest Vista College*
Doug Elrod, *North Central Texas College*
Greg Erianne, *County College of Morris*
Dana Evans, *University of Rio Grande*
J. Alyssa Farnsworth, *Ivy Tech Community College of Indiana*
Bruce Fisher, *Roane State Community College*
Henry Furneaux, *Capital Community College*
Kristine Garner, *University of Arkansas–Fort Smith*
Lori K. Garrett, *Parkland College*
Wanda Goleman, *Northwestern State University*
Emily Gonzalez, *Northern Essex Community College*
Ewa Gorski, *Community College of Baltimore County*
Sylvester Hackworth, *Bishop State Community College*
Monica Hall-Porter, *Lasell College*
Rebecca Harris, *Pitt Community College*
James Hawker, *South Florida Community College*
Marta Heath-Sinclair, *Hawkeye Community College*
Candi Heimgartner, *University of Idaho*
Gary Heisermann, *Salem State University*
Noah Henley, *Rowan–Cabarrus Community College*
Julie Huggins, *Arkansas State University*
Sue Hutchins, *Itasca Community College*
Saiful Islam, *Northern Virginia Community College–Woodbridge*
Jerry Johnson, *Corban University*
Jody Johnson, *Arapahoe Community College*
Jacqueline Jordan, *Clayton State University*
Steve Kash, *Oklahoma City Community College*
Paul Kiser, *Bellarmine University*
Steven Kish, *Zane State College*
Marta Klesath, *North Carolina State University*
Chad Knights, *Northern Virginia Community College–Alexandria*
Megan Knoch, *Indiana University of Pennsylvania*
Louis Kutcher, *University of Cincinnati Blue Ash College*
Tiffany Lamb, *Amarillo College*
Paul Lea, *Northern Virginia Community College–Annandale and Medical Campuses*
Jeffrey Lee, *Essex County College*
Carlos Liachovitzky, *Bronx Community College*
Jerri Lindsey, *Tarrant County College–Northeast*
Christine Maney, *Salem State University*
Jennifer Mansfield-Jones, *University of Louisville*
Bruce Maring, *Daytona State College*
Robert Marino, *Capital Community College*
Darren Mattone, *Muskegon Community College*
Cathy Miller, *Florida State College at Jacksonville*
Liza Mohanty, *Olive–Harvey College*
David Mullaney, *Naugatuck Valley Community College*
Gwen Niksic, *University of Mary*
Zvi Ostrin, *Hostos Community College–CUNY*
Jay O'Sullivan, *University of Tampa*
Debby Palatinus, *Roane State Community College*
Crystal Pietrowicz, *Southern Maine Community College*
Lou Rifici, *Cuyahoga Community College*
Nancy Risner, *Ivy Tech Community College*

Jo Rogers, *University of Cincinnati*
Hope Sasway, *Suffolk County Community College–Grant*
Dee Ann Sato, *Cypress College*
Tracy Schnorr, *Aurora Community College*
Samuel Schwarzlose, *Amarillo College*
Shaumarie Scoggins, *Texas Women's University*
Pamela Siergiej, *Roane State Community College*
Hollis Smith, *Kennebec Valley Community College*
Terry St. John, *Ivy Tech Community College–Richmond*
Claudia Stanescu, *University of Arizona*
Sherry Stewart, *Navarro College*
Bonnie Taylor, *Schoolcraft College*
Stephen Waldow, *Monroe College*
Rachel Willard, *Arapahoe Community College*
Greg Wilson, *Holmes Community College*
Matthew Wood, *Lake Sumter Community College*

INSTRUCTOR CLASS TESTERS OF THE FIRST EDITION

Trinika Addison, *Middle Georgia State College*
Arlene Allam-Assi, *Hostos Community College–CUNY*
Rishika Bajaj, *Hostos Community College–CUNY*
Michael Barnett, *Amarillo College*
Dena Berg, *Tarrant County College–Northwest*
Bharat Bhushan, *Capital Community College*
Nick Butkevich, *Schoolcraft College*
Katherine Butts-Dehms, *Kennebec Valley Community College*
Yavuz Cakir, *Benedict College*
Maria Carles, *Northern Essex Community College*
Joann Chang, *Arizona Western College*
Barbara Coles, *Wake Tech Community College*
Richard Coppings, *Jackson State Community College*
Donna Crapanzano, *Stony Brook University*
Molli Crenshaw, *Texas Christian University*
David Davis, *Middle Georgia State College*
Diane Day, *Clayton State University*
Vivien Diaz-Barrios, *Hostos Community College–CUNY*
Sam Dunlap, *North Central Texas College*
Vyasheslav Dushenkov, *Hostos Community College–CUNY*
Paul Dykes, *Century College*
Yasmin Edwards, *Bronx Community College*
Kurt Elliott, *Northwest Vista College*
J. Alyssa Farnsworth, *Ivy Tech Community College–Muncie*
Jill Feinstein, *Richland Community College*
Bruce Fisher, *Roane State Community College*
Edward Franklin, *Corning Community College*
John Gillen, *Hostos Community College–CUNY*
Wanda Goleman, *Northwestern Louisiana State University*
Emily Gonzalez, *Northern Essex Community College*
Rebecca Harris, *Pitt Community College*
Martha Heath-Sinclair, *Hawkeye Community College*
Jacqueline Jordan, *Clayton State University*
Jeff Keyte, *College of Saint Mary*
Christine Kisiel, *Mount Wachusett Community College*
Greg Klein, *Cincinnati State University*
Eduard Korolyev, *Hostos Community College–CUNY*
Tiffany Lamb, *Amarillo College*
Holly Landrum, *Jackson State Community College*
Damaris Lang, *Hostos Community College–CUNY*
Jerri Lindsey, *Tarrant County College–Northeast*
Sheryl Lumbley, *Cedar Valley College*
Renee McFarlane, *Clayton State University*
Paul Melvin, *Clayton State University*
Zvi Ostrin, *Hostos Community College–CUNY*
Deborah Palatinus, *Roane State Community College*
John Pattillo, *Middle Georgia State College*
Penny Perkins-Johnson, *California State University–San Marcos*
Dawn Poirier, *Dean College*
Faina Riftina, *Hostos Community College–CUNY*
Christine Rigsby, *Middle Georgia State College*
Nancy Risner, *Ivy Tech Community College–Muncie*
Charlene Sayers, *Rutgers University–Camden*
Ralph Schwartz, *Hostos Community College–CUNY*
Samuel Schwarzlose, *Amarillo College*
Shaumarie Scoggins, *Texas Woman's University*
Dustin Scott, *Jackson State Community College*
Dara Lee Shigley, *Ivy Tech Community College–Evansville*
Igor Shiltsov, *Hostos Community College–CUNY*
Jane Slone, *Cedar Valley College*
Jill Stein, *Essex County College*
Julie Trachman, *Hostos Community College–CUNY*
Tiffany Vogler, *Ivy Tech Community College–Southwest*
Heather Wesp, *Montcalm Community College*

It is difficult to produce a publication of this magnitude that is free of mistakes or omissions. Any errors or oversights are my responsibility and do not reflect the work of the editors, reviewers, artists, or production staff. I encourage faculty and students to send their comments or suggestions about the content of this lab manual directly to me at the address given below. I will give all your ideas serious consideration when I prepare for the next edition.

Stephen N. Sarikas
Professor of Biology
Lasell College
1844 Commonwealth Ave.
Newton, MA 02466
ssarikas@lasell.edu

Contents

THE RESPIRATORY SYSTEM

THE DIGESTIVE SYSTEM

THE URINARY SYSTEM

THE REPRODUCTIVE SYSTEMS

CAT DISSECTION EXERCISES

EXERCISE 1

Body Organization and Terminology

LEARNING OUTCOMES

These Learning Outcomes correspond by number to the laboratory activities in this exercise. When you complete the activities, you should be able to:

Activity 1.1 **Describe and demonstrate anatomical position and use anatomical terminology to describe relative positions of structures in the human body.**

Activity 1.2 **Describe and demonstrate the various anatomical planes and sections.**

Activity 1.3 **Summarize functions of each organ system, and list the organs in each.**

Activity 1.4 **Name the anatomical regions of the body.**

Activity 1.5 **Identify the body cavities and the organs that are located in each.**

LABORATORY SUPPLIES

- Human torso model, with dissectible parts
- Various anatomical models of organs and organ systems, with dissectible parts
- Fresh vegetables that are long and cylindrical in shape (e.g., cucumbers or eggplants)
- Small kitchen knives or scalpels
- Human skeleton or skull
- Coloring pencils

PRE-LAB QUIZ

Before you begin, read all the activities in Exercise 1 and the required reading in your textbook that is assigned by your instructor.

1. The locations of body structures are described with reference to the universally accepted position of the human body. This position is referred to as the __________.
2. During this laboratory exercise, you will be completing all the following activities *except*
   a. identifying anatomical planes and sections.
   b. exploring the body cavities.
   c. locating organs in the body and placing them in the correct organ system.
   d. identifying the parts of a typical cell.
3. The orbital cavities, nasal cavity, paranasal sinuses, and the oral cavity are all located in what region of the body? __________
4. During this laboratory exercise, you will use a torso model to complete all the following activities *except*
   a. identifying the four abdominopelvic quadrants.
   b. demonstrating various types of sections that you will observe with a microscope.
   c. identifying anatomical planes.
   d. identifying organs in the various organ systems.
5. During this laboratory exercise, you will be using your own body or your lab partner's body to complete which of the following activities?
   a. describing the relationships of structures using directional terms
   b. identifying organs in the various organ systems
   c. removing parts of organs to identify various sections
   d. identifying the major body cavities
6. The ventral body cavity contains all the following subdivisions *except* the
   a. vertebral cavity.
   b. abdominal cavity.
   c. thoracic cavity.
   d. pelvic cavity.
7. True or False: The femoral region of the body is the thigh. __________
8. True or False: A sagittal plane divides the body into superior and anterior parts. __________
9. Clinicians divide the abdominopelvic region into four quadrants. Anatomists divide this region into __________ smaller segments.
10. True or False: The diaphragm separates the thoracic and abdominopelvic cavities. __________

Activity **1.1**

Using Anatomical Terms to Describe Body Organization

To succeed in this course, you must become familiar with the general organization of the human body and learn the standard anatomical language that is used to describe that organization. Most anatomical names are derived from Latin or Greek words and have remained uniform throughout most of the more than 2000 years during which anatomy has been studied.

A Anatomical Position

Human anatomy is described with reference to the **anatomical position,** a universally accepted standard position for the body. An individual in the anatomical position stands erect with head and eyes directed forward. The upper limbs are by the sides, with the palms facing forward, and the lower limbs are together, with the toes facing forward.

Word Origins

The word *anatomy* is derived from the Greek words *ana* (= "apart") and *tome* (= "a cutting"). Together, the two words mean "a cutting apart." The best way to study the structure of an organism is to dissect it or "cut it apart."

The word *physiology* is also derived from two Greek words: *physis* (= "nature") and *logos* (= "study"). Together, the two words mean "study nature." Physiology is the study of natural processes in the body.

1. Suppose that the accepted anatomical position is to have the palms facing backward. How would this new position change your description of the palms and thumbs?

▶ ______________________________

____

____

2. Based on your answer to the previous question, why do you think it is important to have a universally accepted anatomical position?

▶ ______________________________

____

____

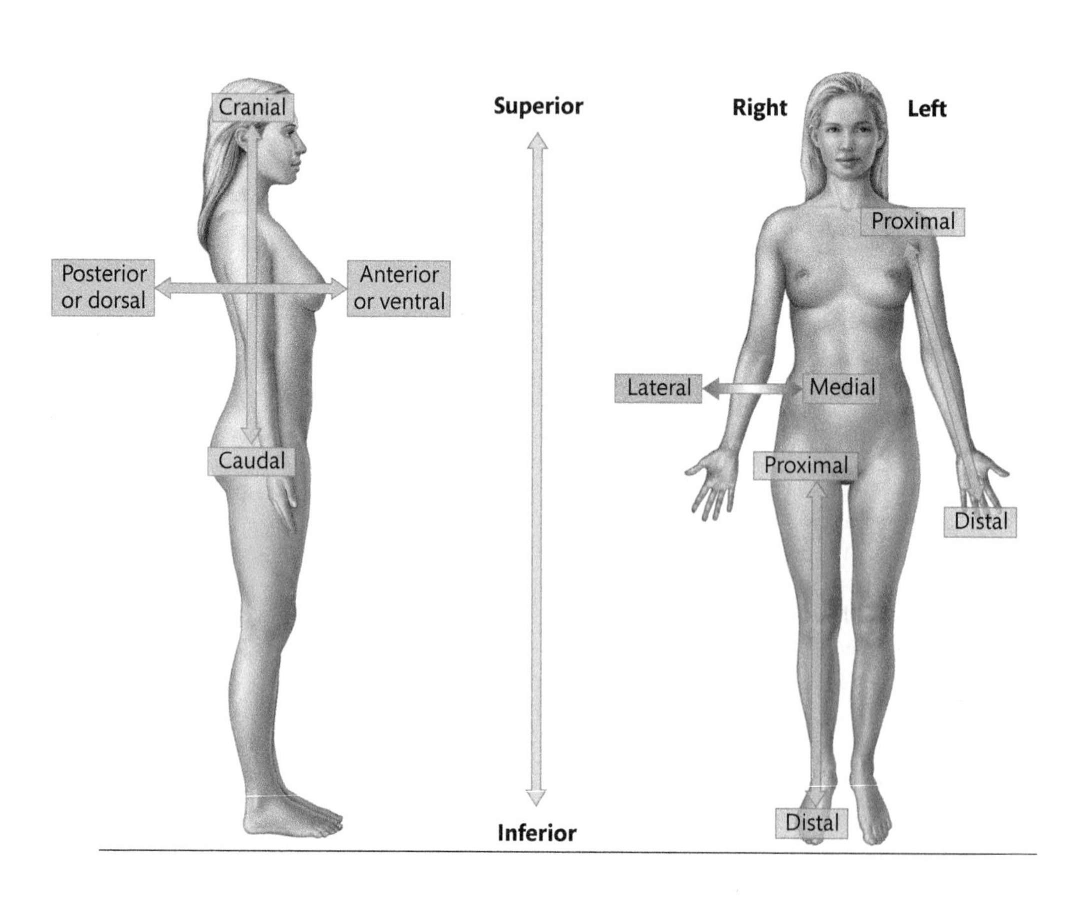

B Anatomical Terms

Several anatomical terms are used to describe the location of one body part in relation to another. These terms are always used to illustrate the relative position of a structure when the body is in the anatomical position. For example, to explain the relative position of the heart to the esophagus, you could state that the heart is **anterior** (closer to the front) to the esophagus. Alternatively, you could say that the esophagus is **posterior** (closer to the back) to the heart. Other terms used to express relative position are described in Table 1.1 and illustrated in the figure on the previous page. Carefully review these terms and be sure that you understand their meanings before you proceed. Make it a habit to periodically review these terms as the course progresses.

1 Ask your lab partner to stand in the anatomical position. Use the directional terms listed in Table 1.1 to describe the relationships of the following structures on your lab partner's body. ▶

a. Left eye to left ear ______________________

b. Thumb to little finger ______________________

c. Right ankle joint to right knee joint ______________________

d. Left elbow joint to right elbow joint ______________________

2 Refer to a torso model. Use directional terms to describe the relationships of the following pairs of internal organs. You will have to "dissect" the torso model to identify some of these structures. ▶

a. Left kidney to spleen ______________________

b. Right lung to right lobe of the liver ______________________

c. Pancreas to stomach ______________________

d. Ascending colon to descending colon ______________________

TABLE 1.1 Anatomical Terms of Relationship and Comparison

	Term	Definition	Example
1.	a. superior (cranial) b. inferior (caudal)	closer to the head closer to the feet	The lungs are *superior* to the stomach. The liver is *inferior* to the heart.
2.	a. anterior (ventral) b. posterior (dorsal)	closer to the front closer to the back	The trachea is *anterior* to the esophagus. The vertebral column is *posterior* to the heart.
3.	a. medial b. lateral	closer to the midline farther from the midline	The nose is *medial* to the cheeks. The spleen is *lateral* to the pancreas.
4.	intermediate	between a more medial and more lateral structure	The clavicle is *intermediate* to the sternum and the shoulder.
5.	a. proximal b. distal	closer to the trunk farther from the trunk	The shoulder is *proximal* to the elbow. The wrist is *distal* to the elbow.
6.	a. superficial (external) b. deep (internal)	closer to or on the surface farther from the surface	The skin is *superficial* to the skeletal muscles. The bones are *deep* to the skin.
7.	a. parietal b. visceral	pertaining to the wall of a body cavity pertaining to the covering of an organ	The membrane lining the thoracic wall is the *parietal* pleura. The membrane that covers the surface of the lungs is the *visceral* pleura.
8.	a. ipsilateral b. contralateral	on the same side of the body on the opposite side of the body	The right lung is *ipsilateral* to the liver. The left arm is *contralateral* to the right lung.

MAKING CONNECTIONS

Often, the terms in Table 1.1 are combined to provide a more specific description of location. For instance, one can say that the heart is **superomedial** to the ascending colon. What does this terminology tell you about the relative positions of the heart and the ascending colon?

▶ ____

____

Activity **1.2**

Defining Anatomical Planes and Sections

A Three types of imaginary planes pass through the body in the anatomical position. Each of the three planes forms a right angle with the other two. The surfaces that are formed by cuts made in the various planes are called **sections.**

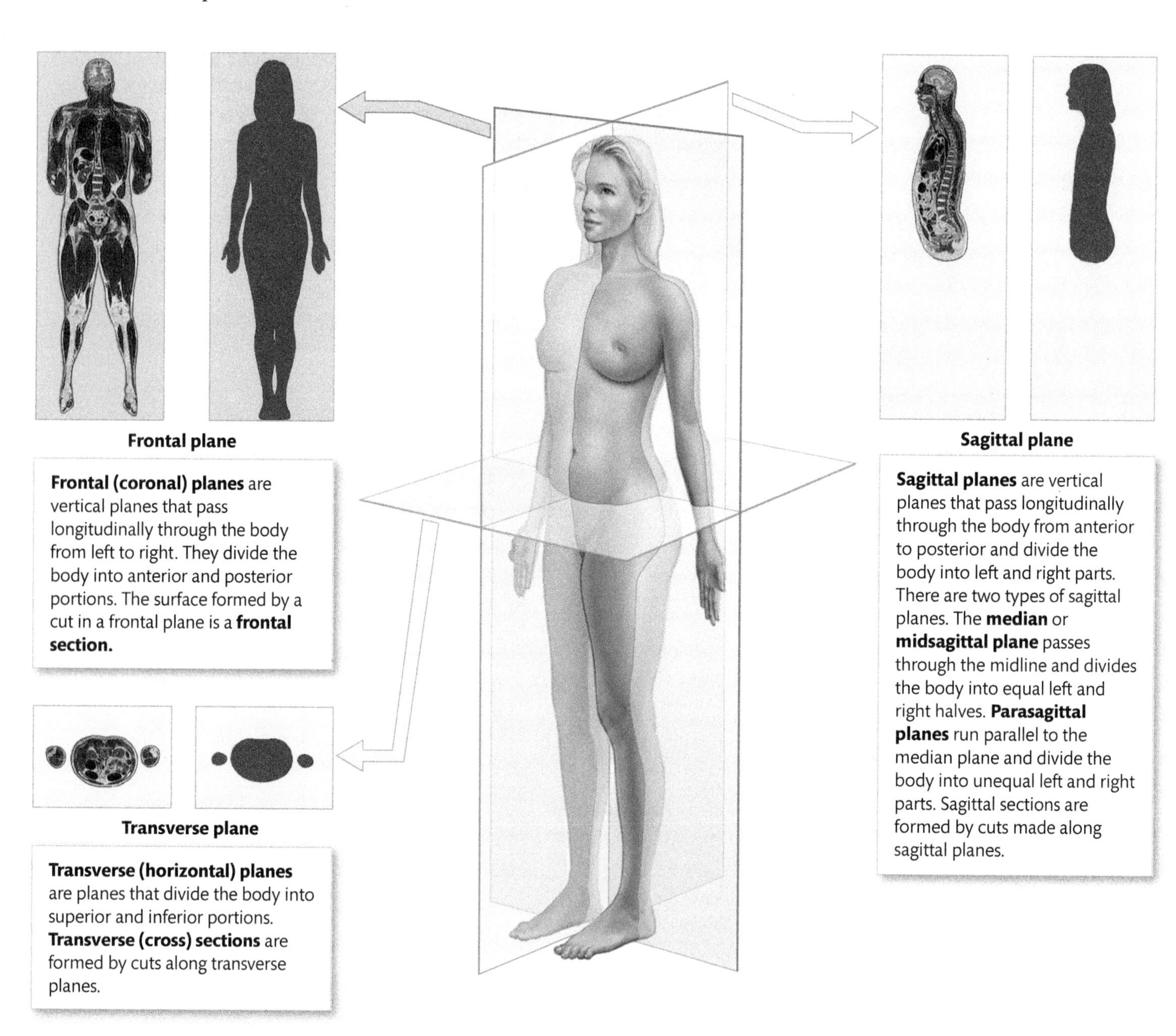

1. Identify the anatomical planes on torso models and on your own or your lab partner's body.

2. Observe several anatomical models with removable parts. As you remove each part, determine what type of section has been made. Remember that you must place the structure in the anatomical position to determine the answer.

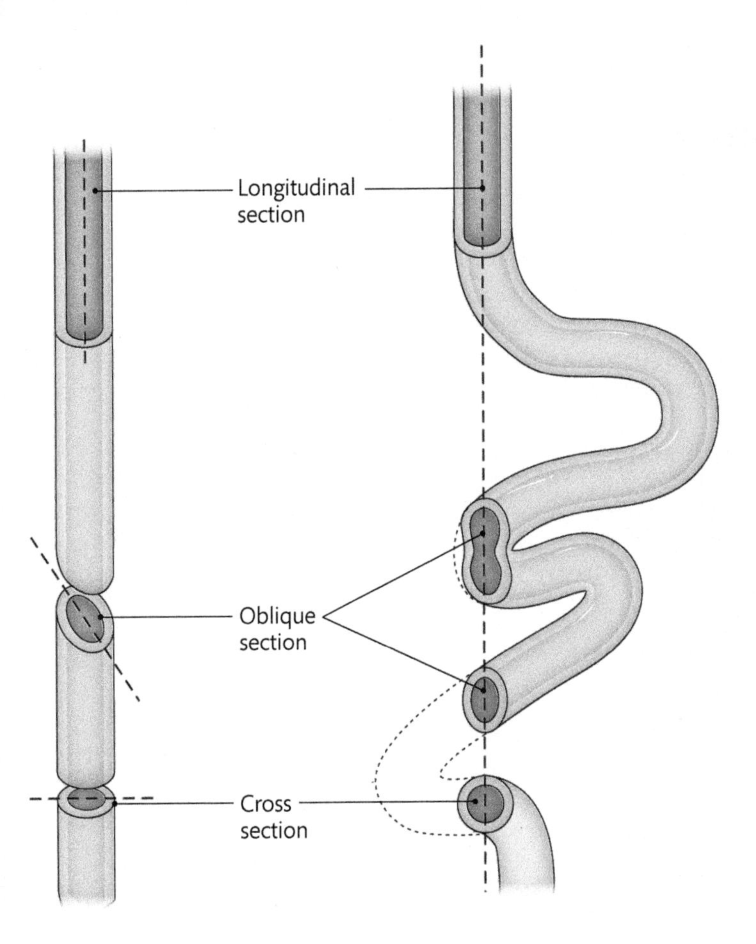

B Sections are typically used to describe cuts made in specific structures rather than the whole body. For example, if you are viewing a microscope slide of the trachea, you may be looking at a cross section (cs) or a longitudinal section (ls). Sometimes, the label on the slide will tell you what type of section you are observing (i.e., cs or ls). In a highly folded structure such as the small intestine, however, you can often find more than one type of section on the same microscope slide.

1. Obtain a vegetable that has a long cylindrical shape (e.g., cucumber, zucchini, or eggplant). Using a knife or scalpel, start near one end of the vegetable and cut, in order, a cross section, a longitudinal section, and an oblique section. Oblique sections are not formed by cuts made along any of the basic anatomical planes as described earlier. Instead, these sections slant or deviate from these planes and intersect them at angles other than 90°.
2. Observe the surfaces that you have produced with these sectional cuts. In the spaces below, draw each section that you have produced.

Cross Section	Longitudinal Section	Oblique Section
▶	▶	▶

MAKING CONNECTIONS

The thoracic cavity contains the heart and lungs. Explain, in a general way, how a view of the thoracic cavity along the midsagittal plane would differ from a view of the thoracic cavity along a transverse plane.

▶ ______________________________

____

____

____

Activity 1.3

Identifying Organs and Organ Systems

An **organ** is a distinct structure that contains at least two, but often all four, types of tissues and carries out specific functions. Most organs are located within body cavities that are closed to the outside. For example, the small intestine contains all four types of tissues and performs the final steps for digesting (breaking down) nutrients into small molecules and absorbing these molecules into the blood or lymph. The small intestine is located within the abdominal cavity.

An **organ system** is a collection of organs that works as a team to complete a common objective. For example, the small intestine is an organ in the digestive system, which also includes the oral cavity, pharynx, esophagus, and most organs in the abdominal cavity. These organs are responsible for ingesting food, digesting and absorbing nutrient molecules, and eliminating undigested wastes.

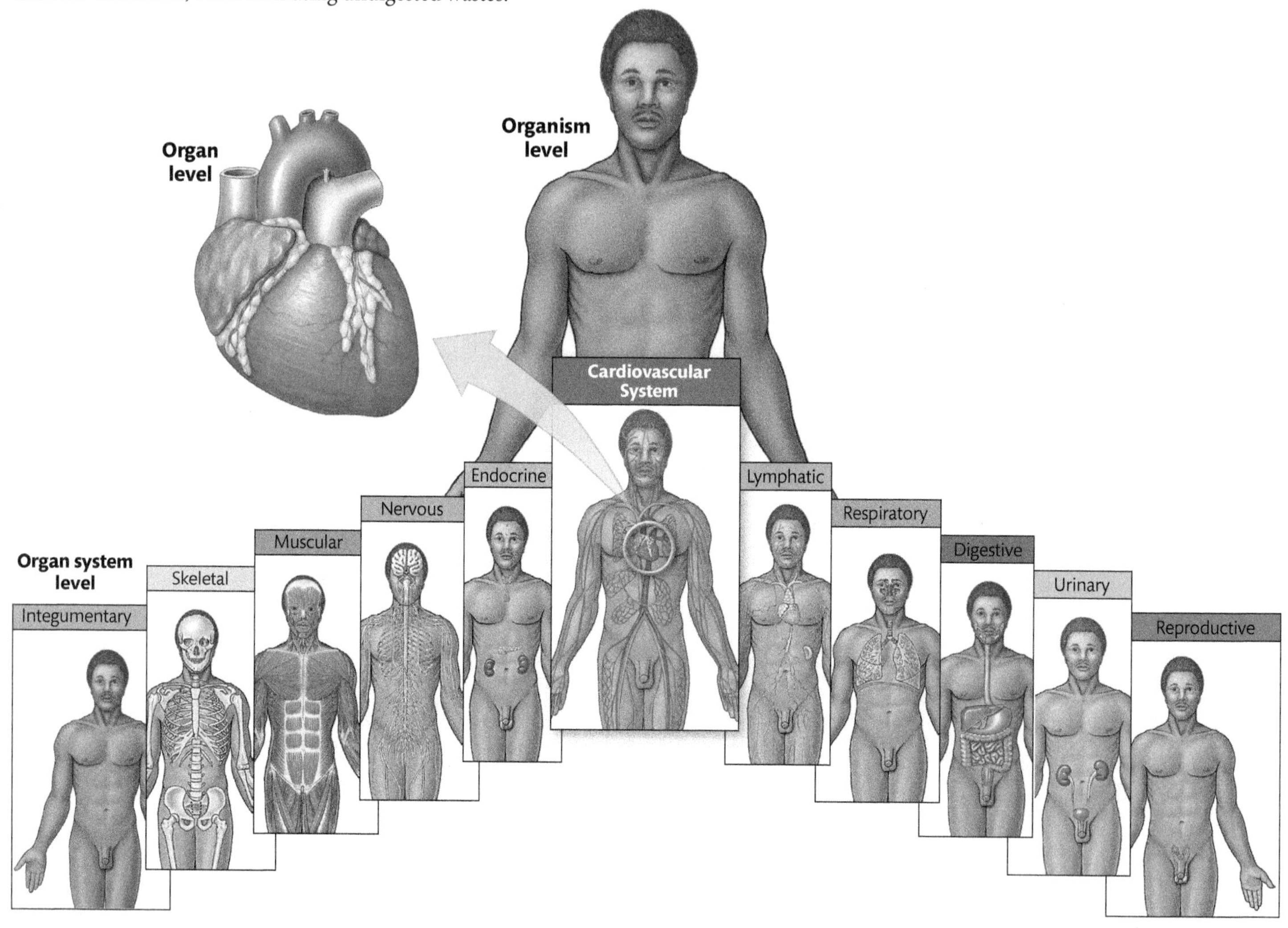

A On a human torso model, observe the anatomical relationships of the internal organs. Notice how adjacent organs are in close contact with one another and that very little unoccupied space remains in the body cavities. Identify all the structures listed below. (Depending on the type of models in your lab, you may not be able to locate all the structures.) Next to each structure, write the organ system to which it belongs. ▶

Aorta ______	Skull ______	Ovaries ______	Tonsils ______
Brain ______	Small intestine ______	Pancreas ______	Trachea ______
Heart ______	Spinal cord ______	Skeletal muscles ______	Urinary bladder ______
Kidneys ______	Spleen ______	Skin ______	Uterus ______
Lungs ______	Testes ______	Liver ______	Stomach ______

B Consult your textbook or have a class discussion to identify the major functions of the 11 organ systems in the human body. Write your answers in Table 1.2. ▶

TABLE 1.2 Organ Systems and Their Major Functions

Organ System		Major Function
	Integumentary system	
	Skeletal system	
	Muscular system	
	Nervous system	
	Endocrine system	
	Cardiovascular system	
	Lymphatic system	
	Respiratory system	
	Digestive system	
	Urinary system	
	Reproductive system	

IN THE CLINIC

Organs and Organ Systems

It is convenient to study the organ systems as discrete entities, but from a functional perspective, each organ system is closely integrated with other systems. Consider the following examples:

- The lymphatic system defends the organs in other systems against infection and plays a pivotal role in tissue repair after an injury.
- The digestive system provides nutrients for cells in all organ systems. These nutrients are transported by the cardiovascular system.

Because the organ systems are so closely connected in function, many diseases present symptoms with a wide range of systemic effects. For example, diabetes mellitus, a disease that is characterized by the inability of cells to take up glucose, forces the body to break down vital proteins and lipids to produce enough energy for metabolism. As a result, many degenerative changes occur throughout the body, leading to myriad medical problems, including blindness, kidney failure, reduced blood flow to the limbs, and heart disease.

As you can see from these examples, during normal function and during periods of disease, the activities of each organ system are influenced and sometimes controlled by the activities of other systems. You should begin to understand and learn to appreciate this close integration of function.

MAKING CONNECTIONS

In this activity, you grouped various organs by organ system. Review your groupings and identify any organs that appear in more than one organ system. Comment on the functional significance of organs having a role in more than one system.

▶ ____

____

____

____

____

____

____

____

____

■ Want more practice? Go to: **MasteringA&P** > Study Area > Menu > Lab Tools > PAL > Anatomical Models > observe torso model or whole organ photos for the following organ systems: Nervous (Central Nervous System only), Endocrine, Cardiovascular, Lymphatic, Respiratory, Digestive, Urinary, and Reproductive

Activity **1.4**

Identifying Anatomical Regions

The body can be divided into two major divisions: the **axial** and the **appendicular.** The **axial division** is the central part of the body and includes the head, neck, and trunk. The **appendicular division** includes the upper and lower limbs. Both the axial and appendicular divisions can be subdivided into numerous smaller regions, each with a specific anatomical and common name. Familiarity with these terms will help you locate and learn the names of other structures later on. For example, the axilla is the region of the body that is commonly referred to as the armpit. The axillary artery and vein travel through this region, and the axillary lymph nodes are also located here.

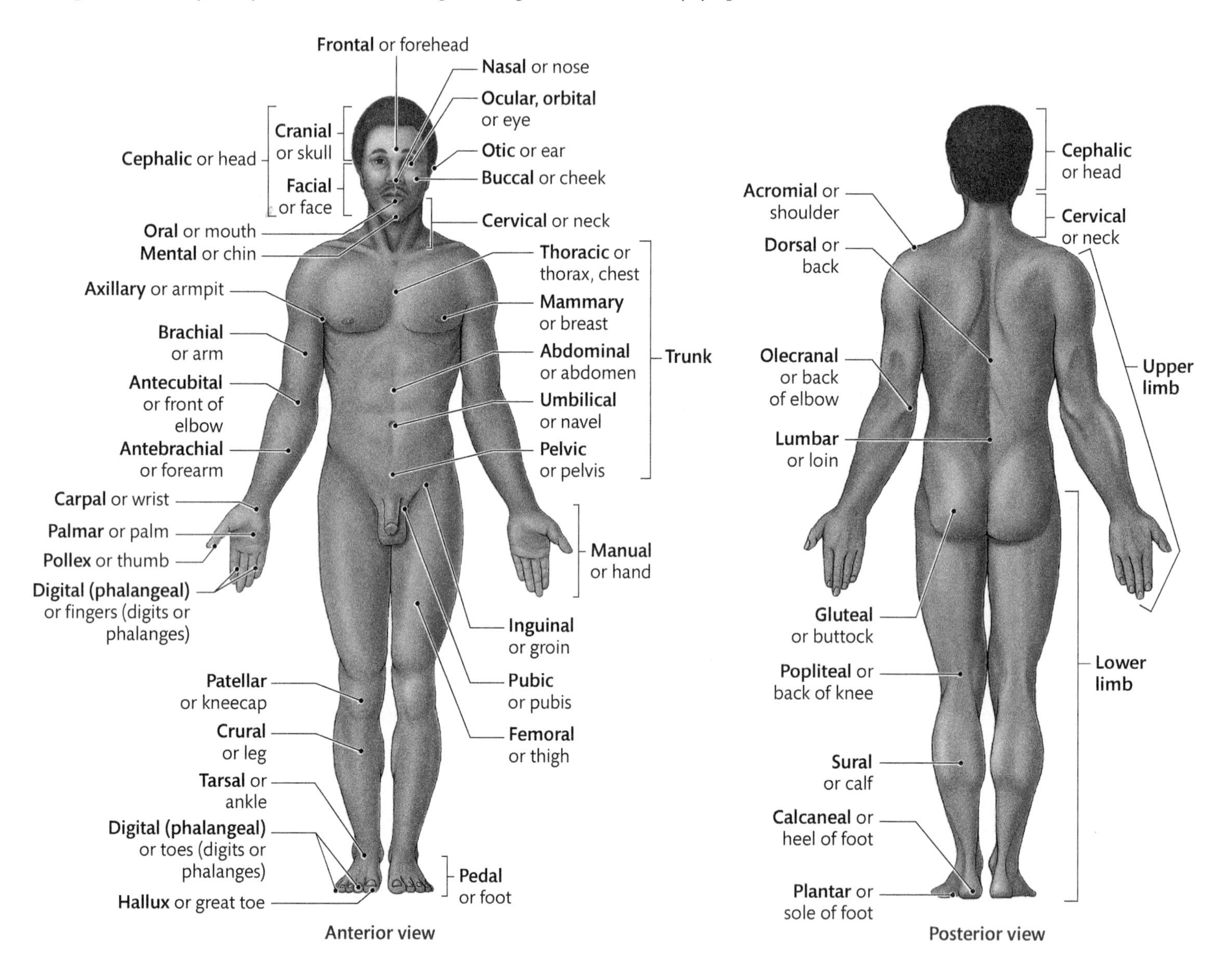

A Identify, by their common names, the anatomical regions in which you are likely to find the following structures: ▶

a. Inguinal canal __________

b. Brachial artery __________

c. Femoral vein __________

d. Facial nerve __________

e. Thoracic vertebrae __________

f. Carpal bones __________

g. Cranial bones __________

h. Popliteal artery __________

B The abdomen and pelvis, together often referred to as the **abdominopelvic** region, can be divided into even smaller segments. Clinicians divide this region into quadrants that are formed by two imaginary, perpendicular lines intersecting at the umbilicus. The four quadrants are the right upper quadrant, left upper quadrant, right lower quadrant, and left lower quadrant. Anatomists usually describe the abdominopelvic area in a more specific manner by dividing it into nine regions.

1. Using a torso model, identify the four abdominopelvic quadrants. Identify two organs, or parts of organs, found within each quadrant and list them in Table 1.3. Use the photos on the right as a reference. ▶
2. Using a torso model, identify the nine abdominopelvic regions. Identify two organs, or parts of organs, found within each region and list them in Table 1.3. Use the photos on the right as a reference. ▶

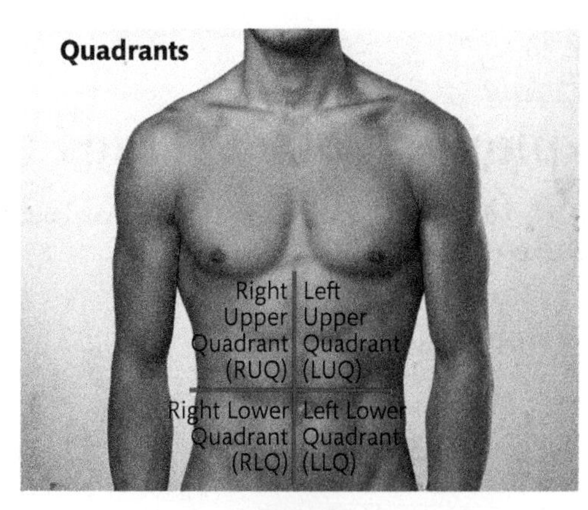

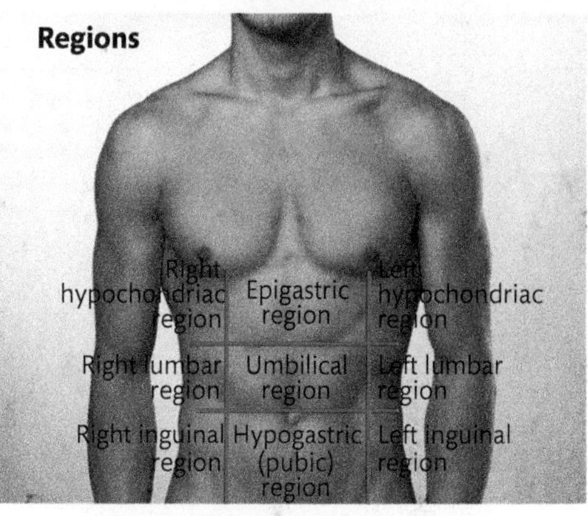

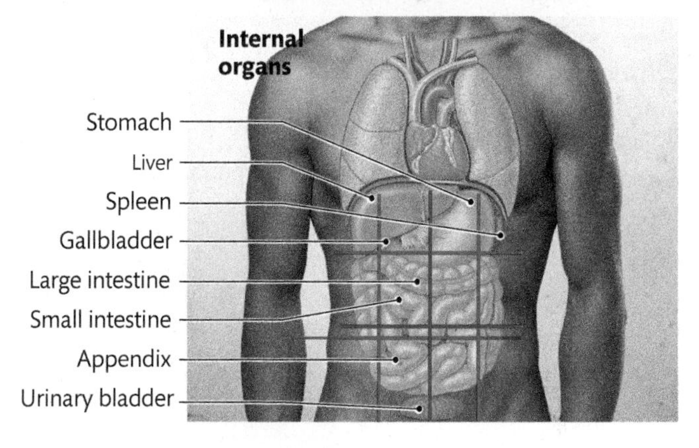

TABLE 1.3 The Abdominopelvic Regions and Underlying Organs

Four Quadrants	Organs
Right upper quadrant	
Left upper quadrant	
Right lower quadrant	
Left lower quadrant	
Nine Regions	**Organs**
Right hypochondriac	
Epigastric	
Left hypochondriac	
Right lumbar	
Umbilical	
Left lumbar	
Right inguinal	
Hypogastric (pubic)	
Left inguinal	

IN THE CLINIC

Abdominopelvic Quadrants

The quadrant system is clinically important because it can be used to identify the general location of underlying organs. For example, the appendix is a wormlike extension attached to the cecum at the origin of the large intestine. The appendix is located in the right lower quadrant. If a patient complains of persistent pain in this region, he or she could have an inflamed appendix, or **appendicitis.**

MAKING CONNECTIONS

Which method of dividing the abdominopelvic region do you find to be more useful from an anatomical perspective? From a clinical perspective? Explain.

▶ ______________________________

____

____

____

____

____

____

Activity 1.5

Exploring Body Cavities

The axial division of the body contains two major body cavities: the **dorsal (posterior) cavity** and the **ventral (anterior) cavity.**

A The Dorsal Cavity

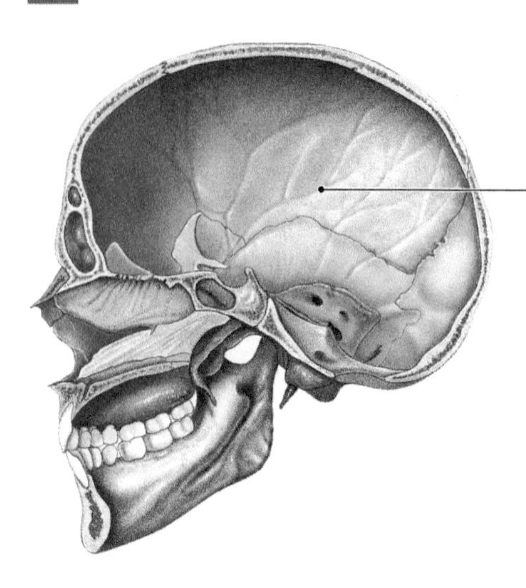

The **dorsal cavity** contains the central nervous system. It has two subdivisions:

1. the **cranial cavity,** which is formed by the cranial bones in the skull.
2. the **vertebral (spinal) cavity,** which is a bony canal formed by consecutive vertebrae in the vertebral column.

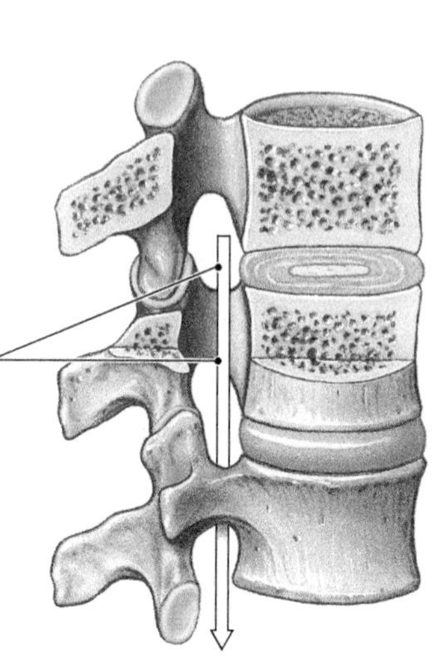

B The Ventral Cavity

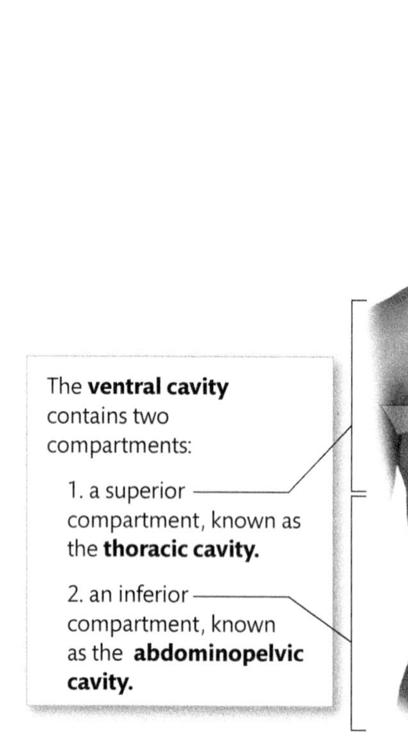

The **ventral cavity** contains two compartments:

1. a superior compartment, known as the **thoracic cavity.**
2. an inferior compartment, known as the **abdominopelvic cavity.**

The **diaphragm** is a muscular partition that separates the thoracic and abdominopelvic cavities.

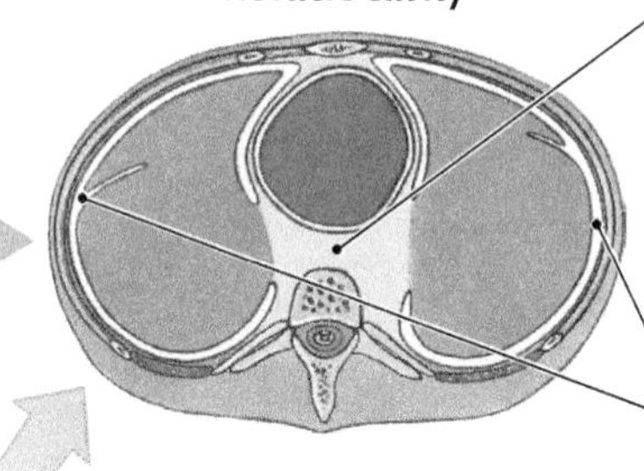

The thoracic cavity has two subdivisons:

1. a centrally located region known as the **mediastinum,** which contains the pericardial cavity surrounding the heart and the thoracic portions of the aorta, trachea, and esophagus.
2. two lateral **pleural cavities,** which surround the lungs.

Abdominopelvic cavity

The abdominopelvic cavity contains organs for digestion, excretion, and reproduction. It has two subdivisions:

1. the superior **abdominal cavity** (light blue).
2. the inferior **pelvic cavity** (dark blue).

The **peritoneal cavity** is the portion of the abdominopelvic cavity that is lined by a membrane called the **peritoneum** (red line).

C Using a torso model, identify the organs associated with each subdivision of the major body cavities and list them in Table 1.4. ▶

TABLE 1.4 The Major Body Cavities

Major Cavity	Subdivisions	Organs
Dorsal (posterior) cavity	Cranial cavity contains	
	Vertebral (spinal) cavity contains	
Ventral (anterior) cavity	Thoracic cavity 1. Pleural cavities surround	
	2. Pericardial cavity surrounds	
	Abdominopelvic cavity 1. Abdominal cavity contains	
	2. Pelvic cavity contains	

D In addition to the major body cavities, a number of smaller functional cavities are located in the head. They include the **orbital cavities** for the eyeballs, the **nasal cavity** and **paranasal sinuses,** and the **oral cavity,** or mouth.

Obtain a human skeleton or skull and identify the orbital, nasal, and oral cavities. On a torso model, identify these same cavities and, if possible, the paranasal sinuses.

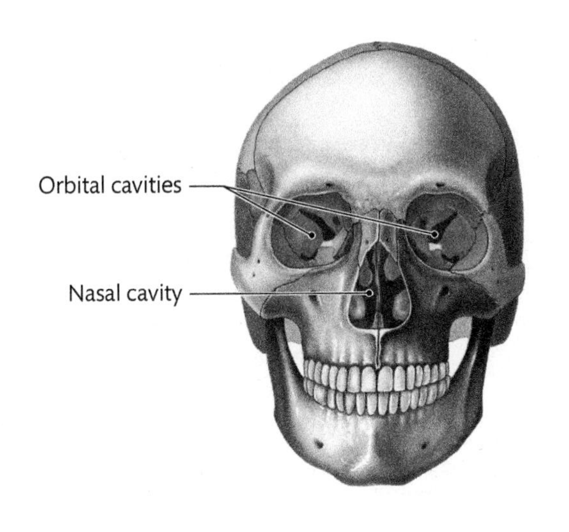

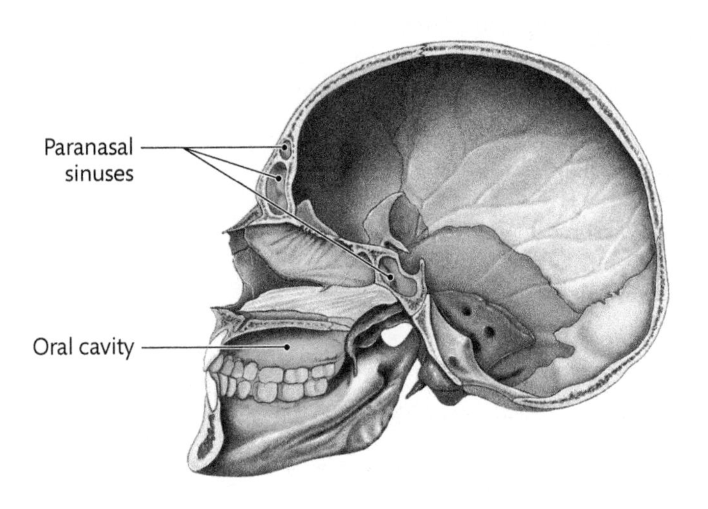

MAKING CONNECTIONS

Examine a skeleton and a human torso model and explain how the skeletal and muscular systems are vital for protecting the organs found within the major body cavities.

▶ ______________________________

____

____

____

____

BEFORE YOU MOVE ON . . .

‹‹ LOOKING BACK

Human **anatomy** is the study of the structure of the human body and the relative relationships among body parts. Human **physiology** is the study of normal function in the human body. To gain a complete understanding of human biology, knowledge of both structure (anatomy) and function (physiology) is essential.

The anatomy and physiology of the human body can be studied from six increasingly complex **levels of organization.** In the exercise just completed, you learned key anatomical terms, identified anatomical regions and body cavities, and studied the internal organs by observing their relative positions to other organs and grouping them into the appropriate organ systems. Your observations were focused on the three highest levels of organization: organ, organ system, and organism. The first in-depth investigation of organs in an organ system will be the integumentary system (Exercise 6). To fully appreciate the unique structure of organs, however, you will first study two lower levels of organization: the cellular level in Exercises 3 and 4 and the tissue level in Exercise 5.

Explain why a solid understanding of the cellular and tissue levels of organization is critical to understanding the structure and function of organs, organ systems, and the complete organism.

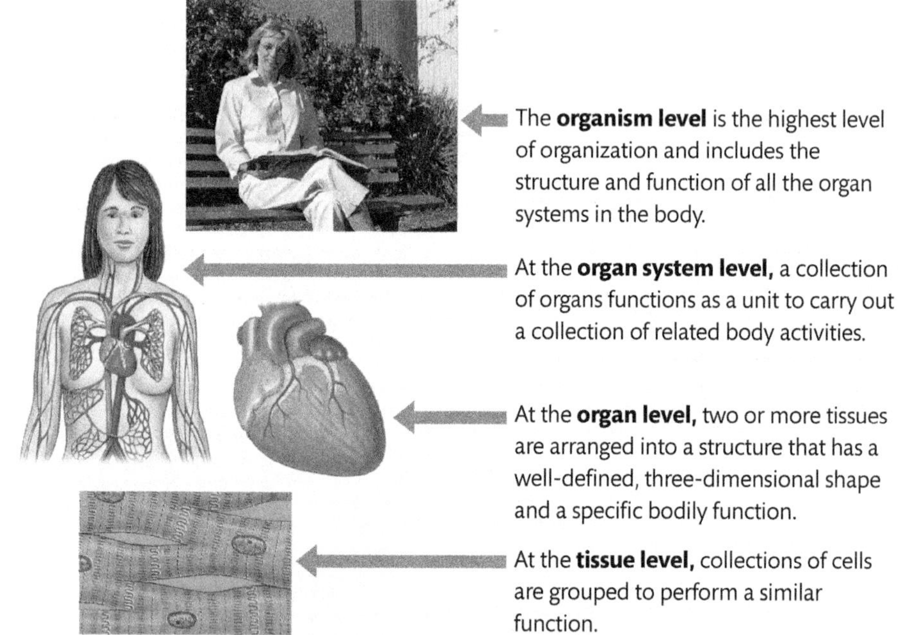

The **organism level** is the highest level of organization and includes the structure and function of all the organ systems in the body.

At the **organ system level,** a collection of organs functions as a unit to carry out a collection of related body activities.

At the **organ level,** two or more tissues are arranged into a structure that has a well-defined, three-dimensional shape and a specific bodily function.

At the **tissue level,** collections of cells are grouped to perform a similar function.

At the **cellular level,** organelles, which are composed of molecules, are organized in a unique way to form cells. The cell represents the fundamental unit of life.

At the **chemical level,** the chemical bonds between atoms give rise to molecules.

LOOKING FORWARD ››

The study of cells and tissues is facilitated by the use of several types of microscopes, including light microscopes, electron microscopes, and fluorescence microscopes. Scientists use microscopes to view images of objects too small for the naked eye to discern. In the following exercise, you will learn how to correctly use a compound light microscope, which employs a series of lenses to bend or refract light waves that pass through an object (e.g., a section of tissue). As a result, the image is magnified, typically 40 to 1000 times.

Name

Lab Section

Date

EXERCISE 1 REVIEW SHEET

Body Organization and Terminology

1. Describe what is meant by the anatomical position. Why is it important to examine the body in this position?

QUESTIONS 2–5: Explain the meaning of the underlined anatomical terms in the following sentences.

2. In the arm, the biceps brachii muscle lies anterior to the brachialis muscle.

3. In the forearm, the radial artery is lateral to the ulnar artery.

4. In females, the uterus is just superior to the urinary bladder; in males, the prostate gland is just inferior to the bladder.

5. In the lower limb, the knee joint is proximal to the ankle joint, but distal to the hip joint.

QUESTIONS 6–8: Identify the body planes that are labeled in the diagram.

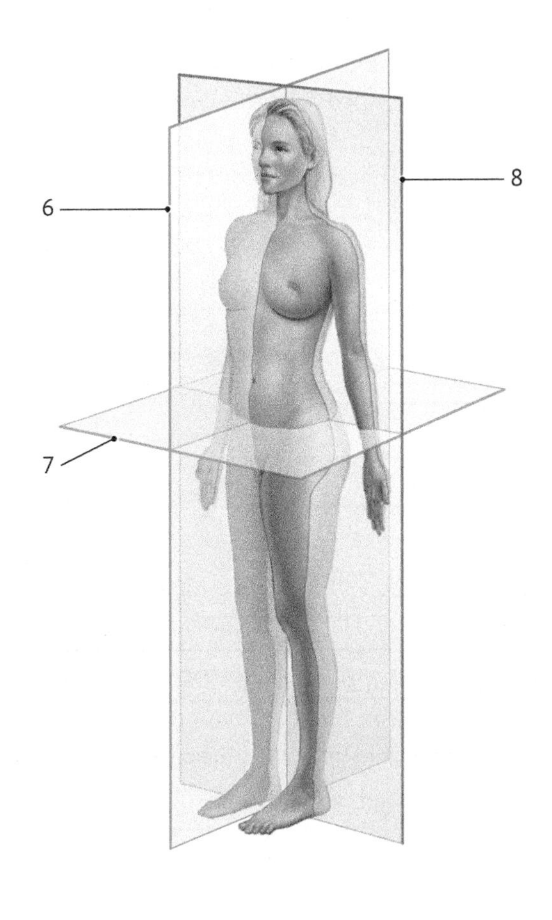

6.

7.

8.

9. What is the difference between a midsagittal plane and a parasagittal plane?

10. Draw dashed lines in the diagram below to demonstrate how longitudinal, cross, and oblique sections would be made. Label the type of section that is made by each line.

QUESTIONS 11–17: Match each organ in column A with the appropriate organ system in column B.

A

11. Trachea ____________

12. Brain ____________

13. Small intestine ____________

14. Prostate gland ____________

15. Kidney ____________

16. Uterus ____________

17. Adrenal (suprarenal) glands ____________

B

a. Urinary system

b. Respiratory system

c. Nervous system

d. Endocrine system

e. Male reproductive system

f. Female reproductive system

g. Digestive system

QUESTIONS 18–23: Match each structure in column A with the body region in which it is found in column B.

A

18. Axillary artery ____________

19. Carpal tunnel ____________

20. Femoral nerve ____________

21. Inguinal canal ____________

22. Tarsal bone ____________

23. Brachial vein ____________

B

a. Arm

b. Ankle

c. Armpit

d. Wrist

e. Groin

f. Thigh

24. Use the following terms to fill in the blank boxes. Some blank boxes have more than one answer.

- Male reproductive system
- Stomach
- Pericardial cavity
- Respiratory system
- Lungs
- Female reproductive system
- Pelvic cavity
- Small intestine
- Cardiovascular system

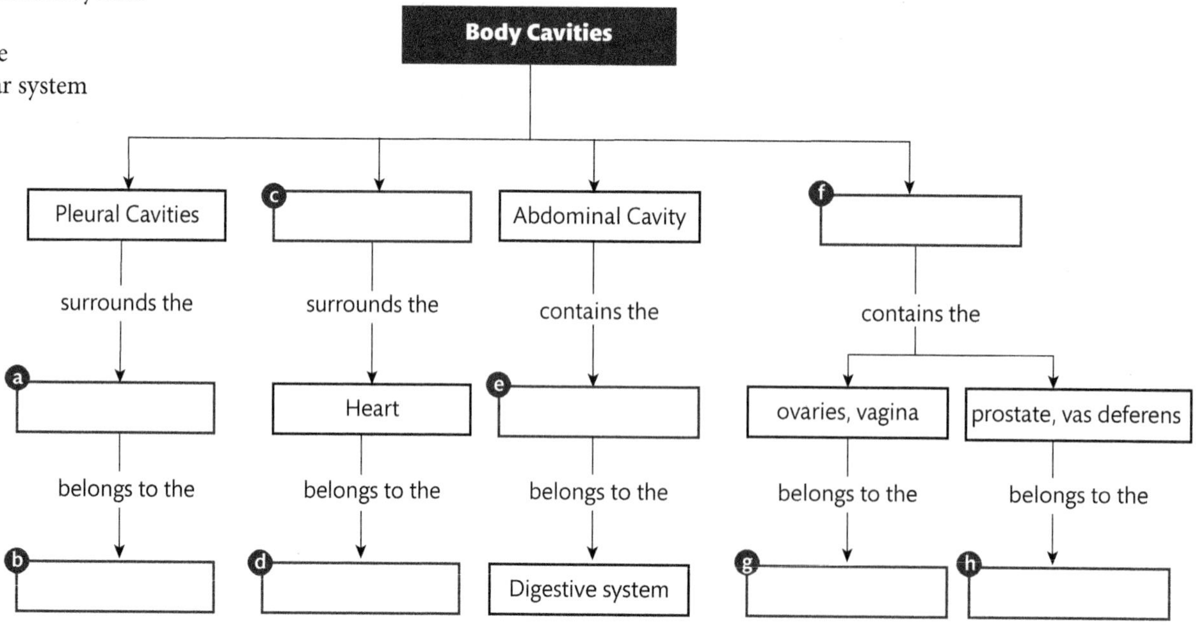

Body Cavities

Pleural Cavities
c
Abdominal Cavity
f

surrounds the
surrounds the
contains the
contains the

a
Heart
e
ovaries, vagina
prostate, vas deferens

belongs to the
belongs to the
belongs to the
belongs to the
belongs to the

b
d
Digestive system
g
h

EXERCISE 2

Care and Use of the Compound Light Microscope

LEARNING OUTCOMES

These Learning Outcomes correspond by number to the laboratory activities in this exercise. When you complete the activities, you should be able to:

Activity 2.1 **Identify the parts of a compound light microscope and explain their functions.**

Activity 2.2 **Demonstrate the proper method for viewing a specimen with the compound microscope.**

Activity 2.3 **Describe the principle of inversion of image.**

Activity 2.4 **Understand the concept of depth of field.**

Activity 2.5 **Measure the diameter of the field of view and estimate the size of structures in a tissue section.**

LABORATORY SUPPLIES

- Compound light microscopes
- Prepared microscope slides of various tissues
- Prepared microscope slides of the letter *e*
- Prepared microscope slides of intersecting colored threads
- Clear millimeter rulers
- Lens paper

PRE-LAB QUIZ

Before you begin, read all the activities in Exercise 2 and the required reading in your textbook that is assigned by your instructor.

1. When carrying a microscope, you should hold it securely with both hands. One hand should be on the ___________, and the other hand should be under the ___________.
   a. arm . . . base
   b. head . . . stage
   c. arm . . . stage
   d. stage . . . base
2. On a compound microscope, where is the light source located?
   a. on the oculars
   b. on the base
   c. on the stage
   d. on the cord
3. True or False: The space between the objective lens and the microscope stage is called the working distance. ___________
4. True or False: When an image is approximately in focus, you can use the coarse adjustment knob to bring it to exact focus. ___________
5. The illuminated area that you view with a microscope is called the ___________.
6. The nosepiece on a microscope is a revolving structure that holds the ___________.
7. The ___________ concentrates the light before it travels through the tissue on the slide.
8. During this laboratory exercise, you will use a prepared slide with the letter *e* to demonstrate
   a. depth of field.
   b. the relationship between total magnification and field diameter.
   c. the working distance.
   d. inversion of image.
9. The thickness of the tissue layer that is in focus is called
   a. resolving power.
   b. image inversion.
   c. depth of field.
   d. field diameter.
10. During this laboratory exercise, you will use a clear millimeter ruler to
    a. estimate the diameter of the field of view.
    b. measure the working distance.
    c. estimate the resolving power of the microscope.
    d. estimate the size of a structure on a tissue section.

Activity 2.1

Learning the Parts of a Light Microscope

The **compound light microscopes** that you will be using to study microscopic anatomy are expensive precision instruments. They can be damaged if they are not handled with prudence and care. When carrying a microscope, hold it firmly in front of you, with one hand on the **arm** and the other hand supporting it under the **base.** Always set it down gently, without sliding it, onto the table.

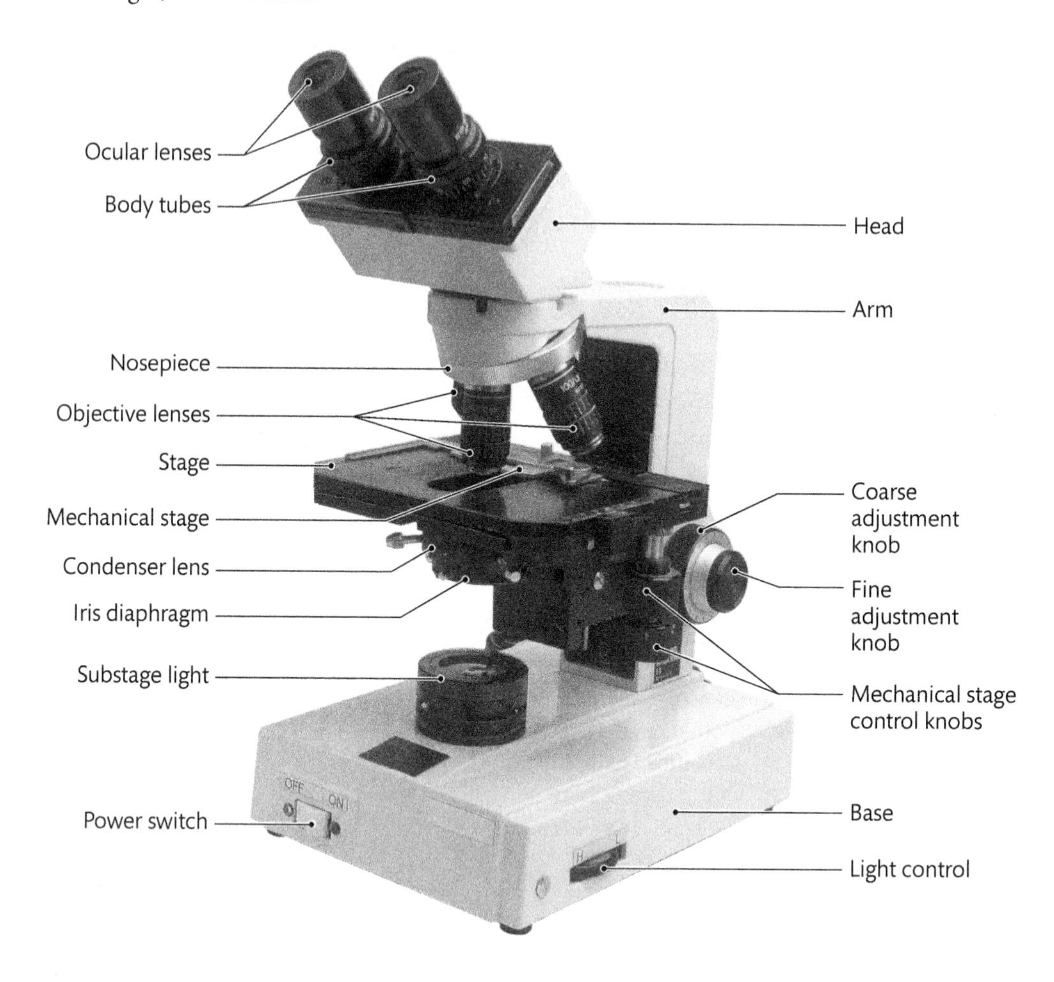

A **simple light microscope** has a magnifying system that uses only one lens, and the **compound light microscope** uses a series of lenses in combination. The microscopes you will be using in the laboratory are **monocular** or **binocular** compound light microscopes. A monocular microscope has only one ocular (eyepiece) lens; thus, only one eye can be used for viewing. A binocular microscope has two ocular lenses, one for each eye.

A Before you use the light microscope for the first time, identify the locations and study the functions of its parts.

1. Plug in the electric cord and switch on the **substage light** built into the base. There may be a light control knob that will allow a range of brightness. If so, always start with a high light intensity and adjust the brightness with the **iris diaphragm.** The lever of the iris diaphragm can be seen under the microscope **stage.** Alternatively, some microscopes have an iris diaphragm composed of a wheel that can be rotated between different-sized openings.

Look through the ocular (eyepiece) lens (or lenses) and move the lever to open and close the diaphragm. Describe what you observe.

▶ ______________________________

____

____

2 The microscope stage is the platform on which a microscope slide is placed. The stage has a central opening or aperture through which light passes to reach the specimen. Typically, the stage is equipped with a clamping device, called a **mechanical stage,** that holds the slide securely and is used to position the object to be viewed directly over the stage aperture. Below the stage, on the side of the microscope, identify the two **mechanical stage control knobs.**

Describe how the mechanical stage moves as you turn each knob.

▶ ____

____

____

____

3 Locate the **condenser lens,** between the stage and the iris diaphragm. After light passes through the aperture (opening) of the iris diaphragm, it travels through the condenser lens. The condenser lens will concentrate the light before it passes through the tissue specimen.

4 Identify the **nosepiece,** located at the base of the **head.** The nosepiece is a revolving structure that holds two to four **objective lenses.** After traveling through the condenser lens and the tissue specimen, light passes through an objective lens and magnifies the image that you see. The magnification of each objective lens, often called the "power," is stamped on its rim. The typical laboratory microscope will be equipped with at least three objective lenses: a low-power (10×) lens, a high-power (40× to 45×) lens, and an oil immersion (100×) lens (× stands for "times"). Some microscopes may also have a scanning lens with a magnification of 4×.

5 Identify the **ocular** or **eyepiece lenses.** As light passes through these lenses, the image is magnified again before it reaches your eyes. For most laboratory microscopes, the ocular lenses are 5×, 10×, or 15×. For the microscopes that you will be using, the magnification is probably 10×. Similar to the objective lenses, you should find the magnification stamped on the rims of the lenses. Some microscopes in your laboratory may be equipped with **pointers.** If present, the pointers are usually attached to the inside of the casing of one ocular lens. Some pointers can be moved by rotating the eyepiece.

6 On each side of the base of the microscope are two knobs, the larger **coarse adjustment knob** and the smaller **fine adjustment knob.** You use these knobs to bring the specimen into focus.

With the lowest-power objective lens in place, describe what you observe when you turn the coarse adjustment knob in a clockwise direction and then in a counterclockwise direction.

▶ ____

____

____

____

____

____

____

____

____

B The **total magnification** of the microscope with a particular objective lens in place can be calculated as follows:

Total Magnification = (Ocular Lens Magnification) × (Objective Lens Magnification)

Suppose that you are viewing a section of lung with a 10× ocular lens and a 40× objective lens. Calculate the total magnification of the section that you are viewing.

▶

IN THE CLINIC

Wearing Eyeglasses While Using a Microscope

Do people who normally wear eyeglasses need to wear them while using a microscope? If you are near- or farsighted, you do not need your eyeglasses because focusing the microscope will correct your vision. However, if you have astigmatism, an image distortion caused by an irregular curvature in the cornea or lens, you should wear your glasses because the microscope cannot make the proper correction.

MAKING CONNECTIONS

The coarse and fine adjustment knobs have similar functions. However, when you turn the fine adjustment knob one complete revolution and observe what occurs, you see no discernible change. Explain why.

▶ ____

____

____

Activity 2.2

Viewing a Specimen with a Compound Microscope

Objects that are magnified by a light microscope will have more clarity of detail only if **resolution** is also increased. Resolution, also referred to as **resolving power,** is the ability to distinguish close objects as separate and distinct. The unaided human eye has a resolving power of about 0.1 millimeter (mm), which means that a person can distinguish two objects that are 0.1 mm apart as distinct entities. If they are less than 0.1 mm apart, they are perceived as being a single object. A good compound microscope can increase resolution to 0.001 mm, so the resolving power is 100 times greater than that of the unaided eye.

A Before you view a slide, the microscope and its objective lens must be in the correct position.

1. Place the microscope on your lab bench so that the ocular lenses are pointing toward you. Switch on the substage light and immediately check the brightness. If the light appears too bright, make an adjustment with the light control knob or the iris diaphragm.

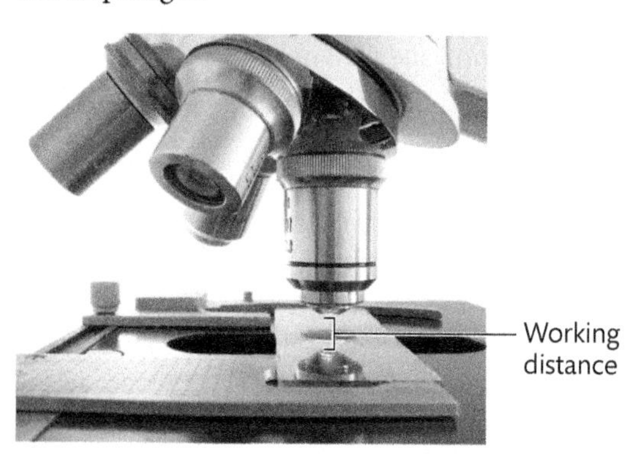

2. Rotate the lowest-power objective lens into position by manually moving the nosepiece until the lens snaps into place. Using the coarse adjustment knob, move the nosepiece up (or move the stage down) to make room for placing a slide in position. The distance between the objective lens and the microscope stage is called the **working distance.** Before you insert or remove a slide, always make sure that the working distance is at its maximum.

3. Clean microscope lenses regularly so that dirt or blurred images will not be confused with cell structures. Always use lens paper for cleaning microscope lenses. Other types of tissues, such as Kimwipes or paper towels, can scratch the lenses. To remove an oily smudge, moisten the lens paper with lens cleaner before cleaning. Avoid touching the lenses with your fingers because oils from your skin can damage the lenses. If you accidentally touch a lens, clean it immediately with lens paper moistened with lens cleaner.

B Obtain a prepared microscope slide of any type of sectioned tissue and place it on the microscope stage.

1. Make sure that the slide is secured in position by the mechanical stage. Adjust the position of the slide with the two mechanical stage control knobs so that the tissue you are about to view is centered over the stage aperture.

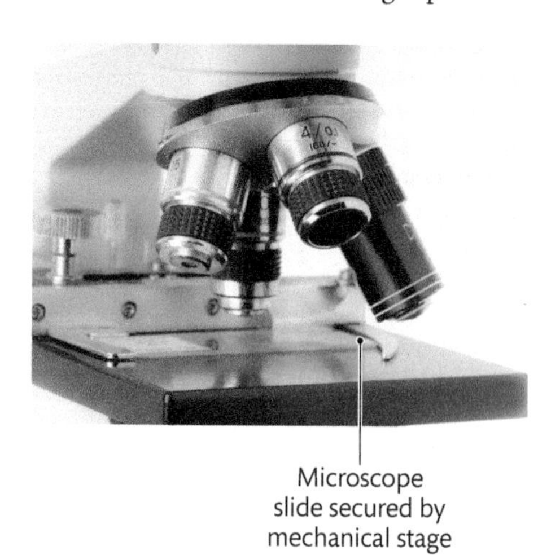

2. Turn the coarse adjustment knob to bring the objective lens and stage as close as possible to each other. Notice that the coarse adjustment knob will stop turning before the objective lens and slide come into contact. That will only occur when a scanning (4×) or low-power (10×) objective lens is in position. Therefore, to guard against damaging a lens or breaking a slide, always begin with the lowest-power objective lens that is on your microscope.

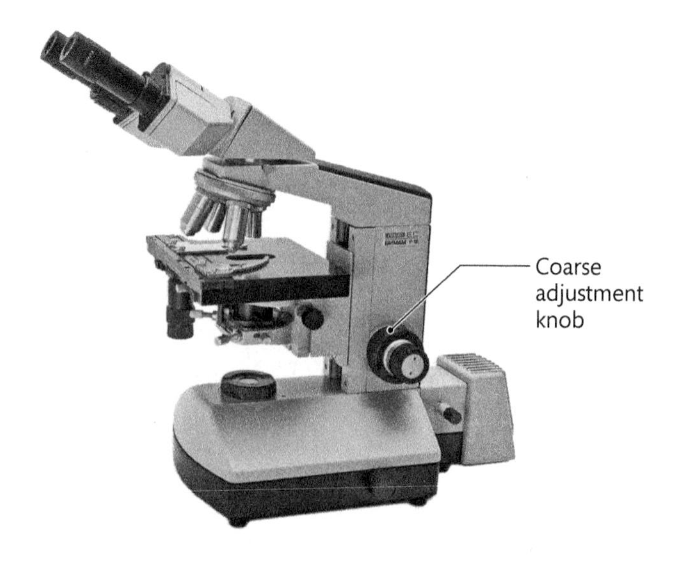

C With a scanning or low-power objective lens and the stage in their correct positions, you can now begin to focus the tissue section.

1 Look into the microscope through the ocular lenses. If you are using a binocular microscope, which has two separate eyepieces, you may have to adjust the ocular lenses by moving them closer together or farther apart. You can do that by turning the adjustment knob, located just below or between the ocular lenses, or by manually moving the lenses closer or farther apart. The adjustment is correct when the images from both eyes fuse into one circular image.

2 The illuminated area that you are viewing is called the **field of view.** Adjust the iris diaphragm so that enough light passes through the tissue section.

3 If your condenser lens is adjustable, keep it just below the stage, as far up as it will go. For most work in the laboratory, the condenser lens should be kept in this position. Experiment with the condenser lens by moving it to various positions under the stage. What changes do you see in the field of view when the condenser lens is moved?

▶ ______________________________

____

4 While looking into the ocular lenses, focus the image by slowly turning the coarse adjustment knob. When the image is approximately in focus, use the fine adjustment knob to bring it to exact focus.

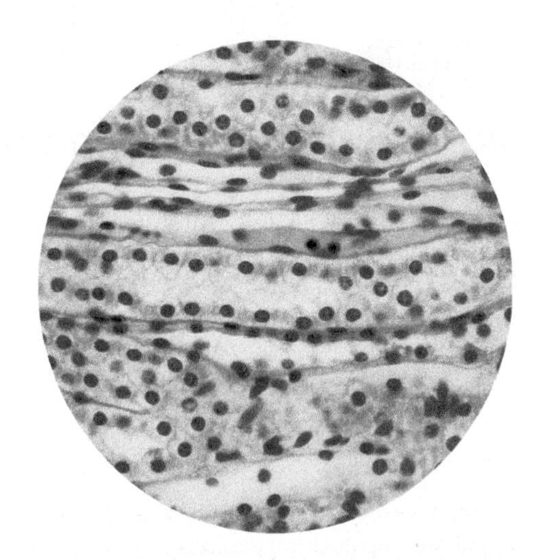

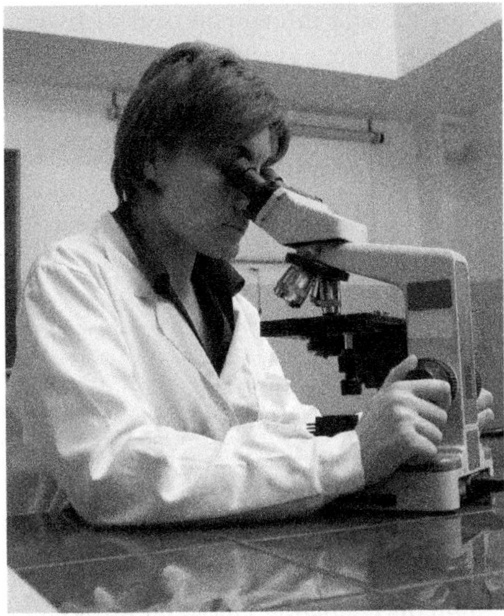

D Select an area on the tissue section that you would like to view at a higher magnification.

1 Move the slide with the mechanical stage control knobs so that the selected area is in the center of the field of view.

2 Rotate the nosepiece so that a higher-magnification (40× to 45×) objective lens is in position. Make sure that the lens will not hit the slide before moving it into position. (The 100× oil immersion lens will not be used regularly for the required microscopic work in this course. If it is used, your laboratory instructor will teach you the correct method for viewing structures with this lens.)

3 Look into the ocular lenses to see if a focusing adjustment must be made. If your microscope is **parfocal,** once the initial focus is made with a low-power objective lens, the image will remain in focus when switching between objective lenses. If your microscope is not parfocal, you will have to make a focusing adjustment with the fine adjustment knob. Do not attempt to focus with the coarse adjustment knob under high power.

4 When you move to a higher magnification, what change occurs in the overall size of the field of view?

▶ ______________________________

____

As you move to a higher magnification, you may have to adjust the iris diaphragm to allow more light to reach the specimen. Explain why.

▶ ______________________________

____

____

____

MAKING CONNECTIONS

During this activity, you were asked to describe any changes in the field of view when switching to a different objective lens. How does a change of objective lens affect the light that passes through the tissue specimen? (*Hint:* Recall the function of the condenser lens.)

▶ ______________________________

____

Activity 2.3

Inversion of Image: Viewing the Letter *e*

A When viewing a tissue section with a microscope, the image that you observe is both **inverted** (turned upside down) and **reversed** (turned from side to side). The following activity demonstrates this principle.

1. With the coarse adjustment knob, maximize the working distance on your microscope.

Place the low-power (10×) objective lens in position.

Place a prepared microscope slide with the letter *e* on the stage in the right-side-up position and centered over the aperture. Do not view the slide through the ocular lens yet. In the space below, draw the letter *e* as viewed with the unaided eye.

▶

2. Position the condenser lens as far up as it will go.

Adjust the iris diaphragm to produce a brightly illuminated field of view.

Focus the letter *e* with the coarse and fine adjustment knobs.

Observe the letter *e* in the field of view. In the space below, draw the *e* as it actually appears under low power.

▶

3. Describe the changes in the orientation of the letter *e* when viewed with a microscope.

▶ ______________________________

____

____

____

____

4. While looking through a microscope, use the mechanical stage control knob to move the slide to the right. In which direction does the *e* move in the field of view?

▶ ______________________________

____

Now use the mechanical stage control knob to move the slide away from you. In which direction does the *e* move in the field of view?

▶ ______________________________

____

MAKING CONNECTIONS

Review the observations that you made in this activity and discuss how they demonstrate the concept of "inversion of image."

▶ ______________________________

____

____

____

____

____

Activity 2.4

Perceiving Depth of Field

Tissue sections on prepared microscope slides are sliced very thinly so that light can pass through them. Despite the translucence of the section, most tissue sections will still have several layers of cells. By careful use of the fine adjustment knob, you can focus on structures located at different levels within the section. Thus, when one layer moves out of focus and becomes blurry, another layer, above or below the first layer, moves into focus and becomes sharp and clear. The term **depth of field** refers to the thickness (depth) of the tissue layer that is currently in focus.

A When you examine prepared slides of various tissues, be aware of the three-dimensional qualities of the structures that you are viewing. By using the fine adjustment knob, you can focus on objects located at different levels within the section.

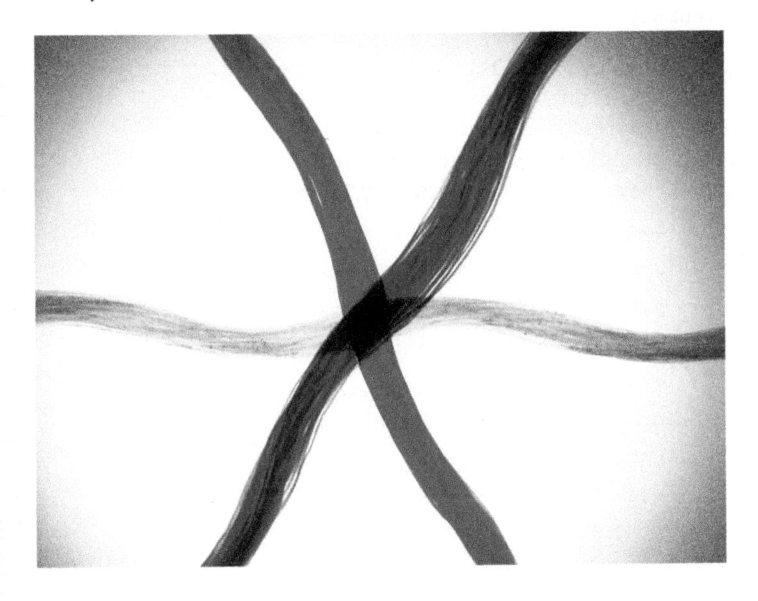

1. Obtain a slide with three different colored threads, all intersecting at a common point.
2. Position the slide so that the intersection point is in the center of the stage aperture.
3. Under low power (4× or 10× objective lens), use the coarse adjustment knob to focus the threads at the point where they all intersect. You should be able to easily identify the different thread colors.
4. Switch to the high-power objective lens (40× to 45×). With the fine adjustment knob, focus on each colored thread separately. Notice that when you are finely focused on one thread, the other two are blurred.
5. After focusing on all three threads, determine the order of colors from top to bottom. ▶
   - Color of top thread ____________
   - Color of middle thread ____________
   - Color of bottom thread ____________

MAKING CONNECTIONS

Speculate on how the depth of field changes (increases or decreases) as the total magnification increases. Explain why.

▶ ____

____

____

____

____

Activity 2.5

Determining the Diameter of the Field of View

When you switch to a higher-power objective lens to increase the total magnification, the diameter of the field of view (field diameter) will decrease proportionately. We can express the relationship between total magnification and field diameter with the following equation:

$$M_1D_1 = M_2D_2$$

where M_1 is the initial total magnification, D_1 is the field diameter at M_1, M_2 is the second total magnification when the objective lens is changed, and D_2 is the field diameter at M_2.

This relationship is useful for predicting the field diameter when the total magnification is increased (or decreased). For example, suppose that at a total magnification of 100×, the diameter of the field of view is 4.0 mm. If you increase the total magnification to 200×, you can estimate the field diameter at the greater magnification as follows:

$$M_1 = 100\times;\ D_1 = 4.0\text{ mm};\ M_2 = 200\times;\ D_2 \text{ is unknown}$$
$$(100\times)(4.0\text{ mm}) = (200\times)(D_2)$$
$$D_2 = (100\times)(4.0\text{ mm})/200\times = 2.0\text{ mm}$$

If the diameter of the field of view is known, you can estimate the size of structures in tissue sections. For example, consider a structure (e.g., a sweat gland in skin) that extends across one-tenth of the diameter of the field of view. If the field diameter is known to be 2.0 mm, the diameter of the object will be one-tenth of 2.0 mm, or 0.2 mm.

A Estimating the Diameter of the Field of View

1. Place a clear millimeter ruler across the aperture on the microscope stage. With the scanning (4×) objective lens in position, focus the ruler lines and make sure that they are crossing the widest portion of the field of view.
2. Align the beginning of a millimeter interval with the left edge of the field of view, as shown in the figure on the left.
3. Estimate the diameter in millimeters (mm).
4. Record the magnification and field diameter in Table 2.1.
5. Repeat steps 1–4 with the low-power (10×) and then the high-power (40× to 45×) objective lenses in position.
6. Record the magnification and field diameter for each lens in Table 2.1. ▶

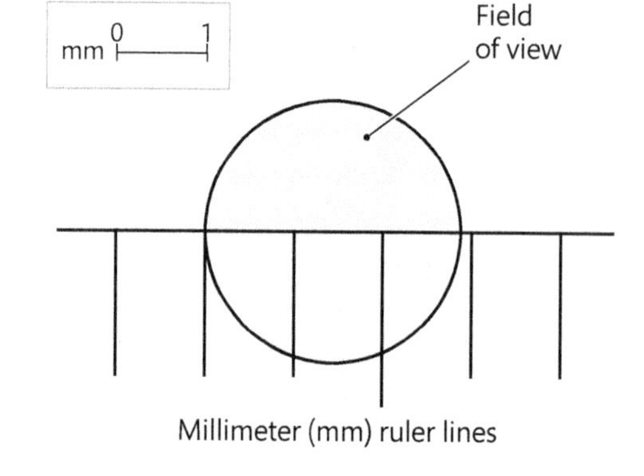

7. How does the diameter change as you increase the magnification of the objective lens?

▶ ______________________________

____

TABLE 2.1 Measuring the Diameter of the Field of View

Type of Objective Lens	Magnification of Objective Lens	Magnification of Ocular Lens	Total Magnification	Diameter of the Field of View (mm)
Scanning				
Low power				
High power				

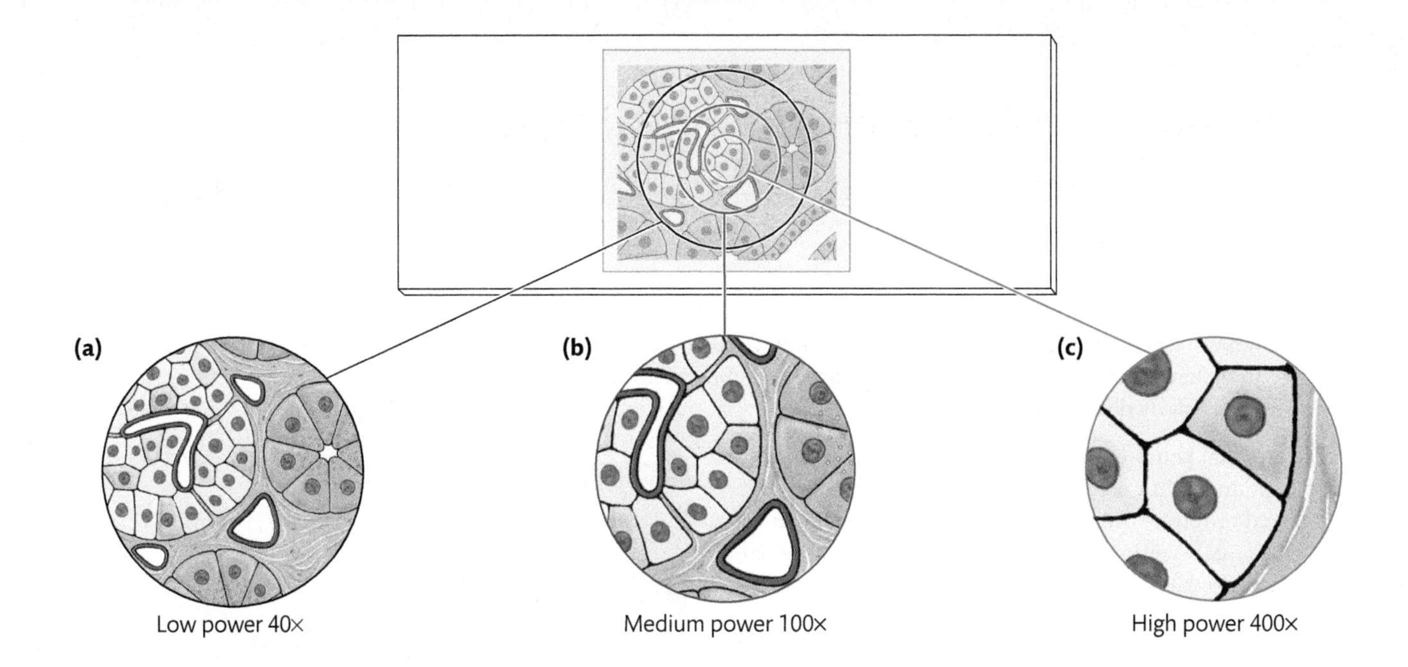

B Estimating the Size of a Structure in the Field of View

1 Place a prepared slide with any type of tissue on the microscope stage and view it with the scanning (4×) objective lens. When you use this objective lens, the total magnification will be 40× as shown in Figure (a), above. Can you explain why?

▶ ______________________________

____

2 Select a specific structure in the tissue that you are viewing. Estimate the proportion of the field diameter across which the structure extends.

3 Use your earlier measurements of the field diameter (Table 2.1) to estimate the structure's size (i.e., diameter of the structure).

Size of structure (mm): ▶ ______________________________

4 Move the slide so that the structure is centered.

5 Switch to the low-power (10×) objective lens and view the structure at a higher magnification. The total magnification is now 100×, as shown in Figure (b), above.

6 Once again, estimate the proportion of the field diameter across which the structure extends.

7 Predict the size of the structure by using the data recorded in Table 2.1. How does this calculation compare with your earlier estimate of structure size at the lower magnification?

▶ ______________________________

____

____

8 View the structure with the high-power (40× or 45×) objective lens so that the total magnification is 400× or 450×, as shown in Figure (c), above, and repeat the previous steps to predict the size of the structure.

How does this calculation compare with your first two estimates of structure size with the scanning and low-power objective lenses?

▶ ______________________________

____

____

MAKING CONNECTIONS

Review your results in Table 2.1. Test the relationship between total magnification and field diameter by inserting your data into the equation $M_1D_1 = M_2D_2$. Do your results support the relationship? Explain.

▶ ______________________________

____

____

____

BEFORE YOU MOVE ON . . .

❮❮ LOOKING BACK

A microscope allows scientists to view tissue and cellular structures that are too small for examination with the naked eye. Clinically, microscopes have become essential diagnostic tools and are used regularly during surgical procedures.

The microscopes that you used in this exercise are monocular or binocular compound light microscopes. The optical lenses in a light microscope bend or refract the light waves that pass through an observed object (e.g., a section of tissue). As a result, the image of the object is magnified, typically 40 to 1000 times.

The light microscope is one of the most important tools you will use to study anatomy and physiology. Speculate on why viewing structures microscopically will improve your knowledge and understanding of anatomy and physiology. Identify some examples that support your speculation.

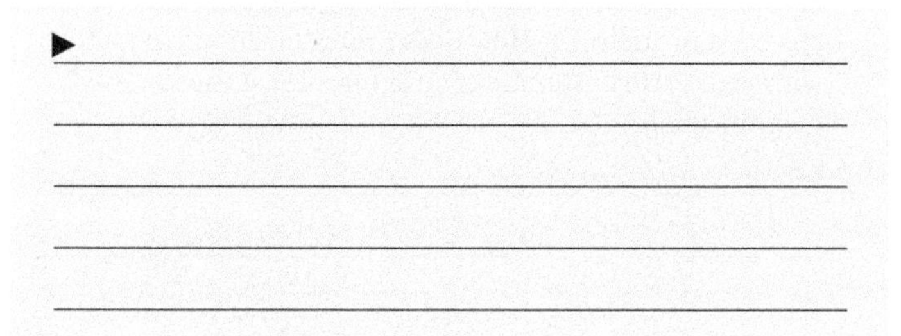

LOOKING FORWARD ❯❯

Microscopes were pivotal in aiding scientists in the discovery of the cell. The term **cell** was first used in 1665 when Robert Hooke was examining a thin slice of cork for a demonstration at the Royal Society of London. He observed evenly spaced rows of boxes that reminded him of the "cells," or living quarters, for monks in a monastery. Hooke did not actually observe living cells because the cork tissue was dead. All that remained was the cell wall, a structure found in plant cells and bacteria. With the advent of the first microscope in 1673, Anton van Leeuwenhoek was the first to identify living cells. Significant advances in cell biology did not occur until the first part of the 19th century, however, when microscopes with stronger magnification and resolving power were developed. In the next exercise, you will use a compound light microscope to examine cell structure and to make comparative observations of different cell types.

Name ________________________

Lab Section ________________________

Date ________________________

EXERCISE **2** **REVIEW SHEET**

Care and Use of the Compound Light Microscope

QUESTIONS 1–8: Identify the labeled structures on the microscope shown below. Select your answers from the following list and write a function for each.

a. Mechanical stage
b. Objective lenses
c. Nosepiece
d. Microscope head
e. Ocular lenses
f. Coarse and fine adjustment knobs
g. Iris diaphragm
h. Substage light

1. ________________________

2. ________________________

3. ________________________

4. ________________________

5. ________________________

6. ________________________

7. ________________________

8. ________________________

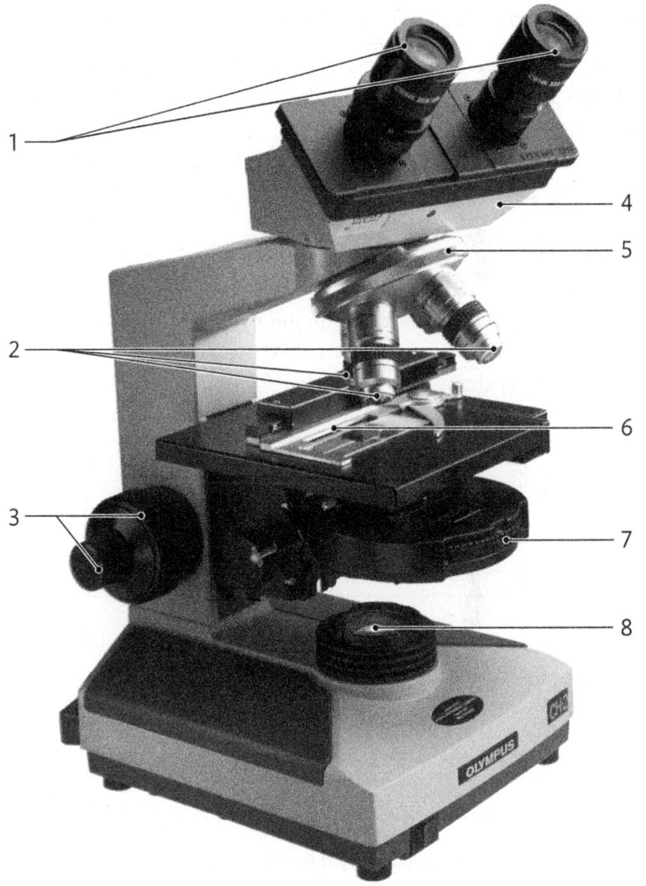

9. What advantage is there to using a microscope that is parfocal?

____

QUESTIONS 10–12: Define the following terms.

10. Resolving power:

____

____

11. Working distance:

____

____

12. Field of view:

____

____

13. Make a comparison of viewing a microscope slide under low magnification and high magnification by identifying benefits and drawbacks for each.

____

____

____

____

____

14. Before placing a slide on the microscope stage, you observe the tissue section with the naked eye and notice that it has the following shape:

Next, you view the slide under low magnification so that the entire tissue section is visible in the field of view. In the space below, draw the tissue section as it will appear under your microscope.

15. Explain what is meant by depth of field.

____

____

16. While viewing a microscope slide with the scanning objective lens, the total magnification is 40× and the diameter of the field of view is 5.0 mm. Switching to the low-power objective lens will increase the total magnification to 100×. Calculate the diameter of the field of view at this higher magnification.

EXERCISE 3

Cell Structure and Cell Division

LEARNING OUTCOMES

These Learning Outcomes correspond by number to the laboratory activities in this exercise. When you complete the activities, you should be able to:

Activity 3.1 **Describe the structure and function of the nucleus and major organelles in a eukaryotic cell.**

Activity 3.2 **Prepare a wet mount of cells derived from your own cheek.**

Activity 3.3 **Compare and contrast light microscopic and electron microscopic observations of cell structure.**

Activity 3.4 **Identify and describe the stages of mitosis.**

LABORATORY SUPPLIES

- Eukaryotic cell model
- Compound light microscopes
- Microscope slides
- Coverslips
- Toothpicks
- 10% methylene blue stain
- Prepared microscope slides of cheek cells
- Prepared microscope slides of various structures
- Electron micrographs of various cell structures
- Prepared slides of whitefish blastula
- Coloring pencils

PRE-LAB QUIZ

Before you begin, read all the activities in Exercise 3 and the required reading in your textbook that is assigned by your instructor.

1. During this laboratory exercise, you will be preparing a wet mount of cells collected from what part of your body?
   a. skin from your arm
   b. hair from your scalp
   c. nail from your finger
   d. inner lining of your cheek
2. During this laboratory exercise, you will be comparing your observations of cell structure using what two types of microscopic techniques?
   a. light microscopy and fluorescence microscopy
   b. light microscopy and electron microscopy
   c. light microscopy and phase contrast microscopy
   d. electron microscopy and fluorescence microscopy
3. The lipid bilayer of a cell membrane is composed of what type of molecules?
   a. proteins
   b. carbohydrates
   c. phospholipids
   d. fatty acids
4. True or False: During the cell cycle, cytokinesis is the period when the cell's cytoplasm divides into two parts. ___________
5. What organelle contains enzymes that can destroy harmful bacteria and viruses?

   ____
6. True or False: The Golgi apparatus produces most of the ATP in a cell. ___________
7. True or False: Chromatid pairs separate during the metaphase stage of mitosis.

   ____
8. Chromatin is composed of what two molecules? ___________________
9. True or False: When you switch to a higher power objective lens to increase the total magnification, the diameter of the field of view also increases. ___________
10. True or False: Whenever body fluids are handled in the laboratory, universal precautions should be followed. ___________

Activity 3.1

Examining Cell Structure

A In your study of anatomy and physiology, you will focus on the structure and function of **eukaryotic cells** because they are the cells found in the human body. Typically, the largest structure inside a eukaryotic cell is the **nucleus,** which contains the cell's DNA and directs all cellular activities.

The **cytoplasm** is the gel-like cell matrix located outside the nucleus. It consists of a fluid portion, the **cytosol,** and various structures called **organelles.** The organelles are divided into two categories: membranous organelles and nonmembranous organelles. **Membranous organelles** are surrounded by a phospholipid membrane, similar in structure to the plasma (cell) membrane. These membrane-bound structures have internal compartments with chemical environments that may be different from the surrounding cytosol. **Nonmembranous organelles** are not surrounded by a membrane and, thus, are in direct contact with the cytosol.

Peroxisomes contain enzymes that neutralize toxins produced by cellular metabolism or are taken in from the outside; they also break down fatty acids.

Lysosomes are filled with digestive (hydrolytic) enzymes. They digest old, worn-out organelles and destroy harmful bacteria, viruses, and toxins.

Centrioles are paired cylindrical structures, arranged at right angles to one another and located at one end of the nucleus in an area of the cytoplasm known as the centrosome. They produce microtubules for the cytoskeleton, form the basal bodies from which the cilia and flagella are produced, and form the mitotic spindle for mitosis.

Microvilli are tiny, fingerlike extensions that increase the surface area along the cell membrane of cells that absorb substances. They contain bundles of microfilaments that are anchored to the terminal web, a filamentous band that runs just below the cell surface.

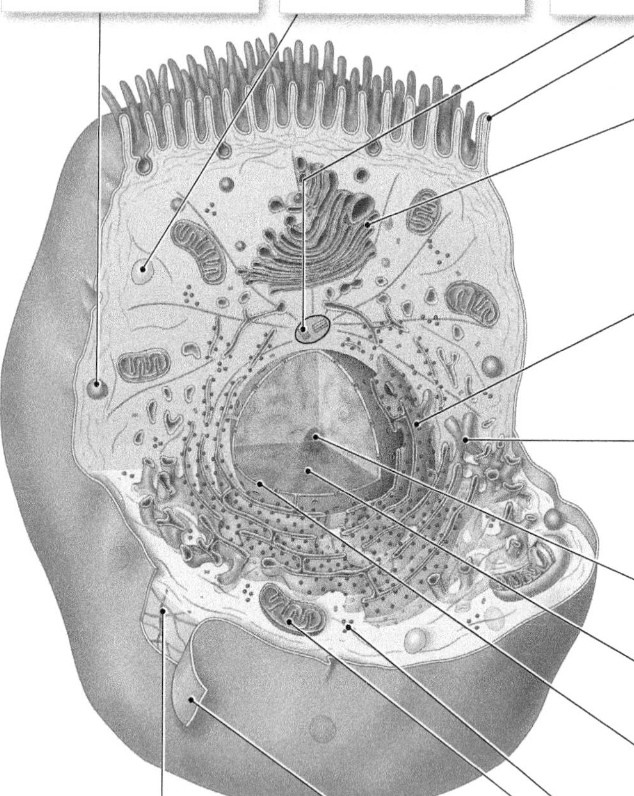

The **Golgi apparatus** is a series of flattened membranous sacs, resembling a stack of pancakes. This organelle modifies and packages proteins that are produced by the rough endoplasmic reticulum.

The **endoplasmic reticulum (ER)** is a highly folded membranous structure that encloses a network of fluid-filled cavities.

The **rough endoplasmic reticulum (RER)** is studded with ribosomes (fixed ribosomes), the sites of protein synthesis. It produces proteins that are secreted by the cell, incorporated into the plasma membrane, or used by lysosomes.

The **smooth endoplasmic reticulum (SER),** which lacks ribosomes, synthesizes lipids and glycogen and detoxifies poisons and various drugs.

The nucleus

The **nucleolus** is suspended in a gel-like matrix called the nucleoplasm. It produces ribosomes.

Chromatin is composed of complex molecules of DNA and protein. DNA is the genetic material of the cell.

The **nuclear envelope** is a double membrane that separates nucleoplasm from cytoplasm. It is dotted with numerous nuclear pores that allow the passage of various substances into and out of the nucleus.

Free ribosomes produce proteins used for cellular metabolism.

The **cytoskeleton** consists of three types of protein filaments: microtubules, intermediate filaments, and microfilaments. They provide strength and flexibility for the cell and support for the various other organelles.

The **plasma (cell) membrane** is composed of phospholipids, proteins, and carbohydrates and acts as a selectively permeable barrier between the cell and its external environment.

Mitochondria are bean-shaped structures surrounded by a double membrane. The inner membrane, which contains numerous inward folds, encloses a gel-like material called the **matrix.** Mitochondria produce most of the cell's energy in the form of ATP.

B Study a model of a typical eukaryotic animal cell or the diagram on the previous page.

1. Identify the important cell structures and become familiar with the basic structure and function of each.

2. In the table below, identify the organelle that matches each example. ▶

> **Word Origins**
>
> The Greek term *karyon* refers to a nucleus, and the prefix *eu-* means "good." A eukaryotic cell, therefore, has a true, well-defined nucleus.

Cell Structure	Example
	White blood cells that fight infections use these organelles to destroy harmful bacteria and viruses.
	Liver cells use this organelle to synthesize blood (plasma) proteins.
	Cells in the small intestine, which absorb nutrients, have these structures to increase surface area along the plasma (cell) membrane.
	Cells that divide regularly, such as the cells in the epidermis of the skin, use these organelles to produce the mitotic spindle prior to cell division.
	Muscle cells, which expend a large amount of energy when they contract, have a large number of these organelles to produce ATP.

C The basic structural component of the plasma (cell) membrane is the **phospholipid bilayer,** which is composed of two layers of phospholipid molecules. At normal body temperature, the membrane has a fluid nature and, therefore, is very flexible. **Cholesterol,** a second lipid component, is interspersed between phospholipids in both layers of the membrane. Cholesterol provides some degree of stability to the phospholipid bilayer structure.

1. Explain why the phospholipid molecules are arranged as a bilayer.

▶ ____________________________________

____

____

2. The plasma membrane has **integral proteins** that are firmly inserted into the lipid bilayer and **peripheral proteins** that rest loosely on the inner surface. In the diagram on the right, proteins are shown in blue. Notice that two integral proteins have channels passing through them. Based on this structural feature, speculate on the function of these proteins.

▶ ____________________________________

____

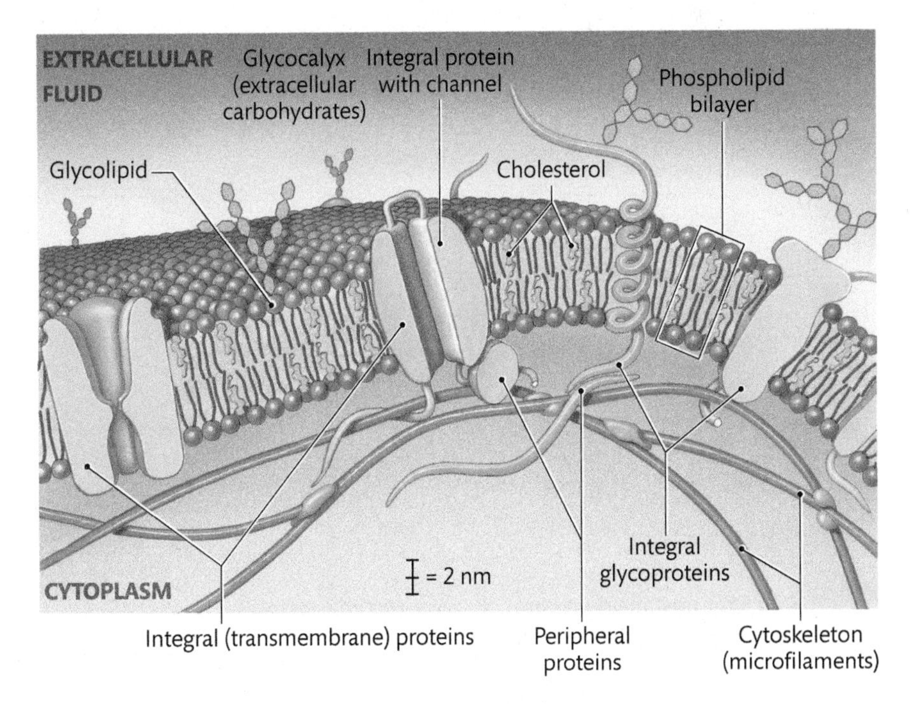

MAKING CONNECTIONS

Why is it important for the plasma membrane to be flexible? What do you think would happen if it became too rigid?

▶ ____

____

____

____

Activity 3.2

Preparing a Wet Mount of Cheek Epithelial Cells

⚠ **Warning:** *Saliva is a body fluid that may contain infectious organisms. Therefore, universal precautions will be followed while you are preparing a wet mount of your cheek cells. Before you begin, carefully review universal precautions, which are described in Appendix A on page A-1 of this lab manual.*

A Observing a Wet Mount of Unstained Cells

1. Gently scrape the inner lining of your cheek with the broad end of a flat toothpick.
2. Stir the toothpick vigorously in a drop of distilled water on a clean microscope slide.
3. Using a new toothpick each time, gently scrape your cheek two or three more times and stir into the water drop.
4. Cover the drop with a coverslip lowered onto the slide at an angle to avoid forming air bubbles.
5. Observe with a light microscope, first on low power and then on high power, and sketch what you observe in the space below. Identify the nucleus, cytoplasm, and plasma (cell) membrane.

▶

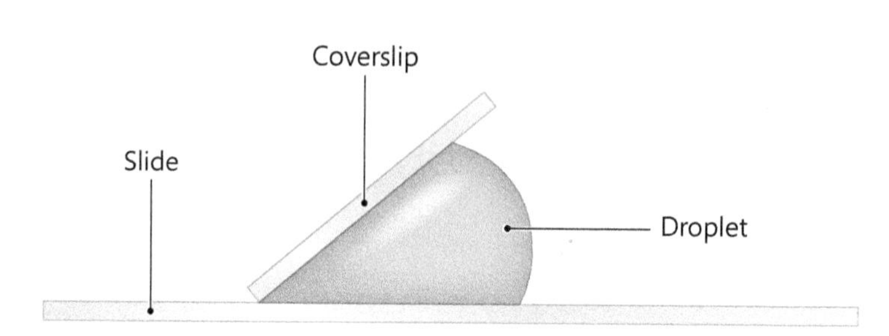

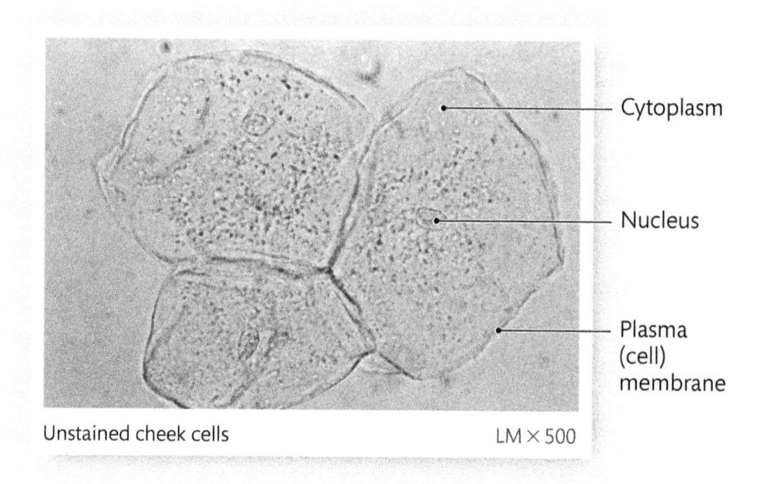

Unstained cheek cells LM × 500

6. In your unstained preparation, were you able to easily identify structures in the cells? Explain.

▶ ______________________________

____

____

____

____

____

____

IN THE CLINIC

Collecting DNA from Cheek Cells

A **buccal swab (buccal smear)** is a technique to collect DNA from the cells that line the mucous membrane of a person's cheek. The buccal swab collection method is more convenient, simpler, and less expensive than blood sampling, and it is relatively noninvasive. To collect the cells for a buccal swab, a sterile swab with a cotton or foam tip is gently rubbed along the inside of the cheek. The sample is then placed into a sterile transport tube or collection envelope and delivered to a testing laboratory. DNA samples collected from cheek cells can be used for forensic investigations at crime scenes, for paternity testing, and for genetic analysis for detecting inherited diseases.

B Observing Cells Treated with a Biological Dye

1. Repeat steps 1 through 3 on the previous page. Before adding the coverslip, stain your preparation with a small drop of 10 percent methylene blue. Be careful not to overstain the slide.

2. Cover the preparation with a coverslip and observe the cells with the microscope. A good preparation will have a faint blue color. If your slide appears dark blue, you have overstained the cells, and it will be difficult to identify structures. If that is the case, you should make a new preparation.

3. Observe the stained cells with a microscope using the low-power objective lens. Switch to the high-power objective lens and sketch what you observe in the space below. Identify the following structures:
   a. Nucleus
   b. Nucleolus
   c. Chromatin
   d. Nuclear membrane
   e. Cytoplasm
   f. Plasma membrane

4. What effect does methylene blue have on the appearance of the cells in your preparation?

▶ ______________________________

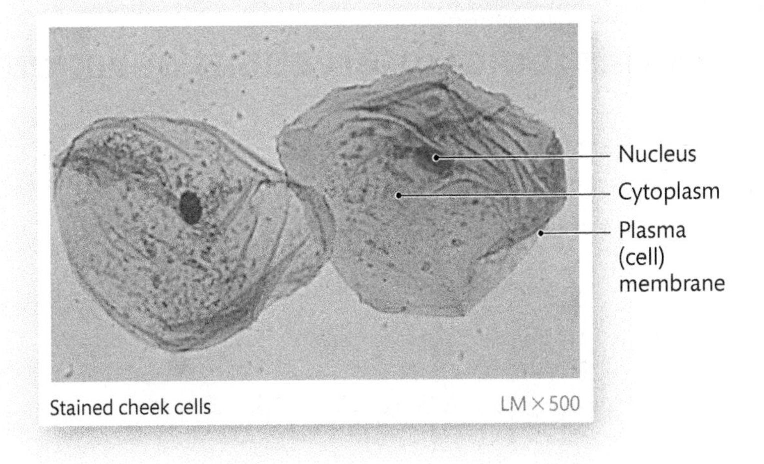

Stained cheek cells LM × 500

5. Compare the observations of your own cheek cells with a prepared slide of the same cell type or with the figure above.

▶ ______________________________

MAKING CONNECTIONS

For a forensic DNA analysis, why is a cheek cell scrape a good source for obtaining a DNA sample from a crime suspect?

▶ ______________________________

Activity 3.3

Microscopic Observations of Various Cell Types

A Light Microscopy

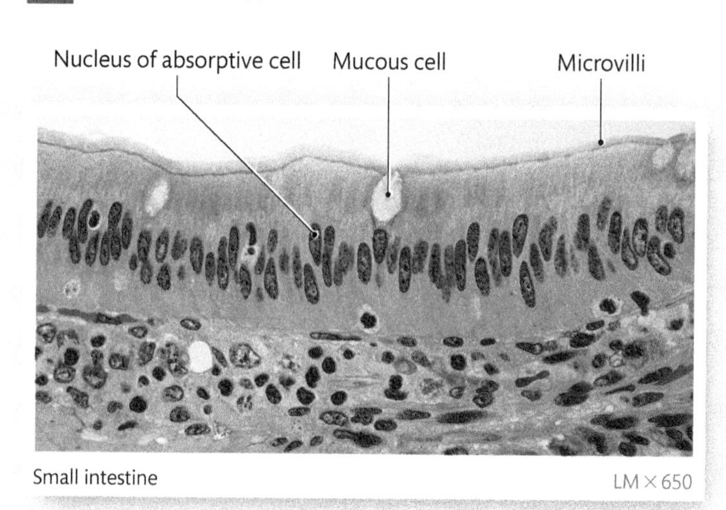

Small intestine LM × 650

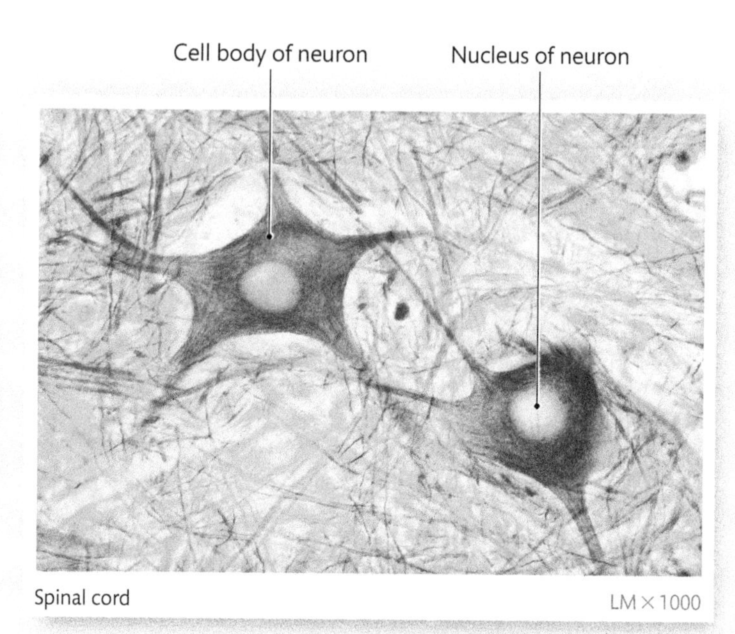

Spinal cord LM × 1000

1 Examine the structure of various cell types by observing several prepared microscope slides. In the spaces below, sketch three different cell types and identify any structures that you recognize. For each drawing, note the tissue or organ observed and the total magnification.

▶ **Cell Type 1**

Tissue/organ: ____________________

Total magnification: ______________

▶ **Cell Type 2**

Tissue/organ: ____________________

Total magnification: ______________

▶ **Cell Type 3**

Tissue/organ: ____________________

Total magnification: ______________

2 Identify one unique characteristic for each cell that you observed. ▶

	Unique Characteristic
Cell Type 1	
Cell Type 2	
Cell Type 3	

B Electron Microscopy

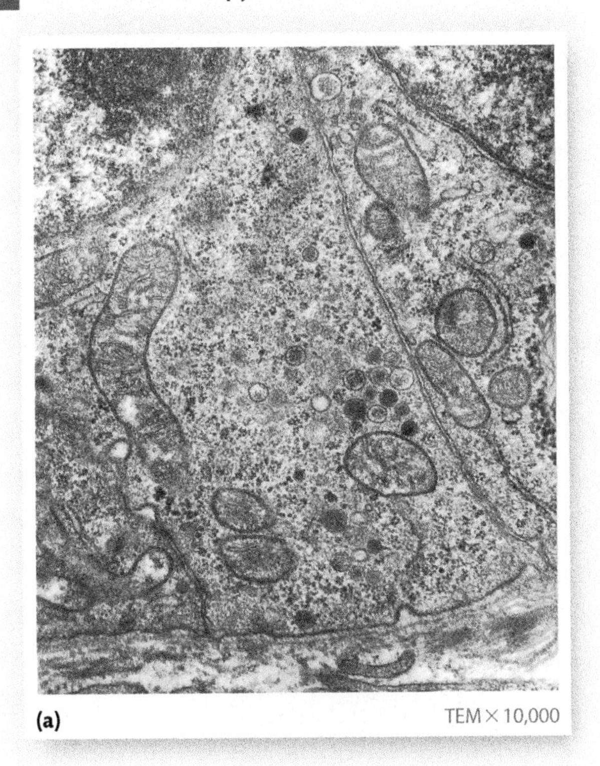

(a) TEM × 10,000

1 The three electron micrographs shown on this page illustrate various organelles found in endocrine cells located in the lung. In the table below, list the organelles that you can identify in each micrograph. ▶

	Organelles
(a)	
(b)	
(c)	

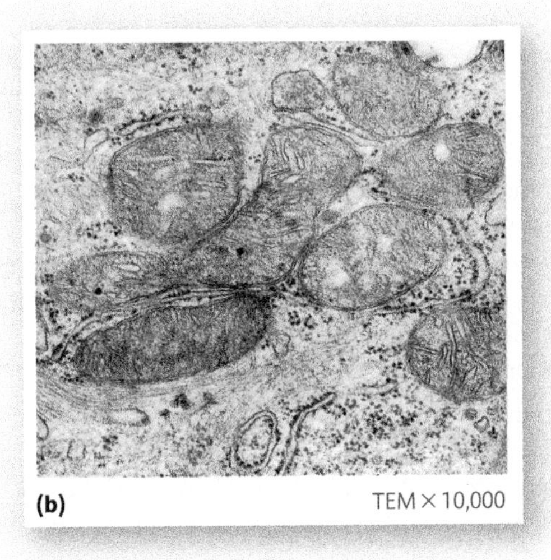

(b) TEM × 10,000

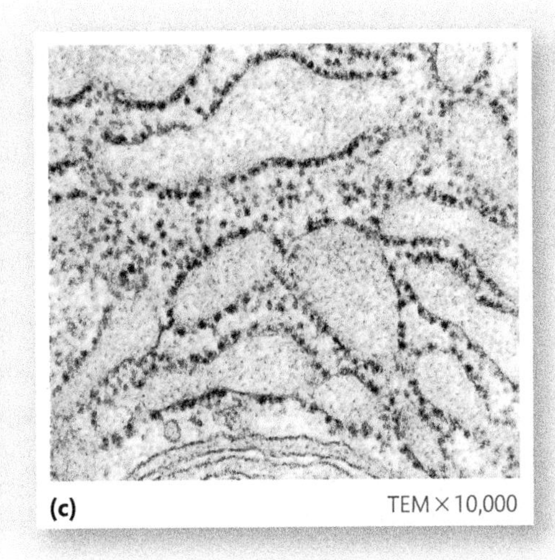

(c) TEM × 10,000

2 Name the two cell structures that are surrounded by a double phospholipid bilayer (a double membrane). ▶

1. ____________________
2. ____________________

3 Examine other electron micrographs of cells that are available in the laboratory and identify the organelles that you observe.

In the spaces below, note the difference in total magnification between these micrographs and your earlier observations with the light microscope. ▶

	Total Magnification	
	Minimum	Maximum
Light microscope		
Electron microscope		

MAKING CONNECTIONS

Compare the amount of detail that you observed through the light microscope and electron microscope. Which structures could you examine in greater depth with the electron microscope?

▶ ____

____

____

____

Activity 3.4

Examining the Stages of Mitosis

A The regularly recurring series of life processes performed by any cell is called the **cell cycle.** The cell cycle is divided into two time periods (see the figure below):

1. **Interphase** is the period when the cell is not dividing. The cell conducts its normal metabolic activities and prepares itself for cell division.
2. **Cell division** is the period during which identical copies of the original cell are produced. Cell division includes two processes:
   - *Mitosis:* the division of a cell's nuclear material
   - *Cytokinesis:* the division of cytoplasmic components

Cell division is analogous to photocopying in that cells divide to produce genetically identical copies of themselves. For one-celled organisms, cell division is a form of reproduction, allowing a population to replace its members and increase its numbers. In multicellular organisms, such as humans, cell division provides growth, wound healing, and remodeling. In the embryo, cell division is the primary mechanism for developing new structures and growth of the individual.

INTERPHASE
S
DNA replication
G_2
Protein synthesis
G_1
Normal cell functions
THE CELL CYCLE
Prophase
Metaphase
Anaphase
Telophase
Cytokinesis
MITOSIS AND CYTOKINESIS
Start
G_0
(nondividing cells)

B Examine a prepared slide that illustrates cell division in the whitefish **blastula.** A blastula is an early embryological stage of the vertebrate embryo. It is a rapidly growing structure, so dividing cells will be abundant. You should be able to identify cells in interphase, the four stages of mitosis, and cytokinesis.

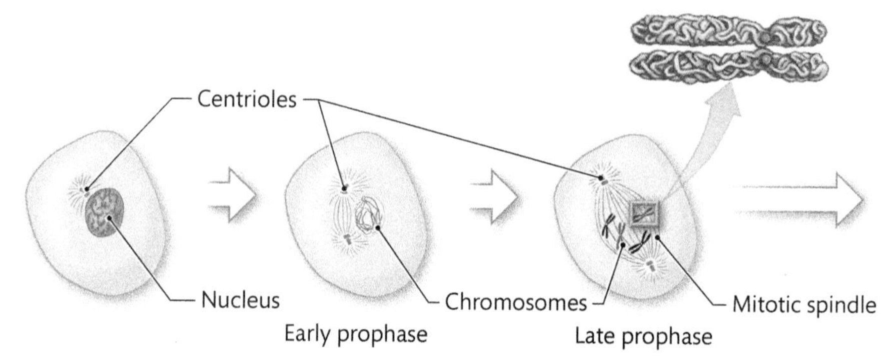

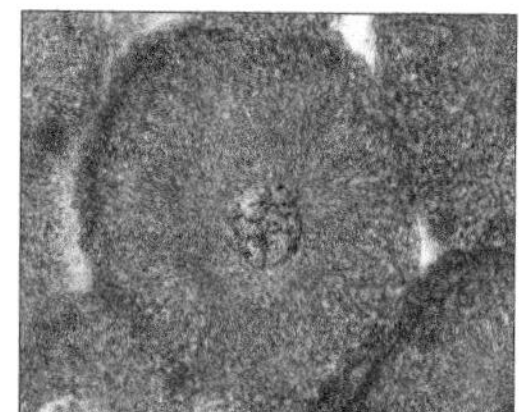
Interphase

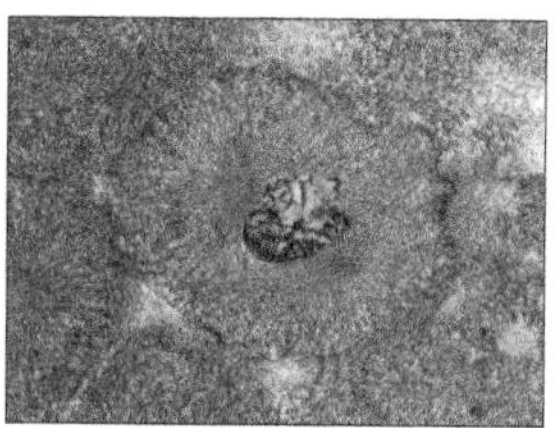
Prophase

1 Identify a cell during interphase. Notice that the chromatin in the nucleus is uncoiled and that chromosomes cannot be identified.

2 During **early prophase,** notice that the chromatin in the nucleus coils and condenses, forming **chromosomes.** Because the DNA has replicated, each chromosome consists of two identical **chromatids,** held together at the **centromere.** At this time, the centrioles begin to separate and produce the microtubules, which will form the **mitotic spindle.** The mitotic spindle will be used to separate the chromatids in each chromosome.

3 During **late prophase,** notice that the nuclear membrane has broken down. The centrioles continue to separate by migrating to opposite ends of the cell. Identify the mitotic spindle, which is now complete.

IN THE CLINIC

When Cells Become Cancerous

Local and genetic controls usually regulate cell division and maintain healthy rates of cell growth and reproduction. **Cancer** refers to a large number of disorders in which mutations disrupt the normal controls of cell division, producing increasingly abnormal cells with unrestricted growth. Typically, cancerous cells do not fully mature, and they stop performing their normal functions. These cells produce **malignant tumors,** and over time, they replace healthy cells and prevent organs from functioning properly. Some cancerous cells may break away or **metastasize** from the primary tumor and spread, often through the bloodstream, to other locations in the body.

4 Locate a cell during **metaphase** and identify the **metaphase plate,** which consists of the chromosomes lined up end to end along the equator of the cell. Some microtubules in the mitotic spindle are attached to **kinetochores,** which are DNA-protein complexes on the centromere of each chromosome.

5 Identify a cell during **anaphase** and notice that the chromatid pairs have separated. The chromatid pairs are being pulled to opposite ends of the cell as the mitotic spindle shortens. Thus, the duplicated chromosomes of the original cell have divided into two identical sets of chromosomes.

6 During **telophase,** notice that a nuclear membrane forms around each new set of chromosomes as nuclei re-form. The chromosomes unravel to form fine threads of chromatin, and the mitotic spindle disappears.

7 During **cytokinesis,** the cytoplasm divides into two approximately equal parts. Notice that two new cells have formed.

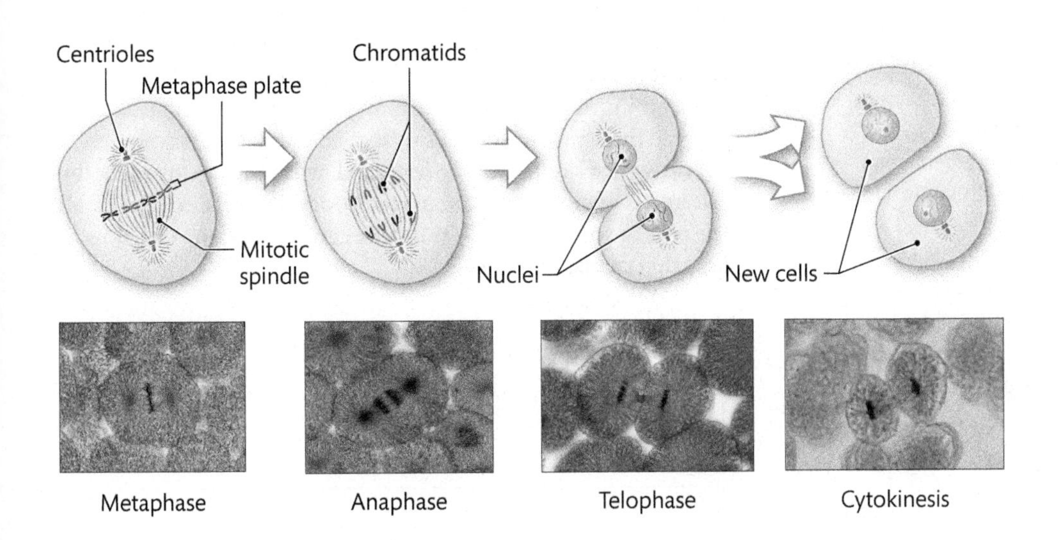

C Sketch what you see in the microscope and label structures (if present) such as chromosomes, mitotic spindles, nuclear membrane, and cell membrane. ▶

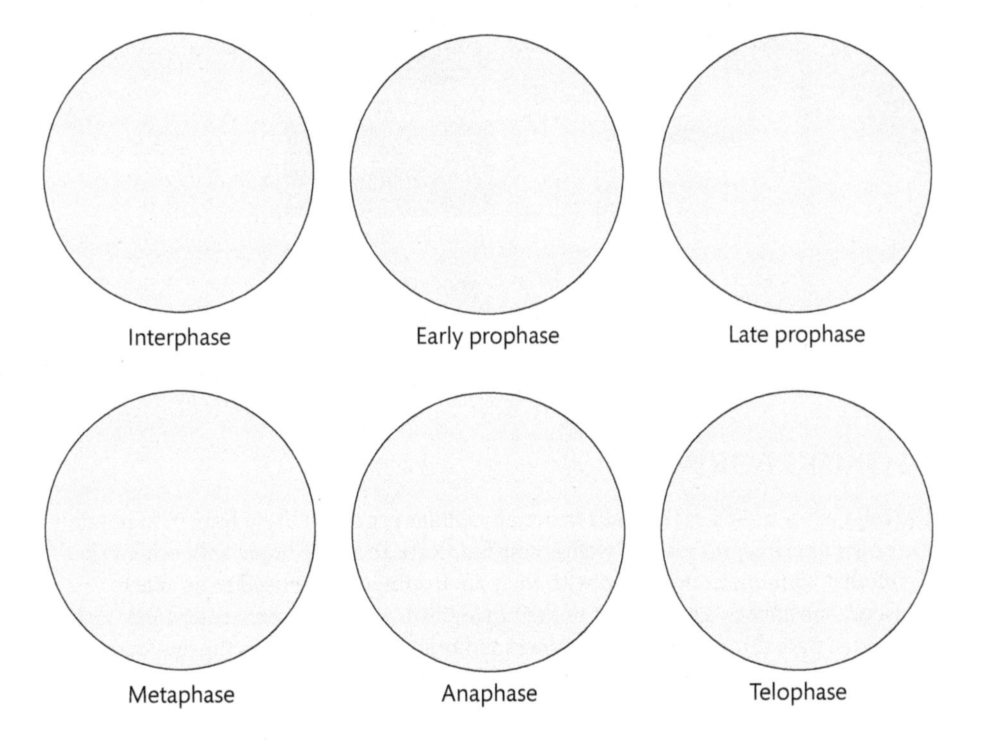

MAKING CONNECTIONS

The cells in the epidermis (superficial layer) of your skin are constantly dividing by mitosis. Explain why.

▶ ________________________________

____

____

____

____

____

____

____

____

____

____

____

____

BEFORE YOU MOVE ON . . .

« LOOKING BACK

At the beginning of the 19th century, medical studies on the human body paved the way for the development of what is known as the **cell theory.** It was during this time that a fundamental understanding of cell structure and function began to emerge. Gradually, the work of various scientists contributed to the modern cell theory, which includes the following concepts:

- Cells are the structural building blocks of all living organisms.
- All cells arise from preexisting cells.
- The cell is the basic unit of life.
- In a multicellular organism, each cell maintains its own metabolism, independent of other cells, yet individual cells depend on other cells for survival.
- The activities of all cells in an organism are essential and highly coordinated.

Different cell types vary greatly in size, shape, and function. You observed this difference when you examined various types of cells in Activity 3.3 of this exercise. For example, muscle cells are strikingly different, in both structure and function, from nerve cells. You will see more cell variability in Exercise 5 when you will learn how groups of similar cells are organized to form tissues. The cell theory, however, tells us that all the cells in our bodies derive from a single cell: the fertilized ovum that forms when the cell nuclei of egg and sperm unite. So, despite their variability, all the cells in your body have the same genetic makeup. Consider how that can be possible.

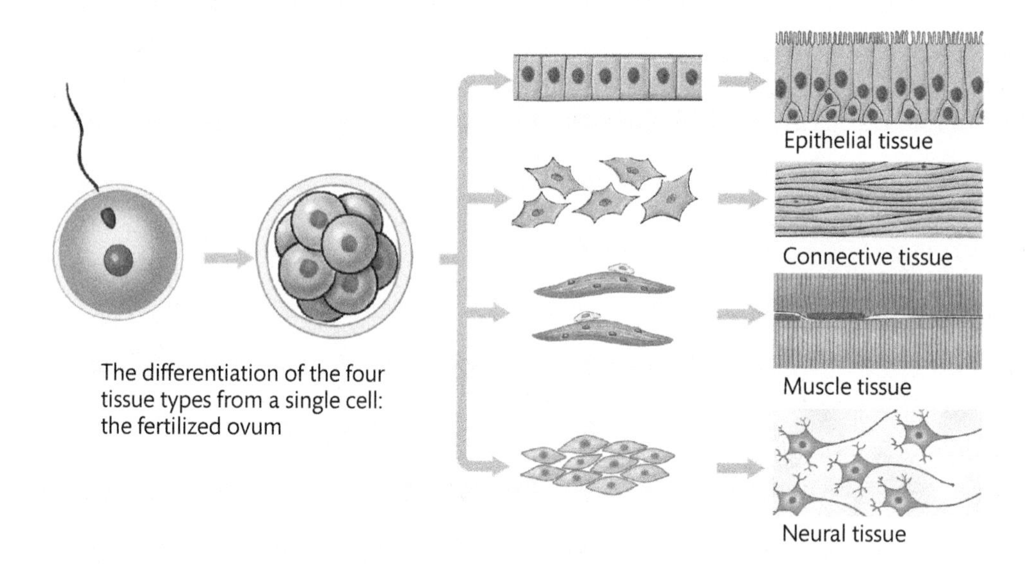

The differentiation of the four tissue types from a single cell: the fertilized ovum

LOOKING FORWARD »

The cell theory indicates that cells in a multicellular organism like a human maintain their own internal processes as well as communicate and coordinate with other cells. To interact with one another and with their environment, cells need to be able to transport substances into and out of their cytoplasm. Any substance entering or exiting a cell must pass through the cell's plasma membrane. Depending on the nature of the substance, there are several different ways in which that can occur. In the next exercise, we will explore some of the mechanisms of membrane transport.

Name ______________________

Lab Section ______________________

Date ______________________

EXERCISE 3 **REVIEW SHEET**

Cell Structure and Cell Division

QUESTIONS 1–8: Identify the labeled structures in the diagram by writing the name next to the appropriate number in column A of the table. Color the structure with the color indicated in column B of the table.

A. Structure	B. Color
1.	green
2.	yellow
3.	red
4.	dark blue
5.	brown
6.	purple
7.	orange
8.	light blue

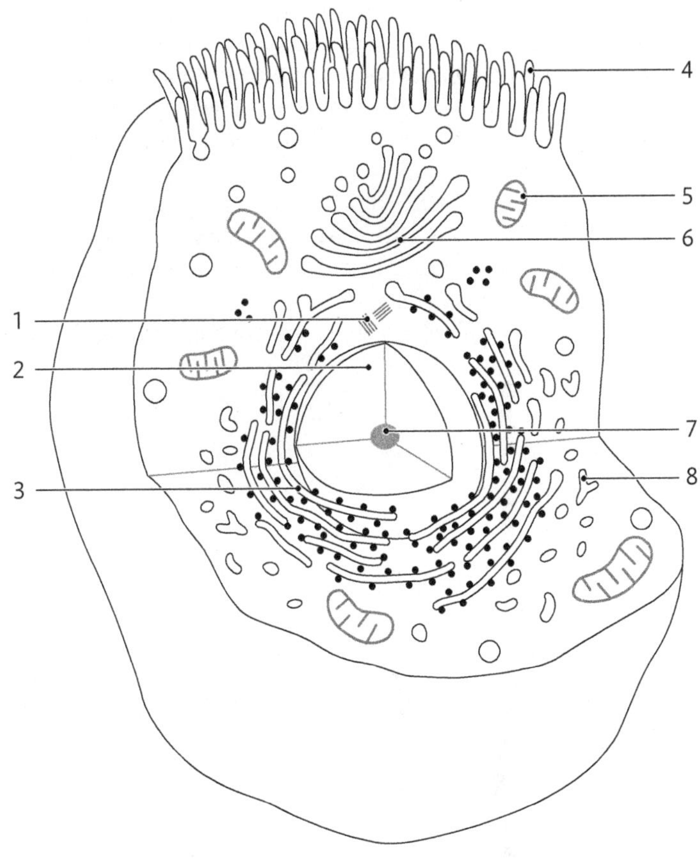

9. When preparing a wet mount of cheek cells, why is it important not to overstain your preparation?

____

____

____

____

____

10. Embryonic stem cells give rise to all cell types in the developing embryo. Would you expect the interphase period of stem cells to be long or short? Explain.

____

____

____

____

____

11. Both of the images below show cell nuclei. Which structures in the nucleus can you identify using an electron microscope that you cannot identify using a light microscope?

____

____

____

____

____

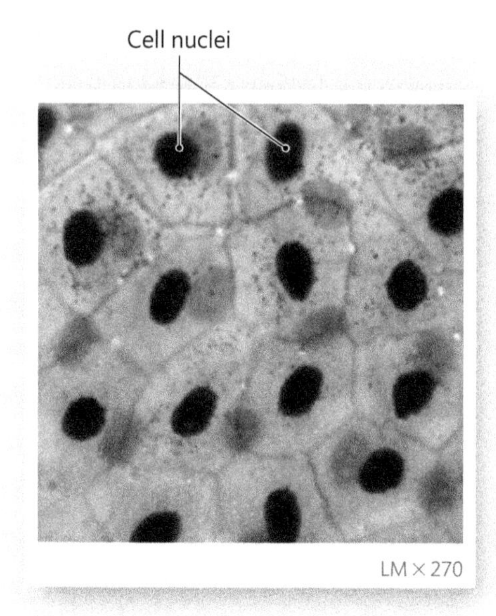

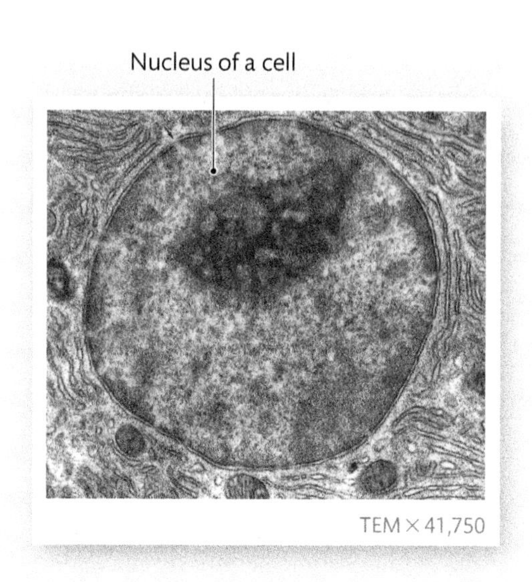

QUESTIONS 12–16: Match the phase of mitosis in column A with the appropriate event in column B. One event in column B will not be used.

A

12. Early prophase ____________________

13. Late prophase ____________________

14. Metaphase ____________________

15. Anaphase ____________________

16. Telophase ____________________

B

a. The chromatid pairs separate and are pulled to opposite ends of the cell.

b. The DNA in the nucleus replicates.

c. The nuclear membrane breaks down.

d. The chromatin molecules in the nucleus become highly condensed, forming chromosomes.

e. A nuclear membrane forms around each new set of chromosomes.

f. The chromosomes line up, end to end, along the equator of the cell.

EXERCISE

4

Membrane Transport

LEARNING OUTCOMES

These Learning Outcomes correspond by number to the laboratory activities in this exercise. When you complete the activities, you should be able to:

Activity 4.1 **Explain how temperature and membrane permeability can affect the rate of diffusion.**

Activity 4.2 **Summarize the fundamental principles of osmosis.**

LABORATORY SUPPLIES

- Plain white paper
- Petri dishes
- Pencils
- Distilled water near freezing (0° to 10°C)
- Distilled water at room temperature (about 23°C)
- Distilled water near boiling (90° to 100°C)
- Methylene blue or toluidine blue dye
- Millimeter rulers
- Pipettes
- Dialysis tubing
- Water
- 10% starch solution
- Iodine potassium iodide (IKI) solution
- Elastic bands or clips
- 500-mL beakers
- Ring stands with clamps
- Thistle tubes with detachable parts
- 25% and 50% molasses solutions
- Marking pens

PRE-LAB QUIZ

Before you begin, read all the activities in Exercise 4 and the required reading in your textbook that is assigned by your instructor.

1. What membrane transport processes will you be examining during this laboratory exercise?
   a. simple diffusion and filtration
   b. simple diffusion and facilitated diffusion
   c. filtration and osmosis
   d. simple diffusion and osmosis
   e. facilitated diffusion and osmosis
2. During this laboratory exercise, you will examine the effect of temperature on diffusion rate by observing the diffusion of a biological dye in what substance?
   a. 10% salt solution
   b. distilled water
   c. 10% glucose solution
3. Under normal conditions, the fluid inside cells (cytoplasmic fluid) and the fluid surrounding cells (extracellular fluid) have equal solute concentrations. If two fluid solutions have similar solute concentrations, they are called
   a. hypotonic solutions.
   b. hypertonic solutions.
   c. isotonic solutions.
4. During this laboratory exercise, to test the effect of membrane permeability on diffusion you will submerge a bag of dialysis tubing, filled with a ________________ solution, in a beaker of water.
5. True or False: Overconsumption of water (water intoxication) can cause the blood and other body fluids to become hypertonic relative to the fluid inside cells. __________

Questions 6–10: Match the membrane transport process in column A with the appropriate description in column B. (You may need to consult your textbook to answer these questions.)

A	B
6. Facilitated diffusion __________	a. Passive movement of substances across a membrane without the use of transport proteins
7. Endocytosis __________	b. The transport of large particles and macromolecules into a cell
8. Ion pump __________	c. Active transport of ions across a membrane
9. Simple diffusion __________	d. The release of substances from a cell by secretory vesicles
10. Exocytosis __________	e. Passive movement of substances across a membrane with the aid of transport proteins

Activity **4.1**

Demonstrating Simple Diffusion

Simple diffusion is a **passive process** that does not require cellular energy (ATP). It is driven by the constant motion of molecules **(kinetic energy)** and the tendency of molecules to move along a **concentration gradient** from an area of high concentration to an area of low concentration.

A The Effect of Temperature on the Rate of Diffusion

An increase in temperature will lead to an increase in kinetic energy. If that occurs, molecules will move faster, and the diffusion rate will accelerate.

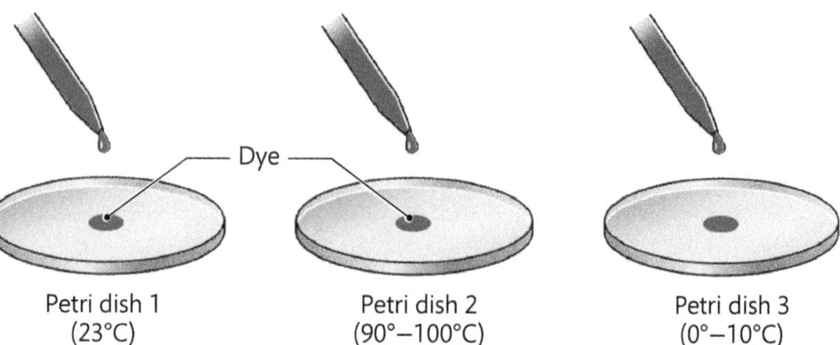

1. Place a piece of plain white paper on the lab bench. With a pencil, mark three dots on the paper, spaced at least 15 cm (6 in.) apart.
2. Place three petri dishes on the paper so that each dish is centered over one of the three dots. Add distilled water to each petri dish until the bottom surface is completely covered according to the following instructions:
   - Petri dish 1: Add water that is at 23°C (room temperature).
   - Petri dish 2: Add water that is at 90° to 100°C.
   - Petri dish 3: Add water that is at 0° to 10°C.

3. **Make a Prediction**

Predict the water temperature at which the diffusion rate will be greatest.

▶ ______________________________

Predict the water temperature at which the diffusion rate will be lowest.

▶ ______________________________

4. Add a drop of blue dye to each petri dish, carefully placing each drop in the center of the dish (over the pencil dot).
5. For each petri dish, observe the diffusion of the blue dye in the water. At 3-minute intervals, compare the relative rates of diffusion for each temperature by measuring the distance (in millimeters) between the origin of the dye and the diffusion front. Record your data in Table 4.1. ▶

TABLE 4.1 Effect of Temperature on Diffusion Rate (Distance from Dye Origin, in mm)

Time (min)	Petri Dish 1 (23°C)	Petri Dish 2 (90°–100°C)	Petri Dish 3 (0°–10°C)
0	0	0	0
3			
6			
9			
12			
15			
18			
21			
24			
27			
30			

6. **Assess the Outcome**

Do your experimental results agree with your earlier prediction? Provide an explanation for the results that you observed.

▶ ______________________________

____

____

____

____

B The Effect of Membrane Permeability on Diffusion

Diffusion across a membrane has two requirements:

1. A concentration gradient must be present.
2. The membrane must be **permeable** to the substance being transported.

Even with a significant difference between the concentrations on each side of a membrane, if the membrane is not permeable to that particular molecule, diffusion cannot occur. The following activity demonstrates this concept.

1. Soak a 10-cm strip of dialysis tubing in water for a few minutes so that it will be easier to open. Seal one end by using a clip or wrapping it with multiple turns of an elastic band to form a tight seal.

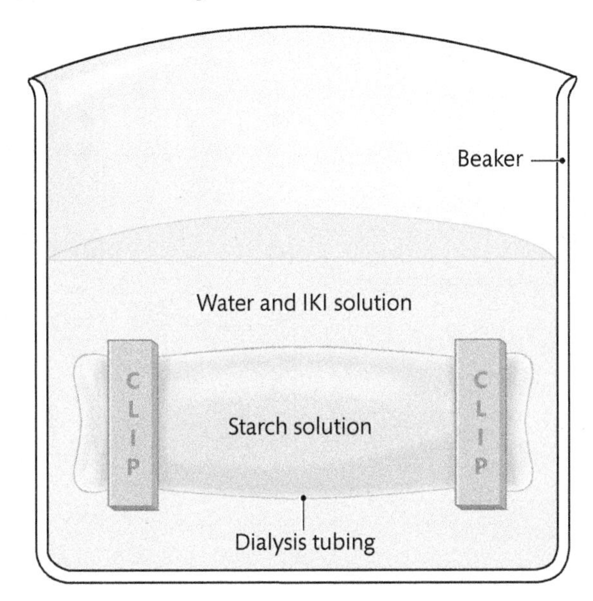

2. Open the other end of the bag and add enough 10% starch solution (starch and distilled water) to fill it to about one-third of its volume. Carefully secure the remaining end with a clip or several turns of an elastic band. Rinse the outside of the bag to remove any starch.

3. Place the bag of starch into a beaker and fill the beaker with warm water (35° to 40°C) until the bag is fully submerged.

4. Add iodine potassium iodide (IKI) solution to the beaker until the water is a medium amber color. The IKI solution is used to test for the presence of starch. The solution will turn dark blue or black if starch is present.

5. Observe the bag after 10 to 20 minutes and record the color of the solutions in the beaker and in the dialysis tubing bag. ▶

What color is the solution in the beaker?

____

What color is the solution in the bag?

____

Based on your results (solution colors), identify the substances (if any) that diffused across the dialysis tubing. Provide an explanation for the results.

____

____

____

____

MAKING CONNECTIONS

In Activity 4.1.A, equilibrium is reached when dye is uniformly distributed throughout the petri dish and there is no net directional movement of dye molecules. Was equilibrium reached at any of the water temperatures that were used? Provide an explanation, based on your observations.

▶ _______________

____

____

____

____

____

____

____

____

____

■ Want more practice? Go to: MasteringA&P > Study Area > Menu > Lab Tools > PhysioEx™ > Exercise 1: Cell Transport Mechanisms and Permeability > Activity 1: Simulating Dialysis (Simple Diffusion)

Activity **4.2**

Observing Osmosis

A A **solute** is a substance that dissolves when added to another substance. The substance that dissolves the solute is a **solvent.** Water is considered the **universal solvent** because it dissolves all other polar substances. For example, our body fluids (e.g., blood, lymph, cytoplasm, and extracellular fluid), which are mostly water, will dissolve various polar compounds such as glucose, proteins, and electrolytes. Solute concentration is measured in **milliosmols (mOsm),** which is the number of dissolved particles per milliliter (mL) of solution.

The terms *hypotonic, hypertonic,* and *isotonic* are descriptive words that can be used to compare the solute concentrations of two solutions: the **hypotonic solution** will have a lower solute concentration, and the **hypertonic solution** will have a higher solute concentration. If two solutions have equal solute concentrations, they are called **isotonic solutions.**

The four solutions in the figure have different solute concentrations, indicated by the varying number of black circles in each beaker. Thus, solution (b) has the greatest solute concentration. Use the terms *hypotonic, hypertonic,* and *isotonic* to make the following comparisons:

(a) (b)

(c) (d)

1. Compare solution (a) with solution (b): ▶
   Solution (a) is

   ____

   Solution (b) is

   ____

2. Compare solution (a) with solution (c): ▶
   Solution (a) is

   ____

   Solution (c) is

   ____

3. Does your description of solution (a) differ in the two comparisons? If so, explain why.
   ▶ ______________________________

4. Which two solutions are isotonic? Explain why.
   ▶ ______________________________

B **Osmosis** refers to the diffusion of water through a **selectively permeable** membrane. The net movement of water across a membrane will be from an area of higher water concentration to an area of lower water concentration. In other words, the net movement of water will be from a solution with a lower total solute concentration (a hypotonic solution) to a solution with a higher total solute concentration (a hypertonic solution).

1. Cut an 8-cm (3-in.) piece of dialysis tubing and separate the two layers by soaking it in tap water. Slide one blade of the scissors between the two layers and cut along the folded margin to form a single-layer membrane.
2. Place the dialysis membrane tightly over the mouth of a thistle tube. Keeping the membrane taut, secure it to the thistle tube with several wrappings of an elastic band.
3. Separate the mouth of the thistle tube from the cylindrical portion. Using a pipette, carefully fill the mouth with a molasses solution until the solution is about to overflow. Half the class should use a 50% molasses solution and the other half a 25% solution.
4. Reconnect the cylindrical portion of the thistle tube to the mouth. Make sure the connection is tightly secured. The meniscus (curved surface) of the molasses solution should rise slightly up the cylinder. If an air bubble is trapped in the thistle tube, the molasses will recede back into the mouth. If that occurs, detach the cylinder and add more molasses to the mouth.

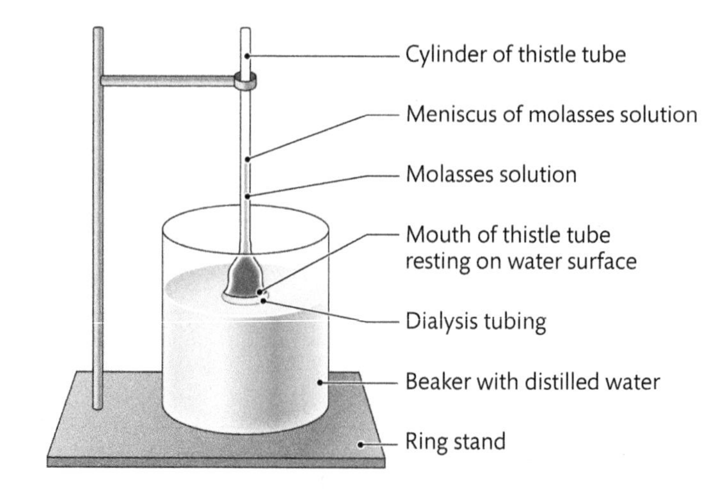

5 Mark the initial level of the meniscus of the molasses solution with a marking pen. Lower the thistle tube into a beaker of distilled water so that the dialysis membrane is submerged. Secure the thistle tube in this position with a ring stand and clamp, being careful not to break the delicate tube (refer to lower figure on previous page). If the water solution becomes discolored, there is a leak in the dialysis tubing. If there is a leak, take apart the assembly and start over.

6 Using your initial mark as your zero point, note the change in the level of the meniscus (in centimeters) every 10 minutes for a 1-hour period. Record your results in Table 4.2. ▶ Obtain results for both the 25% and 50% molasses solutions by sharing your data with other students in the laboratory.

7 For each concentration, what was the total distance that the molasses solution traveled up the thistle tube? ▶

25% molasses: ____________________

50% molasses: ____________________

8 Calculate the rate, in centimeters per minute (cm/min), that each molasses solution rose up the thistle tube. ▶

25% molasses solution: ____________________

50% molasses solution: ____________________

9 Compare the rates at which the two solutions rose up the thistle tube. If there were any differences, explain why.

▶ ____________________

TABLE 4.2 Osmosis Demonstration

	Meniscus Level (cm)	
Time (min)	25% Molasses	50% Molasses
0	0	0
10		
20		
30		
40		
50		
60		

IN THE CLINIC

Water Intoxication

We all know that alcohol consumption can lead to intoxication, but did you know that excessive water consumption can cause similar symptoms? This condition is called water intoxication. As essential as water is to life, overconsumption of it can be fatal. Normally, our kidneys eliminate excess water from the body. However, severe dehydration—such as from prolonged vomiting and diarrhea or excessive sweating—causes the loss not only of water but also of electrolytes (solutes). If a dehydrated person consumes a very large amount of water quickly without replacing electrolytes, the kidneys may not be able to clear the water at the desired rate, and the blood will become hypotonic. As a result, cells in the brain, which are hypertonic to the now diluted blood, will take in too much water by osmosis. If that happens, the brain cells will swell, similar to when the brain swells from serious head trauma, and death can occur.

Word Origins

Osmosis comes from the Greek word *osmos*, which means "a push or impulse." Osmosis is the passage (a "push") of a solvent (e.g., water) through a semipermeable membrane from a hypotonic to a hypertonic solution.

MAKING CONNECTIONS

Under normal conditions, the cells in our body are exposed to isotonic solutions. That is, the solute concentration of the cytoplasm is about the same as the solute concentration of extracellular fluids. Why do you think it is important for the cytoplasm and extracellular fluids to remain isotonic?

▶ ____________________

■ Want more practice? Go to: MasteringA&P > Study Area > Menu > Lab Tools > PhysioEx > Exercise 1: Cell Transport Mechanisms and Permeability > Activity 3: Simulating Osmotic Pressure

BEFORE YOU MOVE ON . . .

‹‹ LOOKING BACK

In this laboratory exercise, you studied two passive membrane transport processes, simple diffusion and osmosis. Passive processes do not require the use of cellular energy (ATP). Rather, the kinetic energy of molecular motion is the fuel that drives these systems.

Other transport mechanisms, called **active processes,** require the cellular energy derived from ATP molecules, most of which is produced by the mitochondria. Active processes transport substances that are too large to pass through the membrane, are not soluble in lipids, or are being transported against a concentration gradient from an area of lower concentration to an area of higher concentration. Substances are transported actively with the assistance of carrier proteins or within vesicles.

Plasma membranes are molecular walls that separate the cytoplasm from the extracellular environment, but there is some communication between the two sides. Membranes are selectively permeable, which means that some substances can travel across the barrier but other substances cannot. This characteristic is vital for the normal function of all cells.

Why do you think a plasma membrane must be selectively permeable? In other words, why can't the membrane be freely permeable, permitting all substances to pass across, or be impermeable, blocking the transport of all substances?

▶ ____

____

____

____

____

____

____

____

____

____

____

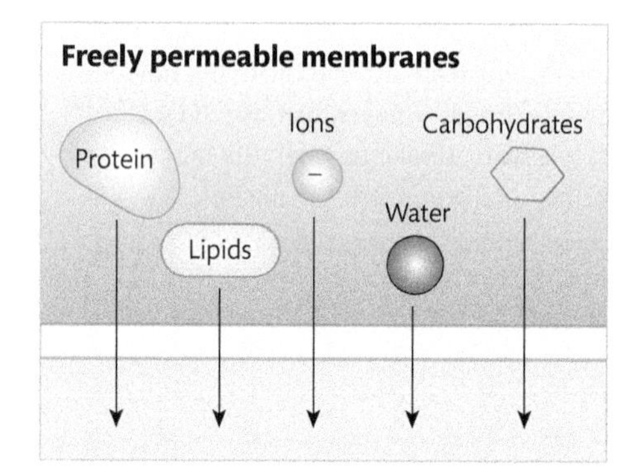

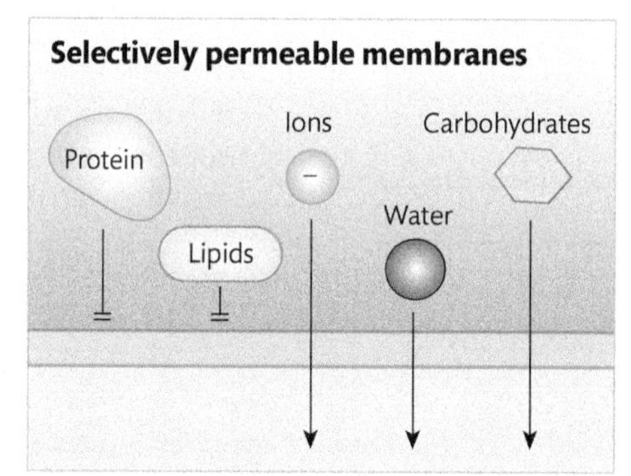

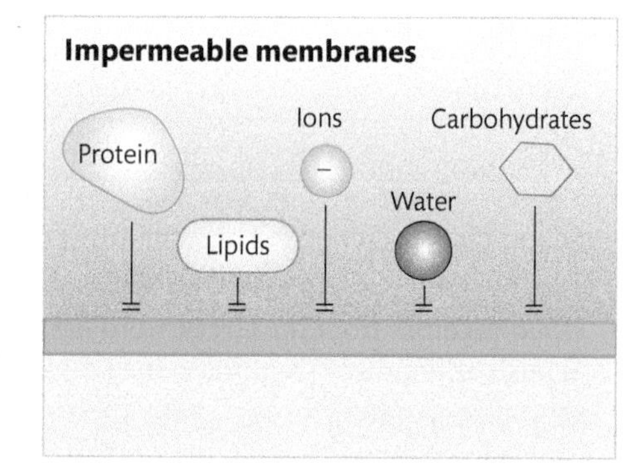

LOOKING FORWARD ››

The transport of substances across plasma membranes enables cells to communicate and work in concert with other cells. A group of cells that work together as a unit to carry out a specialized function is called a tissue. The human body has four major categories of tissues—epithelial tissue, connective tissue, muscle tissue, and nervous tissue—which you will explore in the next exercise.

Membrane Transport

Name ______________________

Lab Section ______________________

Date ______________________

QUESTION 1: With regard to simple diffusion, explain the significance of the following terms.

a. Kinetic energy:

____

____

b. Concentration gradient:

____

____

QUESTIONS 2–3: Explain how the following factors will affect the rate of diffusion across a cell membrane.

2. Temperature:

____

____

____

3. Membrane permeability:

____

____

____

4. Apply the principles of diffusion to explain what will occur after placing a dyed sugar cube in a beaker of water, as illustrated below.

____

____

____

____

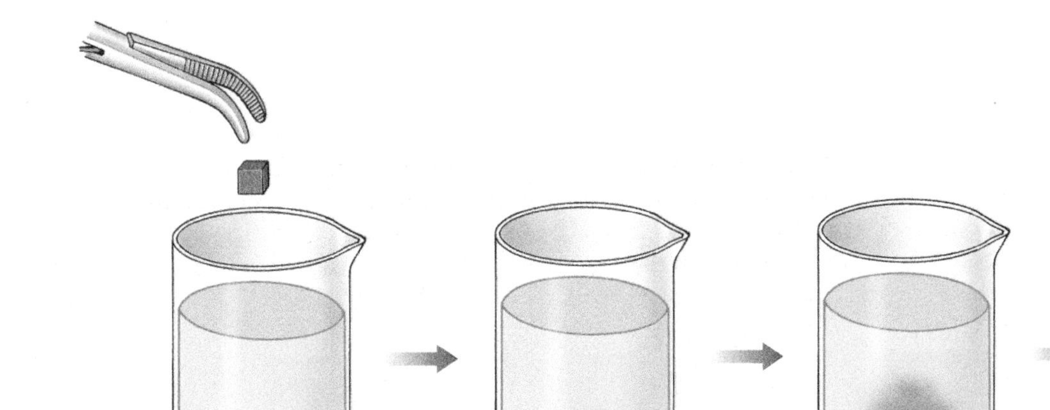

QUESTIONS 5–6: Refer to the diagram on the right to answer the questions.

Substances X and Y may be any substances other than water. Letter size indicates relative concentrations of substances X and Y on each side of the membrane. Therefore, the concentration of X is greater inside the cell than outside, and the concentration of Y is greater outside the cell than inside. The solute concentrations of the extracellular fluid and cytoplasm are measured in milliosmols (mOsm). The membrane is permeable to water.

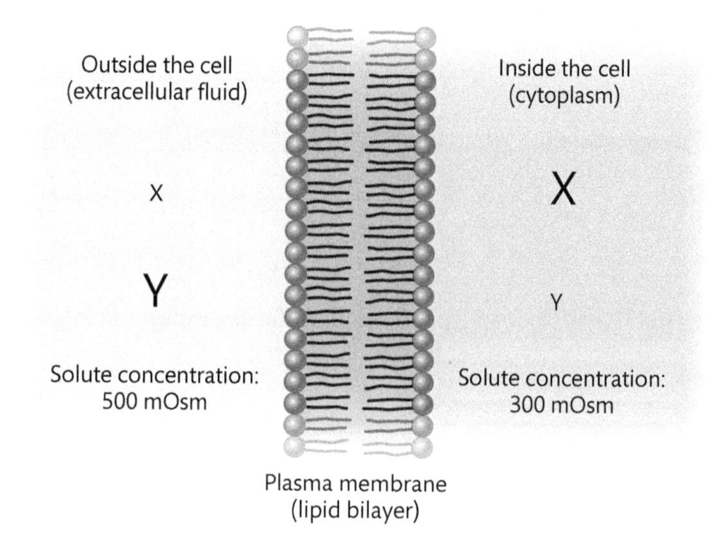

5. If substances X and Y were to diffuse across the cell membrane, in which direction would each substance travel? Explain.

____

____

____

____

6. Given the conditions illustrated in the diagram, the net flow of water across the membrane will be in which direction? Explain.

____

____

____

____

7. The figures below illustrate the effects on red blood cells when they are exposed to isotonic, hypotonic, and hypertonic solutions. The arrows represent the net directional flow of water. Match each figure with the type of solution surrounding the cell.

Figure	Type of Solution (Isotonic, Hypotonic, or Hypertonic)
(a)	
(b)	
(c)	

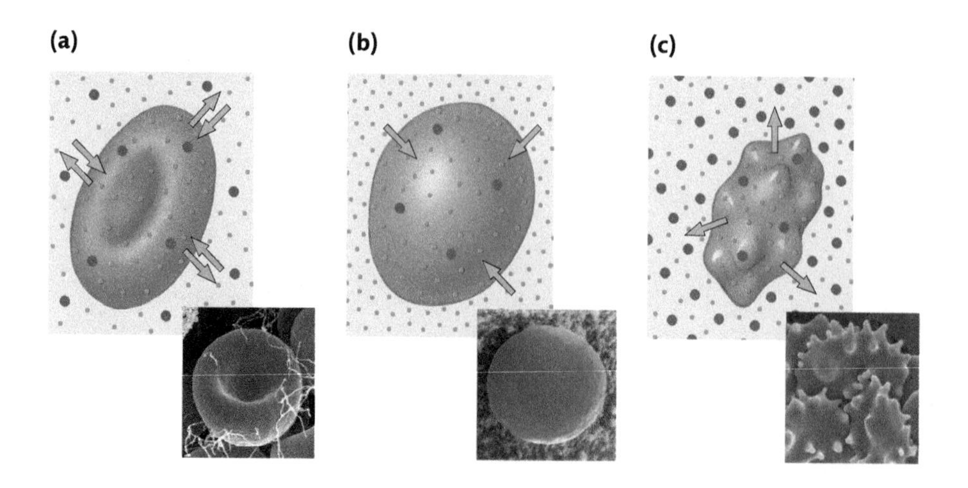

EXERCISE

5

Tissues

LEARNING OUTCOMES

These Learning Outcomes correspond by number to the laboratory activities in this exercise. When you complete the activities, you should be able to:

Activities 5.1–5.3 **Describe the structure, functions, and locations of the various types of epithelial tissue.**

Activities 5.4–5.8 **Discuss the structure, functions, and locations of the various types of connective tissue.**

Activity 5.9 **Compare and contrast the structure, functions, and locations of the three types of muscle tissue.**

Activity 5.10 **Summarize the structure and functions of neurons in nervous (neural) tissue.**

LABORATORY SUPPLIES

- Compound light microscopes
- Colored pencils or markers
- Prepared microscope slides:
  - artery, vein, nerve
  - esophagus
  - human skin (sole)
  - kidney
  - human skin (scalp)
  - urinary bladder
  - small intestine
  - trachea
  - areolar connective tissue
  - adipose tissue
  - reticular tissue
  - dense regular connective tissue
  - elastic connective tissue
  - hyaline cartilage
  - elastic cartilage
  - fibrocartilage
  - bone
  - human blood smear
  - skeletal muscle
  - cardiac muscle
  - smooth muscle
  - ox spinal cord smear

PRE-LAB QUIZ

Before you begin, read all the activities in Exercise 5 and the required reading in your textbook that is assigned by your instructor.

Questions 1–4: Match the type of epithelium in column A with the structure in column B that you will view with a microscope.

A	B
1. Simple columnar ____________	a. Esophagus
2. Stratified squamous ____________	b. Urinary bladder
3. Pseudostratified columnar ____________	c. Small intestine
4. Transitional ____________	d. Trachea

5. True or False: Areolar connective tissue is an example of a loose connective tissue. ____________
6. True or False: Tendons and ligaments are composed of elastic tissue. ____________
7. True or False: Bone and cartilage are examples of dense regular connective tissues. ____________
8. True or False: Erythrocytes and leukocytes are cells found in blood. ____________
9. Specialized cell junctions, called intercalated discs, are found in what type of muscle tissue? ____________
10. What part of a nerve cell (neuron) can generate and transmit electric impulses? ____________

Activity **5.1**

Epithelial Tissue: Microscopic Observations of Squamous Epithelia

A **tissue** is a group of cells that work together as a unit to carry out a specialized function. The body has four types of tissues: epithelial tissue, connective tissue, muscle tissue, and nervous tissue.

Overview of Epithelia

Epithelial tissue (or epithelia; singular = epithelium) covers all body surfaces. It forms the outside surfaces of organs and the inner linings of hollow organs, lines the walls of body cavities and the exterior surface of the body, and is the primary tissue of glands. The base of an epithelium is anchored to connective tissue by a thin band of glycoproteins and protein fibers called the **basal lamina (basement membrane).** Epithelial cells are usually tightly joined by various types of cell junctions. Because of this feature, epithelial tissues are effective in providing protection for the organ they cover.

Epithelia are classified by using two structural criteria:

1. **Number of cell layers.** A **simple epithelium** consists of a single layer of cells, whereas a **stratified epithelium** has several layers.
2. **Shape of the cells.** A **squamous epithelium** contains flattened, irregularly shaped cells. A **cuboidal epithelium** contains cube-shaped cells in which the height and width are about equal. A **columnar epithelium** contains elongated cells (column-shaped) in which the height is much greater than the width.

The types of squamous epithelia are summarized in Table 5.1.

TABLE 5.1 Types of Squamous Epithelia

Type	Description	Examples
Simple squamous	Single layer of flattened cells	Inner lining of blood vessels (endothelium) Lining of body cavities (mesothelium)
Stratified squamous	Several layers of cells; cells near the surface are flattened	Inner lining of esophagus Inner lining of vaginal canal
Stratified squamous, keratinized	Stratified squamous epithelium with a layer of dead cells filled with keratin at the surface	Epidermis of the skin

A Simple Squamous Epithelium

1. Using low magnification, scan a slide of arteries, veins, and nerves in cross section. Locate the circular profile of an artery. Unlike the nerve, the artery and vein will both be hollow structures. The artery will have the thicker wall of the two blood vessels.

2. Switch to high magnification and examine the wall of the artery. The internal space **(lumen)** may contain some brownish-red or pink-staining red blood cells. The arterial wall (surrounding the lumen) is composed of three distinct layers. The innermost layer, appearing as a wavy dark-staining line, is called the tunica intima. Notice that the tunica intima contains a single layer of flattened cells. It is a simple squamous epithelium. All epithelia that line blood vessels are simple squamous and are given the special name **endothelium.** Identify other tissue layers of the arterial wall. The middle layer, the tunica media, consists of smooth muscle, and the outer layer, the tunica externa, contains loose connective tissue.

3. In the space provided, draw your observations of the wall of an artery at high magnification and label the simple squamous epithelium, the muscle layer, and the connective tissue layer.

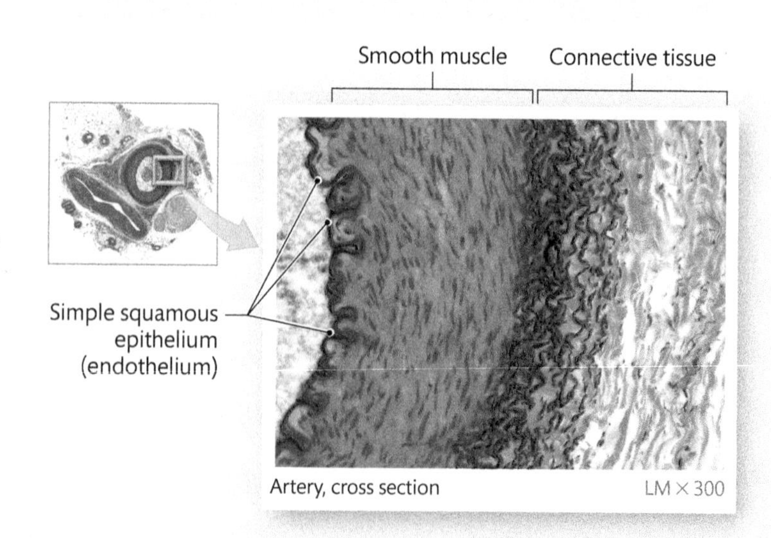

Artery, cross section LM × 300

B Stratified Squamous Epithelium

1. View a microscope slide of the esophagus under low power. Observe that the epithelium lining the lumen (internal space) of the esophagus has many layers of cells.
2. Switch to high magnification and observe that the deeper layers contain cube-shaped cells but the superficial layers consist of flattened cells. That is an example of a stratified squamous epithelium. This type of epithelium provides protection from abrasive and frictional forces, such as when food rubs against the wall of the esophagus.

C Stratified Squamous, Keratinized Epithelium

1. View a microscope slide of thick skin from the sole under low power and identify the epithelium. This epithelium, known as the **epidermis** of the skin, is a stratified squamous type, with a thick layer of dead cells at the surface filled with keratin fibers. The layer of dead cells, known as the **keratinized layer,** serves as a protective barrier that prevents dehydration and infection from airborne pathogens.
2. Deep to the epithelium, identify a layer of connective tissue, known as the **dermis.**

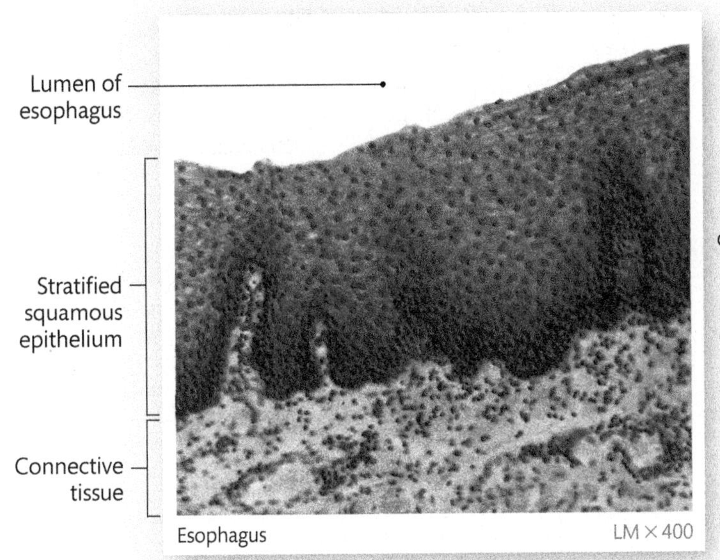

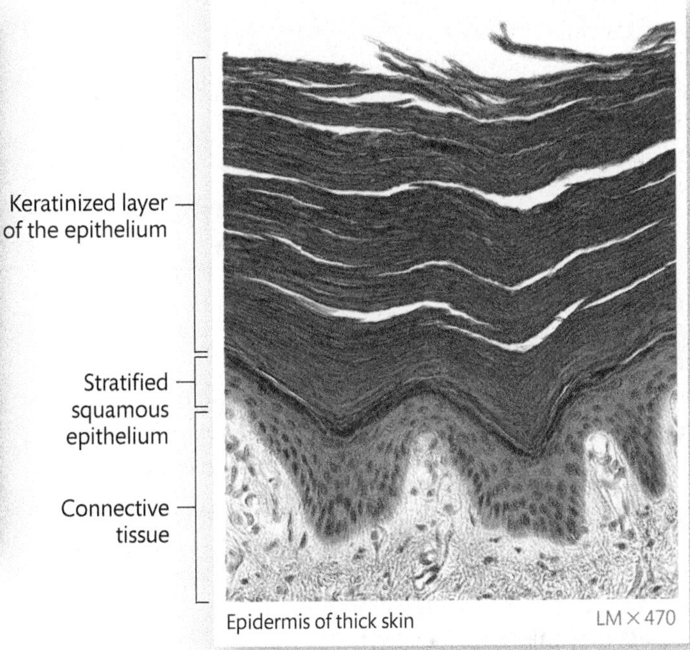

D In the spaces provided, sketch what you see under the microscope.

1. In box (a), draw your observations of the esophagus.
2. In box (b), draw your observations of thick skin.
3. On both drawings, label the stratified squamous epithelium and connective tissue.
4. On the appropriate drawing, label the keratinized layer.

(a) ▶

(b) ▶

MAKING CONNECTIONS

During this activity, you observed that the lumen of a blood vessel, a passageway for blood, is lined by a simple squamous epithelium, which is a very thin cell layer. You also observed that the lumen of the esophagus, a passageway for food to the stomach, is lined by a stratified squamous epithelium, which is much thicker. Why do you think these two structures have epithelia that are so strikingly different?

▶ ______________________________

____

____

____

____

____

____

Activity **5.2**

Epithelial Tissue: Microscopic Observations of Cuboidal Epithelia

Cuboidal epithelia possess cube-shaped cells. Table 5.2 summarizes the three types of cuboidal epithelia:

TABLE 5.2 Types of Cuboidal Epithelia

Type	Description	Examples
Simple cuboidal	Single layer of cube-shaped cells	Kidney (renal) tubules Liver cells
Stratified cuboidal	Two or more layers of cube-shaped cells	Ducts of sweat glands Ducts of mammary glands
Transitional	Appearance of epithelium varies with changing physical conditions: a. relaxed: several layers of cuboidal cells b. stretched: fewer cell layers; cells on surface are flattened	Inner lining of urinary tract (urinary bladder, ureter, and part of the urethra)

A Simple Cuboidal Epithelium

1. Scan a slide of the kidney under low magnification until you locate a field of view that contains numerous **renal tubules,** which are involved in the production of urine.

2. Observe cross sections and longitudinal sections of the tubules at high magnification. Notice that the tubules have a lumen that is lined by a single layer of cube-shaped cells. That is an example of a simple cuboidal epithelium. Also, observe that the tubules are separated by thin layers of connective tissue.

   Simple cuboidal epithelium is found in structures such as the kidney and liver, which are actively involved in the absorption and secretion of various substances.

3. In the space provided, draw your observations and label the simple cuboidal epithelium.

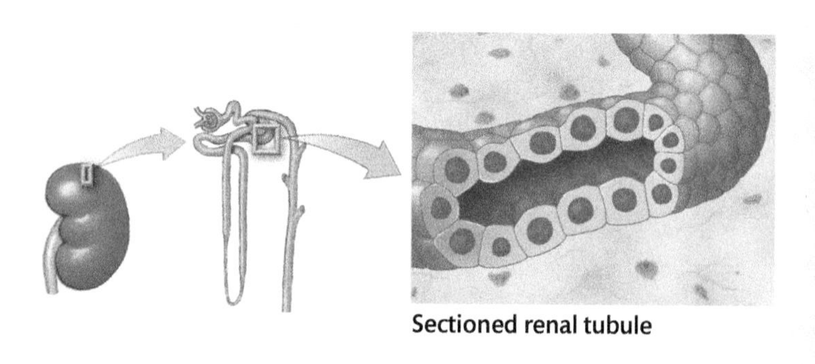

Sectioned renal tubule

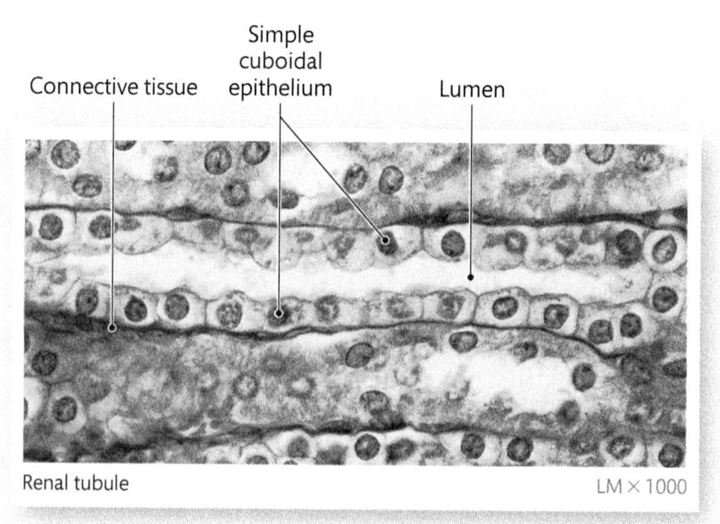

Renal tubule

■ Want more practice? Go to: **MasteringA&P** > Study Area > Menu > Lab Tools > PAL > Histology > Epithelial Tissue

B Stratified Cuboidal Epithelium

1 Under low power, scan a slide of thin skin from the scalp and locate numerous sweat glands scattered throughout the dermis between hair follicles.

2 Examine the sweat glands with the high-power objective lens and locate the darker staining portions, which are the sweat gland ducts. Notice that the lighter staining glandular regions consist of cuboidal epithelial cells. The ducts are lined by a stratified cuboidal epithelium.

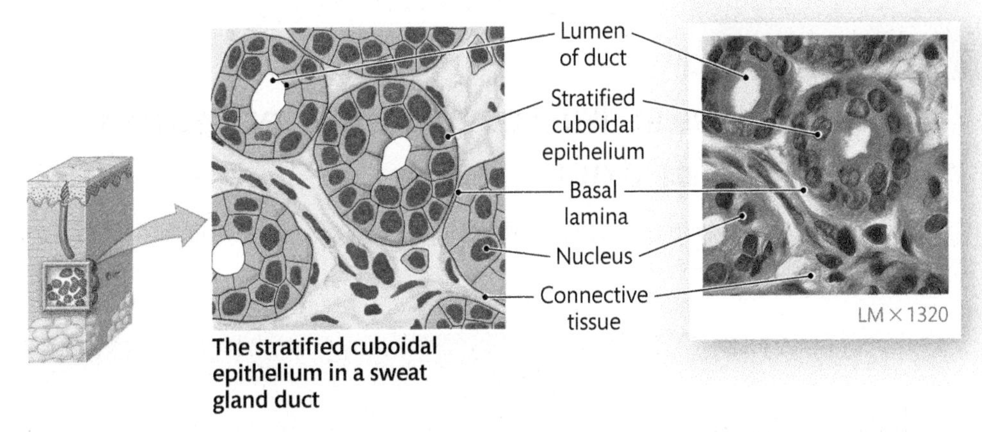

The stratified cuboidal epithelium in a sweat gland duct

C Transitional Epithelium

Transitional epithelium is unique to the urinary tract (urinary bladder, ureters, and urethra).

1 Under low power, view a section of a relaxed bladder and identify the epithelium that lines the lumen. Notice that the bladder wall has numerous folds and the epithelium is relatively thick.

2 Examine another section, showing a stretched bladder, and notice that the folds in the wall have disappeared and the epithelium appears very thin.

3 In the spaces provided below, make two drawings of your observations of transitional epithelium when the bladder wall is (a) relaxed and (b) stretched.

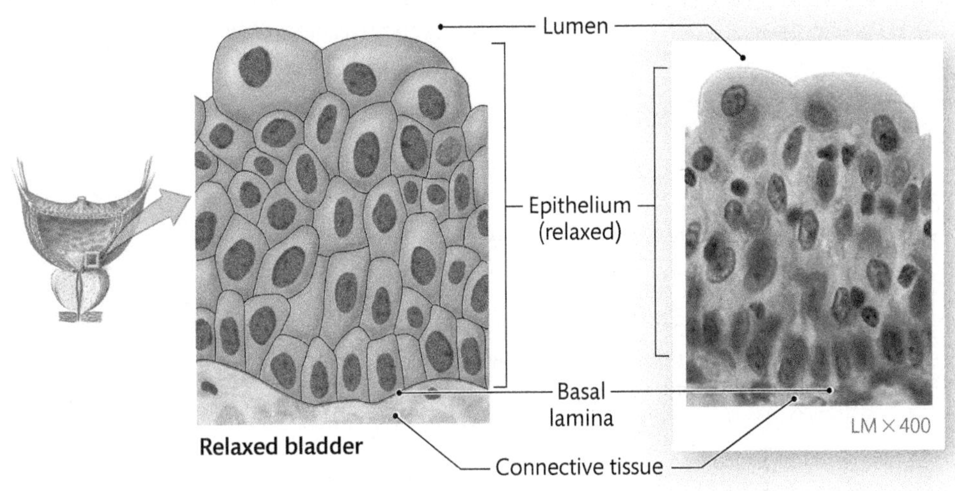

Relaxed bladder

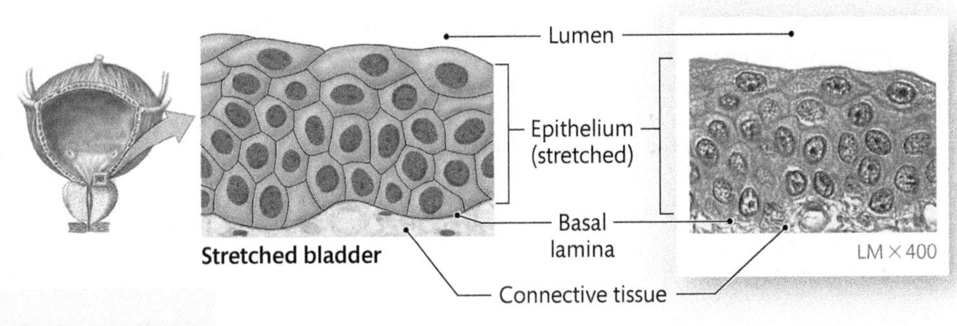

Stretched bladder

▶

(a) relaxed bladder

▶

(b) stretched bladder

MAKING CONNECTIONS

Why do you think it is essential that the epithelium lining the wall of the urinary bladder be composed of several layers of cells rather than a single layer?

▶ ______

Activity **5.3**

Epithelial Tissue: Microscopic Observations of Columnar Epithelia

Columnar epithelia contain cells in which the height is greater than the width. Table 5.3 summarizes the three types of columnar epithelia.

TABLE 5.3 **Types of Columnar Epithelia**

Type	Description	Example
Simple columnar	Single layer of tall (columnar) cells	Inner lining of the gastrointestinal tract (stomach, small and large intestines)
Pseudostratified columnar	Epithelium appears to be stratified, but it really is not because all the cells rest on the basal lamina	Inner lining of trachea and large airways in lung
Stratified columnar (relatively rare, so you will not be required to identify it)	Two or more layers of cells, but only the surface layer is columnar	Large ducts of salivary glands and pancreas

A Simple Columnar Epithelium

Scan a slide of the small intestine under low magnification and locate the epithelium along the surface.

1. Why is this epithelium classified as simple columnar?

▶ ______________________________

____

Notice that all the nuclei are located at about the same level in all the cells. How does this characteristic help you classify this epithelium?

▶ ______________________________

____

____

2. Under high power, observe the darker staining fringe along the surface of the epithelium. This structure, known as the **brush border,** is actually a series of finger-like projections known as **microvilli.** They function to increase the surface area of small intestinal cells to maximize absorption of nutrients.

3. Notice the distinctive, oval-shaped **mucous cells (goblet cells).** The mucus they secrete enhances the digestive process and protects the epithelium from harsh digestive chemicals. Depending on the preparation, these cells may appear clear or stained blue or red.

4. Notice that immediately below the epithelium is a layer of connective tissue. The epithelium is separated from the connective tissue by a **basal lamina.**

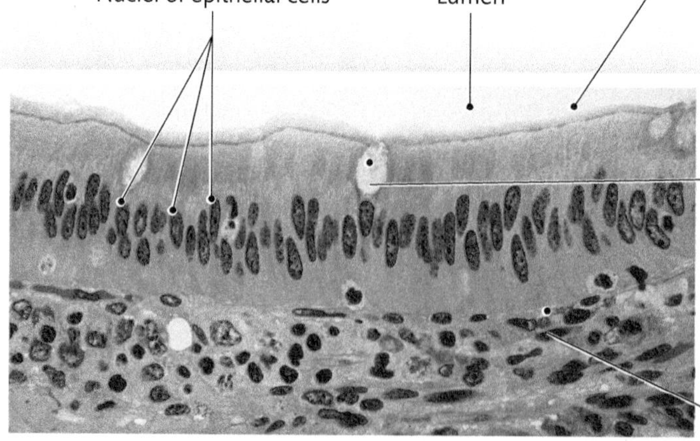

Small intestine, cross section

B Pseudostratified Columnar

View a slide of the trachea under low power. Move the slide so that a region of the epithelium that lines the lumen of the trachea is in the center of the field of view. Switch to high power to observe the epithelium in more detail.

1. Under high power, observe that the nuclei in the epithelial cells are located at different levels, giving the impression that there are several layers of cells. Actually, this epithelium contains only one cell layer because all the cells touch the basal lamina. The nuclei are at different levels because the cells have different shapes and some cells do not reach the surface.

2. Identify **cilia,** thin, hairlike structures projecting from the surface of most epithelial cells. The cilia beat rhythmically to move substances across the surface.

3. This epithelium contains numerous mucous cells (goblet cells) that secrete mucus onto the surface. Identify these cells on the slide. The regions of the cells containing mucus typically appear clear or whitish.

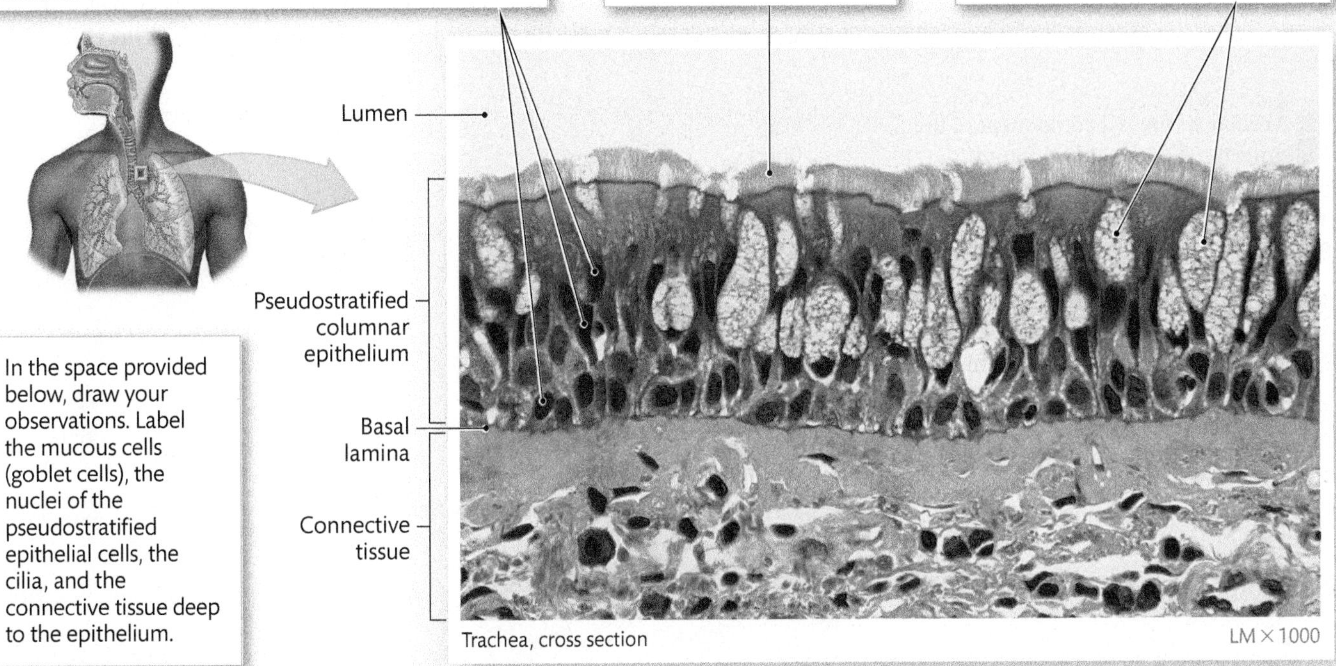

Trachea, cross section

4. In the space provided below, draw your observations. Label the mucous cells (goblet cells), the nuclei of the pseudostratified epithelial cells, the cilia, and the connective tissue deep to the epithelium.

▶

IN THE CLINIC

Mucous Secretions in the Respiratory Tract

The air we breathe contains particulate matter that can promote allergic reactions and cause disease. The mucus secreted by mucous cells (goblet cells) is a sticky substance that traps particles that you breathe in. The beating cilia then move this debris to the throat, where it is swallowed.

MAKING CONNECTIONS

During your observations of epithelia, you were asked to identify the connective tissue that is found just deep to each epithelial layer. Why do you think epithelia and connective tissue are arranged in this way? (*Hint:* Refer to your text to identify the functions of connective tissue.)

▶ ____

____

____

____

____

____

____

____

Activity **5.4**

Connective Tissue: Microscopic Observations of Loose Connective Tissues

Overview of Connective Tissue

Connective tissues are found in all parts of the body and are diverse in both structure and function. They connect and cushion structures, support and protect vital organs, store fat, defend the body from infections, and repair tissue damage. Unlike epithelial cells, connective tissue cells are not packed tightly together. Instead, they are separated by an intercellular material called the **matrix**, which consists of a **ground substance** and various protein fibers.

Depending on the tissue, the consistency of the ground substance varies from liquid to solid and contains complex protein–carbohydrate molecules known as **proteoglycans.** The matrix also contains three types of fibers: collagen, elastic, and reticular. Both collagen and reticular fibers contain the fibrous protein **collagen,** whereas elastic fibers contain the fibrous protein **elastin.** Together, these fibers provide support and resiliency to the tissue.

A **Areolar tissue** is a component of the **superficial fascia** that is found between the skin and muscle. It also forms a thin layer deep to epithelial cells and around capillaries.

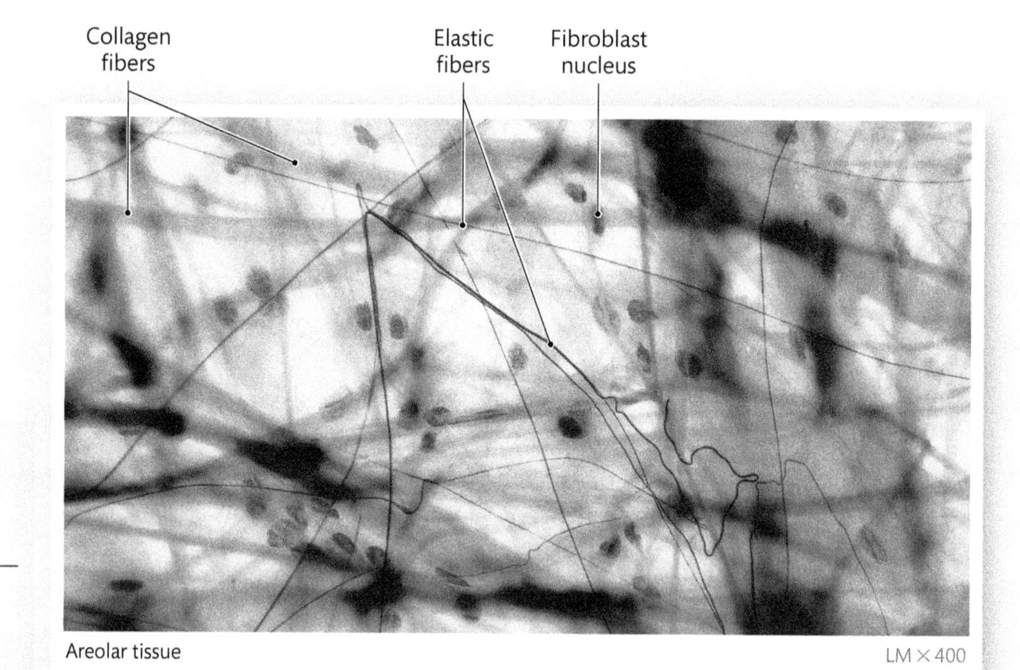

Areolar tissue LM × 400

1. As you view a slide of areolar tissue under low power, notice the loose arrangement of protein fibers and cells. The clear spaces between structures represent the gel-like ground substance in which the cells and fibers are suspended.

2. The thick, pink-staining fibers that you are viewing on your slide are what type of fiber?

▶ ____________________

The very thin, dark-staining (blue-purple) fibers are what type of fiber?

▶ ____________________

3. Most of the cells that you see are **fibroblasts.** They can be identified by the dark-staining nuclei scattered throughout the specimen. The cytoplasm of these cells stains weakly and cannot be easily identified. Refer to your text or class notes and describe the function of these cells.

▶ ____________________

____

Refer to your text, make a list of four other cell types found in areolar connective tissue, and describe the function of each. ▶

Cell Type	Function

4. In the space provided, make a drawing of your observations. Label the collagen fibers, elastic fibers, and the nuclei of fibroblasts.

▶

B **Adipose (fat) tissue** is found in the superficial fascia and is attached to the walls of many internal organs. It is also stored as yellow bone marrow in some bone cavities.

1. Examine a slide of adipose tissue under low power. Notice that the adipose cells, or **adipocytes,** are packed closely together and there is very little intercellular matrix. Most of the volume of adipose cells appears to be empty spaces. In living tissue, these spaces are filled with fat, and the remainder of the cytoplasm and nuclei are squeezed into a thin rim along the periphery.

2. Observe the adipose tissue under high power. At this magnification, you can get a better view of the cytoplasmic rim of the adipose cells. As you scan the slide, find a region along the rim that contains a bulging nucleus.

3. Adipose tissue also serves as **fat pads** by filling spaces around joints. What function do you think these fat pads have?

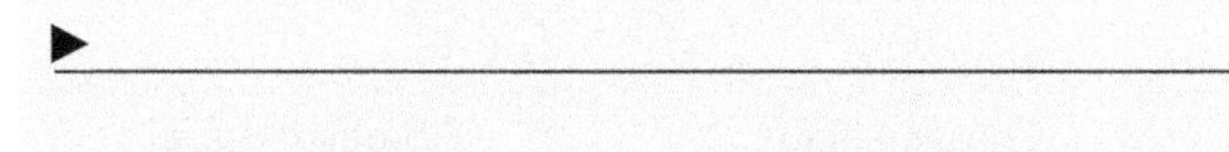

▶ ____

____

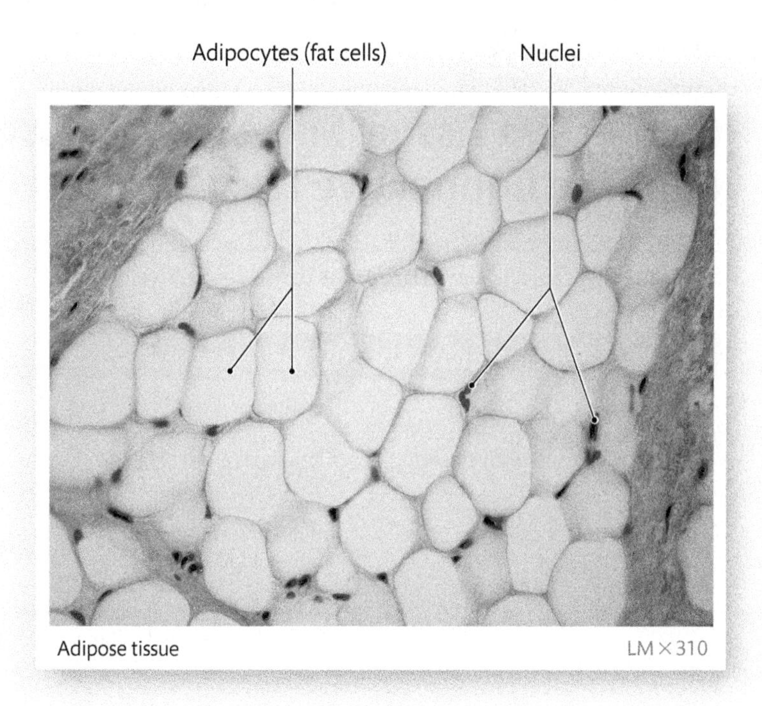

Adipose tissue LM × 310

C **Reticular tissue** forms a supporting framework for organs in the lymphatic system (lymph nodes, thymus, and spleen) as well as in the liver and kidney.

1. Under low power, view a slide that is prepared specifically to demonstrate reticular tissue. Observe the network of black-staining reticular fibers coursing throughout the tissue.

2. By viewing the slide under high power, you can more clearly see the reticular fiber network. The reticular fibers are very thin, but they are made of collagen protein, which gives them strength and durability. You will also see the nuclei of various cells. Many of them are **reticular cells,** which produce the reticular fibers. Other cells include macrophages, fibroblasts, and lymphocytes.

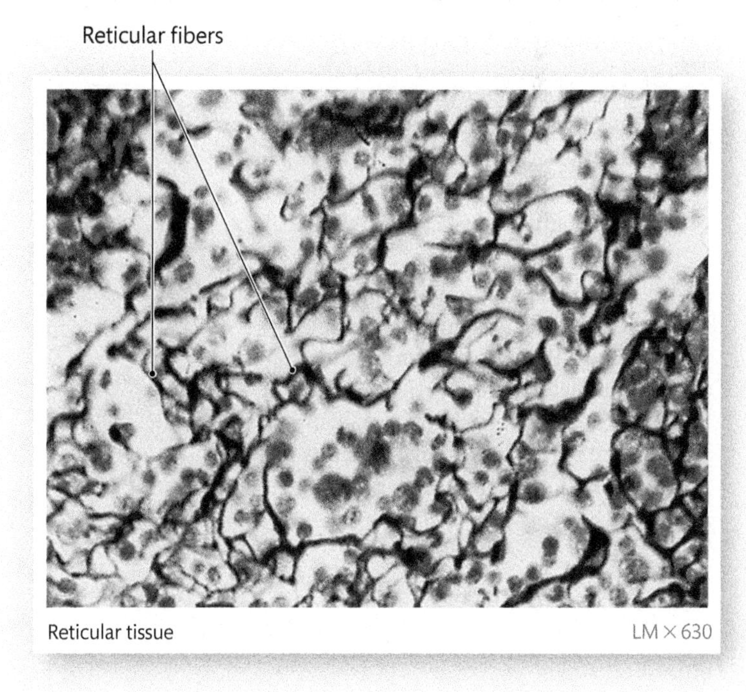

Reticular tissue LM × 630

MAKING CONNECTIONS

Areolar tissue and reticular tissue both have a diverse array of cell types. What does this characteristic tell you about the functions of these connective tissues?

▶ ____

____

____

____

Activity **5.5**

Connective Tissue: Microscopic Observations of Dense Connective Tissues

Dense connective tissues are characterized by the abundance of strong and resilient protein fiber bundles. Table 5.4 summarizes the dense connective tissues.

TABLE 5.4 Types of Dense Connective Tissues

Type	Description	Example
Dense regular connective tissue	Fibroblasts arranged in parallel rows between densely packed bundles of collagen fibers	Tendons, ligaments, and aponeuroses
Dense irregular connective tissue	Irregularly arranged collagen and elastic fibers with intervening fibroblasts	Dermis of skin; capsules around organs; coverings around brain, spinal cord, and nerves
Elastic tissue	Parallel bundles of elastic and collagen fibers with fibroblasts interspersed between them	Elastic ligaments between vertebrae; walls of large arteries

A Dense Regular Connective Tissue

1. View a slide of **dense regular connective tissue** under low power. Notice the large number of collagen fibers, which give it great strength. Blood supply to this tissue is very limited.

2. Center an area of the tissue in the field of view, and then switch to high power. At this magnification, you can see parallel bundles of collagen fibers more clearly. Fibroblasts (dark-staining cells) are arranged in parallel rows between the fibrous bundles.

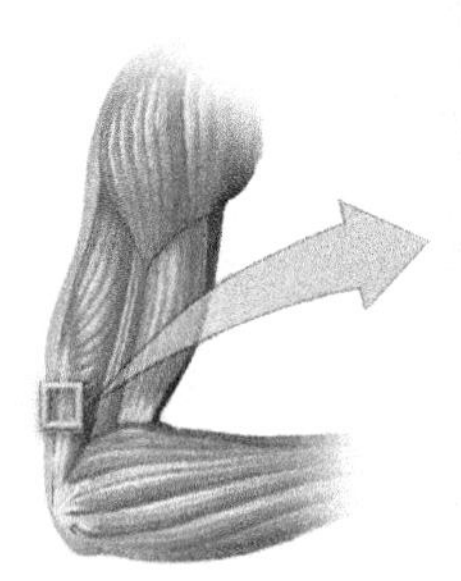

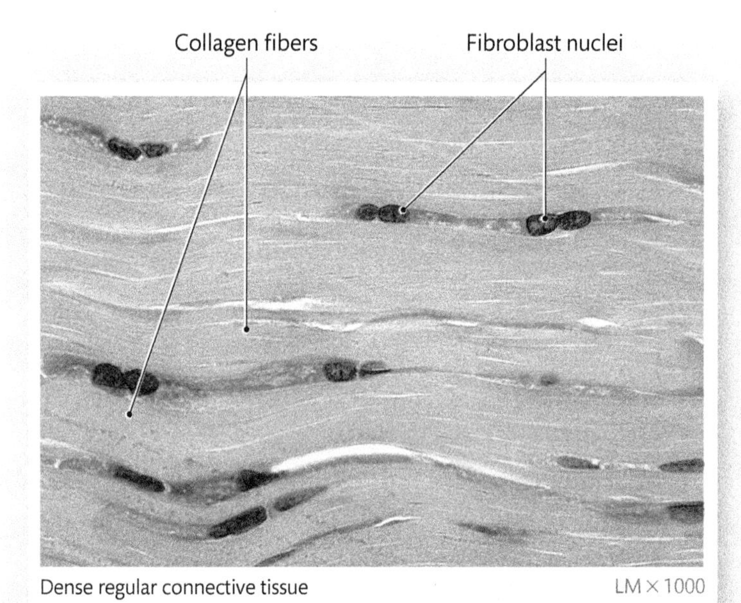

Dense regular connective tissue LM × 1000

3. Explain why this type of tissue is well suited for structures such as tendons and ligaments.

▶ ____

____

____

____

B Dense Irregular Connective Tissue

1. Examine a slide of thick skin under low power. In Activity 5.1, you observed the epidermis. Locate this layer again and identify the type of epithelium.

▶ ______________________

2. Directly below the epidermis is the **dermis** of the skin. The deep region of the dermis is composed of dense irregular connective tissue. This type of connective tissue consists mostly of collagen fibers, although elastic and reticular fibers are also present. How does the arrangement of the fibers in this tissue differ from the arrangement of collagen fibers in dense regular connective tissue?

▶ ______________________

____

____

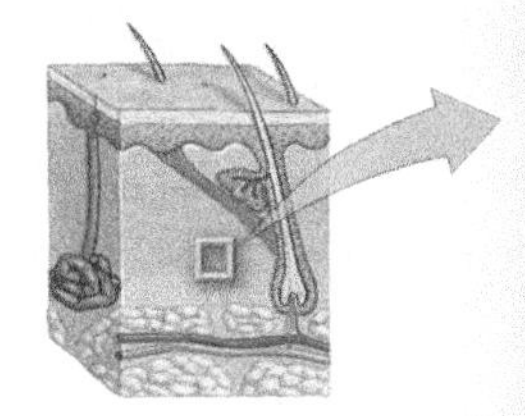

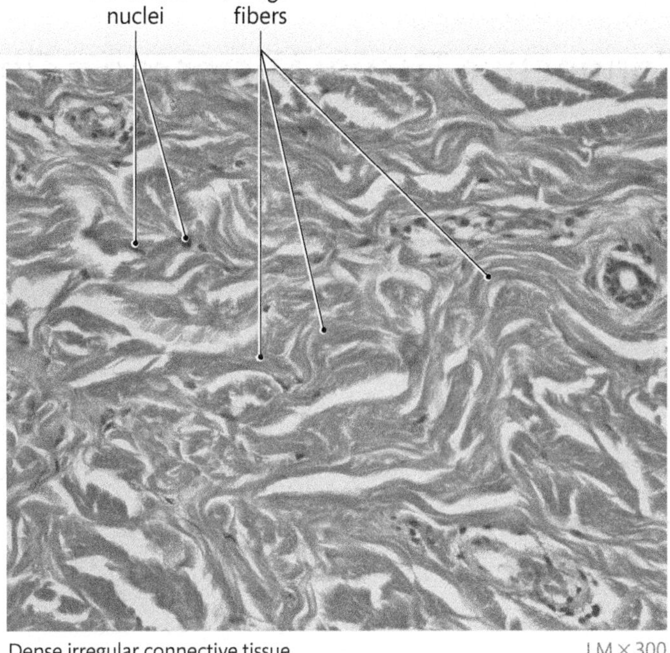

Dense irregular connective tissue LM × 300

C Elastic Tissue

Elastic tissue contains a large amount of elastic fibers and is found in areas that need flexibility.

1. View a slide of elastic tissue, first under low power and then under high power.
2. Note the bundles of elastic fibers, usually dark staining, that predominate in this tissue. Elastic tissue also contains some collagen fibers.

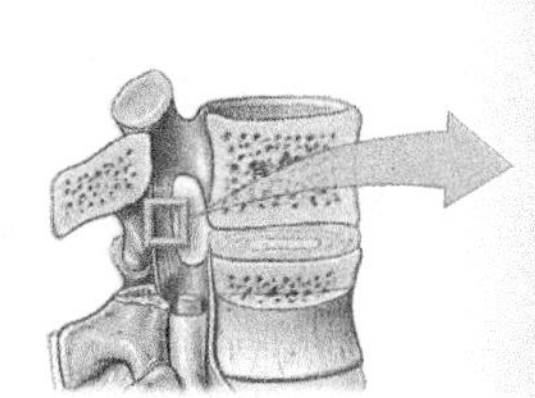

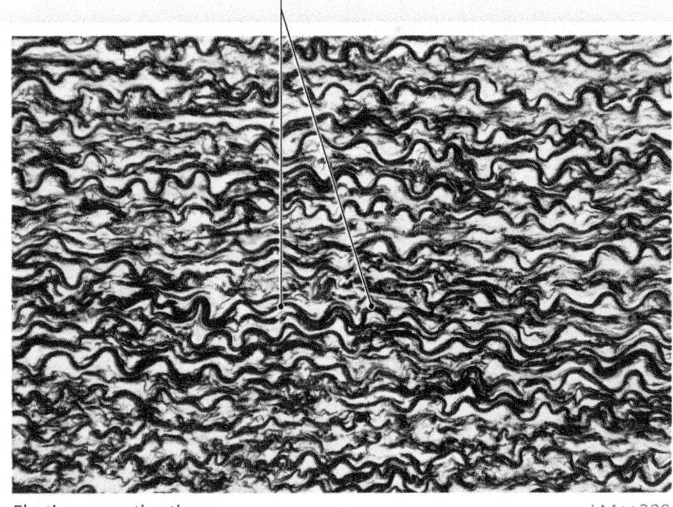

Elastic connective tissue LM × 300

MAKING CONNECTIONS

Why do injuries to ligaments and tendons, such as a sprained ankle or biceps tendonitis, take a long time to completely heal?

▶ ______________________

____

____

____

Activity **5.6**

Connective Tissue: Microscopic Observations of Cartilage

Cartilage has a semisolid matrix that contains a large amount of water, varying amounts of collagen and elastic fibers, and complex polysaccharide molecules, known as **chondroitin sulfates,** which are bound to proteins to form proteoglycans. Cartilage cells, or **chondrocytes,** are located in small cavities called **lacunae** (singular = **lacuna**). The three types of cartilage are summarized in Table 5.5.

TABLE 5.5 Types of Cartilaginous Tissues

Type	Description	Examples
Hyaline cartilage	Matrix contains many collagen fibers and appears smooth and glassy	Cartilage of nose Costal cartilages, connecting the ribs to the sternum Cartilage of the larynx and trachea Articular cartilage at synovial joints Growth plates of developing bones
Elastic cartilage	Matrix contains a large amount of elastic fibers	Epiglottis Auricle (pinna) of the external ear
Fibrocartilage	Rows of lacunae alternate with rows of thick collagen fibers	Intervertebral discs Menisci of the knee joint

A **Hyaline cartilage** provides support in the respiratory tract and other structures, and it allows smooth, low-friction movements at joints.

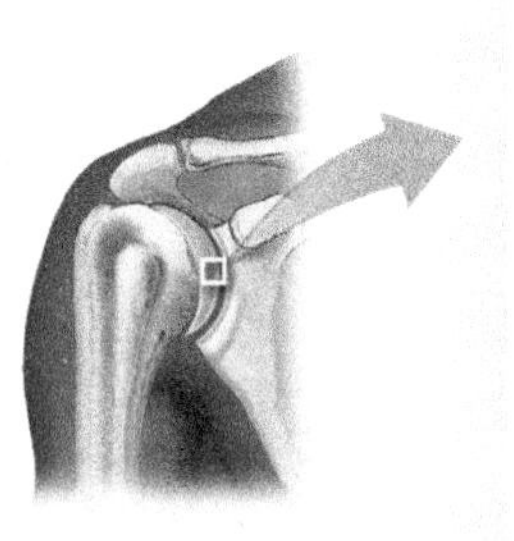

1. View a slide of hyaline cartilage under low power. Identify the lacunae, which are separated from one another by the solid matrix.
2. Switch to high power and identify the chondrocytes within the lacunae.
3. In the space provided, draw your observations. Label the matrix and chondrocytes in the lacunae.

▶

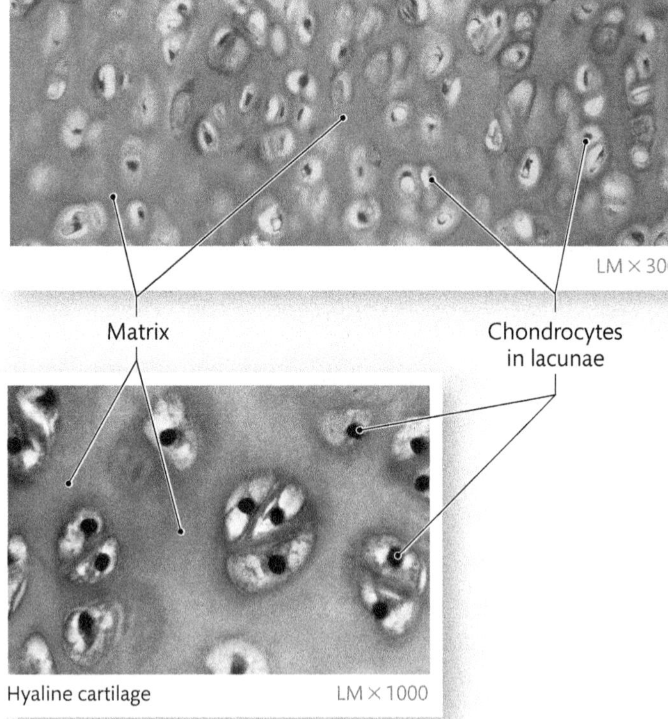

Hyaline cartilage

Word Origins

The term *hyaline* is derived from the Greek word *hyalos,* which means "glass." The name reflects the glassy, smooth quality of the matrix in hyaline cartilage.

B **Elastic cartilage** contains an abundance of elastic fibers.

1 Examine a slide of elastic cartilage under both low and high power. Identify the chondrocytes within the lacunae and notice the dark-staining elastic fibers, which give the matrix a fibrous appearance. Identify the **perichondrium,** which is a layer of dense connective tissue that surrounds cartilage.

2 The abundance of elastic fibers gives this type of cartilage what special characteristic?

▶ 

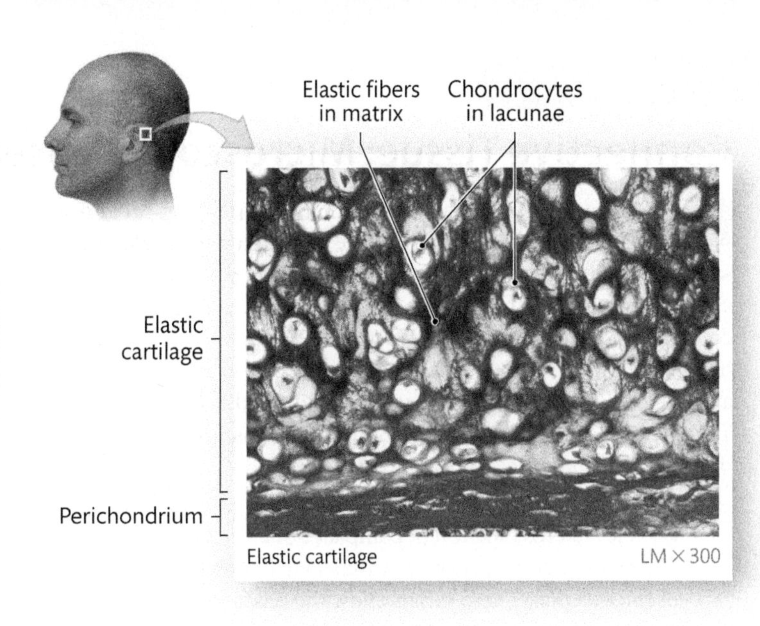

Elastic cartilage LM × 300

C **Fibrocartilage** is located in areas where shock absorption is needed.

1 Examine a slide of fibrocartilage under both low and high power.

2 Identify the rows of collagen fibers, separated by lacunae, which contain chondrocytes.

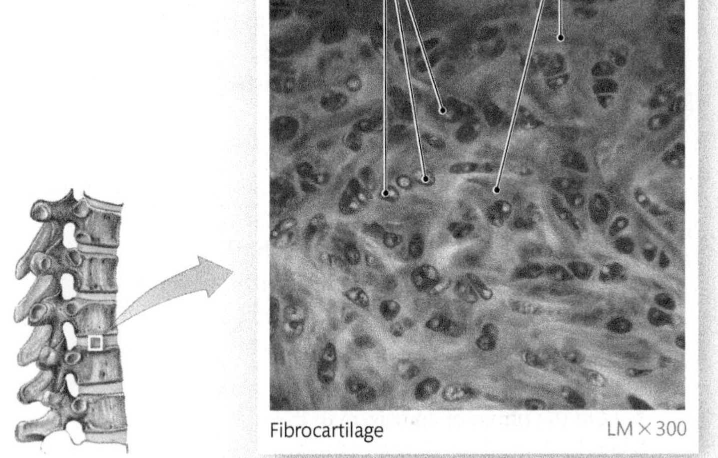

Fibrocartilage LM × 300

Word Origins

The term *lacuna* is derived from the Latin word *lacus,* which refers to a hollow or a lake. In cartilage and bone, a lacuna is a space or cavity in which a cell is located.

IN THE CLINIC

Cartilage Injuries

The matrix in cartilage contains an abundance of water. Because cartilage generally does not have a good blood supply, the water is vital for helping nutrients to diffuse through the matrix. The lack of blood supply in cartilage is starkly contrasted to the rich blood supply found in bone. In fact, damaged cartilage heals much more slowly than a fractured bone. In addition, most bones in the body form from hyaline cartilage. If the cartilage is damaged before the bone is fully grown, the injured bone may stop growing.

MAKING CONNECTIONS

Review the structure and function of the three types of cartilaginous tissue and identify their similarities and differences. ▶

Similarities	Differences

Activity 5.7

Connective Tissue: Microscopic Observations of Bone

A **Bone** is a solid connective tissue that provides support and protection for the body. The bone matrix contains calcium salts such as hydroxyapatite (calcium phosphate combined with calcium hydroxide) and calcium carbonate. The matrix also contains collagen fibers and a variety of ions, including fluoride, magnesium, and sodium. The calcium salts give bone its hardness and rigidity; collagen fibers provide strength and resilience. Together, the calcium and collagen make bone as strong as, but more flexible than, steel or reinforced concrete.

Two types of bone tissue exist in the body. The first type, **compact bone,** is an extremely dense material that forms the hard exterior covering of all bones. The second type, **spongy (cancellous) bone,** fills the interior regions of most bones and forms a thin internal layer along the shafts of long bones. In this activity, we will examine the structure of compact bone.

1. View a cross section of compact bone under low power. Identify the circular **central canals,** located in the centers of the **osteons.** These canals allow the passage of numerous blood vessels and nerves throughout compact bone.
2. The osteon is the basic structural unit of compact bone. In each osteon, identify several layers of bone matrix organized around the central canal. The bone layers are called **bone lamellae** (singular = **lamella**).
3. Locate the lacunae, which appear as small, dark oval or elongated spaces, arranged in a circular pattern between the bone lamellae. In living bone tissue, **osteocytes** (mature bone cells) are located within the lacunae.
4. Switch to high power and identify the thin dark threads that appear to originate from the lacunae and course through the bone matrix. These microscopic canals, the **canaliculi** (singular = **canaliculus**), pass through the lamellae. Cell processes radiating from an osteocyte travel through canaliculi and form cell junctions with the processes of neighboring cells.
5. On the photo of an osteon to the right, label the central canal, bone lamellae, lacunae, and canaliculi. Indicate where the following structures would be in living tissue: (1) an osteocyte; (2) cell processes of an osteocyte; and (3) blood vessels and nerves. ▶

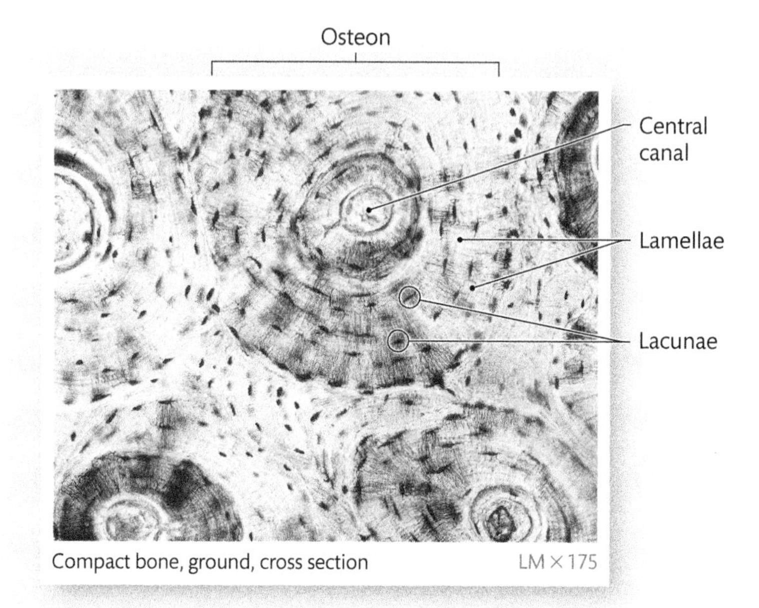

Compact bone, ground, cross section LM × 175

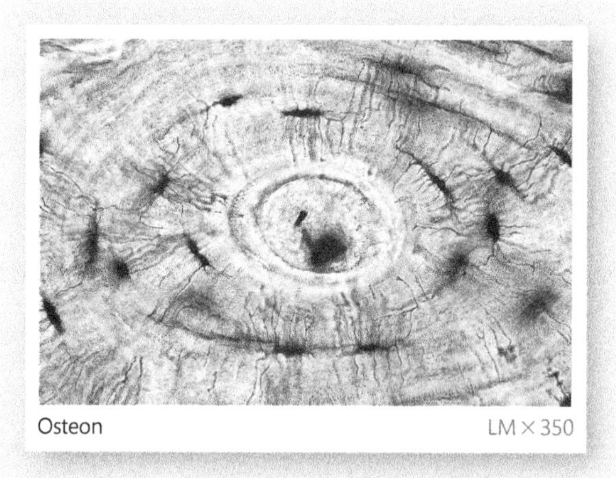

Osteon LM × 350

MAKING CONNECTIONS

Compare the structures of bone and cartilage and identify their similarities and differences. ▶

Similarities	Differences

 ■ Want more practice? Go to: **MasteringA&P** > Study Area > Menu > Lab Tools > PAL > Histology > Connective Tissue

Activity **5.8**

Connective Tissue: Microscopic Observations of Blood

A **Blood** is a connective tissue that has a fluid matrix **(plasma),** which consists of water and numerous dissolved solutes (e.g., proteins, ions, gases, hormones, nutrients). Red blood cells, white blood cells, and platelets, which constitute the **formed elements,** are suspended in the plasma. The various functions of blood include the transportation of substances, the regulation of various blood chemicals, and protection against infections and diseases.

1. View a slide of a human blood smear under low power and notice that the entire field is filled with blood cells. The vast majority of the blood cells are the pinkish-staining **red blood cells,** or **erythrocytes.** As you scan the slide, you should be able to identify the larger **white blood cells,** or **leukocytes.**

2. Move the slide so that some white blood cells are in the center of the field of view. Switch to high power to examine the blood cells more closely. Note that the red blood cells lack a nucleus. They lose their nuclei as they mature so that more space is available to store hemoglobin, the molecule needed to transport oxygen. White blood cells retain their nuclei, which are stained a deep blue or purple. Observe the irregular shapes of the nuclei in many white blood cells.

3. You might also see small cell fragments in the blood smear. These fragments are **platelets,** or **thrombocytes,** which are essential for blood clotting.

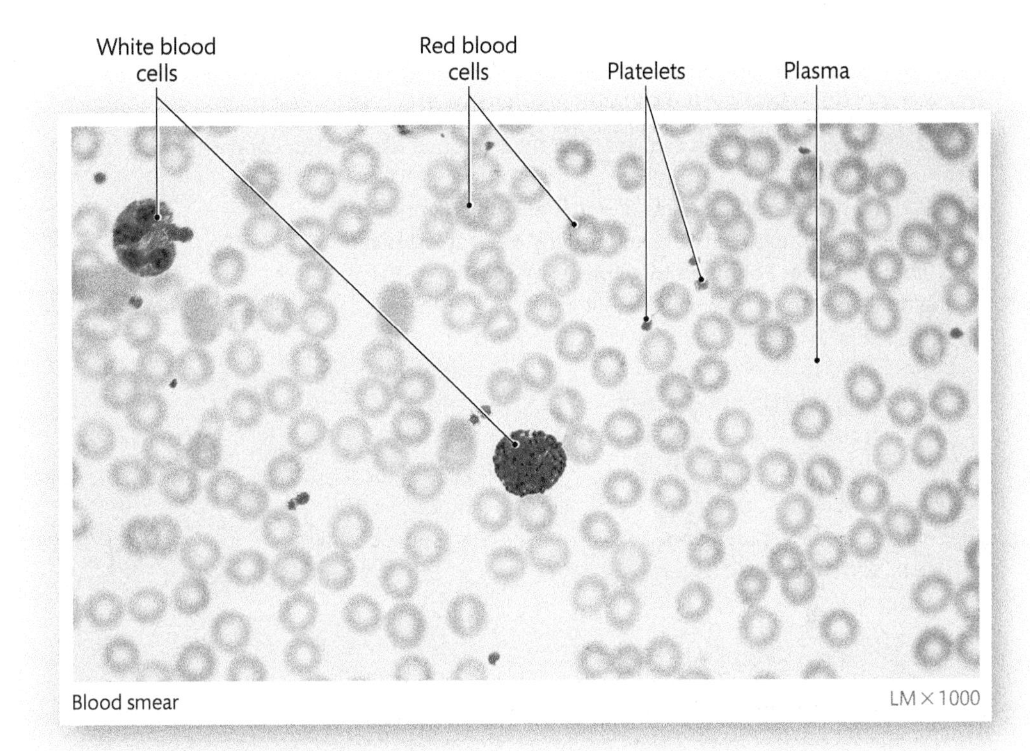

Blood smear

MAKING CONNECTIONS

During Activities 5.3–5.8, you observed connective tissues with diverse structures and functions. Compare the structure (e.g., cell types, protein fibers, ground substance) and function of the following three types of connective tissues: areolar connective tissue, bone, and blood. ▶

Connective Tissue Type	Cell Types	Protein Fiber Types	Consistency of Ground Substance	Function
Areolar connective tissue				
Bone				
Blood				

Activity 5.9

Muscle Tissue: Microscopic Observations of Muscle Tissue

The cells of muscle tissue are called muscle fibers because they have an elongated, fibrous shape. There are three types of muscle tissue—*skeletal muscle, cardiac muscle,* and *smooth muscle*—but the cells in all three types share the following unique characteristics:

- **Excitability.** Muscle fibers possess excitable membranes that receive and respond to stimuli from nerves or hormones.
- **Contractility.** Muscle fibers contain contractile proteins that allow them to change in length, or contract, and thereby generate a force to produce movement.
- **Elasticity.** Muscle fibers can return to their original length after contracting.

A **Skeletal muscles** have tendinous attachments to bone (and, in a few cases, to soft tissue) and usually cross at least one joint. They carry out voluntary movements by pulling on a bone to initiate an action at a joint.

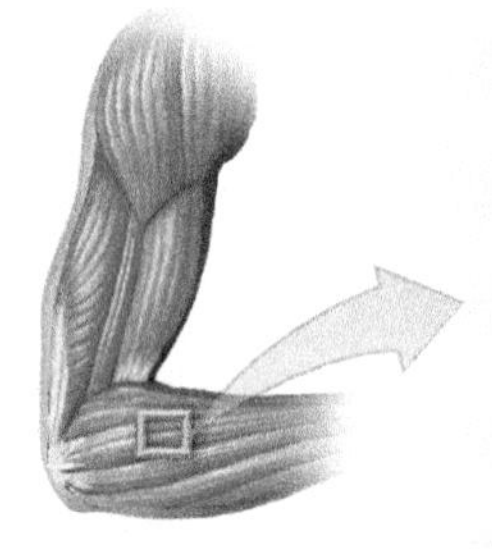

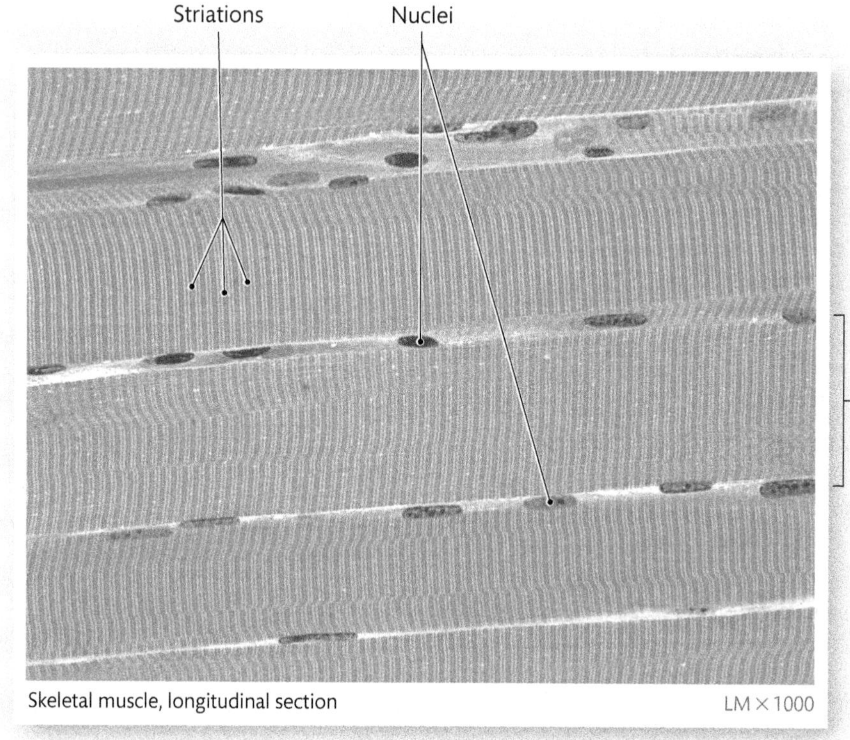

Skeletal muscle, longitudinal section

1. Observe a slide of skeletal muscle under low power. As you scan the slide, identify regions that contain longitudinal sections of muscle fibers.
2. Switch to high power and identify the distinctive **striations,** which will appear as alternating light and dark bands across the width of each muscle fiber. Observe that each muscle cell contains several peripherally located nuclei.
3. In the space provided to the right, draw your observations. Label a muscle fiber, striations, and nuclei.

▶

■ Want more practice? Go to: **MasteringA&P** > Study Area > Menu > Lab Tools > PAL > Histology > Muscle Tissue

B **Cardiac muscle** is located only in the heart wall. It is responsible for the involuntary contractions of the heart chambers (the heartbeat) that pump blood out of the heart.

1 Observe a slide of cardiac muscle under low power. As you scan the slide, move to an area of the field of view that contains longitudinally cut muscle fibers and switch to high power.

2 Notice that cardiac muscle forms a network of branching fibers. Striations are present, but they are less organized than in skeletal muscle. Also unlike skeletal muscle, the nuclei are centrally located. Most cardiac muscle cells have only one nucleus, but you might find a few multinucleated fibers.

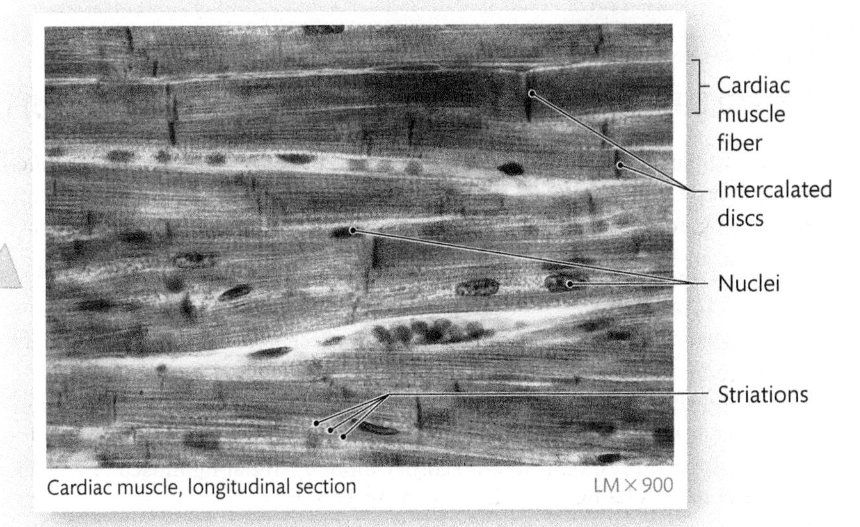

Cardiac muscle, longitudinal section LM × 900

3 Locate the specialized cell junctions known as **intercalated discs.** Under high power, the discs appear as lines traveling across the width of the muscle fibers. They are stained slightly darker than the cytoplasm.

Refer to your text and describe the function of the intercalated discs.

▶ ______________________________

____

____

C **Smooth muscle** is located in the walls of most internal organs. It is responsible for the involuntary movements associated with the function of various organs.

1 Observe a slide of smooth muscle under low power. Find an area of muscle fibers in a longitudinal section and move it to the center of the field of view.

2 Switch to high power. Notice that smooth muscle cells lack striations and contain one centrally located nucleus.

3 Examine an area that contains several layers of cells. Notice that the cells are broad in the middle and taper toward the ends. Also observe that the cells are layered in an overlapping way, similar to the arrangement of bricks in a brick wall.

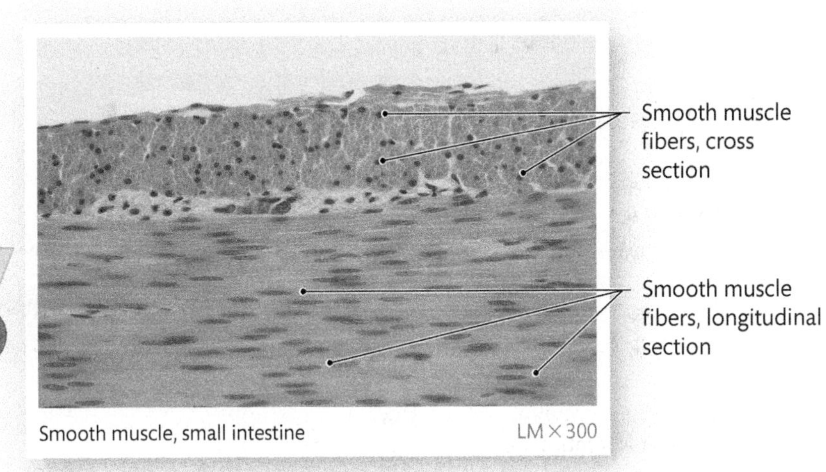

Smooth muscle, small intestine LM × 300

MAKING CONNECTIONS

Compare the structure and function of the three types of muscle tissue.

Muscle Type	Cell Structure	Location	Voluntary/Involuntary Contraction	General Function
Skeletal				
Cardiac				
Smooth				

Activity **5.10**

Nervous Tissue: Microscopic Observations of Neurons

Nervous (neural) tissue, the primary tissue type in the nervous system, provides the body with a means for maintaining homeostasis and ensuring that bodily functions are carried out efficiently. It does so by conducting electric impulses along nerve fibers to target organs. As a result of this electrical activity, the nervous system is able to control and integrate the activities of all the organs and organ systems. Nerve tissue is responsible for three basic functions:

- **Reception.** With the assistance of sensory receptors, it detects changes that occur inside the body and in the surrounding environment.
- **Integration.** It interprets and integrates sensory input by storing the information as memory and producing thoughts.
- **Response.** It responds to this sensory input by initiating muscular contractions or glandular secretions.

A The structural and functional unit of nervous tissue is the **neuron,** or **nerve cell.** More than 99 percent of all nerve cells are **multipolar neurons.** They have a **cell body** that contains a large, round nucleus and most of the cell's organelles. Extending from the cell body are many cell processes; one of these processes is an **axon,** and all the others are **dendrites.**

Other cells in nervous tissue, collectively called **neuroglia,** or **glial cells,** support and protect neurons and other structures. Neuroglia are more numerous but much smaller than neurons.

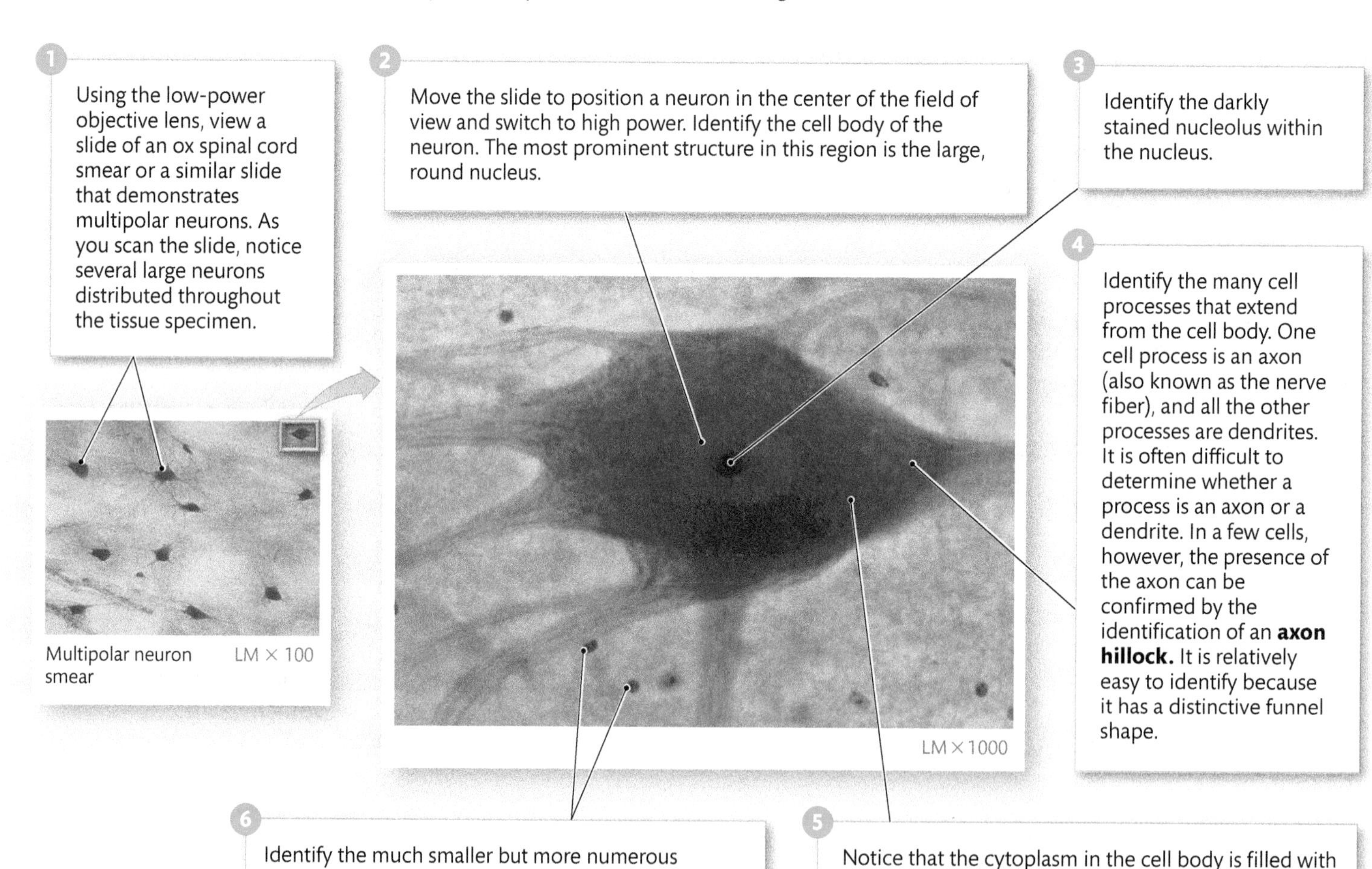

7 In the space provided, draw a region of the spinal cord smear as you view it with the microscope and label the following structures in a neuron: cell body, nucleus, nucleolus, dendrites, axon, axon hillock, and chromatophilic (Nissl) bodies. Also label the glial cells.

▶

Word Origins

The word *glial* is derived from the Greek word *glia,* which means "glue." The name reflects the general function of the glial cells: sustaining and protecting neurons.

The term *chromatophilic* is derived from Greek and means "color loving." The chromatophilic bodies in nerve cells readily stain with biological dyes and typically appear as dark blue or purplish structures.

B The dendrites, along with the cell body, serve as contacts to receive impulses from other neurons. Axons generate electric impulses (action potentials) and transmit them to other cells at special junctions known as **synapses.** Usually, axons give off several collateral branches, each ending with a dilated region called the **synaptic (axon) terminal** or **synaptic knob.** Cells receiving impulses from a neuron could be another neuron, a muscle cell, or a glandular cell.

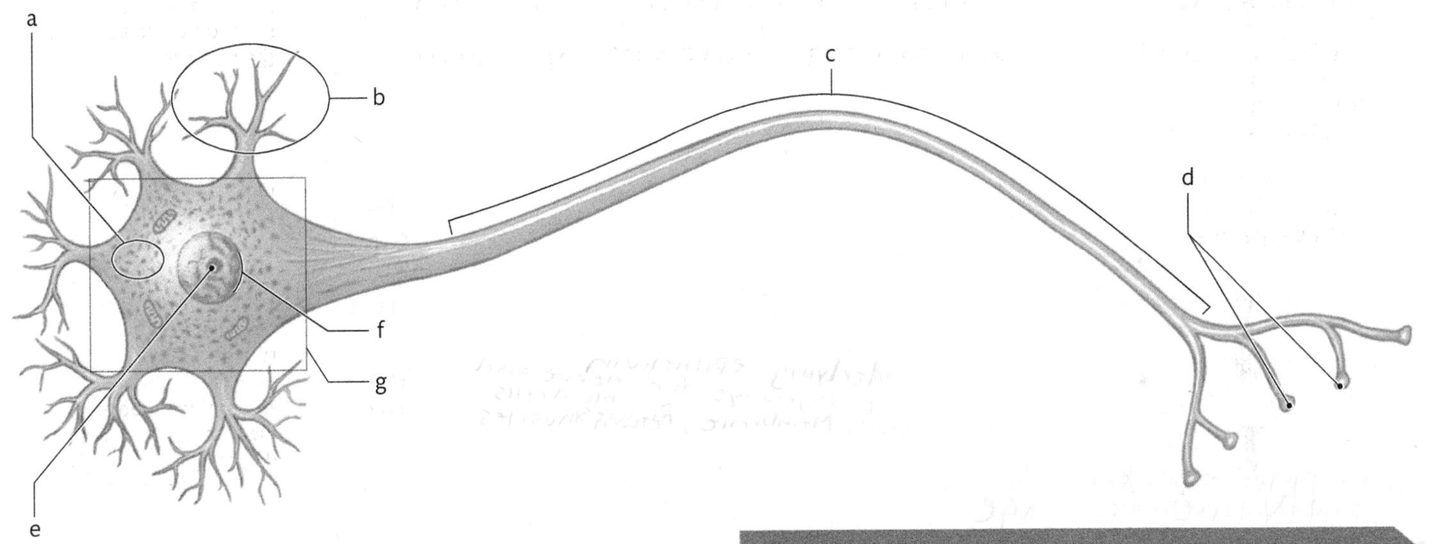

MAKING CONNECTIONS

The cytoplasm of the axon (the axoplasm) contains numerous protein filaments—neurofibrils and neurotubules—that are components of the cytoskeleton. These structures assist in transporting materials from the cell body to the synaptic terminals. Speculate on why this process, known as axoplasmic transport, is a vital activity for sustaining normal function at synapses.

▶ ______________________________

____

____

____

____

1 If available, study a model of the neuron and compare its structure with your microscopic observations.

2 Identify the following structures in the diagram above. ▶

1. cell body	a. ______
2. chromatophilic bodies	b. ______
3. axon	c. ______
4. nucleolus	d. ______
5. dendrites	e. ______
6. nucleus	f. ______
7. synaptic terminals	g. ______

BEFORE YOU MOVE ON . . .

« LOOKING BACK

In this exercise, you were introduced to the biological discipline known as **histology,** which is the microscopic study of tissues. Tissues contain groups of cells that work in concert to perform specific functions. Epithelial tissue provides protection by covering body surfaces with cells that are held firmly together by cell junctions. It also performs other specialized functions depending on the type and location of the epithelium. Connective tissue is highly variable in structure and function. The cells are separated by a matrix that consists of a ground substance and protein fibers. Connective tissues connect structures and provide support and protection for vital organs. They also defend the body from infections and repair damaged tissue. Muscle tissue contains elongated cells, called muscle fibers. Muscle cells are filled with contractile proteins that allow them to contract and produce movement. Nervous tissue contains highly specialized cells, called neurons, that can initiate and transmit electrical impulses to other cells. By conducting electric impulses to target cells, neurons are able to regulate and control body functions.

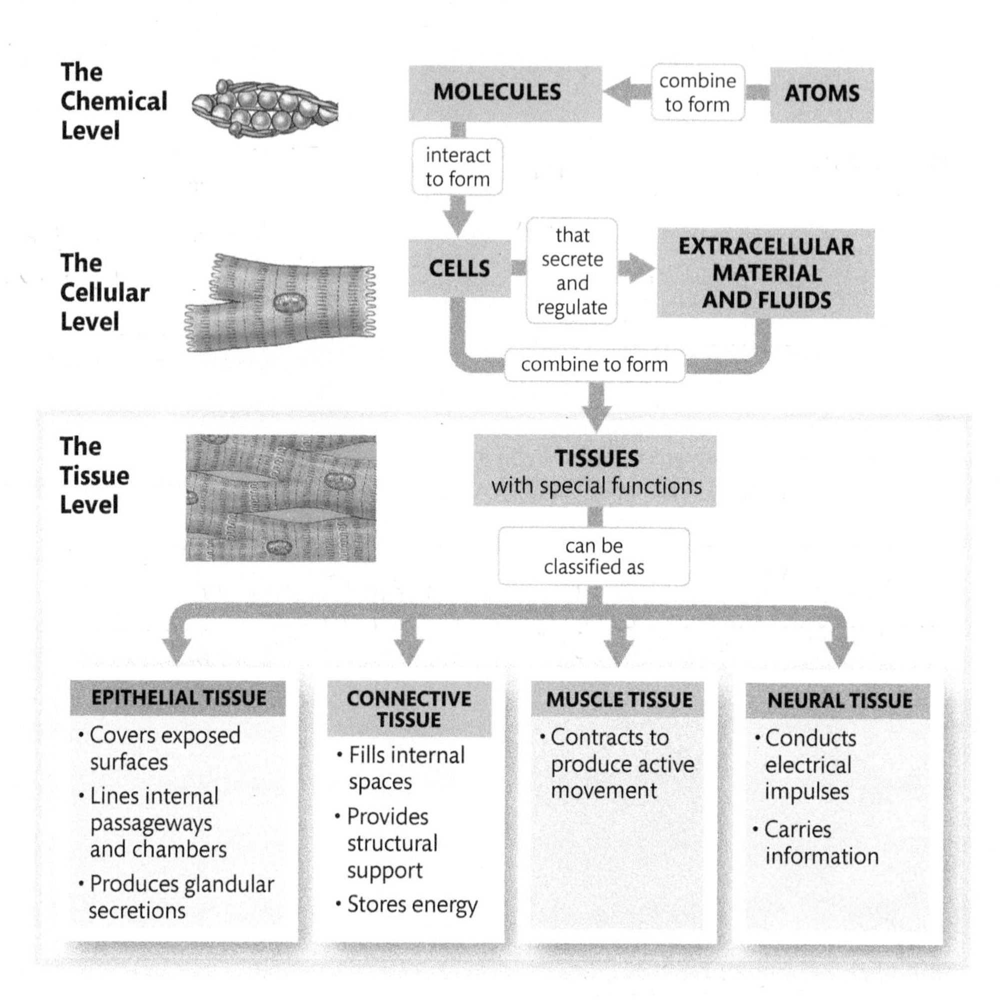

Tissues perform highly specialized and unique functions. For that to be possible, each individual cell must be well coordinated and synchronized with the functions of all the other cells in that tissue. Using muscle and neural tissues as examples, consider why such coordination is necessary.

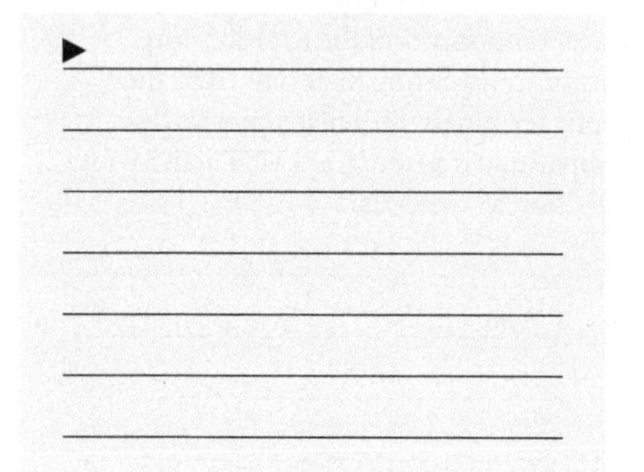

LOOKING FORWARD »

The tissue level is significant because it is the first multicellular level of organization you have studied. As you move forward to the next exercise and beyond, you will begin exploring higher levels of organization: the organ level and the organ system level. An **organ** is a structure that contains two or more of the primary tissue types (and often all four). Each tissue type works collaboratively with the others to perform specific functions. An **organ system** consists of two or more organs that work together in a coordinated manner to perform specific activities (e.g., respiratory activities, digestive activities).

Name Gabriel Vanevenhoven

Lab Section ______

Date 2/13/23

EXERCISE 5 REVIEW SHEET

Tissues

1. Complete the following table.

Type of Epithelium	Structure	Location	Function
Simple squamous	a. Thin Scaly	b. lung / GI	c. Diffusion secretion
d. Pseudostratisfied	e. Multilayered	Trachea	f. Secrection
g. Stratified Squamous	h. Multiple cell layer Top layer of compact dead cells	i. Skin	Protection from abrasion and friction
j. Simple Columnar	One layer of tall cells	k. GI / Uterus	l. Absorbtion Secrection
m. Transitional	n. Multi Layered Round → Flat when stretched	Urinary bladder	o. Filling of Urinary tract
p. Stratisfied Cubodial	q. Two or more Layer of cells surface cells square or round	Skin	r. Secreetion
s. Simple Cubodial	One layer of cube-shaped cells	t. Kidney Liver	u. Absorbtion secrection

2. Identify the epithelial types illustrated in the diagrams to the right.

a. Simple cubodial

b. Stratified Squamous ←(Non Keratized)

c. Pseudostratified Columnar

d. Stratified Columnar

e. Simple Columnar

f. Simple Squamous

(a)

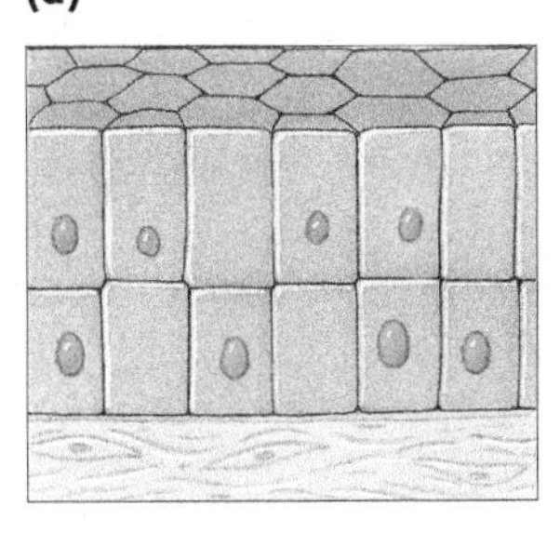

(b)

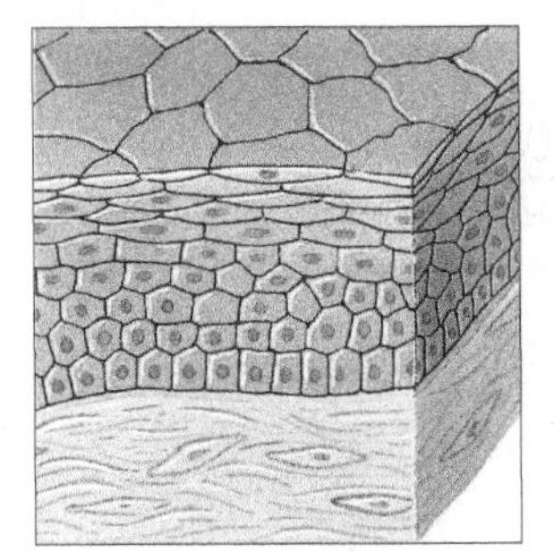

(c)

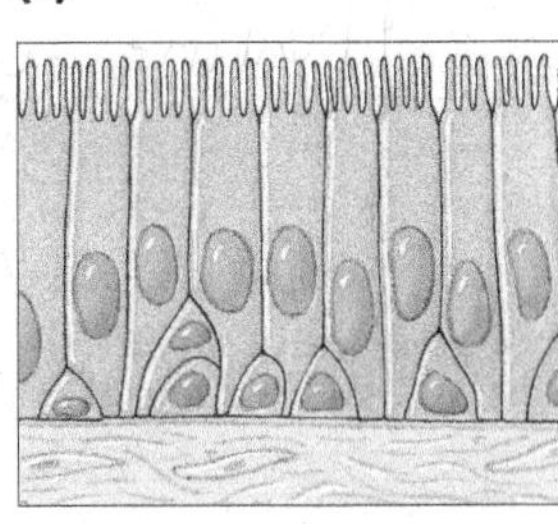

(d)

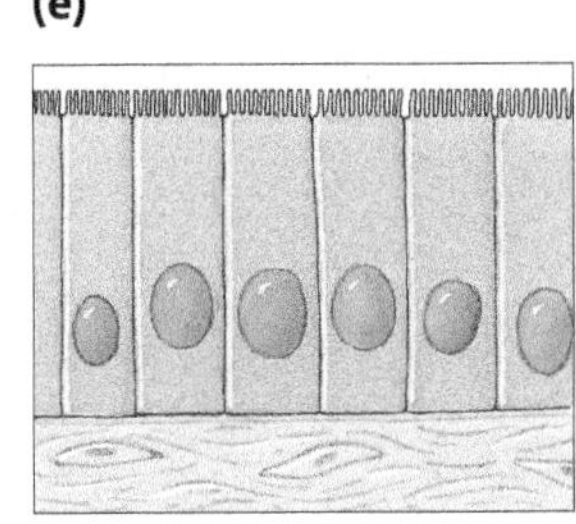

(e)

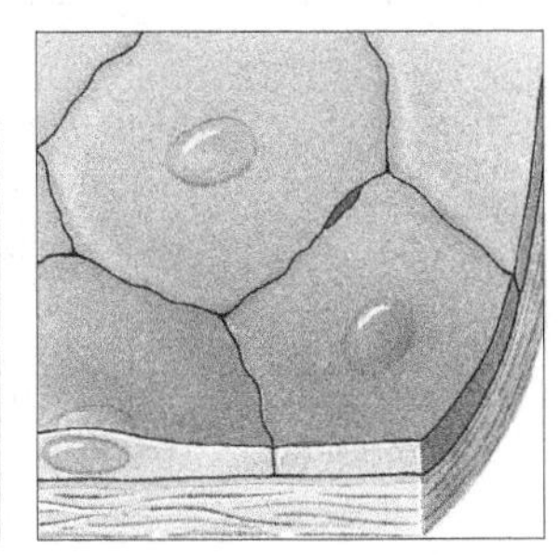

(f)

3. Describe the structure and function of the connective tissue matrix.

The Structure of the connective tissue matrix are cartilages, bones, fibrous tissue, adipose tissue, and blood. The function is the binding of organs, support, physical protection, immune protection, movement, storage, heat production, and transport.

4. Why is it important for the cartilaginous discs located between the vertebrae to be composed of fibrocartilage?

Because the spine is the center for the most movement and impact so the cartilage needs to be able to resist compression but also be able to absorb shock from movement.

5. Complete the following table.

Connective Tissue Type	Location	Function
Adipose tissue	a. Fat, Breast, Surround kidney and eye	b. Energy storage, Insulation, heat production, protect organ
c. Dense Regular Connective Tissue	Tendons and ligaments	d. Lig = Bone → Bone, Tend = Muscle → Bone
e. Fibrocartilage	Intervertebral discs	f. Reisist compression and absorbtion of shock
g. Compact Bone	Skeleton	h. Physical Support, leverage for muscles
Areolar connective tissue	i. Underlying epithelium, Passageways for nerve and blood cells, Serous Membrane, Betwen Muscles	j. Bind epithelium, Passage of nerves, Arena for immune defence
k. Hyaline Cartilage	Articular cartilage at synovial joints	l. Eases Joint movements, Hold air way open, Move vocal chords

QUESTIONS 6–9: Coloring exercises. For each diagram, color the structures with the indicated colors.

6.

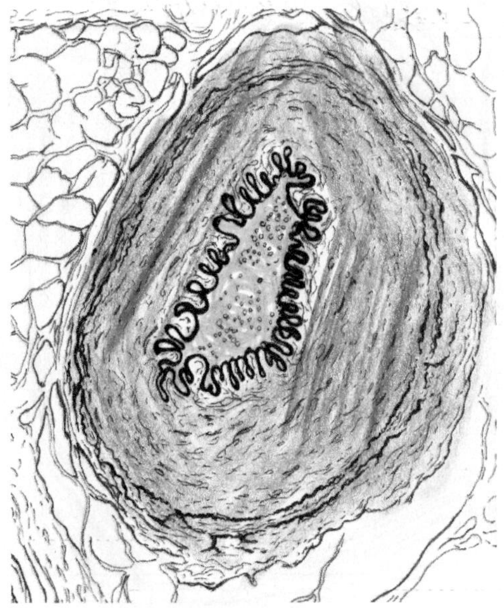

Cross section of an artery

Simple squamous epithelium in the wall of the artery = **green**
Muscle layer = **red**
Connective tissue layer = **yellow**

7.

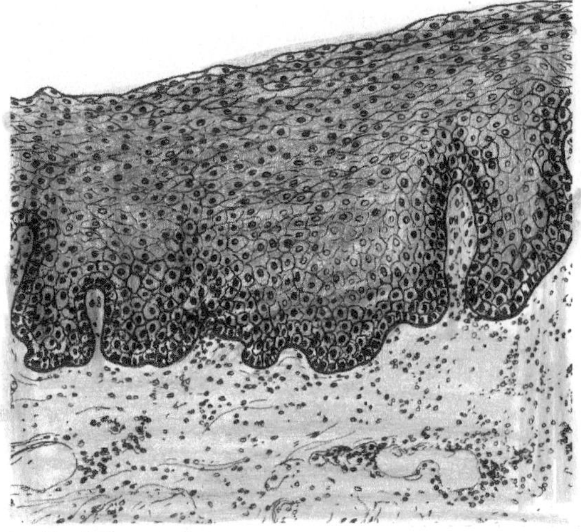

Wall of the esophagus

Epithelium = **red**
What type of epithelium?
Non keritanized Stratified Squamous

Connective tissue = **green**

8.

The skin

Epithelium = **red**
What type of epithelium?
Keratinized Stratified Squamous

Connective tissue = **green**

9.

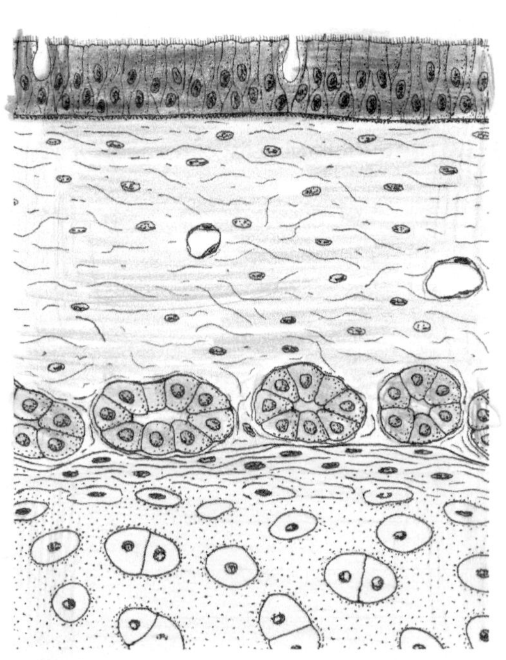

Wall of the trachea

Epithelium covering the surface = **red**
What type of epithelium covers the surface?
Pseudostratified Columnar

Glandular epithelium = **blue**
Dense irregular connective tissue = **green**
Cartilage = **yellow**
What type of cartilage?
Hylaine

10. Identify the type of muscle tissue that is shown in each of the following figures.

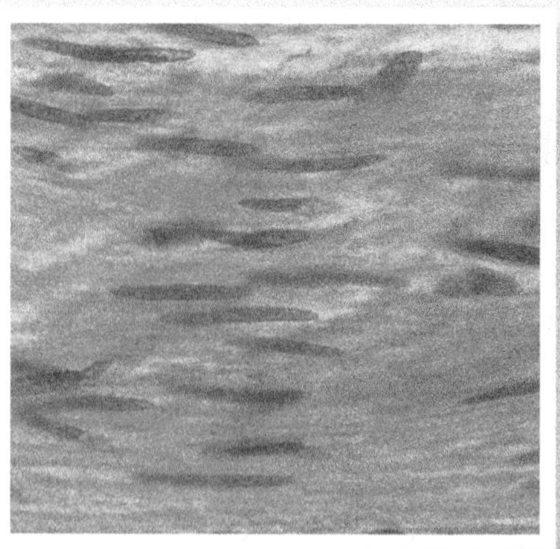

LM × 1100

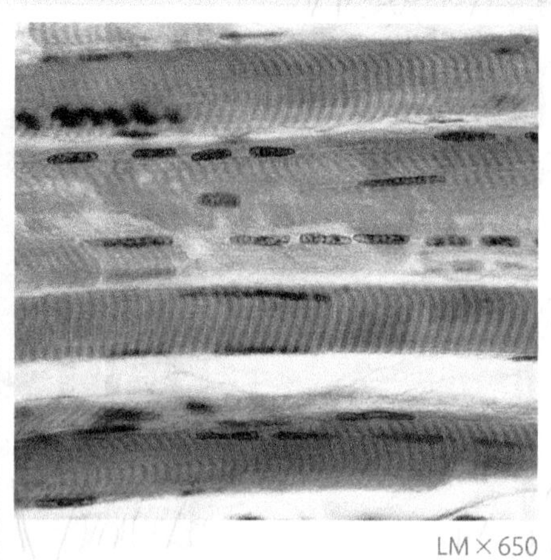

LM × 650

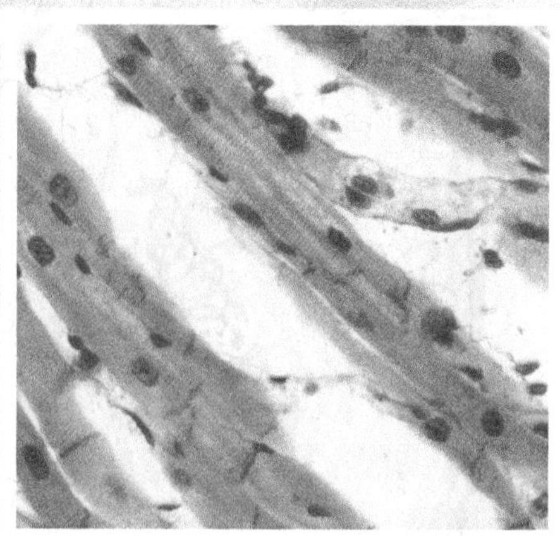

LM × 750

a. Smooth

b. Skeletal

c. Cardiac

11. Explain why bone and blood, which have strikingly diverse structures, are both categorized as connective tissues.

They are both connective tissue because they contribute to the functions of connective tissue. The bones support the body while the blood transports gases, nutrients, wastes, hormones, and blood cells.

QUESTIONS 12–17: Use the following list of terms to identify the labeled structures in the photo.

a. Dendrites
b. Neuroglia
c. Axon hillock
d. Nucleus
e. Cell body
f. Axon

12. Cell Body

13. Nucleous

14. Dendrites

15. Neuroglia

16. Axon

17. Axon Hillock

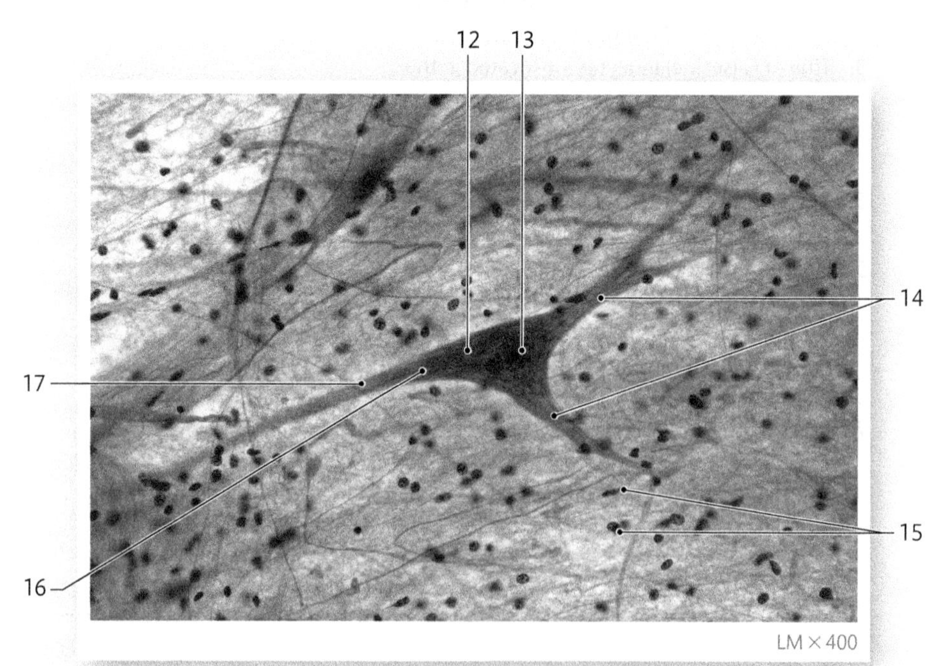

LM × 400

EXERCISE 6

The Integumentary System

LEARNING OUTCOMES

These Learning Outcomes correspond by number to the laboratory activities in this exercise. When you complete the activities, you should be able to:

Activity 6.1 **Describe the organization of the epidermis, dermis, and hypodermis.**

Activity 6.2 **Describe the structure and function of the accessory structures of the skin.**

Activity 6.3 **Compare the three-dimensional organization of the integumentary system with microscopic observations.**

Activity 6.4 **Explain how fingerprints are formed and observe the variation in fingerprint patterns among individuals.**

Activity 6.5 **Describe the structure of nails.**

LABORATORY SUPPLIES

- Compound light miccroscope
- Prepared microscope slides:
  - scalp
  - scalp, pigmented
  - skin, sole of foot
- Anatomical model of skin
- Handheld magnifying glass
- Colored pencils

PRE-LAB QUIZ

Before you begin, read all the activities in Exercise 6 and the required reading in your textbook that is assigned by your instructor.

1. True or False: During this laboratory exercise, you will be viewing a slide of the scalp to observe thick skin. ___________
2. True or False: During this laboratory exercise, you will use a compound light microscope to examine your fingerprints. ___________
3. True or False: Melanocytes are located in the stratum basale of the epidermis.

   ____
4. True or False: Sebum, produced by a sebaceous gland, drains into a hair follicle.

   ____
5. True or False: Merocrine sweat glands do not become functional until puberty.

   ____
6. Which of the following is not an accessory structure of skin?
   a. sweat gland
   b. arrector pili muscle
   c. hair follicle
   d. dermal papilla
7. Lamellated corpuscles are sensory receptors that respond to
   a. light touch.
   b. deep pressure.
   c. pain.
   d. temperature.
8. The epidermis is composed of what type of epithelium? ___________________________

   ____
9. Arrector pili muscles are composed of what type of muscle tissue? ___________________

   ____
10. Similar to a hair shaft, a nail consists of dead cells filled with ______________________.

Activity **6.1**

Examining the Microscopic Structure of Skin

A Examining the Organization and Microscopic Structure of the Skin

In terms of total surface area, the skin (**cutaneous membrane**) is the largest organ of the body. It consists of two tissue regions: the outer epidermis and the inner dermis.

1. View a section of skin from the sole of the foot under low power. Locate the darkly stained **epidermis** at one edge of the section. It consists of a stratified squamous, keratinized epithelium. Most of the cells in the epidermis are called keratinocytes.

2. Locate the **dermis** just deep to the epidermis. The narrow papillary layer, bordering the epidermis, contains areolar connective tissue. The reticular layer comprises the majority of the dermis and consists of dense irregular connective tissue.

3. Deep to the dermis, locate the **hypodermis.** Observe that it is composed largely of areolar connective tissue and adipose tissue. The hypodermis, also called the superficial fascia, is not considered to be a part of the integumentary system, but it contains blood vessels and nerves that supply the skin. In addition, the hypodermis connects the skin to underlying muscles.

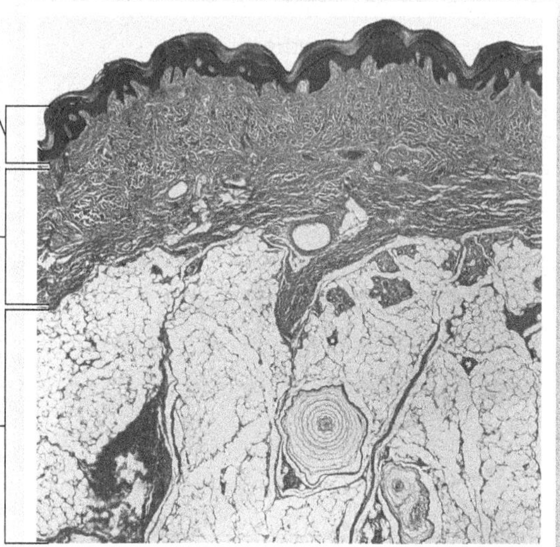

Cutaneous membrane and hypodermis LM × 40

B Epidermis of Thick Skin

In **thick skin,** which covers the palms of the hands and soles of the feet, the epidermis is relatively thick and contains five cell layers: the **stratum basale, stratum spinosum, stratum granulosum, stratum lucidum,** and **stratum corneum.**

1. Switch to high power and observe the epidermis more closely. On conventional microscope slide preparations, it is difficult to identify all the epidermal cell layers. However, the stratum corneum and stratum basale can be located.

2. Locate the **stratum corneum** (keratinized layer), the outermost layer of the epidermis. It contains several layers of dead keratinocytes whose cytoplasm is filled with keratin. The stratum corneum forms a protective, water-resistant covering, but the cells in this layer are eventually shed from the surface. In the skin on the sole, the stratum corneum is very thick.

3. Locate the **stratum basale,** a single layer of cuboidal cells located along the base of the epidermis. These cells regularly divide by mitosis to replace the old keratinocytes that are sloughed off the stratum corneum.

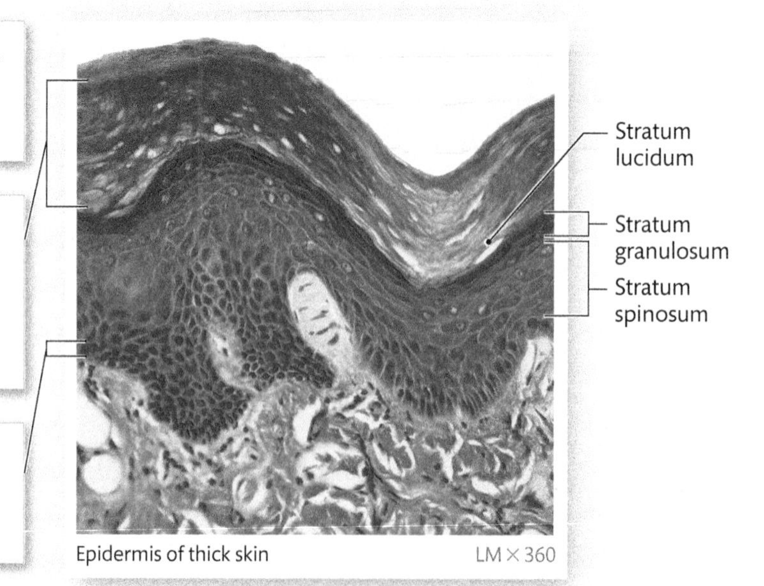

Epidermis of thick skin LM × 360

C Thin Skin

In **thin skin,** which covers all parts of the body except the soles and palms, the stratum lucidum is absent, and the stratum corneum is much thinner than in thick skin.

1. Scan a slide of thin skin from the scalp under low power and identify the three tissue layers that you observed in the previous slide:

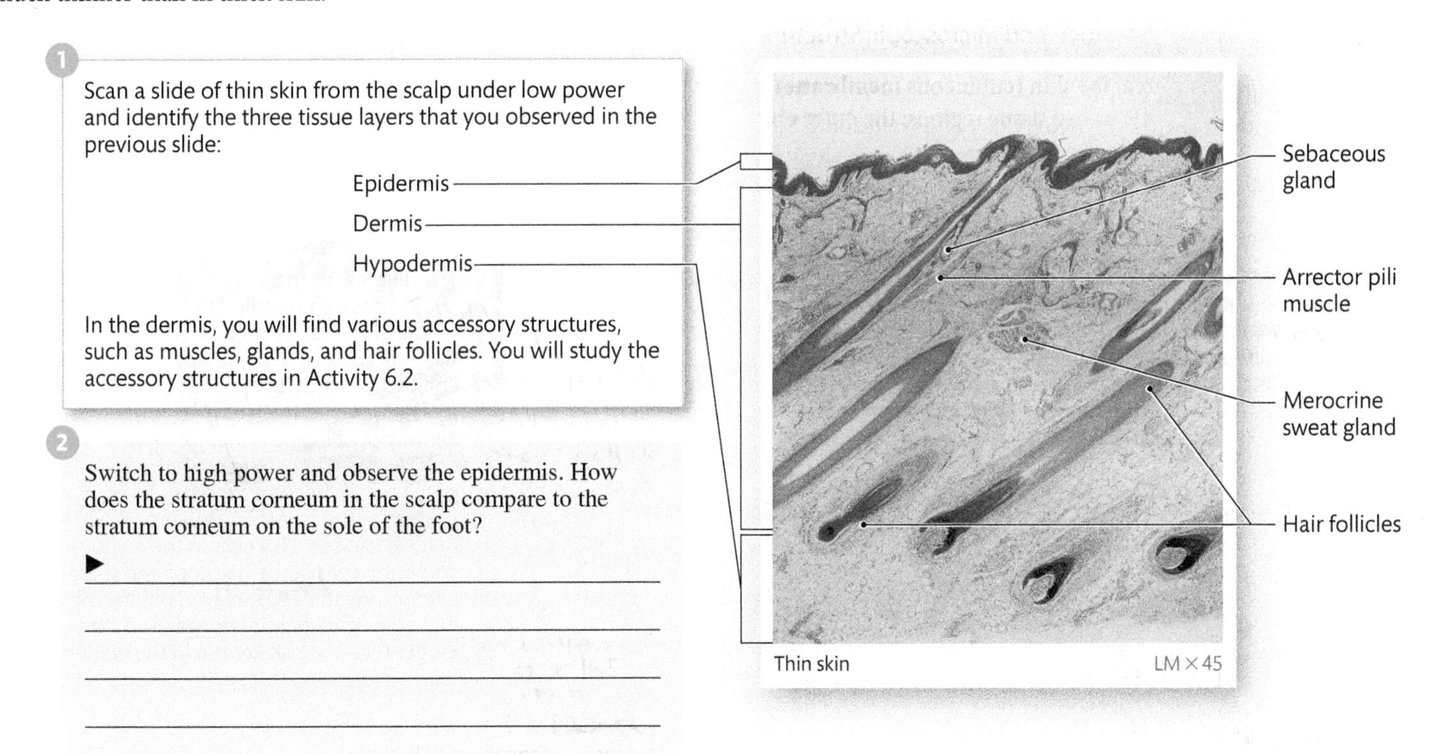

Thin skin LM × 45

In the dermis, you will find various accessory structures, such as muscles, glands, and hair follicles. You will study the accessory structures in Activity 6.2.

2. Switch to high power and observe the epidermis. How does the stratum corneum in the scalp compare to the stratum corneum on the sole of the foot?

▶ ______________________________

____

____

____

3. View a slide of the pigmented scalp under low power. Some cells in the epidermis contain melanin, which is a protein pigment that appears dark brown on your slide. Melanin is produced by cells called melanocytes, which are found deep in the stratum basale. Describe the distribution of the pigment within the epidermis on your slide.

▶ ______________________________

____

____

____

____

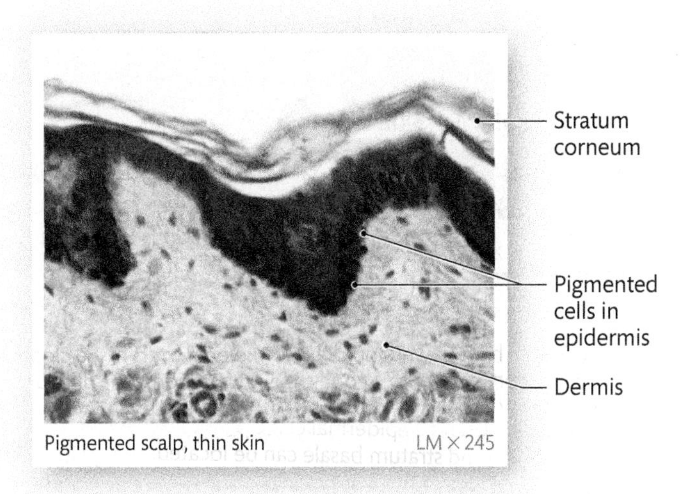

Pigmented scalp, thin skin LM × 245

Word Origins

Dermis is derived from the Greek word *derma,* which means "skin." *Epidermis* and *hypodermis* are derived by adding Greek roots to *-dermis.* The prefix epi- means "upon," and, indeed, the epidermis is located upon the dermis. *Hypo-* means "under"; the hypodermis is deep to the dermis.

MAKING CONNECTIONS

In this activity, you observed that the stratum corneum is much thicker in the sole (thick skin) than in the scalp (thin skin). Explain why you would expect such a difference.

▶ ______________________________

____

____

____

Activity 6.2

Studying the Accessory Structures of Skin

Accessory structures of the skin are well established in the dermis. Portions of some structures, such as hair follicles and the ducts of sweat glands, extend deeply into the hypodermis or superficially into the epidermis. The accessory structures that you will examine in this activity include hair follicles, arrector pili muscles, sebaceous glands, and sweat glands.

A Hair Follicles and Arrector Pili Muscles

1. **Hair follicles** are tubular compartments, each one containing a hair. The follicle walls are continuous with the epidermis and extend deeply into the dermis and hypodermis. Under low power, scan a slide of thin skin from the scalp. Identify numerous hair follicles in the dermis. Locate some follicles that are directly connected to the epidermis. Because of the plane of section, other follicles may not appear to have this connection.

2. The deepest portion of a hair follicle is a knoblike structure called the **hair bulb.** Identify these structures on your slide. The hair bulb contains the hair papilla, an area of connective tissue that contains blood vessels and nerves, and the hair matrix, a region of actively dividing cells that give rise to the hair root. As the hair root approaches the surface of the skin, it becomes the hair shaft. A portion of the hair shaft is that part of the hair that we can see. Like cells in the stratum corneum, cells in the hair root and shaft become keratinized and die. The hard keratin that covers the hair is stronger and lasts longer than the soft keratin in the epidermis.

Epidermis

Hair shaft

Sebaceous gland

Merocrine sweat gland

Hair root

Thin skin LM × 105

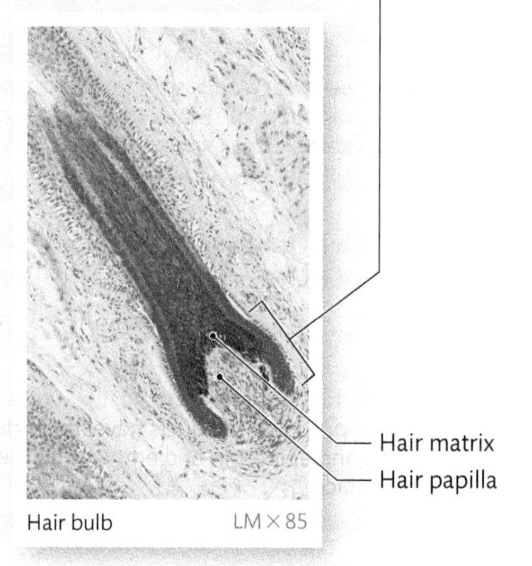

Hair bulb LM × 85

3. Identify **arrector pili muscles.** These structures are smooth muscle fibers attached to the connective tissue sheaths of hair follicles. They pass obliquely toward the papillary layer of the dermis, but because of the plane of section, you may not see the entire muscle. In response to cold temperatures or emotions such as fear or anger, arrector pili muscles contract. This action moves hair follicles to an upright position and dimples the skin to form "goose bumps."

B Sebaceous Glands

Sebaceous glands are found in all regions of the body except the palms of the hands and the soles of the feet. They produce a lipid mixture called **sebum** that lubricates the skin and hair. Sebum may also help protect against bacterial and fungal infections. The glandular cells secrete sebum by rupturing, a process known as holocrine secretion. The sebum empties into a short straight duct, which drains into a hair follicle. In hairless regions of the body, sebaceous glands, called sebaceous follicles, secrete sebum into ducts that run directly to the skin's surface.

1. On a slide of thin skin from the scalp, locate **sebaceous glands** and verify that they are connected to the walls of hair follicles. Due to the plane of section, this connection will not be seen for all glands.

2. Notice that the glandular cells are arranged like clusters of grapes; they are lightly stained or clear with dark-staining nuclei.

Epidermis
Hair follicle

Sebaceous glands LM × 180

C Sweat Glands

There are two types of sweat glands: **merocrine (eccrine)** and **apocrine. Merocrine sweat glands** are vital components of the body's temperature-regulating mechanism. They release sweat via ducts onto the skin surface. When the water in sweat evaporates, excess heat is released, and the body is cooled. **Apocrine sweat glands** are restricted to regions around the nipples, armpits, and anogenital area. They do not become functional until puberty. Their activity increases during periods of stress, pain, or sexual arousal, and their secretions, which are released into hair follicles, have a distinct odor.

1. On a slide of thin skin from the scalp, locate merocrine sweat glands scattered throughout the dermis between hair follicles. (You will not find apocrine sweat glands on this slide.)

2. Identify the **glandular cells,** the regions of the sweat glands that produce and secrete sweat.

3. Identify the sweat gland **ducts,** the darker-staining portions. The ducts travel through the dermis and epidermis and empty onto the skin surface.

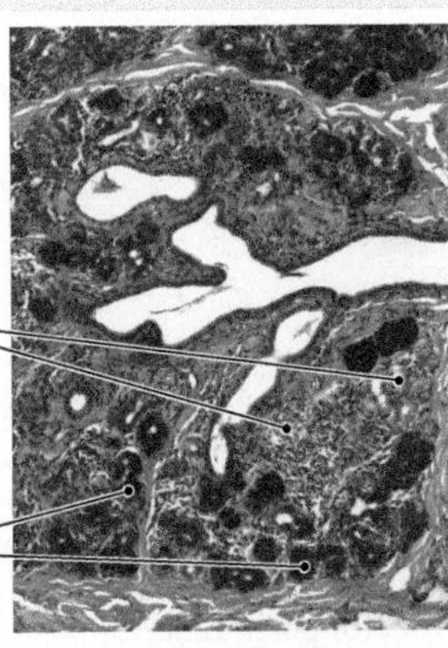

Merocrine sweat glands LM × 70

IN THE CLINIC

Acne

At puberty, under the influence of increasing levels of sex hormones, sebaceous glands enlarge, and sebum secretion increases. The overproduction of sebum can sometimes block sebaceous gland ducts, forming a **whitehead** on the skin's surface. A **blackhead** forms if the sebum is oxidized and darkens. **Acne** infections are caused by bacteria that accumulate in the blocked ducts, leading to inflammation and the formation of pimples or acne lesions on the skin. Treatment for acne depends on the severity of the infection. For mild cases, gentle washing with warm water and mild soap might be sufficient. For severe infections that cause cysts, nodules, and scarring, antibiotic gels and other prescription medications might be required.

MAKING CONNECTIONS

Chemotherapy drugs are strong medications that attack rapidly dividing cancer cells. One of the side effects of these drugs is hair loss, not only on the scalp, but also of pubic and armpit hair, eyelashes, and eyebrows. Why do you think these drugs cause hair loss?

▶ ______

Activity 6.3

Examining an Anatomical Model of the Integumentary System

A Examining a model of the integumentary system allows you to study the spatial arrangement of the accessory structures and sensory receptors in the cutaneous membrane and to compare your observations with previous microscope work.

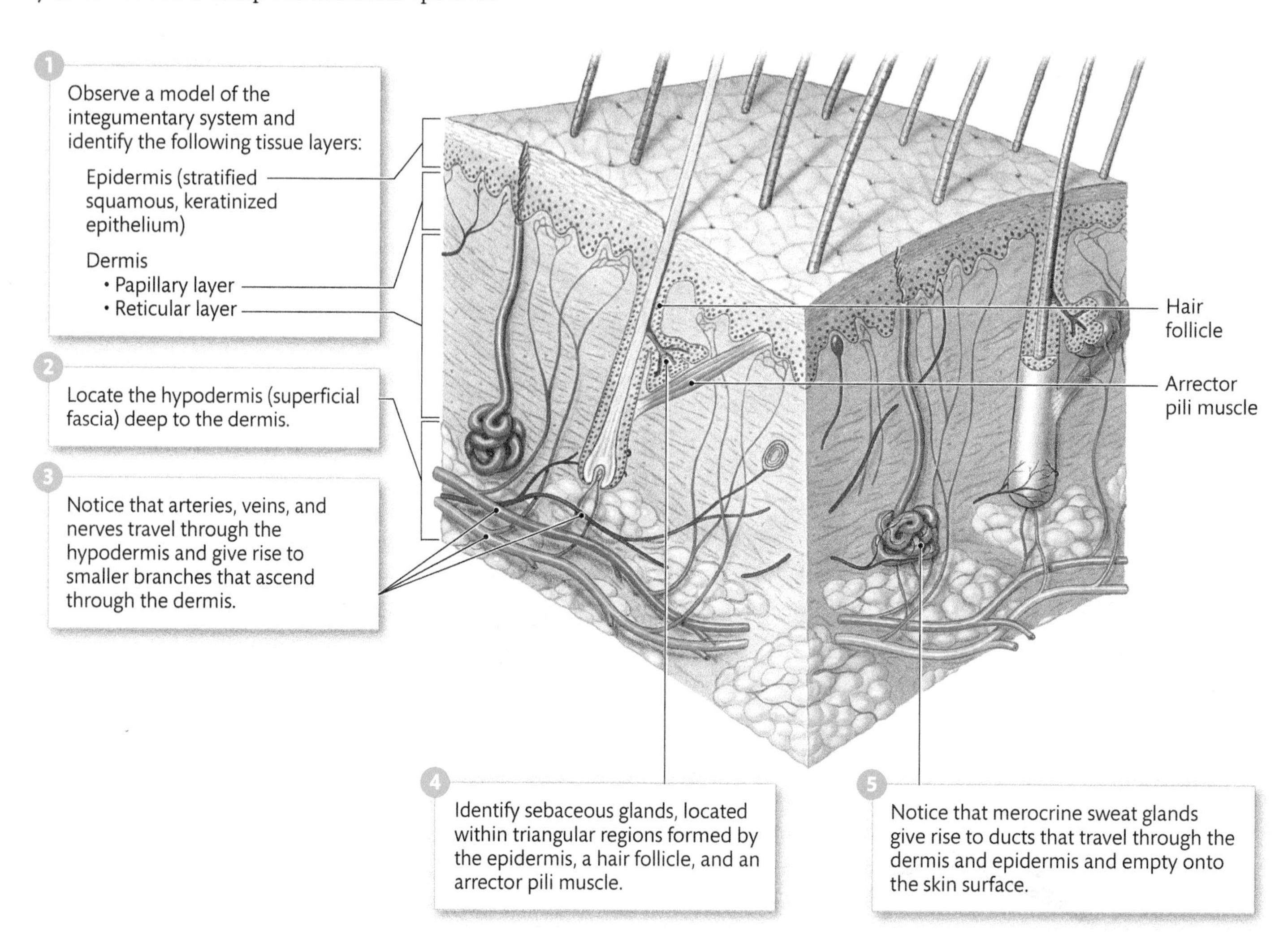

IN THE CLINIC

Malignant Melanoma

Malignant melanoma is a type of skin cancer that causes unregulated reproduction of melanocytes. Because these cells produce the pigment melanin, melanomas are often easily spotted because of their dark coloration. Although these malignancies may grow slowly at first, they become aggressive and, if not caught early, have a high mortality rate. Normally, melanin protects us from genetic damage caused by ultraviolet radiation, which can lead to melanoma and other skin cancers. People who have skin with a high level of melanin have more protection against this damage. The risk of melanoma is about 20 times higher for people of European descent than for people of African descent. Because the occurrence of melanoma in people of African descent is relatively rare, its diagnosis is often delayed. As a result, when the disease is discovered, it is usually at a more advanced stage, and the chances of survival are reduced. Early detection is the key, and that requires frequent self-examination of your skin, including areas that are not exposed to sunlight.

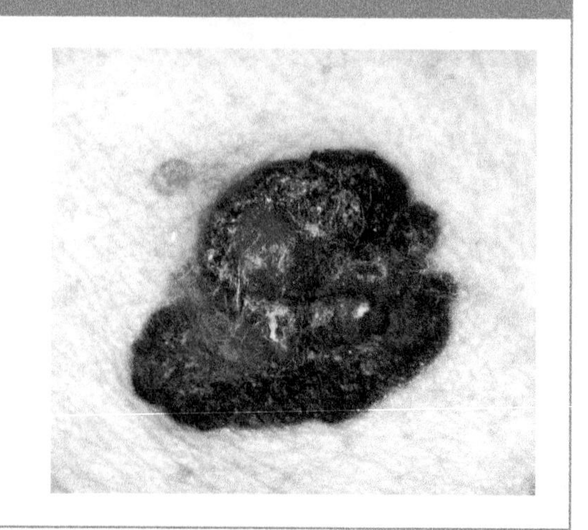

B The epidermis forms an undulating border with the papillary layer, along which dermal papillae project between epidermal ridges.

1. On a model of the integumentary system, identify the **dermal papillae** projecting into the epidermis.

2. Identify the **epidermal ridges** dipping into the papillary layer of the dermis.

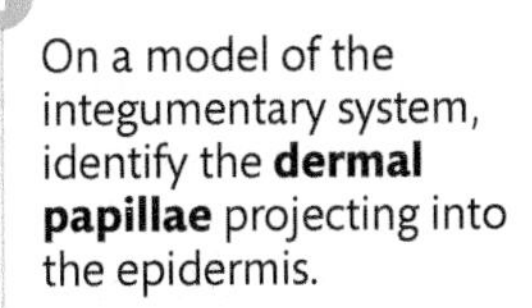

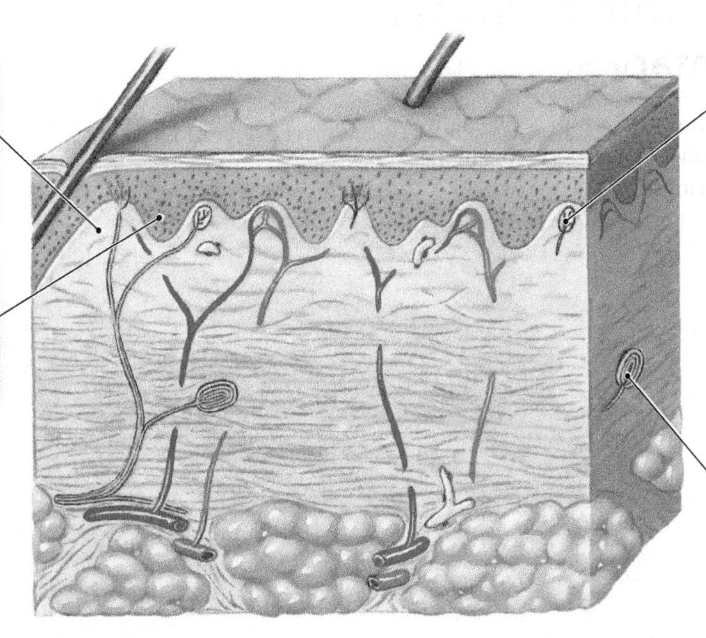

C Skin contains several types of sensory receptors.

1. **Tactile corpuscles** are nerve endings wrapped in a connective tissue sheath. They are sensitive to fine or light touch. Tactile corpuscles are abundant in the lips, nipples, fingertips, palms, soles, and external genitalia. On the model, identify these receptors in the papillary layer of the dermis, immediately deep to the epidermis.

2. **Lamellated corpuscles** are nerve endings enclosed in a relatively large, onion-shaped connective tissue sheath. They respond to deep pressure sensations. Identify these receptors deep in the dermis or in the hypodermis.

3. **Tactile discs** are disc-shaped nerve endings associated with specialized tactile (Merkel) cells found deep in the epidermis. Tactile discs are sensitive to light touch.

4. **Free nerve endings** of sensory neurons extend into the epidermis. They respond to pain and temperature sensations.

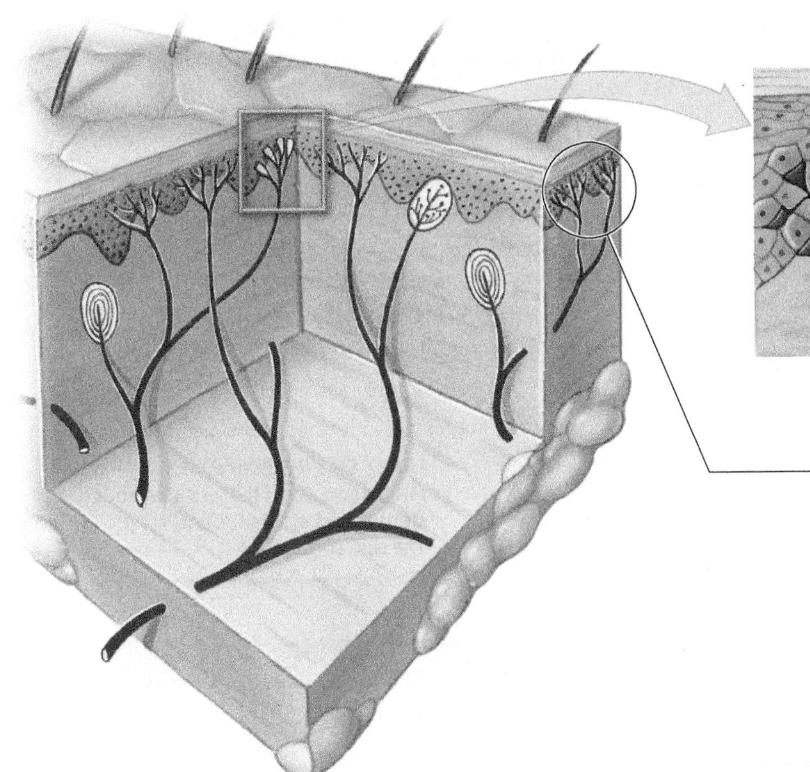

MAKING CONNECTIONS

Why do you think tactile corpuscles are located more superficially in the dermis than lamellated corpuscles?

▶ ______________________________

____

____

____

Activity **6.4**

Examining Fingerprint Patterns

A In each fingertip, the border between the papillary layer and the epidermis produces a distinctive pattern of epidermal ridges that form fingerprints. The idea that fingerprint patterns are unique and specific to each individual is widely accepted in both the scientific and law enforcement communities. Thus, fingerprint impressions, collected at a crime scene, are considered by a court of law to be highly reliable sources of forensic evidence.

1. Observe the pattern of epidermal ridges on the skin that covers your fingertips.

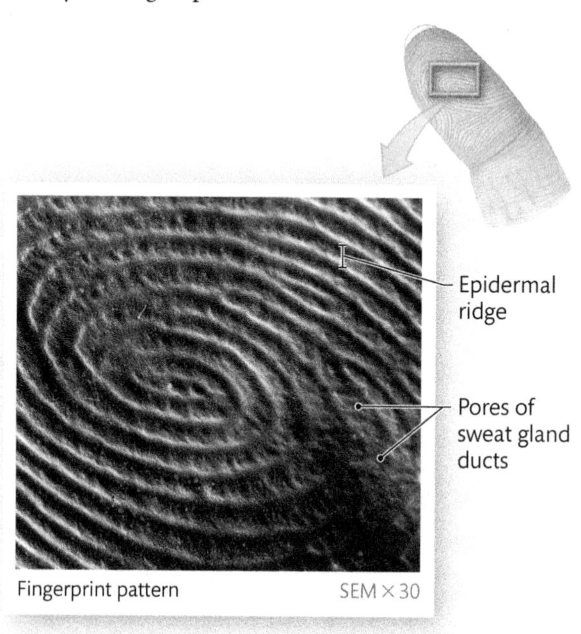

Fingerprint pattern SEM × 30

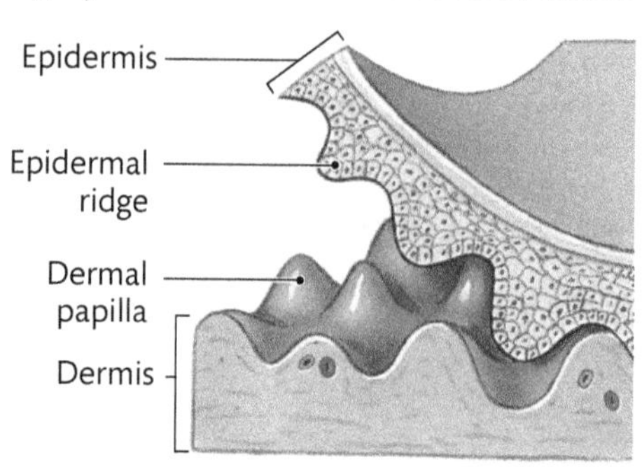

IN THE CLINIC

Using Fingerprints to Detect Drug Use

Forensic scientists are developing a new technique to analyze fingerprints to detect drug use. Fingerprints are routinely identified with iron oxide particles that attach to skin secretions that accumulate on objects touched by a person. In the new technique, iron oxide is combined with antibodies to various chemicals and applied to a fingerprint. A fluorescent dye is added, and if the specific antibody target is present, it will glow. Thus far, antibodies for THC in marijuana, cocaine, nicotine, and methadone are available. Future applications might include the detection of diseases such as cancer, heart disease, and diabetes that produce specific chemicals found in sweat and sebaceous gland secretions.

2. Use a magnifying glass to examine the patterns on two of your fingers more closely. How do the patterns on your fingers compare with each other?

▶ ______________________________

3. Examine the epidermal ridges on the same two fingers of your lab partner's hand. How do the patterns on your fingers compare to your lab partner's?

▶ ______________________________

MAKING CONNECTIONS

What factors might alter a person's fingerprint patterns during his or her lifetime?

▶ ______________________________

Activity 6.5

Examining the Structure of Nails

Similar to hair, nails consist of dead cells filled with hard keratin. They form protective shields that cover the dorsal skin surfaces at the tips of the fingers and toes.

A Inspect the nails on your fingers and identify the following structures:

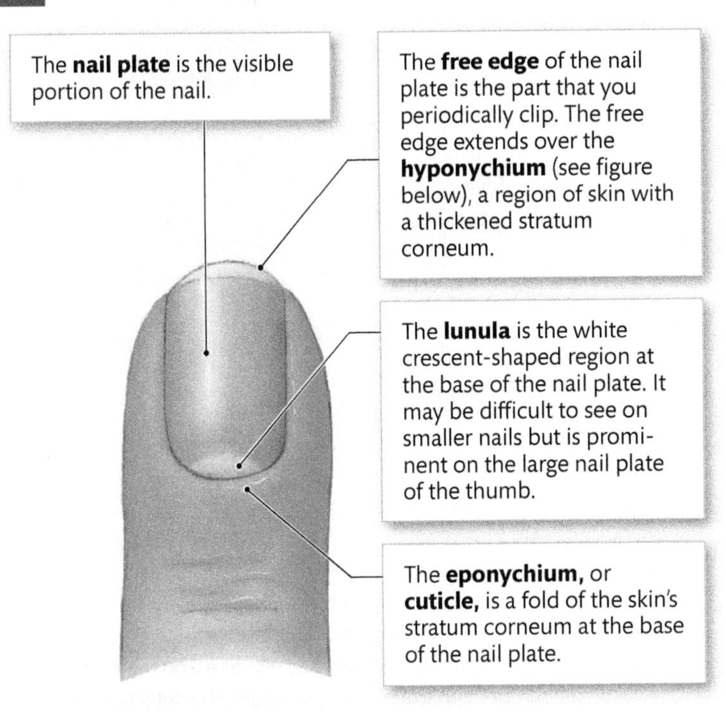

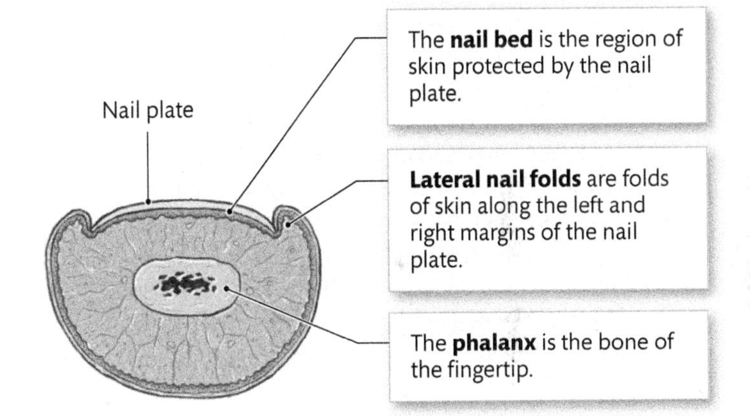

B Deep nail structures cannot be viewed from the surface.

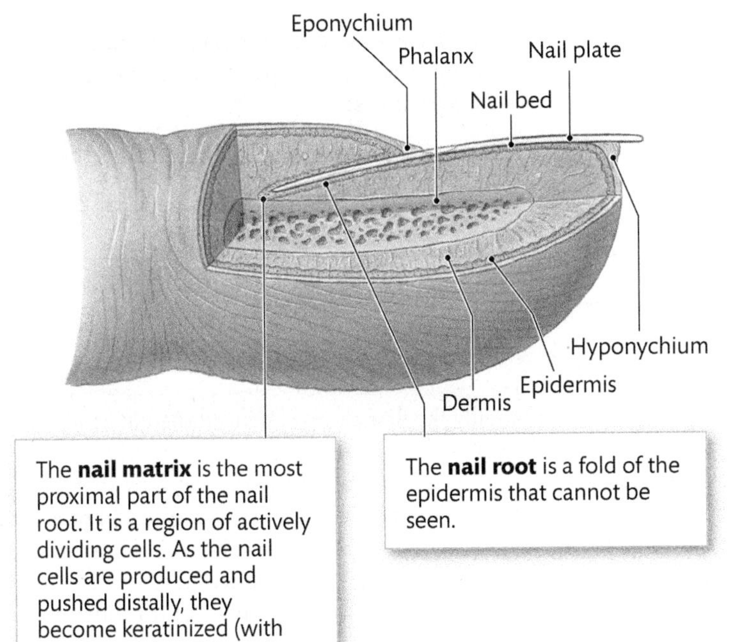

MAKING CONNECTIONS

Like nails, hooves and claws are composed of epidermal cells filled with hard keratin. Animals use hooves for locomotion and use claws for several functions such as climbing, digging, hanging, grasping, and even killing prey. Identify two functions that nails provide for humans.

▶ ______________________________

BEFORE YOU MOVE ON . . .

« LOOKING BACK

In this laboratory exercise, you studied the structure and function of the integumentary system. It marks the first time that you examined the human body from two additional levels of organization: the organ level and the organ system level.

The integumentary system is formed by the cutaneous membrane and the accessory structures. The cutaneous membrane is composed of two tissue regions: the epidermis and the dermis. The epidermis is a surface layer of stratified squamous, keratinized epithelium. The dermis is a connective tissue layer, deep to the epidermis. It is divided into two regions. The thin papillary layer, composed of areolar connective tissue, directly borders the epidermis. The deeper reticular layer is composed of dense irregular connective tissue. The accessory structures include hair follicles, sebaceous glands, arrector pili muscles, sweat glands, sensory receptors, and nails. The accessory structures are well established in the dermis of the skin, but some extend superficially into the epidermis.

Explain why the cutaneous membrane is classified as an organ. Explain why the cutaneous membrane and the embedded accessory structures are classified as an organ system.

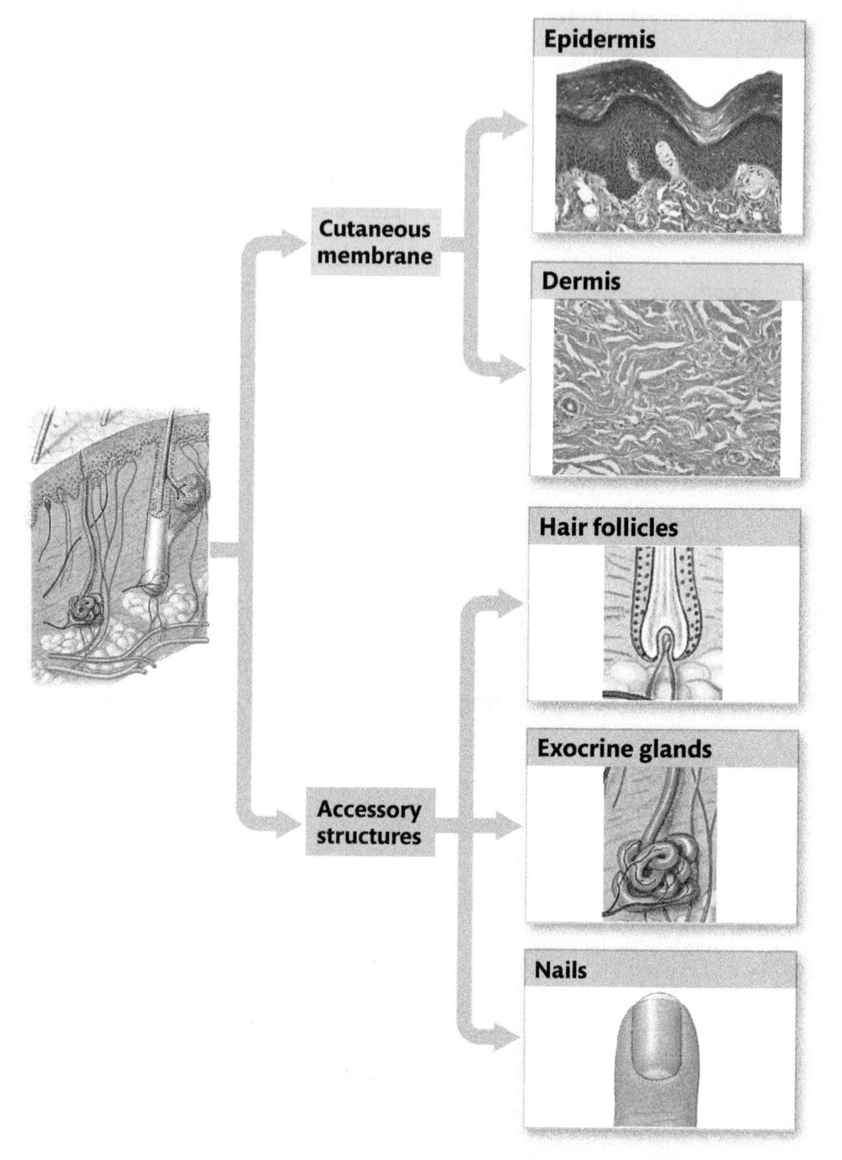

LOOKING FORWARD »

The integumentary system is traditionally the first organ system studied in anatomy and physiology courses because it covers the outer surface of the body and functions as a protective barrier between the external environment and deeper body structures. In the laboratory exercises that follow, you will be studying the structure and function of the organs in the other 10 organ systems in the body, beginning with the skeletal system. This system—in particular, the axial skeleton—is a natural next area of study because it provides the framework that supports and protects most of the other organ systems.

Name ______________________

Lab Section ______________________

Date ______________________

EXERCISE 6 **REVIEW SHEET**

The Integumentary System

1. Complete the following table.

Layer	Structure	Function
Epidermis	a.	b.
Dermis	c.	d.
Hypodermis	e.	f.

2. a. Explain what happens when you get a suntan.

____

____

____

b. If you get a suntan regularly (e.g., every summer), how will it affect your risk of acquiring skin cancer? Explain.

____

____

3. Describe how the contraction of arrector pili muscles affects sebaceous glands and hair follicles.

____

____

____

QUESTIONS 4–13: Match the structure in column A with the correct description in column B.

A	B
4. Melanocyte ______	**a.** Visible portion of a nail
5. Nail matrix ______	**b.** Sensory receptor that responds to deep pressure sensations
6. Tactile corpuscle ______	**c.** Gland that is important for regulating body temperature
7. Merocrine sweat gland ______	**d.** Epidermal cell that produces melanin
8. Keratinocyte ______	**e.** Part of a hair follicle that contains the hair matrix
9. Nail plate ______	**f.** Cell type that comprises the majority of the cells in the epidermis
10. Apocrine sweat gland ______	**g.** Sensory receptor that responds to pain and temperature sensations
11. Hair bulb ______	**h.** Type of gland that is not functional until puberty
12. Lamellated corpuscle ______	**i.** Region of actively dividing cells that form the nail plate
13. Free nerve ending ______	**j.** Sensory receptor that responds to light touch sensations

QUESTIONS 14–15: For each diagram, color the structures with the indicated colors.

14.

Epidermis and dermis of skin

Stratum basale = **green**
Reticular layer = **red**
Papillary layer = **blue**
Stratum corneum = **yellow**

15.

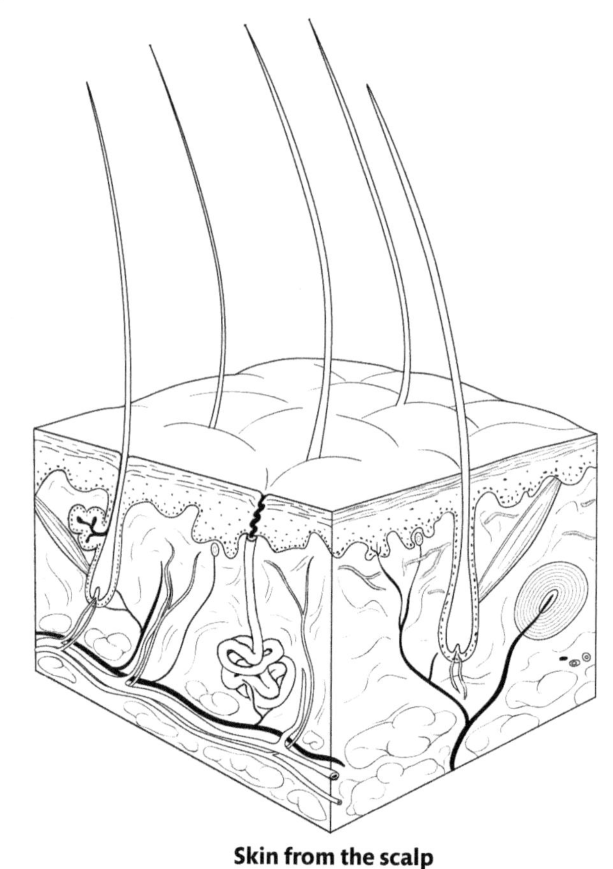

Skin from the scalp

Hypodermis = **gray**
Dermis = **purple**
Epidermis = **blue**
Hair follicle = **green**
Sebaceous gland = **red**
Sweat gland = **brown**

Arrector pili muscle = **orange**
Lamellated corpuscle = **yellow**
Tactile corpuscle = **black**
Hair root = **pink**
Hair shaft = **tan**

QUESTIONS 16–21: In the figure to the right, identify the labeled structures. Select your answers from the following list of structures:

a. Nail plate
b. Eponychium
c. Hyponychium
d. Nail matrix
e. Nail bed
f. Nail root

16. ______________________________

17. ______________________________

18. ______________________________

19. ______________________________

20. ______________________________

21. ______________________________

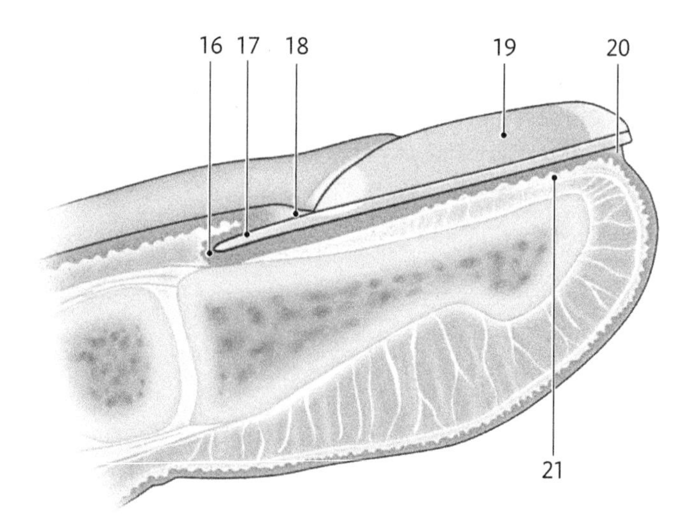

EXERCISE 7

Introduction to the Skeletal System and the Axial Skeleton

LEARNING OUTCOMES

These Learning Outcomes correspond by number to the laboratory activities in this exercise. When you complete the activities, you should be able to:

Activity 7.1	**Classify bones according to their shapes.**
Activity 7.2	**Differentiate between the axial skeleton and appendicular skeleton.**
Activity 7.3	**Describe the microscopic structure of compact and spongy bone.**
Activity 7.4	**Describe the arrangement of compact and spongy bone in the bones of the skeleton.**
Activity 7.5	**Understand the functions of bone markings on bones.**
Activities 7.6–7.9	**Identify the bones of the skull and their bone markings from various views.**
Activity 7.10	**Compare the bones of the fetal skull and the adult skull.**
Activity 7.11	**Recognize the general features of the vertebral column.**
Activities 7.12–7.13	**Compare the unique features of vertebrae from different regions of the vertebral column.**
Activity 7.14	**Describe the structure of the thoracic cage.**

LABORATORY SUPPLIES

- Anatomical model of compact and spongy bone
- Humerus or femur, longitudinal section
- Cranial bones, cross section
- Human skeletons
- Disarticulated bones of the human skeleton
- Human skulls
- Brain models
- Fetal skulls
- Vertebrae from cervical (including C_1 and C_2), thoracic, and lumbar regions
- Sacrum and coccyx
- Ribs
- Colored pencils

PRE-LAB QUIZ

Before you begin, read all the activities in Exercise 7 and the required reading in your textbook that is assigned by your instructor.

1. During this laboratory exercise, you will classify bones according to ______________ and ______________.
2. Consider the following supply list: a skeleton, a heart model, a skull, individual vertebrae, a brain model, and a model of bone tissue. To complete all the activities in this laboratory exercise, which item on the list will *not* be used? ______________
3. True or False: Osteons are the basic functional units of spongy bone. ______________
4. True or False: The carpal bones are classified as short bones. ______________
5. What type of bone has a shaft (diaphysis) and two expanded knoblike ends (epiphyses)? ______________
6. The skull is composed of two distinct groups of bones: the cranial bones and the ______________ bones.
7. The ______________ vertebrae articulate (form joints) with the ribs.
8. In the fetal skull, the developing cranial bones are connected by areas of fibrous connective tissue called ______________.
9. True or False: In a typical vertebra, the pedicles and laminae form the spinous process. ______________
10. True or False: The costal cartilages of vertebrosternal ribs are directly attached to the sternum. ______________

Activity 7.1

Classifying Bones According to Shape

A We can classify bones according to their size and shape.

1. Study a whole skeleton, or a complete collection of disarticulated bones, and notice the wide range of sizes and forms.
2. Identify the bones according to the following shapes:

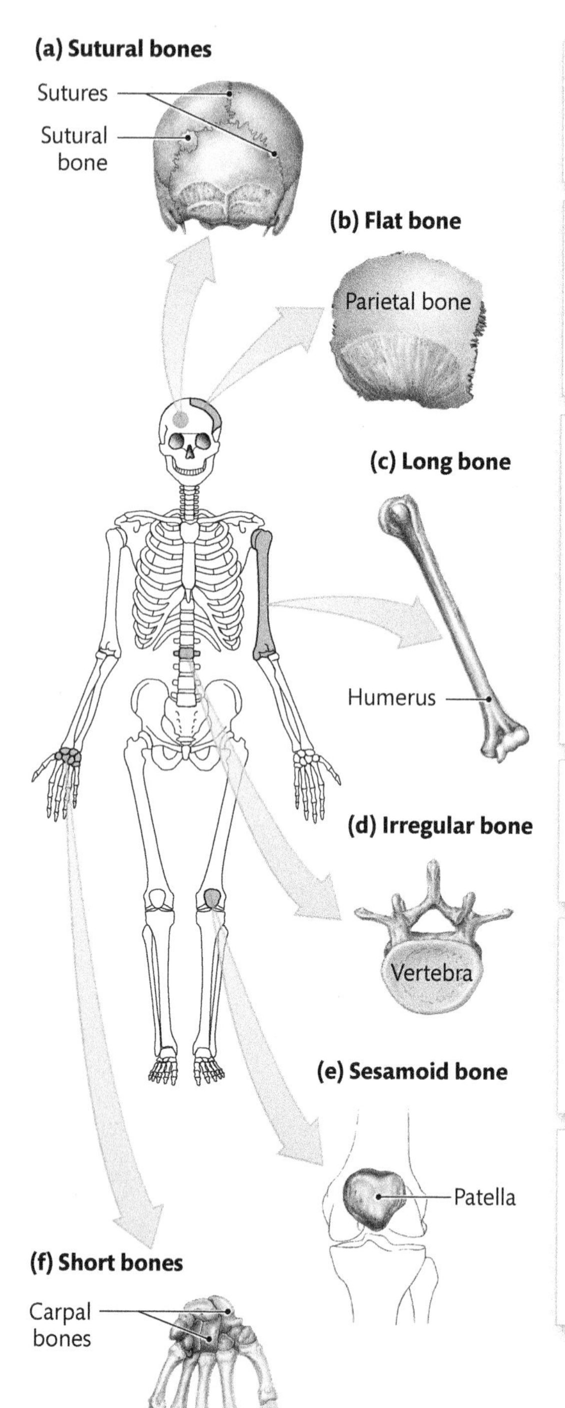

Sutural bones develop between the joints (sutures) of cranial bones (Figure [a]). Their number varies between individuals.

Flat bones are thin, platelike structures. This group includes the **cranial bones** (bones that protect the brain such as the **parietal bone,** Figure [b]), the **sternum,** the **ribs,** and the **scapulae.**

Long bones have extended longitudinal axes so that the length of the bone is much greater than the width. A typical long bone contains an elongated shaft, known as the **diaphysis,** with two expanded, knoblike ends or **epiphyses.** Examples of long bones include the bones in the arm (**humerus,** Figure [c]), forearm (**radius** and **ulna**), thigh (**femur**), and leg (**tibia** and **fibula**).

Irregular bones have a variety of shapes and include the **facial bones** of the skull, the **vertebrae** (Figure [d]), and the **pelvic bones.**

Sesamoid bones are embedded in tendons at articulations. The **patella** (kneecap, Figure [e]) is a sesamoid bone found in all humans. The number of other sesamoid bones varies among individuals.

Short bones are those in which the length and width are about equal, so they appear cube shaped. They include the bones of the wrist (**carpal bones,** Figure [f]) and ankle **(tarsal bones).**

Word Origins

The word *sesamoid* derives from the Greek word *sesamoeides,* which means "like a sesame." Most sesamoid bones are small and roughly resemble the shape of a sesame seed. The patella resembles a small plate or shallow disk, which is what the term *patella* means in Latin.

MAKING CONNECTIONS

The metacarpal bones (in the palms), metatarsal bones (in the soles), and phalanges (in the fingers and toes) are much shorter than most ribs, and yet the bones in the hands and feet are classified as long bones and the ribs as flat bones. Why do you think these bones are classified as they are?

▶ ______________________________

Activity 7.2

Classifying Bones According to Location

Bones can be classified according to their location on the skeleton. Study a whole skeleton, or a complete collection of disarticulated bones.

A The **appendicular skeleton** (unshaded in this figure) includes the bones of the upper and lower limbs (the appendages).

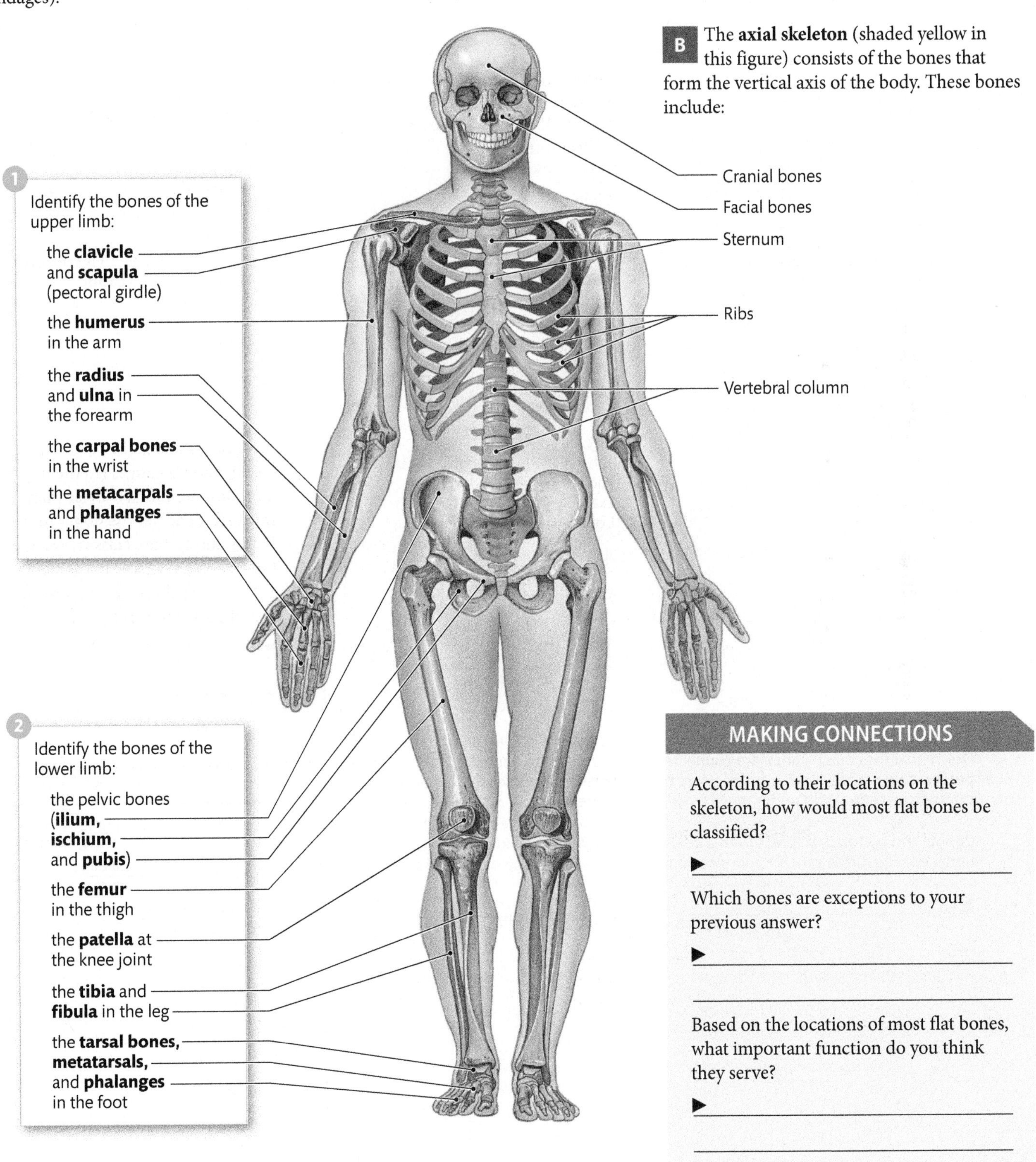

B The **axial skeleton** (shaded yellow in this figure) consists of the bones that form the vertical axis of the body. These bones include:

MAKING CONNECTIONS

According to their locations on the skeleton, how would most flat bones be classified?

▶ ______________________________

Which bones are exceptions to your previous answer?

▶ ______________________________

____

Based on the locations of most flat bones, what important function do you think they serve?

▶ ______________________________

____

Activity 7.3

Examining the Microscopic Structure of Bone

Two types of bone tissue exist in the skeleton: compact and spongy. **Compact bone** is an extremely dense material that forms the hard exterior covering of all bones. **Spongy (cancellous) bone** fills the interior regions of most bones and forms a thin internal layer along the diaphyses of long bones.

A Compact Bone

Examine a three-dimensional model showing a microscopic section of bone tissue. Identify the structures illustrated in the following figures.

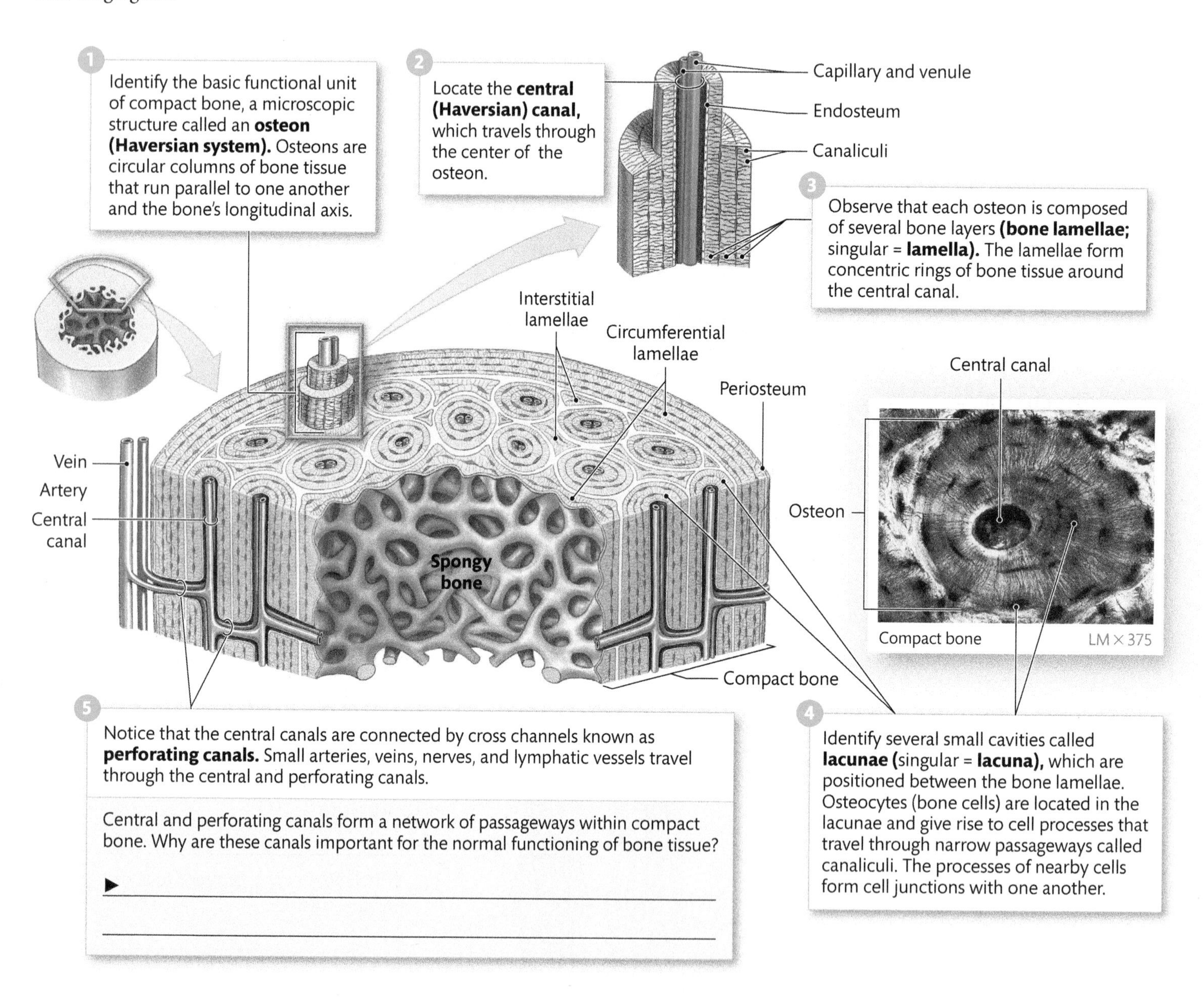

1 Identify the basic functional unit of compact bone, a microscopic structure called an **osteon (Haversian system).** Osteons are circular columns of bone tissue that run parallel to one another and the bone's longitudinal axis.

2 Locate the **central (Haversian) canal,** which travels through the center of the osteon.

3 Observe that each osteon is composed of several bone layers **(bone lamellae;** singular = **lamella).** The lamellae form concentric rings of bone tissue around the central canal.

4 Identify several small cavities called **lacunae** (singular = **lacuna),** which are positioned between the bone lamellae. Osteocytes (bone cells) are located in the lacunae and give rise to cell processes that travel through narrow passageways called canaliculi. The processes of nearby cells form cell junctions with one another.

5 Notice that the central canals are connected by cross channels known as **perforating canals.** Small arteries, veins, nerves, and lymphatic vessels travel through the central and perforating canals.

Central and perforating canals form a network of passageways within compact bone. Why are these canals important for the normal functioning of bone tissue?

▶ ____

____

B Spongy Bone

Unlike compact bone, spongy bone does not have a regular arrangement of osteons. Instead, bone lamellae form an irregular arrangement of interconnecting bony struts called **trabeculae** (singular = **trabecula**) with spaces surrounding the latticework of bony tissue. The spaces are filled with red bone marrow, where blood cells are produced, or yellow bone marrow, where fat is stored. The porous structure of spongy bone makes it suitable for cushioning the impact generated by body movements.

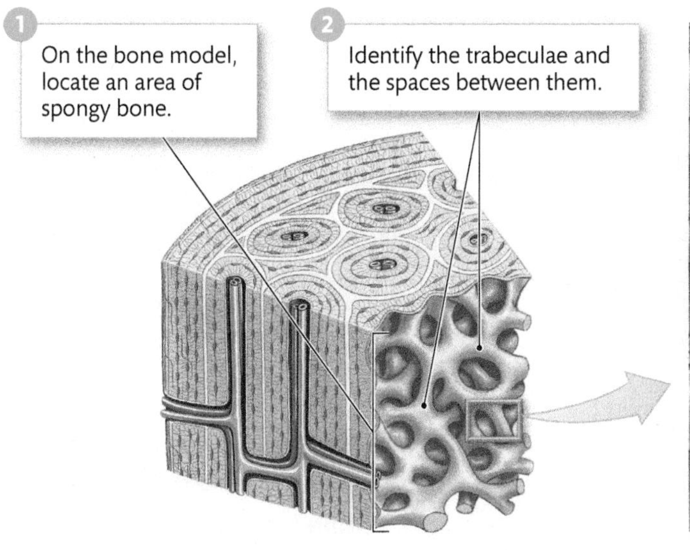

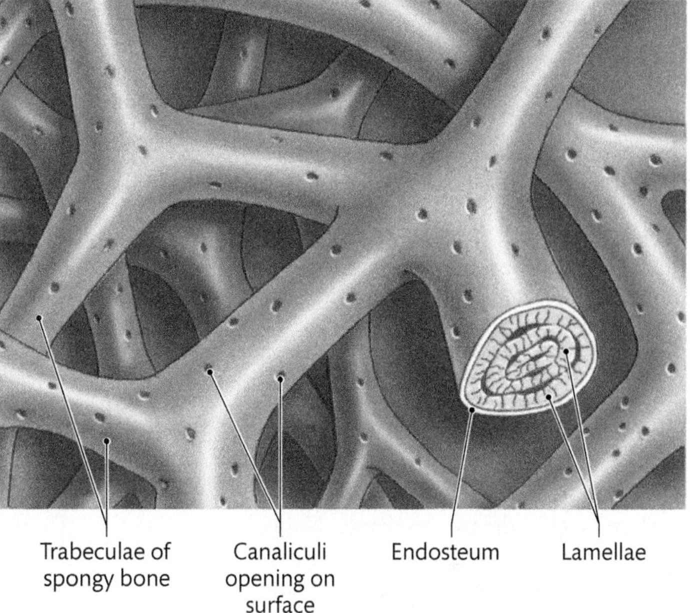

C

During your earlier study of tissues, you learned that bone is a connective tissue with a solid matrix.

Review the structure of the matrix and identify its two main components. Briefly describe the special qualities that each component gives to bone. ▶

Component	Special Quality
1. ________	________
	________
2. ________	________
	________

Word Origins

In Latin, the word *trabecula* means "beam." Spongy bone is composed of interconnecting trabeculae (or beams) of bone tissue.

IN THE CLINIC

Osteoporosis

Part of the aging process involves a decrease in the activity of osteoblasts, which are the cells that deposit new bone matrix. As a consequence, we start to lose some bone mass. If the condition progresses, eventually enough bone mass is lost so that the bone's normal functioning is impaired. This clinical condition is called osteoporosis, which literally means "porous bone." The bones become visibly more porous and, as a result, weaker and more brittle.

MAKING CONNECTIONS

In this activity, you learned that within the canaliculi, the processes of neighboring osteocytes can link together by forming cell junctions. What do you think is the significance of these cell junctions?

▶ ________________________________

____

____

____

____

____

____

Activity 7.4

Examining the Arrangement of Compact and Spongy Bone

A Compact and Spongy Bone in a Long Bone

1 In a long bone, such as the femur, that has been cut along its longitudinal axis in a coronal plane, notice the external layer of solid compact bone. This layer is relatively thick where it surrounds the diaphysis but much thinner around each epiphysis.

Why do you think the compact bone layer is much thicker around the diaphysis compared to the epiphyses?

▶ ______________________________

____

3 The more porous spongy bone predominates at the two epiphyses. Notice the complex network of bony trabeculae. Remember that, in living bone, the spaces between the trabeculae are filled with bone marrow.

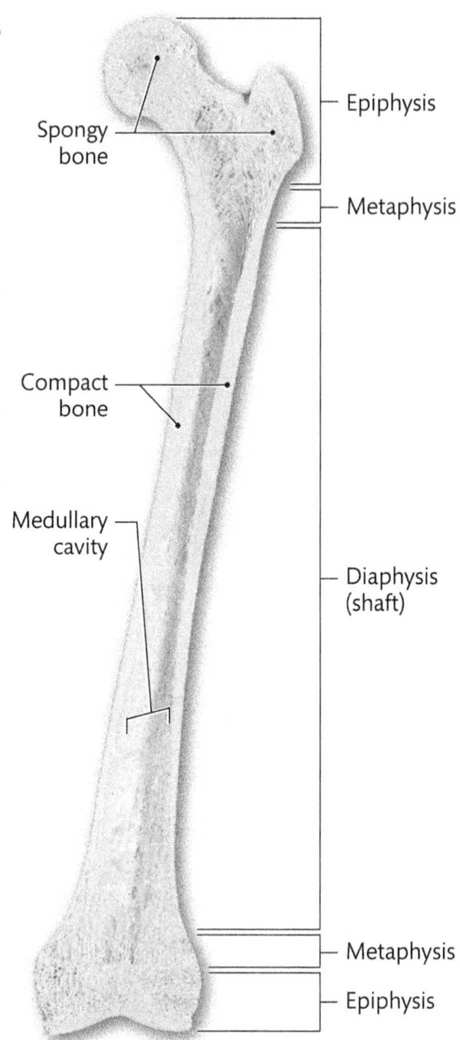

2 Identify the medullary cavity, which is the hollow space inside the diaphysis. During bone development, this cavity is filled with spongy bone and red bone marrow. As the bone matures, the bone is resorbed, and the red bone marrow is converted to yellow bone marrow. Thus, in the adult, the medullary cavity of a typical long bone becomes a storage site for fat.

IN THE CLINIC

Bone Marrow

In the adult skeleton, red marrow is abundant in the spongy bone of the ribs, sternum, cranial bones, bodies of vertebrae, clavicles, scapulae, pelvic bones, and proximal epiphyses of the long bones, such as the femur and humerus. Most other spongy bone is filled with yellow marrow, but during a period of trauma that causes severe blood loss—for instance, a hard blow to the back that causes damage to the spleen—yellow marrow can be converted to red marrow to increase blood cell production.

B Compact and Spongy Bone in a Flat Bone

1 The parietal bone and other cranial bones are flat bones. In a cross section of a flat bone, observe the internal layer of spongy bone called the **diploë.** Notice that the diploë is sandwiched between two thin layers of compact bone.

2 Examine this relationship and remember that the cavities in the diploë are filled with red bone marrow.

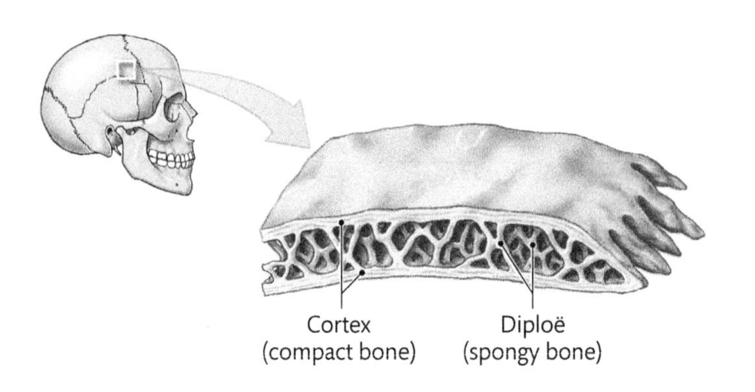

MAKING CONNECTIONS

If bones were composed entirely of compact bone, muscles would have to contract with greater force to produce movements. Explain why.

▶ ______________________________

____

____

____

____

Activity 7.5

Identifying Bone Markings

A All bones possess distinctive markings or landmarks that are designed for specific functions. For example, certain landmarks are used as articulating surfaces to form a joint (an articulation) with another bone on the skeleton. Examples include **condyles, heads, facets,** and **trochleas.** Other bony features, such as **trochanters, tubercles, crests,** and **processes,** provide areas of attachment for tendons and ligaments. In addition, various types of depressions and openings provide passageways for nerves and blood vessels. They include **foramina** (singular = **foramen**), **canals, fissures,** and **grooves.**

1 Identify examples of bone markings found on bones in the skull.

A **meatus** or **canal** is a bony passageway.

A **sinus** is a cavity within a bone.

A **foramen** is a round or oval hole or opening.

A **process** is a bony projection.

A **fissure** is a slit-like opening.

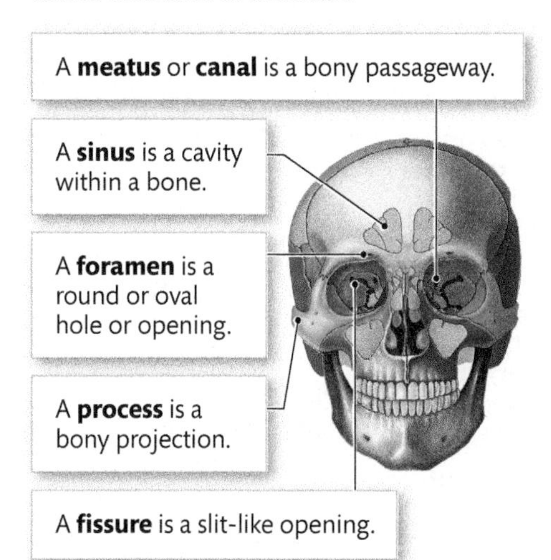

2 Identify examples of bone markings found on the humerus.

A **groove** is a narrow channel.

A **tuberosity** is an elevated projection with a roughened surface.

A **head** is a rounded expansion, connected to a narrow **neck.**

A **tubercle** is a rounded projection or process.

A **trochlea** is a pulley-shaped articular process.

An **epicondyle** is an elevated area, above a condyle.

A **condyle** is a rounded projection.

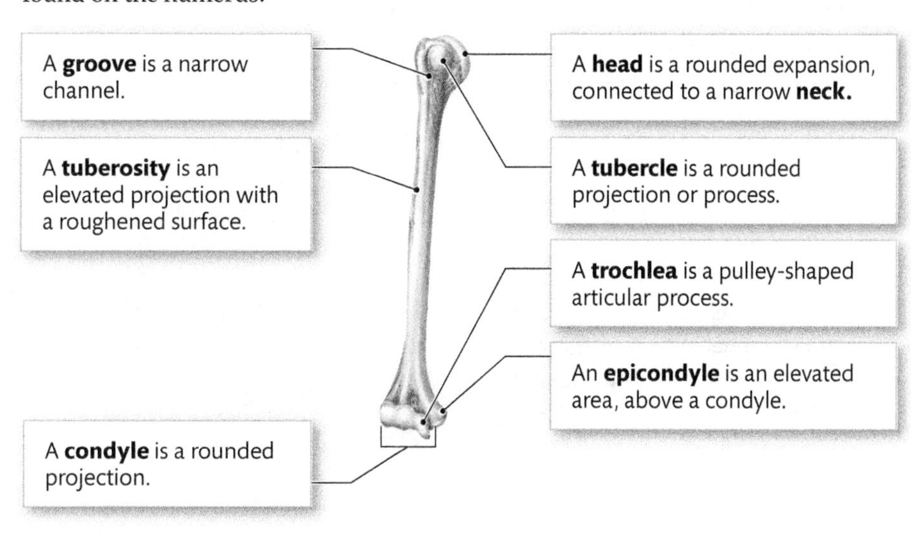

3 Identify examples of bone markings found on the femur.

A **trochanter** is a large irregularly shaped process (on femur only).

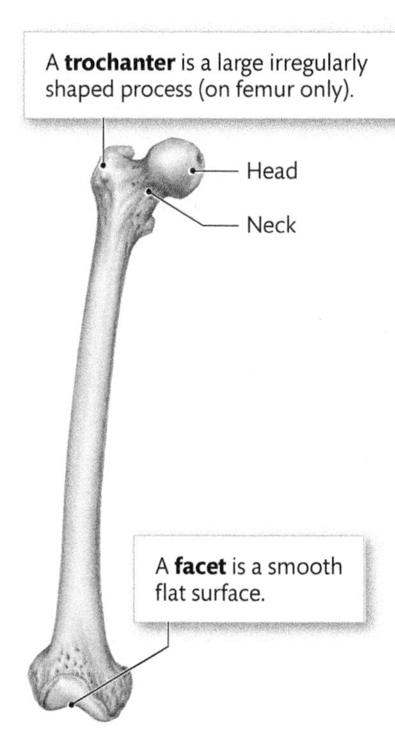

A **facet** is a smooth flat surface.

4 Identify examples of bone markings found on the pelvis.

A **crest** is a prominent narrow ridge.

A **fossa** is a shallow depression.

A **line** is a narrow ridge, less prominent than a crest.

A **spine** is a slender, pointed projection.

A **ramus** is an arm-like projection.

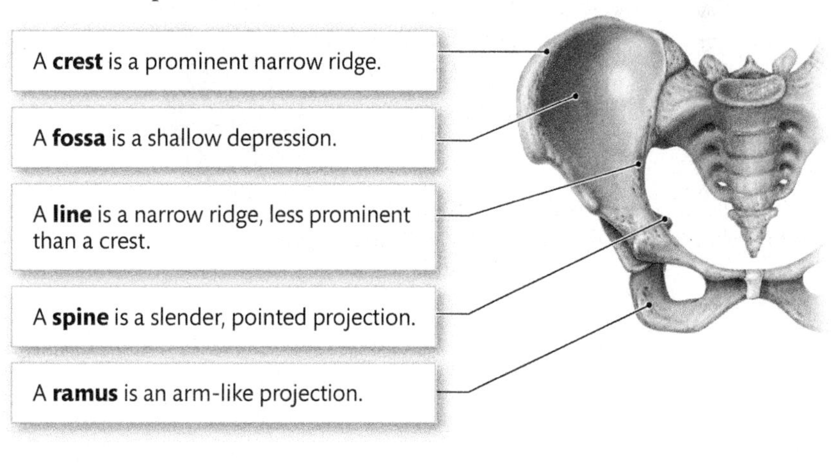

MAKING CONNECTIONS

Compare the heads on a humerus and a femur. How are they similar? How are they different?

▶ ______________________________

____

____

Activity **7.6**

Bones of the Skull: Examining an Anterior View

The skull is composed of two distinct groups of bones: cranial and facial. The eight **cranial bones (neurocranium)** form the walls, roof, and floor of the cranial cavity, where the brain and associated structures are located. Most of these bones are flat or slightly curved. The 14 **facial bones** have irregular shapes. These bones form the face and the walls of orbital and nasal cavities and provide bony sockets for the teeth.

Learning the bones of the skull and their various markings can be quite challenging, so proceed with care and patience. Most of the bones can be identified from more than one view. Consequently, to appreciate the three-dimensional quality of the skull, examine it from several positions.

A From an anterior view, you can identify a combination of facial and cranial bones.

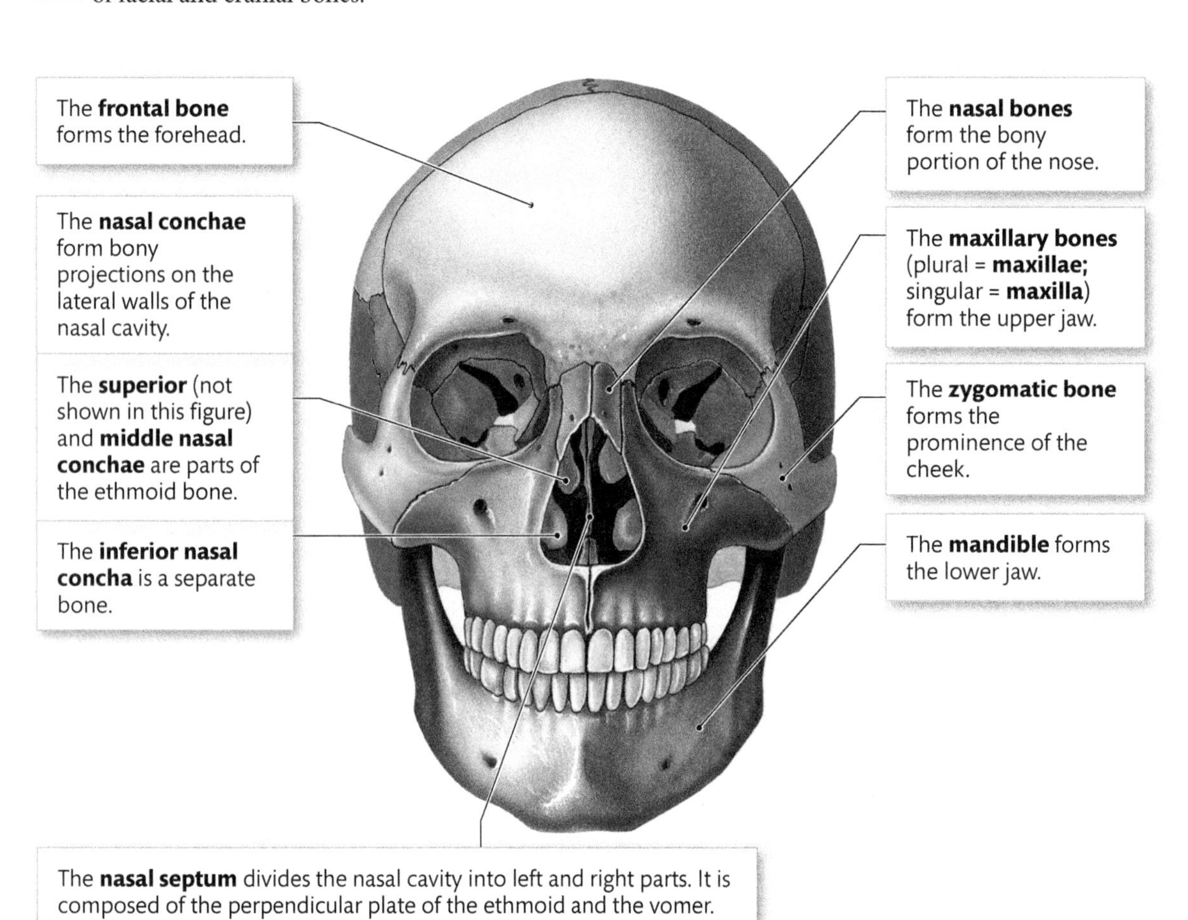

B The **bony orbit** is formed by portions of seven bones: frontal, ethmoid, lacrimal, sphenoid, maxilla, zygomatic, and palatine. On the skull, identify these bones.

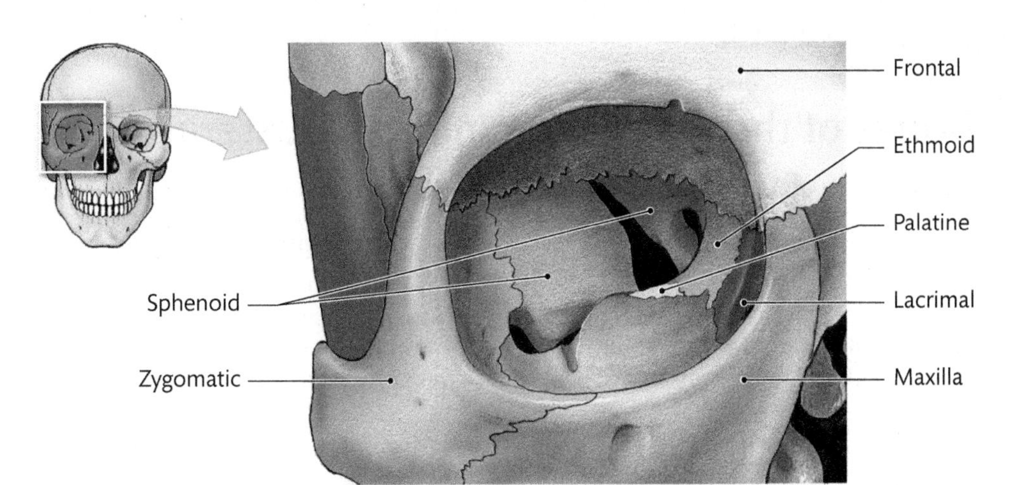

C From an anterior view, identify the following markings on the skull:

- supraorbital foramen
- superior orbital fissure
- optic canal
- infraorbital foramen
- mental foramen

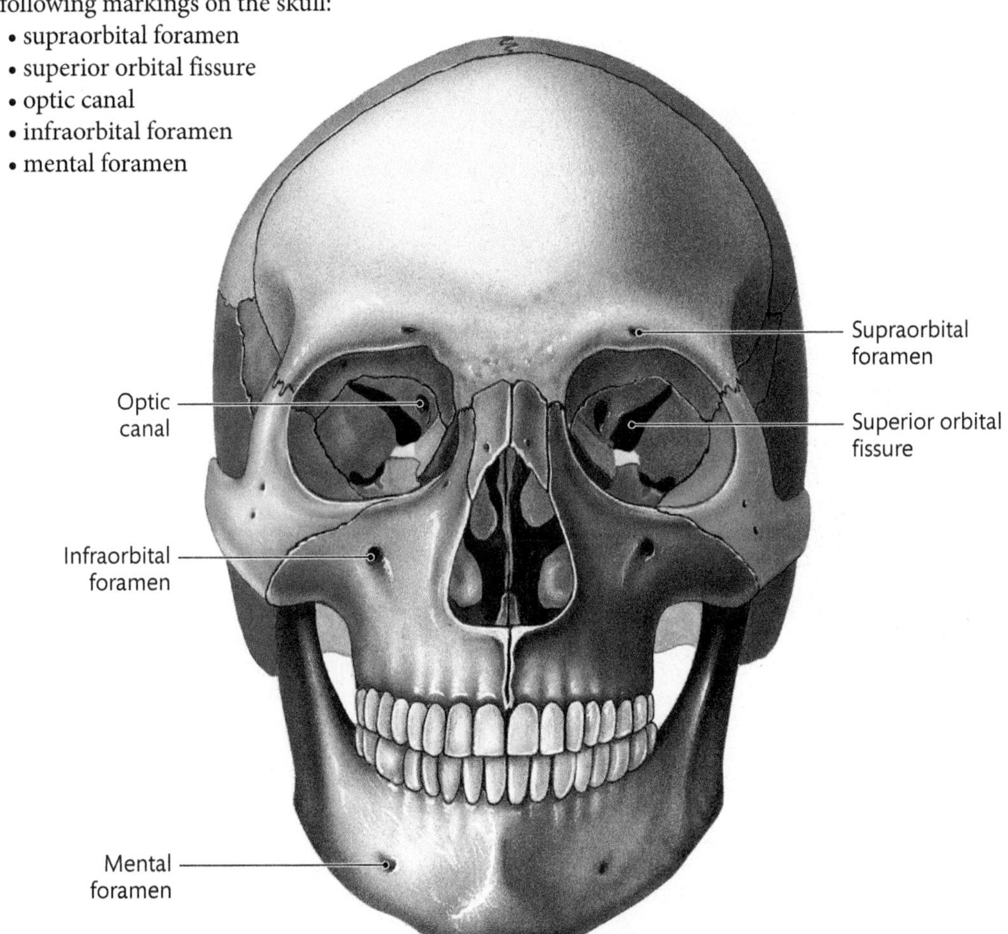

MAKING CONNECTIONS

The walls of the orbits are formed by portions of seven bones. Observe this region of the skull. Can you identify any structural weaknesses in this area?

▶ ______________________________

____

____

____

____

____

Activity 7.7

Bones of the Skull: Examining the Lateral and Posterior Views

From lateral and posterior views of the skull, you can identify many of the bones that form the neurocranium. You can also see the major sutures that hold these bones together.

A From a lateral view, some of the facial bones that you identified in the previous activity can be seen from a different perspective.

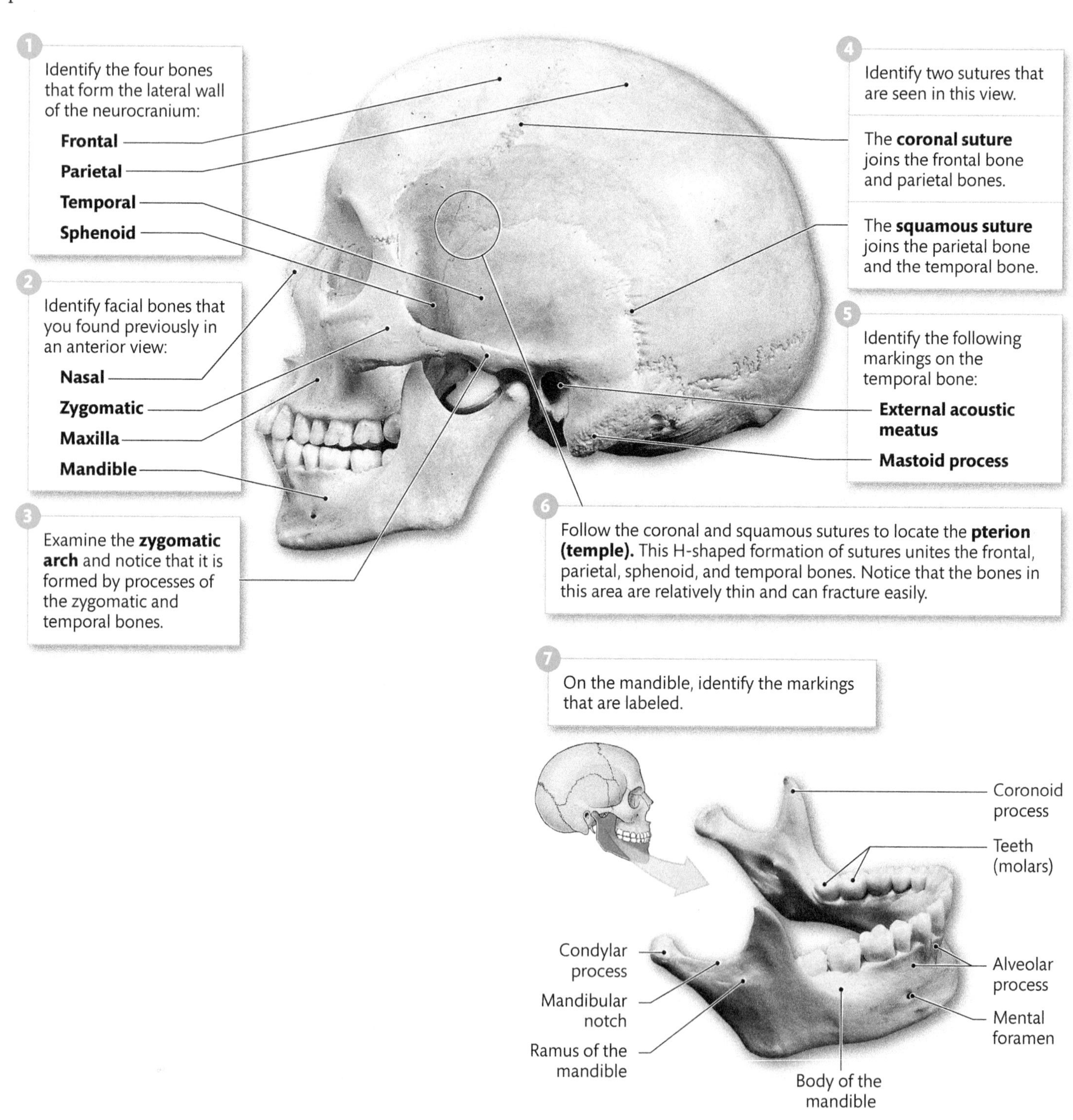

B From the **posterior view** of the skull, observe that the posterior wall of the neurocranium is formed by the **occipital bone** (**occiput** or back of head) and portions of the **parietal bones.**

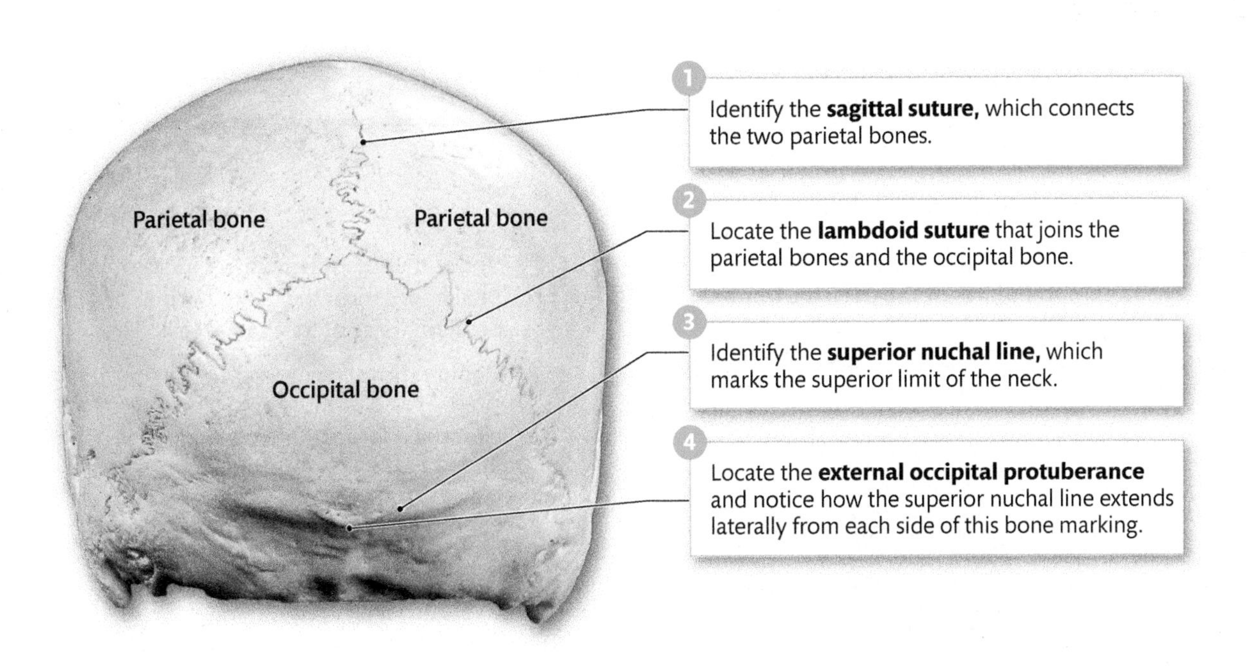

MAKING CONNECTIONS

Observe the suture lines between cranial bones on several skulls in the laboratory. Notice that the line patterns are different on each skull. What do you think accounts for this variability?

▶ ____

____

____

____

____

____

____

____

Activity **7.8**

Examining the Base of the Skull

You can study the base of the skull externally from an inferior view and internally by inspecting the floor of the cranial cavity.

A From an **inferior view** of the skull, you can identify several openings, which lead into the cranial cavity. These openings transmit blood vessels and nerves.

1. Observe that the **hard palate** is formed by the **palatine processes of the maxillary bones** and the **horizontal plates of the palatine bones.**

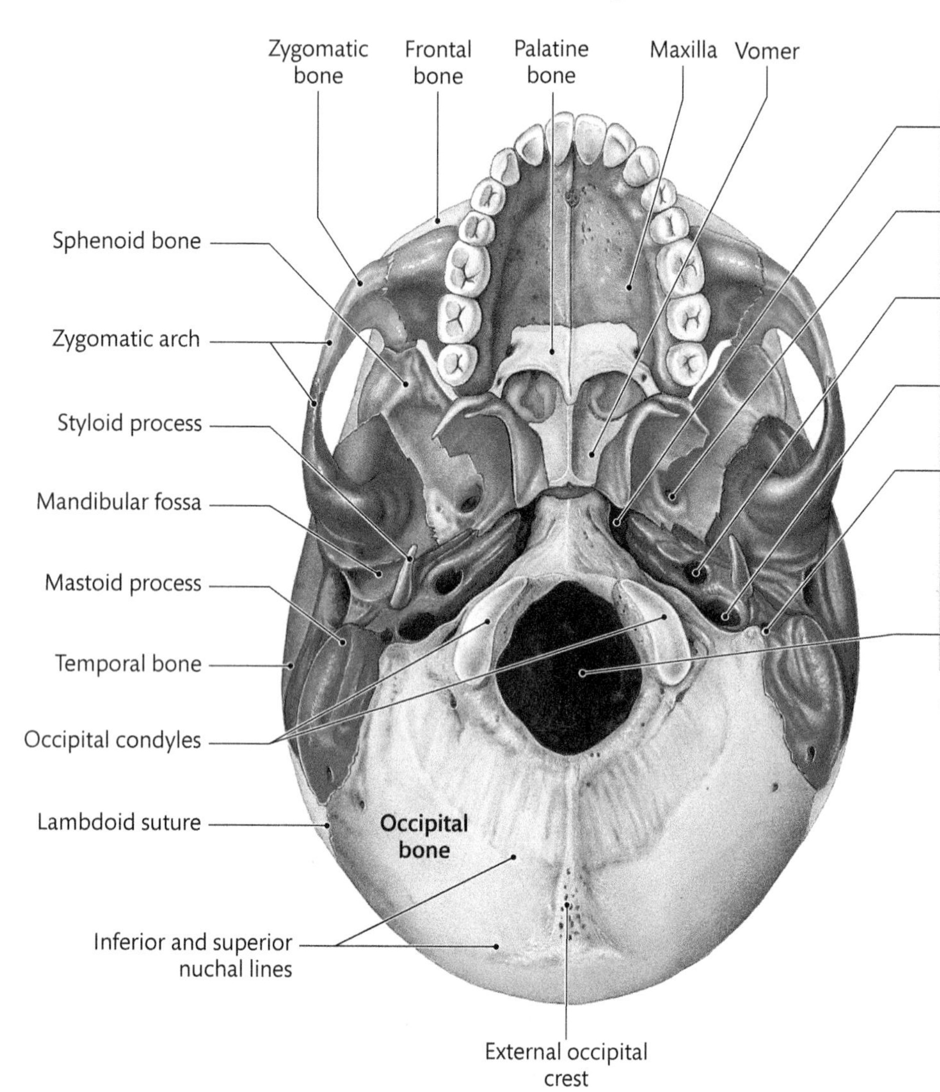

2. Notice that the sphenoid, temporal, and occipital bones form the major portion of the basicranium (base of the cranial cavity).

3. Identify the following openings:

The **foramen lacerum** is located between the temporal and occipital bones.
The **foramen ovale** travels through the sphenoid bone.
The **carotid canal** courses through the temporal bone.
The **jugular foramen** is located between the temporal and occipital bones.
The **stylomastoid foramen** is located between the styloid and mastoid processes in the temporal bone.
The **foramen magnum** is the large opening in the base of the occipital bone where the brain connects to the spinal cord.

4. Identify the following bone markings:
   - Zygomatic arch
   - Styloid process
   - Mandibular fossa
   - Mastoid process
   - Occipital condyle
   - Inferior and superior nuchal lines
   - External occipital crest

B In this view of the floor of the **cranial cavity,** compare the appearance of the sphenoid, temporal, and occipital bones with the same bones as they appear in the inferior view on the previous page.

1 Notice that the floor of the cranial cavity is formed by portions of the frontal, ethmoid, sphenoid, temporal, and occipital bones.

2 Locate the three depressions in the cranial floor: the **anterior, middle,** and **posterior cranial fossae** (singular = **fossa).**

Obtain a model of the brain and place it within the cranial cavity. Verify that:

- The anterior cranial fossa is occupied by the frontal lobes of the brain.
- The middle cranial fossa is occupied by the temporal lobes of the brain.
- The posterior cranial fossa is occupied by the cerebellum and brainstem.

3 In the sphenoid bone, identify the following foramina and canals:

- **Optic canal**
- **Foramen rotundum**
- **Foramen ovale**
- **Foramen spinosum**

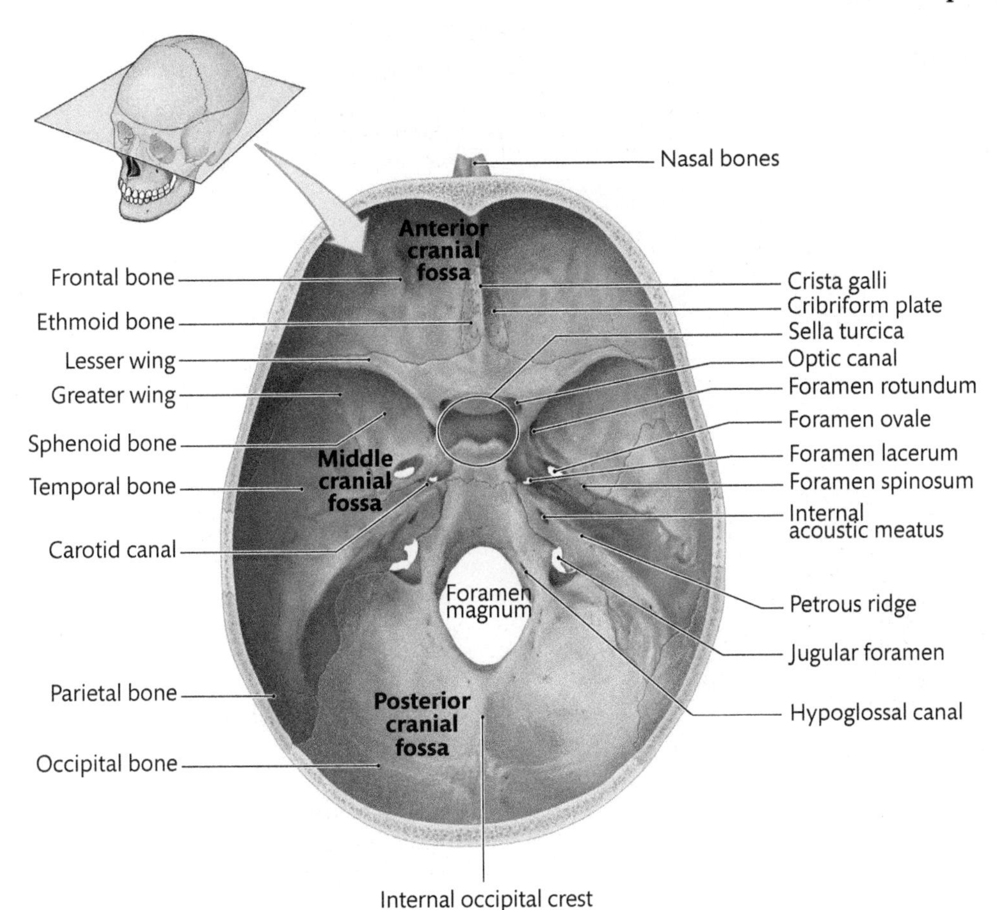

4 In the temporal bone, locate the **carotid canal** and **internal acoustic meatus.**

5 In the occipital bone, find the **hypoglossal canal.**

6 Between the temporal and occipital bones, identify the foramen lacerum and jugular foramen.

7 Identify the bone markings that are listed in Table 7.1.

TABLE 7.1 Bone Markings in the Cranial Cavity

Bone Marking	Bone Location
Crista galli	Ethmoid
Cribriform plate	Ethmoid
Petrous ridge	Temporal
Sella turcica	Sphenoid
Greater wing	Sphenoid
Lesser wing	Sphenoid

MAKING CONNECTIONS

Normally, the two maxillary bones fuse along the midline of the hard palate before birth. If they fail to fuse completely, a **cleft palate** develops. Surgery to correct this condition is usually performed 6 to 12 months after birth. Until that age, what problems do you predict could result from a cleft palate?

▶ ______________________________

____

____

____

Activity 7.9

Examining a Sagittal Section of the Skull

A A **sagittal section** of the skull provides a unique view that is not often studied in the laboratory. Identify the following bones and markings that you can see from this view:

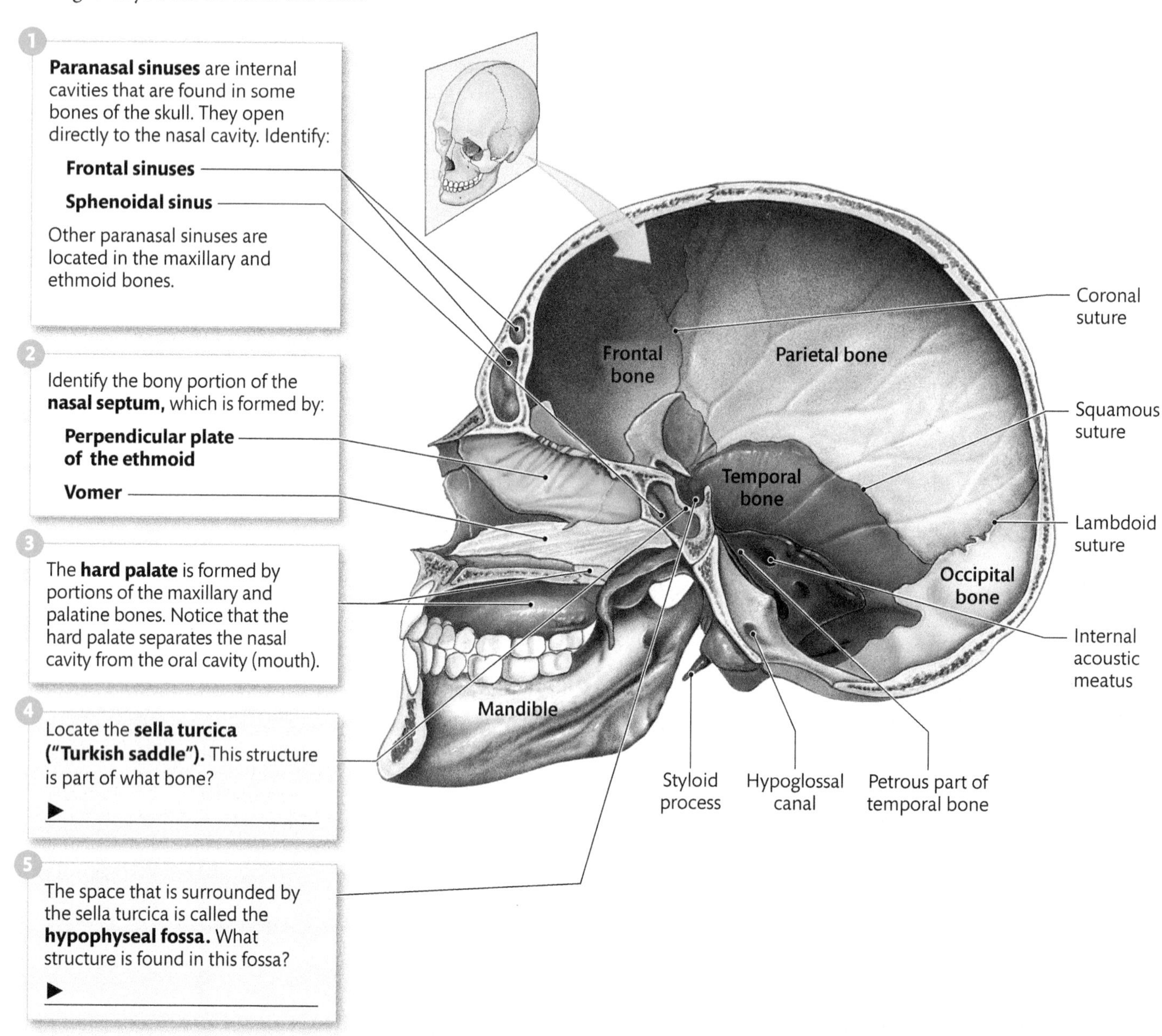

MAKING CONNECTIONS

In the previous four activities, you identified and examined the skull bones from several different positions. Why is it essential that you study the skull from several views?

▶ ______

______

______

______

Activity **7.10**

Examining the Fetal Skull

A In the **fetal skull,** the bones are actively growing and remodeling to accommodate brain growth. Bone markings are smaller and less prominent than those on the adult skull.

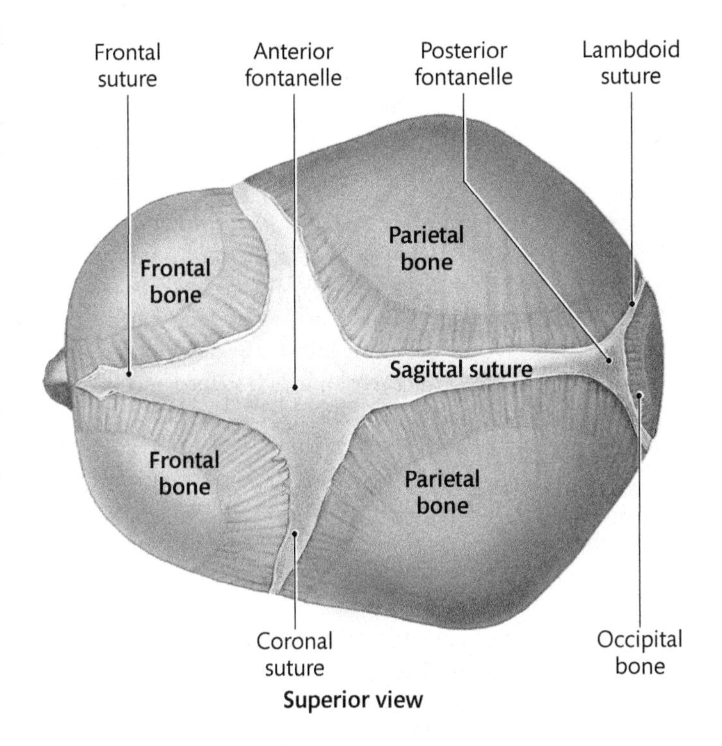

Superior view

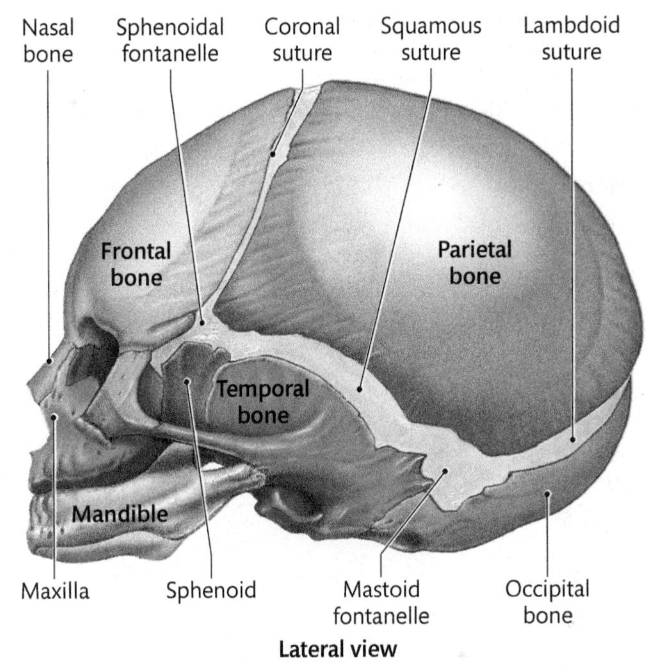

Lateral view

1. On a fetal skull, locate the cranial and facial bones that you identified in the adult skull.
2. Identify the major sutures and notice that they are not fully developed in the fetal skull. The ossification process (bone formation) is incomplete, and adjacent cranial bones are connected by fibrous connective tissue. In some areas, the size of the fibrous connections is relatively large. These regions are known as **fontanelles.** Identify the following fontanelles in the fetal skull:
   - **Anterior (frontal) fontanelle**
   - **Posterior (occipital) fontanelle**
   - **Sphenoidal (anterolateral) fontanelle**
   - **Mastoid (posterolateral) fontanelle**

Word Origins

In French, the word *fontanelle* refers to a fountain or spring. The "soft spots" on a baby's skull are called fontanelles because a pulse can be felt in these areas, especially in the anterior fontanelle.

IN THE CLINIC

The Significance of Fontanelles

The fontanelles provide some flexibility in the head during the birth process. The bones are allowed to slightly slide across each other to reshape the skull as it passes through the birth canal. In addition, the fontanelles also accommodate brain growth in the fetus and infant.

MAKING CONNECTIONS

What potential problems can you predict if the sutures between cranial bones form prematurely after birth?

▶ ______________________________

____

____

____

____

____

____

Activity 7.11

Examining the General Features of the Vertebral Column

A The **vertebral column (spine)** supports the head, neck, and trunk; helps maintain an erect posture while standing and sitting; and protects the spinal cord. During development, the vertebral column is formed by 33 individual **vertebrae** (singular = **vertebra**). As a result of bone fusion, however, the adult spine has only 26 bones and is divided into five regions, as follows: cervical vertebrae (7); thoracic vertebrae (12); lumbar vertebrae (5); fused vertebrae of the sacrum (5); and fused vertebrae of the coccyx (4).

1 Obtain a complete human skeleton and notice the general position of the vertebral column. Identify the five regions of the vertebral column shown below:

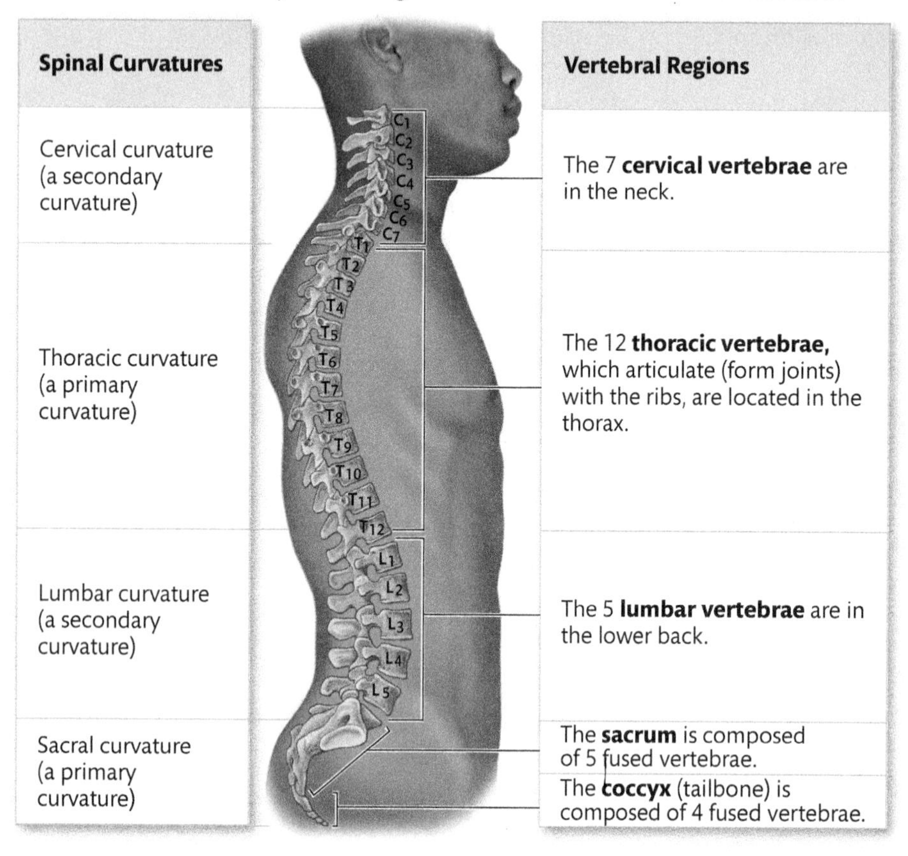

Word Origins

The terms *primary curvature* and *secondary curvature* derive from the structural changes that occur during development of the vertebrae. During fetal life, the initial curvature of the vertebral column is C shaped and concave anteriorly. The thoracic and sacral portions of the vertebral column retain this so-called primary curvature. During infancy and childhood, the cervical and lumbar portions of the vertebral column become progressively concave posteriorly. These secondary curvatures are adaptations to support the head (cervical vertebrae) and the torso (lumbar vertebrae) in an upright position.

2 Observe the four normal curvatures of the adult vertebral column. Notice that in the thoracic and sacral regions, the concave surface of the curvature faces anteriorly (the curve opens to the front). These regions form the **primary curvatures** of the vertebral column. The cervical and lumbar regions curve in the opposite direction (the concave surface faces posteriorly). These regions form the **secondary curvatures** of the vertebral column.

IN THE CLINIC

Abnormal Curvatures of the Vertebral Column

The vertebral column has four normal curvatures, but it can also become abnormally curved. **Kyphosis** (Figure [a]) is an exaggerated thoracic curvature that often gives an individual a hunchback appearance. **Lordosis** (Figure [b]) is an exaggerated lumbar curvature and results in a swayback appearance. **Scoliosis** (Figure [c]) is an abnormal lateral deviation of the vertebral column. Any of these vertebral misalignments can lead to chronic and severe pain. Depending on the severity, they can be treated with physical therapy, back-strengthening exercises, bracing, or surgery.

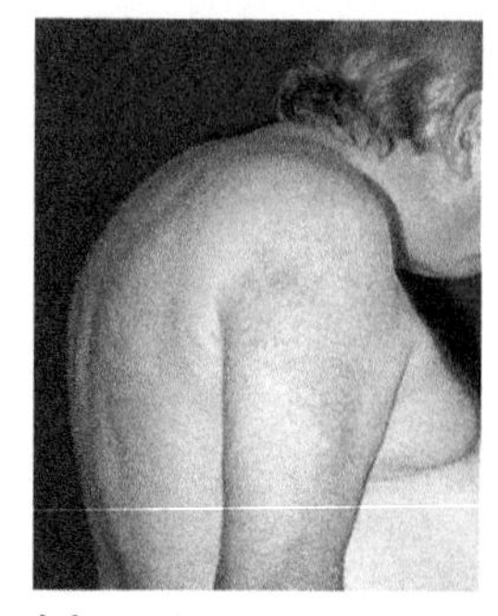

(a) Kyphosis

(b) Lordosis

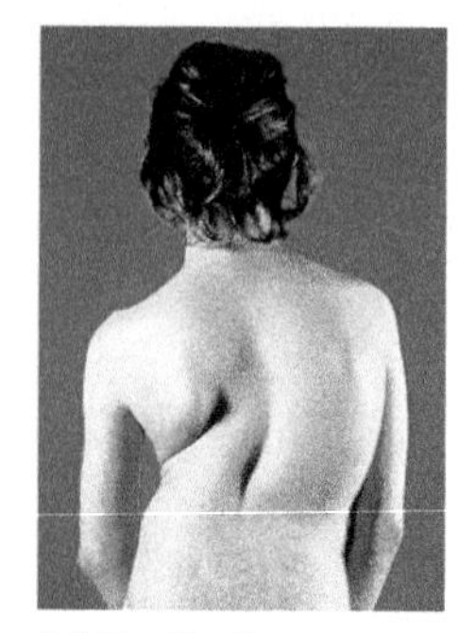

(c) Scoliosis

B A typical vertebra contains a relatively large anterior **vertebral body** and a posterior **vertebral arch.** Several processes extend from the vertebral arch. Identify the following structural features on a vertebra:

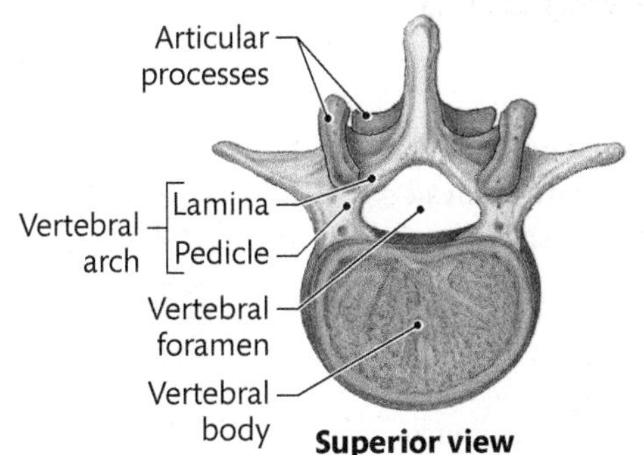

Superior view

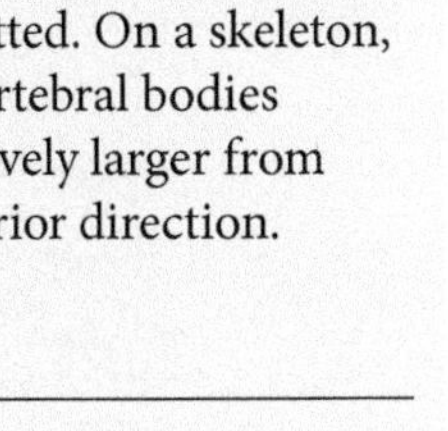

1. The vertebral body is the main anterior bony mass of each vertebra and is the part through which body weight is transmitted. On a skeleton, notice that the vertebral bodies become progressively larger from a superior to inferior direction. Explain why.

▶ ______________________________

____

2. Identify the **pedicles** and **laminae** (singular = **lamina**), which form the vertebral arch. The pedicles project posteriorly from the body and form the lateral walls of the arch. The laminae travel medially from the pedicles to form the roof.

3. The **vertebral foramen** is the opening formed by the posterior surface of the body and the vertebral arch. On a skeleton, notice that the vertebral foramina of all the vertebrae form the **vertebral canal,** through which the spinal cord travels.

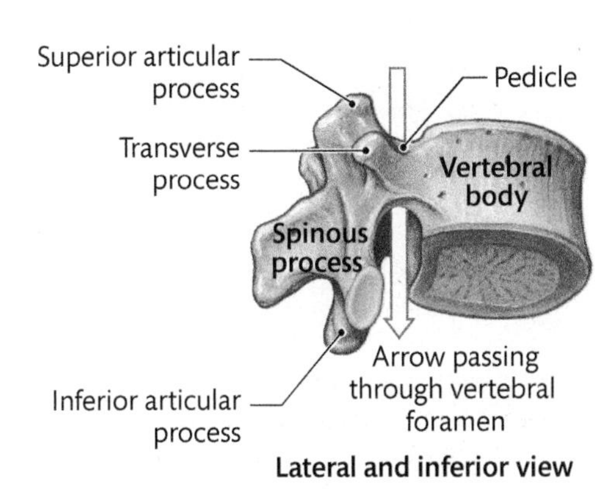

Lateral and inferior view

4. Identify the following bony processes on a typical vertebra:
   - One **spinous process** projects posteriorly from the union of the laminae.
   - Two **transverse processes,** one on each side, project laterally from the union of a pedicle and lamina.
   - Two **inferior articular processes,** one on each side, project inferiorly from a pedicle.
   - Two **superior articular processes,** one on each side, project superiorly from a pedicle.

5. On a skeleton, identify the **intervertebral foramina** between adjacent vertebrae. These openings allow the spinal nerves to exit the vertebral canal.

6. On a skeleton, locate the following articulations (joints) between adjacent vertebrae:
   - **Intervertebral discs** separate the bodies of adjacent vertebrae. The discs are composed primarily of fibrocartilage. They are absent between the first and second cervical vertebrae and between the fused vertebrae in the sacrum and coccyx.
   - **Synovial joints** (joints with a fluid-filled joint capsule) are found between inferior articular processes of the superior vertebra and the superior articular processes of the inferior vertebra. Examine the position of these joints on an intact skeleton. Obtain two vertebrae from the same region and stack them so you can see how the articular processes from adjacent vertebrae form a joint.

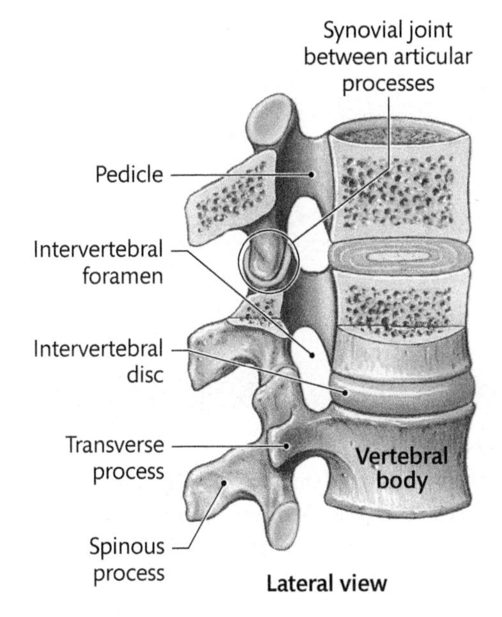

Lateral view

MAKING CONNECTIONS

The vertebral bodies, laminae, and spinous processes of adjacent vertebrae are held together by numerous ligaments. Why are these ligamentous connections important?

▶ ______________________________

____

____

____

____

____

Activity 7.12

Examining the Unique Features of Cervical and Thoracic Vertebrae

A The seven **cervical vertebrae** are the smallest bones in the vertebral column. They are designed to support the head and allow for its movement. On a skeleton, identify the following features of cervical vertebrae.

1. Identify the **transverse foramina,** which are openings on the transverse processes of all cervical vertebrae. On a model of the vertebral column, notice that the transverse foramina, on each side, form a bony canal for the passage of the **vertebral arteries** and **veins.** These blood vessels supply and drain blood to and from the brain.

2. On the third through sixth cervical vertebrae (C_3–C_6), observe that the spinous processes are relatively short and bifurcate at the tip (bifid spinous process).

3. The spinous process of the last cervical vertebra (C_7) is relatively long and does not bifurcate. Place your index finger at the superior margin of your neck, along the midline, and slowly move your finger inferiorly. The first prominent bony protuberance that you feel is the tip of the spinous process of C_7. The seventh cervical vertebra is referred to as the **vertebra prominens.** Why do you think it is given this special name?

▶ ______________________________

____

____

4. The first cervical vertebra (C_1), known as the **atlas,** is a ringlike structure with a short anterior arch and long posterior arch. It lacks a body and a spinous process. Locate the occipital condyles at the base of a skull. Position the atlas so that it articulates with the occipital condyles. This articulation is called the **atlanto-occipital joint.** The "yes" nodding motion of the head occurs at this joint.

5. The second cervical vertebra (C_2) is called the **axis.** Identify the peglike process, called the **dens,** which extends superiorly from the body. The dens protrudes into the opening of the atlas and is held against the anterior arch by the transverse ligament. This arrangement forms the **atlanto-axial joint,** where the "no" rotational motion of the head occurs.

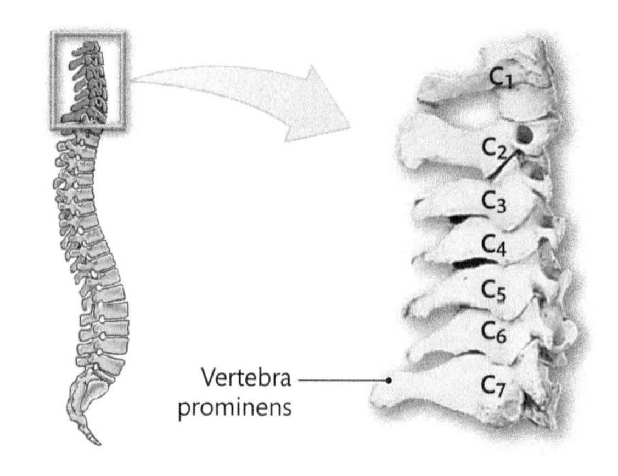

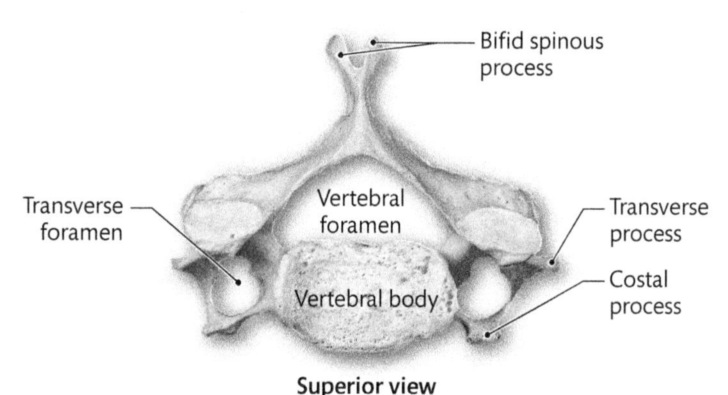

Superior view

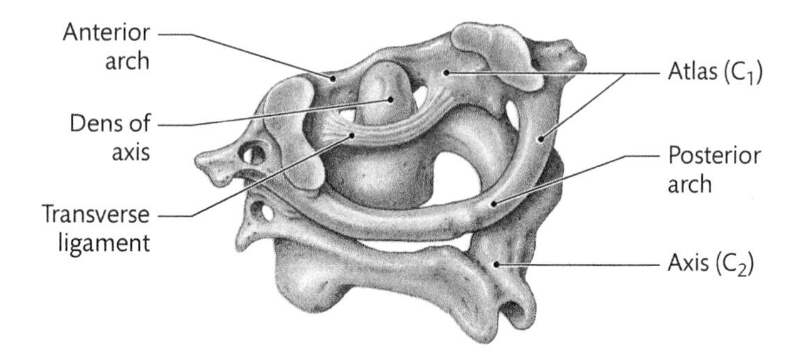

IN THE CLINIC

Whiplash Injuries

Whiplash injuries occur as a result of small tears in the ligament that reinforces the atlanto-axial joint. Tears usually result from abrupt and excessive front-to-back movements, such as when you are hit from the rear in a car accident.

B The 12 **thoracic vertebrae** are easily identified by their articulations with the 12 pairs of ribs. Notice that the vertebrae increase in size as they proceed inferiorly. Identify the thoracic vertebrae on a skeleton.

1 The first four thoracic vertebrae (T_1–T_4) could be identified as cervical vertebrae. Notice that the size of the vertebral bodies and the shape of the spinous processes are similar in appearance to the last cervical vertebra (C_7). One way to distinguish them is to observe the transverse processes. What feature is present in transverse processes of all cervical vertebrae, but missing in thoracic vertebrae?

▶ ______________________________

2 Observe that the last four thoracic vertebrae (T_9–T_{12}) have large vertebral bodies and broad spinous processes, which appear similar to lumbar vertebrae. The middle four thoracic vertebrae (T_5–T_8) demonstrate the "typical" features of thoracic vertebrae.

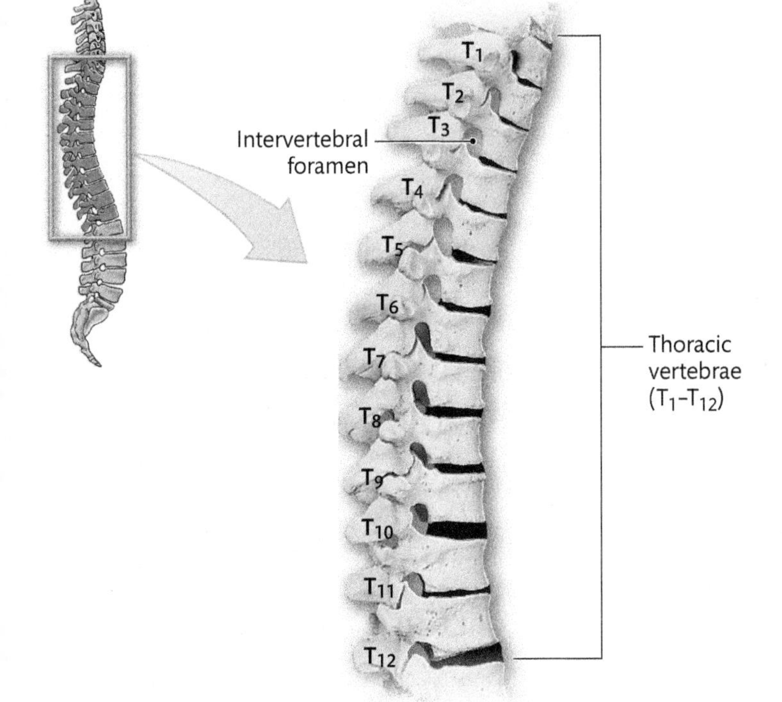

3 In a typical thoracic vertebra, notice that the body has a distinctive heart shape and that the vertebral foramen is circular.

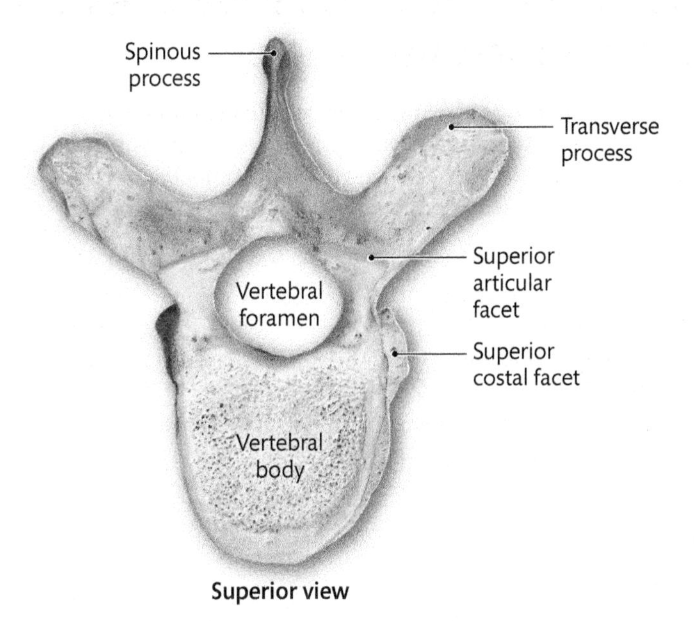

Superior view

4 Notice that the spinous process of a typical thoracic vertebra projects inferiorly. Locate the two flat surfaces on each side of the body and one on the transverse process. These surfaces, called **costal facets,** are found on all thoracic vertebrae. They are articular surfaces for synovial joints with the ribs.

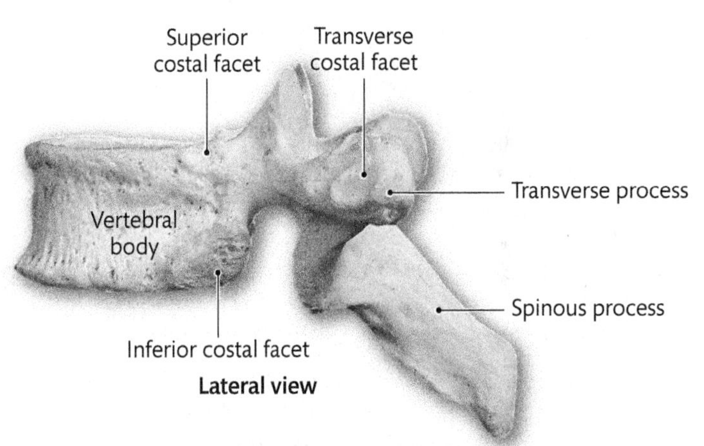

Lateral view

MAKING CONNECTIONS

A chiropractor may perform a spinal manipulation of the cervical vertebrae while treating a patient. With this procedure, there is a slight risk that a blood clot could form, travel to the brain, and cause a stroke. Based on your anatomical knowledge of the cervical vertebrae, explain how that could occur.

▶ ______________________________

____

____

Activity **7.13**

Examining the Unique Features of Lumbar Vertebrae and the Sacrum

A The five **lumbar vertebrae** are located in the lower back, between the thorax and pelvis. They are the largest individual vertebrae and support the most weight. The secondary curvature of these vertebrae is a special adaptation for bipedal locomotion.

1. Identify the lumbar vertebrae on a skeleton.
2. On a typical lumbar vertebra, observe the following unique features:
   - The vertebral body is oval- or kidney-shaped and is relatively large.
   - The spinous process is short and paddle-shaped, and it projects posteriorly.
   - The transverse processes are long and slender.

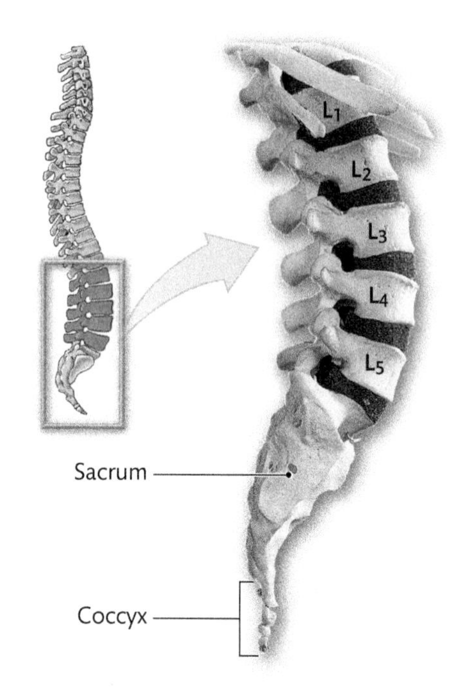

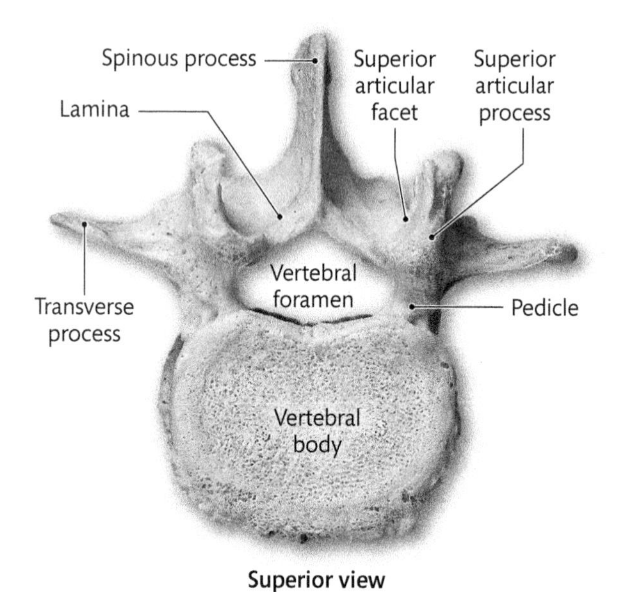

Superior view

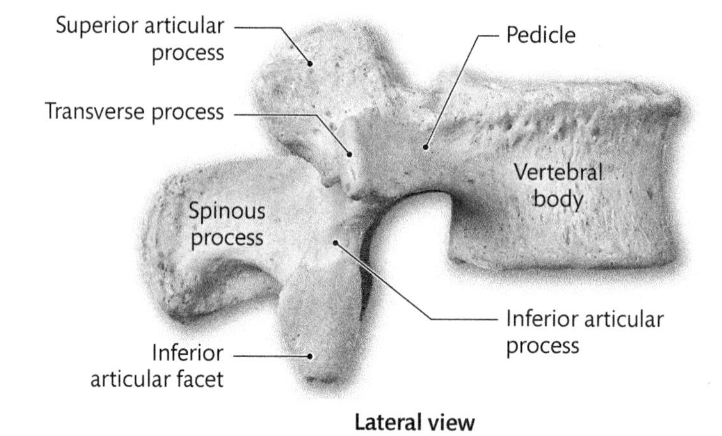

Lateral view

 ■ Want more practice? Go to: **MasteringA&P** > Study Area > Menu > Lab Tools > PAL > Anatomical Models > Axial Skeleton > Vertebral Column

B The **sacrum** develops as five separate vertebrae. Beginning at ages 16 to 18 until the mid-20s, the vertebrae fuse together to form one bone. Identify the sacrum on a skeleton.

1. On the anterior surface of the sacrum, identify the following special features:
   - The four **transverse lines** indicate the areas of fusion between the original bodies of the sacral vertebrae.
   - The **sacral promontory** is a bulge on the anterior margin of the superior surface (the base). It is an important landmark for obstetric pelvic measurements.
   - The inferior end of the sacrum is called the **apex.**
   - The **sacral foramina** represent the original intervertebral foramina. They are openings through which the sacral spinal nerves pass to reach peripheral structures.
   - The **ala** or wing on each side of the base provides a broad area for muscle attachments.
2. On the posterior surface of the sacrum, locate the following structures:
   - The **median sacral crest** is the central bony ridge formed by the fusion of the spinous processes.
   - Along each lateral margin, the **lateral sacral crests** mark the fusion of the transverse processes. Notice that the lateral sacral crests are enlarged and thickened so that they can transmit the weight of the upper body to the ilium and the lower limb.
   - The **sacral canal** is the continuation of the vertebral canal. The roof of the sacral canal is formed by the fusion of the laminae of the sacral vertebrae. The sacral hiatus marks the inferior end of the canal.
   - On each side, the **auricular surface** of the sacrum articulates with the pelvic girdle to form the **sacro-iliac joint.**
3. Observe the relative position of the sacrum with respect to the lumbar vertebrae and the **coccyx** (refer to the figure at the top right of the previous page). An intervertebral disc is located between the body of the fifth lumbar vertebra and the base of the sacrum. In addition, an intervertebral disc is located between the apex of the sacrum and the coccyx.
4. From a lateral view of the sacrum, identify the auricular surface, which articulates with the pelvic bone, and the apex, which articulates with the coccyx.

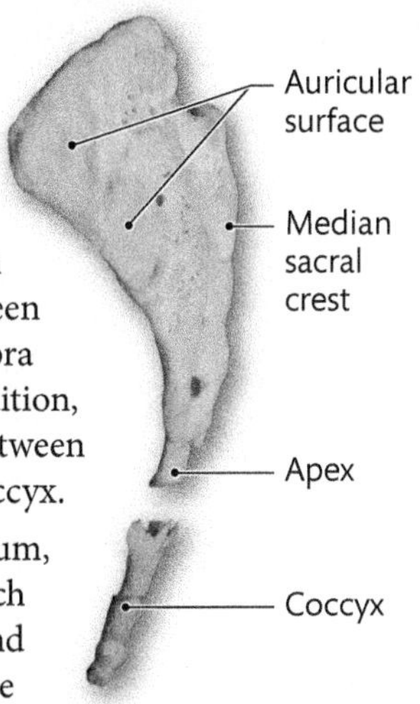

Lateral view

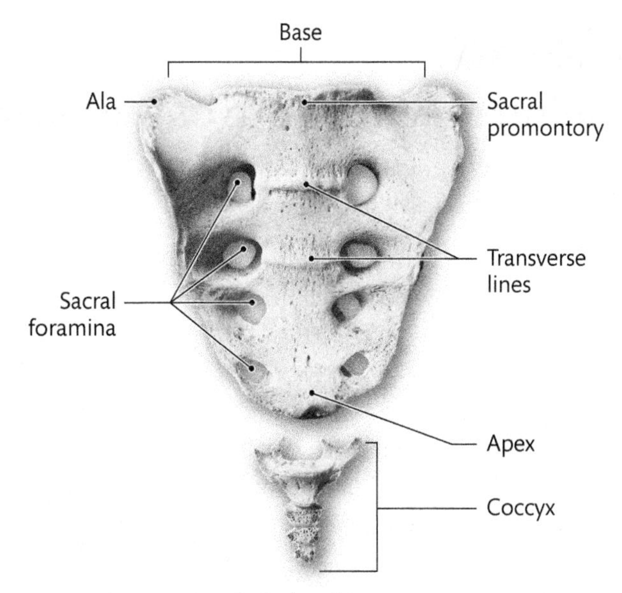

Anterior view

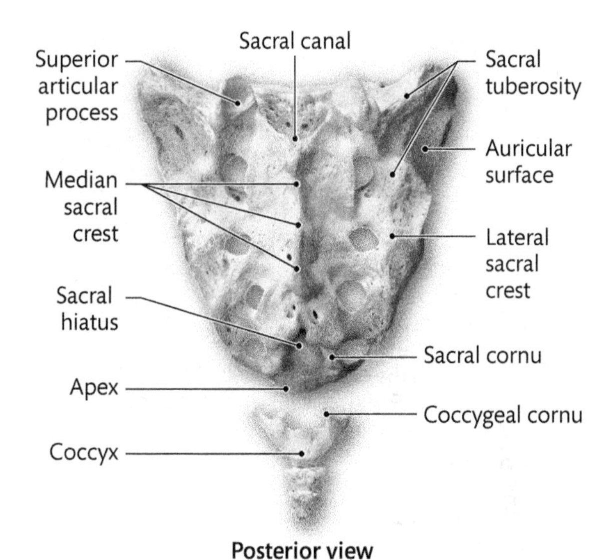

Posterior view

Word Origins

The word *sacrum* is derived from the Latin word *sacer,* which means "holy" or "consecrated." The Romans referred to the sacrum as *os sacrum,* which means "sacred bone." The Romans believed that the sacrum, because of its size and bulk, was the last bone to decay in a corpse. Thus, they thought that this bone might be the site from which the body would form again in the afterlife.

MAKING CONNECTIONS

Speculate on a functional advantage for the sacrum to be one large bone rather than five separate vertebrae.

▶ ____

____

____

____

____

____

____

____

Activity **7.14**

Examining the Structure of the Thoracic Cage

The thoracic cage includes the 12 pairs of ribs and the sternum. These structures surround and protect the lungs, heart, and other structures. In addition, movements of the ribs and sternum are essential for breathing.

A The anterior view of the thoracic cage illustrates the anatomical relationship between the ribs and the sternum.

1 On the whole skeleton, locate the **sternum** on the anterior surface of the thoracic cage and identify its three parts: the manubrium, the body, and the xiphoid process.

The **manubrium** is the broad superior segment. The superior margin of the manubrium is called the jugular notch. List the bones that articulate with the manubrium.

▶ ______________________

____

____

The elongated middle portion is the **body.** Notice that it is the largest portion of the sternum and articulates with the second through seventh pairs of ribs.

The **xiphoid process,** the smallest portion of the sternum, is attached to the inferior end of the body. It remains cartilaginous well into adulthood.

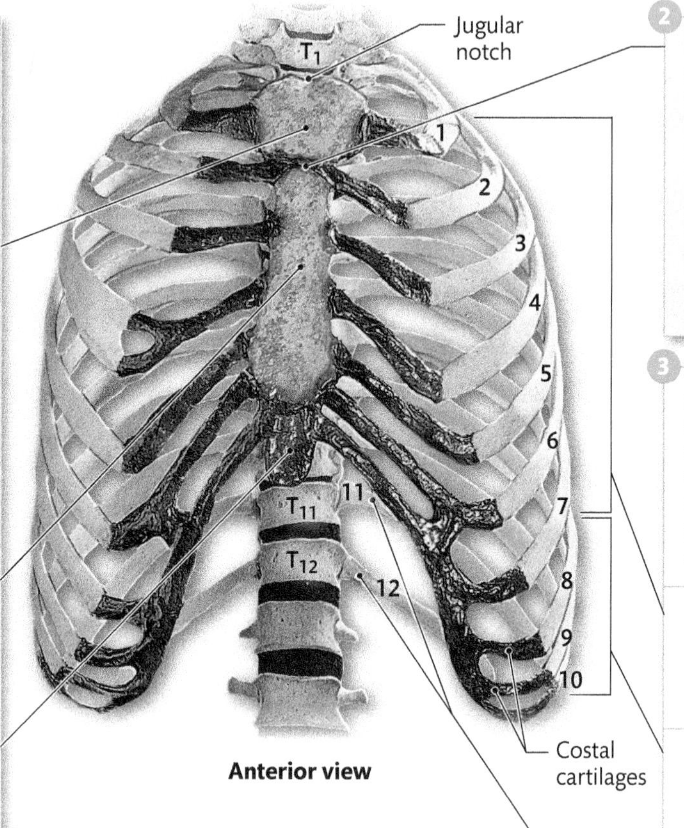

Anterior view

2 Identify the joint between the manubrium and body. This articulation acts as a hinge that allows the body to move anteriorly during inhalation. The angle formed at the joint is the **sternal angle.** Which pair of ribs articulates with the sternum at the sternal angle?

▶ ______________________

3 Anteriorly, notice that the ribs are attached to pieces of cartilage, referred to as **costal cartilages.** The ribs are categorized according to how the costal cartilages are attached to the sternum:

Ribs 1 through 7 are called **vertebrosternal ribs.** Their costal cartilages are directly attached to the sternum.

Ribs 8 through 10 are called **vertebrochondral ribs.** Their costal cartilages are connected to one another and attach to the sternum via the costal cartilage of rib 7. These ribs, therefore, have an indirect attachment to the sternum.

Ribs 11 through 12 are called **floating ribs** because they have no connection to the sternum.

4 Examine the arrangement of the 12 pairs of ribs. How does the length of the ribs change from rib pair 1 through rib pair 7? How does the length change from rib pair 8 through rib pair 12? ▶

Rib Pairs	Change in Length
Rib pairs 1–7	
Rib pairs 8–12	

B The posterior view of the thoracic cage illustrates the anatomical relationship between the ribs and thoracic vertebrae.

1 On a skeleton, notice that most, but not all, ribs articulate with two thoracic vertebrae. Typically, a rib articulates with its numerically corresponding vertebra, the intervertebral disc, and the vertebra superior to it. For example, rib 5 articulates with T_5, T_4, and the intervening disc.

Which ribs articulate with only one thoracic vertebra?

▶ ______________________________

____

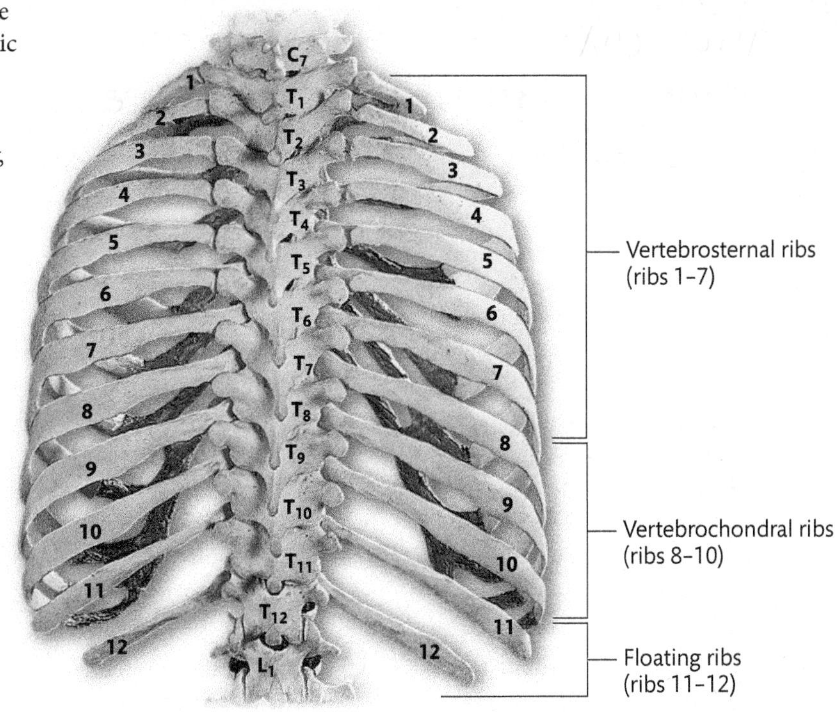

2 Notice the spaces between the ribs, called **intercostal spaces.** Each space contains three layers of **intercostal muscles,** which play an important role in respiration.

C Distinct bone markings are present on the vertebral end of a rib, but are absent on the sternal end.

1 Examine a "typical" rib and differentiate between the vertebral end and the sternal end. At the vertebral end, identify (1) the **head,** which articulates with the body of the numerically corresponding thoracic vertebra, the intervertebral disc, and the body of the superior vertebra; (2) the **tubercle,** which articulates with the transverse process of the numerically corresponding thoracic vertebra; and (3) the **neck,** which is the short length of bone between the head and tubercle.

2 Identify the **shaft,** or **body,** the main part of the rib. Beginning at the vertebral end, the **angle of the rib** is where the shaft curves anteriorly and inferiorly toward the sternum. The **costal groove** runs along the inferior border of the shaft. The groove provides a pathway for intercostal arteries, veins, and nerves.

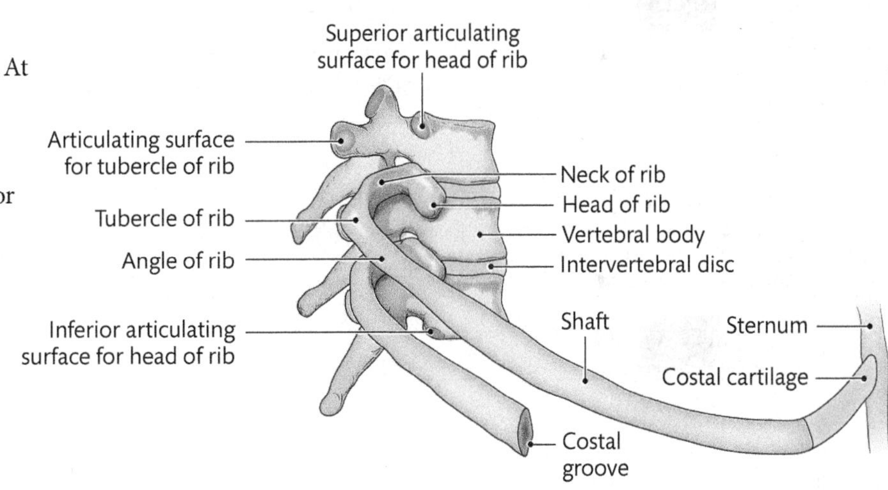

MAKING CONNECTIONS

The 11th and 12th rib pairs are called *floating ribs,* but that term can be misleading. Explain why.

▶ ______________________________

____

____

____

BEFORE YOU MOVE ON . . .

« LOOKING BACK

Whether you are looking at real bones or plastic models, bone tissue appears to be inactive and inert. In living individuals, however, bone is vital, dynamic tissue with many important functions, namely:

1. Providing rigid scaffolding that supports the body
2. Protecting vital organs
3. Providing leverage for skeletal muscles to make body movements possible
4. Producing blood cells and platelets
5. Maintaining normal blood levels of calcium and phosphate ions
6. Storing lipids.

The first three functions might seem obvious to you, but the last three are activities that we do not typically associate with bone tissue. For example, it is difficult to connect the dried calcified remains or plastic models of the bones you are studying with the robust metabolism and high energy that is required to produce blood cells or to regulate ion concentrations in blood.

Review the functions listed above carefully and consider what role the skeletal system has in maintaining homeostasis. The answer may not be obvious, but it will give you a better appreciation of the dynamic nature of bone tissue.

▶ ____

____

____

____

____

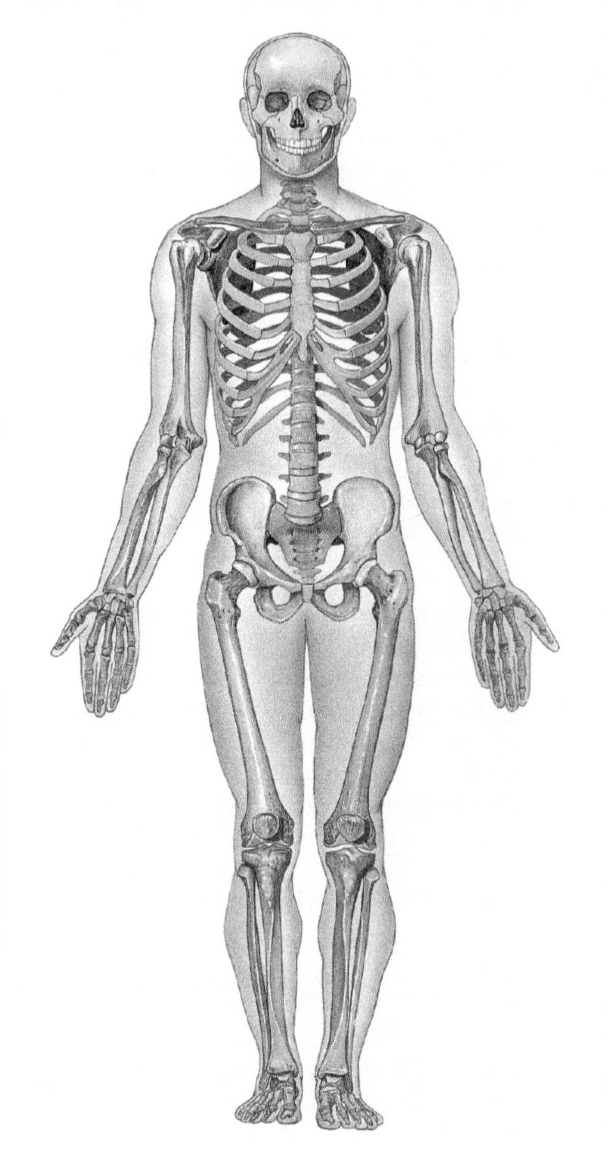

LOOKING FORWARD »

In this laboratory exercise, you were introduced to the bones of the skeleton. You also completed an in-depth examination of the axial skeleton, which includes the bones that form the body's longitudinal axis. In the next exercise, you will study the appendicular skeleton, which includes the bones of the upper and lower limbs.

Name Gabe Vanaenhoven

Lab Section

Date 3/25

EXERCISE 7 REVIEW SHEET

Introduction to the Skeletal System and the Axial Skeleton

1. Complete the following table.

Bone Classification According to Shape	Examples
a. Sesamoid Bone	Patella
b. Long Bone	Humerus c. Femur
Irregular bones	d. Maxilla e. Zygomatic
f. Irregular Bone	Small bone between cranial bones
g. Short Bone	Carpal bones h. Tarsals
Flat bones	i. Scalpula j. Occipital

QUESTIONS 2–11: Match the bone in column A with its correct classsification in column B. Answers may be used more than once.

A

2. Clavicle A

3. Atlas E

4. Parietal D

5. Tibia C

6. Sternum B

7. Carpal bones E

8. Sphenoid D

9. Femur A

10. Sacrum B

11. Humerus

B

a. Axial skeleton, vertebral column

b. Appendicular skeleton, upper limb

c. Axial skeleton, thoracic cage

d. Appendicular skeleton, lower limb

e. Axial skeleton, skull

12. How does the structure of compact bone differ from the structure of spongy bone?

Compact Bone is dense and composed of Osteons where spongy bone is less dense and made up of trabeculae.

13. Describe the arrangement of compact and spongy bone in a typical long bone.

The end contains spongy bone where the outside layer is compact bone.

QUESTIONS 14–19: Match the bone marking in column A with its general function in column B. Answers may be used more than once.

A		B
14. Head of the humerus	A	**a.** Articulating surface at a joint
15. Foramen ovale in the skull	C	**b.** Attachment site for tendons or ligaments
16. Trochanter on the femur	B	**c.** Passageway for nerves or blood vessels
17. Costal facet on a thoracic vertebra	A	
18. Mastoid process on the skull	B	
19. Carotid canal in the skull	C	

QUESTIONS 20–26: Identify the labeled bones in the diagram.

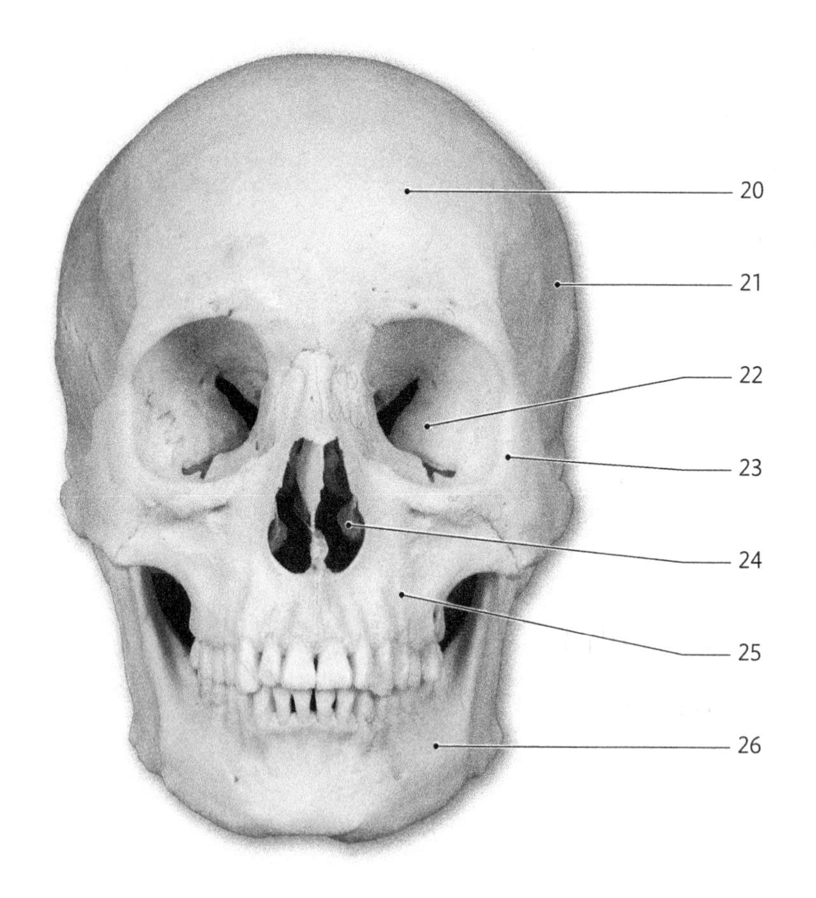

20. Frontal

21. Parietal

22. Sphenoid

23. Zygomatic

24. Inferior Nasal Concha

25. Maxilla

26. Mandible

27. In the following diagram, color the bones with the indicated colors.

- Frontal = **yellow**
- Sphenoid = **purple**
- Zygomatic = **blue**
- Maxillary = **orange**
- Lacrimal = **red**
- Mandible = **gray**
- Parietal = **green**
- Occipital = **orange**
- Temporal = **red**
- Nasal = **purple**

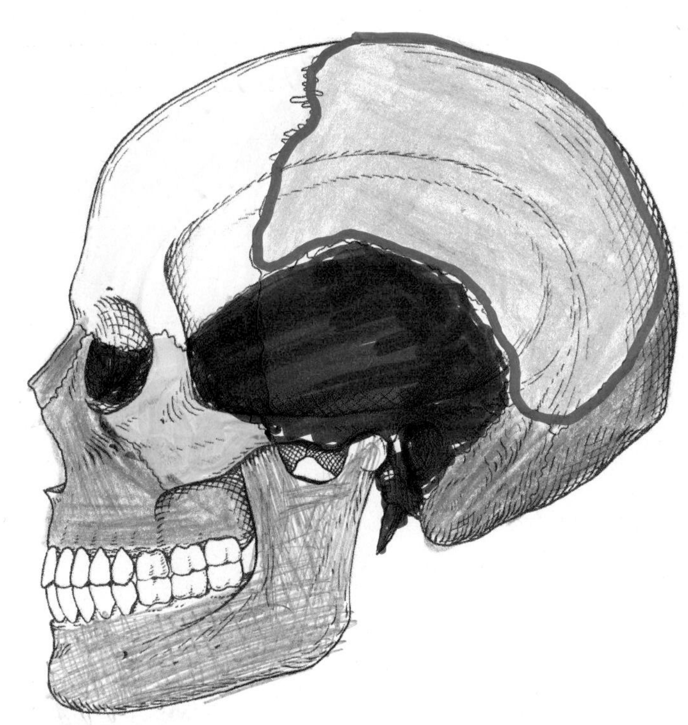

QUESTIONS 28–39: In the following diagram, identify the labeled bony markings. Color the bones with the indicated colors.

- Frontal = **yellow**
- Sphenoid = **purple**
- Temporal = **red**
- Ethmoid = **green**
- Occipital = **orange**
- Parietal = **pink** / Brown

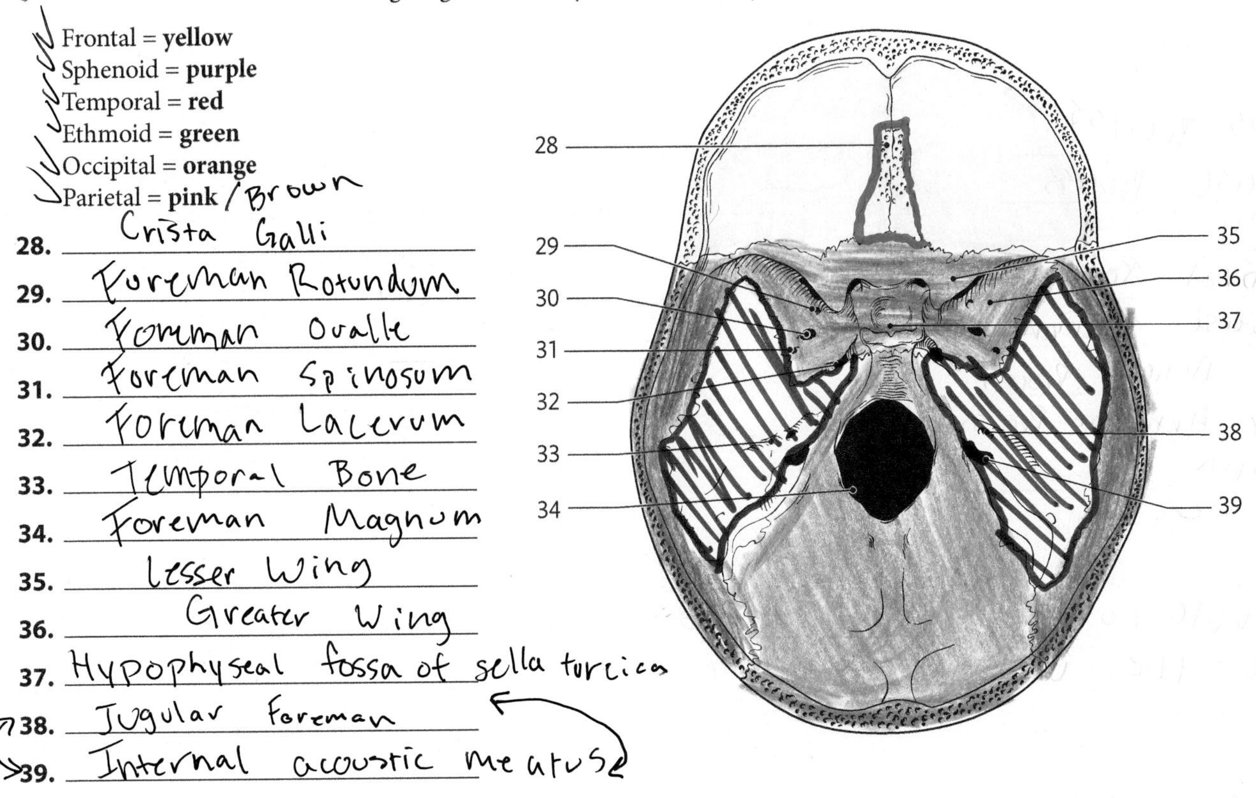

28. Crista Galli

29. Foreman Rotundum

30. Forman Ovalle

31. Foreman Spinosum

32. Foreman Lacerum

33. Temporal Bone

34. Foreman Magnum

35. Lesser Wing

36. Greater Wing

37. Hypophyseal fossa of sella turcica

38. Jugular Foreman

39. Internal acoustic meatuse

QUESTIONS 40–46: Identify the labeled structures in the diagram.

40. Sagittal Suture

41. Parietal Bone

42. Anterior Fontaelle

43. Coronal Suture

44. Frontal Bone

45. Nasal Bone

46. Maxilla

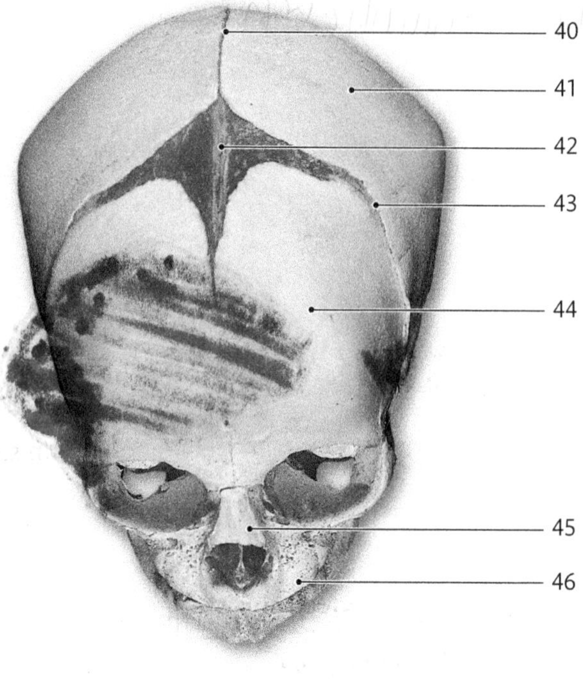

QUESTIONS 47–55: Identify the structures labeled in the following diagrams. Color the bones with the indicated colors.

Cervical vertebra = **green**
Thoracic vertebra = **red**
Lumbar vertebra = **blue**

47. Spinous Process

48. Tranverse Process

49. Body

50. Vertebral Foreman

51. Transverse Foreman

52. Superior Articular Process

53. Inferior Articular Process

54. Lamina

55. Pedicle

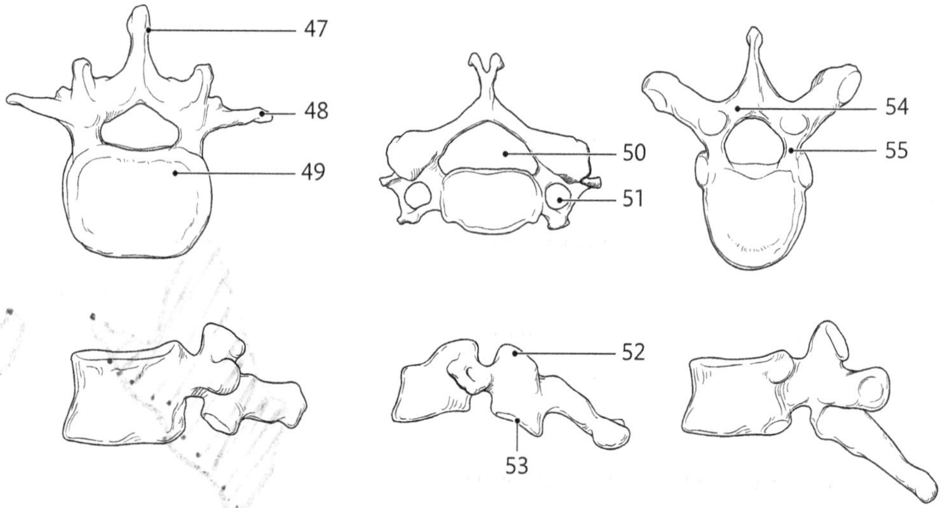

56. Explain the difference between a vertebrosternal rib and a vertebrochondral rib.

The vertebrosternal ribs are attached directly to the sternum where the vertebrochondral ribs are not.

57. Explain how a typical rib articulates with the vertebral column.

Costovertebral joint is connected and the orgin of movement.

EXERCISE

8

The Appendicular Skeleton

LEARNING OUTCOMES

These Learning Outcomes correspond by number to the laboratory activities in this exercise. When you complete the activities, you should be able to:

Activities 8.1–8.3 **Identify the bones of the upper limb and their bone markings and explain how these bones articulate.**

Activities 8.4–8.6 **Identify the bones of the lower limb and their bone markings and explain how these bones articulate.**

LABORATORY SUPPLIES

- Human skeleton
- Model of the knee
- Disarticulated bones of the upper extremity
- Disarticulated bones of the lower extremity
- Colored pencils

PRE-LAB QUIZ

Before you begin, read all the activities in Exercise 8 and the required reading in your textbook that is assigned by your instructor.

1. During this laboratory exercise, you will complete all the following activities *except*
   a. estimate a person's height by measuring the length of the femur.
   b. compare the structure of the male and female bony pelvis.
   c. identify the carpal bones.
   d. observe how the ribs are connected to the sternum.
   e. examine the bones of the pectoral girdle.

Questions 2–5: Select your answers from the following list. For each question, there could be more than one correct answer.

humerus	carpal bones	metacarpals
scapula	clavicle	phalanges
radius	ulna	

2. Identify the bone(s) that form the pectoral girdle. ______________________
3. The shoulder joint is formed by the articulation of what bones? ______________________
4. Identify the bone(s) of the forearm. ______________________
5. Identify the bone(s) of the wrist and hand. ______________________
____

Questions 6–9: Select your answers from the following list. For each question, there could be more than one correct answer.

ilium	femur	fibula	phalanges
ischium	patella	tarsal bones	
pubis	tibia	metatarsals	

6. Identify the bones that form the coxal bone. ______________________
7. Identify the bone(s) of the thigh. ______________________
8. The knee joint is formed by the articulation of what bones? ______________________
9. Identify the bone(s) of the leg. ______________________
____

10. True or False: The transverse arch of the foot is not considered to be a true arch.
______

Activity **8.1**

Upper Limb: Examining the Bones of the Pectoral Girdle

Overview of the Upper Limb Bones

Each upper limb contains four segments. Their locations and the bones found in each are described in Table 8.1.

TABLE 8.1 Upper Limb Segments

Segment	Location	Description
Shoulder	The proximal region of the upper limb that overlaps with portions of the neck, back, and thorax and includes the shoulder joint	Contains one **clavicle** and one **scapula**
Arm	Region between shoulder and elbow joints	Contains one large long bone, the **humerus**
Forearm	Region between elbow and wrist joints	Contains two smaller long bones, the **ulna** and **radius**
Wrist and hand	Region that is distal to forearm	Contains 8 **carpal bones** at the wrist, 5 **metacarpal bones** in the palm, and 14 **phalanges** in the digits (thumb and fingers)

A The **pectoral girdle** forms a bony strut that suspends the upper limbs lateral to the trunk. It consists of a ring of bone formed by the two **clavicles** (collar bones) and **scapulae** (shoulder blades). The ring is completed anteriorly by the manubrium of the sternum, but is incomplete posteriorly. The clavicle and scapula on each side of the body represent one-half of the pectoral girdle.

1 On a skeleton, identify the clavicle and scapula on each side.

2 The pectoral girdle is stabilized almost entirely by muscles that connect it to the sternum, ribs, and vertebrae. There is only one bony articulation between each pectoral girdle and the axial skeleton. Identify this articulation.

▶ ____

____

____

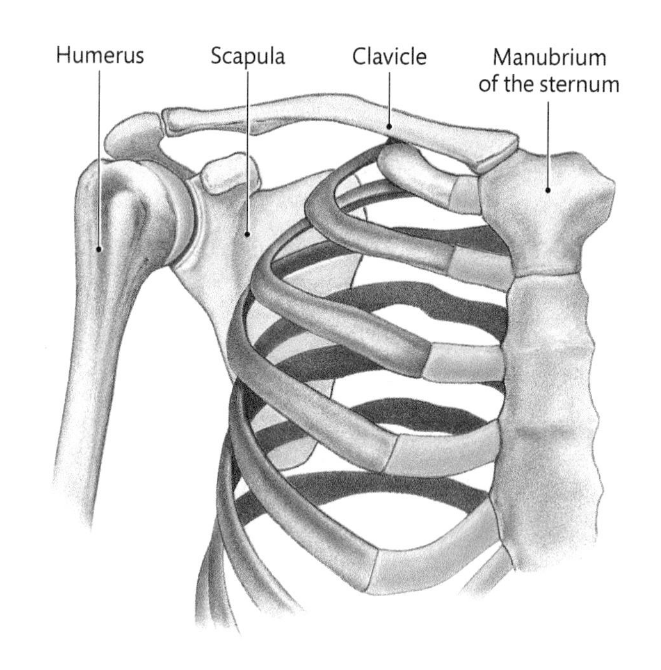

B The clavicle has two distinct curvatures.

1 The medial or ▶ ____________________ end is concave posteriorly and has a pyramidal-shaped end.

2 The lateral or ▶ ____________________ end is concave anteriorly and has a flattened end.

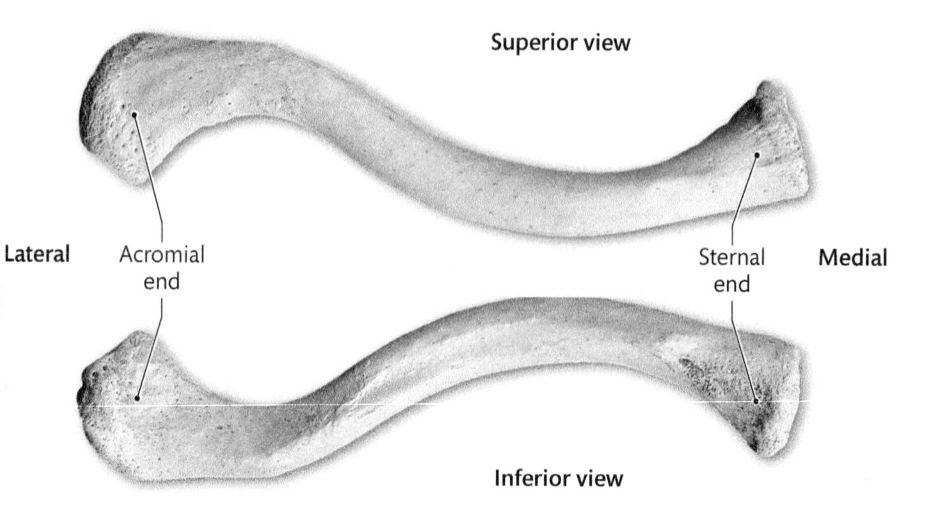

C The scapula is a flat, triangular bone that rests on the posterior thoracic wall.

1 Notice that the **shoulder joint** is formed by the articulation of the glenoid cavity (fossa) of the scapula and the head of the humerus.

2 The shoulder joint is also called the glenohumeral joint. Explain why.

▶ ______________________________

____

____

____

____

____

____

3 On a skeleton or disarticulated scapula, identify the bone markings labeled in the figures on this page.

4 Complete the following table. ▶

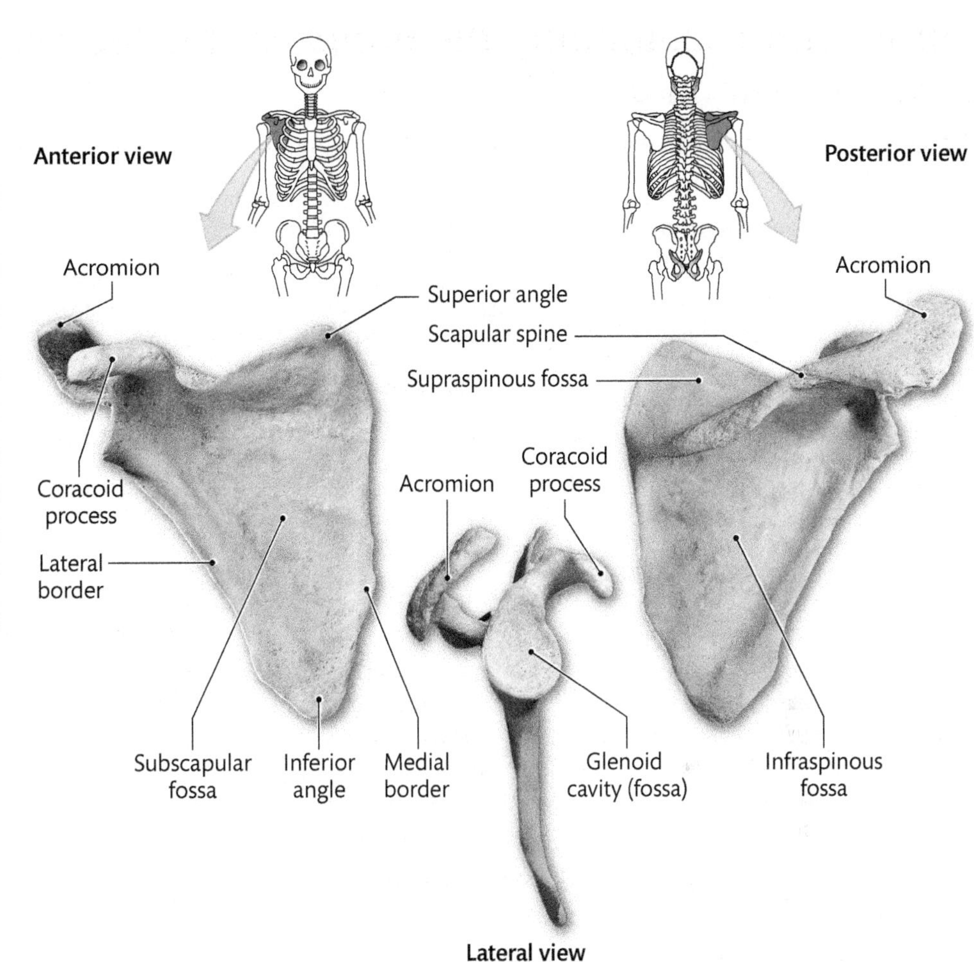

Description	Bone Markings
Bone markings along the anterior aspect of the scapula	
Bone markings along the posterior aspect of the scapula	
Bone marking that provides an articulating surface for the shoulder joint	

MAKING CONNECTIONS

Speculate on the principal function of the clavicle. (*Hint:* Consider the consequences if the clavicle were absent.)

▶ ______________________________

____

____

____

____

____

____

____

____

____

____

____

■ Want more practice? Go to: **MasteringA&P** > Study Area > Menu > Lab Tools > PAL > Anatomical Models > Appendicular Skeleton > Pectoral Girdle

Activity **8.2**

Upper Limb: Examining the Bones of the Arm and Forearm

A The humerus is the only bone found in the arm. It is the longest and largest bone of the upper limb.

1. On a skeleton, locate the humerus in each arm. Notice that it is involved in the formation of two major joints: the shoulder joint and the elbow joint.

2. On a skeleton or disarticulated humerus, identify the bone markings labeled in the figures on this page.

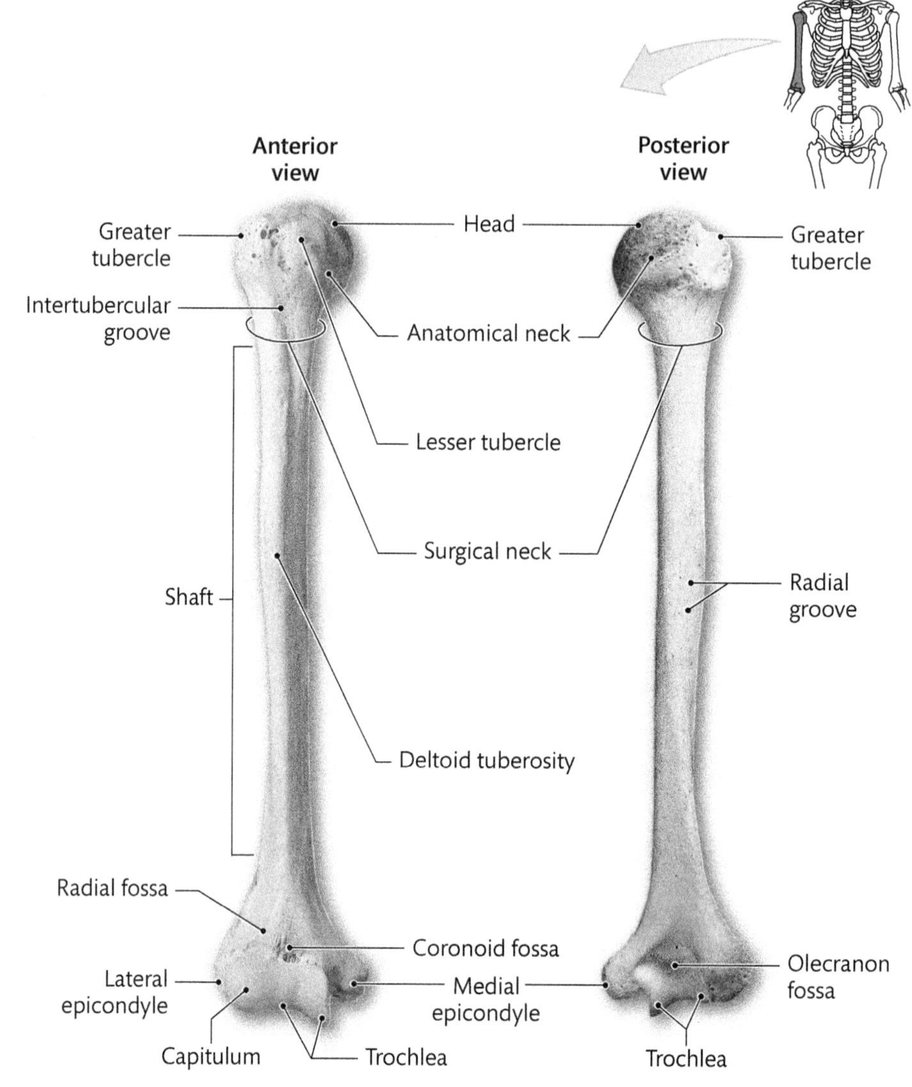

3. For each of the bone markings of the humerus listed in the table, speculate on its function by placing a check mark in the appropriate function column. ▶

Bone Marking	Function		
	Articulating surface at a joint	Attachment site for tendons and ligaments	Passageway for nerves, blood vessels, tendons, or ligaments
Head			
Greater tubercle			
Intertubercular groove			
Medial epicondyle			
Olecranon fossa			
Radial groove			

B The two long bones in the forearm are the ulna and radius. These two bones articulate at their proximal and distal ends to form the **proximal** and **distal radioulnar joints.**

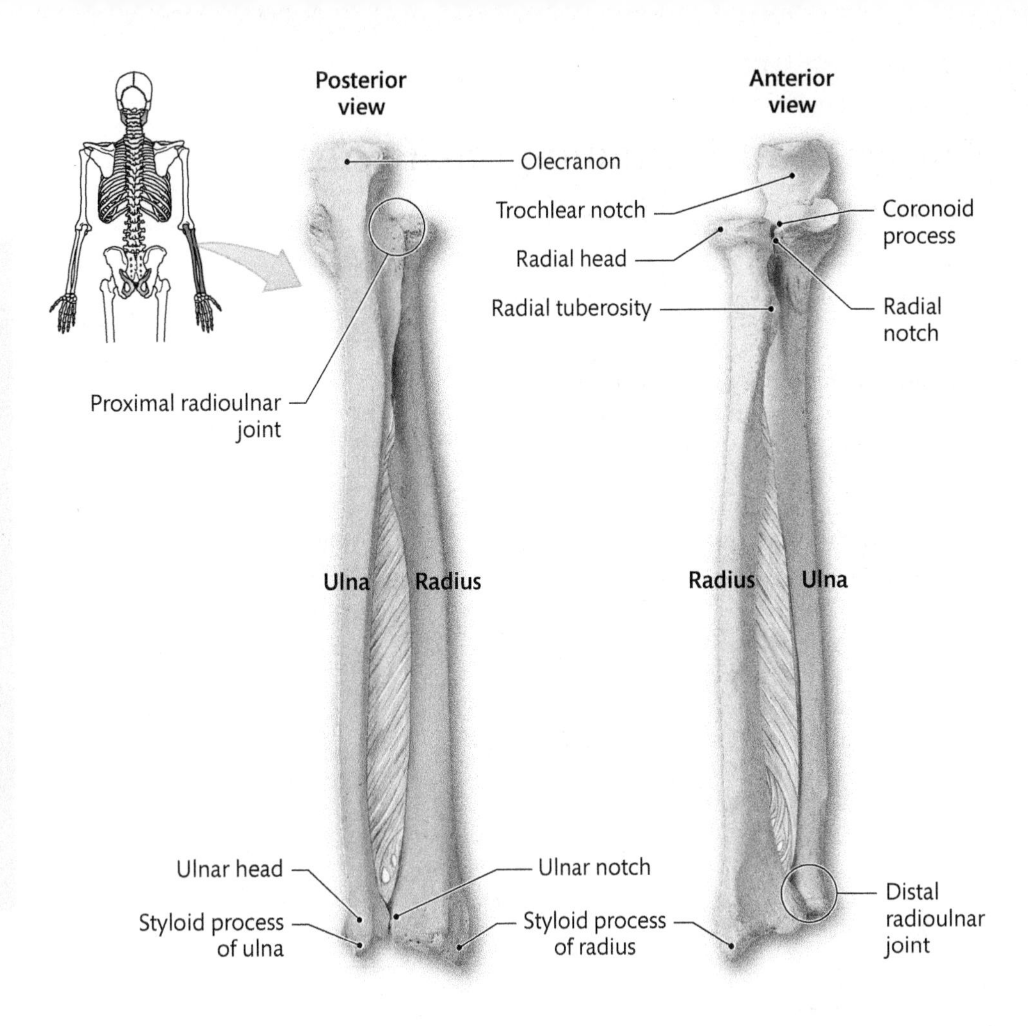

1 Observe the ulna and radius on a skeleton. Which one is the medial bone, and which is the lateral bone? ▶

Medial bone

____

Lateral bone

____

Explain why an understanding of anatomical position is useful for determining your answers.

▶ _______________

____

____

____

____

2 Identify the following bone markings.

Radius:

1. Radial head
2. Radial tuberosity
3. Ulnar notch
4. Styloid process of radius

Ulna:

1. Olecranon
2. Trochlear notch
3. Coronoid process
4. Radial notch
5. Ulnar head
6. Styloid process of ulna

3 Both the radius and ulna articulate with the humerus at the elbow joint. Identify the bone markings that articulate between the humerus and radius and between the humerus and ulna at the elbow joint. ▶

Articulating Bones	Bone Markings
Humerus and radius	Humerus: Radius:
Humerus and ulna	Humerus: Ulna:

MAKING CONNECTIONS

A long bone consists of a diaphysis (the shaft) and two epiphyses (the knoblike ends). Examine a humerus, a radius, and an ulna. For each bone, list the bone markings that form the proximal epiphysis and the distal epiphysis. ▶

Bone	Proximal Epiphysis	Distal Epiphysis
Humerus		
Radius		
Ulna		

Activity **8.3**

Upper Limb: Examining the Bones of the Wrist and Hand

A The carpal bones are the short bones positioned between the radius and ulna in the forearm and the metacarpal bones in the hand.

1. Examine the eight carpal bones at the wrist and notice that they are arranged in two rows of four (proximal and distal). As a unit, carpal bones are shaped so that the anterior (palmar) surface is concave and the posterior (dorsal) surface is convex.

2. At the wrist joint, notice that the inferior surface of the radius articulates with the proximal row of carpal bones. Although it might appear otherwise, the ulna does not contribute an articulating surface at the wrist.

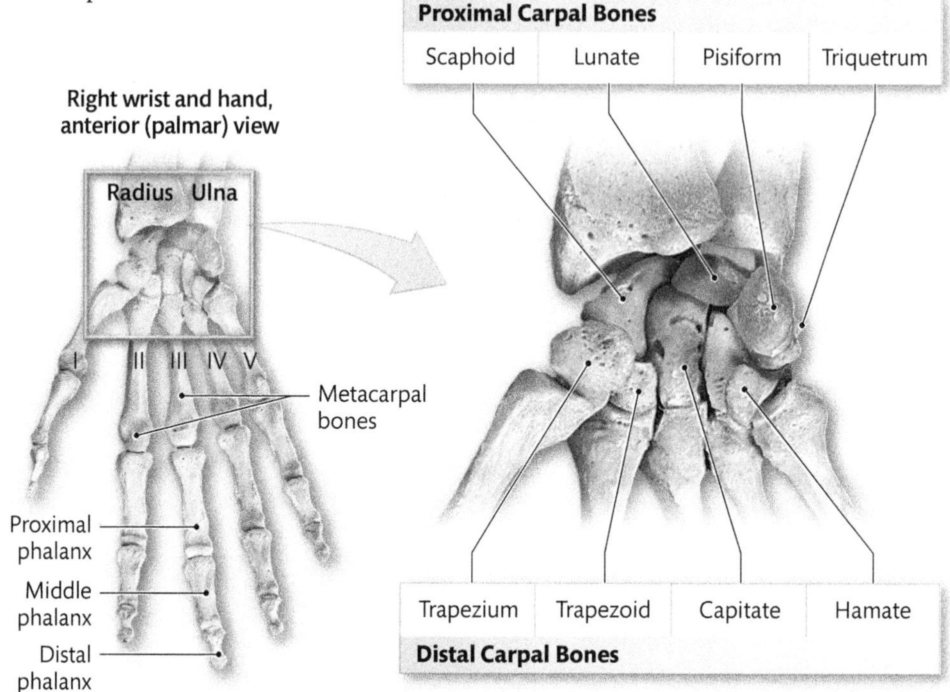

3. In the space below, draw and label the eight carpal bones, positioning them in sequence along the proximal and distal rows from lateral to medial.

▶

Word Origins

The appearances of the carpal bones can help you remember their names. For instance, in the proximal row, the lunate (from the Latin *luna,* moon) is a comma-shaped bone resembling a crescent moon. The largest and most central carpal, located in the distal row, is the capitate. Think of its position and size as making it the "head" bone because *caput* means "head" in Latin. Finally, next to the capitate is the curved hamate, whose name derives from the Latin word *hamatus,* which means "hooked."

B In the hand, the metacarpal bones are located in the palm; the phalanges (singular = phalanx) are the bones of the digits (thumb and fingers).

1. Locate the five metacarpal bones that travel through the palm of the hand. The proximal ends (bases) of these miniature long bones articulate with the carpal bones. Distally, the heads articulate with the proximal phalanges of the digits to form the **metacarpophalangeal (MP) joints** (the knuckles). The metacarpals are numbered by Roman numerals I through V, from lateral to medial.

2. Notice that, similar to the metacarpal bones, the phalanges are long bones with a proximal base and a distal head. The thumb contains only two phalanges; all the other digits contain three. The articulations between phalanges are called **interphalangeal (IP) joints.**

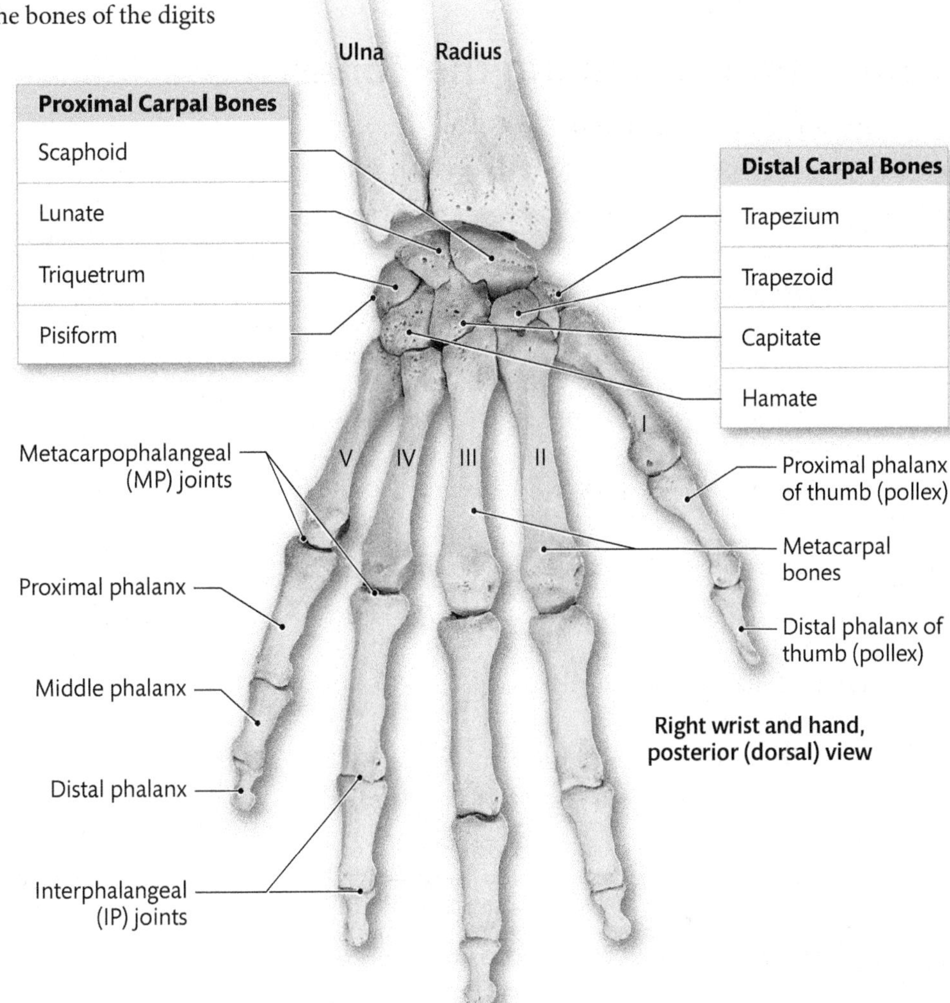

Right wrist and hand, posterior (dorsal) view

IN THE CLINIC

Carpal Bone Fractures

The scaphoid is the most common carpal bone to be fractured, accounting for 70 percent of all cases. Fractures of the capitate are rare because it fits snugly between other bones and is in a protected position. Fractures are usually caused by a hard blow to the dorsum (posterior or dorsal surface) of the hand or if the hand is used for protection from a fall, causing excessive extension of the wrist. Carpal fractures are not always clearly demonstrated on x-ray images; bone and CT scans may be required to confirm a diagnosis. Most carpal bone fractures can be treated by placing the wrist in a cast for about six weeks followed by physical therapy. More severe fractures may require surgery, during which the broken bone is held together with pins or plates prior to immobilizing in a cast.

MAKING CONNECTIONS

The carpal bones are held together by ligaments that restrict their mobility to gliding movements. From a functional standpoint, why do you think this arrangement is necessary?

▶ ______________________________

____

____

____

____

____

____

____

■ Want more practice? Go to: **MasteringA&P** > Study Area > Menu > Lab Tools > PAL > Anatomical Models > Appendicular Skeleton > Upper Limb

Activity **8.4**

Lower Limb: Examining the Bones of the Pelvic Girdle

Overview of the Lower Limb Bones

Each lower limb contains four segments. Their locations and the bones found in each are described in Table 8.2.

TABLE 8.2 **Lower Limb Segments**

Segment	Location	Description
Hip	Extends from superior margin of the **coxal bone (hip bone)** to the hip joint	The hip bone is formed by the fusion of three bones: the **ilium, ischium,** and **pubis**
Thigh	Region between hip and knee joints	Contains the **femur,** a massive long bone, and the **patella,** a sesamoid bone that protects the knee anteriorly
Leg	Region between knee and ankle joints	Contains two smaller long bones, the medial **tibia** and lateral **fibula**
Foot	Region that is distal to leg	Contains 7 **tarsal bones,** 5 **metatarsal bones,** and 14 **phalanges**

A The **pelvic girdle** consists of the right and left coxal bones.

1. On a skeleton, identify the three fused bones that form the coxal bone:
   - The ilium (plural = ilia) is the largest and most superior bone. The expansive and flat medial and lateral surfaces on this bone serve as attachment sites for muscles.
   - The ischium (plural = ischia) forms the posteroinferior portion of the coxal bone. It contains the ischial tuberosity, which helps support the body while sitting.
   - The pubis (pubic bone) forms the anteroinferior portion of the coxal bone.

2. Notice that the ilium, ischium, and pubis join together on the lateral surface of the coxal bone to form a deep bony socket known as the **acetabulum.** The acetabulum articulates with the head of the femur to form the hip joint.

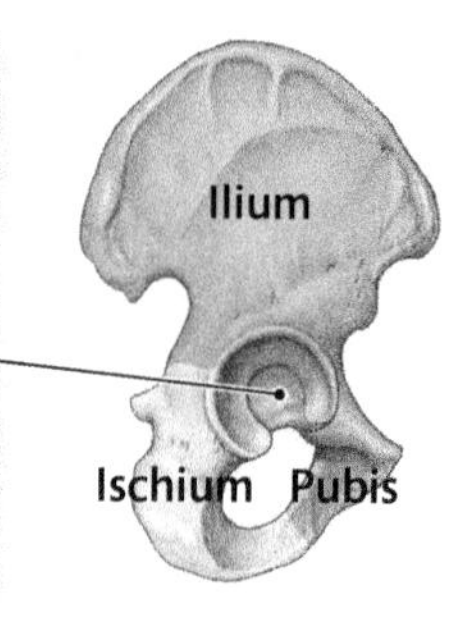

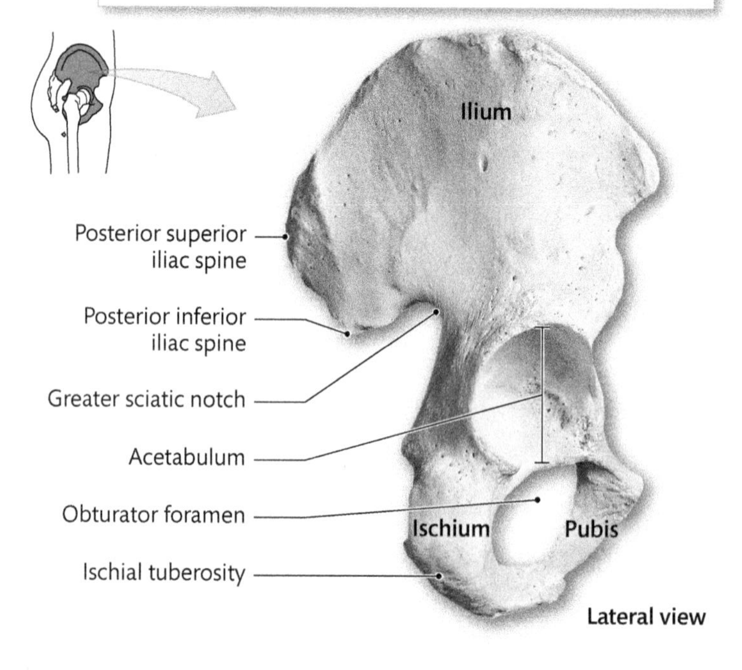

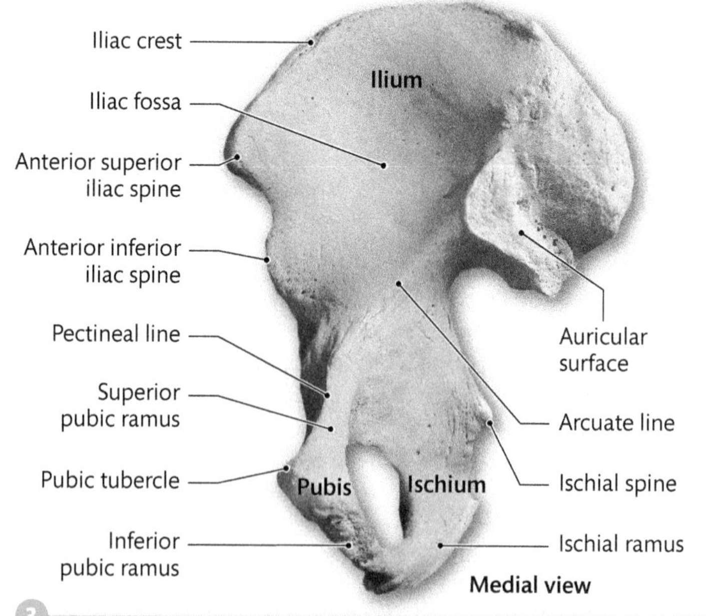

3. On a skeleton or disarticulated pelvic bones, identify the bone markings labeled in the figures above.

■ Want more practice? Go to: **MasteringA&P** > Study Area > Menu > Lab Tools > PAL > Anatomical Models > Appendicular Skeleton > Pelvic Girdle

B The pelvic girdle, sacrum, and coccyx form the **bony pelvis.**

1 Observe that the auricular surfaces of the sacrum articulate with the auricular surfaces of the two ilia to form the **sacro-iliac joints.**

2 Anteriorly, observe that the two pubic bones are joined by a disc of fibrocartilage. This articulation is called the **pubic symphysis.**

3 Identify the **pubic arch.** It is the angle formed by the two pubic bones at the pubic symphysis.

Sacrum

4 Identify the bones in the bony pelvis that belong to the appendicular skeleton and those that belong to the axial skeleton. ▶

Skeletal Region	Bones
Appendicular	
Axial	

C The two regions of the bony pelvis are the false (greater) pelvis and the true (lesser) pelvis.

1 Identify the **pelvic brim.** From either side of the sacrum, it is a circular margin that extends along the arcuate and pectineal lines to the superior margin of the pubic symphysis.

2 Locate the **false pelvis.** It is the region superior to the pelvic brim and contains the organs in the inferior portion of the abdominal cavity.

3 Find the **true pelvis,** which is the region inferior to the pelvic brim (shown in purple) and contains the organs of the pelvic cavity.

4 Identify the **pelvic outlet.** This is the opening leading out of the pelvic cavity, formed by the inferior margins of the bony pelvis.

List the structures that form the inferior margins (connected by the double arrows) of the pelvic outlet.

▶ ______________________

____

____

____

5 Observe the **pelvic inlet.** It is the opening leading into the pelvic cavity. It is formed by the pelvic brim.

MAKING CONNECTIONS

Compare the bony pelvis of a male with that of a female and describe the anatomical differences in the following features. ▶

Anatomical Feature	Male-Female Comparison
Pelvic inlet and outlet	
Curvature of the sacrum	
Angle of the pubic arch	
Lateral projection of ilia	

Speculate on a reason for these differences.

▶ ____

____

Activity 8.5

Lower Limb: Examining the Bones of the Thigh and Leg

A The femur is the long bone of the thigh. It is the heaviest, largest, and longest bone of the body.

1 On a skeleton, notice that the femur contributes to the formation of two major joints: the hip and the knee.

2 Why do you think the femur is such a large and massive bone?

▶ ______________________________

____

____

____

____

____

3 A person's height can be estimated by measuring the length of the femur and multiplying that value by four. Measure the length of a disarticulated femur and estimate the height of the individual. ▶

1. Length of femur (cm)

   ____

2. Estimated height (cm)

   ____

3. Estimated height (ft)

   ____

   (2.54 cm = 1 in)

4 On a skeleton or disarticulated femur, identify the bone markings labeled in the figures on the right.

5 For each bone marking listed in the table, speculate on its function by placing a check mark in the appropriate function column. ▶

	Function	
Bone Marking	**Articulating surface at a joint**	**Attachment site for tendons and ligaments**
Femoral head		
Greater trochanter		
Linea aspera		
Medial condyle		
Lateral epicondyle		

B The patella is a sesamoid bone that is located within the tendon of the quadriceps femoris muscles, a group of four muscles in the anterior thigh.

1. Locate the attachment sites for the quadriceps tendon (a) and the patellar ligament (b) on the anterior surface of the patella.

2. On the posterior surface of the patella, identify the region that articulates with the patellar surface on the femur.

(a)

(b)

Anterior view **Posterior view**

C The tibia and fibula are the two long bones in the leg.

1. On a skeleton, locate the tibia and fibula. The more massive tibia (the shin bone) is located anteromedially. The fibula is lateral and roughly parallel to the tibia. Both bones provide attachment sites for muscles and stabilize the ankle, but only the tibia is weight bearing. Notice that the fibula and tibia articulate at their proximal and distal ends to form the **proximal** and **distal tibiofibular joints.**

2. Identify the following bone markings:

Tibia:

1. Medial condyle
2. Lateral condyle
3. Tibial tuberosity
4. Medial malleolus

Fibula:

1. Head of fibula
2. Lateral malleolus

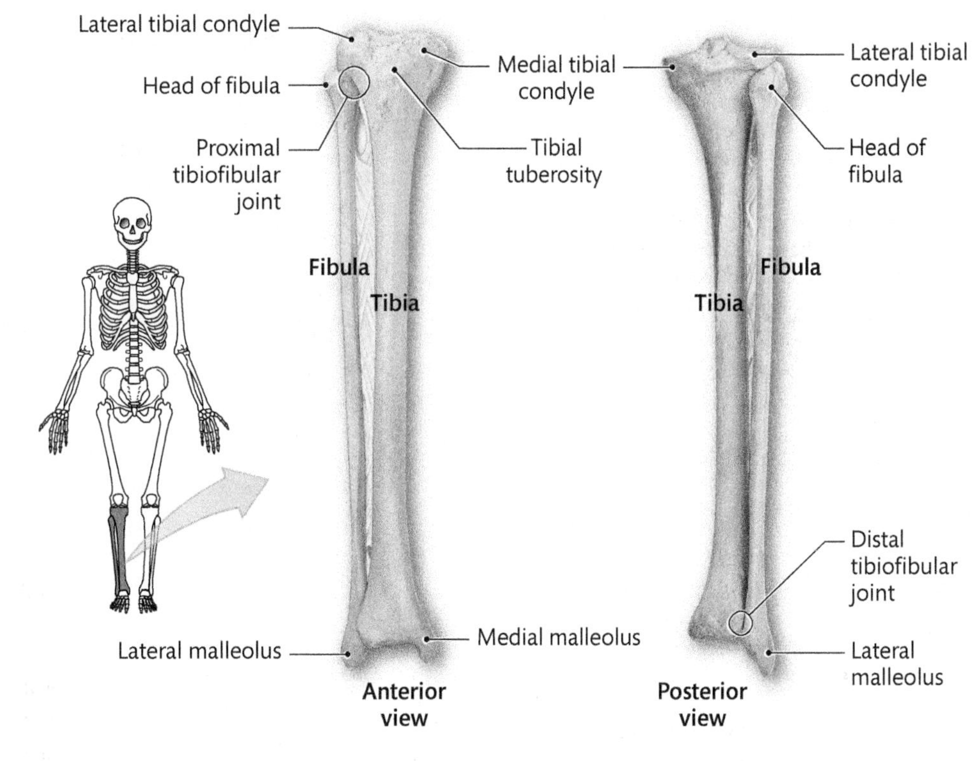

3. Notice that the fibula plays no role in forming the knee joint. Identify the bone markings that form the lateral and medial articulations between the femur and tibia at the knee joint. ▶

Articulation	Bone Markings
Lateral articulation	
Medial articulation	

MAKING CONNECTIONS

Speculate on the function of the patella at the knee joint.

▶ ______________________________

____

____

____

____

____

■ Want more practice? Go to: **MasteringA&P** > Study Area > Menu > Lab Tools > PAL > Anatomical Models > Appendicular Skeleton > Lower Limb

Activity **8.6**

Lower Limb: Examining the Bones of the Foot

A The bones of the foot include 7 tarsal bones, 5 metatarsal bones, and 14 phalanges.

1. Identify the **talus.** It is the most superior bone of the foot. It articulates with the tibia and fibula to form the ankle joint and receives the body's weight.

2. Locate the **calcaneus,** which is the largest tarsal bone. It is positioned inferior to the talus and extends posteriorly to form the heel. Because of its location, the calcaneus transmits most of the body's weight from the talus to the ground while standing.

3. On the lateral side of the foot, find the **cuboid,** which articulates with the calcaneus, posteriorly.

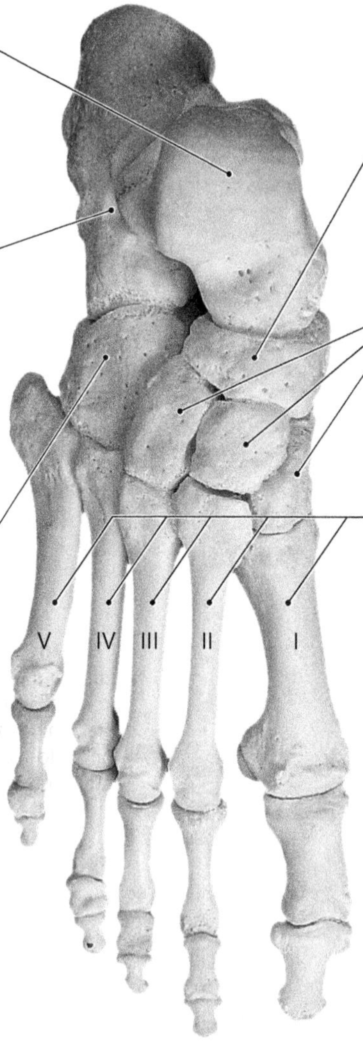

4. On the medial side of the foot, locate the **navicular,** which articulates with the talus, posteriorly, and the three cuneiform bones, anteriorly.

5. Anterior to the navicular and medial to the cuboid are the three **cuneiforms.** Identify these bones. From medial to lateral, they are the **medial cuneiform, intermediate cuneiform,** and **lateral cuneiform.**

6. Identify the five long **metatarsals** and notice their position in your own foot. From medial to lateral, the metatarsals are numbered by Roman numerals I through V.

How is the numbering of the metatarsals in the foot similar to the numbering of the metacarpals in the hand? How is it different? ▶

Similarities:

____

____

Differences:

____

____

V IV III II I

Dorsal view

7. Examine the phalanges in the toes. How does their arrangement compare to that of the phalanges in the fingers?

▶ ______________________________

____

____

____

____

____

Word Origins

Two tarsal bones are named according to their shape. *Cuboid* is derived from the Greek word *kybos,* which means "cube." The cuboid bone is cube shaped. The navicular bone is shaped roughly like a boat or barge; its name is derived from the Latin word *navicularis,* which means "related to shipping."

The other tarsal bones have names that reflect their position in the foot. *Talus* is the Latin word for "ankle." The talus articulates with the tibia and fibula to form the ankle joint. *Calcaneus* is derived from the Latin word *calcaneum,* which means "heel." The calcaneus is the largest tarsal bone and forms the heel of the foot. *Cuneiform* originates from the Latin word *cuneus,* which means "wedge." Notice that the three cuneiform bones are wedged between the navicular, cuboid, and first three metatarsal bones.

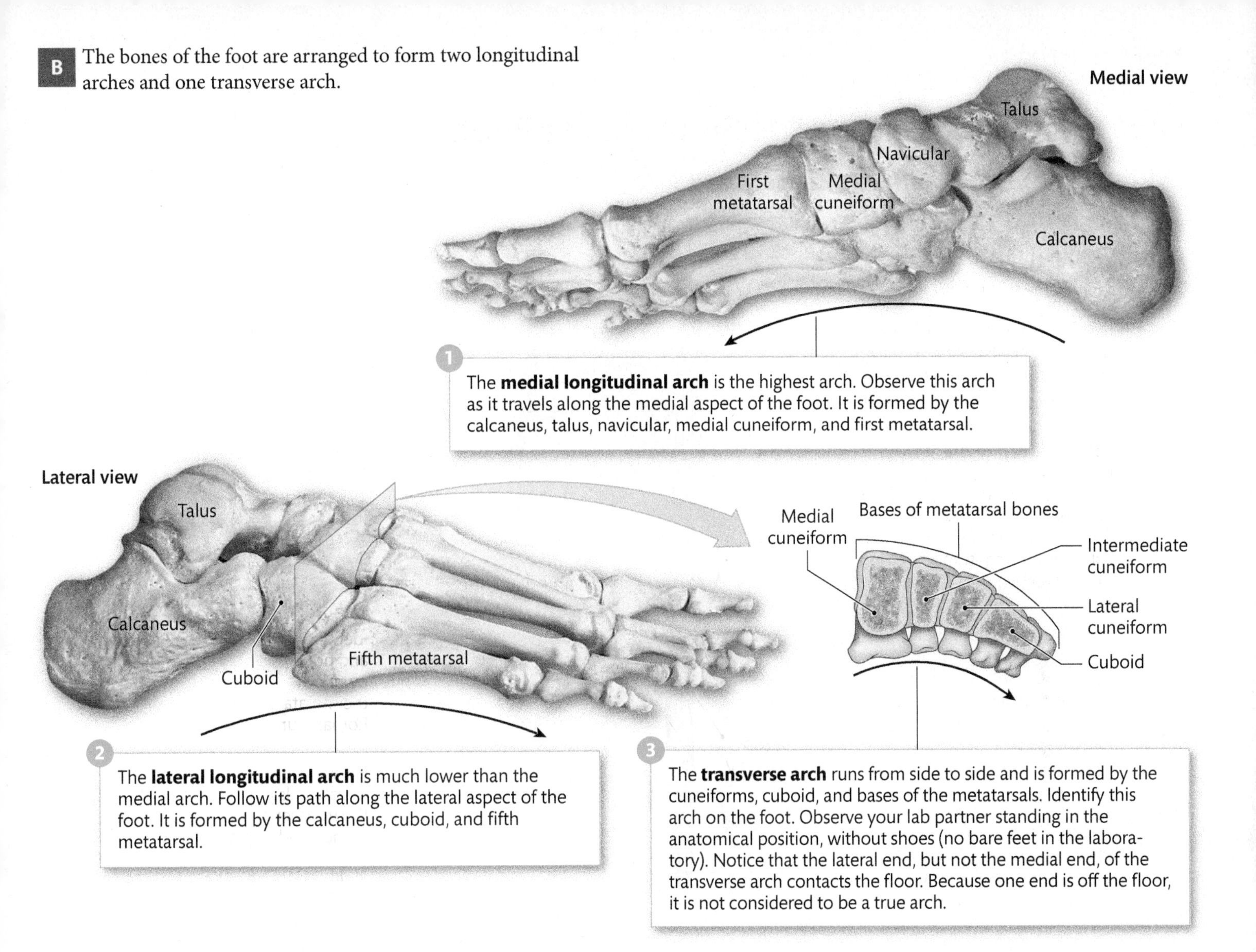

IN THE CLINIC

Abnormal Arches

Pes cavus, more commonly called **high arches,** is a condition in which the medial longitudinal arches are higher than normal. When a person with high arches stands, excessive stress is placed on the metatarsals. Over time, this stress could cause bone and nerve damage. Mild cases of pes cavus may not require any treatment. Wearing supportive shoes or using prescription orthotic arch supports can provide relief. In severe cases, surgery to reduce the height of the arch may be required.

Flatfeet is a condition in which the medial longitudinal arch is lower than normal. It can be caused by lack of arch development during childhood or injury to the ligaments or tendons that support the arches. Often, flatfeet has no associated symptoms or pain, and treatment is not necessary. In other cases, it can cause stress and misalignment of the ankles and knees, resulting in swelling and pain at these joints. Orthotic devices and stretching exercises may help to relieve the pain. In very severe cases, surgery to repair damaged ligaments and tendons may be necessary.

MAKING CONNECTIONS

Speculate on the function of the arches in the foot, particularly during walking and running.

▶ ____

____

____

____

____

____

____

____

____

____

BEFORE YOU MOVE ON . . .

‹‹ LOOKING BACK

The appendicular skeleton comprises the bones of the upper and lower limbs (the appendages). In this laboratory exercise, you learned that each upper limb includes a clavicle and scapula (1/2 pectoral girdle) and the bones of the arm, forearm, wrist, and hand. Each lower limb includes a coxal (hip) bone (1/2 pelvic girdle) and the bones of the thigh, leg, and foot. You also observed that the organization of the bones in the upper limb is comparable to those in the lower limb. For example, the arm and thigh each contain one large long bone; the forearm and leg each contain two smaller long bones that are roughly parallel.

Despite these similarities, there are important structural and functional differences between the upper and lower limbs.

Consider these questions: ▶

1. **Why are the bones of the lower limb larger than those of the upper limb?**

____

____

____

____

____

____

2. **What is the fundamental difference in function between the foot and the hand?**

____

____

____

____

____

____

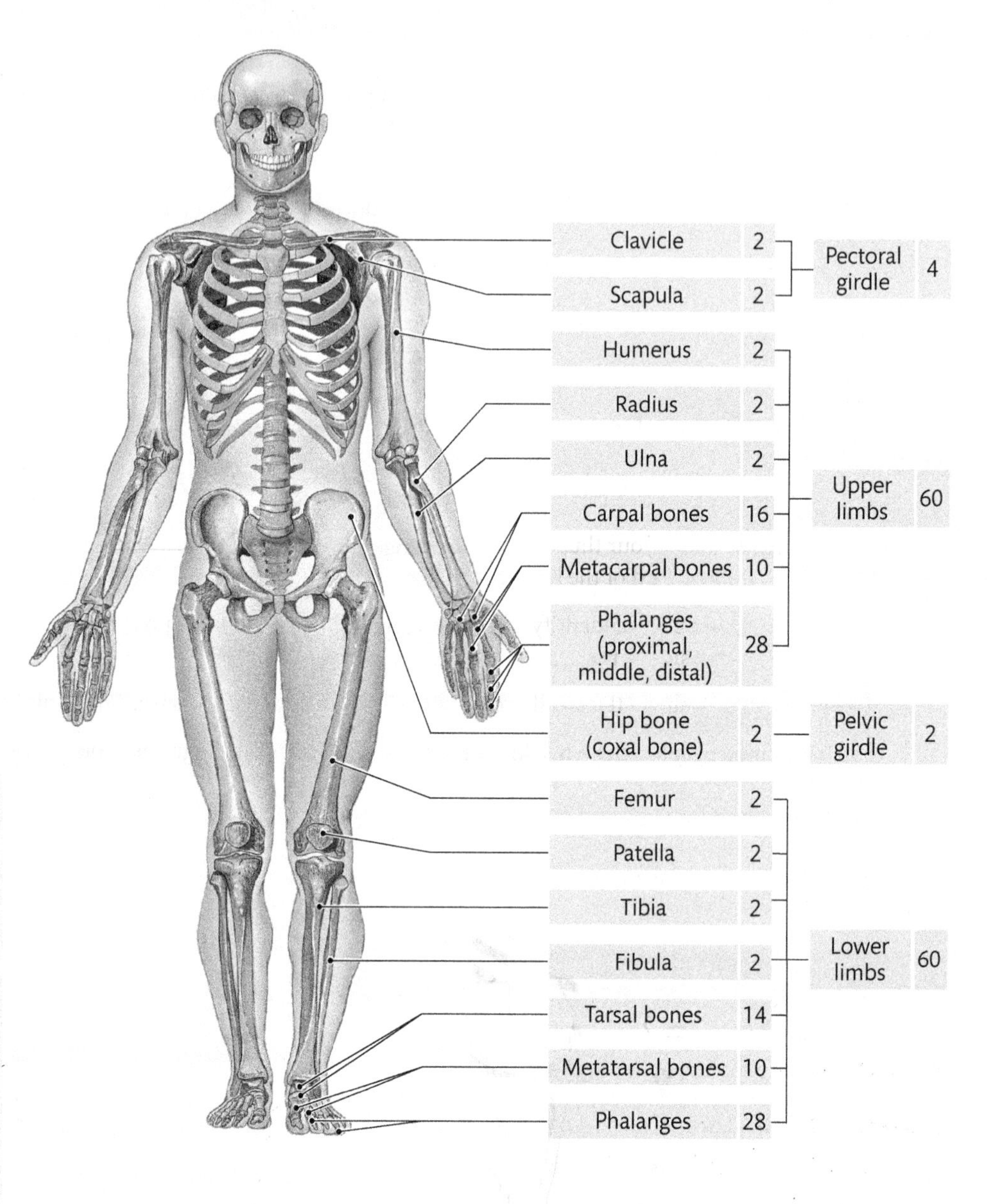

LOOKING FORWARD ››

Be aware that the skeletal system not only includes all the bones of the body, but also the cartilage, tendons, and ligaments associated with the articulations (joints). Tendons and ligaments are both composed of dense regular connective tissue. At a joint, a tendon connects a muscle to a bone; a ligament connects one bone to another bone. In the next laboratory exercise (Laboratory Exercise 9), you will study articulations. Think of articulations as the functional junctions between bones. They bind various parts of the skeletal system together, are locations on the body where movement occurs, allow bone growth and development, and permit parts of the skeleton to change shape.

Name ______________________

Lab Section ______________________

Date ______________________

EXERCISE 8 **REVIEW SHEET**

The Appendicular Skeleton

QUESTIONS 1–5: Identify the following bone markings by palpating (feeling through the skin) them on yourself or your lab partner. Select your answers from the following list. Answers may be used once or not at all.

a. radial head
b. medial epicondyle
c. scapular spine
d. coracoid process
e. greater tubercle
f. styloid process of ulna
g. acromion
h. styloid process of the radius
i. lateral epicondyle
j. lesser tubercle
k. ulnar head
l. head of the humerus

1. The prominent bony ridge that you can palpate across the posterior surface of the scapula is the ___C___.

2. Place your index finger on the manubrium of your sternum and move it laterally in either direction along the S-shaped path of the clavicle. The lateral end of the clavicle articulates with the ___G___ of the scapula.

3. Near the elbow joint, use your thumb and index finger to palpate the ___B___ and ___I___, the two prominent projections at the distal end of the humerus.

4. With your forearm extended, identify a shallow lateral depression on the posterior surface of the elbow. The hard surface that you can feel within this depression is the ___K___.

5. Near the wrist joint, palpate the small projection at the distal end of the ulna. This projection is the ___f___.

QUESTIONS 6–16: Identify the bone markings labeled in the following diagrams. Color the bones with the indicated colors.

Clavicle = **green**
Scapula = **red**
Humerus = **blue**

6. Supraspinous Fossa
7. Spine
8. Coracoid Process
9. Acromion Process
10. Head of Humerus
11. Greater Tubercle
12. Anatomical Neck
13. Intertubercle Sulcus
14. Lesser Tubercle
15. Sub Scapular Fossa
16. Infraspinous Fossa

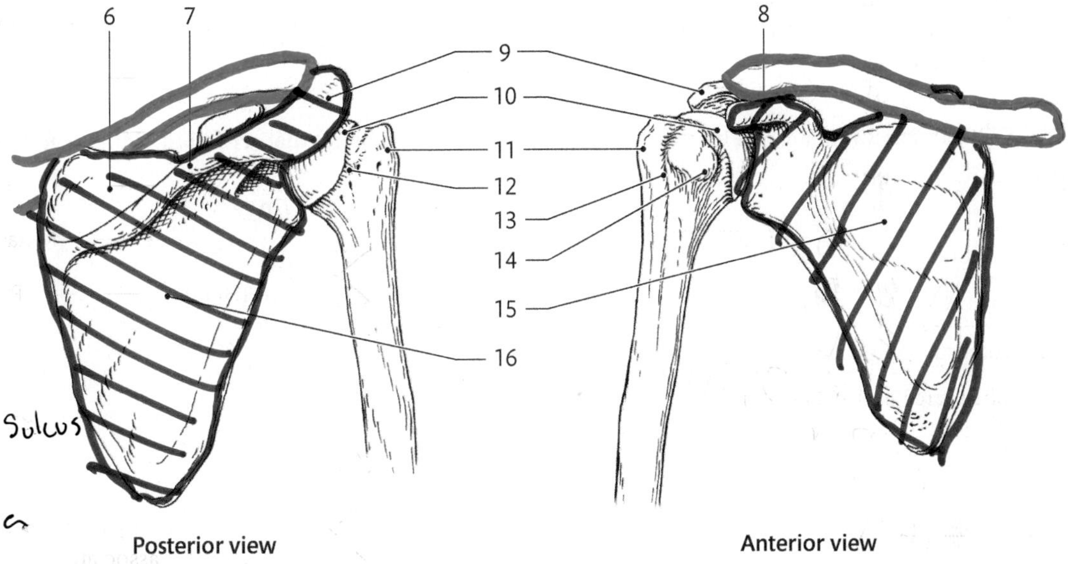

6 7 8 9 10 11 12 13 14 15 16

Posterior view

Anterior view

QUESTIONS 17–25: Identify the bone markings labeled in the following diagrams. Color the bones with the indicated colors.

Humerus = **blue**
Radius = **red**
Ulna = **yellow**
Carpals = **green**
Metacarpals = **brown**
Phalanges = **orange**

17. Capitheluim
18. ~~[illegible]~~ Head of Radius
19. Radial Tuberosity
20. Styloid process of radius
21. Trochlea
22. Medial Epicondyle
23. Olecranon
24. Styloid process of ulna
25. Lateral Epicondyle

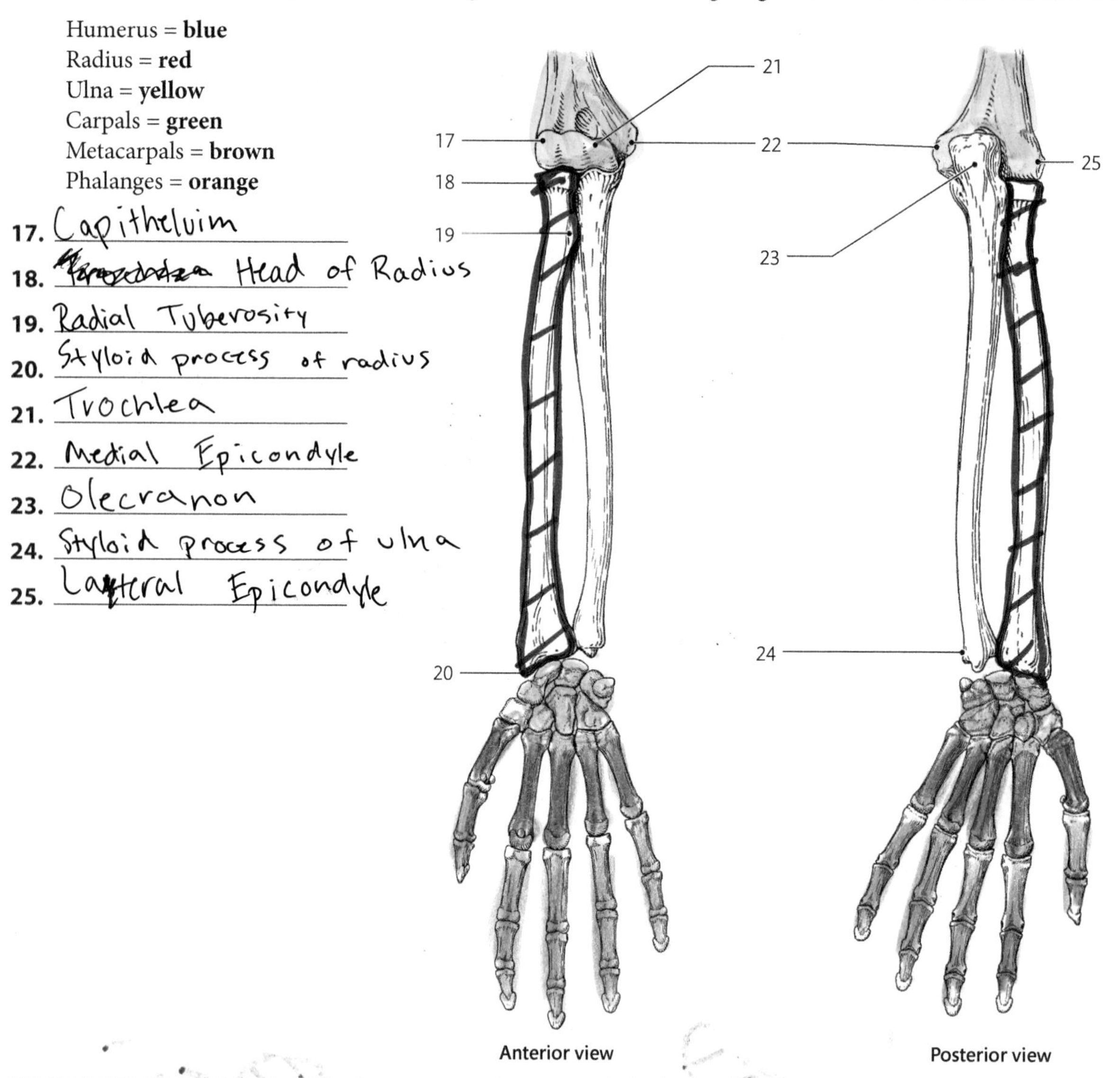

QUESTIONS 26–33: Match the body region in column A with the bones that are found in that region in column B. Some questions have more than one answer.

A

26. Pectoral girdle J, C
27. Arm H, ~~[illegible]~~, ~~[illegible]~~,
28. Forearm E, L
29. Wrist and hand F, B, M
30. Pelvic girdle D, K
31. Thigh O
32. Leg ~~[illegible]~~, A, I
33. Foot N, G

B

a. Fibula
b. Metacarpal bones
c. Clavicle
d. Ilium
e. Radius
f. Hamate
g. Calcaneus
h. Humerus
i. Tibia
j. Scapula
k. Pubis
l. Ulna
m. Triquetrum
n. Cuboid
o. Femur

QUESTIONS 34–39: Identify the following bone markings by palpating them on yourself or your lab partner. Select your answers from the following list. Answers may be used once or not at all.

- **a.** patella
- **b.** greater trochanter
- **c.** lateral malleolus
- **d.** iliac crest
- **e.** medial condyle
- **f.** tibial tuberosity
- **g.** lesser trochanter
- **h.** ischial tuberosity
- **i.** lateral condyle
- **j.** medial malleolus
- **k.** patellar surface
- **l.** head of the fibula

34. Press gently with your fingers along your waistline to feel the ___D___, which forms the superior border of the bony pelvis.

35. While standing, palpate the shallow depression on the posterolateral aspect of your buttock. From this position, move your fingers anteriorly to feel the ___B___, the massive process that projects laterally from the proximal femur.

36. Sit in a chair and lift up slightly on one side. The ___H___ can be felt along the inferior margin of the buttock.

37. Sit in a chair and place one hand over your knee and use your fingers to feel the ___E___ and ___I___ of the femur on each side. As you are doing so, the palm of your hand is in contact with the ___K___, which rests on the anterior surface of the knee joint.

38. Feel the prominent bump on the anterior surface of the tibia slightly inferior to the knee. This bump is the ___F___, which marks the point of attachment for the patellar ligament.

39. At the distal end of the leg, palpate the two prominent bumps on either side of the ankle. These projections are the ___J___ of the tibia and the ___C___ of the fibula.

QUESTIONS 40–51: Identify the bone markings labeled in the following diagrams. Color the bones with the indicated colors.

Ilium = **green**
Ischium = **red**
Pubis = **yellow**

40. Posterior Inferior Iliac Spine
41. Greater Sciatic Notch
42. Lesser Sciatic Notch
43. Ischium
44. Illium
45. Anterion Superior Iliac Spine
46. Anterior Inferior Iliac Spine
47. Acetabulum
48. Superior Ramus of Pubis
49. Obturator Foremen
50. Auricular Surface
51. Ischial Spine

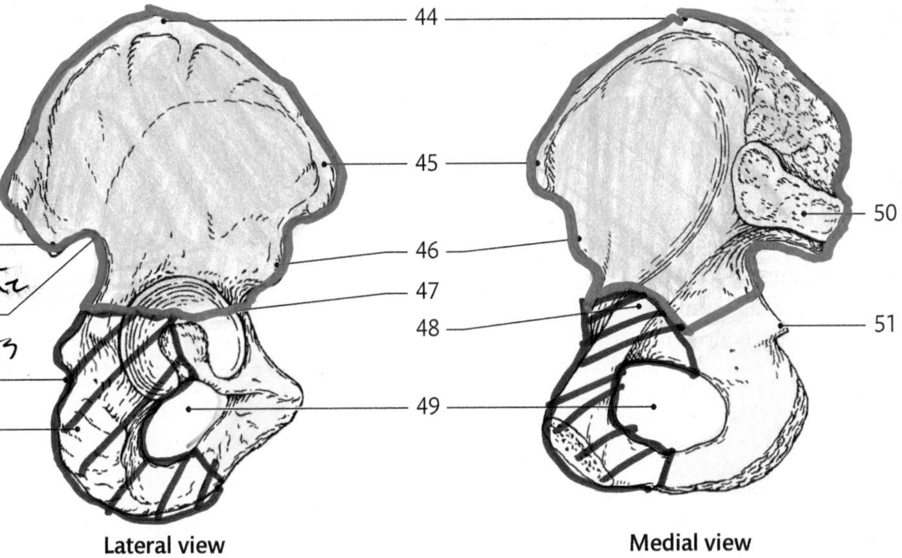

Lateral view

Medial view

QUESTIONS 52–64: Identify the bone markings labeled in the following diagrams. Color the bones with the indicated colors.

Femur = **green**
Tibia = **red**
Fibula = **yellow**

52. Neck
53. Inter chanteric line
54. Lateral epi condyle
55. Patellar Surface
56. Lateral Condyle
57. Head
58. Head of fibula
59. Lesser Trochanter
60. Medial epicondyle
61. Medial Condyle
62. Tibial Tuberosity
63. Medial Malleolus
64. Lateral Malleolus

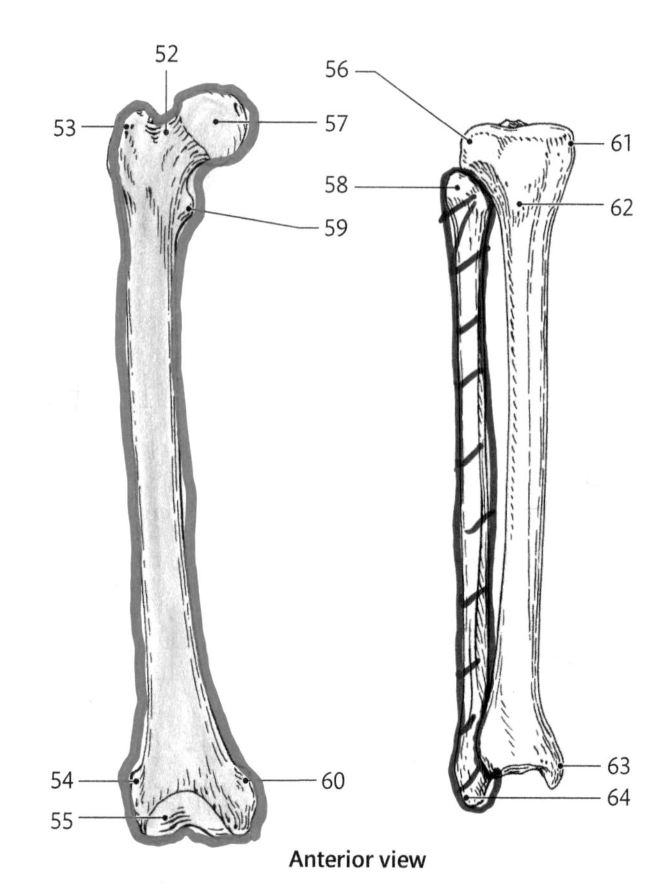

Anterior view

65. Color the bones with the indicated colors.

Calcaneus = **red**
Talus = **green**
Navicular = **yellow**
Cuboid = **blue**
Cuneiforms = **orange**
Metatarsals = **purple**
Phalanges = **brown**

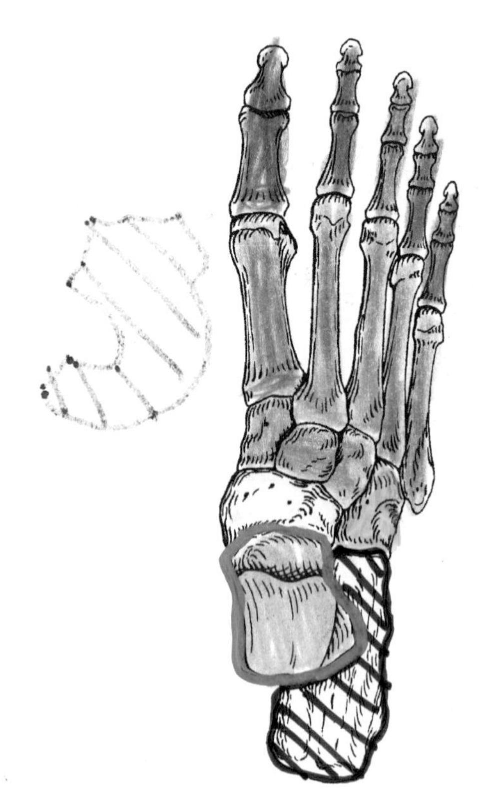
Dorsal view

EXERCISE

9

Articulations

LEARNING OUTCOMES

These Learning Outcomes correspond by number to the laboratory activities in this exercise. When you complete the activities, you should be able to:

Activity 9.1 Describe and provide examples of the different types of fibrous joints.

Activity 9.2 Describe and provide examples of the different types of cartilaginous joints.

Activities 9.3–9.4 Demonstrate the types of movements at synovial joints.

Activities 9.5–9.7 Provide examples of the six types of synovial joints and compare the movements possible at each type.

Activity 9.8 Discuss the structure and function of synovial joints in the upper limb.

Activity 9.9 Discuss the structure and function of synovial joints in the lower limb.

LABORATORY SUPPLIES

- Human skull
- Human fetal skull
- Human skeleton, intact
- Human skeleton, disarticulated
- Model of the vertebral column
- Model of intervertebral disc
- Model of herniated disc
- Longitudinal section (coronal plane) of adult humerus or femur
- Joint models:
  - shoulder joint
  - elbow joint with proximal radioulnar joint
  - hip joint
  - knee joint

PRE-LAB QUIZ

Before you begin, read all the activities in Exercise 9 and the required reading in your textbook that is assigned by your instructor.

1. During this laboratory exercise, you will be completing all the following activities *except*
   a. examining the types of movements that occur at synovial joints.
   b. identifying the major sutures between cranial bones.
   c. examining the microscopic structure of the elbow and knee joints.
   d. identifying various types of joints on a skeleton.
   e. identifying ligaments at synovial joints in the upper and lower limbs.
2. In this laboratory exercise, you will examine and identify the ligaments associated with all the following joints *except* the
   a. knee joint.
   b. elbow joint.
   c. shoulder joint.
   d. wrist joint.
   e. hip joint.

Questions 3–7: Identify each of the following articulations as a fibrous joint, cartilaginous joint, or synovial joint.

3. A tooth anchored to a bony socket ____________
4. Ankle joint ____________
5. Symphysis pubis ____________
6. A suture between two skull bones ____________
7. Sacroiliac joint ____________
8. True or False: The carpometacarpal joint of the thumb is an example of a ball and socket joint. ________
9. True or False: Pronation and supination are rotational movements that occur at the proximal and distal radioulnar joints. ________
10. True or False: The epiphyseal (growth) plate in a long bone is an example of a syndesmosis joint. ________

Activity 9.1

Examining the Structure of Fibrous Joints

The bones that form a **fibrous joint** are held firmly together by fibrous connective tissue in which collagen fibers predominate. A joint cavity is not present, and little or no movement occurs between the bones. The three types of fibrous joints in the human body are **sutures, syndesmoses** (singular = **syndesmosis**), and **gomphoses** (singular = **gomphosis**).

A Sutures are very tight articulations between adjacent bones. This type of joint is only found in the skull. In the adult, the connective tissue fibers that connect the bones become completely ossified. Thus, the bones are fused together and no movement occurs between them.

1. On the adult skull, identify the major sutures between cranial bones.

The **coronal suture** is the joint between the frontal bone, anteriorly, and the two parietal bones, posteriorly.	The **squamous suture** on each side of the skull is the joint between the parietal bone, superiorly, and the temporal bone, inferiorly.	The **lambdoid suture** is the joint between the occipital bone, posteriorly, and the two parietal bones, anteriorly.	The **sagittal suture** is the joint on the superior surface of the skull, between the two parietal bones.

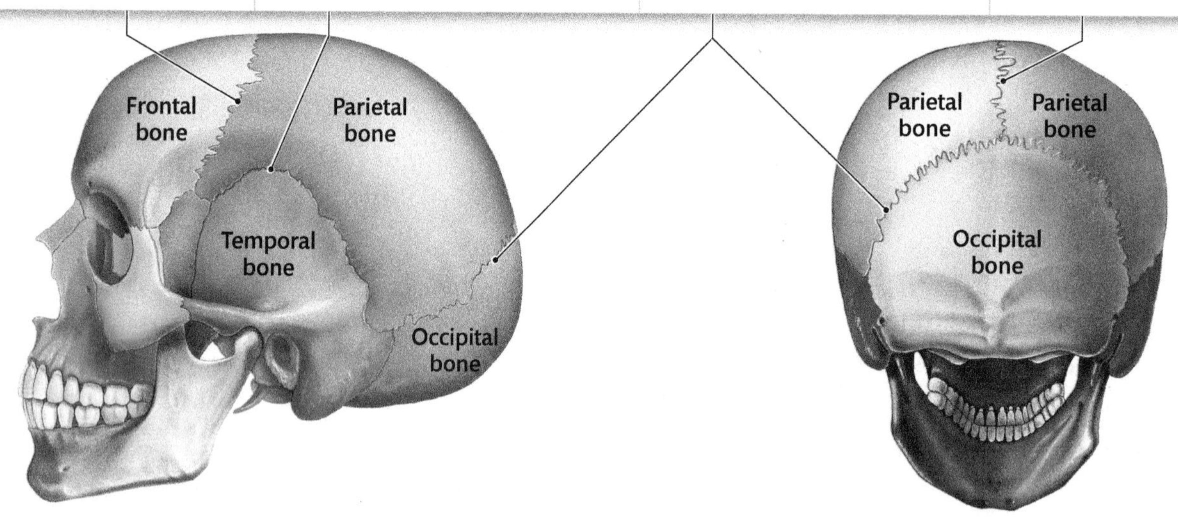

2. In the fetus and infant, ossification is not complete, and the articulating bones at the sutures are held together by areas of connective tissue fibers. The largest regions of connective tissue are called fontanelles, or "soft spots." Because the sutures are not fully developed, there is some movement between the bones. Identify the major fontanelles in the fetal skull: ▶

The **anterior (frontal) fontanelle** is located at the junction of the ________________ and ________________ sutures.	The **posterior (occipital) fontanelle** is located at the junction of the ________________ and ________________ sutures.	The two **anterolateral (sphenoidal) fontanelles** are each located at the junction of the ________________ and ________________ sutures.	The two **posterolateral (mastoid) fontanelles** are each located at the junction of the ________________ and ________________ sutures.

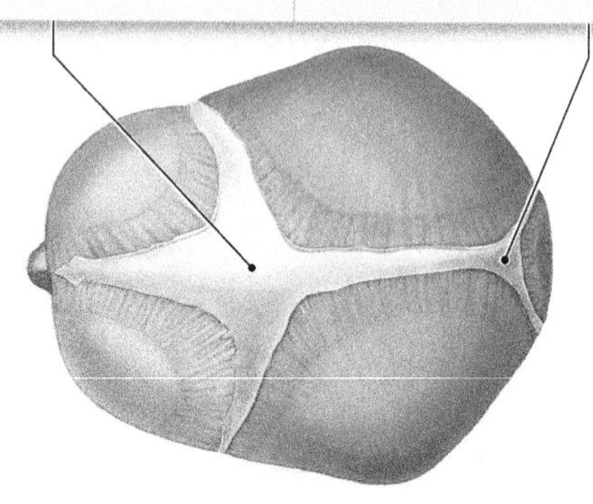

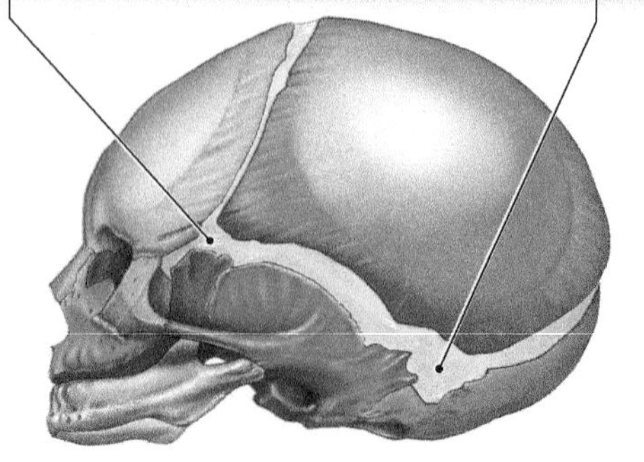

B In a syndesmosis, the bones are held together by strong, fibrous connective tissue. The articulating surfaces of the bones may be either relatively small and held together by cordlike ligaments or broad and held together by fibrous sheets called interosseous membranes. The movement between bones can vary, but usually is quite limited.

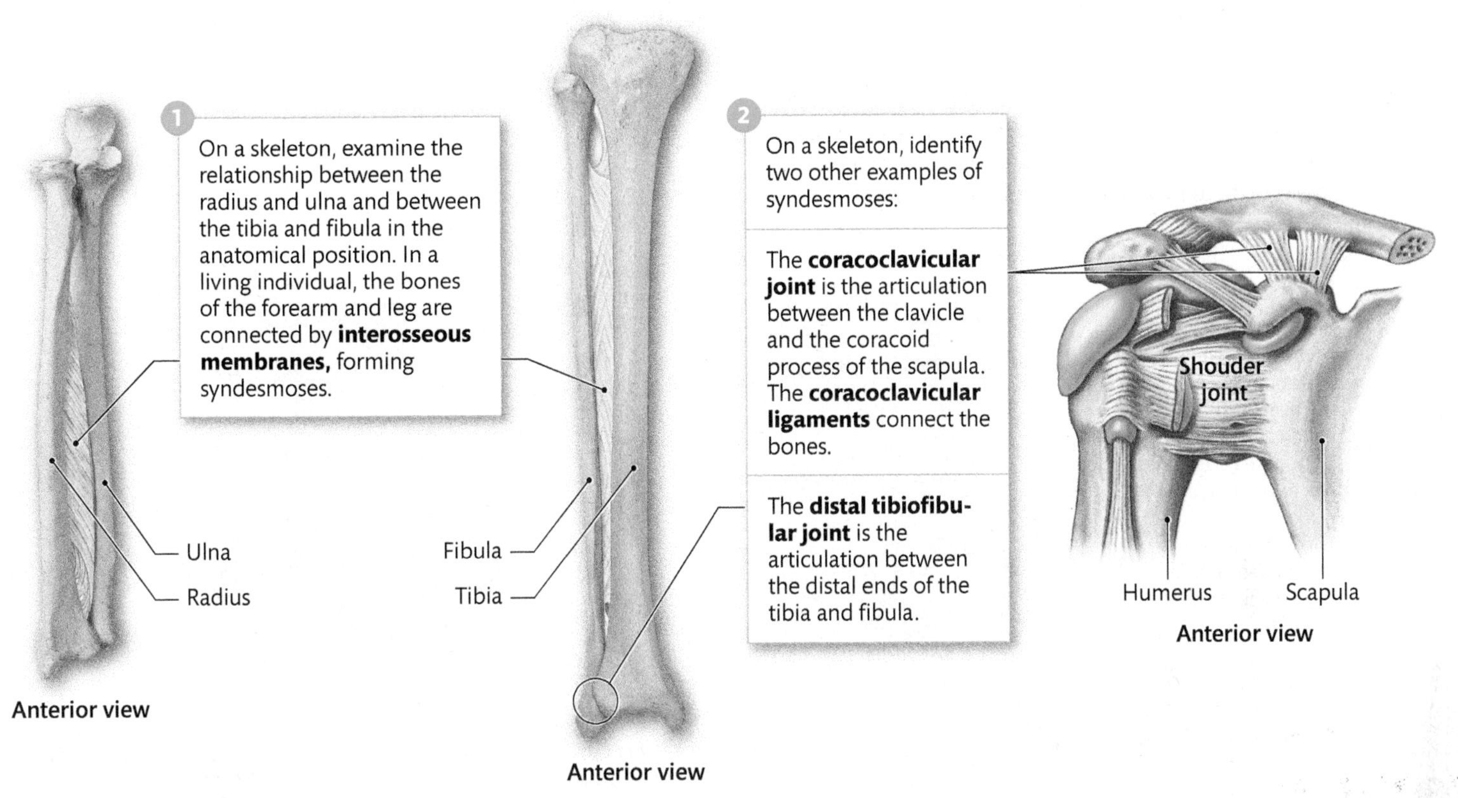

C A gomphosis is a unique peg-and-socket joint at which no movement occurs. The only gomphoses are the articulations between the permanent teeth and the maxilla (upper teeth) and mandible (lower teeth).

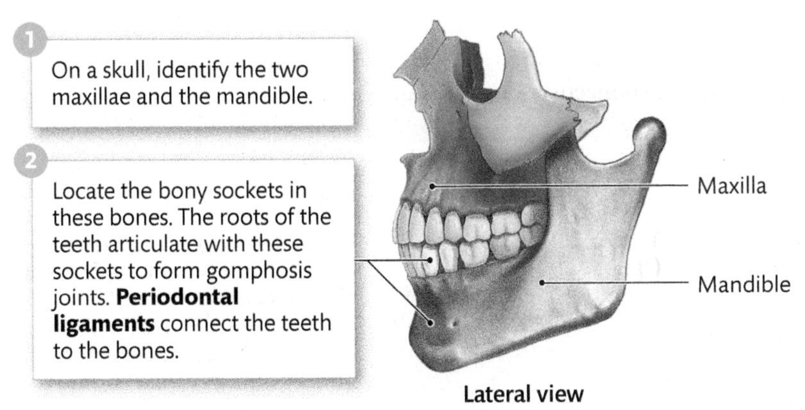

MAKING CONNECTIONS

Review the connections of the coracoclavicular ligaments at the coracoclavicular joint. What role do you think these ligaments have in stabilizing the pectoral girdle?

▶ ______________________________

____

____

____

____

____

____

____

____

____

____

Word Origins

Gomphosis is derived from the Greek word *gomphos,* which means a "bolt" or "nail." Think of the teeth as bolts inserted into the bones.

Activity **9.2**

Examining the Structure of Cartilaginous Joints

In **cartilaginous joints,** the articulating bones are held together by either a plate of hyaline cartilage or a fibrocartilage disc. Similar to fibrous joints, cartilaginous joints lack a joint cavity. The two types of cartilaginous joints in the human body are **symphyses** (singular = **symphysis**) and **synchondroses** (singular = **synchondrosis**).

A At a symphysis, the articulating surfaces of the bones are covered with hyaline cartilage, and a disc of shock-absorbing fibrocartilage is sandwiched between the bones, holding them together. Movement at these joints is limited.

1 Inspect the vertebral column on a skeleton and notice that the bodies of adjacent vertebrae are separated by **intervertebral discs,** which are symphyses. Each intervertebral disc has two parts:

The outer region, composed of fibrocartilage, is the **anulus fibrosus.**

The central core of the disc, a gelatinous mass, is the **nucleus pulposus.**

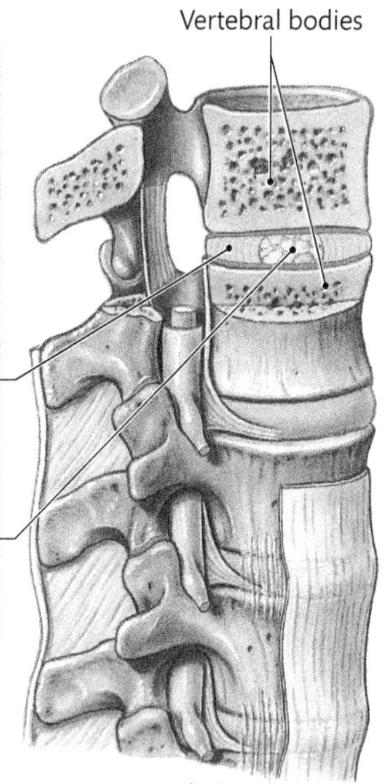

Lateral view

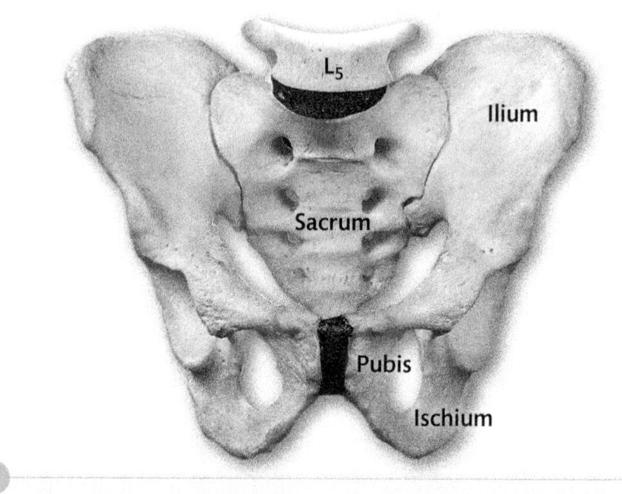

2 On a whole skeleton or bony pelvis, identify the articulation between the two pubic bones. This joint, the **symphysis pubis (pubic symphysis),** is another example of a symphysis. What type of cartilage is found at this joint?

▶ ______________________________

____

IN THE CLINIC

Herniated Disc

A **herniated disc** is caused by deterioration of the posterior and lateral aspects of the anulus fibrosus. As a result, the nucleus pulposus is displaced posterolaterally from its normal position, causing tissue inflammation and compression of the emerging spinal nerve (see figure on right). In most cases, herniated discs occur between lumbar vertebrae and may cause severe low back and leg pain. Only 10 percent of people with herniated discs require surgery. With physical therapy and anti-inflammatory medication, recovery typically occurs within 6 months. The inflammation gradually dissipates over time, and the herniated disc heals naturally as the displaced gel of the nucleus pulposus is broken down and resorbed.

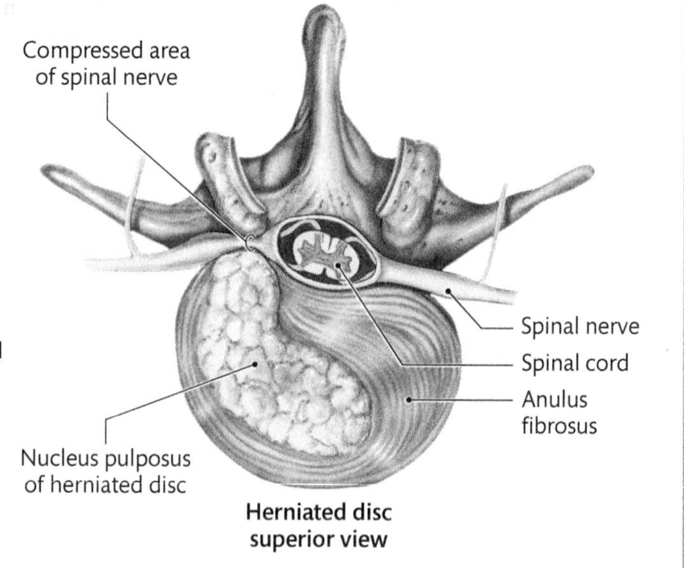

Herniated disc superior view

B At a synchondrosis, a plate of hyaline cartilage unites the bones. Its primary function is to allow bone growth; movement does not occur at these articulations.

1 In a developing long bone, growth occurs at the **epiphyseal plate,** which is an example of a synchondrosis. Why are epiphyseal plates classified as joints?

▶ ______________________

____

____

____

____

Epiphyseal plates are temporary joints. Explain why.

▶ ______________________

____

____

____

____

____

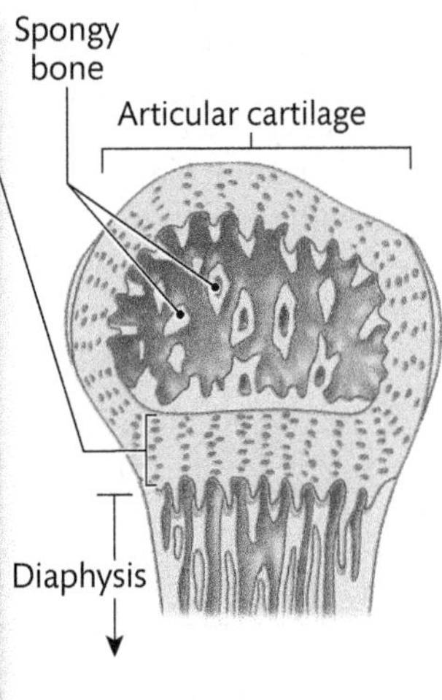

2 Inspect an adult long bone that has been cut along its longitudinal axis in the coronal plane. At the proximal end of the bone, observe the border between the diaphysis and epiphysis. In this region, a fine line of ossified tissue travels horizontally across the bone. This **epiphyseal line** marks the area where the epiphyseal plate was located during active bone growth.

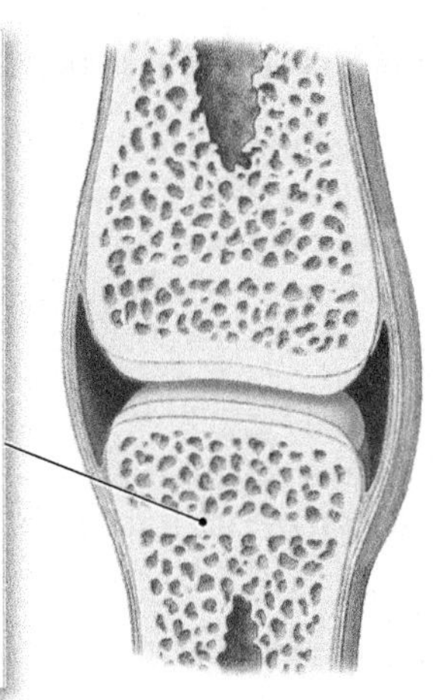

Word Origins

Synchondrosis is derived from the Greek word *sunkhondrosis,* which means "together by cartilage." A synchondrosis joint is the joining of two bones by cartilage.

3 On a skeleton, identify the articulation between the first rib and the manubrium of the sternum. This joint is unique because it is the only permanent synchondrosis in the human body. All other ribs that articulate with the sternum form synovial joints.

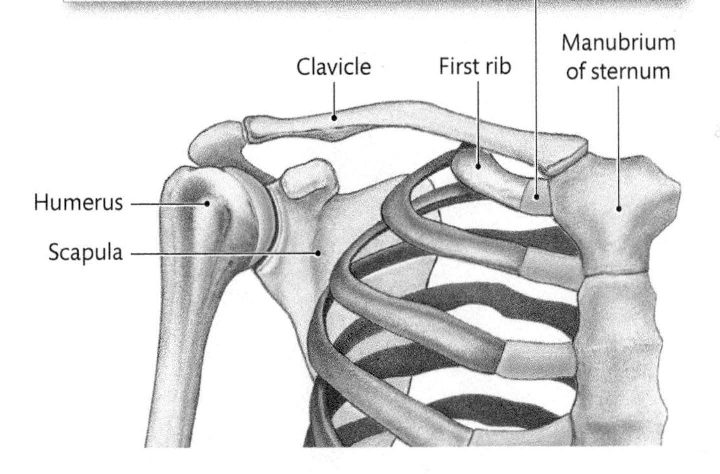

MAKING CONNECTIONS

Very limited movement occurs at one intervertebral disc between two vertebrae. However, if you bend forward, backward, or laterally at the waist, a great deal of movement occurs along the vertebral column. How do you account for this movement?

▶ ______________________

____

____

____

____

____

____

____

____

____

____

____

Activity 9.3

Understanding the Types of Movements at Synovial Joints: Gliding and Angular Movements

Most articulations in the body are synovial joints, which are defined by the presence of a fluid-filled cavity between the articulating surfaces. Synovial joints are capable of a variety of movements in several planes, and specific terms are used to describe these movements. It is essential that you become familiar with these terms before you begin to analyze and compare movements at specific joints. Study the following terms of movement. Work with your lab partner by quizzing each other on the performance of each movement.

A

Gliding movements occur when articulating surfaces of two bones move back and forth or side to side.

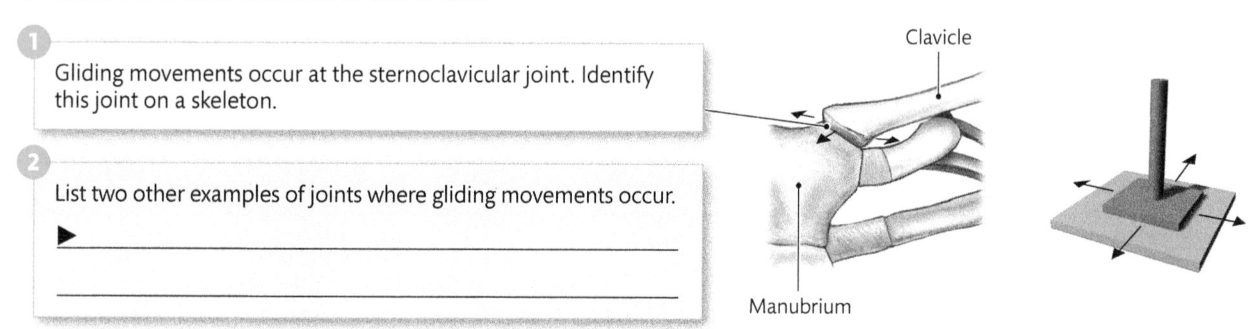

1. Gliding movements occur at the sternoclavicular joint. Identify this joint on a skeleton.

2. List two other examples of joints where gliding movements occur.

▶ ______________________________

____

B

During **flexion** and **extension,** the angle between articulating bones changes.

1. From the anatomical position, perform flexion of the forearm. How does the angle between the arm and forearm change?

▶ ______________________________

____

____

2. Return to the anatomical position by extending the forearm. How does the arm-forearm angle change?

▶ ______________________________

____

____

3. Identify two other joints where flexion and extension occur.

▶ ______________________________

____

C **Abduction** and **adduction** are angular movements that occur only at joints in the limbs.

1. From the anatomical position, abduct your arm by moving it away from your torso.

2. Return to the anatomical position by adducting your arm. At what joint did abduction and adduction occur?

   ▶ ______________________________

3. Perform abduction and adduction at the hip and wrist joints. For all these examples of abduction and adduction, in what body plane did the movements occur?

   ▶ ______________________________

   ____

Abduction
Adduction
Abduction
Adduction
Abduction
Adduction

D **Circumduction** is a circular motion that results from a combination of angular movements; the distal end of the part being moved describes a circle.

1. Perform circumduction at the shoulder joint.

2. Identify one other joint where circumduction can occur.

   ▶ ______________________________

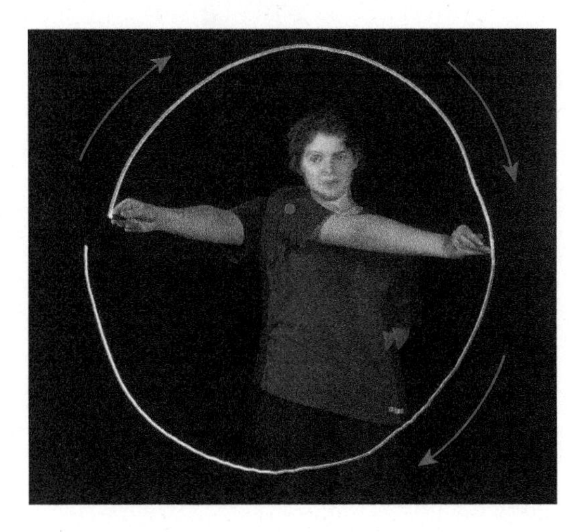

MAKING CONNECTIONS

Slowly perform circumduction at the shoulder. Identify the various angular movements that are included in this action and list them in the space below.

▶ ______________________________

____

____

____

____

____

____

____

____

____

____

Activity 9.4

Understanding the Types of Movements at Synovial Joints: Rotational and Special Movements

A **Rotational movements** occur around the long axis of a bone.

1 **Left** and **right rotation** refer to the turning of a body part to the right or left around the midline of the body. Rotation of the head to signify "no" is an example.

2 **Medial (internal) rotation** occurs when the anterior surface of the moving part is brought toward the median plane. **Lateral (external) rotation** occurs when the anterior surface is brought away from the median plane.

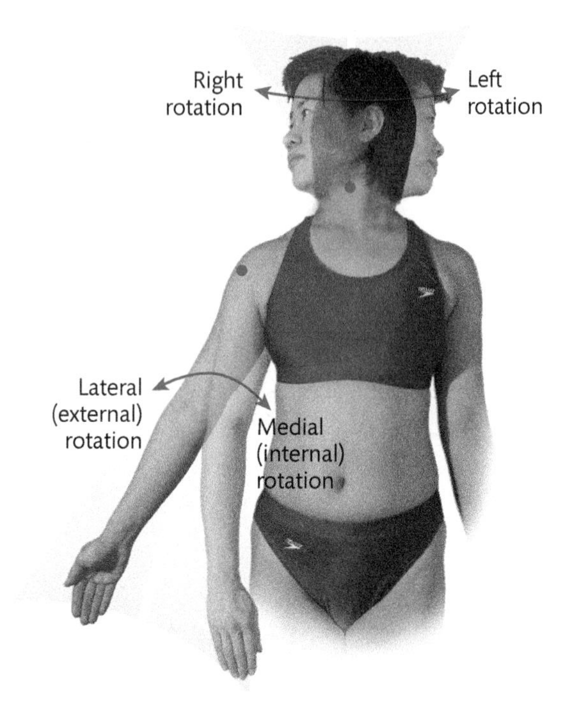

3 Perform medial and lateral rotation of the arm. At what joint do these actions occur?

▶ ______________________________

Identify another joint where these movements occur.

▶ ______________________________

4 **Pronation** and **supination** specifically apply to the rotation of the radius around the ulna at the proximal and distal radioulnar joints. Identify these joints on a skeleton.

5 From the anatomical position, pronate the forearm by rotating it medially so that the palm of the hand faces posteriorly. Supinate the forearm by rotating it laterally so that the palm faces anteriorly. Notice that you have returned to the anatomical position.

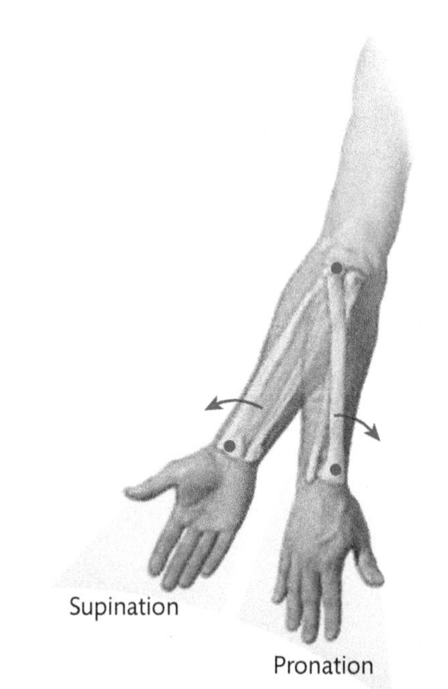

6 Describe the relationship between the radius and ulna when: ▶

a. the forearm is pronated:

____

____

____

____

b. the forearm is supinated:

____

____

____

____

B **Special movements** are unique actions that occur at specific joints. Perform the following movements on your own body.

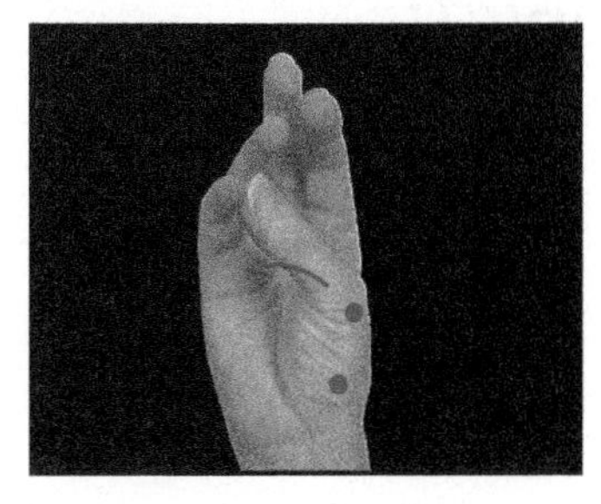

1 **Opposition** occurs when the thumb is brought over to touch another digit. During **reposition,** the thumb is brought back to anatomical position. Perform these actions with your thumb. During opposition, movement occurs at the carpometacarpal and metacarpophalangeal joints of the thumb.

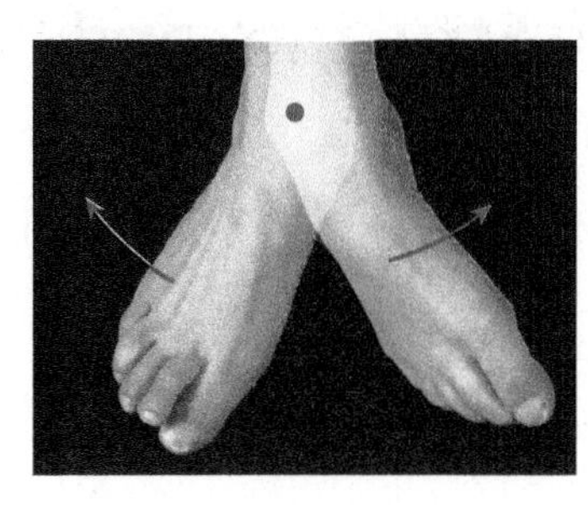

2 **Eversion** is an action that moves the sole of the foot away from the median plane. **Inversion** is an action that moves the sole of the foot toward the median plane. Perform these actions with your foot. At what joints do eversion and inversion occur?

▶ ______________________________

____

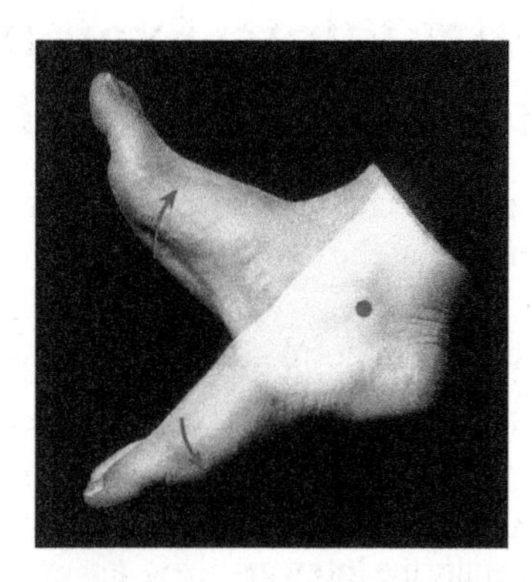

3 **Dorsiflexion** is a bending action that elevates the soles, such as when you stand on your heels. **Plantar flexion** is a bending action that elevates the heels, such as when you stand on your toes. Stand in the anatomical position and perform dorsiflexion and plantar flexion. At what joint do these actions occur?

▶ ______________________________

____

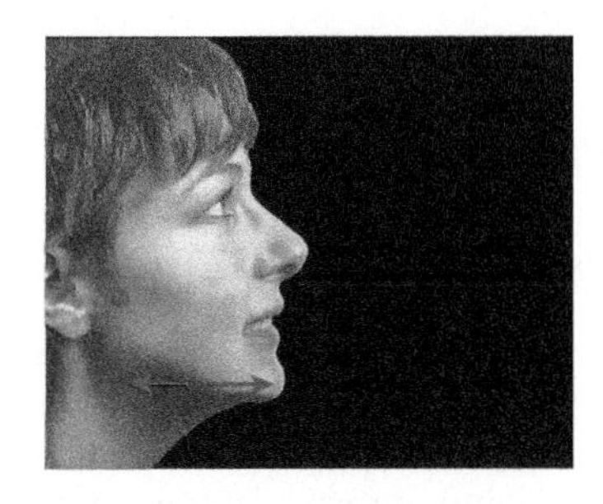

4 During **protraction,** the mandible moves anteriorly. During **retraction,** the mandible moves posteriorly. Place your index finger on the temporomandibular joint, just anterior to the external acoustic meatus. Protract and retract your mandible, and you will detect movement at the joint.

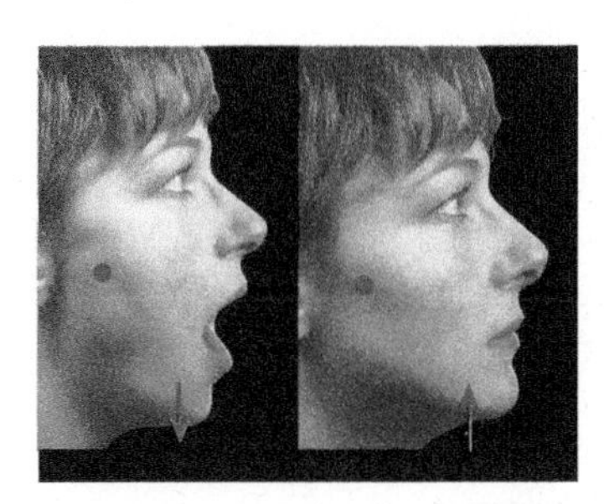

5 **Depression** lowers the mandible inferiorly. **Elevation** raises the mandible superiorly. Once again, place your finger on the temporomandibular joint and feel the movement when you lower and raise your mandible.

MAKING CONNECTIONS

Opposition and reposition of the thumb are critical for performing fine-motor skills with your hand. Explain why. (*Hint:* Try to write your name with a pen without using your thumb.)

▶ ______________________________

____

____

____

Activity 9.5

Synovial Joints: Examining the Structure and Function of Gliding and Hinge Joints

There are six types of synovial joints. Each type is categorized by the shape of its articulating surfaces and the range of motion at the joint.

A The bones that form **gliding joints** have small, flat articulating surfaces.

1. On a skeleton, identify the **intercarpal joints** at the wrist.

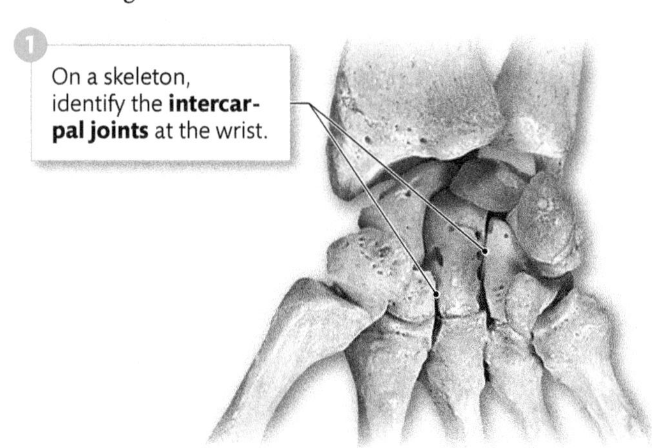

2. Rub your palms together, first side to side and then back and forth. This motion resembles the **sliding movements** at your intercarpal joints. Movement at the joints will be much more restrictive than movement when the palms are rubbed together. Because these movements do not occur along a specific body plane or around an axis, gliding joints are referred to as **nonaxial joints.**

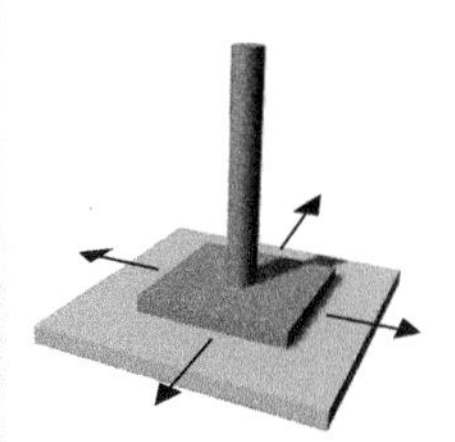

3. On a skeleton, identify the following gliding joints. For each joint, list the bones that form the articulaton. ▶

Acromioclavicular joint

____

Sternoclavicular joint

____

Sacroiliac joint

____

Joints between articular processes of adjacent vertebrae

____

Vertebrocostal joints

____

Intertarsal joints

____

Word Origins

The names of many synovial joints are derived from the bones that form the articulation. Consider the following examples:

- The *intercarpal* joints are articulations between carpal bones in the wrist.
- The *metacarpophalangeal* joints are articulations between the metacarpal bones and the proximal phalanges in the hand. They are also called the knuckle joints.
- The proximal and distal *radioulnar* joints are articulations in the forearm between the radius and ulna.

There are exceptions, but this general rule will make it easier to learn the names and location of many joints.

B At **hinge joints**, one articulating surface is convex, and the other is concave.

1 On a skeleton, identify the **elbow joint.** Observe the articulating surfaces on disarticulated bones:

Between the **capitulum** on the humerus and the **radial head** on the radius.

Between the **trochlea** on the humerus and the **trochlear notch** on the ulna.

Humerus

Ulna

Radius

2 Notice that the capitulum and trochlea on the humerus are convex, and the radial head on the radius and trochlear notch on the ulna are concave.

3 Flexion and extension are the two principal movements at hinge joints. Flex and extend your forearm and notice that these movements occur along a single plane or axis. Because movement is restricted to one plane, hinge joints are a type of **uniaxial joint.**

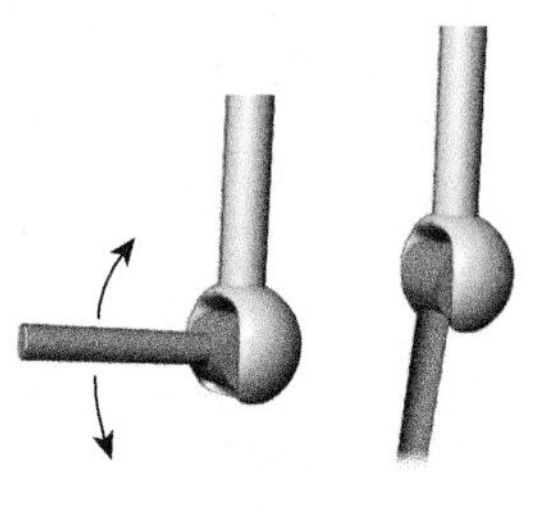

4 On a skeleton, identify the following hinge joints:

Interphalangeal joints

Knee joints

Ankle joints

Move these joints on your body. Based on the movements that you were able to perform, are all these joints uniaxial? Explain.

▶

MAKING CONNECTIONS

Describe the following actions and identify the joint at which they occur. ▶

a. Flexion of the leg:

b. Inversion of the foot:

c. Extension of the arm:

Activity 9.6

Synovial Joints: Examining the Structure and Function of Pivot and Ellipsoid Joints

A At **pivot joints**, a rounded surface of one bone fits into a shallow depression of another bone.

1 On a skeleton, identify the proximal and distal radioulnar joints. Observe the articulations for these joints:

At the proximal joint, the head of the radius articulates with the radial notch on the ulna.

At the distal joint, the head of the ulna articulates with the ulnar notch on the radius.

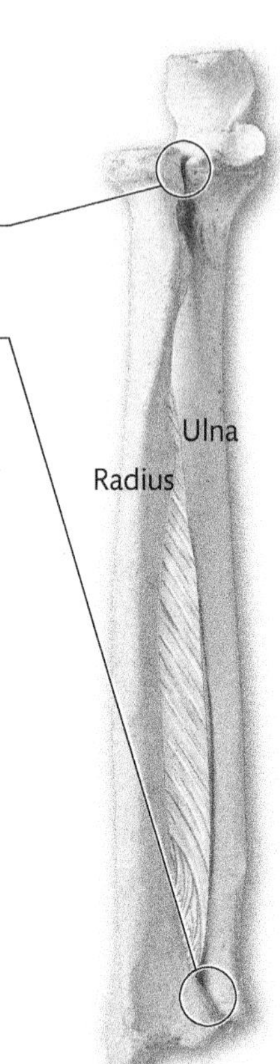

2 At the radioulnar joints, the rotational movements are pronation and supination of the forearm. Perform these actions on the right forearm. Notice that movement occurs at the two radioulnar joints, and as a result, the entire radius rotates around the ulna.

3 The actions of pivot joints are restricted to rotational movements around a central axis. They are classified as uniaxial joints. Explain why.

▶ ______________________________

____

____

____

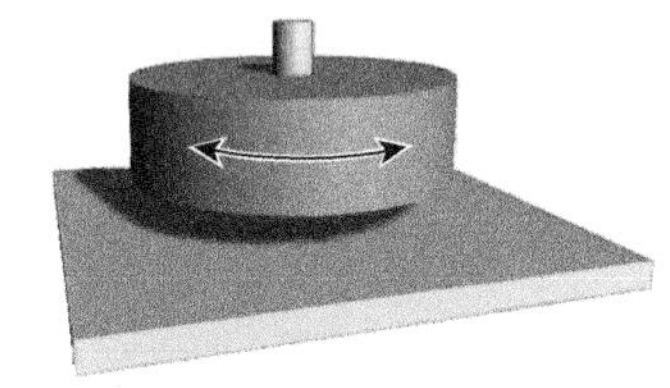

4 On a skeleton, identify the **atlantoaxial joint.** This joint is an articulation between the first two cervical vertebrae, the atlas (C_1) and the axis (C_2). Notice how the dens of the axis articulates with a shallow depression on the anterior arch of the atlas. The atlantoaxial joint is another example of a pivot joint. Explain why.

▶ ______________________________

____

____

____

____

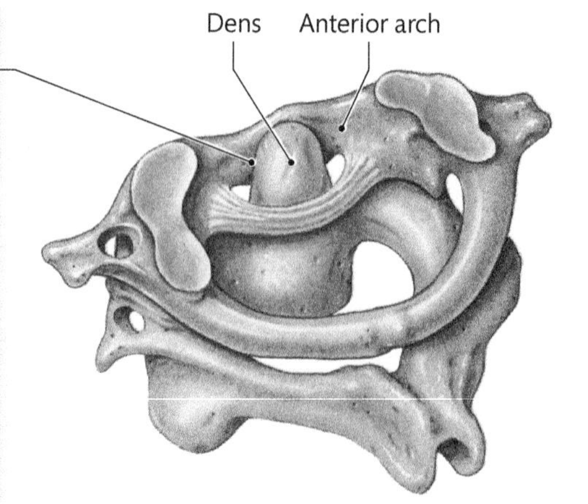

B At **ellipsoid joints**, an oval-shaped convex surface articulates with a shallow elliptical cavity.

1. On a skeleton, identify a **wrist joint.** At this joint, a condyle, formed by the scaphoid and lunate bones, articulates with an elliptical cavity, formed by the distal end of the radius. Notice that the ulna does not contribute an articulating surface to the wrist joint.

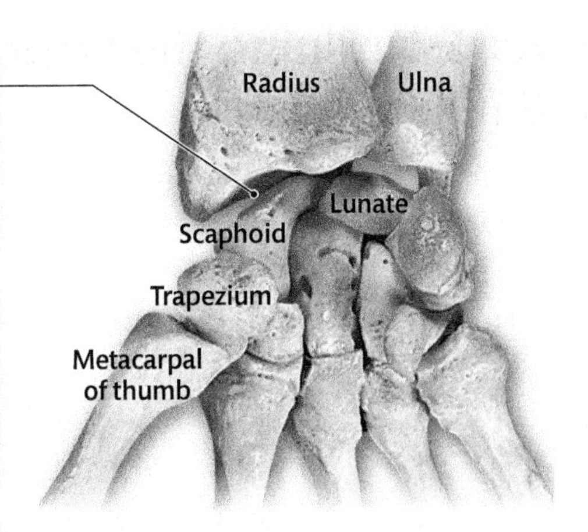

2. With your body in anatomical position, bend your right wrist so that your hand moves anteriorly. This movement is called flexion of the hand. To extend your hand, move your hand posteriorly back to the anatomical position. Notice that these two movements occur in the same plane.

3. With your body in anatomical position, bend your right wrist so that your hand moves laterally. This movement is called abduction of the hand. To adduct your hand, bend your wrist to move your hand medially. Notice that these two movements occur in a plane that is perpendicular to the plane in which flexion and extension occur. Because movements pass through two different planes, the wrist and other ellipsoid joints are classified as **biaxial.**

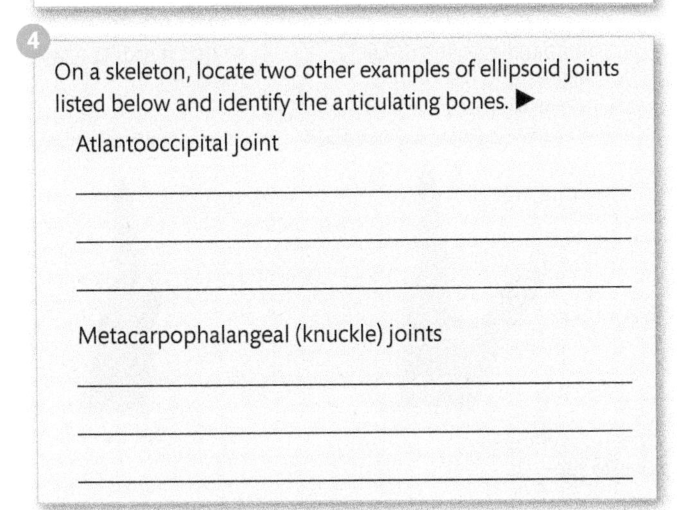

4. On a skeleton, locate two other examples of ellipsoid joints listed below and identify the articulating bones. ▶

Atlantooccipital joint

____

____

____

Metacarpophalangeal (knuckle) joints

____

____

____

MAKING CONNECTIONS

Flexion and extension of the fingers can occur at the metacarpophalangeal (knuckle) joints and the interphalangeal joints. Abduction and adduction of the fingers (moving the fingers apart and bringing them back together) occur at the metacarpophalangeal joints only. Based on the structure of these two types of joints, provide an explanation for the difference in function.

▶ ____________________________________

Activity 9.7

Synovial Joints: Examining the Structure and Function of Saddle and Ball and Socket Joints

A At **saddle joints**, each articulating surface has a convex and a concave region. The shape of each surface resembles a saddle.

1. On a skeleton, identify a **carpometacarpal joint** of the thumb. This joint is formed by the articulation between the trapezium and the metacarpal bone of the thumb.

2. The primary movements that occur at saddle joints are flexion and extension, and abduction and adduction. To demonstrate these movements, position your right hand with the palm directly in front of your face. Perform the following movements at the carpometacarpal joint of the thumb:

Flexion of the thumb: Bend the thumb so that it passes directly across the palmar surface of your hand.

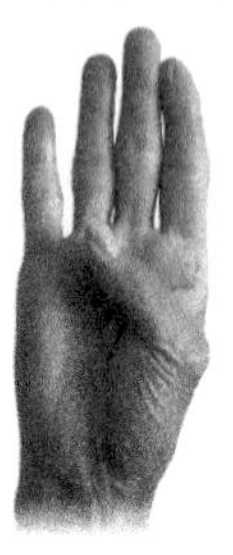

Extension of the thumb: Bend the thumb so that it moves away from the hand, but in the same plane as flexion (like a hitchhiker).

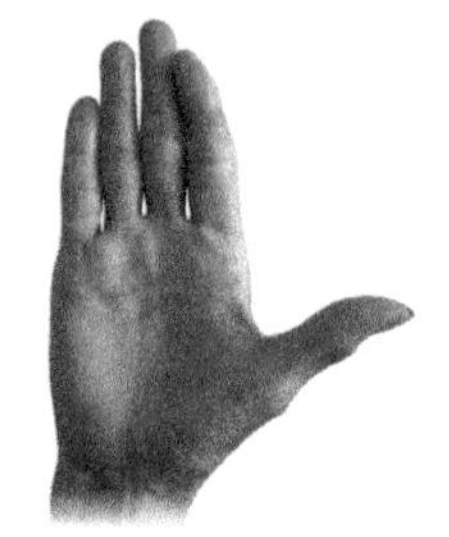

Abduction of the thumb: With the thumb positioned tightly against the palm and parallel to the index finger, bend it so that it moves toward your face.

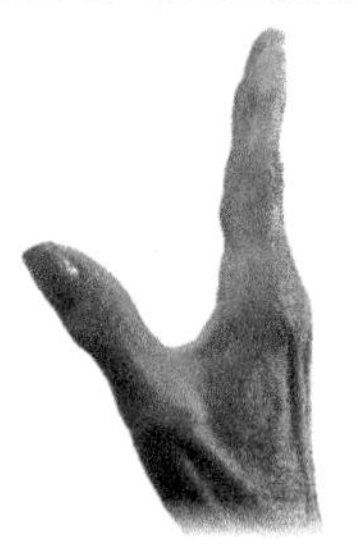

Adduction of the thumb: With the thumb in the abducted position, bend it so that it moves toward the palm.

Notice that flexion and extension occur in a plane that is perpendicular to the plane in which abduction and adduction occur.

3. Saddle joints are biaxial. The shape of the articular surfaces at the saddle joint for the thumb allows for much greater freedom of movement compared to other biaxial joints, however. Flexion and abduction are combined with slight medial rotation so that the thumb can be brought into contact with the palmar surface of the pinky finger. This action is called **opposition.** Bringing the thumb back to the anatomical position is called reposition. Perform these movements with your own thumb.

Want more practice? Go to: **MasteringA&P** > Study Area > Menu > Animations & Videos > ***A&PFlix*** > Group Muscle Actions and Joints >

- Unit 1: Muscles that act on the shoulder joint and humerus > Movement at the glenohumeral joint: An overview
- Unit 4: Muscles that act on the hip joint and femur > Movement at the hip joint: An overview

B At **ball and socket joints,** a rounded head articulates with a cuplike concavity.

1 On a skeleton, identify the **shoulder joint.** At this joint, the head of the humerus articulates with the glenoid cavity.

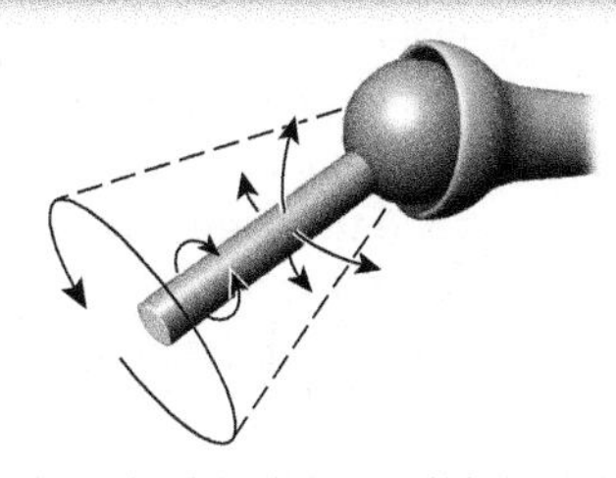

2 Ball-and-socket joints are **triaxial,** allowing movements in three planes: flexion and extension in the first plane, abduction and adduction in the second, and medial and lateral rotation in the third. Circumduction can also be performed at these joints. From the anatomical position, perform the following movements at the shoulder joint:

- Flexion of the arm
- Extension of the arm
- Abduction of the arm
- Adduction of the arm
- Medial rotation of the arm
- Lateral rotation of the arm
- Circumduction of the arm

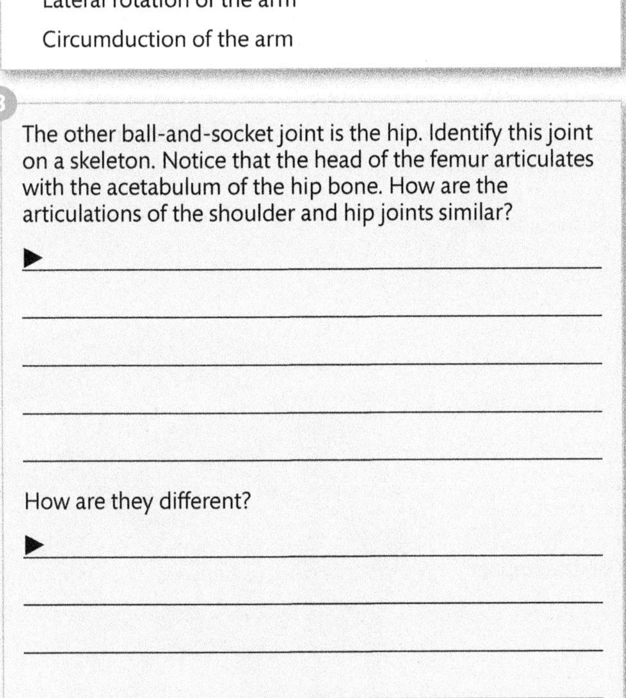

3 The other ball-and-socket joint is the hip. Identify this joint on a skeleton. Notice that the head of the femur articulates with the acetabulum of the hip bone. How are the articulations of the shoulder and hip joints similar?

▶ ______________________________

____

____

____

____

How are they different?

▶ ______________________________

____

____

____

____

4 At the hip joint, perform flexion/extension, abduction/adduction, and medial/lateral rotation of the thigh. How does the range of motion at the hip compare with that of the shoulder?

▶ ______________________________

____

____

____

MAKING CONNECTIONS

List two activities that involve circumduction of the arm. ▶

1. ______________________________

____

____

2. ______________________________

____

____

Activity **9.8**

Synovial Joints: Examining the Structure of Synovial Joints in the Upper Limb

Synovial joints are defined by the presence of a fluid-filled **joint cavity.** The articulating bony surfaces are covered by a smooth layer of hyaline cartilage, known as the **articular cartilage,** which absorbs shock and reduces friction. A **joint capsule,** composed of strong fibrous connective tissue, is attached to each bone in the joint and encloses the joint cavity. Lining the inside of the joint capsule is a thin layer of loose connective tissue called the **synovial membrane.** It produces **synovial fluid** that is secreted into the joint cavity to provide lubrication and shock absorption during movement. Accessory structures that strengthen and support synovial joints include **ligaments, bursae** (singular = **bursa**), **menisci** (singular = **meniscus**), and **fat pads.**

A The **shoulder joint** has the greatest range of motion of any joint in the body.

1 On a skeleton, examine the shoulder joint and identify the surfaces that are covered by articular cartilage.

▶ ______________________________

____

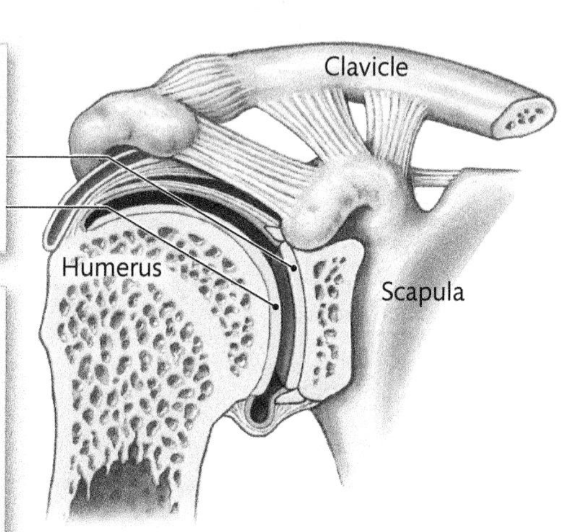

Frontal section

2 Cut out a strip of paper that is 5 cm long and 0.5 cm wide. On a skeleton or disarticulated scapula, wrap the paper strip around the glenoid cavity so that the edge extends about 3 mm beyond the rim. The paper strip approximates the position of the fibrocartilaginous ring called the **glenoid labrum.** What do you think is the function of this structure?

▶ ______________________________

____

3 On a shoulder model, identify the following ligaments that support the shoulder. ▶

The **acromioclavicular ligament** supports the superior aspect of the shoulder. This ligament is actually a part of the acromioclavicular (AC) joint. How would you classify the AC joint?

____

The **coracoclavicular ligaments** help stabilize the clavicle. What type of fibrous joint do these ligaments form?

____

The **coracoacromial ligament** connects what two bone markings on the scapula?

____

____

The **coracohumeral ligament** supports the superior aspect of the shoulder. To what structures does this ligament connect?

____

____

The **glenohumeral ligaments** support the anterior aspect of the shoulder. Identify their bony connections.

____

____

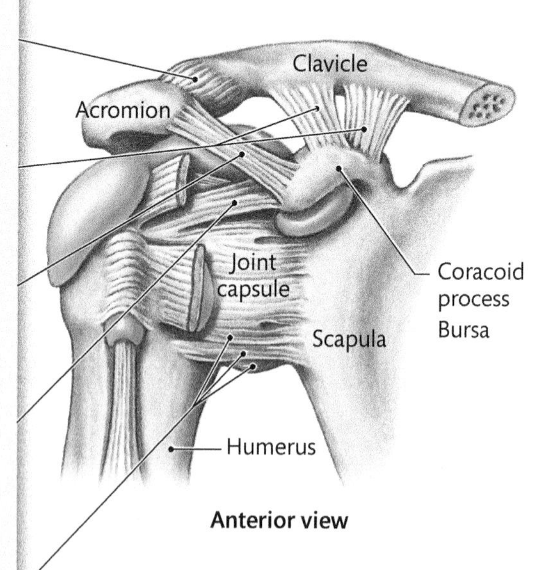

Anterior view

B At the **elbow joint,** the shapes of the articulating surfaces and the strong collateral ligaments allow flexion and extension but prevent other movements.

1. On a skeleton, examine the bony components of the elbow joint. The elbow joint shares a common joint cavity with what other joint?

▶ ______________________________

2. On a model of the elbow joint, identify the **annular ligament.** Notice how the ligament attaches to the anterior and posterior margins of the radial notch and wraps around the radial head. Speculate on the function of this ligament.

▶ ______________________________

____

____

3. On a model of the elbow joint, identify the two triangular-shaped collateral ligaments: ▶

On the medial side, describe the attachments of the **medial (ulnar) collateral ligament.**

____

____

____

____

____

On the lateral side, the base of the **lateral (radial) collateral ligament** blends and becomes continuous with the annular ligament. The apex is attached to what bone marking on the humerus?

____

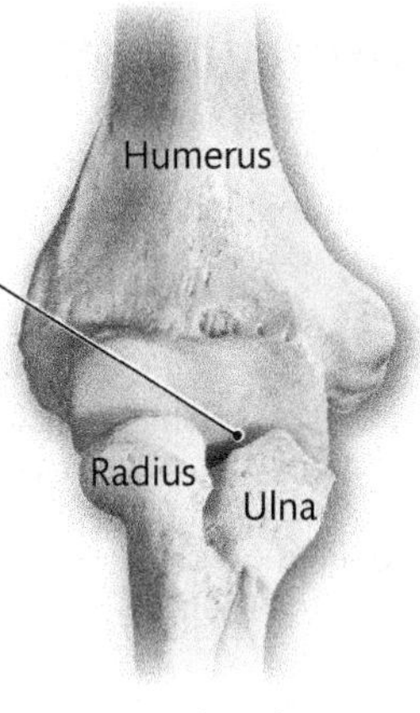

Humerus
Radius
Ulna

Anterior view

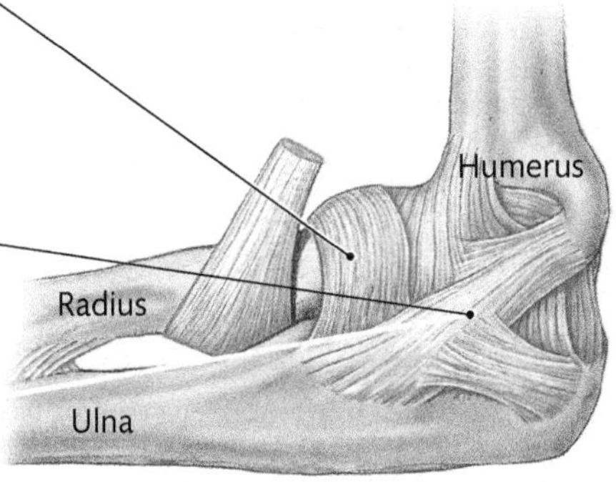

Humerus
Radius
Ulna

Medial view

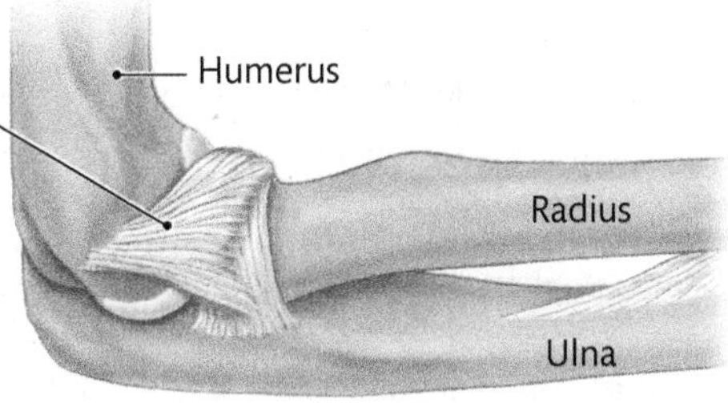

Humerus
Radius
Ulna

Lateral view

Word Origins

The term *annular* is derived from the Latin word *anulus,* which means "a ring." Annular means "shaped like a ring." The *annular ligament* is a ringlike structure that wraps around the radial head at the proximal radioulnar joint.

MAKING CONNECTIONS

Bursae are flattened sacs lined with a synovial membrane and filled with synovial fluid. They are located at strategic locations where tendons, ligaments, muscles, or skin rub against bone or neighboring soft tissue structures (see the blue structures in the bottom figure on page 144). Speculate on the function of bursae at synovial joints.

▶ ______________________________

____

____

____

Want more practice? Go to: **MasteringA&P** > Study Area > Menu > Lab Tools > PAL >
- Anatomical Models > Joints
- Human Cadaver > Joints

Activity **9.9**

Synovial Joints: Examining the Structure of Synovial Joints in the Lower Limb

A The structure of the **hip joint** provides durability and stability for supporting and transmitting all the weight of the upper body.

1. On a model of the hip joint, identify the **iliofemoral ligament.** It is a Y-shaped ligament that strengthens the anterior aspect of the hip joint. Proximally, it is attached to the ilium at the anterior inferior iliac spine and the rim of the acetabulum. Distally, it is attached to the femur at the intertrochanteric line.

2. Identify another anterior ligament, the **pubofemoral ligament.** Proximally, this structure is attached to the pubic part of the acetabular rim. It travels distally and blends with the medial portion of the iliofemoral ligament.

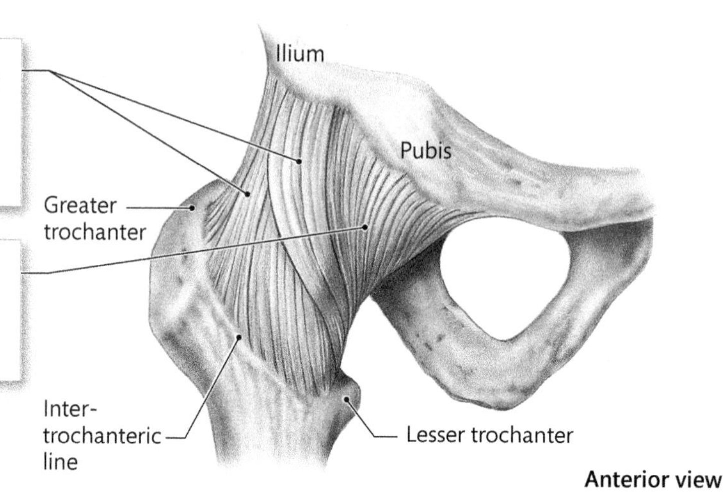

Anterior view

3. Locate the **ischiofemoral ligament,** which reinforces the posterior aspect of the joint. Its attachments include the ischial portion of the acetabular rim and the neck of the femur, medial to the base of the greater trochanter.

4. Inside the hip joint, a fat pad rests along the floor of the acetabulum, and a horseshoe-shaped fibrocartilage pad, the **acetabular labrum,** forms an incomplete ring around the periphery. What do you think is the function of these structures?

▶ ______________________________

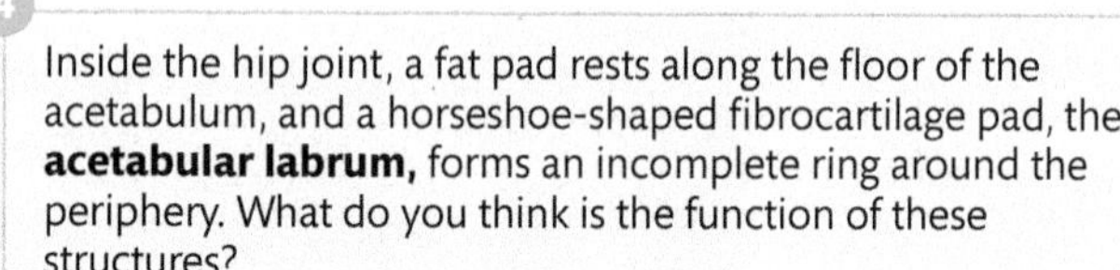

____

____

____

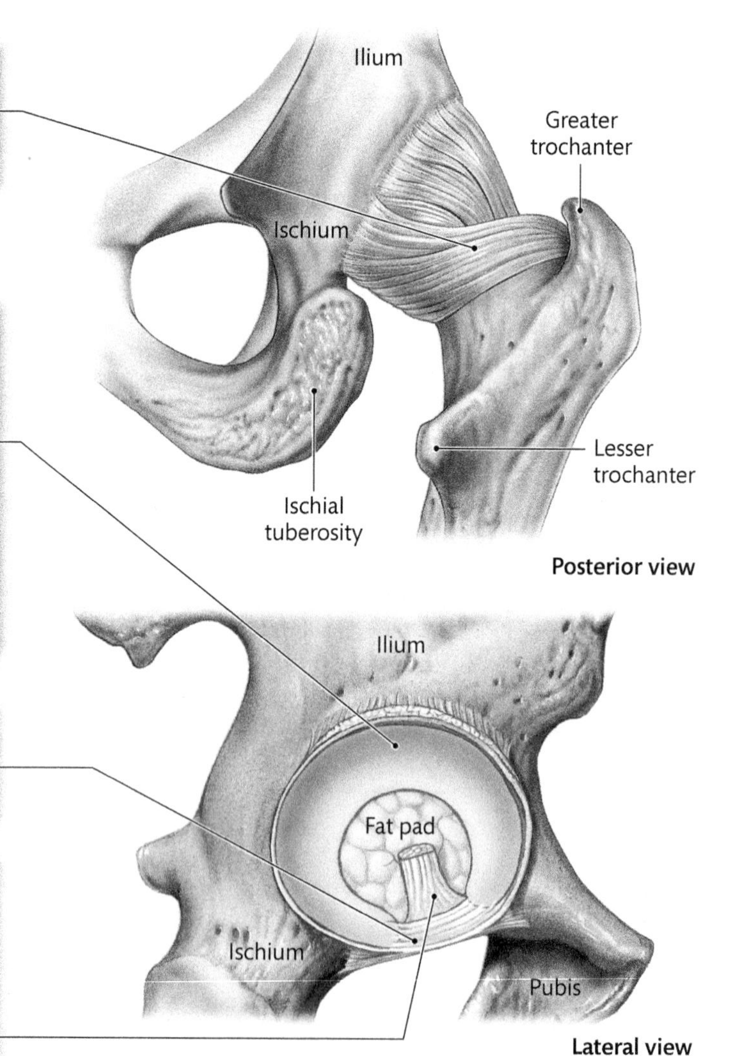

Posterior view

Lateral view

5. On a skeleton, identify the **acetabular notch,** a small gap along the inferior margin of the acetabulum. The **transverse acetabular ligament** stretches across the acetabular notch.

6. The **ligament of the femoral head** extends from the transverse acetabular ligament to the fovea capitis, a small depression on the femoral head, (ligament is not attached to the femoral head in the figure). Speculate on the function of this ligament.

 ______________________________

____

B The **knee joint** is not a typical hinge joint. Although the principal movements are flexion and extension of the leg, slight medial rotation and lateral rotation are also possible. Therefore, the knee is often referred to as a modified hinge joint.

1. On a knee model, identify the **quadriceps tendon** that attaches the quadriceps femoris muscle to the patella.

2. Notice that the **patellar ligament** is a continuation of the quadriceps tendon. It extends from the patella to the tibial tuberosity. What is the difference between a tendon and a ligament?

▶ ______________________________

____

____

3. Along the lateral side of the knee, identify the **fibular collateral ligament.** Identify its attachments to the femur and fibula. ▶

Femur: ______________________________

Fibula: ______________________________

Patella

Fibula Tibia

Anterior view

5. Along the medial side of the knee, find the **tibial collateral ligament.** Identify its attachments to the femur and tibia. ▶

Femur: ______________________________

Tibia: ______________________________

6. Move the patella and flex the knee to expose the **anterior cruciate liagment (ACL)** and the **posterior cruciate ligament (PCL).** Notice that these ligaments are found within the joint capsule and cross each other. Both ligaments connect the tibia to the femur, but they are named according to their attachments to the intercondylar (between the condyles) area of the tibia.

4. Along the posterior aspect of the knee, the **popliteal ligaments** connect the femur to the medial tibial condyle and the fibular head. Why are these ligaments important?

▶ ______________________________

____

____

____

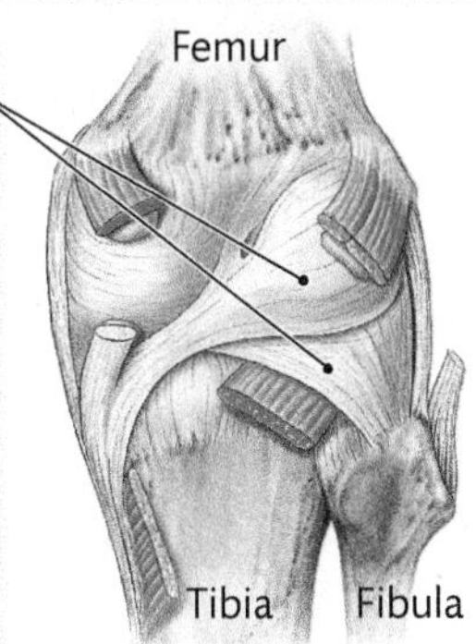

Posterior view

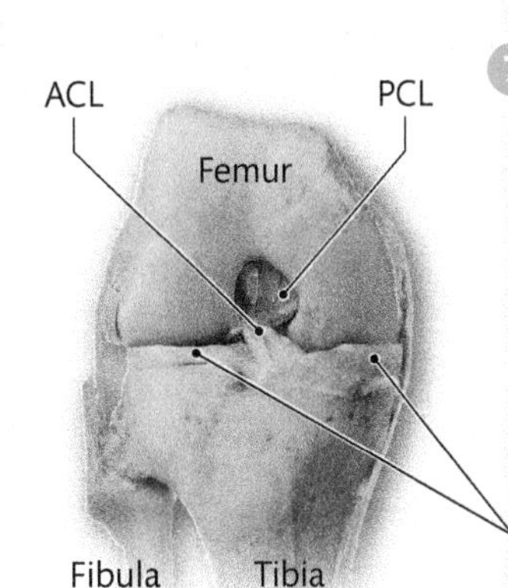

Deep anterior view, flexed

7. With the knee in the flexed position, identify the **lateral meniscus** and the **medial meniscus.** The two menisci are C-shaped fibrocartilage plates that rest on the articular surface of the tibia. Observe how the menisci form a deeper articulating surface on the tibia for the condyles of the femur. This arrangement increases the stability of the knee. In addition, the menisci have an important shock-absorbing function.

IN THE CLINIC

Knee Injuries

The knee is a complex joint that supports and transmits all your body weight. It is under a lot of stress, and despite its stabilizing specializations, the knee can be injured. An abrupt movement that causes the knee to buckle medially can tear the medial collateral ligament, and that injury, in turn, often tears the medial meniscus, which is attached to it. A knee injury generally known as "torn cartilage" occurs when cartilage fragments are dislodged from one or both menisci. In addition, the cruciate ligaments can be injured if the knee is moved too forcefully in an anterior or posterior direction.

MAKING CONNECTIONS

Speculate on the functions of the anterior and posterior cruciate ligaments.

▶ ______________________________

____

____

____

____

____

Want more practice? Go to: **MasteringA&P** > Study Area > Menu > Lab Tools > PAL >

- Anatomical Models > Joints
- Human Cadaver > Joints

BEFORE YOU MOVE ON . . .

« LOOKING BACK

In this laboratory exercise, you studied the joints in the body according to their structural classification (type of articulation). You learned that joints can be placed in one of three structural groups:

1. **Fibrous joints:** Articulating bones are held together by fibrous connective tissue.
2. **Cartilaginous joints:** Articulating bones are held together by cartilage.
3. **Synovial joints:** A connective tissue capsule encloses a fluid-filled cavity between the articulating bones.

You also examined the types of movements that can occur at joints, particularly at synovial joints. Accordingly, articulations are also classified according to functional characteristics (degree of movement):

1. **Synarthroses** (singular = **synarthrosis**): joints where no movement occurs
2. **Amphiarthroses** (singular = **amphiarthrosis**): joints that have limited movement
3. **Diarthroses** (singular = **diarthrosis**): joints that are freely movable

Realize that synarthroses and amphiarthroses can be either fibrous or cartilaginous joints. All diarthroses are synovial joints.

There are distinct differences in the range of motion and level of stability among different types of joints. If a joint has a wide range of motion, how do you think this feature will affect the joint's stability?

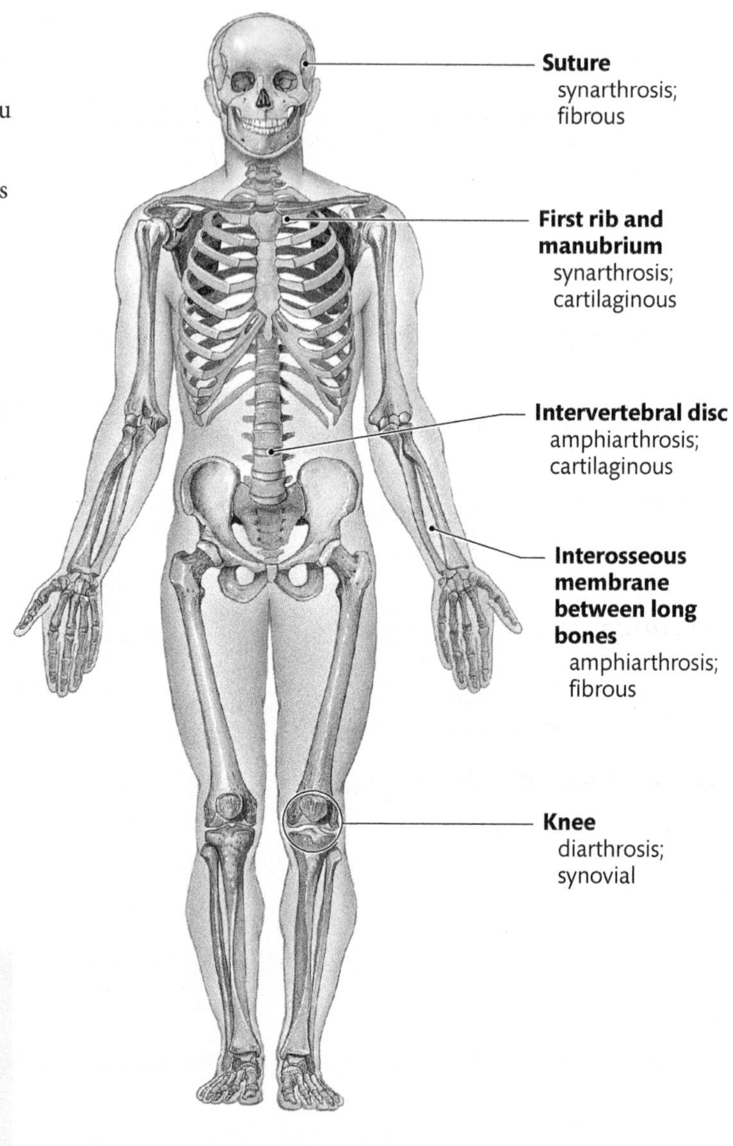

LOOKING FORWARD »

As you move forward to study the muscular system (Exercises 10–13), you will see that muscle contractions provide the force for producing movements at joints. Skeletal muscles carry out voluntary movements by pulling on a bone to initiate an action at a joint. They also have a fundamental role in maintaining posture and stabilizing joints.

Because of their elongated, fibrous shape, muscle cells are usually referred to as muscle fibers. In the next exercise, you will explore the microscopic structure of skeletal muscle fibers and investigate the mechanisms involved in their contraction.

Name Gabe Vanaenhoven

Lab Section

Date

EXERCISE 9 REVIEW SHEET

Articulations

1. Complete the following table.

Classification of Fibrous and Cartilaginous Joints

Category	Type	Structure	Degree of Movement
a.	Suture	b.	c.
d.	Symphysis	e.	f.
Cartilaginous	g.	h.	No movement
Fibrous	i.	j.	Little movement
k.	l.	Periodontal ligaments anchor teeth to bones.	m.

QUESTIONS 2–10: Match the term in column A with the correct description in column B.

A

2. Articular cartilage c

3. Interosseous membrane f

4. Synovial membrane i

5. Bursa d

6. Fontanelle G

7. Meniscus a

8. Intervertebral disc h

9. Epiphyseal plate b

10. Periodontal ligament e

B

a. C-shaped plate of fibrocartilage that provides shock absorption at the knee joint

b. Synchondrosis where bone growth occurs

c. Layer of hyaline cartilage that covers articulating bony surfaces at synovial joints

d. Fluid-filled sac that provides cushioning and reduces friction at a synovial joint

e. Structure that anchors a tooth to its bony socket

f. Sheet of fibrous connective tissue that connects the long bones in the forearm or leg

g. Region of connective tissue between cranial bones in the fetal skull, where ossification is not complete

h. Shock-absorbing fibrocartilage located between the bodies of adjacent vertebrae

i. Structure that produces the fluid found in the joint capsule of a synovial joint

11. Complete the following table by providing two examples of a uniaxial joint, a biaxial joint, and a triaxial joint. For each example, identify the bone markings that serve as articulating surfaces and list the movements (flexion, extension, abduction, adduction, etc.) that occur.

Type of Joint	Examples	Articulating Surfaces	Movements
Uniaxial joints	a. Ankle	b. Tibia Medial and Lateral Malleolus	c. Dorsiflexion
	d. Knee	e. Medial + Lateral femoral condyles	f. Flexion + Extention
Biaxial joints	g. Condylar Joint (Phalangirs)	h. Metacarpals	i. Adduction Abduction Flexion, Extension
	j. Saddle Joint (Thumb)	k. Carpals	l. Adduction Abduction Flexion Extension
Triaxial joints	m. Shoulder (Ball n socket)	n. Clavical Scalpula	o. Flexion Extension Adduction Abduction Rotation
	p. Hip (Ball n socket)	q. Femur	r. Flexion Extension Adduction Abduction Rotation

12. Complete the following table by identifying the type of movement (flexion, extension, abduction, adduction, etc.) that is described and the joint where the movement is occurring.

Description of Movement	Type of Movement	Joint Where Movement Is Occurring
Moving the thigh toward the torso, in the coronal plane	a. Flexion	b. Ball and Socket
Rotating forearm and hand laterally so that palm faces anteriorly	c. Lateral Rotation external	d. Pivot
A bending action that elevates the soles, such as when you stand on your heels	e. Dorsiflexion	f. Synovial
Bringing the thumb over to touch another digit	g. Opposition	h. Saddle
Anterior bending of the forearm that decreases the angle between the arm and forearm	i. Flexion	j. Hinge

QUESTIONS 13–15: Identify the labeled ligaments in the diagram below.

13. Acromioclavicular Ligament

14. Coracoacromial Ligament

15. Coracoclavicular Ligament

13 14 15 Clavicle

Acromion process

Head of humerus

Coracoid process

Frontal section

QUESTIONS 16–18: Identify the labeled ligaments in the diagram below.

16. Radial Collateral Ligament

17. Anular Lig. of radius

18. Ulnar Collateral Ligament

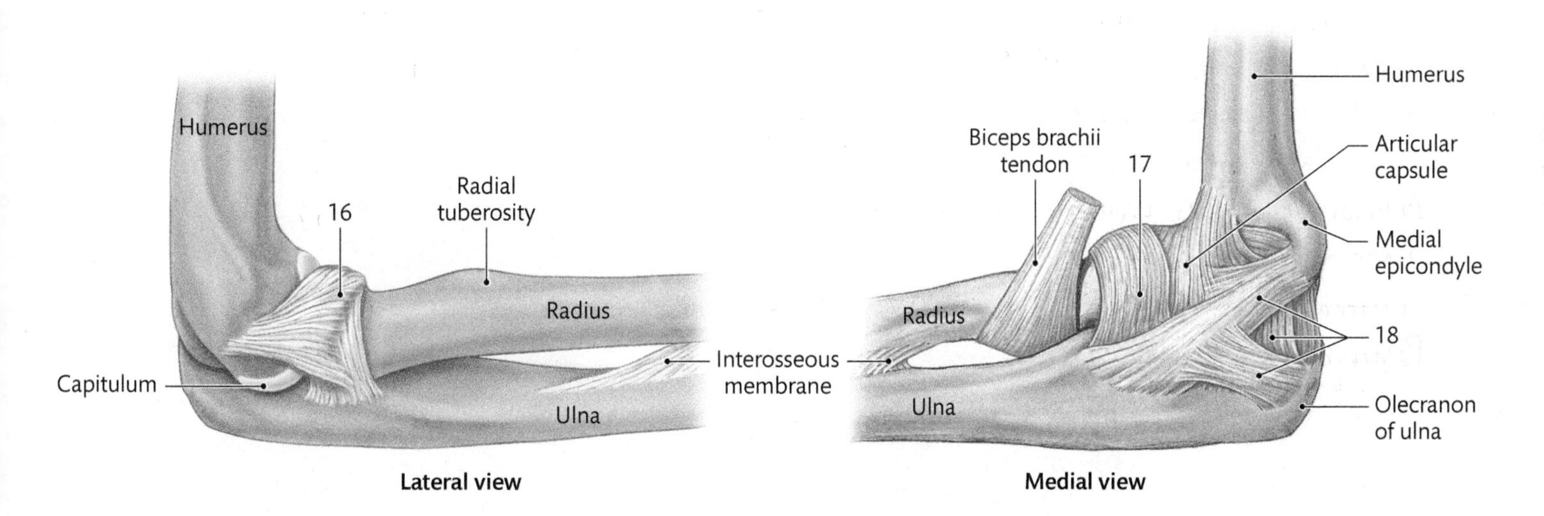

Lateral view **Medial view**

QUESTIONS 19–21: Identify the labeled ligaments in the diagram below.

19. Ilifemoral Ligament
20. Pubofemoral Ligament
21. Ischiofemoral Ligament

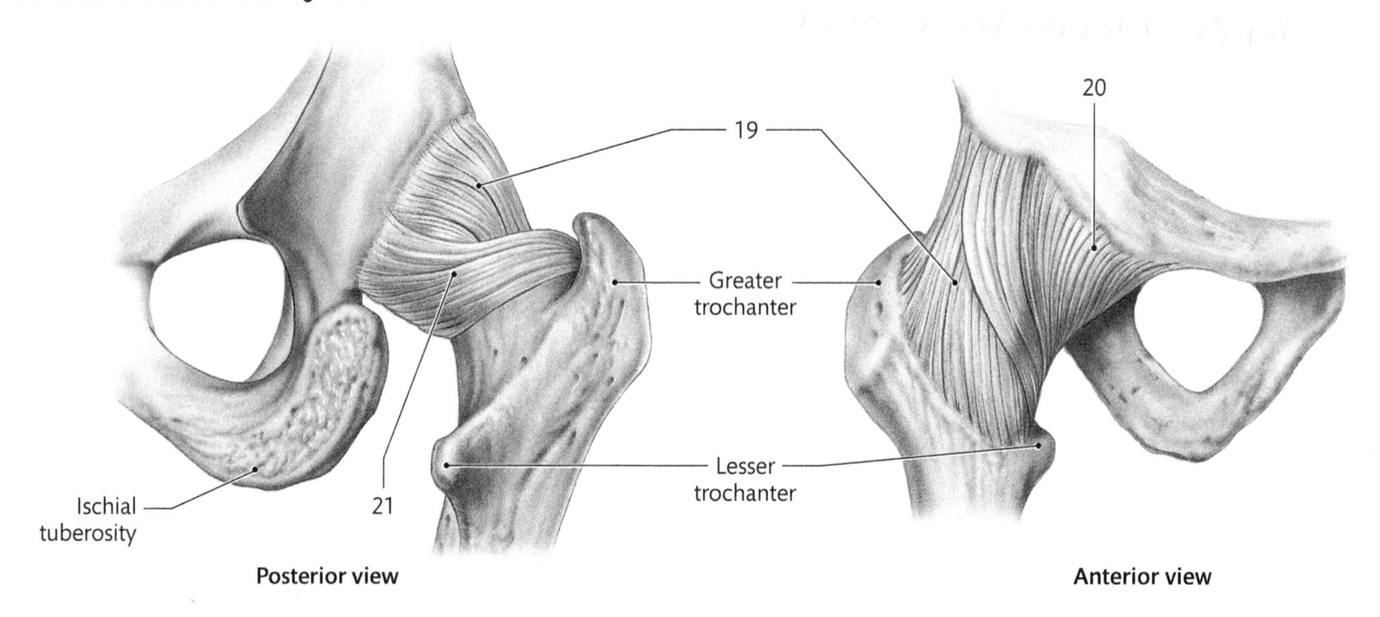

QUESTIONS 22–29: Identify the labeled ligaments/tendons in the diagram below.

22. Quadricept Tendon
23. Medial Collateral Ligament
24. Medial Meniscus Lig
25. Patellar Ligament
26. Anterior Cruciate Ligament
27. Lateral Collarteral Lig
28. Lateral Menisus Lig
29. Posterior Cruciate Ligament

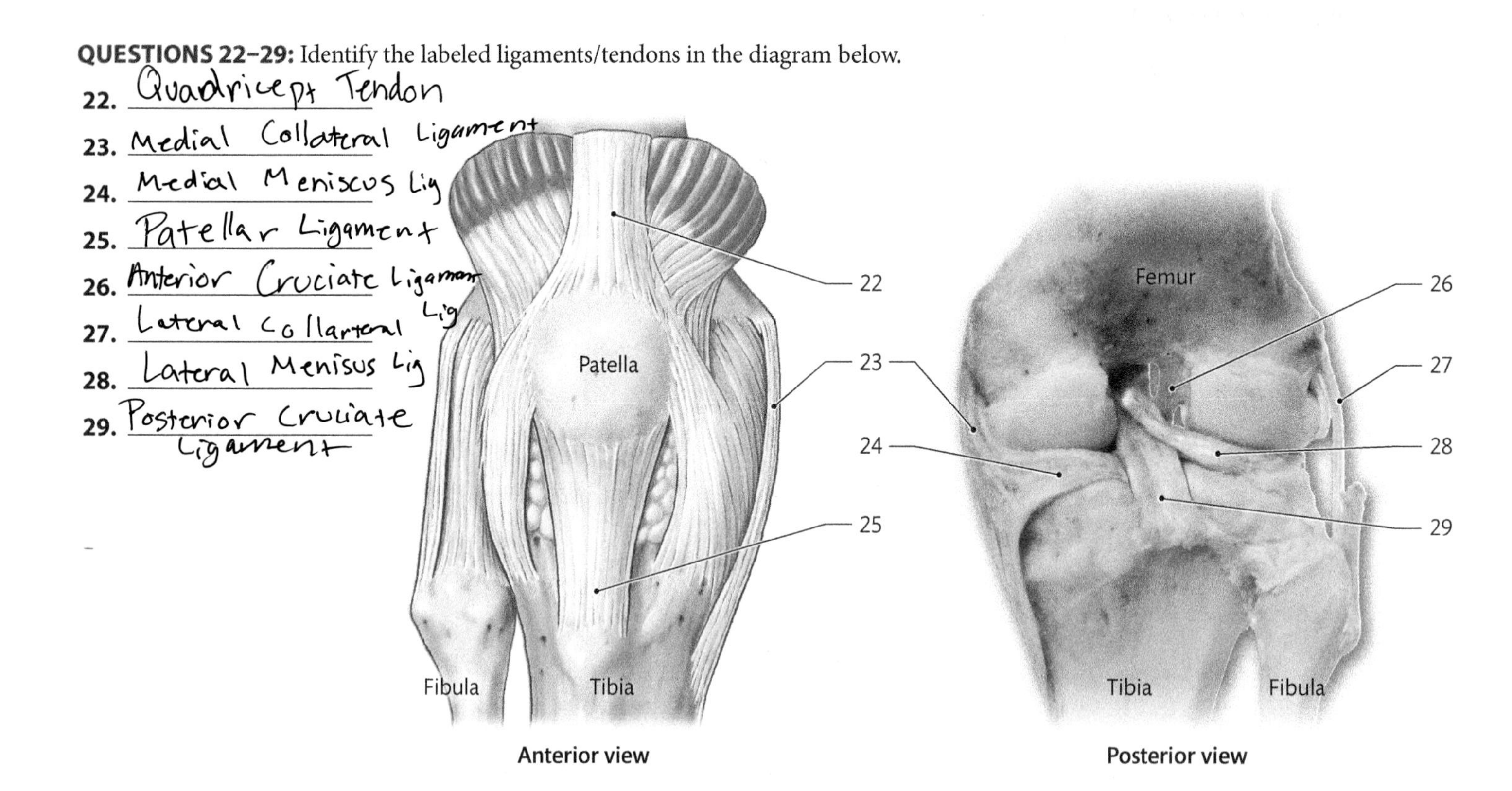

EXERCISE

10

Introduction to Skeletal Muscle

LEARNING OUTCOMES

These Learning Outcomes correspond by number to the laboratory activities in this exercise. When you complete the activities, you should be able to:

Activity 10.1 **Identify the principal microscopic features of skeletal muscle.**

Activity 10.2 **Describe the structure of a neuromuscular junction and discuss the sequence of events that occur there.**

Activity 10.3 **Predict how ATP and various ions affect skeletal muscle contraction and test your prediction.**

LABORATORY SUPPLIES

- Prepared microscope slides of skeletal muscle
- Compound light microscopes
- Dissecting microscopes
- Colored pencils
- Three-dimensional model of a neuromuscular junction
- Prepared microscope slides of a neuromuscular junction
- Three-dimensional model of a skeletal muscle fiber
- Glycerinated skeletal muscle preparation
- 0.25% ATP in distilled water
- Ion solution: 0.05M potassium chloride (KCl) and 0.001M magnesium chloride ($MgCl_2$) in distilled water
- 50% glycerol in distilled water
- Microscope slides and coverslips
- Forceps and dissecting needles

PRE-LAB QUIZ

Before you begin, read all the activities in Exercise 10 and the required reading in your textbook that is assigned by your instructor.

1. During this laboratory exercise, you will complete all the following activities *except*
   a. observing the contraction of skeletal muscle under a dissecting microscope.
   b. examining the microscopic structure of skeletal muscle.
   c. studying the events at a neuromuscular junction.
   d. learning how muscles can be classified according to their functions.
   e. observing the structure of skeletal muscle fibers on a three-dimensional model.
2. True or False: During one activity in this laboratory exercise, you will be using skeletal muscle fibers that have been incubated in a saline solution. ________
3. List the two models that you will use to study skeletal muscle in this laboratory exercise.

   ____

   ____
4. True or False: The striations in skeletal muscle are formed by the arrangement of thick and thin filaments. ________
5. True or False: In skeletal muscle cells, the sarcoplasmic reticulum is continuous with the sarcolemma. ________
6. The interval between two Z lines is called a ________________.
7. At a neuromuscular junction, the space between the plasma membrane of the synaptic terminal and the motor end plate is the ________________.
8. An action potential (electric impulse) is transferred from a motor neuron to a muscle cell at a ________________.
9. True or False: Like most other cells in the body, skeletal muscle cells have only one nucleus. ________
10. True or False: The influx of sodium ions generates an action potential at the motor end plate. ________

Activity **10.1**

Examining the Microscopic Structure of Skeletal Muscle Fibers

A **Skeletal muscle fibers** possess two unique anatomical features:

- Multiple nuclei located along the periphery of the fibers.
- Alternating light and dark bands (striations) produced by the overlapping arrangement of the thick and thin filaments.

1 Scan a slide of **skeletal muscle** under low power with a compound light microscope. Your slide will likely contain areas of muscle fibers in both longitudinal section and cross section.

2 Switch to high power and observe the longitudinal and cross sections more closely.

3 In the cross-sectional view, identify the darkly stained nuclei arranged around the periphery of each muscle fiber.

Why do you think the nuclei in skeletal muscle fibers are located along the periphery of the cell and not in the center?

▶ ______________________________

____

____

____

____

____

____

____

____

____

4 In the longitudinal view, identify the distinctive **striations.** They will appear as alternating light and dark bands across the width of each muscle fiber. The striations are formed by the arrangement of the thick and thin contractile protein filaments, within long cylindrical structures called myofibrils. The lighter I bands contain only thin filaments. The darker A bands contain overlapping thin and thick filaments. In the longitudinal section, notice also the peripheral location of the nuclei.

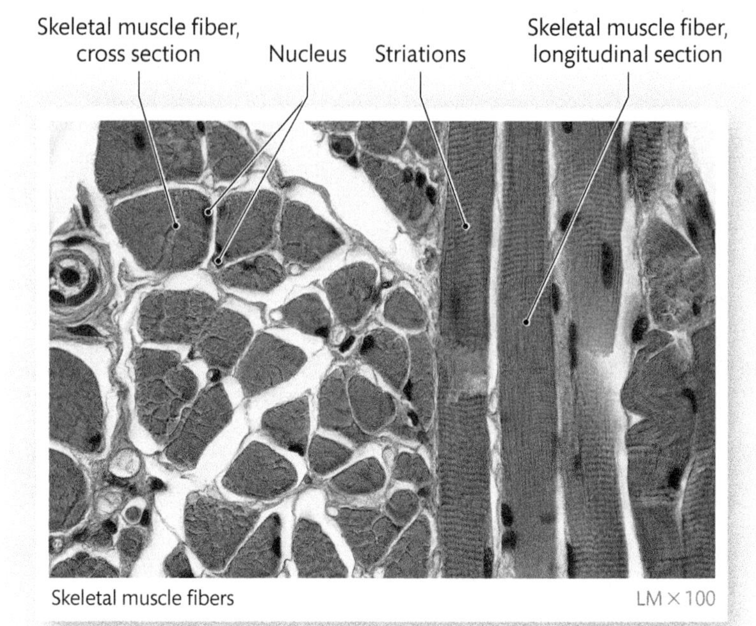

Skeletal muscle fibers

Word Origins

The prefix *myo-* has its origins from the Greek word *mys,* for "muscle." Thus, any word with the prefix *myo-* is related to muscle. Myofilaments are muscle filaments, and myofibrils are bundles of myofilaments.

The prefix *sarco* is derived from the Greek word for "flesh," *sarx.* Any word that begins with *sarco* refers to "muscular substance" or "resemblance to flesh." Thus, the sarcolemma is the cell membrane, the sarcoplasm is the cytoplasm, and the sarcoplasmic reticulum is the endoplasmic reticulum of a muscle cell.

IN THE CLINIC

Delayed Onset Muscle Soreness

Many people experience muscle pain 24 to 36 hours after strenuous physical activity such as shoveling snow, chopping wood, or a new exercise routine. This condition is called **delayed onset muscle soreness (DOMS).** The pain associated with DOMS is believed to be the result of damage to cell membranes caused by small tears in the muscle tissue or minor injury to the tendons that attach muscle to bone. The soreness typically disappears in a few days, but any treatment that increases blood flow to the area, such as light massage, will accelerate the recovery.

■ Want more practice? Go to: **MasteringA&P** > Study Area > Menu > Lab Tools > PAL > Histology > Muscle Tissue (skeletal muscle only)

B Skeletal muscle cells contain the contractile proteins **myosin,** which is the main component of thick filaments, and **actin,** which is the main component of thin filaments. Collectively, the thick and thin filaments are often referred to as the **myofilaments** of the muscle cell. The myofilaments are packaged into parallel cylindrical bundles, called **myofibrils,** that extend the entire length of the muscle fiber and account for most of the cell's volume. The unique arrangement of actin and myosin in the myofibrils yields a pattern of alternating light and dark bands (striations) throughout the cytoplasm. The membrane structures of a skeletal muscle fiber—**transverse (T) tubules, sarcoplasmic reticulum,** and **cisternae**—surround the thick and thin filaments in the myofibrils. When electrical impulses called action potentials travel along the membranes of T tubules, they promote the release of calcium ions from the sarcoplasmic reticulum and cisternae into the **sarcoplasm** (muscle cell cytoplasm). These actions cause the muscle fiber to contract.

1. On a three-dimensional model of a skeletal muscle fiber, identify the following structures:

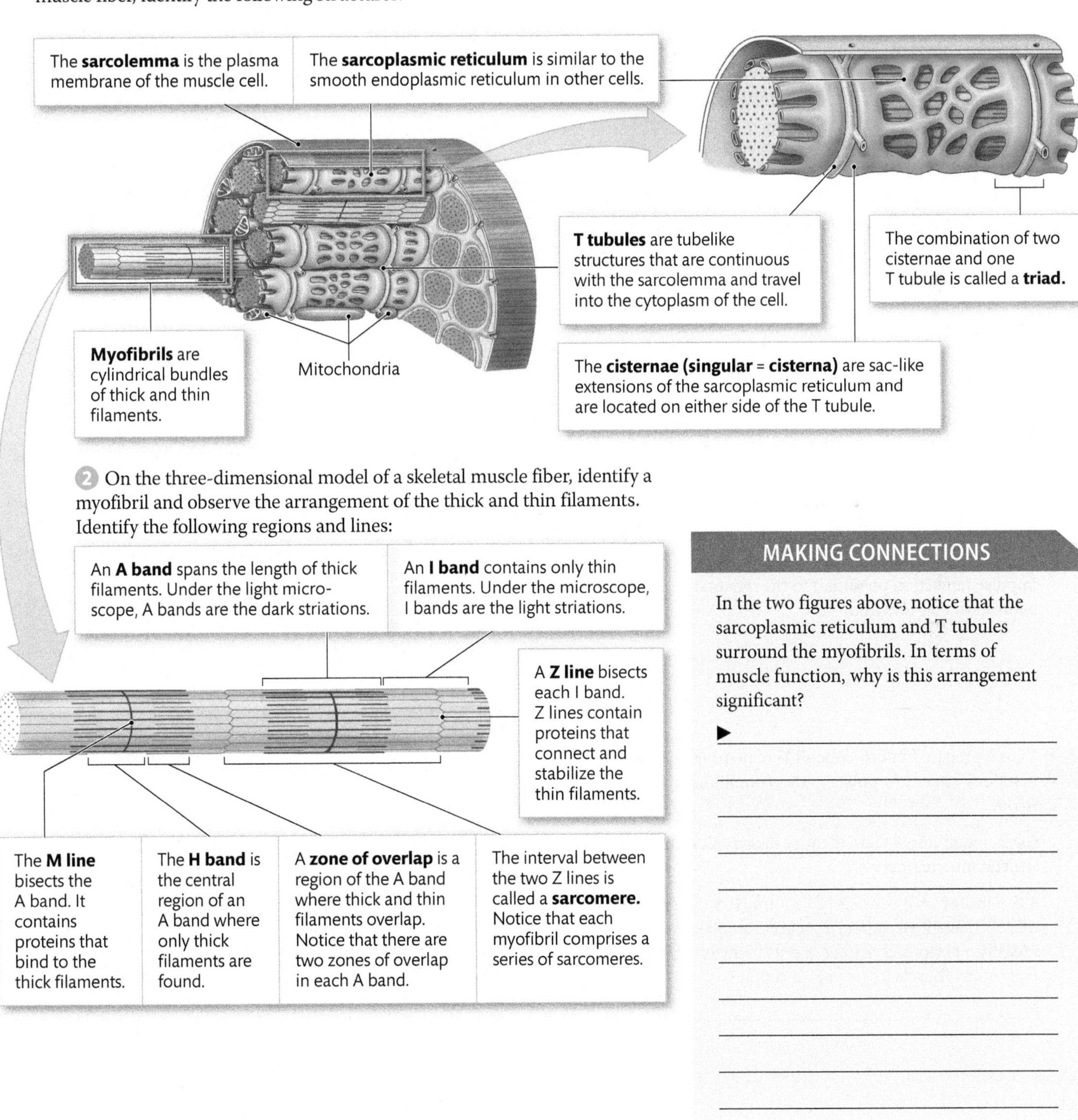

2. On the three-dimensional model of a skeletal muscle fiber, identify a myofibril and observe the arrangement of the thick and thin filaments. Identify the following regions and lines:

MAKING CONNECTIONS

In the two figures above, notice that the sarcoplasmic reticulum and T tubules surround the myofibrils. In terms of muscle function, why is this arrangement significant?

▶ ______________________________

____

____

____

____

____

____

____

____

____

Activity 10.2

Examining the Neuromuscular Junction at a Skeletal Muscle Cell

A Skeletal muscle contractions are controlled by **motor neurons** that originate in the central nervous system (brain or spinal cord). These neurons have long processes called **axons** that travel along spinal and cranial nerves. The end of each axon (the synaptic terminal) interacts with a muscle fiber at a **neuromuscular junction.**

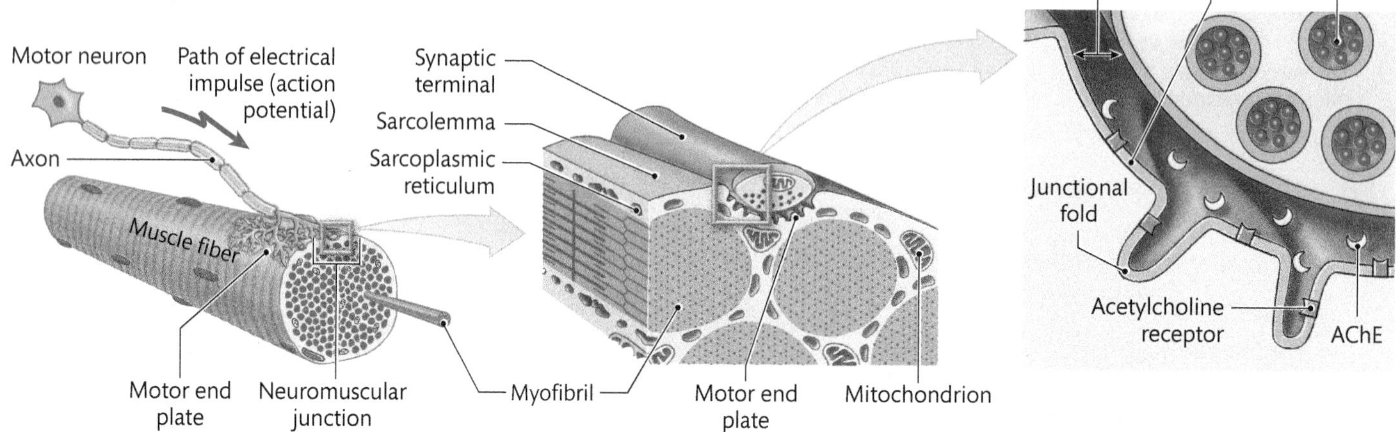

1 On a model of a neuromuscular junction, review the organization of a skeletal muscle fiber by identifying the following structures:

- sarcolemma
- myofibrils in the sarcoplasm
- I bands, A bands, and Z lines in the myofibrils.

Identify the motor nerve fiber (axon) as it approaches the neuromuscular junction. The axon transmits an electric impulse (an action potential) to the synaptic terminals at the neuromuscular junction.

2 At the neuromuscular junction, identify the following structures:

- The **synaptic terminal** is the expanded knob at the end of the nerve fiber. It contains numerous secretory vesicles filled with the **neurotransmitter, acetylcholine (ACh)**.
- The **motor end plate** is the length of the sarcolemma associated with the neuromuscular junction. It contains receptors that bind to acetylcholine. Notice the **junctional folds** along the motor end plate membrane. The folds increase surface area and, therefore, the number of ACh receptors along the motor end plate.
- The **synaptic cleft** is the narrow space between the plasma membrane of the synaptic terminal and the motor end plate. The enzyme, **acetylcholinesterase (AChE),** which breaks down ACh, is found in this space.

B View a prepared microscope slide of neuromuscular junctions under low power with a compound light microscope.

1 In longitudinal view, identify the striations in several skeletal muscle fibers.

2 A **motor unit** consists of a motor nerve fiber and all the muscle fibers it stimulates at neuromuscular junctions. Identify a motor nerve fiber and its many branches. How many muscle fibers does it supply?

▶ ______________________________

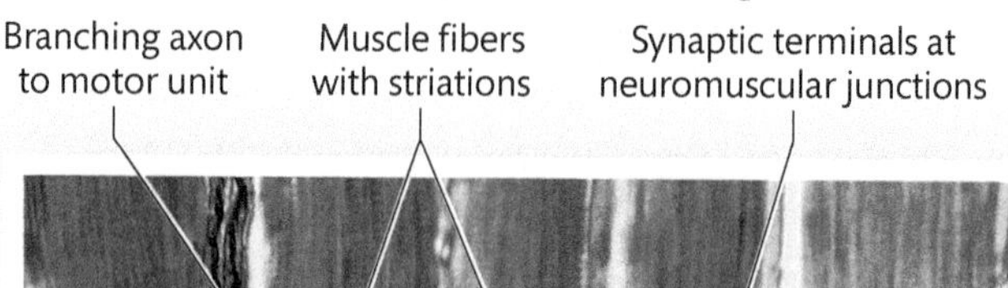

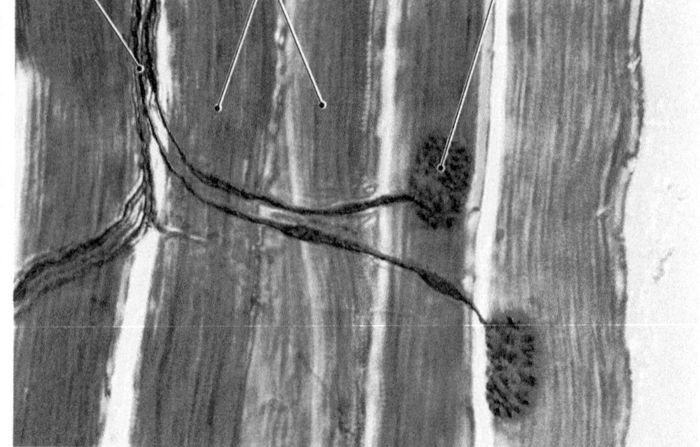

A motor unit

C The four figures below illustrate the activities that occur at the neuromuscular junction. Carefully review the activities depicted in each figure and then answer the questions in Table 10.1. ▶

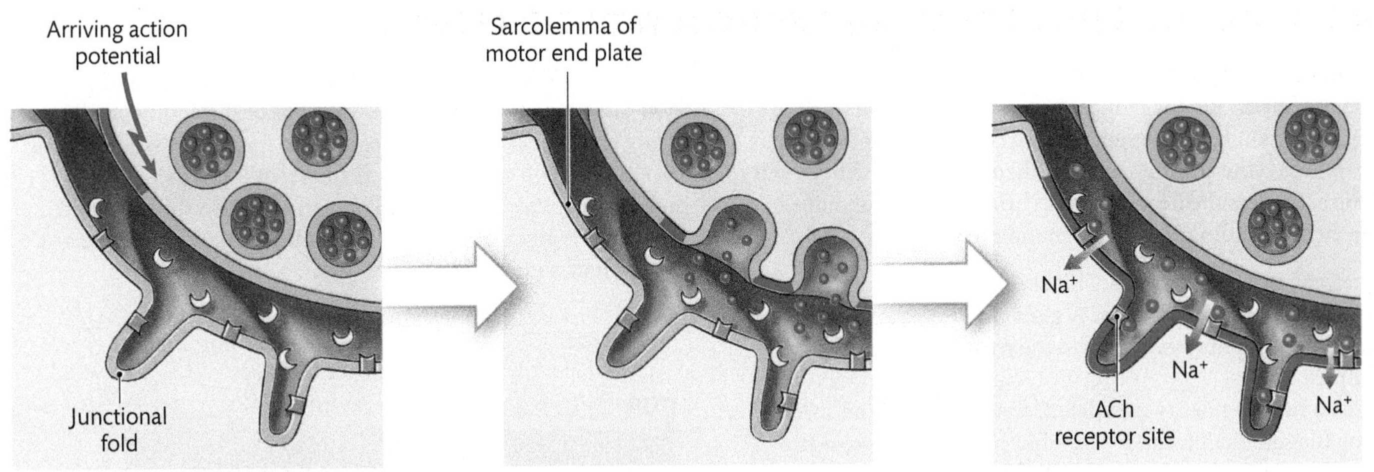

1. The action potential arrives at the synaptic terminal.

2. Acetylcholine (ACh) is released from secretory vesicles into the synaptic cleft.

3. ACh diffuses across the synaptic cleft and binds to receptors on the motor end plate. The binding of ACh initiates the flow of sodium ions into the muscle cell.

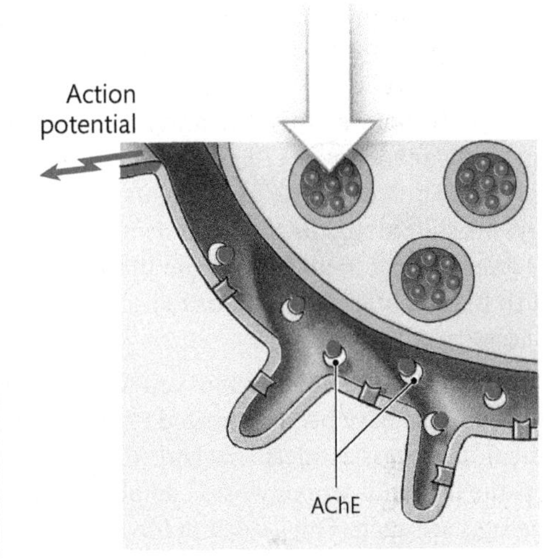

4. The influx of sodium ions generates an action potential at the motor end plate. The enzyme, acetylcholinesterase (AChE), binds to ACh and breaks it down.

TABLE 10.1 Sequence of Events at the Neuromuscular Junction

Events	Questions	Answers
Arrival of action potential	What is an action potential?	
Release of ACh	What transport process is used to release ACh?	
ACh diffuses across the synaptic cleft	What is the function of ACh receptors?	
Generation of action potential at motor end plate	What do you think would happen if AChE did not break down ACh?	

MAKING CONNECTIONS

The muscular system disease **myasthenia gravis** occurs when the body's immune system mistakenly attacks the ACh receptors on the motor end plate. Comment on how myasthenia gravis will affect the normal sequence of events at neuromuscular junctions.

▶ ____________________

____

____

Activity **10.3**

Observing the Contraction of Skeletal Muscle Fibers

A When a muscle fiber contracts, the lengths of the myofilaments do not change. Instead, the thin and thick filaments slide along each other, causing changes in the lengths of the sarcomeres **(the sliding filament theory).** A muscle fiber will not contract unless the heads on cross bridges of myosin molecules are attached to binding sites on actin molecules.

- At rest, actin binding sites are blocked by double stranded proteins called **tropomyosin.** A globular protein, **troponin,** is attached to tropomyosin to form the **troponin-tropomyosin complex.**
- When a muscle fiber is stimulated, an action potential spreads along the sarcolemma and T tubules, causing the release of calcium ions from the sarcoplasmic reticulum and cisternae into the surrounding sarcoplasm. These actions, known as **excitation contraction coupling,** must occur for a muscle fiber to contract.
- The contraction cycle begins when calcium ions bind to troponin, inducing a shift in the position of the troponin-tropomyosin complex. This action exposes the binding sites, and the heads on myosin cross bridges can now attach to them.
- Using cellular energy (ATP), the myosin heads pivot **(power stroke)** and pull on the actin filaments so that the thin filaments slide along the thick filaments.
- The cross bridges then detach from the actin filaments and return to their original positions to begin a new contraction cycle.

To observe muscle contraction, you will be using skeletal muscle fibers that have been incubated in a glycerol solution. The glycerination process extracts ions and ATP from the cells. It also disrupts the troponin-tropomyosin complex so that the actin binding sites are open. Thus, calcium ions are not required for contraction to proceed. Potassium and magnesium ions, however, are needed for the normal function of ATPase, the enzyme that is used to extract energy from ATP.

Make a Prediction

You will be observing contraction of muscle fibers when they are exposed to three conditions:

- only ATP
- only potassium and magnesium ions
- ATP and ions.

Before you begin, predict which condition will result in the greatest contraction of the muscle fibers and which will result in the least contraction.

▶

B Place a sample of glycerinated skeletal muscle fibers under a dissecting microscope. Use forceps or dissecting needles to gently tease the muscle fibers apart into four groups. Each group should have two or three muscle fibers, or if possible, a single muscle fiber. Place each group on a microscope slide and add a drop of 50% glycerol to prevent dehydration. You should have a total of four slides.

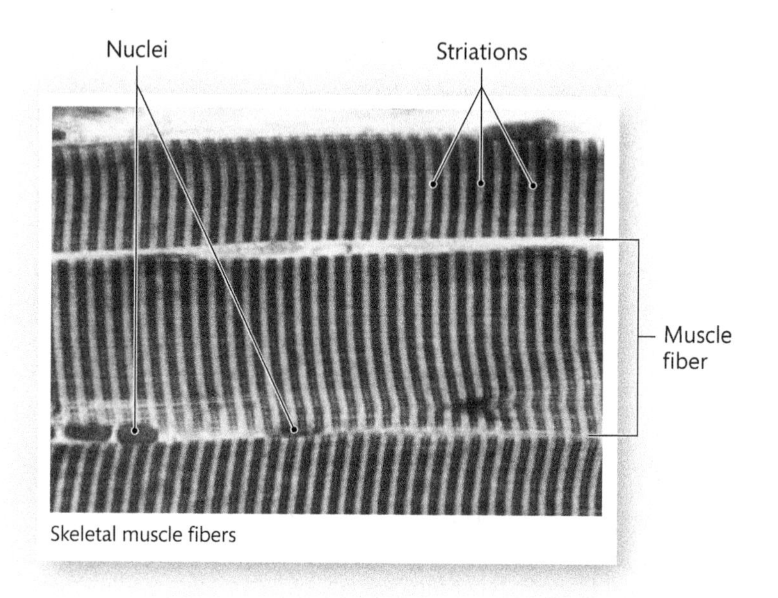

Skeletal muscle fibers

Perform the following procedures:

1. *Slide 1:* Place a coverslip over the sample and observe the fibers with a compound light microscope. Begin at low power and then switch to high power. Identify the striations and the multiple nuclei in each fiber. Draw your observations in the space below.

▶

2 *Slide 2:* Without a coverslip, observe the relaxed fibers with a dissecting microscope and measure their length with a millimeter ruler. Record the results in Table 10.2. Keep the fibers under the dissecting microscope and add a drop of 0.25% ATP solution. Observe any changes that occur. After 45 seconds, measure the length of the fibers and record the results in Table 10.2. ▶

3 *Slide 3:* Repeat the procedure for slide 2, but use an ion solution (a mixture of 0.05M potassium chloride [KCl] and 0.001M magnesium chloride [$MgCl_2$]) in place of the ATP solution. Record the lengths of the muscle fibers before and after exposure to the ion solution in Table 10.2. ▶

4 *Slide 4:* Repeat the procedure for slide 2, but this time use a solution containing both ATP and ions. Record the length of the muscle fibers before and after exposure to the ATP-ion solution in Table 10.2.

5 For each of the above procedures, calculate the net change and percent change in muscle fiber length. Record the results in Table 10.2. ▶

TABLE 10.2 Effect of ATP and Ions on Skeletal Muscle Contraction

Solution	Length of Relaxed Muscle Fibers (mm)	Length of Muscle Fibers after Exposure to Solution (mm)	Net Change in Muscle Fiber Length (mm)*	% Change in Muscle Fiber Length**
ATP (Slide 2)				
Ions (KCl + $MgCl_2$) (Slide 3)				
ATP + ions (Slide 4)				

*Net change = Length of relaxed muscle fibers − Length of muscle fibers after exposure to solution.
** % change = Net change ÷ Length of relaxed muscle × 100.

6 Assess the Outcome

How do your data in Table 10.2 compare to your predictions?

▶ ______________________________

____

____

IN THE CLINIC

Botulism Poisoning

Botulism poisoning is caused by the consumption of canned or smoked food contaminated by the bacteria *Clostridium botulinum*. These bacteria produce a potent toxin that prevents the release of acetylcholine at synaptic terminals. Symptoms typically begin to appear 18 to 36 hours after the initial exposure, but can occur as early as six hours after eating contaminated food. The first symptoms to appear typically affect the eyes, such as dilated pupils, blurred and double vision, and droopy eyelids. Muscle weakness or paralysis can lead to difficulty with speaking, standing, or walking. The primary concern with botulism is the potential for respiratory muscle paralysis, leading to death by suffocation. The mortality rate in the United States is only 10 percent, however. Most people recover gradually with the appropriate supportive care, which could include bed rest, medications, and careful observation.

MAKING CONNECTIONS

Based on your observations, what conclusion can you make about the chemical requirements for muscle contraction?

▶ ______________________________

____

____

____

____

____

____

____

____

BEFORE YOU MOVE ON . . .

« LOOKING BACK

In this laboratory exercise, you examined the unique microscopic structure of skeletal muscle fibers. You observed the alternating light and dark bands (striations) that travel across the width of each muscle fiber. You learned that the thick and thin filaments (myofilaments) are arranged into cylindrical bundles, called myofibrils and that the T tubules, sarcoplasmic reticulum, and cisternae surround the myofibrils. You also studied the sequence of activities that occur at neuromuscular junctions and observed muscle fiber contraction under various conditions.

All muscle cells (skeletal, cardiac, and smooth) are called muscle fibers because they have an elongated, fibrous shape. Skeletal muscle cells are unique, however, because unlike most cells, which have a single nucleus, they are multinucleated (possess several nuclei).

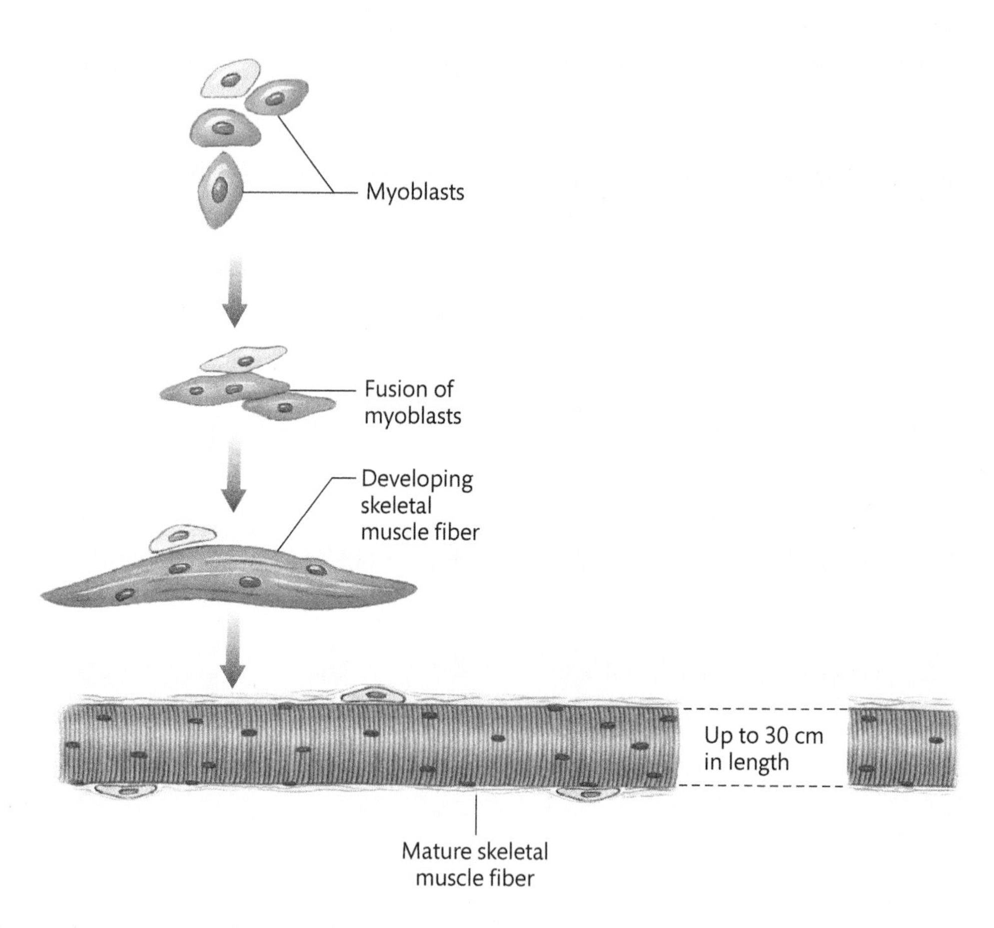

Why is it that mature skeletal muscle fibers are multinucleated cells? (*Hint:* In your textbook, review how groups of embryonic cells called myoblasts differentiate into a mature skeletal muscle fiber.)

► ____

____

____

____

LOOKING FORWARD »

Skeletal muscles are attached to bones by tendons and initiate actions at joints when they contract. Skeletal muscles are divided into two groups—axial and appendicular—according to their locations on the body and the actions they produce. The **axial muscles** include the muscles of the head, neck, and trunk. In the next exercise, you will study the structure and function of the axial muscles, and you will learn why these muscles are vitally important for supporting the head and vertebral column and for allowing us to maintain erect posture.

Name ____________________

Lab Section ____________________

Date ____________________

EXERCISE 10 REVIEW QUESTIONS

Introduction to Skeletal Muscle

1. What is the structural basis for the alternating light and dark striations in skeletal muscle fibers?

____

____

____

____

2. Describe the band(s), line(s), and zone(s) that are found in a single sarcomere.

____

____

____

____

____

QUESTIONS 3–9: Identify the labeled structures in the following diagram.

3. ____________________

4. ____________________

5. ____________________

6. ____________________

7. ____________________

8. ____________________

9. ____________________

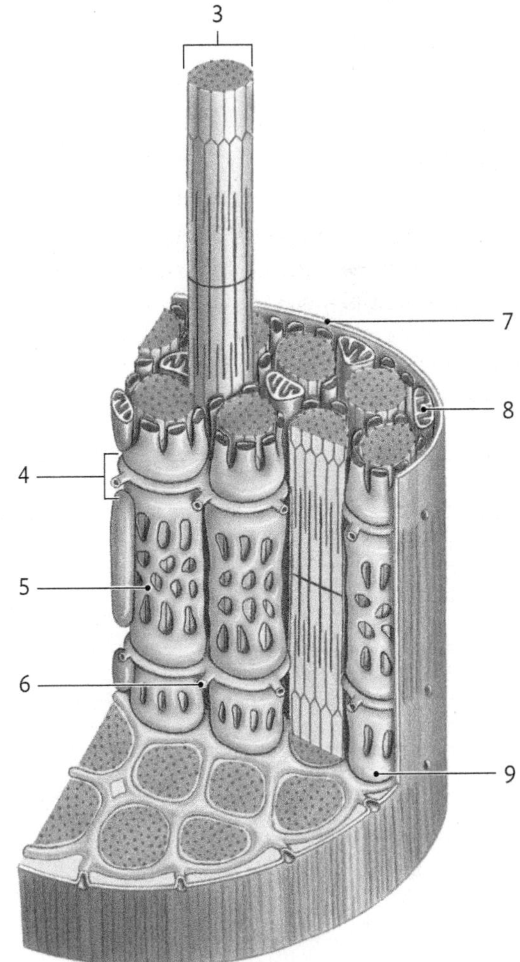

10. Review the two figures below and compare the condition of the sarcomere at rest and after contraction:

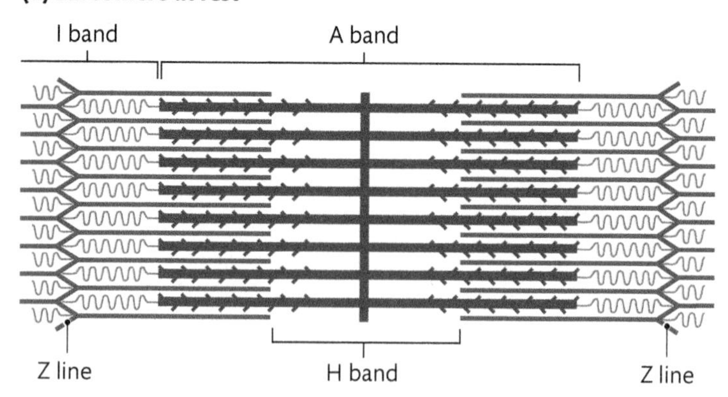

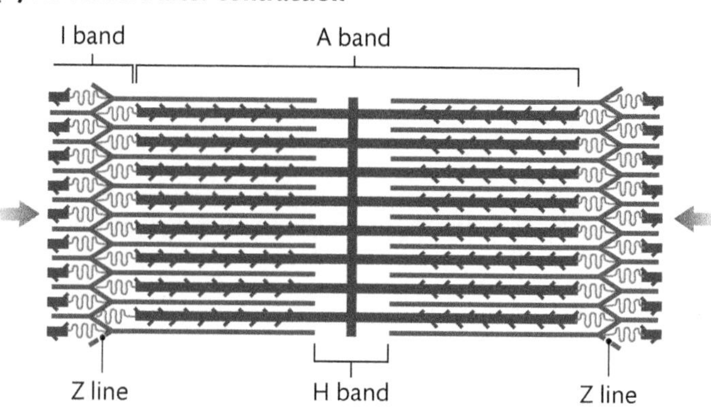

a. Which band(s) change in length during contraction?

____

b. Which band(s) do not change in length during contraction?

____

11. Fill in the blank boxes in the concept map at right by using the terms in the list below.

- T tubules
- Acetylcholine
- Cisternae
- Sodium ions
- Sarcoplasm

Excitation-contraction coupling

At a neuromuscular junction, synaptic vesicles release

(a) ______________

which binds to receptors on the

motor end plate

and initiates an influx of

(b) ______________

An action potential spreads along the sarcolemma and

(c) ______________

causing the release of calcium ions from the

sarcoplasmic reticulum *and* (d) ______________

12. Describe the role that each of the following has in the contraction of a muscle fiber according to the sliding filament theory.

a. The power stroke:

____

____

____

____

____

b. Calcium ions:

____

____

____

____

c. Troponin-tropomyosin complex:

____

____

____

____

EXERCISE 11

Muscles of the Head, Neck, and Trunk

LEARNING OUTCOMES

These Learning Outcomes correspond by number to the laboratory activities in this exercise. When you complete the activities, you should be able to:

Activity 11.1 **Explain how skeletal muscles produce movement by using bones as levers.**

Identify and explain the functions of the following muscles:

Activity 11.2 **The head: Muscles of facial expression and mastication.**

Activity 11.3 **The head: Extrinsic eye muscles, extrinsic tongue muscles, and muscles of the pharynx.**

Activity 11.4 **Muscles of the neck.**

Activity 11.5 **The trunk: Muscles of the thorax.**

Activity 11.6 **The trunk: Muscles of the abdominal wall.**

Activity 11.7 **The trunk: Deep back muscles.**

Activity 11.8 **The trunk: Muscles of the pelvic floor.**

LABORATORY SUPPLIES

- Skulls
- Skeletons
- Hand mirrors
- Anatomical models, charts, and atlases of human musculature
- Pipe cleaners or some other type of flexible probe
- Metric ruler

PRE-LAB QUIZ

Before you begin, read all the activities in Exercise 11 and the required reading in your textbook that is assigned by your instructor.

1. During this laboratory exercise, you will studying all the following muscles *except*
   a. muscles of the pharynx.
   b. infrahyoid muscles.
   c. muscles of the shoulder.
   d. muscle of mastication.
   e. muscles of the abdominal wall.
2. True or False: The sternocleidomastoid separates the anterior and posterior triangles in the neck. ___________
3. True or False: When a muscle contracts to cause an action at a joint, the origin is the moving attachment of the muscle, and the insertion is the stationary attachment. ___________
4. Most of which group of muscles originate on bone and insert on the skin? ___________
5. The _______________ muscles perform voluntary eye movements.
6. True or False: The rectus abdominis is the primary respiratory muscle. ___________
7. True or False: There are three layers of abdominal oblique muscles. ___________
8. True or False: The deep back (erector spinae) muscles flex the vertebral column. ___________
9. Muscles of the pelvic floor are located in a region known as the _______________.
10. The suprahyoid muscles connect the hyoid bone to the _______________.

Activity **11.1**

Understanding How Skeletal Muscles Produce Movement

A A **lever** is a rigid object that moves on a fixed point, called a **fulcrum,** when a force is applied. In the body, when a muscle contracts, it applies a force that causes a bone to move at a joint. The bone acts as a lever, and the joint serves as a fulcrum.

1 Most skeletal muscles are attached to bones by dense regular connective tissue in the form of cordlike **tendons** or membranous sheets called **aponeuroses** (singular = **aponeurosis**). With few exceptions, muscles cross one or more joints along their paths. When a muscle contracts, it causes an action at the joint(s) that it crosses. When an action occurs, one bony attachment, the **origin,** remains fixed or stationary, whereas the other attachment, the **insertion,** moves.

Which attachment—origin or insertion—acts like a lever? Explain why.

▶ ____

____

____

2 The **masseter** is a muscle of mastication (chewing). It is attached to the lateral surface of the mandible and the zygomatic arch. The muscle elevates the mandible.

Which attachment is the origin, and which is the insertion? ▶

Origin: ____________________________________

Insertion: __________________________________

3 On the diagram on the right, draw an arrow to identify the joint that acts as a fulcrum when the mandible is elevated. ▶

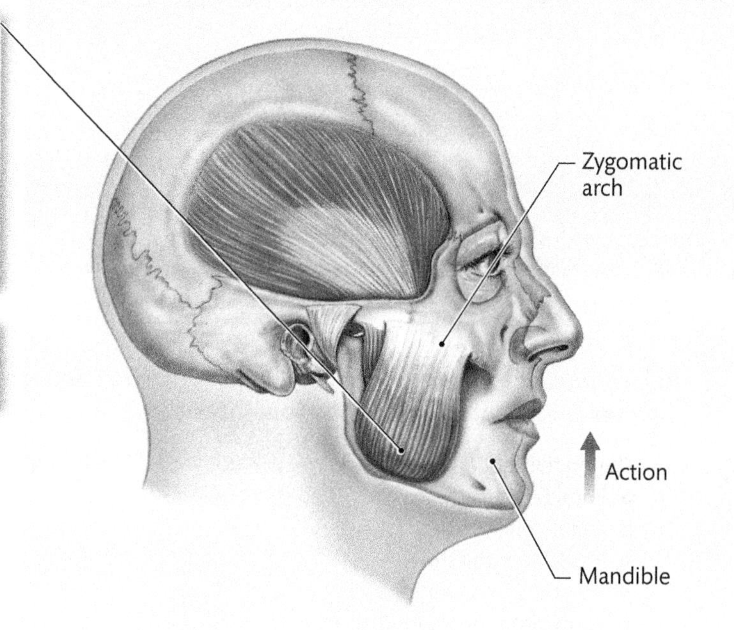

4 The gastrocnemius is a large superficial muscle in the posterior leg (calf). It is attached to the distal end of the femur and to the calcaneus. One of its functions is plantar flexion of the foot, which allows you to stand on your toes. Observe this action on your lab partner and use the information to complete the following table. ▶

Origin of gastrocnemius	
Insertion of gastrocnemius	
Structure acting as a lever	
Structure acting as a fulcrum	

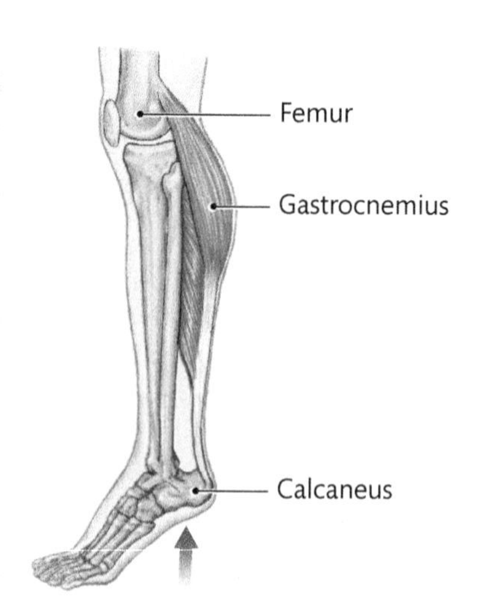

B Muscles can be classified according to their functions.

1 An **agonist,** or prime mover, is a muscle that directly brings about a specific action. For example, the **biceps brachii** is an agonist for flexion of the forearm.

2 An **antagonist** muscle directly opposes a specific action. The **triceps brachii** extends the forearm and is an antagonist muscle that opposes flexion.

3 If extension of the forearm is the action that you are considering, the roles of the biceps brachii and triceps brachii are reversed. Explain why.

▶ ______________________________

____

____

____

____

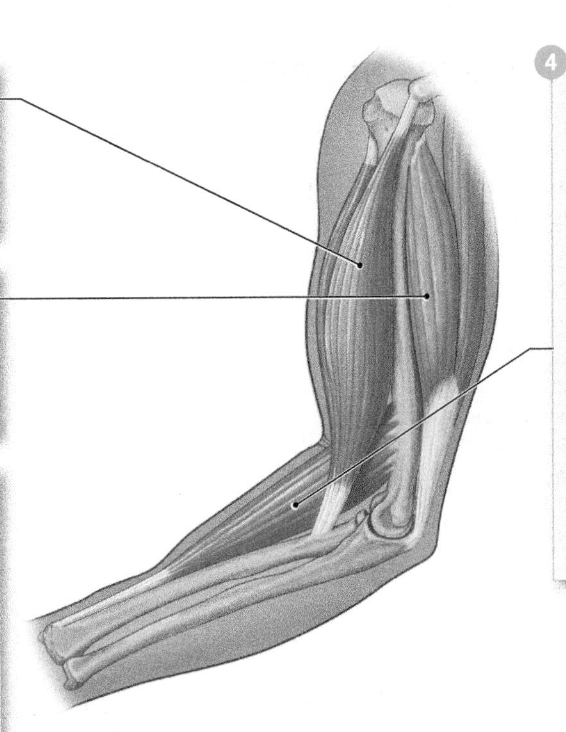

4 Muscles can also act as **synergists** ("working together") by promoting or assisting a specific action or by reducing unnecessary movements while the action is performed. The **brachioradialis** flexes the forearm and stabilizes the elbow joint. For which muscle, the biceps brachii or the triceps brachii, does the brachioradialis act as a synergist? Explain.

▶ ______________________________

____

____

____

MAKING CONNECTIONS

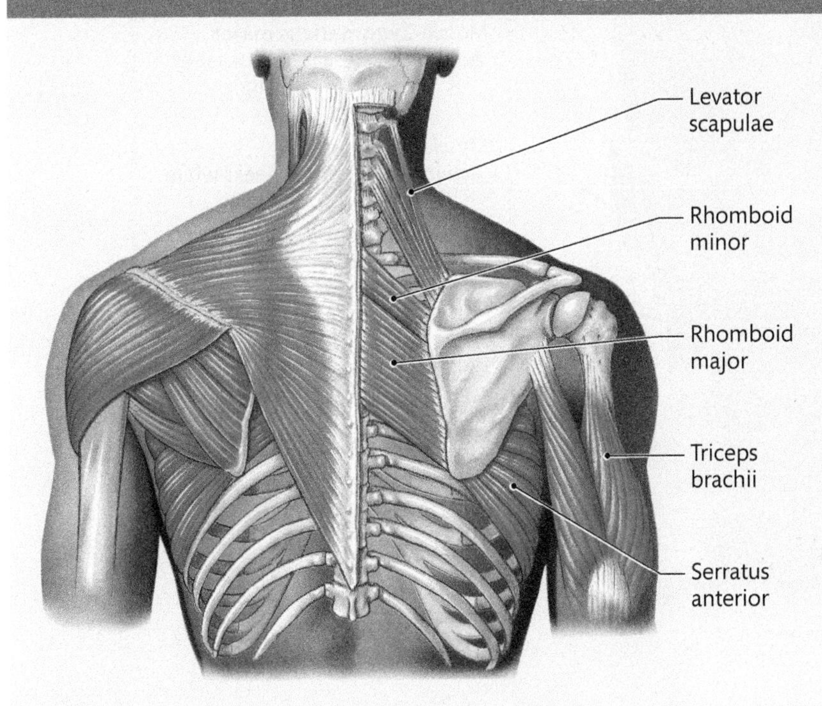

When the triceps brachii extends the arm, the levator scapulae, rhomboids, and serratus anterior serve as synergists. Speculate on the reason why. (*Hint:* All the synergist muscles attach the scapula to the axial skeleton.)

▶ ______________________________

____

____

____

____

____

____

____

____

Activity **11.2**

The Head: Examining Muscles of Facial Expression and Mastication

A **Muscles of facial expression** allow us to move our facial features. They are located deep to the skin of the face, neck, and scalp. Most originate on bone and insert into the skin.

1. Obtain a skull and review the arrangement of the facial bones. Recall that the bones articulate at sutures, which are immovable joints. Thus, rather than initiating movement at a joint, the muscles of facial expression move the skin on the head and neck. These actions change facial expressions as moods change. Identify the four facial bones that are labeled on the diagram to the right. ▶

a. ______________________________

b. ______________________________

c. ______________________________

d. ______________________________

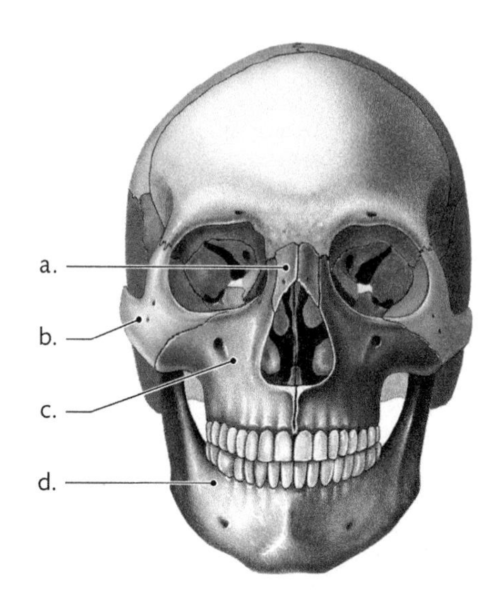

2. On an anatomical model, identify the following muscles of facial expression. Examine the action of each by watching yourself in a mirror or by observing your lab partner.

Muscle: **Frontalis**
Action: Raises eyebrows; wrinkles forehead.

Muscle: **Procerus**
Action: Depresses eyebrow; wrinkles skin over bridge of nose.

Muscle: **Nasalis**
Action: Compresses and dilates nostrils.

Muscle: **Orbicularis oculi**
Action: Closes eyelid during squinting and blinking.

Muscle: **Levator labii superioris**
Action: Elevates upper lip.

Muscle: **Orbicularis oris**
Action: Closes and protrudes lips while speaking, kissing, and whistling.

Muscle: **Platysma**
Action: Tenses skin of the neck; depresses the mandible.

Muscle: **Mentalis** (**cut**)
Action: Protrudes lip; wrinkles skin on chin.

Muscle: **Zygomaticus minor**
Action: Elevates upper lip.

Muscle: **Zygomaticus major**
Action: Elevates lateral corner of mouth for smiling.

Muscle: **Buccinator**
Action: Compresses cheeks while whistling or blowing.

Muscle: **Depressor anguli oris**
Action: Depresses angle of mouth.

Muscle: **Depressor labii inferioris**
Action: Depresses lower lip.

Want more practice? Go to: **MasteringA&P** > Study Area > Menu > Lab Tools > PAL >
■ Anatomical Models > Muscular System > Head and Neck
■ Human Cadaver > Muscular System > Head and Neck

B The four muscles of mastication are involved in chewing food.

1 The muscles of mastication act on the **temporomandibular joint (TMJ),** which is formed by the articulation between the condylar process on the mandible and the mandibular fossa on the temporal bone. Identify the TMJ on the skull.

The TMJ is a synovial joint. What type of synovial joint is it?

▶ ______________________________

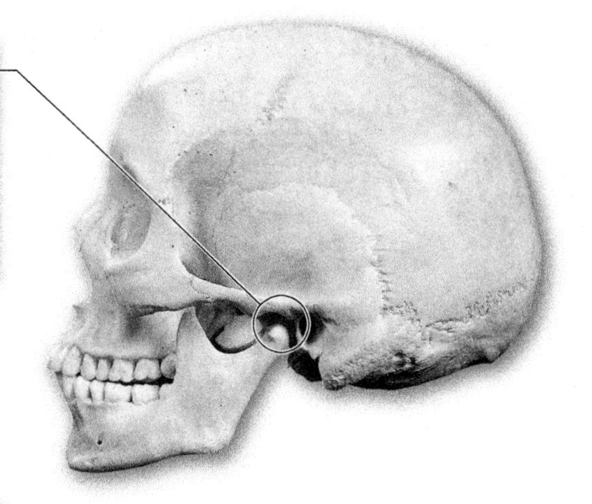

2 On an anatomical model, identify the four muscles of masticaton. On a skull, identify the origin and insertion for each muscle.

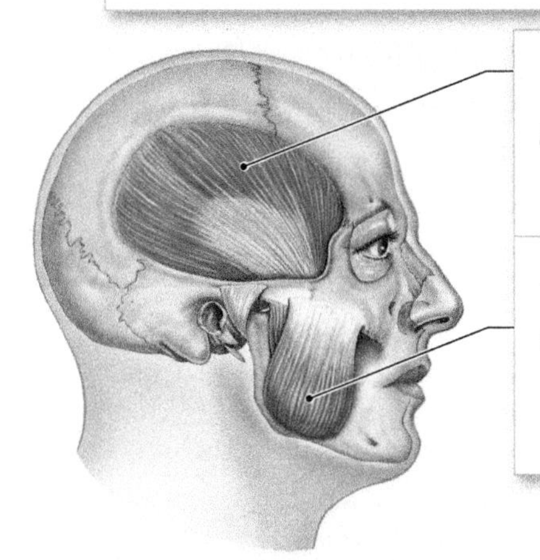

Muscle: **Temporalis**
Origin: Temporal bone
Insertion: Coronoid process of mandible

Muscle: **Masseter**
Origin: Zygomatic arch
Insertion: Lateral surface of mandibular ramus and angle

Muscle: **Lateral pterygoid**
Origin: Lateral pterygoid plate
Insertion: Neck of mandible

Muscle: **Medial pterygoid**
Origin: Lateral pterygoid plate
Insertion: Medial surface of mandibular ramus and angle

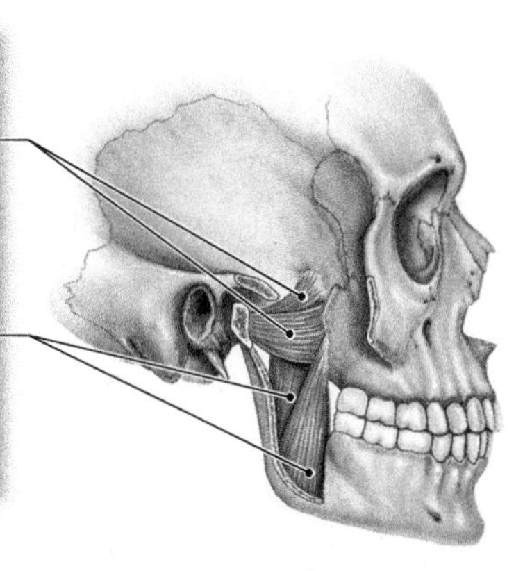

3 On a skull, arrange a pipe cleaner along the path of each muscle of mastication by connecting its origin and insertion. Based on the observed path that each muscle takes, speculate on its action in the table below. ▶

Muscle	Action
Temporalis	
Masseter	
Lateral pterygoid	
Medial pterygoid	

MAKING CONNECTIONS

Bell's palsy is a viral infection that causes an inflammation of the facial nerves and temporary paralysis of muscles of facial expression. Speculate on what might occur if Bell's palsy affected the left facial nerve of an individual but the right facial nerve functioned normally.

▶ ______________________________

Activity **11.3**

The Head: Examining Extrinsic Eye Muscles, Extrinsic Tongue Muscles, and Muscles of the Pharynx

A The six **extrinsic eye muscles** are responsible for voluntary eye movements.

1. On a skull, identify the two bony orbits. Observe the opening of the **optic canal** along the posterior wall of each orbit. The optic nerve passes through the canal to reach the eye.

2. All but one of the extrinsic eye muscles originate from a tendinous ring that surrounds this opening. From their origins, these small muscles insert onto the outer layer of the eyeball. On a model or a skull, locate the position of the tendinous ring.

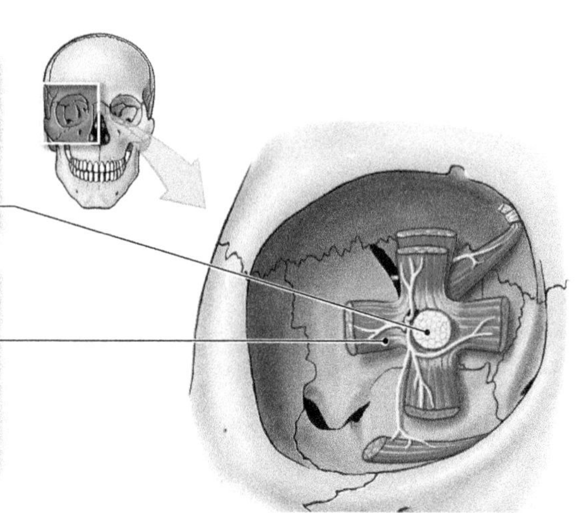

3. On an anatomical model, identify the six extrinsic muscles of the eye. Perform the action that is described for each.

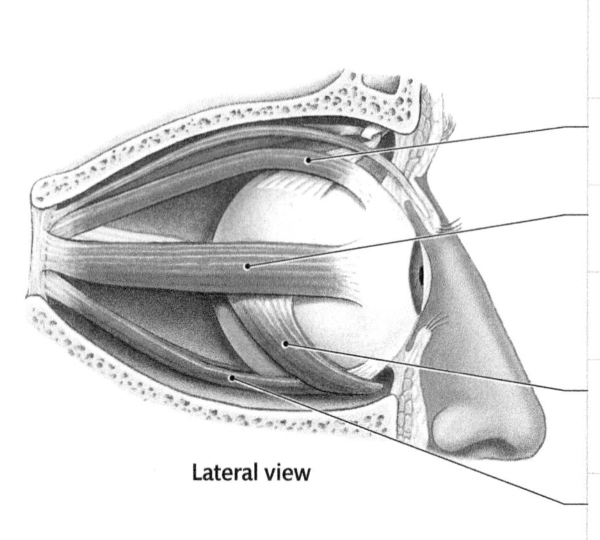

Lateral view

Muscle: **Superior oblique**
Action: Rotates eyeball inferiorly and laterally.

Muscle: **Superior rectus**
Action: Rotates eyeball superiorly.

Muscle: **Lateral rectus**
Action: Rotates eyeball laterally.

Muscle: **Medial rectus**
Action: Rotates eyeball medially.

Muscle: **Inferior oblique**
Action: Rotates eyeball superiorly and laterally.

Muscle: **Inferior rectus**
Action: Rotates eyeball inferiorly.

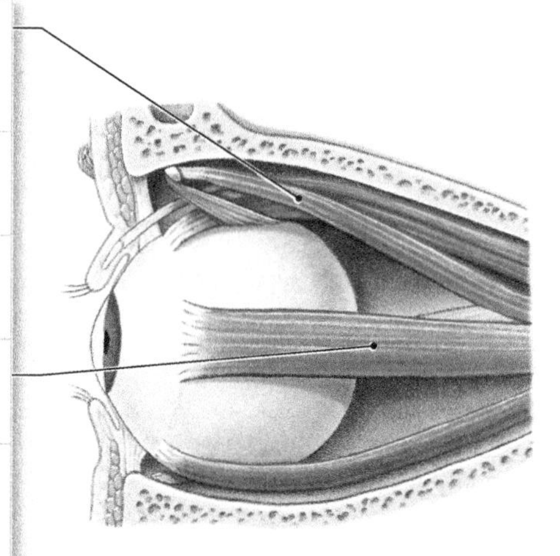

Medial view

4. Which extrinsic eye muscle does not originate from the tendinous ring around the opening of the optic canal?

▶ ______________________________

Where does this muscle originate?

▶ ______________________________

____

5. Observe the unique path of the superior oblique. This muscle runs along the superior surface of the eyeball, angling medially toward the nose. At this position, the muscle's tendon travels through a pulley-like structure called the trochlea before curving laterally to insert on the superior surface of the eye.

B The four **extrinsic tongue muscles** are involved in voluntary tongue movements during chewing, swallowing, and speaking. The muscles are illustrated in the diagram below.

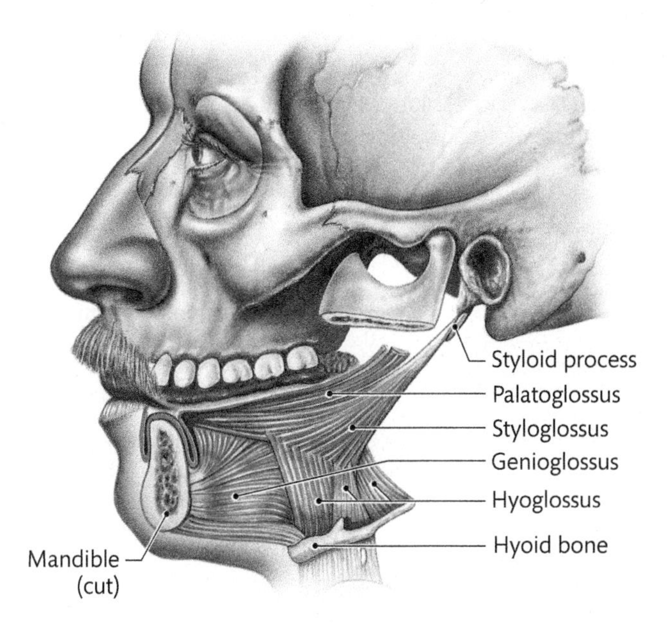

1 The names of the extrinsic tongue muscles provide clues for identifying their attachment sites. For example, *glossus* refers to the tongue. The terms *palato, stylo, genio,* and *hyo* refer to the soft palate, styloid process, mandible, and hyoid bone, respectively. Use this information to identify the origins and insertions of the extrinsic tongue muscles in the table below. ▶

Muscle	Origin	Insertion
Palatoglossus		
Styloglossus		
Genioglossus		
Hyoglossus		

2 On a skeleton, hold a pipe cleaner to each muscle's origin and extend it to where the tongue would be. The various tracks of the pipe cleaners approximate the paths of the four extrinsic tongue muscles.

C The muscles of the pharynx are necessary for swallowing.

1 On an anatomical model, identify the muscles of the pharynx. Notice that many of these muscles support the larynx by suspending it from the base of the skull. They are organized into three groups:

Muscles of the soft palate include **levator veli palatini** and **tensor veli palatini.** They elevate and tighten the soft palate during swallowing.

Laryngeal elevators elevate the larynx during swallowing.

Pharyngeal constrictors include the **superior, middle,** and **inferior pharyngeal constrictors.** They form a circular layer of muscle around the wall of the pharynx.

Thyroid cartilage

2 Observe your lab partner while he or she is swallowing. Notice that during the swallowing process, the larynx is elevated toward the pharynx. You can observe this action by identifying the thyroid cartilage along the anterior aspect of the neck and watching it move superiorly. The act of swallowing also involves the actions of muscles that elevate the soft palate and constrict the pharynx.

MAKING CONNECTIONS

Identify at least three daily activities that require normal function of the extrinsic tongue muscles.

▶ ______________________________

____

____

____

____

____

Activity **11.4**

Examining Muscles of the Neck

A The most prominent muscle in the neck is the sternocleidomastoid, which is a relatively long and superficial straplike muscle on each side of the neck. It is an important anatomical landmark because it separates the neck on each side into two triangular-shaped regions known as the **anterior** and **posterior triangles.**

1. Using an anatomical model, identify the **sternocleidomastoid.**

2. On a skeleton, identify the two origins of the sternocleidomastoid: the manubrium of the sternum and the medial end of the clavicle.

   Identify the muscle's insertion on the mastoid process of the temporal bone.

Temporalis
Frontalis
Masseter
Posterior triangle
Anterior triangle

Lateral view

3. Sit in a relaxed position, with your head facing forward. Gently place your fingers on the sternocleidomastoid muscle on the right side of your neck. As you turn your head to the left and look up, feel this muscle contract. In this position, palpate the entire length of the sternocleidomastoid from the mastoid process at the base of the skull to the manubrium and clavicle. Describe the action of the sternocleidomastoid.

   ▶ ______________________________

   ____

   ____

B The three scalene muscles are found in the posterior triangle of the neck.

1. On a model identify the **anterior,** **middle,** and **posterior scalene muscles.**

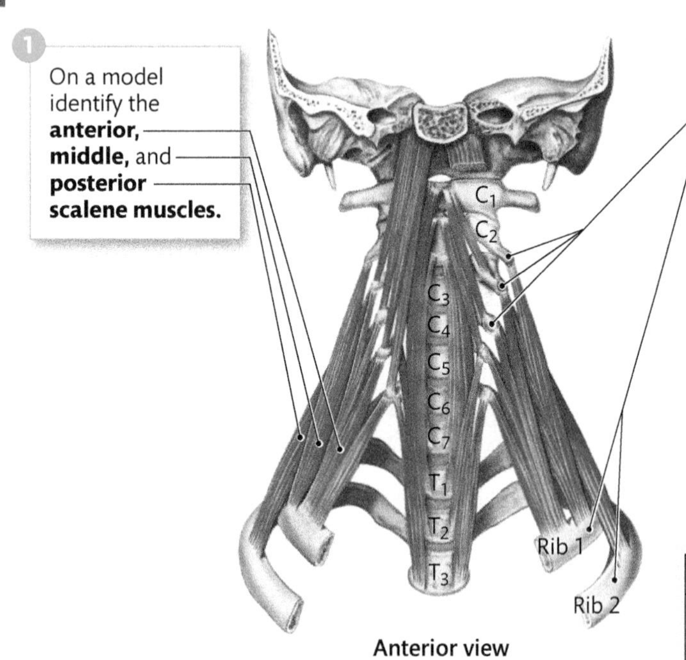

Anterior view

2. On a skeleton, identify the attachments of the scalenes. They extend from the transverse processes of cervical vertebrae to ribs 1 and 2. Connect the attachments with pipe cleaners to demonstrate the paths of these muscles.

3. Place your fingers on the right side of your neck just posterior to the sternocleidomastoid, in the posterior triangle. Feel your scalenes contract as you flex your neck to the right.

   During forced inhalation, your ribs are elevated. With your fingers in the same position, breathe in deeply and feel your scalenes contract.

   For each action, identify the attachment that served as the origin and the attachment that served as the insertion in the table below. ▶

Action	Origin	Insertion
Flex the neck		
Elevate the ribs		

C Muscles of the anterior triangle include two groups of muscles.

- The **suprahyoid muscles** are superior to the hyoid bone and connect it to the skull.
- The **infrahyoid muscles** are inferior to the hyoid bone and connect it to the sternum, clavicle, and scapula.

1 On an anatomical model, identify the muscles found in the anterior triangle. On a skeleton, identify the origin and insertion for each muscle.

IN THE CLINIC

Injury to the Sternocleidomastoid

The nerve that stimulates the sternocleidomastoid branches through the posterior triangle of the neck. It is vulnerable to injury during surgical procedures in this area (e.g., a biopsy of lymph nodes to check for cancer). Damage to this nerve can lead to weakness or loss of muscle function. Symptoms include difficulty in rotating the head to the opposite side or laterally flexing the neck to the same side.

Suprahyoid Muscles

Muscle: **Geniohyoid**
Origin: Medial surface of mandible at chin
Insertion: Hyoid bone

Muscle: **Mylohyoid**
Origin: Medial surface of mandible
Insertion: Hyoid bone

Muscle: **Stylohyoid**
Origin: Styloid process of temporal bone
Insertion: Hyoid bone

Muscle: **Digastric**
Origin of posterior belly: Mastoid region of temporal bone
Origin of anterior belly: Inferior margin of mandible
Insertion: Hyoid bone

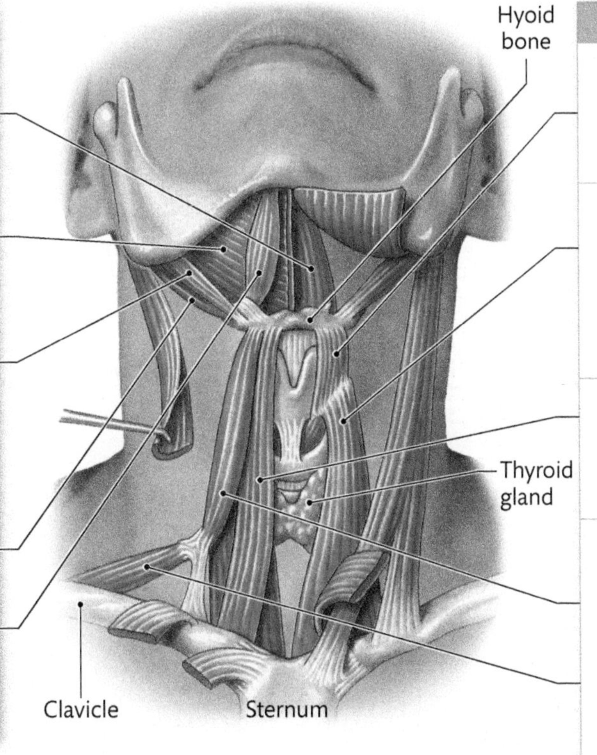

Infrahyoid Muscles

Muscle: **Thyrohyoid**
Origin: Thyroid cartilage of larynx
Insertion: Hyoid bone

Muscle: **Sternothyroid**
Origin: Manubrium of sternum and first costal cartilage
Insertion: Thyroid cartilage of larynx

Muscle: **Sternohyoid**
Origin: Clavicle and manubrium of sternum
Insertion: Hyoid bone

Muscle: **Omohyoid**
Origin of superior belly: By fascia, from clavicle and first rib
Origin of inferior belly: Superior border of scapula
Insertion: Hyoid bone

2 On a skeleton, use a pipe cleaner to connect the origin of each suprahyoid muscle with its insertion. Visualize each muscle's path. As a group, what common function do you think these muscles have during swallowing and speaking?

▶ ____________________

____

Repeat the procedure, described above, for the infrahyoid muscles. What would be their common function during swallowing and speaking?

▶ ____________________

____

MAKING CONNECTIONS

Based on the muscular anatomy of the anterior triangle, explain why the hyoid bone and larynx move simultaneously during swallowing and speaking.

▶ ____________________

____

____

____

____

____

____

Activity 11.5

The Trunk: Examining Muscles of the Thorax

A The intercostal spaces (between the ribs) contain three layers of **intercostal muscles,** which play an important role in respiration (breathing).

1. On a skeleton, identify the **intercostal spaces.** The intercostal muscles that fill these spaces attach to adjacent ribs.

2. Along the lateral aspect of one intercostal space, position a pipe cleaner so that it begins at the superior rib and travels inferiorly and anteriorly to the inferior rib. This pipe cleaner represents the **external intercostal muscle** (red arrow). In each intercostal space, this muscle originates on the superior rib and inserts on the inferior rib.

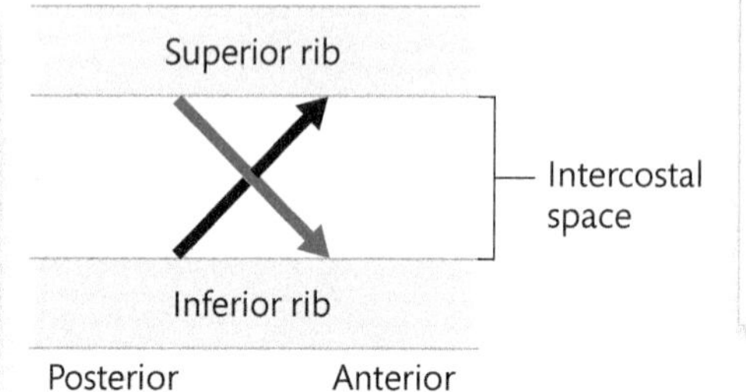

3. In the same intercostal space, place a second pipe cleaner so that it begins at the inferior rib and travels superiorly and anteriorly to the superior rib. The second pipe cleaner represents the **internal intercostal muscle** (black arrow). In each intercostal space, this muscle originates on the inferior rib and inserts on the superior rib. Notice that the internal intercostal muscles are roughly perpendicular to the external intercostal muscles.

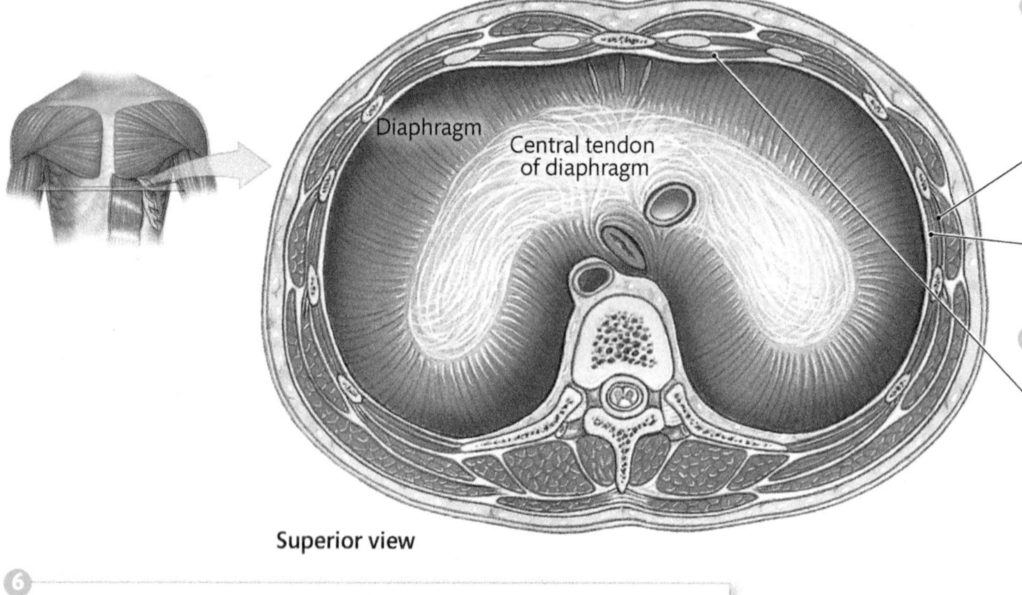

Superior view

4. On a model that shows the thoracic wall in cross section, identify the first two layers of intercostal muscles.

As a group, the **external intercostal muscles** elevate the ribs.

As a group, the **internal intercostal muscles** depress the ribs.

5. The third and deepest layer of intercostal muscles is divided into three groups:

Transversus thoracis muscles, anteriorly

Innermost intercostal muscles, laterally (not shown)

Subcostal muscles, posteriorly (not shown)

These muscles assist the internal intercostals in depressing the ribs.

6. The intercostal muscles have an important role in respiration. To investigate this function, place fingers from each hand on your ribs:

- Inhale quietly. You will feel the external intercostal muscles elevating your ribs.
- Exhale quietly. You will feel your ribs being depressed. Quiet exhalation occurs passively (without the assistance of muscle activity).
- Breathe normally for a few cycles, and then exhale forcefully. You will feel your ribs being depressed even more by the internal intercostal muscles. The transversus thoracis, innermost intercostal, and subcostal muscles are synergists for rib depression.

B The **diaphragm** is a musculotendinous partition that separates the thoracic and abdominopelvic cavities. It has a concave surface that faces the abdominal cavity (see the figure below) and a convex surface that faces the thoracic cavity and forms two domes (see the bottom figure on the previous page). The diaphragm is the primary respiratory muscle.

Word Origins

Diaphragm is derived from Greek and means "partition or wall." The diaphragm forms a wall between the thoracic and abdominopelvic cavities.

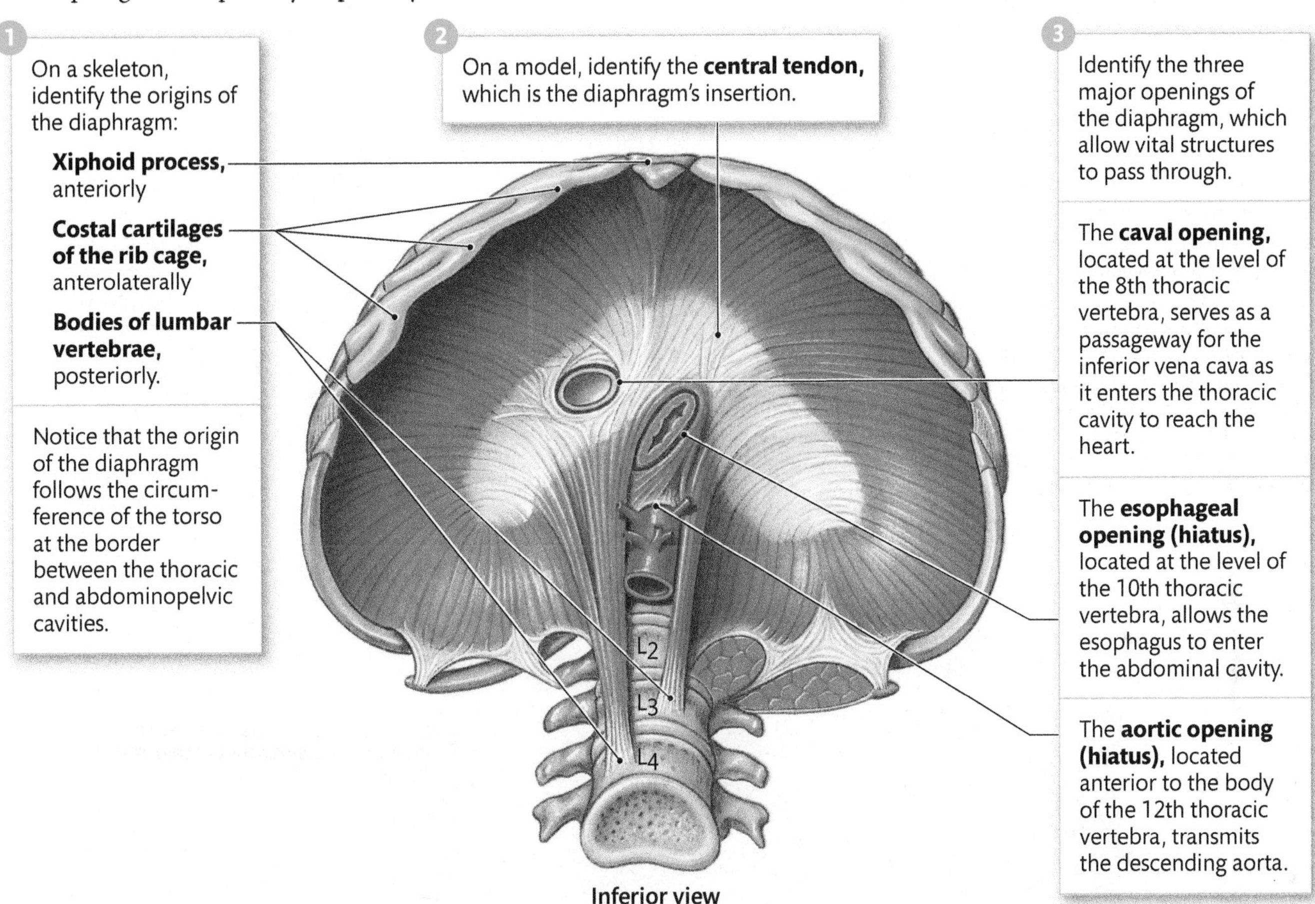

Inferior view

IN THE CLINIC

Diaphragmatic Hernia

A congenital diaphragmatic hernia occurs about once in every 2200 births. It is typically caused by a structural defect in the posterolateral aspect of the diaphragm, usually on the left side. As a result, abdominal contents (e.g., portions of the small intestine) are displaced into the thoracic cavity. The condition can inhibit normal lung development and lead to life-threatening breathing difficulties. Surgery is required to place the abdominal organs in their proper position and to repair the damage in the diaphragm.

MAKING CONNECTIONS

Unlike the caval and esophageal openings, the aortic opening does not pass through the diaphragm. Instead, it is a passageway that is posterior to the diaphragm. Explain why this distinction is significant.

▶ ______________________________

____

____

____

____

____

____

Want more practice? Go to: **MasteringA&P** > Study Area > Menu > Lab Tools > PAL >

- Anatomical Models > Muscular System > Trunk
- Human Cadaver > Muscular System > Trunk

Activity **11.6**

The Trunk: Examining Muscles of the Abdominal Wall

A The muscles of the abdominal wall form a strong, multilayered enclosure that protects vulnerable abdominal organs from injury. Included in this group are the long, vertical rectus abdominis muscles and three pairs of broad, flat oblique muscles.

1 On a torso model, identify the **rectus abdominis** muscles, which cover the anterior abdominal wall.

Origin: Pubic symphysis and pubic crest
Insertion: Xiphoid process of the sternum and costal cartilages of ribs 5 through 7.

2 Identify the **rectus sheath,** which is formed by the aponeuroses of the three oblique muscles. The rectus sheath encloses the rectus abdominis muscles.

3 The **linea alba** is formed by the interlacing fibers of the rectus sheath and separates the two rectus abdominis muscles.

4 Three layers of oblique muscles cover the antero-lateral aspect of the abdominal wall. Notice that the muscle fibers in each layer travel in different directions. From superficial to deep, they are:

Muscle: **External oblique**
Origin: External surfaces of inferior eight ribs
Insertion: Linea alba, pubic tubercle, iliac crest

Muscle: **Internal oblique**
Origin: Thoracolumbar (lumbodorsal) fascia, iliac crest
Insertion: Linea alba, inferior three or four ribs, pubic crest

Muscle: **Transversus abdominis**
Origin: Thoracolumbar (lumbodorsal) fascia, costal cartilages of inferior six ribs, iliac crest
Insertion: Linea alba, pubis

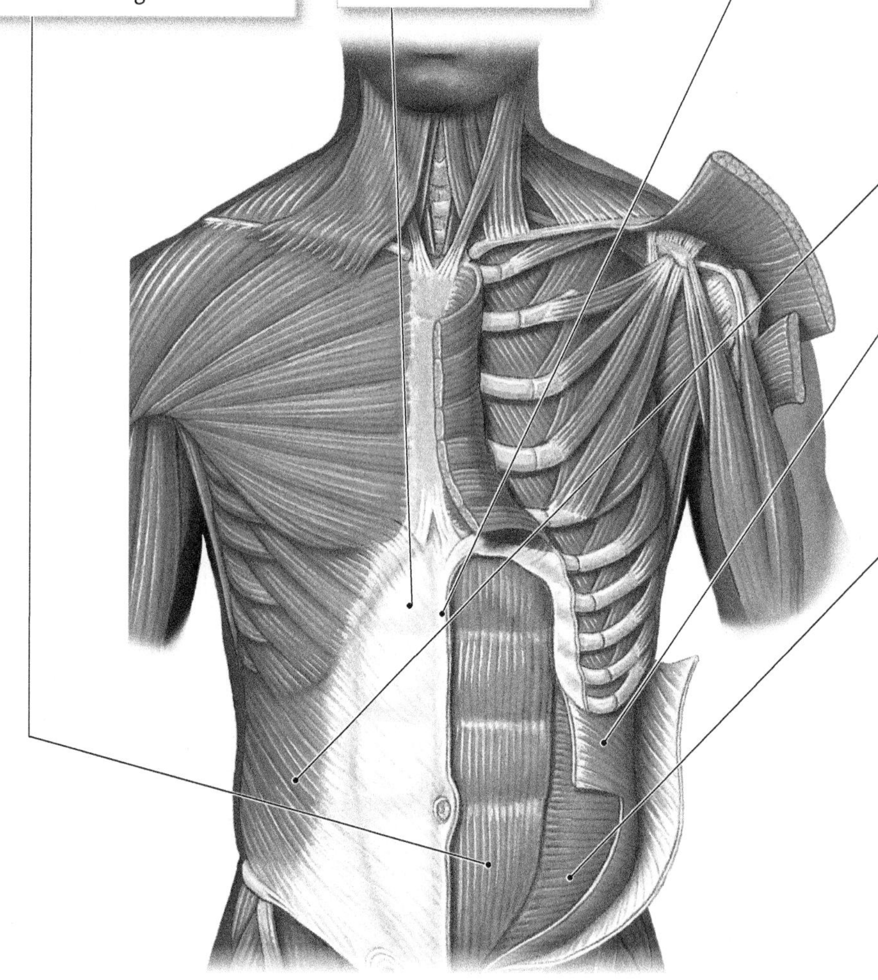

Anterior view

Want more practice? Go to: **MasteringA&P** > Study Area > Menu > Lab Tools > PAL >
- Anatomical Models > Muscular System > Trunk
- Human Cadaver > Muscular System > Trunk

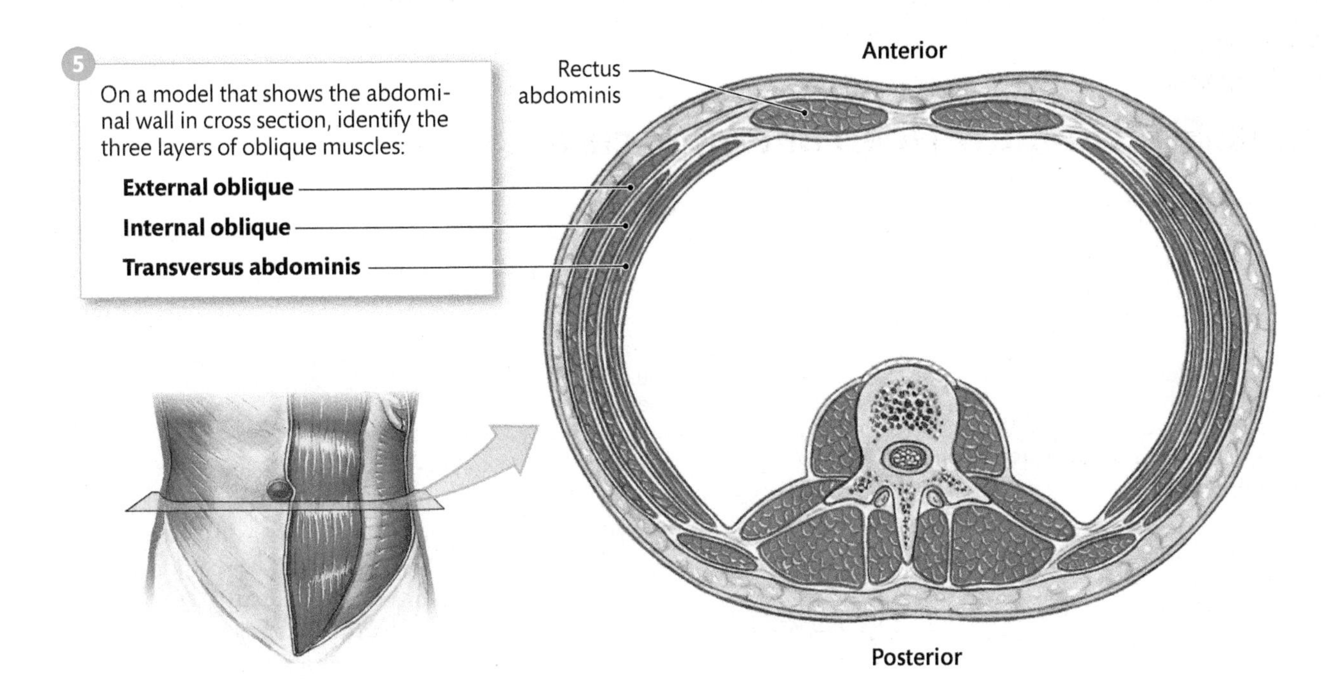

B The abdominal muscles move the vertebral column in several directions and are important respiratory muscles. Perform the following movements while your lab partner observes:

1. From the anatomical position, flex your vertebral column by bending forward at the waist. The rectus abdominis is the prime mover for **flexion of the vertebral column** in the thoracic and lumbar regions.
2. Laterally flex your vertebral column by bending to the right and left at the waist. Then, rotate your body at the waist. **Lateral flexion and rotation of the torso** are actions of the external and internal oblique muscles.
3. Sit or stand quietly with your fingers resting on the anterior wall of your abdomen. Breathe normally for a few cycles, and then exhale forcefully. During the forced expiration, you will feel your rectus abdominis contract. Contractions of the rectus abdominis and the other abdominal wall muscles push abdominal organs up against the diaphragm, which increases internal thoracic pressure and helps force air out of the lungs.

Word Origins

The names of the abdominal muscles provide clues about their orientation. *Rectus* comes from Latin and means "straight." The muscle fibers of rectus abdominis travel in a straight, vertical path. *Oblique* also derives from Latin and means "diagonal" or "slanting." The fibers of the external and internal oblique muscles travel diagonally across the anterolateral abdominal wall. *Transversus* derives from Latin and French and means "to turn across." The fibers of transversus abdominis travel horizontally across the abdominal wall.

MAKING CONNECTIONS

Speculate on the significance of the muscle fibers in the three oblique muscle layers traveling in different directions.

▶ ____

____

____

____

____

____

____

____

____

____

Activity **11.7**

The Trunk: Studying the Deep Back Muscles

A The **deep back muscles** extend, rotate, and laterally flex the vertebral column and head, and they have an important role in maintaining normal posture. They include the splenius muscles, the erector spinae muscles, and the transversospinalis muscles.

1. The **splenius muscles** originate from the spinous processes of vertebrae C_7 through T_6 and the ligamentum nuchae, a fibrous cord that connects the spinous processes of the cervical vertebrae, and insert on the transverse processes of C_1 to C_3, mastoid process, and occipital bone. The splenius muscles include:

2. The **erector spinae muscles** are comprised of three muscular bands that travel vertically from the sacrum to the posterior surface of the skull, along each side of the vertebral column. From lateral to medial, the three muscles are:

3. The **transversospinalis muscles** include a number of short muscles that are located deep to the erector spinae in the grooves formed by the spinous and transverse processes of vertebrae. The transversospinalis consists of the following muscle groups:

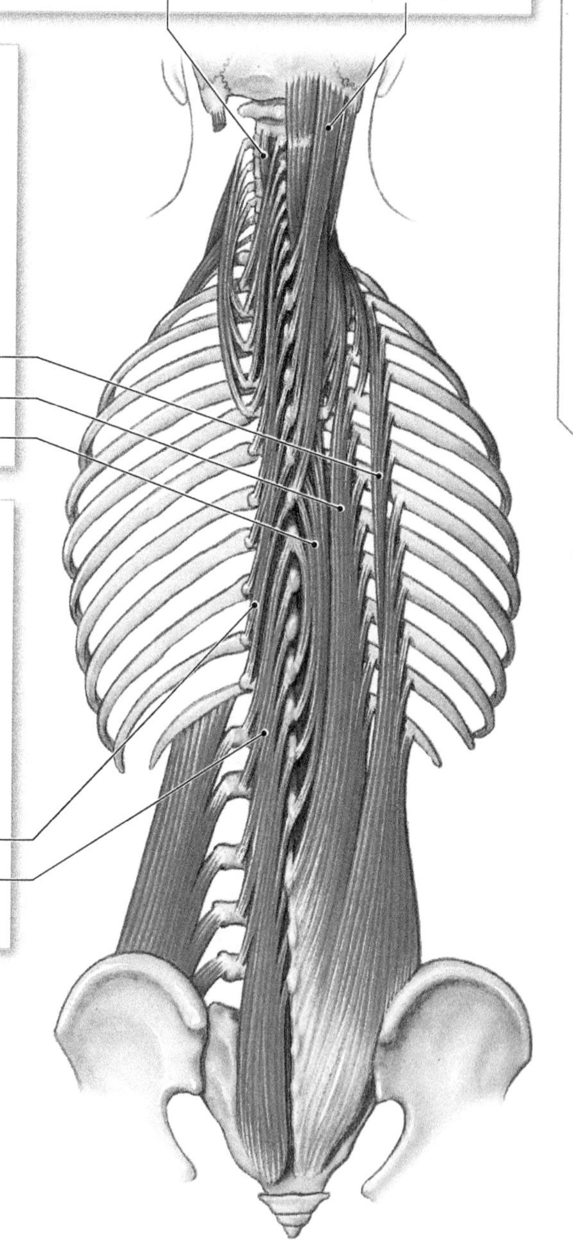

IN THE CLINIC

Back Pain

Sudden or excessive rotation or extension of the vertebral column can strain back muscles. Back injuries usually occur in the lumbar region (lower back), perhaps while participating in sports or incorrectly lifting heavy objects. The disproportionate force tears muscle fibers in the erector spinae and damages associated back ligaments. Often, the injured muscles go into spasm, causing pain and loss of function. Most back pain can be treated by applying ice and taking nonsteroidal anti-inflammatory drugs, followed by physical therapy. In addition, clinical evidence has shown that chiropractic and acupuncture treatments are effective in relieving some types of back pain.

4. Eighty percent of Americans will experience back pain at least once in their lives. It is one of the most common reasons people see a doctor and a major medical concern for college students who carry heavy backpacks to their classes. Identify at least two ways that back injuries can be prevented.

▶ ______________________________

____

____

____

____

____

____

B Most anatomical models do not provide good views of the deep back muscles. They can, however, be demonstrated by identifying their attachments on a skeleton.

1 On a skeleton, identify the posterior aspect of the sacrum, iliac crest, spinous processes of T_{11} through T_{12}, and all five lumbar vertebrae. These skeletal structures are the bony origins of the erector spinae muscles.

2 Link pipe cleaners together to make chains of different lengths. Position the ends of these chains on the posterior surface of the sacrum and extend them vertically to the various insertions of the erector spinae:

- **Iliocostalis:** angles of the ribs or transverse processes of C_4 through C_6 (red arrow in diagram projects to a rib angle).
- **Longissimus:** transverse processes of thoracic and cervical vertebrae, tubercles of the inferior nine ribs, or the mastoid process (yellow arrow projects to mastoid process).
- **Spinalis:** spinous processes of thoracic or cervical vertebrae (black arrow projects to a thoracic vertebra).

3 Examine the connections of the **transversospinalis.** On a skeleton, place one end of a pipe cleaner on the transverse process of any vertebra and extend it superiorly by one to three vertebrae to a spinous process (blue arrows). Notice that the transversospinalis muscles are within the bony groove formed by the serial arrangement of transverse and spinous processes.

4 Bend forward (flex) at the waist. As you slowly straighten (extend) the spine, concentrate on feeling the pull in your back, along the vertebrae. You are feeling the collective actions of the deep back muscles.

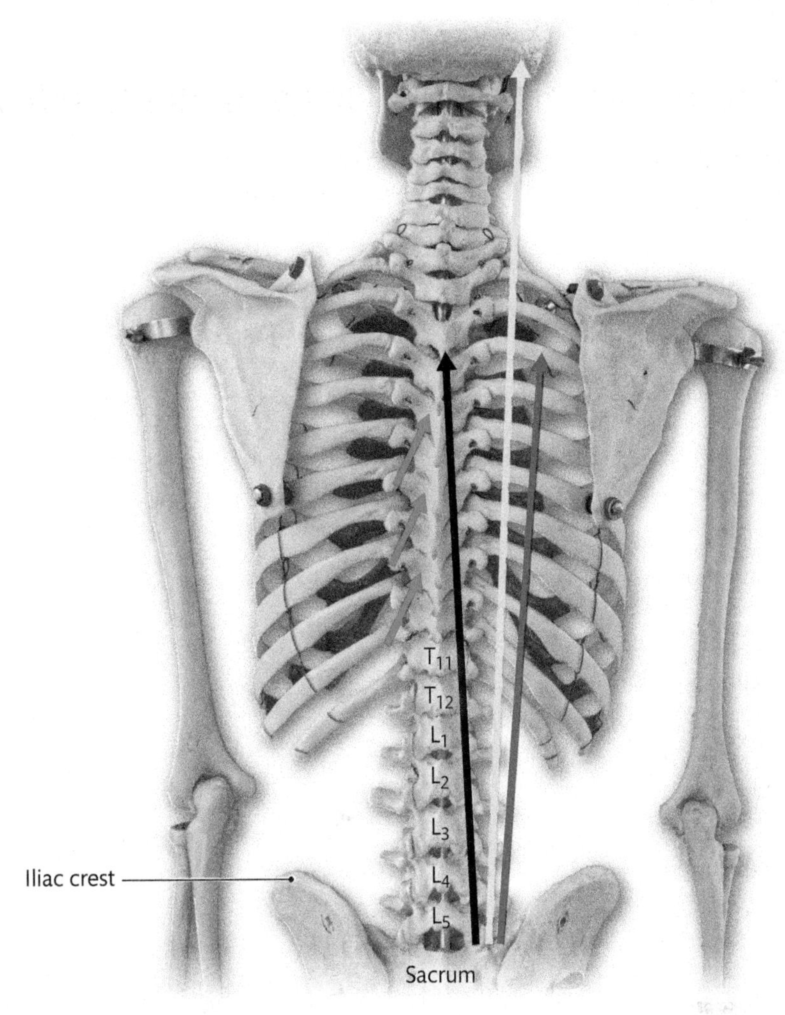

Posterior view

MAKING CONNECTIONS

Acting together, the erector spinae muscles are the primary extensors of the vertebral column and the head. The transversospinalis muscles are synergists for this action. Identify a muscle that acts as an antagonist for this action and explain why it is.

▶ ______________________________

Activity **11.8**

The Trunk: Examining Muscles of the Pelvic Floor

A The urogenital and pelvic diaphragms form a funnel-shaped muscular layer that supports the organs in the pelvic cavity and closes the outlet of the bony pelvis.

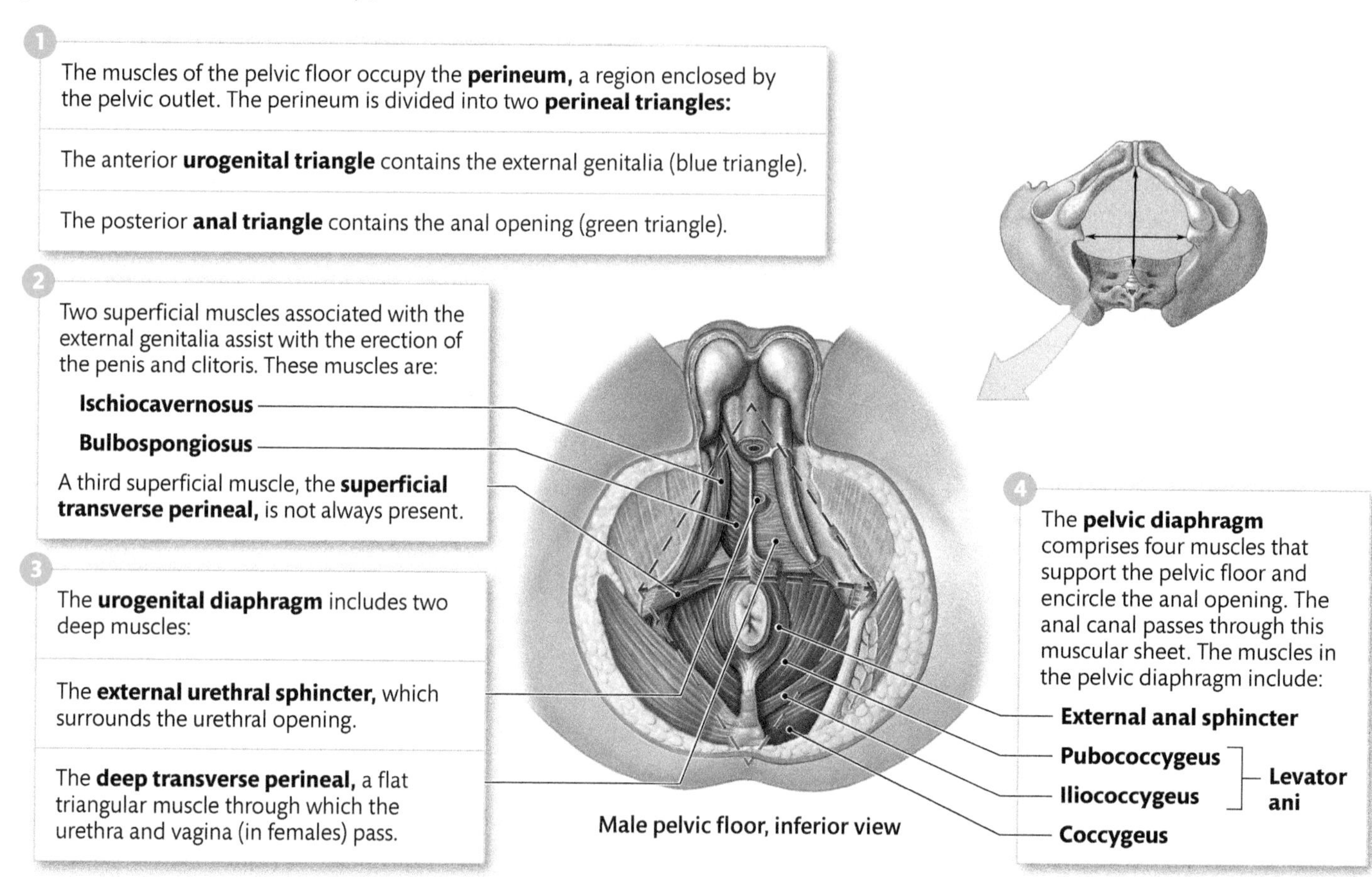

Male pelvic floor, inferior view

B In this diagram of the female pelvic floor, identify the same muscles that were described above in the male.

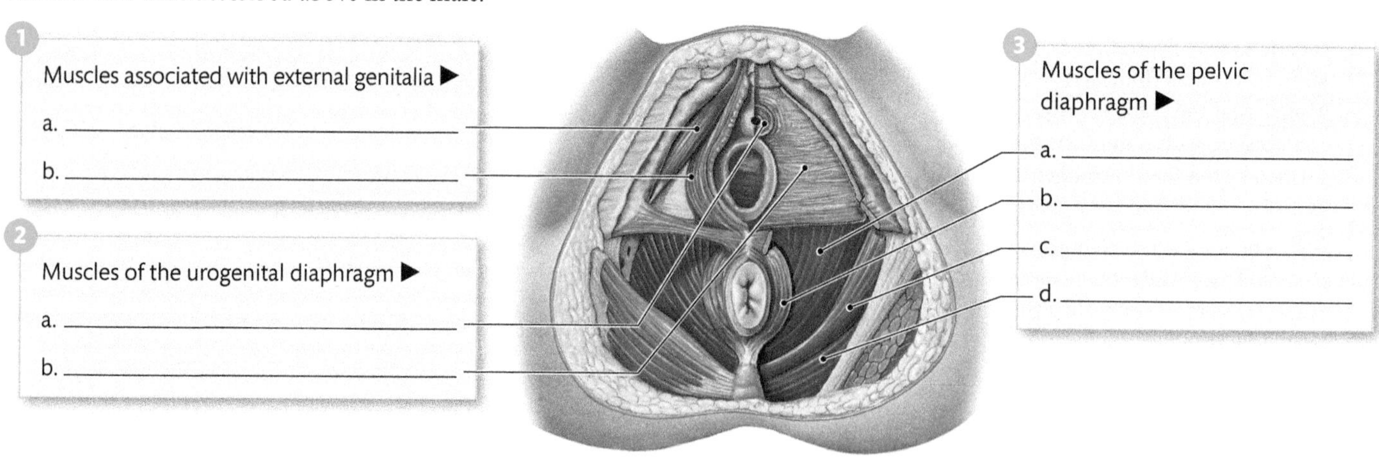

Female pelvic floor, inferior view

C Most anatomical models do not provide good views of the pelvic floor muscles, but some of these muscles can be demonstrated by identifying their attachments on the bony pelvis.

1. On a skeleton, identify the bony margins of the diamond-shaped perineum:
   - Two ischial tuberosities
   - Pubic symphysis
   - Coccyx

   Next, identify the ischial spines.

2. Position a ruler to connect the ischial tuberosities. Doing so divides the perineum into two perineal triangles.
   - Urogenital triangle (anterior)
   - Anal triangle (posterior)

3. Demonstrate the positions of the following muscles by connecting their attachments with pipe cleaners:

   Urogenital diaphragm
   - **Deep transverse perineal:** Position a pipe cleaner so that it connects the two ischial rami (purple line).

   Pelvic diaphragm
   - **Pubococcygeus:** Extend a pipe cleaner from the internal margins of the pubis to the coccyx (green arrow).
   - **Iliococcygeus:** Position a pipe cleaner so that it connects the ischial spine and coccyx (red arrow).
   - **Coccygeus:** Connect the ischial spine and the inferior border of the sacrum with a pipe cleaner (blue arrow).

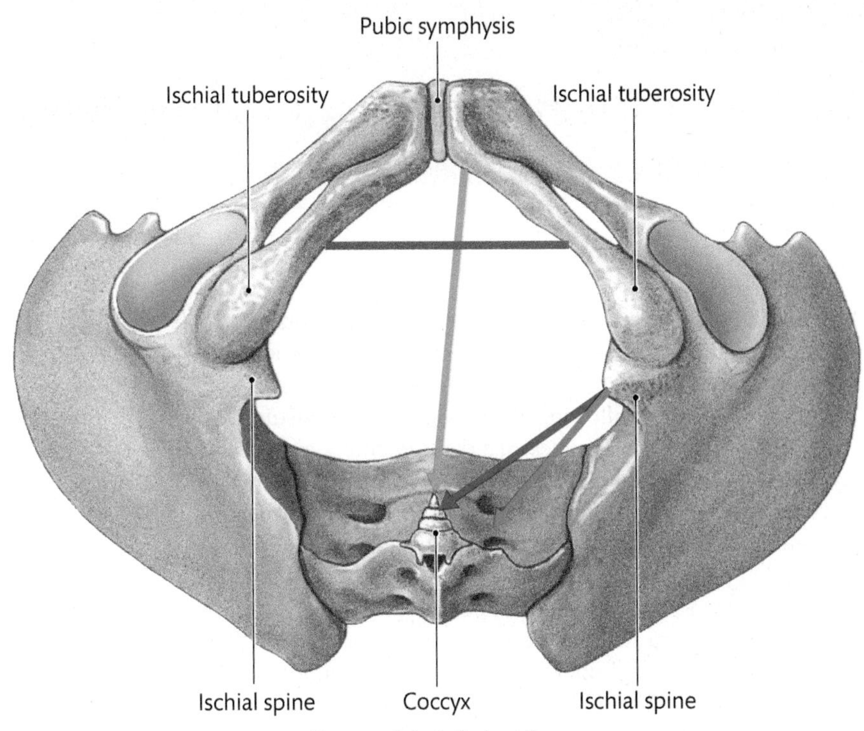

Bony pelvis, inferior view

MAKING CONNECTIONS

Speculate on complications that could develop if the levator ani muscles (pubococcygeus and iliococcygeus) were damaged or weakened.

▶ ____

BEFORE YOU MOVE ON . . .

« LOOKING BACK

The muscles of the head, neck, and trunk—the axial muscles—include about 60 percent of all skeletal muscles in the body. In general, they are important for supporting the head and vertebral column and allowing us to maintain erect posture. In this laboratory exercise, you learned that axial muscles in specific regions have other more specific functions, many that we might take for granted. For example, muscles in the head allow us to move our eyes and make facial expressions that convey emotions such as happiness, surprise, grief, and sadness. Other muscles in the head allow us to chew and manipulate food, and muscles in the neck allow us to swallow the food we have just chewed. Some trunk muscles have an important respiratory function, and others support urinary and reproductive organs in the pelvic cavity.

Make a list of all the organ systems that are functionally connected to the axial muscles. Then, consider how your quality of life would be affected if any of these muscles were not functioning properly. For instance, how would your life be different if you could not make facial expressions?

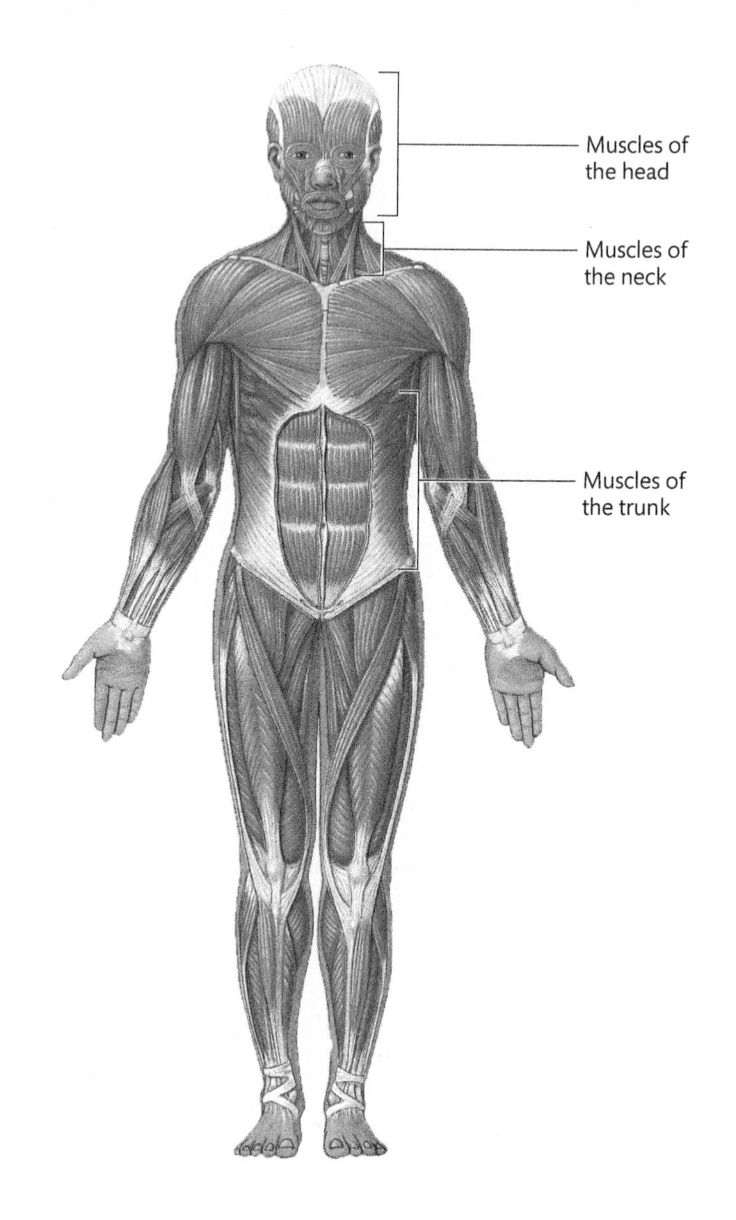

LOOKING FORWARD »

In the next exercise, you will study the structure and function of the **appendicular muscles,** which include the muscles of the upper and lower limbs. The muscles of the upper limb are essential for our ability to carry out fine motor skills. The muscles of the lower limb are important for locomotion, support, and erect posture. Some appendicular muscles that act at the shoulder and hip joints originate on bones of the axial skeleton and insert on bones of the appendicular skeleton. When you study these muscles, you will observe the structural and functional connections between the axial and appendicular regions of the body.

Name ______________________

Lab Section ______________________

Date ______________________

EXERCISE 11 REVIEW SHEET

Muscles of the Head, Neck, and Trunk

1. The rectus abdominis covers the anterior abdominal wall. Its superior attachment includes the xiphoid process and costal cartilages of ribs 5 through 7; its inferior attachment includes the pubic symphysis and pubic crest. The muscle depresses the ribs and flexes the vertebral column. Which attachment (superior or inferior) is the origin, and which is the insertion? Explain.

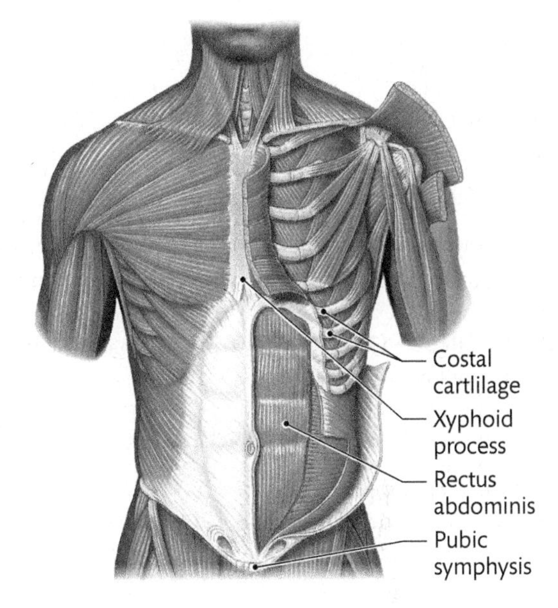

2. When the hyoid bone is elevated, the suprahyoid muscles are the agonists. What muscles act as antagonists? Explain.

3. The splenius capitis is a deep muscle on the posterior surface of the neck. It is attached to the occipital bone and to spinous processes of cervical and thoracic vertebrae. One of its functions is to extend the head. Use this information to complete the following table.

Origin of splenius capitis	
Insertion of splenius capitis	
Structure acting as a lever	
Structure acting as the fulcrum	

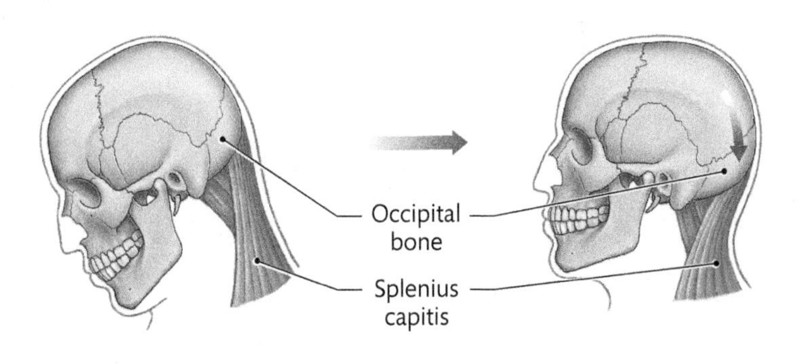

QUESTIONS 4–8: Identify the muscle of facial expression that performs the action that is described. Choose your answers from the following list. Answers may be used once or not at all.

a. frontalis
b. orbicularis oculi
c. buccinator
d. platysma
e. zygomaticus major
f. depressor anguli oris
g. orbicularis oris
h. mentalis

4. When you smile, the ______________________ is the primary muscle that elevates the corners of your mouth.

5. Close one eye by blinking or squinting. This movement is initiated by the ______________________, a circular muscle that surrounds the eye.

6. When you frown, the ______________________ is the primary muscle that depresses the corners of your mouth.

7. As you grit your teeth, palpate the skin along the anterior surface of your neck. Notice that the skin is very tense. This response is caused by the contraction of the ______________________.

8. Compress your cheeks by whistling or blowing. This action is produced by the ______________________, which is the relatively large muscle of facial expression that can be palpated in your cheek wall.

QUESTIONS 9–17: Identify the muscles that are labeled in the following diagram. Select your answers from the following list. Answers may be used once or not at all.

a. Levator labii superioris
b. Frontalis
c. Orbicularis oris
d. Omohyoid
e. Sternocleidomastoid
f. Masseter
g. Buccinator
h. Orbicularis oculi
i. Zygomaticus major
j. Temporalis
k. Nasalis
l. Zygomaticus minor

9. ______________________

10. ______________________

11. ______________________

12. ______________________

13. ______________________

14. ______________________

15. ______________________

16. ______________________

17. ______________________

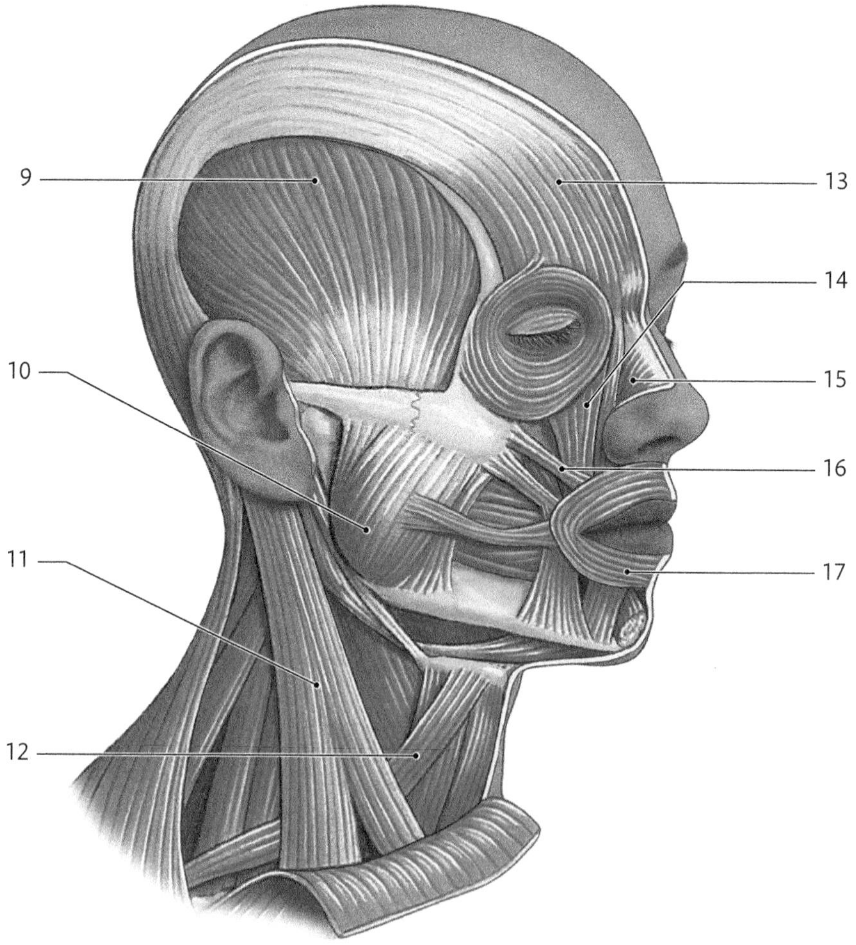

18. The extrinsic muscles of the tongue are supplied by the hypoglossal nerves. If a person's right hypoglossal nerve is not functioning, what do you think would happen if she stuck her tongue out? Explain.

____

____

____

____

QUESTIONS 19–24: Identify the muscles that are labeled in the following diagram. Select your answers from the following list. Answers may be used once or not at all.

a. Thyrohyoid
b. Stylohyoid
c. Mylohyoid
d. Digastric
e. Omohyoid
f. Geniohyoid
g. Sternohyoid
h. Sternothyroid

19. ____________________

20. ____________________

21. ____________________

22. ____________________

23. ____________________

24. ____________________

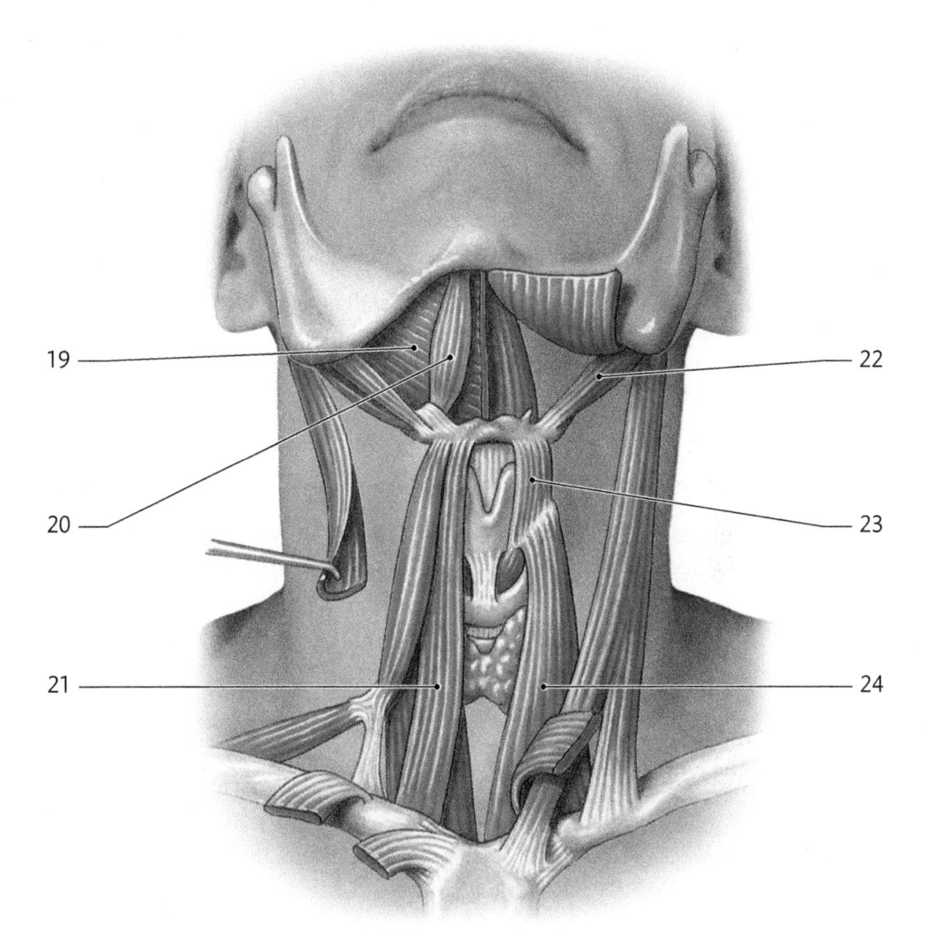

QUESTIONS 25–31: Identify the structures that are labeled in the following diagram. Select your answers from the following list. Answers may be used once or not at all.

a. Internal intercostal muscle
b. Erector spinae muscles
c. Transversus thoracis muscle
d. External intercostal muscle
e. Central tendon of the diaphragm
f. Aortic opening
g. Caval opening
h. Esophageal opening

25. ______________________

26. ______________________

27. ______________________

28. ______________________

29. ______________________

30. ______________________

31. ______________________

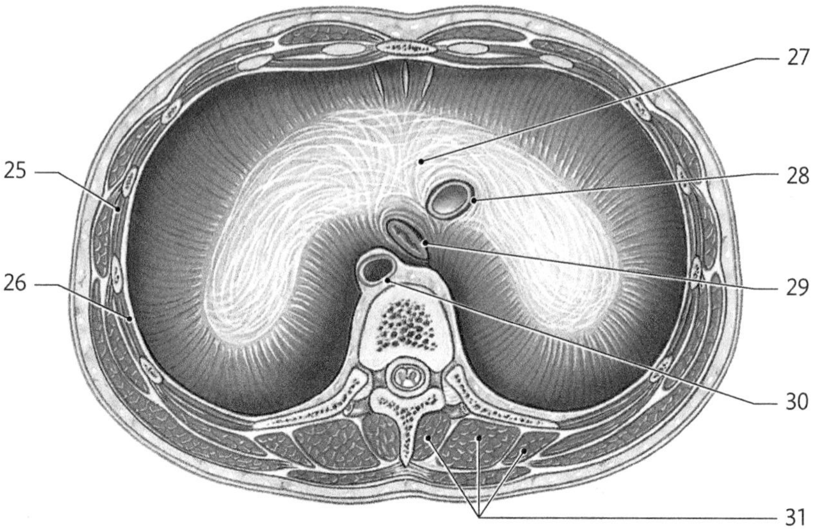

32. The external and internal oblique muscles laterally flex the vertebral column to the same side. Which muscles would act as antagonists for lateral flexion of the vertebral column to the left side? Explain.

____

____

____

33. Why is it important to strengthen deep back muscles as well as abdominal wall muscles? (*Hint:* Consider the negative impact of having strong deep back muscles but weak abdominal muscles.)

____

____

____

34. Complete the following table.

Muscle(s) of the Pelvic Floor	Function
Bulbospongiosus, ischiospongiosus	a.
b.	Supports the pelvic floor
External anal sphincter	c.
d.	Opens and closes urethral opening

EXERCISE

12

Muscles of the Upper and Lower Limbs

LEARNING OUTCOMES

These Learning Outcomes correspond by number to the laboratory activities in this exercise. When you complete the activities, you should be able to:

Identify the following muscles and explain their actions:

Activity 12.1 Muscles of the shoulder that move the scapula.

Activity 12.2 Muscles of the shoulder that move the arm.

Activity 12.3 Muscles of the arm.

Activity 12.4 Muscles of the anterior forearm.

Activity 12.5 Muscles of the posterior forearm.

Activity 12.6 Intrinsic muscles of the hand.

Activity 12.7 Muscles of the gluteal region.

Activity 12.8 Muscles of the anterior thigh.

Activity 12.9 Muscles of the medial thigh.

Activity 12.10 Muscles of the posterior thigh.

Activity 12.11 Muscles of the anterior and lateral leg.

Activity 12.12 Muscles of the posterior leg.

Activity 12.13 Intrinsic muscles of the foot.

Identify the following muscles and associated structures by palpation:

Activity 12.14 Surface anatomy of upper limb muscles.

Activity 12.15 Surface anatomy of lower limb muscles.

LABORATORY SUPPLIES

- Anatomical models, charts, and atlases of human musculature
- Human skeletons
- Pipe cleaners or some other type of flexible probe
- Metric ruler

PRE-LAB QUIZ

Before you begin, read all the activities in Exercise 12 and the required reading in your textbook that is assigned by your instructor.

1. During this laboratory exercise, you will use ____________ to trace the paths of muscles on a skeleton.
2. True or False: The levator scapulae and rhomboids connect the scapula to the vertebral column. __________
3. Rotator cuff muscles connect the scapula to the greater and lesser tubercles of the __________.
4. True or False: The coracobrachialis is in the posterior compartment of the arm. __________
5. The ____________________ on the humerus is called the common flexor origin.
6. All muscles with *pollicis* in their names will act on the ____________________.
7. True or False: Both the piriformis and obturator internus muscles lie partly within the pelvic cavity and partly within the gluteal region. __________
8. True or False: The quadriceps femoris muscles are located in the posterior thigh. __________
9. The groin muscles are located in what region of the thigh? ____________________
10. The extensor digitorum longus muscle is located in the __________ compartment of the leg.

Activity **12.1**

Muscles of the Upper Limb: Examining Muscles of the Shoulder That Move the Scapula

The upper limb muscles act at the shoulder, elbow, wrist, and hand. Muscles of the shoulder can be divided into two groups: muscles that move the shoulder (discussed in this activity) and muscles that move the arm (see Activity 12.2).

A Three deep muscles on the anterior thoracic wall—pectoralis minor, subclavius, and serratus anterior—help stabilize the pectoral girdle during movements at the shoulder.

1 On a model, identify the following shoulder muscles:

Muscle: **Pectoralis minor**	*Muscle:* **Subclavius**	*Muscle:* **Serratus anterior**
Origin: Ribs 3–5	*Origin:* First rib	*Origin:* Ribs 1–8
Insertion: Coracoid process of scapula	*Insertion:* Inferior surface of clavicle	*Insertion:* Medial border of scapula

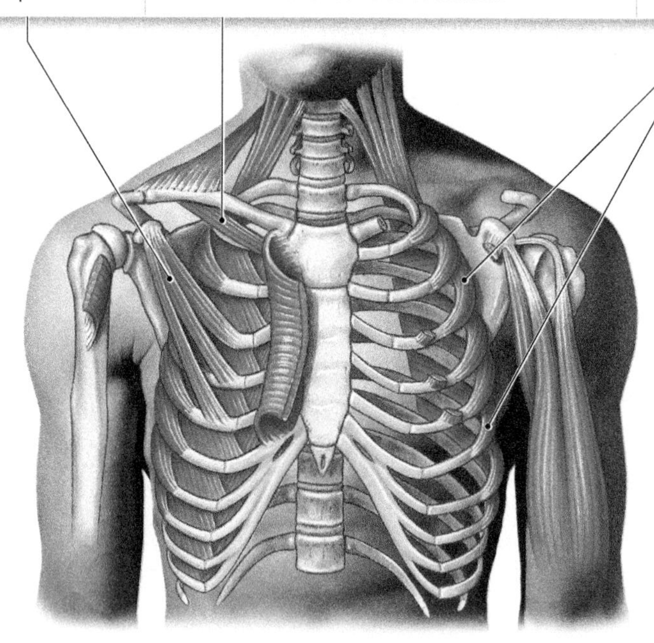

Anterior view

2 On a skeleton, use pipe cleaners to connect the origins and insertions of the pectoralis minor, subclavius, and serratus anterior. Visualize the paths of the muscles by observing the positions of the pipe cleaners. Answer the following questions: ▶

Will the pectoralis minor:

Elevate or depress the scapula?

____

Inferiorly or superiorly rotate the scapula?

____

Will the subclavius:

Elevate or depress the clavicle?

____

Will the serratus anterior:

Protract or retract the scapula?

____

Inferiorly or superiorly rotate the scapula?

____

Word Origins

The origin of serratus anterior on the ribs resembles the cutting edge of a serrated knife. It is the basis for the muscle's name, *serratus.*

IN THE CLINIC

Winged Scapula

The serratus anterior keeps the scapula in position against the thoracic wall. If this muscle is paralyzed, the medial border of the scapula moves laterally and posteriorly, especially when the individual leans on the hands against a wall. This gives the scapula the appearance of a wing; thus, this condition is called a winged scapula. This injury can cause significant pain and loss of strength and flexibility at the shoulder. Physical therapy treatments include range of motion and strengthening exercises, with the patient wearing a brace to keep the scapula against the thoracic wall. If paralysis is permanent, corrective surgery might be required.

B Four posterior shoulder muscles connect the scapula to bones on the axial skeleton. The trapezius is a broad superficial muscle that spreads across the posterior thoracic wall. The levator scapulae, rhomboid minor, and rhomboid major are deep to the trapezius.

1 On a model, identify the following muscles:

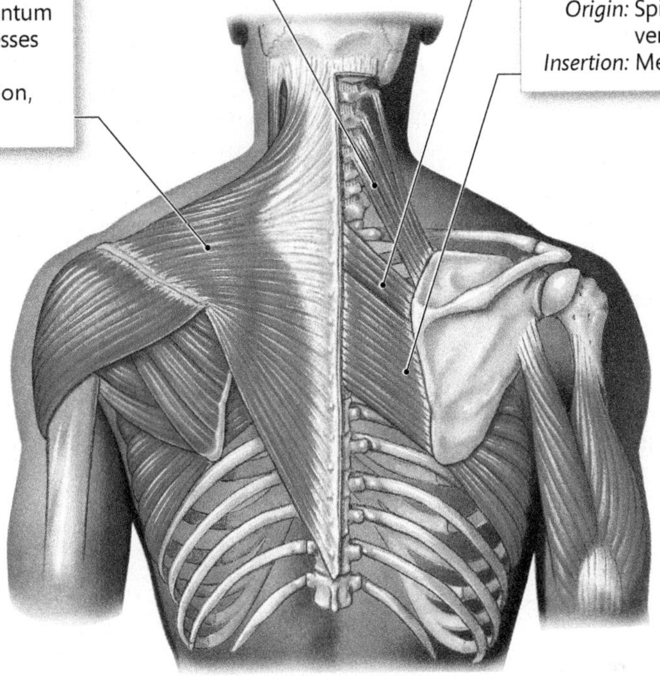

Muscle: **Levator scapulae**
Origin: Transverse processes of vertebrae C_1–C_4
Insertion: Medial border of scapula, near superior angle

Muscle: **Rhomboid minor**
Origin: Ligamentum nuchae, spinous processes of vertebrae C_7–T_1
Insertion: Medial border of scapula

Muscle: **Trapezius**
Origin: Occipital bone, ligamentum nuchae, spinous processes of thoracic vertebrae
Insertion: Scapular spine, acromion, lateral third of clavicle

Muscle: **Rhomboid major**
Origin: Spinous processes of vertebrae T_2–T_5
Insertion: Medial border of scapula

Posterior view

2 Because the trapezius has a broad origin, the muscle can be divided into three functional regions: superior, middle, and inferior. On a skeleton, use pipe cleaners to illustrate the paths of the three regions and answer the following questions about their actions: ▶

Will the superior region elevate or depress the scapula?

____

Will the middle region protract or retract the scapula?

____

Will the inferior region elevate or depress the scapula?

____

3 On a skeleton, use pipe cleaners to connect the origins and insertions of the levator scapulae and the two rhomboids. Notice that all three muscles connect the vertebral column to the medial border of the scapulae. Answer the following questions: ▶

Will all three muscles elevate or depress the scapula?

____

Will the two rhomboids protract or retract the scapula?

____

MAKING CONNECTIONS

Explain how the stability of the scapula would be affected if you were to lose the function of the pectoralis minor, serratus anterior, and rhomboids.

▶ _______________

____

____

____

____

____

____

Want more practice? Go to: **MasteringA&P** > Study Area > Menu >
- Lab Tools > PAL > Anatomical Models > Muscular System > Upper Limb
- Animations & Videos > *A&PFlix* > Group Muscle Actions and Joints > Unit 1: Muscles that act on the shoulder joint and humerus

Activity 12.2

Muscles of the Upper Limb: Examining Muscles of the Shoulder That Move the Arm

A The pectoralis major is a large superficial muscle on the anterior thoracic wall that moves the arm in several planes.

1. On a model, identify the **pectoralis major** on the anterior thoracic wall.

2. On a skeleton, identify the attachments of the pectoralis major. The muscle originates from the medial end of the clavicle, anterior surface of the sternum, and the costal cartilages of ribs 1 through 6. From this large attachment site, the muscle passes laterally and converges at its insertion on the humerus, lateral to the intertubercular groove.

3. Identify the muscle that lies deep to the pectoralis major:

▶ ______________________

____

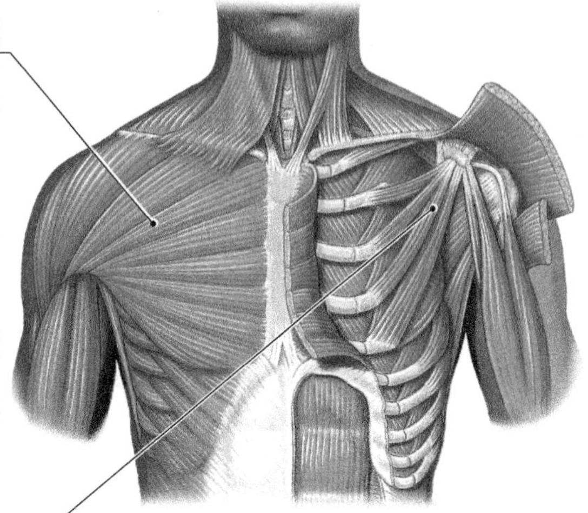

Anterior view

4. On a skeleton, place a pipe cleaner between the origin of pectoralis major on the sternum and its insertion on the humerus. Does the pectoralis major cross the shoulder joint?

▶ ______________________

5. The pipe cleaner illustrates the approximate pathway of the pectoralis major. Identify its actions: ▶

Does the muscle extend or flex the arm?

____

Does the muscle medially or laterally rotate the arm?

____

Does the muscle adduct or abduct the arm?

____

B On the posterior body wall, the latissimus dorsi connects the axial skeleton to the humerus. It is the largest muscle that moves the arm. Teres major is a much smaller muscle but performs the same actions as latissismus dorsi.

1. On a model, identify the latissimus dorsi and teres major muscles:

Muscle: **Teres major**
Origin: Scapula, near inferior angle
Insertion: Medial lip of intertubercular groove on humerus

Muscle: **Latissimus dorsi**
Origin: Spinous processes of vertebrae T_6–T_{12}, inferior five ribs, thoracolumbar (lumbodorsal) fascia
Insertion: Floor of intertubercular groove on humerus

Thoracolumbar fascia

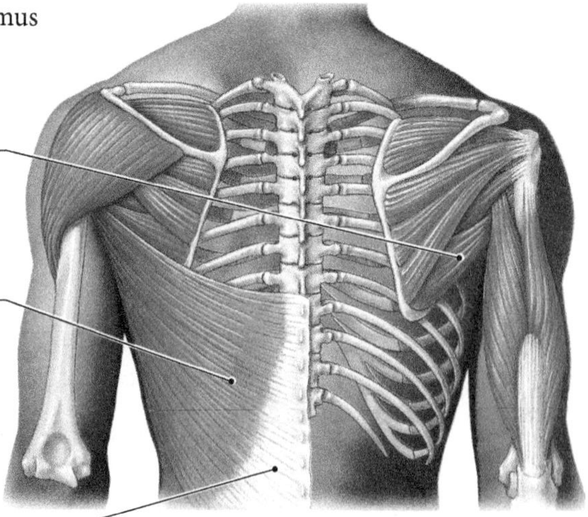

Posterior view

2. On a skeleton, use pipe cleaners to connect the origins and insertions of the teres major and latissimus dorsi. Notice that both muscles pass medial to the humeral shaft before inserting on the intertubercular groove. These muscles extend, medially rotate, and adduct the arm. Why do you think they perform the same actions?

▶ ______________________

____

____

____

Want more practice? Go to: **MasteringA&P** > Study Area > Menu >
- Lab Tools > PAL > Human Cadaver > Muscular System > Upper Limb
- Animations & Videos > *A&PFlix* > Origins, Insertions, Actions, Innervations > Trunk

C The deltoid is a large muscle that drapes over the shoulder, where it performs multiple actions.

1 On a model, identify the **deltoid muscle** covering the shoulder joint. The muscle inserts on the deltoid tuberosity of the humerus. Its origin is similar to the insertion of the trapezius. Describe the origin of the deltoid muscle.

▶ ____

____

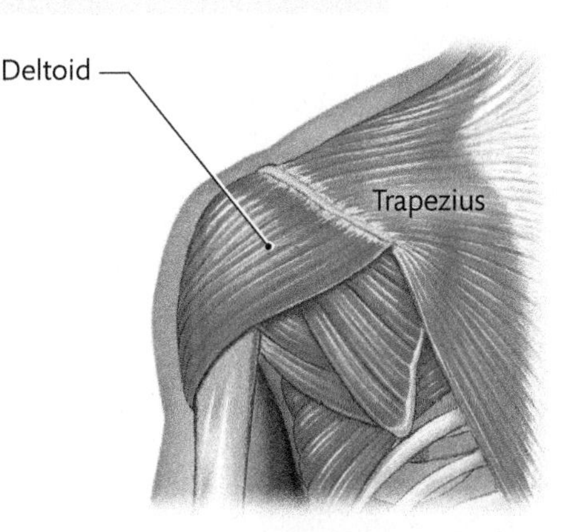

Right shoulder, anterior view **Left shoulder, posterior view**

2 The deltoid is divided into three functional regions: anterior, middle, and posterior. On a skeleton, use pipe cleaners to illustrate the paths of the three regions and answer the following questions: ▶

Will the anterior region:

Medially or laterally rotate the arm?

____

Extend or flex the arm?

____

Will the middle region:

Adduct or abduct the arm?

____

Will the posterior region:

Medially or laterally rotate the arm?

____

Extend or flex the arm?

____

Acting together, will the anterior and posterior regions adduct or abduct the arm?

____

D The **rotator cuff muscles**—supraspinatus, infraspinatus, teres minor, and subscapularis—connect the scapula to the greater and lesser tubercles of the humerus.

1 On a model, identify the four rotator cuff muscles:

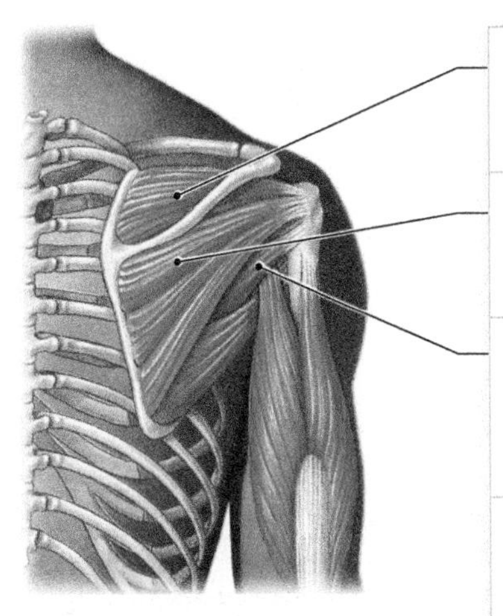

Right shoulder, posterior view

Muscle: **Supraspinatus**
Origin: Supraspinous fossa of scapula
Insertion: Greater tubercle of humerus

Muscle: **Infraspinatus**
Origin: Infraspinous fossa of scapula
Insertion: Greater tubercle of humerus

Muscle: **Teres minor**
Origin: Posterior aspect of scapula along lateral margin
Insertion: Greater tubercle of humerus

Muscle: **Subscapularis**
Origin: Subscapular fossa of scapula
Insertion: Lesser tubercle of humerus

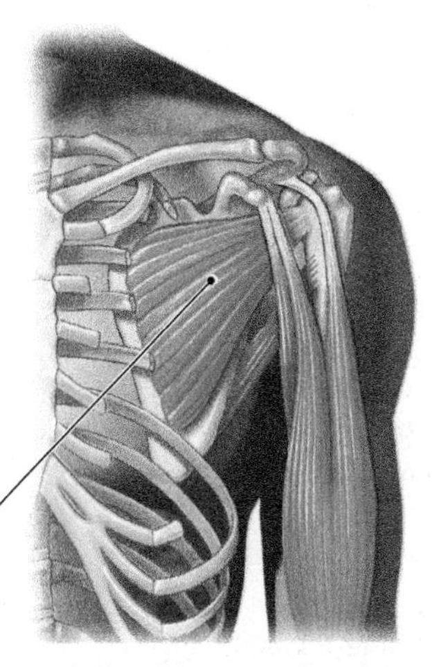

Left shoulder, anterior view

2 On a skeleton, use pipe cleaners to illustrate the paths of the rotator cuff muscles and answer the following questions about their actions: ▶

Will supraspinatus abduct or adduct the arm?

____

Will infraspinatus and teres minor medially or laterally rotate the arm?

____

Will subscapularis medially or laterally rotate the arm?

____

MAKING CONNECTIONS

From a functional standpoint, explain why the deltoid is an extremely important muscle.

▶ ____

____

____

____

Activity **12.3**

Muscles of the Upper Limb: Examining Muscles of the Arm

A Muscles of the arm are divided into anterior and posterior compartments. The **anterior compartment** muscles are flexors of the arm and forearm (the **flexor compartment**).

1 On a model, identify the anterior compartment muscles of the arm. The biceps brachii is the most superficial muscle. The coracobrachialis is a slender muscle deep and medial to the biceps brachii. The brachialis is deep to the biceps brachii.

2 Run pipe cleaners between the origins and insertions of the anterior arm muscles and notice the following:

- The coracobrachialis crosses only the shoulder joint.
- The brachialis crosses only the elbow joint.
- The biceps brachii crosses the shoulder, elbow, and proximal radioulnar joints.
- The long head of the biceps brachii travels along the intertubercular groove and through the joint cavity of the shoulder to reach its origin on the superior margin of the glenoid cavity of the scapula.

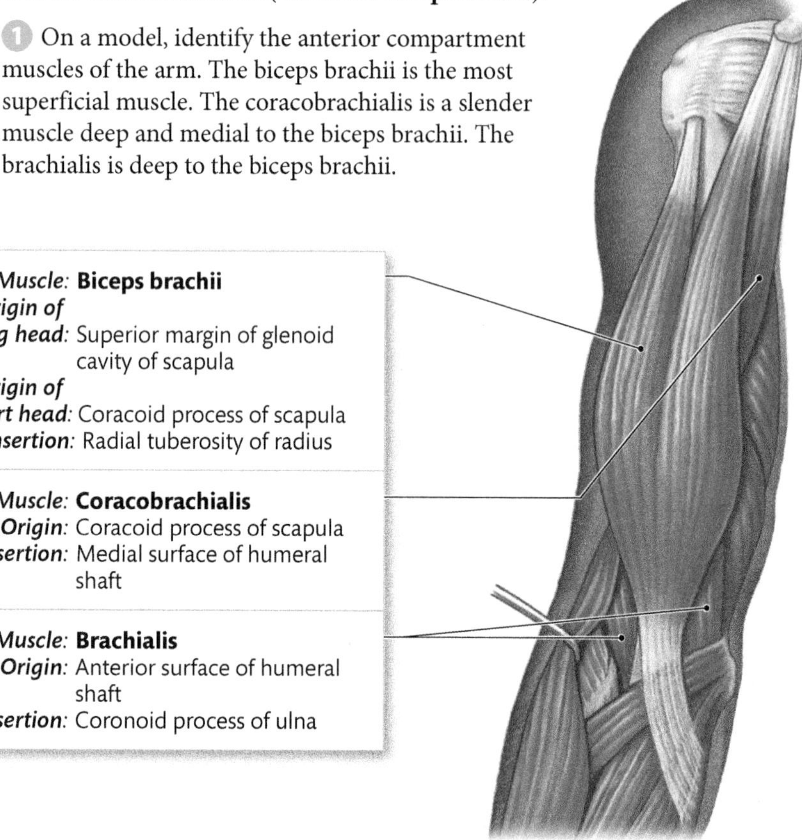

Anterior view

3 The biceps brachii and brachialis are agonists for flexion of the forearm. Biceps brachii is also an agonist for supination of the forearm. Place your hand on the anterior surface of your arm while flexing or supinating your forearm. What change do you feel in the biceps brachii while performing these movements?

▶ ______________________________

____

____

The biceps brachii and coracobrachialis are synergists for flexion of the arm. Coracobrachialis is also a synergist for adduction of the arm. Identify a muscle that is an agonist for: ▶

Flexion of the arm:

____

Adduction of the arm:

____

B The triceps brachii, the only muscle in the **posterior compartment,** extends the arm and forearm (the **extensor compartment**).

1 On a model, identify the three heads of the triceps brachii:

Muscle: **Triceps brachii**
Origin of long head: Inferior margin of glenoid cavity
Origin of lateral head: Proximal end of posterior and lateral humeral shaft
Origin of medial head: Distal end of posterior humeral shaft (not shown; deep to long and lateral heads)
Insertion: Olecranon process of ulna

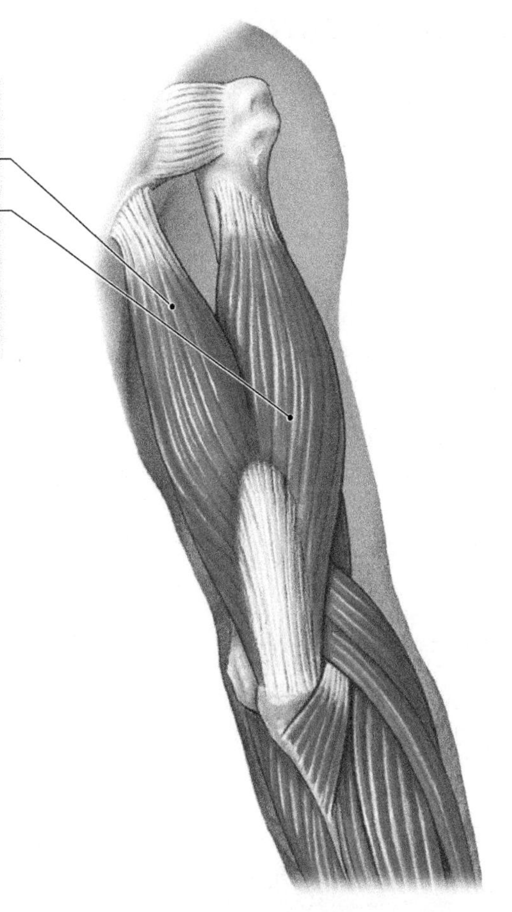

Posterior view

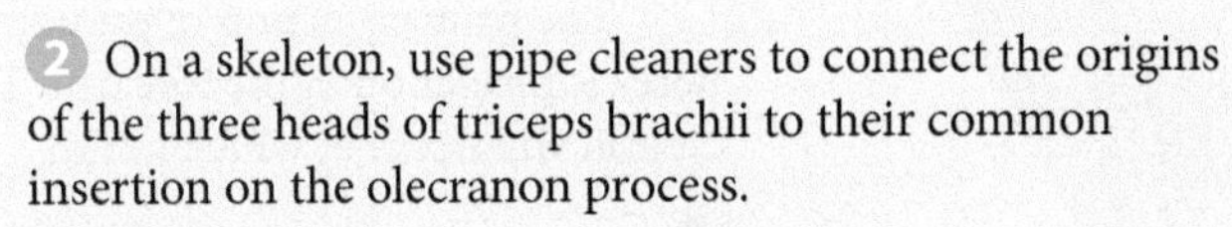

2 On a skeleton, use pipe cleaners to connect the origins of the three heads of triceps brachii to their common insertion on the olecranon process.

Which head(s) crosses the shoulder joint?

▶ ____________________

Which head(s) crosses the elbow joint?

▶ ____________________

3 The triceps brachii is an agonist for extension of the forearm. Which muscles are antagonists for this movement?

▶ ____________________

____

The triceps brachii is a synergist for extension and adduction of the arm.

Which head(s) can perform these movements?

▶ ____________________

Identify a muscle that is an agonist for extension of the arm.

▶ ____________________

Identify a muscle that is an antagonist for extension of the arm.

▶ ____________________

Which anterior arm muscle also adducts the arm?

▶ ____________________

MAKING CONNECTIONS

Compare the following pairs of actions by completing the table. ▶

Actions	Joint Where Action Occurs	Bone(s) That Move
Flexion and extension of the arm		
Flexion and extension of the forearm		

Want more practice? Go to: **MasteringA&P** > Study Area > Menu >
- Lab Tools > PAL > Anatomical Models > Muscular System > Upper Limb
- Lab Tools > PAL > Human Cadaver > Muscular System > Upper Limb
- Animations & Videos > ***A&PFlix*** > Origins, Insertions, Actions, Innervations > Upper Limb
- Animations & Videos > ***A&PFlix*** > Group Muscle Actions and Joints > Unit 2: Muscles that act on the elbow joint and forearm

Activity **12.4**

Muscles of the Upper Limb: Examining Muscles of the Anterior Forearm

Similar to the muscles of the arm, the muscles of the forearm are divided into the anterior compartment (discussed in this activity) and the posterior compartment (see Activity 12.5).

The anterior compartment muscles are divided into superficial, intermediate, and deep layers. The principal actions of these muscles are flexion of the hand and fingers and pronation of the forearm.

A The **superficial layer** of the anterior compartment contains four muscles.

1 The superficial muscles have a common origin at the medial epicondyle of the humerus. This bone marking is called the **common flexor origin.** On a model, identify the four superficial muscles:

Muscle: **Pronator teres**
Additional origin: Coronoid process of ulna
Insertion: Lateral surface of radial shaft

Muscle: **Flexor carpi radialis**
Insertion: Bases of metacarpal bones 2 and 3

Muscle: **Palmaris longus**
Insertion: Flexor retinaculum

Muscle: **Flexor carpi ulnaris**
Additional origin: Olecranon of ulna
Insertion: Medial carpal bones; base of 5th metacarpal

Flexor retinaculum

Anterior view

2 On a skeleton, place a pipe cleaner on the common flexor origin and extend it to the insertion of the pronator teres. Notice that this muscle crosses the elbow and proximal radioulnar joints.

The primary action of pronator teres is pronation of the forearm. This action will occur at which joint?

▶ ______________________

3 Place three pipe cleaners on the common flexor origin. Extend the first pipe cleaner to the insertion of flexor carpi radialis, the second to the insertion of palmaris longus, and the third to the insertion of flexor carpi ulnaris. Examine the positions of the three pipe cleaners. Notice that all these muscles cross the elbow and wrist joints and, therefore, will function at both locations. Their primary action will be flexion at the wrist joint, however.

In addition to wrist flexion: ▶

Will flexor carpi radialis abduct or adduct the wrist?

____

Will flexor carpi ulnaris abduct or adduct the wrist?

____

B The **intermediate layer** contains one muscle that also originates from the medial epicondyle.

1 On a model, identify the **flexor digitorum superficialis** in the intermediate layer.

Additional origin: Coronoid process of ulna, anterior surface of radius
Insertion: Middle phalanges of medial four digits.

2 On a skeleton, position a series of pipe cleaners so they begin at the common flexor origin and extend to the middle phalanx of the index finger. Observe the path of the pipe cleaners and notice that the flexor digitorum superficialis crosses several joints. Identify the muscle's main actions.

▶ ______________________

____

Identify the joints where the actions occur.

▶ ______________________

____

Anterior view

Want more practice? Go to: **MasteringA&P** > Study Area > Menu >
■ Lab Tools > PAL > Anatomical Models > Muscular System > Upper Limb
■ Animations & Videos > *A&PFlix* > Group Muscle Actions and Joints > Unit 3: Muscles that act on the wrist and fingers

C The **deep layer** of the anterior compartment contains three muscles.

1 Two of the three deep muscles have long tendons extending to insertion sites on the thumb and fingers. On a model, identify the following muscles:

Muscle: **Flexor digitorum profundus**
Origin: Medial and anterior surfaces of ulna
Insertion: Distal phalanges of medial four digits

Muscle: **Flexor pollicis longus**
Origin: Anterior surface of radius; interosseous membrane
Insertion: Distal phalanx of thumb

Muscle: **Pronator quadratus**
Origin: Anterior surface of distal ulna
Insertion: Anterior surface of distal radius

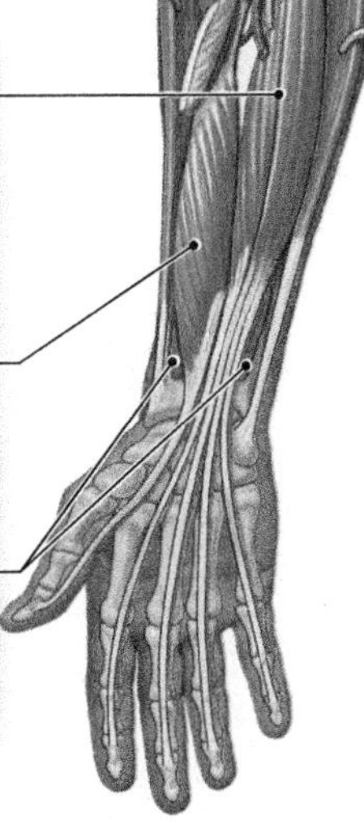

Anterior view

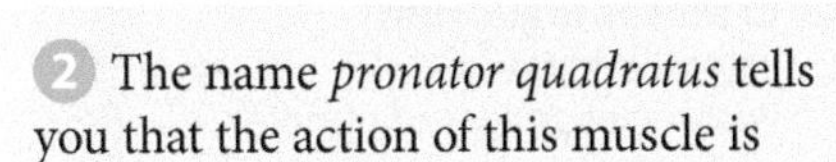

2 The name *pronator quadratus* tells you that the action of this muscle is

▶ ______________________

3 On a skeleton, position a series of pipe cleaners so that they connect the origin of flexor digitorum profundus with its insertion on the index finger. Use additional pipe cleaners to connect the origin and insertion of flexor pollicis longus. Observe the path of the pipe cleaners and explain why these muscles do not act at the elbow joint.

▶ ______________________

Complete the following table. ▶

Muscle	Main Actions	Joints Where Actions Occur
Flexor digitorum profundus		
Flexor pollicis longus		

Word Origins

Carpi refers to the wrist; any muscle with *carpi* in its name will act at the wrist joint.

Pollicis refers to the thumb; any muscle with *pollicis* in its name will act on the thumb.

D The **carpal tunnel** is a passageway for the long tendons of anterior compartment muscles that insert on the hand.

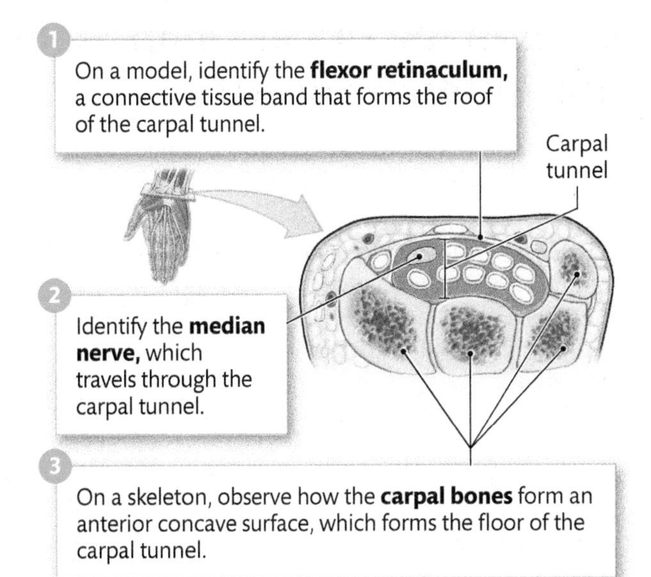

1 On a model, identify the **flexor retinaculum,** a connective tissue band that forms the roof of the carpal tunnel.

2 Identify the **median nerve,** which travels through the carpal tunnel.

3 On a skeleton, observe how the **carpal bones** form an anterior concave surface, which forms the floor of the carpal tunnel.

IN THE CLINIC

Carpal Tunnel Syndrome

The median nerve passes through the carpal tunnel and supplies the skin over roughly the lateral half of the hand and several small hand muscles. Individuals who perform tasks involving repeated flexion of the wrists and fingers, such as typing and texting, may develop carpal tunnel syndrome. This condition is characterized by inflammation and swelling of the tendons passing through the carpal tunnel, which compress the median nerve.

Symptoms of carpal tunnel syndrome include tingling, numbness, or pain in the wrist, thumb, and lateral three fingers and difficulty opposing and flexing the thumb. The partial loss of thumb function impairs fine motor activities. Usually, rest and anti-inflammatory drugs can diminish the symptoms. In severe cases, it may be necessary to surgically cut the flexor retinaculum to relieve pressure (carpal tunnel release).

MAKING CONNECTIONS

If you were to lose all function of flexor carpi radialis, flexor carpi ulnaris, and palmaris longus, would you still be able to flex your wrist? Explain.

▶ ______________________

Activity **12.5**

Muscles of the Upper Limb: Examining Muscles of the Posterior Forearm

The posterior compartment muscles are divided into superficial and deep layers. The principal actions of these muscles are extension of the hand and fingers and supination of the forearm.

A The **superficial layer** of the posterior compartment contains seven muscles. Five of these muscles originate at the lateral epicondyle of the humerus, which is called the **common extensor origin.**

1. On a model, identify the seven muscles in the superficial layer:

Muscle: **Anconeus**
Origin: Lateral epicondyle
Insertion: Olecranon of ulna

Muscle: **Extensor carpi ulnaris**
Origin: Lateral epicondyle; posterior surface of ulna
Insertion: Metacarpal of fifth digit

Muscle: **Extensor digiti minimi**
Origin: Lateral epicondyle
Insertion: Posterior surfaces of middle and distal phalanges of fifth digit

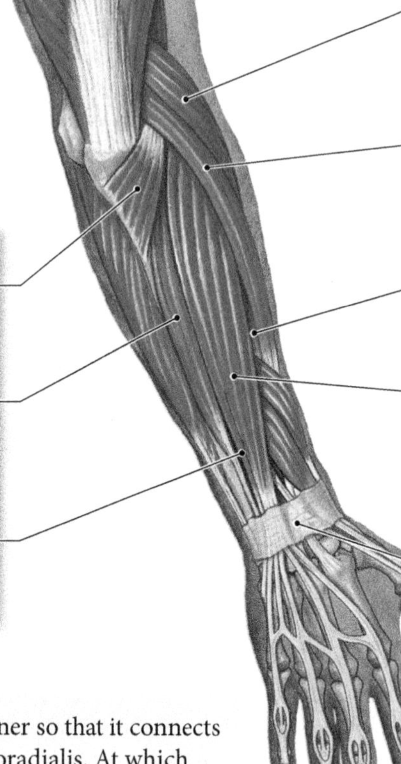

Muscle: **Brachioradialis**
Origin: Lateral supracondylar ridge of humerus
Insertion: Distal end of radius

Muscle: **Extensor carpi radialis longus**
Origin: Lateral supracondylar ridge of humerus
Insertion: Metacarpal of second digit

Muscle: **Extensor carpi radialis brevis**
Origin: Lateral epicondyle
Insertion: Metacarpal of third digit

Muscle: **Extensor digitorum**
Origin: Lateral epicondyle
Insertion: Posterior surfaces of middle and distal phalanges of medial four digits

Extensor retinaculum

Posterior view

2. On a skeleton, position a pipe cleaner so that it connects the origin and insertion of the brachioradialis. At which joint will this muscle act?

▶ ____________________

Brachioradialis is unique because it functions like an anterior compartment muscle. What do you think is the action of this muscle?

▶ ____________________

3. Position pipe cleaners so that they connect the origins and insertions of extensor carpi radialis longus and brevis and extensor carpi ulnaris. These muscles will act at the elbow and wrist joints. Explain why.

▶ ____________________

The primary action of these muscles is extension of the hand.

In addition to hand extension: ▶

Will extensor carpi radialis longus and brevis abduct or adduct the wrist?

____

Will extensor carpi ulnaris abduct or adduct the wrist?

____

4. Position a series of pipe cleaners so that they connect the origin of extensor digitorum to its insertion on the second digit. Identify four joints that the muscle crosses.

▶ ____________________

____

Identify the main actions of the extensor digitorum.

▶ ____________________

____

Identify the joints where these actions occur.

▶ ____________________

____

Based on its name, what do you think is the main action of the extensor digiti minimi?

▶ ____________________

____

Want more practice? Go to: **MasteringA&P** > Study Area > Menu >
- Lab Tools > PAL > Human Cadaver > Muscular System > Upper Limb
- Animations & Videos > *A&PFlix* > Origins, Insertions, Actions, Innervations > Upper Limb

B The **deep layer** of the posterior compartment contains five muscles.

1 Four deep muscles originate on the posterior surface of the ulna or radius and extend the thumb or index finger. On a model, identify the deep muscles in the posterior forearm:

Muscle: **Supinator**
Origin: Lateral epicondyle of humerus; proximal ulna
Insertion: Posterior, lateral, and anterior surfaces of proximal radius

Muscle: **Extensor pollicis longus**
Origin: Posterior surface of ulna; interosseous membrane
Insertion: Distal phalanx of thumb

Muscle: **Abductor pollicis longus**
Origin: Posterior surfaces of radius and ulna; interosseous membrane
Insertion: Metacarpal of thumb

Muscle: **Extensor pollicis brevis**
Origin: Posterior surface of radius; interosseous membrane
Insertion: Proximal phalanx of thumb

Muscle: **Extensor indicis**
Origin: Posterior surface of ulna; interosseous membrane
Insertion: Posterior surfaces of middle and distal phalanges of second digit

Posterior view

2 Along with the supinator, what other muscle supinates the forearm?

▶ ______________________________

Identify two muscles that are antagonists to forearm supination.

▶ ______________________________

3 On a skeleton, use pipe cleaners to approximate the positions of the three pollicis muscles and the extensor indicis. All these muscles are synergists for extension of the hand.

Using their names as clues, identify the main actions of these muscles: ▶

Muscle	Main Action
Extensor pollicis longus	
Extensor pollicis brevis	
Abductor pollicis longus	
Extensor indicis	

C The long tendons of posterior compartment muscles travel through tunnels to reach their insertions on the hand, just as the tendons of anterior compartment muscles travel through the carpal tunnel.

1 On a skeleton, position a pipe cleaner across the posterior surfaces of the distal radius and ulna. The pipe cleaner approximates the position of the **extensor retinaculum,** a thick band of connective tissue that forms the roof of several small tunnels that transmit the tendons of the posterior forearm muscles.

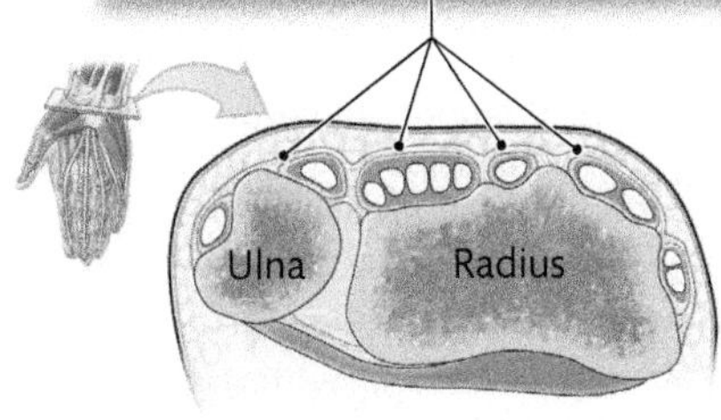

2 The distal ends of the **ulna** and **radius** form the floor of these tunnels. Speculate on how muscle function would be affected if the tendons did not travel through these tunnels.

▶ ______________________________

____

____

MAKING CONNECTIONS

The extensor carpi radialis longus and extensor carpi radialis brevis abduct the wrist; the extensor carpi ulnaris adducts the wrist. If you were to lose the function of all these muscles, abduction and adduction of the wrist would still be possible. Explain why.

▶ ______________________________

____

____

____

____

____

____

Activity **12.6**

Muscles of the Upper Limb: Examining the Intrinsic Muscles of the Hand

The intrinsic muscles of the hand perform fine motor skills. Without them, activities such as writing with a pen, holding a cell phone and texting, grasping the handle of a coffee mug, buttoning a shirt, or snapping your fingers would not be possible.

A The four lumbricals are superficial muscles that travel through the palm of the hand. They are unique because all their attachments are to soft tissue rather than bone.

1 On a model, identify the **lumbricals:**

Origin: Tendons of flexor digitorum profundus of digits 2–5
Insertion: Tendons of extensor digitorum of digits 2–5

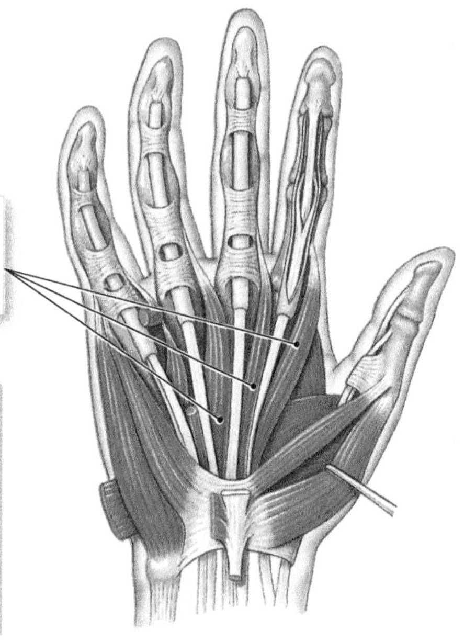

Anterior (palmar) view

Word Origins

Lumbrical derives from Latin and means "earthworm." Each lumbrical is a small, slender muscle that does indeed look like a worm. *Interossei,* also from Latin, means "between the bones." The name reflects the position of these muscles between the metacarpal bones.

2 The lumbricals flex the fingers at the metacarpophalangeal (knuckle) joints and extend the fingers at the interphalangeal joints. On a model, examine the path of these muscles and speculate why they act as flexors at one joint and extensors at another.

▶ ______________________________

____

____

____

____

____

____

B Four dorsal interossei and three or four palmar interossei fill the spaces between the metacarpal bones.

1 On a model, identify the four **palmar interossei:**

Origin: Sides of metacarpals of all digits except the third
Insertion: Bases of proximal phalanges of all digits except the third

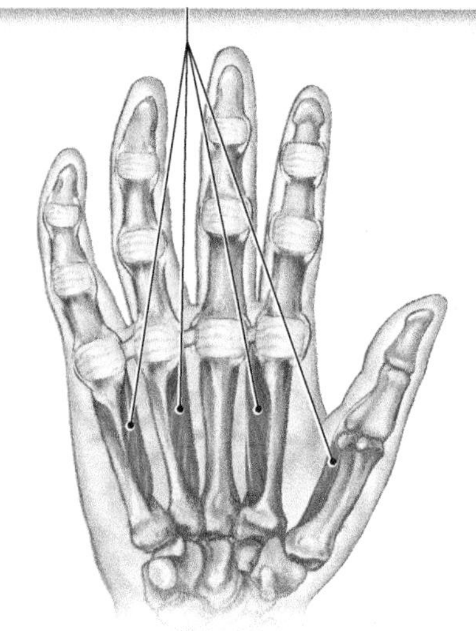

Anterior (palmar) view

2 On a model, identify the four **dorsal interossei:**

Origin: Sides of adjacent metacarpals
Insertion: Bases of proximal phalanges of the second, third, and fourth digits

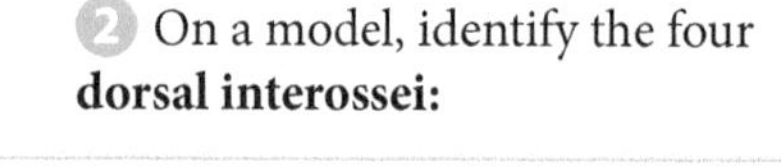

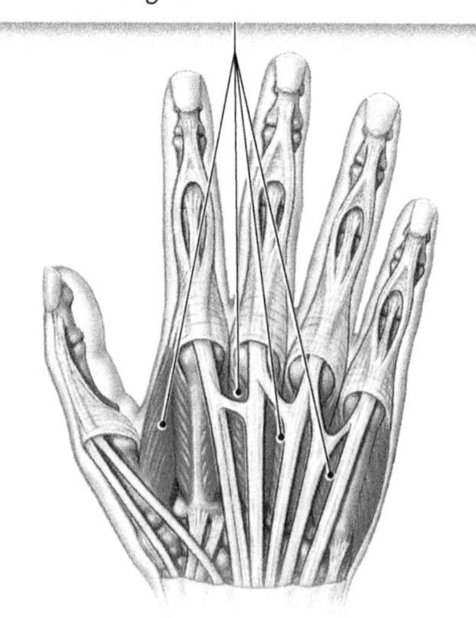

Posterior view

3 Similar to the lumbricals, the interossei flex the fingers at the metacarpophalangeal joints and extend the fingers at the interphalangeal joints. Identify a muscle that will be an antagonist to these muscles at the ▶

metacarpophalangeal joints:

____

interphalangeal joints:

____

4 The dorsal interossei abduct the fingers, and the palmar interossei adduct the fingers. Hold your palm in front of your face and imagine a vertical midline running through the middle finger. Abduct your fingers by moving them away from the midline; adduct your fingers by moving them toward the midline.

C The **thenar eminence** is the rounded mound of soft tissue at the base of the thumb. The four muscles of the thenar eminence are divided into a superficial group and a deep group. These muscles act exclusively on the thumb.

1 On a model, identify the superficial muscles of the thenar eminence: ▶

Muscle: **Flexor pollicis brevis**
Origin: Lateral carpal bones, flexor retinaculum
Insertion: Proximal phalanx of the thumb

Muscle: **Abductor pollicis brevis**
Origin: Lateral carpal bones, flexor retinaculum
Insertion: Proximal phalanx of the thumb

2 On a model, identify the deep muscles of the thenar eminence: ▶

Muscle: **Adductor pollicis**
Origin: Second and third metacarpals and capitate bone
Insertion: Proximal phalanx of the thumb

Muscle: **Opponens pollicis**
Origin: Trapezium and flexor retinaculum
Insertion: Metacarpal bone of the thumb

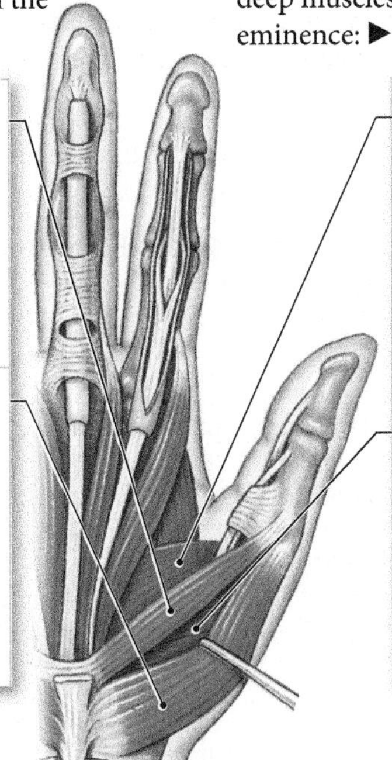

Anterior (palmar) view

3 Using their names as clues, identify the main actions of the thenar muscles: ▶

Muscle	Main Action
Flexor pollicis brevis	
Abductor pollicis brevis	
Adductor pollicis	
Opponens pollicis	

4 From anatomical position, perform opposition by rotating the thumb medially so that it touches the little finger. To demonstrate the importance of thumb opposition, grasp a plastic bottle as you normally would. Now, put the bottle down and grasp it again without using your thumb. Compare the two actions describing how effectively you were able to hold the bottle.

▶ ______________________________

____

____

D The **hypothenar eminence** is the rounded mound of soft tissue at the base of the little finger. There are three muscles of the hypothenar eminence: two superficial muscles and one deep muscle. They act exclusiviely on the little finger.

1 On a model, identify the two superficial muscles of the hypothenar eminence: ▶

Muscle: **Flexor digiti minimi brevis**
Origin: Hamate bone and flexor retinaculum
Insertion: Proximal phalanx of the fifth digit

Muscle: **Abductor digiti minimi**
Origin: Pisiform bone
Insertion: Proximal phalanx of the fifth digit

2 Identify the deep muscle of the hypothenar eminence: ▶

Muscle: **Opponens digiti minimi**
Origin: Hamate bone and flexor retinaculum
Insertion: Metacarpal bone of the fifth digit

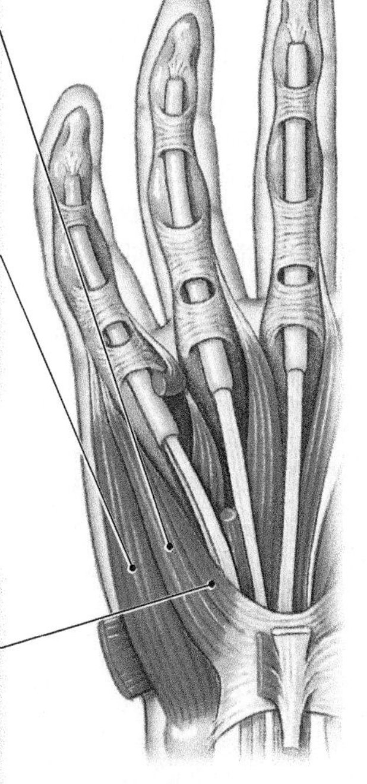

Anterior (palmar) view

3 The hypothenar muscles abduct, flex, and oppose the little finger. Try to oppose your little finger. Compare the range of motion of this action with thumb opposition.

▶ ______________________________

____

MAKING CONNECTIONS

The median nerve supplies the intrinsic hand muscles that oppose, abduct, and flex the thumb. The ulnar nerve supplies the intrinsic hand muscles that oppose the little finger, abduct and adduct the fingers, and adduct the thumb. Consider two possible hand injuries: damage to the median nerve and damage to the ulnar nerve. Which injury would be more serious? Explain:

▶ ______________________________

____

____

____

Muscles of the Lower Limb: Examining the Muscles of the Gluteal Region

The muscles of the lower limb act at the hip, knee, and ankle joints as well as the joints in the foot. The **gluteal region** extends from the iliac crest, superiorly, to the inferior border of the gluteus maximus, inferiorly. The muscles in this region are divided into superficial and deep groups.

A The superficial group includes the three gluteal muscles (gluteus maximus, gluteus medius, and gluteus minimus), which form a mass of muscle tissue known as the **buttock,** and tensor fasciae latae. These muscles are important extensors, abductors, and rotators of the thigh. They also stabilize the pelvis and thigh while standing erect.

1 On a model, identify the following superficial gluteal muscles:

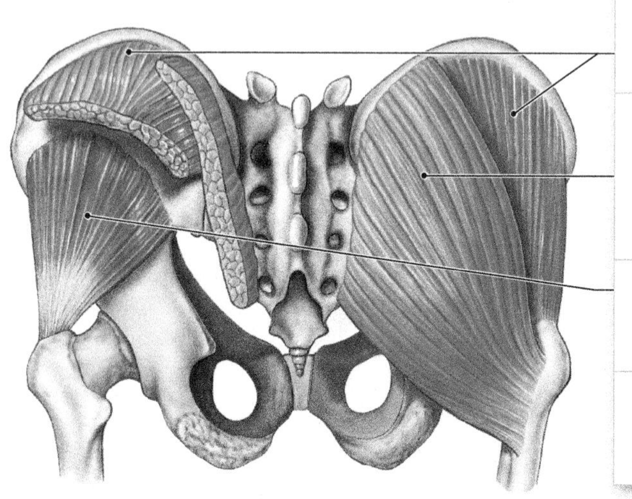

Posterior view

Muscle:	**Gluteus medius**
Origin:	Lateral surface of ilium
Insertion:	Greater trochanter of femur
Muscle:	**Gluteus maximus**
Origin:	Lateral surface of ilium, dorsal surface of sacrum and coccyx
Insertion:	Iliotibial tract, gluteal tuberosity of femur
Muscle:	**Gluteus minimus**
Origin:	Lateral surface of ilium
Insertion:	Greater trochanter of femur
Muscle:	**Tensor fasciae latae**
Origin:	Anterior superior iliac spine
Insertion:	Iliotibial tract

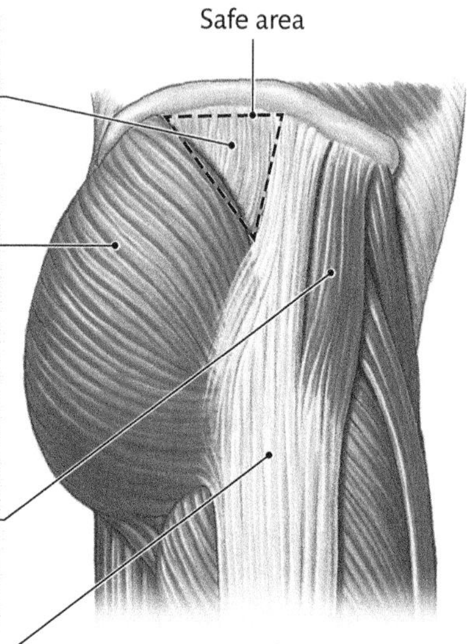

Lateral view

2 On a skeleton, use pipe cleaners to connect the attachments of the gluteus maximus between the sacrum and gluteal tuberosity. Observe the path of the muscle and predict its actions: ▶

Will gluteus maximus extend or flex the thigh?

____

Will gluteus maximus medially or laterally rotate the thigh?

____

3 The gluteus medius and gluteus minimus have similar attachments and functions. Extend pipe cleaners between the lateral surface of the ilium and the superior and anterior surfaces of the greater trochanter and speculate whether these muscles will: ▶

Adduct or abduct the thigh?

____

Medially or laterally rotate the thigh?

____

4 The gluteus medius and gluteus minimus are important muscles for locomotion because they stabilize the pelvis during walking and running. Stand in anatomical position and palpate the left superolateral corner of the gluteal region (the "safe area"; see "In the Clinic" on the next page). Raise your right foot off the ground. You can feel the left gluteus medius contracting. When the right limb is raised, the unsupported right side of the pelvis tends to tilt inferiorly. The left gluteal muscles steady the pelvis by pulling the left side inferiorly. This action prevents the right side from sagging, allowing the right foot to clear the ground while walking.

5 On a model, observe the path of tensor fasciae latae. Notice that this muscle inserts on the **iliotibial tract,** a thick band of deep fascia that travels along the lateral thigh between the iliac crest and the tibia. When tensor fasciae latae contracts, it tenses the iliotibial tract, which laterally supports the knee. Speculate whether this muscle will: ▶

Flex or extend the thigh:

____

Adduct or abduct the thigh:

____

Medially or laterally rotate the thigh:

____

B The deep group of gluteal muscles are the piriformis, superior gemellus, obturator internus, inferior gemellus, and quadratus femoris. The piriformis abducts the thigh; the other muscles in this group laterally rotate the thigh.

1. On a model, identify the deep gluteal muscles:

Muscle: **Piriformis**
Origin: Anterior surface of sacrum
Insertion: Greater trochanter of femur

Muscle: **Superior gemellus**
Origin: Ischial spine
Insertion: Greater trochanter of femur

Muscle: **Obturator internus**
Origin: Internal margin of oburator foramen
Insertion: Greater trochanter of femur

Muscle: **Inferior gemellus**
Origin: Ischial tuberosity
Insertion: Greater trochanter of femur

Muscle: **Quadratus femoris**
Origin: Ischial tuberosity
Insertion: Intertrochanteric crest of femur

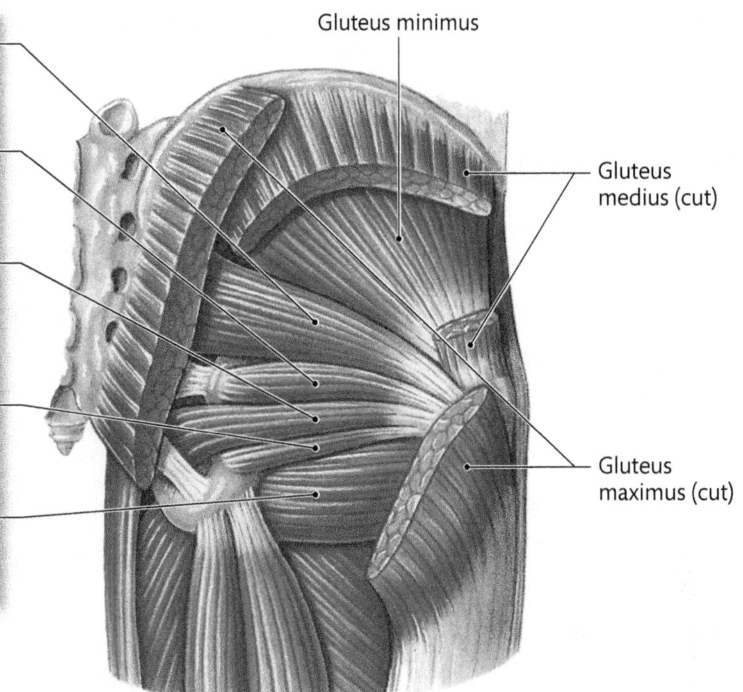

Posterior view

2. On a skeleton, place one end of a pipe cleaner on the anterior surface of the sacrum. Extend it laterally through the greater sciatic notch so that the other end reaches the superior tip of the greater trochanter. This spot marks the position of the piriformis. Notice that the piriformis is partly within the pelvic cavity and partly within the gluteal region.

3. Position one end of a pipe cleaner on the internal surface of bone rimming the obturator foramen. Extend the pipe cleaner laterally, through the lesser sciatic notch, and bend it so that the other end reaches the greater trochanter. The pipe cleaner marks the position of the obturator internus. Observe the following:

- The obturator internus lies partly within the pelvic cavity and partly within the gluteal region.
- To insert on the greater trochanter, the obturator internus must turn anteriorly shortly after it passes through the lesser sciatic notch.

IN THE CLINIC

The Safe Area

Most of the gluteus medius lies deep to the gluteus maximus. In the superolateral part of the buttock, however, the gluteus medius extends beyond the lateral border of the gluteus maximus. The exposed portion of the gluteus medius is called the safe area because intramuscular injections can be made in this region without injury to the sciatic nerve.

MAKING CONNECTIONS

Explain how the gluteal muscles, which act at the hip joint, are similar to the rotator cuff muscles, which act at the shoulder joint.

▶ ______________________________

____

____

____

____

____

____

____

Want more practice? Go to: **MasteringA&P** > Study Area > Menu >
- Lab Tools > PAL > Anatomical Models > Muscular System > Lower Limb
- Animations & Videos > ***A&PFlix*** > Group Muscle Actions and Joints > Unit 4: Muscles that act on the hip joint and femur

Activity 12.8

Muscles of the Lower Limb: Examining the Muscles of the Anterior Thigh

The **anterior thigh** is dominated by four muscles, known collectively as the quadriceps femoris. Other muscles in the anterior thigh include the iliopsoas and sartorius.

A The iliopsoas is formed by two separate muscles, the psoas major and iliacus, which originate in the abdominopelvic cavity and insert in the anterior thigh.

1 On a model, identify the two muscles that make up the iliopsoas:

Muscle: **Psoas major**
Origin: Bodies of vertebrae T_{12}–L_5
Insertion: Lesser trochanter of femur

Muscle: **Iliacus**
Origin: Iliac fossa
Insertion: Lesser trochanter of femur

Inguinal ligament

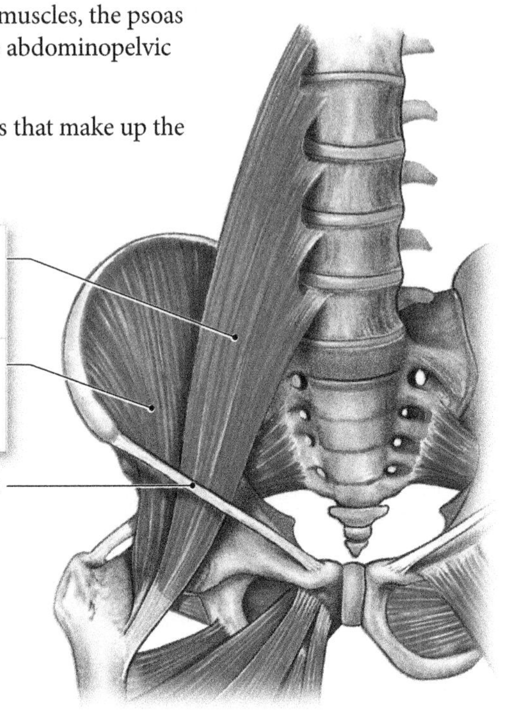

Anterior view

2 On a skeleton, use a pipe cleaner to connect the anterior superior iliac spine and the pubic tubercle. The pipe cleaner illustrates the position of the **inguinal ligament,** which forms the superior border of the anterior thigh.

3 Place the end of one pipe cleaner on the body of a lumbar vertebra and a second on the iliac fossa. Allow both pipe cleaners to pass inferiorly and converge on the lesser trochanter. The pipe cleaners represent the paths of the psoas major and iliacus. Both muscles leave the abdominopelvic cavity and enter the anterior thigh by passing deep to the inguinal ligament.

4 Walk across the room. The iliopsoas is the agonist for flexing the thigh to lift the lower limb off the floor as you walk.

Aside from walking or running, name another activity during which flexion of the thigh is important.

▶ ______________________________

Identify a gluteal muscle that is a synergist for the iliopsoas.

▶ ______________________________

Identify a gluteal muscle that is an antagonist of the iliopsoas.

▶ ______________________________

Want more practice? Go to: MasteringA&P > Study Area > Menu >
- Lab Tools > PAL > Anatomical Models > Muscular System > Lower Limb
- Lab Tools > PAL > Human Cadaver > Muscular System > Lower Limb
- Animations & Videos > A&PFlix > Origins, Insertions, Actions, Innervations > Lower Limb
- Animations & Videos > A&PFlix > Group Muscle Actions and Joints > Unit 4: Muscles that act on the hip joint and femur
- Animations & Videos > A&PFlix > Group Muscle Actions and Joints > Unit 5: Muscles that act on the knee joint and lower leg

B The four muscles that make up the **quadriceps femoris**—rectus femoris, vastus lateralis, vastus medialis, and vastus intermedius—cover nearly all the anterior, lateral, and medial aspects of the femur.

1. On a model, identify the four muscles in the quadriceps femoris group:

Muscle: **Rectus femoris**
Origin: Anterior inferior iliac spine
Insertion: Patella by quadriceps tendon and tibial tuberosity by patellar ligament

Muscle: **Vastus lateralis**
Origin: Inferior to greater trochanter and linea aspera
Insertion: Patella by quadriceps tendon and tibial tuberosity by patellar ligament

Muscle: **Vastus medialis**
Origin: Intertrochanteric line and linea aspera
Insertion: Patella by quadriceps tendon and tibial tuberosity by patellar ligament

Muscle: **Vastus intermedius** (not visible; deep to rectus femoris and vastus lateralis)
Origin: Anterior and lateral surfaces of femoral shaft
Insertion: Patella by quadriceps tendon and tibial tuberosity by patellar ligament

Anterior view

2. Sit on a lab stool with your legs and feet hanging freely and place your hand over the right quadriceps femoris muscles. Extend your leg so that it is elevated to a horizontal position. Feel the muscles contract as you raise your leg. The quadriceps femoris muscles are the prime movers for leg extension.

3. The rectus femoris is a synergist for hip flexion, but the other three muscles in the quadriceps group are not. Explain why.

▶ ______________________________

____

____

____

____

____

C Sartorius is the most superficial muscle in the anterior thigh.

1. On a model, identify **sartorius** and trace its path along the anterior thigh. From its origin at the anterior superior iliac spine, the muscle passes inferomedially across the thigh to its insertion on the medial surface of the proximal end of the tibia. Notice that it crosses both the hip and knee joints.

2. The various actions of sartorius are required for the crossed-legged sitting position. Sit on a stool with both feet resting on the floor. Now, lift your right foot and cross your lower limbs so that the right leg rests on the left thigh. Identify the various actions of your right lower limb as you assume this position.

At the hip, identify three actions:

▶ ______________________________

____

____

At the knee, identify one action:

▶ ______________________________

Do you think sartorius is an agonist or synergist for these movements? Explain.

▶ ______________________________

____

____

MAKING CONNECTIONS

The iliopsoas can also flex the trunk when sitting up from a supine position. In this example, the origins and insertions of the muscles are reversed. Can you explain why?

▶ ______________________________

____

____

Activity **12.9**

Muscles of the Lower Limb: Examining the Muscles of the Medial Thigh

The **medial thigh muscles** are commonly referred to as the groin muscles. Most of these muscles originate on the pubic bone or ischium and insert along the posterior aspect of the femur. They function as synergists while walking and play a role in maintaining erect posture.

A All three superficial muscles in the medial thigh originate from the pubic bone.

1 On a model, identify the three superficial muscles:

Muscle: **Pectineus**
Origin: Superior ramus of pubic bone
Insertion: Pectineal line of femur

Muscle: **Adductor longus**
Origin: Body of pubic bone
Insertion: Linea aspera of femur

Muscle: **Gracilis**
Origin: Body and inferior ramus of pubic bone
Insertion: Medial surface of proximal end of tibia

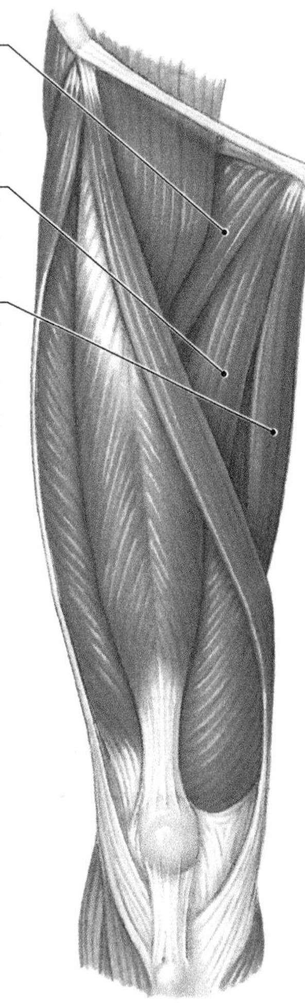

Anterior view

2 Connect the origins and insertions of pectineus and adductor longus with pipe cleaners. Observe the path of these muscles; notice that from their origins, the muscles travel posteriorly and inferiorly to reach their insertions.

Predict whether these muscles will: ▶

Adduct or abduct the thigh:

____

Extend or flex the thigh:

____

3 Use pipe cleaners to connect the origin and insertion of gracilis. Identify the joints that the muscle crosses. Predict what its actions will be at each joint. ▶

	Joint Name	Function(s) of Gracilis at Joint
Joint 1		
Joint 2		

IN THE CLINIC

Groin Pull

A groin pull (groin injury) refers to some degree of damage to one or more medial thigh muscles. Typically, the injury presents as a strain, stretch, or tear of the muscles near their origins on the pelvis. Groin injuries often occur among athletes who play sports that require quick running starts, such as sprinting, kicking a ball in soccer, stealing bases in baseball, running downfield to catch a pass in football, or executing a fast break in basketball. Groin pulls can be treated by taking anti-inflammatory medications, icing the injured area, and rest, followed by physical therapy to improve flexibility and strength.

Want more practice? Go to: **MasteringA&P** > Study Area > Menu >

- Lab Tools > PAL > Anatomical Models > Muscular System > Lower Limb
- Lab Tools > PAL > Human Cadaver > Muscular System > Lower Limb
- Animations & Videos > *A&PFlix* > Origins, Insertions, Actions, Innervations > Lower Limb
- Animations & Videos > *A&PFlix* > Group Muscle Actions and Joints > Unit 4: Muscles that act on the hip joint and femur

B All three deep muscles in the medial thigh insert on the femur.

1 On a model, identify the three deep mucles:

Muscle: **Obturator externus**
Origin: External margin of obturator foramen
Insertion: Trochanteric fossa of femur

Muscle: **Adductor brevis**
Origin: Body and inferior ramus of pubic bone
Insertion: Pectineal line and linea aspera of femur

Muscle: **Adductor magnus**
Origin of adductor part: Inferior ramus of pubic bone; ramus of ischium
Origin of hamstring part: Ischial tuberosity
Insertion of adductor part: Linea aspera and medial supracondylar line of femur
Insertion of hamstring part: Adductor tubercle of femur

Anterior view

2 Connect the origin and insertion of the obturator externus with a pipe cleaner. Observe the muscle's path and predict its action.

▶ ______________________

____

List two other muscles that have the same action as obturator externus.

▶ ______________________

____

3 Sit on a lab stool with your feet hanging freely and your thighs separated. With your hands on your medial thighs, attempt to draw them together against the resistance of the seat. As you perform this action (adduction of the thighs), you will feel your medial thigh muscles contract.

Identify the gluteal muscles that are antagonists for thigh adduction.

▶ ______________________

____

____

____

Word Origins

The medial thigh muscles include adductor brevis, adductor longus, and adductor magnus. In Latin, *brevis, longus,* and *magnus* mean "short," "long," and "large," respectively. These terms are used regularly to compare muscles that are located in the same region and share a common action.

MAKING CONNECTIONS

You are at a nightclub in which the main attraction is a mechanical bull. You decide to try your luck with "bull riding." Although it is a wild ride, you surprise yourself by staying on for 10 seconds. Shortly thereafter, you develop tightness and pain in your medial thighs. Explain why you are experiencing this discomfort.

▶ ______________________

____

____

____

____

____

____

____

Activity **12.10**

Muscles of the Lower Extremity: Examining the Muscles of the Posterior Thigh

The three long muscles in the posterior thigh are called the "hamstrings." These muscles share a common origin on the ischial tuberosity, deep to the gluteus maximus.

A Laterally, the biceps femoris has two heads of origin. Medially, the superficial semitendinosus travels parallel to the deeper semimembranosus.

① On a model, identify the three muscles of the posterior thigh:

Muscle: **Biceps femoris**
Origin of long head: Ischial tuberosity
Origin of short head: Linea aspera and lateral supracondylar line
Insertion: Lateral side of fibular head

Muscle: **Semitendinosus**
Origin: Ischial tuberosity
Insertion: Medial surface of proximal tibia

Muscle: **Semimembranosus**
Origin: Ischial tuberosity
Insertion: Posterior surface of medial condyle of tibia

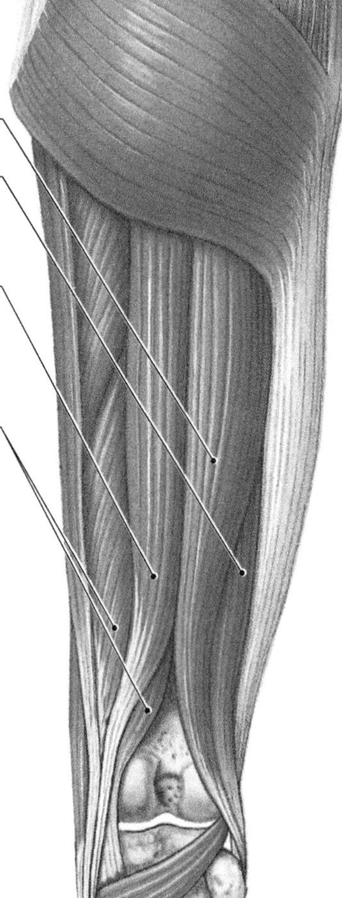

Posterior view

Word Origins

The name *hamstrings* comes from the practice of butchers in 18th-century England who inserted hooks into these muscles to hang the carcasses of pigs in their shops.

② On a skeleton, use pipe cleaners to connect the origins and insertions of the posterior thigh muscles. Examine the paths of the muscles. Identify the joints that each muscle crosses and predict the action of the muscle at each joint. ▶

Muscle	Joints Crossed by Muscle	Action of Muscle at Joint
Semimem-branosus		
Semiten-dinosus		
Biceps femoris, long head		
Biceps femoris, short head		

Is there a muscle in this group that differs structurally and functionally from all the others? Explain.

▶ ____

____

____

____

____

____

____

Want more practice? Go to: **MasteringA&P** > Study Area > Menu >

- Lab Tools > PAL > Anatomical Models > Muscular System > Lower Limb
- Lab Tools > PAL > Human Cadaver > Muscular System > Lower Limb
- Animations & Videos > ***A&PFlix*** > Origins, Insertions, Actions, Innervations > Lower Limb
- Animations & Videos > ***A&PFlix*** > Group Muscle Actions and Joints > Unit 5: Muscles that act on the knee joint and lower leg

B The length of hamstring muscles varies and can change depending on a person's exercise routine.

1. From the anatomical position, bend forward slowly (without bouncing) at the waist and attempt to touch your toes with the knees extended.
2. Some people cannot touch their toes because their hamstrings are too short. Others can place their palms flat on the floor because their hamstrings are very long. Compare your ability to touch your toes with five other students in the class: ▶

Name	Ability to Touch Toes (Yes/No/ Touch Palms to Floor)

3. Stretching the hamstrings on a regular basis will lengthen the muscles. If you are unable to touch your toes, begin to do exercises that will lengthen the hamstring muscles. After a few weeks of regular, gentle stretching, try touching your toes again. You should notice a difference. What do you think are the benefits of having longer hamstring muscles?

▶ ______________________________

____

____

IN THE CLINIC

Hamstring Injuries

Hamstring injuries are common among individuals who participate in sports that involve running, jumping, or quick starts and stops. Hamstring pulls can vary in severity, ranging from small tears of a few muscle fibers to an avulsion (forcible separation) of tendinous attachments from the ischial tuberosity. These injuries can be painful and slow to heal. Initial treatment includes rest, icing, and anti-inflammatory medication to reduce pain and swelling. This regimen would be followed by physical therapy to improve range of motion, flexibility, and strength. Severe injuries may require surgical repair.

MAKING CONNECTIONS

When the thighs and legs are flexed (e.g., when you are seated), the semimembranosus and semitendinosus can extend the trunk (e.g., when you stand from a seated position). Describe the origins and insertions of these muscles during trunk extension.

▶ ______________________________

____

____

____

____

Activity **12.11**

Muscles of the Lower Limb: Examining the Muscles of the Anterior and Lateral Leg

The leg is divided into three muscular compartments: anterior and lateral (discussed in this activity) and posterior (see Activity 12.12). The muscles in each compartment have similar functions.

A The **anterior compartment** of the leg contains three muscles that originate on the tibia or fibula and insert on bones in the foot. The anterior compartment muscles have long tendons that travel through tunnels, similar to the carpal tunnel at the wrist. The floors of these tunnels are formed by the distal ends of the tibia and fibula, the interosseous membrane that connects those bones, and the tarsal bones of the foot. The roofs are formed by thick bands of connective tissue known as the **superior extensor retinaculum** and **inferior extensor retinaculum.**

1 On a model, identify the muscles in the anterior compartment of the leg:

- *Muscle:* **Tibialis anterior**
- *Origin:* Lateral condyle and lateral surface of tibia; interosseous membrane
- *Insertion:* Medial cuneiform bone and first metatarsal of foot

- *Muscle:* **Extensor digitorum longus**
- *Origin:* Lateral condyle of tibia; medial surface of fibula; interosseous membrane
- *Insertion:* Middle and distal phalanges of digits 2–5

- *Muscle:* **Extensor hallucis longus**
- *Origin:* Anterior surface of fibula; interosseous membrane
- *Insertion:* Distal phalanx of great toe

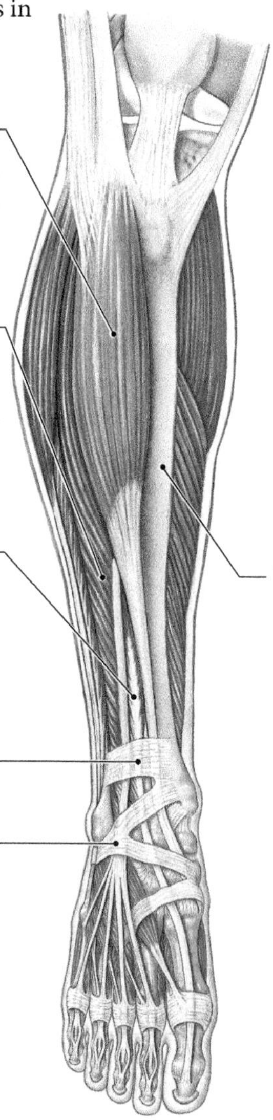

Tibia

Superior extensor retinaculum

Inferior extensor retinaculum

Anterior view

2 On a skeleton, use pipe cleaners to connect the origins and insertions of the anterior compartment muscles. Determine the joints that each muscle crosses and place a check mark in the appropriate boxes in the table. ▶

Muscle	Knee	Ankle	Intertarsal Joints	Interphalangeal Joints
Tibialis anterior				
Extensor digitorum longus				
Extensor hallucis longus				

3 Observe the paths of these muscles as illustrated by the pipe cleaners. Answer the following questions: ▶

All three muscles have a common action. Do they plantar flex or dorsiflex the foot?

____

Does tibialis anterior invert or evert the foot?

____

Identify another action of extensor digitorum longus.

____

Identify another action of extensor hallucis longus.

____

Word Origins

Hallucis comes from the Latin word *hallux,* which means "great toe." Any muscle with *hallucis* in its name will act on the great toe.

Want more practice? Go to: **MasteringA&P** > Study Area > Menu >
- Lab Tools > PAL > Anatomical Models > Muscular System > Lower Limb
- Lab Tools > PAL > Human Cadaver > Muscular System > Lower Limb
- Animations & Videos > A&PFlix > Origins, Insertions, Actions, Innervations > Lower Limb
- Animations & Videos > A&PFlix > Group Muscle Actions and Joints > Unit 6: Muscles that act on the ankle joint

B The two muscles in the **lateral compartment** of the leg originate on the lateral side of the fibula and insert on bones in the foot. Similar to the anterior compartment muscles, the inserting tendons travel through tunnels, formed by bone and retinacula, to reach their insertions.

1 On a model, identify the muscles in the lateral compartment of the leg:

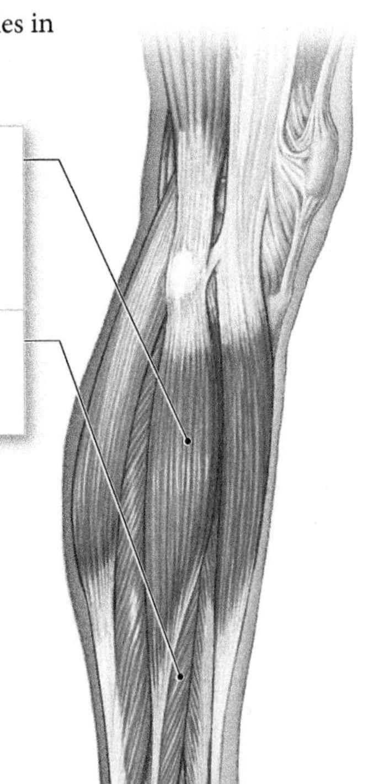

Lateral view

2 Use a chain of 2–3 pipe cleaners to trace the path of fibularis longus:

- Place one end of the pipe cleaner chain on the fibular head.
- Extend the pipe cleaners inferiorly. Pass them lateral to the calcaneus and posterior to the lateral malleolus, and then bend them anteriorly toward the cuboid.
- At the cuboid, curve the chain onto the sole of the foot so it travels medially.
- Finally, on the medial side of the foot, rest the free end of the chain on the medial cuneiform.
- Observe how the tendon supports the lateral longitudinal and transverse arches of the foot.

3 Use a pipe cleaner to trace the path of fibularis brevis:

- Set one end of the pipe cleaner along the the lateral surface of the fibula at its distal end.
- As before, pass the pipe cleaner lateral to the calcaneus and posterior to the lateral malleolus.
- As the pipe cleaner bends anteriorly toward the cuboid, place the free end on the base of the metatarsal of the little toe.

4 Observe the paths of the fibularis muscles, represented by the pipe cleaners. Predict the primary action of these muscles.

▶ ______________________________

____

____

IN THE CLINIC

Foot Drop

The tibialis anterior is a strong dorsiflexor of the foot, particularly when the foot is elevated while walking. If the tibialis anterior is impaired, the foot remains plantar flexed and drags along the ground. This condition, called **foot drop,** can be temporary or permanent, depending on the cause. The most common treatments are the use of orthotics to support the foot and physical therapy to increase muscle strength and improve gait.

MAKING CONNECTIONS

If the two muscles in the lateral leg compartment were paralyzed, the foot would remain in an abnormal inverted position. Explain why.

▶ ______________________________

____

____

____

____

____

____

Activity **12.12**

Muscles of the Lower Limb: Examining the Muscles of the Posterior Leg

The **posterior compartment** of the leg contains seven muscles arranged in three layers: superficial, intermediate, and deep.

A The one muscle in the superficial layer and the three muscles in the intermediate layer are commonly referred to as "calf muscles." Three of the four muscles have a common insertion on the posterior surface of the calcaneus.

1. On a model, identify the muscles in the superficial and intermediate layers:

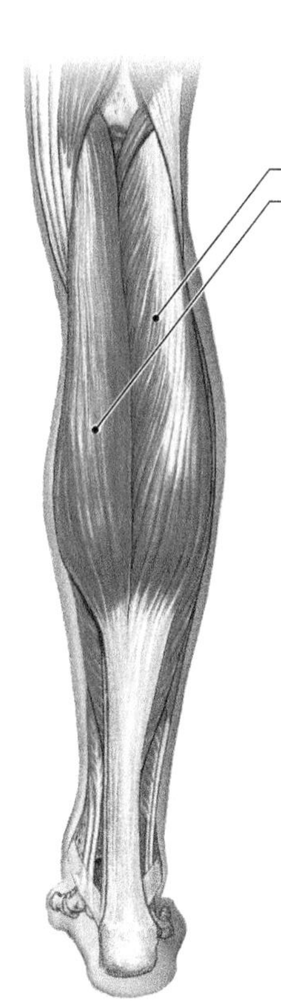

Superficial layer, posterior view

Muscle: **Gastrocnemius**
Origin of lateral head: Lateral condyle of femur
Origin of medial head: Medial condyle of femur
Insertion: Calcaneus, by the calcaneal (Achilles) tendon

Muscle: **Plantaris**
Origin: Lateral supracondylar line of femur
Insertion: Calcaneus

Muscle: **Popliteus**
Origin: Lateral condyle of femur
Insertion: Posterior surface of proximal tibia

Muscle: **Soleus**
Origin: Head and posterior surface of fibula; medial surface of tibia
Insertion: Calcaneus, by the calcaneal (Achilles) tendon

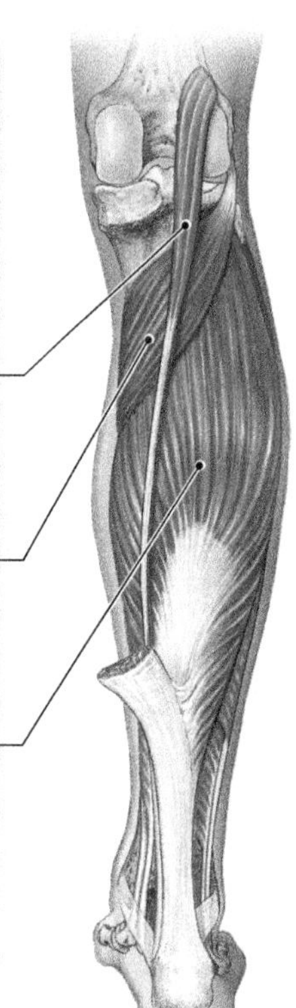

Intermediate layer, posterior view

2. On a skeleton, use pipe cleaners to connect the origins and insertions of gastrocnemius and soleus. Identify the joints that are crossed by: ▶

Gastrocnemius: ______________________

Soleus: ______________________

3. Assume a crouched position, as if you are in starting blocks and about to start a 100-m race. Push off with the back leg to start the "race."

⚠ **Warning:** *Perform the "starting block" activity only if your instructor agrees and there is enough space available.*

During the push-off, your gastrocnemius and soleus are performing what action at the foot?

▶ ______________________

What other action can gastrocnemius perform but soleus cannot?

▶ ______________________

4. Sit on a lab stool and extend your right leg so that it is parallel to the floor. In this position, your leg is fully extended and "locked." Now, flex your right leg so that it rests on the floor or hangs freely. When you perform this movement, the popliteus "unlocks" the knee by rotating the leg medially.

IN THE CLINIC

The Plantaris Muscle

The plantaris muscle is absent in 7 to 10 percent of the population. Functionally, the muscle has little importance, but its long tendon is often used for reconstructive surgery to repair ligaments and tendons in other areas of the body.

Word Origins

Gastrocnemius comes from two Greek words: the word *gaster* means "stomach," and *kneme* means "leg." Taken literally, gastrocnemius means "stomach of the leg."

Soleus derives from the Latin word *solea,* meaning "sole fish." The soleus muscle resembles this fish species in appearance.

Want more practice? Go to: **MasteringA&P** > Study Area > Menu > Lab Tools > PAL >
- Anatomical Models > Muscular System > Lower Limb
- Human Cadaver > Muscular System > Lower Limb

B The three deep muscles in the posterior compartment have long tendons of insertion that reach the foot by passing through a tunnel formed by bone and a thick band of connective tissue **(flexor retinaculum)**.

1 On a model, identify the deep muscles of the **posterior compartment:**

Muscle: **Tibialis posterior**
Origin: Posterior surfaces of tibia and fibula; interosseous membrane
Insertion: Navicular, cuboid, and cuneiform bones; metatarsals 1–3

Muscle: **Flexor digitorum longus**
Origin: Posterior surface of tibia
Insertion: Distal phalanges of digits 2–5

Muscle: **Flexor hallucis longus**
Origin: Posterior surface of fibula; interosseous membrane
Insertion: Distal phalanx of great toe

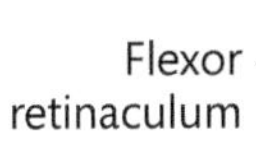

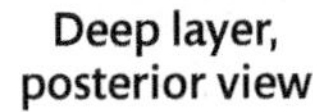

Deep layer, posterior view

2 On a skeleton, use pipe cleaners to connect the origins and insertions of the three deep muscles. For all three muscles, the pipe cleaners will pass posterior to the medial malleolus and onto the medial side of the calcaneus. From there, they will curve onto the sole of the foot to reach their insertions. If your instructor allows you, gently tape the pipe cleaners to their bony attachments so that you can observe the relative positions of the muscles. Notice that the tendons of flexor digitorum longus and flexor hallucis longus cross just before they turn onto the sole of the foot.

3 Observe the paths of all the tendons along the sole of the foot. Which foot arch do they help to support?

▶ ______________________________

4 Sit on a stool so that your legs can hang freely and plantar flex your feet. Notice that when you plantar flex the feet, the toes are in the flexed position. The flexor hallucis longus (big toe) and flexor digitorum longus (lateral four toes) are agonists for toe flexion and synergists for plantar flexion.

Identify a muscle that is an agonist for plantar flexion.

▶ ______________________________

Identify a muscle that is an antagonist for plantar flexion.

▶ ______________________________

5 An important action of tibialis posterior is foot inversion. Identify two muscles that are antagonists of tibialis posterior.

▶ ______________________________

____

IN THE CLINIC

Calcaneal Tendonitis

Calcaneal (Achilles) tendonitis is a common running injury. It is caused by tissue inflammation, leading to small tears in the calcaneal tendon. The injury causes pain while walking and is extremely slow to heal. Individuals with a history of calcaneal tendonitis are vulnerable to a much more severe injury, a ruptured calcaneal tendon. The tendon can rupture if a plantar-flexed foot is suddenly dorsiflexed. Typically, there is a sudden pain in the calf and an audible snap as the tendon breaks. A fleshy lump forms in the calf due to the shortening of the gastrocnemius and soleus, and a gap can be felt where the tendon once was. When the calcaneal tendon is ruptured, the individual cannot use the limb. The foot can be dorsiflexed, but plantar flexion is difficult.

MAKING CONNECTIONS

In this activity, you learned that tibialis posterior inverts the foot. If you were to lose the function of this muscle, would foot inversion still be possible? Explain.

▶ ______________________________

____

____

____

____

____

____

Activity 12.13

Muscles of the Lower Limb: Examining the Intrinsic Foot Muscles

The intrinsic foot muscles are organized into four layers along the plantar surface; the first layer is the most superficial, and the fourth layer is the deepest. Structurally, there are many parallels between the musculature of the foot and the hand. The foot cannot perform the very specific, fine motor skills characteristic of the hand, however.

A All the muscles in the first layer originate on the calcaneus.

1 On a model, identify the three muscles of the first layer:

Abductor hallucis travels medially and inserts on the proximal phalanx of the great toe.

Abductor digiti minimi travels laterally and inserts on the proximal phalanx of the little toe.

Flexor digitorum brevis travels between these two muscles and inserts on the middle phalanges of the four lateral toes.

Plantar view

2 The names of the three muscles in the first layer suggest that they abduct and flex the toes but they are not very efficient at performing these actions. What do you think is a more important function of these muscles?

▶ ______________________________

____

____

____

____

____

____

B All the muscles in the second layer insert on the tendons of other muscles.

1 On a model, identify the muscles of the second layer:

The **lumbricals** originate on the tendons of flexor digitorum longus and insert on the tendons of extensor digitorum longus.

Quadratus plantae originates on the calcaneus and inserts on the tendon of flexor digitorum longus.

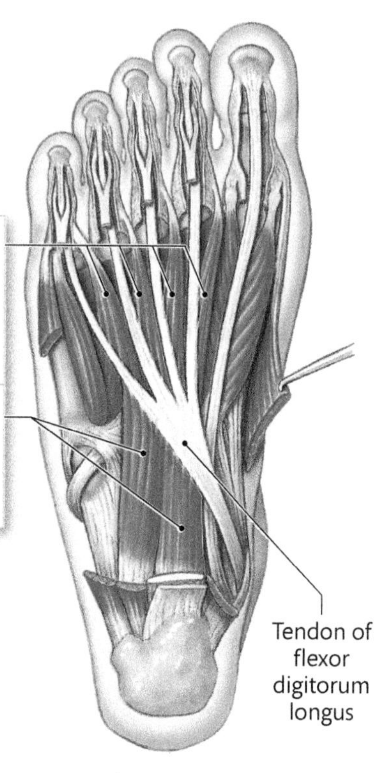

Plantar view

2 The function of the lumbricals in the foot is similar to their counterparts in the hand. Use this information to determine the function of these muscles.

▶ ______________________________

____

____

3 Complete the following table. ▶

Muscle	Function
Quadratus plantae	Synergist for flexion of digits 2–5 in the foot
	Agonist for flexion of digits 2–5 in the foot
	Antagonist for flexion of digits 2–5 in the foot

C The muscles in the third layer act on either the great toe or little toe.

1. On a model, identify the muscles of the third layer:

Flexor hallucis brevis originates from the cuboid and lateral cuneiform and inserts on the proximal phalanx of the great toe.

Flexor digiti minimi brevis originates at the base of the fifth metatarsal and inserts on the proximal phalanx of the same toe.

Adductor hallucis (not shown) has two heads of origin on the second through fourth metatarsals. The two heads converge and insert on the lateral side of the proximal phalanx of the great toe.

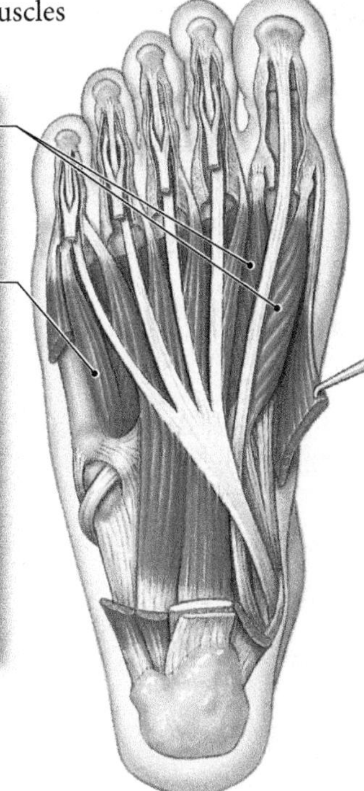

Plantar view

2. Complete the following table. ▶

Muscle	Function
Flexor hallucis brevis	Synergist for flexion of the great toe
	Agonist for flexion of the great toe
	Antagonist for flexion of the great toe

D The interossei muscles occupy the fourth layer.

1. Four **dorsal interossei** and three **plantar interossei** fill the spaces between the metatarsal bones. The muscles originate on these bones and run between them, inserting on the proximal phalanges of the lateral four digits. On a model, identify the dorsal interossei (plantar interossei are not shown).

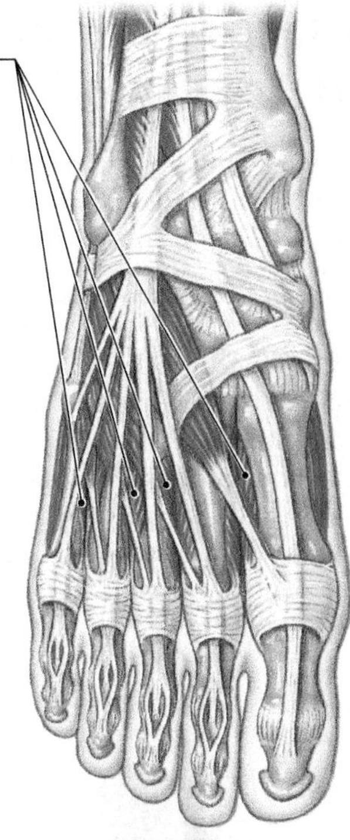

Dorsal view

2. The dorsal and plantar interossei weakly abduct and adduct the toes. Speculate on another more important action of these muscles.

▶ ____________________

____

____

____

____

____

____

MAKING CONNECTIONS

The hand is capable of finer motor skills than the foot partly due to functional differences between the thumb and the great toe. From a functional standpoint, explain the difference and why it is significant.

▶ ____________________

____

____

Want more practice? Go to: **MasteringA&P** > Study Area > Menu > Lab Tools > PAL >
- Anatomical Models > Muscular System > Lower Limb
- Human Cadaver > Muscular System > Lower Limb

Activity **12.14**

Surface Anatomy: Upper Limb Muscles and Associated Structures

Surface anatomy is the study of internal structures in relation to features on the surface of the body. We can identify internal structures by feeling them through the skin **(palpation).**

A Surface Anatomy of the Shoulder Muscles

1 Abduct your arm so that it is parallel to the floor and palpate the soft mound of tissue that forms the anterior border of the axilla (armpit). You are feeling the **anterior axillary fold,** which is produced by **pectoralis major** as it travels to its insertion on the humerus.

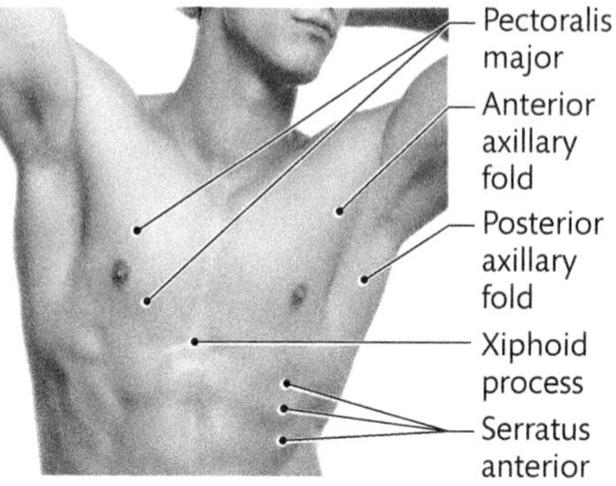

Anterior view

2 Keep your arm parallel to the floor and palpate the soft tissue that forms the posterior border of the axilla. This is the **posterior axillary fold,** which includes **latissimus dorsi** and **teres major.** Palpate the posterior axillary fold along its course to verify that these muscles also insert on the humerus. You can palpate the bulk of latissimus dorsi in the lumbar region of your back.

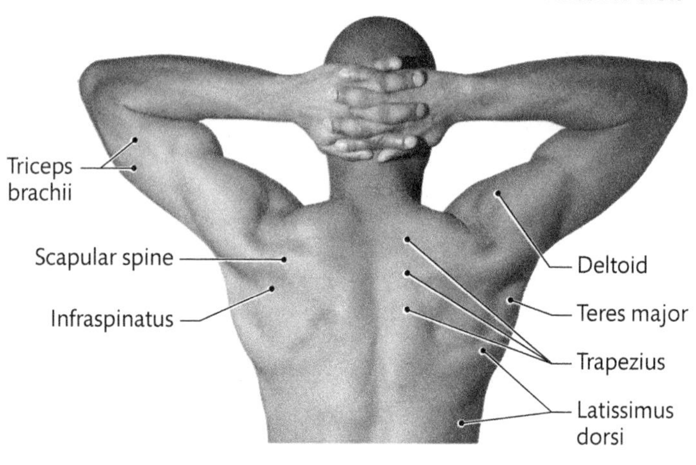

Posterior view

3 Place your hand on the posterior surface of your neck. From this position, palpate the **trapezius** as it sweeps across the posterior neck toward the shoulder.

4 Locate the position of the **scapular spine.** Palpate the **infraspinatus** in the region inferior to the spine.

5 Palpate the **deltoid** as it drapes over the shoulder joint. Follow the muscle's path to its insertion on the deltoid tuberosity.

B Surface Anatomy of Muscles in the Arm

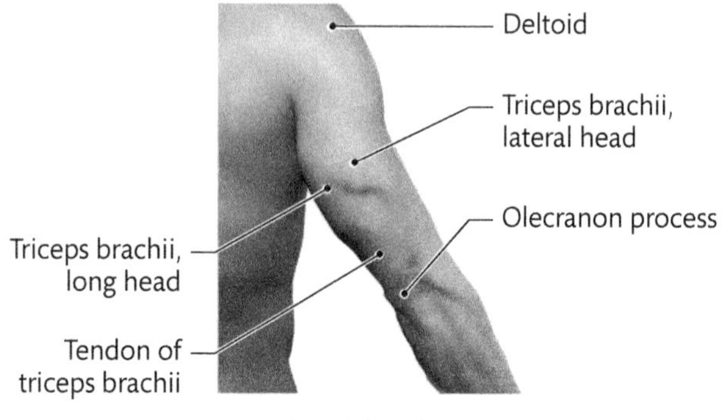

Posterior view

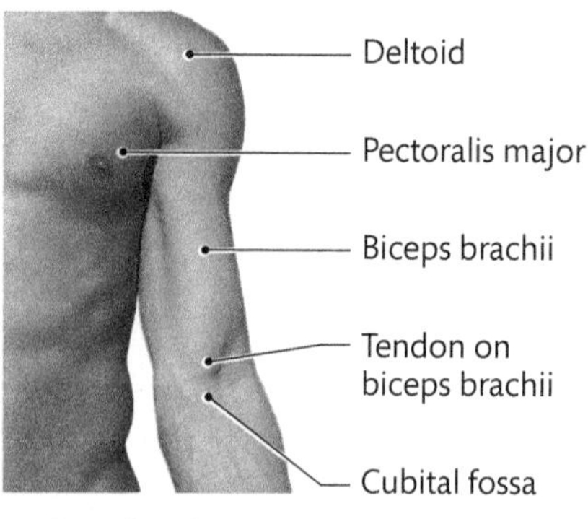

Anterior view

1 Extend your forearm and feel the contraction of **triceps brachii** on the posterior surface of the arm. Move your fingers distally to palpate the **triceps brachii tendon.** Follow the tendon to its insertion on the **olecranon** of the ulna.

2 Flex your forearm and palpate the anterior surface of your arm to feel the contraction of the **biceps brachii.** Keep your forearm flexed and palpate the **biceps tendon** as it travels through the **cubital fossa,** which is the shallow depression anterior to the elbow joint.

C Surface Anatomy of Muscles in the Forearm

1 Place your left thumb on the posterior surface of your right arm, just proximal to the elbow joint, and rest your fingers on the anterior forearm so that they are directed toward the wrist. The relative positions of the four superficial muscles of the anterior forearm are revealed by the placement of your fingers:

- Index finger = **pronator teres**
- Middle finger = **flexor carpi radialis**
- Ring finger = **palmaris longus**
- Little finger = **flexor carpi ulnaris**

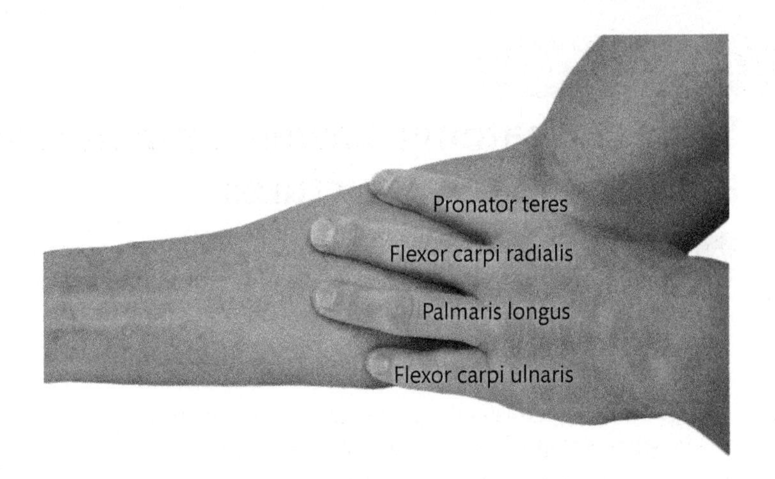

2 By flexing your fingers and hand, you can identify the tendons of several flexor muscles as they approach the wrist. From lateral to medial, identify the **tendons of flexor carpi radialis, palmaris longus, flexor digitorum superficialis,** and **flexor carpi ulnaris.** The palmaris longus tendon has a central position at the wrist and will likely be the most prominent. About 15 to 20 percent of the population lack this tendon on at least one side, however. Survey your class to see if anyone is missing the palmaris longus on one or both sides. The loss of palmaris longus has no effect on function.

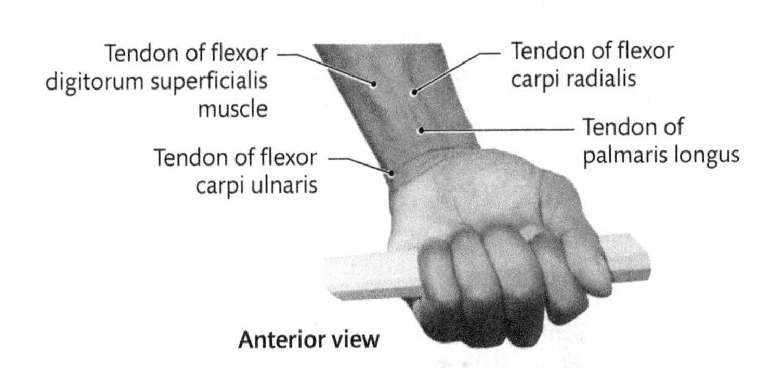

3 Extend your fingers and palpate the four tendons of **extensor digitorum** as they travel across the hand. Trace the tendons to their insertions on the fingers.

4 Extend your thumb as far as possible. The depression that appears at the base of this digit is known as the **anatomical snuff box.** The name comes from the 18th-century practice of placing powdered tobacco (snuff) in the depression and sniffing it into the nostrils. The anatomical snuffbox is bordered by the **tendon of extensor pollicis brevis,** laterally, and the more prominent **tendon of extensor pollicis longus,** medially. Palpate these tendons and trace them to their insertions on the thumb.

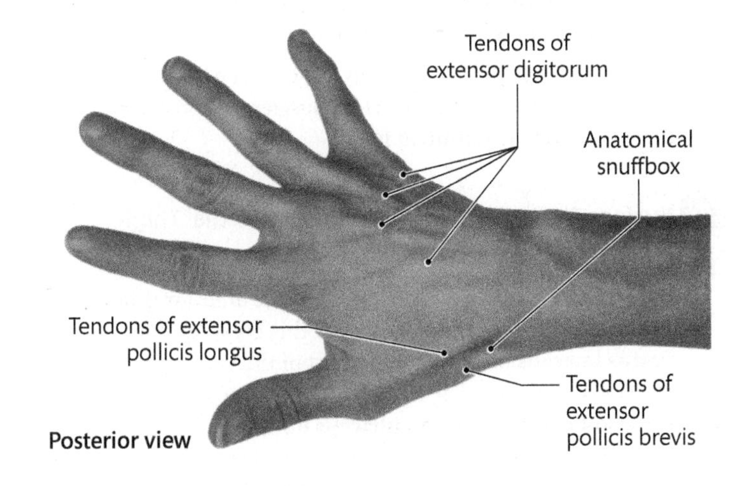

MAKING CONNECTIONS

What benefits can be derived from studying the surface anatomy of the muscular system?

▶ ____

____

____

____

Activity **12.15**

Surface Anatomy: Lower Limb Muscles and Associated Structures

A Surface Anatomy of the Gluteal Region

1 The **gluteus maximus** is the largest muscle in the buttock and can be easily palpated. Feel the muscle contracting when you extend your thigh.

2 In the superolateral region of the buttock, palpate the portion of **gluteus medius** that is not covered by gluteus maximus. You can feel the muscle contracting when you abduct your thigh.

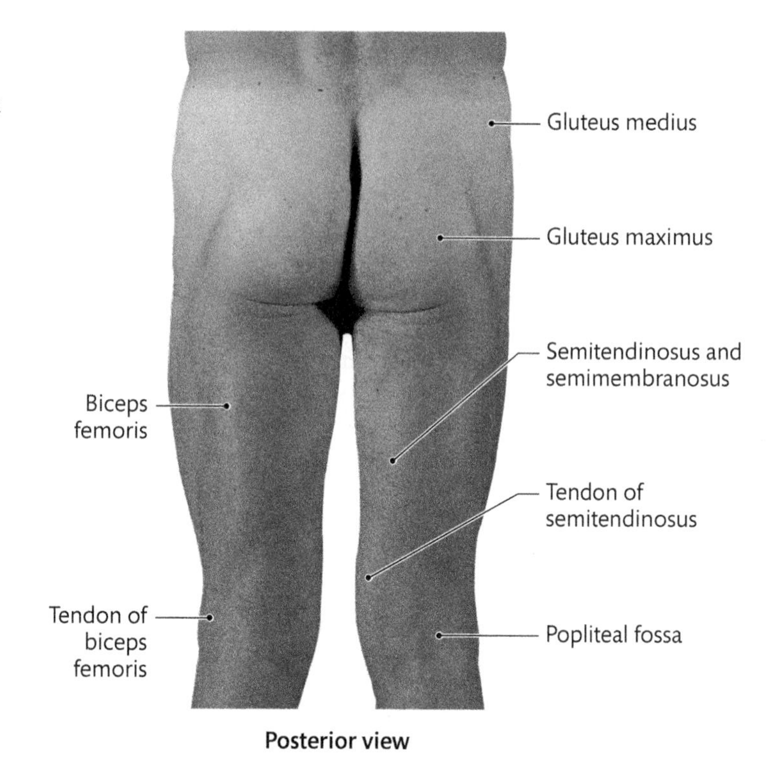

Posterior view

B Surface Anatomy of the Posterior Thigh

1 As you flex your knee, palpate your posterior thigh or hamstring muscles as they contract.

- The **biceps femoris** is the lateral muscle. Move your fingers inferiorly along the muscle until you can feel the **biceps femoris tendon.** Follow the tendon to its insertion on the head of the fibula.
- The **semitendinosus** and **semimembranosus** are the two medial muscles. As you move your hand inferiorly, you will be able to palpate the tendons of these muscles along the medial aspect of the knee.

2 The depression that you can feel between the lateral and medial tendons of the hamstring muscles and posterior to the knee is called the **popliteal fossa.**

C Surface Anatomy of the Anterior and Medial Thigh

1 Extend your leg and feel the contractions of the **quadriceps femoris muscle group** in the anterior thigh. Palpate three of the four muscles in this complex: the **vastus lateralis** on the lateral aspect of the thigh, the **vastus medialis** on the medial aspect, and the **rectus femoris** in the middle. Palpate the **quadriceps tendon** as it connects to the **patella.** As you flex and extend your leg, feel the **patellar ligament** between the patella and **tibial tuberosity.**

2 Flex your thigh and identify a shallow depression, called the **femoral triangle,** in the anterosuperior region of your thigh. Palpate the muscular borders of the femoral triangle: **Sartorius** travels diagonally across the anterior thigh and forms the lateral border. **Adductor longus,** one of the large adductor muscles in the medial thigh, forms the medial border.

3 Along the medial aspect of the thigh, palpate the **gracilis** as it travels inferiorly to its insertion on the tibia.

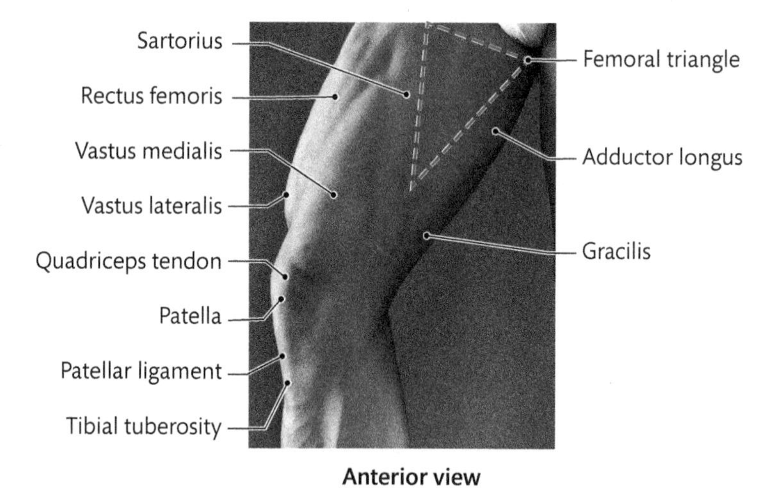

Anterior view

D Surface Anatomy of the Leg and Foot

1 On the leg, locate the subcutaneous **anterior border of the tibia** (the shin). Move your fingers just lateral to the bone and palpate the **tibialis anterior.** You can feel the muscle contracting when you dorsiflex the foot.

2 Extend your toes and observe the dorsum of your foot. Identify the **tendons of extensor digitorum longus and extensor hallucis longus.** Palpate the tendons as they pass to their insertions on the phalanges.

3 From a seated position, extend your leg and plantar flex your foot (alternatively, you can stand on your tiptoes). Palpate the lateral and medial heads of **gastrocnemius.** This muscle makes up the bulk of the muscle mass in the posterior leg or calf. The **soleus** lies deep to gastrocnemius and extends more laterally. You can best feel this muscle along the lateral margins of the posterior leg.

4 Move your hand distally along the calf muscles until you can feel the **calcaneal (Achilles) tendon.** Follow the tendon to its attachment on the posterior surface of the **calcaneus.**

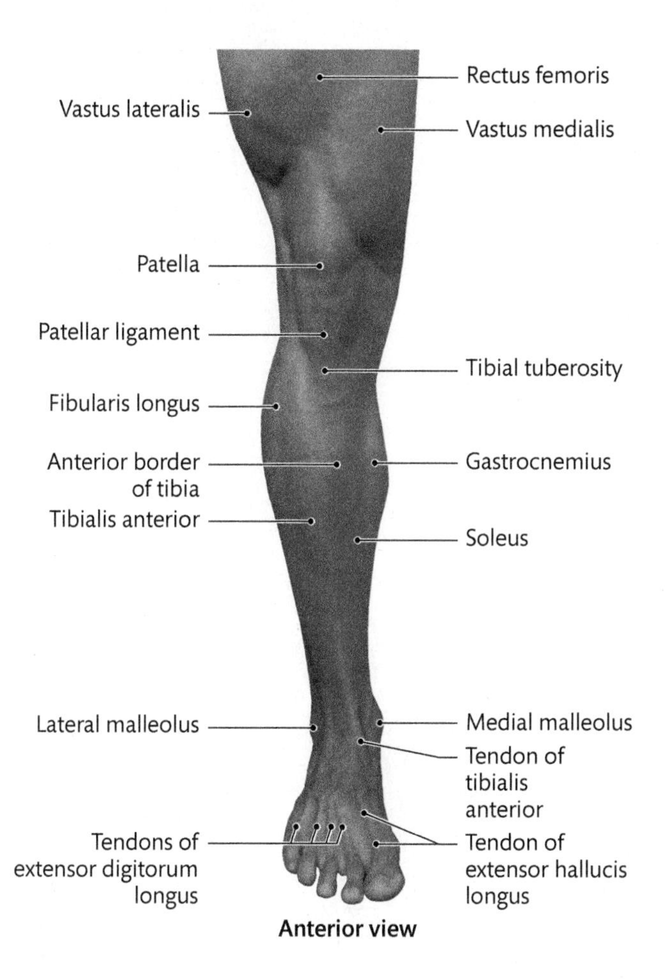

Anterior view

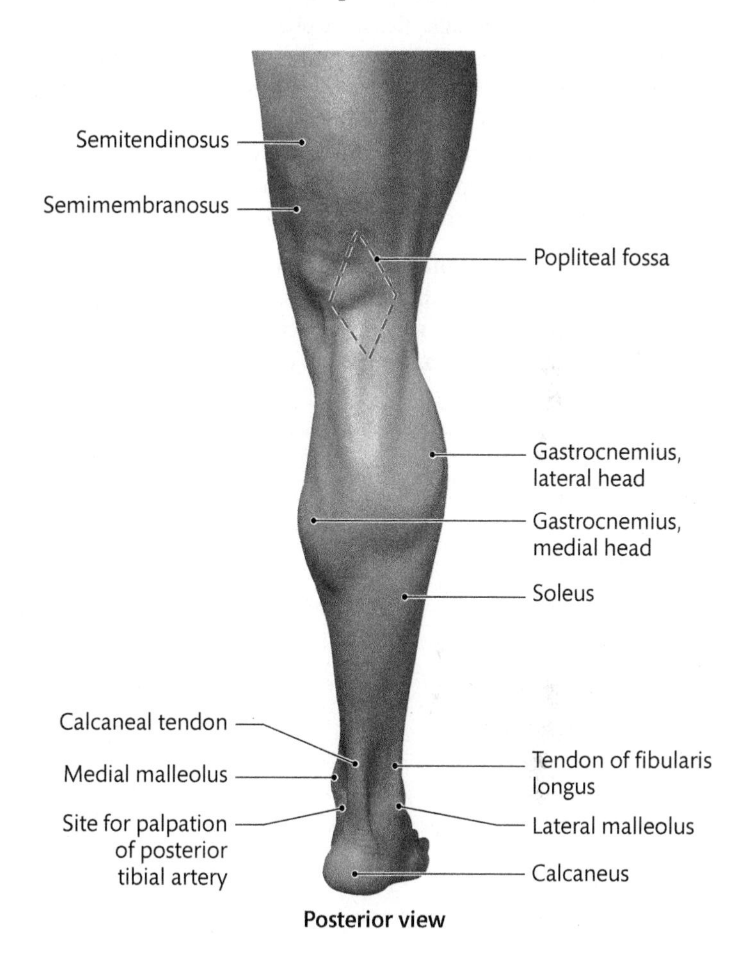

Posterior view

MAKING CONNECTIONS

If you wanted to study the surface anatomy of the muscular system, you would likely select a well-trained athlete as a subject. Other than bodybuilding, from what other sport could you choose a good subject? Explain why.

▶ ______________________________

____

____

____

____

BEFORE YOU MOVE ON . . .

‹‹ LOOKING BACK

As a general rule, a muscle will perform an action at a joint that it crosses. Many of the muscles that you studied in this exercise cross two or more joints. For example, the biceps brachii in the arm crosses the shoulder and elbow joints. In the forearm, several muscles—such as flexor carpi radialis, flexor carpi ulnaris, and extensor carpi ulnaris—cross the elbow and wrist joints. Other forearm muscles with very long tendons originate on the radius or ulna and cross the wrist joint and several joints in the hand to reach their insertions. Examples include flexor digitorum profundus, flexor pollicis longus, extensor pollicis longus, and extensor indicis. In the lower limb, several muscles in the thigh and leg—such as rectus femoris, the hamstring muscles, gastrocnemius, and extensor digitorum longus—also cross multiple joints.

While studying a muscle that crosses two or more joints, you observed that the muscle could act as an agonist for the action at one joint and as a synergist for the action at another joint. Speculate on why it is functionally significant to have muscles that can perform actions at more than one joint.

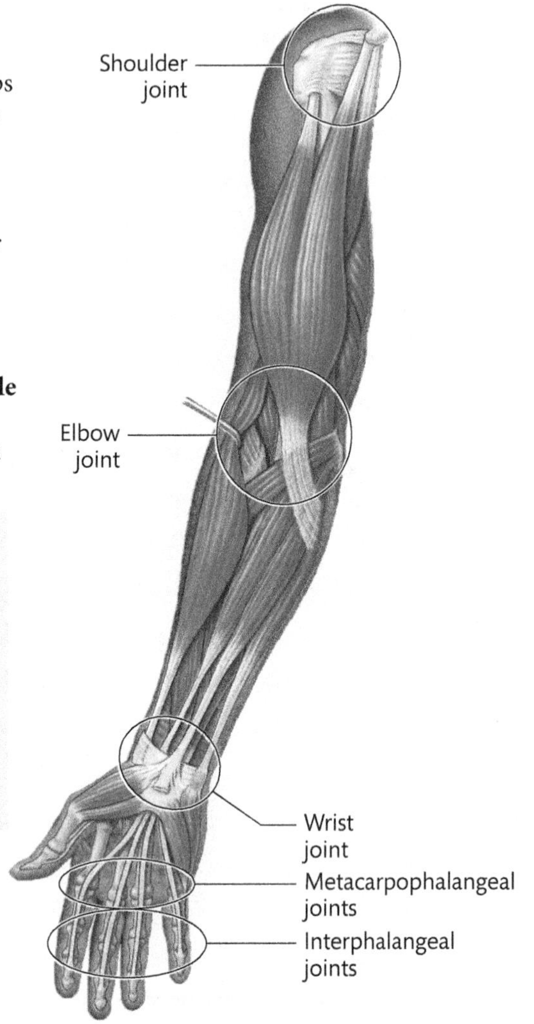

LOOKING FORWARD ››

When a muscle contracts, it exerts a force, called muscle tension, on an object. The object in turn exerts an opposing force, known as the load, which resists the muscle tension. For example, to lift your anatomy and physiology textbook off a desk, muscles in your upper limb contract, and the force generated (muscle tension) will lift the book. The weight of the book (the load), however, will oppose the tension produced by the contracting muscles. To lift the book off the desk, muscle tension must be greater than the load. In the next exercise, you will examine the relationship between muscle tension and load when muscles contract.

Name ______________________

Lab Section ______________________

Date ______________________

EXERCISE 12 REVIEW SHEET

Muscles of the Upper and Lower Limbs

1. Pectoralis minor, serratus anterior, and the rhomboids connect the scapula to bones on the axial skeleton. Would the loss of function of these muscles have any effect on the function of the deltoid or rotator cuff muscles, which act on the arm? Explain.

____

____

____

____

____

____

____

QUESTIONS 2–7: Match the function with the correct muscle in the diagram. Answers may be used once or not at all.

a. Muscle that has the same function as teres major
b. The trapezius and this muscle can elevate and retract the scapula
c. Rotator cuff muscle that medially rotates the arm
d. Rotator cuff muscle that laterally rotates the arm
e. Muscle that abducts, adducts, medially and laterally rotates, extends, and flexes the arm
f. Rotator cuff muscle that abducts the arm
g. Inferior region of this muscle depresses the scapula

2. ______________________

3. ______________________

4. ______________________

5. ______________________

6. ______________________

7. ______________________

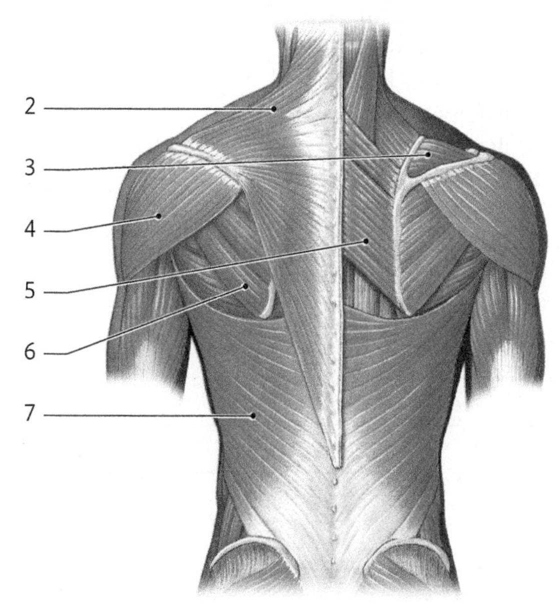

Posterior view

8. For the following actions, identify a muscle that would act as an agonist, an antagonist, and a synergist.

Action	Agonist	Antagonist	Synergist
Flexion of the forearm	a.	b.	c.
Abduction of the arm	d.	e.	f.
Pronation of the forearm	g.	h.	i.
Medial rotation of the arm	j.	k.	l.

9. The brachialis does not act at the shoulder, and the coracobrachialis does not act at the elbow joint. Explain why.

____

____

10. Explain why the force of gravity will allow you to extend your forearm even if the triceps brachii is not functioning.

____

____

QUESTIONS 11–22: Identify the labeled muscles. Select your answer from the list below. Answers may be used once or not at all.

a. Pronator teres
b. Extensor carpi radialis longus
c. Flexor carpi ulnaris
d. Flexor digiti minimi brevis
e. Brachialis
f. Triceps brachii
g. Flexor carpi radialis
h. Biceps brachii
i. Extensor carpi radialis brevis
j. Brachioradialis
k. Abductor pollicis brevis
l. Extensor digitorum
m. Flexor pollicis brevis
n. Extensor carpi ulnaris
o. Palmaris longus

11. ______________________

12. ______________________

13. ______________________

14. ______________________

15. ______________________

16. ______________________

17. ______________________

18. ______________________

19. ______________________

20. ______________________

21. ______________________

22. ______________________

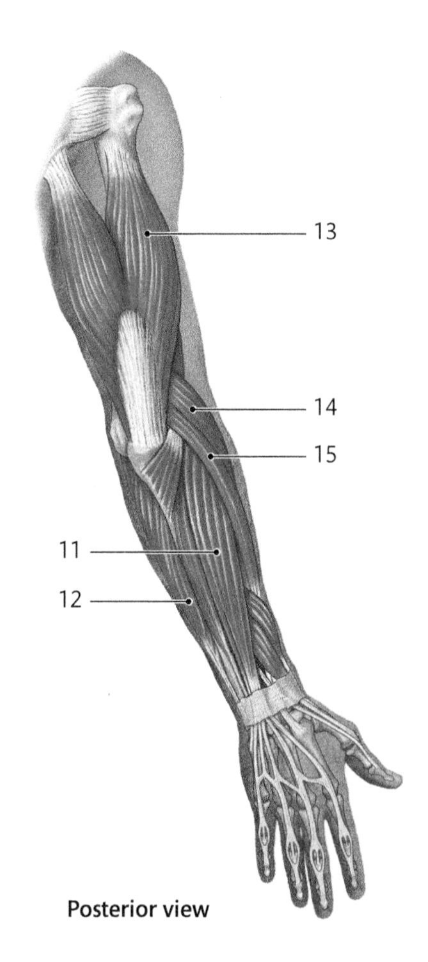

Posterior view

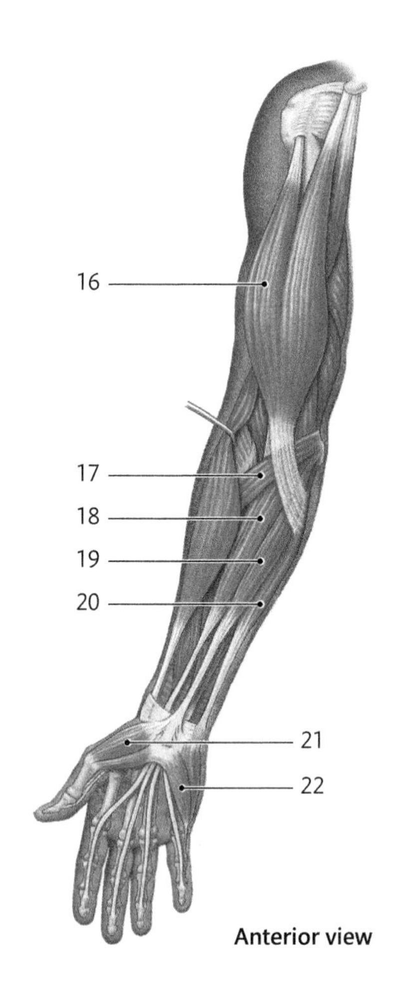

Anterior view

23. Explain why anatomists describe the adductor magnus as a muscle composed of two parts: an adductor part and a hamstring part.

____

____

____

____

24. Complete the following table.

Action	Agonist (1 example)	Antagonist (1 example)	Synergist (1 example)
Abduction of the thigh	a.	b.	Piriformis
c.	d.	Quadriceps femoris	Sartorius
e.	Tibialis anterior	f.	Extensor hallucis longus
g.	h.	Gluteus maximus	Tensor fasciae latae

QUESTIONS 25–33: Match the description with the correct muscle in the diagram. Answers may be used once or not at all.

a. Muscle that is the medial part of the quadriceps femoris group
b. Muscle that has two heads of origin
c. Muscle that has an origin in the abdominopelvic cavity
d. Muscle that has the same origin as tensor fasciae latae
e. Muscle that abducts the thigh
f. Only muscle in the medial thigh that crosses the knee joint
g. Only muscle in the quadriceps femoris group that can flex the thigh
h. Medial hamstring muscle
i. Muscle that adducts and flexes the thigh
j. Muscle that extends and laterally rotates the thigh

25. ______________________

26. ______________________

27. ______________________

28. ______________________

29. ______________________

30. ______________________

31. ______________________

32. ______________________

33. ______________________

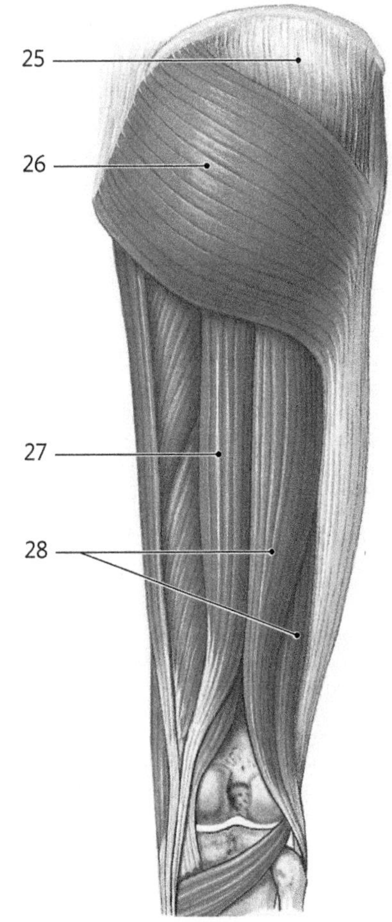
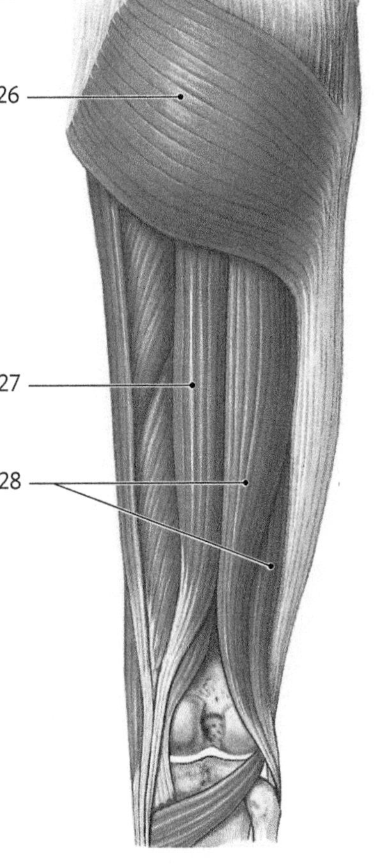

Posterior view

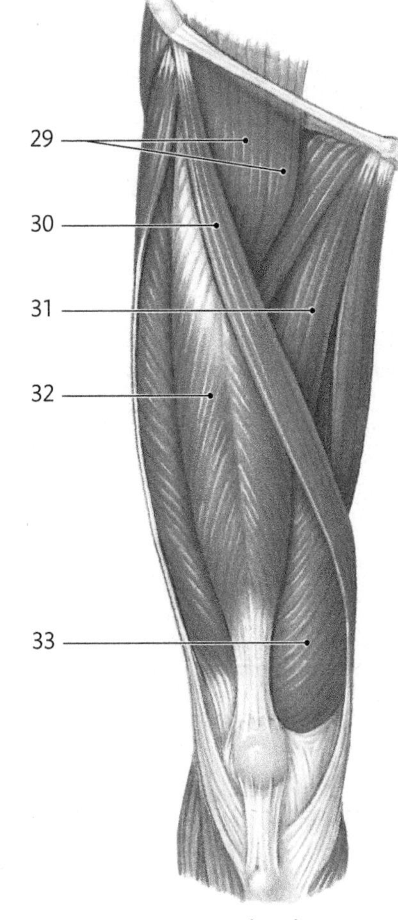

Anterior view

QUESTIONS 34–37: Identify the labeled muscles. Select your answer from the list below. Answers may be used once or not at all.

a. Fibularis longus
b. Tibialis posterior
c. Tibialis anterior
d. Gastrocnemius
e. Fibularis brevis
f. Soleus

34. ______________________________

35. ______________________________

36. ______________________________

37. ______________________________

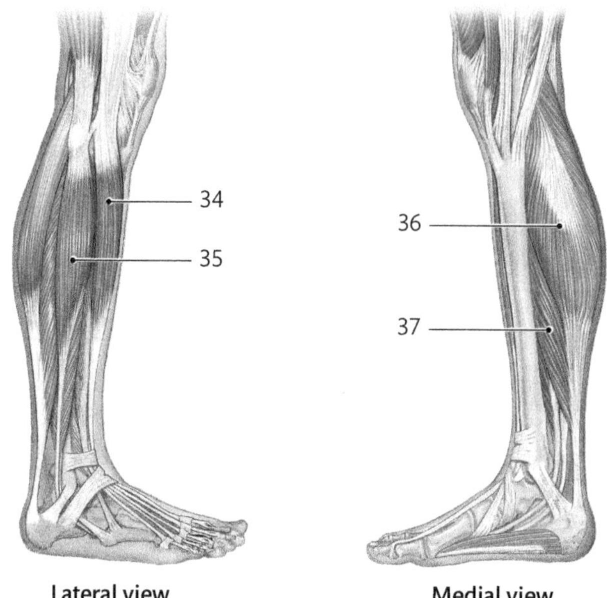

QUESTIONS 38–41: The vast majority of skeletal muscles are named according to one or more distinguishing features that describe their structure or function. Complete the following table by providing one example of a muscle whose name is derived from each listed criterion.

Criterion		Example
38. Number of origins	biceps = two origins	a.
	triceps = three origins	b.
	quadriceps = four origins	c.
39. Primary action of muscle	flexor = flexion of body part	a.
	extensor = extension of body part	b.
	abductor = abduction of body part	c.
	adductor = adduction of body part	d.
	supinator = supination of body part	e.
	pronator = pronation of body part	f.
40. Shape of muscle	serratus = serrated edge	a.
	rhombus = diamond shaped	b.
	quadratus = square shaped	c.
41. Size of muscle	maximus = largest	a.
	minimus = smallest	b.
	longus = long	c.
	brevis = short	d.
	major = larger	e.
	minor = smaller	f.

EXERCISE 13

Physiology of the Muscular System

LEARNING OUTCOMES

These Learning Outcomes correspond by number to the laboratory activities in this exercise. When you complete the activities, you should be able to:

Activity 13.1 **Explain the difference between isotonic and isometric muscle contractions and the interactions between muscles that contract concentrically and eccentrically.**

Activity 13.2 **Describe a motor unit and explain the functional significance of motor unit recruitment.**

Activity 13.3 **Analyze EMG recordings of motor unit recruitment.**

Activity 13.4 **Discuss the relationship between muscle load and muscle contraction speed.**

Activity 13.5 **Explain how muscle fatigue affects normal muscle activity.**

Activity 13.6 **Evaluate the range of motion of muscles at various joints.**

LABORATORY SUPPLIES

- Computer with Windows or Mac operating system
- Biopac Student Lab System (MP35, MP36, or MP45)
- Electrode lead set (SS2L)
- Disposable electrodes (EL503)
- Biopac electrode gel and abrasive pad
- Dumbbells of various weights or similar objects that are easy to hold
- Goniometers

PRE-LAB QUIZ

Before you begin, read all the activities in Exercise 13 and the required reading in your textbook that is assigned by your instructor.

1. During this laboratory exercise, dumbbells will *not* be used to complete which of the following activities?
   a. demonstrating muscle fatigue
   b. demonstrating isotonic and isometric contractions
   c. examining the relationship between resistance and contraction speed
2. During this laboratory exercise, what activity will you perform to demonstrate an isometric contraction? ______________________
3. During this laboratory exercise, what instrument will you use to measure range of motion? ______________________
4. Identify two joints at which you will be measuring range of motion during this laboratory exercise. ______________________
5. True or False: During an eccentric muscle contraction, muscle tension will exceed the load. __________
6. True or False: Muscle length will shorten during an isometric contraction. __________
7. True or False: It takes a longer period of time for a muscle to exceed the load of a heavier object than a lighter object. __________
8. The minimum stimulus required to initiate a muscle contraction is called the ______________________.
9. When muscle fatigue occurs, what metabolic waste product accumulates in muscle cells? ______________________
10. When additional ______________________ are activated, the force and speed of muscle contraction increases.

Activity 13.1

Demonstrating Isotonic and Isometric Contractions

Muscle tension is the force produced by a contracting muscle when it acts on an object. The **load**, which is generated by the object, is a force that opposes the muscle tension. The type of muscle contraction depends on the relation between the muscle tension and load.

A Isotonic Muscle Contractions: Concentric and Eccentric

Nearly all muscular activities involve **isotonic contractions** during which the length of the muscle will change when muscle tension rises. There are two types of isotonic contractions. (1) During a **concentric contraction**, muscle tension exceeds load. The muscle shortens when it contracts and performs work by moving an object from one position to another. One example is the contraction of the biceps brachii when you lift a glass of water off a table. (2) During an **eccentric contraction,** the load exceeds muscle tension, and the muscle lengthens when it contracts. If you place the glass back on the table, the biceps brachii is still contracting, but it is lengthening and acting as a brake to prevent you from dropping the glass.

1. Stand in an upright position with your feet flat on the floor.
2. Slowly and carefully lower your torso by bending your knees until your thighs are nearly parallel to the floor.
3. Remain in the crouched position for a few seconds. Then, slowly elevate your torso, returning to the upright position.
4. During the knee bend, identify the thigh muscles that are performing the following: ▶

   a. Concentric isotonic contraction:

   ____

   ____

   b. Eccentric isotonic contraction:

   ____

5. As you return to an upright position, identify the thigh muscles that are performing the following: ▶

   a. Concentric isotonic contraction:

   ____

   b. Eccentric isotonic contraction:

   ____

   ____

IN THE CLINIC

Tetanus

A period of smooth, sustained muscle contraction is known as **tetanus.** The disease tetanus has no relationship to normal muscle function but gets its name from the primary effect on its victim: sustained and forceful contractions of skeletal muscles. The disease tetanus is caused by the bacteria *Clostridium tetani*, which grow rapidly in anaerobic (no oxygen) environments such as a deep puncture wound from a nail. Once the *Clostridium* bacteria are established, they multiply rapidly and release a powerful toxin that inhibits the regulation of motor neuron activity. The outcome is sustained and abnormally forceful contractions of skeletal muscles. One common early symptom is difficulty in opening the mouth. For this reason, the disease is sometimes called **lockjaw.**

Symptoms of tetanus, such as difficulty swallowing and muscle spasms and stiffness, usually appear about 2 weeks after the initial infection. People can recover from the disease, but those with severe cases may die from respiratory failure if the diaphragm is affected. In the United States and other developed countries, tetanus is not considered a serious public health problem because proper immunization practices (tetanus shots) prevent infection. Booster shots are required every 10 years.

B Isometric Muscle Contractions

During an **isometric contraction,** muscle tension is unable to overcome the load exerted by the object. As tension increases, muscle length remains constant, and the object does not move.

If you attempt to push a parked automobile with its emergency brake engaged, you can feel muscles in your upper and lower limbs tighten as they contract, but the car will not budge. Your effort will be in vain because your muscles are contracting isometrically. That is, the load generated by the car is greater than the tension produced by your muscles.

1. Place the palms of your hands flat against a wall and attempt to push the wall away from you.

2. Identify the muscles that are contracting when you push against a wall and list the action of each muscle. ▶

Muscle	Action

3. Explain why the muscles you identified in step 2 are contracting isometrically.

▶ ______________________________

____

____

MAKING CONNECTIONS

At what point during a knee bend, which you performed during this activity, are your thigh muscles contracting isometrically? Explain your answer.

▶ ______________________________

____

____

____

____

____

____

____

Word Origins

Isotonic comes from Greek and means "equal tension" (*isos,* "equal"; *tonos,* "tension"). The name refers to the state of a contracting muscle as it shortens to overcome a constant load (tension), such as lifting a bottle of water off a table.

Isometric also has Greek origins and means "of equal dimensions" (*isos,* "equal"; *metron,* "measure"). The name refers to the condition of a muscle that contracts but does not change in length.

Activity 13.2

Electromyography Using the Biopac Student Laboratory System: Recording Motor Unit Recruitment Data

Skeletal muscles are controlled by motor nerve fibers (motor axons) that originate in the brain or spinal cord and travel along some cranial and all spinal nerves to reach their destinations. When a nerve fiber penetrates a muscle, it divides into numerous terminal branches. Each branch forms a neuromuscular junction with one muscle fiber. Each nerve fiber, its terminal branches, and all the muscle fibers it controls represent a **motor unit.** The contraction of muscle fibers in a motor unit is determined by the motor unit's **threshold stimulus,** which is the minimum stimulus strength required to initiate a contraction. If the threshold stimulus is reached, all the muscle fibers in a motor unit contract simultaneously. A whole muscle contains a mosaic of motor units with varying threshold stimuli.

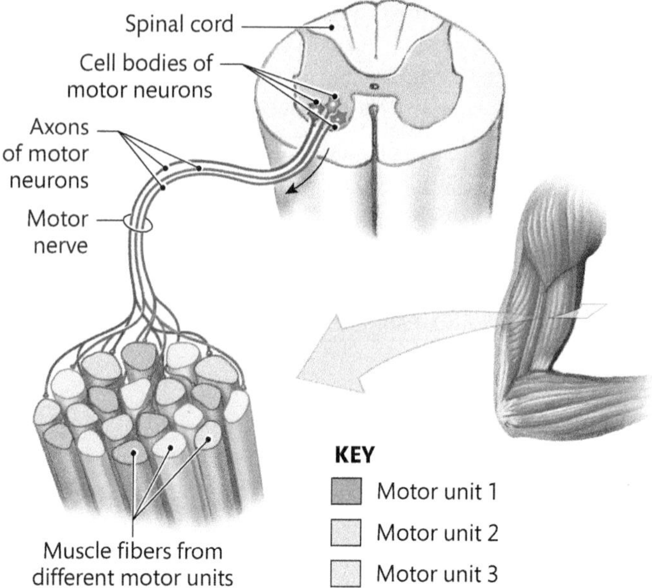

The strength of a muscle contraction depends on the number of motor units stimulated. If the muscle contracts slowly and with little force, relatively few motor units have been activated. If the muscle receives more frequent nerve impulses, additional motor units are activated, and the muscle contracts faster and with more power. The activation of increasing numbers of motor units is called **motor unit recruitment.**

If electrodes are placed on the skin, the changes in the cutaneous electrical activity, caused by the electric impulses generated when a muscle contracts, can be detected and recorded by a technique called **electromyography.** The recording that is made is called an **electromyogram (EMG).** You will use the Biopac Student Laboratory System to record the EMG generated by your lab partner's forearm as he or she clenches a fist. By comparing the force recordings with the EMG, you will be able to correlate motor unit recruitment with strength of muscle contraction.

A Setup and Calibration

1. Turn the computer on. Make sure the Biopac MP35 unit is off.
2. Have your lab partner remove all jewelry, watches, or other metal objects from his or her body. Position three electrodes on your lab partner's dominant forearm as shown in the figure on the right. Attach electrode leads (SS2L) as illustrated. In the same manner, position three electrodes on your lab partner's nondominant forearm, but do not attach electrode leads at this time.

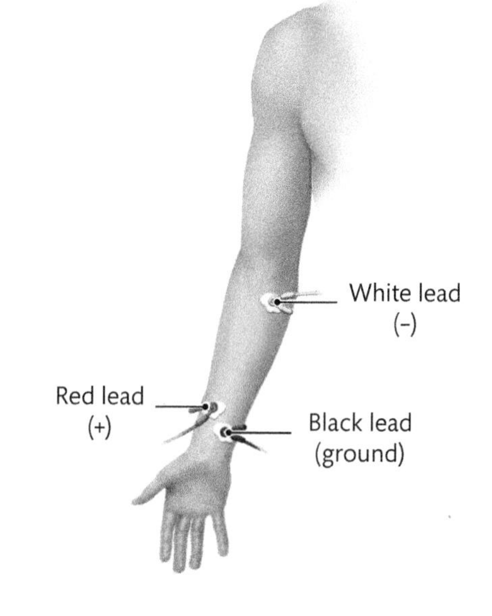

3. Attach the electrode leads (SS2L) to Channel 3 on the front of the Biopac unit (MP35, MP36, or MP45).
4. Turn on the Biopac unit. Start the Biopac Student Laboratory system on the computer, choose lesson 1 (LO1-EMG-1), and click **OK.** When prompted, enter a unique file name and click **OK.**
5. Click **Calibrate,** read the dialog box, and click **OK** when you are ready to begin.
6. Wait 2 seconds and have your lab partner clench the fist of his or her dominant forearm as forcefully as possible for about 5 seconds and then relax. This procedure tells the computer what a maximum clench looks like. The calibration procedure will automatically stop after 8 seconds.
7. Your recording will show a zero baseline and positive signals representing a maximum clench. Your recording should look like the figure shown on the right. If it does not, check the electrodes and connections and click **Re-do Calibration.**

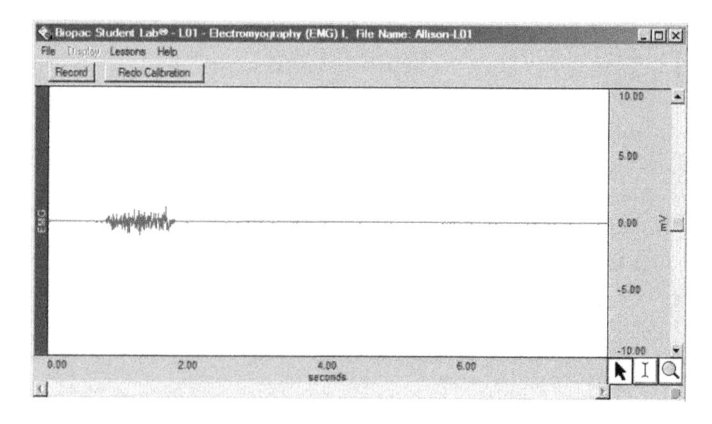

■ Want more practice? Go to: MasteringA&P > Study Area > Menu > Lab Tools > PhysioEx > Exercise 2: Skeletal Muscle Physiology > Activity 2: The Effect of Stimulus Voltage on Skeletal Muscle Contraction

Recording Data

In this part of the activity, you will make two EMG recordings. In Segment 1, your lab partner will make a series of four fist clenches using the dominant forearm. The force of each clench will increase in equal amounts, and the last clench will be at maximum tension. Segment 2 will be a similar series of fist clenches using the nondominant forearm. Read this entire section first so that you will be familiar with the activities before you begin recording data.

B Segment 1: Dominant Forearm

1. Click **Record** to begin the recording. Have your lab partner perform a series of four fist clenches, increasing the force equally after each clench. The fourth clench should be at maximum force. Each clench should be held for 2 seconds, followed by a 2-second relaxation period before the next 2-second clench. To keep track of time, notice that each vertical line on the screen is a 2-second interval.

2. When you complete the series of clenches (Segment 1), click **Suspend** to pause the recording. The resulting data should resemble that shown in the top figure on the right. If not, click **Redo** and repeat step 1.

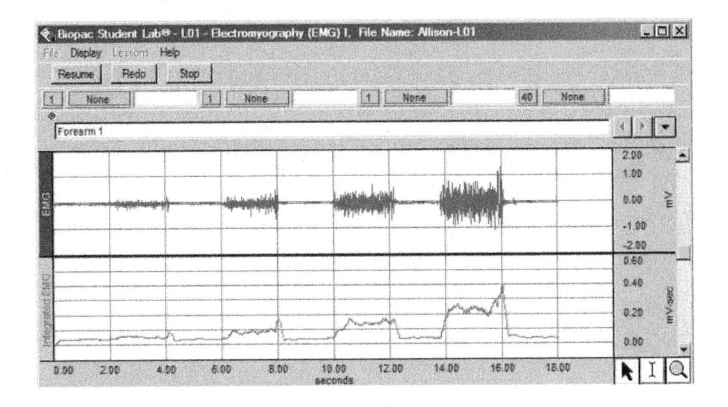

C Segment 2: Nondominant Forearm

1. Disconnect the electrode leads from your lab partner's dominant forearm and reconnect them in the same manner to the nondominant forearm.

2. Click **Resume** to begin Segment 2. A marker will be automatically inserted to identify the beginning of the segment.

3. Repeat the same cycle of clenches that your lab partner performed in Segment 1 for the dominant forearm. Each clench should be for 2 seconds followed by a 2-second relaxation. The clenches should increase equally in strength until maximum tension is reached on the fourth clench.

4. Click **Suspend** to pause the recording. The resulting data should appear similar to that illustrated in the bottom figure on the right. If not, repeat steps 2 and 3.

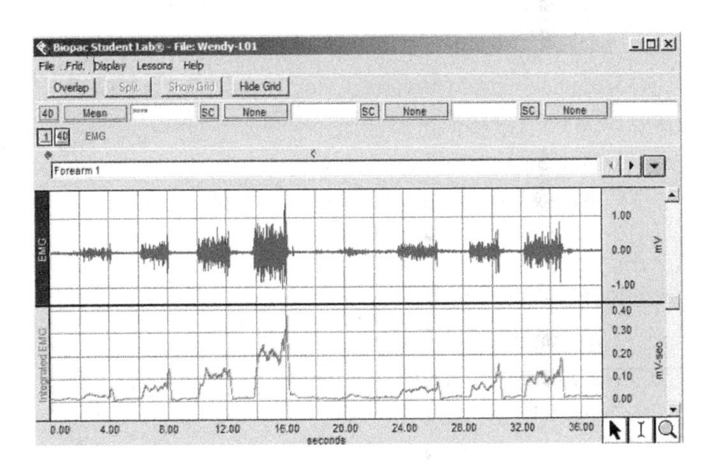

5. Click **Stop** when you are finished. A pop-up window will appear asking if you are sure you want to stop recording. Clicking **Yes** will end your recording session. All your data will be automatically saved.

6. Gently peel the electrodes off your lab partner's forearm and wash with soap and water to remove the electrode gel residue from his or her skin.

MAKING CONNECTIONS

Identify sources of experimental error when measuring fist clench force.

▶ ____

____

____

____

Activity 13.3

Electromyography Using the Biopac Student Laboratory System: Data Analysis of Motor Unit Recruitment Recordings

A Data Analysis

You will measure the force generated in the dominant and nondominant forearms for each fist clench. Then, for both forearms, you will calculate the percent increase in the clench force between the weakest and strongest clench.

1. Enter the **Review Saved Data** mode and select Lesson 1 (L01-EMG-1)

2. Notice the Channel Number designations along the top of the screen:

CH 3	Raw EMG
CH 40	Integrated EMG

Raw EMG is a recording of the electrical activity detected by the electrodes. Integrated EMG is like a "contour" tracing of the shape and intensity of the Raw EMG. It reduces some of the "noise" that makes taking measurements from the raw signal difficult.

3. Set up your display window for optimal viewing of the data from Segment 1 (forearm 1, dominant). You can use the following tools to help you adjust the data window:

Autoscale	Zoom previous
Horizontal	Horizontal (Time) scroll bar
Autoscale	Vertical (Adjustment) scroll bar
Waveforms	Overlap button
Zoom tool	Split button

4. Notice the measurement box settings along the top of the data window:

CH 3	Min
CH 3	Max
CH 3	p-p
CH 40	Mean

Each measurement box has three sections: channel number, measurement type, and result. The first two sections have drop-down menus. They are activated when you click on them.

5. Using the I-beam cursor tool, highlight the "plateau" of the first clench. The plateau should be a relatively flat region of the tracing in the middle of the clench, representing the "hold" period of the clench-hold-release cycle (see the figure at the top of this page). Record the measurement data in Table 13.1 (Cluster Number 1, Forearm 1 [Dominant]). ▶

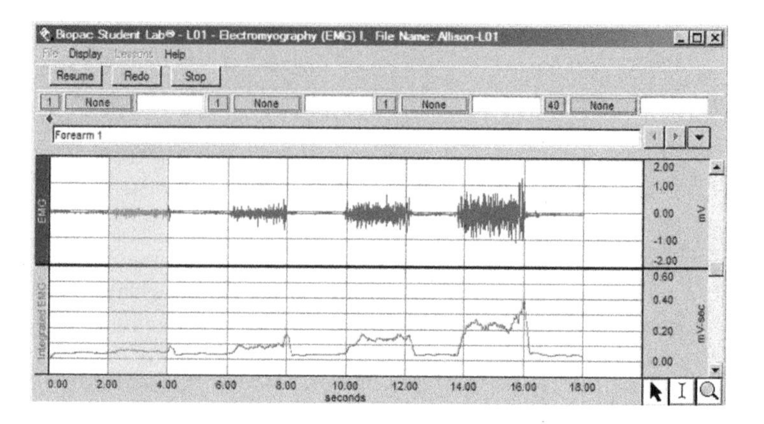

6. Use the I-beam cursor to highlight each successive plateau region for the other EMG clusters in segment 1. Record the data for clusters 2, 3, and 4 in Table 13.1. ▶

7. Use the I-beam cursor to highlight the plateau regions for the four clenches in Segment 2 (forearm 2, nondominant). Record the data for all four clusters in Table 13.1. ▶

8. For the dominant and nondominant forearms, use the mean measurement in Table 13.1 to calculate the percent increase in the clench force between the weakest clench and the strongest clench. ▶

1. Dominant forearm: Percent increase ________
2. Nondominant forearm: Percent increase ________

MAKING CONNECTIONS

Describe the relationship between force and EMG activity. Explain how it relates to motor unit recruitment.

▶ ____

____

____

____

____

TABLE 13.1 Motor Unit Recruitment: Comparing Dominant and Nondominant Forearms

	Forearm 1 (Dominant)				Forearm 2 (Nondominant)			
Number	Min (CH 3 min)	Max (CH 3 max)	p-p (CH 3 p-p)	Mean (CH 40 mean)	Min (CH 3 min)	Max (CH 3 max)	p-p (CH 3 p-p)	Mean (CH 40 mean)
1								
2								
3								
4								

Activity **13.4**

Examining Resistance versus Contraction Speed

A The speed at which a muscle contracts is inversely related to the resistance or load. When a muscle begins to contract, tension increases gradually. To move an object, muscle tension must exceed the load, and it will take longer for tension to exceed the load of a heavier object than a lighter object. Thus, you can lift a light object more quickly than a heavy object.

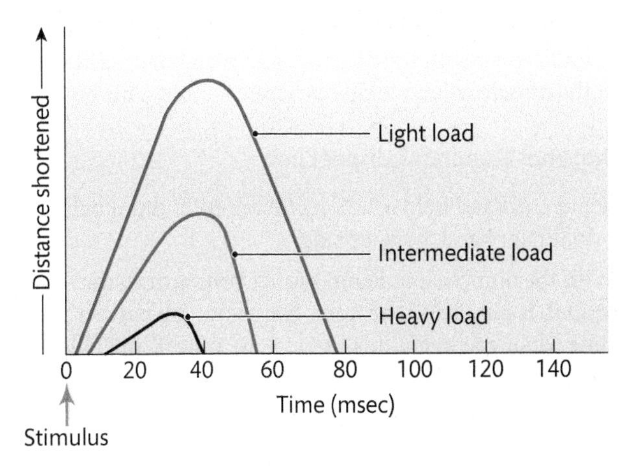

1. Obtain three or four dumbbells (or other objects that are easy to hold) of various weights.

2. Rest the forearm of your dominant side flat on the laboratory bench with your palm directed superiorly.
3. From this position, grasp the lightest dumbbell (or other object) and lift it with moderate force by flexing your forearm. Do not use maximum force and keep your elbow on the bench surface as you lift.
4. Repeat steps 2 and 3 with the other dumbbells, in ascending weight order. Be sure you apply the same level of force with each lift as you used in the previous step and keep your arm and forearm in the same positions.

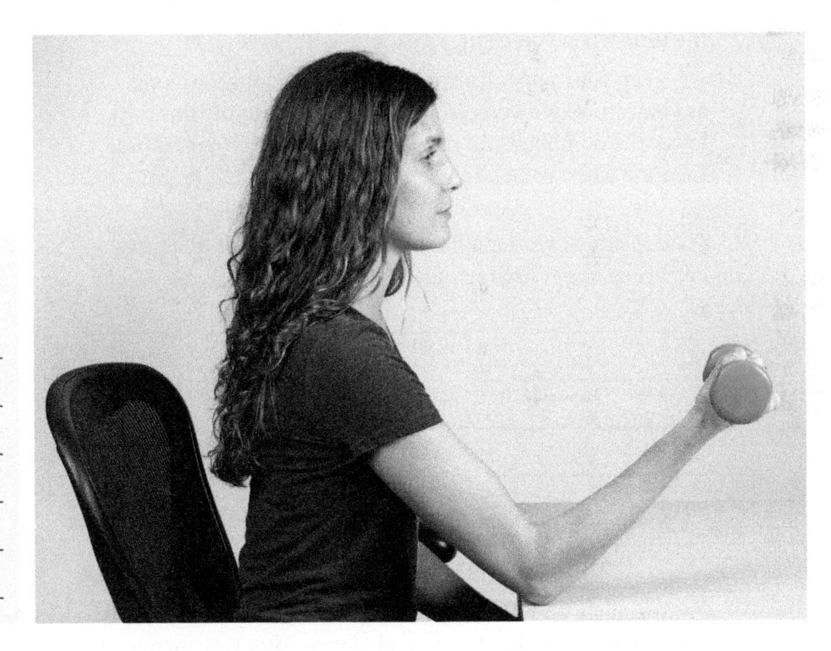

5. Describe your results in terms of speed of contraction versus increasing load (resistance).

▶ ______________________________

____

____

____

____

____

MAKING CONNECTIONS

Explain why your thigh and leg muscles will contract more slowly while running up a steep hill compared to on a level surface.

▶ ______________________________

____

____

____

____

■ Want more practice? Go to: **Mastering A&P** > Study Area > Menu > Lab Tools > **PhysioEx** > Exercise 2: Skeletal Muscle Physiology > Activity 7: Isotonic Contractions and the Load-Velocity Relationship

Activity 13.5

Demonstrating Muscle Fatigue

Muscle fibers can contract as long as enough energy is available. Over time, however, the metabolic production of energy cannot keep up with the demand, and metabolic wastes, such as lactic acid, accumulate in the sarcoplasm. As a result, the muscle will no longer generate a force. This breakdown in function is called **muscle fatigue.**

A Testing Your Dominant Upper Limb

1 Stand erect and hold a 2.25-kg (about 5-lb) dumbbell in your dominant hand, by your side.

2 With the dumbbell in hand, abduct your arm so that your upper limb is parallel to the floor. Have your lab partner note the time when you begin this exercise as "time 0" in Table 13.2. ▶

3 Keep your upper limb parallel to the floor for as long as you can.

4 When your upper limb begins to sag and you are laboring to keep it parallel to the floor, change your position by moving your arm anteriorly so that it is projecting forward but is still parallel to the floor. Have your lab partner record the time when you change position in Table 13.2. ▶

5 Keep your upper limb parallel to the floor for as long as you can. When you can no longer maintain the position, lower the dumbbell and rest. Have your lab partner record the time when you end the exercise in Table 13.2. ▶

6 Explain why, after a period of time, you were unable to keep your upper limb parallel to the floor.

▶ ____________________

7 Speculate on why you were able to hold the dumbbell in position for an additional period of time after you changed the position of your upper limb.

▶ ____________________

■ Want more practice? Go to: MasteringA&P > Study Area > Menu > Lab Tools > PhysioEx > Exercise 2: Skeletal Muscle Physiology > Activity 5: Fatigue in Isolated Skeletal Muscle

TABLE 13.2 Muscle Fatigue: Comparing Dominant and Nondominant Upper Limbs			
	Start Time of Exercise (sec)	Time of Position Change (min:sec)	End Time of Exercise (min:sec)
Dominant side	0		
Nondominant side	0		

B Testing Your Nondominant Upper Limb

1 Repeat steps 1 through 5 described on the previous page using your nondominant upper limb. Have your lab partner record the times in Table 13.2. ▶

2 Did you note any differences between your dominant and nondominant upper limbs? Provide an explanation for these differences, if any.

▶ ______

______

______

______

______

______

IN THE CLINIC

Oxygen Debt

Virtually all the energy (ATP) used by muscles at rest or during light to moderate exercise is derived from aerobic respiration, a process that occurs in the mitochondria and requires oxygen. As long as oxygen is available, aerobic metabolism will supply muscle cells with enough ATP to function normally. If activity becomes more intense and continues for an extended period, however, a point will be reached when oxygen supply will not keep pace with demand. If that occurs, muscle metabolism will switch to an anaerobic process called glycolysis.

Aerobic respiration produces much more ATP than glycolysis, but it is rather slow because it involves a long series of biochemical steps. Glycolysis is fast and efficient, but produces only a small amount of ATP. It serves as a "quick fix" when muscle cells need energy immediately, such as when running a 100-meter sprint. If you rely too heavily on glycolysis for your energy needs, however, there is a price to pay. The primary waste product of glycolysis is lactic acid. If enough oxygen is available, muscle cells are able to convert lactic acid to pyruvic acid, which can be used to produce ATP. Lactic acid that is released into the bloodstream is transported to the liver, where it is converted to glucose.

During intense muscle activity, there is an oxygen deficiency. As a result, lactic acid accumulates in muscle cells, and muscle fatigue sets in. The oxygen deficiency leads to what is called the oxygen debt, which is the amount of oxygen required for muscles to recover from a period of intense activity. Recovery includes converting lactic acid to pyruvic acid (in muscle cells) or glucose (in the liver) and replenishing energy reserves (ATP and creatine phosphate) in muscle cells.

MAKING CONNECTIONS

One factor that causes muscle fatigue is the accumulation of lactic acid in the sarcoplasm of muscle cells. How will that impair normal muscle function? (*Hint:* Consider the pH inside muscle cells.)

▶ ______

______

______

______

______

______

______

______

Activity 13.6

Testing Muscle Flexibility

Although muscle contractions are typically evaluated by the forces they generate, muscle flexibility is also important. Muscle flexibility refers to the range of motion (ROM) at a particular joint. The degree of flexibility depends on the potential of muscles and tendons to lengthen when movement occurs at a joint. Two types of flexibility are static and dynamic. **Static flexibility** measures ROM without considering the degree of difficulty in performing a movement. **Dynamic flexibility** also evaluates ROM, but resistance to motion is a consideration, and the ease or difficulty of movement is assessed. Dynamic flexibility is more important than static flexibility for evaluating athletic ability or the physical condition of a joint (i.e., evaluating a joint for arthritis), but it is difficult to accurately measure without specialized equipment.

In this activity, you will use an instrument called a **goniometer** to measure static flexibility. A goniometer is used like a protractor to measure the angle formed by a moving body part at a joint during a ROM evaluation. It has two arms, one stationary and one moving, that are joined by a central pivot (the fulcrum). To measure the ROM of a joint, the fulcrum of the goniometer is placed over the joint, and the two arms are placed in line with the moving body part. When the subject moves, the stationary arm remains in its original position; the moving arm sweeps across an arc that reflects the movement of the body part. The angle formed by the two arms of the goniometer is the range of motion.

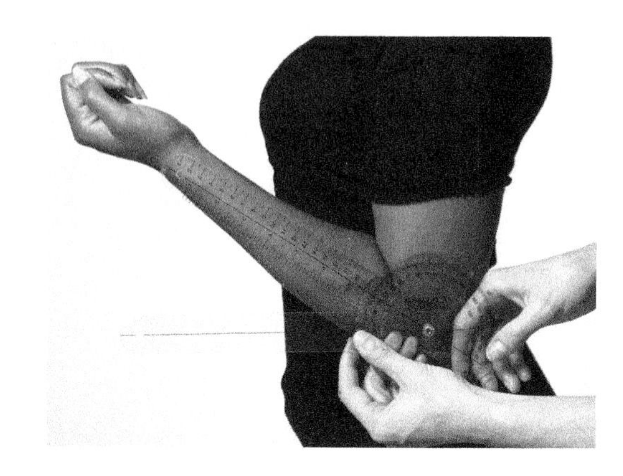

Measure the ROM with a goniometer at the shoulder, elbow, hip, and knee on both sides of the body. If a goniometer is not available, have your lab partner observe you during the ROM evaluation and make visual comparisons with the appropriate diagram.

A Shoulder Abduction and Adduction

1. From anatomical position, abduct your arm until it is parallel to the floor. This is your start position.

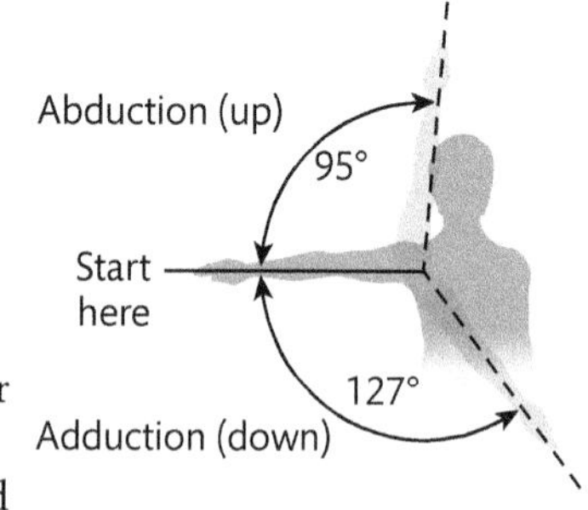

2. Have your lab partner position the fulcrum of the goniometer over the anterior aspect of your shoulder; the two goniometer arms should be aligned with the long axis of your arm.
3. Abduct your arm as much as possible so that your hand is above your head.
4. Have your lab partner adjust the moving arm of the goniometer and measure the angle produced. Record the result in Table 13.3. ▶
5. Return your arm and the goniometer to the start position and then adduct your arm as much as possible so that it crosses anterior to your torso.
6. Your lab partner should measure and record the adduction angle in Table 13.3. ▶

B Elbow Flexion and Extension

1. From anatomical position, flex one forearm so that it is perpendicular (90°) to your arm. This is your start position.

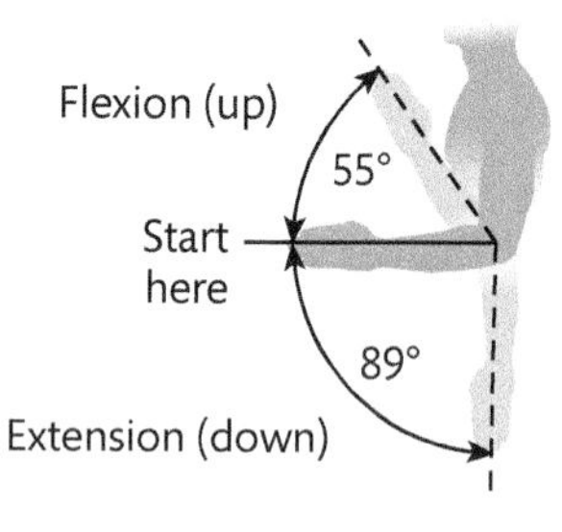

2. Place the fulcrum of the goniometer over the lateral aspect of the elbow with the arms parallel to the long axis of the forearm.
3. Flex your forearm as much as possible and have your lab partner measure the angle with the goniometer and record the result in Table 13.3. ▶
4. Return to the start position and then extend your forearm as much as possible. Have your lab partner measure the extension angle with the goniometer. Record the result in Table 13.3. ▶

TABLE 13.3 Flexibility Tests: Measuring Range of Motion (ROM)						
			Measured ROM (mark with "x" for visual estimation)		Rating (below average/average/above average)	
Joint	Movement	Average ROM	Right	Left	Right	Left
Shoulder	Abduction	95°				
	Adduction	127°				
Elbow	Flexion	55°				
	Extension	89°				
Hip	Abduction	45°				
	Adduction	26°				
Knee	Flexion	140°				

C Hip Abduction and Adduction

1. Stand erect with your feet flat on the floor. Have your lab partner place the fulcrum of the goniometer over the anterior aspect of your hip joint with the arms aligned with the long axis of your thigh.

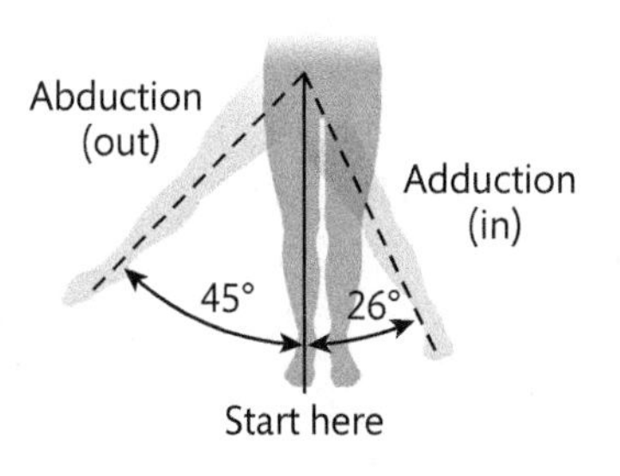

2. Abduct your thigh as much as possible by moving your lower limb away from the midline of the body in a coronal plane.
3. Your lab partner should measure the abduction angle with the goniometer and record the result in Table 13.3. ▶
4. Adduct your thigh back to the start position. From here, adduct your thigh even more by swinging the lower limb as much as possible to the contralateral side of your body. Have your lab partner measure and record the adduction angle. ▶

D Knee Flexion

1. Stand erect with your feet flat on the floor. Have your lab partner place the fulcrum of the goniometer over the lateral aspect of your knee joint with the arms aligned with the long axis of your leg.
2. Flex your leg as much as possible. Your lab partner should measure and record the flexion angle. ▶

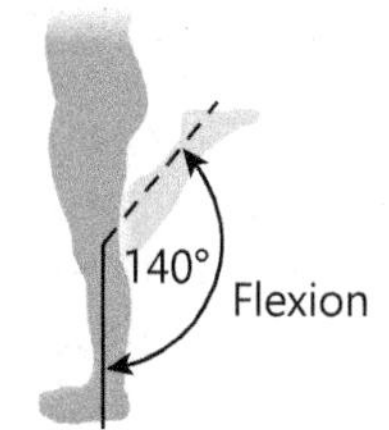

MAKING CONNECTIONS

Speculate on how below-average range of motion could have a negative effect on athletic ability. Provide an example by discussing the role of range of motion in one sport.

▶ ______________________________

BEFORE YOU MOVE ON . . .

« LOOKING BACK

In this laboratory exercise, you studied the relationship between muscle tension and load. You performed various activities, such as deep knee bends and lifting and lowering dumbbells on a lab bench, to demonstrate the two types of isotonic contractions. You learned that most muscular activities are a combination of both types:

- During a concentric contraction, muscle length shortens; tension is greater than the load; and the object, upon which the muscle acts, is moved.
- During an eccentric contraction, muscle length increases, and the tension is less than the load. Muscles that contract eccentrically often function to control the actions of antagonist muscles that are contracting concentrically. For example, when lifting a dumbbell off a lab bench, the triceps brachii contracts eccentrically to control the action of the biceps brachii when the forearm is flexed. When the dumbbell is returned to the lab bench, the biceps brachii contracts eccentrically to control the action of the triceps brachii when the forearm is extended.

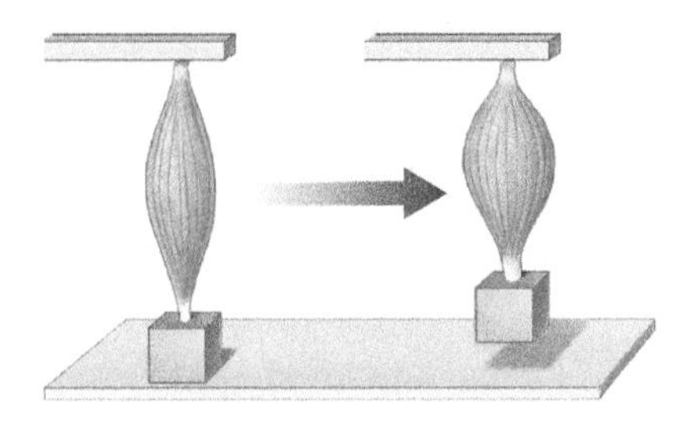

Concentric contraction

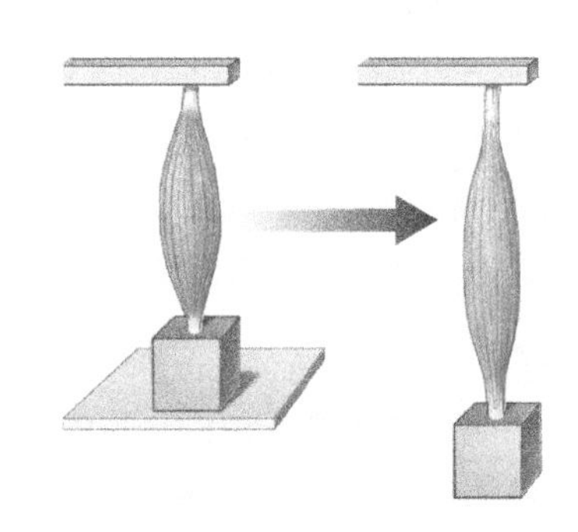

Eccentric contraction

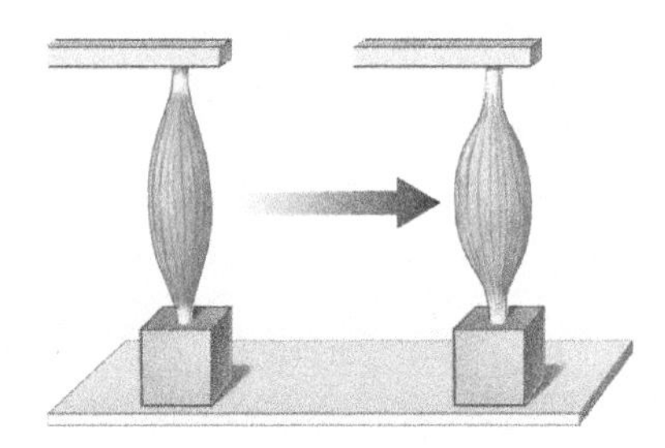

Isometric contraction

You also learned that when you perform isometric contractions, such as pushing against a wall, muscle tension increases, but there is no change in muscle length. Muscle tension does not exceed the load, and the object does not move.

Although isometric contractions do not produce any meaningful work (i.e., an object is not moved), they have other important functions. Speculate on their significance. Can you think of activities in which isometric contractions play a key role?

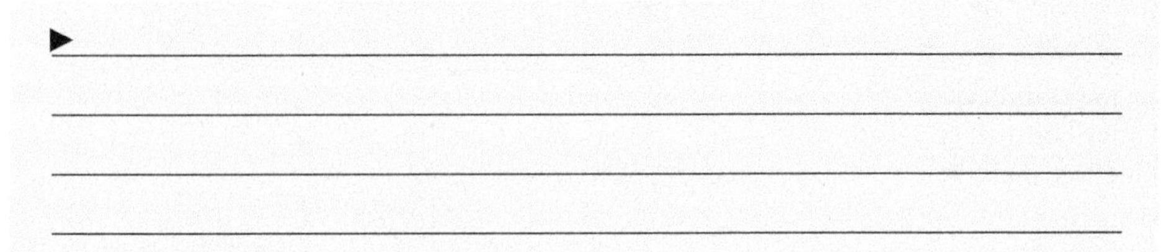

LOOKING FORWARD »

All the voluntary movements of skeletal muscles and the involuntary movements of cardiac and smooth muscle are regulated and controlled by motor neurons from the brain and spinal cord. Neurons (nerve cells) have the ability to conduct electric impulses along nerve fibers to various target organs. As a result of this electrical activity, the nervous system is able to control and integrate the activities of all the organs and organ systems. In the next laboratory exercise, you will study the microscopic structure of neurons and other components of nerve tissue in various nervous system structures. You will also learn how neurons interact with target cells at special junctions known as synapses.

Name ______________________

Lab Section ______________________

Date ______________________

EXERCISE 13 REVIEW SHEET

Physiology of the Muscular System

1. From a functional standpoint, discuss the difference between isotonic and isometric muscle contractions.

____

____

____

____

2. The graph on the right compares speed of muscle contraction with load. Which object will you be able to lift most quickly, and which will you lift most slowly? Explain your answer by discussing the relationship between load and speed of contraction.

____

____

____

____

____

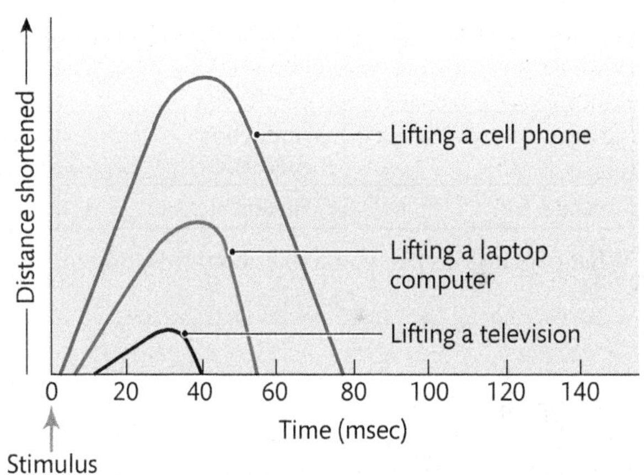

3. The figure on the right shows the organization of three small motor units. Notice that the muscle fibers in each motor unit are dispersed throughout a muscle and intermingle with muscle fibers of the other motor units. What do you think is the functional significance of such an arrangement?

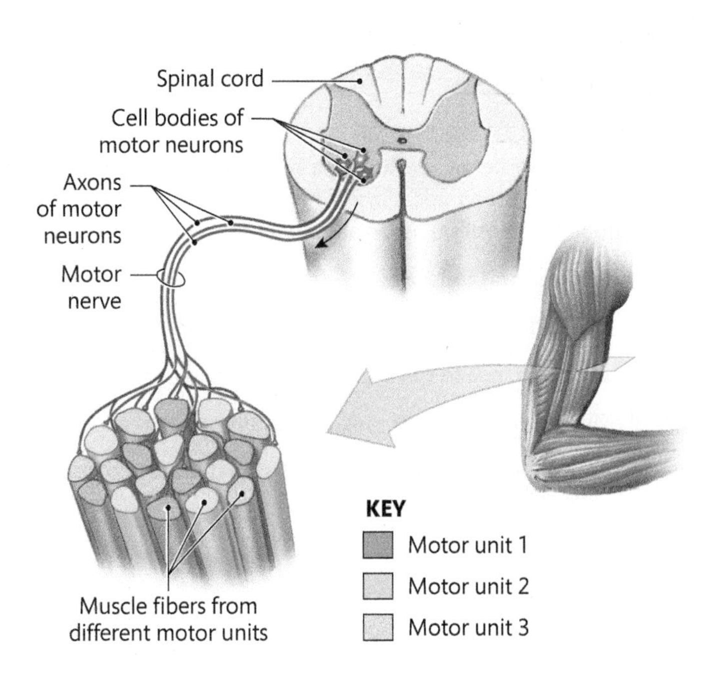

QUESTIONS 4–8: Match the term in column A with the appropriate description in column B.

A		B
4. Threshold stimulus	______	**a.** Condition of a muscle characterized by its inability to generate a force
5. Motor unit	______	**b.** Minimum stimulus required to initiate a muscle fiber contraction
6. Motor unit recruitment	______	**c.** Activation of increasing numbers of motor units in a muscle
7. Muscle flexibility	______	**d.** Motor nerve fiber and all the muscle fibers it controls
8. Muscle fatigue	______	**e.** Range of motion at a joint

9. The following data were collected by measuring the range of motion (ROM) at various joints with a goniometer:

			Measured ROM (mark with "x" for visual estimation)	
Joint	Movement	Average ROM	Right	Left
Shoulder	Abduction Adduction	95° 127°	87° 119°	86° 120°
Elbow	Flexion Extension	55° 89°	48° 85°	49° 86°
Hip	Abduction Adduction	45° 26°	55° 30°	55° 32°
Knee	Flexion	140°	145°	145°

What conclusions can you make about this individual's flexibility?

____

____

____

10. Identify sources of experimental error when measuring the range of motion with a goniometer.

____

____

____

11. To demonstrate muscle fatigue, a student held an 8-lb dumbbell in her hand and abducted her upper limb so that it was parallel to the floor. After 45 seconds, she was having trouble holding that position, so she rotated her hand 90°. By doing so, she was able to hold the dumbbell in position for an additional 20 seconds. Which of the following statements best explains why the student could do that?

a. The load was reduced.

b. Different motor units were contracting while others were relaxing.

c. The resistance to contraction was reduced.

d. Muscles began to contract isotonically rather than isometrically.

e. The change in position provided a greater range of motion at the shoulder.

EXERCISE

14

Nervous Tissue

LEARNING OUTCOMES

These Learning Outcomes correspond by number to the laboratory activities in this exercise. When you complete the activities, you should be able to:

Activity 14.1 **Discuss the structure and function of the different types of neurons and neuroglia.**

Activity 14.2 **Identify important microscopic structures in the cerebrum.**

Activity 14.3 **Identify important microscopic structures in the cerebellum.**

Activity 14.4 **Explain the microscopic organization of the spinal cord.**

Activity 14.5 **Discuss how ventral and dorsal spinal roots give rise to a spinal nerve.**

Activity 14.6 **Describe the organization of a peripheral nerve.**

Activity 14.7 **Demonstrate how action potentials are propagated along myelinated and unmyelinated axons.**

LABORATORY SUPPLIES

- Compound light microscopes
- Prepared microscope slides of the following nerve tissue:
  - spinal cord smear
  - cerebrum
  - cerebellum
  - spinal cord cross section
  - spinal cord cross section with dorsal and ventral roots
  - peripheral nerve cross section
- Stopwatches
- Colored pencils

PRE-LAB QUIZZ

Before you begin, read all the activities in Exercise 14 and the required reading in your textbook that is assigned by your instructor.

1. During this laboratory exercise, you will be viewing microscope slides of all the following structures *except*
   a. the brainstem.
   b. the cerebrum.
   c. the spinal cord.
   d. the cerebellum.
2. During this laboratory exercise, the class will be divided into two groups to demonstrate the functional significance of what structure? ______________________
3. True or False: Neuroglial cells are found in the central nervous system but not in the peripheral nervous system. ____________
4. In the cerebellum, the large multipolar neurons located in the middle layer of the cortex are called ______________________.
5. True or False: Similar to the cerebrum and cerebellum, the spinal cord contains a surface layer of gray matter. ____________
6. True or False: A peripheral nerve contains thousands of nerve fibers. All these fibers are surrounded by a myelin sheath. ____________
7. True or False: A ventral root contains all motor nerve fibers; a dorsal root contains all sensory nerve fibers. ____________
8. True or False: The white matter in the cerebral cortex is called the arbor vitae.

   ____
9. What type of neuron contains two cell processes? ______________________
10. True or False: In the gray matter of the brain and spinal cord, there are many more neuroglia cells than neurons. ____________

Activity **14.1**

Overview of Cell Types in Nervous Tissue

A The structural and functional unit of the nervous system is the **neuron,** or **nerve cell.** A typical neuron has a **cell body** that contains a large, round nucleus with a well-defined nucleolus. In the surrounding cytoplasm, the most prominent organelles include mitochondria, the Golgi apparatus, and **Nissl (chromatophilic) bodies,** which are clusters of rough endoplasmic reticulum (RER) and free ribosomes. Extending from the cell body are two types of processes: **dendrites** and **axons.** Dendrites, along with the cell body, serve as contacts to receive impulses from other neurons. Typically a neuron has only one axon (nerve fiber), which transmits impulses to other cells. Neurons can be classified into four types according to the number of cell processes they possess and their functions (Table 14.1).

TABLE 14.1 The Four Types of Neurons

Cell Type	Structure	Function	Locations
Multipolar	Many cell processes: one axon, multiple dendrites	Motor neurons for voluntary and involuntary movement; interneurons that form fiber tract links between brain and spinal cord regions	Cerebral cortex; cerebellar cortex; gray matter in spinal cord
Bipolar	Two cell processes: one axon, one dendrite	Sensory neurons for vision, hearing, equilibrium, and smell	Retina of the eye; inner ear; olfactory epithelium in nasal cavity
Unipolar	One cell process that divides into two branches: both branches are axons; receptive endings of peripheral branch are dendrites	Sensory neurons for general sensations (touch, pressure, temperature, pain, proprioception)	Sensory ganglia for spinal nerves; most cranial nerve ganglia
Anaxonic	Many cell processes: all are dendrites	Function is unknown	Various regions in the brain

1 Identify the type of neuron that is illustrated in each figure. Provide a rationale for each identification. ▶

Neuron	Type	Rationale
A		
B		
C		
D		

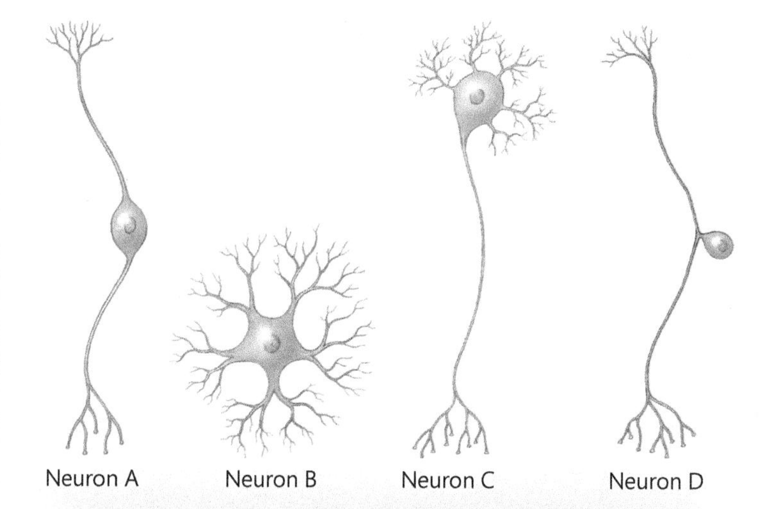

2 Which type of neuron is shown in the photo? Explain.

▶ ______________________________

____

____

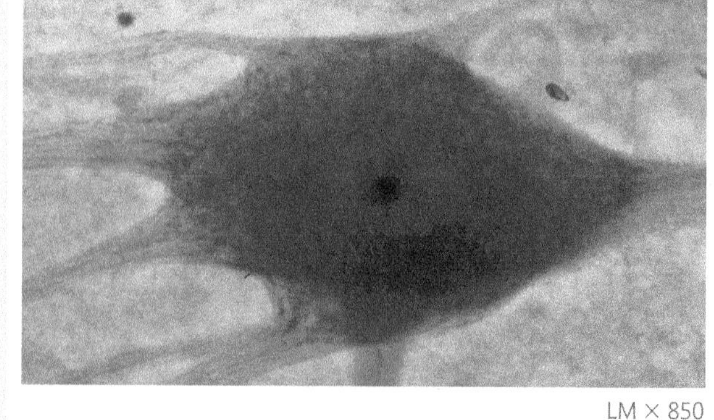
LM × 850

B **Neuroglia**, or **glial cells**, perform a variety of functions that protect and support neurons and other structures. They have relatively small cell bodies from which cytoplasmic processes extend. The number of processes and the complexity of their branching patterns vary among the different cell types. Neuroglia are found in both the central and peripheral nervous systems. The specific functions of each type are described in Table 14.2.

1. View a slide of a spinal cord smear under low power and identify the large multipolar neurons scattered throughout the field of view.
2. Center a neuron in the field of view and switch to high power. Identify the cell body, nucleus, and cell processes (axon and dendrites) of the neuron.
3. Identify the nuclei of neuroglia surrounding the neuron. They appear as small darkly stained circular structures. The cell bodies cannot be clearly identified. Notice that glial cells far outnumber neurons.

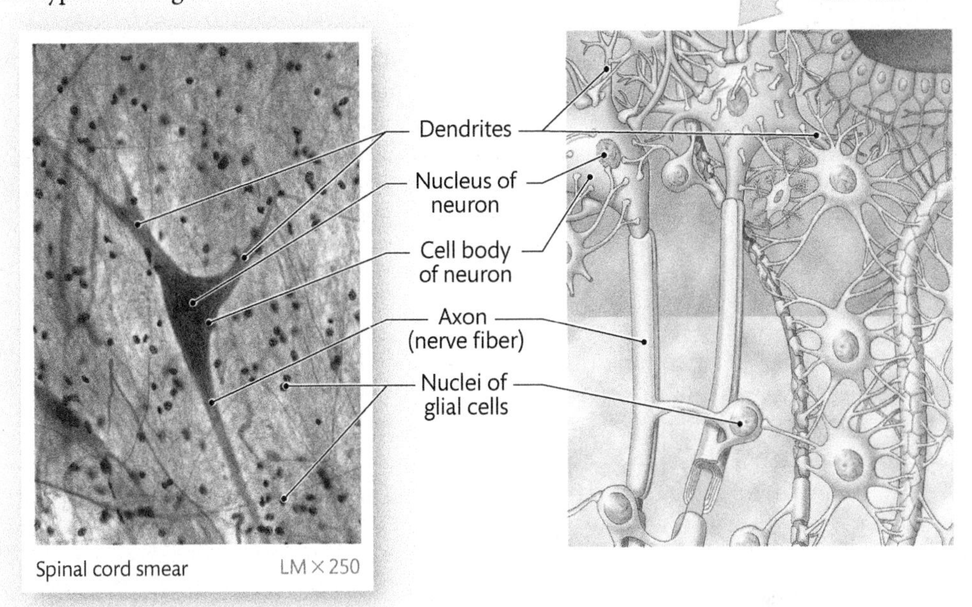

Spinal cord smear

TABLE 14.2 **Neuroglial Cell Types**	
Cell Type	**Function**
A. Neuroglia in the CNS	
1. Astrocytes	Cell processes form a supporting network that connects neurons to blood vessels; help form the blood–brain barrier, which prevents the passage of potentially harmful substances from the blood to brain tissue; and help regulate levels of oxygen, carbon dioxide, and nutrients.
2. Microglia (microglial cells)	Act as phagocytes; protect the CNS from disease-causing microorganisms and clear away cellular debris.
3. Ependymal cells (ependyma)	Modified epithelial cells that line the ventricles of the brain and central canal of the spinal cord; facilitate the circulation of cerebrospinal fluid in the ventricles.
4. Oligodendrocytes	Produce the myelin sheath around axons in the CNS.
B. Neuroglia in the PNS	
1. Satellite cells	Surround neuron cell bodies in peripheral ganglia and regulate levels of oxygen, carbon dioxide, and nutrients.
2. Schwann cells	Produce the myelin sheath around axons in the PNS.

MAKING CONNECTIONS

Recall that centrioles are organelles that make the spindle apparatus that separates chromosomes during mitosis. Most neurons in the CNS do not have centrioles. Explain why.

▶ ______________________________

____

Activity 14.2

Central Nervous System: Examining the Microscopic Structure of the Cerebrum

A The cerebrum of the brain consists of **two cerebral hemispheres.** Similar to most regions of the brain, each cerebral hemisphere has two types of nervous tissue: **gray matter** and **white matter.** The gray matter consists of neuron cell bodies, dendrites, unmyelinated axons, and neuroglia. It is located along the surface **(cerebral cortex)** and in deeper regions **(nuclei).** Immediately deep to the cerebral cortex is the white matter, which contains myelinated axons and neuroglia. Bundles of axons in the white matter form fiber tracts that connect various brain regions.

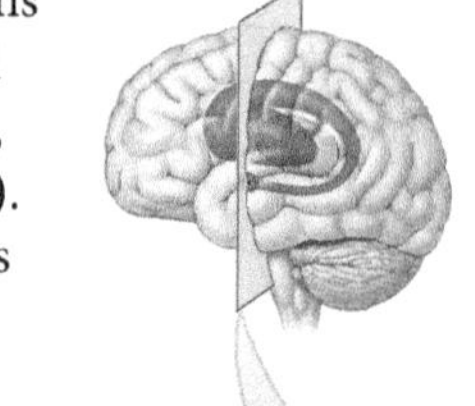

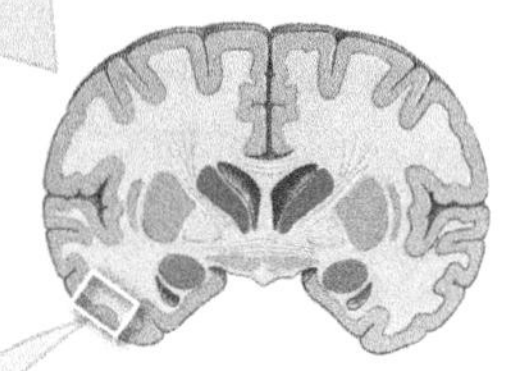

Coronal section of the cerebrum

1. View a slide of the cerebrum under low power. As you scan the slide, notice two distinct regions:

Along the surface is the layer of gray matter known as the **cerebral cortex.**

Deep to the cerebral cortex is the **white matter.**

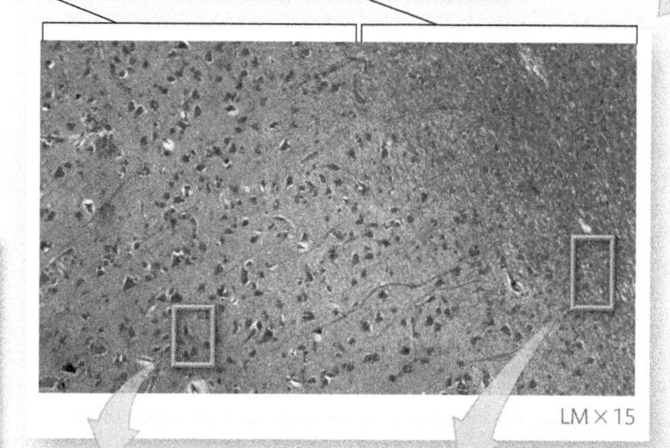

LM × 15

2. Center a region of gray matter in the field of view and switch to high power. In the gray matter, you will see darkly stained cells throughout the field of view.

Identify the relatively large cell bodies of **multipolar neurons.**

Locate the nuclei of the much smaller **neuroglia,** whose cell bodies are difficult to see. Notice that the neuroglia outnumber the neurons.

3. Examine one multipolar neuron and identify the following structures in the cell body:
   - the **nucleus**
   - a distinct **nucleolus** inside the nucleus
   - **Nissl bodies** distributed throughout the cytoplasm

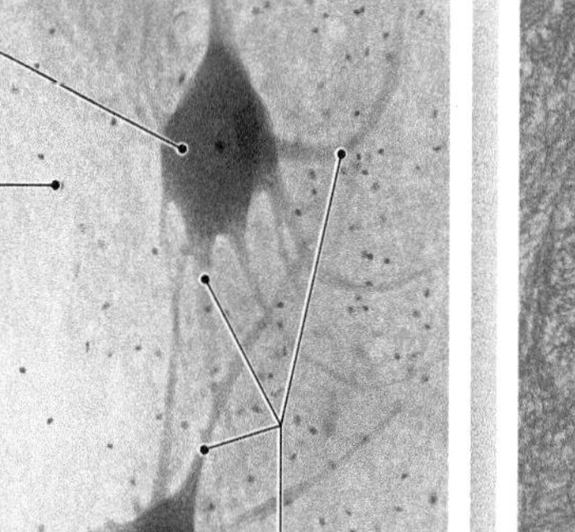

Cerebral cortex LM × 135

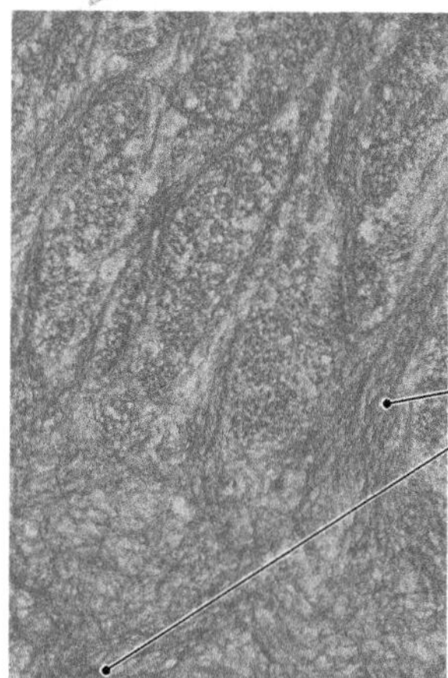

White matter LM × 140

5. Move the slide to a deeper area where very few cell bodies are present. This area is the white matter, which contains mostly **myelinated nerve fibers.**

4. Identify the many **cell processes** that extend from the cell body. One cell process is the axon, and all others are dendrites.

6. In the white matter, you should be able to identify the nuclei of neuroglia. Which neuroglial type would you expect to find in white matter? Explain.

▶ ______________________________

____

____

7

In the space provided, draw a region of the cerebrum as you viewed it with the microscope and label the following structures:

- Gray matter in the cerebral cortex
- White matter deep to the cerebral cortex
- Multipolar neurons
- Nuclei of neuroglia (in both gray matter and white matter)
- Myelinated axons in white matter

▶

B In the cerebral cortex, the cell processes of multipolar neurons form a complex, three-dimensional network.

1. Under low power, view the slide of the cerebrum stained with silver. The slide has been prepared to highlight neurons by staining them black against a light background. As you scan the slide, notice the black-stained network of neuron cell processes. In the white matter, you will see only nerve fibers, but if you move to an area of gray matter, you will see the cell bodies and numerous cell processes of **multipolar neurons.**

2. Place a region of gray matter containing several neurons in the center of the field of view and switch to high power. At this magnification, you can clearly see cell processes extending from all sides of the cell bodies.

3. Slowly turn the fine adjustment knob of the microscope. Notice that the cells that you were viewing move out of focus, but others at different depths of field come into focus. You can truly appreciate the three-dimensional quality of the structure you are viewing.

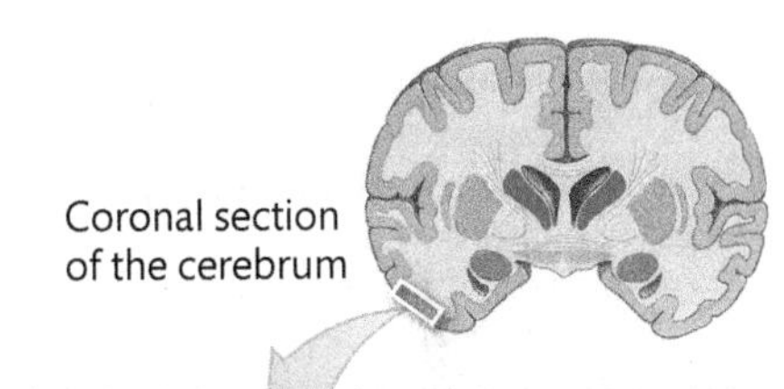
Coronal section of the cerebrum

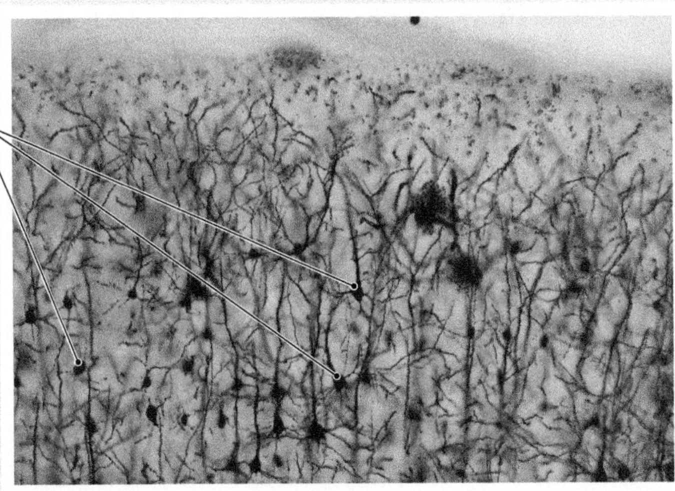
Cerebral cortex

MAKING CONNECTIONS

From your observations of the cerebrum using the light microscope, you likely noticed that virtually all the neurons are multipolar. In fact, more than 99 percent of all neurons are this type. Explain why. (*Hint:* Consider the general function of these neurons.)

▶ ______________________________

____

____

____

Activity **14.3**

Central Nervous System: Examining the Microscopic Structure of the Cerebellum

In the cerebellum of the brain, the arrangement of nervous tissue is similar to what is seen in the cerebrum. The cerebellum consists of two **cerebellar hemispheres.** Each hemisphere contains a surface layer of gray matter, the **cerebellar cortex,** and a deep region of white matter called the **arbor vitae.** Deep to the arbor vitae are additional regions of gray matter known as the **cerebellar nuclei.**

Cerebrum

Midsagittal section of brain

Cerebellum

A The cerebellar cortex contains three distinct layers.

1 View a slide of the cerebellum under low power. As you scan the slide, identify the three layers of the cerebellar cortex:

The relatively thick superficial **molecular layer** contains interneurons that interconnect the other neurons in the cerebellar cortex.

The thin **Purkinje cell layer** is occupied by large multipolar neurons whose axons serve as a major output channel, transmitting information from the cortex to deeper cerebellar nuclei.

The deep **granular layer** includes neurons that receive most of the information coming to the cerebellum from other brain areas.

2 Deep to the granular layer, identify the region of white matter known as the **arbor vitae.** Notice that this white matter is deep to the cerebellar cortex. This pattern is similar to what you observed in the cerebrum.

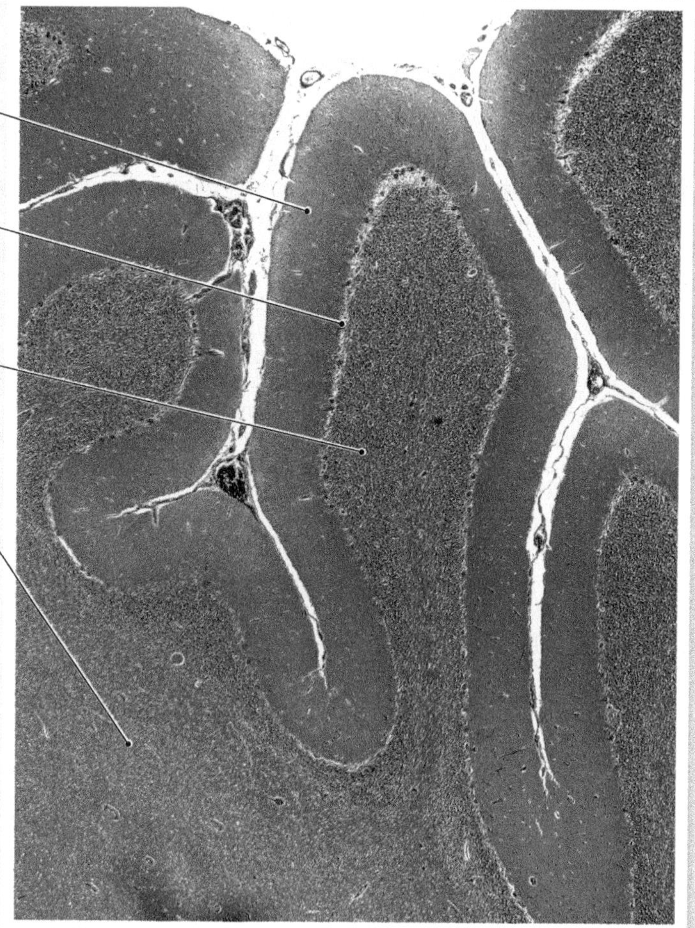

Cerebellum LM × 70

B The distinguishing structural feature in the cerebellum is the single row of large, flask-shaped Purkinje cells in the Purkinje cell layer.

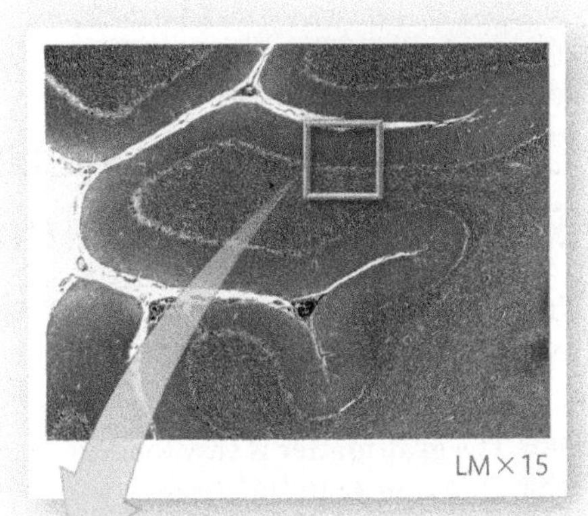

LM × 15

1. Switch to high power. Within the molecular and granular layers, identify the **cell bodies** of small multipolar neurons. On some cells, you will see the initial segments of cell processes (axons and dendrites) extending from the cell body.

2. Identify the large multipolar **Purkinje cells** in the middle layer of the cortex.

3. In the space below, draw a region of the cerebellum that you observed and label the following structures:

   Cerebellar cortex
   - Molecular layer
   - Purkinje cell layer
   - Granular layer

   Arbor vitae (white matter)

Cerebellar cortex LM × 175

▶

MAKING CONNECTIONS

Compare the microscopic structures of the cerebellum and cerebrum. Identify similarities and differences (e.g., cell types and shapes, layers, or structural patterns of nervous tissue).

Similarities:

▶ ____________________

____

____

____

Differences:

▶ ____________________

____

____

____

Activity **14.4**

Central Nervous System: Examining the Microscopic Structure of the Spinal Cord

Unlike the cerebrum and cerebellum in the brain, the spinal cord lacks a surface (cortical) layer of gray matter. Instead, all the gray matter is deep and completely surrounded by the more superficial white matter.

A The **gray matter** is easy to identify in cross section because of its unique butterfly shape.

1. View the slide of the spinal cord in cross section under low power. At this magnification, you can clearly identify the butterfly-shaped gray matter, surrounded by peripheral regions of white matter.

2. Carefully observe the gray matter. On each side, identify the following regions:

   The **posterior horn** receives sensory fibers from spinal nerves.

   The **lateral horn** in thoracic and lumbar spinal cord levels contains the cell bodies of autonomic (involuntary) motor neurons.

   The **gray commissure** is a narrow band of tissue that connects the gray matter on each side.

   The **central canal,** which contains cerebrospinal fluid, passes through the gray commissure.

   The **anterior horn** contains the cell bodies of somatic (voluntary) motor neurons.

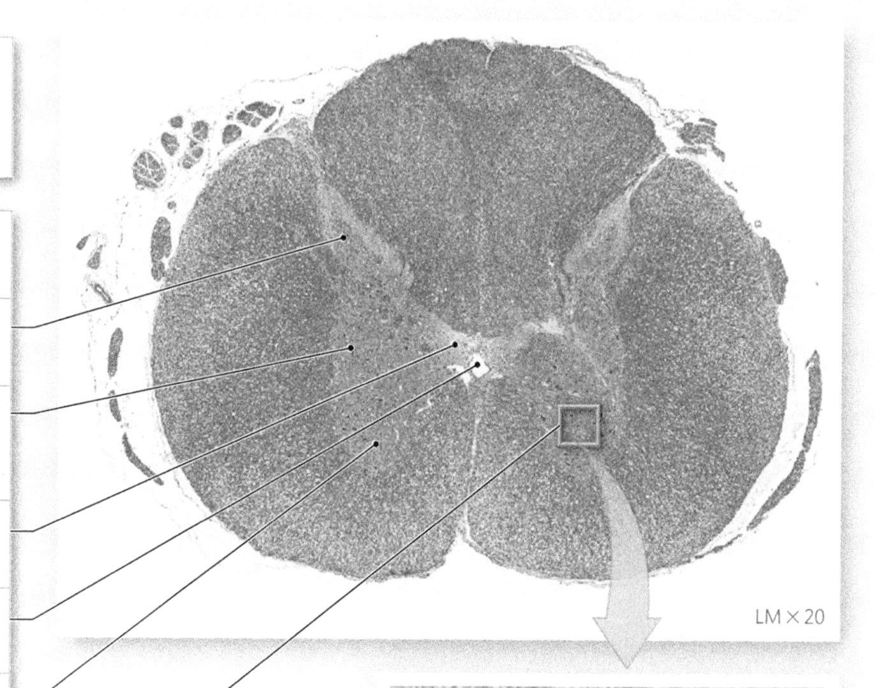

LM × 20

3. Center a region of the anterior horn in the field of view and switch to high power.

4. Identify the cell bodies of multipolar somatic motor neurons. These cells are also called **lower motor neurons.** They give rise to axons that travel in spinal nerves to reach skeletal muscle.

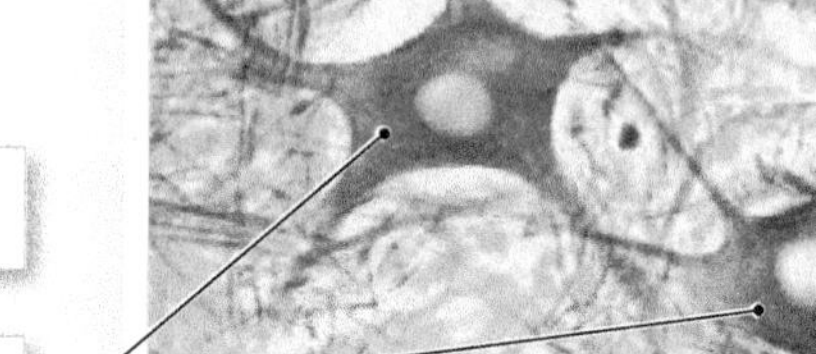

Spinal cord: anterior horn LM × 330

B On each side of the spinal cord, the white matter is divided into three distinct columns (**funiculi;** singular = **funiculus**) of tissue. Each is named according to its relative position. The white matter columns contain two types of fiber tracts that connect the spinal cord and the brain: The **ascending tracts** transmit sensory information from sensory receptors, and the **descending tracts** convey motor information to effectors (muscles and glands).

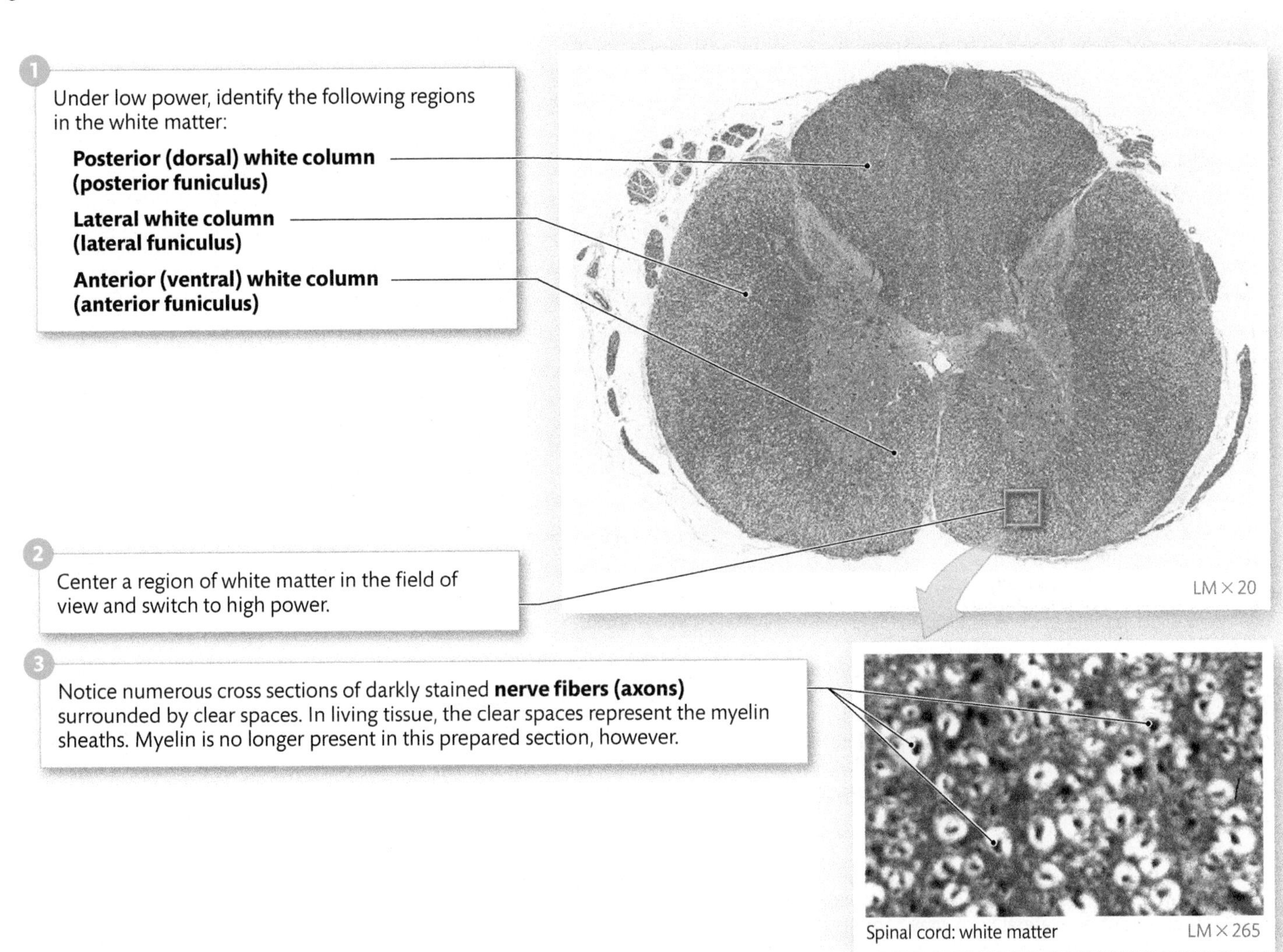

Spinal cord: white matter

MAKING CONNECTIONS

The axons of some motor neurons that begin in the spinal cord and supply skeletal muscles can be longer than 1 meter (3 to 4 feet). Explain how that could be possible.

▶ ____

____

____

____

____

Activity **14.5**

Peripheral Nervous System: Examining the Microscopic Structure of Spinal Nerve Roots

The peripheral nervous system includes all nervous tissue outside the central nervous system. In the PNS, each spinal nerve is formed by the union of a **ventral root** and a **dorsal root.** The ventral root contains motor nerve fibers traveling from the anterior and lateral horns of the spinal cord to a spinal nerve. The dorsal root contains sensory nerve fibers traveling from a spinal nerve to the posterior horn of the spinal cord.

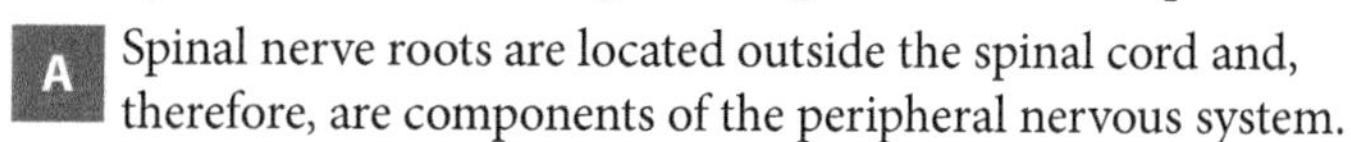

A Spinal nerve roots are located outside the spinal cord and, therefore, are components of the peripheral nervous system.

B The dorsal root ganglia contain cell bodies of unipolar neurons.

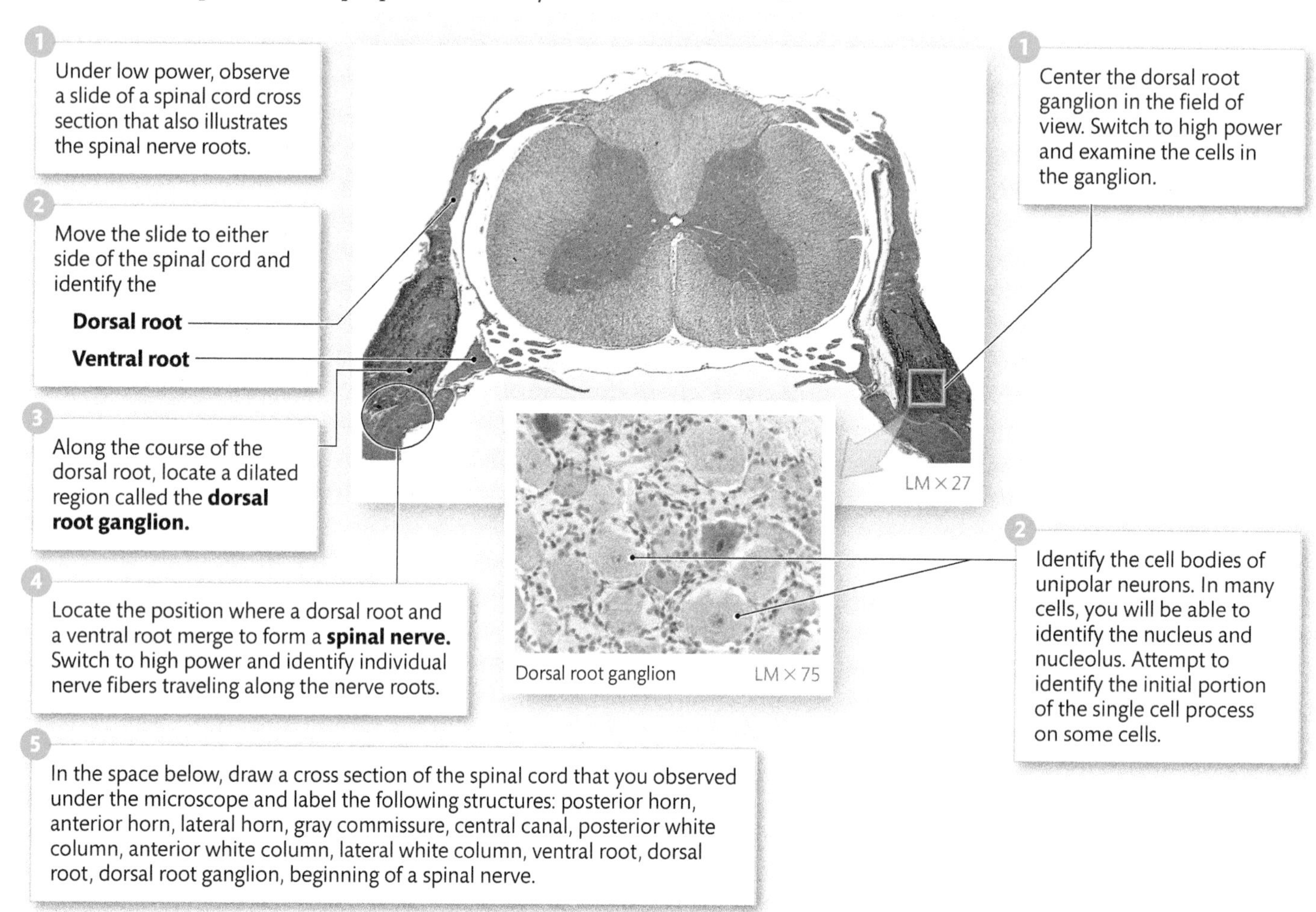

5 In the space below, draw a cross section of the spinal cord that you observed under the microscope and label the following structures: posterior horn, anterior horn, lateral horn, gray commissure, central canal, posterior white column, anterior white column, lateral white column, ventral root, dorsal root, dorsal root ganglion, beginning of a spinal nerve.

▶

MAKING CONNECTIONS

Explain what would happen if all the ventral roots that supply the spinal nerves to the lower limbs were severed.

▶ ______________________________

____

____

____

____

____

■ Want more practice? Go to: **MasteringA&P** > Study Area > Menu > Lab Tools > PAL > Histology > Nervous Tissue

Activity **14.6**

Peripheral Nervous System: Examining the Microscopic Structure of Peripheral Nerves

A A typical peripheral nerve (cranial or spinal nerve) contains thousands of nerve fibers that are organized into bundles (fascicles) and enclosed by connective tissue coverings. Some fibers are surrounded by a myelin sheath, whereas others are unmyelinated.

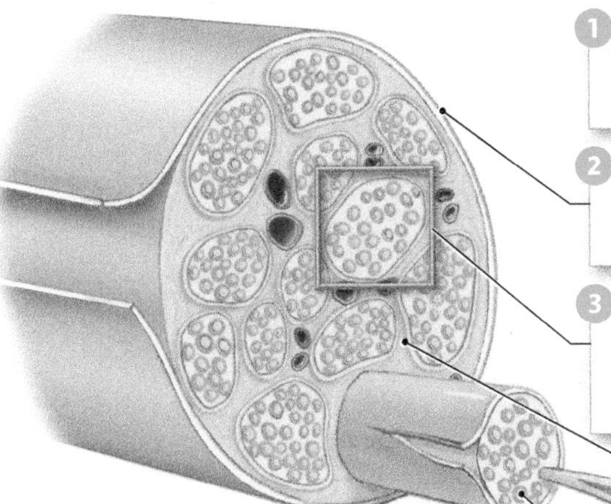

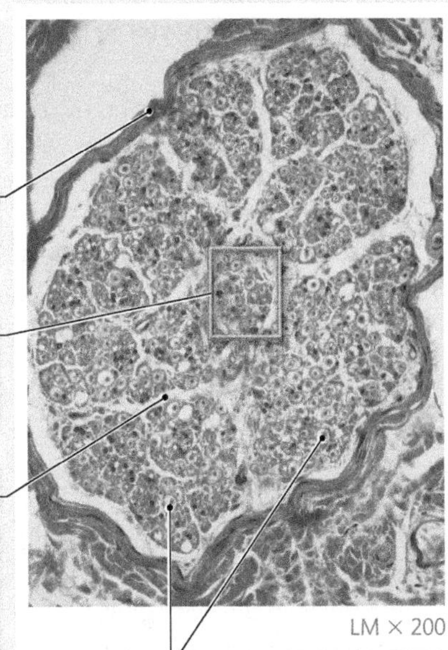

1. View a cross section of a peripheral nerve under low power.
2. Identify the **epineurium,** the layer of connective tissue that surrounds the entire nerve.
3. Identify several **nerve fascicles (nerve bundles).** Each fascicle contains numerous nerve fibers (axons).
4. Identify the **perineurium,** a connective tissue layer that surrounds each fascicle.
5. In a fascicle, identify individual **nerve fibers (axons).** The fibers will appear as darkly stained circles. Myelinated nerve fibers are surrounded by a clear space where the sheath is located. Due to the special preparation of the tissue, the myelin sheath is no longer present. The endoneurium is a connective tissue layer that covers each nerve fiber. It will be difficult to see on your slide.
6. In the space below, draw a cross section of a peripheral nerve and label the following structures: whole nerve, nerve bundle (fascicle), nerve fiber (axon), epineurium, perineurium, endoneurium.

▶

MAKING CONNECTIONS

How do the levels of organization in a peripheral nerve compare to the levels of organization in a skeletal muscle?

▶ ______________________________

Activity 14.7

Demonstrating the Significance of the Myelin Sheath

A Some axons are surrounded by a **myelin sheath,** which consists of several circular layers of fused cell membrane. Because of its high lipid content, the myelin sheath provides an insulating covering for nerve fibers, much like the insulation around electric wires in your home. The myelin sheath is produced by two types of neuroglia: **Schwann cells** in the peripheral nervous system and **oligodendrocytes** in the central nervous system.

Axons that are surrounded by myelin are **myelinated fibers;** those without myelin are **unmyelinated fibers.** At regular intervals along a myelinated axon are small gaps in the myelin sheath. These interruptions are called **nodes (nodes of Ranvier).** The myelinated segments between the nodes are called **internodes.**

Use your textbook to help you complete the first two steps in this activity.

1 In the two diagrams on the right, label the following structures with the appropriate letter: ▶

a. Neuron cell bodies
b. Nuclei in neuron cell bodies
c. Internodes in myelinated axons
d. Nodes in myelinated axons
e. Unmyelinated axon (central nervous system only)
f. Schwann cell nucleus
g. Oligodendrocyte nucleus

2 Examine the myelinated axons in the diagrams on the right. Is there a difference in the way Schwann cells and oligodendocytes myelinate axons? Explain.

▶ ______________________________

Peripheral nervous system

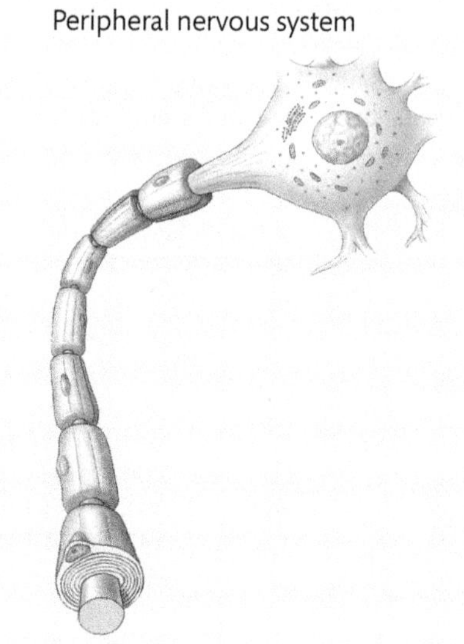

Central nervous system

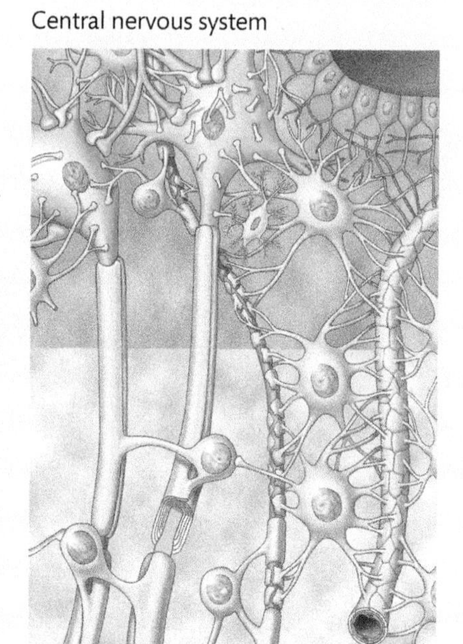

IN THE CLINIC

Multiple Sclerosis

Multiple sclerosis (MS) is a neurological disease that typically occurs in young adults. It is caused by the destruction of the myelin sheath around motor and sensory axons in the brain and spinal cord. The result is a progressive loss of motor and sensory function. Typical symptoms include muscle paralysis, impaired vision, loss of balance, slurred speech, and skin numbness. Multiple sclerosis is an autoimmune disease, which means that the affected individual's own immune system mistakenly attacks and destroys the myelin sheath. Recent investigations suggest that the immune system confuses the myelin sheath proteins for viral proteins with similar amino acid sequences and attacks them. Treatment with interferons—antiviral proteins produced by cells in the immune system—can ease the symptoms of MS, but at present there is no cure.

■ Want more practice? Go to: **MasteringA&P** > Study Area > Menu > Lab Tools > **PhysioEx** > Exercise 3: Neurophysiology of Nerve Impulses > Activity 7: The Action Potential: Conduction Velocity

B The presence or absence of myelin will influence the way action potentials are propagated. Action potentials can be generated along the entire length of an unmyelinated axon. Myelinated axons generate action potentials only at the nodes, where myelin is missing. The insulating property of the myelin sheath inhibits the movement of sodium and potassium ions across the membrane, and this movement creates the action potential. Thus, action potentials are not produced along the myelinated segments of the axon. Instead, they are transmitted by skipping from node to node. This type of transmission, known as **saltatory propagation,** is much faster than the **continuous propagation** that occurs in an unmyelinated axon.

1. Organize two groups of students. The first group, "axon A," should include 9 or 10 students; the second group, "axon B," should include 4 or 5 students. Two additional students, with stopwatches, will serve as timers.

2. Arrange the groups to form two parallel rows, each about 20 feet long. The first person in each group represents the initial segment of an axon. They should stand side by side at the zero mark. The last person in each group represents a synaptic terminal. They should stand side by side at the 20-foot mark. All other members in both groups should position themselves so there is equal distance between them. The first person in each group should hold a pencil.

3. **Make a Prediction**

Before you begin, predict which group (axon A or axon B) will be the first to pass the pencil the entire length of the "axon." Provide an explanation to support your prediction.

▶ ______________________________

____

____

____

____

4. When your instructor gives the command to start, pass the pencil down the line to the end. Members of axon A should be able to pass the pencil without walking toward the next person. Members of axon B will have to walk to the next person.

5. After completing the first pass of the pencil, verify your results by repeating the exercise two more times. Record the results in Table 14.3. ▶

TABLE 14.3 Action Potential Propagation Experiment

"Axon"	Time (sec)		
	Trial 1	Trial 2	Trial 3
A			
B			

6. **Assess the Outcome**

a. Based on the results of this activity, which group represents a myelinated axon? Did you correctly predict that this "axon" would be faster? Explain.

▶ ______________________________

____

b. What function does the passing of the pencil represent?

▶ ______________________________

c. Do the individuals in the myelinated axon represent internodes or nodes? Explain.

▶ ______________________________

____

____

MAKING CONNECTIONS

After reading In the Clinic on the previous page, you learned that multiple sclerosis is caused by damage to the myelin sheath.

a. How does multiple sclerosis affect the transmission of action potentials?

▶ ______________________________

____

b. To demonstrate the effects of multiple sclerosis, would you add or subtract members of the myelinated axon group? Explain.

▶ ______________________________

____

BEFORE YOU MOVE ON . . .

« LOOKING BACK

In this laboratory exercise, you examined the structure of a typical neuron and compared the structure and function of the four types of nerve cells. You also studied the structural organization of the cerebrum and cerebellum in the brain, the spinal cord, and a peripheral nerve.

Neurons have the ability to conduct electric impulses along their axons. They can also transmit impulses to target cells at special junctions called synapses. This highly specialized function allows the nervous system to regulate and maintain homeostasis. For example, when a sensory receptor responds to a change in the environment, sensory information is transmitted along nerve fibers to the central nervous system (CNS), where the information is analyzed. The CNS then sends a motor command to adjust the activity of an effector organ. This process allows the nervous system to monitor all bodily functions.

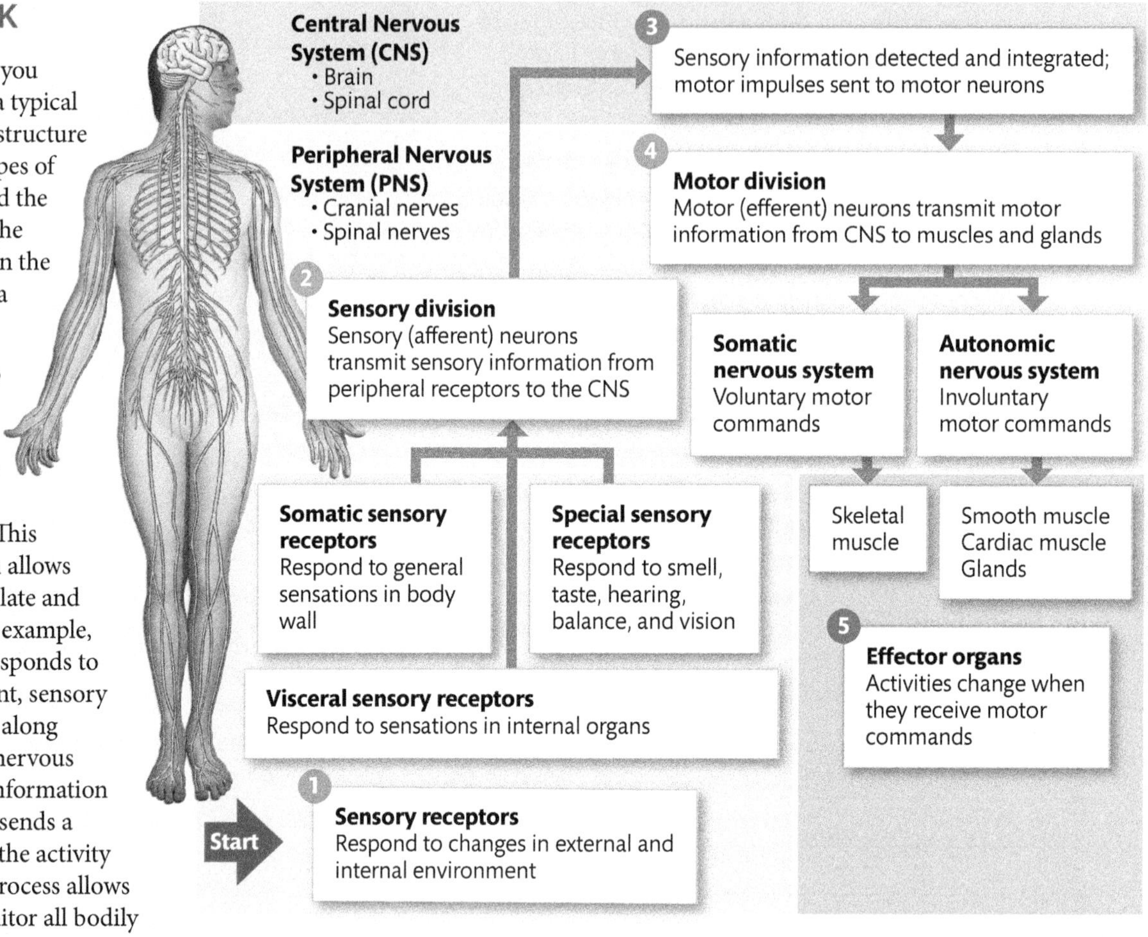

Central Nervous System (CNS)
- Brain
- Spinal cord

Peripheral Nervous System (PNS)
- Cranial nerves
- Spinal nerves

Start

1 **Sensory receptors** Respond to changes in external and internal environment

Somatic sensory receptors Respond to general sensations in body wall

Special sensory receptors Respond to smell, taste, hearing, balance, and vision

Visceral sensory receptors Respond to sensations in internal organs

2 **Sensory division** Sensory (afferent) neurons transmit sensory information from peripheral receptors to the CNS

3 Sensory information detected and integrated; motor impulses sent to motor neurons

4 **Motor division** Motor (efferent) neurons transmit motor information from CNS to muscles and glands

Somatic nervous system Voluntary motor commands

Autonomic nervous system Involuntary motor commands

Skeletal muscle

Smooth muscle
Cardiac muscle
Glands

5 **Effector organs** Activities change when they receive motor commands

Provide examples of how the nervous system can influence and control the function of two organ systems.

▶ ____

____

____

LOOKING FORWARD »

The nervous system, which contains organs composed primarily of nervous tissue, is divided into two structural divisions. The **central nervous system (CNS)** includes the brain and spinal cord. It is the control center for all nervous system function. Sensory information is delivered to the CNS, where it is detected and integrated. In response to sensory inputs, the CNS transmits motor impulses to muscles and glands. The **peripheral nervous system (PNS)** comprises all the nerves that connect the brain and spinal cord to muscles, glands, and receptors. Nerves connected to the brain are called **cranial nerves;** those connected to the spinal cord are **spinal nerves.** In Exercise 15, you will study the structure and function of the brain and cranial nerves; in Exercise 16, you will examine the spinal cord and spinal nerves.

Name ______________________________

Lab Section ______________________________

Date ______________________________

EXERCISE 14 REVIEW SHEET

Nervous Tissue

1. Complete the following table.

Types of Neurons

Unipolar/Bipolar/Multipolar/ Anaxonic	Sensory/Motor/Interneuron/ Unknown	# Axons (one/several/none)	# Dendrites (one/several/none)	Location
a.	b.	c.	d.	Dorsal root ganglia
e.	Interneuron	f.	g.	h.
Multipolar	i.	j.	k.	l.
m.	n.	o.	p.	Retina
q.	Unknown	r.	s.	

QUESTIONS 2–11: Match the structure in column A with the appropriate description in column B.

A

2. Axon ______________________________

3. Synaptic cleft ______________________________

4. Epineurium ______________________________

5. Nissl bodies ______________________________

6. Synaptic terminals ______________________________

7. Perineurium ______________________________

8. Node ______________________________

9. Endoneurium ______________________________

10. Dendrite ______________________________

11. Arbor vitae ______________________________

B

a. Connective tissue layer that surrounds each axon

b. Contains secretory vesicles filled with a neurotransmitter

c. Neuron cell process that receives impulses from other neurons

d. Narrow space that separates the presynaptic and postsynaptic membranes

e. Region of white matter deep to the cerebellar cortex

f. Neuron cell process that transmits electric impulses

g. Connective tissue layer that surrounds an entire nerve

h. Short unmyelinated segment along a myelinated axon

i. Rough endoplasmic reticulum in a neuron

j. Connective tissue layer that surrounds a nerve fascicle

12. Explain the difference between white matter and gray matter.

____

____

____

____

____

QUESTIONS 13–18: Match the neuroglial cell type in column A with the appropriate function in column B.

A	B
13. Ependymal cells ______	**a.** Help form the blood–brain barrier
14. Astrocytes ______	**b.** Produce the myelin sheath in the PNS
15. Schwann cells ______	**c.** Line the walls of ventricles in the brain
16. Microglia ______	**d.** Surround and protect neuron cell bodies in peripheral ganglia
17. Oligodendrocytes ______	**e.** Protect the CNS from disease-causing microorganisms
18. Satellite cells ______	**f.** Produce the myelin sheath in the CNS

19. In the following diagram, identify the structures by using the colors indicated.

Anterior horn = **dark green**
Posterior horn = **red**
Lateral horn = **yellow**
Gray commissure = **black**
Lateral white column = **dark blue**
Anterior white column = **purple**
Posterior white column = **brown**
Dorsal root ganglion = **pink**
Dorsal root = **light green**
Ventral root = **light blue**

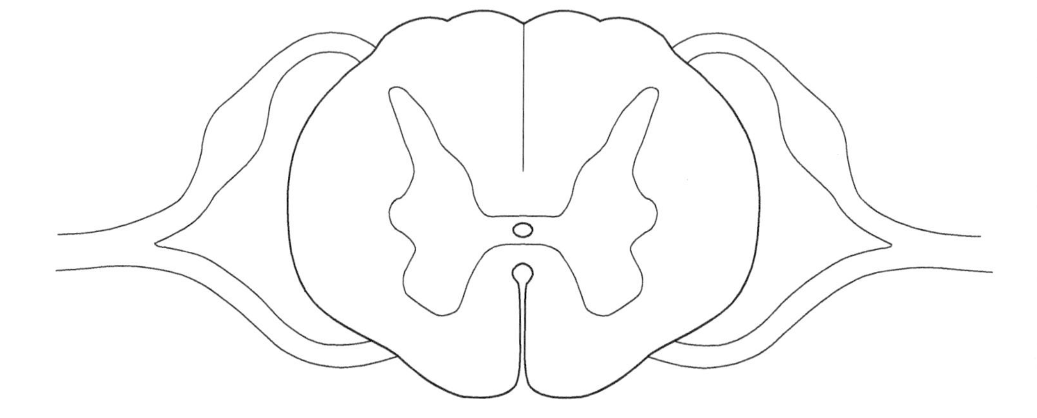

20. What is a myelin sheath? How does the myelin sheath affect the conduction of impulses along an axon?

____

____

____

____

____

____

EXERCISE

15

The Brain and Cranial Nerves

LEARNING OUTCOMES

These Learning Outcomes correspond by number to the laboratory activities in this exercise. When you complete the activities, you should be able to:

Activities	15.1–15.4	Dissect and identify the major anatomical structures on a sheep brain and compare its structure with that of a human brain.
Activity	15.5	Identify the brain ventricles and describe the flow of cerebrospinal fluid.
Activity	15.6	Examine and compare the wave patterns produced by an electrocardiogram (EEG).
Activity	15.7	Locate the origins of the cranial nerves on the brain.
Activities	15.8–15.10	Evaluate the functions of the cranial nerves.
Activity	15.11	Identify the openings in the skull through which the cranial nerves exit the cranial cavity.

LABORATORY SUPPLIES

- Models of the human brain
- Preserved sheep brains
- Dissecting trays
- Dissecting tools
- Dissecting gloves
- Protective eyewear
- Face masks (optional)
- Computer with Windows or Mac operating system
- Biopac Student Lab System (MP35, MP36, or MP45)
- Electrode lead set (SS2L)
- Disposable electrodes (EL503)
- Biopac electrode gel and abrasive pad
- Human skulls
- Eye models
- Model of a midsagittal section of the head
- Pipe cleaners
- Small flashlights
- Cotton swabs
- Samples of coffee, pepper, and aromatic spices
- Stool with wheels
- 10% table salt (NaCl) solution
- 10% sugar (sucrose) solution
- 10% quinine solution
- Colored pencils

PRE-LAB QUIZ

Before you begin, read all the activities in Exercise 15 and the required reading in your textbook that is assigned by your instructor.

1. During this laboratory exercise, you will be dissecting a brain from what animal? ______________________
2. The outermost meningeal layer that surrounds the brain and spinal cord is the ______________________.
3. True or False: In both the human and the sheep brain, the cerebellum is the most prominent structure. __________
4. What type of section is the classic view for examining the internal anatomy of the brain? ______________________
5. True or False: When the internal anatomy of the brain is viewed from a coronal section, the white matter, which is located deep to the cerebral cortex, can be identified. __________
6. The aqueduct of the midbrain is a narrow channel that connects what two brain ventricles? ______________________
7. True or False: There are 10 pairs of cranial nerves. __________
8. All of the following cranial nerves innervate the extrinsic eye muscles *except* the
   a. oculomotor nerve (III).
   b. optic nerve (II).
   c. abducens nerve (VI).
   d. trochlear nerve (IV).
9. During this laboratory exercise, you will identify salty and sweet tastes to test the sensory function of which cranial nerve?
   a. facial nerve (VII)
   b. trigeminal nerve (V)
   c. vestibulocochlear nerve (VIII)
   d. accessory nerve (XI)
10. During this laboratory exercise, you will insert ______________________ into various skull openings to identify the destinations of the cranial nerves.

Activity **15.1**

Dissecting the Sheep Meninges

The sheep brain is anatomically similar to the human brain, so it serves as an excellent model for dissection. At various stages in the dissection, you will be asked to compare structures in the sheep brain and human brain. Have anatomical models of the human brain available for this purpose.

The brain and spinal cord are covered by three connective tissue layers: **dura mater, arachnoid mater,** and **pia mater.** These layers are known collectively as the **meninges.** Along with the cranial bones, vertebrae, and cerebrospinal fluid, the meninges protect the brain and spinal cord from trauma and injury.

A The dura mater is the tough, opaque outer covering of the brain.

1 On a sheep brain, identify the **dura mater** (the dura mater may have already been removed during commercial preparation).

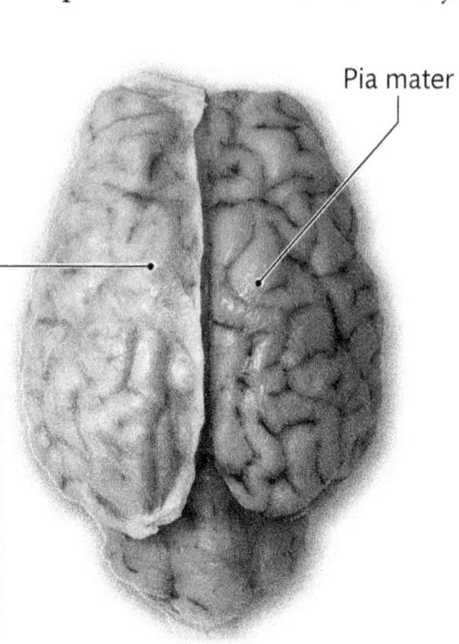

2 Carefully lift the dura mater away from the surface of the brain with forceps and cut the membrane with a pair of scissors. Begin at the posterior end of the brain and cut just to the left or right of the midsagittal plane along the superior surface. Proceed cautiously to avoid damaging underlying structures.

5 The dura mater has two layers:

An outer **endosteal layer** is fused to the periosteum of the cranial bones.

An inner **meningeal layer** forms the brain covering.

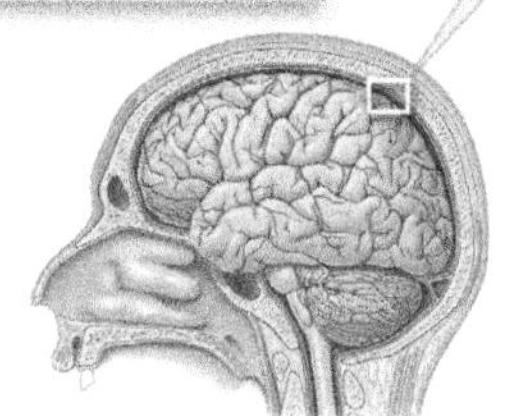

3 Folds of the dura mater form partitions between major brain structures. As you remove the dura mater, identify three **dural folds.** Compare their positions on the sheep brain with a human model.

The **falx cerebri** folds into a deep groove called the longitudinal fissure and separates the cerebral hemispheres.

The **tentorium cerebelli** separates the cerebrum and cerebellum.

The **falx cerebelli** separates the cerebellar hemispheres.

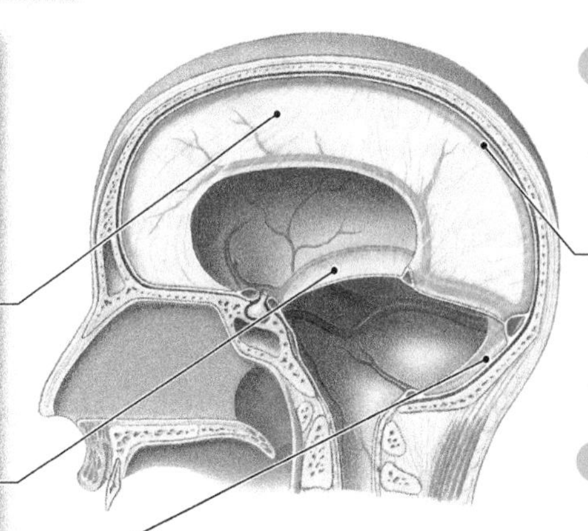

6 You cannot identify the two layers of the dura mater except at specific locations where they separate to form **dural sinuses.**

Along the midsagittal plane, verify that the two layers of the dura mater separate to form the **superior sagittal sinus.** The dural sinuses drain venous blood from the brain to the internal jugular vein.

4 Once you have identified the dural folds, you will have to make at least one cut across each of them with your scissors. Doing so will allow easier removal of the dura mater. What benefits do you think the dural folds provide to the brain?

▶ ______________________________

____

____

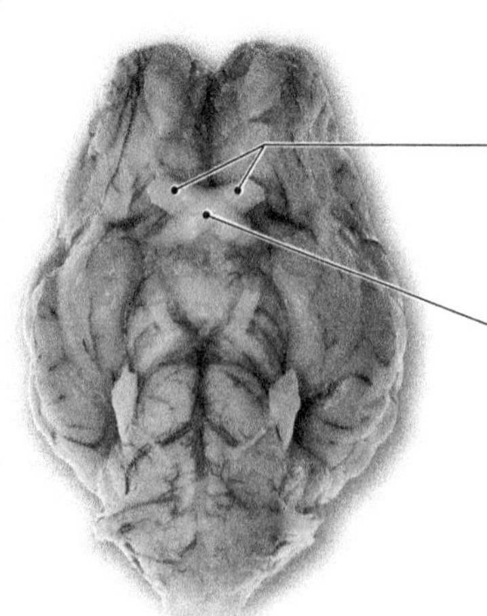

7 It is difficult to remove the dura mater from the inferior surface of the brain without damaging the origins of the cranial nerves. Do not be concerned if that occurs. Do attempt to save the following structures, however:

Two **optic nerves** (cranial nerve II)

The **optic chiasm,** where some nerve fibers from each optic nerve cross over to the opposite side

B The arachnoid mater lies deep to the dura mater. It is separated from the pia mater by the subarachnoid space.

1. Identify the **arachnoid mater** on a model or diagram of the human brain. You may see remnants of the arachnoid mater on your sheep brain, but this layer does not preserve well.

2. Locate the **subarachnoid space** deep to the arachnoid mater. This space is filled with cerebrospinal fluid, which provides a protective, shock-absorbing layer for the brain. Traveling through the space are major blood vessels that supply the brain.

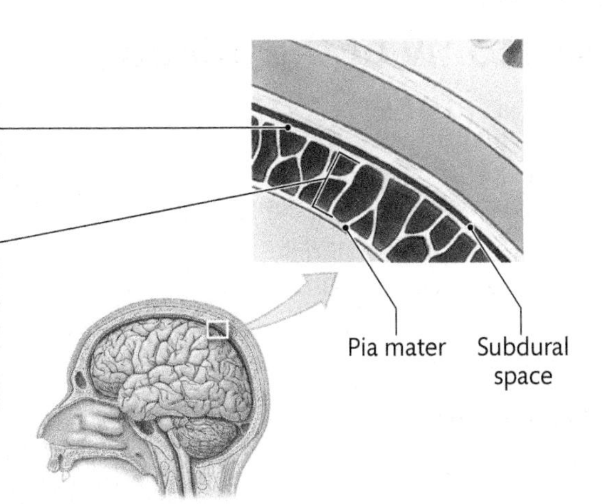

Word Origins

Arachnoid derives from Greek and means "like a spider." The middle meningeal layer is called the arachnoid mater because slender filamentous strands that traverse the subarachnoid space and connect the arachnoid mater to the deeper pia mater resemble a spider's web.

IN THE CLINIC

Epidural and Subdural Hematomas

The dura mater is associated with two potential spaces: the **epidural space** between the dura mater and cranial bones and the **subdural space** between the dura mater and arachnoid mater. The term *potential space* means that a true space does not really exist, but one could form as a result of injury (e.g., trauma such as a blow to the head).

Epidural and **subdural hematomas** can form when meningeal blood vessels are torn by trauma. In an epidural hematoma, blood collects in the epidural space when meningeal arteries, traveling along the outer dural layer, rupture. The pterion (the temple) is a common location for an epidural hematoma because the anterior branch of the middle meningeal artery travels superficially through this region.

A subdural hematoma develops when blood from damaged meningeal veins collects in the subdural space. They typically form when a blow to the head shakes the brain inside the skull.

Epidural hematomas are more dangerous than subdural hematomas because arterial blood pressure is much greater than venous blood pressure. Thus, an epidural hematoma spreads quickly over a relatively short period of time. In both cases, however, an expanding hematoma may exert fatal pressure on the brain unless it is promptly recognized and surgically treated.

C The pia mater is a thin, delicate layer that follows the complicated pattern of convolutions and grooves along the surface of the cerebrum.

1. With the dura mater removed, notice that the surface of your sheep brain has a glossy appearance. This gloss is due to the presence of the **pia mater,** a thin, translucent layer that is closely attached to the surface of the brain (see top figure on previous page).

2. Carefully lift a small portion of the pia mater away from the surface of the brain with a pair of forceps. Notice that removing the pia mater exposes the surface of the brain, which is not glossy.

MAKING CONNECTIONS

Identify three special features of the meninges around the sheep brain that are similar to the meninges around the human brain.

▶ ______________________________

____

____

____

____

____

____

____

____

____

____

Want more practice? Go to: **MasteringA&P** > Study Area > Menu > Lab Tools > PAL >
- Cat > Nervous System > Sheep brain photos (external surface views)
- Human Cadaver > Nervous System > Central Nervous System > Brain photos (midsagittal sections)

Activity **15.2**

Surface Anatomy of the Sheep Brain: Examining Surface Structures

You will begin your study of brain anatomy by identifying some of the prominent external structures. This activity will give you a good overview of the brain's structure before you make any incisions to examine the internal anatomy.

A In both the human and sheep brains, the cerebrum is the largest and most prominent structure.

1. On the sheep brain, identify the cerebrum, which consists of two **cerebral hemispheres.**

Left cerebral hemisphere

Right cerebral hemisphere

2. In each hemisphere, identify the following lobes:

The **frontal lobe** is located at the anterior end of the cerebrum.

The **parietal lobe** is just posterior to the frontal lobe.

The **temporal lobe** is located along the lateral aspect of each hemisphere.

The **occipital lobe** is located at the posterior tip of the cerebrum.

3. Identify the **longitudinal fissure,** which separates the two hemispheres.

4. On a human brain model, identify the frontal, parietal, temporal, and occipital lobes. Identify each cerebral lobe in the diagram on the right. ▶

Lobe A ____________

Lobe B ____________

Lobe C ____________

Lobe D ____________

5. Identify a deep cerebral lobe known as the **insula.** The insula is deep to what other lobe?

▶ ____________

6. Inspect the surface (cerebral cortex) of each cerebral hemisphere. Notice the complex folding pattern that forms ridges of tissue known as **gyri** (singular = **gyrus**). Each gyrus is separated by a groove. A shallow groove is a **sulcus** (plural = **sulci**), and a deep groove is a **fissure.** On a brain model, identify the following major grooves and the lobes that are separated by them.

The **parieto-occipital sulcus** separates what two lobes?

▶ ____________

The **central sulcus** separates what two lobes?

▶ ____________

The **lateral fissure** separates what three lobes?

▶ ____________

Want more practice? Go to: **MasteringA&P** > Study Area > Menu > Lab Tools > PAL >

- Cat > Nervous System > Sheep brain photos (external surface views)
- Human Cadaver > Nervous System > Central Nervous System > Brain photos (external surface views)

B The cerebellum is located just posterior to the cerebrum. It is separated from the cerebrum by the **transverse fissure**. When the cerebrum and cerebellum are moved apart at the transverse fissure, the midbrain is exposed.

1 On the sheep brain, identify the **cerebellum,** just posterior to the cerebrum. Notice that the cerebellum, like the cerebrum, is divided into two hemispheres, the cerebellar hemispheres.

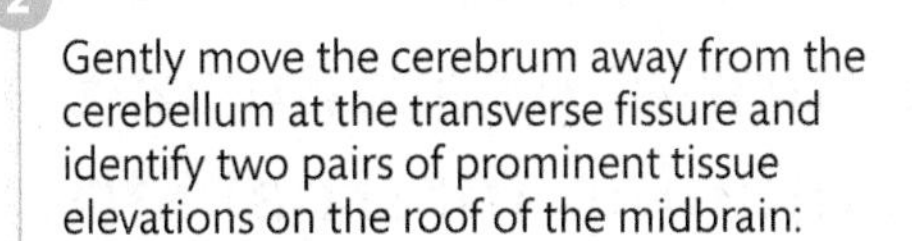

2 Gently move the cerebrum away from the cerebellum at the transverse fissure and identify two pairs of prominent tissue elevations on the roof of the midbrain:

The anterior pair is the **superior colliculi** (singular = **colliculus**).

The posterior pair is the **inferior colliculi.**

Collectively, all four of these tissue elevations are called the **corpora quadrigemina.**

C The pituitary gland and brainstem can be identified on the the inferior surface of the brain.

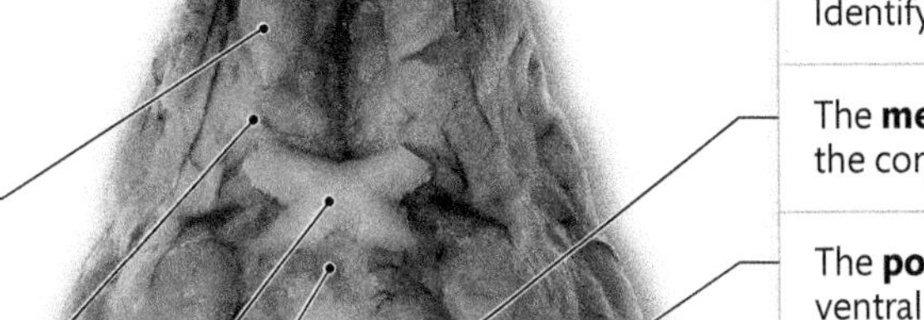

1 Observe the inferior surface of the sheep brain and identify the following structures:

The **olfactory bulbs** are relatively large, ovoid masses on the inferior surface of the frontal lobe.

The **olfactory tracts** leave the olfactory bulbs and pass posteriorly to the olfactory areas in the temporal lobes.

The **optic chiasm,** where the two optic nerves converge, is anterior to the pituitary gland.

The **pituitary gland** (removed in figure) is connected to the hypothalamus by the **infundibulum.**

2 Identify the three parts of the **brainstem:**

The **mesencephalon** or **midbrain** is where the corpora quadrigemina are located.

The **pons** is an elevated mass of tissue on the ventral surface.

The **medulla oblongata** is the column of tissue that extends posteriorly from the pons and is directly continuous with the spinal cord.

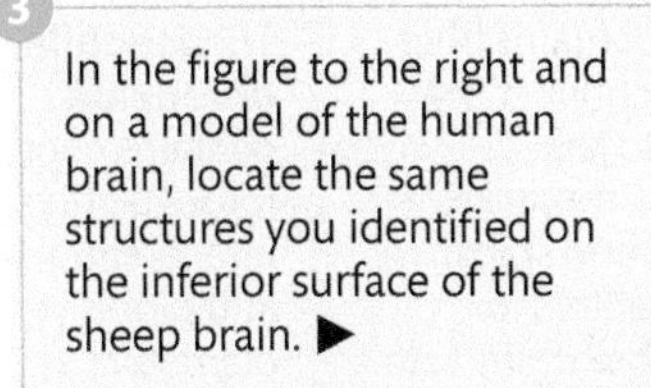

3 In the figure to the right and on a model of the human brain, locate the same structures you identified on the inferior surface of the sheep brain. ▶

1. ____________________
2. ____________________
3. ____________________
4. ____________________
5. ____________________
6. ____________________
7. ____________________

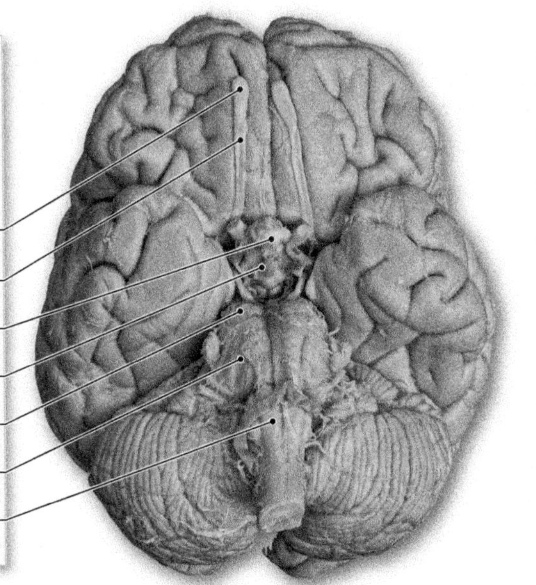

MAKING CONNECTIONS

During this activity, you observed that the cerebral cortex is characterized by a complex series of gyri. Speculate on the functional advantage for this structural pattern.

▶ ____

____

____

____

____

____

____

____

____

Sectional Anatomy of the Sheep Brain: Examining Internal Structures from a Midsagittal Section

By studying sectional views, you will be able to identify most external brain features as well as several deep structures that cannot be seen from the surface.

A The **midsagittal section** is a classic view for studying brain anatomy. It provides a clear anterior-to-posterior view of most major structures and demonstrates the dominance in size of the cerebrum over other brain regions.

1. Place the sheep brain on the dissecting tray with the dorsal surface up.
2. Using a scalpel or small knife, make an incision along the midsagittal plane of the brain. Cut along the longitudinal fissure in an anterior-to-posterior direction, staying as close to the midline as possible.
3. When you reach the posterior end of the cerebrum, continue your midsagittal incision through the cerebellum and along the brainstem.
4. From this view, identify the following structures:

The **cerebrum** interprets sensory impulses, controls voluntary motor activity, regulates muscle tone, stores information in memory, and develops emotional and intellectual thoughts.

The **cingulate gyrus** is a part of the limbic system where emotions and other related behaviors are regulated.

The **pineal body** secretes melatonin, a hormone that assists the hypothalamus in the regulation of sleep/wake cycles.

The **thalamus** acts as a relay station for sensory impulses going to the cerebral cortex.

The **corpus callosum** contains myelinated nerve fiber tracts that connect the two cerebral hemispheres.

The **fornix** is a fiber tract that connects the mamillary body to the hippocampus. You will not be able to identify the hippocampus because it is located deep within the temporal lobe.

The **cerebellum** regulates motor activity so that muscle contractions are smooth, coordinated, and timely.

The **hypothalamus** regulates the autonomic nervous system and pituitary gland. It contains control centers for body temperature and food and water intake.

The **mamillary body** is a region in the hypothalamus. A component of the limbic system, it serves as a control center for motor reflexes associated with eating.

The **midbrain** contains the superior colliculi, which control reflexes related to visual stimuli, and the inferior colliculi, which control reflexes related to auditory stimuli.

The **pons** contains fiber tracts that connect the cerebrum to the cerebellum, medulla oblongata, and spinal cord and centers for regulating breathing rate.

The **medulla oblongata** contains centers for regulating heart rate, breathing rate, and blood vessel diameter.

Word Origins

In a midsagittal section, the highly branched pattern of the cerebellar white matter can be clearly seen (see the figure to the right). Because it resembles the branches of a tree, the white matter in the cerebellum is often referred to as the ***arbor vitae,*** which is the Latin term for "tree of life."

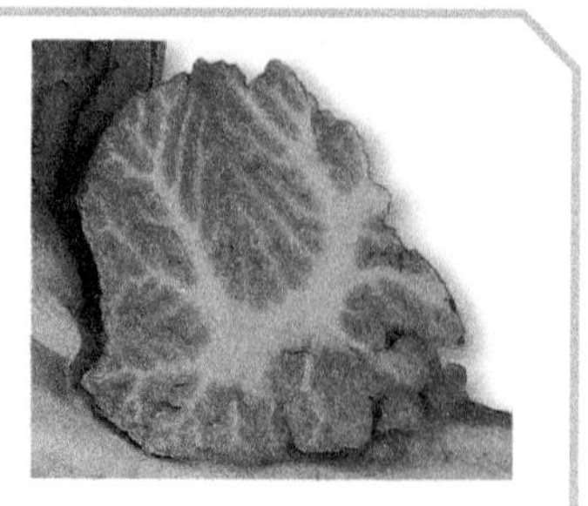

B Midsagittal sections of the human and sheep brains are strikingly similar. Compare your dissection with a human brain model.

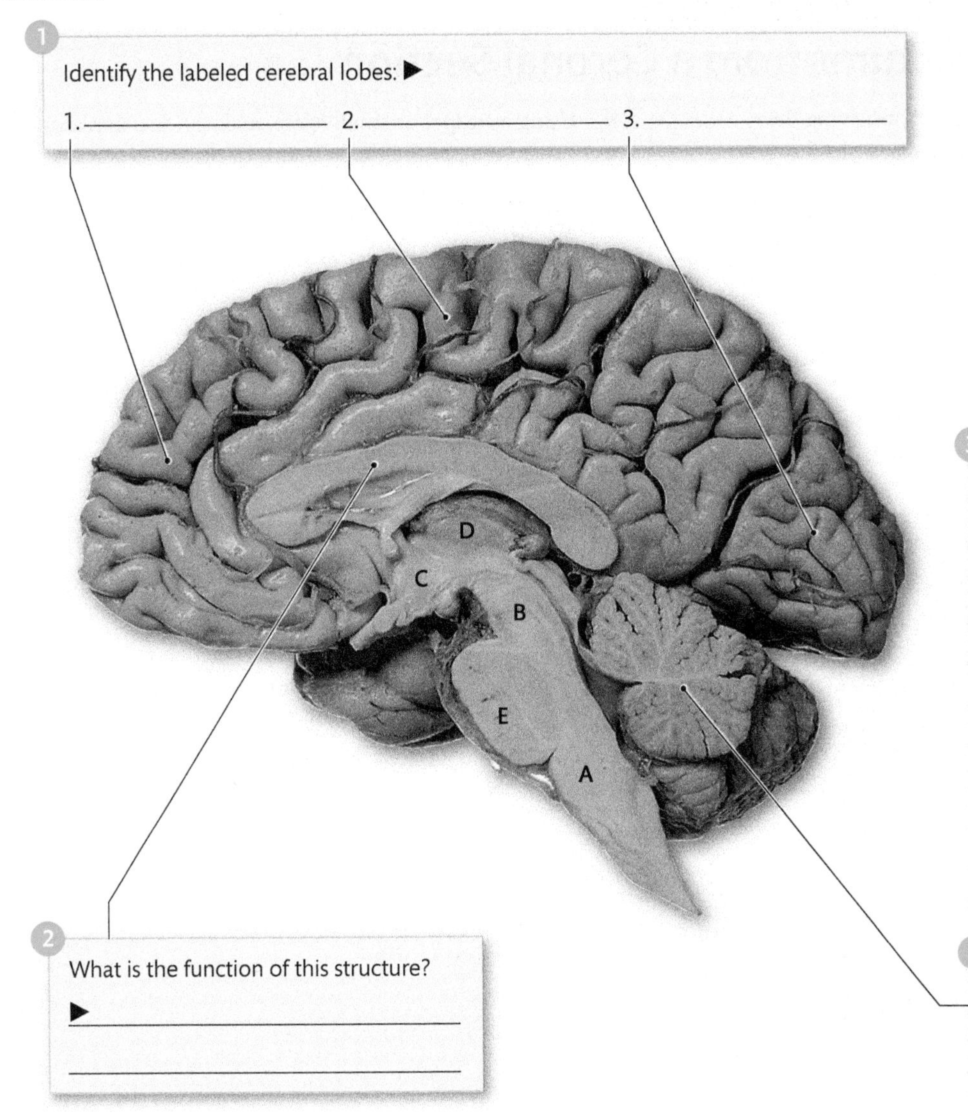

Word Origins

Colliculus derives from Latin and means "mound." The collective name for the inferior and superior colliculi, *corpora quadrigemina,* is also Latin and means, literally, "bodies of quadruplets." More loosely, it translates to "four bodies"; it refers to the four mounds (bodies) of tissue that make up the inferior and superior colliculi.

3 Identify structures A–E in the figure. ▶

Structure A ______

Structure B ______

Structure C ______

Structure D ______

Structure E ______

Which of the five structures, above, form the brainstem?

▶ ______

4 This structure is a component of which brain region?

▶ ______

IN THE CLINIC

Concussions

A concussion is a traumatic brain injury caused by a blow to the head or a fall that causes the brain to shake inside the skull. Some people may briefly lose consciousness after the injury but others will not. Other possible symptoms include temporary memory loss, drowsiness, headache, difficulty concentrating, dizziness, double or blurred vision, and sensitivity to light and noise. Individuals who participate in contact sports have an increased risk of getting a concussion. Although a single concussion is not considered to be life threatening, people who have experienced multiple concussions could develop a degenerative brain disease known as Chronic Traumatic Encephalopathy (CTE).

The best way to recover from a concussion is to rest, both physically and mentally. All physical activity such as sports or physical labor should stop. Mental activities such as using a computer or smartphone, reading, or doing schoolwork should also be avoided. The recovery period should be carefully monitored by a physician, and normal activities should not resume until all concussion symptoms are gone.

MAKING CONNECTIONS

Identify one structural difference between the human and sheep brain.

▶ ______

Want more practice? Go to: **MasteringA&P** > Study Area > Menu > Lab Tools > PAL >

- Cat > Nervous System > Sheep brain photos (midsagittal sections)
- Anatomical Models > Nervous System > Central Nervous System > Brain photos (midsagittal sections)
- Human Cadaver > Nervous System > Central Nervous System > Brain photos (midsagittal sections)

Activity **15.4**

Sectional Anatomy of the Sheep Brain: Examining Internal Structures from a Coronal Section

A A **coronal section** is particularly useful for observing the organization of the gray matter and white matter in the cerebrum. From this view, the position of the surface gray matter in the cerebral cortex in relation to the white matter and deep gray matter nuclei (i.e., the basal nuclei) is clearly seen.

1. Using a scalpel or small knife, make an incision along a coronal plane of the sheep brain. Begin your incision by cutting through the parietal lobe. Be sure that your cut passes through the temporal lobes.
2. Examine your coronal section of the brain and identify the following structures:

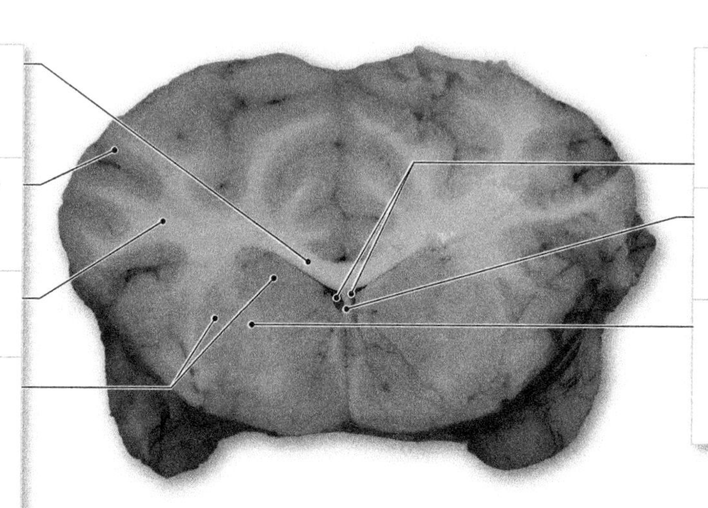

The **corpus callosum** is a fiber tract that travels between the two cerebral hemispheres.

The **cerebral cortex** is a thin layer of gray matter on the surface of the cerebrum.

The **white matter** is deep to the cerebral cortex.

The **basal nuclei** are regions of gray matter deep within the cerebrum. They regulate the intensity and degree of voluntary movements by making adjustments in muscle tone.

The two **lateral ventricles** (see Activity 15.5), deep within the cerebrum, contain cerebrospinal fluid.

The **septum pellucidum** is a partition that separates the two lateral ventricles.

The **internal capsule** is a band of nerve fiber tracts (white matter) that travels between the thalamus and basal nuclei.

IN THE CLINIC

Parkinson's Disease

Neurons located in a midbrain nucleus called the **substantia nigra** project nerve fibers to the basal nuclei, where they release the neurotransmitter **dopamine.** Damage to the substantia nigra reduces dopamine levels and basal nuclei activity. The result is a gradual systemic increase in muscle tone, and symptoms of **Parkinson's disease** can develop: muscle tremors at rest and difficulty initiating and controlling voluntary movements. While walking, people with Parkinson's disease tend to shuffle and bend forward. It is not clear how Parkinson's develops, but recent research suggests that exposure to pesticides might increase the risk. Treatment for Parkinson's disease includes administration of L-dopa, a precursor of dopamine, in combination with other drugs. Recent studies using transplants of fetal substantia nigra tissue have shown promising results. The use of fetal tissue for this purpose is highly controversial in the United States, however, and whether research will progress in this direction remains questionable.

■ Want more practice? Go to: **MasteringA&P** > Study Area > Menu > Lab Tools > PAL > Human Cadaver > Nervous System > Central Nervous System > Brain photos (coronal and transverse sections)

B Compare your coronal section dissection with a similar view on a human brain model.

1 Identify structures A–H in the figure. ▶

Structure A ________________________

Structure B ________________________

Structure C ________________________

Structure D ________________________

Structure E ________________________

Structure F ________________________

Structure G ________________________

Structure H ________________________

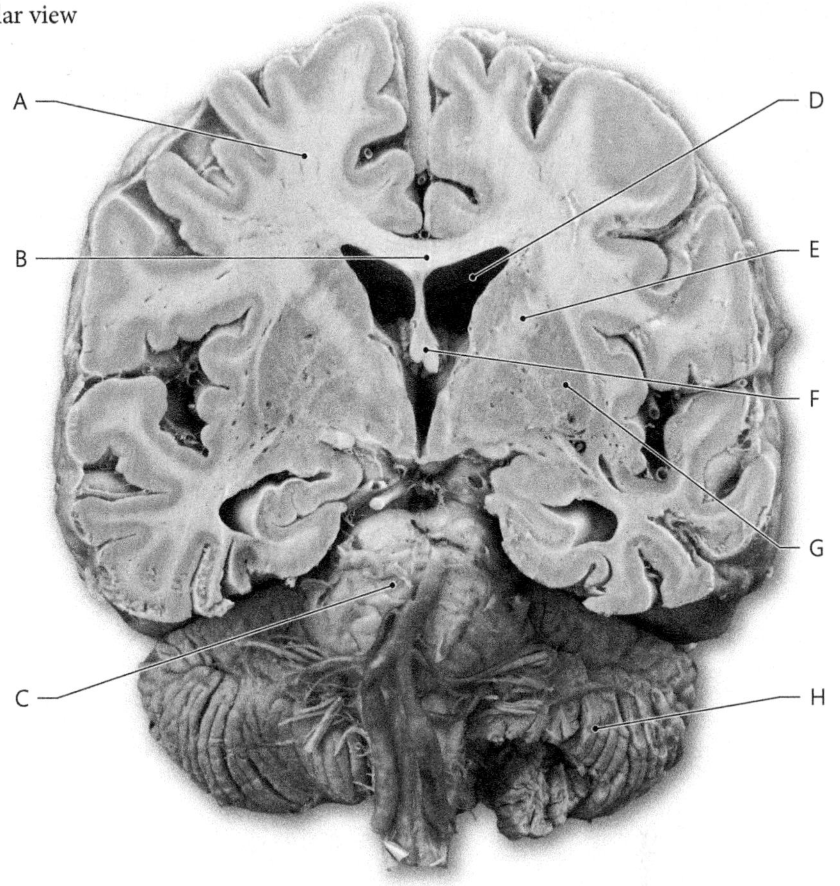

A
B
C
D
E
F
G
H

2 Perform the following simple test:

- Have your lab partner sit with his or her eyes closed and palms facing up.
- Lightly touch one of your lab partner's fingers with a pencil or pen.
- With the other hand, ask your lab partner to touch the corresponding finger with the thumb. (*Example: If you touch the right index finger, your lab partner must touch his or her left index finger with his or her left thumb).*

3 To perform this test successfully, your lab partner must be able to coordinate the actions of both hands by passing information between the two cerebral hemispheres. What structure allows you to do that? Explain.

▶ ________________________

MAKING CONNECTIONS

What structures can you see in a coronal section of the brain that you cannot identify in a midsagittal section?

▶ ________________________

Activity **15.5**

Examining Brain Ventricles and the Flow of Cerebrospinal Fluid

A The **ventricles** are a series of interconnected deep cavities that are derived from the internal space of the embryonic **neural tube.** The walls of the ventricles are lined by **ependymal cells,** a type of neuroglia, and the lumen is filled with **cerebrospinal fluid (CSF).** To appreciate the three-dimensional qualities of the ventricles, examine them from a variety of views and sections.

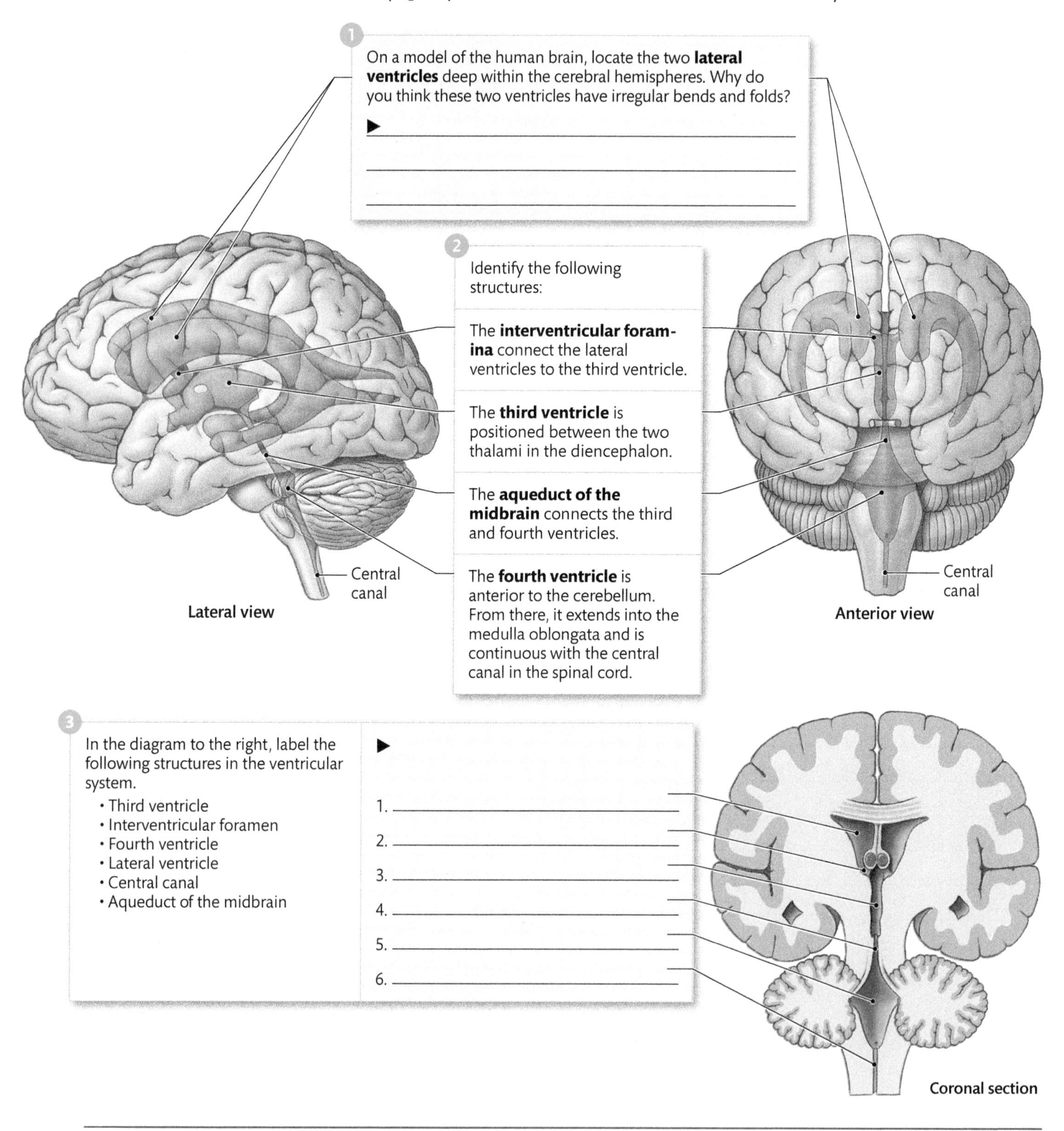

Want more practice? Go to: **MasteringA&P** > Study Area > Menu > Lab Tools > PAL >

- Anatomical Models > Nervous System > Central Nervous System > Brain photos (ventricles)
- Human Cadaver > Nervous System > Central Nervous System > Brain photos (midsagittal and coronal sections)

B Cerebrospinal fluid (CSF) flows through the brain ventricles and subarachnoid space.

1 On a model of the brain, locate the roofs of the brain ventricles, where the **choroid plexuses** are located. The choroid plexuses produce CSF. Each plexus is composed of capillary networks that are in close contact with modified ependymal cells. CSF forms when blood plasma filters through the walls of the capillaries, travels through the ependymal cells, and empties into the ventricles. CSF flows through the ventricles and into the central canal of the spinal cord.

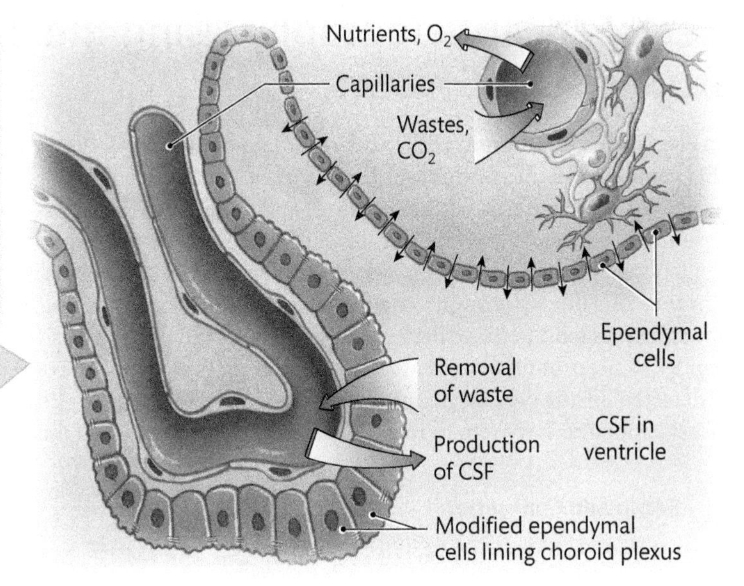

Dura mater
Superior sagittal sinus
Cranium
Arachnoid granulation
CSF movement
Subdural space
Arachnoid mater
Subarachnoid space
Pia mater
Cerebral cortex

2 On a brain model, identify the **fourth ventricle.** CSF enters the subarachnoid space by passing through openings in the roof of this ventricle. Because the subarachnoid space is filled with cerebrospinal fluid, the brain and spinal cord are surrounded by an aqueous (mostly water) liquid layer. Explain why this arrangement protects the central nervous system.

▶ ______________________

____

____

____

____

____

____

____

____

____

3 Cerebrospinal fluid is drained by the **arachnoid granulations.** The granulations are extensions of the arachnoid mater that project into the dural sinuses. CSF filters through the granulations and drains into venous blood by entering one of the dural sinuses.

MAKING CONNECTIONS

List all structures, in the correct order, that a drop of CSF encounters in its pathway. Start with the choroid plexus along the roof of the right lateral ventricle and end with the central canal of the spinal cord.

▶ ______________________

____

____

Beginning once again at the choroid plexus in the right lateral ventricle, trace the pathway that a drop of CSF would take to reach the subarachnoid space.

▶ ______________________

____

____

Activity **15.6**

Electroencephalography Using the Biopac Student Lab System

An **electroencephalogram (EEG)** is a recording of the brain's electrical activity. The wave patterns produced by an EEG generally fall into one of four basic types, each characterized by a particular combination of frequency (the number of wave patterns per time period) and amplitude (the size of the waves). **Alpha waves** (low-frequency, high-amplitude) are detected in healthy adults who are awake but resting, with their eyes closed. Higher-frequency, lower-amplitude **beta waves** are identified in individuals who are experiencing a stressful situation or concentrating on performing a specific task. They also occur during rapid eye movement (REM) or deep sleep, when the eyes tend to move back and forth. **Theta** and **delta waves** typically occur during periods of sleep other than REM sleep, but they may also occur if a brain tumor or some other disorder is present. Theta waves can also be detected during periods of intense frustration in otherwise healthy individuals. In this activity, you will use the Biopac Student Lab System to record your lab partner's EEG and then examine the four EEG waveforms.

A Setup and Calibration

1. Turn the computer on but keep the Biopac unit (MP35, MP36, or MP45) off.
2. Position three electrodes on your lab partner's scalp as shown in the figure on the right. Part your lab partner's hair tightly and apply a drop of electrode gel to the exposed scalp. Firmly press the disposable electrode onto the gel. Attach the red, white, and black electrode leads (SS2L) as illustrated. After the electrodes are attached, a swim cap or self-adhering wrap can be used to help keep them in place.

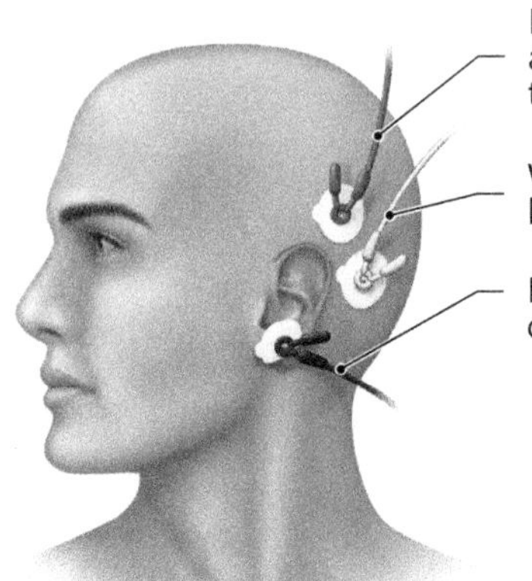

3. Instruct your lab partner to sit comfortably in a chair and minimize movement throughout the experiment. The room should be quiet so that your lab partner can relax.
4. Attach the electrode leads (SS2L) to Channel 1 and turn on the Biopac unit.
5. Start the Biopac Student Lab System, choose Lesson 3 (L03 EEG-1), and click **OK**. When prompted, enter a unique file name and click **OK**.
6. Your lab partner should be relaxed with his or her eyes closed. Click on the Calibrate button and wait for the computer to adjust the signal. Calibration automatically stops after 8 seconds. If the recording is relatively flat without any large fluctuations or spikes (see the figure to the right), you are ready to record. If not, check the electrodes and connections and click **Redo Calibration.**

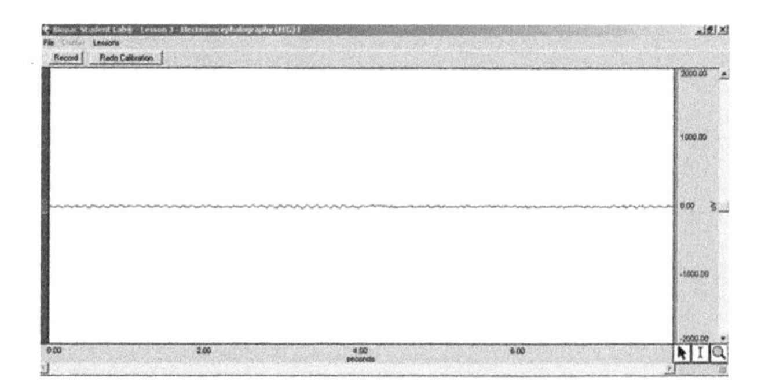

B Recording Data

You will record three segments of data in this exercise: 20 seconds with your lab partner's eyes closed, 20 seconds with eyes open, and 20 seconds with eyes reclosed. Read this entire section first so that you will be familiar with the activities before you begin recording data.

1. Have your lab partner sit quietly and breathe slowly with his or her eyes closed. Click on the **Record** button.
2. After 20 seconds, ask your lab partner to open his or her eyes and try not to blink. Immediately press the F4 button to insert a marker.
3. Record for an additional 20 seconds with the eyes open and not blinking. After 20 seconds, ask your lab partner to close his or her eyes. Immediately press the F5 button to insert a marker.
4. Record for an additional 20 seconds and click **Suspend.**
5. Your raw EEG recording should resemble what is shown in the figure to the right. If it does, click **Done.** If it does not, click **Redo.**

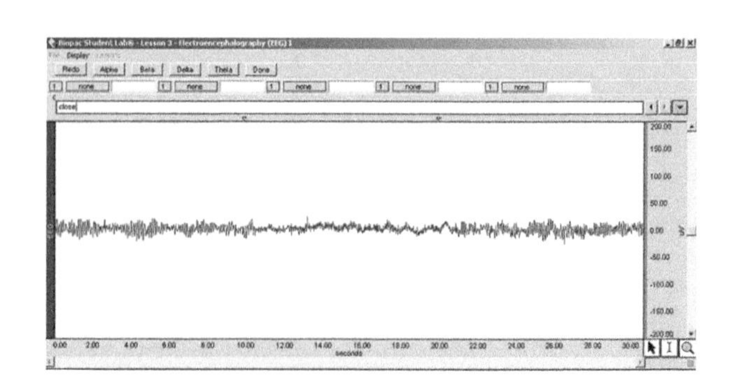

TABLE 15.1	Channel Designations and Measurement Box Settings		
Channel	**Designations**	**Measurements for Box Settings**	
CH 1	EEG (hidden)		
CH 40	Alpha	CH 40	Stddev
CH 41	Beta	CH 41	Stddev
CH 42	Delta	CH 42	Stddev
CH 43	Theta	CH 43	Stddev
		SC	Freq

Note: **Stddev** is the standard deviation, which measures the variability in the wave amplitude at each data point.

Freq is the frequency, which converts the time segment of the selected area to cycles/sec.

C Data Analysis

1. Enter the **Review Saved Data** Mode. Select the appropriate data file for Lesson 3, which ends in L-03.
2. Review the channel (CH) designations and the measurement box settings. They are listed in Table 15.1.
3. Using the I-beam cursor tool, select the region corresponding to the first "eyes closed" segment (from time 0 to the first marker) as shown in the figure below and record the standard deviations for all four waveforms in Table 15.2. ▶

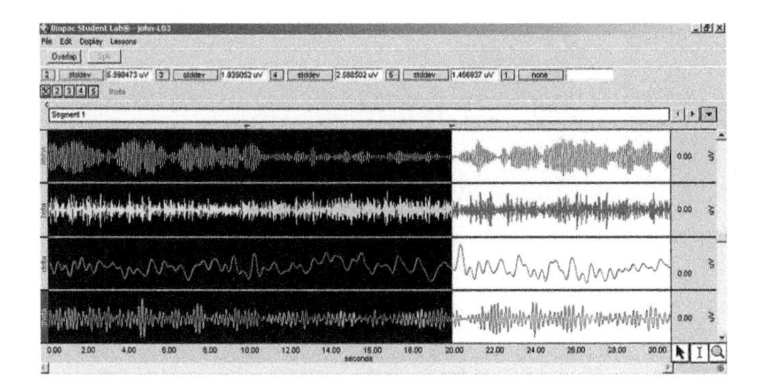

4. Repeat step 3 for the region corresponding to the "eyes open" segment (from the first marker to the second) and the region corresponding to the "eyes reclosed" segment (from the second marker to the end). Record the standard deviations for all four waveforms in Table 15.2. ▶

TABLE 15.2	EEG Amplitude Measurements (Stddev. uV) in Each Segment			
Waveform	**Channel**	**Eyes Closed**	**Eyes Open**	**Eyes Reclosed**
Alpha	CH 40			
Beta	CH 41			
Delta	CH 42			
Theta	CH 43			

5. What changes do you observe in the amplitude of the alpha and beta waves when the eyes are closed and when they are open?

▶ ______________________________

____

____

6. Using the zoom cursor, zoom in on a 3- to 4-second region of the "Eyes Closed" segment. Then, using the I-beam cursor, select a single peak-to-peak cycle from the alpha wave as shown in the figure below (top waveform). Record the frequency in Table 15.3. Take two additional single-cycle measurements from the alpha wave and record the frequencies in Table 15.3. Calculate the mean frequency and record the result in Table 15.3. ▶

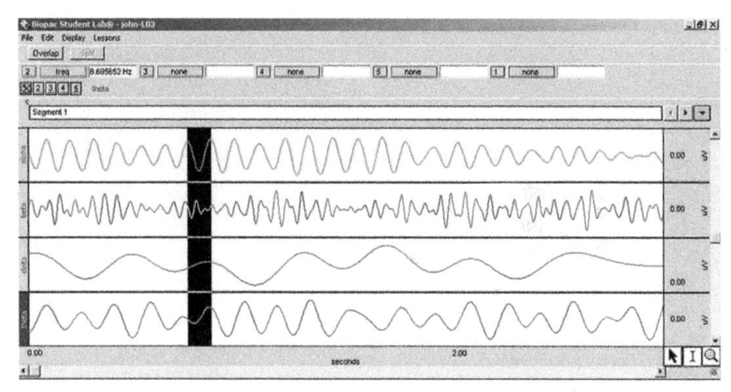

7. Repeat step 6 for three cycles from each of the beta, delta, and theta waves. Calculate the mean frequencies and record the data in Table 15.3. ▶

TABLE 15.3	EEG Frequency Measurements (Hz) in the Eyes Closed Segment					
Waveform	**Channel**	**Cycle 1**	**Cycle 2**	**Cycle 3**	**Mean**	**Normal**
Alpha	CH 40					8–13
Beta	CH 41					13–30
Delta	CH 42					1–5
Theta	CH 43					4–8

MAKING CONNECTIONS

How do the frequencies of the four waves you measured compare to the normal ranges presented in Table 15.3?

▶ ______________________________

____

____

____

____

Activity 15.7

Identifying the Origins of the Cranial Nerves

The 12 pairs of cranial nerves have special names that describe their general distribution or functions. They are also identified by Roman numerals in the order of their origins, from anterior to posterior.

A The **olfactory nerves (cranial nerves I)** orginate in the nasal cavity and form synapses with neurons in the olfactory bulbs.

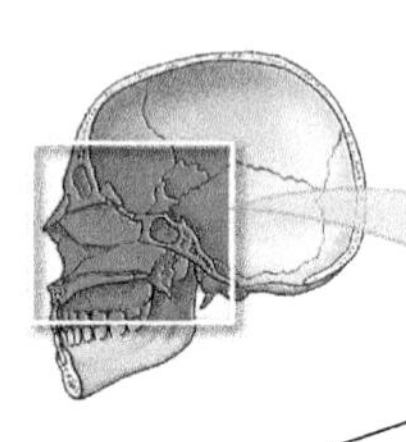

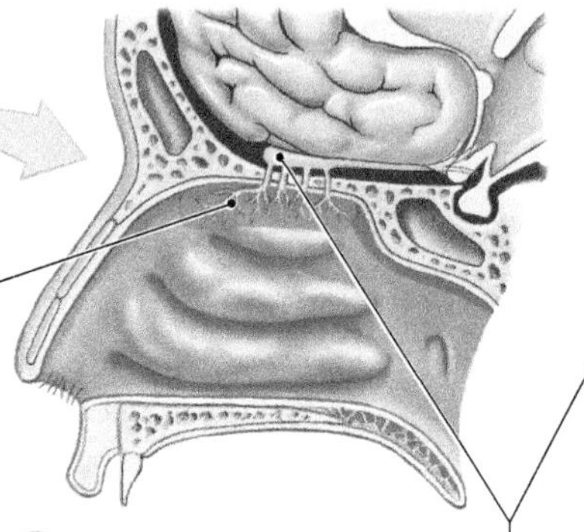

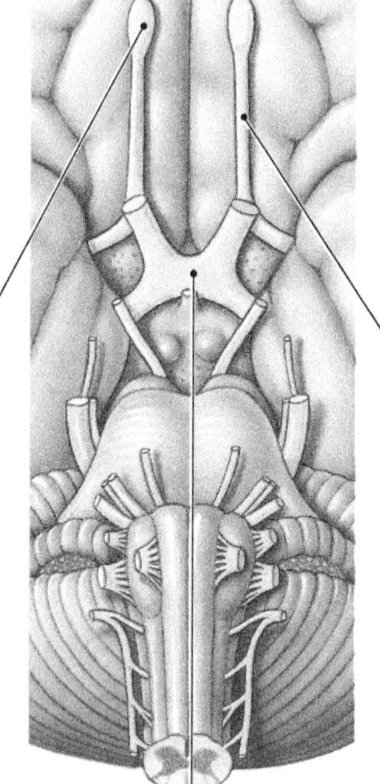

1. On a model of a midsagittal section of the head, locate the superior region of the nasal cavity. This region contains the olfactory epithelium, where the **olfactory nerves** originate.

2. On a model of the brain, identify the **olfactory bulbs** on the inferior surface of the frontal lobes. The nerve fibers of the olfactory nerves pass through small holes in the cribriform plate of the ethmoid bone and synapse with the neurons in the olfactory bulbs.

3. From the olfactory bulbs, nerve fibers travel within the **olfactory tracts** to reach the olfactory areas in the temporal lobes. Identify the olfactory tracts on the brain model.

B The **optic nerves (cranial nerves II)** originate from the retinas of the eyes and travel to the optic chiasm.

1. On a model of the eye, identify the **optic nerve** as it emerges from the posterior wall of the eyeball.

2. On a model of the brain, observe the two optic nerves converging at the **optic chiasm.** At this location, some fibers of each nerve cross over to the contralateral side before they reach the visual areas in the occipital lobes.

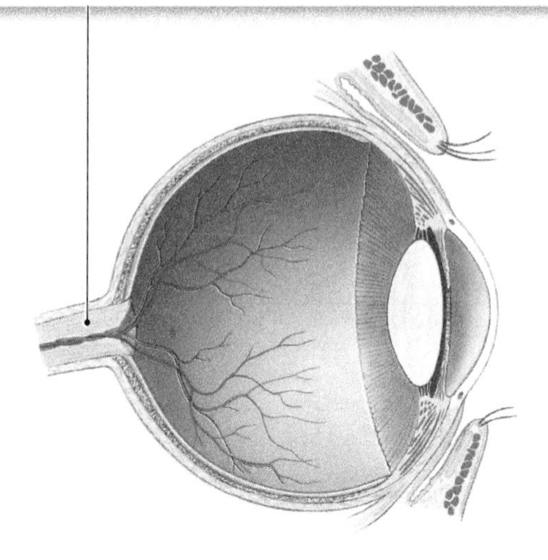

C Cranial nerves III through XII emerge from various regions of the brainstem. They arise in ascending order, from anterior to posterior.

1 On a model of the brain, identify the cranial nerves that originate from the midbrain:

Oculomotor nerve (cranial nerve III)

Trochlear nerve (cranial nerve IV)

2 Identify the cranial nerves that originate from the pons:

Trigeminal nerve (cranial nerve V)

Abducens nerve (cranial nerve VI)

Facial nerve (cranial nerve VII)

3 Locate the **vestibulocochlear nerve (cranial nerve VIII),** which originates on the pons-medulla border.

4 Identify the final four pairs of cranial nerves that originate from the medulla oblongata:

Glossopharyngeal nerve (cranial nerve IX)

Vagus nerve (cranial nerve X)

Accessory nerve (cranial nerve XI)

Hypoglossal nerve (cranial nerve XII)

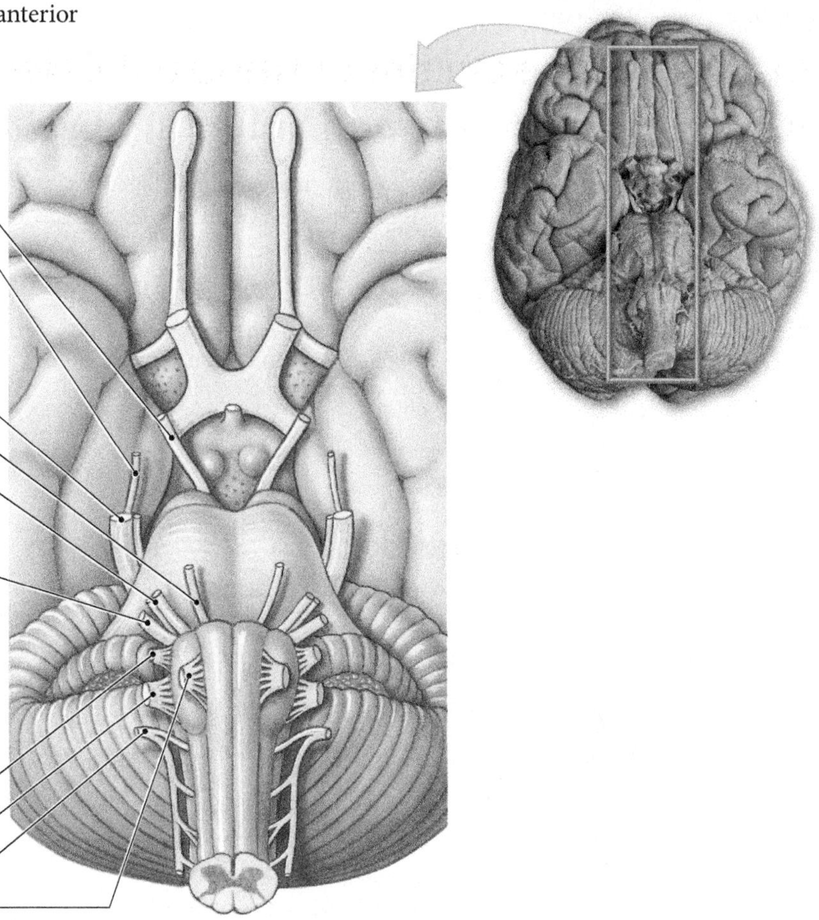

MAKING CONNECTIONS

Fractures in the base of the skull can often damage one or more cranial nerves. Why would the cranial nerves be vulnerable to these injuries?

▶ ______________________________

Want more practice? Go to: **MasteringA&P** > Study Area > Menu > Lab Tools > PAL >

- Anatomical Models > Nervous System > Central Nervous System > Brain photos (inferior view and cranial nerves)
- Human Cadaver > Nervous System > Central Nervous System > Brain photos (inferior view and cranial nerves)

Activity **15.8**

Cranial Nerve Function: Testing the Cranial Nerves for Olfaction, Vision, and Voluntary Eye Movements

We can evaluate cranial nerve function by performing a simple battery of tests. Working with your laboratory partner, carry out the following analysis of cranial nerve function. For many of these tests, only a portion of the nerve's function will be evaluated.

A The **olfactory nerve (I)** begins in the nasal cavity and contains sensory fibers for olfaction (smell).

1. Have your lab partner close both eyes and one nostril.
2. Bring samples of coffee, pepper, and other spices close to the open nostril, one at a time, and ask your partner to identify them.

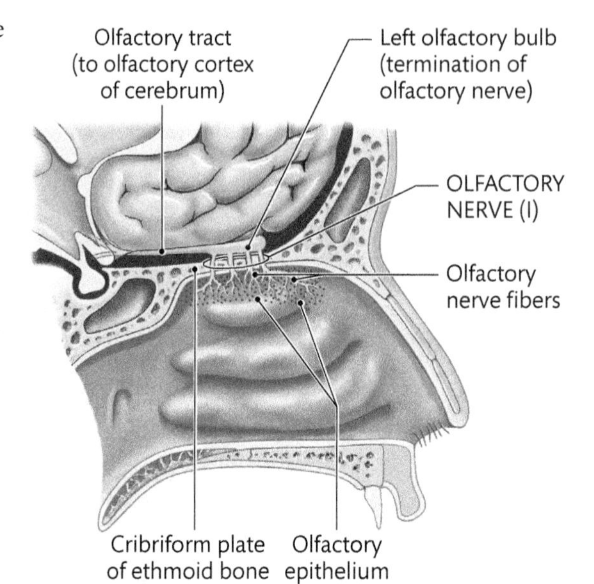

3. Perform the same test with the other nostril, but pass the substances in a different order.
4. Correct identification of each substance indicates normal function of the olfactory nerve. Record the results in Table 15.4. ▶

B The **optic nerve (II)** begins in the retina of the eyes and is responsible for vision. The left and right visual fields overlap in the center. The combined visual field is divided into quadrants indicated by the colored regions in the figure on the right.

1. To test visual fields, have your lab partner stand in front of you and look directly at your face.
2. Without moving his or her eyes, ask your lab partner to state how many fingers you are showing as you move them into each quadrant of the combined visual field.
3. Correctly reporting the number of fingers being shown indicates normal optic nerve function. Record your results in Table 15.4. ▶

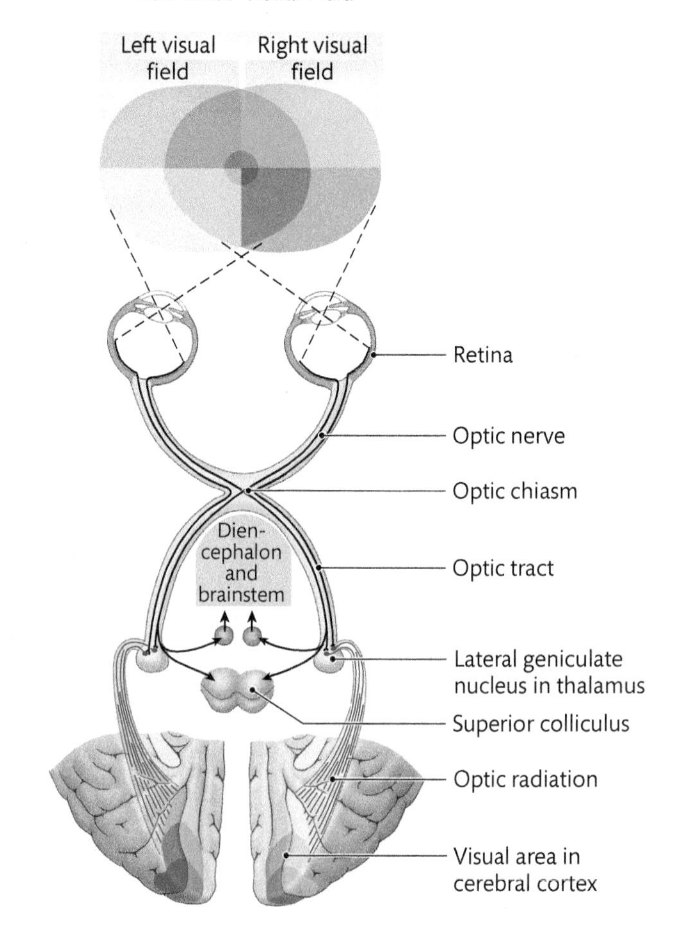

C The **oculomotor nerve (III), trochlear nerve (IV),** and **abducens nerve (VI)** innervate the six extrinsic muscles that perform voluntary eye movements. The muscles innervated by each nerve are as follows:

- Oculomotor: inferior oblique, superior rectus, inferior rectus, medial rectus
- Trochlear: superior oblique
- Abducens: lateral rectus

1 Have your lab partner sit comfortably on a laboratory seat, head facing forward and one eye covered.

2 Place your finger about 30 cm (1 foot) in front of your partner's open eye and move it in all directions. With the head still, your partner should be able to follow your finger with normal eye movements.

3 Perform the same test with the other eye. Record your results in Table 15.4. ▶

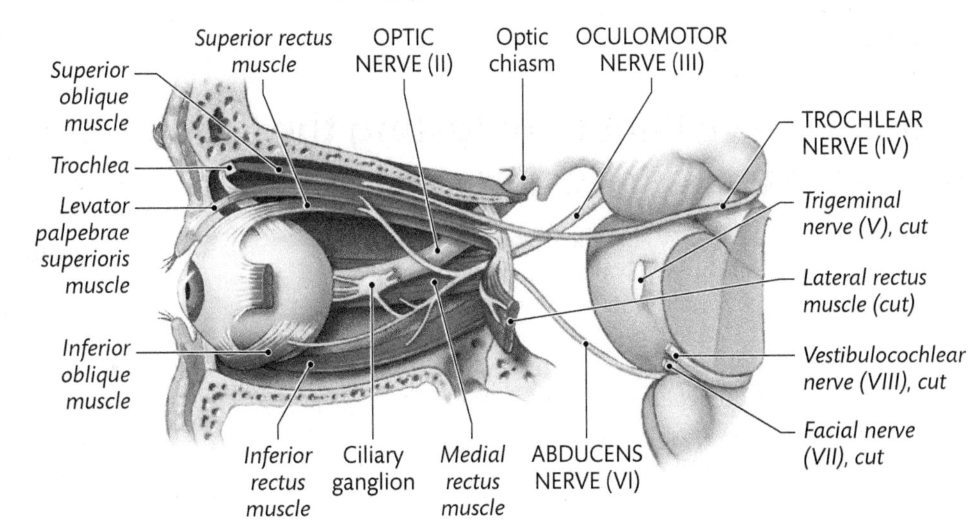

D The oculomotor nerve (III) also contains autonomic (parasympathetic) nerve fibers that cause the constrictor muscles in the iris to contract. This contraction decreases the diameter of the pupil, which reduces the amount of light entering the eye.

1 Shine a flashlight at an angle (not directly) into your lab partner's eyes and observe the change in the diameter of the pupils.

2 Remove the light and observe the change in pupil diameter. Record your results in Table 15.4. ▶

TABLE 15.4 Functional Assessment of Cranial Nerves for Olfaction, Vision, and Voluntary Eye Movements

Cranial Nerve	Test Performed	Results	Assessment of Nerve Function
Olfactory (I)			
Optic (II)			
Oculomotor (III), trochlear (IV), and abducens (VI)			
Oculomotor (III) autonomic function			

MAKING CONNECTIONS

Three cranial nerves innervate the extrinsic muscles of the eye. The greatest impairment of eye movement would result from damage to which of these cranial nerves? Explain.

▶ ______________________________

____

____

____

Activity **15.9**

Cranial Nerve Function: Testing the Cranial Nerves to the Face and Inner Ear

A The **trigeminal nerve (V)** has three branches:

- The **ophthalmic branch** contains sensory fibers from the cornea, skin of the nose, forehead, and anterior scalp.
- The **maxillary branch** contains sensory fibers from the mucous membrane in the nasal cavity, upper teeth and gums, palate, upper lip, and skin of the cheek.
- The **mandibular branch** contains sensory fibers from the lower teeth, gums and lips, anterior two-thirds of the tongue, and skin of the chin and lower jaw. It also supplies voluntary motor fibers to the muscles of mastication (chewing).

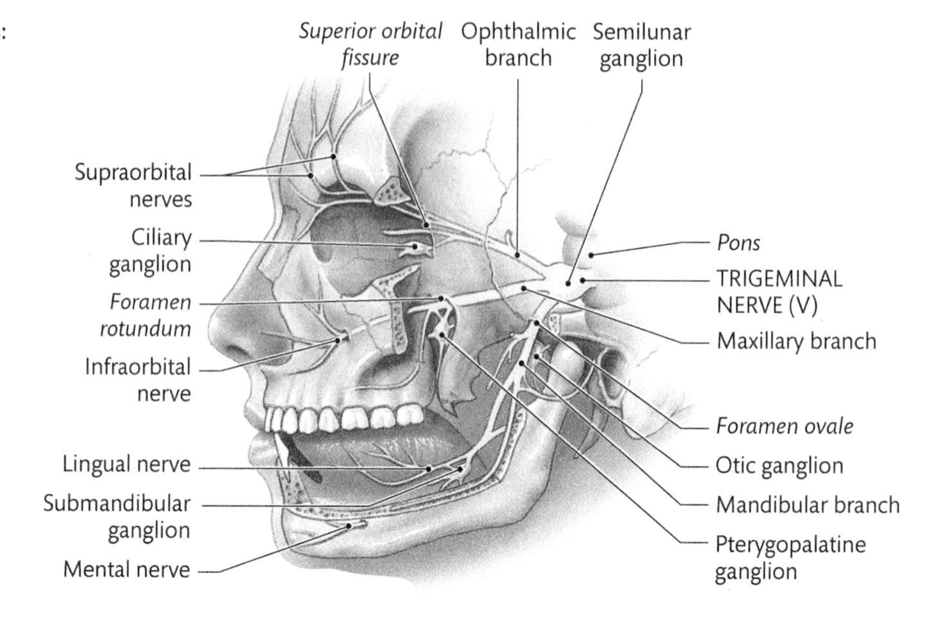

1 To test sensory function:

- Have your partner close his or her eyes.
- Gently press one fingertip on the mandibular, maxillary, and ophthalmic regions of your partner's face.
- Ask your partner to detect where each sensation originates.
- Record your results in Table 15.5. ▶

2 To test motor function:

- Place your hand under your partner's mandible. Ask your lab partner to open his or her mouth against your resistance.
- The ability to depress the mandible indicates normal motor function of the muscles of mastication.
- Record your results in Table 15.5. ▶

B The **facial nerve (VII)** has both sensory and motor components. Its sensory fibers transmit taste sensations from the anterior two-thirds of the tongue to the brain. Voluntary motor fibers control the actions of the muscles of facial expression, and autonomic motor fibers regulate secretions from the lacrimal (tear) glands and salivary glands.

1 To test sensory function:

- Have your partner sit with closed eyes and open mouth, tongue slightly protruding.
- Dip separate cotton swabs in a 10% table salt solution and a 10% sucrose solution.
- Gently place the swabs onto the tip of your partner's tongue in any order and ask him or her to identify each taste.
- If taste sensation is normal, your partner should be able to differentiate between salty and sugary tastes.
- Record your results in Table 15.5. ▶

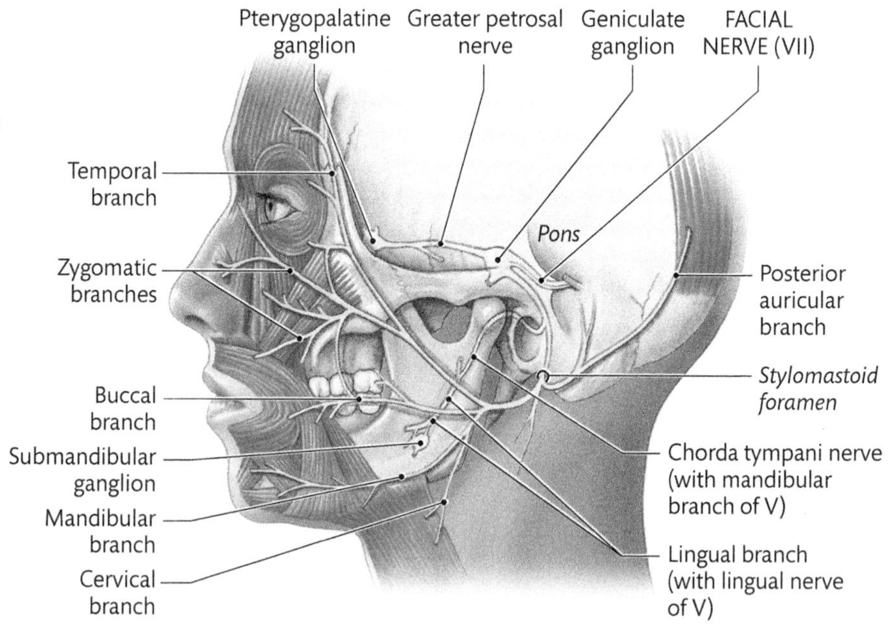

2 To test motor function:

- Have your lab partner perform various facial expressions (e.g., smiling, puffing the cheeks, raising the eyebrows).
- Check to see if normal function occurs on both sides of the face. For example, when smiling, are both corners of the mouth elevated?
- Record your results in Table 15.5. ▶

C The **vestibulocochlear nerve (VIII)** transmits sensory information from receptors in the inner ear to the brain. It has two branches:

- The **cochlear branch** is responsible for our ability to hear sounds.
- The **vestibular branch** allows us to detect movements that are linear (along a straight line) and angular (turning). This function is part of our equilibrium (balance) sense.

1. To assess vestibular function, have your lab partner sit, with eyes closed and ears plugged, on a stool with wheels.
2. Slowly roll the stool, forward or backward, and ask your partner to identify the direction of movement.
3. Next, slowly move the stool forward for a short distance and then turn it to the right or left. Ask your partner to identify the direction of the turn.
4. Record your results in Table 15.5. ▶

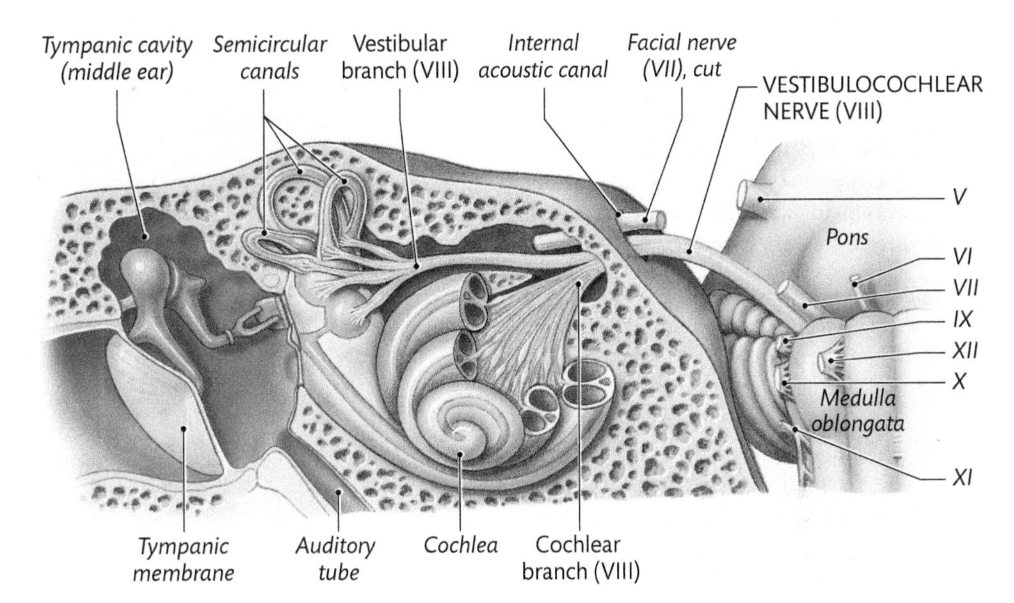

TABLE 15.5 Functional Assessment of Cranial Nerves to the Face and Inner Ear

Cranial Nerve	Test Performed	Results	Assessment of Nerve Function
Trigeminal (V) sensory function			
Trigeminal (V) motor function			
Facial (VII) sensory function			
Facial (VII) motor function			
Vestibulocochlear (VIII) function			

MAKING CONNECTIONS

You have dental caries (tooth decay causing a cavity) in your lower left second molar. Your dentist injects an anesthetic to eliminate any possible pain during treatment. Which cranial nerve is affected by the anesthetic? Explain.

▶ ______________________________

____

____

Activity **15.10**

Cranial Nerve Function: Testing the Cranial Nerves to the Tongue, Pharynx, and Internal Organs

A The **glossopharyngeal nerve (IX)** is both sensory and motor. It supplies sensory neurons for taste to the posterior third of the tongue and general sensory neurons to the pharynx, posterior tongue, and tympanic membrane (eardrum). Its motor component includes voluntary motor fibers to one pharyngeal muscle (stylopharyngeus) that has a role in swallowing and autonomic motor fibers to the parotid salivary glands.

1. To test the nerve's taste function:
   - Dip a cotton swab in a 10% quinine solution.
   - Gently place the swab on the posterior third of your lab partner's tongue.
   - The ability to taste the bitter flavor of quinine indicates normal function.
2. Record your results in Table 15.6. ▶

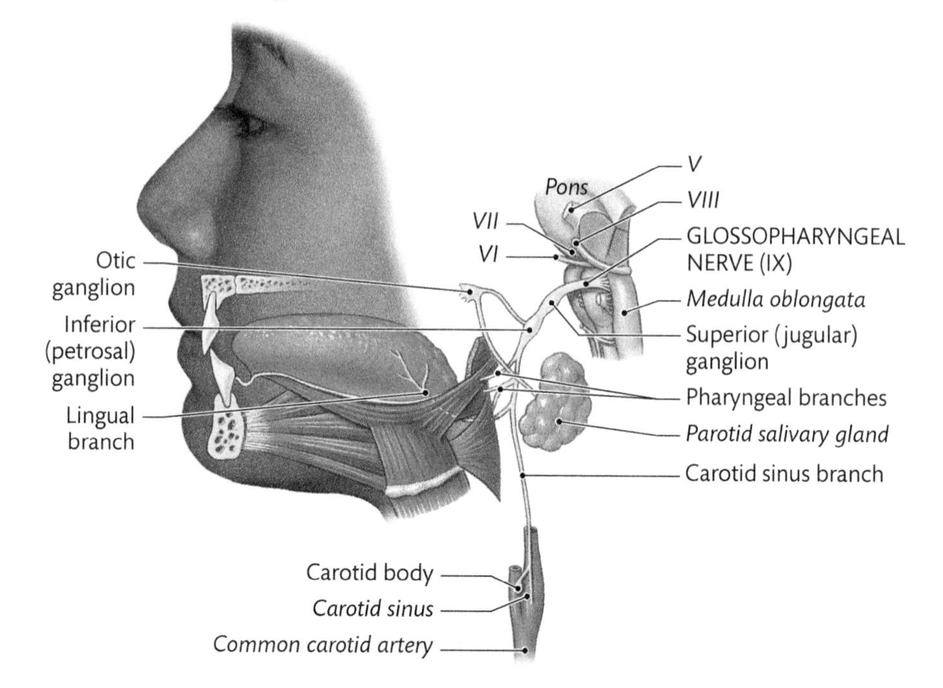

B The **vagus nerve (X)** supplies sensory neurons to a portion of the pharynx, external acoustic meatus, diaphragm, and internal organs in the thoracic and abdominopelvic cavities. Its motor component includes voluntary motor fibers to the palate and pharynx and autonomic motor fibers to the internal organs in the thoracic and abdominal cavities.

1. To test the function of the vagus nerve, have your lab partner drink some water. The swallowing reflex while drinking indicates normal nerve function.
2. Record your results in Table 15.6. ▶
3. Demonstrating the swallowing reflex is also a test for what other cranial nerve? Explain.

▶

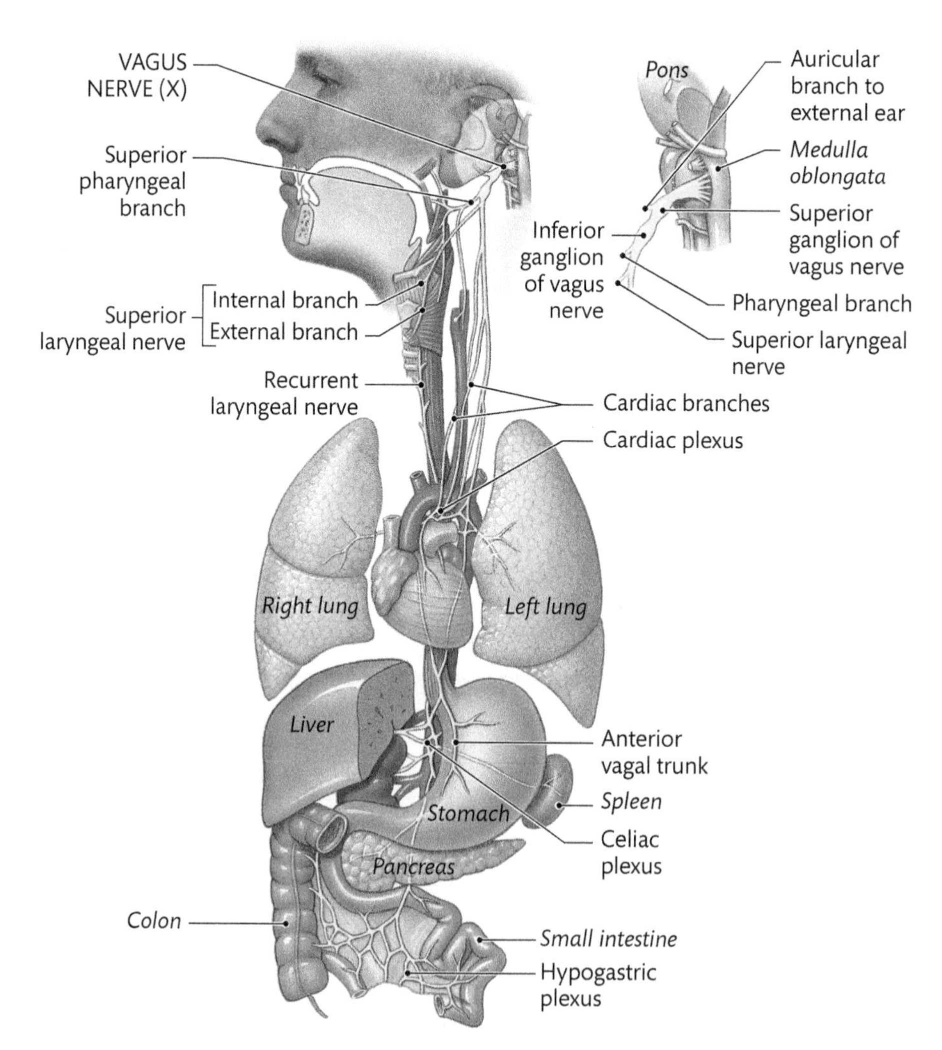

C The **accessory nerve (XI)** is entirely motor and innervates the trapezius and sternocleidomastoid muscles.

1. Ask your laboratory partner to shrug his or her shoulders. This action, which requires elevation of the scapulae, is performed by the trapezius muscles.
2. Have your partner turn (rotate) his or her head to the right and look up at the same time. As the action is being performed, feel (by palpation) the contraction of the sternocleidomastoid on the left side.
3. Record your results in Table 15.6. ▶

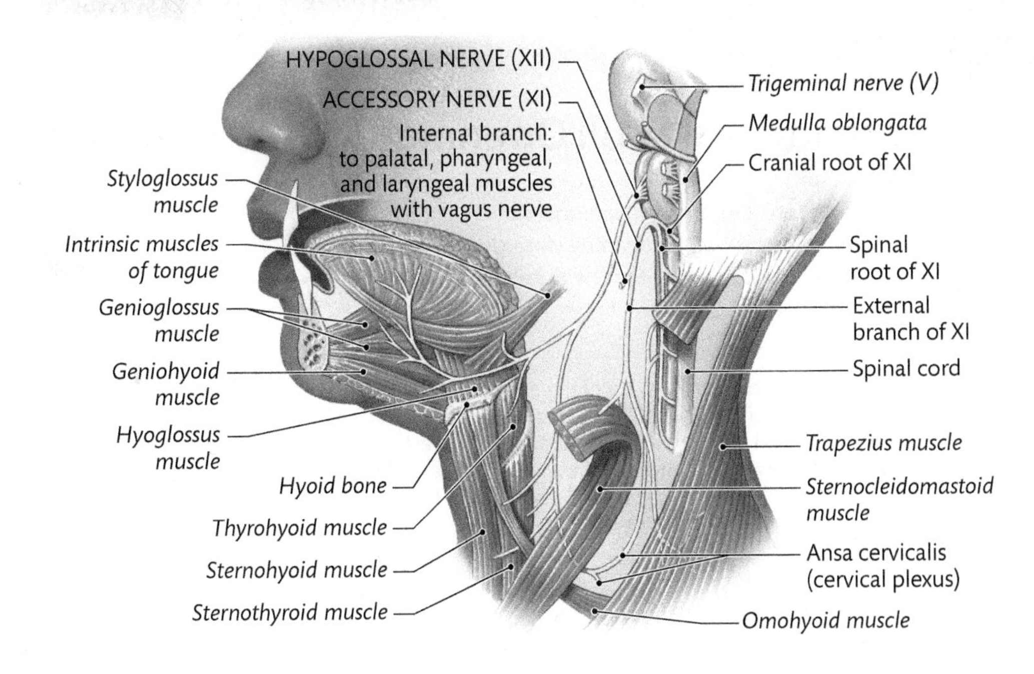

D The **hypoglossal nerve (XII)** is also entirely motor and innervates the muscles that move the tongue.

1. Have your laboratory partner protrude his or her tongue.
   - If both hypoglossal nerves are functioning normally, the tongue will move straight out when protruded.
   - If the tongue deviates to one side, damage to the hypoglossal nerve on that side could be indicated.
2. Record your results in Table 15.6. ▶

TABLE 15.6 Functional Assessment of Cranial Nerves to the Tongue, Pharynx, and Internal Organs

Cranial Nerve	Test Performed	Results	Assessment of Nerve Function
Glossopharyngeal (IX)			
Vagus (X)			
Accessory (XI)			
Hypoglossal (XII)			

MAKING CONNECTIONS

Sensory fibers from the glossopharyngeal nerve (IX) supply taste receptors on the tongue. If these sensory fibers are severed, taste sensation will not be completely lost. Explain why.

▶ ______________________________

____

____

Activity 15.11

Identifying the Skull Openings for the Cranial Nerves

A To reach the structures that they innervate, the cranial nerves must first pass through an opening (foramen, fissure, or canal) in the skull. The best way to view these openings is to look inside the cranial cavity.

1. Identify the **cribriform plate** of the ethmoid bone and notice the numerous small holes that dot its surface. These openings lead to the roof of the nasal cavity. Fibers of the olfactory nerve travel through these holes to reach the olfactory bulbs. Within the cranial cavity, the olfactory bulbs rest on the cribriform plate.

2. There are seven additional openings that transmit cranial nerves. On a skull, identify these passageways, which are labeled in the diagram.

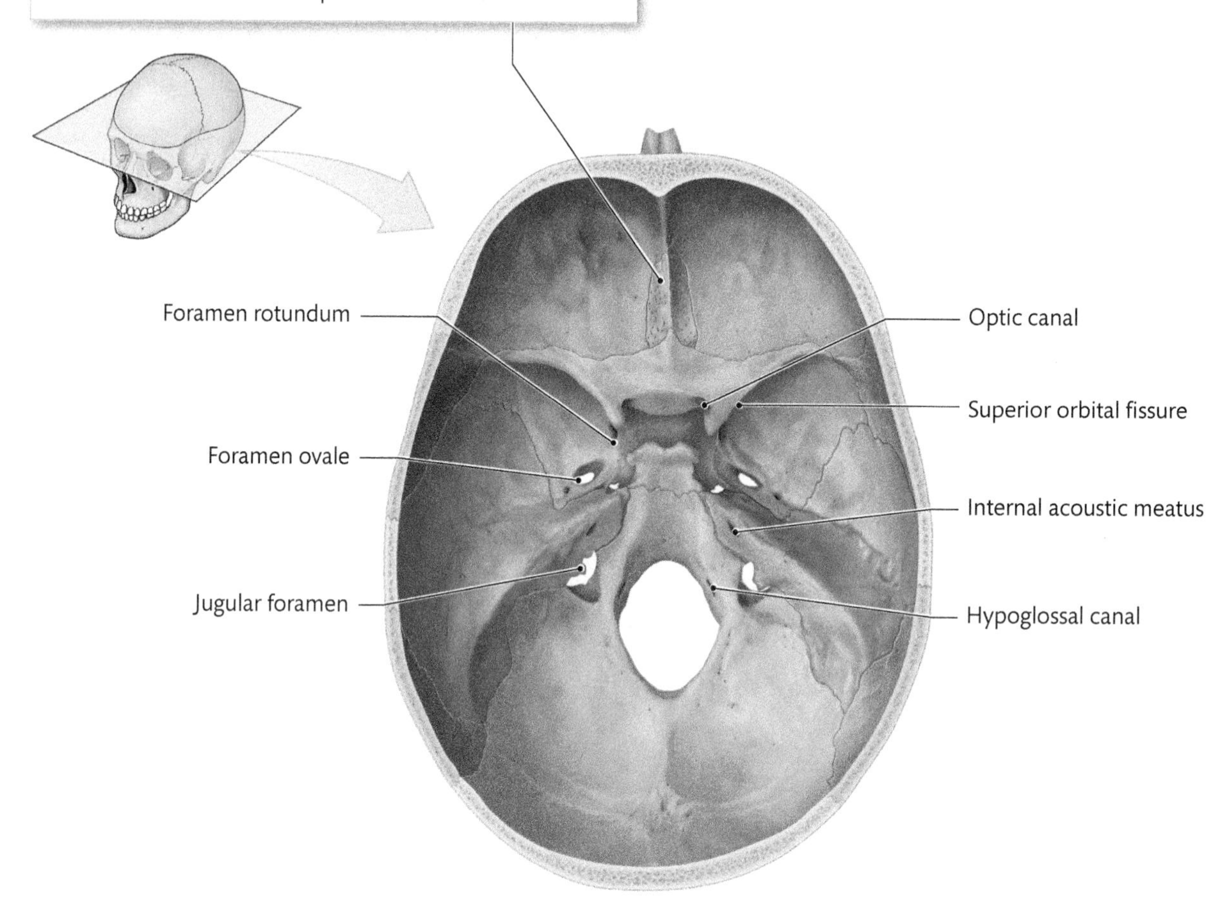

■ Want more practice? Go to: **MasteringA&P** > Study Area > Menu > Lab Tools > PAL > Anatomical Models > Axial Skeleton > Skull (cranial cavity and inferior view)

B Each skull opening leads to a particular region of the head or neck. If you know the destination of the opening, you can predict which cranial nerve(s) are being transmitted.

1. From the cranial cavity, insert a pipe cleaner through the optic canal.
2. Notice that the pipe cleaner exits the optic canal by entering the bony orbit where the eyeball is located.
3. You can use this information to predict that the nerve passing through the optic canal is the optic nerve. This prediction is recorded for you in the first row of Table 15.7. ▶
4. Insert a pipe cleaner into each of the other openings listed in Table 15.7. ▶
5. Identify the position of the pipe cleaner at the exit of each opening. Record your answer in the second column of Table 15.7. ▶
6. Predict which cranial nerve(s) pass through each opening. Record your prediction in the third column of Table 15.7. ▶

C The facial nerve (VII) has a rather complex pathway. It enters the internal acoustic meatus. From this passageway, branches of the nerve travel through canals in the temporal bone (these canals are deep and cannot be identified) to reach the lacrimal (tear) gland, the mucous membrane in the nasal cavity, and salivary glands in the oral cavity.

1. Another branch of the facial nerve exits the skull through the **stylomastoid foramen.** On the base of a skull, position a pipe cleaner at this opening.

2. Bend the pipe cleaner anteriorly.
   The pipe cleaner is directed toward which region of the skull?

   ▶ ______________________

   What do you think is the function of this branch of the facial nerve?

   ▶ ______________________

   ____

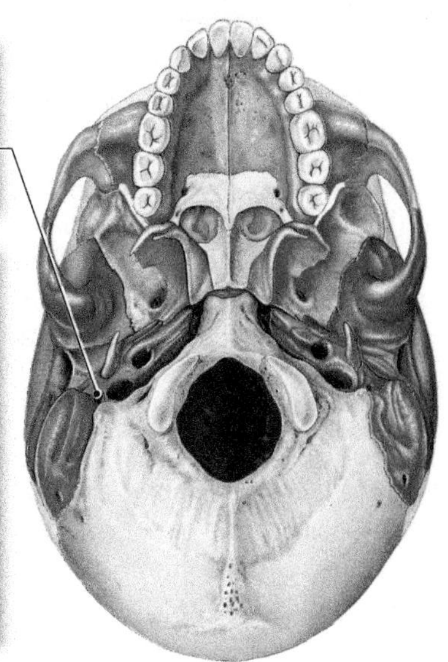

TABLE 15.7 Identifying the Destinations of the Cranial Nerves

Bony Opening	Position of Pipe Cleaner as It Exits the Opening	Cranial Nerves That Travel Through Opening
Optic canal	Bony orbit	Optic nerve
Superior orbital fissure		
Foramen rotundum		
Foramen ovale		
Internal acoustic meatus		
Jugular foramen		
Hypoglossal canal		

MAKING CONNECTIONS

The superior orbital fissure, the internal acoustic meatus, and the jugular foramen all transmit two or more cranial nerves. Explain why.

▶ ______________________

____

____

BEFORE YOU MOVE ON . . .

‹‹ LOOKING BACK

In this laboratory exercise, you studied the major structures of the brain, including the meninges, surface features, internal structures, and the ventricles. You also studied the cranial nerves and tested their functions. The central nervous system is amazingly complex both structurally and functionally. In the embryo, it develops from a single cylindrical structure called the **neural tube.** The cephalic (toward the head) end of the tube begins to grow at a faster rate than the caudal end, and after four weeks of development the three **primary brain vesicles** (brain regions) have formed. A week later, two of these vesicles subdivide, giving rise to the five **secondary brain vesicles.** All the brain structures that you studied in this laboratory exercise are derived from the five brain vesicles.

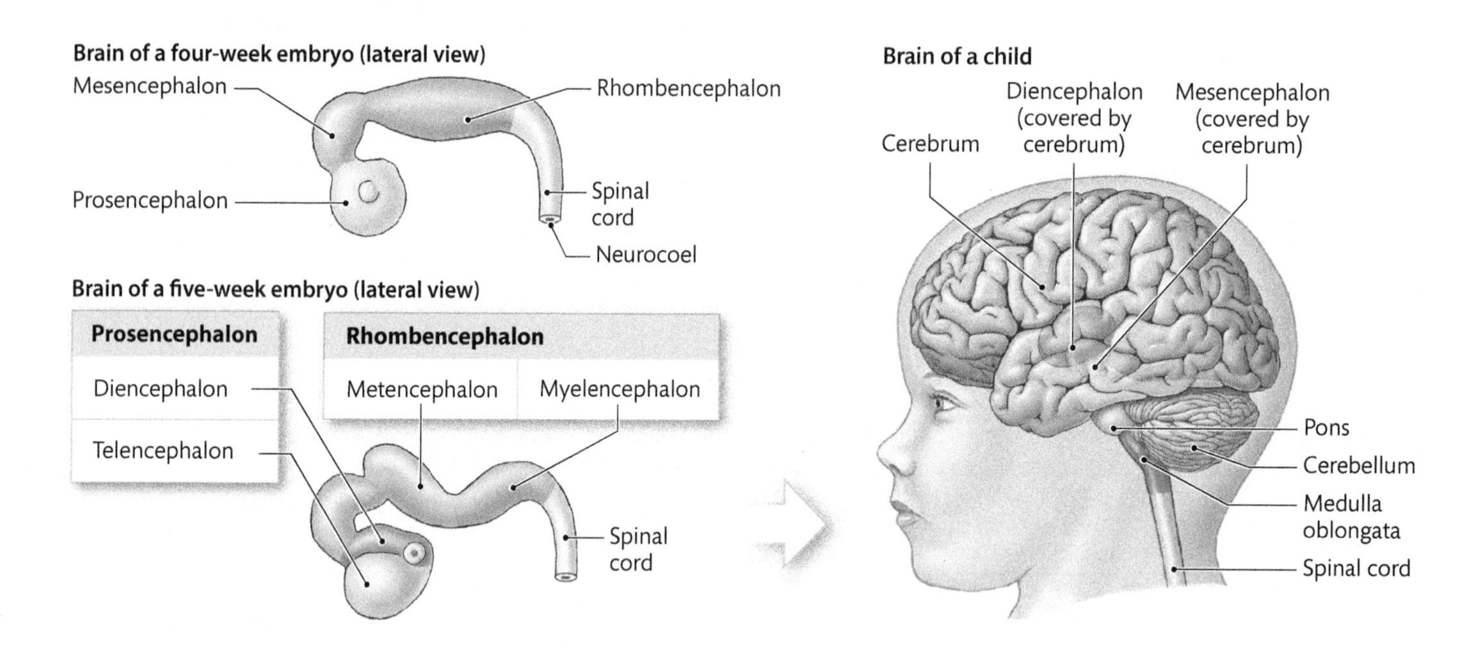

In the forebrain, the **telencephalon** becomes the cerebrum, and the **diencephalon** includes the thalamus, hypothalamus, and pineal gland. The midbrain or **mesencephalon** contains the superior and inferior colliculi. In the hindbrain, the pons and cerebellum are located in the **metencephalon,** and the medulla oblongata is in the **myelencephalon.**

A remarkable developmental feature of the primate brain—in particular the human brain—is the rapid growth of the cerebrum. In the adult human brain, the cerebrum comprises about 67 percent of the brain's total mass and covers most of the other structures. Why do you think the cerebrum is so large in the human brain?

▶ ____

____

____

LOOKING FORWARD ››

The spinal cord is a direct continuation of the medulla oblongata in the brain. It develops from the caudal (toward the tail) end of the neural tube and begins at the foramen magnum on the base of the skull. In the next exercise, you will examine the gross anatomical structure of the spinal cord and the organization of spinal nerves. You will also conduct tests to evaluate the sensory function of the spinal cord.

Name ______________________

Lab Section ______________________

Date ______________________

EXERCISE 15 REVIEW SHEET

The Brain and Cranial Nerves

QUESTIONS 1–5: Match the descriptions with the labeled structures in the figure. Labeled structures may be used once or not at all.

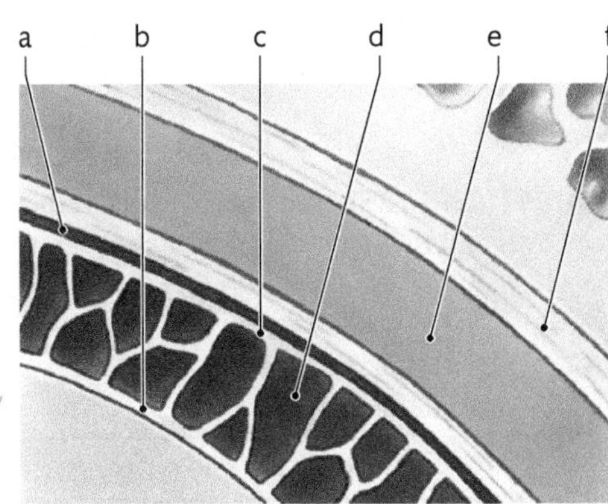

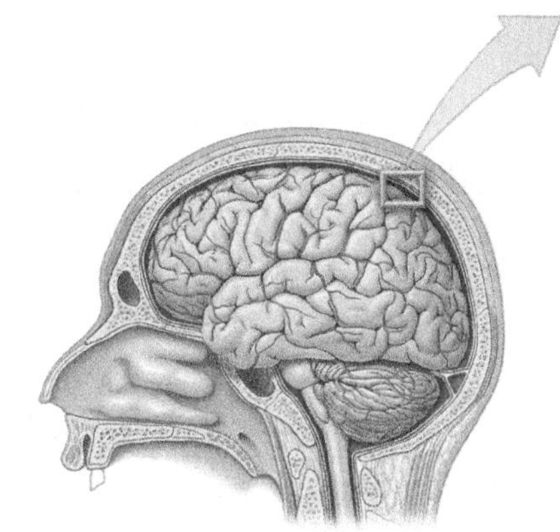

1. This space is filled with cerebrospinal fluid. ______________________
2. This meningeal layer is the pia mater. ______________________
3. Hematomas around the brain can occur when damaged meningeal blood vessels leak blood into this space. ______________________
4. This structure drains venous blood from the brain. ______________________
5. This meningeal layer is the arachnoid mater. ______________________

QUESTIONS 6–10: Match the brain region in column A with the appropriate structures in column B. (Use all the structures in column B; some regions in column A will have multiple answers.)

A

6. Telencephalon ______________________
7. Diencephalon ______________________
8. Mesencephalon ______________________
9. Metencephalon ______________________
10. Myelencephalon ______________________

B

a. Third ventricle

b. Corpora quadrigemina

c. Frontal lobe

d. Cerebellar cortex

e. Mamillary body

f. Regulatory center for heart rate

g. Basal nuclei

h. Corpus callosum

11. In the following diagram, identify the structures by labeling with the color that is indicated.

Temporal lobe = **green**
Frontal lobe = **red**
Occipital lobe = **blue**
Parietal lobe = **yellow**
Pons = **brown**
Cerebellum = **purple**
Medulla oblongata = **pink**

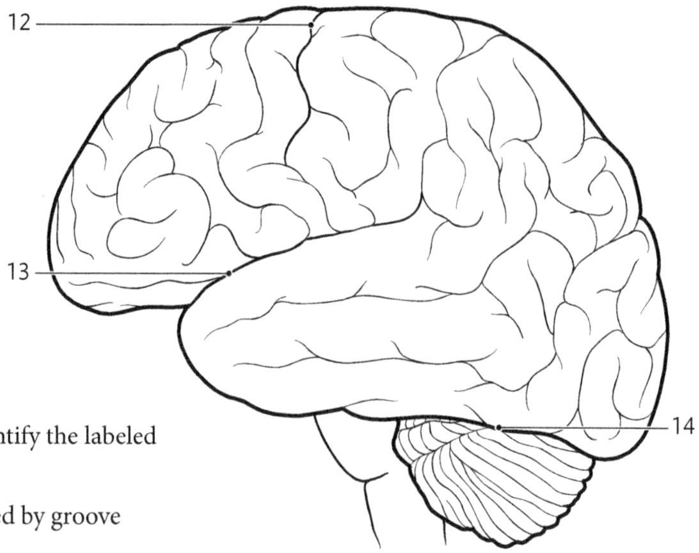

QUESTIONS 12–14: In the same diagram, identify the labeled grooves and the structures that they separate.

	Groove	Structures separated by groove
12.	______________	______________________________
13.	______________	______________________________
14.	______________	______________________________

15. The overaccumulation of cerebrospinal fluid in the brain ventricles or subarachnoid space is called hydrocephalus ("water on the brain"). Hydrocephalus can be caused by a blockage or reduction of normal drainage of CSF from the brain ventricles at the arachnoid granulations. What potential problem could develop if this condition is not treated promptly?

16. How would brain function be affected by lesions (tissue damage) in the following structures?

a. Basal nuclei:

b. Inferior colliculi:

c. Food intake center in hypothalamus:

QUESTION 17: Complete the following table.

Major Structures in the Adult Brain

Brain Region	Structure	Function
Telencephalon (anterior forebrain)	a.	b.
c.	d.	Controls reflexes related to auditory stimuli
e.	Pons	f.
g.	h.	Secretes melatonin, a hormone that assists the hypothalamus in the regulation of sleep/wake cycles
Myelencephalon (posterior hindbrain)	i.	j.
k.	Hypothalamus	l.
m.	n.	Regulates skeletal muscle contractions so that they are smooth and well coordinated
o.	p.	Acts as relay station for sensory impulses to the cerebral cortex
q.	Superior colliculi	r.

QUESTIONS 18–24: All the labeled structures in the photo are cranial nerves except for one. Name the labeled structures and identify the one that is not a cranial nerve.

18. ______________________

19. ______________________

20. ______________________

21. ______________________

22. ______________________

23. ______________________

24. Which of the structures that you identified in Questions 18–23 is not a cranial nerve? ______________

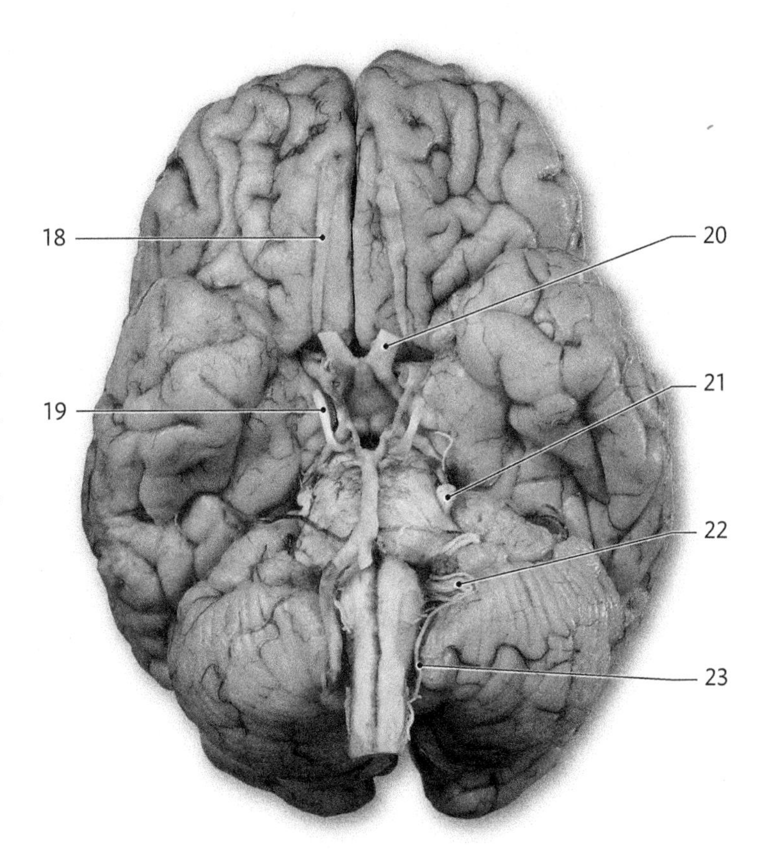

QUESTIONS 25–29: Match the descriptions of the cranial nerves with the correct skull opening that is labeled in the figure. (Answers may be used once or not at all.)

25. This cranial nerve supplies voluntary motor nerve fibers to the muscles of mastication. ____________

26. This cranial nerve supplies autonomic motor fibers to the heart. ____________

27. This cranial nerve contains sensory fibers for vision. ____________

28. This cranial nerve contains sensory fibers from the upper teeth and gums. ____________

29. This cranial nerve supplies voluntary motor fibers to the tongue. ____________

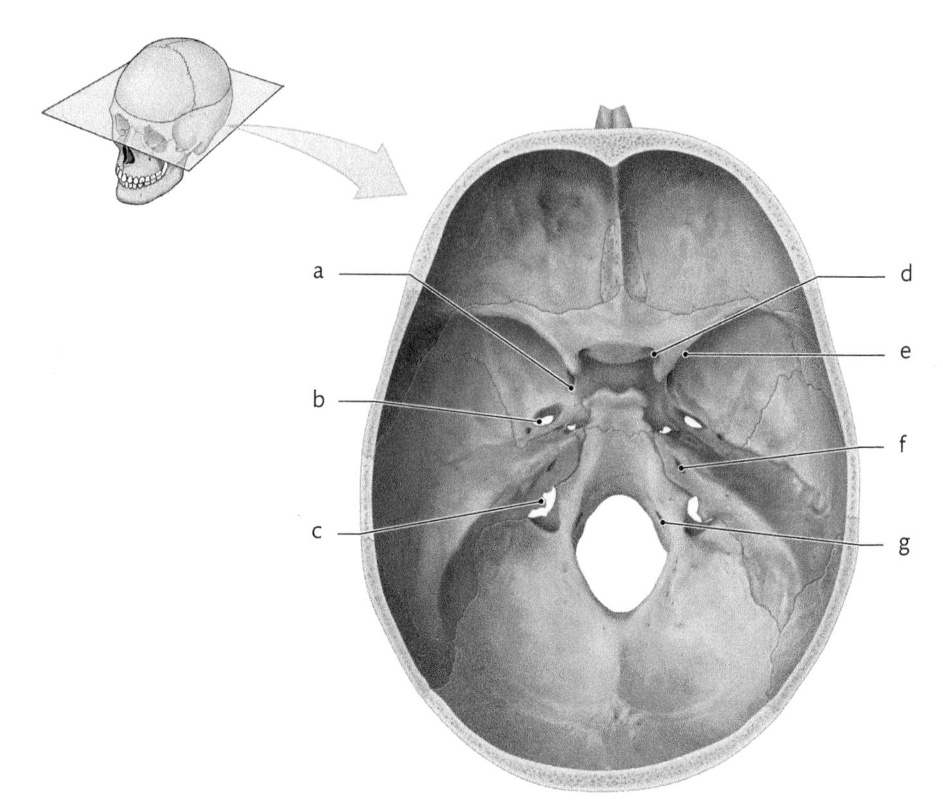

QUESTIONS 30–34: Assuming an injury to the cranial nerves in column A, match each nerve with the correct possible symptom in column B.

A	B
30. Facial nerve (VII) ____________	**a.** Dizziness
31. Accessory nerve (XI) ____________	**b.** Partial loss of smell
32. Olfactory nerve (I) ____________	**c.** Loss of sensation in the skin covering the cheek
33. Vestibulocohlear nerve (VIII) ____________	**d.** Loss of taste sensation on the anterior two-thirds of the tongue
34. Maxillary branch of trigeminal nerve (V) ____________	**e.** Difficulty shrugging the shoulders

EXERCISE 16

The Spinal Cord and Spinal Nerves

LEARNING OUTCOMES

These Learning Outcomes correspond by number to the laboratory activities in this exercise. When you complete the activities, you should be able to:

Activity 16.1 **Describe the organization of the spinal meninges.**

Activity 16.2 **Explain the gross anatomical structure of the spinal cord and describe the formation and organization of spinal nerves.**

Activity 16.3 **Assess spinal nerve function using a two-point discrimination test.**

Activity 16.4 **Examine spinal nerve function by testing general sensory function in dermatomes.**

LABORATORY SUPPLIES

- Models of the entire spinal cord and vertebral column
- Models of a spinal cord cross section showing its relationship with the spinal meninges and vertebrae
- Diagrams of the spinal cord and spinal nerves from anatomical atlases
- Brain models
- Measuring compasses with a millimeter scale
- Tuning forks

PRE-LAB QUIZ

Before you begin, read all the activities in Exercise 16 and the required reading in your textbook that is assigned by your instructor.

1. During this laboratory exercise, what piece of equipment will you use to test sensory function in dermatomes? ____________
2. During this laboratory exercise, what piece of equipment will you use for the two-point discrimination test? ____________
3. True or False: The primary sensory area is located in the frontal lobe of the cerebrum.

   ______
4. True or False: In a two-point discrimination test, the greater the distance between two pins at which two distinct sensations can be felt, the greater the density of skin receptors.

   ______
5. The body parts map in the primary sensory area is called the ____________.
6. True or False: A spinal nerve begins where the two nerve roots unite. ______
7. Spinal nerves are classified as mixed nerves because ____________

   ____
8. True or False: The spinal cord ends between the T_{12} and L_1 vertebrae. ______
9. True or False: If two pins touch your skin simultaneously and you detect only one sensation, the pins touched the skin in the same receptive field. ______
10. True or False: There is very little variability in the distribution of dermatomes among individuals. ______

Activity **16.1**

Identifying Spinal Meninges and Associated Structures

A The spinal cord is protected by the same three meningeal layers as the brain: the dura mater, arachnoid mater, and pia mater. These layers are continuous with the cranial meninges at the foramen magnum of the skull.

1 On models of the spinal cord, observe that the basic organization of the dura mater, arachnoid mater, and pia mater is similar to that of the brain.

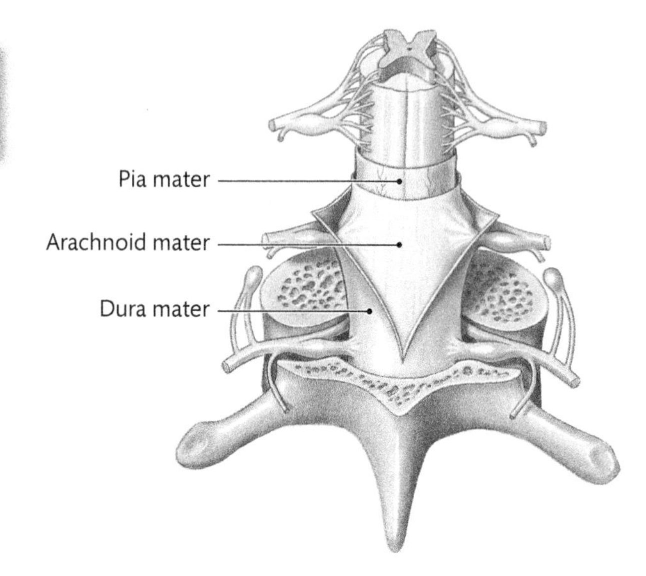

2 In the diagram to the right, label the three meningeal layers: ▶

1. ______________________

2. ______________________

3. ______________________

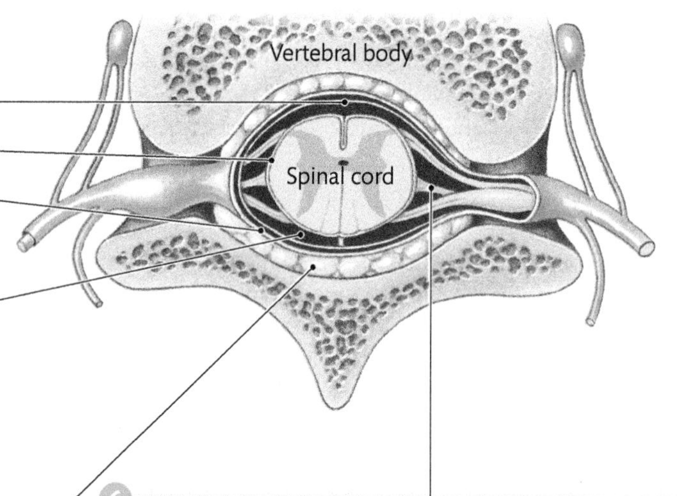

3 On a model, identify the fluid-filled **subarachnoid space.**

4 Notice the following special structural features on the spinal dura mater that differ from the cranial dura mater:

Around the spinal cord, there is only one layer of dura mater. It is a continuation of the inner dural layer around the brain.

The spinal dura mater is not attached to the inner surfaces of vertebrae as the cranial dura mater is to the cranial bones. Instead, a true space, called the **epidural space,** exists between the dura mater and the vertebrae. Blood vessels and nerves travel through this space.

5 The spinal epidural space is filled with fat tissue. What function do you think this fat tissue serves?

▶ ______________________

____

____

6 Identify the **denticulate ligaments,** which are lateral extensions of the pia mater that connect the spinal cord along its entire length to the dura mater. Speculate on the function of these structures.

▶ ______________________

____

____

____

IN THE CLINIC

Epidural Block

Anesthetics can be injected into the epidural space, resulting in a temporary loss of function (both sensory and motor) to the spinal nerves at the area of injection. This procedure, known as an **epidural block,** is difficult to do in most areas of the spinal cord because the epidural space is quite narrow. To reduce the pain of labor and childbirth, an epidural block can be performed, with relatively little risk by injecting anesthetics into the epidural space between the L_3 and L_4 vertebrae.

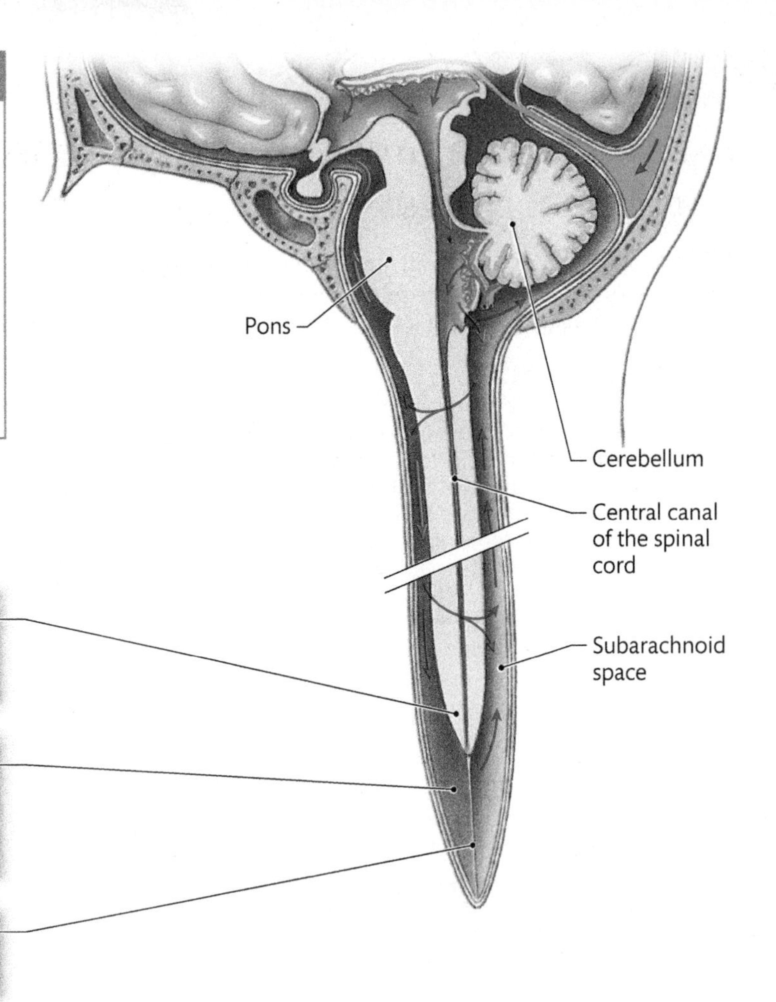

B The subarachnoid space is filled with cerebrospinal fluid and extends beyond the termination of the spinal cord.

1. On a model, identify the cone-shaped termination of the spinal cord called the **conus medullaris.** The spinal cord ends between vertebrae L_1 and L_2.

2. The dura mater and arachnoid mater extend beyond the conus medullaris to between vertebrae S_2 and S_3. Thus, the **subarachnoid space ends as a blind sac** filled with cerebrospinal fluid.

3. The **filum terminale** is a thin strand of pia mater that extends inferiorly from the conus medullaris to the coccyx.

Word Origins

Filum terminale derives from Latin and means "terminal thread." The name aptly describes this structure: a thin strand of pia mater that extends from the termination of the spinal cord to the coccyx.

Denticulate comes from the Latin word *dentatus,* which means "toothed." The denticulate ligaments resemble rows of small teeth on each side of the spinal cord.

MAKING CONNECTIONS

A **lumbar puncture (spinal tap)** is a procedure during which a needle is inserted between adjacent lumbar vertebrae into the subarachnoid space to collect a sample of cerebrospinal fluid. Explain why a lumbar puncture is done inferior to the second lumbar vertebra.

▶ ______________________________

____

____

____

____

____

____

____

■ Want more practice? Go to: **MasteringA&P** > Study Area > Menu > Lab Tools > PAL > Anatomical Models > Nervous System > Central Nervous System > Spinal cord photos

Activity **16.2**

Examining the Gross Anatomy of the Spinal Cord and Spinal Nerves

A The spinal cord serves as an information pathway that connects the brain and the spinal nerves. It is also an integration center for spinal reflexes. The spinal cord is divided into 31 segments: 8 cervical, 12 thoracic, 5 lumbar, 5 sacral, and 1 coccygeal. Each spinal cord segment gives rise to a pair of spinal nerves, which are numbered in the figure. Throughout its course, the spinal cord is enclosed and protected by the **vertebral canal.**

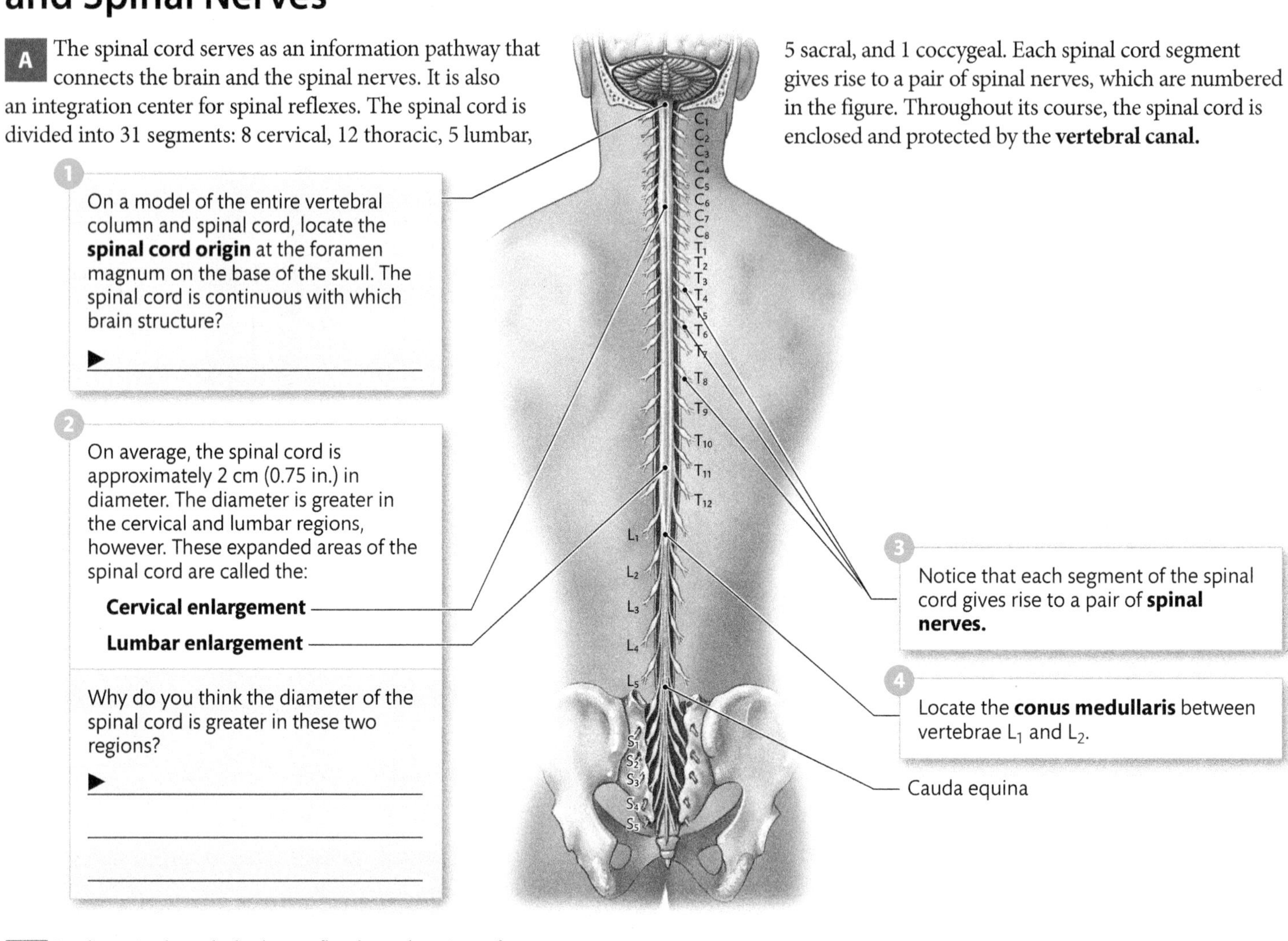

1. On a model of the entire vertebral column and spinal cord, locate the **spinal cord origin** at the foramen magnum on the base of the skull. The spinal cord is continuous with which brain structure?

   ▶ ______________________________

2. On average, the spinal cord is approximately 2 cm (0.75 in.) in diameter. The diameter is greater in the cervical and lumbar regions, however. These expanded areas of the spinal cord are called the:

   Cervical enlargement

   Lumbar enlargement

   Why do you think the diameter of the spinal cord is greater in these two regions?

   ▶ ______________________________

   ____

   ____

3. Notice that each segment of the spinal cord gives rise to a pair of **spinal nerves.**

4. Locate the **conus medullaris** between vertebrae L_1 and L_2.

B In the spinal cord, the butterfly-shaped region of gray matter is deep to the white matter.

1. On a model of a spinal cord cross section, identify the following regions of gray matter:

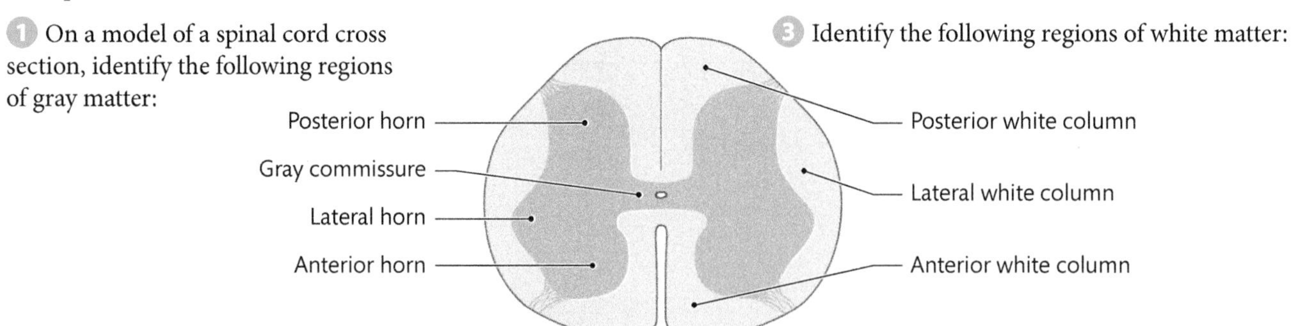

3. Identify the following regions of white matter:

2. Use your textbook to complete the following table. ▶

Gray Matter Region	Function
Posterior horn	
Lateral horn	
Anterior horn	

4. What is the function of the white columns? (*Hint*: The white columns contain myelinated nerve fibers.)

▶ ______________________________

____

____

____

C The spinal nerves are identified according to the spinal cord segment from which they originate. They are all classified as **mixed nerves** because they contain both **sensory (afferent)** and **motor (efferent)** nerve fibers.

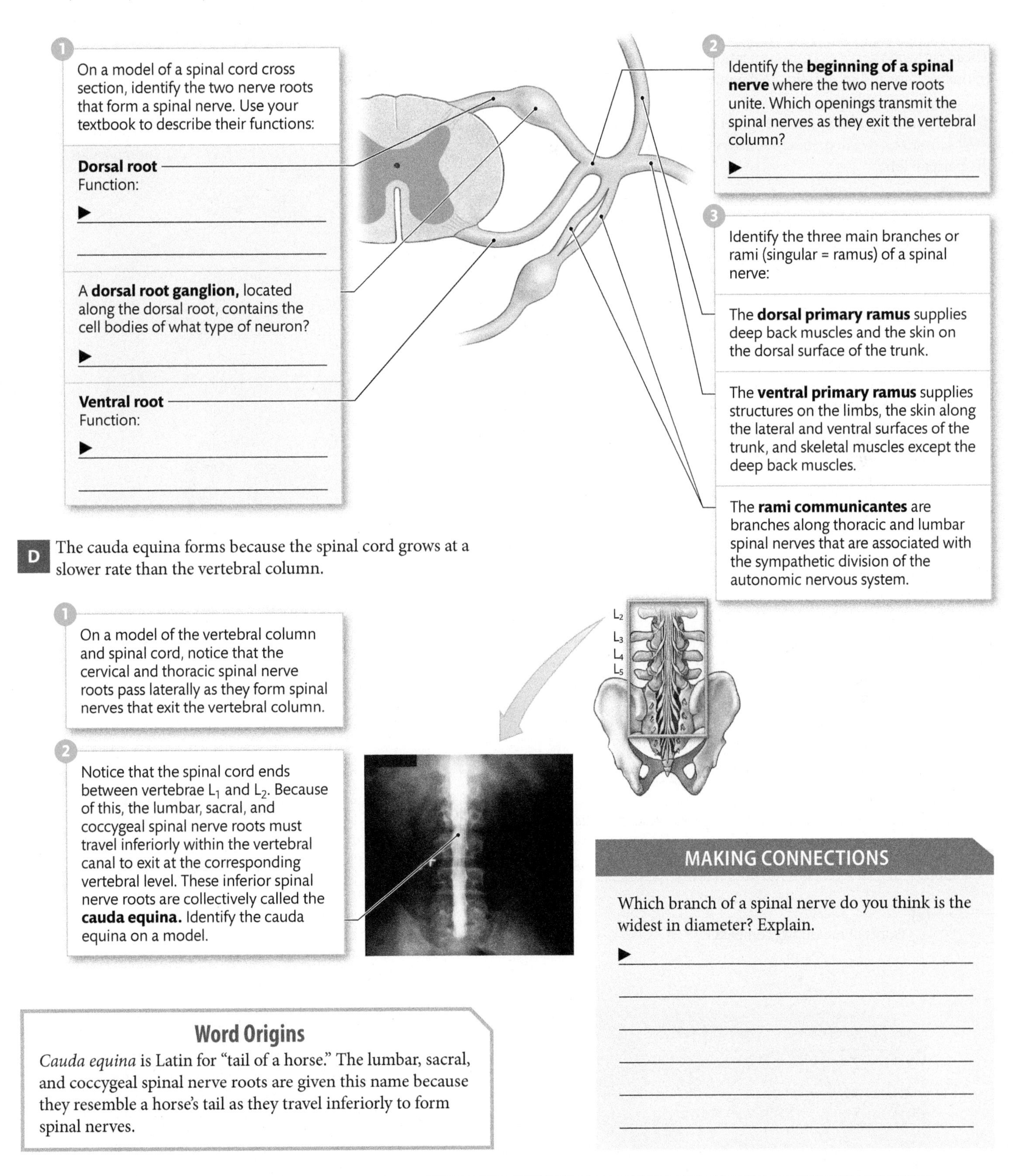

1 On a model of a spinal cord cross section, identify the two nerve roots that form a spinal nerve. Use your textbook to describe their functions:

Dorsal root
Function:

▶ ______________________________

____

A **dorsal root ganglion,** located along the dorsal root, contains the cell bodies of what type of neuron?

▶ ______________________________

Ventral root
Function:

▶ ______________________________

____

2 Identify the **beginning of a spinal nerve** where the two nerve roots unite. Which openings transmit the spinal nerves as they exit the vertebral column?

▶ ______________________________

3 Identify the three main branches or rami (singular = ramus) of a spinal nerve:

The **dorsal primary ramus** supplies deep back muscles and the skin on the dorsal surface of the trunk.

The **ventral primary ramus** supplies structures on the limbs, the skin along the lateral and ventral surfaces of the trunk, and skeletal muscles except the deep back muscles.

The **rami communicantes** are branches along thoracic and lumbar spinal nerves that are associated with the sympathetic division of the autonomic nervous system.

D The cauda equina forms because the spinal cord grows at a slower rate than the vertebral column.

1 On a model of the vertebral column and spinal cord, notice that the cervical and thoracic spinal nerve roots pass laterally as they form spinal nerves that exit the vertebral column.

2 Notice that the spinal cord ends between vertebrae L_1 and L_2. Because of this, the lumbar, sacral, and coccygeal spinal nerve roots must travel inferiorly within the vertebral canal to exit at the corresponding vertebral level. These inferior spinal nerve roots are collectively called the **cauda equina.** Identify the cauda equina on a model.

MAKING CONNECTIONS

Which branch of a spinal nerve do you think is the widest in diameter? Explain.

▶ ______________________________

____

____

____

____

____

Word Origins

Cauda equina is Latin for "tail of a horse." The lumbar, sacral, and coccygeal spinal nerve roots are given this name because they resemble a horse's tail as they travel inferiorly to form spinal nerves.

Want more practice? Go to: **MasteringA&P** > Study Area > Menu > Lab Tools > PAL >

- Anatomical Models > Nervous System > Central Nervous System > Spinal cord photos
- Human Cadaver > Nervous System > Central Nervous System > Spinal cord photos

Activity **16.3**

Using a Two-Point Discrimination Test

The white matter in the spinal cord contains **ascending sensory pathways** and **descending motor pathways.** These pathways serve as communication links between spinal nerves and the sensory and motor areas in the brain. In this activity and in Activity 16.4, you will assess the spinal cord's sensory function.

A Ascending pathways in the spinal cord transmit general sensations to the **primary sensory area** in the **postcentral gyrus** of the parietal lobe.

1. On a model of the brain, locate the postcentral gyrus. The cerebral cortex in this gyrus is organized into areas that detect sensations from specific body regions. A body parts map called the **sensory homunculus** compares the relative sizes of these areas.
2. In a typical sensory pathway, **first-order neurons** transmit sensory information from a receptor to the spinal cord or brainstem, where they synapse on a second-order neuron. Label the arrow that represents a first-order neuron in the diagram on the right. ▶
3. **Second-order neurons** cross over to the opposite side and travel to the thalamus, where they synapse on a third-order neuron. Label the arrow that represents a second-order neuron in the diagram on the right. ▶
4. **Third-order neurons** travel from the thalamus to the cerebral cortex. They synapse on neurons in the primary sensory area. Label the arrow that represents a third-order neuron in the diagram on the right. ▶

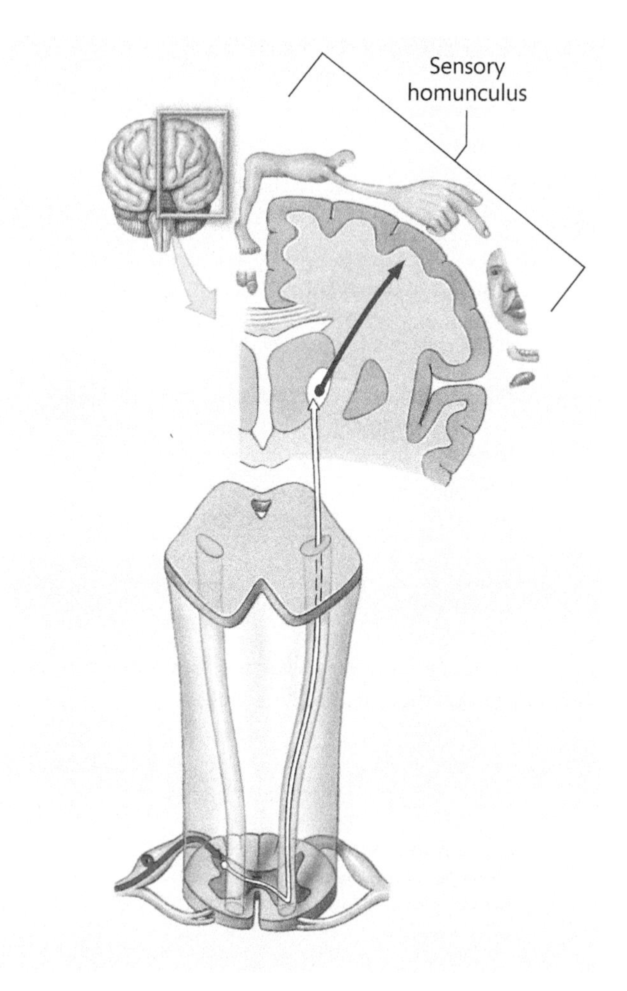

B A **receptive field** is an area on the body that is monitored by a single sensory neuron. The smaller the receptive field, the more accurate you can be in localizing the origin of a stimulus. The **two-point discrimination test** assesses spinal nerve function and compares the relative density of receptors in the skin of selected body regions. During this test, the two points of a measuring compass are gently pressed onto the skin, as shown below. Depending on the distance between the two compass points and the density of receptors, you may feel one or two distinct sensations.

1. Which figure illustrates a two-point discrimination test during which only one sensation is perceived?

▶ ______________________________

2. Which figure illustrates a test during which two distinct sensations are perceived?

▶ ______________________________

3. Explain the answers you gave in steps 1 and 2.

▶ ______________________________

____

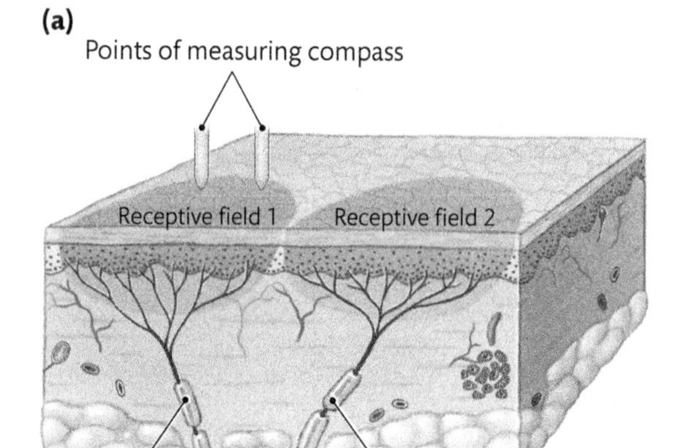

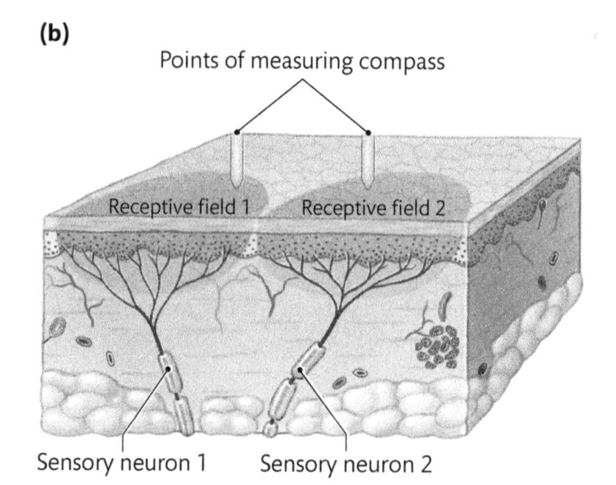

C The relative densities of skin receptors on various skin regions can be determined by performing a series of two-point discrimination tests. The distance between the compass points is inversely related to the density of receptors. Accordingly, the shorter the distance between the two compass points at which two distinct sensations can be felt, the greater the density of skin receptors.

1 Make a Prediction

You will be performing two-point discrimination tests on the following skin regions: (1) tip of the index finger, (2) dorsal surface of the hand, (3) posterior surface of the neck, (4) side of the nose, and (5) anterior surface of the arm. Before you begin, examine the sensory homunculus in the figure on the previous page and predict the relative densities of skin receptors in these five regions. Rank them on a scale of 1 to 5, with 1 being the highest density and 5 being the lowest. Record your rankings in the "Predicted" column in Table 16.1. ▶

TABLE 16.1 Two-Point Discrimination Test

Area of Skin	Distance between Compass Points at Which Two-Point Discrimination Can Be Made (mm)	Relative Density of Sensory Receptors (ranked 1 to 5)	
		Predicted	Actual
Tip of the index finger			
Dorsal surface of the hand			
Posterior surface of the neck			
Side of the nose			
Anterior surface of the arm			

2 Arrange the two points of a measuring compass so that they are 1 mm apart.

3 With your laboratory partner's eyes closed, gently place the points of the compass on the tip of his or her index finger.

4 Your laboratory partner's eyes should remain closed as you increase the distance between the two compass points slightly and place them on the same area of the finger.

5 Repeat step 4. Each time, increase the distance between the points by small increments until your laboratory partner feels two distinct points on the skin.

6 In Table 16.1, record the distance at which two-point discrimination can be made. ▶

7 Return the compass points to their original positions (1 mm apart) and repeat steps 2 through 6 on the other skin regions listed in Table 16.1. Record your results in the table. ▶

8 Review your test results and rank the five skin regions in terms of relative density of sensory receptors (1 = highest density, 5 = lowest density). Record your rankings in the "Actual" column in Table 16.1. ▶

9 Assess the Outcome

Compare your predictions of relative receptor densities with your actual experimental results. How accurate were your predictions? Explain.

▶ ____________________

____

____

MAKING CONNECTIONS

Review your results of the two-point discrimination test. Speculate on why the sensory homunculus is arranged disproportionately. For instance, why is so much of the sensory area devoted to the hands and fingers compared to other, larger regions of the body?

▶ ____________________

____

____

Activity **16.4**

Testing General Sensory Function in Dermatomes

The surface of the body can be divided into sensory cutaneous regions, each innervated by a single pair of spinal nerves. These cutaneous regions are called **dermatomes.**

A The figure to the right illustrates a generalized organization of dermatomes on the human body. Adjacent dermatomes, particularly those on the trunk, may overlap, and their distribution varies greatly among individuals. Some skin areas are typically within the defined dermatome for a specific spinal nerve, however. The function of the spinal nerve can be tested by asking the subject if he or she feels the vibration of a tuning fork or the pinch of a needle (pin-prick test) in a selected skin area.

1. Use the dermatome map on the right to identify a region on the lower limb that can be used to test the function of spinal nerve S_2.

▶ ______________________________

____

____

2. A pin-prick test on the skin covering the dorsal surface of the big toe will test the function of which spinal nerve?

▶ ______________________________

____

____

3. Why do you think it is difficult to assess the function of individual thoracic spinal nerves?

▶ ______________________________

____

____

____

____

____

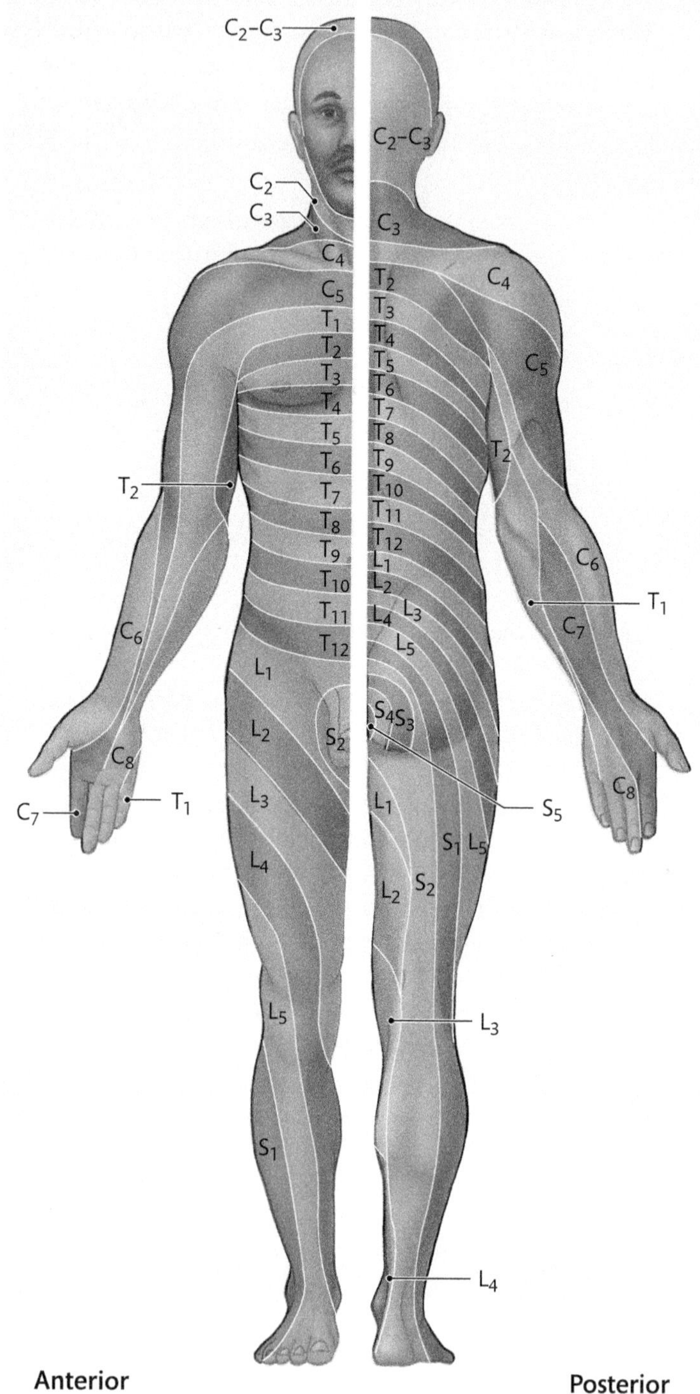

B The peripheral nerves derived from the **brachial plexus** will be used as a model for testing the function of individual spinal nerves. The brachial plexus is formed by the ventral rami of spinal nerves C_5 through T_1. It gives rise to the peripheral nerves that innervate structures in the upper limb.

1. Strike the prongs of a tuning fork against the edge of your lab bench so that their vibration is loud enough to be easily heard.
2. On your lab partner, quickly but gently press the handle of the tuning fork on the dermatomal skin area described in Table 16.2 for spinal nerve C_5.
3. The ability to feel the vibration generated by the tuning fork indicates normal function of the spinal nerve. Record your result in Table 16.2. ▶
4. Repeat steps 1 through 3 for the other dermatomal areas listed in Table 16.2. ▶

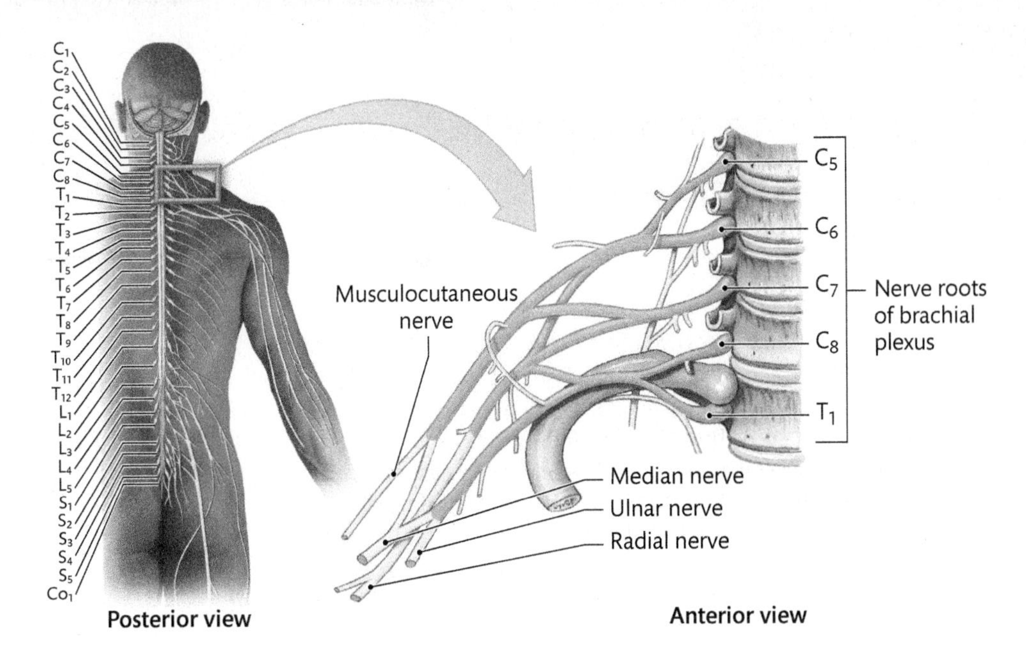

TABLE 16.2 Functional Assessment for Spinal Nerves of the Brachial Plexus

Spinal Nerve	Dermatomal Area	Result of Test
C_5	Skin covering anterior and lateral aspects of deltoid muscle	
C_6	Skin covering the thenar eminence in the hand	
C_7	Skin covering the anterior surface of the distal end of the index finger	
C_8	Skin covering the hypothenar eminence in the hand	
T_1	Skin covering the superomedial aspect of the arm	

MAKING CONNECTIONS

The median and radial nerves contain nerve fibers of all the spinal nerves in the brachial plexus (C_5–T_1). If an assessment of spinal nerves in the brachial plexus indicated a loss of function in spinal nerve C_8, would the function of the median and radial nerves be significantly affected? Explain.

▶ ______________________________

____

____

____

____

____

BEFORE YOU MOVE ON ...

« LOOKING BACK

Along its course, the spinal cord gives rise to 31 pairs of spinal nerves. Each spinal nerve is formed by the union of two nerve roots. The dorsal root transmits sensory fibers from the spinal nerve to the posterior horn of the spinal cord. The ventral root transmits motor fibers from the spinal cord to a spinal nerve. Each spinal nerve has three main branches (rami): the dorsal primary ramus, the ventral primary ramus, and the rami communicantes. The ventral primary rami of many spinal nerves form **plexuses** or nerve networks. The four spinal nerve plexuses are summarized in the following table:

Plexus	Spinal Nerve	Distribution
Cervical plexus	C_1–C_5	Portions of the head, neck, and shoulders
Brachial plexus	C_5–T_1	All structures in the upper limb
Lumbar plexus	T_{12}–L_4	Abdominal wall, external genitalia, part of lower limb
Sacral plexus	L_4–S_4	Buttocks, perineum, part of lower limb

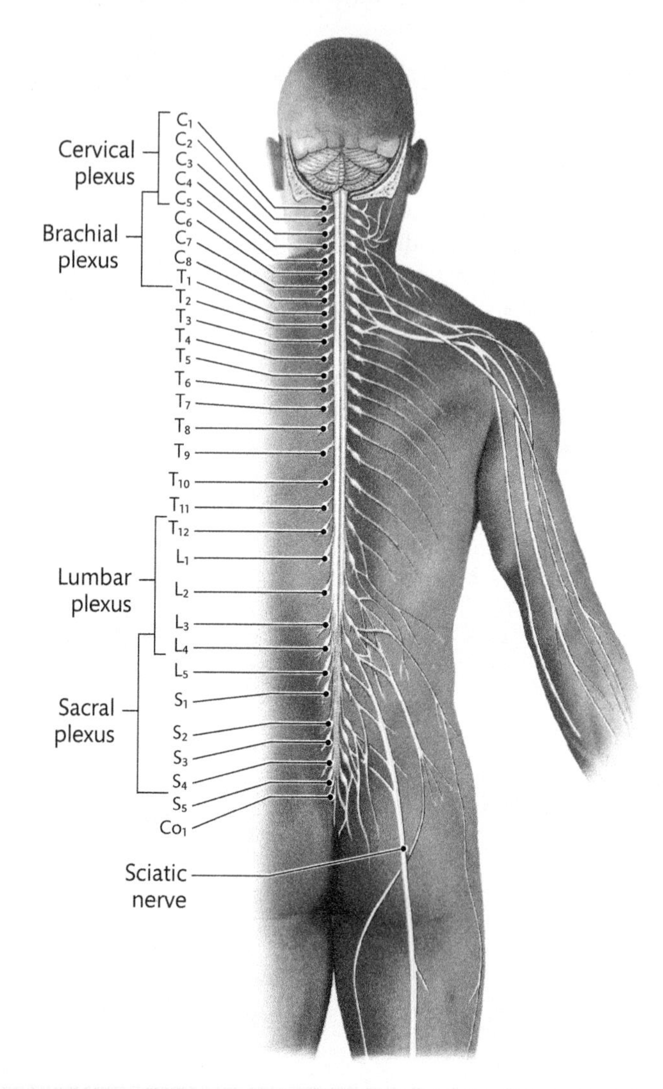

Each plexus gives rise to peripheral nerves that contain contributions from two or more spinal nerves. For example, the **sciatic nerve,** which supplies structures in the lower limb, originates from the sacral plexus. It contains components from spinal nerves L_4 to S_3.

Is there any functional advantage to having peripheral nerves with multiple spinal nerve contributions? Notice that most thoracic spinal nerves do not participate in the formation of plexuses. Can you think of a reason why?

LOOKING FORWARD »

In this laboratory exercise, you assessed the spinal cord's sensory function by completing two-point discrimination tests and evaluating the sensory function in dermatomes. In the next exercise, you will examine various types of reflexes, which require a combination of sensory and motor function and have an important role in maintaining homeostasis.

Name ______________________

Lab Section ______________________

Date ______________________

EXERCISE 16 REVIEW SHEET

The Spinal Cord and Spinal Nerves

1. Explain how the arrangement of the dura mater around the spinal cord differs from its arrangement around the brain.

____

____

____

____

____

QUESTIONS 2–10: Identify the labeled structures in the diagram below.

2. ______________________

3. ______________________

4. ______________________

5. ______________________

6. ______________________

7. ______________________

8. ______________________

9. ______________________

10. ______________________

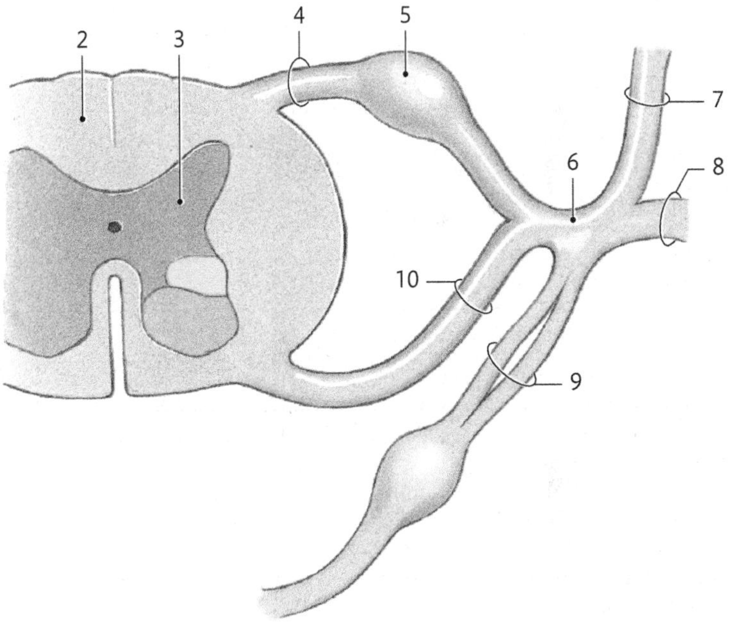

11. What is the cauda equina? Explain why it forms.

____

____

____

____

____

12. The results of a two-point discrimination test on four skin areas are given in the table. Complete the table by ranking the skin areas 1 through 4 according to the relative densities of sensory receptors (1 = highest density, 4 = lowest density). In the figure of the sensory homunculus, identify the four labeled skin regions, A, B, C, and D, based on your ranking of sensory receptor density.

Two-Point Discrimination Test

Area of Skin	Distance between Compass Points at Which Two-Point Discrimination Can Be Made (mm)	Relative Density of Sensory Receptors (ranked 1 to 4)
A	5.0 mm	
B	3.2 mm	
C	8.6 mm	
D	1.5 mm	

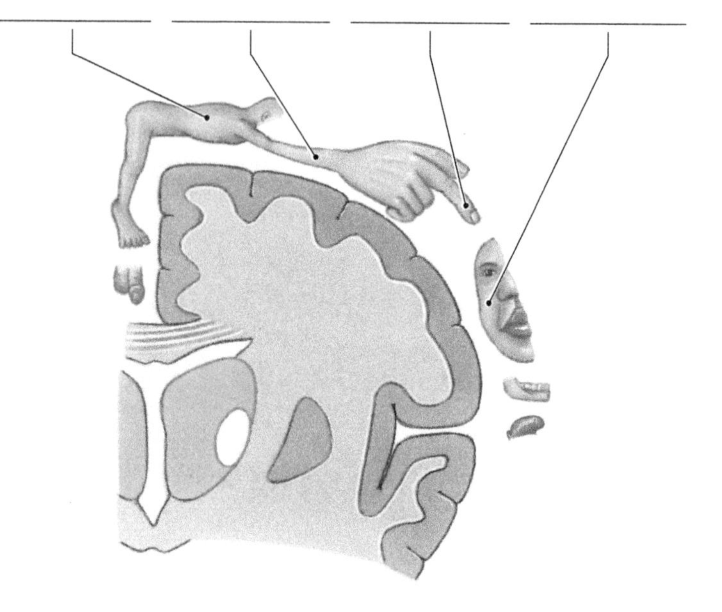

Provide a rationale for your answers.

13. The following peripheral nerves are derived from the lumbar plexus:

Peripheral Nerve	Spinal Nerve Contributions
Ilioinguinal nerve	L_1
Genitofemoral nerve	L_1, L_2
Lateral femoral cutaneous nerve	L_2, L_3
Femoral nerve	L_2, L_3, L_4

A functional assessment of the spinal nerves that form the lumbar plexus indicated a functional loss of spinal nerve L_2 on the right side. How would this condition affect the nerves listed in the table? Would each nerve be equally affected? Explain your answer.

EXERCISE

17

Human Reflex Physiology

LEARNING OUTCOMES

These Learning Outcomes correspond by number to the laboratory activities in this exercise. When you complete the activities, you should be able to:

Activities 17.1–17.3 Assess control of skeletal muscles in coordinated movement and equilibrium (balance).

Activities 17.4–17.5 Perform spinal reflex tests for the upper and lower limbs.

LABORATORY SUPPLIES

- Reflex hammers
- Chalk
- Labeling or masking tape

PRE-LAB QUIZ

Before you begin, read all the activities in Exercise 17 and the required reading in your textbook that is assigned by your instructor.

1. During this laboratory exercise, what piece of equipment will you use to perform spinal reflex tests? ____________
2. During this laboratory exercise, you will analyze your lab partner's coordination and balance. These activities will test the function of what brain region?
   a. cerebrum
   b. thalamus
   c. cerebellum
   d. hypothalamus
3. ____________ is a general sensory awareness of position and movement of your own body parts.
4. During this laboratory exercise, you will assess coordination by
   a. walking in a straight line, heel to toe.
   b. touching the tip of your nose.
   c. standing erect, adjacent to a vertical line.
   d. both (a) and (c).
5. During this laboratory exercise, you will assess balance by
   a. walking in a straight line, heel to toe.
   b. standing erect, adjacent to a vertical line.
   c. moving your arms forward and bringing your fingertips together.
   d. both (a) and (b).
6. Stretch reflexes involve the activity of what type of sensory receptors? ____________
7. The normal response for the ____________ reflex test is a slight extension of the leg.
8. The normal response for both the plantar reflex and the ____________ reflex is plantar flexion of the foot.
9. True or False: Stretch reflexes are important because they protect muscles from damage caused by overextension. ____________
10. True or False: The biceps reflex test assesses the function of the C_5 and C_6 segments of the spinal cord. ____________

Activity 17.1

Assessing Coordination

The cerebellum receives the following inputs:

- **Motor commands** from the cerebrum
- Stimuli for **equilibrium (balance)** from the inner ear
- Stimuli for **vision** from the eyes
- Stimuli for **proprioception** (body position in space) from receptors called **proprioceptors** in muscles and joints.

After analyzing this input, the cerebellum sends information that fine-tunes the motor activity back to the cerebrum.

Normal cerebellar function gives us the ability to maintain balance and posture and allows us to perform highly specialized motor functions such as dancing, riding a bicycle, and climbing a ladder.

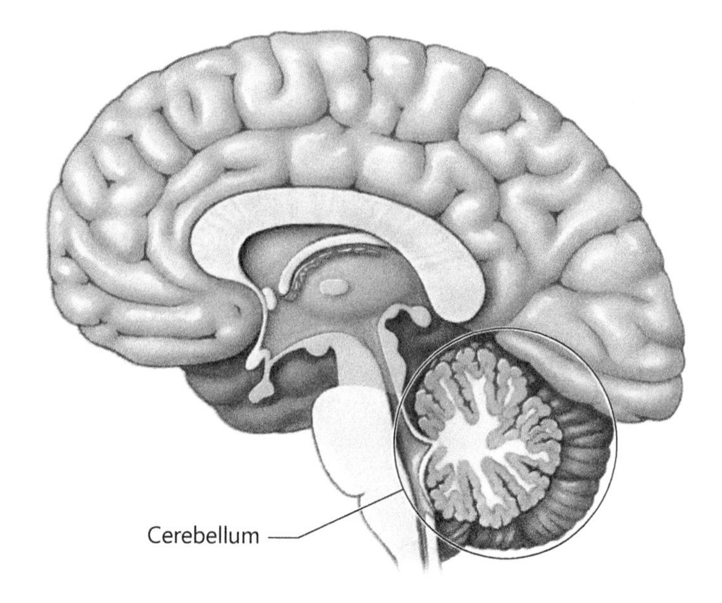

A Assessing Coordination: Test 1

1. Stand in the anatomical position with your eyes open.
2. Move your right hand to your face and touch the tip of your nose with your index finger. Repeat this action with your left hand.
3. Repeat steps 1 and 2 with your eyes closed.
4. In the space below, record and analyze your test results. How accurate are you in touching the tip of your nose when your eyes are open versus closed? If you missed the tip of your nose, record the position where your finger made contact with your face and its distance from the tip of the nose.

▶ ______________________________

____

____

____

____

Word Origins

Proprioception derives from two Latin words: *proprius* ("one's own") and *capio* ("to take") Proprioception is a general sensory awareness of the position and movement of one's own body parts.

B Assessing Coordination: Test 2

1. Stand in the anatomical position with your eyes open.
2. Abduct your arms so that they are parallel to the floor.
3. With each index finger extended, move your arms forward and bring your fingertips together.
4. Repeat steps 1 through 3 with your eyes closed.
5. In the space provided, record and analyze your test results. How accurate are you in touching your fingertips with your eyes open versus closed? If your fingertips do not make contact, record the distance between your fingers as they pass by each other.

▶ ______________________________

____

____

____

____

IN THE CLINIC

Ataxia

Damage to the cerebellum, due to trauma or stroke, can result in **ataxia,** a condition characterized by uncoordinated, shaky movements and an inability to maintain balance. A person with ataxia may not be able to stand or sit in an erect postion. Voluntary movements are performed very slowly and require a great deal of concentration. In addition, timing and coordination of movements are impaired so that an attempt to grab an object (e.g., reaching for a coffee cup or pen) often results in moving the hand beyond the target.

MAKING CONNECTIONS

When your eyes are closed: ▶

a. Is it more or less difficult to perform the coordination tests in this activity? Explain.

____

____

____

____

b. What other sensory stimuli do you rely on to successfully complete these tests?

____

____

____

____

Activity **17.2**

Assessing Balance

Several drugs, including alcohol, impair cerebellar function temporarily. For this reason, police officers ask drivers who are suspected of driving under the influence of alcohol or drugs to walk in a straight line.

A Assessing Balance: Test 1

1. With your eyes open, walk slowly, heel to toe, in a straight line without losing your balance. Have your lab partner record in Table 17.1 the time it takes you to walk 6 m (20 ft). ▶
2. Repeat step 1 several times, walking faster each time. Record the time after each turn in Table 17.1. Stop the test when you can no longer maintain your balance. ▶
3. Repeat steps 1 and 2 with your eyes closed. Record the time of each turn in Table 17.1. ▶

(Important: For safety, two people should remain at the sides of the subject to act as spotters while he or she performs this activity.)

4. Compare your results with the results of other students in the laboratory.

▶ ______________________________

____

____

____

____

____

____

TABLE 17.1 Assessing Balance by Walking along a Straight Line

Trial #	Time (sec)	
	Walking with Eyes Open	Walking with Eyes Closed
1		
2		
3		
4		
5		

B Assessing Balance: Test 2

1. Draw a vertical line on a chalkboard or whiteboard or mark a vertical line on a wall in the laboratory with a piece of tape.

2. Stand erect, adjacent to the chalk or tape line, with your upper limbs by your sides and your eyes open. Remain standing for 2 minutes.

3. Your laboratory partner will time the exercise and record any body movements that you make during the test. Use the chalk or tape line as a reference position for recording movements.

▶ ______________________________

____

____

____

____

4. When the 2-minute period is completed, sit for a short time. After you are rested and comfortable, stand up again next to the vertical reference line.

5. **Make a Prediction**

For this test, you will stand for 2 minutes, but this time you will close your eyes the whole time. Before you begin, predict how these results will compare with the results of the previous test with your eyes open.

▶ ______________________________

____

____

____

____

6. Conduct the test with your eyes closed. Your laboratory partner should note any body movements that you make during the test.

▶ ______________________________

____

____

____

____

7. **Assess the Outcome**

Compare the results from the two trials. In particular, contrast the number and type of movements that you made and note any evidence of a loss of balance while your eyes were open or closed. Do your results support or refute your hypothesis?

▶ ______________________________

____

____

____

____

MAKING CONNECTIONS

Do you believe that your ability to perceive and interpret visual stimuli is more important for performing the coordinated movements in Activity 17.1 or for maintaining balance in this activity? Explain.

▶ ______________________________

____

____

____

____

____

Activity 17.3

Assessing Coordinated Movement and Balance

Complex and highly specialized actions require a combination of balance and well-coordinated muscle contractions.

A Assessing Coordinated Movement and Balance: Test 1

1. Stand upright, with your upper limbs by your sides and your eyes open.
2. When you are relaxed, raise your right leg off the ground and touch your heel to the anterior surface (the shin) of your left leg.
3. Move your heel inferiorly (toward the floor), but try to keep it in contact with your other leg.
4. During the test, your lab partner should evaluate your performance by observing how fluid your movements are and how well you maintain your balance. Record and analyze the results in Table 17.2. ▶
5. Repeat steps 1 through 4 with your other leg. Record and analyze the results in Table 17.2. ▶

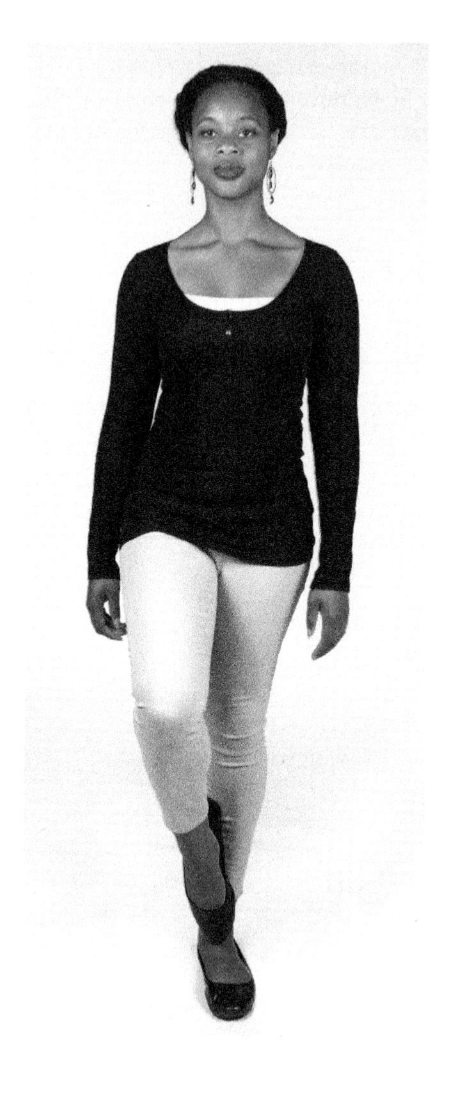

TABLE 17.2 Assessing Coordinated Movement and Balance

Test	Right Leg		Left Leg	
	Results	Analysis	Results	Analysis
Test 1: Placing heel on shin				
Test 2: Moving foot in a figure-eight pattern				

B Assessing Coordinated Movement and Balance: Test 2

1. Stand upright, with your upper limbs by your sides and your eyes open.
2. When you are relaxed, slightly raise your right foot off the ground and move your entire lower limb so that your foot traces a figure-eight pattern that is parallel to the floor. Attempt to do 10 figure-eight patterns before lowering your foot.
3. During the test, your lab partner should evaluate your performance by observing how fluid your movements are and how well you maintain your balance. Record and analyze the results in Table 17.2. ▶
4. Repeat steps 1 through 3 with your left foot. Record and analyze the results in Table 17.2. ▶
5. Obtain results and analyses of Test 1 and Test 2 from two other students in the laboratory and record the data in Table 17.3. ▶

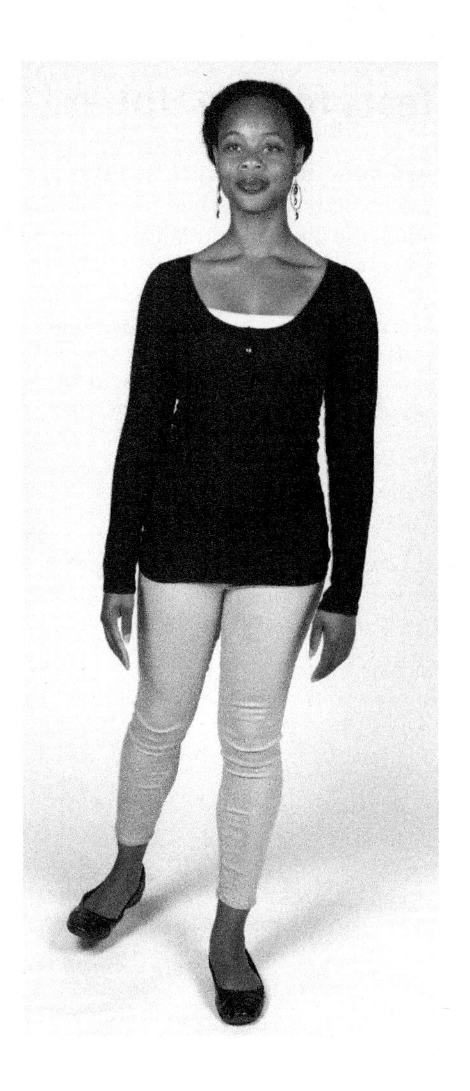

TABLE 17.3 Assessing Coordinated Movement and Balance: Data from Two Other Students

Test 1: Placing heel on shin		
Student Name	**Results**	**Analysis**
Test 2: Moving foot in a figure-eight pattern		
Student Name	**Results**	**Analysis**

6. Compare your results and analyses of the two tests with the results and analyses of the other two students.

▶ ______________________________

____

____

____

____

MAKING CONNECTIONS

Discuss why well-coordinated muscle contractions, balance, and posture are important for performing complex actions (e.g., dancing, running, climbing).

▶ ______________________________

____

____

____

____

____

____

____

Activity 17.4

Reflex Tests for the Upper Limb

A **reflex** is an automatic, involuntary response to a change that occurs inside or outside the body. Many reflexes play a critical role in protecting the body by regulating homeostasis. The neuronal pathway of a reflex, known as a **reflex arc,** contains the five elements illustrated in the figure.

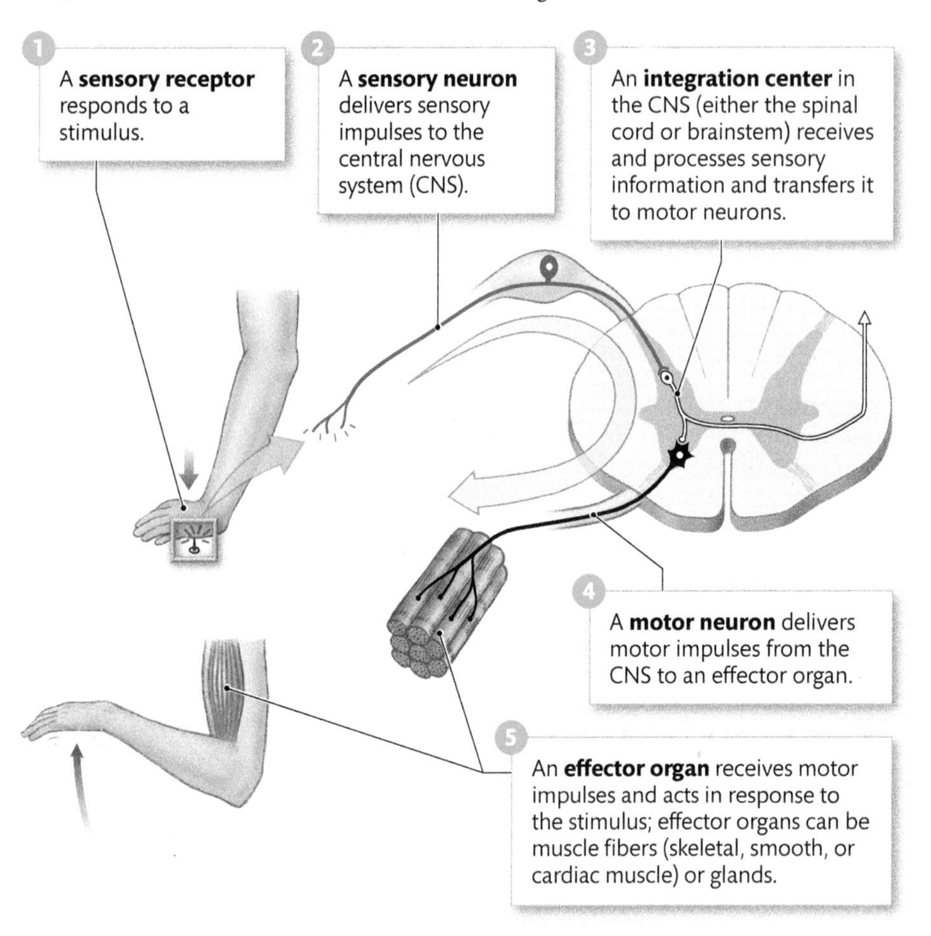

The spinal reflexes that you will observe in the laboratory are examples of **stretch reflexes.** They involve the activity of receptors called **muscle spindles** that are embedded between muscle fascicles in skeletal muscles. When a muscle stretches, muscle spindles lengthen. This lengthening triggers a reflex muscle contraction, which prevents excessive stretching of the muscle.

A The **biceps reflex test** assesses the function of the C_5 and C_6 segments of the spinal cord.

1. While your lab partner is sitting comfortably, identify his or her biceps brachii tendon in the right arm. The tendon can be located more easily when the forearm is flexed. The arm should not be resting on a table or any other surface.
2. Palpate the tendon as it passes through the cubital fossa (the shallow depression anterior to the elbow joint).
3. Position your finger or thumb over the tendon and stabilize the elbow with the rest of your hand.
4. Gently tap your finger or thumb with the pointed end of a rubber reflex hammer. A normal response will be a contraction of the biceps brachii muscle; you should notice a slight flexion of the forearm.
5. Repeat steps 1 through 4 on the left arm. Record your results in Table 17.4. ▶

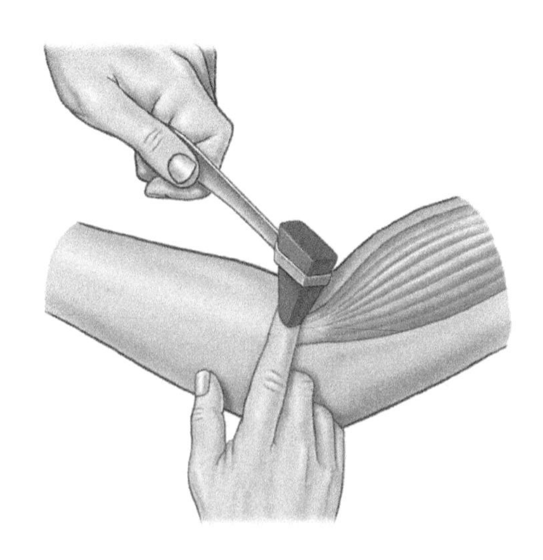

B The **triceps reflex test** assesses the function of the C_7 and C_8 segments of the spinal cord.

1. While your lab partner is sitting comfortably, palpate his or her triceps brachii muscle in the posterior right arm. The arm should not be resting on a table or any other surface. Follow the muscle inferiorly until you can feel the triceps tendon just superior to the elbow joint.
2. Gently tap the triceps tendon with the pointed end of the reflex hammer. A normal response will be a contraction of the triceps brachii muscle; you should notice a slight extension of the forearm.
3. Repeat steps 1 and 2 on the left side. Record your results in Table 17.4. ▶

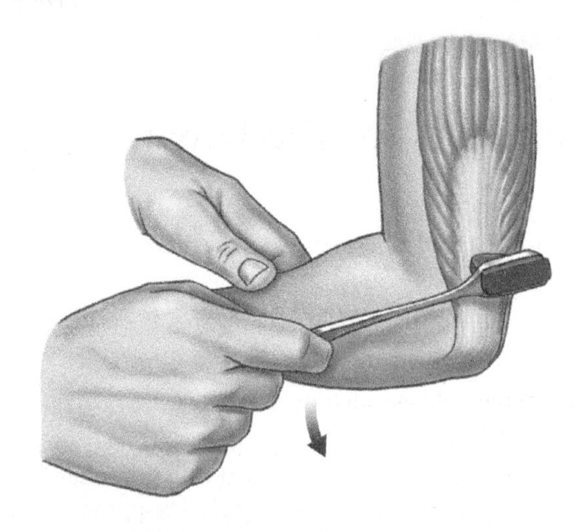

TABLE 17.4 Reflex Tests for the Upper Limb

	Test Results	
Reflex Test	**Right Arm**	**Left Arm**
Biceps reflex test		
Triceps reflex test		

IN THE CLINIC

Stretch Reflexes

Stretch reflexes play a critical role in reducing the risk of muscle damage due to overextension. They are also important for maintaining normal upright posture. Accordingly, they are very active in the muscles of the upper and lower limbs. Health professionals often conduct stretch reflex tests to check normal function of the spinal nerves. For example, an absent or weak biceps reflex could indicate a lack of sensation or reduced motor output along spinal nerve C_5 or C_6. Similar results for the triceps reflex would indicate a sensory or motor loss along spinal nerve C_7 or C_8.

MAKING CONNECTIONS

Unlike smooth and cardiac muscle, skeletal muscle is voluntary, and we have conscious control of the movements produced by its contractions. During reflexes that involve skeletal muscles, are the movements voluntary or involuntary? Explain.

▶ ____

____

____

____

____

____

____

Activity **17.5**

Reflex Tests for the Lower Limb

A The **patellar reflex test** evaluates the function of the L_2, L_3, and L_4 levels of the spinal cord.

1. Have your lab partner sit on the laboratory bench with his or her legs hanging freely.
2. On the right leg, identify the patellar ligament just inferior to the patella.
3. Gently tap the patellar ligament with the pointed end of the rubber reflex hammer. A normal response will be a contraction of the quadriceps femoris muscles; you should notice a slight extension of the leg.
4. Repeat steps 1 through 3 on the left leg. Record your results in Table 17.5. ▶

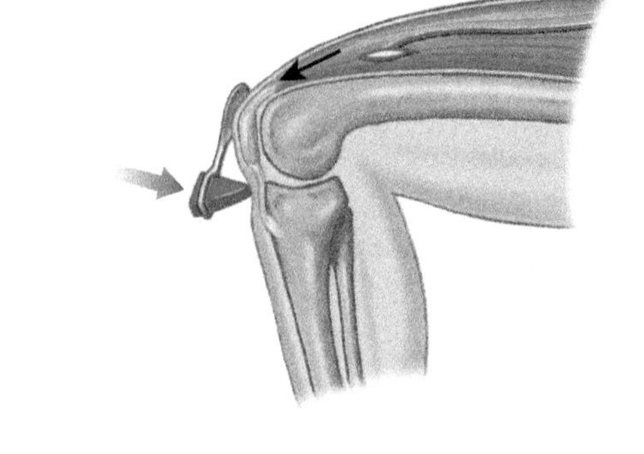

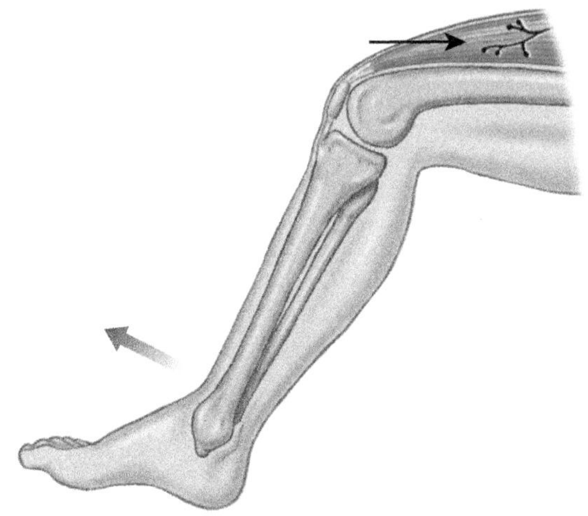

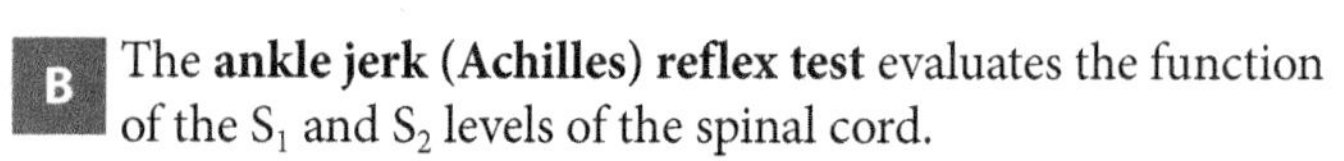

B The **ankle jerk (Achilles) reflex test** evaluates the function of the S_1 and S_2 levels of the spinal cord.

1. Have your lab partner sit on the laboratory bench with his or her legs hanging freely.
2. On the right leg, identify the calcaneal (Achilles) tendon as it passes posterior to the ankle joint.
3. Gently tap the calcaneal tendon with the broad side of the reflex hammer. A normal response will be a contraction of the gastrocnemius and soleus muscles; you should notice a slight plantar flexion of the foot.
4. Repeat steps 1 through 3 on the left leg. Record your results in Table 17.5. ▶

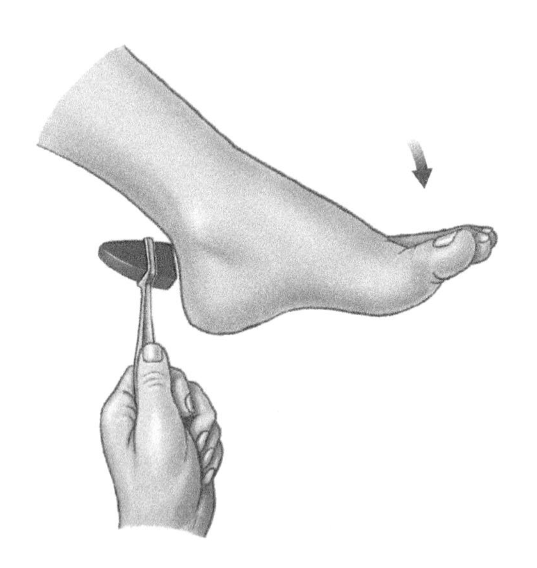

C The **plantar reflex test,** known also as the **negative Babinski reflex,** examines the function of the S_1 and S_2 and, to a lesser extent, the L_4 and L_5 levels of the spinal cord.

1. Have your lab partner sit on the laboratory bench with his or her legs hanging freely and shoes removed.
2. On the right foot, gently run the handle of the reflex hammer from the heel along the lateral margin of the foot to the tarsometatarsal joints; then move across the plantar surface toward the great toe.
3. A normal response is plantar flexion of the foot. An abnormal response is extension of the great toe and abduction (spreading) of the other toes.
4. Repeat steps 1 through 3 on the left foot. Record your results in Table 17.5. ▶

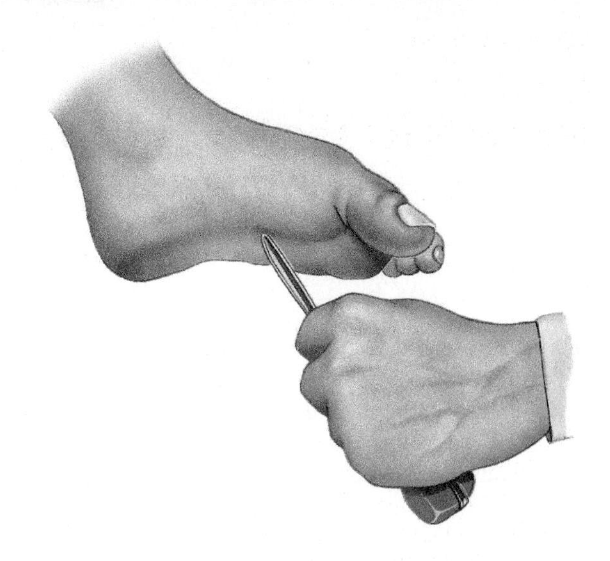

TABLE 17.5 Reflex Tests for the Lower Limb

	Test Results	
Reflex Test	**Right Lower Limb**	**Left Lower Limb**
Patellar reflex test		
Ankle jerk reflex test		
Plantar reflex test		

IN THE CLINIC

Babinski Sign

An infant's nerve fibers are not yet completely myelinated. For this reason, babies commonly show an abnormal plantar reflex—spreading of the toes—called a positive Babinski sign. In older children and adults, however, a positive Babinski sign can indicate damage to the descending (motor) tracts in the spinal cord.

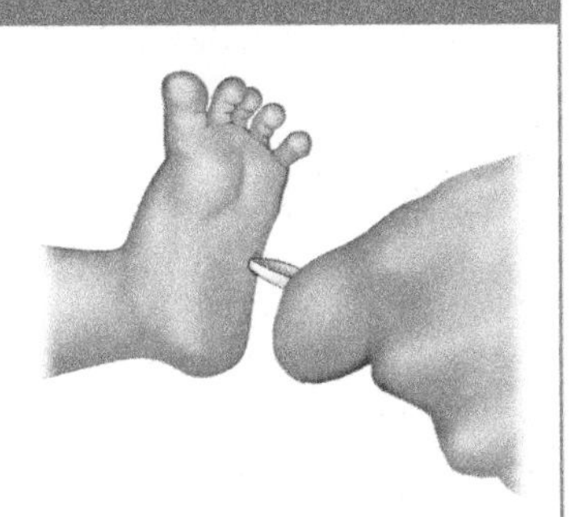

MAKING CONNECTIONS

Amyotrophic lateral sclerosis (ALS), commonly known as Lou Gehrig's disease, is a progressive disorder that leads to the destruction of motor neurons in the spinal cord. Explain how ALS would affect the stretch reflex tests that you conducted in this activity.

▶ ______________________________

____

____

____

____

BEFORE YOU MOVE ON . . .

❮❮ LOOKING BACK

In this laboratory exercise, you examined cerebellar function by performing a number of tests to assess coordination and balance. The cerebellum receives motor commands from the cerebrum and sensory stimuli for equilibrium, vision, and proprioception. After analyzing this information, the cerebellum sends information back to the cerebrum. This information adjusts the voluntary motor activity initiated by the **primary motor area.** Normal cerebellar function is vital for allowing us to maintain balance and posture and to carry out all motor functions gracefully and efficiently.

You also conducted a number of stretch reflex tests to evaluate the function of spinal nerves. A stretch reflex that is absent or weak could mean that sensory or motor function at a particular spinal cord level is impaired.

The motor pathways contain two types of motor neurons. The cell bodies of **upper motor neurons** are located in the primary motor area. Along their pathways, they cross over to the opposite side, either in the brainstem or spinal cord. Axons of upper motor neurons synapse with **lower motor neurons,** which have axons that travel in cranial and spinal nerves and innervate skeletal muscles.

The primary motor area can be divided into body regions represented by a **motor homunculus.** This body parts map correlates a cortical region with the body part it controls. The body regions in the motor homunculus are proportioned differently from the body itself. For instance, your hand is much smaller than your torso, but the cortical region that controls your hand is relatively large whereas the cortical region that controls your torso is relatively small.

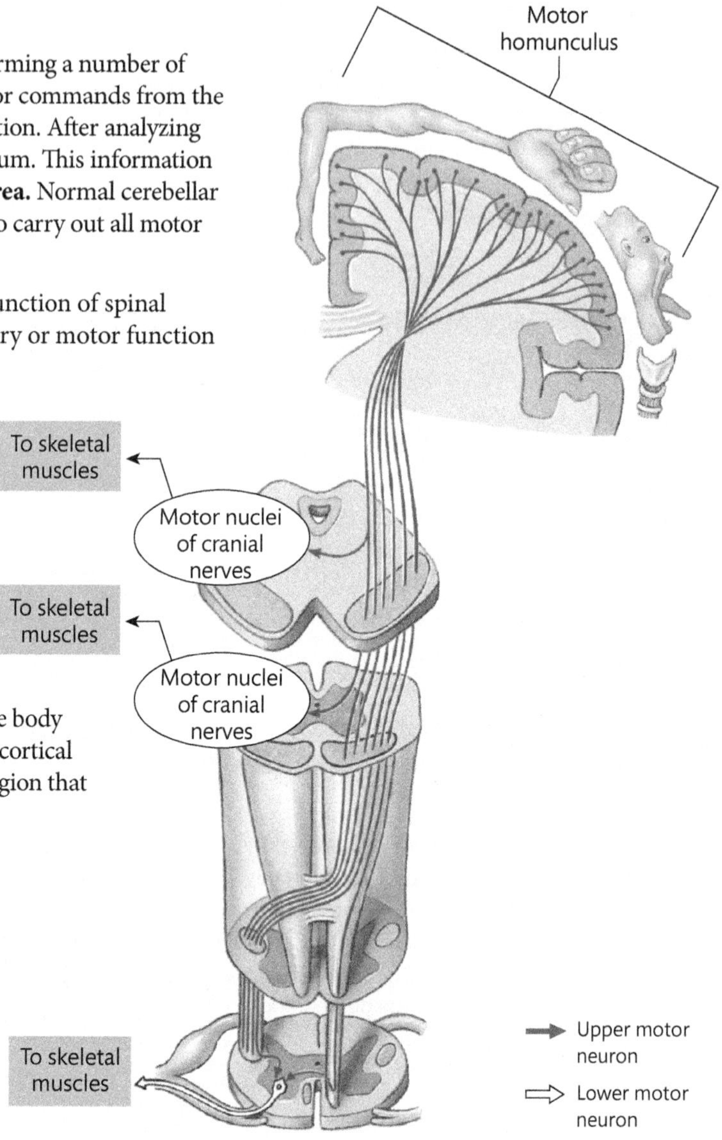

Why do you think so much of the primary motor area is devoted to controlling our hands?

▶ ____

____

____

____

When the spinal cord is damaged, the motor pathways to spinal cord levels inferior to the injury will be blocked, but reflex movements may still occur. How is that possible?

▶ ____

____

____

____

____

____

Motor homunculus

To skeletal muscles

Motor nuclei of cranial nerves

To skeletal muscles

Motor nuclei of cranial nerves

To skeletal muscles

Upper motor neuron

Lower motor neuron

LOOKING FORWARD ❯❯

In this and other previous laboratory exercises (Exercise 6, "The Integumentary System"; Exercise 15, "The Brain and Cranial Nerves"; and Exercise 16, "The Spinal Cord and Spinal Nerves"), the role of general sensory receptors was examined. In the next exercise, you will study the special senses, which include **olfaction** (smell), **gustation** (taste), **vision, hearing,** and **equilibrium** (body orientation). Receptors for these sensations are localized to specific regions in the head and have a complex structure. You will see that special sensory information must travel along complex pathways to reach the central nervous system.

Name ______________________

Lab Section ______________________

Date ______________________

EXERCISE 17 REVIEW SHEET

Human Reflex Physiology

QUESTIONS 1–3: Define the following terms.

1. Primary motor area:

____

2. Motor homunculus:

____

3. Proprioception:

____

4. Describe the neural inputs that the cerebellum receives and the neural outputs that it sends out.

____

5. One evening as you are reading a book by your favorite author, there is a power outage in your neighborhood. Discuss the importance of proprioception for allowing you to navigate through your house in the dark.

____

QUESTIONS 6–10: Match the descriptions of reflex arc components with the appropriate structure labeled in the diagram.

6. This structure contracts as a result of a stimulus.

____

7. Located in the CNS, this structure receives sensory impulses and sends out motor impulses.

____

8. This cell delivers nerve impulses to an effector organ.

____

9. This structure responds to a stimulus. _______________

10. This cell delivers nerve impulses to the CNS.

____

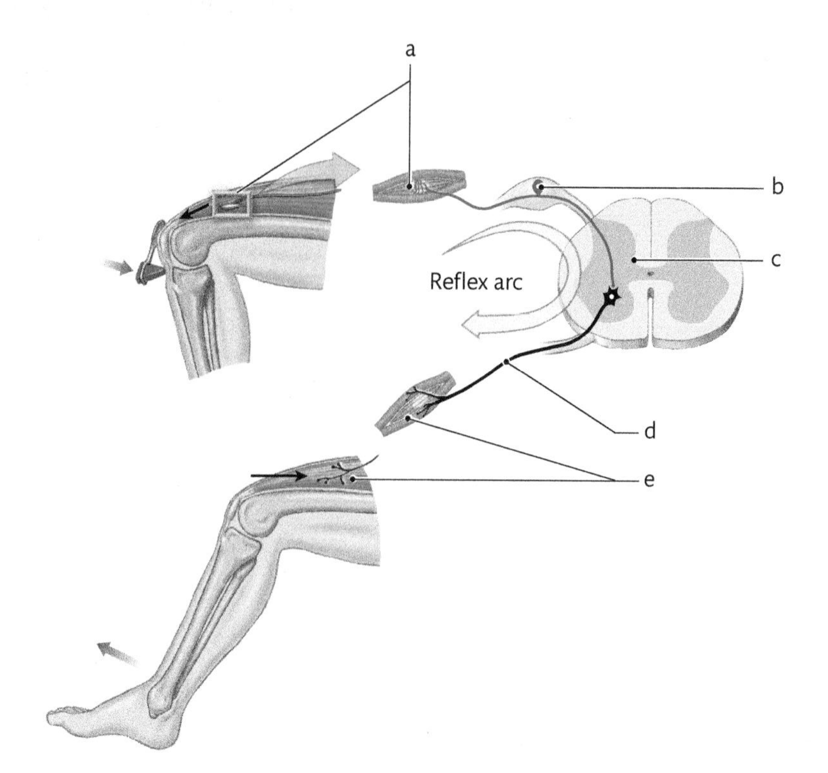

11. Transverse myelitis is a CNS disease that is caused by inflammation of the spinal cord. It can affect one or several levels of the spinal cord. How could this disease affect the results of the stretch reflex tests that you performed in this laboratory exercise? Explain.

____

____

____

____

EXERCISE 18

Special Senses

LEARNING OUTCOMES

These Learning Outcomes correspond by number to the laboratory activities in this exercise. When you complete the activities, you should be able to:

Activity 18.1	Describe the gross and microscopic anatomy of olfactory structures.
Activity 18.2	Test and explain olfactory adaptation.
Activity 18.3	Describe the gross and microscopic anatomy of gustatory structures.
Activity 18.4	Test gustatory sensations.
Activity 18.5	Identify and explain the function of accessory eye structures.
Activity 18.6	Explain the gross anatomy of the eye.
Activity 18.7	Dissect a cow eye.
Activity 18.8	Describe the microscopic structure of the retina.
Activities 18.9–18.12	Perform tests to assess vision.
Activity 18.13	Identify the gross anatomical structures of the ear.
Activity 18.14	Examine the microscopic structure of the cochlea.
Activities 18.15–18.16	Perform tests to assess hearing.

LABORATORY SUPPLIES

- Anatomical models of the following:
  - midsagittal section of the head
  - human brain
  - human eye
  - human ear
- Human skulls
- Compound light microscopes
- Prepared microscope slides of the following:
  - olfactory epithelium
  - tongue with taste buds
  - retina
  - cochlea
- Stopwatch
- Laboratory timer
- Vials of wintergreen and peppermint oils (or other substances with strong, distinct odors)
- Cotton swabs
- Paper towels
- Granulated sugar (sucrose)
- 10% granulated sugar solution
- Phenylthiocarbamide (PTC) test strips
- Cubes of various food items (food types to remain unknown to students)
- Pipe cleaners or slender probes
- Dissecting tools, gloves, and trays
- Protective eyewear
- Face masks (optional)
- Fresh or preserved cow eyes
- Dissecting pins
- Metric rulers
- Pencils
- Flashlights
- Snellen charts
- Astigmatism charts
- Ishihara color plates
- Tuning forks
- Colored pencils

PRE-LAB QUIZ

Before you begin, read all the activities in Exercise 18 and the required reading in your textbook that is assigned by your instructor.

1. True or False: Adaptation refers to increasing sensitivity to a stimulus over time.
   ____
2. Axons of the olfactory nerve reach the olfactory bulb by passing through narrow openings in what bone structure? ____________________________
3. True or False: Of the three types of papillae on the surface of the tongue, only the filiform papillae contain taste buds. ____________
4. The area of the retina that lacks photoreceptive cells is called the ____________________.
5. What structure contains smooth muscle that can dilate or constrict the pupil?
   ____
6. What structure contains smooth muscle that can change the shape of the lens?
   ____
7. During this laboratory exercise, what piece of equipment will you use to test visual acuity?
   ____
8. During this laboratory exercise, you will study the microscopic structure of the ____________________ in the eye and of the ____________________ in the ear.
9. During this laboratory exercise, what piece of equipment will you use for the Weber and Rinne hearing tests? ____________________________
10. True or False: The spiral organ is innervated by the cochlear division of the vestibulocochlear nerve. ____________

Activity **18.1**

Olfaction: Examining the Anatomy of Olfactory Structures

A **Gross Anatomy:** Olfactory structures are associated with the superior concha, perpendicular plate of the ethmoid, and cribriform plate.

1

On a model of a midsagittal section of the head, identify the nasal cavity and the following structures:

The inferior surface of the **cribriform plate**

Superior nasal concha

These structures and the superior portion of the **perpendicular plate** (area shaded in purple) are covered by the mucous membrane that contains the olfactory epithelium.

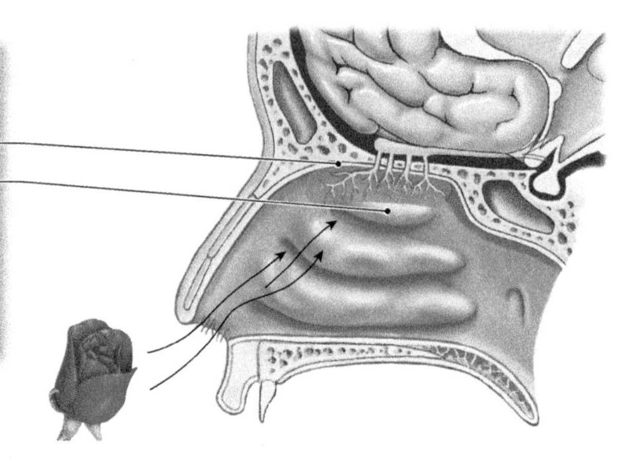

2

Inside the cranial cavity of a skull, identify the numerous olfactory foramina, which are narrow openings that pass through the **cribriform plate.** These passageways transmit the unmyelinated axons of the olfactory nerve (cranial nerve I). The axons originate from the cell bodies of olfactory receptor cells in the olfactory epithelium (see top figure).

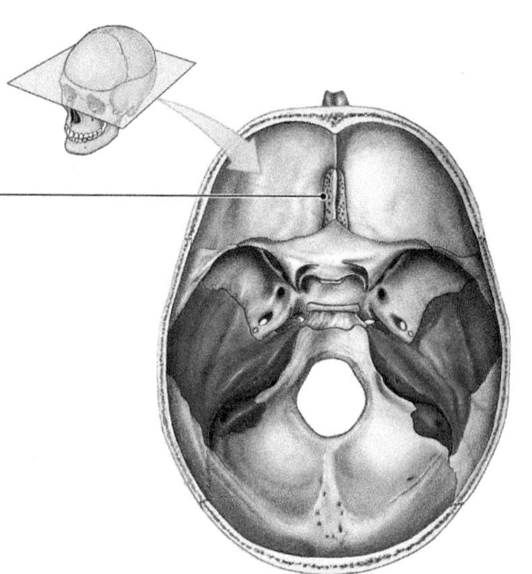

3

On a model of the brain, identify the two **olfactory bulbs** on the inferior surface of the frontal lobes. The olfactory bulbs are regions of gray matter where olfactory nerve axons synapse with multipolar neurons.

4

Identify the **olfactory tracts** that project posteriorly from the olfactory bulbs, sending axons to the olfactory area in the temporal lobe.

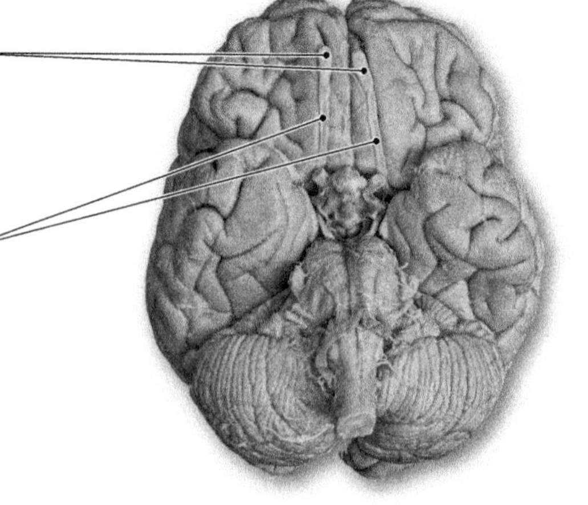

5

Place the brain into the cranial cavity of a skull (do not expect an exact fit). Upon what structure do the olfactory bulbs rest?

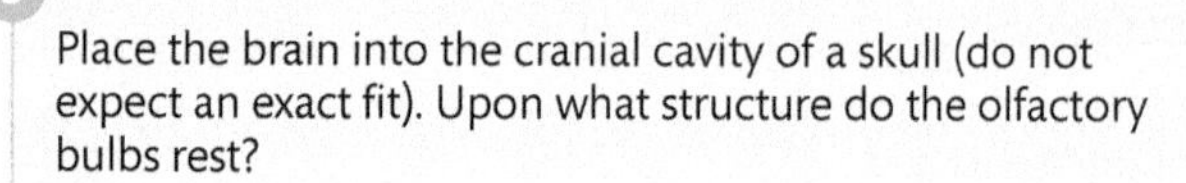

B **Microscopic Anatomy:** Sensory receptors for olfaction are located in the olfactory epithelium that covers the superior portion of the nasal cavity.

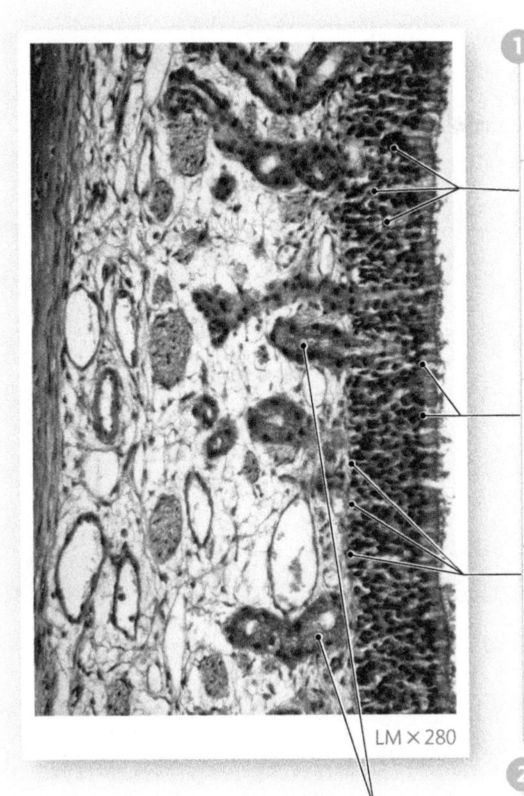
LM × 280

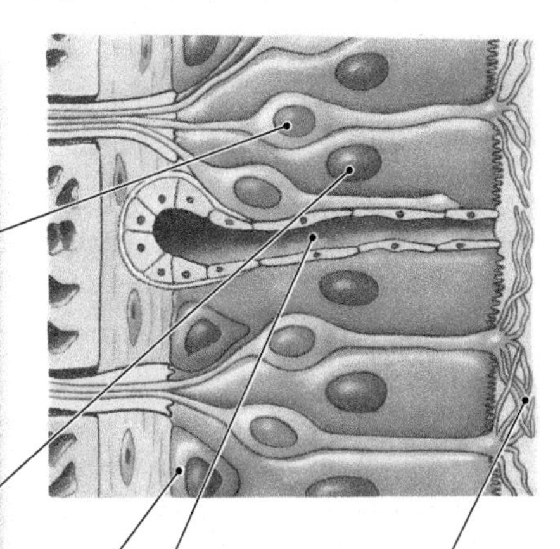

1 View a slide of the olfactory epithelium under low power and identify three cell types:

Olfactory receptor cells are bipolar neurons that are stimulated by chemical substances (odorants) in the nasal cavity. They are columnar cells that extend from the base to the surface of the epithelium. Their nuclei are centrally located in the epithelium.

Supporting cells are columnar epithelial cells that surround the olfactory receptors. Their nuclei are positioned closer to the epithelial surface.

Basal cells are interspersed between the bases of the supporting cells and divide regularly to produce new olfactory receptor cells. Their nuclei are located along the base of the epithelium.

2 Identify the lamina propria, deep to the olfactory epithelium. This connective tissue area contains **olfactory glands** that secrete mucus onto the surface of the olfactory epithelium.

3 Switch to high power and identify **cilia,** which are located along the surface of the olfactory receptor cells. Chemical odorants that enter the nasal cavity stimulate olfactory receptor cells by binding to the receptors along the membranes of the cilia.

MAKING CONNECTIONS

Why would sniffing (forceful inhalation through the nose), rather than inhaling normally, tend to increase the stimulation of olfactory receptor cells?

▶ ____

____

____

____

____

Activity **18.2**

Olfaction: Testing Olfactory Adaptation

A **Adaptation** refers to a decreasing sensitivity to a stimulus over time. For olfaction, adaptation means that your sensitivity to an odor diminishes during continual stimulation.

1. Place several drops of peppermint oil (or any other substance with a strong, distinct odor) on a cotton swab.
2. Sit with both eyes closed and squeeze your left nostril shut.
3. Hold the cotton swab under your open right nostril. Inhale normally through your nose but exhale through your mouth.
4. Have your lab partner time how long it takes for olfactory adaptation to occur. Adaptation will occur when the odor is significantly diminished or disappears. Record the result in Table 18.1. ▶
5. Test a second substance, such as wintergreen oil, with the same nostril and record the adaptation time in Table 18.1. ▶
6. Repeat the adaptation tests for peppermint and wintergreen oils for the left nostril and record the results in Table 18.1. ▶

Word Origins

Olfaction comes from the Latin word *olfacio* ("to smell"). Structures with *olfaction* or *olfactory* in their names have functions related to the sense of smell.

TABLE 18.1 Olfactory Adaptation

Odorant Substance	Adaptation Time (sec)	
	Right Nostril	Left Nostril

7 Compare your results with those of other students in your laboratory. For each substance, describe the variation in adaptation time among individuals. ▶

1. Type of odorant: ______________________________

____

____

____

____

2. Type of odorant: ______________________________

____

____

____

____

IN THE CLINIC

Anosmia

Anosmia is the complete loss of olfactory sensation. In most cases, this condition is temporary and is caused by nasal congestion due to a cold or sinus infection. Anosmia could also develop as a result of damage to olfactory nerves after surgery or trauma, nasal polyps (noncancerous growths) that block the nasal cavity, exposure to toxic chemicals, or cocaine abuse. Your sense of smell can also diminish with age. The most obvious sign of anosmia is that odors of familiar things, like food or flowers, can no longer be detected. This loss could place people in dangerous situations if, for example, they cannot smell a gas leak or smoke from a fire.

If anosmia is caused by a cold or infection, it may be treated with decongestants, antibiotics, or other medications. If polyps are found in the nasal cavity, surgery may be required to restore olfaction. Sometimes, smell sensation will return without treatment, but if aging is the underlying cause, it may not be treatable.

MAKING CONNECTIONS

Explain why the olfactory adaptation time for one substance can be relatively short, but for another substance it can be much longer.

▶ ____

____

____

____

____

____

____

____

____

____

Activity **18.3**

Gustation: Examining the Anatomy of Gustatory Structures

A **Gross Anatomy:** Gustatory structures are located in the oral cavity and pharynx.

1. On a midsagittal section of the head, locate the **oral cavity.** Notice that the hard palate separates the oral cavity from the nasal cavity.

2. Identify the **tongue** within the oral cavity.

3. Locate the **pharynx,** which is positioned posterior to the nasal and oral cavities.

4. Locate the **epiglottis,** a plate of cartilage covered by a mucous membrane. When you swallow while eating or drinking, the epiglottis covers the entrance to the larynx. Why is this action important?

▶ ______________________________

____

____

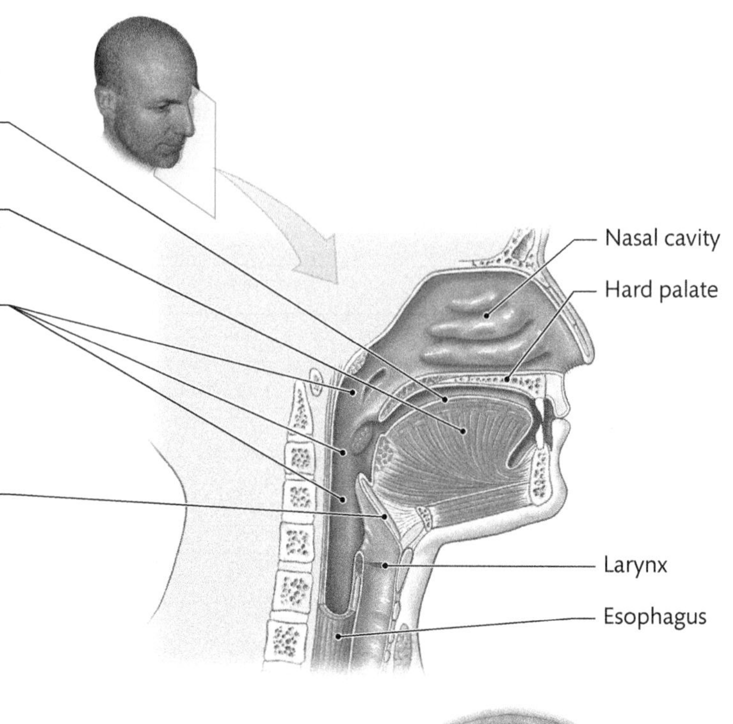

5. Sensory receptors for gustation are located in taste buds, primarily on the superior surface of the tongue but also on the soft palate, pharynx, and epiglottis. On the tongue, taste buds are located in the epithelium of surface projections known as **papillae** (singular = **papilla**). Look in a mirror and stick your tongue out to observe the three types of papillae:

Nine to twelve **circumvallate papillae** are aligned in a V shape on the posterior aspect of the tongue. Each one contains more than 100 taste buds.

Fungiform papillae are scattered over the entire anterior two-thirds of the tongue. Each fungiform papilla contains between 5 and 10 taste buds.

Filiform papillae are also distributed over the anterior two-thirds of the tongue interspersed among the fungiform papillae. Although they are more numerous, the filiform papillae lack taste buds.

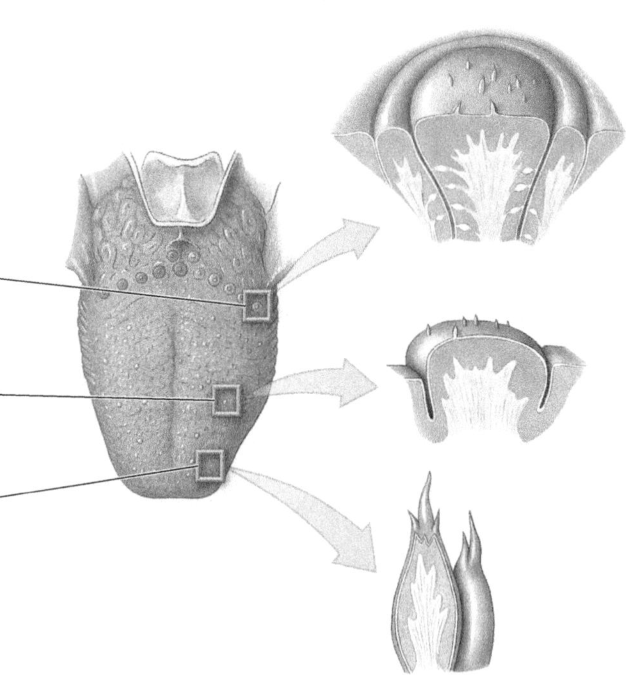

Want more practice? Go to: **MasteringA&P** > Study Area > Menu > Lab Tools > PAL >
- Human Cadaver > Nervous System > Special Senses > Head and tongue photos
- Histology > Special Senses > Taste bud photo

B **Microscopic Anatomy:** Taste buds are ovoid structures that resemble a flower bud. On the tongue, they are embedded in the surrounding epithelium, but at the apex, a taste pore opens to the oral cavity. Taste buds are composed of three different cell types. The basal cells divide regularly to produce new transitional (supporting) cells. Transitonal cells, in turn, differentiate into gustatory cells, which have a life span of about 10 days.

1 Under low power, observe a microscope slide of the tongue that is specially prepared to illustrate taste buds.

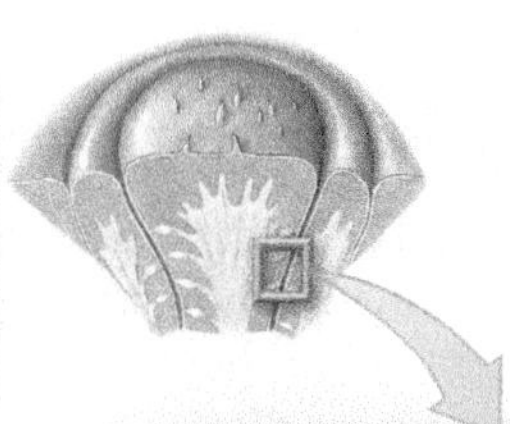

2 Locate **taste buds** embedded in the epithelium covering the papillae. In a typical slide, they will appear as oval-shaped cell clusters, which are lightly stained and contrast sharply with the darker-staining epithelial cells.

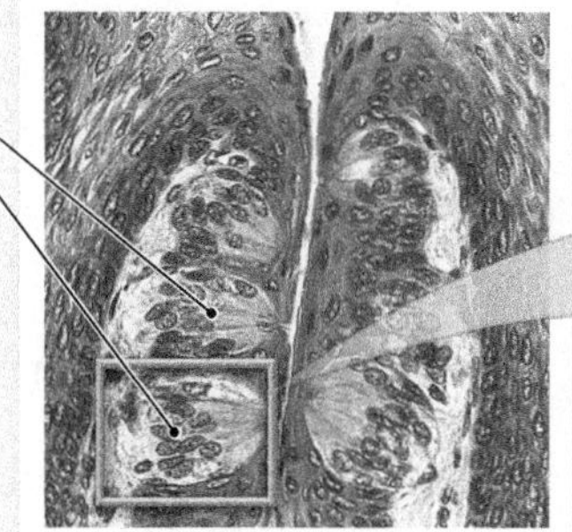

Taste buds LM × 100

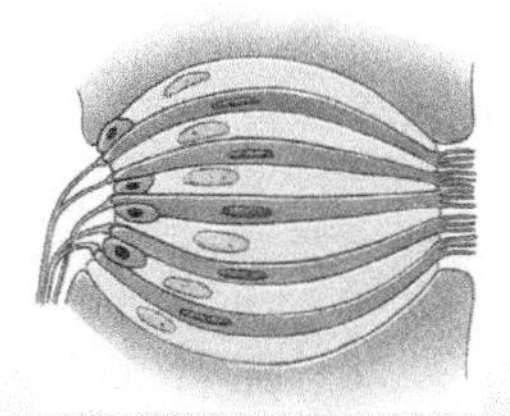

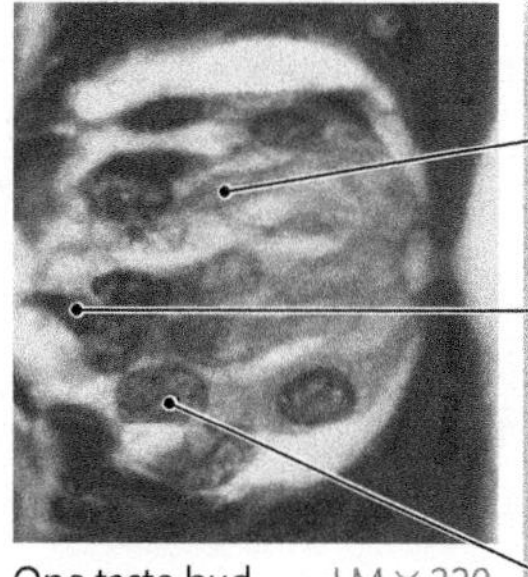

One taste bud LM × 220

3 Switch to high power and observe a taste bud more closely. Depending on the quality of your slide, you may be able to identify the nuclei of cells in the taste bud.

The nuclei of **gustatory cells** are lightly stained.

The nuclei of **basal cells** are darkly stained and are located near the base of the taste bud.

The nuclei of **transitional cells** are also stained darkly but are more centrally positioned.

Word Origins

Gustation is derived from the Latin word *gustacio* ("to taste"). Structures with *gustation* or *gustatory* in their names have functions related to the sense of taste.

MAKING CONNECTIONS

Taste buds are embedded in the epithelium of circumvallate and fungiform papillae. What type of papillae were you most likely observing in your microscope slide of the tongue? Explain. (*Hint:* Qualitatively assess the population density of taste buds on your slide.)

▶ ______________________________

____

____

____

____

Activity **18.4**

Gustation: Testing Gustatory Sensations

⚠ **Warning:** *Before you begin this activity, inform your instructor if you have any allergies to food or food flavorings. Some of the food items in this activity will be unknown to the food taster, so it is important to give your instructor this information before you begin. If you do have an allergy to any food item that will be used, do not perform this activity.*

A Assessing Taste Bud Stimulation

1. Dry the superior surface of your tongue with a clean paper towel. Place a few crystals of granulated sugar on the tip of your tongue. Have your lab partner note the time.

2. Alert your lab partner by raising your hand as soon as you can taste the sugar. Record the time of stimulation (sec):

▶ ______________________________

3. Rinse your mouth with water and dry your tongue with a clean paper towel.

4. **Make a Prediction**

Before you continue, predict whether it will take more or less time to taste the sugar in solution compared to the solid crystals.

▶ ______________________________

Repeat steps 1 and 2 using a cotton swab dipped in a 10% granulated sugar solution instead of the sugar crystals.

Record the time of stimulation (sec).

▶ ______________________________

5. **Assess the Outcome**

Did it take a longer, shorter, or the same length of time to detect the sweet taste of the sugar solution compared to the sugar crystals? Explain your results.

▶ ______________________________

____

____

____

____

B Investigating the Link between Taste and Inheritance

1. The ability to detect the bitter taste of phenylthiocarbamide (PTC) is an inherited trait that is present in about 75 percent of the population. Predict the number of PTC tasters and nontasters in your class and record your predictions in Table 18.2. ▶
   - Predicted PTC tasters = 0.75 × total class population
   - Predicted PTC nontasters = 0.25 × total class population

2. Place a paper strip flavored with PTC on your tongue. Chew the paper to mix the PTC with your saliva. Do not swallow the paper.

3. Can you detect the bitter taste of PTC?

▶ ______________________________

4. Tally the actual numbers and calculate the percentages of PTC tasters and nontasters in your class. Record your results in Table 18.2. ▶
   - % PTC tasters = (number of PTC tasters / total class population) × 100
   - % PTC nontasters = (number of PTC nontasters / total class population) × 100

5. How do your predicted results compare with the actual results? If they are different, suggest a reason why.

▶ ______________________________

____

____

TABLE 18.2 Percentage of Phenylthiocarbamide (PTC) Tasters in the Laboratory Class

	General Population (%)	Predicted Class Population		Actual Class Population	
		Number	Percentage	Number	Percentage
PTC tasters	75%		75%		
PTC nontasters	25%		25%		

C Examining the Effect of Olfaction on Gustation

1 Obtain small cubes of three to five different types of food that have been selected by your lab instructor. All the cubes should be the same size. Throughout this experiment, do not reveal the identity of the food types to your lab partner (the food taster). Wear gloves when handling the food cubes.

2 Have your lab partner dry his or her tongue with a paper towel and then sit with the eyes closed and both nostrils pinched shut. Place the first food cube on your lab partner's dry tongue.

3 Ask your lab partner to identify the food item in the following sequence of conditions:

- Immediately after placing the cube on the tongue, with eyes and nostrils closed

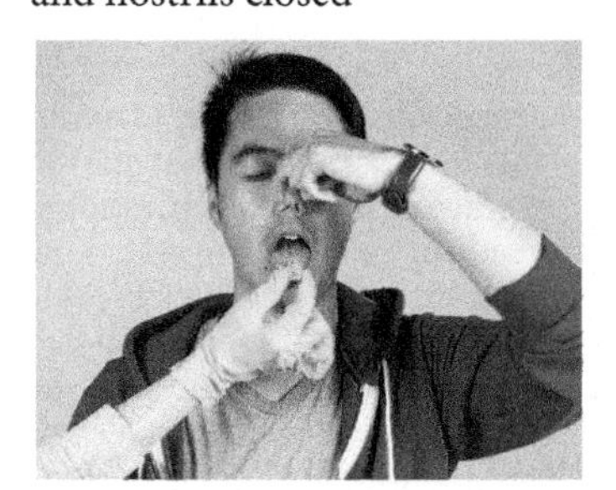

- After chewing the cube, with eyes and nostrils remaining closed

- After opening the nostrils, but with eyes still closed

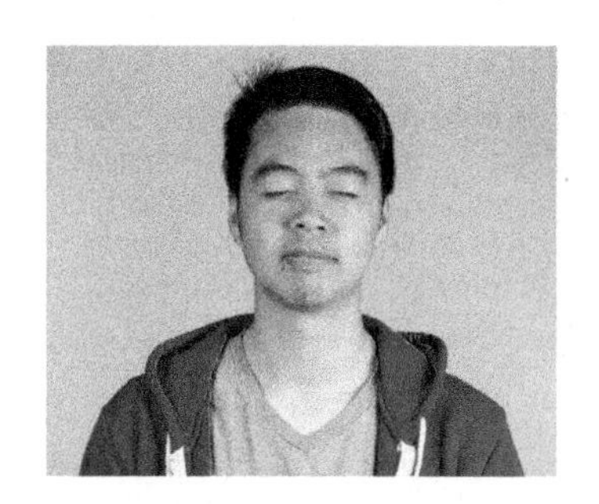

4 Record your results in Table 18.3 using the following protocol: ▶

- Place one plus (+) sign to indicate the condition under which the food was first tasted even if you cannot identify the type of food you are tasting.
- If the food was not tasted, place a minus (−) sign under the appropriate condition.
- If, after tasting the food, the taste was enhanced under subsequent conditions, place two or three plus signs in the appropriate space.

5 Repeat steps 2 through 4 for the other foods. Be sure that your lab partner rinses his or her mouth with water and dries the tongue with a clean paper towel between each taste test. Record the results in Table 18.3. ▶

TABLE 18.3 Identification of Food Items

	Conditions for Identification		
Food Item	On Tongue before Chewing; Nostrils and Eyes Closed	Chewing with Nostrils and Eyes Closed	Chewing with Nostrils Open and Eyes Closed

MAKING CONNECTIONS

Based on the data that you collected in this activity, does your sense of smell have any effect on your sense of taste? Explain.

▶ ____

____

____

____

____

Activity **18.5**

Vision: Examining the Anatomy of Accessory Eye Structures

A Accessory eye structures provide protection from airborne particles and dust, lubricate the surface of the eye, and allow light to reach the retina. Observe your lab partner's face or look at yourself in a mirror and identify the following accessory structures of the eye:

1. The **eyebrows** (not shown) are bands of thick hair that arch across the superior margins of the orbits and protect the eyeballs from perspiration and direct sunlight.

2. The **eyelids (palpebrae)** are fleshy coverings composed mostly of skeletal muscle and skin. Blinking the eyelids moistens the eye surface. Each eyelid contains modified sebaceous glands called tarsal (Meibomian) glands. The secretions from these glands prevent the eyelids from sticking together while blinking.

3. The **conjunctiva** is a thin, protective mucous membrane that lines the inner surface of the eyelids **(palpebral conjunctiva)** and curves onto the anterior surface of the eye **(ocular conjunctiva).** Look in a mirror and observe the shiny surface of your eyes. The shiny layer is the ocular conjunctiva.

4. The **palpebral fissure** is the space between the two eyelids. Through the fissure, you can see only the anterior one-sixth of the eyeball.

5. The two eyelids converge medially at the **medial canthus.**

6. The **lacrimal caruncle** is the small reddish mass of soft tissue at the medial canthus. Glands in this structure produce mucus that often accumulates while sleeping.

7. The two eyelids converge laterally at the **lateral canthus.**

8. **Eyelashes** are the thick hairs emerging from the margins of both eyelids that filter out particulate matter in the air.

IN THE CLINIC

Fracture of the Bony Orbit

A traumatic injury to the eye can cause a "blowout" fracture to the bony orbit. The bones most vulnerable to injury are the very thin ethmoid and lacrimal bones along the medial orbital wall and the maxilla on the inferior wall. Fractures to these bones may also damage adjacent paranasal sinuses, tear the extrinsic eye muscles, cause bleeding in the anterior chamber of the eye, and impair vision. Most blowout fractures will heal without surgery. Anti-inflammatory drugs can be administered to reduce swelling, and antibiotics are given to prevent infection. Surgery, if required, usually involves the insertion of an orbital implant.

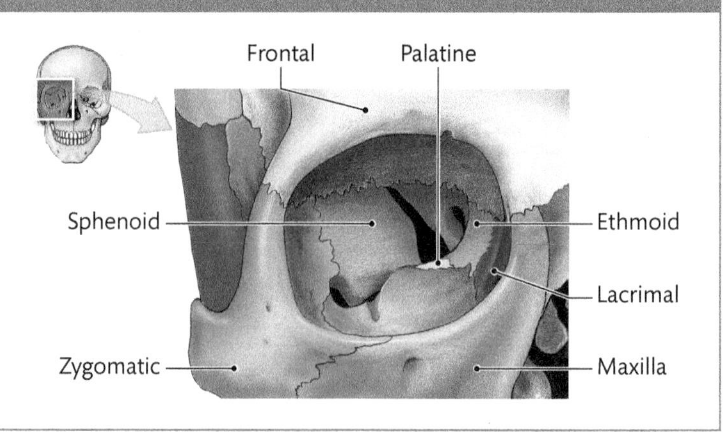

B The lacrimal gland and its accessory structures are known as the **lacrimal apparatus.** The lacrimal gland produces tears, consisting of a watery, alkaline fluid. After they are secreted, tears move across the surface of the eye and drain into the nasal cavity. Tears contain an antibacterial enzyme called **lysozyme.** Thus, in addition to cleaning and lubricating the surface of the eyeball, tears protect the eye from many bacterial infections.

1. On a skull, locate a small depression in the frontal bone, just inside the superolateral margin of the orbital cavity. This depression marks the location of the **lacrimal gland.**

2. On a model of the eye, locate the lacrimal gland and the **lacrimal ducts,** which transport tears to the surface of the eyeball, beneath the upper eyelid.

3. On your lab partner or yourself, once again locate the lacrimal caruncle at the medial canthus. When you blink, tears are swept across the surface of the eye to the lacrimal caruncle, where they drain into two small pores called **lacrimal puncta** (singular = **punctum**).

4. From the lacrimal puncta, tears flow through two small canals to reach the **lacrimal sac.** Identify this structure on a model.

5. On a skull, locate the lacrimal bone on the medial wall of the orbit. A small groove traveling along the bone, known as the lacrimal fossa, marks the location of the lacrimal sac.

6. On a skull, locate the inferior meatus, the space inferior to the inferior nasal concha in the nasal cavity. The **nasal lacrimal duct** drains tears from the lacrimal sac. The duct passes through a canal in the maxilla and empties into the inferior meatus.

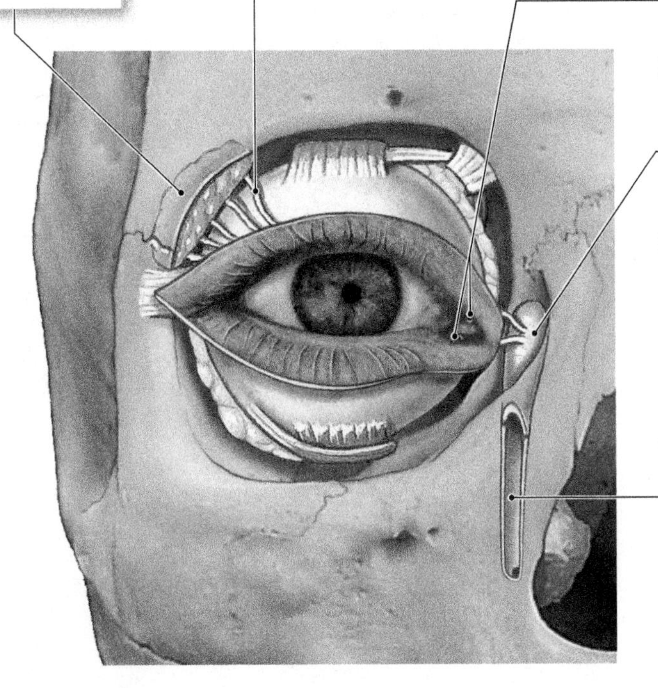

IN THE CLINIC

Conjunctivitis

Conjunctivitis—damage and swelling of the conjunctiva—can be caused by a bacterial or viral infection, allergic reaction, or chemical exposure. The surface of the eye becomes red due to dilated blood vessels, giving conjunctivitis its common name of **pinkeye.** Some forms of conjuctivitis are highly contagious. Treatment depends on the cause and could include antibiotics, allergy medications, or flushing the eye with water.

MAKING CONNECTIONS

Recent studies have linked the use of contact lenses to reduced lacrimal gland secretions. How could this condition affect normal eye function?

▶ ______________________________

____

____

____

____

Activity **18.6**

Vision: Gross Anatomy of the Eye

The **eyes** are spherical structures, approximately 2.5 cm (1 in.) in diameter, that function as the visual sense organs of the body. They are located within the **bony orbits** of the skull, where they are cushioned by a protective layer of fat. The eyes are supplied by the **optic nerves (cranial nerves II),** which travel through the optic canals in the sphenoid bone to reach the **optic chiasm.** At the optic chiasm, some fibers from each nerve cross over to the opposite side.

A Each eye is composed of three layers (tunics) of tissue: the outer fibrous tunic, the middle vascular tunic (uvea), and the inner neural tunic or retina.

1 On a model of the eye, identify the fibrous tunic. Notice that the fibrous tunic has two parts:

The vast majority is a tough, opaque membrane known as the **sclera,** or white of the eye. Consisting of tightly bound elastic and collagen fibers, the sclera protects and supports the eyeball.

The smaller anterior portion of the fibrous tunic is the **cornea.** Unlike the sclera, the cornea is delicate, lacks blood vessels, and is transparent to allow light to enter the eyeball.

2 Remove a portion of the fibrous tunic to reveal the vascular tunic. Observe that this middle layer is composed mostly of the **choroid,** a thin layer of tissue that contains numerous small blood vessels. The choroid is densely populated with melanocytes. The melanin produced by these cells prevents incoming light waves from being reflected back out of the eye.

3 Identify the two anterior extensions of the choroid:

The **iris** is composed of pigmented cells that are responsible for eye color and two layers of smooth muscle that surround a central opening known as the pupil. Contraction of smooth muscle in the iris dilates or constricts the pupil, regulating the amount of light entering the eye.

The **ciliary body** is a ring of tissue, composed mostly of smooth muscle, that surrounds the lens. Suspensory ligaments connect the ciliary body to the lens. The ciliary body and the suspensory ligaments change the shape of the lens while focusing on near or distant objects.

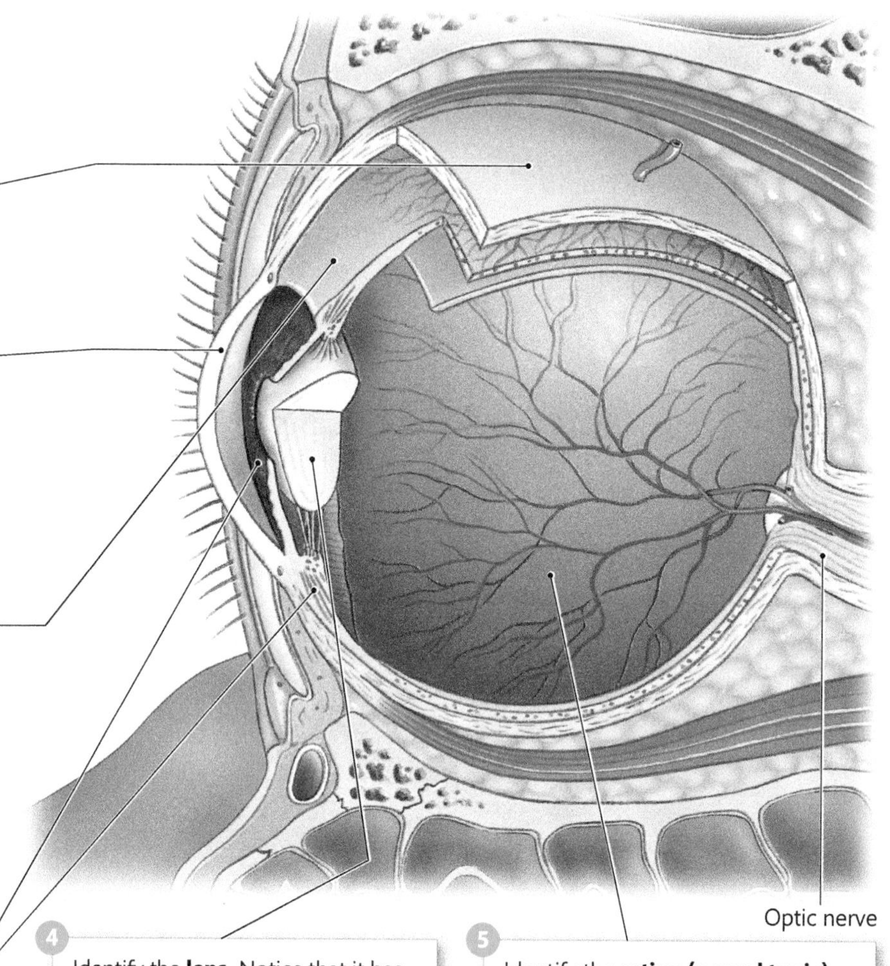

4 Identify the **lens.** Notice that it has a biconvex shape and, like the cornea, is transparent. The lens is composed of several tightly packed layers of fibrous proteins, arranged like the layers of an onion. It refracts (bends) incoming light waves and focuses them onto the retina.

5 Identify the **retina (neural tunic).** The retina contains three layers of nerve cells: the photoreceptor cells (rods and cones), bipolar cells, and ganglion cells. Axons of the ganglion cells form the optic nerve.

■ Want more practice? Go to: **MasteringA&P** > Study Area > Menu > Lab Tools > PAL > Anatomical Models > Nervous System > Special Senses > Eye photos

B The macula lutea and the optic disc are highly specialized areas on the retina.

1. Identify the **macula lutea** along the posterior surface of the retina. In the center of the macula lutea is a small region known as the **fovea centralis.** These two structures contain the highest concentrations of cone cells. Thus, our vision is sharpest when we focus images on these retinal regions. The fovea centralis is unique because it contains only cones and is the area of keenest vision.

2. Identify the **optic disc** medial to the macula lutea. This location marks where the optic nerve exits the eyeball. The optic disc is also called the **blind spot** because it lacks photoreceptor cells. Therefore, light that is focused on this region cannot be seen.

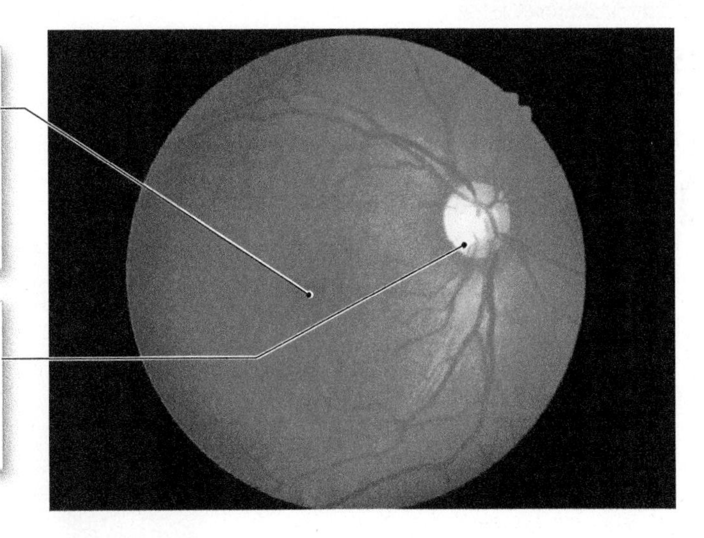

C The interior of the eyeball is divided into the anterior cavity and the posterior cavity.

1. On a model of the eye, identify the **anterior cavity,** which extends from the cornea, anteriorly, to the lens, posteriorly. It is filled with a watery fluid called **aqueous humor.** Notice that the anterior cavity is subdivided into two smaller spaces:

   The **anterior chamber** is the space between the cornea and iris.

   The **posterior chamber** is the space between the iris and lens.

2. Identify the larger **posterior cavity (vitreous chamber),** which extends from the lens, anteriorly, to the retina, posteriorly. It is filled with a jelly-like substance called the **vitreous humor,** which helps maintain normal intraocular pressure and holds the retina firmly against the choroid.

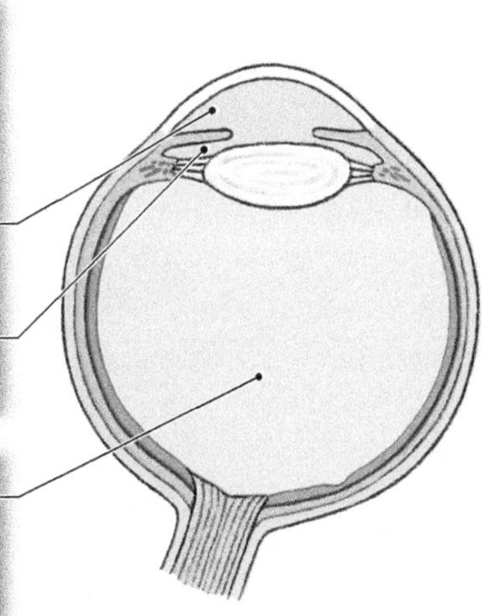

IN THE CLINIC

Glaucoma

In the anterior cavity, aqueous humor is produced by the ciliary body in the posterior chamber and drained by veins in the wall of the anterior chamber. If normal drainage is blocked, intraocular pressure can increase and damage the retina and optic nerve. This condition, called **glaucoma,** can be controlled in its early stages by administering eyedrops that enhance the rate of fluid drainage.

MAKING CONNECTIONS

Macular degeneration is the progressive deterioration of the macula lutea. It usually occurs in adults over 60 years of age. How do you think this condition will affect a person's vision?

▶ ____

____

____

____

Activity **18.7**

Vision: Dissection of the Cow Eye

The cow eye is stucturally similar to the human eye and, therefore, is a good model for anatomical study.

A External Anatomy of the Cow Eye

1. Place a preserved or fresh cow eye on a dissecting tray. Notice the thick layer of fat on the external surface of the eye. Carefully remove the fat, without damaging other structures. What function do you think this fat serves?

   ▶ ______________________

   ____

   ____

2. Identify the **sclera,** or white of the eye, and notice that it makes up most of the external fibrous tunic.

3. Identify the **optic nerve (cranial nerve II)** as it exits from the posterior aspect of the eyeball. It will appear as a solid white cord.

4. Locate the **cornea,** which is the anterior portion of the fibrous tunic. Normally, the cornea is transparent, but if your cow eye is preserved, it will be opaque.

5. On the sclera, locate the attachments for the **extrinsic muscles** of the eye.

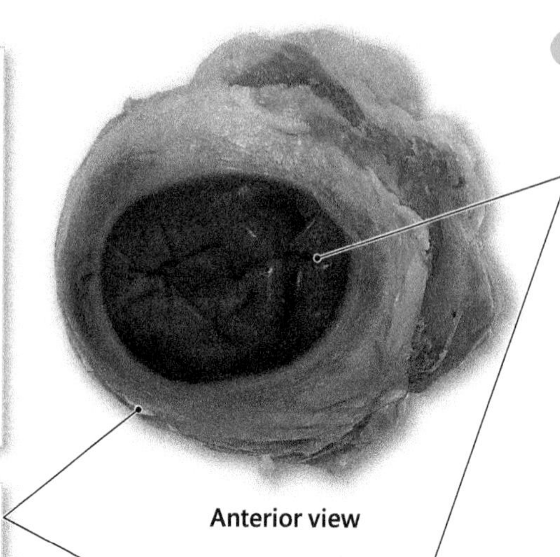

Anterior view

Anterolateral view

IN THE CLINIC

Cataract

A **cataract** occurs when the lens thickens and becomes less flexible than normal. As a result, the normally transparent lens becomes clouded, and vision is blurred. Cataracts can be caused by exposure to chemicals or ultraviolet radiation, trauma, or aging. They can be corrected by removing the damaged lens and replacing it with an artificial implant.

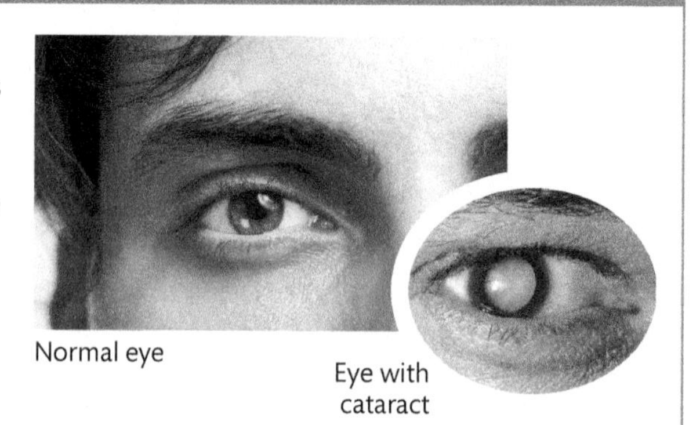

Normal eye

Eye with cataract

B Internal Anatomy of the Cow Eye

Hold the posterior portion of the eyeball securely on the dissecting tray. With a scalpel, make an incision about 6 mm (0.25 in.) posterior to the cornea. The sclera is a relatively thick, fibrous layer, so be sure to apply enough pressure to cut through the wall.

Warning: *Wear safety glasses to protect your eyes. Aqueous humor from the anterior cavity could squirt out during this procedure.*

2. Insert scissors into the initial incision and make a circular cut around the cornea, always remaining about 6 mm (0.25 in.) posterior to the corneal margin.

3. Identify the following structures in the anterior portion of the eyeball:

The centrally located **lens** is a biconvex disc. Like the cornea, it is normally transparent, but will be opaque if your cow eye is preserved.

The **iris** is heavily pigmented and will probably appear very dark or black. The opening through the center of the iris is the pupil.

The **ciliary body** appears as a thick black ring around the lens.

Gently lift one side of the lens with a blunt probe and identify the delicate **suspensory ligaments** (not shown). Notice that the suspensory ligaments connect the lens to the ciliary body.

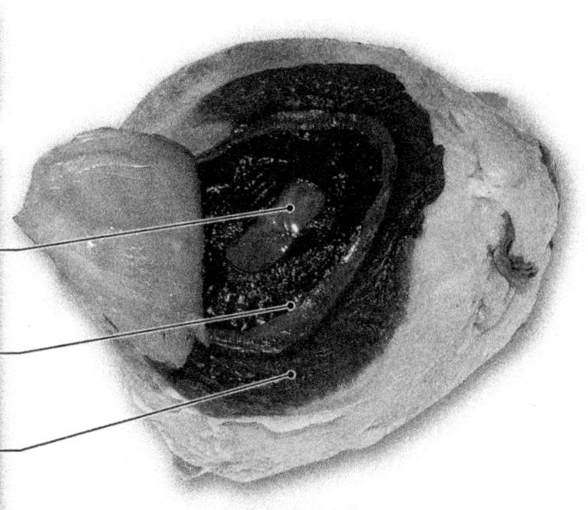

Anterior view

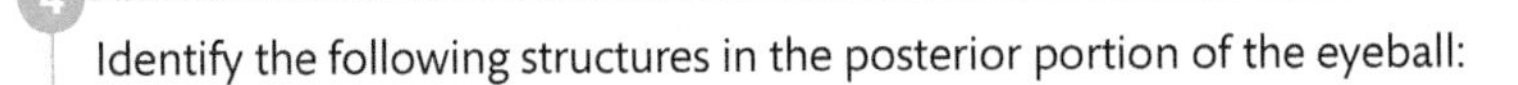

4. Identify the following structures in the posterior portion of the eyeball:

The **vitreous humor** (not shown) is the thick, gel-like substance that fills the posterior cavity.

Remove the vitreous humor and identify the **retina,** which appears as a whitish or yellowish layer.

Identify the **optic disc (blind spot).** Notice that the optic disc is the location where the optic nerve exits the eye.

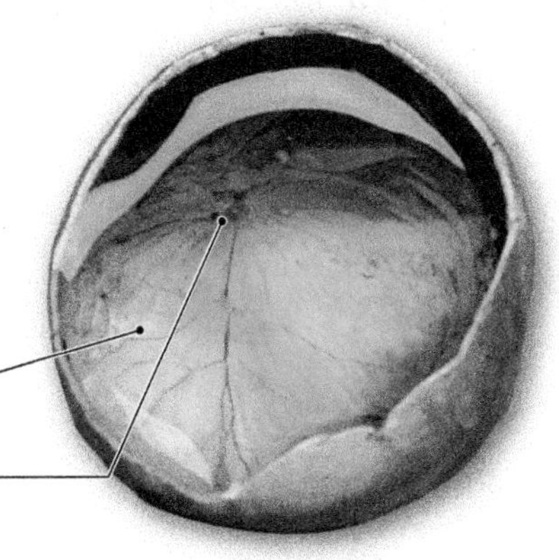

Anterior view

5. Remove a portion of the retina to expose the deeply pigmented **choroid.** The choroid appears iridescent due to the presence of a membrane called the **tapetum lucidum.** This unique structure reflects some light back onto the retina, thus improving vision at night or in dim light. The tapetum lucidum is not found in the human eye but is found in many nocturnal vertebrates such as raccoons, opossums, skunks, rabbits, and domestic dogs and cats.

MAKING CONNECTIONS

Compare the anatomy of the human eye to the cow eye and identify any differences in their structures.

▶ ______________________________

____

____

Activity **18.8**

Vision: Examining the Retina

A The sensory receptors for vision are located in the retina, which consists of an outer pigmented layer and an inner nervous layer. In the pigmented layer, **melanocytes** produce the pigment melanin, which prevents reflection or scattering of light waves when they reach the retina.

The nervous layer contains three types of neurons, each in their own cellular layer.

- Photoreceptor cells form the outermost layer. The two types of photoreceptor cells, rods and cones, are sensitive to light stimuli. Rods are important for our ability to see during low-light conditions and for peripheral vision. Cones are important for visual acuity (viewing objects in sharp focus) and color vision.
- Bipolar cells are located in the middle layer and synapse with the photoreceptor cells.
- Ganglion cells, the innermost layer, synapse with the bipolar cells. The axons of the ganglion cells form the **optic nerve**.

1 View a microscope slide of the retina with the low-power objective lens. If possible, adjust your slide so that your field of view is similar to the figure below. Thus, the choroid and sclera will be at the top of your field of view.

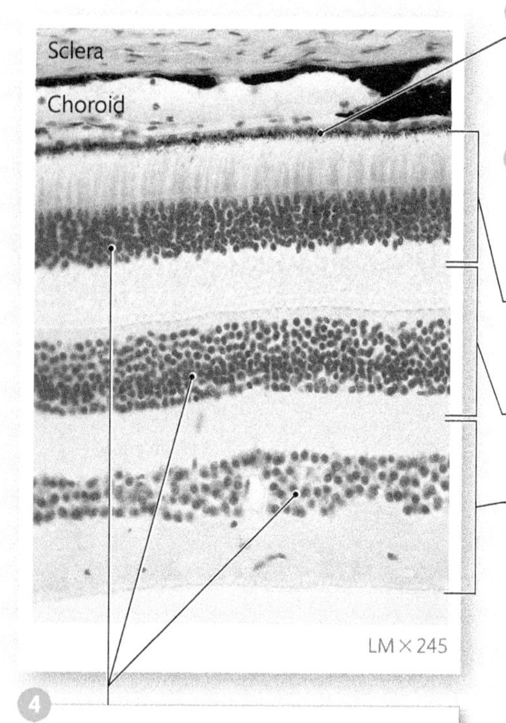

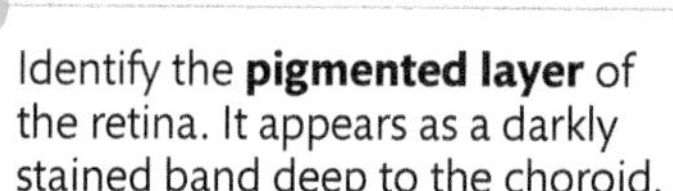

2 Identify the **pigmented layer** of the retina. It appears as a darkly stained band deep to the choroid.

3 Locate the **nervous layer** of the retina, which contains the three bands of nerve cells:

The **photoreceptor cell layer,** consisting of the rods and cones, is just deep to the pigmented layer.

The middle layer contains **bipolar cells.**

The third layer is a band of **ganglion cells.**

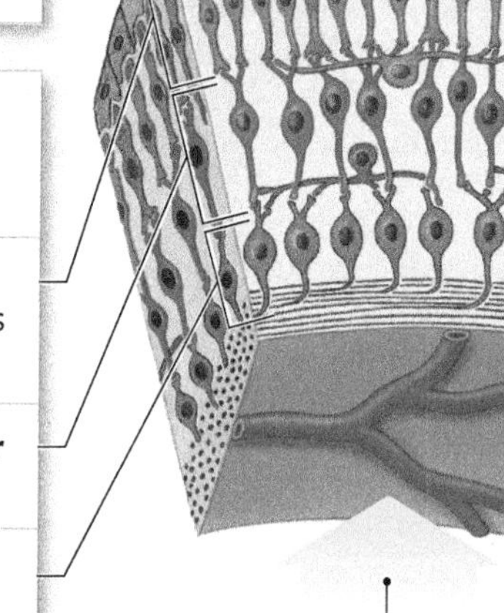

4 You can identify the three cell types in the nervous layer by locating their nuclei, which are organized into three distinct bands.

5 Realize that light first contacts the retina at the ganglion cell layer. Light must pass through these cells and the bipolar cells to reach and stimulate the **rods and cones** in the photorecepetor cell layer.

Want more practice? Go to: **MasteringA&P** > Study Area > Menu > Lab Tools > PAL > Histology > Special Senses > Retina photos

B The **optic disc** is also called the **blind spot** because light that is focused on this region cannot be seen. You can demonstrate the blind spot by performing the following activity.

1. Hold the figure on the right about 46 cm (18 in.) from your eyes.
2. Close your left eye and focus your right eye on the plus (+) sign.
3. Move the figure slowly toward your face until the black dot disappears.
4. Explain why the dot becomes invisible.

▶ ______________________________

____

____

____

____

____

____

+ ●

IN THE CLINIC

Detached Retina

A detached retina is a separation between the pigmented and nervous layers of the retina. It can be caused by a blow to the head or progressive degeneration of the retina due to disease or old age. Fluid can accumulate between the separated layers. If this occurs, the nervous layer bulges forward, causing distorted or loss of vision in the affected visual field. A detached retina is an emergency, but can be repaired with prompt laser surgery or cryosurgery (applying extreme cold to the damaged area).

MAKING CONNECTIONS

In the human eye, the distribution of rods and cones varies. The concentration of rods increases as you move from the center to the periphery of the retina. On the other hand, the concentration of cones increases as you move from the periphery to the center where the macula lutea and fovea centralis are located. If you compare the functions of rods and cones, how do you account for the different distribution patterns of these cells?

▶ ______________________________

____

____

____

____

____

____

Activity 18.9

Performing Visual Tests: Binocular Vision and Depth Perception

At the **optic chiasm,** some fibers from each optic nerve cross over to the other side. Thus, the visual area in each cerebral hemisphere receives and interprets visual information from both eyes. Alone, each eye has a different view, but together, the left and right visual fields overlap, providing humans (and other mammals with eyes toward the front of the face) with **binocular vision.** The visual cortex translates the information from overlapping visual fields into **three-dimensional vision** and the ability to locate objects in space **(depth perception).**

A Demonstrating Binocular Vision

1. Hold a pencil vertically, at arm's length, in front of your face. Focus on the pencil with both eyes. Make sure that there is a sharp contrast between the pencil and the background of your visual field so that you can see the pencil clearly.
2. Close your right eye completely but continue to look at the pencil with your left eye. Then, close your left eye and open your right, always focusing on the pencil.
3. Repeat step 2 several times. Do you notice any change in the relative position of the pencil when you switch back and forth between the visual fields of your left and right eyes? Provide an explanation for your results.

▶ ____

____

____

____

____

____

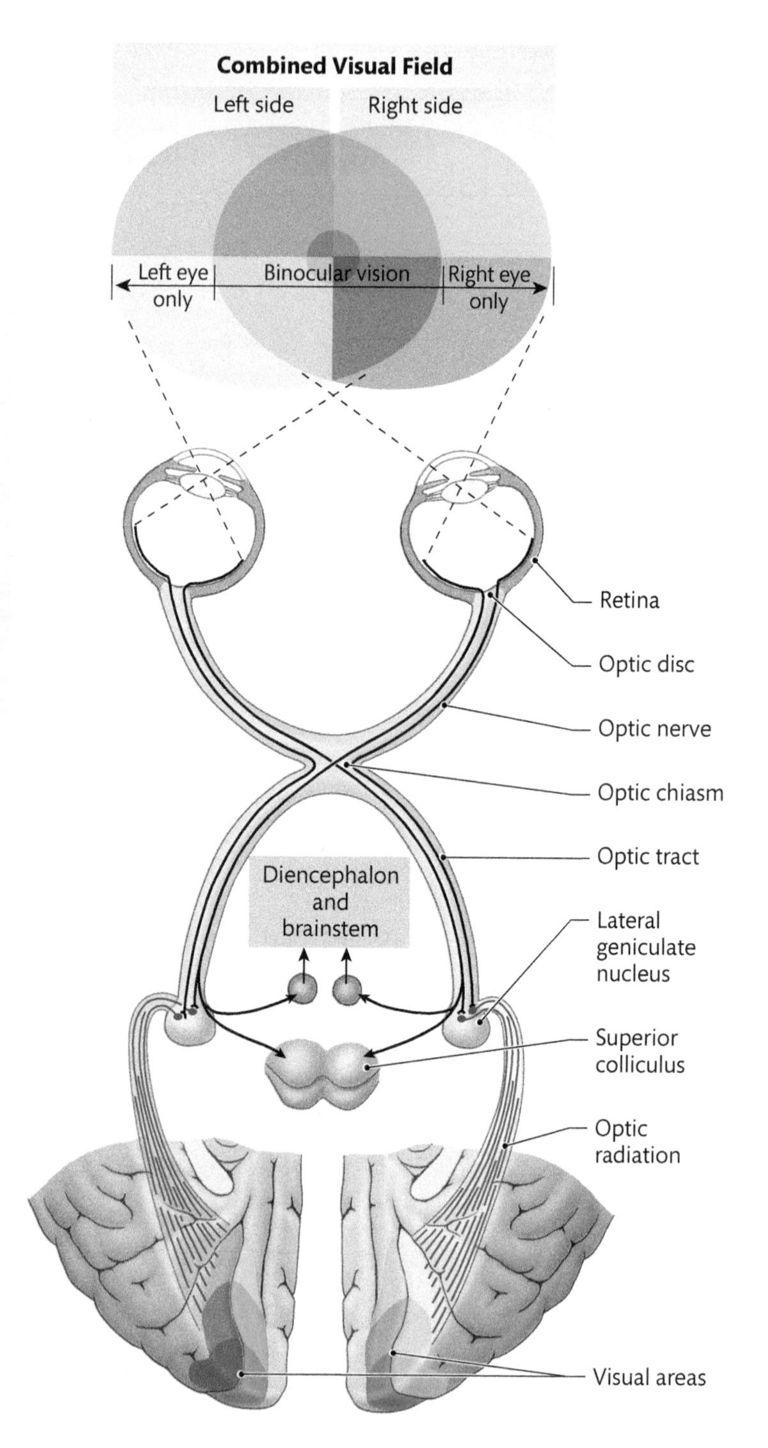

B Demonstrating Depth Perception

1 Hold a pencil vertically, at arm's length, in front of your face. If you are left-handed, hold the pencil with your right hand; if you are right-handed, hold the pencil with your left hand.

2 With both eyes open, quickly place the tip of the index finger of your free hand on the top of the pencil. Repeat this action two or three times.

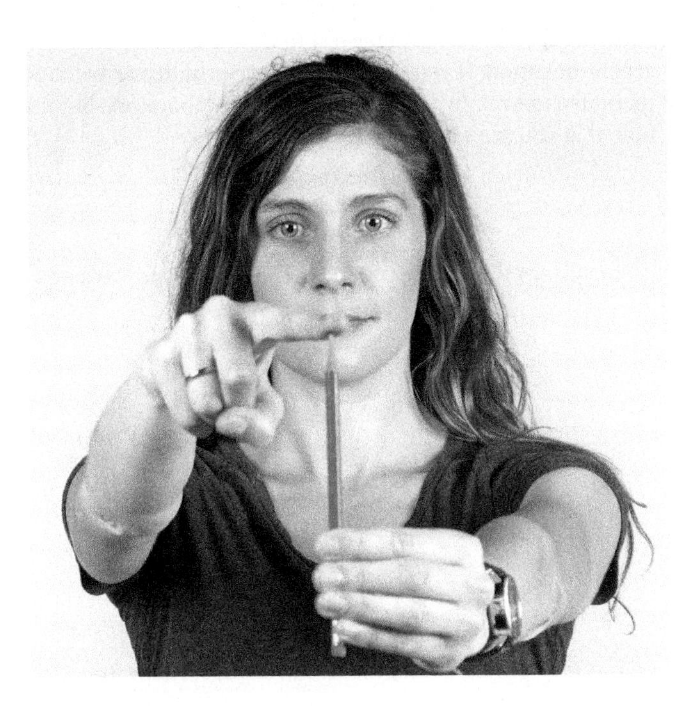

3 Close your left eye and again place the tip of your index finger on the top of the pencil. As before, perform this action quickly, not slowly and thoughtfully. Repeat this action two or three times.

4 Repeat step 3 with your right eye closed and left eye open.

5 Compare the results for each test by answering the following questions: ▶

1. Was it more difficult or less difficult to place the tip of your finger on the pencil when only one eye was open compared to when both eyes were open?

____

____

2. How accurate were you in placing the tip of your finger on the pencil when both eyes were open compared to when only one eye was open?

____

____

____

3. Did you notice any difference in your ability to accurately perform this task with only the right eye open compared to only the left eye open?

____

____

____

MAKING CONNECTIONS

Review your results for the demonstration of depth perception. Explain any differences in the degree of difficulty or accuracy that you noted. If you did not notice any differences, explain that result.

▶ ______________________________

____

____

____

____

____

Activity **18.10**

Performing Visual Tests: Near-Point Accommodation

A The lens is surrounded by the ciliary body, a ring of tissue composed mostly of smooth muscle. Suspensory ligaments connect the ciliary body to the lens. Relaxation and contraction of the smooth muscle in the ciliary body change the shape of the lens to focus on distant or near objects. For example, when focusing on distant objects, the muscle is relaxed, and the lens is flat; when focusing on near objects, the muscle contracts, and the lens is more spherical. Thus, the lens exhibits some degree of elasticity. As a person ages, the lens becomes less elastic than before, and focusing on close objects becomes more difficult. This condition is called **presbyopia.** We can test the elasticity of the lens by measuring the **near-point accommodation,** the distance from the eyes at which an object begins to be blurred or distorted. Perform the following simple test to measure your near-point accommodation. If you wear glasses, perform this test without them. Do not remove contact lenses in the laboratory, but make a note that you are wearing them.

1. **Make a Prediction**

   Before you begin, refer to Table 18.4 and predict the near-point accommodation for you and your lab partner. Record your predictions in Table 18.5. ▶

2. Hold a dissecting pin at arm's length in front of your eyes. Close your right eye and slowly move the pin toward your face until the image just begins to become blurred or distorted.

3. Hold the pin at this position and have your lab partner measure the distance, in centimeters, between the pin and your left eye. Repeat the procedure to confirm your results and record the data in Table 18.5. ▶

4. Repeat the procedure for the other eye. Record the data in Table 18.5. ▶

For Distant Vision: Ciliary Muscle Relaxed, Lens Flattened

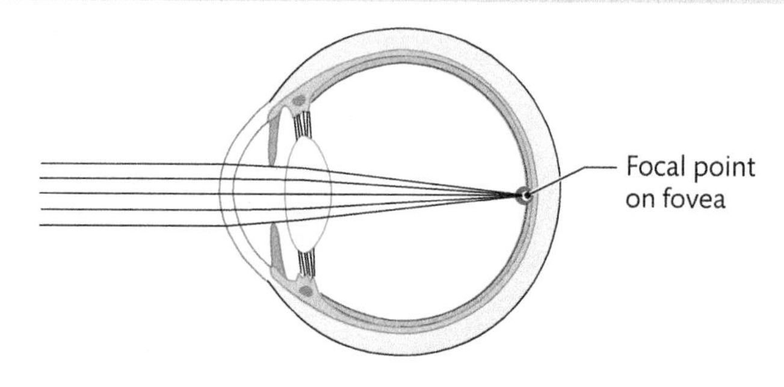

For Close Vision: Ciliary Muscle Contracted, Lens Rounded

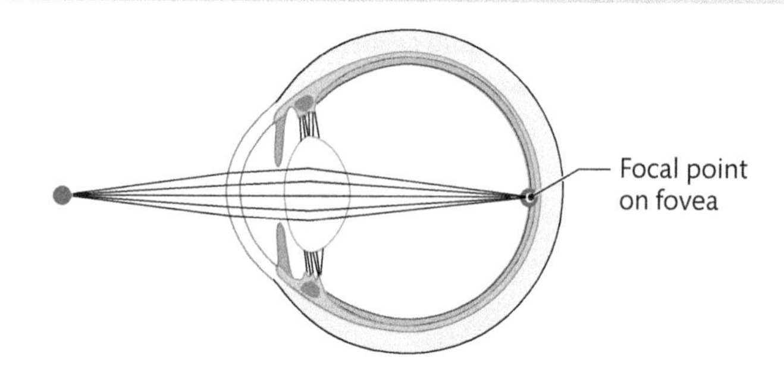

TABLE 18.4 **Near-Point Accommodation Distances at Various Ages**

Age (yr)	Accommodation Distance (cm)
10	7.5
20	9.0
30	11.5
40	17.2
50	52.5
60	83.3

TABLE 18.5 Measuring Near-Point Accommodation

Subject	Age	Near-Point Accommodation (cm)		
		Predicted Value	Actual Value, Left Eye	Actual Value, Right Eye

5 Assess the Outcome

Compare your results with those of your lab partner and one or two other students in the laboratory and consider the following: ▶

- For each subject, was the actual near-point accommodation about the same for both eyes? Explain.

- For each subject, compare the actual near-point accommodation (for both eyes) with the predicted value and decide whether it is normal for your age.

Word Origins

Presbyopia derives from two Greek words: *presbys* ("old man") and *ops* ("eye"). Thus, presbyopia means "eyes of an old man" or "old vision."

MAKING CONNECTIONS

Explain why many people, as they age, require reading glasses.

▶

Activity **18.11**

Performing Visual Tests: Visual Acuity

A **Visual acuity** refers to an individual's ability to see objects clearly at various distances (sharpness of vision). It is generally tested by using a **Snellen chart.** The chart contains rows of letters printed in progressively smaller sizes on a white background. Individuals with normal vision should be able to clearly see the letters of a given size from a specific distance. That distance is printed at the end of each line. You can test your own visual acuity by performing the following test. If you wear glasses, perform the test without the glasses and then repeat it with your glasses. Do not remove contact lenses in the laboratory, but make a note that you are wearing them.

1. Attach a Snellen chart to a wall in the laboratory.
2. Stand 20 ft from the chart. Cover your left eye and read each row of letters from top to bottom until you can no longer read them accurately. Have your lab partner check your accuracy.
3. Check the last line that you read accurately. If "20/20" is printed at the end of the line, you have normal vision **(emmetropia).** If your vision is normal, the objects you are viewing are focused on your retina. Record your results in Table 18.6. ▶

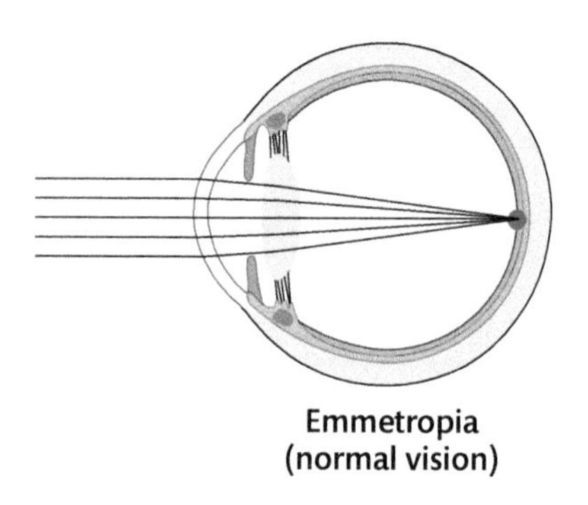

Emmetropia (normal vision)

TABLE 18.6 Testing Visual Acuity Using a Snellen Chart

	Visual Acuity	
	Without Glasses	With Glasses
Right eye		
Left eye		
Both eyes		

4 If the ratio of the last line you read is less than one, you have difficulty seeing distant objects; you are **nearsighted** or **myopic.** A ratio of 20/40, for example, indicates that objects you can clearly see at 20 ft are clearly seen by a person with normal vision at 40 ft. If you are nearsighted, the objects you are viewing are focused in front of your retina. Prescription eyewear that spreads light waves apart before they reach the lens will refocus the image on the retina. How does the shape of the eyeball differ in myopia compared to emmetropia?

▶ ______________________________

____

____

____

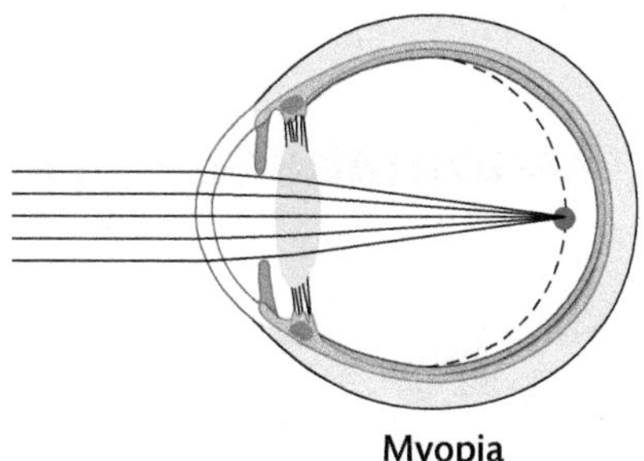
Myopia (nearsighted)

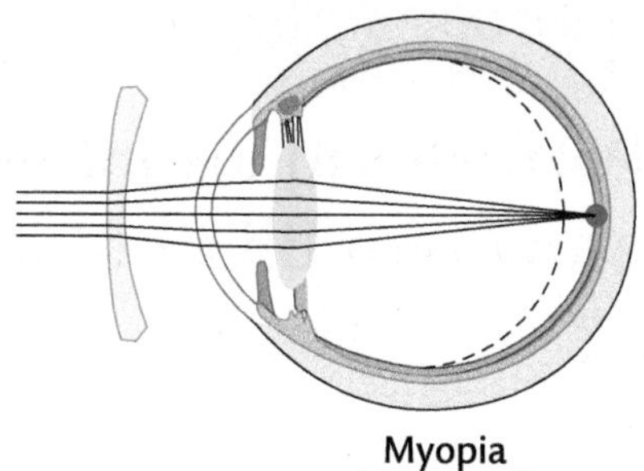
Myopia (corrected)

5 A ratio greater than one suggests that your vision is better than normal. What does it mean if your vision is 20/15?

▶ ______________________________

____

____

____

6 Repeat the test for the other eye and then with both eyes. Record your results in Table 18.6. ▶

IN THE CLINIC

Lasik Eye Surgery

Lasik eye surgery uses a laser to reshape the curvature of the cornea. The technique is designed to correct the refraction of light by the cornea so that the lens can properly focus an image onto the retina. Lasik surgery is becoming an increasingly acceptable alternative to wearing glasses or contact lenses, although it can have complications.

MAKING CONNECTIONS

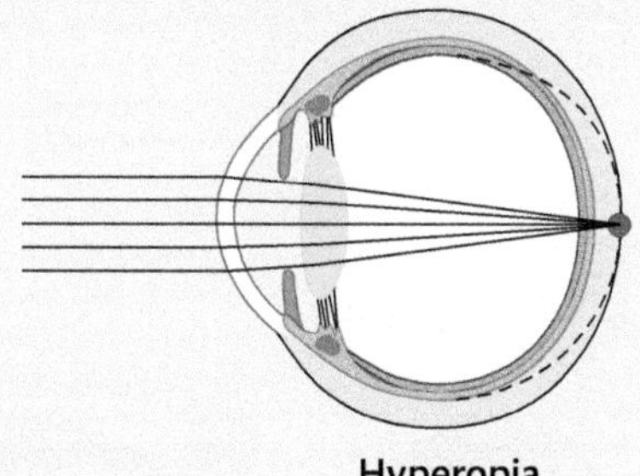
Hyperopia (farsighted)

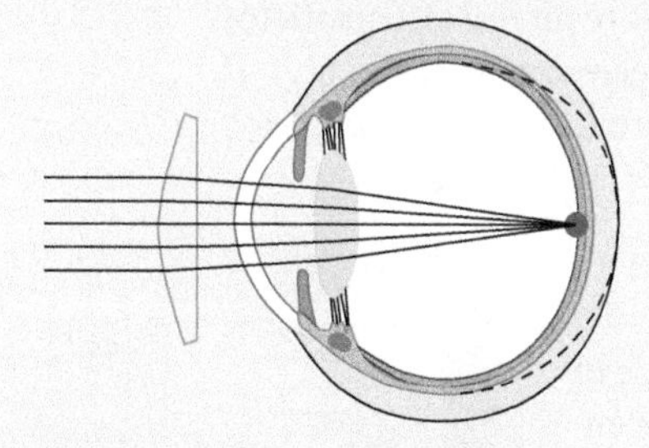
Hyperopia (corrected)

The figure above illustrates hyperopia, in which the lens cannot focus close objects on the retina. People with this condition are said to be farsighted because they are able to clearly see distant objects. Make the following comparisons between hyperopia and myopia: ▶

	Hyperopia	Myopia
Shape of the eyeball		
Where objects are focused in relation to the retina		
How prescription eyewear corrects the problem		

Activity **18.12**

Performing Visual Tests: Astigmatism, Pupillary Reflex, and Color Blindness

A If there are defects in the curvature of the lens or cornea, light entering the eye will not be focused at a single point on the retina. This condition, known as **astigmatism,** can cause blurred vision. You can test for astigmatism in your own eyes by completing the following test. If you wear glasses, perform the test without them and then repeat the test with them. Do not remove contact lenses in the laboratory, but make a note that you are wearing them.

1. With your left eye closed, view the astigmatism chart on the right. Focus on the center of the chart.
2. If all the radiating lines are in sharp focus, there is no astigmatism. If some of the lines are blurred or appear lighter than others, some astigmatism is apparent. Record your result in Table 18.7 with a (+) sign if you have astigmatism or a (–) sign if there is no apparent astigmatism. ▶
3. Repeat the test with your left eye and record the result in Table 18.7. ▶

TABLE 18.7 Astigmatism Test

	Astigmatism	
	Without Glasses (+/–)	With Glasses (+/–)
Right eye		
Left eye		

B The **pupillary reflex** refers to the reflex contraction of constrictor and dilator muscles in the iris that control the diameter of the pupil. When light intensity increases, constrictor muscles contract, and pupil diameter decreases. When light intensity decreases, dilator muscles contract to increase pupil diameter.

1. Have your lab partner close his or her left eye and hold a piece of cardboard or index card on the side of the nose to serve as a barrier between the eyes.
2. Shine a flashlight, at an angle (not directly), into the open right eye for 5 to 10 seconds at a distance of 20 cm (about 8 in.). In Table 18.8, describe any change that occurs in the size of the right pupil. ▶

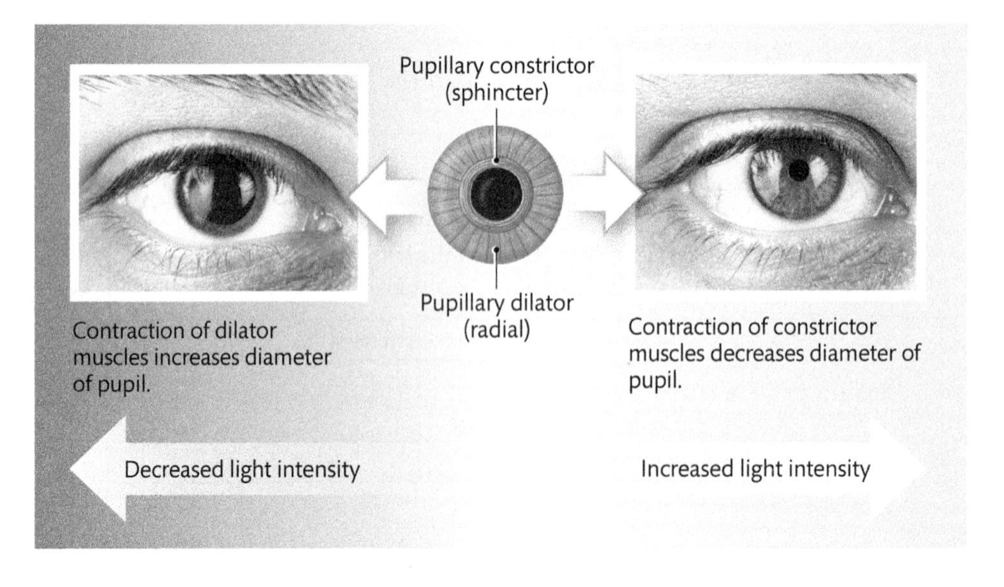

3 Wait 5 minutes and repeat step 2 with the left eye open and the right eye protected from the light. Record your result in Table 18.8. ▶

4 Wait 5 minutes and shine a flashlight, at an angle, on the face with both eyes open. Record your result in Table 18.8. ▶

5 After the pupils of both eyes have returned to normal size, have your lab partner move into a dark room (without windows, if possible) for 5 minutes.

6 When your lab partner emerges from the dark room, immediately observe the condition of his or her pupils. Describe any change that you observe.

▶ ______________________________________

____

____

TABLE 18.8 Pupillary Reflex

Test	Result after Increasing Light Intensity
Right eye open, left eye closed	
Left eye open, right eye closed	
Both eyes open	

C In the retina, three types of cones absorb light at specific wavelengths: Red cones absorb red light, blue cones absorb blue light, and green cones absorb green light. The lack of one or more types of cones, or having abnormal cone pigments, can cause **color blindness.**

1 Have your lab partner hold **Ishihara color plates** about 80 cm (30 in.) in front of you. Be sure that you are viewing them in a brightly lit room.

2 Within 2 or 3 seconds, report what you see in the plates. Have your lab partner record your responses and compare them with those in the plate book. Inability to correctly identify the figures in the plates may indicate some degree of color blindness.

3 Repeat this procedure to test for color blindness in your lab partner.

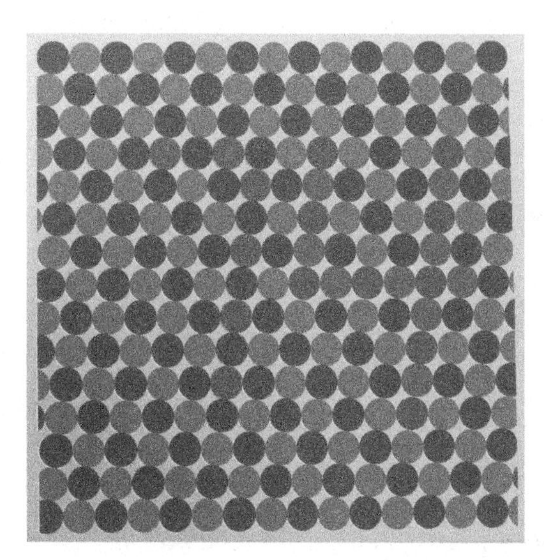

MAKING CONNECTIONS

How would the diameter of the pupil change during a "fight-or-flight" (sympathetic) response? Explain why.

▶ ______________________________________

____

____

____

____

Activity **18.13**

Hearing: Examining the Gross Anatomy of the Ear

A The ear is a sensory organ for **hearing** and **equilibrium** (balance). It is divided into three regions: the **external ear**, **middle ear (tympanic cavity),** and **inner ear.** The inner ear consists of two compartments: the **bony labyrinth** and the **membranous labyrinth.** The bony labyrinth is a maze of bony passageways that course through the temporal bone. The membranous labyrinth is a network of membranous capsules contained within the bony labyrinth. Both compartments are filled with fluid; the bony labyrinth contains **perilymph** and the membranous labyrinth contains **endolymph.** Thus, the inner ear is a "tube within a tube" with the inner membranous labyrinth floating in the fluid contained in the bony labyrinth. The fluids in the inner ear provide a medium for conducting sound waves (sense of hearing) and detecting changes in body position and movement (sense of equilibrium).

1 On a skull, look inside the cranial cavity and identify the **petrous portions of the temporal bones.** The middle and inner ears are located inside chambers formed by these bony regions.

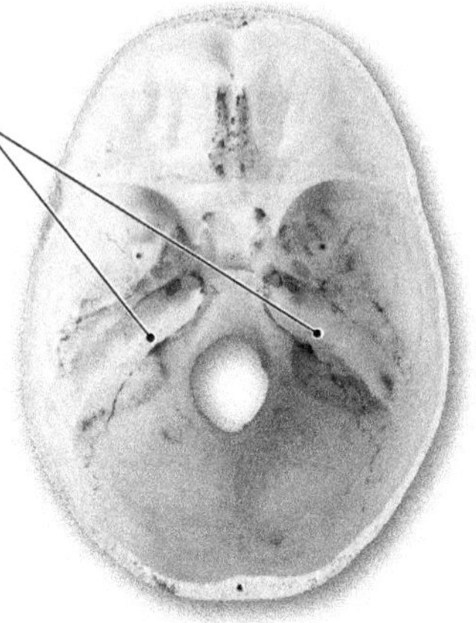

2 On a model of the ear, identify its three divisions:

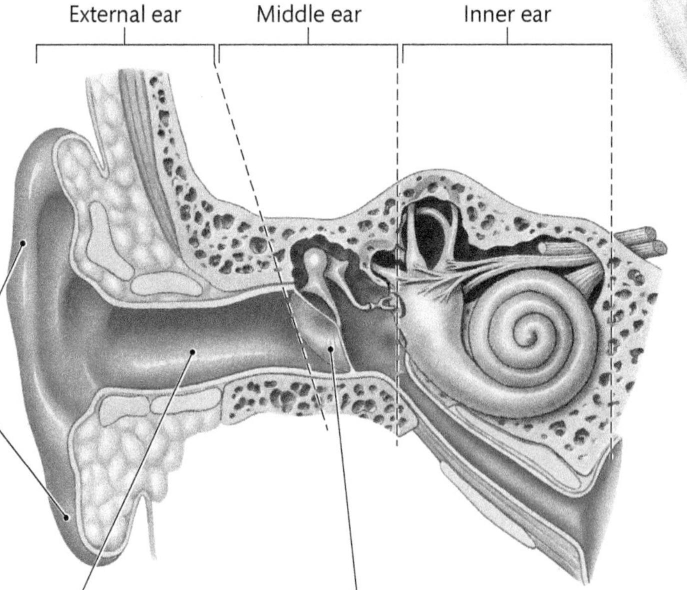

B The external ear includes the auricle (pinna) and external acoustic meatus. The tympanic membrane (eardrum) is positioned at the end of the external acoustic meatus.

1 On a model, identify the **auricle.** On yourself or a lab partner, verify that the auricle consists of:

- a cartilagenous rim known as the **helix**
- an inferior fleshy **lobule** or **earlobe**

What do you think is the function of the auricle?

▶ ______________________

____

2 Observe that the **external acoustic meatus** travels through the temporal bone, connecting the auricle to the tympanic membrane. The external acoustic meatus is lined by skin containing hair follicles, sebaceous glands, and ceruminous (wax) glands. The wax glands are modified apocrine sweat glands that produce cerumen, a waxy substance that keeps the tympanic membrane soft and waterproof and traps foreign particles that enter the external acoustic meatus.

3 Identify the **tympanic membrane.** It is a thin, translucent membrane of fibrous connective tissue that forms a partition between the external and middle ears. Sound waves traveling through the external acoustic meatus cause the tympanic membrane to vibrate. This action transfers the sound waves to the bony ossicles in the middle ear.

If ceruminous glands secrete too much cerumen, will the function of the tympanic membrane be affected? Explain.

▶ ______________________

____

____

____

Want more practice? Go to: **MasteringA&P** > Study Area > Menu > Lab Tools > PAL >
- Anatomical Models > Nervous System > Special Senses > Ear photos
- Human Cadaver > Nervous System > Special Senses > Ear photos

C The middle ear (tympanic cavity) is an air-filled chamber lined by a mucous membrane. It contains the three auditory ossicles.

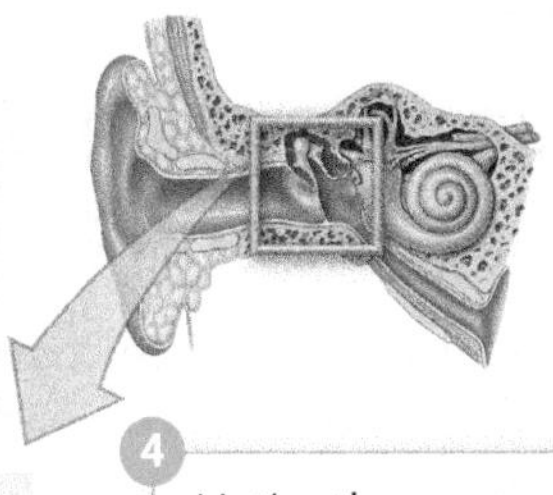

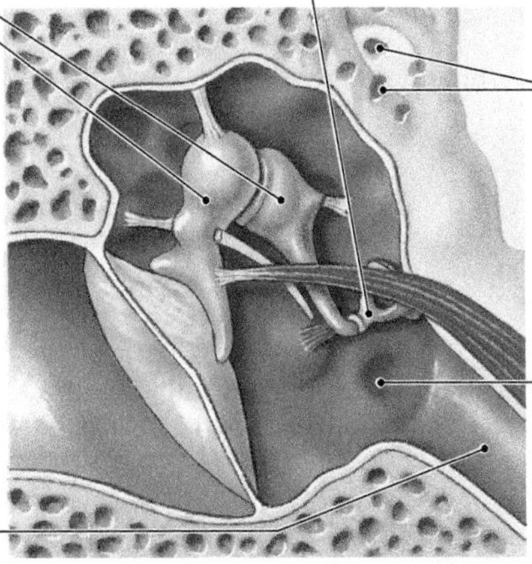

1. On a model, locate the **middle ear** and identify the three **auditory ossicles:**
   - **Malleus** (attached to tympanic membrane)
   - **Incus**
   - **Stapes** (attached to oval window)

2. Notice that these small bones connect the tympanic membrane to the **oval window** on the wall of the inner ear. The auditory ossicles transmit and amplify sound waves across the tympanic cavity to the oval window.

3. Identify the **pharyngotympanic (auditory** or **eustachian) tube** along the anterior wall of the tympanic cavity. Notice that this tube provides a direct link between the middle ear and the nasopharynx (superior region of the throat).

4. Notice the numerous cavities throughout the petrous portion of the temporal bone. These spaces are the **mastoid air cells.** They are directly connected to the posterior wall of the tympanic cavity by several tubular connections.

5. Locate the **round window** near the entrance to the auditory tube at the base of the cochlea.

D The inner ear, or **labyrinth,** contains the vestibule, semicircular canals, and cochlea.

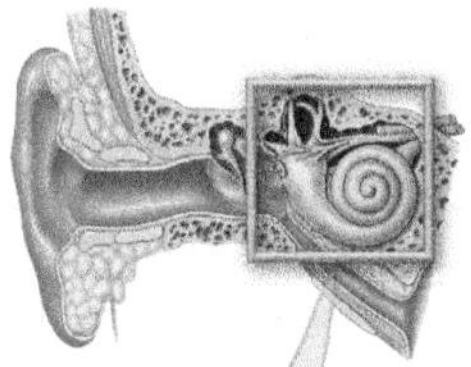

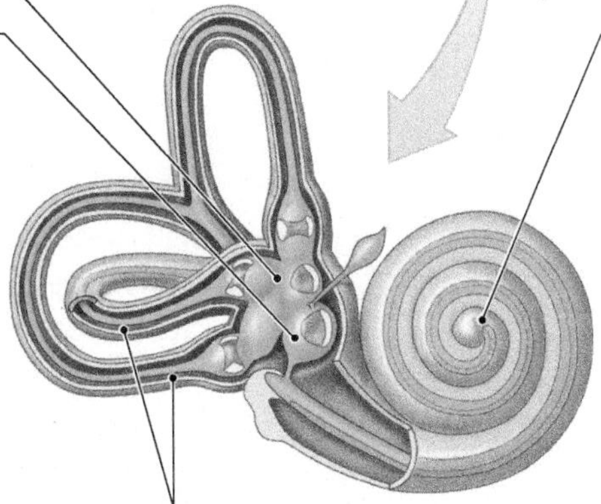

1. Identify the **vestibule,** which is the central area of the inner ear. It is subdivided into two regions:
   - **Utricle**
   - **Saccule**

   The vestibule contains sensory receptors, called **maculae** (singular = macula), that detect changes in linear movements and respond to gravitational forces (static equilibrium). List two examples of body position changes that would be detected by the maculae.

   ▶ ______________________________

   ____

2. Identify the **semicircular canals,** which consist of three semicircular tubes connected to the posterior wall of the vestibule. The semicircular canals lie at right angles to one another along the three planes of movement. Thus, there is an anterior, a posterior, and a lateral semicircular canal. At one end of each semicircular canal, there is an expanded region called an **ampulla.** Inside each ampulla is a region called the **crista ampullaris,** which contains sensory receptors that detect changes in angular acceleration or deceleration (dynamic equilibrium). List two examples of body postion changes that would be detected by receptors in a crista ampullaris.

   ▶ ______________________________

   ____

3. Identify the **cochlea,** a snail-shaped structure that coils 2-1/2 times around a central axis of bone. It is attached to the anterior wall of the vestibule. The cochlea contains the sensory receptors for hearing, known as the spiral organ or the organ of Corti.

MAKING CONNECTIONS

Bacterial or viral infections that originate in the nose or throat can spread to the middle ear and mastoid air cells. Explain how that is possible.

▶ ______________________________

____

____

____

____

Activity **18.14**

Hearing: Examining the Microscopic Anatomy of the Cochlea

A The cochlea is a bony snail-shaped structure. It contains three chambers—vestibular duct, cochlear duct, and tympanic duct—that spiral for 2-1/2 turns.

1. View a slide of the cochlea, in cross section, under low power. Identify a cross section showing the three cochlear chambers.

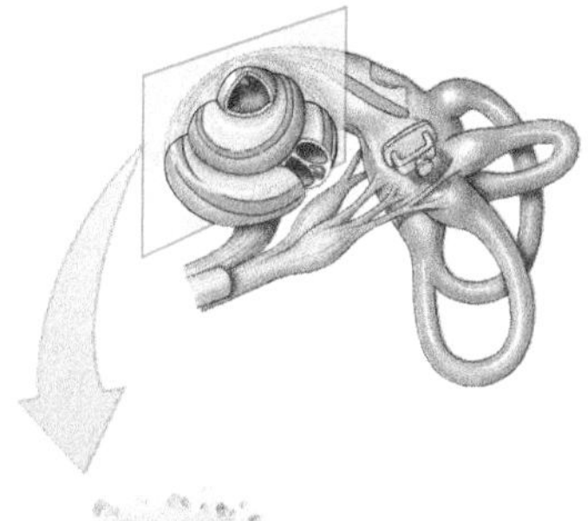

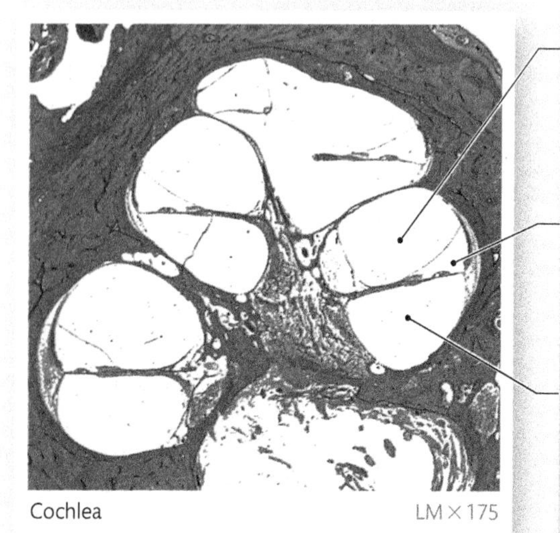

Cochlea LM × 175

The **vestibular duct** is part of the bony labyrinth and is filled with perilymph. It begins at the oval window and spirals to the apex of the cochlea.

The **cochlear duct,** the middle chamber, is a part of the membranous labyrinth and is filled with endolymph. The spiral organ is located inside the cochlear duct.

The **tympanic duct,** part of the bony labyrinth, is filled with perilymph. The duct begins at the apex of the cochlea, where it is continuous with the vestibular duct. It spirals to the base and ends at the round window.

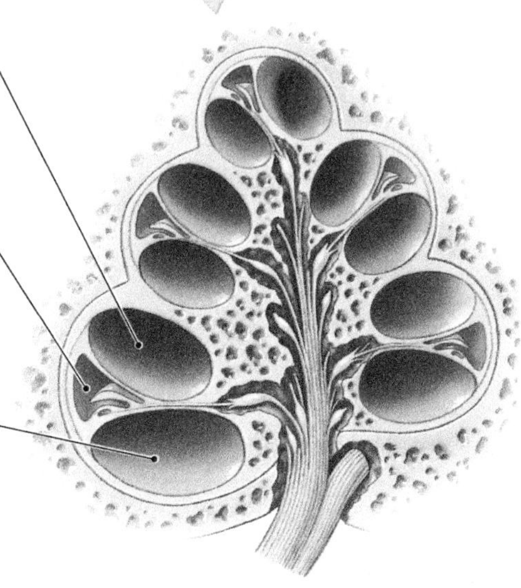

IN THE CLINIC

Altitude and the Tympanic Membrane

Equal air pressure on both sides of the tympanic membrane is functionally important for hearing. When atmospheric pressure is reduced, such as when traveling to higher elevations, the tympanic membrane bulges outward because the auditory tube collapses and the pressure inside the tympanic cavity is greater than the pressure outside. The bulging may be painful and can impair hearing because the membrane is less flexible than usual. Yawning or swallowing will open the auditory tube and equalize the air pressure on both sides of the tympanic membrane.

B The cochlear duct contains the sensory receptor for hearing, the spiral organ (organ of Corti).

1 Adjust the position of the slide so that the **spiral organ** in the cochlear duct is in the center of your field of view. Switch to high power and observe the structures in the cochlear duct.

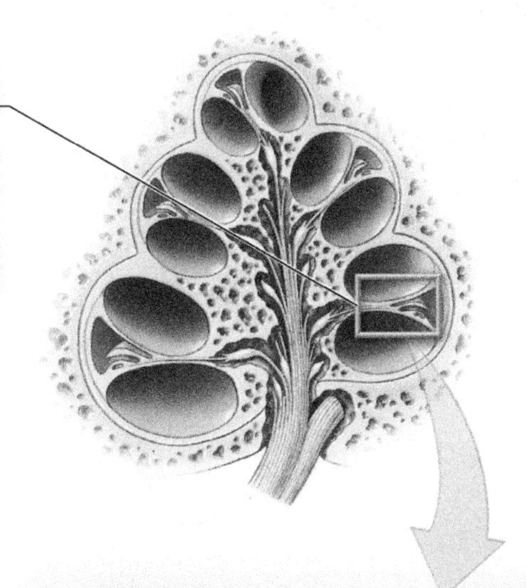

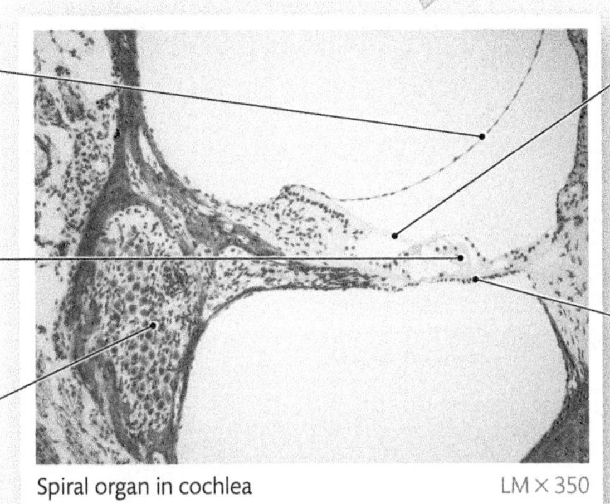

Spiral organ in cochlea LM × 350

2 Identify the **vestibular membrane,** which separates the cochlear duct from the vestibular duct.

3 Locate the **hair cells** (receptor cells) in the spiral organ.

4 Find the **spiral ganglion.** It contains neuron cell bodies of the cochlear division of the vestibulocochlear nerve (cranial nerve VIII), which innervates the hair cells.

5 Observe the **tectorial membrane,** which arches over the hair cells. The stereocilia of the hair cells are in contact with this membrane.

6 The **basilar membrane** separates the cochlear duct from the tympanic duct. Hair cells of the spiral organ rest on the basilar membrane.

MAKING CONNECTIONS

Sound waves traveling through the vestibular duct distort the basilar membrane. Movement of the basilar membrane pushes hair cells against the tectorial membrane. Stimulated hair cells transmit sensory information to the cochlear branch of the vestibulocochlear nerve. Different regions of the basilar membrane are sensitive to specific frequencies of sound waves: high frequencies at the base of the cochlea, low frequencies at the apex, and intermediate frequencies in the middle sections. Why do you think this arrangement is important for our ability to hear different sounds?

▶ ______________________________

____

____

____

Activity **18.15**

Performing Hearing Tests: Hearing Acuity and Localizing Sound

Perform the following hearing tests in a quiet room or in an area of the laboratory where as much background sound as possible can be eliminated.

A **Hearing acuity** refers to an individual's ability to hear sounds clearly at various distances.

1. With your index finger, squeeze shut the entrance to the external auditory canal of your left ear. (*Warning:* Do not insert your finger into the canal.)
2. While sitting quietly, have your lab partner hold a laboratory timer with an audible ticking sound very close to your right ear.
3. Have your lab partner slowly move the timer away from your ear until you no longer hear it ticking.
4. Measure the distance (in centimeters) at which you no longer hear the sound. Record the result in Table 18.9. ▶
5. Repeat steps 1 through 4 for your left ear.

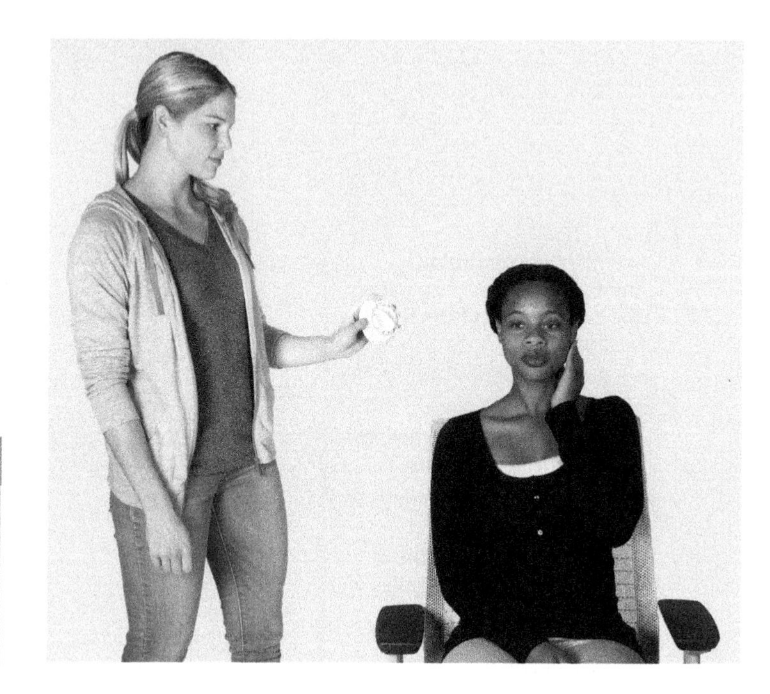

TABLE 18.9 Hearing Acuity Test

	Distance at which Ticking Becomes Inaudible (cm)
Right ear	
Left ear	

6. What conclusion can you make if the distance at which the ticking sound becomes inaudible: ▶
   - is the same for both ears?

   ____

   ____

   ____

   ____

   - is greater in one ear than the other?

   ____

   ____

   ____

   ____

B **Localizing Sound:** When an individual has a hearing loss in one ear, sounds coming from some locations may be heard better than others. For example, if you have better hearing acuity in your right ear than your left ear, a sound originating closer to your right ear will be heard more clearly than one originating closer to your left ear. In such a case, you might prefer to use a telephone with your right ear instead of your left.

1. Sit quietly with your eyes closed. Have your lab partner hold a laboratory timer with an audible ticking sound close to one of your ears so that you can hear it.
2. Have your partner move the timer to various positions around your head so that it can always be heard.
3. After each move, indicate the new position of the timer to your lab partner.

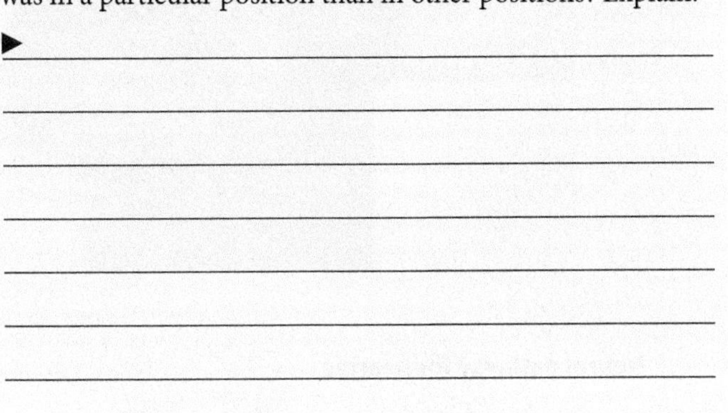

4. Was it more difficult to localize the sound when the timer was in a particular position than in other positions? Explain.

▶ ______________________________

MAKING CONNECTIONS

Do your results indicate the possibility of better hearing acuity in one ear over the other, or is your hearing acuity about the same in both ears? Explain.

▶ ______________________________

Activity **18.16**

Performing Hearing Tests: Weber Test and Rinne Test

A The **Weber test** is used to determine whether an individual has **sensorineural deafness** or **conduction deafness.** Sensorineural deafness is the result of damage along the neural pathway for hearing. In conduction deafness, transmission of sound waves is disturbed in the external acoustic meatus, at the tympanic membrane, or along the auditory ossicles.

1. Strike the prongs of a tuning fork against the edge of your laboratory bench.
2. Place the end of the tuning fork handle on your forehead, midway between each ear, and listen for the tone of the fork.
3. Record your results in Table 18.10 as positive (+) or negative (−) for the following possible outcomes: ▶
   - If you hear the tone of the fork equally well in both ears, you have equal hearing (possibly, equal loss of hearing) in both ears.
   - If you have sensorineural deafness, you will detect the tone in the normal ear only. You will be unable to hear the tone in the affected ear because the neural pathway for hearing is damaged and will be unable to send information to the auditory area in the temporal lobe.
   - If you have conduction deafness, you will hear the tone in both ears, but it will be louder in the affected ear. In the normal ear, the cochlea receives background noise from the environment as well as vibrations from the tuning fork. The affected ear receives only the tuning fork vibrations, so they will sound louder.

Neural pathway for hearing

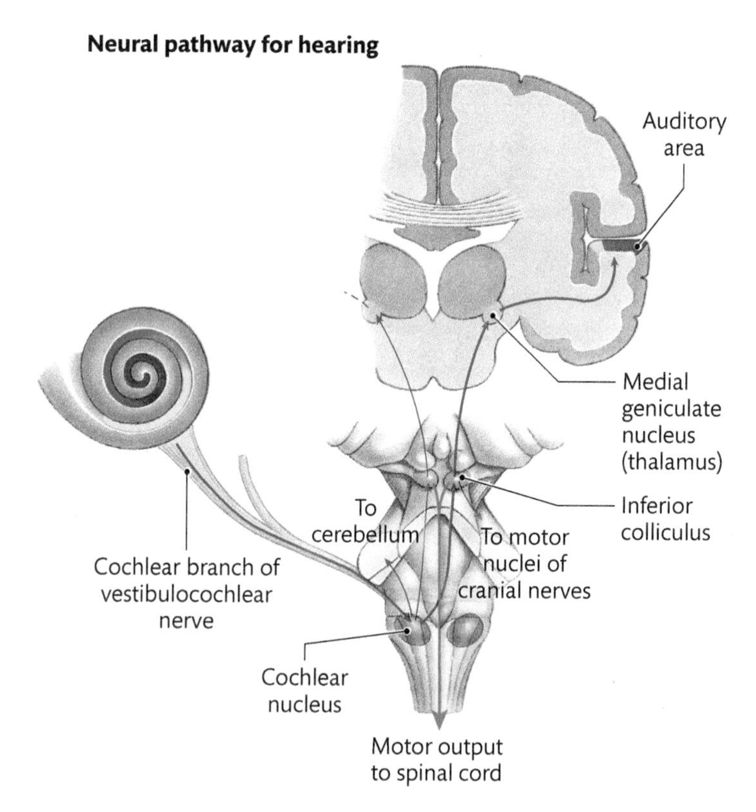

TABLE 18.10 Weber Test Results (+/−)

	Normal Hearing	Possible Sensorineural Deafness	Possible Conduction Deafness
Right ear			
Left ear			

B The **Rinne test** is used to determine whether sound conduction along the auditory ossicles is impaired. Abnormal function of the auditory ossicles may indicate conduction deafness.

1. Strike the prongs of a tuning fork against the edge of your laboratory bench.
2. Place the end of the tuning fork handle on the mastoid process on the right side of your skull. You should be able to hear the sound of the vibrating prongs by bone conduction.
3. When you can no longer hear the sound, place the tuning fork prongs near the entrance to the external acoustic meatus of your right ear.
4. If you can hear the sound again, this time by air conduction, your hearing is normal. If you cannot hear the sound, conduction deafness is possible. Record your results in Table 18.11 as positive (+) if hearing is normal or negative (−) if there is a possibility of conduction deafness. ▶
5. Strike the tuning fork again, but this time place the prongs near the entrance to the external auditory meatus of your right ear so that you can first hear the sound by air conduction.
6. When you can no longer hear the sound, place the handle of the tuning fork on your right mastoid process to hear the sound by bone conduction.
7. If you cannot hear the sound by bone conduction, your hearing is normal. If you can hear the sound, conduction deafness in your right ear is a possibility. Record your results in Table 18.11 as positive (+) if hearing is normal or negative (−) if there is a possible conductive hearing loss. ▶
8. Repeat steps 1 through 7 for your left ear.

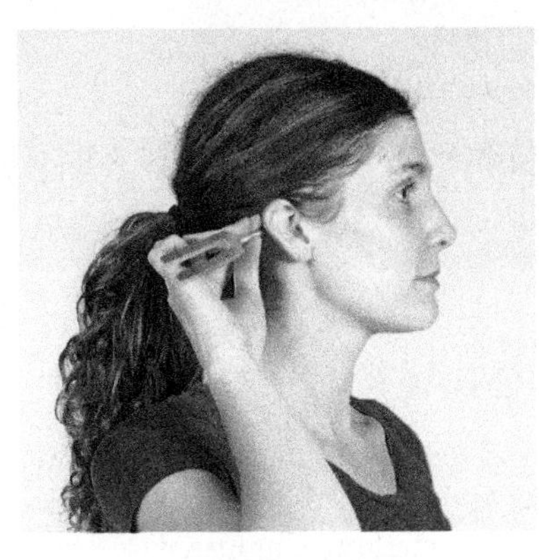

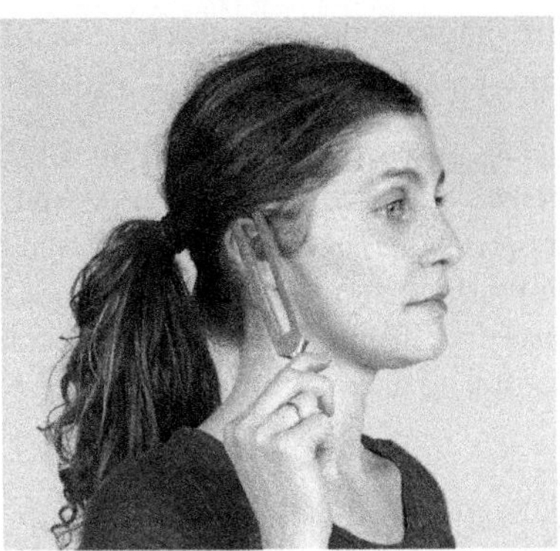

TABLE 18.11 Rinne Test Results (+/−)

	Tuning Fork on Mastoid Process First and Moved to External Acoustic Meatus	Tuning Fork Near External Acoustic Meatus First and Moved to Mastoid Process
Right ear		
Left ear		

MAKING CONNECTIONS

On a scale of 1 to 5, with 1 being "poor" and 5 being "excellent," how would you rate your hearing ability? Based on the results of the hearing tests that you completed in the last two activities, provide supporting evidence for your rating by writing an analysis that describes the general condition of your ability to hear. ▶

	Poor				Excellent
Hearing rating	1	2	3	4	5

Analysis:

____

____

____

____

____

____

BEFORE YOU MOVE ON . . .

‹‹ LOOKING BACK

In this laboratory exercise, you studied the anatomy and physiology of sensory receptors for the five special senses: olfaction (smell), gustation (taste), vision, hearing, and equilibrium (body orientation). Compared to general sensations, perception of special sensations requires larger and more complex receptors and more intricate neural pathways to reach the central nervous system.

Sensory receptors are the peripheral extensions of the nervous system. It is through our senses that awareness of our environment is possible. One concept you should keep in mind is that our perceptions of the world are based on awareness derived not from only one sensation, but rather from a mosaic of multiple sensory inputs. For example, we appreciate the taste and aroma of food, but we also respond to its visual presentation and perhaps to the sounds of its preparation.

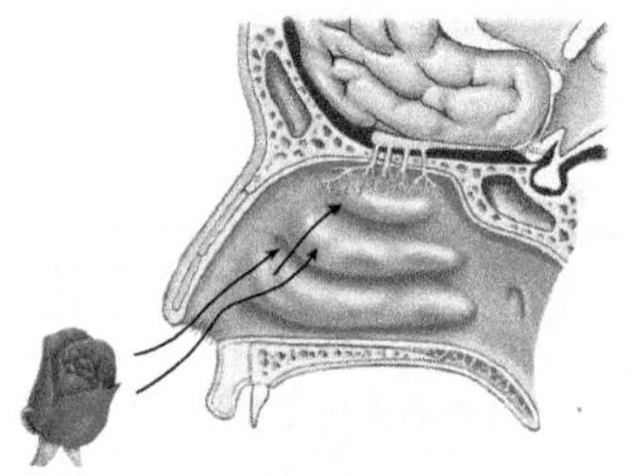

Olfaction

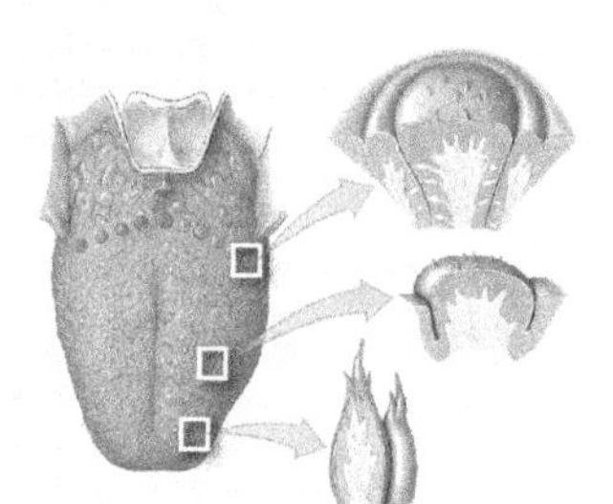

Gustation

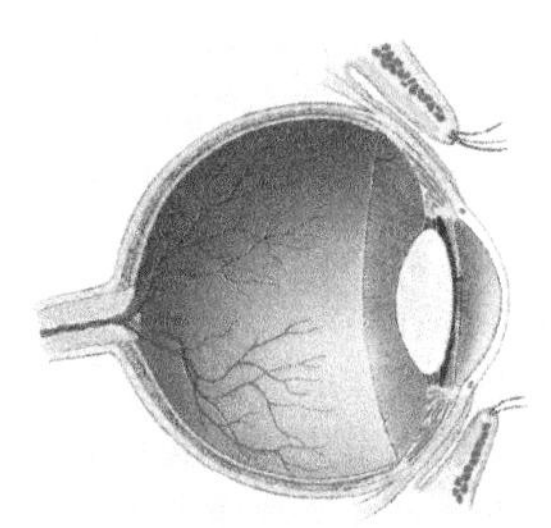

Vision

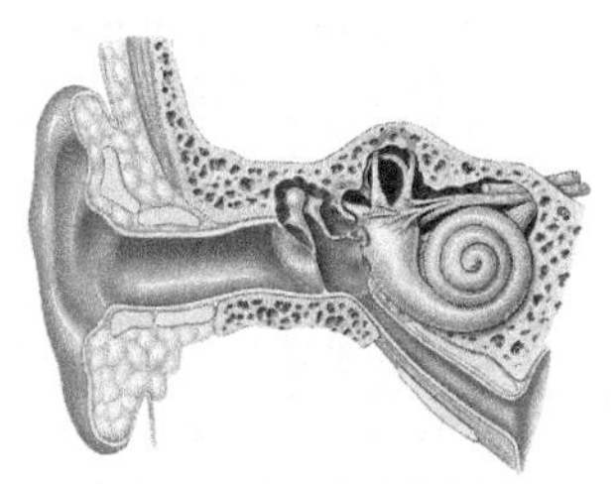

Hearing and equilibrium

Identify other everyday experiences that you can recognize and perceive by overlapping sensory inputs.

▶ ____

____

____

____

____

LOOKING FORWARD ››

This laboratory exercise is the last in a series of five exercises (Laboratory Exercises 14–18) during which you studied the anatomy and physiology of the nervous system. In the next exercise, you will study the structure and function of the endocrine system, which consists of a diverse collection of organs and tissues that contain endocrine glands. These glands secrete chemicals known as hormones into the bloodstream, which are transported to target cells at some distant location.

The endocrine system operates in conjunction with the nervous system to maintain homeostasis and ensure that bodily functions are carried out efficiently. This functional relationship is sometimes expressed as a **neuroendocrine effect** in which nerve impulses can affect the release of hormones, and, in turn, hormones can regulate the transmission of nerve impulses.

Name ____________________

Lab Section ____________________

Date ____________________

EXERCISE 18 REVIEW SHEET

Special Senses

QUESTIONS 1–12: Match the structure in column A with the appropriate description in column B.

A

1. Cochlea ____________
2. Retina ____________
3. Cribriform plate ____________
4. External acoustic meatus ____________
5. Cornea ____________
6. Fovea centralis ____________
7. Lens ____________
8. Iris ____________
9. Taste buds ____________
10. Olfactory bulb ____________
11. Vestibule ____________
12. Auditory tube ____________

B

a. The innermost layer of the eye; contains the photoreceptor cells

b. The sensory receptors for gustation

c. Contains the sensory receptors for static equilibrium

d. Region of the retina that contains only cone cells

e. Contains numerous small openings that transmit the axons of the olfactory nerve

f. Structure that contains the spiral organ

g. Connects the auricle to the tympanic membrane

h. Structure posterior to the iris that focuses light onto the retina

i. Structure in which the axons of the olfactory nerve synapse with other neurons

j. Contains two layers of smooth muscle that constrict and dilate the pupil

k. Connects the middle ear to the nasopharynx

l. The transparent portion of the fibrous tunic, which allows light to enter the eyeball

13. Complete the following table:

Special Senses

Sense	Receptor	Location
a.	b.	Mucous membrane covering superior nasal concha
c.	Retina	d.
e.	f.	Semicircular canals
Gustation	g.	h.
Hearing	i.	j.
k.	Maculae	l.

14. You arrive at home just before dinner, and the smell of a chocolate cake baking in the oven fills the air. As the cake is baking, the chocolate odor diminishes and is replaced by the smell of tomato sauce simmering on the stove. Provide an explanation for this scenario.

____

____

____

____

15. The results in the table below were obtained from an examination to test the effects of olfaction on gustation. Explain what the results mean.

Effects of Olfaction on Gustation

Food Item	Conditions for Identification		
	On Tongue before Chewing; Eyes and Nostrils Closed	Chewing with Eyes and Nostrils Closed	Chewing with Nostrils Open and Eyes Closed
Chocolate	−	+	++
Black licorice	+	++	+++
Almond	−	−	+
Pineapple	−	+	++

One minus (−) sign means that the food item could not be tasted.
One plus (+) sign means that the food item could be tasted.
Multiple plus (++) signs mean that the taste was enhanced.

____

____

____

____

16. Describe three ways that the accessory eye structures protect the eye.

____

____

____

____

17. Albinism is a genetic condition characterized by the lack of melanin production.

a. Identify the eye structures that this condition will affect.

____

____

b. Will vision be affected? Explain.

____

____

____

____

18. A patient's visual acuity was tested using a Snellen chart. The results indicated that her vision was 20/50. What does this result mean?

____

____

____

____

19. Following eye surgery, a man was required to wear an eye patch over his right eye. What effect will this patch have on the man's vision?

____

____

____

____

20. During a test for near-point accommodation, Subject A had an accommodation distance of 15 cm in both eyes, and Subject B had an accommodation distance of 30 cm. Which one is likely the younger person? Explain.

____

____

____

____

21. Explain why many vertebrates, including cows and domestic dogs and cats, have much better night vision than humans.

____

____

____

____

22. Describe the tests you would use to assess a patient's hearing acuity. What results would you expect if the patient's ability to hear and localize sounds was suspected to be below normal?

____

____

____

____

23. During a Weber test, the arm of a tuning fork was placed on the patient's forehead. She was able to hear the tone in the left ear, but not in the right ear. Explain what these results indicate.

____

____

____

____

24. In the following diagram, identify the structures by using the colors indicated.

Lens = **yellow**
Iris = **dark blue**
Sclera = **green**
Choroid = **red**
Retina = **orange**
Ciliary body = **brown**
Cornea = **purple**
Optic nerve = **light blue**

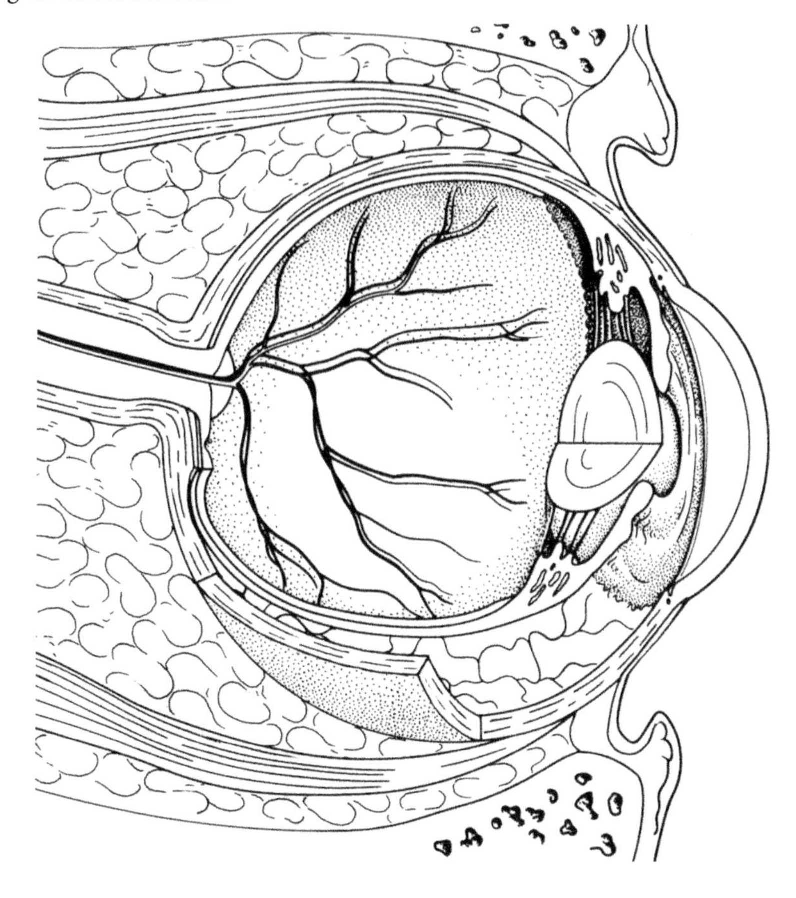

25. In the following diagram, identify the structures by using the colors indicated.

Semicircular canals = **red**
Cochlea = **dark green**
External acoustic meatus = **dark blue**
Vestibule = **yellow**
Auditory ossicles = **brown**
Vestibulocochlear nerve (VIII):
  a. Cochlear division = **purple**
  b. Vestibular division = **light blue**
Auditory tube = **orange**
Tympanic membrane = **pink**
Auricle (Pinna) = **light green**

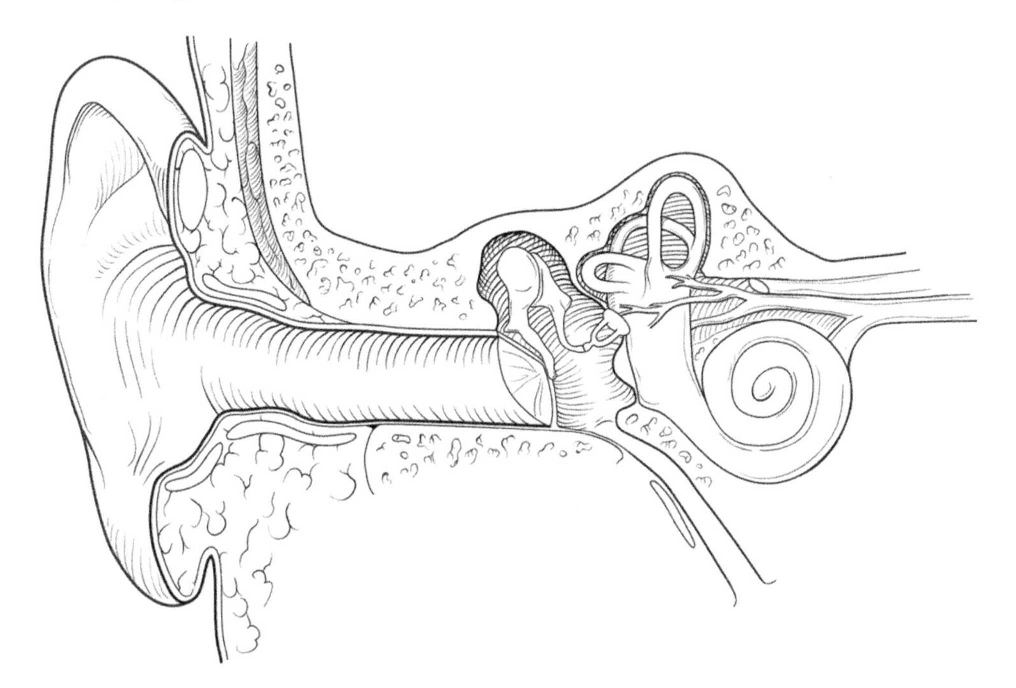

EXERCISE 19

The Endocrine System

LEARNING OUTCOMES

These Learning Outcomes correspond by number to the laboratory activities in this exercise. When you complete the activities, you should be able to:

Activities 19.1–19.3 **Describe the gross anatomical structure and the functions of endocrine glands located in the head, neck, thoracic cavity, and abdominopelvic cavity.**

Activities 19.4–19.6 **Describe the microscopic anatomy of various endocrine glands.**

LABORATORY SUPPLIES

- Anatomical models:
  - human brain
  - human torso
  - female pelvis
  - male pelvis
- Human skulls
- Compound light microscopes
- Prepared microscope slides:
  - pituitary gland
  - thyroid gland
  - parathyroid gland
  - pancreas
  - adrenal gland

PRE-LAB QUIZ

Before you begin, read all the activities in Exercise 19 and the required reading in your textbook that is assigned by your instructor.

1. During this laboratory exercise, you will view microscope slides of all the following structures *except* the
   a. pituitary gland.
   b. adrenal gland.
   c. kidney.
   d. pancreas.
   e. thyroid gland.
2. During this laboratory exercise, you will identify all the following endocrine organs in the head *except* the
   a. pineal gland.
   b. thymus.
   c. hypothalamus.
   d. pituitary gland.
3. During this laboratory exercise, you will identify endocrine organs in the head by viewing a ________________ section of a human brain model.
4. True or False: The infundibulum connects the pituitary to the hypothalamus. ________
5. The ________________ covers a portion of the heart and extends into the base of the neck.
6. True or False: The inner portion of the adrenal gland is called the adrenal cortex. ________
7. You will be viewing slides of an organ that contains islets of Langerhans. What organ will you be viewing? ________________
8. The posterior lobe of the pituitary contains axons of neuron cell bodies located in the ________________.
9. When you view a slide of the parathyroid, you likely see tissue from what other structure? ________________
10. You will be viewing slides of an organ that contains three cellular regions in its cortex. What organ will you be viewing? ________________

Activity **19.1**

Examining the Gross Anatomy of the Endocrine System: Endocrine Organs in the Head

A Three endocrine organs are located in the brain.

1. On a model of a midsagittal section of a human brain, identify the **pineal gland,** located along the roof of the third ventricle. The pineal gland secretes melatonin. Use your textbook to identify the function of this hormone.

▶ ______________________

____

____

2. Locate the **hypothalamus,** just inferior to the thalamus. It produces a number of releasing hormones that increase, and inhibiting hormones that reduce the production and secretion of hormones in the anterior pituitary. The hypothalamus also produces antidiuretic hormone (ADH) and oxytocin. Use your textbook to identify the functions of these hormones. ▶

ADH: ______________________

____

Oxytocin: ______________________

____

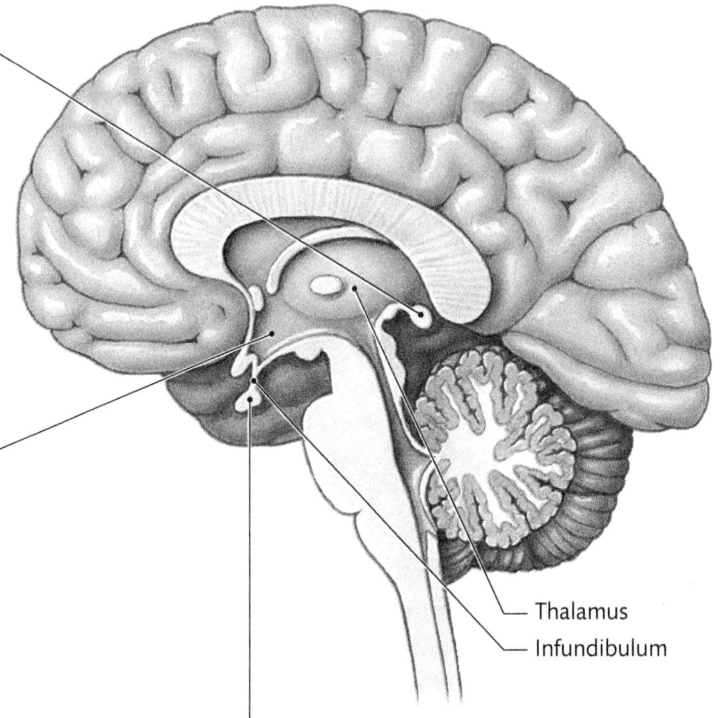

3. Locate the **pituitary gland,** or **hypophysis,** which is directly connected to the hypothalamus by a stalk of tissue called the infundibulum.

Thalamus

Infundibulum

Word Origins

Hormone comes from the Greek word *hormao,* which means "to provoke" or "set in motion." Hormones released by endocrine glands influence their target organs by "setting in motion" or promoting a specific function.

IN THE CLINIC

Seasonal Affective Disorder

Melatonin production and secretion increase during periods of darkness and decrease during periods of light. This fluctuation in activity is believed to be the underlying cause of **seasonal affective disorder (SAD).** Symptoms include unusual changes in mood, sleeping pattern, and appetite in some people living at high latitudes where periods of darkness are quite long during winter months. Symptoms for SAD can be managed by getting enough sleep, eating a healthy diet, and exercising regularly. During fall and winter, daily 30-minute light therapy using a light source that mimics sunlight may also be an effective treatment.

Want more practice? Go to: MasteringA&P > Study Area > Menu > Lab Tools > PAL >
- Anatomical Models > Endocrine System
- Human Cadaver > Endocrine System

B The pituitary gland consists of an **anterior lobe** and a **posterior lobe.**

1. On a skull, identify the **sella turcica** along the floor of the cranial cavity. The pituitary gland rests in this bony depression.

2. On a model of the brain, locate the pituitary gland. Identify the following structures:

The **infundibulum** is a stalk of nervous tissue that connects the pituitary to the hypothalamus.

The **anterior lobe,** or **adenohypophysis,** contains endocrine cells that produce seven hormones.

The **posterior lobe,** or **neurohypophysis,** is an extension of the brain. Oxytocin and ADH, produced in the hypothalamus, are transported along axons that pass through the infundibulum to the posterior lobe, where they are stored and eventually released.

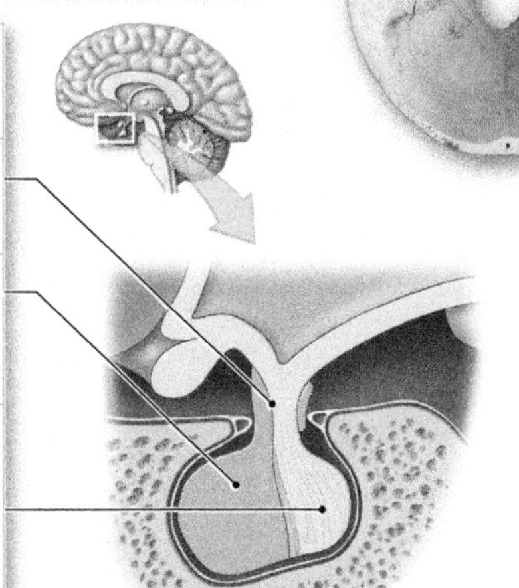

3. The hormones produced by the anterior pituitary are listed in Table 19.1. Refer to your textbook and complete the table by listing the target organ for each hormone and the hormone's effect. ▶

TABLE 19.1 Hormones Produced by the Anterior Pituitary

Hormone	Target	Effect
Thyroid-stimulating hormone (TSH)		
Adrenocorticotropic hormone (ACTH)		
Gonadotropins a. Follicle-stimulating hormone (FSH) b. Luteinizing hormone (LH)		
Prolactin		
Growth hormone (GH)		
Melanocyte-stimulating hormone (MSH)		

MAKING CONNECTIONS

A pituitary tumor can have widespread effects on endocrine system function. Explain why.

▶ ____

____

____

Activity **19.2**

Examining the Gross Anatomy of the Endocrine System: Endocrine Organs in the Neck and Thoracic Cavity

A Endocrine organs in the neck produce hormones that regulate cell metabolism and control calcium ion concentration in body fluids.

1

On a torso or head and neck model, locate the **thyroid gland,** just inferior to the thyroid cartilage. The thyroid gland includes:

Left and right lobes on each side of the trachea.

The **isthmus of the thyroid,** which travels across the anterior surface of the trachea and connects the two lobes.

The thyroid gland produces the hormones thyroxine (T_4) and triiodothyronine (T_3), which regulate cell metabolism, general growth and development, and the normal development and maturation of the nervous system. It also produces calcitonin, which reduces the levels of calcium ions in body fluids.

2

Hyperthyroidism is a thyroid disorder that results in overproduction of thyroid hormones (T_4 and T_3). A person with hyperthyroidism might have a higher-than-normal body temperature and experience weight loss despite a good appetite. Explain why.

▶ ______________________________

____

____

____

____

3

Identify two pairs of pea-size **parathyroid glands** embedded on the posterior surfaces of the thyroid gland lobes. They produce parathyroid hormone (PTH), which opposes the action of calcitonin by increasing the concentration of calcium ions in body fluids.

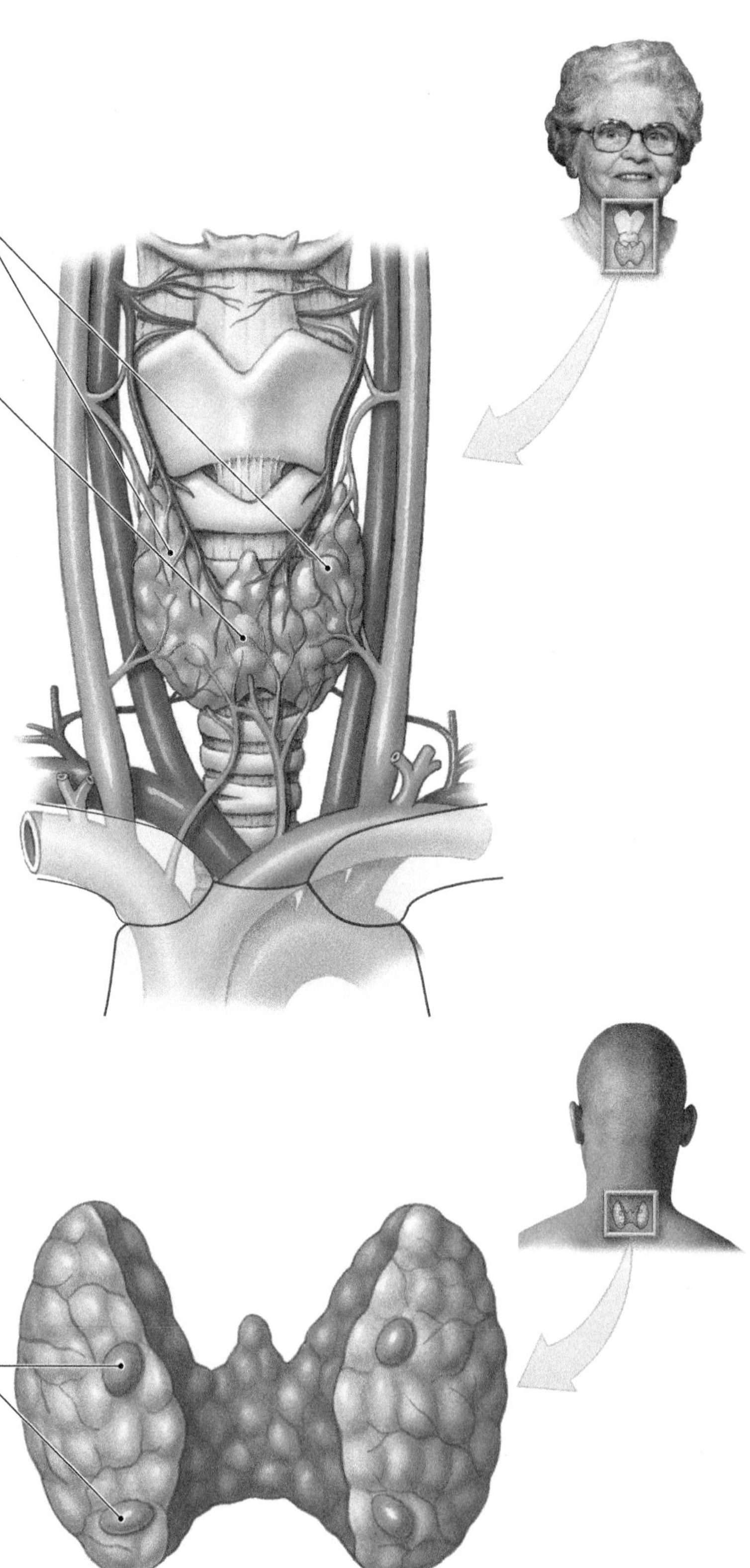

B The thoracic cavity contains two organs that produce and secrete hormones, but have other primary functions.

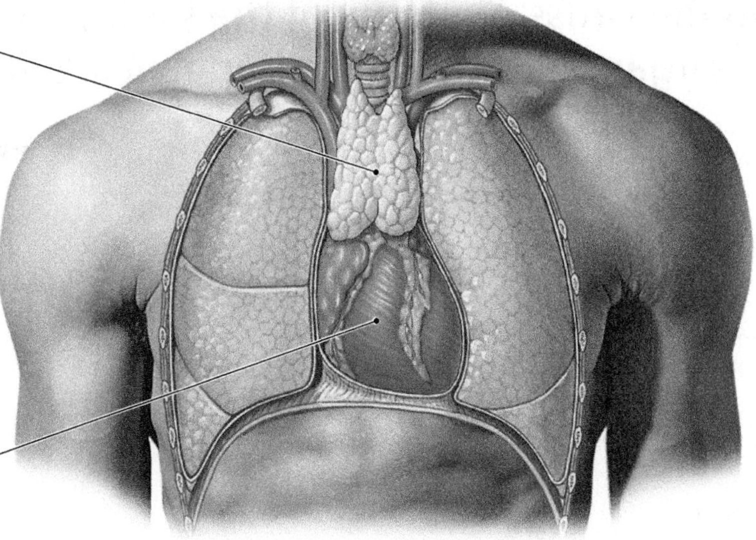

1. Remove the anterior body wall from a torso model to expose the thoracic cavity. Identify the **thymus gland** just posterior to the sternum. Observe how it covers the superior portion of the heart and extends superiorly into the base of the neck. The thymus produces a group of hormones called thymosins that promote the maturation of T-lymphocytes, white blood cells that coordinate the body's immune response.

2. Notice that the **heart** is located in the central region of the thoracic cavity, known as the mediastinum. If blood pressure rises above normal, cardiac muscle cells in the heart wall secrete natriuretic peptides. These hormones act on the kidneys to promote the loss of sodium ions and water. This action will lower blood pressure back to normal. Why do you think a loss of water will cause blood pressure to decline?

▶ ______________________________

____

____

____

____

____

____

____

Word Origins

Natriuretic peptides promote **natriuresis,** or the excretion of sodium in the urine. *Natriuresis* originates from two Greek words: *natrium,* meaning "sodium" (the chemical symbol for sodium is Na), and *ouron,* meaning "urine."

MAKING CONNECTIONS

Calcitonin and PTH regulate the concentration of calcium ions in body fluids. Why do you think it is important to maintain calcium within a normal range?

▶ ______________________________

____

____

____

____

Want more practice? Go to: **MasteringA&P** > Study Area > Menu > Lab Tools > PAL >

- Anatomical Models > Endocrine System
- Human Cadaver > Endocrine System

Activity **19.3**

Examining the Gross Anatomy of the Endocrine System: Endocrine Organs in the Abdominopelvic Cavity

A Digestive organs in the abdominopeivic cavity also have endocrine functions.

1. On a torso model, identify the **stomach** and **small intestine** in the abdominopelvic cavity. These organs produce several hormones that are important for regulating digestive activities.

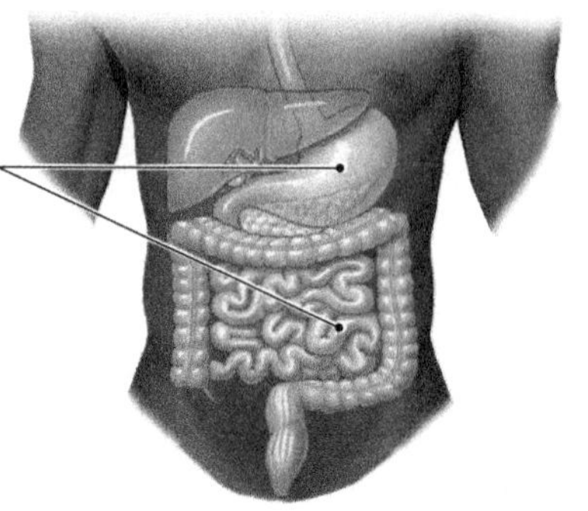

2. Remove the digestive organs from the abdominopelvic cavity to expose the structures along the posterior wall. Locate the elongated pancreas that stretches across the posterior body wall between the duodenum (first part of the small intestine) and the spleen.

The **head of the pancreas** is nestled within the C-shaped curvature of the duodenum on the right side.

The **body of the pancreas** is the main portion of the organ.

The **tail of the pancreas** extends to the left toward the spleen.

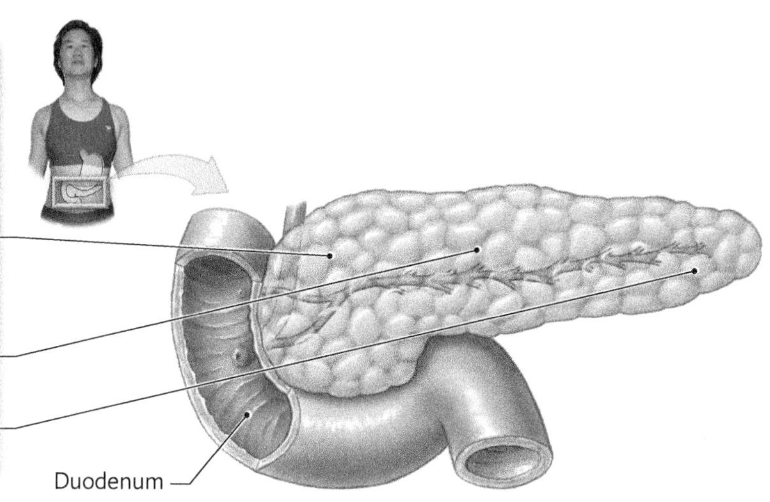

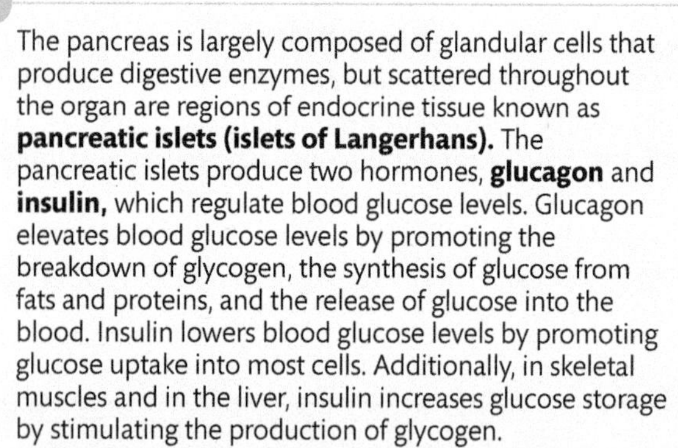

3. The pancreas is largely composed of glandular cells that produce digestive enzymes, but scattered throughout the organ are regions of endocrine tissue known as **pancreatic islets (islets of Langerhans).** The pancreatic islets produce two hormones, **glucagon** and **insulin,** which regulate blood glucose levels. Glucagon elevates blood glucose levels by promoting the breakdown of glycogen, the synthesis of glucose from fats and proteins, and the release of glucose into the blood. Insulin lowers blood glucose levels by promoting glucose uptake into most cells. Additionally, in skeletal muscles and in the liver, insulin increases glucose storage by stimulating the production of glycogen.

If insulin secretion decreased, how would blood glucose levels change? Explain.

▶ ______________________________

____

____

____

____

____

IN THE CLINIC

Diabetes Mellitus

There are two main types of diabetes mellitus. **Type 1 diabetes** accounts for 5 percent to 10 percent of all cases in the United States. It usually develops in children or young adults and destroys the pancreatic cells that produce insulin. It can be treated by daily administration of insulin, supplemented by a carefully monitored dietary plan.

Type 2 diabetes is far more common, making up 90 percent to 95 percent of all cases. A strong correlation exists between type 2 diabetes and obesity. People with type 2 diabetes produce normal amounts of insulin but cannot utilize the hormone effectively. This condition could be due to the production of defective insulin molecules or the lack of insulin receptors on target cells. Careful dietary control, weight reduction, and other lifestyle changes (e.g., regular exercise) are the best treatments for this form of the disease.

Diabetes is a long-term, progressive disorder that has potentially serious systemic effects. It can contribute to blindness, heart disease, stroke, kidney failure, nerve damage, and circulatory problems resulting in limb amputations. It is one of the leading causes of death in the United States.

B The adrenal glands produce and secrete several types of hormones. Although the kidneys are mostly involved with waste removal, they also have endocrine functions.

1 Along the posterior wall of the abdominopelvic cavity, locate the pyramid-shaped **adrenal glands** resting on the superior margins of the kidneys.

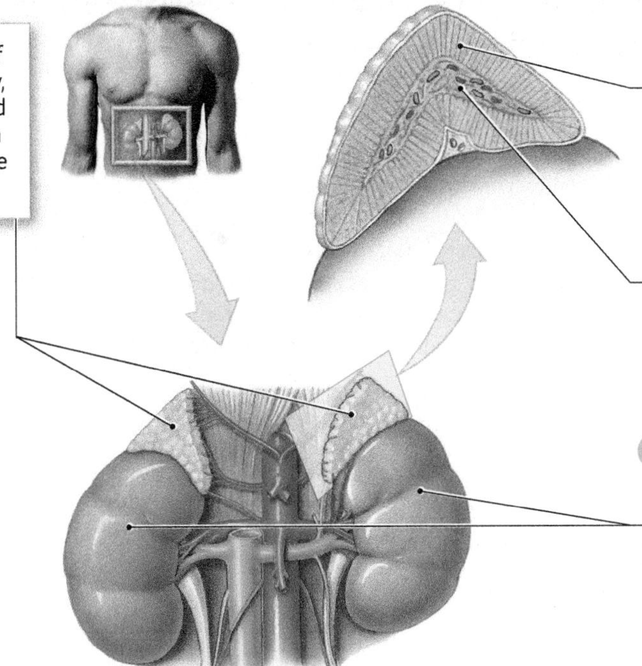

2 Remove the anterior portion of one adrenal gland and observe its internal structure.

The outer region is the **adrenal cortex.** It produces hormones that regulate sodium and potassium levels in body fluids and glucose metabolism. This region also produces androgens (male sex hormones).

The inner region is the **adrenal medulla.** Cells in this region release epinephrine and norepinephrine in response to sympathetic nervous system activation.

3 Under the influence of parathyroid hormone, the **kidneys** release a hormone called calcitriol, which acts on the small intestine to increase absorption of calcium and phosphate. The kidneys also release erythropoietin (EPO), which stimulates red blood cell production in bone marrow.

C The sex hormones control the development and maturation of sex cells (egg and sperm), maintain accessory sex organs, and support secondary sex characteristics.

1 On a female model of the pelvis, locate the **ovaries** along the lateral wall of the pelvic cavity. The ovaries are the primary sex organs in females. They produce female sex hormones called estrogens.

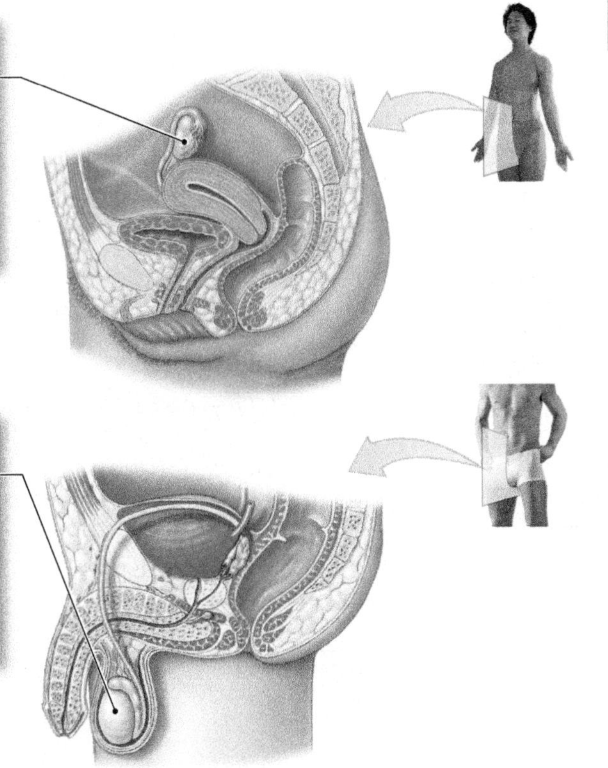

2 On a male model of the pelvis, identify the **testes** in the scrotum. The testes are the primary sex organs in males. They produce male sex hormones (androgens), of which testosterone is the most important.

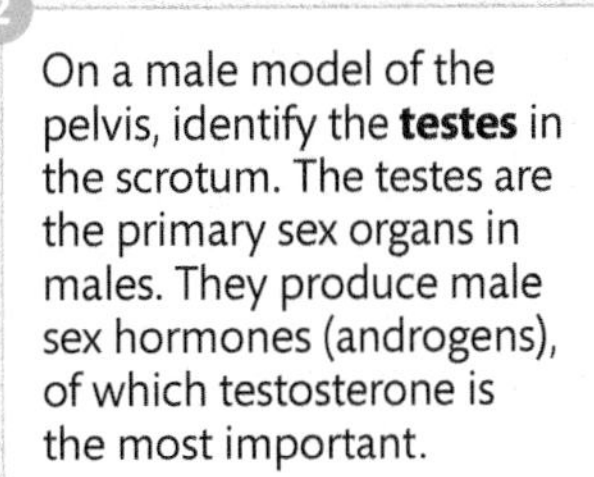

MAKING CONNECTIONS

Endocrine glands are surrounded by an extensive network of blood capillaries. Suggest a reason why this anatomical relationship is significant.

▶ ______________________________

____

____

____

____

____

____

____

____

Want more practice? Go to: **MasteringA&P** > Study Area > Menu > Lab Tools > PAL >

- Anatomical Models > Endocrine System
- Human Cadaver > Endocrine System

Activity **19.4**

Examining the Microscopic Anatomy of the Endocrine Organs: Pituitary Gland

The cells of endocrine glands possess the following common features:

- The cells are usually cuboidal or polyhedral (many sides), with large, spherical nuclei.
- With the exception of the hypothalamus, all endocrine cells are derived from epithelial tissue.
- The cells are typically arranged in clusters, small islands (islets), or cords.
- Endocrine cells form glands that lack a system of ducts. Hormones are secreted directly into the surrounding tissue spaces and eventually enter into the blood circulation.
- Endocrine cells have an extensive blood supply, and they all have at least one surface that is directly adjacent to a capillary.

As you study the microscopic anatomy of the various endocrine organs, be aware of these similarities as well as the unique features that characterize each structure.

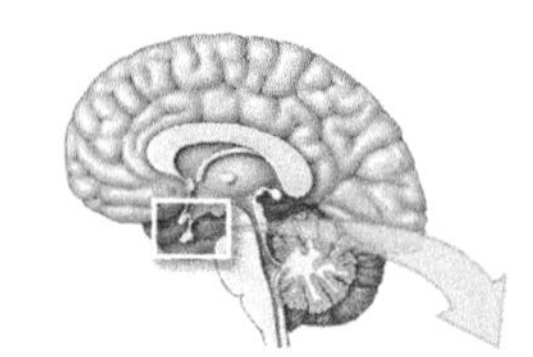

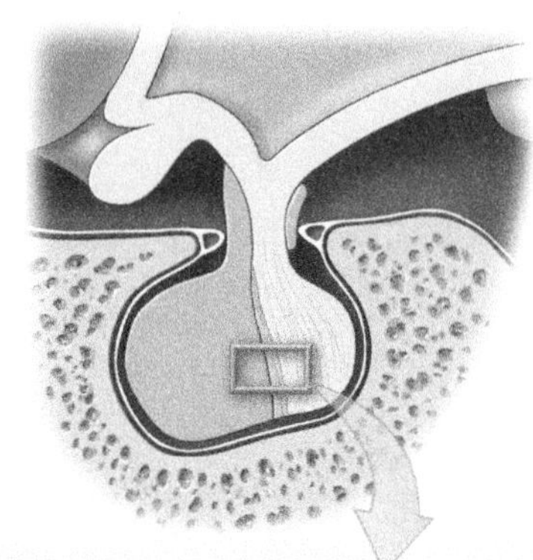

A Under the microscope, the anterior and posterior lobes of the pituitary gland are distinguished by their varied staining qualities.

1. View a slide of the pituitary gland with the scanning or low-power objective lens. Adjacent brain and bone tissue may also be present.
   Which brain structure would you expect to see on your slide?
   ▶ ______________________
   Which skull bone would you expect to be present?
   ▶ ______________________

2. Center the pituitary gland on the slide and identify:
   the darker-staining **anterior lobe (adenohypophysis)**
   the lighter-staining **posterior lobe (neurohypophysis)**

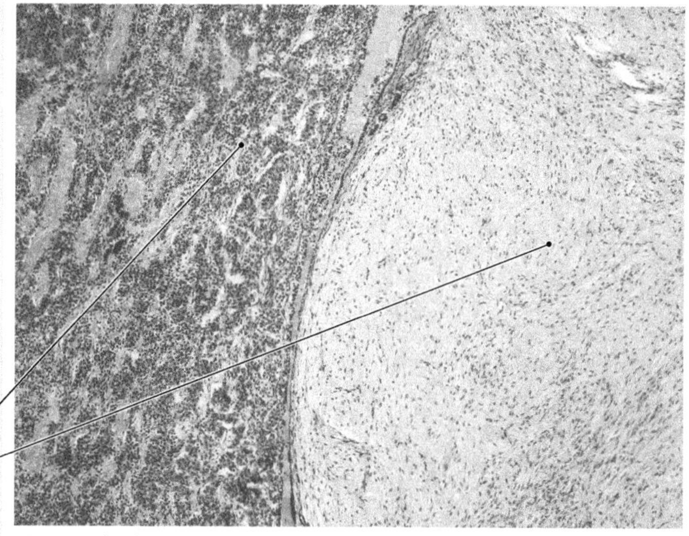

Pituitary gland LM × 70

Word Origins

Pituitary is derived from the Latin word *pituita,* which means "phlegm" or "thick mucus secretion." Renaissance anatomist Andreas Vesalius gave the pituitary its name because he mistakenly thought it produced a mucous secretion related to the throat. When scientists discovered the true function of the pituitary 200 years later, they gave it a new name, the *hypophysis,* which is the Greek word for "undergrowth." Hypophysis is a better name because it describes the gland's position suspended from the inferior surface of the hypothalamus.

B The diverse staining features in the pituitary reflect the structural differences between the anterior and posterior lobes.

1 Observe the anterior lobe of the pituitary with high power and identify the following regions:

The **anterior lobe (adenohypophysis)** consists of glandular epithelial cells arranged in cords or clusters. Notice that these cells have a cuboidal shape and possess well-defined nuclei. As you scan the slide, you will see that the cells vary considerably in their staining properties. What do you think is the reason for this variability?

▶ ____

____

____

____

The **pars intermedia** is a narrow band of tissue in the anterior lobe along the border with the posterior lobe. In the fetus, young children, and pregnant women, the cells in this region produce melanocyte-stimulating hormone. In most adults, this region is inactive.

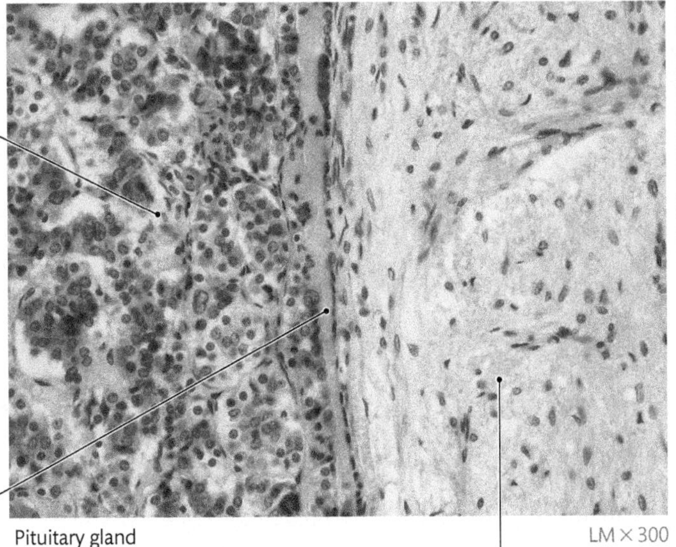

Pituitary gland LM × 300

2 View the **posterior lobe** with high power and notice that most of this region contains axons, which originate from neuron cell bodies in the hypothalamus. Antidiuretic hormone and oxytocin, produced in the hypothalamus, travel along these axons and are released from axon terminals in the posterior lobe.

IN THE CLINIC

Growth Hormone Abnormalities

Growth hormone (GH), secreted by the anterior lobe of the pituitary gland, promotes protein synthesis in virtually all cells. It is particularly important for the growth and development of muscle, cartilage, and bone. Inadequate production (hyposecretion) of GH before puberty leads to a condition called **pituitary growth failure (pituitary dwarfism).** People with this disorder have normal body proportions but abnormally short bones due to reduced activity at the epiphyseal plates. Pituitary dwarfism can be successfully treated before puberty by administering synthetic GH.

Excessive secretion (hypersecretion) of GH can also cause problems. **Gigantism** results from the hypersecretion of GH before bone fusion. Individuals with gigantism have normal body proportions but excessively long limbs and can reach heights up to 8.5 ft. Hypersecretion of GH after bone fusion causes **acromegaly.** In this condition, bones cannot lengthen, but instead become thicker and denser, particularly in the face, hands, and feet. Both gigantism and acromegaly are usually caused by a tumor in the anterior lobe and can be treated by its surgical removal.

MAKING CONNECTIONS

The anterior lobe of the pituitary is an endocrine gland, but the posterior lobe is not. Explain why.

▶ ____

____

____

____

____

____

____

Activity **19.5**

Examining the Microscopic Anatomy of the Endocrine Organs: Thyroid and Parathyroid Glands

A The thyroid gland has a distinctive structure, consisting of numerous **thyroid follicles** of various sizes.

1. View a slide of the thyroid gland with the scanning or low-power objective lens. Identify the **thyroid follicles,** which can be seen throughout the field of view.

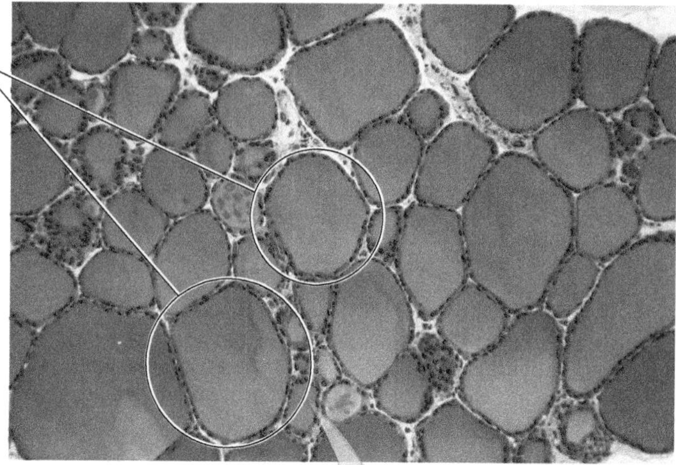

Thyroid gland LM × 260

2. Use the high-power objective lens to examine the thyroid follicles more closely. Notice that each follicle is surrounded by a single layer of cuboidal **follicle cells.**

3. Identify the **follicle cavities,** which contain a lightly staining material known as colloid. Follicle cells produce a globular protein known as thyroglobulin, which is stored in the colloid and later used to synthesize the thyroxine (T_4) and triiodothyronine (T_3). Both hormones are derivatives of the amino acid, tyrosine, combined with iodide ions.

4. In the regions of connective tissue between the follicles, identify the **parafollicular cells (C cells),** which produce calcitonin. They usually appear in small clusters and are characterized by their pale or lightly stained cytoplasm and large nuclei.

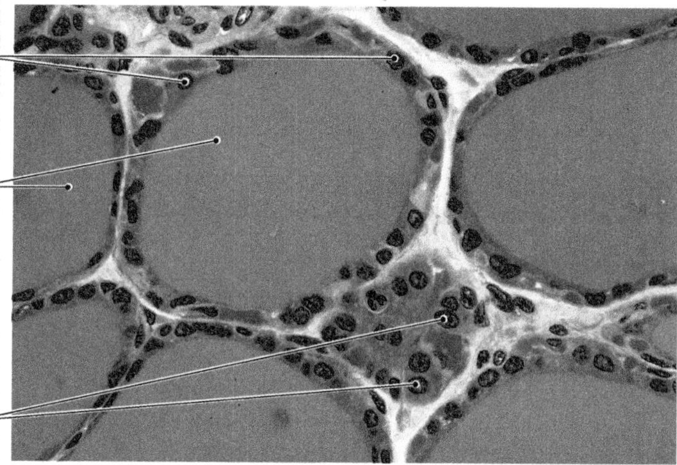

Thyroid gland LM × 600

5. In the space provided, make a drawing of your observations. Label the follicle cells, follicle cavities with colloid, and parafollicular cells.

▶

B Because the parathyroid glands are embedded in the posterior wall of the thyroid gland, prepared microscope slides will often display tissue from both structures.

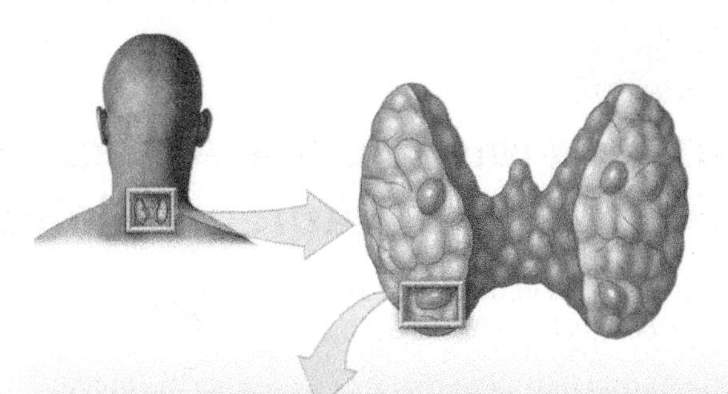

1 View a slide of the **parathyroid gland** with the scanning or low-power objective lens. Locate the parathyroid gland and the surrounding follicles from the thyroid gland.

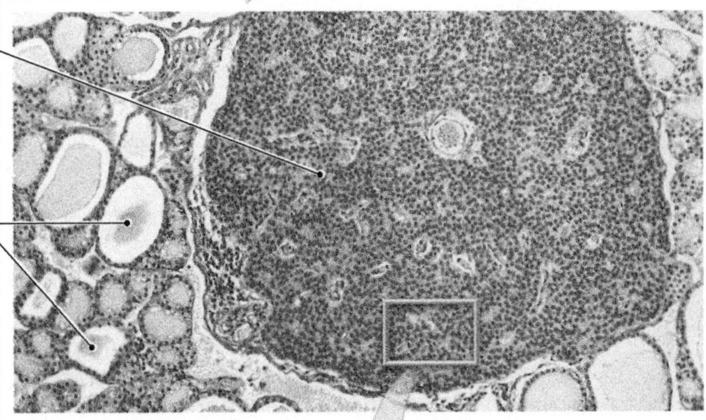

Parathyroid and thyroid glands LM × 100

2 Center an area of parathyroid tissue and switch to high power. The darkly stained cells that fill the field are **parathyroid (chief) cells,** which produce parathyroid hormone (PTH).

3 The **oxyphil cells,** a second cell type, are found only in human parathyroid glands. If you are viewing a human parathyroid, you will see these cells occurring singly or in small clusters. They are larger, stain lighter, and are far fewer than the parathyroid cells. The function of the oxyphil cells is unknown.

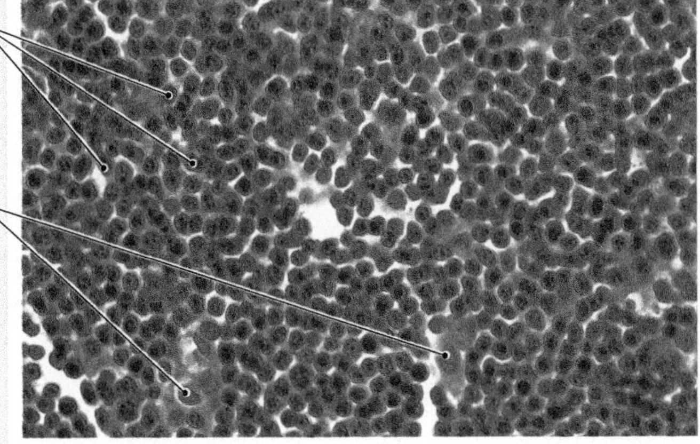

Parathyroid gland LM × 660

4 Calcitonin and PTH have opposite effects on calcium ion levels in body fluids. Both hormones influence the function of osteoclasts, which break down bone tissue. Describe the effects that calcitonin and PTH have on osteoclast activity (*Hint:* Review the functions of calcitonin and PTH in Activity 19.2.).

▶ ______________________________________

____

____

____

____

____

MAKING CONNECTIONS

One treatment for thyroid cancer is to administer radioactive iodine after surgery to kill any remaining thyroid cancer cells. Why is radioactive iodine used?

▶ ______________________________________

____

____

____

____

____

Activity **19.6**

Examining the Microscopic Anatomy of the Endocrine Organs: Pancreas and Adrenal Gland

A The endocrine portion of the pancreas includes the pancreatic islets, which are scattered throughout the organ.

1. View a slide of the pancreas with the low-power objective lens and identify the **pancreatic acini** (singular = **acinus).** Each acinus contains a cluster of cuboidal cells (pancreatic acinar cells), arranged around a central lumen. The acinar cells are nonendocrine glandular cells that produce digestive enzymes. Observe that these darkly stained cells compose the vast majority of the pancreas.

2. Scattered among the pancreatic acini, identify the islands of lighter-staining cells. They are the **pancreatic islets** or **islets of Langerhans,** which are the endocrine portion of the pancreas. The pancreatic islets contain four cell types. Each type is responsible for producing a specific hormone, as follows:
   - Alpha cells produce glucagon.
   - Beta cells produce insulin.
   - Delta cells produce somatostatin.
   - F cells produce pancreatic polypeptide (PP).

   On your slides, you will probably be unable to identify the different cell types.

3. In the space provided, draw your observations. Label the pancreatic acini and pancreatic islets.

▶

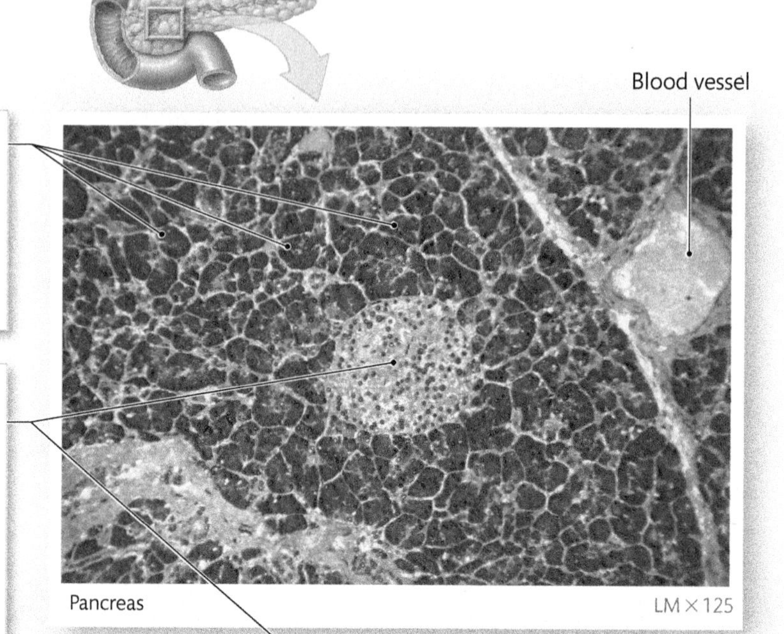

Pancreas LM × 125

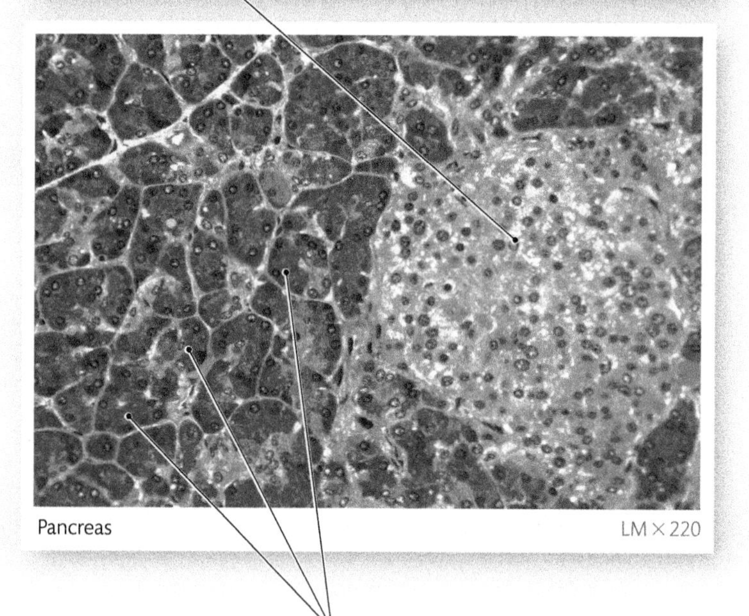

Pancreas LM × 220

B The adrenal cortex produces more than 24 steroid hormones, collectively called adrenocortical steroids. Table 19.2 summarizes the hormones produced by the adrenal cortex.

TABLE 19.2 Hormones Produced by the Adrenal Cortex

Cortical Region	Hormones	Function
Zona glomerulosa	Mineralocorticoids (ex. aldosterone)	Act on the kidneys to conserve sodium and water
Zona fasciculata	Glucocorticoids (ex. cortisol)	Act on many cells to conserve glucose by utilizing fatty acids and proteins as an energy source (glucose-sparing effect)
Zona reticularis	Androgens (male sex hormones)	Stimulate growth of axillary and pubic hair before puberty

1 View a slide of the adrenal gland with the scanning or low-power objective lens and identify its two major regions:

the outer **adrenal cortex**

the inner **adrenal medulla**

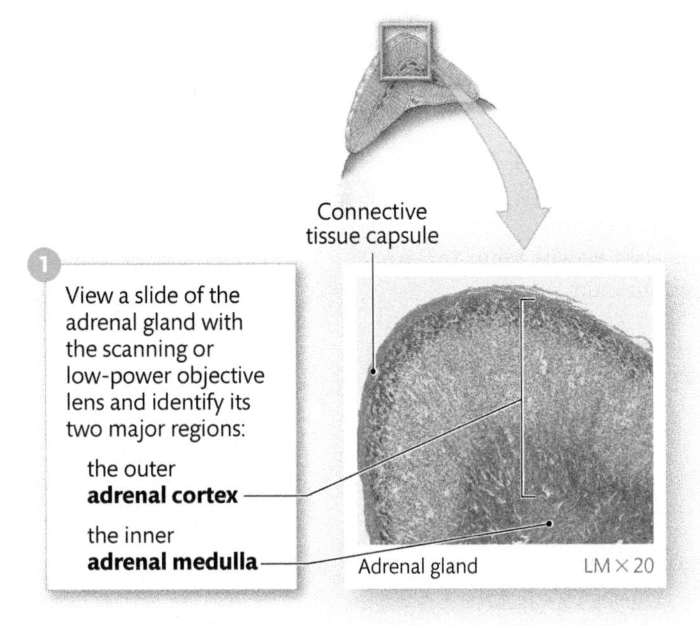

Adrenal gland LM × 20

2 Center the adrenal cortex and switch to high power. Identify the three cortical regions:

Zona glomerulosa (outermost layer): Covered by a connective tissue capsule

Zona fasciculata (middle layer): Cells are larger and more lightly stained than those in the zona glomerulosa

Zona reticularis (innermost layer): Borders the adrenal medulla

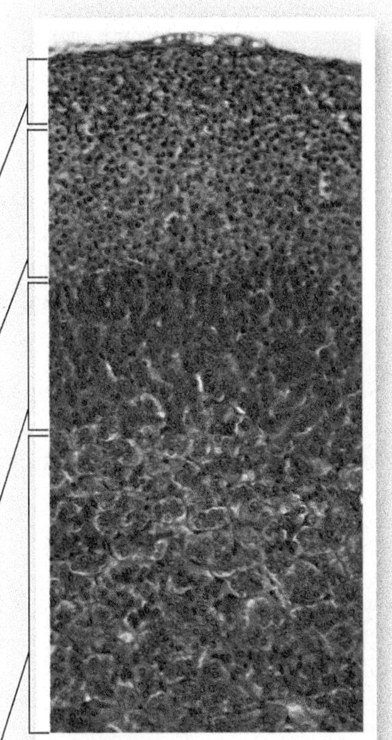
Adrenal gland LM × 50

3 Switch back to low power, locate the **adrenal medulla,** and center this region in the field of view. With high power, observe the cells in this area. An extensive network of capillaries, which travel between the cells, can also be identified. Preganglionic sympathetic nerve fibers stimulate the endocrine cells in the adrenal medulla to produce and secrete epinephrine and norepinephrine. This action initiates the fight-or-flight response.

4 In each region of the adrenal cortex, the endocrine cells are arranged in a unique way. Describe and compare the cellular arrangement in each region. ▶

a. Zona glomerulosa: ______________________________

____

b. Zona fasciculata: ______________________________

____

c. Zona reticularis: ______________________________

____

MAKING CONNECTIONS

If a person becomes dehydrated, in which region of the adrenal cortex will hormone production and secretion increase? Explain.

▶ ______________________________

____

____

____

____

BEFORE YOU MOVE ON . . .

« LOOKING BACK

In this laboratory exercise, you learned that there are many organs in the body that have an endocrine function. For several organs, such as the pituitary and pineal glands in the brain, the thyroid and parathyroid glands in the neck, and the pancreatic islets and adrenal glands in the abdominopelvic cavity, the primary function is the production and secretion of hormones. Many other organs, such as the heart, thymus, small intestine, and reproductive organs, also produce hormones but have other primary functions. All hormones are secreted into blood capillaries and are transported via the blood circulation to distant target organs. Thus, all endocrine glands will have an extensive blood supply, and all the cells in an endocrine gland will have at least one surface directly adjacent to a capillary.

Both the endocrine system and the nervous system are important for maintaining homeostasis and regulating metabolic activity. Both systems use chemical messengers, hormones and neurotransmitters, that bind to receptors on cells at target organs. Some of these chemicals, such as norepinephrine, can function as a hormone and a neurotransmitter. The hormonal effect on a target organ is typically slow acting with long duration, whereas the neuronal effect tends to be fast acting with short duration.

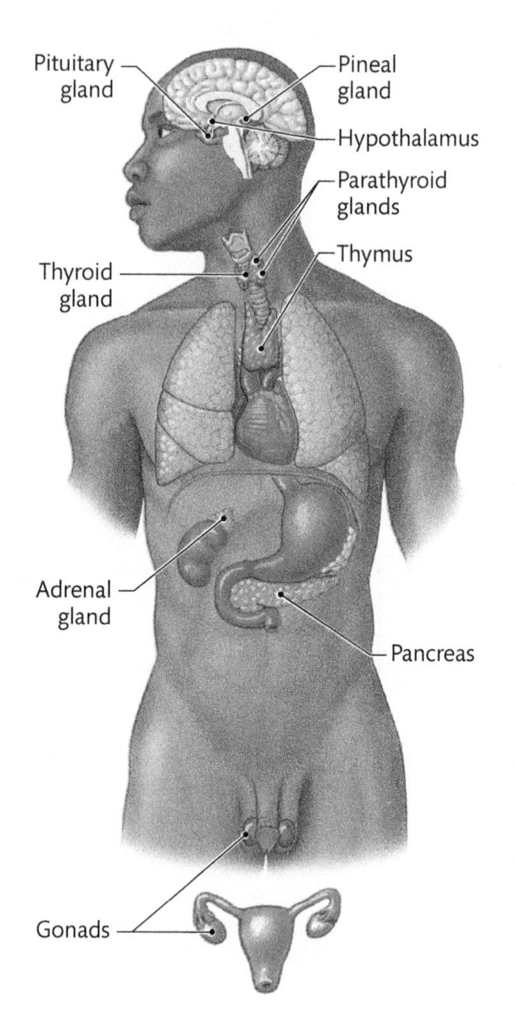

The combined effect of the endocrine and nervous systems on a target organ is referred to as a neuroendocrine effect. Why is it important for the endocrine and nervous systems to have a close functional relationship?

► ______________________________

____

____

____

____

____

LOOKING FORWARD »

Hormones from endocrine glands are distributed to their target organs in the blood circulation. The cardiovascular system, which includes the blood, heart, and blood vessels, will be studied in Exercises 20 through 23. In the next exercise (Exercise 20), you will examine the structure and function of blood cells.

Name ______________________________

Lab Section ______________________________

Date ______________________________

EXERCISE **19** **REVIEW SHEET**

The Endocrine System

QUESTIONS 1–12: Identify the labeled endocrine structures in the diagram on the right. For each structure, list the hormone(s) it produces.

Structure	Hormone(s) Produced
1.	
2.	
3.	
4.	
5.	
6.	
7.	
8.	
9.	
10.	
11.	
12a.	
12b.	

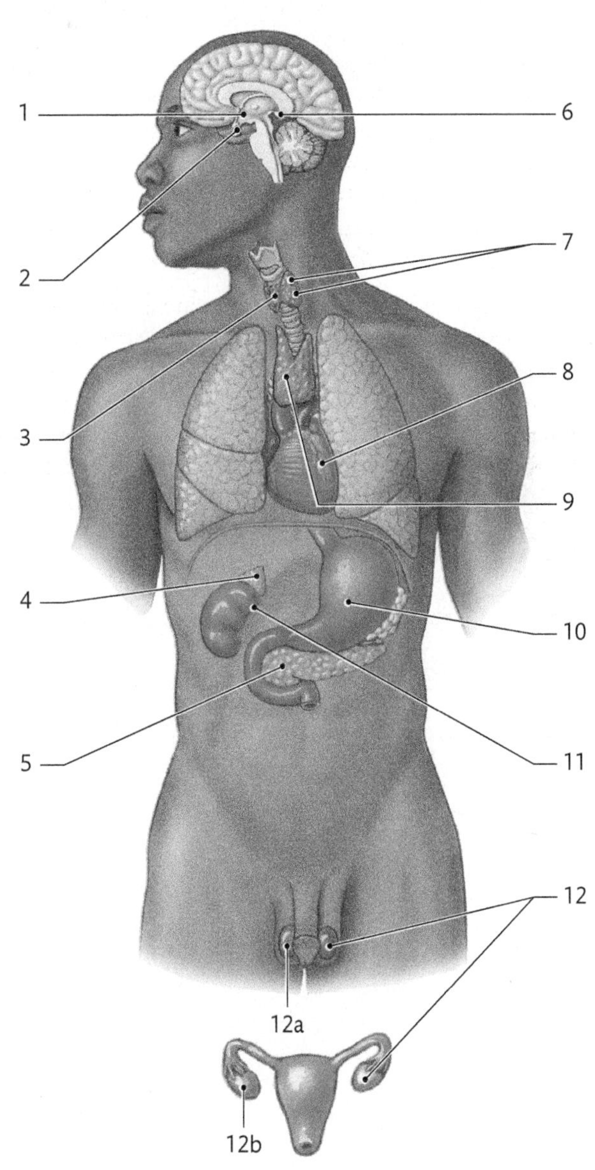

13. Explain how the hypothalamus influences the function of the anterior lobe of the pituitary gland.

____

____

____

14. Identify the endocrine organs that are illustrated in the following photos.

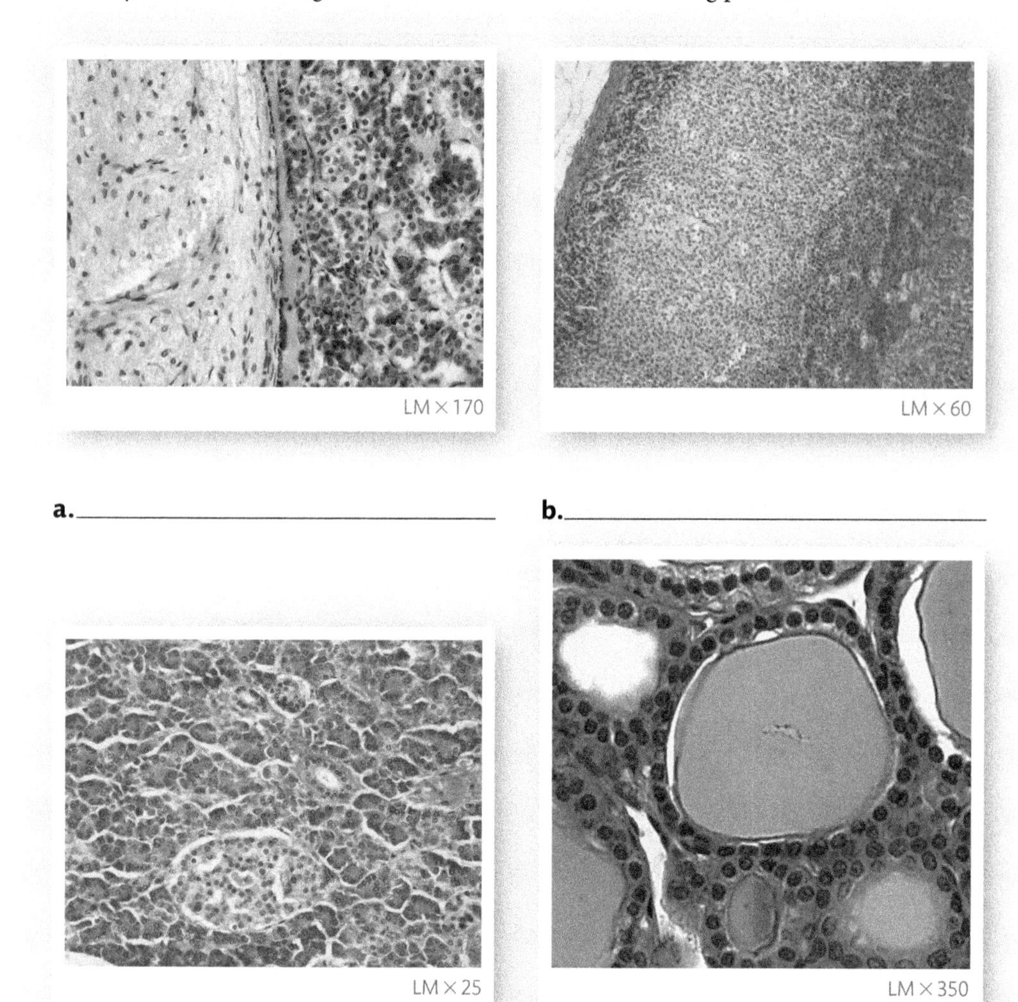

a. ______________________ **b.** ______________________

c. ______________________ **d.** ______________________

15. Complete the following table.

Hormone	Organ That Produces Hormone	Hormone Function
Insulin	a.	b.
c.	Adrenal medulla	d.
T_3 and T_4	e.	f.
g.	h.	Elevates blood glucose levels
Oxytocin	i.	j.
k.	l.	Promotes secretion of thyroid hormones
Aldosterone	m.	n.
o.	p.	Promotes secretion of glucocorticoids
PTH	q.	r.

EXERCISE

20

Blood Cells

LEARNING OUTCOMES

These Learning Outcomes correspond by number to the laboratory activities in this exercise. When you complete the activities, you should be able to:

Activity 20.1 **Identify red blood cells, white blood cells, and platelets in a human blood smear and describe their functions.**

Activity 20.2 **Identify the five types of white blood cells in a human blood smear.**

Activity 20.3 **Perform a differential white blood cell count.**

Activity 20.4 **Determine your ABO and Rh blood type, using universal precautions to collect a blood sample and discard wastes.**

LABORATORY SUPPLIES

- Compound light microscopes
- Prepared microscope slides of human blood smears
- Clean microscope slides
- Gloves
- Face masks
- Protective eyewear
- Wax labeling pencils
- Sterile blood lancets
- Sterile alcohol pads
- Blood mixing sticks
- Warming tray
- Paper towels
- Anti-A, anti-B, and anti-D (anti-Rh) blood typing solutions
- Containers for the disposal of biohazardous wastes
- Simulated blood typing kits (an alternative to using human blood)

PRE-LAB QUIZ

Before you begin, read all the activities in Exercise 20 and the required reading in your textbook that is assigned by your instructor.

1. During this laboratory exercise, you will complete all the following activities *except*
   a. determine your blood type.
   b. perform a differential white blood cell count.
   c. measure the pH of your blood.
   d. identify the types of white blood cells.
   e. compare the structure of red blood cells and white blood cells.
2. You are viewing a blood smear and identify a white blood cell with a bilobed nucleus and bright red granules in the cytoplasm. What type of white blood cell are you viewing?

   ____
3. True or False: When you are viewing a normal blood smear, you are likely to see more neutrophils than any other type of white blood cell. __________
4. The ______________________ are white blood cells that can differentiate into macrophages.
5. What term best describes the shape of a red blood cell? ______________________
6. What component of the formed elements has an important role in the formation of a blood clot? ______________________
7. When you are doing a differential white blood cell count, the cell that you are least likely to identify is a ______________________.
8. True or False: If a person has type AB blood, it means that surface antigens A and B are not present on the cell membranes of red blood cells. __________
9. True or False: If compatible blood types are mixed, agglutination will occur. __________
10. True or False: Universal precautions should always be used when carrying out laboratory activities that expose you to human body fluids. __________

Activity **20.1**

Identifying Blood Cells

Blood consists of various blood cells and cell fragments **(formed elements)** suspended in a fluid matrix **(blood plasma).** Plasma, which makes up 55 percent of total blood volume, is 90 percent water, but it also includes a wide variety of dissolved substances, including gases, nutrients, hormones, waste products, ions, and proteins. The formed elements make up the remaining 45 percent of total blood volume.

A Red blood cells (RBCs) or erythrocytes make up 99.9 percent of the formed elements. Thus, when you examine a normal human blood smear, the vast majority of the cells that you observe will be red blood cells. Scan the slide carefully to locate the other formed elements: white blood cells (WBCs) and platelets.

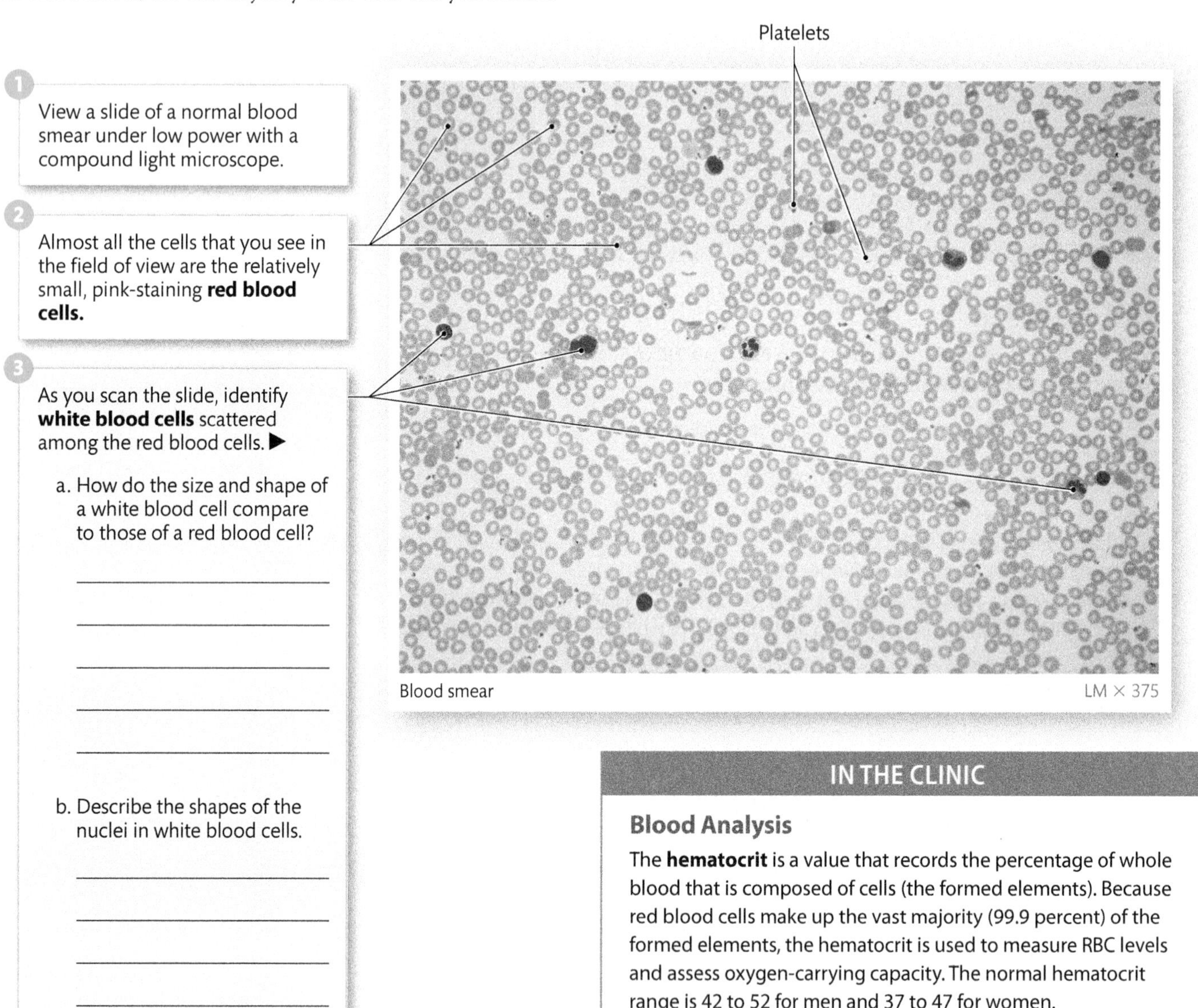

Blood smear LM × 375

Word Origins

Erythrocyte is derived from two Greek words: *erythros,* meaning "red," and *kytos,* meaning "cell." Thus, red blood cells are called erythrocytes. *Leukocyte* is also derived from Greek and means "white cell" (*leukos* means "white").

IN THE CLINIC

Blood Analysis

The **hematocrit** is a value that records the percentage of whole blood that is composed of cells (the formed elements). Because red blood cells make up the vast majority (99.9 percent) of the formed elements, the hematocrit is used to measure RBC levels and assess oxygen-carrying capacity. The normal hematocrit range is 42 to 52 for men and 37 to 47 for women.

Another diagnostic measure for oxygen-carrying capacity is **hemoglobin (Hb) concentration** in red blood cells. Hb concentration is measured in grams per deciliter (g/dL). Typically, a decline in hematocrit will also cause a decline in Hb concentration. Normal Hb concentrations are 14 to 18 g/dL in men and 12 to 16 g/dL in women.

Anemia is a condition characterized by a reduction in oxygen-carrying capacity and a resulting decline in oxygen transport to cells and tissues. Lower-than-normal hematocrit or Hb concentration is a warning sign for anemia.

B Red blood cells are flattened biconcave discs that lack nuclei and most organelles. Most of their cytoplasm is filled with the protein **hemoglobin,** which transports oxygen and carbon dioxide.

1 Examine **red blood cells** more closely under high magnification. They are uniquely shaped as biconcave discs with a relatively thin central region and thick peripheral region. Many resemble doughnuts or Life Saver candies because the thin central region stains more lightly than the thick peripheral region. Notice that mature RBCs lack nuclei.

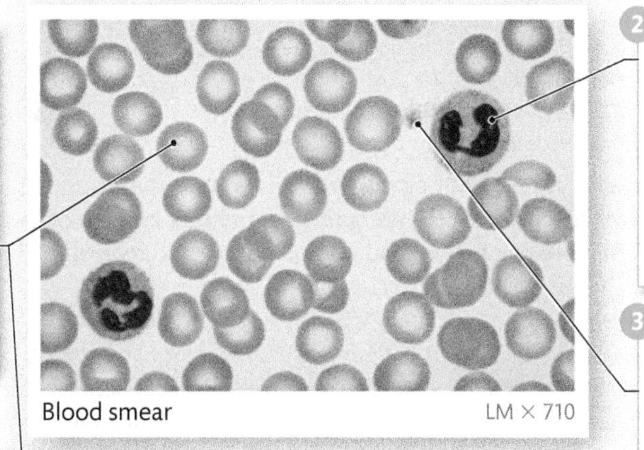
Blood smear LM × 710

2 Identify **white blood cells** in your field of view. Most WBCs are much larger than RBCs. They contain distinct nuclei that usually stain a deep blue or purple. White blood cells play essential roles in defending the body against infection.

3 Identify **platelets,** which are cytoplasmic fragments derived from cells called megakaryocytes. Platelets have a critical role in blood clotting and the repair of damaged blood vessels.

Colorized SEM × 6400

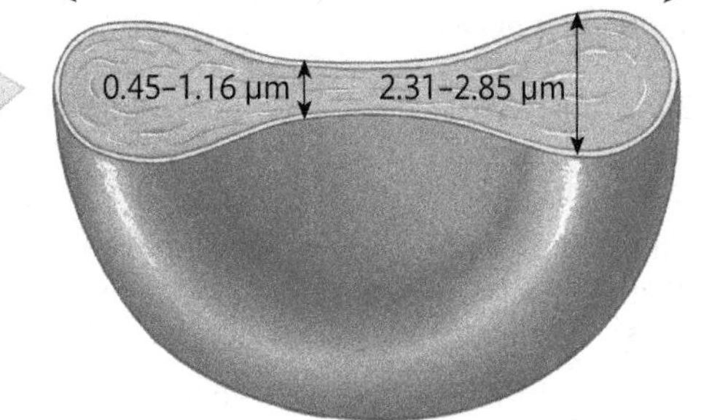

MAKING CONNECTIONS

Hemoglobin is composed of four globular protein subunits. Each subunit contains a **heme** pigment molecule (see figure at right). The vast majority of oxygen that is transported in the blood is bound to iron in the heme molecules. Given that another gas, carbon monoxide, has a much stronger affinity than oxygen to the same binding sites on iron, describe how carbon monoxide poisoning would occur.

▶ ____

____

____

____

____

____

____

____

____

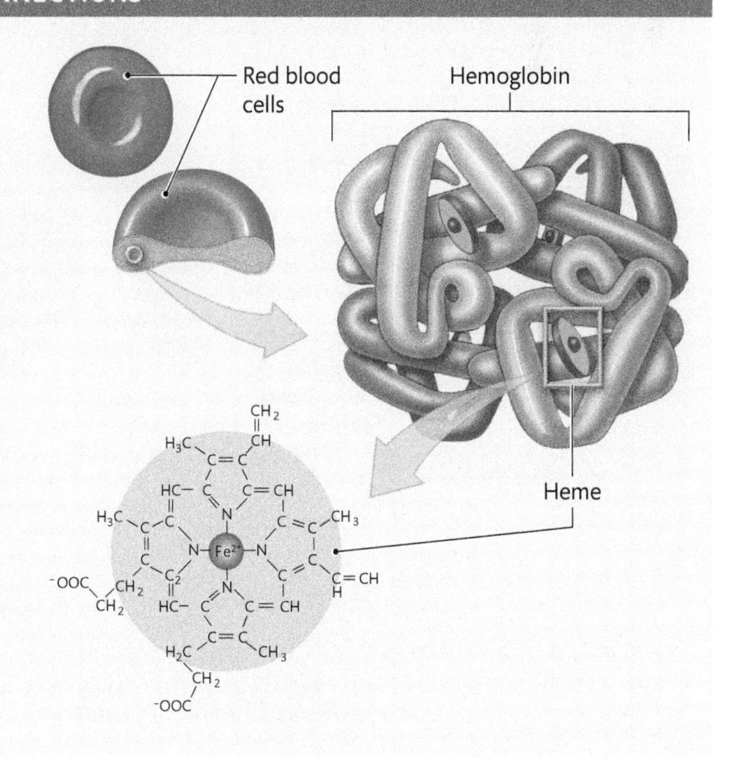

Activity **20.2**

Identifying White Blood Cell Types

White blood cells are categorized into two major groups:

- The **granulocytes** include neutrophils, eosinophils, and basophils.
- The **agranulocytes** include monocytes and lymphocytes.

White blood cells play vital roles in defending the body against infections and promoting inflammation in response to tissue damage and allergies. When these cells are activated, they are capable of migrating from blood vessels to surrounding tissues by squeezing between adjacent cells in capillary walls. This process is called **diapedesis.**

You have already discovered that white blood cells are not nearly as abundant as red blood cells. They make up only 1 percent of all blood cells (6000–9000 cells/μL). They are easy to identify because of their large size and distinctive nuclei, however. Using the high-power objective lens, carefully scan a normal human blood smear. When you identify a white blood cell, use immersion oil and the oil immersion lens to observe the cell at a high magnification. Refer to the descriptions and photographs of the white blood cells, below and on the next page, to help you identify each type.

A Granulocytes have distinct granules in their cytoplasm and unique bilobed or multilobed nuclei.

1. **Neutrophils** are the most abundant white blood cells and the easiest to identify; they make up 50 to 70 percent of all white blood cells (1800–7300 cells/μL). They are about twice the size of red blood cells (10–14 μm in diameter). Neutrophils have distinctive mutlilobed nuclei and pale-staining granules in their cytoplasm. These cells act as phagocytes by engulfing and destroying pathogens and releasing cytotoxic enzymes.

2. **Eosinophils** are difficult to identify on a normal blood smear because they make up only 2 to 4 percent of all white blood cells (0–700 cells/μL). They are the same size as neutrophils and have bilobed or multilobed nuclei. Their cytoplasm is filled with bright red or orange granules. Eosinophils are phagocytic cells that attack parasitic organisms. They also diminish the effects of allergies and inflammation by releasing antihistamines.

3. **Basophils** are similar in size to neutrophils and eosinophils. They have bilobed nuclei that are usually hidden by their dark blue or purple cytoplasmic granules. Basophils enhance inflammation and tissue repair by releasing histamine and heparin. They are quite rare, making up less than 1 percent of all white blood cells (0–150 cells/μL). If you think that you have found one of these cells, verify the identification with your instructor.

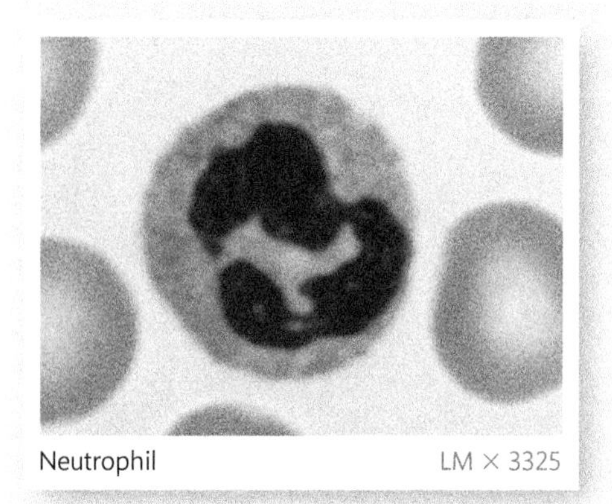

Neutrophil LM × 3325

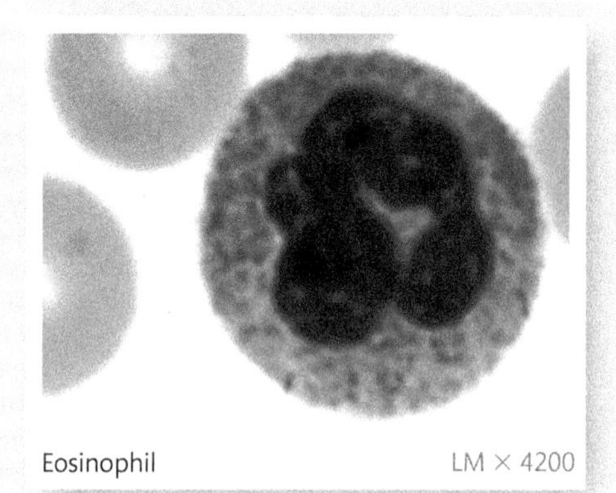

Eosinophil LM × 4200

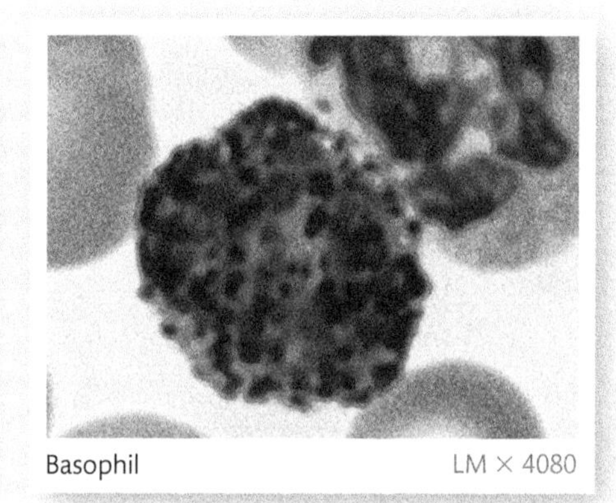

Basophil LM × 4080

B Agranulocytes are noted for their absence of cytoplasmic granules.

1 **Monocytes** are the largest white blood cells, ranging from 14 to 24 μm in diameter. They make up 2 to 8 percent of all white blood cells (200–950 cells/μL). Most monocytes have kidney-shaped nuclei with a deep indentation on one side. These cells differentiate into macrophages. They attack and destroy bacteria and viruses by phagocytosis.

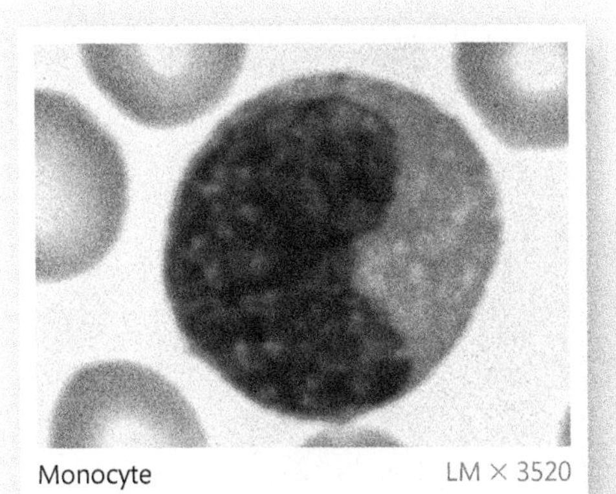
Monocyte LM × 3520

2 **Lymphocytes** are relatively abundant in a normal blood smear, making up 20 to 30 percent of all white blood cells (1500–4000 cells/μL). They range in diameter from 5 to 17 μm and are often classified as small (5–8 μm), medium (9–12 μm), and large (13–17 μm). Their nuclei are round and occupy the vast majority of the cell volume, leaving only a narrow rim of cytoplasm around the periphery. Lymphocytes regulate the immune response by attacking cells directly or by producing antibodies.

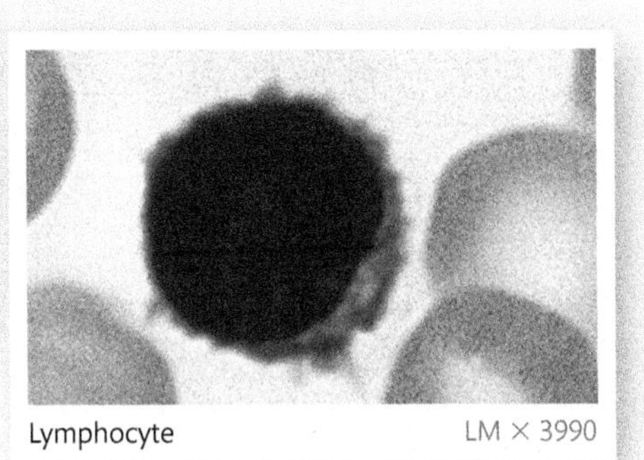
Lymphocyte LM × 3990

3 In the space provided, draw and label each white blood cell type that you observe.

▶

Word Origins

Diapedesis is derived from Greek and means "leaping through." White blood cells can pass or "leap through" the walls of blood vessels to enter surrounding tissues, where they destroy disease-causing organisms and ingest the remains of dead cells.

MAKING CONNECTIONS

Over-the-counter antihistamine drugs relieve the pain and discomfort of inflammation due to allergies, cold, and fever. Overuse of these drugs could prolong your symptoms, however. Speculate on a reason for this effect. (*Hint:* Consider the function of basophils and eosinophils.)

▶ ______________________________

____

____

____

____

____

Activity 20.3

Performing a Differential White Blood Cell Count

Under normal conditions, blood contains a certain percentage of each white blood cell type. Any deviation from the normal percentage ranges could indicate an abnormal condition such as a bacterial, viral, or parasitic infection.

A Before you begin your differential white blood cell count, review the identification of the cell types.

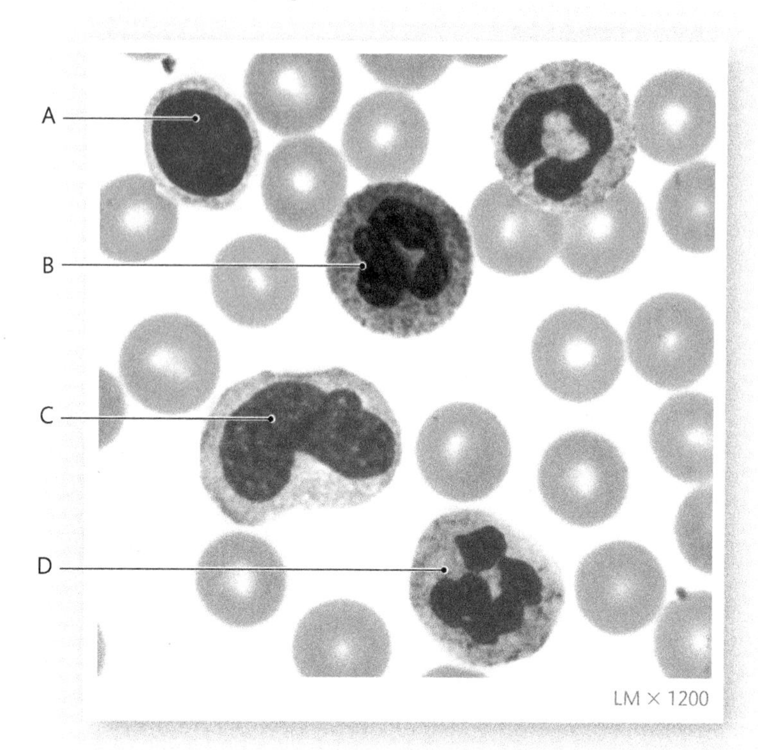

LM × 1200

1 In the figure to the right, which white blood cell type is not shown?

▶ ______________________________

____

2 Identify the white blood cells labeled A through D. ▶

Cell A ______________________________

Cell B ______________________________

Cell C ______________________________

Cell D ______________________________

3 The unlabeled white blood cell is the same type as which other cell in the figure?

▶ ______________________________

IN THE CLINIC

Leukemia

Leukemia is a blood cancer that involves the uncontrolled propagation of abnormal white blood cells in the bone marrow (see figure on right). Leukemias can be classified according to the type of white blood cell involved. **Myeloid leukemias** involve granulocytes; **lymphoid leukemias** involve lymphocytes. For both categories, the disease can advance quickly (acute) or slowly (chronic). As leukemia progresses, abnormal white blood cells gradually replace normal cells, impairing bone marrow function. Production of red blood cells, normal white blood cells, and platelets declines significantly, leading to anemia, infections, and reduced blood clotting. One option to fight leukemia is a bone marrow transplant. This procedure exposes the patient to a massive dose of radiation or chemotherapy to kill abnormal and cancerous cells in the bone marrow. This exposure also kills normal cells, so the individual is left highly susceptible to infections that could be fatal. Next, the patient is given healthy bone marrow tissue that, it is hoped, will generate new populations of normal blood cells. Compatibility of blood and tissue types is a critical factor in transplant operations. If tissue rejection occurs, the donor's lymphocytes could attack and destroy the recipient's tissues, a condition that could be fatal.

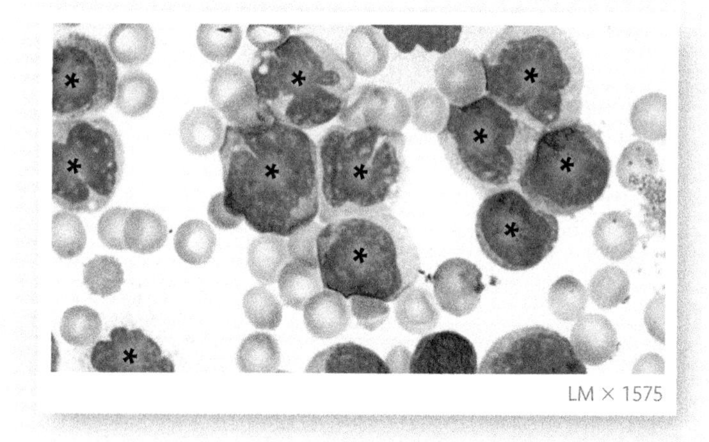
LM × 1575

4 Read the In the Clinic box on the left and answer the following question:

In the figure above, abnormal WBCs are marked with a black asterisk. What type of leukemia, myeloid or lymphoid, is shown? Explain.

▶ ______________________________

____

____

B A **differential white blood cell count** is a laboratory test that determines the percentage of each white blood cell type. Although this procedure can be completed quickly using computers, the manual method described here will yield similar results.

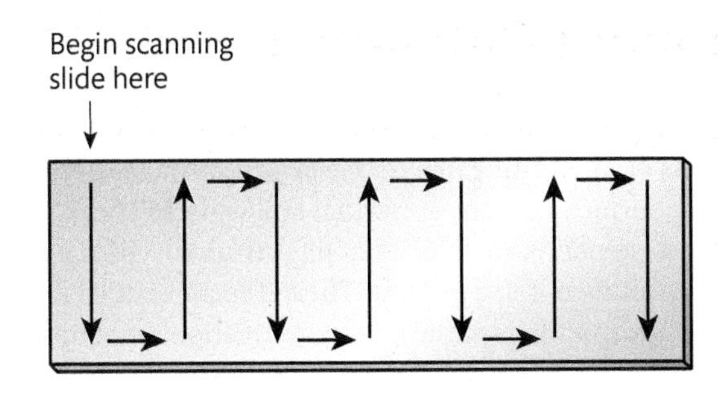

1 **Make a Prediction**
To complete your differential white blood cell count, you will identify at least 100 white blood cells. Before you begin, use the information in Table 20.1 to predict the number of each white blood cell type that you expect to find. Record your predictions in the "Predicted # of Cells" column in Table 20.1. ▶

2 Position a slide of a human blood smear on your microscope stage so that the upper-left margin of the blood smear is in your field of view.

3 Focus the image under low power and then switch to high power and observe the upper-left region of the blood smear. Find an area in which the blood cells are dispersed evenly throughout the field of view. For the most accurate results, use the oil immersion lens, although you can complete this activity with the highest dry objective lens.

4 Beginning at the upper-left region of the blood smear, carefully scan the entire slide in the back-and-forth pattern illustrated in the figure. When you observe a white blood cell, identify the type and record it in the "Observed # of Cells" column in Table 20.1. ▶

5 After you have identified at least 100 white blood cells, convert the number of each cell type to a percentage of the total, using the following equation. Record your results in the right-hand column of Table 20.1. ▶

Percentage (%) = (# cells observed/total # counted) × 100

TABLE 20.1 **Differential White Blood Cell Count**

Cell Type	% of All White Blood Cells	Predicted # of Cells	Observed # of Cells	% of Total White Blood Cells
Neutrophil	50–70			
Eosinophil	2–4			
Basophil	< 1			
Monocyte	2–8			
Lymphocyte	20–30			
Total	100			100

6 **Assess the Outcome**
Do your predicted results agree with your observed results? Provide an explanation for your observations.

▶ ______________________________

____

____

____

____

MAKING CONNECTIONS

Infectious mononucleosis is a disease believed to be caused by an infection with the Epstein-Barr virus. It is characterized by fever, sore throat, swollen lymph nodes, and an enlarged spleen. Discuss how a differential white blood cell count from an individual with this disease would differ from a normal count.

▶ ______________________________

____

____

____

Activity **20.4**

Determining Your Blood Type

Blood type is a genetically determined trait. It is based on the presence of specific glycoprotein molecules, called **surface antigens** or **agglutinogens,** on the cell membranes of red blood cells. An individual's immune system recognizes these surface antigens as normal and will not attack them as a foreign substance.

Blood plasma contains **antibodies** or **agglutinins,** which are each genetically programmed to react with a specific surface antigen if it is present. Thus, if incompatible types of blood are mixed, an antigen–antibody reaction will occur, and the result will be agglutination, or clumping of red blood cells.

A In humans, there are more than 50 blood groups, but most do not cause significant reactions when different types are mixed. In this activity, you will study two blood groups, ABO and Rh, that cause significant antigenic reactions.

1. The presence or absence of surface antigens is the key factor in determining the compatibility of a blood transfusion. For example, type A blood contains surface antigen A and anti-B antibodies. A type A recipient can receive any blood type that lacks surface antigen B, which could react with anti-B antibodies in the type A blood. Thus, a type A recipient is compatible with type A or type O donors. Which donor blood types are compatible with a type B recipient? Explain.

▶ ______________________________

____

____

2. Which blood type is considered to be a **universal donor?**

▶ ______________________________

Which blood type is considered to be a **universal recipient?**

▶ ______________________________

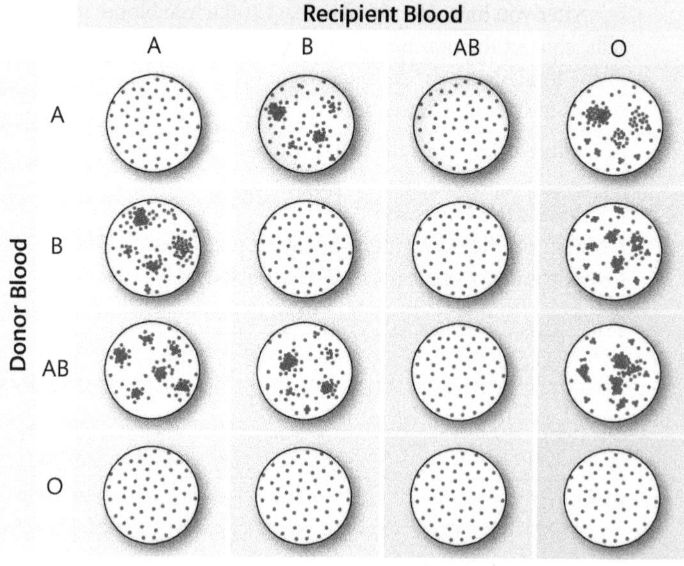

B Blood type can be determined by observing the presence or absence of agglutination when your blood sample is exposed to anti-A, anti-B, and anti-D (anti-Rh) antibodies.

⚠ **Warning:** *Blood may contain infectious organisms. Before you begin this activity, carefully review* ***universal precautions*** *described in Appendix A.*

1. **Make a Prediction**
Before you begin typing your blood, review the data in Table 20.2, which lists the distribution of blood types in the United States, and predict your blood type.

Predicted blood type:

▶ ______________________________

TABLE 20.2 Incidence of Blood Types in the United States (%)

Population Group	O	A	B	AB	Rh(+)
Caucasian	45	40	11	4	85
African American	49	27	20	4	95
Korean American	32	28	30	10	100
Japanese American	31	39	21	10	100
Chinese American	42	27	25	6	100
Native American	79	16	4	1	100
Hawaiian	46	46	5	3	100

2 Obtain a sterile glass microscope slide. With a wax pencil, draw a line that divides the slide into left and right sides. Label the left side "A" and the right side "B." Obtain a second sterile glass microscope slide and label it "Rh."

3 Place a paper towel on your work space. This towel will be used to place blood-collecting instruments prior to disposal.

4 Wash your hands thoroughly with warm water and soap and dry them completely with a paper towel. You will be collecting a blood sample from yourself. Put on protective eyewear and a face mask. Place a surgical glove on the hand that will hold the lancet, but not on the hand from which you will collect the sample. Use a sterile alcohol pad to clean the tip of your index finger on all sides. Place the used pad on the paper towel.

5 Open a sterile blood lancet to expose the sharp tip. Quickly jab the lancet tip into your fingertip, and then place the lancet on the paper towel. Never use the same lancet more than once.

6 Squeeze a drop of blood onto each half of the slide labeled A and B, and a third drop onto the slide labeled Rh. Add a drop of anti-A serum to blood sample A. Do not touch the blood samples with the dropper. In the same way, add a drop of anti-B serum to blood sample B and a drop of anti-D serum to blood sample Rh.

7 Mix the blood samples and antisera with clean mixing sticks. Use a new, clean stick to mix each sample. Place the used mixing sticks on the paper towel. Place both slides on a warming tray. Gently agitate the samples back and forth for 2 minutes.

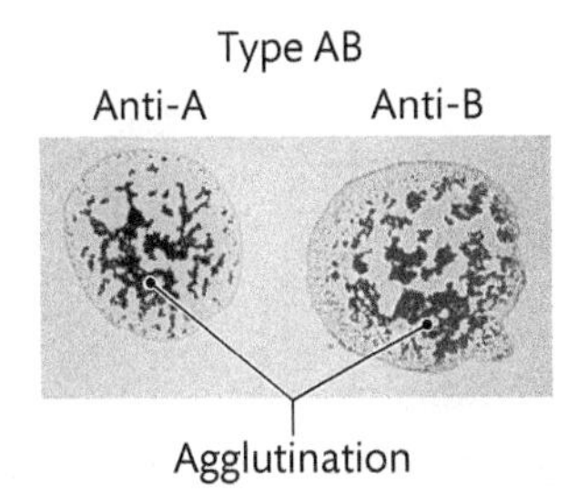

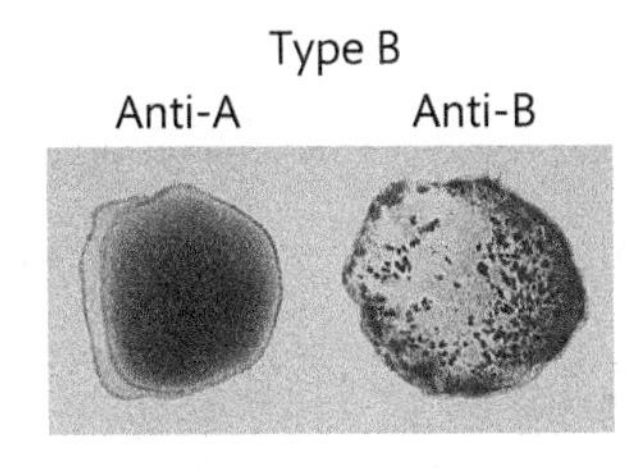

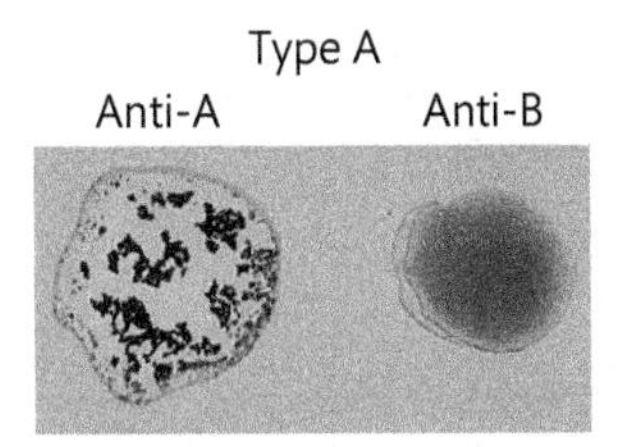

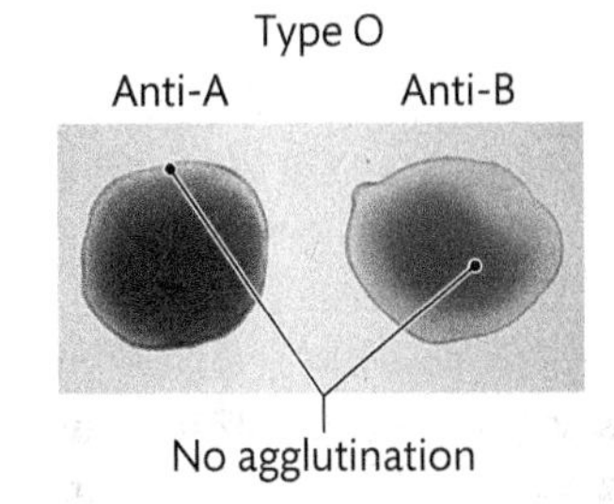

8 Examine the blood samples for evidence of agglutination. Agglutination will occur when antibodies in the antiserum react with the corresponding surface antigen on red blood cells. Record your ABO blood type:

▶ ______________________

9 Rh+ blood will agglutinate when exposed to anti-D serum, but Rh– blood will not. Rh agglutination is often difficult to observe with the unaided eye. If that is the case, use a microscope to observe the reaction. Record your Rh status:

▶ ______________________

10 **Assess the Outcome**

What is your blood type?

▶ ______________________

Did your result agree with your prediction? Explain.

▶ ______________________

____

____

MAKING CONNECTIONS

While typing the blood of a patient who is about to undergo surgery, the medical technician determines that agglutination occurs when the blood is exposed to both anti-A and anti-B antibodies. Agglutination does not occur when the blood is exposed to anti-D antibodies. What is the patient's blood type? Explain.

▶ ____

____

____

BEFORE YOU MOVE ON . . .

« LOOKING BACK

Blood is a component of the **cardiovascular system.** It performs a variety of functions that fall into five categories:

- *Transportation of substances:* Blood delivers oxygen and nutrients, needed for metabolic activities, to all body tissues. It transports carbon dioxide to the lungs and cellular wastes to the kidneys for elimination. Hormones from the endocrine glands are distributed by blood to various target organs.
- *Regulatory activities:* Blood proteins act as buffers that function to maintain stable blood pH levels. Diffusion between blood capillaries and interstitial fluids stabilizes blood concentrations of calcium, potassium, and other important ions.
- *Prevention of fluid loss:* Platelets and various blood proteins protect the body from excessive blood loss by repairing damaged blood vessels and forming blood clots.
- *Defensive activities:* White blood cells, antibodies, and various blood proteins protect the body from infections caused by bacteria, viruses, and other pathogens.
- *Regulation of body temperature:* Blood distributes heat throughout the body and shunts excess heat to the skin's surface for elimination.

Explain why each of the five functions, described above, is important for maintaining homeostasis.

▶ ____

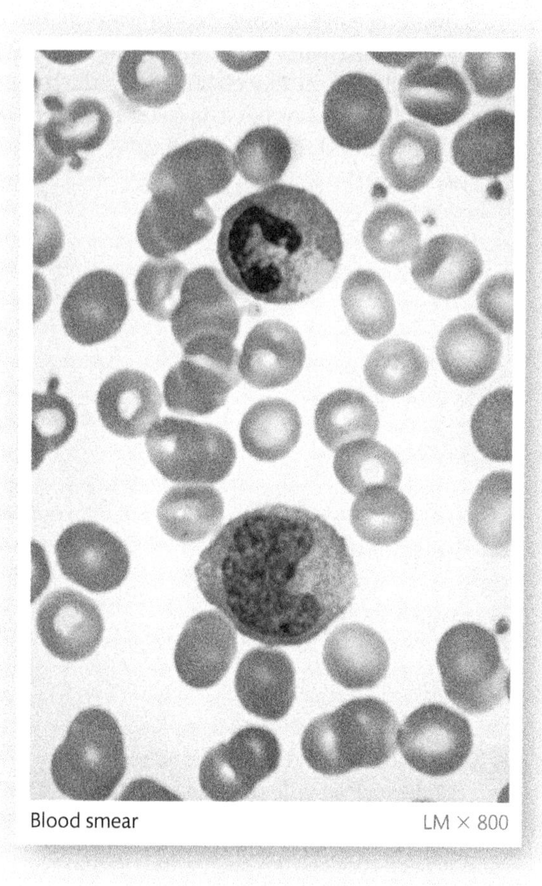

Blood smear LM × 800

LOOKING FORWARD »

Blood is distributed to all parts of the body by the pumping action of the heart. In the next laboratory exercise, you will examine the external and internal anatomy of the heart, identify the major blood vessels of the coronary circulation, and understand how blood flows through the heart chambers.

Name ______________________

Lab Section ______________________

Date ______________________

EXERCISE 20 REVIEW SHEET

Blood Cells

1. Complete the table by identifying the blood cell that is shown in each figure and describing its function.

	Cell Type	Function
	a.	b.
	c.	d.
	e.	f.
	g.	h.
	i.	j.
	k.	l.

2. A mature red blood cell lacks a nucleus. How does this characteristic explain why red blood cells have a life span of only 120 days?

____

____

____

3. White blood cells are able to move from blood plasma to surrounding tissues. What is the functional significance of this ability?

____

____

____

QUESTIONS 4–9: Match the term in column A with the correct description in column B.

A	B
4. Agglutinins ______	**a.** The percentage of whole blood that is composed of blood cells
5. Diapedesis ______	**b.** The clumping of red blood cells that occurs when incompatible blood types are mixed
6. Hematocrit ______	**c.** Cell fragments involved in blood clotting and the repair of damaged blood vessels
7. Heme ______	**d.** The iron-containing pigment molecule in hemoglobin that binds to oxygen
8. Platelets ______	**e.** Antibodies that react to specific surface antigens on red blood cells
9. Agglutination ______	**f.** The migration of white blood cells from capillaries to surrounding tissues

10. A patient's blood test revealed an elevated level of neutrophils, a condition called neutrophilia. Suggest a possible cause for this result.

____

____

____

11. If a woman has type A blood, what donor blood type(s) can she safely receive during a blood transfusion? Explain.

____

____

____

12. The results of a blood typing test are shown below. What is the person's blood type? Explain.

____

____

____

____

____

____

Anti-A　　Anti-B　　Anti-D

13. In your textbook or on the Internet, research the condition called hemolytic disease of the newborn (HDN). Describe HDN and explain why it is important for a couple to know their blood types before having children.

____

____

____

____

EXERCISE 21

Gross Anatomy of the Heart

LEARNING OUTCOMES

These Learning Outcomes correspond by number to the laboratory activities in this exercise. When you complete the activities, you should be able to:

Activity 21.1	**Describe the arrangement of the pericardium that surrounds the heart.**
Activity 21.2	**Identify important structures on the surface of the heart.**
Activity 21.3	**Describe the internal structure of the heart chambers.**
Activity 21.4	**Identify and describe the structure of the heart valves.**
Activity 21.5	**Identify the tissue layers of the heart wall.**
Activity 21.6	**Describe the coronary circulation.**
Activity 21.7	**Describe the flow of blood through the heart.**
Activity 21.8	**Identify the position of the heart in the thoracic cavity by palpation.**
Activities 21.9–21.10	**Dissect a sheep heart and compare its external and internal anatomy to that of the human heart.**

LABORATORY SUPPLIES

- Human heart models
- Plastic bags
- Masking tape
- Cotton or cheesecloth
- Torso model
- Preserved sheep hearts
- Dissecting trays
- Dissecting tools
- Dissecting gloves
- Protective eyewear
- Colored pencils

PRE-LAB QUIZ

Before you begin, read all the activities in Exercise 21 and the required reading in your textbook that is assigned by your instructor.

1. During this laboratory exercise, you will complete all the following activities *except*
   a. examine the external and internal anatomy of the heart.
   b. identify the different components of the pericardium.
   c. study the stages of the cardiac cycle.
   d. examine the coronary circulation.
2. During this laboratory exercise, you will dissect a heart from what animal? ______________
3. The heart is located in a region of the thoracic cavity known as the ______________.
4. True or False: The superior and inferior venae cavae return blood to the heart by draining into the left atrium. __________
5. True or False: The epicardium and endocardium are both serous membranes. __________
6. The right and left coronary arteries are branches of the ______________.
7. True or False: Most cardiac veins drain directly into the right atrium. __________
8. True or False: Deoxygenated blood in the pulmonary circulation is transported to the lungs, where it is oxygenated. __________
9. The inferior border of the heart rests on what structure? ______________
10. During your heart dissection, you will study the internal anatomy by cutting the heart in what sectional plane? ______________

Activity **21.1**

Examining the Organization of the Pericardium

The heart is enclosed by a membranous sac called the **pericardium.** This structure consists of two parts. The outer **fibrous pericardium** is a tough, connective tissue layer. The inner **serous pericardium** is a delicate serous membrane that forms a double-layered sac around the heart.

A You can demonstrate the organization of the pericardium by pushing your fist into a plastic bag.

1. Close off the open end of a large, clear plastic bag with masking tape.

2. Push your fist into the wall of the plastic bag as shown in the diagram to the right. Notice that your fist is covered by two layers of plastic with a space between the two layers.

3. Your fist represents the heart.

4. The plastic bag represents the two layers of the **serous pericardium:**

   The inner plastic layer corresponds to the **visceral pericardium,** which is in contact with the heart wall.

   The outer plastic layer corresponds to the **parietal pericardium.**

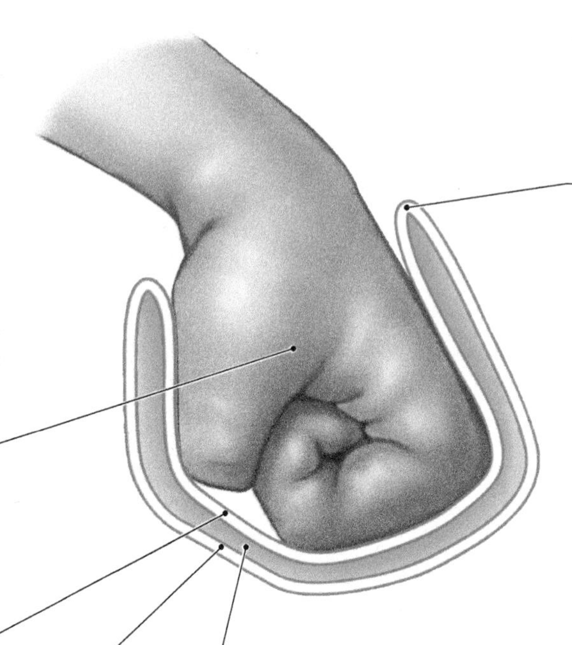

5. Your wrist represents the superior portion of the heart (the base of the heart). Notice that the two layers of plastic are continuous with each other at this position. In other words, the visceral pericardium is continuous with the parietal pericardium where the great blood vessels are connected to the heart.

6. The space between the two layers of plastic represents the **pericardial cavity.**

7. Wrap a layer of cotton or cheesecloth over the outer plastic layer (the parietal pericardium). This layer represents the **fibrous pericardium,** which is fused to connective tissue that surrounds adjacent structures. On a torso model, identify three structures that are adjacent to the heart. Write the names of the structures below.

▶ ______________________________

____

____

B The heart is surrounded by the pericardial cavity.

1 Label the following structures in the diagram to the right. ▶

- Parietal pericardium
- Visceral pericardium
- Pericardial cavity
- Location where the visceral pericardium is continuous with the parietal pericardium
- Fibrous pericardium

2 Within the inset, label the following: ▶

- Parietal pericardium
- Fibrous pericardium
- Pericardial cavity

3 What important function does the fibrous pericardium serve?

▶ ________________________________

____

____

____

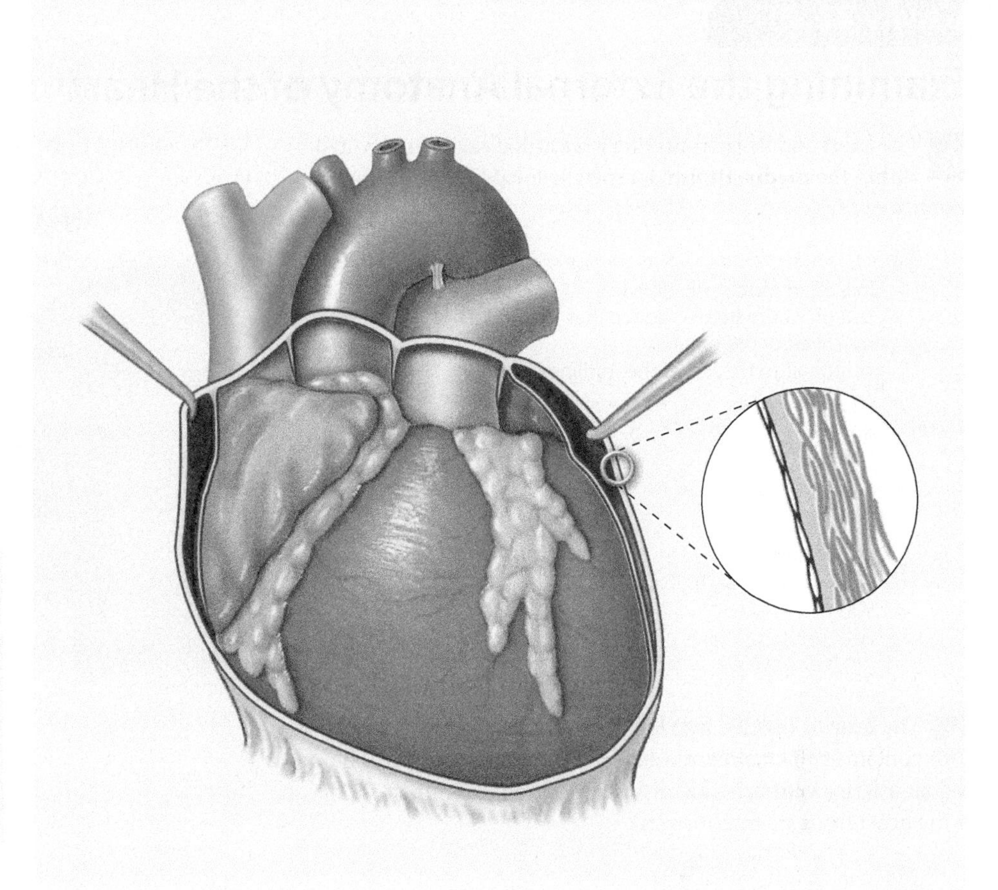

IN THE CLINIC

Pericarditis

Inflammation of the pericardial membranes, known as **pericarditis,** increases the friction between the two membranes and causes an overproduction of fluid. As the fluid accumulates in the pericardial cavity, the normal movements of the heart wall are inhibited, leading to a condition called **cardiac tamponade.** Pericarditis is most often caused by a viral infection. It can also develop after a heart attack or heart surgery. Most cases can be successfully treated with over-the-counter or prescription anti-inflammatory drugs. For severe or recurrent cases, **pericardiocentesis** may be performed. During this procedure, a needle is inserted into the pericardial cavity to remove excess fluid.

MAKING CONNECTIONS

Explain why the heart is surrounded by, but not within, the pericardial cavity.

▶ ____

____

____

____

____

Activity **21.2**

Examining the External Anatomy of the Heart

A The heart and its surrounding pericardial cavity are located within the **mediastinum,** a centrally located area within the thoracic cavity.

1. On a torso model, observe the position of the **heart.** Notice that two-thirds of the organ is positioned to the left of the midline.

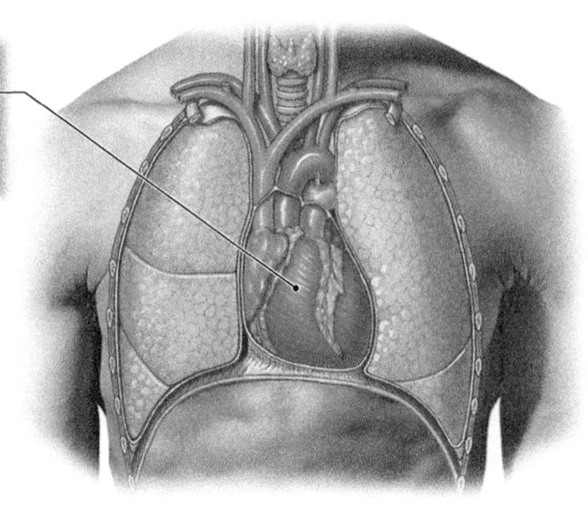

2. Identify the structures that border the heart: ▶

a. Anteriorly ______________________

b. Laterally ______________________

c. Inferiorly ______________________

3. Remove the heart and identify the structures that border the heart posteriorly.

▶ ______________________

____

____

B The heart is divided into left and right sides. Each side contains two chambers: a superior **atrium** that receives blood and an inferior **ventricle** that discharges blood. Examine a model of the heart from an anterior view.

1. Identify three of the four heart chambers:

right atrium and its medial appendage, the **right auricle**

right ventricle

left ventricle

left atrium (best observed from a posterior view)

2. Locate the **apex** of the heart. It is formed by the inferior tip of the left ventricle. Label the apex on the diagram. ▶

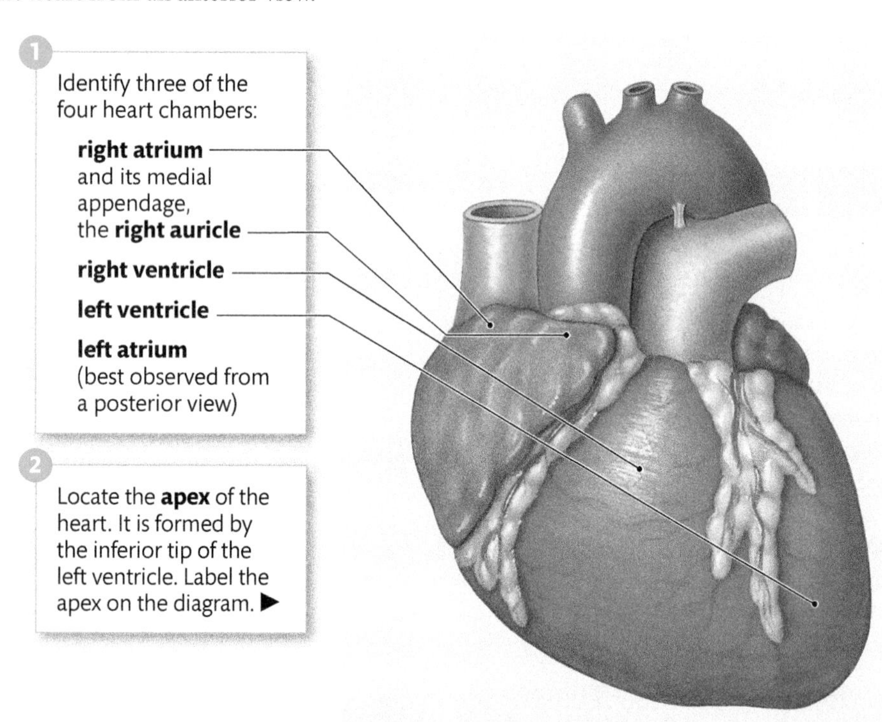

Anterior view

3. Identify the great blood vessels that are connected to the base of the heart and label them on the diagram. ▶

The **superior vena cava** empties into the right atrium.

The **ascending aorta** receives blood from the left ventricle and gives rise to the arch of the aorta.

The **pulmonary trunk** is located anterior to the ascending aorta and receives blood from the right ventricle.

4. Identify two major sulci (grooves) on the anterior surface and label them on the diagram. ▶

The **coronary sulcus** divides the atria, superiorly, from the ventricles, inferiorly.

The **anterior interventricular sulcus** travels toward the apex from the coronary sulcus and forms a border between the left and right ventricles.

Want more practice? Go to: **MasteringA&P** > Study Area > Menu > Lab Tools > PAL >

- Anatomical Models > Cardiovascular System > Heart
- Human Cadaver > Cardiovascular System > Heart

C The left atrium and the base of the heart can best be identified from a posterior view.

1 List the four heart chambers and label them in the diagram. ▶

1. ______________________
2. ______________________
3. ______________________
4. ______________________

2 The **base** of the heart is at the heart's posterior and superior aspects.

3 Identify two major sulci (grooves) on the posterior surface and label them on the diagram. ▶

From the anterior surface, the **coronary sulcus** continues along the posterior surface between the atria and ventricles. Notice that it forms a complete circle around the heart.

The **posterior interventricular sulcus,** like the anterior interventricular sulcus, descends toward the apex from the coronary sulcus and forms a border between the left and right ventricles.

Posterior view

D The great blood vessels are attached to the heart at the base. You can identify these structures from a posterior view.

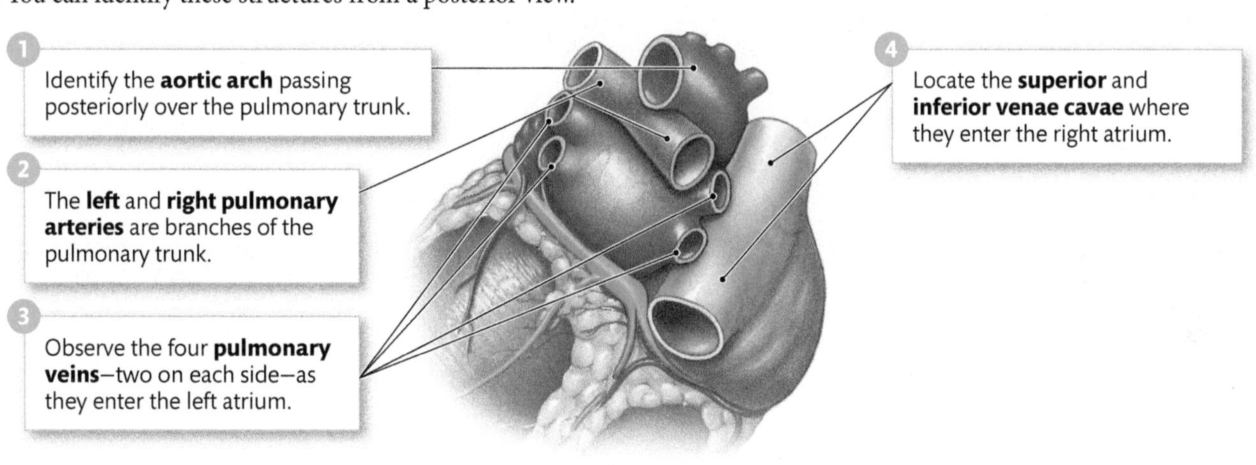

Word Origins

Auricle is derived from *auricular*, the Latin word for "external ear." Early anatomists gave auricles that name because they resembled the external ear.

MAKING CONNECTIONS

From an external view, what features could you use to distinguish the right atrium from the left atrium?

▶ ____

____

From an external view, what features could you use to distinguish the right ventricle from the left ventricle?

▶ ____

____

Activity **21.3**

Examining the Internal Anatomy of the Heart: The Heart Chambers

A The two atria are thin-walled chambers that receive blood from the great veins.

1 On a heart model, open the heart wall to expose the internal structures. Identify the two superior chambers:

Left atrium ________

Right atrium ________

2 With one hand, place your thumb on one side and your index finger on the other side of the wall that separates the two atria. Your fingers are holding the interatrial septum. Identify the **fossa ovalis,** an oval depression along the interatrial septum within the right atrium.

3 In the right atrium, identify the following regions:

The anterior wall, including the auricle, has a rough surface formed by the **pectinate muscles.**

The posterior wall lacks pectinate muscles and is smooth. Locate the **openings for the superior vena cava and inferior vena cava** along this portion of the atrial wall.

4 At the inferior end of the right atrium, identify the **right atrioventricular (AV) orifice,** the opening that leads into the right ventricle.

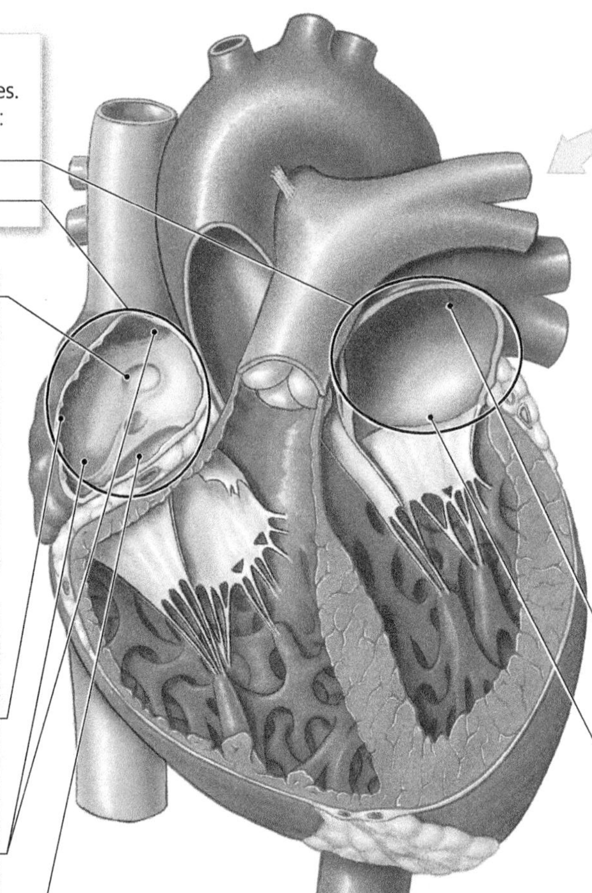

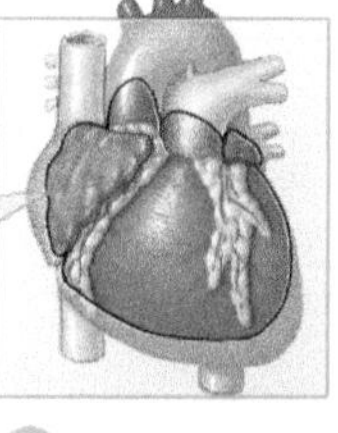

5 Examine the internal structure of the left atrium. How is it similar to the right atrium?

▶ ________

____

____

____

____

6 Identify the **openings for the pulmonary veins.**

7 Identify the **left atrioventricular (AV) orifice,** at the inferior end of the left atrium, leading into the left ventricle.

B The two ventricles are thick-walled chambers that pump blood into the great arteries.

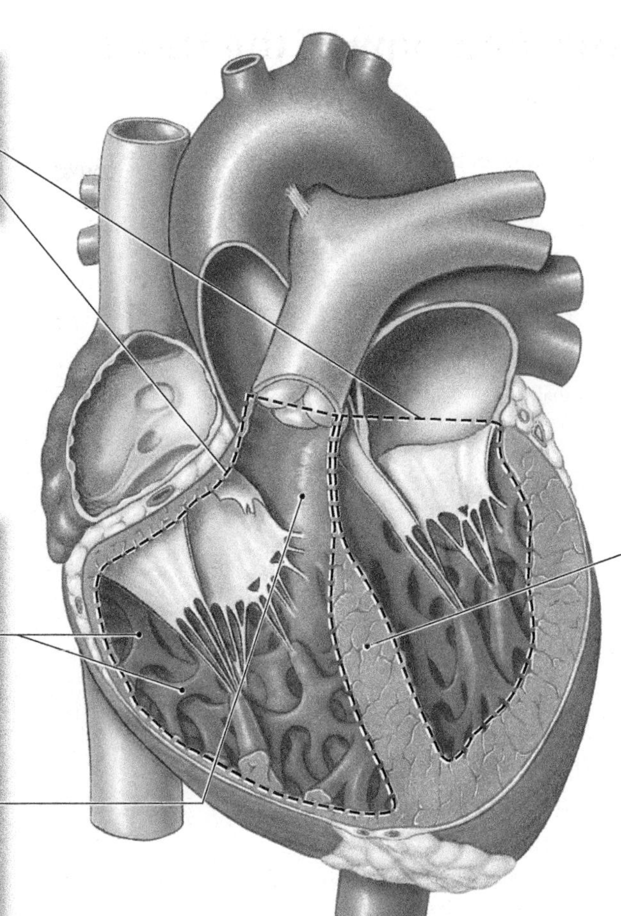

1 Identify the two inferior heart chambers:

Left ventricle ____________

Right ventricle ____________

2 Within the right ventricle, identify the following regions:

The inferior portion receives blood from the right atrium. Its walls are covered by an irregular network of muscular elevations called the **trabeculae carneae.**

Superiorly, the right ventricle narrows into a cone-shaped chamber, the **conus arteriosus,** which leads to the pulmonary trunk. The wall of the conus arteriosus is smooth and lacks trabeculae carneae.

3 Inside the left ventricle, the **aortic vestibule** is the smooth-walled, superior region that leads to the aorta. It is similar to the conus arteriosus in the right ventricle. Describe another structural similarity between the two ventricles.

▶ ____________________________________

____

____

____

4 Place the thumb and index finger of one hand on either side of the wall that separates the two ventricles. This structure is the **interventricular septum.** Notice that this wall is much thicker than the interatrial septum.

5 On the surface of the heart, what two sulci form the anterior and posterior margins of the interventricular septum?

▶ ____________________________________

____

____

____

IN THE CLINIC

Atrial Septal Defect

Inside the right atrium, along the interatrial septum, there is an oval depression called the fossa ovalis. This depression marks the site of the **foramen ovale,** an opening that connects the atria in the fetal heart. The foramen ovale has a valve that allows blood to travel from the right atrium to the left atrium but not in the reverse direction. This specialization in the fetal circulation allows most of the oxygen-rich blood from the placenta to bypass the lungs and pulmonary circulation and pass directly to other vital organs via the systemic circulation. At birth, the foramen ovale closes when the valve fuses with the interatrial septum. Incomplete closure of the foramen ovale, called an **atrial septal defect,** allows oxygen-rich blood in the left atrium to mix with oxygen-poor blood in the right atrium. This malformation can be repaired surgically to prevent the two blood supplies from blending.

MAKING CONNECTIONS

Speculate on the function of the trabeculae carneae.

▶ ____________________________________

____

____

____

____

____

Want more practice? Go to: **MasteringA&P** > Study Area > Menu > Lab Tools > PAL >

- Anatomical Models > Cardiovascular System > Heart
- Human Cadaver > Cardiovascular System > Heart

Activity **21.4**

Examining the Internal Anatomy of the Heart: The Heart Valves

A The two **semilunar valves** regulate the flow of blood between the ventricles and the great arteries. These valves are controlled by changes in pressure. They open when the ventricular pressure is greater than the arterial pressure and the ventricles are pumping blood into the arteries. They close when the ventricular pressure drops below the arterial pressure and the pumping action is complete.

1 Identify the **pulmonary semilunar valve** at the junction of the right ventricle and the pulmonary trunk. Notice that the valve has three crescent-shaped cusps. The cusps are extensions of the pulmonary trunk wall.

2 Identify the **aortic semilunar valve** (not shown) at the junction of the left ventricle and the ascending aorta. How is its structure similar to that of the pulmonary semilunar valve?

▶ ______________________________

____

____

____

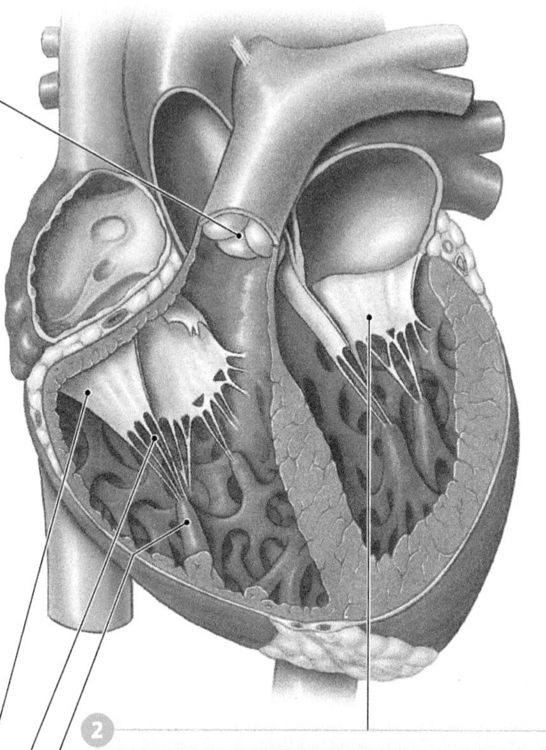

B The two **atrioventricular (AV) valves** are positioned between the atria and ventricles at the atrioventricular orifices. These valves open when the atrial pressure is greater than the ventricular pressure and blood flows from the atria to the ventricles. They close when the ventricular pressure rises above the atrial pressure.

1 The **right AV valve** is also called the **tricuspid valve** because of its three cusps. Identify the following structures on this valve:

The **cusps** extend into the ventricular chambers.

The **chordae tendineae** are fibrous cords that connect the inferior free margins of the cusps to papillary muscles.

The **papillary muscles** are located on the ventricular wall.

2 Identify the **left AV (bicuspid** or **mitral) valve** on the left side of the heart. How is this valve similar to the tricuspid valve?

▶ ______________________________

____

____

How is it different?

▶ ______________________________

____

____

MAKING CONNECTIONS

When the AV valves close, blood cannot flow backward into the atria. Speculate on how the chordae tendineae and papillary muscles function to close the valves.

▶ ______________________________

____

____

Activity **21.5**

Examining the Internal Anatomy of the Heart: The Heart Wall

A The heart wall consists of three tissue layers: the endocardium, myocardium, and epicardium.

1 Observe the left ventricular wall on a heart model and identify the three layers of the heart wall:

The inner **endocardium** is a simple squamous epithelium with a thin supporting layer of areolar connective tissue. It lines the internal walls of the heart chambers. The cusps of the AV valves are extensions of the endocardium.

The middle **myocardium** is the thickest layer. It is composed primarily of cardiac muscle fibers, separated by connective tissue containing capillaries and nerves. Contractions of cardiac muscle fibers are responsible for the pumping action of the heart.

The outermost **epicardium** is a serous membrane of connective tissue and a simple squamous epithelium. This layer is also called the *visceral pericardium*.

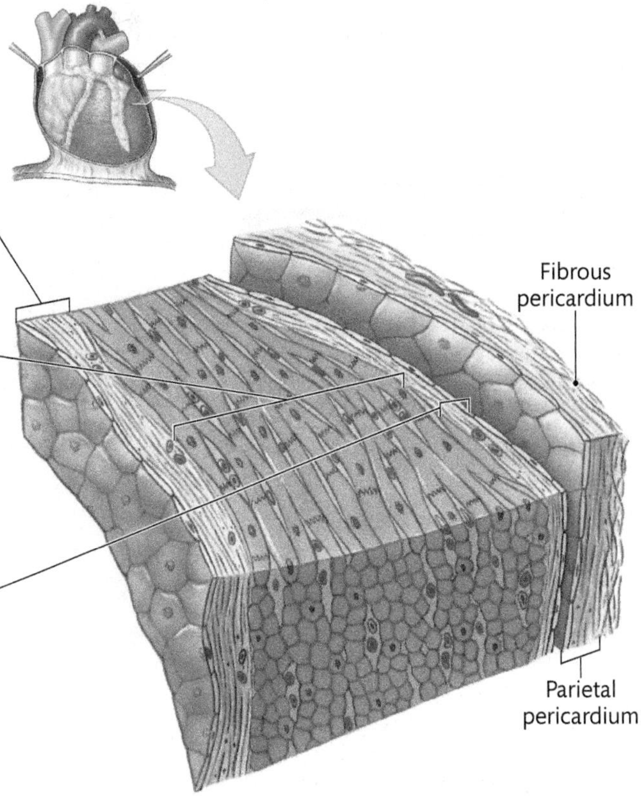

Fibrous pericardium

Parietal pericardium

2 Notice that the walls of the ventricles are much thicker than the walls of the atria. Explain why.

▶ ______________________________

____

____

____

____

3 Observe that the left ventricular wall is thicker than the right ventricular wall. Explain why.

▶ ______________________________

____

____

____

____

MAKING CONNECTIONS

What is **endocarditis**? Explain why this condition could impede normal blood flow through the heart.

▶ ______________________________

____

____

____

____

____

____

____

Activity **21.6**

Identifying Blood Vessels of the Coronary Circulation

Like any organ, the heart must have an adequate blood supply that delivers oxygen and nutrients to cells and carries away carbon dioxide and other metabolic wastes.

A The right and left **coronary arteries** branch directly off the ascending aorta. These arteries and their branches deliver blood to all regions of the heart.

1. Locate the **right coronary artery** as it travels to the right along the coronary sulcus.

2. Just before the right coronary artery curves around to the posterior surface, it gives off a branch called the **right marginal artery,** which descends along the lateral margin of the right ventricle.

3. Identify the **left coronary artery.** It branches off the ascending aorta just superior to the aortic semilunar valve and passes along the coronary sulcus, posterior to the pulmonary trunk.

4. The left coronary artery gives off two main branches:

   The **circumflex artery** travels along the coronary sulcus. It curves around the left side and continues onto the posterior surface.

   The **anterior interventricular artery** descends toward the apex along the anterior interventricular sulcus.

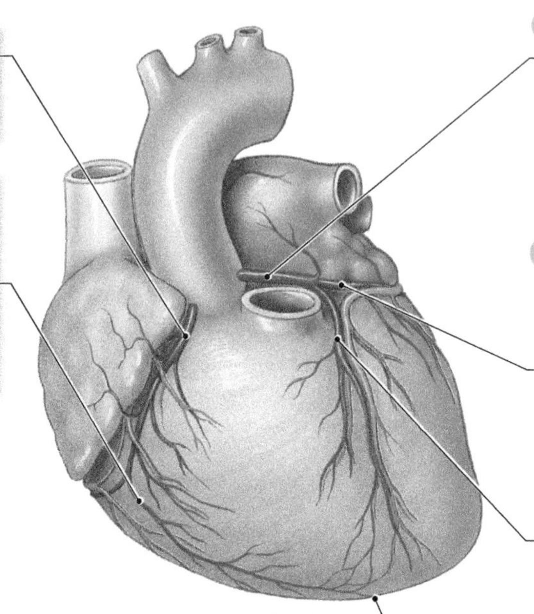

Anterior view

B On the posterior surface of the heart, the circumflex and right coronary arteries give off two large branches.

1. Identify the **left marginal artery** as it branches off the circumflex artery. It descends along the posterolateral wall of the left ventricle.

2. Find the **posterior interventricular artery.** It branches off the right coronary artery and descends toward the apex along the posterior interventricular sulcus.

3. Locate two examples of **anastomoses** (singular = **anastomosis:** a natural connection between two blood vessels) in the coronary circulation.

   Near the apex, branches of the anterior and posterior interventricular arteries come together.

   Along the coronary sulcus on the posterior surface, branches of the circumflex and right coronary arteries merge.

   What do you think is the functional significance of these arterial connections?

   ▶ ________________________________

   ____

   ____

   ____

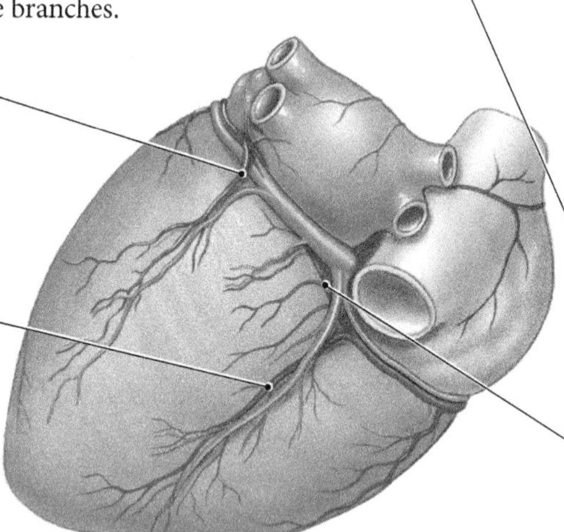

Posterior view

C **Cardiac veins** drain blood from the heart wall. Most of these veins drain into the **coronary sinus,** a large dilated sac that empties into the right atrium.

1. On the anterior surface of the heart, identify the **great cardiac vein.** It ascends along the anterior interventricular sulcus, running alongside the anterior interventricular artery. At the coronary sulcus, it travels with the circumflex artery to the posterior surface, where it drains into the coronary sinus.

2. Find the **small cardiac vein** as it travels alongside the right marginal artery toward the coronary sulcus. Along the coronary sulcus, it travels with the right coronary artery to the posterior surface and drains into the coronary sinus.

3. The **anterior cardiac veins** are small veins that travel a short distance along the anterior surface of the right ventricle. They drain directly into the right atrium.

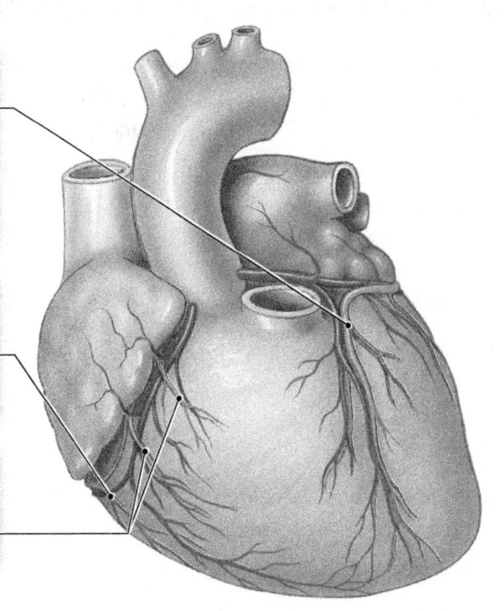

Anterior view

4. On the posterior surface of the heart, identify two other veins that drain into the coronary sinus:

   The **posterior cardiac vein** travels with the left marginal artery along the left ventricular wall.

   The **middle cardiac vein** ascends along the posterior interventricular sulcus, traveling with the posterior interventricular artery.

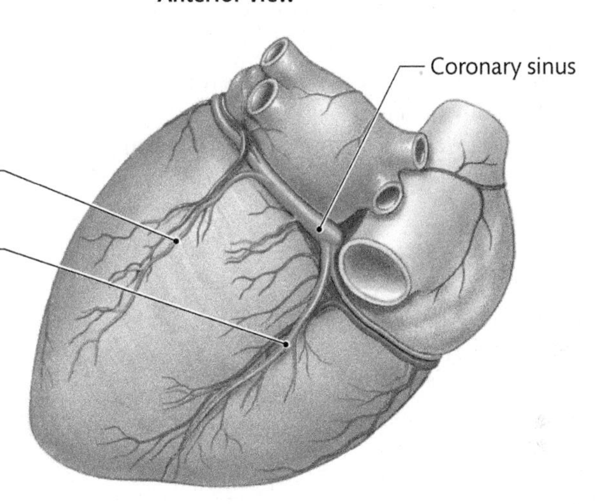

Posterior view

MAKING CONNECTIONS

By observing the paths of the coronary arteries and their branches, predict the regions of the heart wall that will be supplied by the following blood vessels: ▶

Anterior interventricular artery: ______________________________

Circumflex artery: ______________________________

Posterior interventricular artery: ______________________________

Right marginal artery: ______________________________

Left marginal artery: ______________________________

Activity 21.7

Tracing Blood Flow through the Heart Chambers

The heart is a two-sided muscular pump that regulates two separate loops for the circulation of blood. The right side of the heart controls the pulmonary circulation. The left side of the heart controls the systemic circulation. The flow of blood through pulmonary and systemic circuits occurs simultaneously.

A **Pulmonary circulation:** Deoxygenated (oxygen-poor) blood is transported to the lungs, where it is oxygenated (gains oxygen). On a model of the heart, trace the pathway of blood through the pulmonary circulation by reviewing the following steps. Label the steps in the diagram.

1. Deoxygenated blood enters the right atrium via the venae cavae, coronary sinus, and anterior cardiac veins.
2. Blood passes through the tricuspid valve and enters the right ventricle.
3. The right ventricle pumps blood through the pulmonary semilunar valve into the pulmonary trunk.
4. The pulmonary arteries and their branches send blood to the lungs to be oxygenated.
5. The pulmonary veins transport the oxygenated (oxygen-rich) blood to the left atrium.

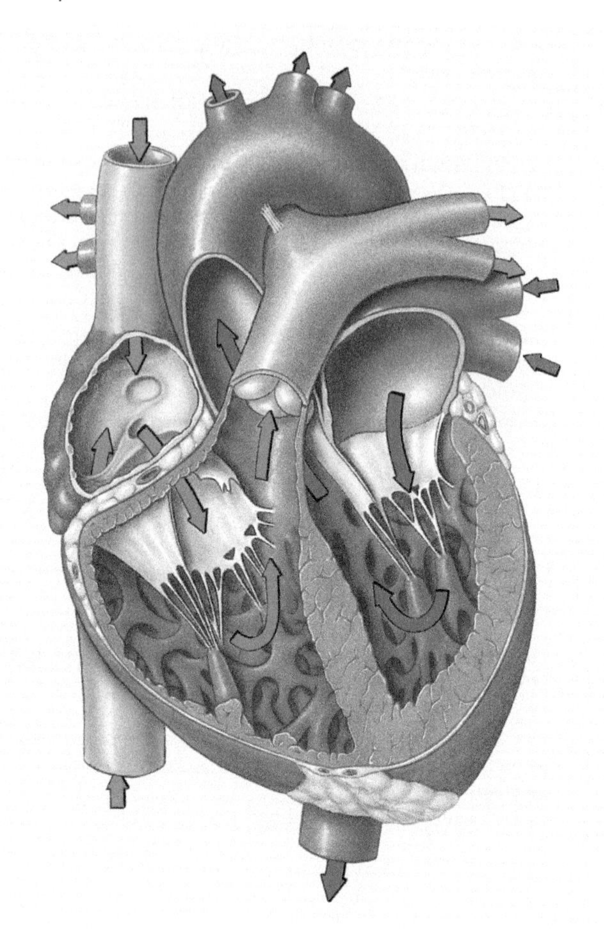

B **Systemic circulation:** Oxygenated blood is transported throughout the body to deliver oxygen to all cells and tissues. On a model of the heart, trace the pathway of blood through the systemic circulation by reviewing the following steps. Label the steps in the diagram. ▶

6. Oxygenated blood passes through the bicuspid (mitral) valve and enters the left ventricle.
7. The left ventricle pumps blood through the aortic semilunar valve into the aorta.
8. The aorta and its branches send oxygenated blood to:

   8a the head, neck, and upper limbs via branches of the aortic arch.

   8b the thorax, abdomen, pelvis, and lower limbs via branches of the descending aorta.
9. Deoxygenated blood returns to the right atrium via the superior and inferior venae cavae.

MAKING CONNECTIONS

If the left ventricle is damaged and cannot pump its normal volume of blood into the aorta, it will lag behind the pace of the right ventricle, and eventually blood will back up in the pulmonary circulation. This condition is called **congestive heart failure.** Explain how restrictions in blood flow from the left ventricle into the systemic circulation can have a negative effect on blood flow in the pulmonary circuit.

▶ ______________________________

____

____

____

____

Activity **21.8**

Surface Anatomy: Identifying the Position of the Heart

A If you can locate the positions of your ribs and intervening intercostal spaces, you will be able to identify the position of your heart along the anterior thoracic wall.

1. Palpate the **jugular notch,** which is the slightly curved superior margin of the manubrium.

2. From the jugular notch, move your finger inferiorly until you can palpate the **sternal angle,** a slender horizontal ridge that marks the articulation between the manubrium and the body of the sternum.

3. The costal cartilages of the second ribs attach to the sternum at the sternal angle. If you move your finger laterally, in either direction, from the sternal angle, you will palpate the **second costal cartilage** and the **second rib.**

4. By moving your finger inferiorly from the second rib, you will palpate an alternating series of ribs and intercostal spaces. For example, just inferior to the second rib is the second intercostal space, followed by the third rib and third intercostal space, and so on.

B Identify the following cardiac landmarks and borders by palpating their positions on your own body and identifying them on a skeleton.

1. The **base** of the heart extends from the inferior border of the second costal cartilage on the left side to the superior border of the third costal cartilage on the right.

2. The **left border of the heart** extends from the left second costal cartilage to the apex of the heart in the left fifth intercostal space.

3. The **apex** of the heart is located in the left fifth intercostal space, 7 to 9 cm (2.7–3.5 in.) to the left of the median plane.

4. The **inferior border of the heart** rests on the diaphragm. It extends from the sixth costal cartilage on the right side to the apex of the heart on the left side.

5. The **right border of the heart** is a curved margin that extends from the third costal cartilage to the sixth costal cartilage on the right side.

MAKING CONNECTIONS

Examine a heart model and determine which heart chamber(s) form the: ▶

Base of the heart: ______________________

Apex of the heart: ______________________

Left border of the heart: ______________________

Right border of the heart: ______________________

Inferior border of the heart: ______________________

Activity 21.9

Dissecting the Sheep Heart: External Anatomy

The sheep heart is remarkably similar to the human heart and is an excellent model for studying cardiac structure. As you dissect, have models, illustrations, or photographs of the human heart available so that you can make structural comparisons.

A To study the external anatomy of the heart, the pericardial sac must first be removed.

1. Identify the **pericardial sac** if it is present. This structure includes the **fibrous pericardium** and the **parietal pericardium.** The fibrous pericardium is a thick outer layer of fibrous connective tissue and fat that encloses the heart. The parietal pericardium is a thin serous membrane attached to the inner wall of the fibrous pericardium.
2. Carefully remove the pericardial sac to expose the entire heart. From the apex of the heart, cut along the pericardial sac toward the base and detach its attachments to the great vessels.
3. Notice that the heart wall is covered by a thin, translucent membrane. With forceps, lift a portion of this membrane off the heart's surface. This structure is the **visceral pericardium.** Recall that the visceral pericardium and **epicardium** are the same structure.

B Hold the sheep heart so that the anterior surface faces you.

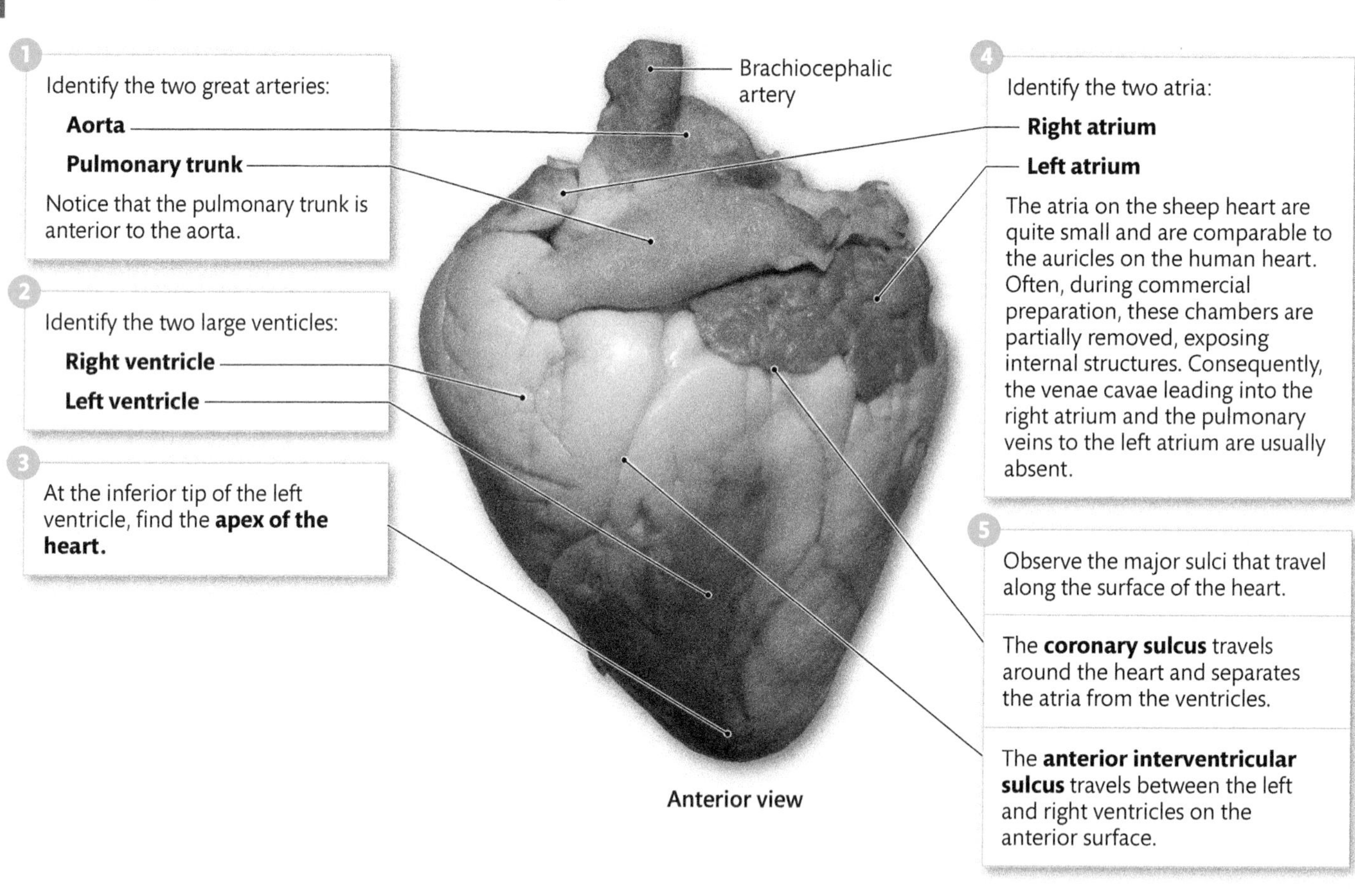

Anterior view

C Hold the sheep heart so that the posterior surface faces you.

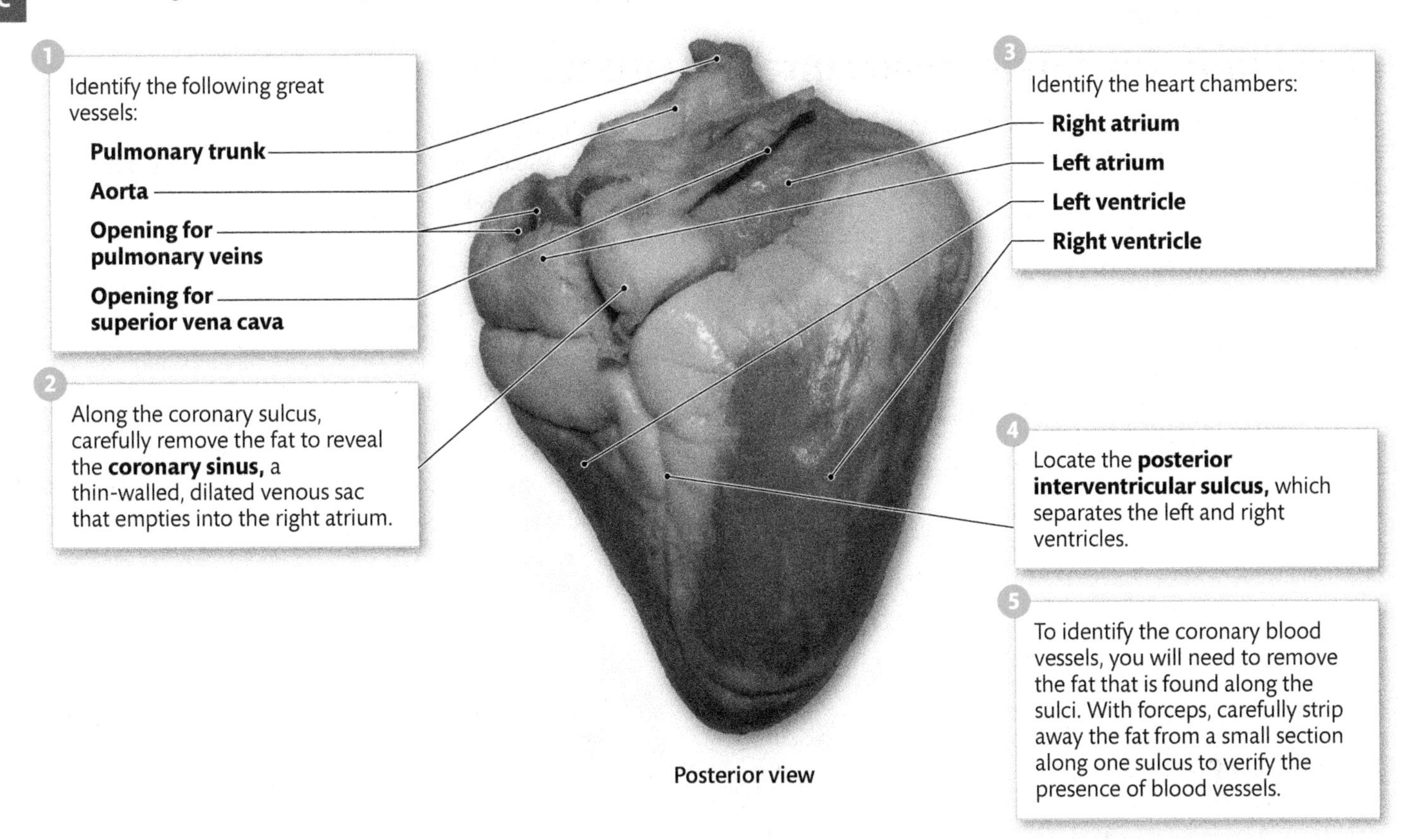

Posterior view

MAKING CONNECTIONS

Summarize similarities and differences in external anatomy between the sheep heart and the human heart. ▶

Similarities:

____

____

____

____

Differences:

____

____

____

____

Activity **21.10**

Dissecting the Sheep Heart: Internal Anatomy

A To study the internal anatomy, you will cut the heart in the coronal (frontal) plane to expose the interior of the heart chambers.

1. To expose the interior of the right atrium and ventricle:
   - Insert the blunt end of a pair of scissors into the superior vena cava. If the superior vena cava is not present, insert the scissors into the opening where it would drain into the right atrium.
   - Cut along the lateral margin of the right atrium and right ventricle to the apex. Be sure to cut through the entire thickness of the ventricular wall, but avoid damaging internal structures.
2. To expose the interior of the left atrium and ventricle:
   - Make a small incision in the lateral wall of the left atrium with a scalpel or the sharp end of a pair of scissors.
   - Insert the blunt end of the scissors into the incision and cut along the lateral margin of the left atrium and left ventricle to the apex.
3. Identify the interatrial septum that separates the left and right atria and the interventricular septum that divides the two ventricles. Beginning at the apex, cut through the interventricular septum with a scalpel or scissors. Continue cutting through the interatrial septum until you have completed a coronal section of the heart as shown in the figure below.

B You can view many internal structures in the coronal section.

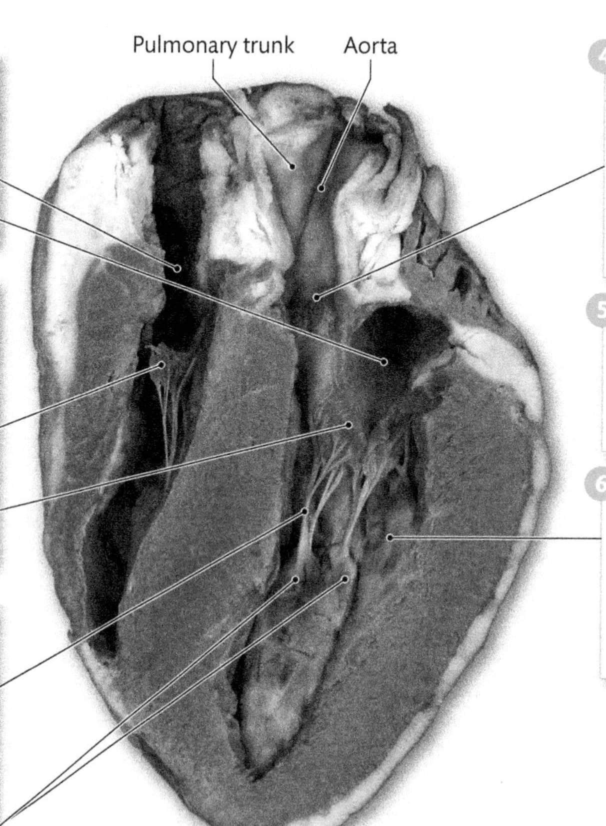

1. Identify the two atrial chambers. Along their anterior walls, locate the pectinate muscles.
   Right atrium
   Left atrium

2. The atrioventricular (AV) valves are located between the atria and ventricles.
   The **tricuspid valve,** with three cusps, is on the right side.
   The **bicuspid (mitral) valve,** with two cusps, is on the left.

3. For each valve, observe that:
   Chordae tendineae are attached to the free margins of the cusps.
   The chordae tendineae extend inferiorly and are attached to **papillary muscles.**

4. From the severed free margin of the aorta, cut along its walls toward the left ventricle until you reach the **aortic semilunar valve.** Observe that the valve is composed of three crescent-shaped cusps.

5. Cut along the wall of the pulmonary trunk until you reach the **pulmonary semilunar valve** (not shown).

6. The **trabeculae carneae** are muscular elevations along the walls of both ventricles. Notice that they are found predominantly in the inferior portions of these chambers.

C The heart wall is dominated by the myocardium, the thick middle layer of cardiac muscle.

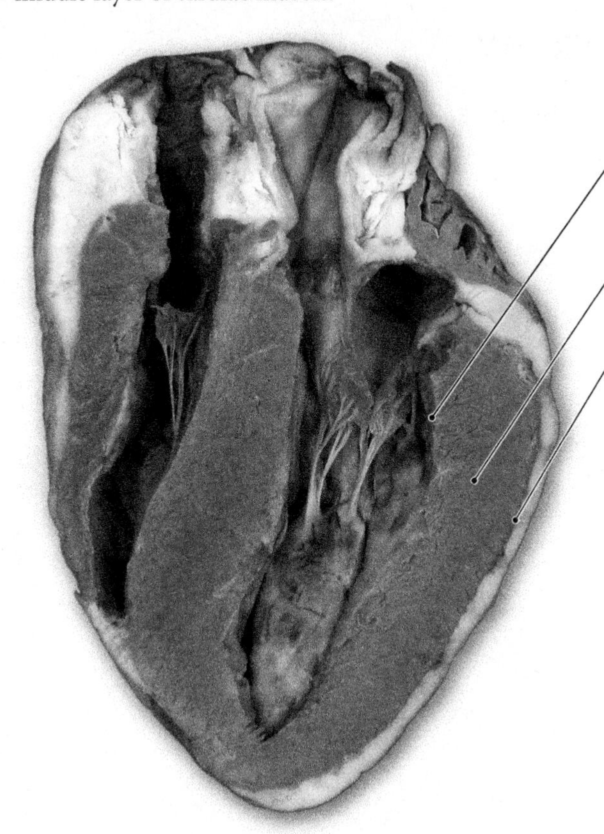

1 Observe the left ventricular wall and identify the three layers of the heart wall:

The inner **endocardium** is a thin serous membrane. It appears as a smooth, shiny surface lining the internal walls of the heart chambers.

The middle **myocardium** is the thickest layer. It is composed primarily of cardiac muscle fibers.

The outer **epicardium,** which is the **visceral pericardium,** is also a serous membrane.

2 Examine the relative thickness of the walls surrounding the heart chambers. Observe that the atrial walls are much thinner than the ventricular walls.

3 Compare the two ventricular walls and observe that the left ventricular wall is thicker than the right ventricular wall.

Word Origins

The bicuspid valve is often referred to as the **mitral valve.** That is because when it is closed, the cusps resemble a *miter,* a tall, pointed headdress worn by bishops and other members of the clergy.

Semilunar means "half moon." The name refers to the half-moon shape of the cusps in the semilunar valves.

MAKING CONNECTIONS

A ventricular septal defect, one of the most common congenital heart defects, refers to one or more holes in the interventricular septum. Blood pressure in the left ventricle is higher than in the right ventricle. How will this condition affect the normal flow of blood through the heart?

▶ ____

____

____

____

____

____

____

____

BEFORE YOU MOVE ON . . .

« LOOKING BACK

The heart is a two-sided, double-pumping organ. The left side (the left pump) sends blood to all tissues and cells, where oxygen and nutrients are delivered and metabolic wastes are taken away. The right side (the right pump) sends blood to the lungs, where oxygen stores in red blood cells are replenished and carbon dioxide, a metabolic waste, is released.

Every day, your heart beats about 100,000 times and pumps 7000 to 9000 liters of blood. By any standard, it is an arduous workload, but the heart's ability to maintain this level of activity for decades, without stopping, is nothing short of remarkable. What special features allow the heart to work so efficiently for such a long period of time?

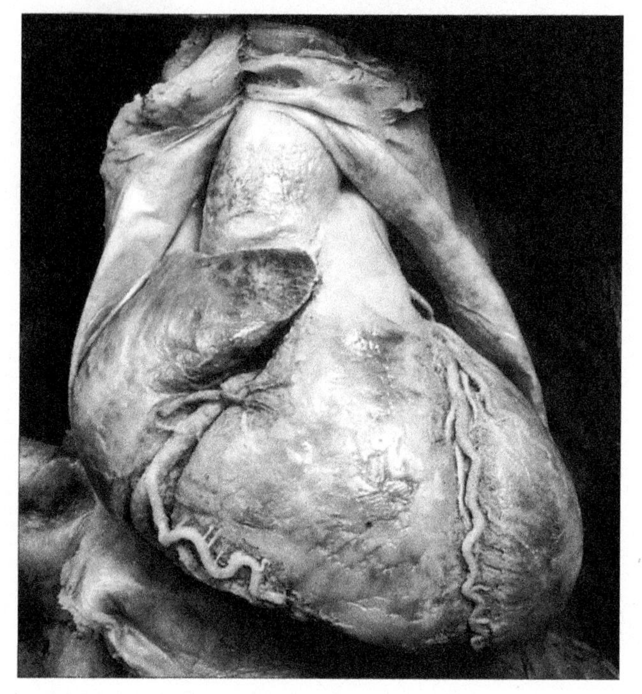

LOOKING FORWARD »

The left and right ventricles pump blood into the great arteries, the aorta and the pulmonary trunk, respectively. Arteries transport blood away from the heart and deliver it to capillaries, where gas and nutrient exchange occurs between blood and cells. Capillaries drain blood into veins, which transport blood toward the heart. In the next laboratory exercise, you will study the microscopic anatomy of arteries, capillaries, and veins and identify the major blood vessels in all regions of the body.

Name ______________________

Lab Section ______________________

Date ______________________

EXERCISE 21 REVIEW SHEET

Gross Anatomy of the Heart

1. The apex of the heart is formed by the
   a. right atrium.
   b. left atrium.
   c. right ventricle.
   d. left ventricle.

2. Which heart groove travels between the atria and the ventricles?
   a. anterior interventricular sulcus
   b. posterior interventricular sulcus
   c. coronary sulcus
   d. both (a) and (b)
   e. (a), (b), and (c)

3. The epicardium and the ______________________ are the same structure.

4. The ______________________ artery forms an anastomosis with the right coronary artery.

5. The adult heart structure that marks the location of an opening between the two atria in the fetal heart is called the ______________________.

QUESTIONS 6–10: Answer the following questions by selecting the correct labeled structure. Answers may be used once or not at all.

6. This structure pumps deoxygenated blood into the pulmonary trunk. ______________

7. The pulmonary veins deliver oxygenated blood to this structure. ______________

8. This structure delivers deoxygenated blood to the right atrium. ______________

9. This structure pumps oxygenated blood into the aorta. ______________

10. This structure and its branches deliver deoxygenated blood to the lungs. ______________

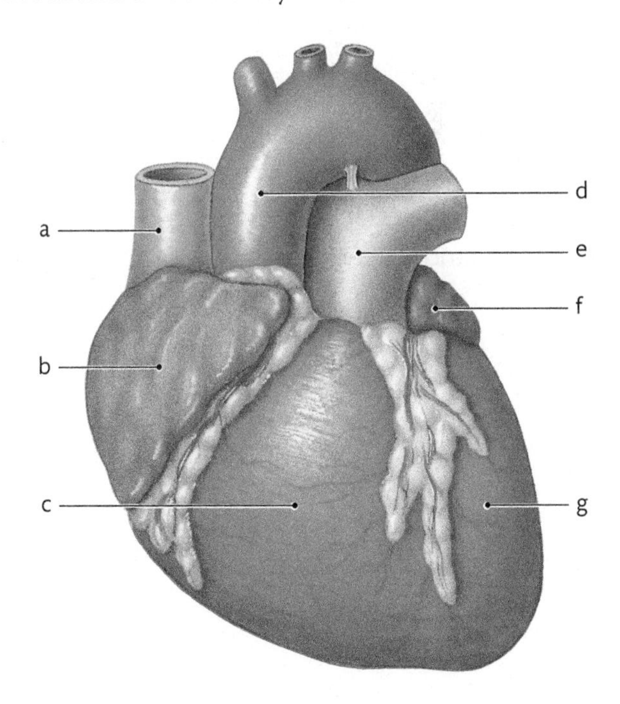

11. Each heart valve is located at the junction of an atrium and ventricle, or a ventricle and great artery. Pressure differences on either side of the valves regulate their opening and closing. Use these concepts to complete the following table.

Heart Valve	The Valve Is Located between the ________ and ________	When the Valve Is Open, the Pressure Is Greater on the ________ Side	When the Valve Is Closed, the Pressure Is Greater on the ________ Side
Biscuspid valve	a.	atrial	b.
c.	right atrium; right ventricle	d.	ventricular
e.	f.	g.	pulmonary trunk
h.	left ventricle; aorta	i.	j.

12. Complete the following table.

Artery	Vessel from Which Artery Branches	Sulcus in Which Artery Travels	Vein That Travels with the Artery
a.	Ascending aorta	b.	Small cardiac vein
Anterior interventricular artery	c.	d.	e.
f.	Left coronary artery	Coronary sulcus	g.
h.	i.	Posterior interventricular sulcus	j.

QUESTIONS 13–19: Match the following terms with the correct labeled structure.

13. Apex ________

14. Right atrium ________

15. Pulmonary trunk ________

16. Left ventricle ________

17. Aorta ________

18. Left atrium ________

19. Right ventricle ________

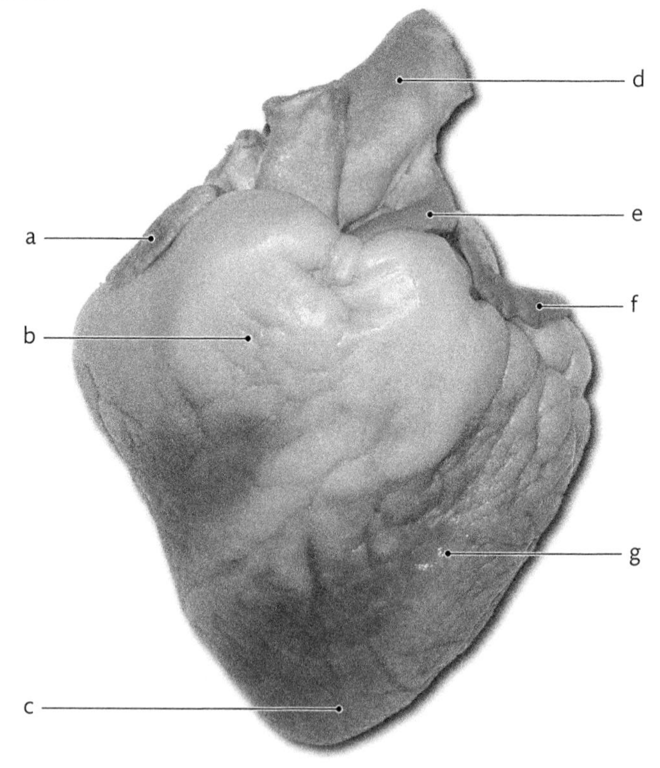

20. In the following diagram, color the structures with the indicated colors.

Right atrium = **yellow**
Left ventricle = **gray**
Aorta = **red**
Left atrium = **dark green**
Pulmonary trunk = **dark blue**
Superior vena cava = **purple**
Right ventricle = **orange**
Inferior vena cava = **pink**
Coronary sinus = **light blue**
Pulmonary arteries = **brown**
Pulmonary veins = **light green**

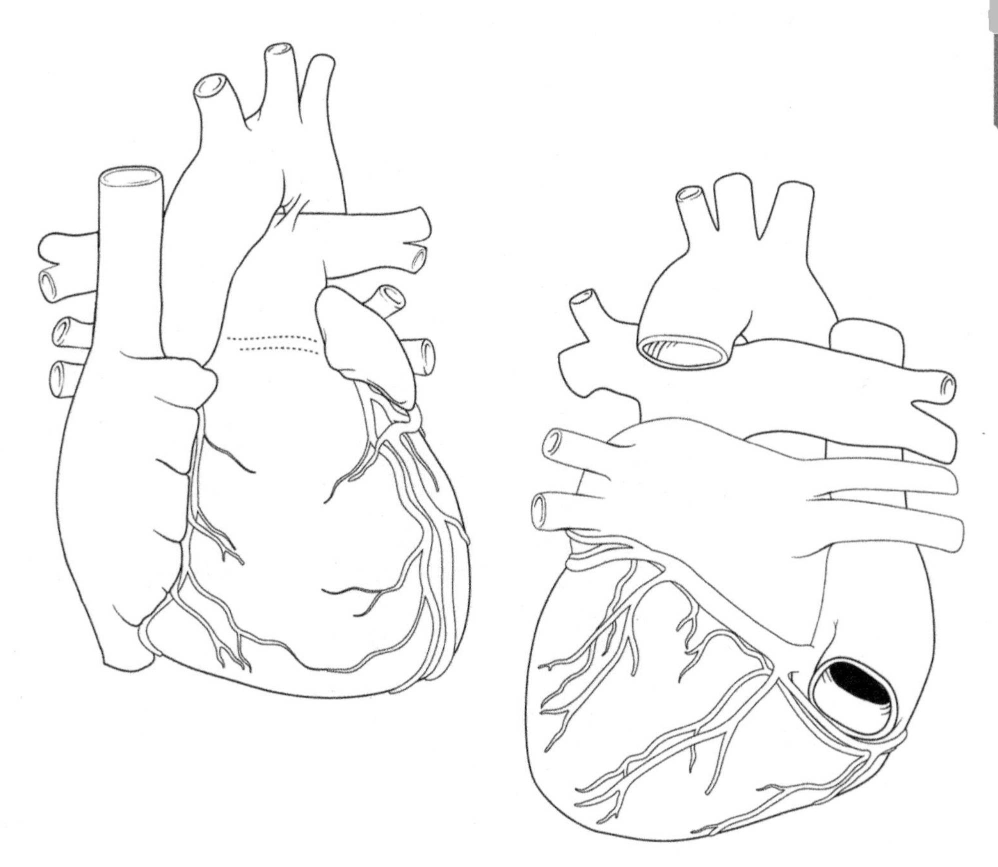

QUESTIONS 21–25: On the photo of the thoracic cage, identify the locations of the following cardiac landmarks. Label all the landmarks that you identify.

21. Draw a line to show the position of the base of the heart.

22. Draw a line to show the position of the left border of the heart.

23. Draw a line to show the position of the right border of the heart.

24. Draw a line to show the position of the inferior border of the heart.

25. Use an arrow to identify the position of the apex.

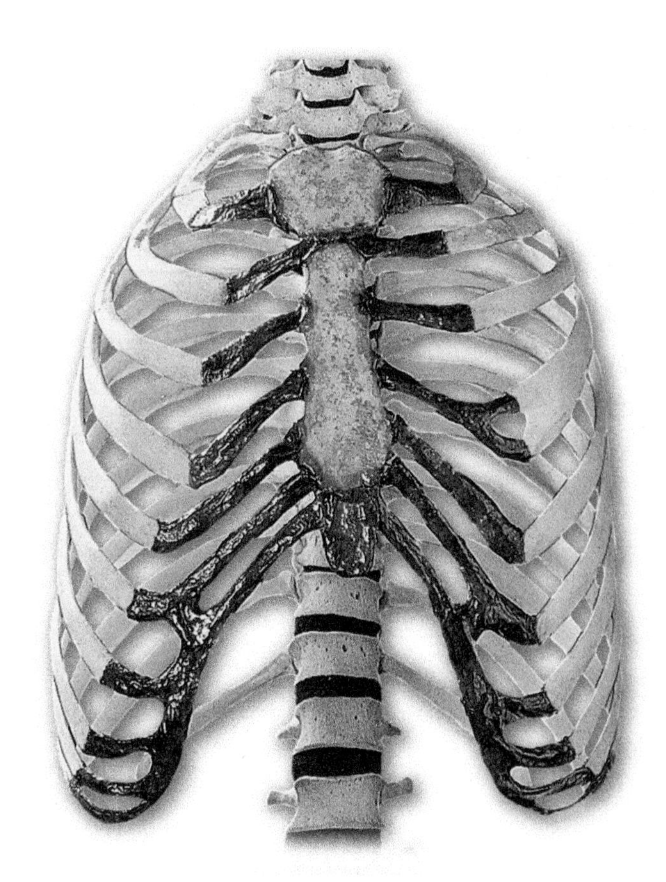

QUESTIONS 26–35: In the following photo, identify the labeled structures. Select your answers from the list in column B. Answers may be used once or not at all.

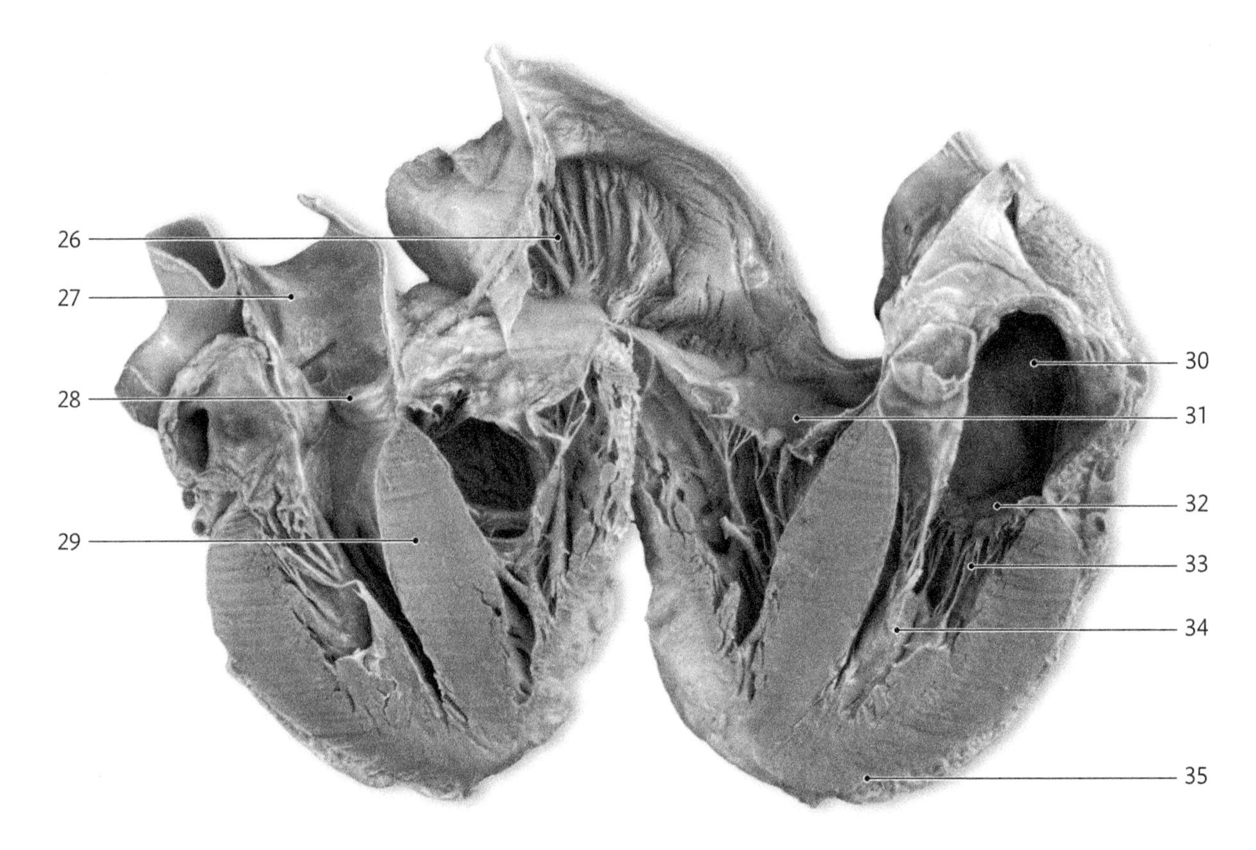

A	B
26. ______________	a. Apex of the heart
27. ______________	b. Cusp of tricuspid valve
28. ______________	c. Right atrium
29. ______________	d. Chordae tendineae
30. ______________	e. Aortic semilunar valve
31. ______________	f. Trabeculae carneae
32. ______________	g. Papillary muscle
33. ______________	h. Interventricular septum
34. ______________	i. Pectinate muscles
35. ______________	j. Cusp of bicuspid valve
	k. Ascending aorta
	l. Left atrium

EXERCISE

22

Anatomy of Blood Vessels

LEARNING OUTCOMES

These Learning Outcomes correspond by number to the laboratory activities in this exercise. When you complete the activities, you should be able to:

Activities 22.1–22.2	Compare the structure and function of arteries, veins, and capillaries.
Activity 22.3	Identify the major arteries and veins in the pulmonary circulation.
Activities 22.4–22.9	Identify the major arteries and veins in the systemic circulation.
Activity 22.10	Identify superficial veins by palpation and surface observation.

LABORATORY SUPPLIES

- Compound light microscopes
- Prepared microscope slides of arteries, veins, and capillaries
- Anatomical models that illustrate major blood vessels:
  - heart
  - brain
  - torso or whole body
  - head and neck
  - vertebral column
  - upper limb
  - lower limb

PRE-LAB QUIZ

Before you begin, read all the activities in Exercise 22 and the required reading in your textbook that is assigned by your instructor.

1. During this laboratory exercise, you will complete all the following activities *except*
   a. examine the microscopic structure of blood vessels.
   b. identify the major arteries in the coronary circulation.
   c. identify superficial veins on your body.
   d. identify the major arteries and veins in the systemic circulation.
2. Blood is supplied to the brain by way of the two internal carotid arteries and the two ______________________ arteries.
3. True or False: The superior vena cava is formed by the union of the two brachiocephalic veins. __________
4. True or False: The azygos system of veins drains blood from abdominal structures. __________
5. The ______________________ is the middle tissue layer in a blood vessel wall.
6. What type of blood vessels are sites where nutrients and wastes are exchanged between blood and body cells? ______________________
7. True or False: In the pulmonary circulation, the pulmonary arteries deliver oxygenated blood to the heart. __________
8. The ______________________ arteries supply blood to the kidneys.
9. The two terminal branches of the brachial artery are the ______________________ and ______________________ arteries.
10. True or False: The great saphenous vein drains into the femoral vein in the anterior thigh. __________

Activity **22.1**

Examining the Microscopic Structure of Blood Vessels: Arteries

Blood vessel walls have three distinct tissue layers. The **tunica intima (interna)** is the innermost layer. It contains a simple squamous epithelium, referred to as the **endothelium,** which lines the lumen (internal space) of the blood vessel; a basal lamina; and a thin layer of loose connective tissue. The **tunica media** is the middle layer. It is usually the thickest layer and is composed of smooth muscle and elastic fibers. The **tunica externa (adventitia)** is the outermost layer. It is composed of connective tissue with numerous elastic and collagen fibers and is continuous with the connective tissue of adjacent structures.

A The arterial system contains three types of arteries: elastic arteries, muscular arteries, and arterioles. **Elastic arteries** include the great arteries (aorta and pulmonary trunk) and some of their primary branches. Compared to other arteries, they have the largest luminal diameters, but their walls are relatively thin.

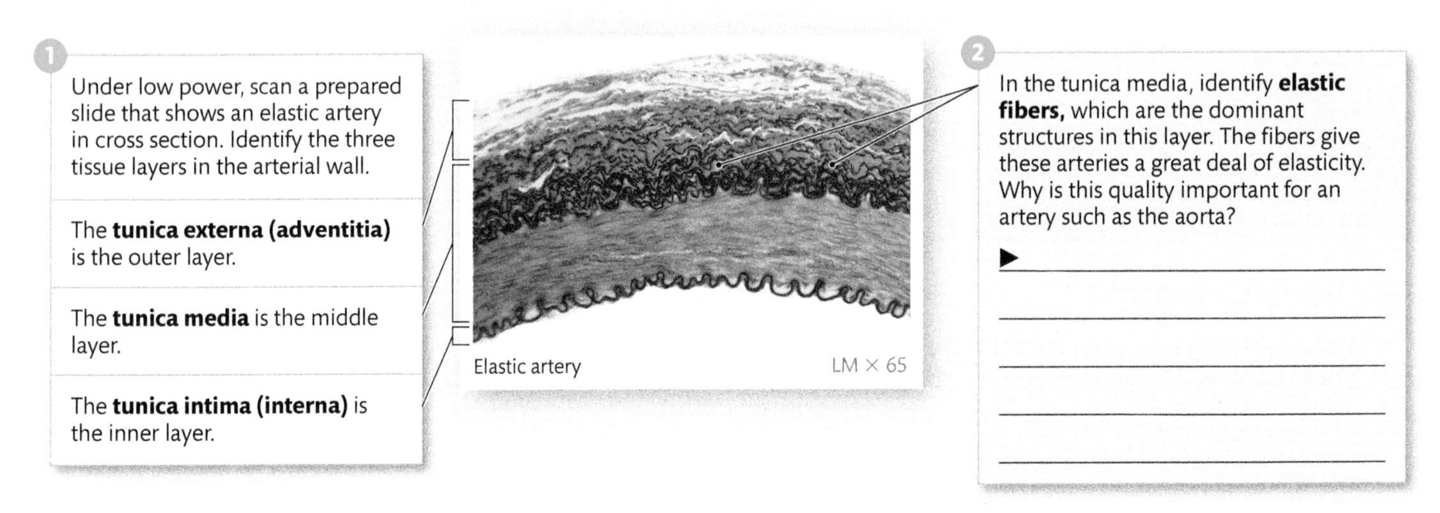

B **Muscular arteries** deliver blood to specific body regions or organs. For example, the brachial artery supplies the arm, and the renal artery supplies the kidney. Proportionately, muscular arteries have the thickest walls of all blood vessels. The tunica media contains mostly smooth muscle and relatively few (compared to elastic arteries) elastic fibers.

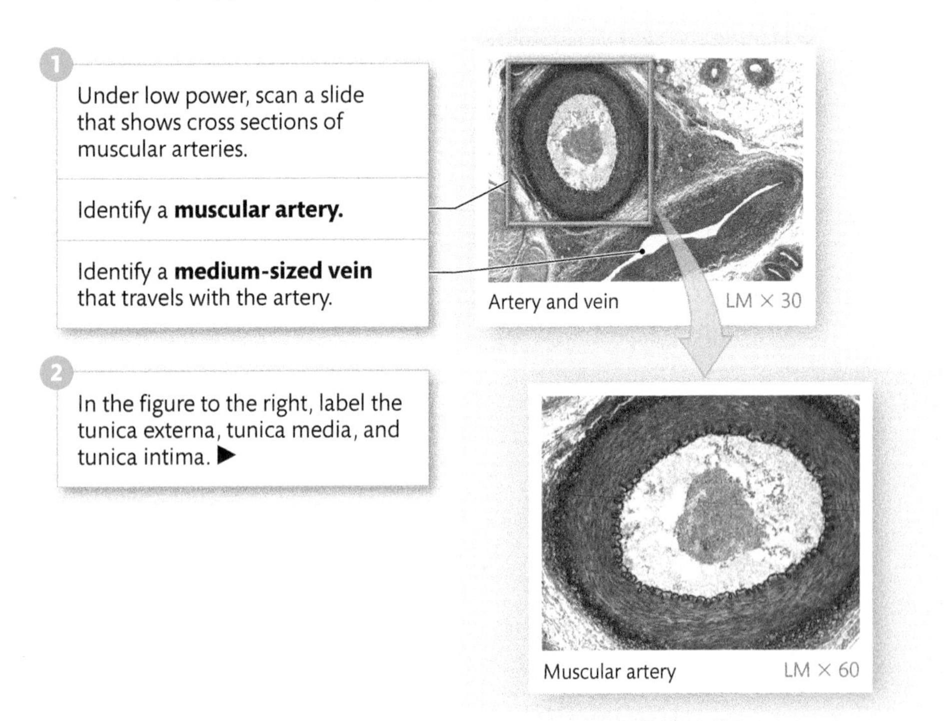

IN THE CLINIC

Atherosclerosis

Atherosclerosis is the accumulation of fatty deposits in the tunica intima of arteries. This condition is usually accompanied by damage and subsequent calcification (deposition of calcium salts) of the tunica media. The progressive accumulation of lipids narrows the lumen and reduces blood flow. Atherosclerosis of the coronary arteries can lead to coronary artery disease and heart attack, and atherosclerosis in arteries supplying the brain can lead to a stroke.

C **Arterioles** are small arteries that deliver blood to capillary beds. The walls of arterioles are very thin but contain all three tissue layers. The tunica media consists mostly of smooth muscle with very few elastic fibers. As arterioles get closer and closer to capillary beds, their walls become progressively thinner. The smallest arterioles contain only an endothelium and a single layer of smooth muscle fibers in the tunica media.

1. Under low power, scan a slide that shows cross sections of arteries and corresponding veins. Switch to high power and find examples of **arterioles.**

2. Notice that an arteriole has a small diameter and the **tunica media** is composed of one to five layers of smooth muscle.

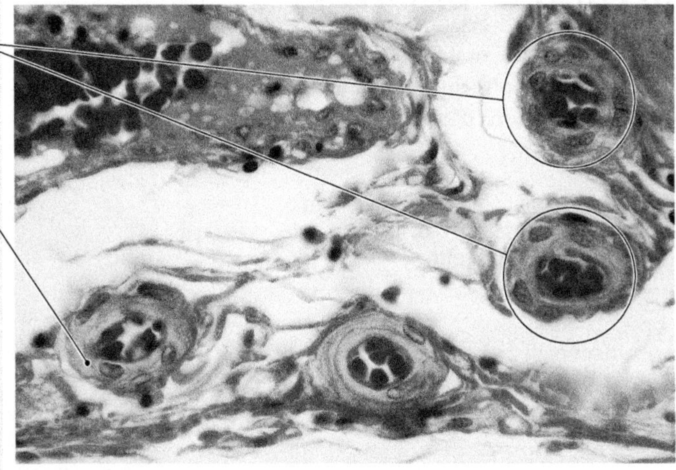

Arterioles LM × 600

IN THE CLINIC

Vasoconstriction and Vasodilation

The activity of smooth muscle in the tunica media of muscular arteries and arterioles is influenced by sympathetic nerve fibers and various hormones. When smooth muscle fibers contract, the vessel diameter decreases. This process is known as **vasoconstriction.** When the muscle fibers relax, the diameter increases. This process is called **vasodilation.** In muscular arteries, vasodilation and vasoconstriction help regulate the volume of blood flowing to a particular structure. In arterioles, vasodilation and vasoconstriction are directly linked to changes in blood pressure. When arterioles dilate, resistance to blood flow declines, and less pressure is required to move blood forward. When arterioles constrict, resistance increases, and more pressure is required to force blood through the vessels. Thus, changes in the resistance to blood flow in arterioles, referred to as **peripheral resistance,** can have a noticeable effect on an individual's blood pressure.

MAKING CONNECTIONS

During exercise, what changes would you expect to occur in the muscular arteries that supply digestive organs and the muscular arteries that supply skeletal muscles?

▶ ____

____

____

____

____

____

____

Examining the Microscopic Structure of Blood Vessels: Veins and Capillaries

A Veins possess the same three tissue layers as arteries. Many veins have valves that prevent the backflow of blood. The valves are continuations of the tunica intima and are similar to the cardiac semilunar valves, both structurally and functionally. Valves are most abundant in the veins where blood must flow back to the heart against the force of gravity.

1. Under low power, scan a slide that shows cross sections of arteries and veins. Identify a **vein.**

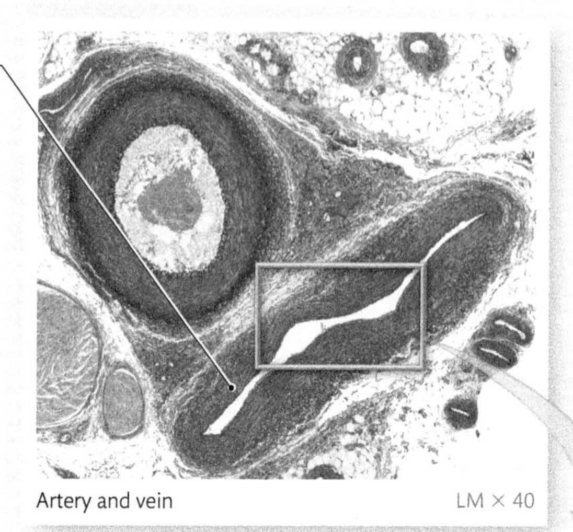

Artery and vein LM × 40

2. Under high power, identify the three tissue layers in the wall of a vein. List the layers and label them in the lower figure to the right. ▶

____

____

____

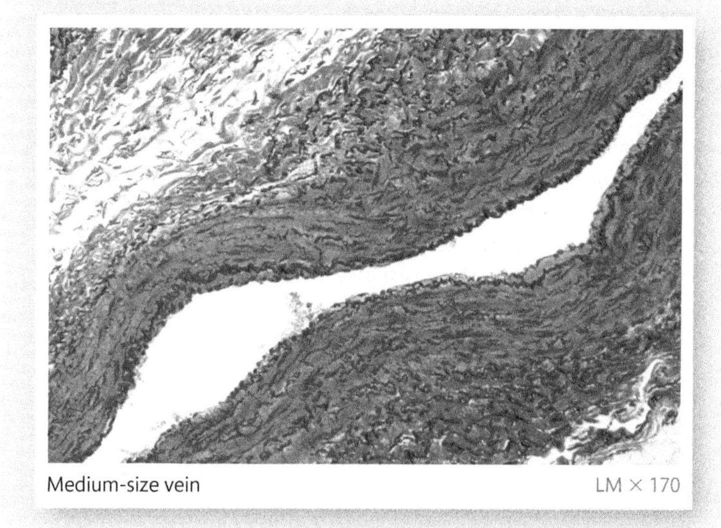

Medium-size vein LM × 170

3. Describe the structural similarities and differences between a muscular artery and a medium-sized vein. ▶

Similarities:

____

____

____

Differences:

____

____

____

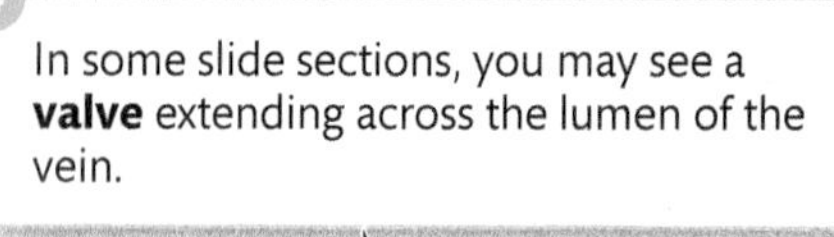

4. In some slide sections, you may see a **valve** extending across the lumen of the vein.

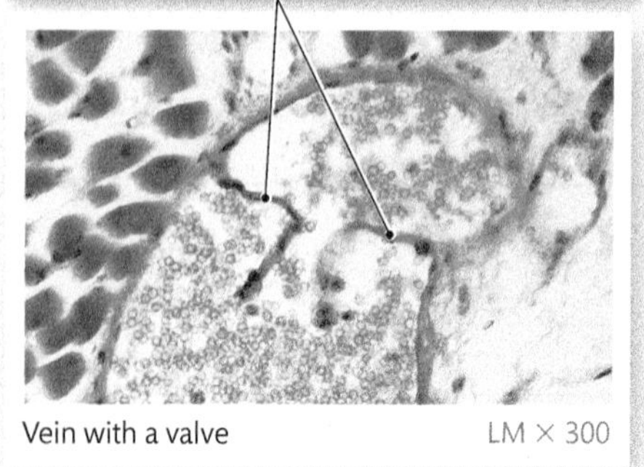

Vein with a valve LM × 300

IN THE CLINIC

Varicose Veins

In the lower limbs and other regions where blood must flow back to the heart against gravity, the veins have numerous valves to prevent backflow. If the valves do not work properly, blood tends to flow in the reverse direction and becomes stagnant in some locations. The slowdown in venous return creates areas where blood pools and vessels dilate. This condition, referred to as **varicose veins,** can be caused by a number of factors, including genetic inheritance, pregnancy, abdominal tumors, obesity, and standing or sitting for long periods of time.

B Capillaries, the smallest blood vessels, connect the arterial and venous circulatory networks. They are sites where nutrients and wastes are exchanged between the blood and body cells. The walls of capillaries are very thin, consisting only of an endothelium and a basal lamina in the tunica intima; the tunica media and tunica externa are absent. Some capillaries are more permeable than others and thus allow more substances to be exchanged. Based on differences in permeability, there are three categories of capillaries:

Continuous capillaries are the least permeable because the endothelial lining is uninterrupted and the cells are sometimes held together by tight junctions.

Fenestrated capillaries are more permeable than the continuous variety because their endothelial cells contain pores (fenestrae) that are covered by a very thin basal lamina.

Sinusoids are the most permeable capillaries. Their endothelial lining is highly irregular and loosely arranged, with many pores and spaces between cells.

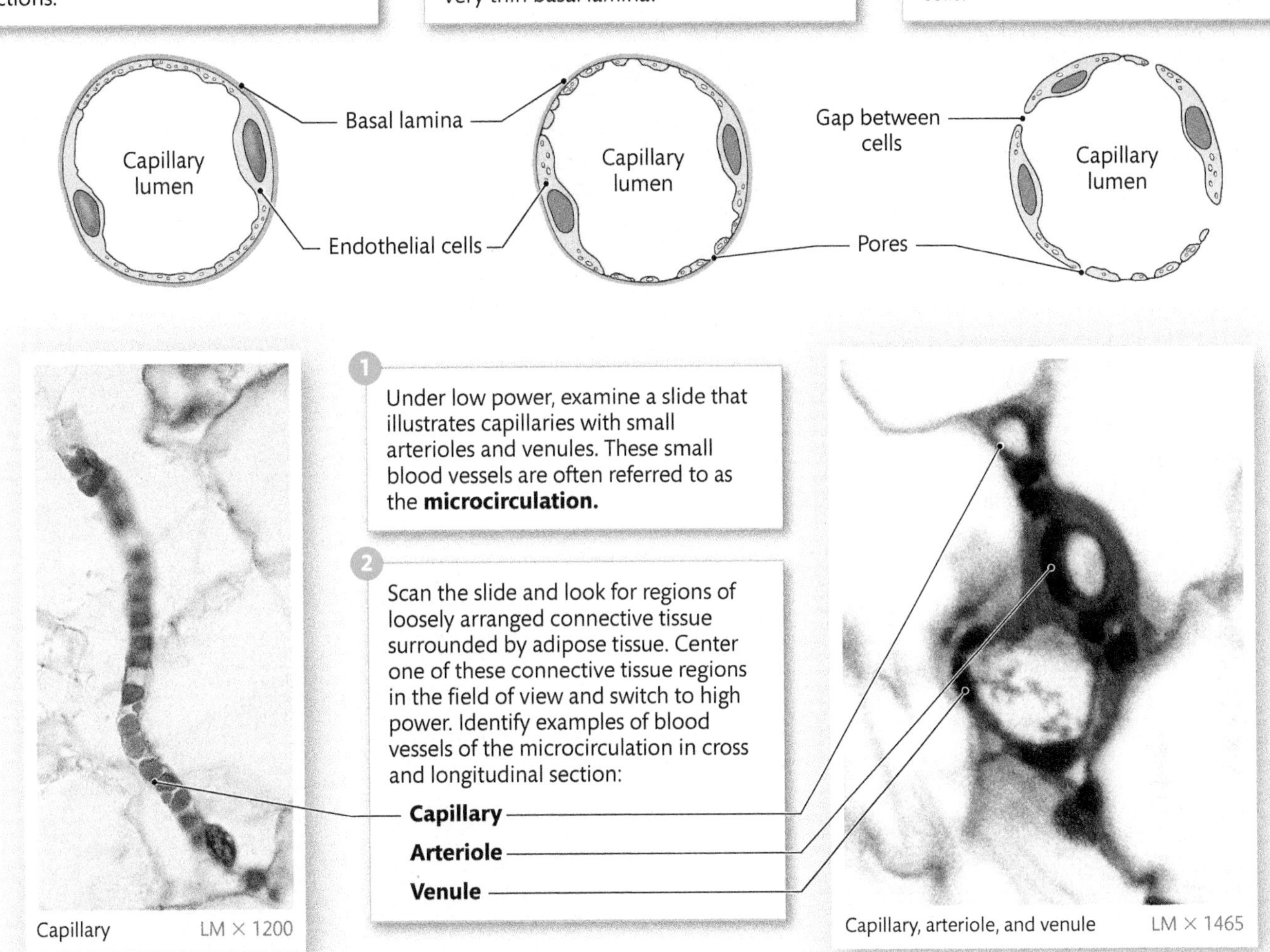

Capillary LM × 1200

Capillary, arteriole, and venule LM × 1465

1. Under low power, examine a slide that illustrates capillaries with small arterioles and venules. These small blood vessels are often referred to as the **microcirculation.**

2. Scan the slide and look for regions of loosely arranged connective tissue surrounded by adipose tissue. Center one of these connective tissue regions in the field of view and switch to high power. Identify examples of blood vessels of the microcirculation in cross and longitudinal section:
   - **Capillary**
   - **Arteriole**
   - **Venule**

MAKING CONNECTIONS

Continuous capillaries are located in the brain, and fenestrated capillaries are found in endocrine organs. Why do you think there is a difference in the type of capillaries found in these structures?

▶ ______________________________

____

____

Activity **22.3**

Identifying Blood Vessels in the Pulmonary Circulation

The **pulmonary circulation,** which is driven by the pumping action of the right side of the heart, delivers deoxygenated blood to pulmonary capillary networks in the lungs and returns oxygenated blood back to the heart.

A At the beginning of the pulmonary circulation, the right ventricle pumps blood into the pulmonary trunk.

1. On the heart model, identify the following structures and label them on the diagram. ▶
   - **Right ventricle**
   - **Pulmonary trunk**

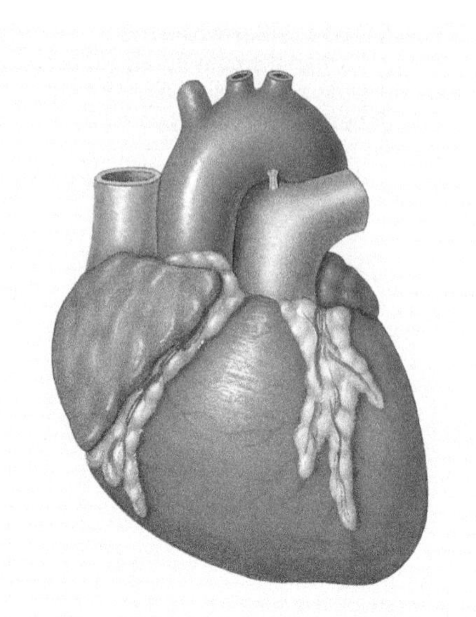

2. The pulmonary trunk gives rise to the **right** and **left pulmonary arteries.** The pulmonary arteries and their branches deliver blood to the lungs for oxygenation. Identify the pulmonary arteries on the heart model. Label the left pulmonary artery on the diagram. ▶

3. The beginning of the right pulmonary artery cannot be seen because it travels posterior to what structure?

▶ ______________________________

B Blood flows into the **pulmonary arteries** and their branches to reach the pulmonary capillaries.

1. On a model of the thorax, identify the **right** and **left pulmonary arteries** branching off the pulmonary trunk.

2. Identify several **branches of the pulmonary arteries** traveling through the lungs. These blood vessels deliver blood to pulmonary capillary beds.

3. **Pulmonary capillaries** surround tiny air sacs (alveoli; singular = alveolus), where gas exchange occurs. How does the composition of blood change as a result of gas exchange?

▶ ______________________________

____

____

____

____

____

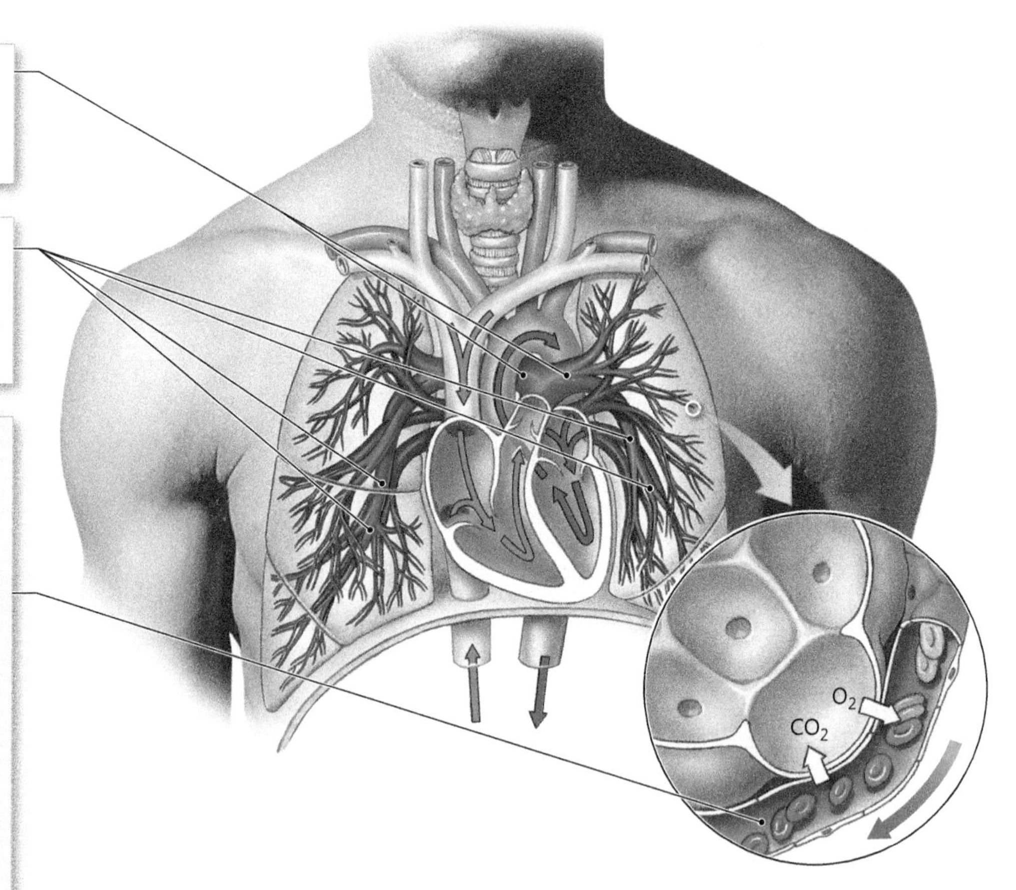

C Oxygenated blood returns to the heart via the right and left **pulmonary veins.**

1. On a thorax model, identify **venules** and **small veins** that deliver blood to the pulmonary veins.

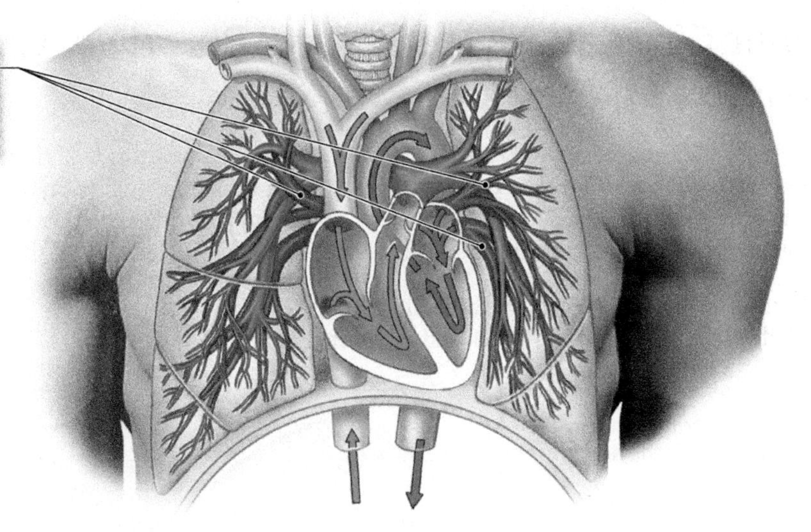

2. On a heart model, identify the **pulmonary veins.** Notice that there are two right pulmonary veins and two left pulmonary veins. Label the pulmonary veins in the diagram. ▶

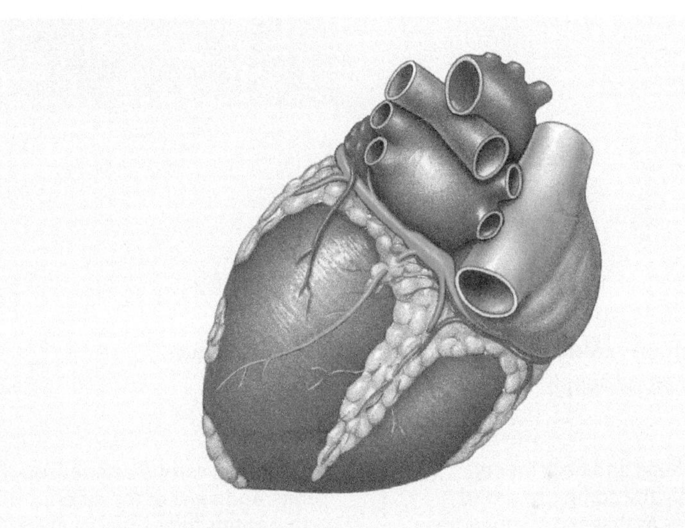

3. Which heart chamber receives blood from the pulmonary veins?

▶ ______________________________

IN THE CLINIC

Pulmonary Embolism

A **pulmonary embolism** is a blood clot that blocks a branch of the pulmonary artery. It is caused by a blood clot that forms elsewhere in the body, usually in a lower limb, called a **deep vein thrombosis.** A piece of the thrombosis breaks away from the main clot, travels back to the right side of the heart, and enters the pulmonary circulation. Most embolisms are small and not life threatening, but large clots can block a main branch of a pulmonary artery and can be deadly. Symptoms include shortness of breath, severe chest pain, heart palpitations, and coughing up bloody sputum. Treatments include anticoagulant drugs such as heparin or Coumadin. Surgery may be required to remove a large clot, particularly if drug treatment is not effective.

MAKING CONNECTIONS

Anemia can develop if a person's red blood cell count or hemoglobin level is lower than normal. How will this condition affect gas exchange in the lungs?

▶ ______________________________

____

____

____

____

____

____

____

Activity **22.4**

Systemic Circulation: Identifying Blood Vessels in the Head and Neck

The **systemic circulation,** which is driven by the pumping action of the left side of the heart, delivers oxygenated blood to the aorta. The aorta and its branches distribute blood to all body tissues. In the systemic capillary beds, oxygen and nutrients are delivered to cells and tissues, and carbon dioxide and other metabolic wastes are picked up. The systemic venous system returns the deoxygenated blood to the right side of the heart.

A The arch of the aorta gives off three arterial branches that distribute blood to the head and neck and to the upper limbs.

1 On a torso or thoracic cavity model, identify the **aortic arch.**

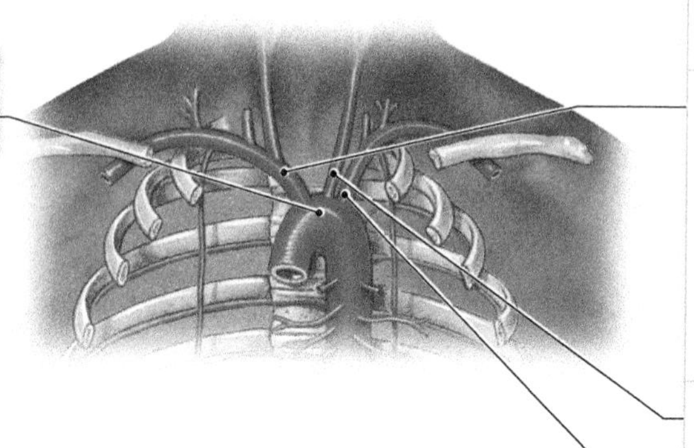

2 Identify the three arterial branches coming off the aortic arch:

The **brachiocephalic trunk** travels a short distance and then bifurcates (splits into two branches) to form the right common carotid artery, which supplies the head and neck on the right side, and the right subclavian artery, which supplies the neck and upper limb on the right side.

The **left common carotid artery** supplies the head and neck on the left side.

The **left subclavian artery** supplies the neck and upper limb on the left side.

B The common carotid arteries and their branches distribute the main blood supply to the head and neck.

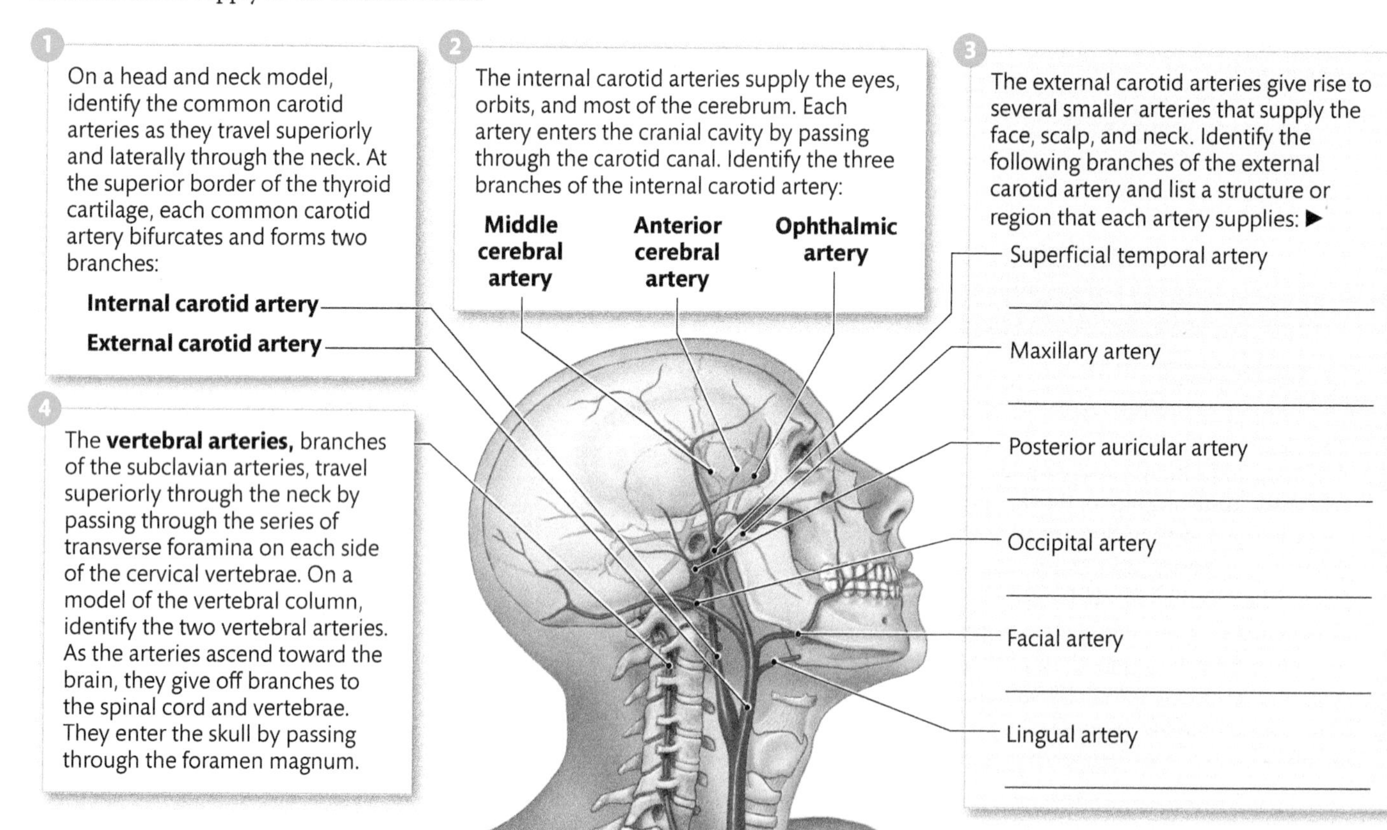

1 On a head and neck model, identify the common carotid arteries as they travel superiorly and laterally through the neck. At the superior border of the thyroid cartilage, each common carotid artery bifurcates and forms two branches:

Internal carotid artery

External carotid artery

2 The internal carotid arteries supply the eyes, orbits, and most of the cerebrum. Each artery enters the cranial cavity by passing through the carotid canal. Identify the three branches of the internal carotid artery:

Middle cerebral artery | **Anterior cerebral artery** | **Ophthalmic artery**

3 The external carotid arteries give rise to several smaller arteries that supply the face, scalp, and neck. Identify the following branches of the external carotid artery and list a structure or region that each artery supplies: ▶

Superficial temporal artery ______________________

Maxillary artery ______________________

Posterior auricular artery ______________________

Occipital artery ______________________

Facial artery ______________________

Lingual artery ______________________

4 The **vertebral arteries,** branches of the subclavian arteries, travel superiorly through the neck by passing through the series of transverse foramina on each side of the cervical vertebrae. On a model of the vertebral column, identify the two vertebral arteries. As the arteries ascend toward the brain, they give off branches to the spinal cord and vertebrae. They enter the skull by passing through the foramen magnum.

C Veins that drain blood from the head and neck, thorax, and upper limbs empty into the superior vena cava.

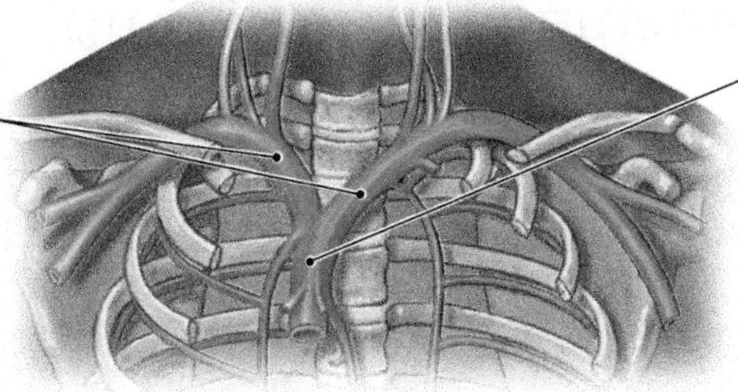

1. On a torso model, identify the **right** and **left brachiocephalic veins.** Notice that the left vein is longer than the right vein. Explain why.

▶ ______________________________

____

____

____

____

____

2. Identify the **superior vena cava.** It is formed by the union of the right and left brachiocephalic veins. The superior vena cava empties into which heart chamber?

▶ ______________________________

D The internal and external jugular veins drain blood from the head and neck.

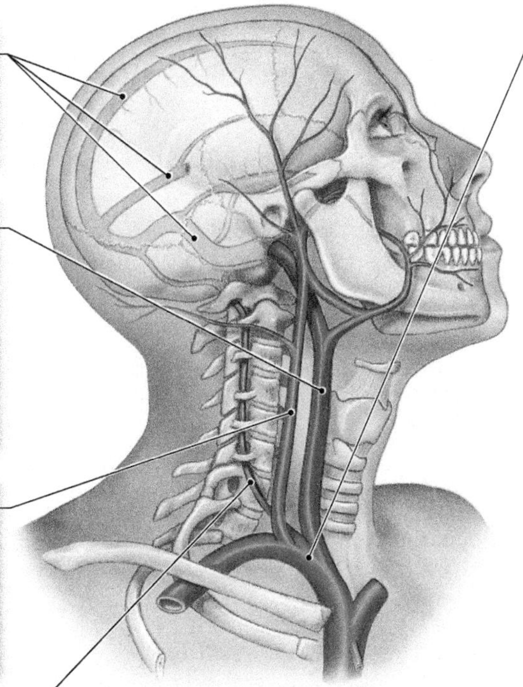

1. On a model of the brain in the cranial cavity, identify the network of **dural sinuses** that drains blood from the brain.

2. On a torso or head and neck model, identify the **internal jugular veins** as they descend through the neck. Notice that they run parallel to the common carotid arteries. The internal jugular veins receive blood from the dural sinuses.

3. Locate the **external jugular veins.** These veins descend through the neck, superficial to the sternocleidomastoid muscles. They drain blood from the face, scalp, and neck.

4. On a model of the vertebral column, identify the vertebral arteries, described earlier. **Vertebral veins,** which are usually not demonstrated on models, travel alongside these arteries. These veins drain blood from the brain, posterior skull bones, and cervical vertebrae.

5. On a torso model, identify the **subclavian veins.** On each side of the body, the internal jugular vein and subclavian vein merge to form what vein?

▶ ______________________________

List two veins that drain into the subclavian veins.

▶ ______________________________

____

MAKING CONNECTIONS

All the major arteries in the systemic circulation have a deep position in the body. Explain why.

▶ ______________________________

____

____

____

Want more practice? Go to: **MasteringA&P** > Study Area > Menu > Lab Tools > PAL > Anatomical Models > Cardiovascular System >

- Arteries
- Veins

Activity **22.5**

Systemic Circulation: Examining the Blood Supply to the Brain

A The blood supply to the brain is derived from branches of the internal carotid arteries and the vertebral arteries.

1. On a brain model, identify the two branches of each internal carotid artery that supply the cerebrum.

   Each **anterior cerebral artery** supplies the medial side of a cerebral hemisphere.

   Each **middle cerebral artery** supplies the lateral sides of the temporal and parietal lobes of a cerebral hemisphere.

2. Locate the **vertebral arteries** traveling along the anterolateral aspects of the brainstem. These arteries give off branches to the medulla oblongata and the cerebellum.

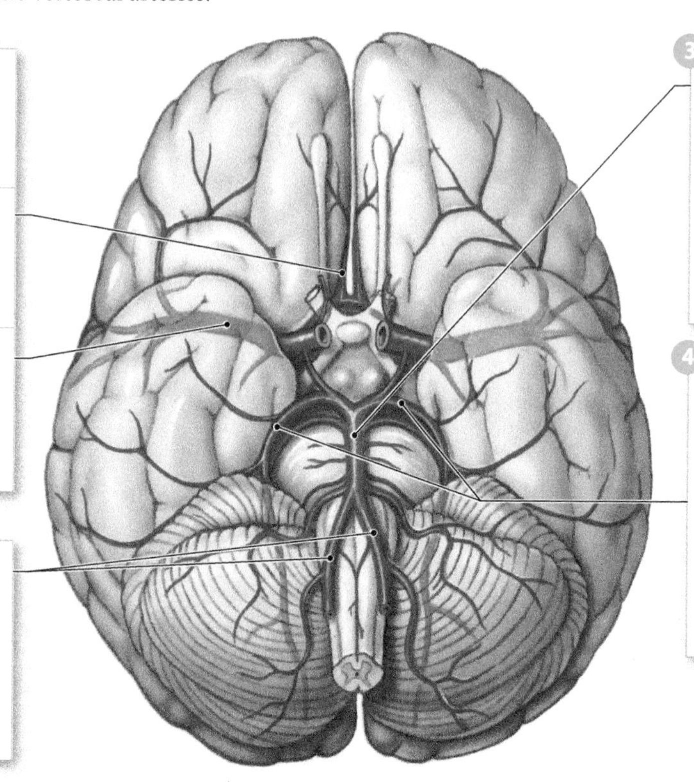

3. At the medulla–pons border, the vertebral arteries merge to form the **basilar artery.** Identify this artery as it travels along the anterior surface of the brainstem and supplies branches to the pons and cerebellum.

4. At the border between the pons and midbrain, identify the two **posterior cerebral arteries,** which branch off the end of the basilar artery. These two blood vessels supply the occipital lobes and portions of the temporal lobes.

B The **cerebral arterial circle (circle of Willis)** interconnects the arteries that supply blood to the brain.

1. On the ventral surface of the brain, identify the **cerebral arterial circle.** Notice that it surrounds the optic chiasm and the pituitary gland.

2. Identify the blood vessels that connect the arteries supplying blood to the brain and complete the cerebral arterial circle:

   The **anterior communicating artery** connects the two anterior cerebral arteries.

   The two **posterior communicating arteries** connect the internal carotid arteries with the posterior cerebral arteries on each side.

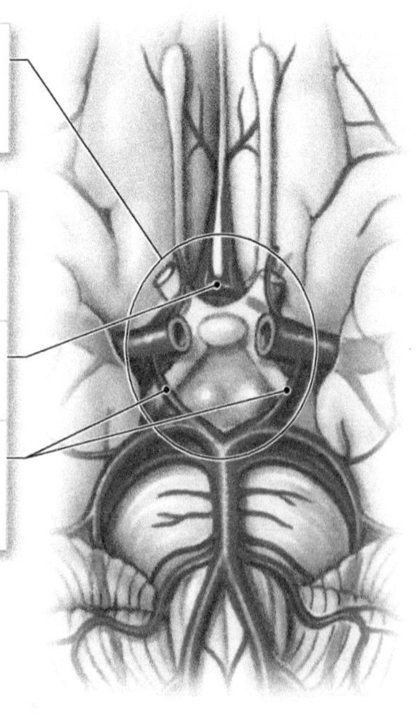

MAKING CONNECTIONS

The cerebral arterial circle is an example of an arterial anastomosis: a natural communication between branches of two or more arteries with different origins. Why do you think anastomoses are important?

▶ ______________________________

____

____

____

____

____

____

____

____

____

Activity **22.6**

Systemic Circulation: Identifying Blood Vessels in the Thoracic Cavity

A The organs in the thoracic cavity and much of the thoracic wall are supplied by branches of the thoracic division of the descending aorta.

1 On a whole body or torso model, identify the **thoracic division of the descending aorta (thoracic aorta)** as it travels along the posterior wall of the thoracic cavity. Describe the relationship of the thoracic aorta with the vertebral column.

▶ ______________________

____

____

____

____

____

____

____

2 The following visceral branches of the thoracic aorta supply the organs in the thoracic cavity (may not be shown on your model). List the structures that they supply. ▶

Bronchial arteries supply the

____

Esophageal arteries supply the

____

Mediastinal arteries supply the

____

Pericardial arteries supply the

____

3 Identify the series of paired arteries that branch off the thoracic aorta and travel within the intercostal spaces. These vessels are the **posterior intercostal arteries.** What structures do they supply?

▶ ______________________

____

4 Locate the paired **superior phrenic arteries.** What structure do these arteries supply?

▶ ______________________

B The **azygos system of veins** is a highly variable system of blood vessels that drains blood from most thoracic structures.

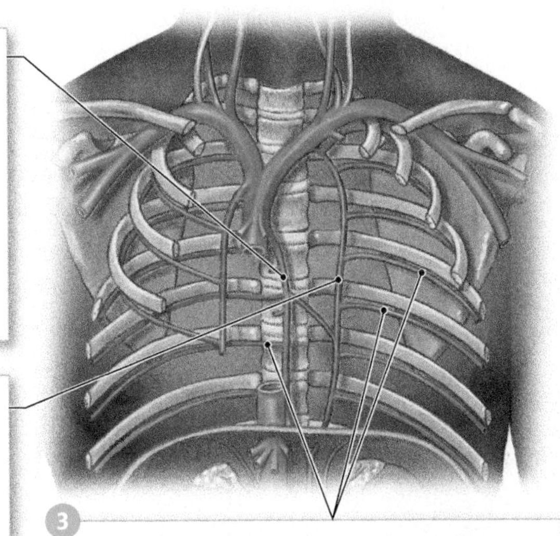

1 On a torso model, identify the **azygos vein.** Observe how the vein ascends along the anterior surface of the vertebral column, just to the right of the midline, and empties into the superior vena cava.

2 Identify the **hemiazygos vein** as it ascends along the left side of the vertebral column. It is connected to the azygos vein by one or more communicating veins.

3 Identify the **posterior intercostal veins** as they travel through the intercostal spaces. Verify that most of these veins drain into the azygos vein on the right side and the hemiazygos vein on the left side.

MAKING CONNECTIONS

In addition to draining the thoracic wall, the azygos system of veins also receives blood from veins that drain internal thoracic structures. Two examples are the **pericardial veins** and the **bronchial veins.** ▶

a. What is the functional difference between pericardial veins and cardiac veins?

____

____

____

b. What is the functional difference between bronchial veins and pulmonary veins?

____

____

____

Activity **22.7**

Systemic Circulation: Identifying Blood Vessels in the Abdominopelvic Cavity

A At the level of the T_{12} vertebra, the descending aorta enters the abdominopelvic cavity by passing through the aortic opening posterior to the diaphragm. The abdominal aorta gives off three major unpaired branches and several paired branches.

1 On a torso model, locate the abdominal aorta. Along its anterior surface, identify the three unpaired arterial branches that supply abdominal organs:

The **celiac trunk** arises from the aorta just inferior to the diaphragm.

The **superior mesenteric artery** originates about 2.5 cm (1 in.) inferior to the celiac trunk.

The **inferior mesenteric artery** arises about 3 to 4 cm (1.2 to 1.6 in.) superior to the termination of the abdominal aorta.

2 A small single branch, the **median sacral artery,** originates on the posterior surface of the abdominal aorta, near its termination. Identify this artery as it travels inferiorly to supply the sacrum and coccyx.

3 In the diagram below, label the three major unpaired branches of the abdominal aorta. List the structures that each supplies in the table below. ▶

4 Identify the paired branches of the abdominal aorta.

The **inferior phrenic arteries** arise just superior to the celiac trunk. They supply the inferior surface of the diaphragm.

The **adrenal arteries** supply the adrenal glands, located on the superior poles of the kidneys.

The **renal arteries** arise just inferior to the superior mesenteric artery. They supply the kidneys.

The **gonadal (ovarian** or **testicular) arteries** branch off the aorta between the superior and inferior mesenteric arteries. They supply the primary sex organs.

Along the posterolateral surface of the aorta, identify several pairs of **lumbar arteries,** which supply the lumbar vertebrae and abdominal wall.

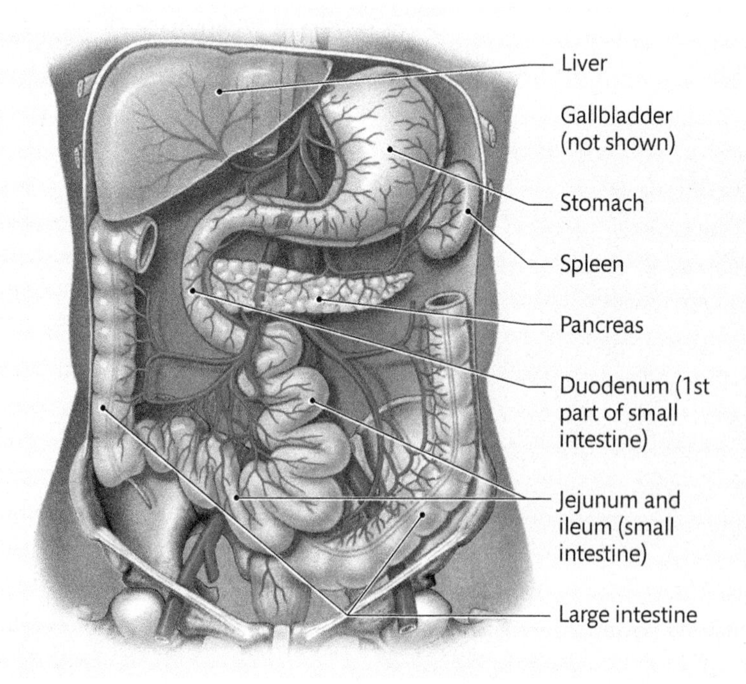

Branch	Structures Supplied
Celiac trunk	
Superior mesenteric artery	
Inferior mesenteric artery	

B Veins that drain blood from the abdominopelvic cavity and lower limbs empty into the inferior vena cava.

1 On a model of the torso or abdominal cavity, identify the **inferior vena cava** as it ascends to the right of the aorta along the posterior abdominal wall.

2 Observe that the inferior vena cava is formed by the union of the two **common iliac veins.**

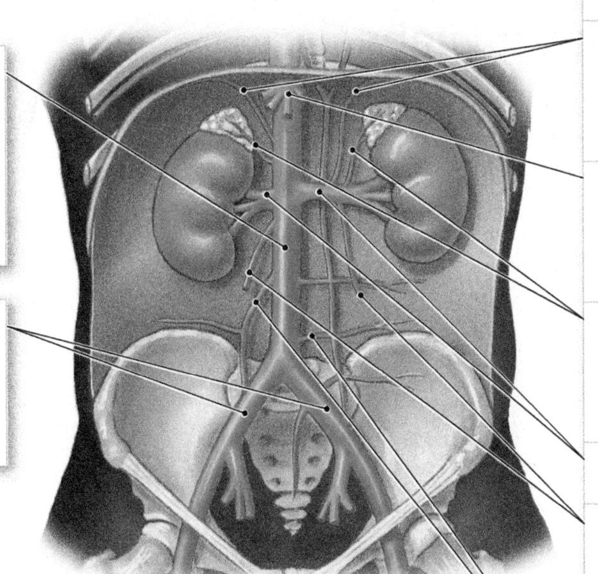

3 Most of the veins emptying into the inferior vena cava travel with their corresponding arteries. Identify these veins on a torso model.

The **phrenic veins** drain the inferior surface of the diaphragm. The right vein empties directly into the inferior vena cava. The left vein usually drains into the left renal vein.

The **hepatic veins** empty directly into the inferior vena cava. They drain blood from the liver sinusoids of the hepatic portal system, which is described below.

The **adrenal veins** drain blood from the adrenal glands. Notice that the right vein empties directly into the inferior vena cava. The left vein usually drains into the left renal vein.

The **renal veins** drain blood from the kidneys.

The **gonadal veins** drain the ovaries **(ovarian veins)** in the female and the testes **(testicular veins)** in the male. The right gonadal vein empties into the inferior vena cava, but the left gonadal vein often drains into the left renal vein.

Several pairs of **lumbar veins** empty into the inferior vena cava in the lumbar region. These veins drain the posterior abdominal wall.

C A **portal system** is a modified portion of the systemic circulation in which blood passes through an extra capillary bed before entering the veins that return blood to the heart. The **hepatic portal system** drains blood from the capillaries of the digestive organs into a set of veins leading into the hepatic portal vein. The hepatic portal vein directs blood into the liver sinusoids, which serve as the second capillary bed. Blood percolates through the sinusoids and drains into the hepatic veins and inferior vena cava for its return to the heart.

1 On the model, identify the **hepatic portal vein.**

2 Identify the following branches that drain into the hepatic portal vein.

Gastric veins drain blood from the stomach.

The **splenic vein** drains blood from the spleen and portions of the stomach and pancreas.

The **superior mesenteric vein** drains blood from the small intestine and the first (proximal) half of the large intestine.

The **inferior mesenteric vein** drains blood from the second (distal) half of the large intestine.

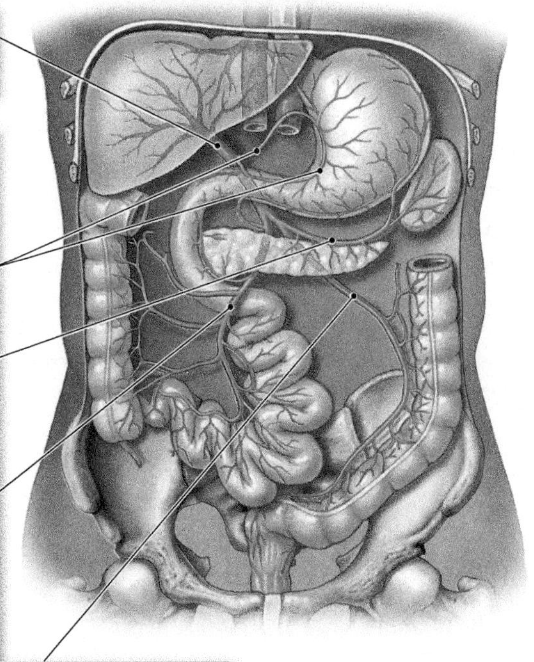

MAKING CONNECTIONS

Why do you think blood from the capillaries of digestive organs travels to the liver before returning to the heart? (*Hint:* Look ahead in your textbook to review the functions of the liver.)

▶ ______________________________

____

____

____

____

____

____

____

____

____

____

Activity **22.8**

Systemic Circulation: Identifying Blood Vessels in the Thoracic Wall and Upper Limbs

A The anterior thoracic wall and upper limbs are supplied by blood vessels that arise from the subclavian arteries.

1. On a torso model, identify the **subclavian artery** on one side of the body. Notice that this artery passes between the clavicle and the first rib.

2. Remove the anterior body wall from a torso model. On its posterior surface, identify the **internal thoracic artery** traveling just lateral to the sternum. Notice that this artery is a branch of the subclavian artery.

3. Each internal thoracic artery gives rise to a series of **anterior intercostal arteries.** Identify these blood vessels traveling through the intercostal spaces.

4. Identify the **axillary artery.** It begins at the lateral border of the first rib as a direct continuation of the subclavian artery. As it travels through the axilla, the axillary artery gives off several branches that supply the shoulder and thoracic wall.

5. On a model of the upper limb, find the **brachial artery.** Locate its origin at the inferior border of the teres major and verify that it is a continuation of the axillary artery.

6. Identify the **deep brachial artery,** an important branch of the brachial artery that travels to the posterior region of the arm.

7. Locate the brachial artery as it passes through the cubital fossa (the region anterior to the elbow joint). Near the head of the radius, the brachial artery divides into its two terminal branches.

   The **ulnar artery** travels along the medial side of the forearm.

   The **radial artery** passes along the lateral side.

8. Identify the two palmar arches:

   Superficial palmar arch

   Deep palmar arch

   The palmar arches arise from the ulnar and radial arteries as they enter the hand.

9. Identify structures or regions of the body that are supplied by the arteries listed in the table below. ▶

Arteries in the Upper Limb

Artery	Structures / Body Regions Supplied
Anterior intercostal	
Brachial	
Deep brachial	
Ulnar	
Radial	
Palmar arches	

B The anterior thoracic wall and upper limbs are drained by veins that empty into the brachiocephalic veins.

1. On a torso or upper limb model, identify the **axillary vein.** Notice that this blood vessel travels through the axilla, medial to the axillary artery.

2. Observe the axillary vein as it travels toward the base of the neck and passes between the first rib and the clavicle. At this position, the axillary vein becomes the **subclavian vein.**

3. In the upper limb, identify the following deep veins (if not shown on your model, you can determine their approximate position by locating the corresponding artery).

   The **brachial vein** travels through the arm and drains into the axillary vein.

   The radial and ulnar veins are the deep veins in the forearm. They arise from venous arches in the hand and travel superiorly through the forearm. Anterior to the elbow, they merge to form the brachial vein.

   The **radial vein** travels laterally in the forearm.

   The **ulnar vein** travels medially.

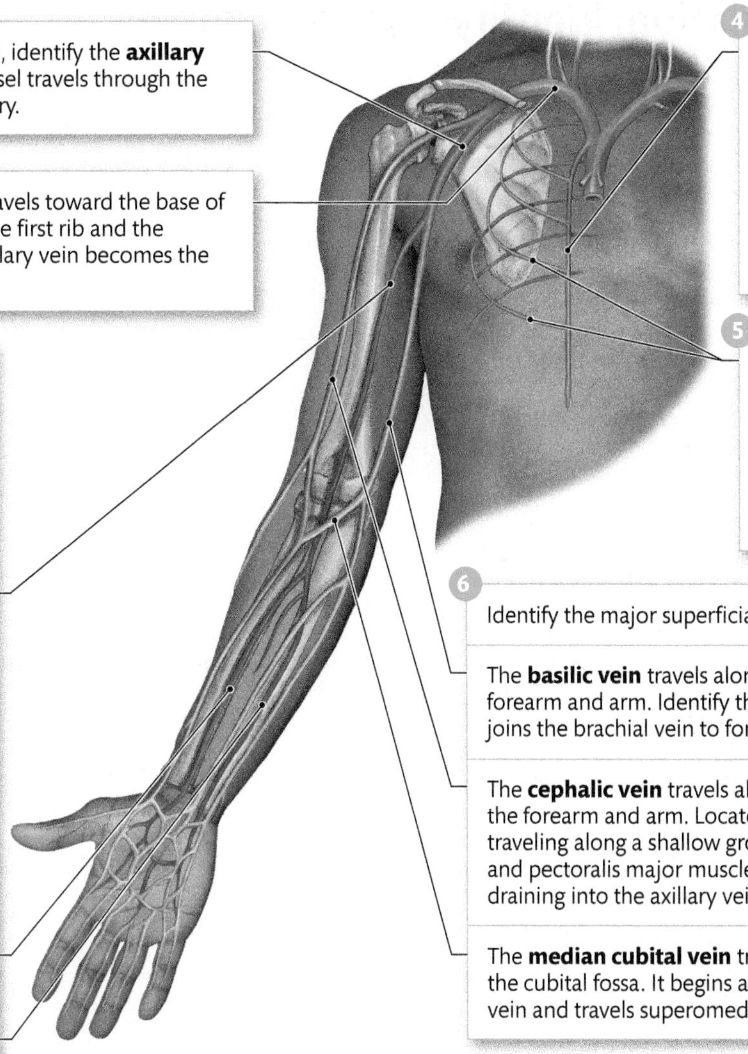

4. Remove the anterior body wall from a torso model. On its posterior surface, identify the **internal thoracic vein** traveling laterally to the sternum. Notice that this vein empties into the brachiocephalic vein.

5. Each internal thoracic vein receives blood from a series of **anterior intercostal veins.** Identify these blood vessels as they travel through the intercostal spaces.

6. Identify the major superficial veins in the upper limb.

   The **basilic vein** travels along the medial side of the forearm and arm. Identify this blood vessel where it joins the brachial vein to form the axillary vein.

   The **cephalic vein** travels along the lateral aspect of the forearm and arm. Locate its terminal portion traveling along a shallow groove, between the deltoid and pectoralis major muscles. The vein ends by draining into the axillary vein.

   The **median cubital vein** travels obliquely through the cubital fossa. It begins as a branch of the cephalic vein and travels superomedially to join the basilic vein.

MAKING CONNECTIONS

While studying for an important anatomy and physiology exam, you might support your head with your forearm and hand. This position flexes your forearm for an extended period of time, partially occluding your brachial artery and thus reducing the main blood supply to your forearm and hands. Refer to the figure on the right and describe the collateral pathways around your elbow that will deliver an adequate amount of blood to your forearm and hand.

▶ ____

____

____

____

____

____

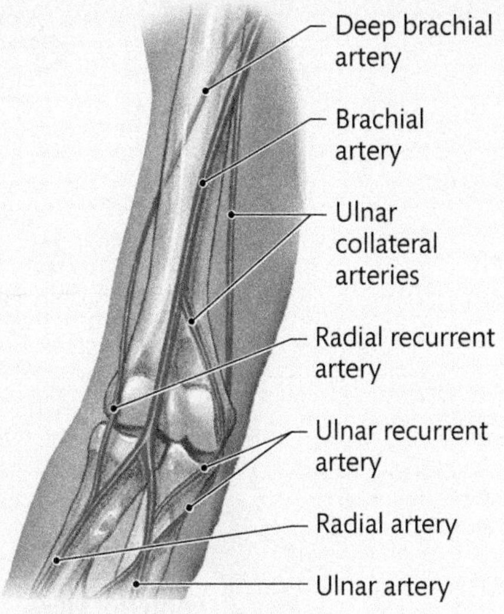

Want more practice? Go to: **MasteringA&P** > Study Area > Menu > Lab Tools > PAL > Anatomical Models > Cardiovascular System >
- Arteries
- Veins

Activity **22.9**

Systemic Circulation: Identifying Blood Vessels in the Lower Limbs

A The pelvis and lower limbs are supplied by blood vessels that arise from the common iliac arteries.

1. On a torso model, locate the two terminal branches of the abdominal aorta: the **left** and **right common iliac arteries.**

2. On one side of the torso model, trace the common iliac artery to its termination, where it bifurcates to form two branches:
   - **External iliac artery**
   - **Internal iliac artery**

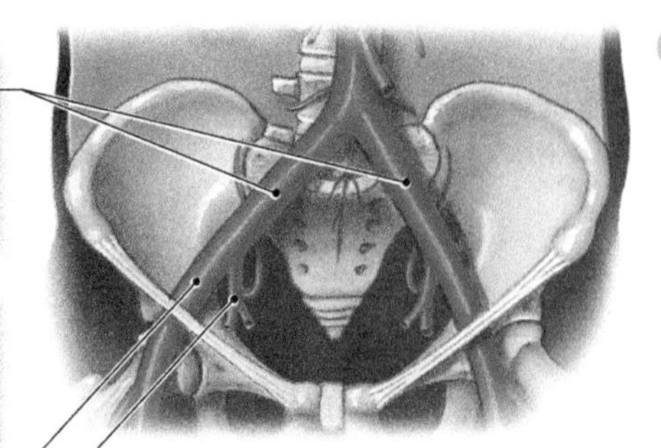

3. Consult you textbook and list three structures that are supplied by branches of the internal iliac arteries.
   - ▶ ______________________
   - ______________________
   - ______________________

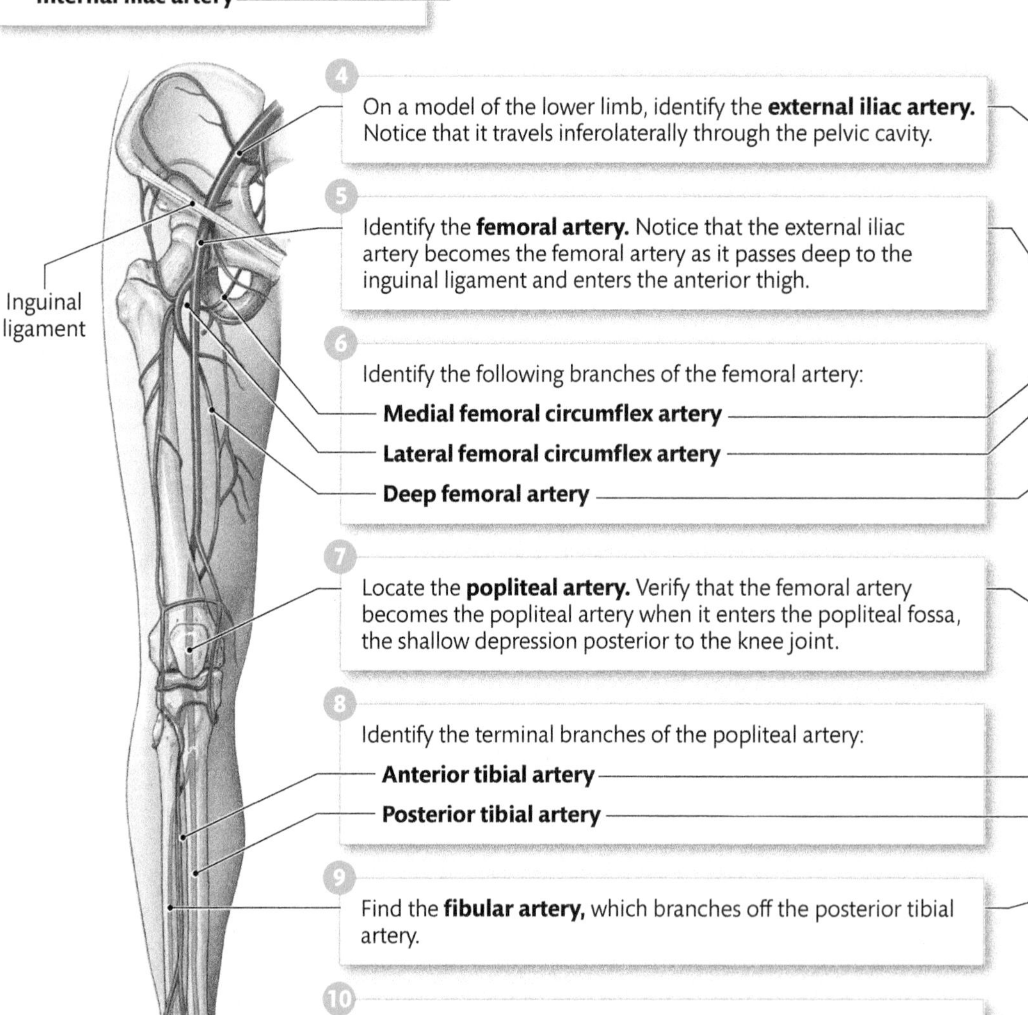

4. On a model of the lower limb, identify the **external iliac artery.** Notice that it travels inferolaterally through the pelvic cavity.

5. Identify the **femoral artery.** Notice that the external iliac artery becomes the femoral artery as it passes deep to the inguinal ligament and enters the anterior thigh.

6. Identify the following branches of the femoral artery:
   - **Medial femoral circumflex artery**
   - **Lateral femoral circumflex artery**
   - **Deep femoral artery**

7. Locate the **popliteal artery.** Verify that the femoral artery becomes the popliteal artery when it enters the popliteal fossa, the shallow depression posterior to the knee joint.

8. Identify the terminal branches of the popliteal artery:
   - **Anterior tibial artery**
   - **Posterior tibial artery**

9. Find the **fibular artery,** which branches off the posterior tibial artery.

10. On the dorsum of the foot, notice that the anterior tibial artery becomes the **dorsalis pedis artery.**

Want more practice? Go to: **MasteringA&P** > Study Area > Menu > Lab Tools > PAL >
- Anatomical Models > Cardiovascular System > Arteries
- Human Cadaver > Cardiovascular System > Blood Vessels

B The pelvis and lower limbs are drained by veins that empty into the common iliac veins.

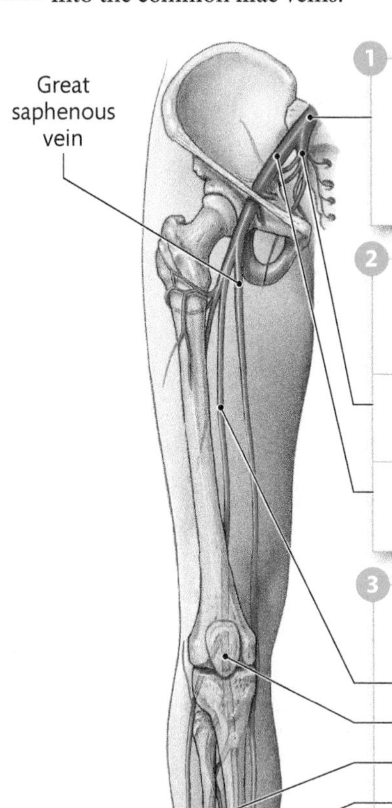

1 On a torso model, identify the two **common iliac veins,** which merge to form the inferior vena cava. Blood returning to the heart from the pelvis and lower limbs flows into these vessels before draining into the inferior vena cava.

2 The common iliac veins are formed by the union of the internal and external iliac veins. Identify these blood vessels on a torso model.

The **internal iliac veins** drain blood from structures along the pelvic wall and pelvic cavity.

The **external iliac veins** drain blood from the lower limbs.

3 On a model of the lower limb, identify the following deep veins:

- The **femoral vein** in the thigh
- The **popliteal vein** in the popliteal fossa
- The **anterior tibial vein** in the anterior leg
- The **posterior tibial vein** in the posterior leg
- The **fibular vein** in the lateral leg

These veins may not be shown on the anatomical models in your laboratory, but you can determine their locations by identifying the corresponding arteries.

4 The **great saphenous vein,** which is the longest vein in the body, is the main superficial vein in the lower limb. It originates on the medial side of the foot, and ascends along the medial aspect of the leg and thigh. At its termination in the anterior thigh, the great saphenous vein empties into the femoral vein. On a torso model, identify the great saphenous vein where it joins the femoral vein in the anterior thigh.

MAKING CONNECTIONS

Identify a structure or body region that is supplied by the following arteries: ▶

a. Femoral circumflex arteries:

____

b. Deep femoral artery:

____

c. Anterior tibial artery:

____

d. Posterior tibial artery:

____

e. Fibular artery:

____

f. Dorsalis pedis artery:

____

IN THE CLINIC

Vein Grafts

Because the great saphenous vein is long and easily accessible, it is often used for vein graft surgery to bypass obstructions in other blood vessels. For example, during **coronary bypass surgery,** a section of the great saphenous vein is attached before and after an obstruction in a coronary artery, thus establishing a circulatory pathway that travels around the blockage. Removal of the great saphenous vein does not produce significant problems with blood drainage because numerous additional veins will take over its function.

Activity 22.10

Surface Anatomy: Identifying Superficial Veins

You can identify many superficial veins, particularly in the neck and limbs, as they travel through the superficial fascia, deep to the skin.

A The external jugular vein is a prominent superficial vein in the neck.

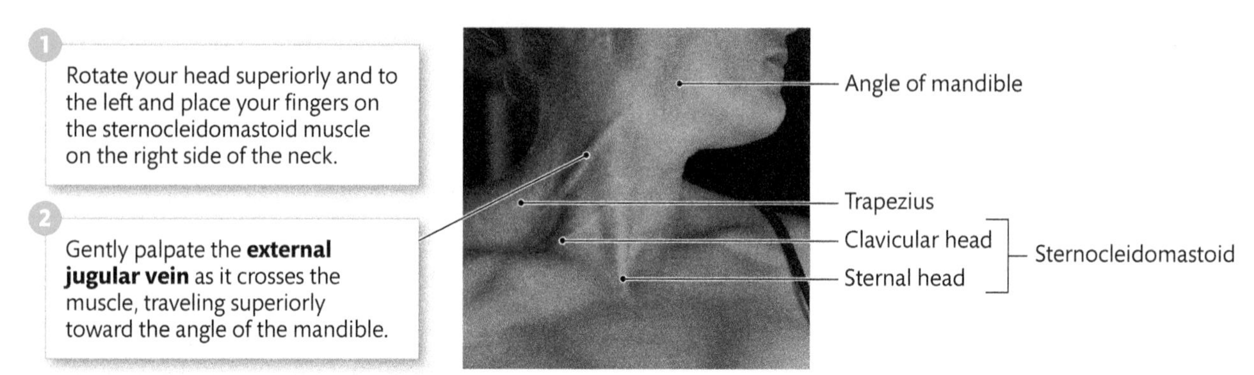

B You can identify superficial veins in the upper limb as they travel through the cubital fossa.

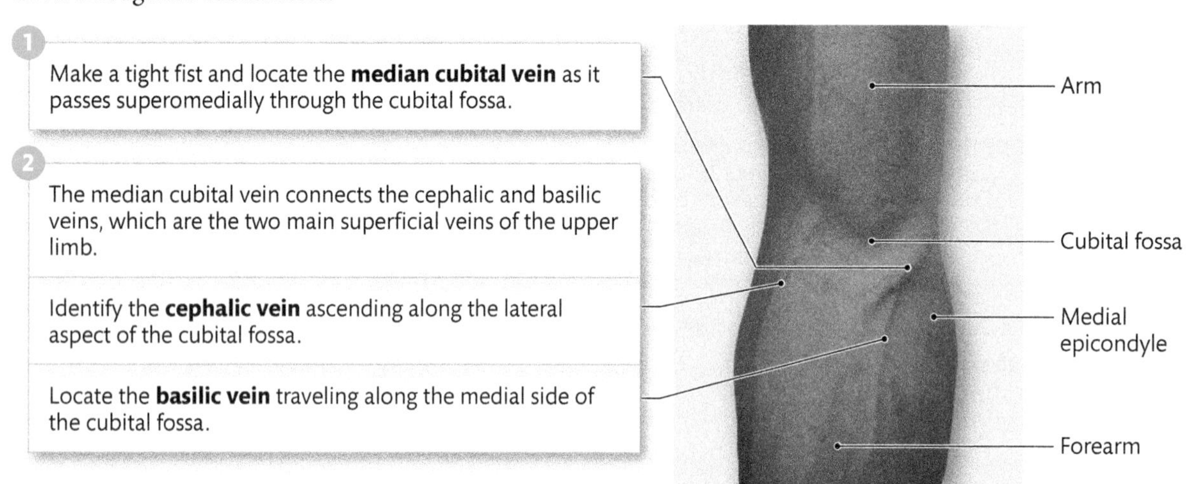

C The dorsal venous network drains blood from the hand. These veins empty into the deep and superficial veins of the forearm.

1. Place the palm of your hand flat on your laboratory table and spread your fingers apart.

2. On the dorsal surface of your hand, observe the superficial veins that form the **dorsal venous network.**

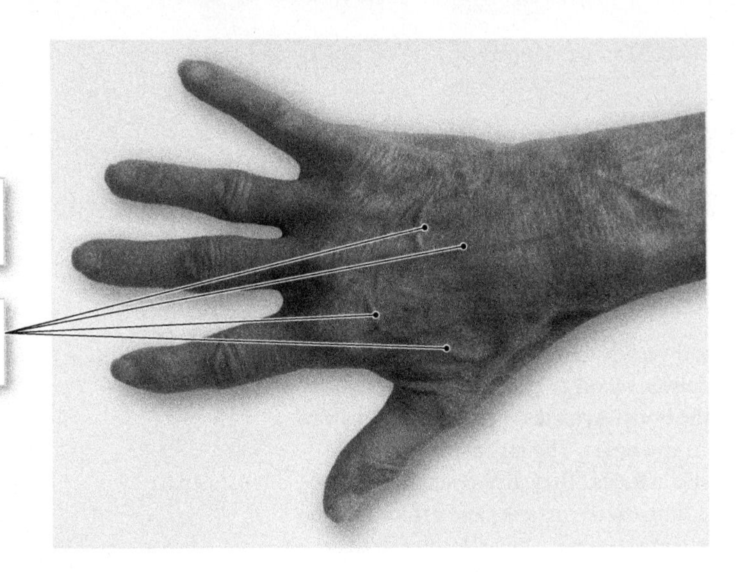

D The dorsal venous arch drains blood from the foot and empties into the deep and superficial veins of the leg.

1. Sit comfortably on a laboratory stool so that your feet are hanging freely. Remove your shoe and sock from one foot. Do not allow your bare foot to contact the laboratory floor.

2. Palpate the **dorsal venous arch** as it curves around the dorsal surface of the foot.

3. Locate the **great saphenous vein** near its origin, anterior to the medial malleolus. You should be able to palpate the vein as it travels subcutaneously at this location.

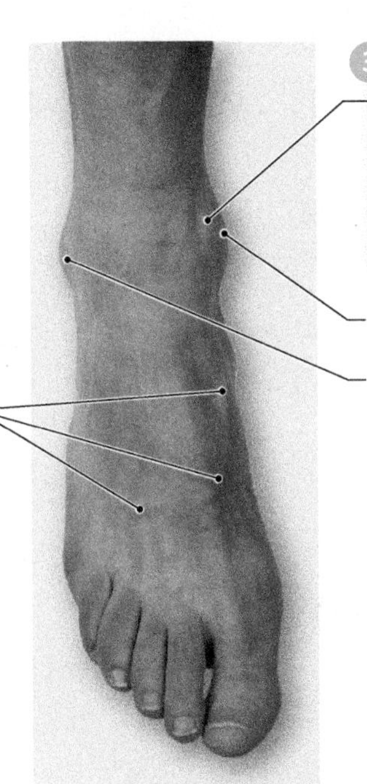

IN THE CLINIC

Venipuncture

The superficial veins of the upper limb are easily accessible and are therefore routinely used for drawing blood or injecting solutions **(venipuncture)**. For example, if you give a blood sample during a medical examination, your median cubital vein is often the blood vessel used. If you have just finished running a marathon on a hot day and are suffering from dehydration, you will be taken to the medical tent where a solution of fluids and electrolytes may be slowly injected into your cephalic vein, basilic vein, or dorsal venous network. This process is called **intravenous feeding.**

MAKING CONNECTIONS

The pattern of superficial veins can vary among individuals. How much variation exists in your class? What do you think is the significance of this variation?

▶ ______________________________

____

____

____

____

____

BEFORE YOU MOVE ON . . .

‹‹ LOOKING BACK

Blood vessels form an extensive network to deliver blood to all the cells and tissues in the body. **Arteries** transport blood away from the heart. The largest arteries are elastic arteries. They give rise to muscular (medium-sized) arteries and arterioles (the smallest arteries). The arterioles deliver blood to **capillaries,** where gas and nutrient exchange occurs between blood and cells.

Veins are blood vessels that transport blood toward the heart. Venules are the smallest veins that directly receive the blood that flows out of capillary beds. Venules converge to form small veins, and small veins in turn give rise to larger and larger veins. Ultimately, blood returns to the right atrium via the two great veins, the superior and inferior venae cavae, and the left atrium via the pulmonary veins.

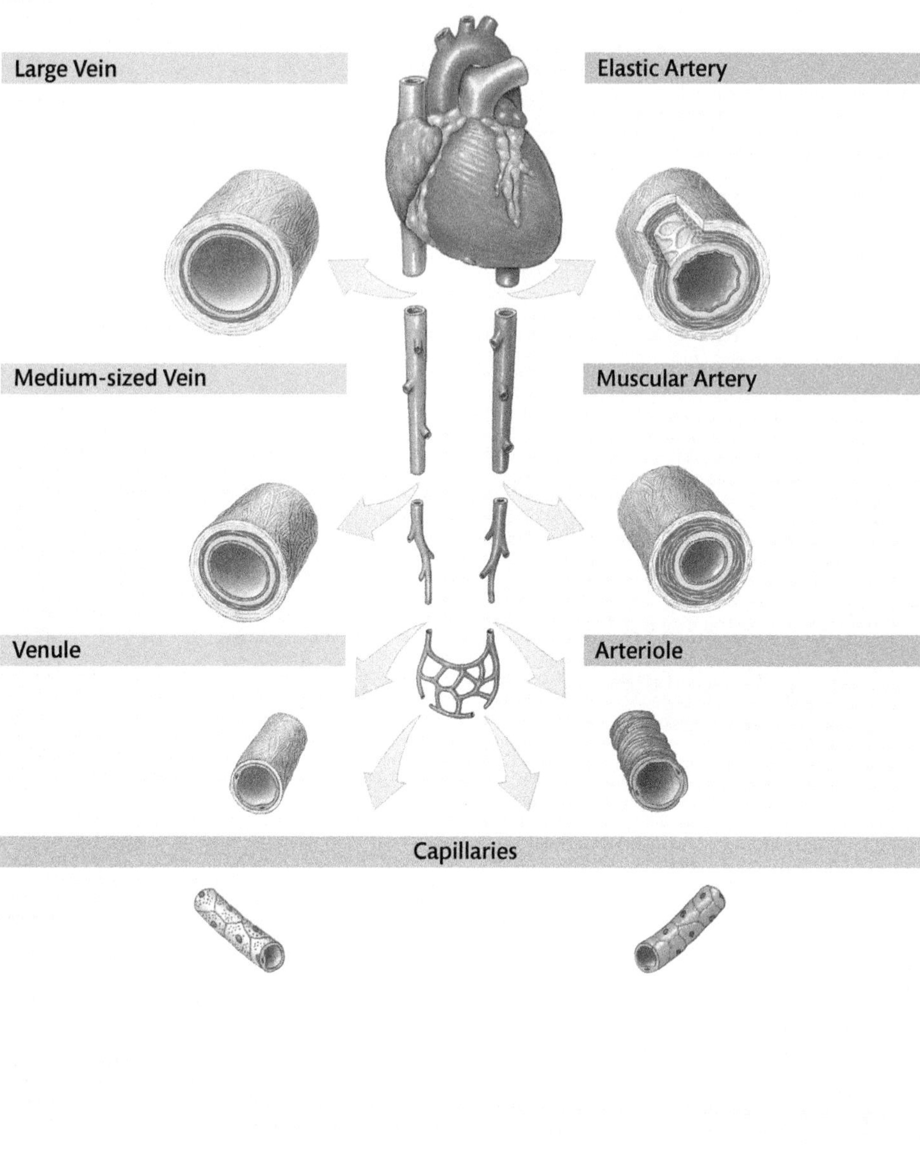

Compared to other types of blood vessels, capillaries have the largest total cross-sectional area and lowest blood flow velocity. Why are these two characteristics significant for capillaries to function as exchange vessels?

LOOKING FORWARD ››

In Exercises 20 (blood) and 21 (heart) and in this exercise (blood vessels), you studied the anatomy of the cardiovascular system. In the next laboratory exercise, you will investigate the physiology of the cardiovascular system. You will learn how the pumping action during the cardiac cycle establishes a pressure gradient that moves blood through the vast network of blood vessels, listen for heart sounds that are produced when the heart valves close during the cardiac cycle, learn how to measure blood pressure, and identify the best locations to feel your pulse. In addition, you will observe how a period of exercise affects blood pressure and pulse rate.

Name ______________________

Lab Section ______________________

Date ______________________

EXERCISE 22 REVIEW SHEET

Anatomy of Blood Vessels

1. Describe the main structural difference between elastic and muscular arteries. What is the functional significance of this difference?

2. Describe and compare the two main circulatory pathways: the pulmonary circulation and the systemic circulation.

3. Compare and contrast the structure and function of the three main types of capillaries by completing the following table.

Capillary Type	Structure of Capillary Wall	Functional Significance of Capillary Structure
Continuous	a.	b.
Fenestrated	c.	d.
Sinusoid	e.	f.

4. What is the functional difference between pulmonary arteries and bronchial arteries?

____

____

____

5. The primary blood flow to the posterior cerebral arteries travels through the vertebral and basilar arteries. Describe the collateral circulation to these arteries if atherosclerosis reduces blood flow through the basilar artery.

____

____

____

____

6. In this exercise, you examined the collateral circulation to the brain and around the elbow. Using the models in the laboratory, identify and describe two other examples of collateral circulation.

a. Example 1:

____

____

____

b. Example 2:

____

____

____

QUESTIONS 7–11: Match the term in column A with the appropriate description in column B.

A

7. Endothelium ____________

8. Microcirculation ____________

9. Anastomosis ____________

10. Sinusoids ____________

11. Portal circulation ____________

B

a. The most permeable capillaries, located in the liver

b. The single layer of squamous cells that lines the lumen of a blood vessel

c. The merging of two or more blood vessels with different origins

d. A specialized part of the systemic circulation in which blood passes through an additional capillary bed before returning to the heart

e. Small-diameter blood vessels, such as arterioles, capillaries, and venules

12. In addition to the superficial veins in the upper limb, a good location for intravenous feeding is near the origin of the great saphenous vein. Explain why.

____

____

____

QUESTIONS 13–44: Identify the labeled arteries in the figure.

13. ______________________________

14. ______________________________

15. ______________________________

16. ______________________________

17. ______________________________

18. ______________________________

19. ______________________________

20. ______________________________

21. ______________________________

22. ______________________________

23. ______________________________

24. ______________________________

25. ______________________________

26. ______________________________

27. ______________________________

28. ______________________________

29. ______________________________

30. ______________________________

31. ______________________________

32. ______________________________

33. ______________________________

34. ______________________________

35. ______________________________

36. ______________________________

37. ______________________________

38. ______________________________

39. ______________________________

40. ______________________________

41. ______________________________

42. ______________________________

43. ______________________________

44. ______________________________

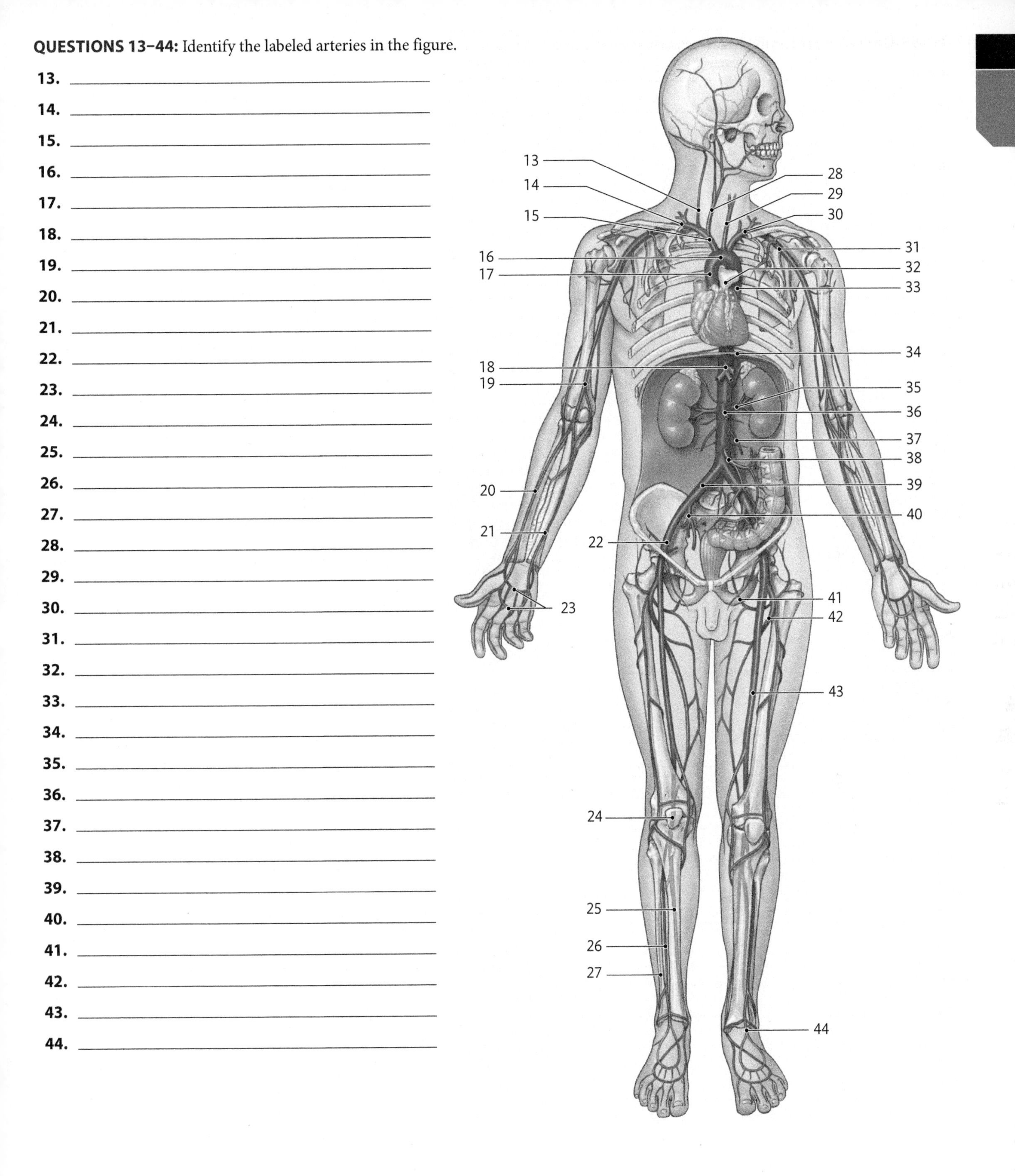

QUESTIONS 45–75: Identify the labeled veins in the figure.

45. ______________________________

46. ______________________________

47. ______________________________

48. ______________________________

49. ______________________________

50. ______________________________

51. ______________________________

52. ______________________________

53. ______________________________

54. ______________________________

55. ______________________________

56. ______________________________

57. ______________________________

58. ______________________________

59. ______________________________

60. ______________________________

61. ______________________________

62. ______________________________

63. ______________________________

64. ______________________________

65. ______________________________

66. ______________________________

67. ______________________________

68. ______________________________

69. ______________________________

70. ______________________________

71. ______________________________

72. ______________________________

73. ______________________________

74. ______________________________

75. ______________________________

45
46
61
47
62
48
49
50
63
64
51
52
65
66
53
67
54
68
55
69
56
70
71
72
73
57
58
59
74
75
60

Superficial veins
Deep veins

EXERCISE

23

Cardiovascular Physiology

LEARNING OUTCOMES

These Learning Outcomes correspond by number to the laboratory activities in this exercise. When you complete the activities, you should be able to:

Activity 23.1	Use a stethoscope to listen to the heart sounds at the auscultation areas.
Activity 23.2	Measure blood pressure and calculate the pulse pressure and mean arterial pressure (MAP).
Activity 23.3	Measure the pulse rate at various locations on the body.
Activity 23.4	Examine the effect of exercise on blood pressure and pulse rate.
Activities 23.5–23.6	Record, measure, and analyze the electrical activity of the heart under various conditions.

LABORATORY SUPPLIES

- Stethoscopes
- Alcohol swabs
- Sphygmomanometers
- Stopwatches or digital timers
- Stationary cycle, elliptical machine, or treadmill
- Computer with Windows or Mac operating system
- Biopac Student Lab System (MP35, MP36, or MP45)
- Electrode lead set (SS2L)
- Disposable electrodes (EL503)
- Biopac electrode gel and abrasive pad

PRE-LAB QUIZ

Before you begin, read all the activities in Exercise 23 and the required reading in your textbook that is assigned by your instructor.

1. During this laboratory exercise, which of the following activities will you complete?
   a. examining the effect of exercise on blood pressure and pulse
   b. listening for heart sounds
   c. measuring your pulse rate
   d. measuring your blood pressure
   e. all the above
2. True or False: Blood pressure is measured in units of measure called millimeters of mercury. ________
3. True or False: During the cardiac cycle, the closing of the semilunar valves produces the first heart sound. ________
4. True or False: The auscultation areas for the tricuspid and bicuspid valves are located in the right fifth intercostal space. ________
5. The ______________ pressure measures the resistance to blood flow in the arteries.
6. The average pressure that moves blood through the systemic circulation is the ______________ pressure.
7. You can feel your pulse in all the following locations *except* the
   a. popliteal fossa.
   b. anterior surface of the lateral wrist.
   c. femoral triangle.
   d. posterior thigh.
   e. temporal area of the head.
8. In Activity 23.4 of this laboratory exercise, you will be asked to predict the effect that exercise will have on all the following *except*
   a. mean arterial pressure (MAP).
   b. diastolic pressure.
   c. capillary pressure.
   d. systolic pressure.
   e. pulse pressure.
9. True or False: Ventricular ejection is the period during the heart cycle when the ventricles pump blood into the great vessels. ________
10. True or False: Blood pressure in the aorta and other elastic arteries is smooth or constant, with little or no fluctuation. ________

Activity **23.1**

Listening for Heart Sounds

A During the **cardiac cycle,** the opening and closing of the heart valves directs the flow of blood from the atria to the ventricles and from the ventricles to the great vessels. The closing of the valves produces two distinctive heart sounds, "lubb-dupp," that you can hear with a stethoscope.

1 The cardiac cycle begins when a small volume of blood is pumped into each ventricle during **atrial systole.** The AV valves are open. At this time, is the atrial blood pressure greater or less than the ventricular blood pressure?

▶ ______________________

2 When atrial systole ends and **ventricular systole** begins, the atrioventricular (AV) valves close, causing vibrations that produce the **first heart sound ("lubb").** Describe the blood pressure changes that occur in the atria and ventricles when the AV valves close.

▶ ______________________

____

3 During the second part of ventricular systole, the semilunar valves open, and the ventricles pump blood into the great vessels. This is known as **ventricular ejection.** Describe the blood pressure differences between the ventricles and great arteries when the semilunar valves are open.

▶ ______________________

____

____

____

____

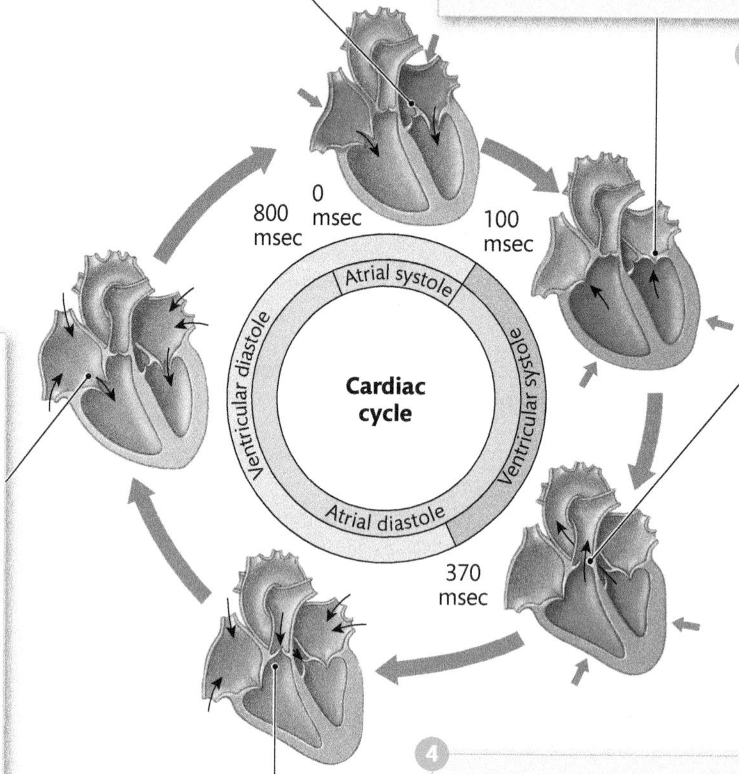

0 msec

100 msec

370 msec

800 msec

Atrial systole

Ventricular systole

Atrial diastole

Ventricular diastole

Cardiac cycle

5 At the end of the cardiac cycle, **the AV valves open,** and passive filling of the ventricles begins. Describe the blood pressure differences between the ventricles and atria, which allow the AV valves to open.

▶ ______________________

____

____

____

____

4 At the beginning of **ventricular diastole,** the semilunar valves close. Vibrations, generated by the closing of these valves, create the **second heart sound ("dupp").** At this time, all heart chambers are relaxed. Describe the blood pressure changes that occur in the ventricles and great arteries when the semilunar valves close.

▶ ______________________

____

____

IN THE CLINIC

Mitral Valve Prolapse

Incomplete closure of the AV valves can allow regurgitation or backflow of blood into the atria. This backflow can cause an abnormal gurgling sound known as a **heart murmur.** On the left side of the heart, incomplete closure of the bicuspid (mitral) valve is called a **mitral valve prolapse.** Minor prolapses are fairly common. Most people live with them and do not experience adverse effects. A major prolapse, possibly caused by ruptured chordae tendineae or severe damage to the cusps, can have serious if not life-threatening consequences, however.

B The best locations to hear **heart sounds** are the **auscultation areas for the heart** on the anterior thoracic wall. There are four auscultation areas. Each is named after the heart valve that can best be heard at that location. Locate the four auscultation areas on your body:

1 The **aortic semilunar area** is located in the second intercostal space, just to the right of the sternum.

Locate the second intercostal space by palpating the superior margin of the manubrium.

Move your finger inferiorly until you can feel the junction between the manubrium and the body of the sternum.

Move your finger laterally to the right of the sternum and feel the costal cartilage of the second rib. The second intercostal space is just inferior to this rib.

The aortic semilunar area is located in this space, just lateral to the sternum.

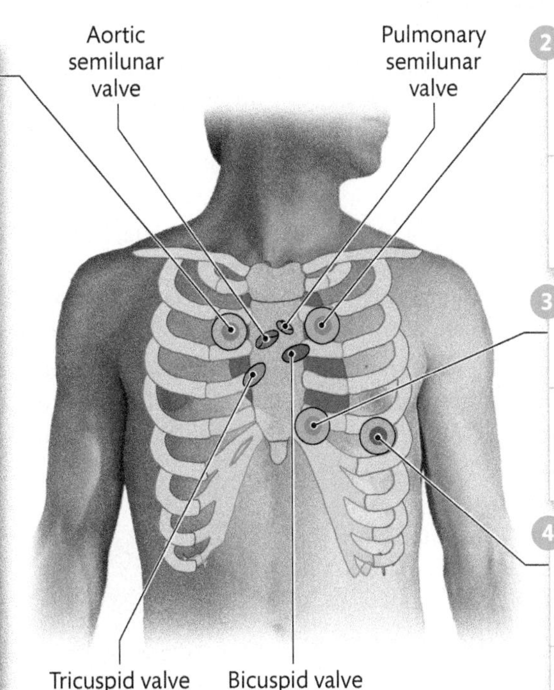

2 The **pulmonary semilunar area** is located in the second intercostal space, just to the left of the sternum.

Follow the guidelines for locating the aortic semilunar area, but this time, move to the left side of the sternum.

3 The **tricuspid area** is located in the left fifth intercostal space.

Palpate the inferior end of the sternum. The tricuspid area is just to the left of that point.

4 The **bicuspid area** is located in the left fifth intercostal space, where the apex of the heart is found.

Find the inferior end of the sternum.

Move your finger about 7 cm (2.75 in.) to the left. You will feel the fifth intercostal space between the fifth and sixth ribs.

C You can use a stethoscope to hear heart sounds at the auscultation areas.

1. Sterilize the eartips of a stethoscope with an alcohol swab before placing them in your ears.
2. Hold the chest piece between your index and middle fingers. Holding it with your fingertips may produce excess background noise.
3. Place the chest piece of the stethoscope on the bicuspid area and listen. Can you hear both heart sounds ("lubb-dupp")? Can you hear one sound better than the other? Record your answers in Table 23.1. ▶
4. Listen to the heart sounds at the other auscultation areas. Answer the same two questions that were asked for the biscuspid area in step 3. Record your answers in Table 23.1. ▶

TABLE 23.1 Listening for Heart Sounds

Auscultation Area	Can You Hear Both Heart Sounds?	Can You Hear One Sound Better?
Biscuspid area		
Tricuspid area		
Aortic semilunar area		
Pulmonary semilunar area		

MAKING CONNECTIONS

During a physical examination, why does the doctor listen to the heart sounds at all four auscultation areas?

▶ ______________________________

____

____

____

Activity **23.2**

Measuring Blood Pressure

Blood pressure is the force exerted by blood on the walls of blood vessels. It is a function of the pumping action of the heart and the resistance to blood flow. Blood flows throughout the circulatory pathways due to the existence of a pressure gradient that allows blood to move from areas of high pressure to areas of low pressure. In large elastic arteries, blood pressure fluctuates between a maximum and minimum value corresponding to the cardiac cycle.

- During ventricular systole, ventricular ejection stretches the wall of the aorta, and the pressure inside the aorta reaches a peak. This maximum level is called the **systolic pressure.**
- During ventricular diastole, the aortic semilunar valve closes, and the elastic fibers in the wall of the aorta recoil to force blood forward. Aortic pressure declines to a minimum level called the **diastolic pressure.**

Thus, blood pressure in the aorta is not smooth or constant, but pulsatile in nature. This characteristic is also true for other elastic arteries, but it diminishes in the smaller arteries and arterioles as the number of elastic fibers in the vessel walls decreases. Blood flowing through capillaries and veins travels under relatively low pressure with little or no fluctuation.

A You will use a **sphygmomanometer** to measure blood pressure. A sphygmomanometer consists of a rubber cuff with two attached rubber tubes. At the end of one tube is a compressible hand bulb; at the end of the other tube is a dial that records pressure in units called **millimeters of mercury (mm Hg).** When blood pressure is measured, it is the arterial blood pressure in the systemic circulation that is recorded.

Because the pressure in the arteries is pulsatile, both systolic and diastolic pressures are measured. A blood pressure of 120/80 means:

- Systolic pressure = 120 mm Hg (the force exerted by the left ventricle when it pumps blood into the aorta)
- Diastolic pressure = 80 mm Hg (the resistance to blood flow in the arteries)

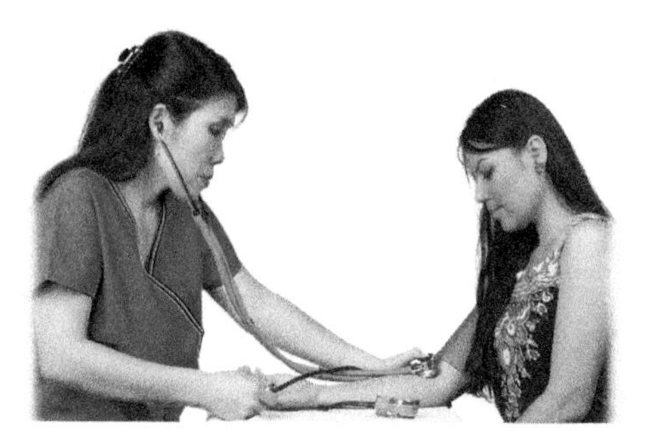

Word Origins

A millimeter of mercury (mm Hg) is the force that causes a fluid, with a density equal to that of mercury, to rise one millimeter up a tube. If the pressure in a blood vessel is 95 mm Hg, for example, the force exerted by the blood will cause a column of fluid to rise 95 millimeters.

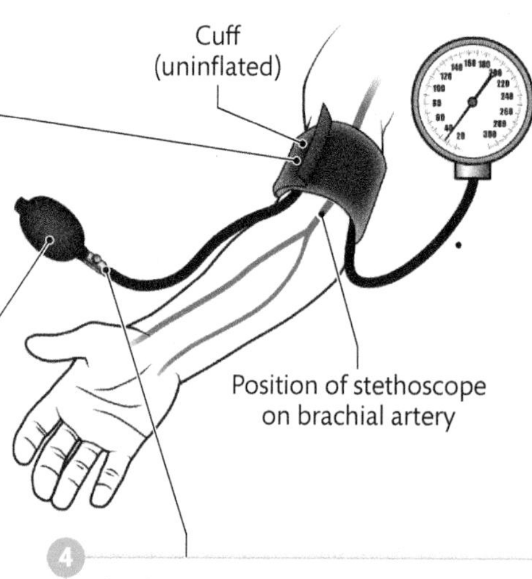

1. Have your lab partner sit quietly. Wrap the **cuff of a sphygmomanometer** around his or her arm and position a stethoscope over the brachial artery, distal to the cuff.

2. Make sure the valve at the base of the **hand bulb** is closed. Then, squeeze the hand bulb to inflate the cuff until the cuff pressure is greater than the pressure in the brachial artery. For most people, a cuff pressure of 160 mm Hg will be sufficient.

3. The **walls of the brachial artery** are now fully compressed, preventing blood from flowing to the forearm. You can verify this compression by listening to the brachial artery as you squeeze the cuff. When you hear no sound, compression is complete.

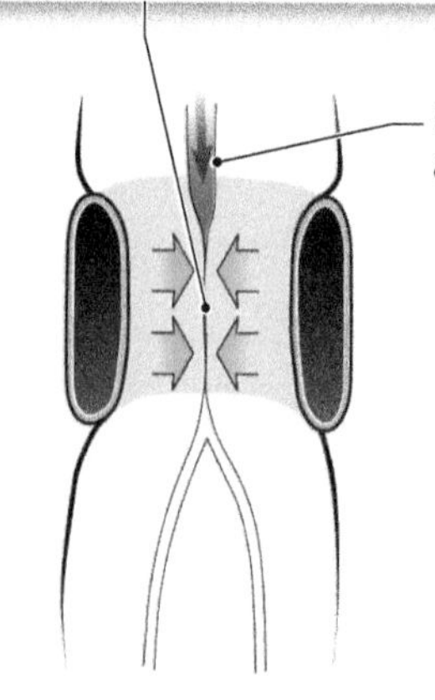

4. Slowly deflate the cuff while listening to the brachial artery with the stethoscope. The cuff can be deflated by opening the **valve at the base of the hand bulb.** Deflate the cuff at a rate of 2 to 3 mm Hg per second.

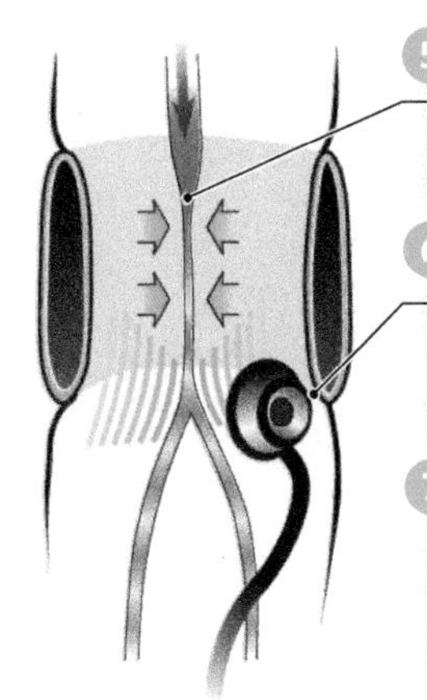

5. As the cuff pressure slowly decreases, it will eventually become less than the arterial pressure. At this point, the artery is partially open and blood will begin to pass through.

6. Because the blood flow is turbulent at this time, you will hear thumping sounds. These **Korotkoff sounds** correspond to the systolic blood pressure. Make a mental note of the pressure reading at the time that you begin to hear Korotkoff sounds.

7. Continue to slowly deflate the cuff. As you do so, the constriction in the artery is reduced and blood flow becomes less turbulent. As a result, the thumping sounds will become faint and eventually disappear. The pressure at the time when the thumping sounds stop corresponds to the diastolic pressure. Make a mental note of the pressure reading at this time.

8. Record your partner's blood pressure in Table 23.2. ▶

9. With your lab partner's assistance, repeat steps 1 through 7 to determine your own blood pressure. Record this value in Table 23.2. ▶

B The difference between the systolic and diastolic pressures is the **pulse pressure.**

Pulse pressure = systolic pressure − diastolic pressure

1. Calculate the pulse pressure for you and your lab partner.
2. Record the pulse pressures in Table 23.2. ▶

C The average pressure that drives blood through the systemic circulation is the **mean arterial pressure (MAP)**. MAP is crucial for maintaining a steady blood flow from the heart to the capillaries. It is equal to the diastolic pressure plus one-third of the pulse pressure and is represented by the following equation:

MAP = diastolic pressure + 1/3 pulse pressure

1. Calculate the mean arterial pressure for you and your lab partner.
2. Record the mean arterial pressures in Table 23.2. ▶

TABLE 23.2 Blood Pressure Readings

Subject	Blood Pressure (mm Hg)	Pulse Pressure (mm Hg)	Mean Arterial Pressure (mm Hg)
Population average*	120/80	40	93.3
Your lab partner			
Yourself			

*Keep in mind that these readings are average values. You should expect variation in the population.

MAKING CONNECTIONS

You measured arterial blood pressure by wrapping the sphygmomanometer cuff around the brachial artery. Why do you think the brachial artery is used for blood pressure measurements?

▶ ______________________________

____

____

____

____

Activity **23.3**

Measuring the Pulse Rate

In the systemic circulation, when the left ventricle pumps blood into the aorta, the pulse pressure generates pressure waves. These waves travel along other elastic arteries whose walls expand and recoil at a frequency that corresponds to the heartbeat. The rhythmic expansion and recoil of the arteries is known as the **pulse.** You can feel pulses at various locations on the body. They tend to diminish in smaller arteries and are absent in capillaries and veins.

A On the forearm, you can feel the pulse at the wrist.

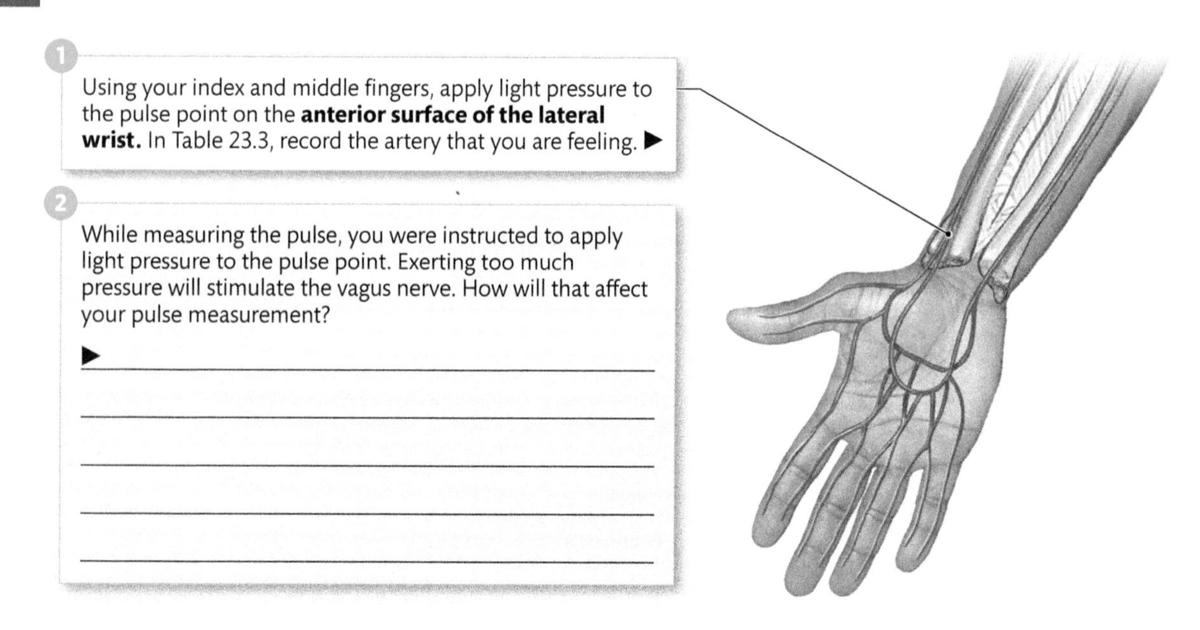

1. Using your index and middle fingers, apply light pressure to the pulse point on the **anterior surface of the lateral wrist.** In Table 23.3, record the artery that you are feeling. ▶

2. While measuring the pulse, you were instructed to apply light pressure to the pulse point. Exerting too much pressure will stimulate the vagus nerve. How will that affect your pulse measurement?

▶ ____

____

____

____

____

B On the lower limb, you can feel a pulse at your thigh and popliteal fossa.

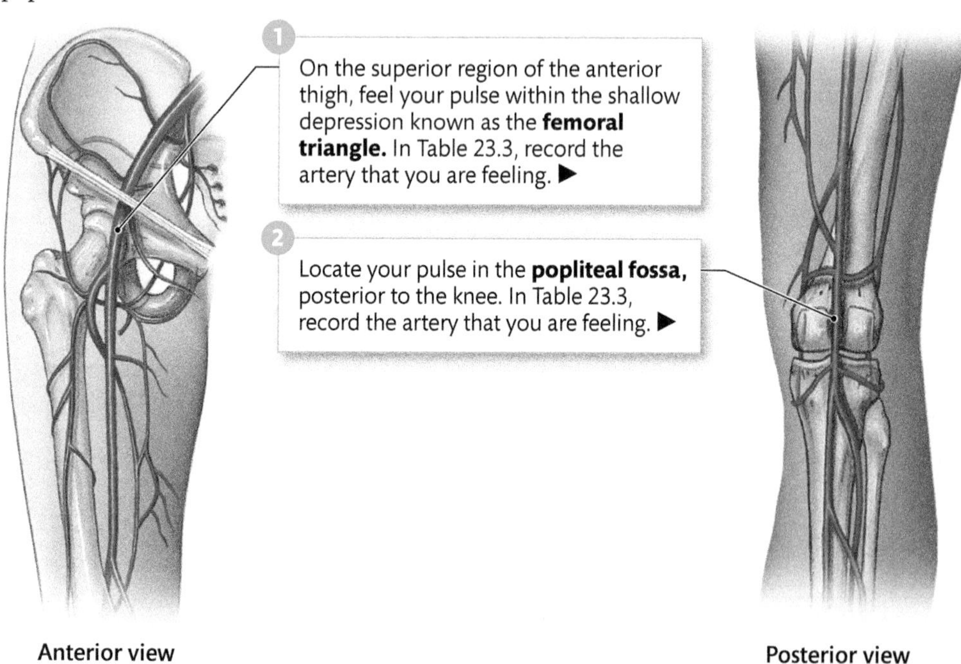

1. On the superior region of the anterior thigh, feel your pulse within the shallow depression known as the **femoral triangle.** In Table 23.3, record the artery that you are feeling. ▶

2. Locate your pulse in the **popliteal fossa,** posterior to the knee. In Table 23.3, record the artery that you are feeling. ▶

Anterior view

Posterior view

■ Want more practice? Go to: **MasteringA&P** > Study Area > Menu > Lab Tools > **PhysioEx**™ > Exercise 6: Cardiovascular Physiology > Activity 2: Examining the Effect of Vagus Nerve Stimulation

C You can feel a pulse at two locations in the head and neck.

1. Feel your pulse on the side of the head in the **temporal area.** In Table 23.3, record the artery that you are feeling. ▶

2. Locate the pulse point in the **superior region of the neck,** about 1 cm inferomedial to the jaw. In Table 23.3, record the artery that you are feeling. ▶

TABLE 23.3 Pulse Points

Body Region	Artery at Pulse Point
Anterior surface of lateral wrist	
Femoral triangle	
Popliteal fossa	
Temporal area of the head	
Superior neck	

D You can measure your resting pulse at any of the pulse points on the body.

1. Sit quietly for 3 minutes and then take your resting pulse at any pulse point. The most accurate measurement would be to count the number of beats for 1 full minute, but a resting pulse can usually be measured accurately by counting for 10 seconds and multiplying by 6.

2. Record your resting pulse:

▶ ______________________ beats/min

MAKING CONNECTIONS

The average resting pulse is 70 to 80 beats per minute, but there is considerable variation in the population. For example, it is not unusual for well-trained athletes to have resting pulse rates as low as 40 to 50 beats per minute. (The resting pulse rates of some world-class marathon runners are 35 to 40 beats per minute!) Why do athletes tend to have slower resting pulse rates than nonathletes?

▶ ____

____

____

____

____

____

Activity 23.4

Examining the Effect of Exercise on Blood Pressure and Pulse Rate

A Temporary elevations in blood pressure and pulse rate are normal when the body is adapting to changing conditions that occur during physical exercise. In this activity, you will measure blood pressure and pulse rate before, during, and after exercise. Try to work in teams of three: One person will be the test subject, one person will measure blood pressure, and one person will measure pulse rate.

⚠ **Warning:** *Anyone who is physically or medically unable to perform physical exercise should not participate as a test subject.*

1. *Preexercise:* After the test subject has sat quietly for 3 minutes, measure his or her resting blood pressure and pulse rate. Keep the sphygmomanometer and stethoscope in position after measuring the blood pressure. Record these results in Table 23.4. ▶

2. **Make a Prediction**

Before you begin, predict what effect exercise will have on blood pressure, pulse pressure, and mean arterial pressure (MAP): increase, decrease, or no change. ▶

- Effect on blood pressure:
  systolic pressure
  ____
  diastolic pressure
  ____
- Effect on pulse pressure
  ____
- Effect on MAP
  ____

Predict how many minutes after exercise it will take for the test subject's blood pressure and pulse rate to return to resting levels. ▶

- Return to resting blood pressure in __________ minutes.
- Return to resting pulse rate in __________ minutes.

3. *Exercise period:* Have the test subject exercise for 15 minutes on a stationary bicycle, treadmill, or elliptical machine. If exercise equipment is not available, another form of exercise, such as running in place, can be substituted. Keep the sphygmomanometer and stethoscope in position during the entire exercise period.

4. At 3-minute intervals during the exercise period and again immediately after exercise ends (15 minutes), measure the test subject's blood pressure and pulse rate. Measure the pulse by counting beats for 10 seconds and multiplying by 6. For each blood pressure measurement, calculate the pulse pressure and MAP. Record all results in Table 23.4. ▶

5. *Postexercise:* One minute after exercise ends, measure the test subject's blood pressure and pulse rate. Take additional measurements at 3-minute intervals until blood pressure and pulse rate return to resting levels. Measure the pulse by counting beats for 10 seconds and multiplying by 6. For each blood pressure measurement, calculate the pulse pressure and MAP. Record all the results in Table 23.4. ▶

Want more practice? Go to: MasteringA&P > Study Area > Menu > Lab Tools > PhysioEx > Exercise 6: Cardiovascular Physiology >

- Activity 3: Examining the Effect of Temperature on Heart Rate
- Activity 4: Examining the Effects of Chemical Modifiers on Heart Rate

TABLE 23.4 Effect of Exercise on Blood Pressure and Pulse Rate

Stage	Time	Blood Pressure (mm Hg)	Pulse Pressure (mm Hg)	Mean Arterial Pressure (mm Hg)	Pulse (beats/minute)
Preexercise	At rest				
Exercise period	3 minutes				
	6 minutes				
	9 minutes				
	12 minutes				
	15 minutes				
Postexercise	1 minute				
	4 minutes				
	7 minutes				
	10 minutes				
	13 minutes				
	16 minutes				

B Assess the Outcome

1. How do the actual changes in systolic pressure, diastolic pressure, pulse pressure, and MAP compare to your predictions? Explain.

▶ ______________________________

2. The length of time it takes for blood pressure and pulse to recover to a resting level is an indication of physical fitness and cardiovascular efficiency. As a general rule, the shorter the recovery time, the better the fitness. How does your test subject compare with other members of the class?

▶ ______________________________

MAKING CONNECTIONS

During and immediately after exercise, counting the pulse for 10 seconds and multiplying by 6 is more accurate than counting for a full minute. Explain why.

▶ ______________________________

Electrocardiography Using the Biopac Student Lab System: Recording the Electrical Activity of the Heart under Various Conditions

The specialized cardiac muscle cells in the cardiac conduction system generate action potentials without inputs from nerves or hormones. The electrical activity occuring along the heart wall spreads through the rest of the body and creates changes in the electrical potential of the skin. These electrical changes can be detected by electrodes on the skin and recorded in a procedure known as **electrocardiography.** From the resulting **electrocardiogram (ECG),** it is possible to examine the heart's mechanical activity.

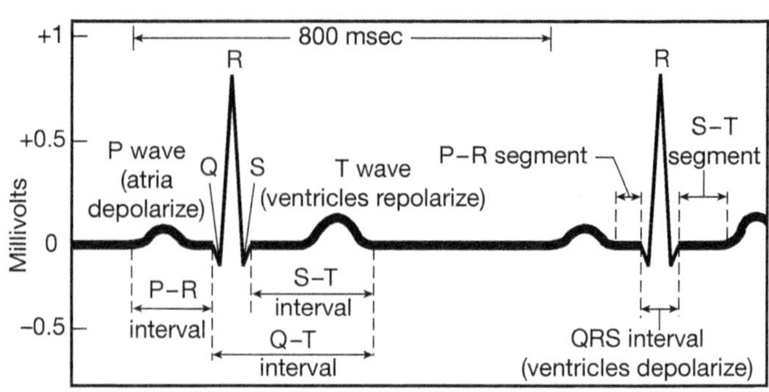

The normal ECG tracing is a flat baseline interrupted by a series of waves. In a single cardiac cycle, the **P wave** indicates the depolarization of the atria just prior to the beginning of atrial contraction or systole. The **QRS complex** represents the depolarization of the ventricles, which precedes ventricular systole. The **T wave** results from ventricular repolarization, which occurs before ventricular relaxation or diastole. The wave associated with repolarization of the atria is hidden by the much larger QRS complex.

In this activity, you will record the ECG of your lab partner under varying conditions, correlate electrical and mechanical events of the heart, and observe the changes in heart rate associated with body position and breathing pattern.

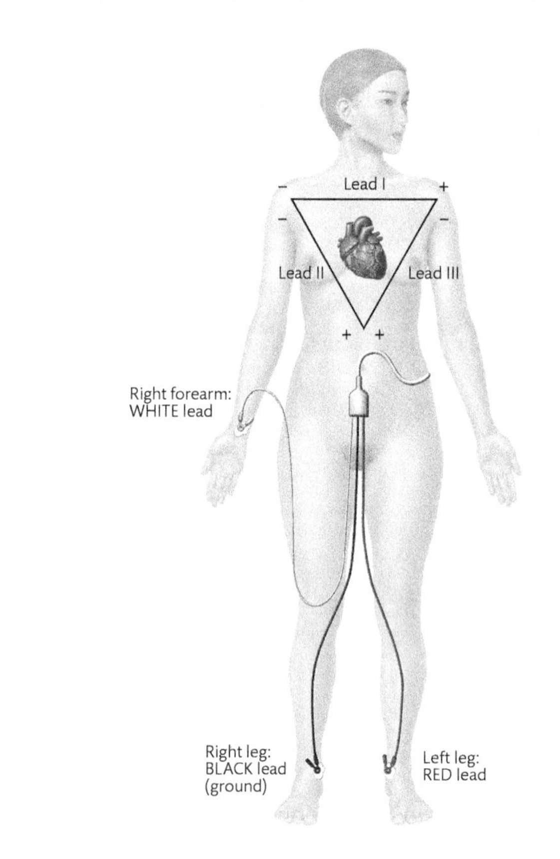

A Setup and Calibration

1. Instruct your lab partner to remove all jewelry, watches, and other metal objects. Ask your partner to lie down and relax.
2. Attach electrode leads (SS2L) to Channel 2 of the Biopac unit (MP35, MP36, or MP45) and then turn the unit on.
3. Attach three electrodes to your lab partner as shown in the figure on the right. Attach the white electrode lead to the anterior surface of the right wrist, the red lead to the medial aspect of the left leg just superior to the ankle, and the black lead to the medial aspect of the right leg just superior to the ankle. Use a small amount of gel on the skin where each electrode will be placed.
4. Start the Biopac Student Lab System, choose Lesson 5 (L05-ECG-1), and click **OK.** When prompted, enter a unique filename and click **OK.**
5. Click on the **Calibrate** button and wait while the computer adjusts for optimal recording. The calibration procedure will automatically stop after 8 seconds. If there is a small ECG waveform with a relatively flat baseline (see the figure to the right), calibration was successful. If not, click on the **Redo Calibration** button, check all connections, and repeat the calibration.

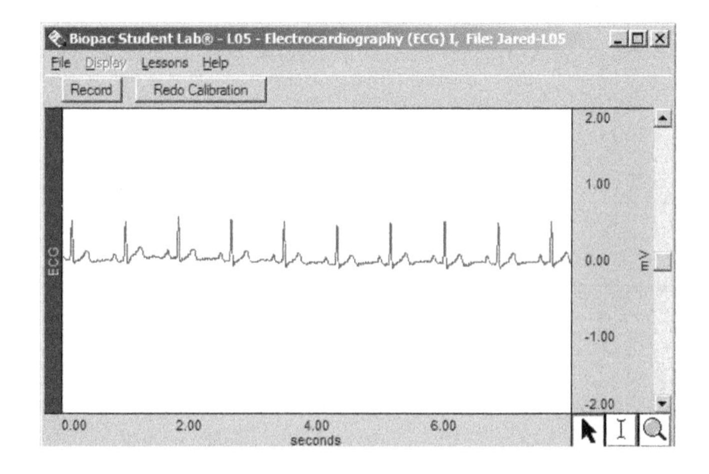

B Recording Data

In this activity, you will record the ECG from your lab partner under four conditions: lying down, after sitting up, while breathing deeply, and after exercise. In each case, it is important that the actual recording be done while your partner is physically still. Electrical activity from muscular movements, like that recorded in an EMG, can corrupt the ECG signal.

1. With your lab partner lying comfortably, click on the **Record** button. Wait 20 seconds and then click on the **Suspend** button. If the data do not appear correct (see the upper graph in the figure below) or have significant baseline drift (an extended period without a wave pattern), click on the **Redo** button and repeat this recording segment.

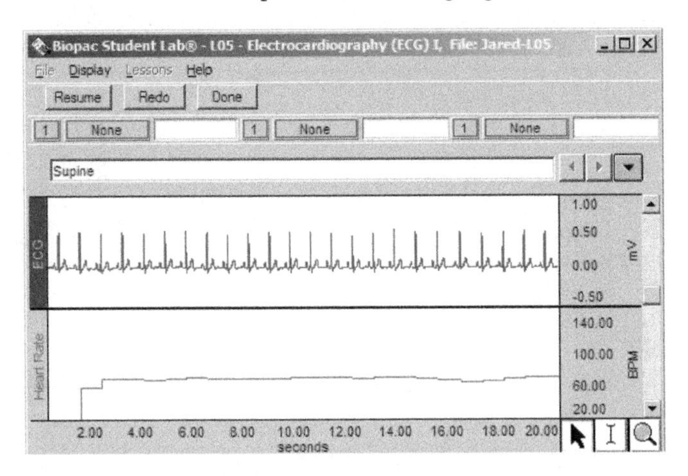

2. Instruct your lab partner to quickly sit upright. Immediately click on the **Resume** button as soon as he or she is sitting fully upright. A marker labeled **Seated** will be automatically inserted into the recording. Record for 20 seconds and then click on the **Suspend** button. If the data do not appear correct (see the upper graph in the figure below), click on the **Redo** button and repeat this recording segment.

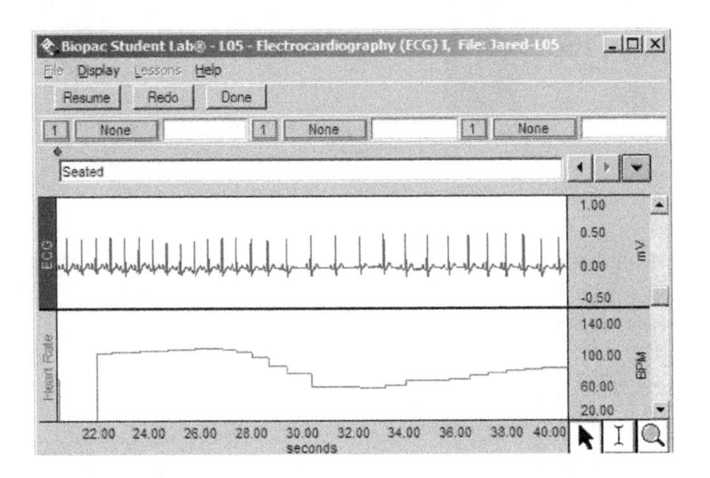

3. With your partner remaining seated, click on the **Resume** button. A marker labeled **Deep breathing** will be automatically inserted into the recording. Record for 20 seconds and then have your lab partner take five long, slow, deep breaths. You should insert markers at the beginning of each inhalation and at the subsequent exhalation. Markers are inserted by pressing the **Esc** key (Mac) or **F9** key (PC).

4. Click on the **Suspend** button and inspect the data. Deep breathing may cause some baseline drift. As long as it is not excessive, you do not need to redo the recording. If the data do not appear correct (see the upper graph in the figure below), click on the **Redo** button and repeat this recording segment.

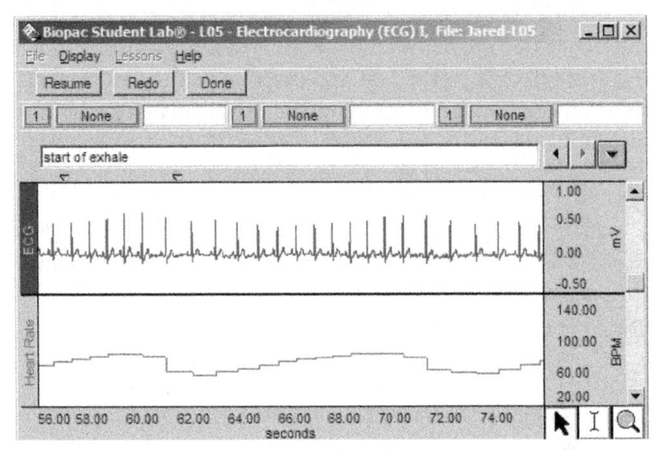

5. Have your lab partner perform an exercise, such as pushups, jumping jacks, or running in place to raise the heart rate. The exercise period should last for about 3 minutes.

6. Immediately after exercising, have your lab partner quickly sit down. Immediately click on the **Resume** button to capture the ECG while he or she is recovering from exercise. Record for 60 seconds and click on the **Suspend** button. Inspect the data. Some baseline drift is normal. As long as it is not excessive, click the **Done** button. If the data do not appear correct (see the upper graph in the figure below), click the **Redo** button and repeat this segment.

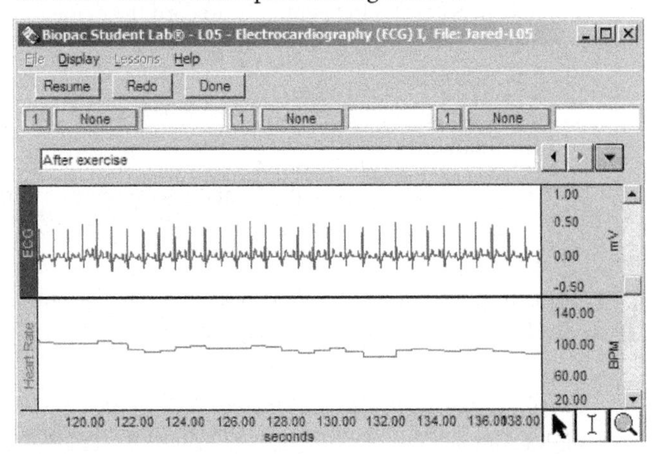

MAKING CONNECTIONS

In a normal ECG, why do you think the size of the R wave in the QRS complex is greater than the size of the P wave?

▶ ____

____

Activity 23.6

Electrocardiography Using the Biopac Student Lab System: Data Analysis of the Electrocardiogram (ECG) Recordings

A Data Analysis

In this activity, you will measure the duration and amplitude of various components of an ECG. The duration is calculated using the **delta T** measurement. This measurement computes the difference in time between the ending and beginning points of the area selected by the I-beam cursor tool. The amplitude (in millivolts, mV) of an ECG component is calculated using the **delta** measurement.

1. Enter the **Review Saved Data** mode and select the appropriate data file for Lesson 5, which ends in -LO5.
2. Use the drop down menu bars to set the three channel 2 measurement boxes as follows:

   Ch 2 delta T

   Ch 2 delta

   Ch 2 BPM (beats per minute)
3. Using the top figure on the right as a guide, use the I-beam cursor tool to select a P wave from one cardiac cycle in Segment 1 (resting, lying down). The bottom figure on the right illustrates the selection of a P wave. To accurately select a P wave on the ECG, you will need to use the magnifying glass tool to zoom in on the desired heart cycle. Then, if necessary, from the Display menu, use Autoscale waveforms to scale and position the ECG for optimal analysis. Once you make a selection, the duration (delta T) and the amplitude (delta) of the P wave will be computed. Record these values under the column labeled Cycle 1 in Table 23.5.
4. Select two additional P waves from two additional complete heart cycles in Segment 1. Record the delta T and delta values in the appropriate columns in Table 23.5 and calculate the mean.
5. Repeat steps 3 and 4 for the following components of the cardiac cycle and record the results in Table 23.5. For the intervals and segments, only delta T will be calculated.
   - PR interval
   - PR segment
   - QRS complex
   - QT interval
   - ST segment
   - T wave

Note: Accurately interpreting ECGs requires significant training and practice. A trained health care professional is best able to determine what abnormalities are due to normal variation, experimental noise, and medical conditions. Do not be alarmed if your ECG is different from those illustrated or from the normal values in the table.

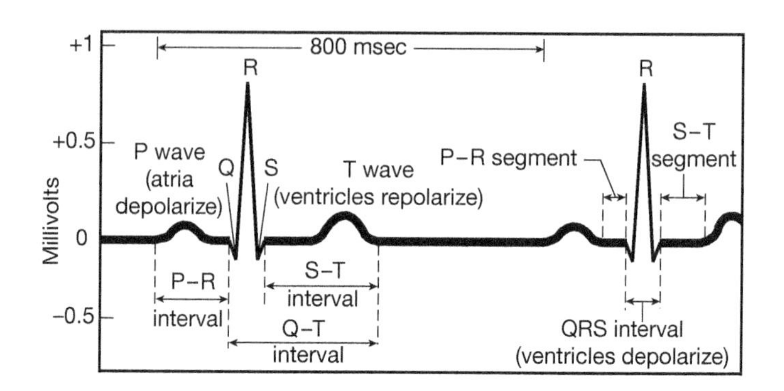

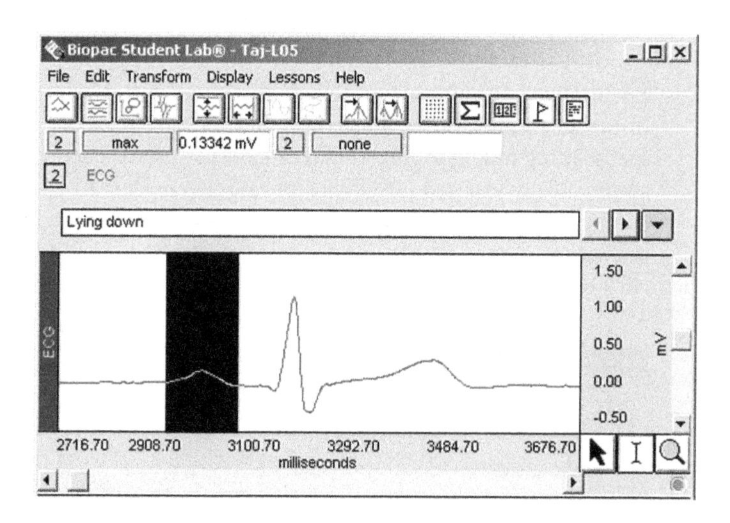

TABLE 23.5 Components of the ECG in Segment 1 (resting, lying down)					
Component	Normal Range	Cycle 1	Cycle 2	Cycle 3	Mean
Duration of P wave (delta T)	0.06–0.12 sec				
Amplitude of P wave (delta)	0.1–0.3 mV				
Duration of PR interval (delta T)	0.12–0.20 sec				
Duration of PR segment (delta T)	0.06–0.12 sec				
Duration of QRS complex (delta T)	0.06–0.10 sec				
Amplitude of QRS complex (delta)	0.8–1.2 mV				
Duration of the QT interval (delta T)	0.36–0.44 sec				
Duration of the ST segment (delta T)	0.12 sec				
Duration of the T wave (delta T)	0.12–0.16 sec				
Amplitude of the T wave (delta)	0.3 mV				

6 In Segment 1, use the I-beam cursor tool to select a single cardiac cycle from R wave peak to R wave peak (see figure on the right). The computer will measure the duration of the complete cycle and convert this measurement to beats per minute (BPM). Record the duration (delta T) and heart rate (BPM) in Table 23.6. Repeat for two additional cycles in Segment 1 and calculate the mean.

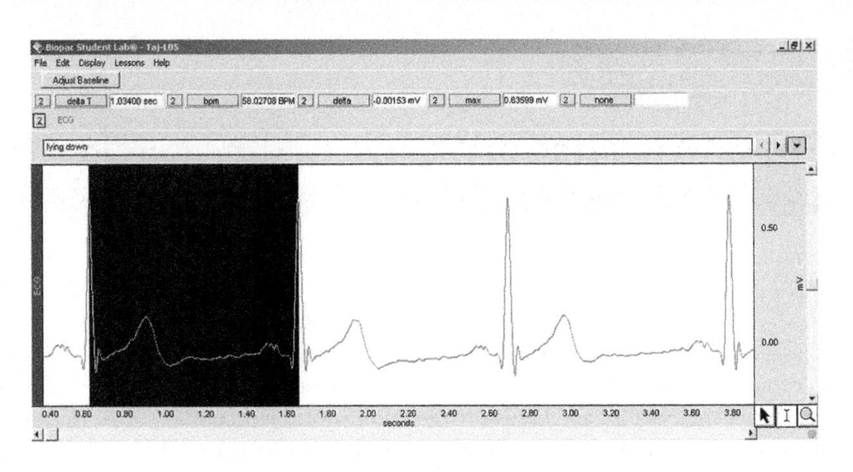

7 Repeat step 6 for each of the remaining three recording segments and record your results in Table 23.6. In Segment 3 (deep breathing), select three cycles that occurred during inhalations and then repeat three cycles that occurred during exhalations.

8 Measure changes in the duration of ventricular systole and diastole that occur during exercise. Ventricular systole corresponds to the QT interval, which is the period from the Q wave to the end of the T wave. Ventricular diastole, then, is measured from the end of the T wave to the subsequent R wave.

a. Transfer the data you recorded in Table 23.5 for the QT interval into the appropriate cells of Table 23.7 for Segment 1 (resting, lying down).
b. Scroll to the ECG waveform from Segment 1. Using the I-beam cursor tool, select the region corresponding to ventricular diastole (from the end of T to next R) and record the delta T measurement. Repeat for two additional cycles.
c. Scroll to Segment 4 (after exercise) and measure the duration of ventricular diastole and systole from three cardiac cycles. Record the results in Table 23.7 and calculate the means.

TABLE 23.6 Changes in Heart Rate and Duration of the Cardiac Cycle

Segment	Measurement	Cardiac Cycle			Mean	Range
		1	2	3		
1–Resting, lying down	Delta T					
	BPM					
2–Sitting up	Delta T					
	BPM					
3–Seated, inhalation	Delta T					
	BPM					
3–Seated, exhalation	Delta T					
	BPM					
4–After exercise	Delta T					
	BPM					

TABLE 23.7 Changes in Duration (delta T) of Ventricular Systole and Diastole

Segment	Measurement	Cardiac Cycle			Mean
		1	2	3	
1–Resting, lying down	QT interval (ventricular systole)				
	End of T to subsequent R (Ventricular diastole)				
4–After exercise	QT interval (ventricular systole)				
	End of T to subsequent R (Ventricular diastole)				

9 How does the duration (delta T) of ventricular systole and diastole change from resting to after exercise? ▶

____

____

____

MAKING CONNECTIONS

How does the heart rate (BPM) change during each of the four experimental conditions? Describe the physiological mechanisms causing these changes.

▶ ______________________________

____

____

____

BEFORE YOU MOVE ON . . .

‹‹ LOOKING BACK

The cardiac cycle refers to the series of events—relaxation and contraction—that occur during one heartbeat. A period of relaxation in a heart chamber is called **diastole (atrial diastole, ventricular diastole).** A period of contraction in a heart chamber is called **systole (atrial systole, ventricular systole).** As blood flows through the heart during the cardiac cycle, the valves open and close at specific times. The vibrations that are produced when the valves close create the heart sounds that you heard with a stethoscope.

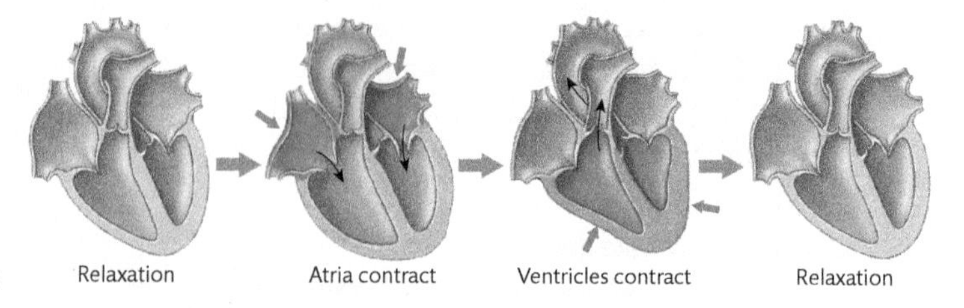

The pumping action of the ventricles creates pressure waves that allow blood to flow along a pressure gradient. These pressure waves can be detected at several locations as rhythmic pulses that parallel the heartbeat.

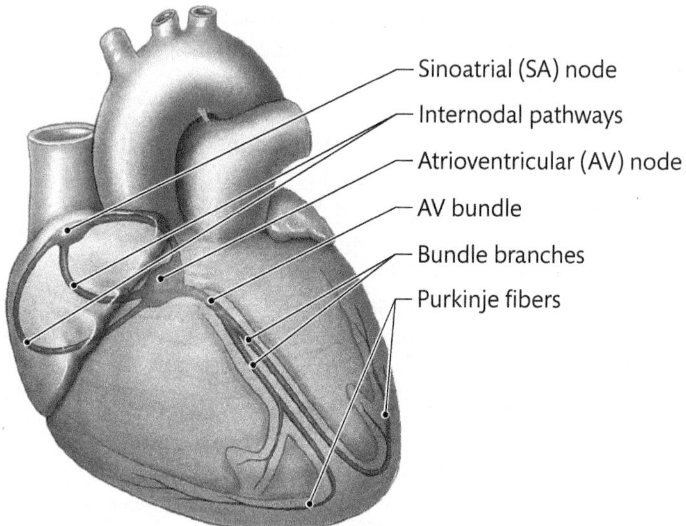

The cardiac cycle is regulated by a network of specialized cardiac muscle cells known as the **cardiac conduction system.** The **sinoatrial (SA) node**, located in the posterior wall of the right atrium, serves as the heart's pacemaker. It spontaneously generates action potentials (electric impulses) without neural or hormonal stimulation. From the SA node, action potentials travel through the cardiac conduction system. As a result, impulses spread across the walls of the atria (atrial systole) and then to the ventricles (ventricular systole).

Why is it important for cardiac muscle to generate action potentials and contract on its own? Without this ability, would heart transplants be possible?

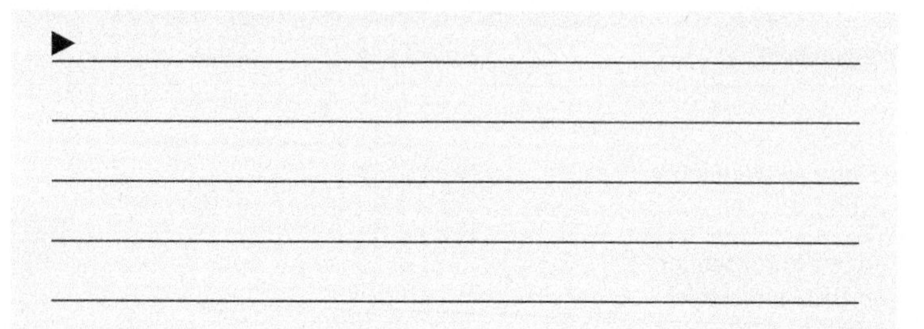

LOOKING FORWARD ››

The cardiovascular system and the lymphatic system, which you will study in the next laboratory exercise, have many functional connections. The white blood cells, which fight infections and promote tissue repair, are produced in red bone marrow before being distributed by the cardiovascular system to lymphatic organs such as lymph nodes and the spleen. Antibodies, which are produced in lymphatic organs, are transported in the blood. In addition, fluids that leak out of blood capillaries are taken up by lymphatic capillaries and recycled back to the blood by the lymphatic circulatory system.

Name ____________________

Lab Section ____________________

Date ____________________

EXERCISE 23 REVIEW SHEET

Cardiovascular Physiology

1. Why is it important for the walls of large arteries to have an abundant supply of elastic fibers?

____

____

____

QUESTIONS 2–9: Match the term in column A with the appropriate description in column B.

A

2. Systolic pressure ____________

3. Sphygmomanometer ____________

4. Heart sounds ____________

5. Diastolic pressure ____________

6. Cardiac cycle ____________

7. Stethoscope ____________

8. Korotkoff sounds ____________

9. Pulse ____________

B

a. Instrument used to measure blood pressure

b. The minimum pressure in the aorta during the cardiac cycle

c. Sounds of turbulent blood flow that can be heard while measuring blood pressure

d. Instrument used for hearing heart sounds

e. The maximum pressure in the aorta during the cardiac cycle

f. Series of events that occur in the heart during one heartbeat

g. Sounds that are produced when the heart valves close

h. Cyclic expansion and contraction of the arteries

10. A woman's blood pressure at rest is 125/75. After 30 minutes of exercise, her blood pressure is 190/80.

a. Calculate her pulse pressure and mean arterial pressure (MAP) at rest and after exercise.

b. How did her pulse pressure and MAP change after exercise?

____

____

c. During exercise, why do you think there is a large increase in systolic pressure, but very little change in diastolic pressure?

____

____

____

____

____

QUESTIONS 11–16: The following graph demonstrates the blood pressure changes in the left ventricle (red), left atrium (blue), and aorta (black) during the cardiac cycle. Match each event in column A with the correct arrow in column B. Answers may be used more than once.

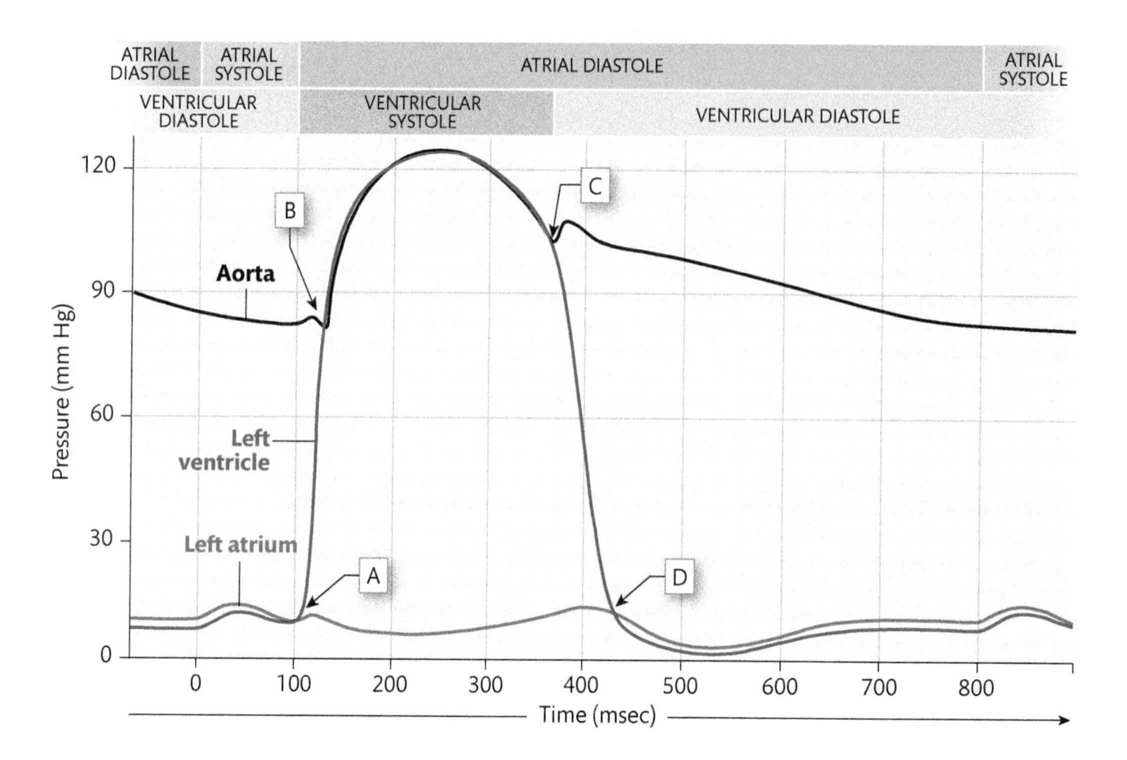

A	B
11. Aortic valve closes ______	**a.** Arrow A
12. First heart sound occurs ______	**b.** Arrow B
13. Left AV valve opens ______	**c.** Arrow C
14. Left AV valve closes ______	**d.** Arrow D
15. Aortic valve opens ______	
16. Second heart sound occurs ______	

17. Arteriosclerosis decreases the elasticity of arterial walls and increases arterial resistance to blood flow. Suppose there are two patients. One has a resting blood pressure of 115/70. The other has a resting blood pressure of 140/100.

a. Calculate the pulse pressure and MAP of each individual.

b. Which patient do you think has arteriosclerosis? Explain.

______

______

c. For the person with arteriosclerosis, how must the pulse pressure change to maintain normal blood flow?

______

______

______

EXERCISE 24

The Lymphatic System

LEARNING OUTCOMES

These Learning Outcomes correspond by number to the laboratory activities in this exercise. When you complete the activities, you should be able to:

Activities 24.1–24.2	Describe the structure, anatomical relationships, and function of lymphatic system structures.
Activity 24.3	Compare the structure and function of different types of lymphatic vessels.
Activities 24.4–24.5	Describe the microscopic structure of lymphatic organs and nodules.

LABORATORY SUPPLIES

- Anatomical models:
  - midsagittal section of the head and neck
  - human torso
- Compound light microscopes
- Prepared microscope slides:
  - thymus
  - spleen
  - lymph node
  - tonsils
  - small intestine (ileum)
  - lymphatic vessels

PRE-LAB QUIZ

Before you begin, read all the activities in Exercise 24 and the required reading in your textbook that is assigned by your instructor.

1. During this laboratory exercise, you will study the light microscopic structure of all the following structures *except*
   a. the spleen.
   b. lymphatic nodules in the lung.
   c. a tonsil.
   d. a lymph node.
   e. the thymus.
2. The primary cells for fighting infections are ______________________.
3. True or False: Lymph nodes are located along the course of lymphatic vessels. __________
4. True or False: Unlike lymphatic organs, lymphatic nodules are surrounded by a connective tissue capsule. __________
5. The jugular trunks drain lymph from the
   a. head and neck.
   b. abdominal cavity.
   c. lower limbs.
   d. upper limbs.
   e. thoracic cavity.
6. True or False: The thoracic duct drains lymph from the upper right quadrant of the body. __________
7. In the spleen, the network of blood sinusoids is called the ______________________.
8. During this laboratory exercise, you will view microscope slides of lymphatic nodules in what abdominal organ? ________________
9. True or False: The region of a lymphatic nodule where new lymphocytes are produced is called the germinal center. __________
10. True or False: When activated, T lymphocytes can differentiate into plasma cells, which produce antibodies. __________

Activity **24.1**

Examining the Gross Anatomical Structure of Lymphatic Organs

Lymphoid tissue is a specialized loose connective tissue in lymphatic structures. It contains an extensive network of **reticular fibers** that provide a supporting framework, and dense aggregations of **lymphocytes,** the primary cells involved in fighting off infection and disease. **Lymphatic organs** are organized areas of lymphoid tissue surrounded by a connective tissue capsule. These structures include the thymus, spleen, and lymph nodes.

A The thymus gland is a bilobed organ located in the mediastinum.

1. Remove the anterior body wall on a torso model to expose the organs in the thoracic and abdominopelvic cavities.

2. Identify the **thymus gland.** Observe how it covers the superior portion of the heart and the great vessels and extends superiorly into the base of the neck, where it blankets the inferior end of the trachea.

3. The thymus produces a group of hormones called **thymic hormones (thymosins).** Refer to your textbook and identify the function of these hormones.

▶ ______________________________

____

____

____

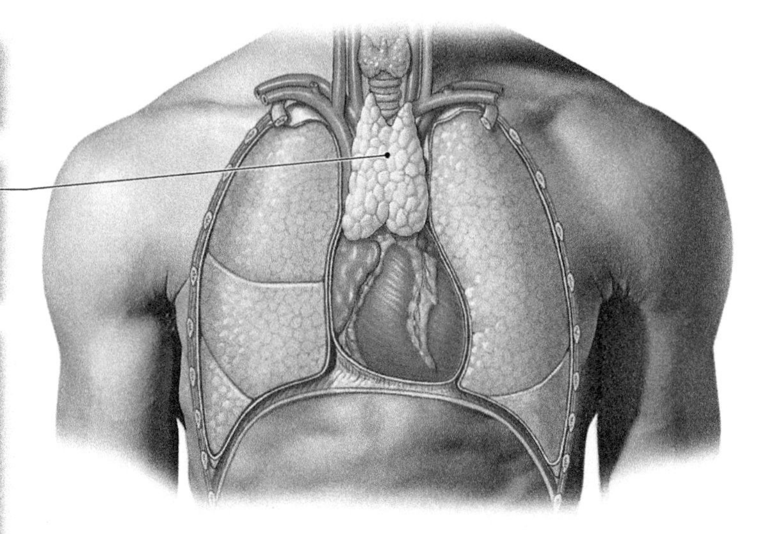

Word Origins

Thymus is the Latin word for sweetbread. If you go to a restaurant and order sweetbreads from the menu, you will probably be eating the thymus gland of a lamb, cow, or pig.

B The spleen is in the upper-left quadrant of the abdominal cavity.

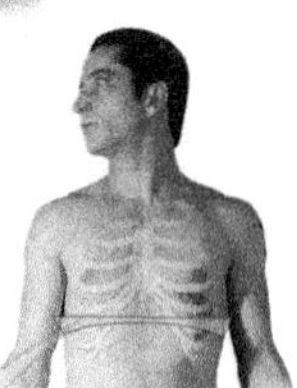

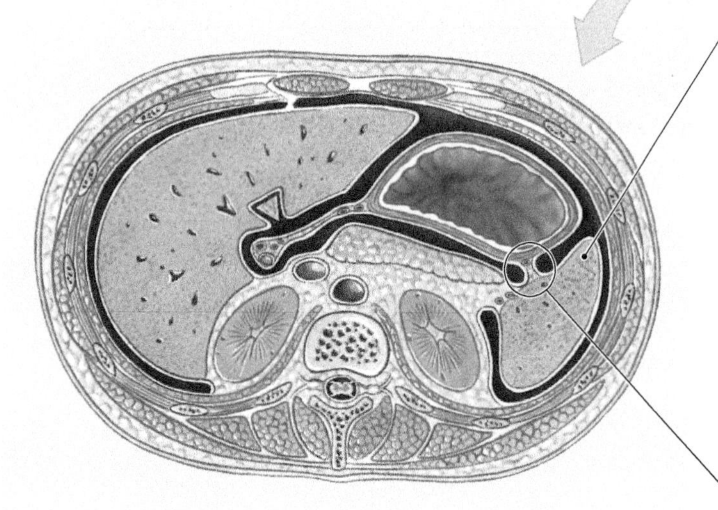

1. On a torso model, remove the stomach to expose the anterior surface of the **spleen** in the abdominopelvic cavity. Label the following organs on the diagram to the left and describe their anatomical relationship to the spleen: ▶

   Stomach

   ____

   Pancreas

   ____

   Left kidney

   ____

2. Locate the **hilum** of the spleen, which is the region where blood vessels, lymphatic vessels, and nerves enter and exit the organ.

C Lymph nodes are found deep and superficially throughout the body, but are aggregated into clusters in specific body regions such as the axilla and groin.

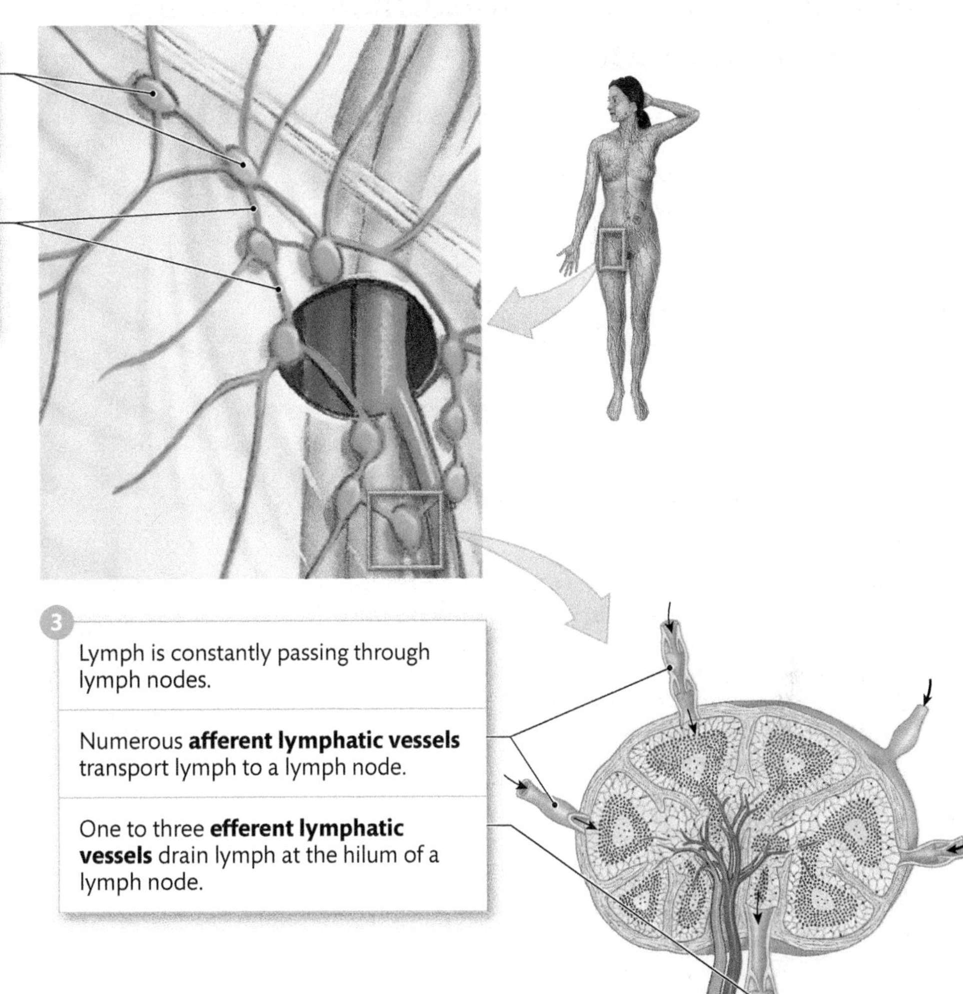

IN THE CLINIC

HIV Infection

The primary target for the **human immunodeficiency virus (HIV),** the virus that causes AIDS, is a type of lymphocyte called a **helper T cell.** These cells are needed to stimulate all other immune functions. As HIV disease progresses to AIDS, the helper T-cell population declines, and the body's immune system is suppressed. As a result, the body's ability to fight off infections (opportunistic infections) is compromised. AIDS-related deaths occur when the body succumbs to infections that a suppressed immune system can no longer control.

Lymphadenopathy is enlargement or swelling of lymph nodes. It can be caused by a number of diseases, and swelling normally declines when the illness ends. For people who are HIV positive, however, lymph nodes can remain swollen for several months, even if there are no other indications or symptoms of a disease.

MAKING CONNECTIONS

If cancer cells from a primary tumor spread to neighboring lymph nodes, the risk of a secondary tumor (metastasis) forming in another body region or organ increases significantly. Explain why.

▶ ______________________________________

____

____

____

____

____

Want more practice? Go to: **MasteringA&P** > Study Area > Menu > Lab Tools > PAL >
- Anatomical Models > Lymphatic System
- Human Cadavar > Lymphatic System

Activity **24.2**

Examining the Gross Anatomical Structure of Lymphatic Nodules

Lymphatic **nodules** are concentrated regions of lymphoid tissue that lack a connective tissue capsule.

A The tonsils are regions of lymphoid tissue located where the oral and nasal cavities connect to the pharynx.

1. On a midsagittal section of the head and neck, locate the single **pharyngeal tonsil (adenoid)** along the posterior wall of the superior region of the pharynx, known as the nasopharynx.

2. There are two **palatine tonsils.** Identify one along the lateral wall of the pharynx near the posterior end of the palate.

3. There are two **lingual tonsils,** located bilaterally at the base of the tongue. Locate one on the model.

4. Why do you think the pharyngeal region contains such a large amount of lymphoid tissue?

▶ ______________________________

____

____

____

____

____

____

IN THE CLINIC

Tonsillectomy

A **tonsillectomy** is the surgical removal of inflamed and infected tonsils **(tonsillitis).** The palatine tonsils are most commonly removed, but the pharyngeal tonsil may be removed as well. Children are more vulnerable to tonsillitis than adults because the tonsils are most active before puberty. Tonsillectomy was once a common procedure, but today it is done less frequently. It is recommended if the tonsils become frequently infected (more than seven occurrences per year) and do not respond to medication or if they become large enough to make breathing and swallowing difficult.

Want more practice? Go to: **MasteringA&P** > Study Area > Menu > Lab Tools > PAL >

- Anatomical Models > Lymphatic System
- Human Cadavar > Lymphatic System

B Lymphoid nodules, found in the mucous membranes of internal organs in the thoracic and abdominopelvic cavities, are called mucosa-associated lymphoid tissue (MALT).

1. On a torso model, locate the junction between the small and large intestines in the lower-right quadrant of the abdominopelvic cavity.

2. The first part of the large intestine is called the cecum. Identify the **appendix,** the wormlike appendage extending from the wall of the cecum. The appendix contains lymphoid nodules, which are structurally similar to the tonsils.

3. Lymphoid nodules are scattered throughout the mucous membranes of the following structures:

   Small intestine

   Large intestine

   Bronchial passageways in the lungs

   Observe the positions of these structures on a torso model and follow the entire pathways of the digestive and respiratory tracts. Based on your observations, can you explain why it is important to have lymphoid tissue in digestive and respiratory structures?

   ▶ ______________________________

   ____

   ____

   ____

   ____

4. Identify two other organ systems that contain mucosa-associated lymphoid tissue. (*Hint:* What other organ systems have tracts with pathways that are similar to the digestive and respiratory tracts?)

   ▶ ______________________________

   ____

   ____

   ____

MAKING CONNECTIONS

How do the tonsils illustrate a functional connection between the lymphatic system and the respiratory system?

▶ ______________________________

____

____

____

Activity **24.3**

Examining the Gross Anatomical Structure of the Lymphatic Circulation

Fluid is constantly moving in and out of blood capillary beds. More fluid moves out than in, however, and it would accumulate in the surrounding connective tissue (interstitial tissue) if there were no proper drainage. **Lymphatic capillaries** are found in most tissues and form an interlacing network with blood capillaries. Excess fluid in the interstitial tissue, known as **intersitial fluid,** drains into lymphatic capillaries to become **lymph.** Lymphatic capillaries mark the beginning of the **lymphatic circulation,** which transports lymph through a series of vessels and returns it to the blood circulation by draining into the subclavian veins.

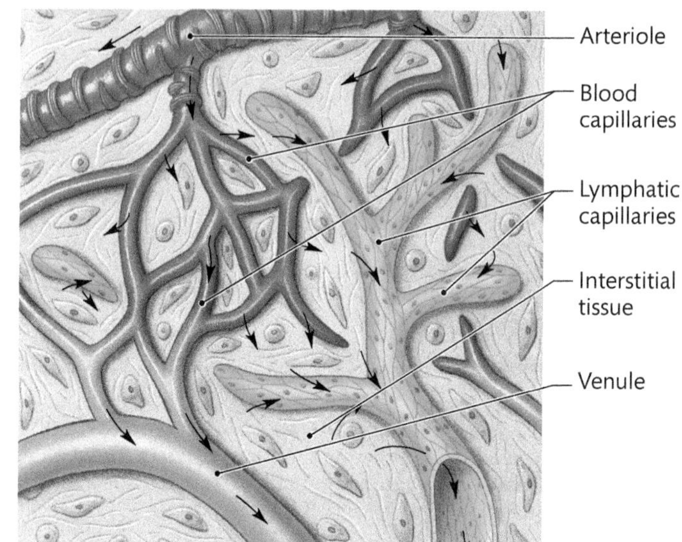

Word Origins

Lymph is derived from the Latin word *lympha* ("clear spring water"). Typically, lymph is a clear fluid, although in the small intestine, it has a milky appearance due to the high fat content.

A Lymphatic capillaries drain excess fluid released from blood capillaries and empty into larger **lymphatic vessels.**

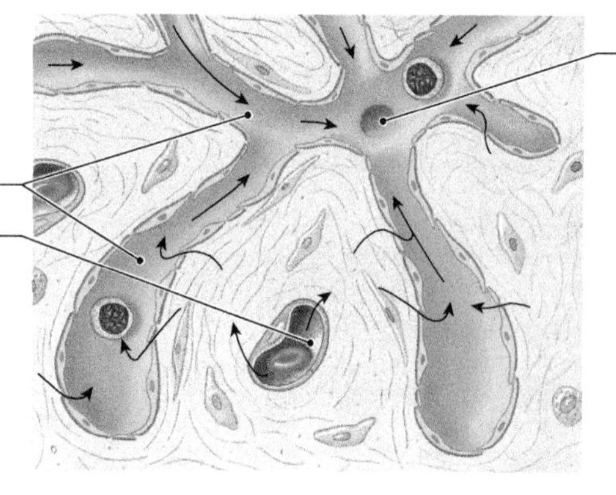

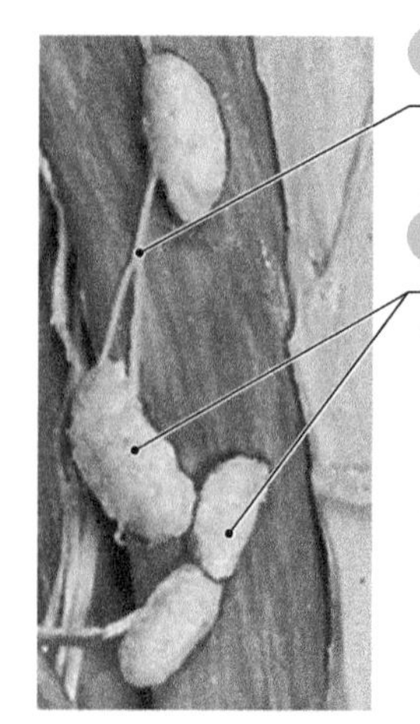

1 On a torso model, identify networks of **lymphatic vessels** in the axillary or inguinal regions.

2 As you observed earlier, **lymph nodes** are distributed along the course of these vessels. Why do you think there are large numbers of lymph nodes in the axillary and inguinal regions?

▶ ______________________________________

____

____

____

____

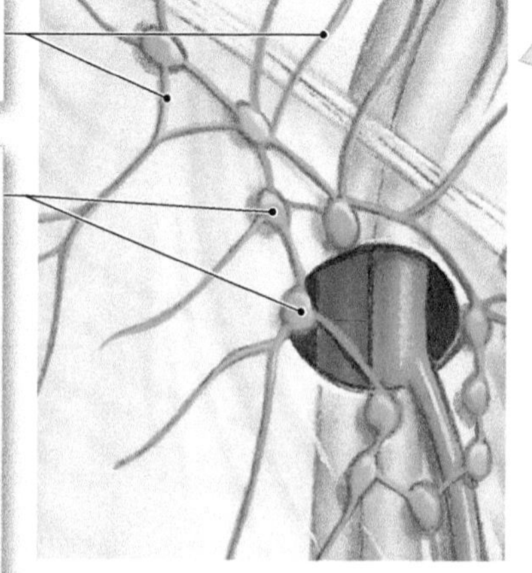

B **Lymphatic trunks** are positioned at strategic locations throughout the body. They are formed by the union of several lymphatic vessels and drain the lymph from a specific region of the body. The lymphatic trunks empty into one of two **lymphatic ducts,** which drain lymph into the venous circulation.

Major Lymphatic Trunks		
Lymphatic Trunk	**Regions Drained**	**Structures Drained**
Right and left jugular trunks	Head and neck	
Right and left subclavian trunks	Upper limbs	
Right and left bronchomediastinal trunks	Anterior thoracic wall; thoracic cavity	
Intestinal trunk	Abdominal cavity	
Right and left lumbar trunks	Abdominal and pelvic walls; pelvic cavity; lower limbs	

1 Review the regions that are drained by the lymphatic trunks listed in the table to the right. Complete the table by naming three structures in each region that will be drained by its lymphatic trunk(s). ▶

2 On each side of a torso model, find the location where the subclavian and internal jugular veins join to form the brachiocephalic vein.

On the right side, the **right lymphatic duct** empties into the bloodstream near this venous junction.

The thoracic duct has a similar drainage pattern on the left side.

3 On a torso model, identify the abdominal aorta. The **thoracic duct** ascends along the left side of the vertebral column adjacent to the aorta.

4 On a torso model or skeleton, identify the L_2 vertebra. The thoracic duct begins at this level as a dilated sac called the **cisterna chyli.**

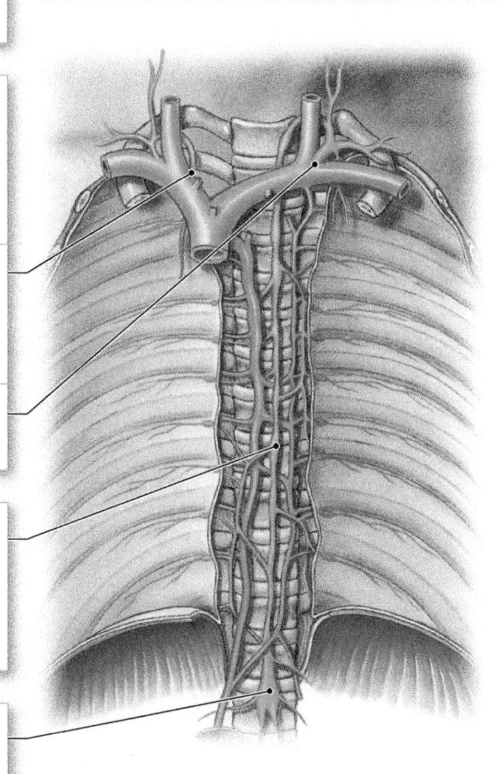

5 The right lymphatic duct drains lymph from the upper-right quadrant of the body (light green area in the figure on the right). The much larger thoracic duct drains lymph from all other body regions (dark green area). Based on this drainage pattern, list the lymphatic trunks that drain into each duct in the table below. ▶

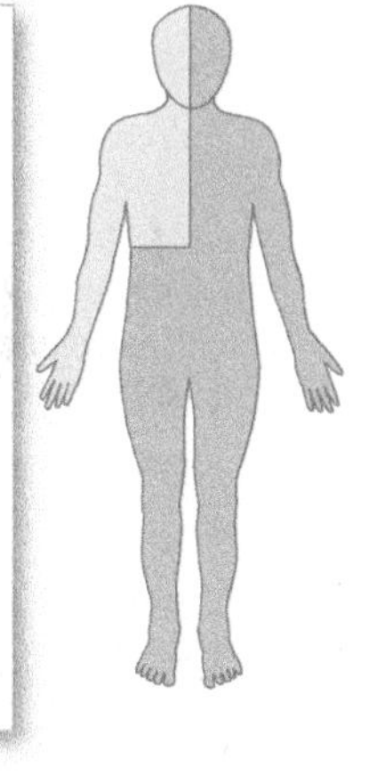

Lymphatic Duct	Trunks That Drain into Each Duct
Right lymphatic duct	
Thoracic duct	

MAKING CONNECTIONS

Filarial elephantiasis is a tropical disease caused by a parasitic worm that is transmitted by mosquitoes. Once inside the body, the parasite obstructs normal drainage of interstitial fluid into lymphatic capillaries. What do you think will be the outcome of this infection? Explain.

▶ ______________________________

Examining the Light Microscopic Structure of the Thymus and Spleen

Lymphoid tissue contains a variety of cell types that are largely involved in protecting the body from infection and disease. Lymphocytes are the principal cells in lymphoid tissue. **T lymphocytes (T cells)** are responsible for the **cell-mediated immune response.** They directly attack and destroy foreign microorganisms by phagocytosis or by releasing chemicals. One type of T lymphocyte, the **helper T cell,** activates all immune activity. **B lymphocytes (B cells)** initiate the **humoral immune response** by differentiating into **plasma cells,** which produce **antibodies.** As they circulate through the blood and lymph, antibodies bind to and destroy foreign **antigens** (antigen–antibody reaction), such as pathogens, and certain foods and drugs. Other cells in lymphoid tissue include **macrophages,** which attack foreign cells by phagocytosis and activate T lymphocytes; **fibroblasts,** which produce the reticular fibers; and **dendritic cells,** which also play a role in T-lymphocyte activation.

A The **thymus gland** is most active during childhood. After puberty, it gradually gets smaller as glandular tissue is replaced by fibrous tissue and fat. By middle age, the thymus has very little glandular function.

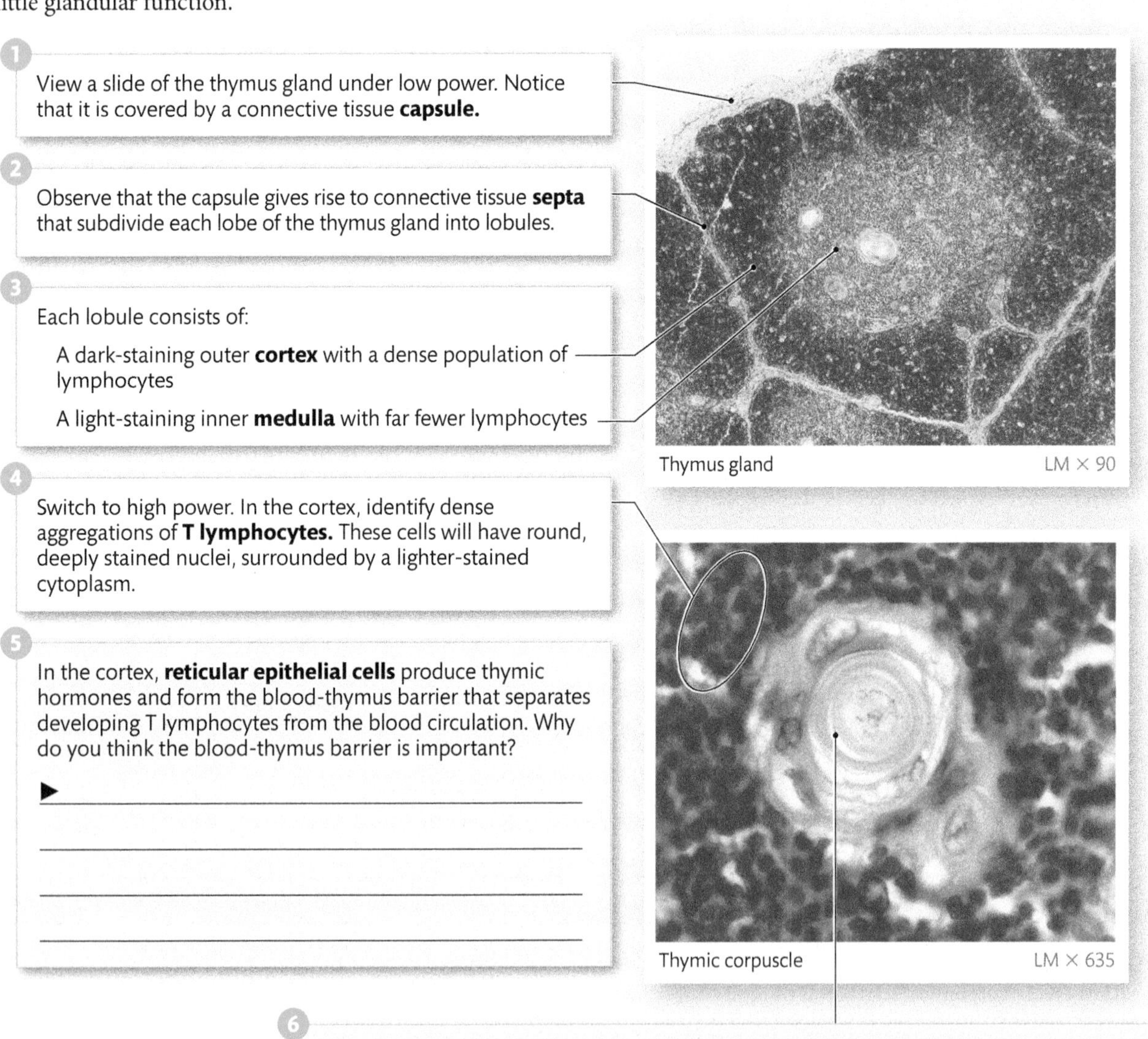

1. View a slide of the thymus gland under low power. Notice that it is covered by a connective tissue **capsule.**

2. Observe that the capsule gives rise to connective tissue **septa** that subdivide each lobe of the thymus gland into lobules.

3. Each lobule consists of:

   A dark-staining outer **cortex** with a dense population of lymphocytes

   A light-staining inner **medulla** with far fewer lymphocytes

Thymus gland LM × 90

4. Switch to high power. In the cortex, identify dense aggregations of **T lymphocytes.** These cells will have round, deeply stained nuclei, surrounded by a lighter-stained cytoplasm.

5. In the cortex, **reticular epithelial cells** produce thymic hormones and form the blood-thymus barrier that separates developing T lymphocytes from the blood circulation. Why do you think the blood-thymus barrier is important?

   ▶ ____________________

   ____

   ____

   ____

Thymic corpuscle LM × 635

6. In the medulla, reticular epithelial cells form clusters called **thymic corpuscles,** which are unique to the thymus. The blood-thymus barrier does not exist in the medulla, and mature T lymphocytes leave by entering a blood or lymphatic vessel. Where do the T lymphocytes go after leaving the thymus?

   ▶ ____________________

B The spleen is the largest lymphatic organ in the body.

1

View a slide of the spleen under low power.

Like the thymus, the spleen is surrounded by a thin connective tissue **capsule.**

The capsule gives rise to **trabeculae (septa)** that divide the organ into lobules.

2

Identify the two functional tissue components of the spleen:

Regions of **white pulp** are compact masses of lymphocytes that are scattered throughout the spleen.

Most of the spleen consists of **red pulp,** which completely surrounds the areas of white pulp. Red pulp includes a vast network of sinusoids that stores red blood cells. It also contains numerous lymphocytes, monocytes, dendritic cells, and macrophages.

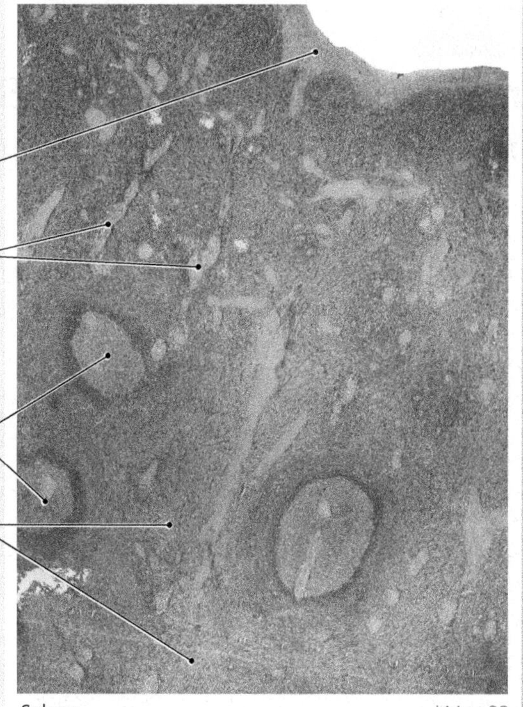

Spleen LM × 32

3

Move the slide so that an area of white pulp is in the center of the field of view. Switch to the high-power objective lens to examine the region more closely.

Dense clusters of **lymphocytes** can be seen around the periphery of white pulp.

In the center are the lighter-staining **germinal centers,** where new lymphocytes are produced.

Identify a **central artery,** which is a branch of the splenic artery. These blood vessels are found in areas of white pulp.

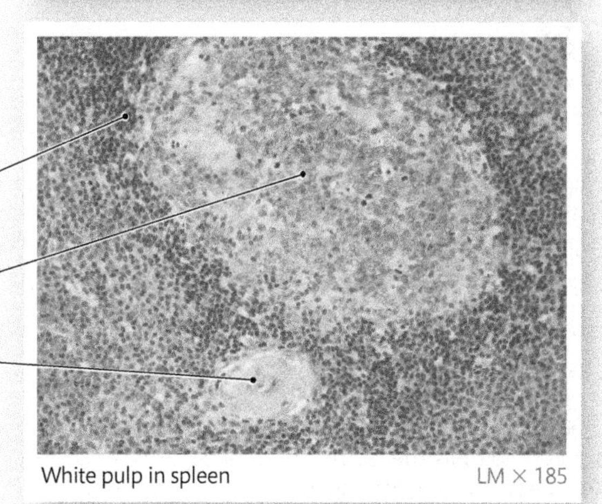

White pulp in spleen LM × 185

4

Refer to your textbook and list two functions of the spleen. ▶

Function #1: ______________________

____

____

____

Function #2: ______________________

____

____

____

MAKING CONNECTIONS

How would the adult immune system be affected if the thymus were not functioning normally during childhood?

▶ ____

____

____

____

____

Want more practice? Go to: **MasteringA&P** > Study Area > Menu > Lab Tools >

- **PAL** > Histology > Lymphatic System
- **PhysioEx** > Exercise 12: Serological Testing > Activity 2: Comparing Samples with Ouchterlony Double Diffusion

Activity **24.5**

Examining the Light Microscopic Structure of Lymph Nodes and Lymphatic Nodules

A When lymph enters a lymph node, it travels through a network of lymphatic sinusoids. In the sinusoids, nearly all foreign substances are recognized, and immune responses are initiated to neutralize them. Thus, lymph nodes are able to remove disease-causing microorganisms before the lymph returns to the blood.

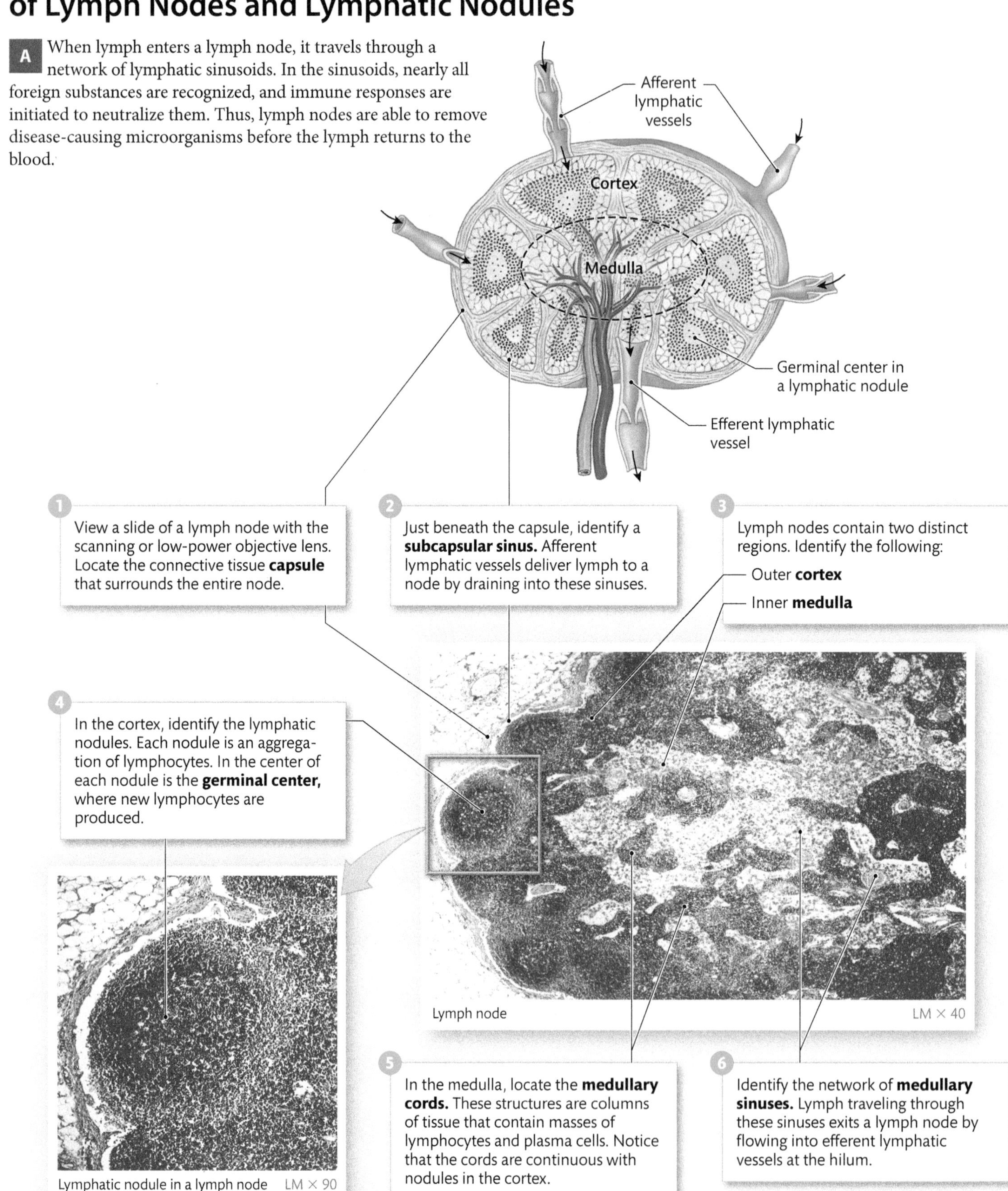

Lymph node LM × 40

Lymphatic nodule in a lymph node LM × 90

B Tonsils are lymphatic nodules that protect the upper respiratory and digestive tracts from infections.

1. View a slide of a tonsil under low power. Notice that unlike lymph nodes or the spleen, the tonsil lacks a connective tissue capsule.

2. Identify the numerous **lymphatic nodules** embedded in the mucous membrane of the pharyngeal wall.

Tonsil LM × 33

3. Switch to high power and examine the nodules more closely. Notice that they contain:
   - Outer rings of densely packed lymphocytes
   - Centrally located germinal centers

Lymphatic nodules in a tonsil LM × 68

C Aggregated lymphoid nodules, or Peyer's patches, are located in the ileum of the small intestine.

1. View a slide of the ileum (last segment of the small intestine) under low power.

2. Scan the slide and identify **aggregated lymphoid nodules** in the mucous membrane.

3. Examine the nodules more closely. Are they surrounded by a connective tissue capsule?

   ▶ ______________________

   Do they contain germinal centers?

   ▶ ______________________

Aggregated lymphoid nodules in the ileum LM × 43

MAKING CONNECTIONS

Identify similarities and differences in the microscopic structure of lymph nodes, tonsils, and aggregated lymphoid nodules. Focus your attention on the arrangement and structure of the lymphoid tissue in each structure. ▶

Similarities:

____

____

____

____

Differences:

____

____

____

____

BEFORE YOU MOVE ON . . .

‹‹ LOOKING BACK

The lymphatic system consists of an extensive network of lymphatic vessels that return lymph back to the blood and various lymphoid organs and lymphoid nodules that act to defend the body against infections and disease. The main cell type is the lymphocyte. These cells, found in both the lymphatic and cardiovascular systems, initiate specific immune responses against foreign particles called antigens. The immune response includes antibody production and a direct cellular or chemical attack.

The efficiency and specificity of the body's immune response makes tissue typing key to the success of organ transplant surgery. Even when the tissue match between donor and recipient is very close, a transplanted organ is recognized as foreign tissue, and it will induce an immune response. To reduce the risk of tissue rejection, patients who receive an organ transplant must take drugs that suppress the immune system. With a lifetime of taking immunosuppressive drugs ahead, what future health risks do organ transplant recipients face?

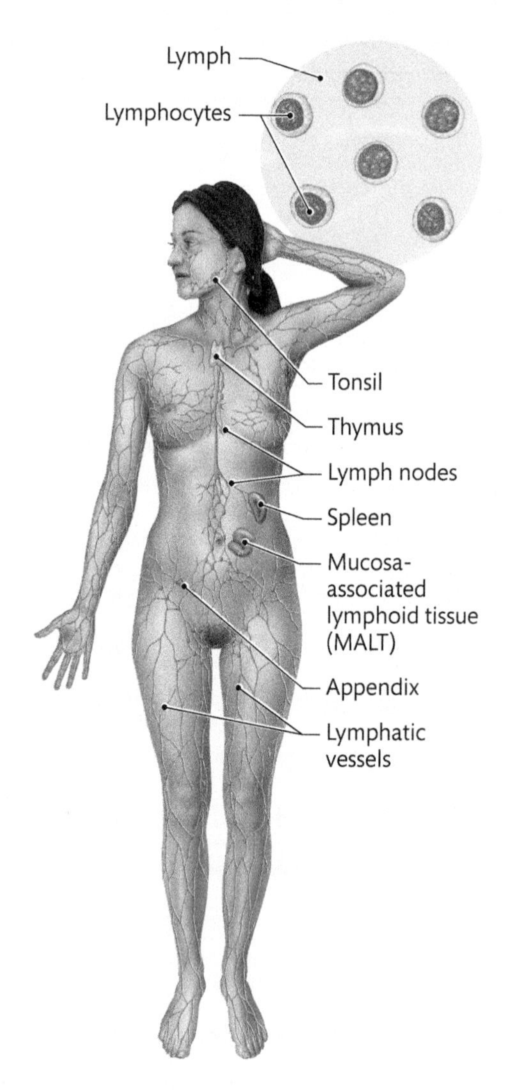

LOOKING FORWARD ››

Lymphoid nodules are located in the mucous membranes of many internal organs. These nodules, known as mucosa-associated lymphoid tissue (MALT), are abundant along the mucous membranes of the respiratory tract and are vitally important for destroying potentially harmful particles in the air that we inhale. In the next laboratory exercise, you will study the gross anatomy and microscopic structure of the organs in the respiratory system.

Name ______________________

Lab Section ______________________

Date ______________________

EXERCISE **24** **REVIEW SHEET**

The Lymphatic System

1. What is the main structural difference between a lymphatic organ and a lymphatic nodule?

2. During a football game, a quarterback sustained a hard blow to the back and fell hard to the ground. As a result, his spleen ruptured and was removed during emergency surgery. The player fully recovered, but his doctor told him that he would have to take extra precautions, such as annual flu vaccinations and frequent hand washing, to prevent infections. Explain why.

3. A woman's right axillary lymph nodes were removed during a mastectomy. The operation was successful, but over the next 5 years, she developed significant edema in her right arm, forearm, and hand. Explain why.

4. Explain why the actions of the thymus demonstrate a close functional connection between the lymphatic system and the endocrine system.

5. Explain the difference between the cell-mediated immune response and the humoral immune response.

6. Identify the lymphatic structures that are shown in the following photos.

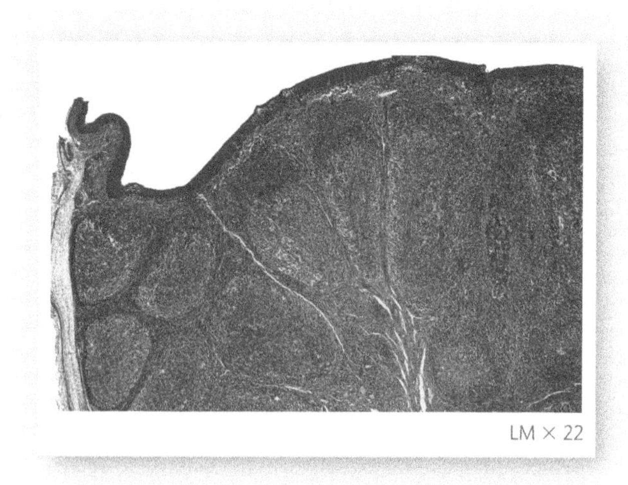
LM × 22

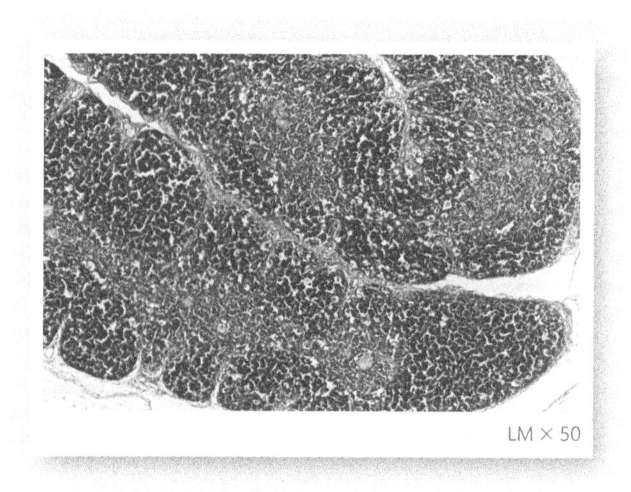
LM × 50

a. ______________________ b. ______________________

7. Identify the structures indicated by the arrows in the following photos.

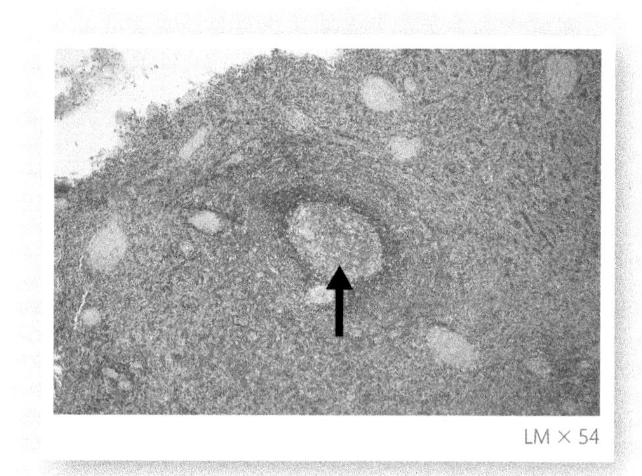
LM × 54

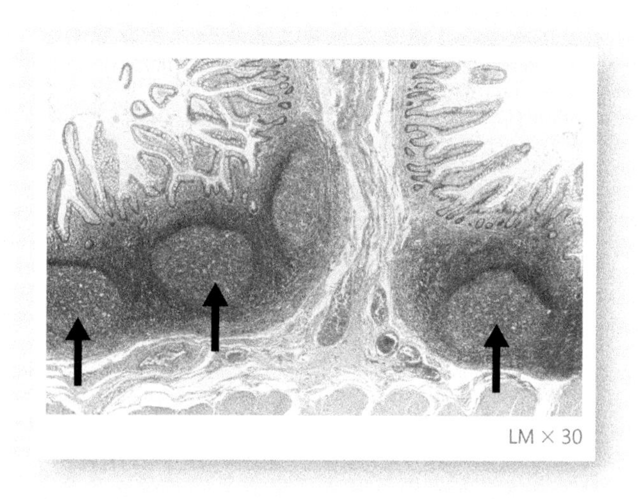
LM × 30

a. ______________________ b. ______________________

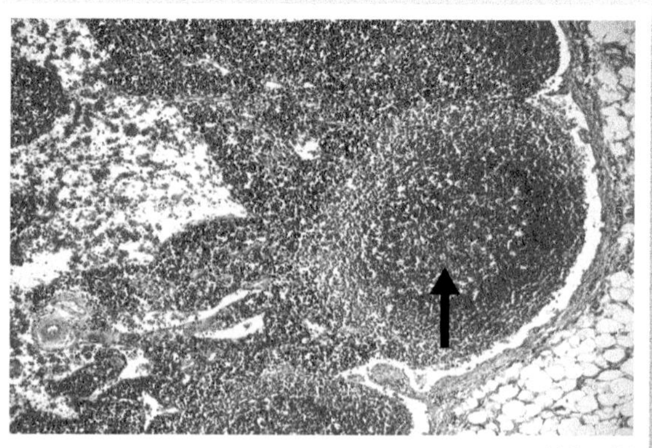
LM × 54

c. ______________________

8. The photo below is a cadaver dissection showing an anterior view of the upper abdominal cavity. Identify the organ indicated by the arrow and describe its function.

____

____

____

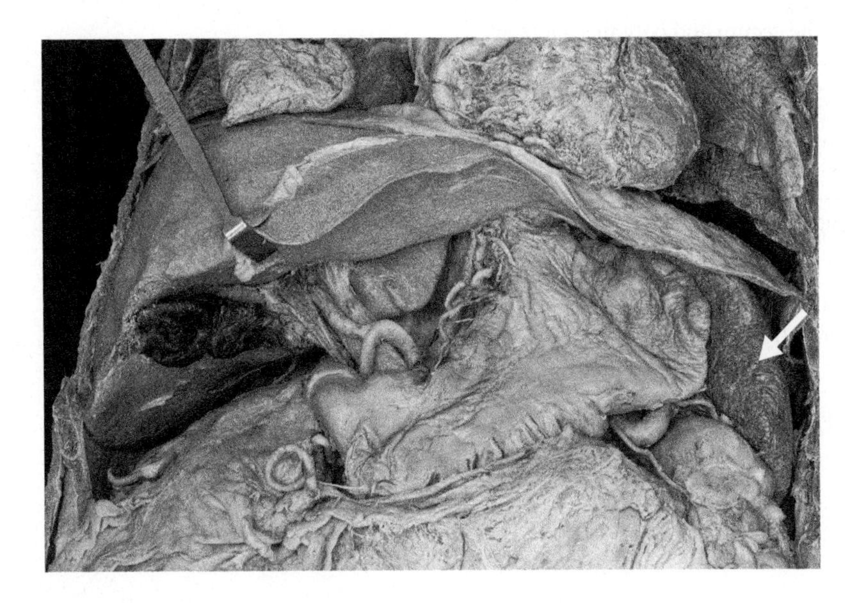

QUESTIONS 9–18: Match the term in column A with the appropriate description in column B.

A	B
9. Mucosa-associated lymphoid tissue ________	**a.** Lymphatic nodules located in the pharynx
10. White pulp ________	**b.** Cluster of reticular epithelial cells in the thymus gland
11. Medulla ________	**c.** Lymph exits a lymph node by draining into this structure
12. Afferent lymphatic vessel ________	**d.** Sinusoids are located in this region of the spleen
13. Germinal center ________	**e.** Inner region of the thymus gland or a lymph node
14. Thymic corpuscle ________	**f.** Region of a lymphatic nodule where new lymphocytes are produced
15. Red pulp ________	**g.** Lymphatic nodules located in the small and large intestines
16. Cortex ________	**h.** Structure that transports lymph to a lymph node
17. Efferent lymphatic vessel ________	**i.** Outer region of the thymus gland or a lymph node
18. Tonsils ________	**j.** Densely packed regions of lymphocytes in the spleen

QUESTIONS 19–27: Match the labeled structures (a–i) in the diagram with the appropriate description. Each answer should include the correct letter and name of the structure.

19. Drains lymph from the upper-right quadrant of the body.

____

20. Sac-like structure that is the origin of the thoracic duct.

____

21. Extends from the wall of the cecum and contains aggregated lymphoid nodules.

____

22. Functions as a blood filter.

____

23. Drains lymph from all regions of the body except the upper-right quadrant.

____

24. Site of T-lymphocyte maturation.

____

25. Lymphatic nodules that protect the body from infections in the pharynx.

____

26. Lymph is filtered as it passes through the sinusoids of these structures.

____

27. Lymphatic nodules located in the small intestine.

____

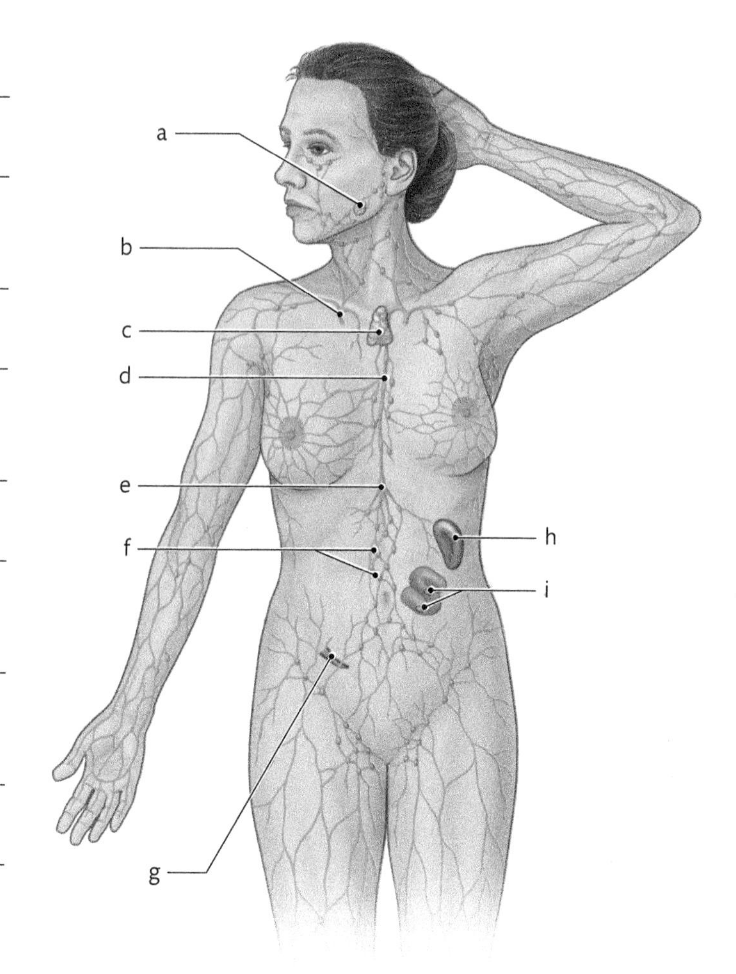

EXERCISE

25

Anatomy of the Respiratory System

LEARNING OUTCOMES

These Learning Outcomes correspond by number to the laboratory activities in this exercise. When you complete the activities, you should be able to:

Activities 25.1–25.3	**Describe the structure and function of the nose, nasal cavity, pharynx, larynx, and trachea.**
Activity 25.4	**Identify respiratory structures in the neck by palpation.**
Activity 25.5	**Describe the structure and function of the lungs.**
Activities 25.6–25.7	**Describe the microscopic structure of the trachea and lungs.**

LABORATORY SUPPLIES

- Human skull
- Anatomical models:
  - torso
  - head and neck
  - larynx and trachea
  - bronchial tree
  - right and left lungs
- Prepared microscope slides:
  - trachea
  - lung

PRE-LAB QUIZ

Before you begin, read all the activities in Exercise 25 and the required reading in your textbook that is assigned by your instructor.

1. During this laboratory exercise, you will study the light microscopic structure of what two structures? ________________
2. True or False: The right lung contains two lobes, and the left lung contains three lobes. ________
3. The inferior border of the cricoid cartilage marks the border between what two structures? ________________
4. True or False: The cartilaginous rings in the trachea cover all sides of the tracheal wall except the posterior surface. ________
5. The perpendicular plate of the ethmoid and what other bone form the bony nasal septum? ________________
6. The posterior tip of the soft palate is called the ________________.
7. During this laboratory exercise, by palpating along the surface of the neck, you will be able to identify all the following structures *except* the
   a. thyroid cartilage.
   b. arytenoid cartilages.
   c. hyoid bone.
   d. cricoid cartilage.
   c. first cartilaginous ring of the trachea.
8. True or False: The thyroid cartilage forms a complete ring, but the cricoid cartilage is open posteriorly. ________
9. True or False: Cartilaginous plates are present in the wall of a bronchus, but they are absent in the wall of a bronchiole. ________
10. The respiratory epithelium lining the trachea contains ciliated columnar cells and ________________ cells.

Activity **25.1**

Gross Anatomy: Examining the Structure of the Nose and Nasal Cavity

A The **nose** is the only externally visible component of the respiratory system. It is covered by skin and supported by cartilage and bone.

1. On a skull, identify the bones that make up the bony portion of the nose.

   Superiorly, the nasal bridge is formed by:

   A portion of the **frontal bone**

   The two **nasal bones**

   Processes of the **maxillae** make up the lateral walls.

2. 

   On yourself, palpate the:

   Nasal bone bridge

   Cartilaginous portions of your nose

   Notice that the nasal cartilages extend laterally from the bridge of the nose. What function do you think they have?

   ▶ ______________________________

   ____

   ____

   ____

   ____

   ____

3. The **nostrils (external nares)** allow air to enter the nasal cavity. Internal hairs line these passageways. What do you think is the function of these hairs?

   ▶ ______________________________

   ____

   ____

   ____

   ____

   ____

B The **nasal cavity** is the open space posterior to the nose. It is divided into right and left halves by a bony and cartilaginous partition called the **nasal septum.**

1. On a skull, identify the **perpendicular plate of the ethmoid,** which forms the larger superior portion of the bony nasal septum.

2. Locate the **vomer.** It forms the smaller inferior portion of the bony nasal septum.

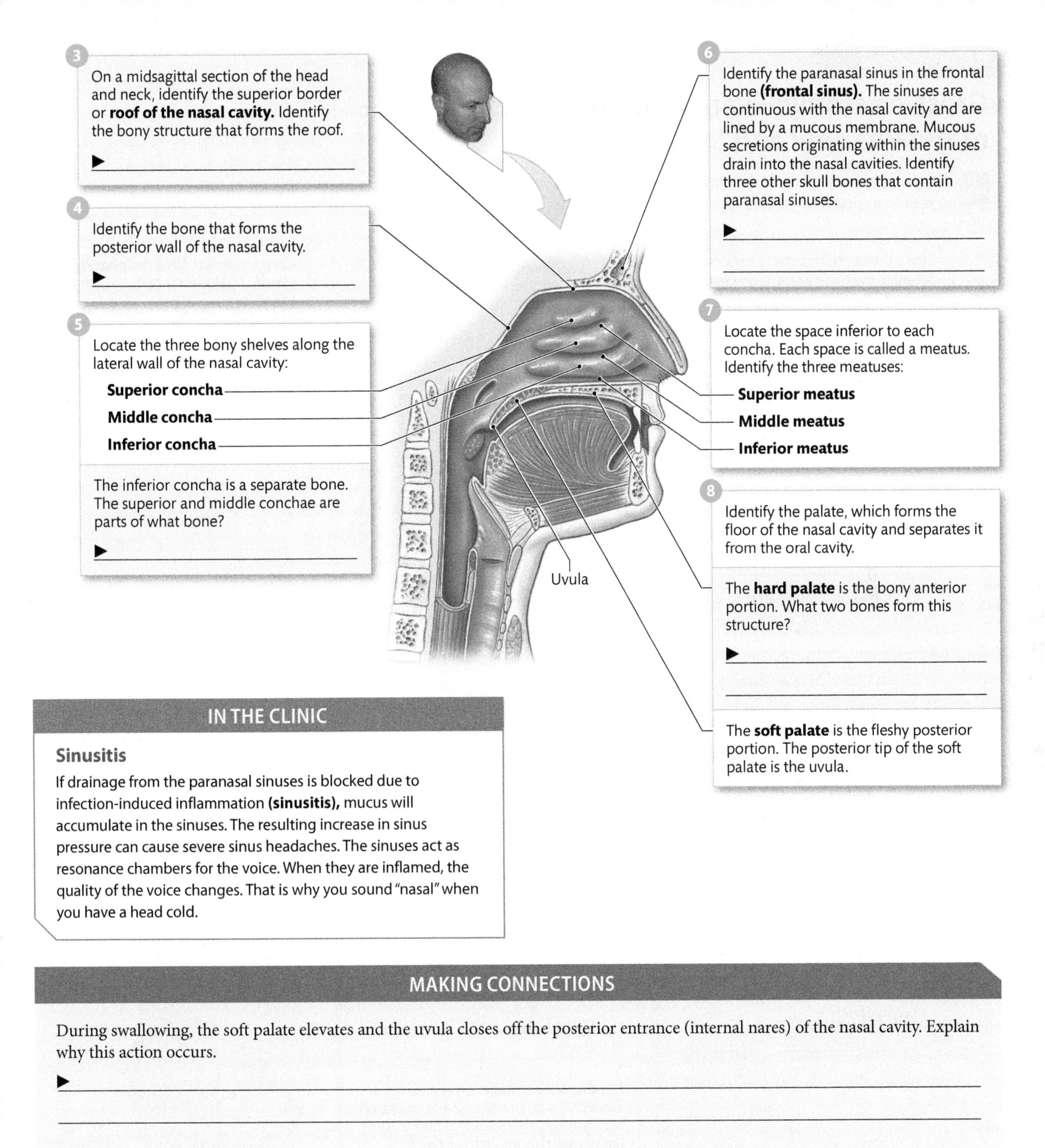

MAKING CONNECTIONS

During swallowing, the soft palate elevates and the uvula closes off the posterior entrance (internal nares) of the nasal cavity. Explain why this action occurs.

▶

Want more practice? Go to: **MasteringA&P** > Study Area > Menu > Lab Tools > PAL >

- Anatomical Models > Respiratory System
- Human Cadaver > Respiratory System

Activity **25.2**

Gross Anatomy: Examining the Structure of the Pharynx and Larynx

A The **pharynx,** or throat, is the cavity posterior to the nasal and oral cavities. It serves as a passageway for air to the larynx and food to the esophagus.

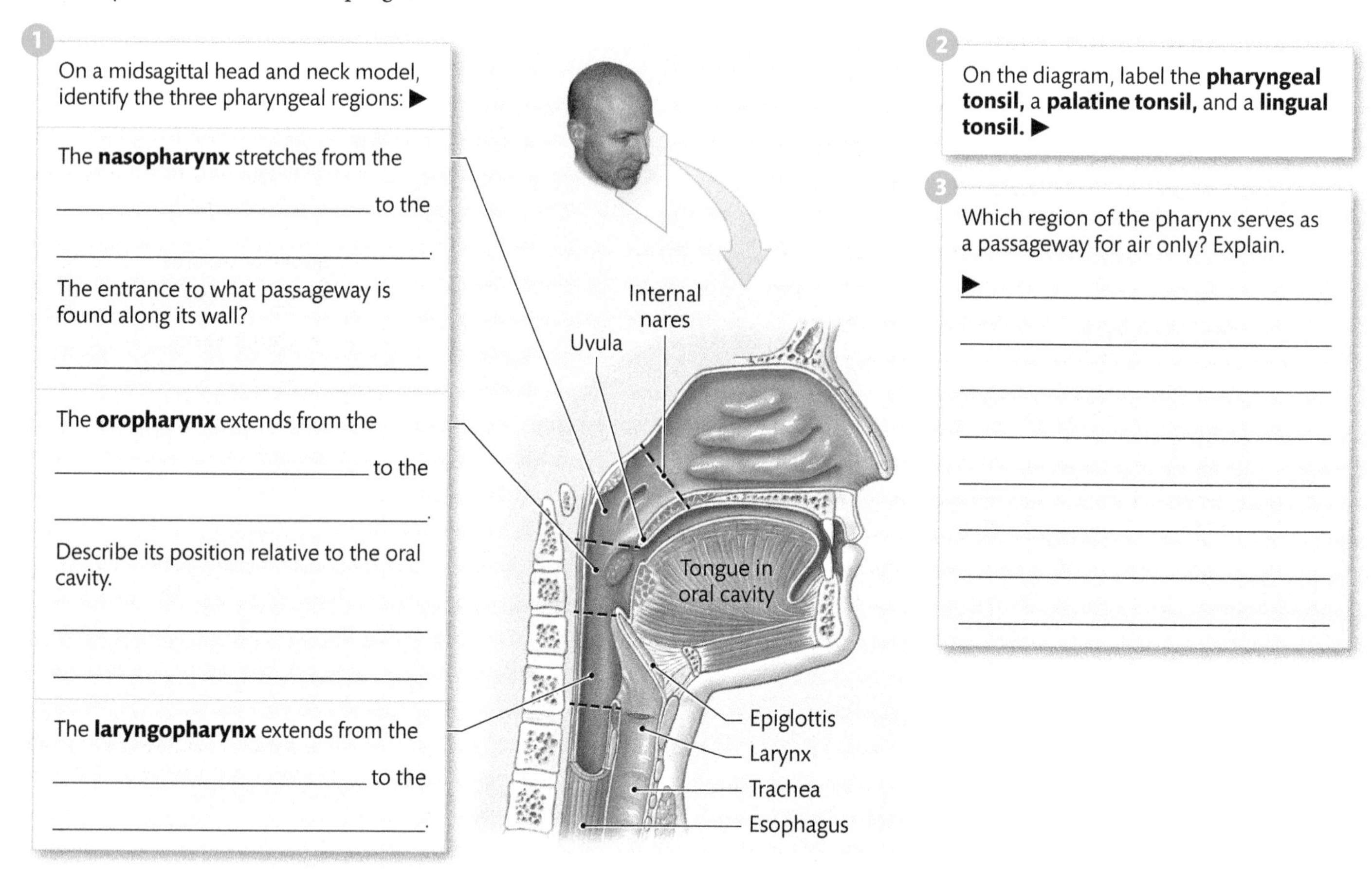

1 On a midsagittal head and neck model, identify the three pharyngeal regions: ▶

The **nasopharynx** stretches from the

______________________ to the

______________________.

The entrance to what passageway is found along its wall?

____

The **oropharynx** extends from the

______________________ to the

______________________.

Describe its position relative to the oral cavity.

____

____

The **laryngopharynx** extends from the

______________________ to the

______________________.

2 On the diagram, label the **pharyngeal tonsil,** a **palatine tonsil,** and a **lingual tonsil.** ▶

3 Which region of the pharynx serves as a passageway for air only? Explain.

▶ ______________________

____

____

____

____

____

____

B The wall of the **larynx** is composed of several pieces of cartilage connected by ligaments and muscles. It provides a passageway for air to enter and exit the trachea during pulmonary ventilation, and it contains the vocal cords, which are used for sound production.

1 Identify the larynx on a head and neck model. Describe its anatomical relationship with the: ▶

Laryngopharynx:

____

____

____

Trachea:

____

____

____

2 On a model of the larynx, observe its anterior surface and identify the following structures:

The **hyoid bone** is positioned superior to the larynx.

The **thyroid cartilage** forms an anterior protective wall for the larynx. It consists of two cartilaginous plates that fuse to form the laryngeal prominence, or Adam's apple. The thyroid cartilage is attached to the hyoid bone by connective tissue.

The **cricoid cartilage** forms the inferior border of the larynx. It is attached by connective tissue to the thyroid cartilage, superiorly, and the first tracheal cartilage, inferiorly.

Epiglottis

Laryngeal prominence

Anterior view

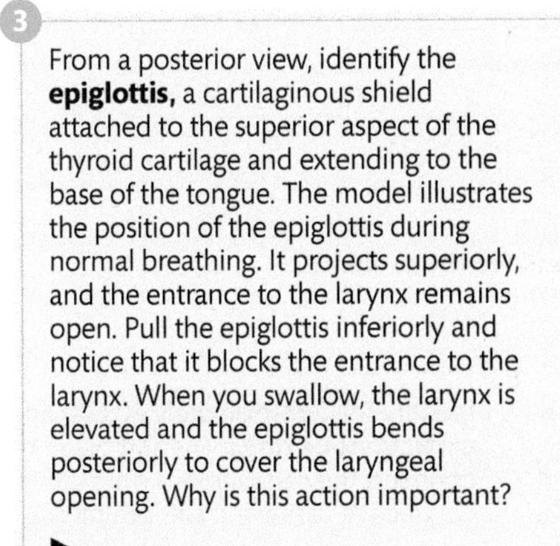

3 From a posterior view, identify the **epiglottis,** a cartilaginous shield attached to the superior aspect of the thyroid cartilage and extending to the base of the tongue. The model illustrates the position of the epiglottis during normal breathing. It projects superiorly, and the entrance to the larynx remains open. Pull the epiglottis inferiorly and notice that it blocks the entrance to the larynx. When you swallow, the larynx is elevated and the epiglottis bends posteriorly to cover the laryngeal opening. Why is this action important?

▶ ______________________________

____

4 Notice the arrangement of the thyroid and cricoid cartilages.

The **thyroid cartilage** is open posteriorly.

The **cricoid cartilage** forms a complete ring.

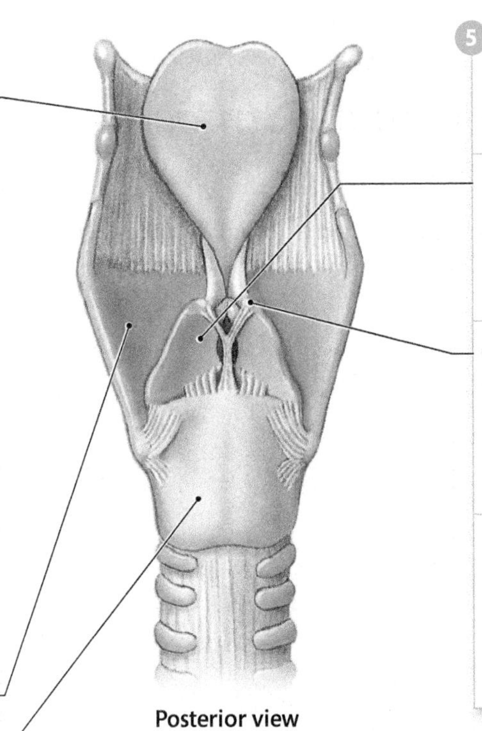

Posterior view

5 Identify three pairs of small cartilages located along the posterior and lateral walls of the larynx:

The paired pyramid-shaped **arytenoid cartilages** rest on the superior margin of the cricoid cartilage. Along with the thyroid cartilage, they provide attachment sites for the vocal cords.

The cone-shaped **corniculate cartilages** are connected to the superior tips of the arytenoid cartilages. They serve as attachments for muscles that regulate the tension of the vocal cords during speech.

The cylinder-shaped **cuneiform cartilages** (not shown) are embedded in the mucous membrane that lines the larynx. During sound production, they stiffen the soft tissue along the lateral laryngeal wall.

C Inside the larynx, two pairs of tissue folds—the vestibular folds and the vocal folds—are attached to the thyroid cartilage anteriorly and the arytenoid cartilages posteriorly.

1 Identify the superior **vestibular ligaments,** which are surrounded by the vestibular folds. These structures are not involved in sound production.

2 Locate the vocal cords. Each vocal cord includes a band of elastic tissue, the **vocal ligament,** which is covered by a fold of epithelium, the vocal fold.

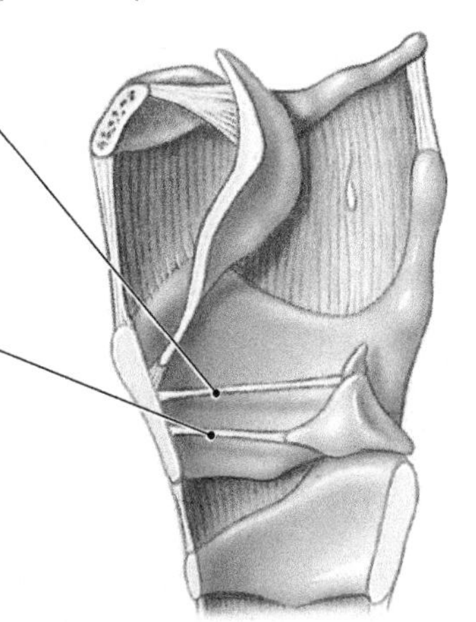

Sagittal section

3 Identify the opening between the vocal cords, known as the **glottis.** Sound is produced when air travels through the glottis and vibrates the vocal folds.

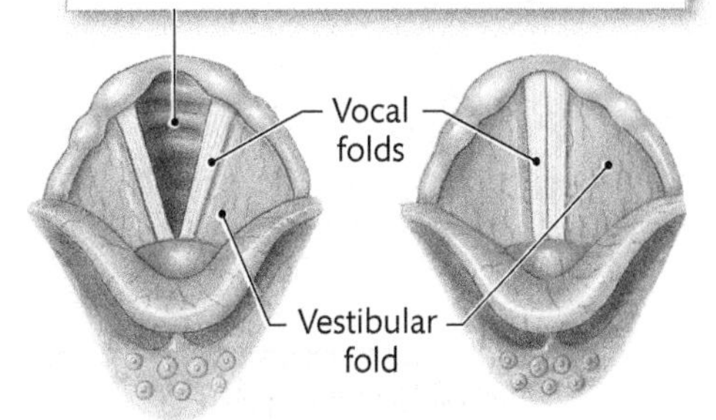

Superior view

MAKING CONNECTIONS

Sounds are amplified by rotational movements of the arytenoid cartilages, which open and close the glottis. Loud sounds, such as yelling, are produced when the cords are widely separated; soft sounds, such as whispering, are produced when the cords are separated slightly. Explain why.

▶ ______________________________

____

____

Want more practice? Go to: **MasteringA&P** > Study Area > Menu > Lab Tools > PAL >

- Anatomical Models > Respiratory System
- Human Cadaver > Respiratory System

Activity **25.3**

Gross Anatomy: Examining the Structure of the Trachea

A The **trachea,** commonly referred to as the windpipe, is a tubular structure that delivers air to the bronchial tree. It is approximately 12 cm (4.5 in.) long and 2.5 cm (1 in.) in diameter.

1. On a model of the torso or thoracic cavity, identify the **trachea.** Observe that it begins at the level of the C_6 vertebra, where it is directly continuous with the larynx, and ends at the level of T_5, where it divides to form the right and left primary bronchi.

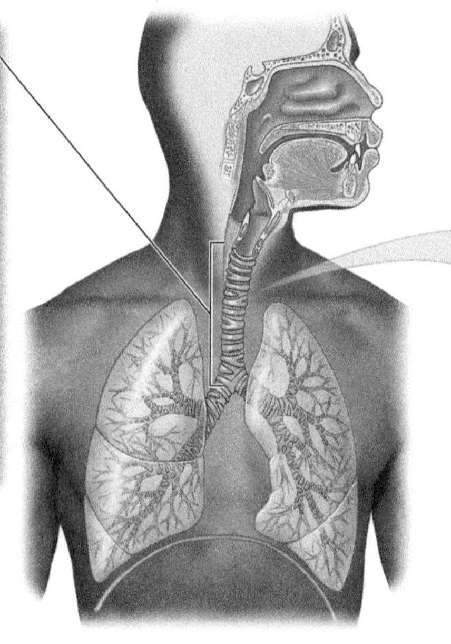

2. Identify the approximately 20 C-shaped **cartilaginous rings.** The soft tissue that connects the cartilaginous rings contains connective tissue and smooth muscle.

3. Identify the last cartilage, the **carina,** where the trachea divides. The mucous membrane that covers the carina contains many sensory receptors. If stimulated by dust particles, these receptors initiate a strong cough reflex to prevent foreign debris from entering the bronchial tree.

B The esophagus travels along the posterior surface of the trachea.

1. On the torso model, identify the esophagus as it passes posterior to the trachea.

2. Along the posterior tracheal surface, identify the band of smooth muscle known as the **trachealis muscle.** Notice that this muscle connects the open ends of the tracheal rings.

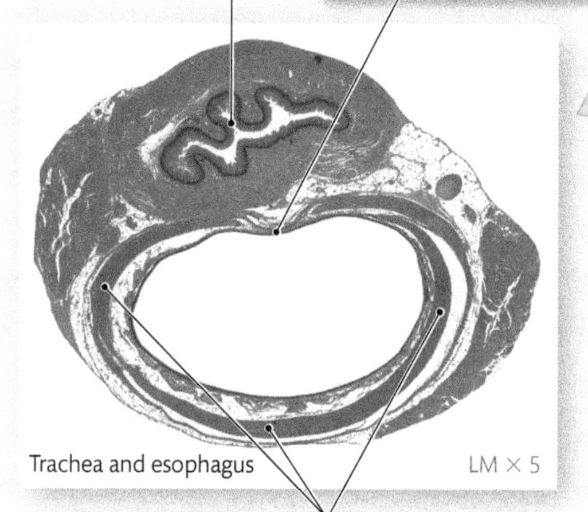

Trachea and esophagus LM × 5

3. Notice that each **tracheal cartilage** covers the anterior and lateral aspects of the tracheal wall, but not the posterior aspect.

Word Origins

Carina is a Latin word that means "the keel of a boat." The carina at the inferior end of the trachea is a cartilaginous ridge that resembles a keel.

MAKING CONNECTIONS

Why do you think cartilage is missing along the posterior wall of the trachea?

▶ ______________________________

____

____

____

____

Want more practice? Go to: **MasteringA&P** > Study Area > Menu > Lab Tools > PAL >

- Anatomical Models > Respiratory System
- Human Cadaver > Respiratory System

Activity 25.4

Surface Anatomy: Respiratory Structures in the Neck

A You can identify two laryngeal cartilages and other respiratory structures by palpation as they travel through the neck.

1. Along the anterior midline of your neck, about 4.0 to 4.5 cm (about 1.5 in.) inferior to your mandible, gently palpate the **laryngeal prominence** of your thyroid cartilage. Keep your finger on the prominence and feel it move when you swallow.

2. The laryngeal prominence is larger in males than in females. Verify this difference by observing and comparing the laryngeal prominence on both men and women in your class. Why do you think there is a difference?

▶ ______________________________

____

3. From the laryngeal prominence, slowly move your finger superiorly until you can feel the **hyoid bone.** Similar to the laryngeal prominence, you can feel the hyoid bone move by pressing gently and swallowing.

4. Place your finger on your laryngeal prominence again, and slowly move it inferiorly along the midline of the neck. As your finger moves off the thyroid cartilage, you will feel a soft tissue gap where the **cricothyroid ligament** is located.

Thyroid gland

5. Continue moving your finger inferiorly until you feel the hard surface of the **cricoid cartilage.** The cricoid cartilage is positioned at the level of the 6th cervical vertebra. At this level and posterior to the cartilage, the pharynx ends and the esophagus begins.

6. The inferior margin of the cricoid cartilage is the boundary between the larynx and the trachea. From the cricoid cartilage, move your finger inferiorly until you can feel the **1st cartilaginous ring of the trachea.**

MAKING CONNECTIONS

Explain why the surface position of the laryngeal prominence is an important landmark for identifying other respiratory structures in the neck.

▶ ______________________________

____

____

____

____

Activity **25.5**

Gross Anatomy: Examining the Structure of the Lungs

A The lungs are pyramid- or cone-shaped organs located in the left and right sides of the thoracic cavity. They are lined by a double serous membrane known as the **pleural membrane.** The **pleural cavity** is the narrow, fluid-filled space between the two pleural membranes.

1 On a model of the thoracic cavity, observe the position of the **lungs** in relation to the heart. The heart is centrally located within an area called the mediastinum. The lungs are positioned on either side of the heart.

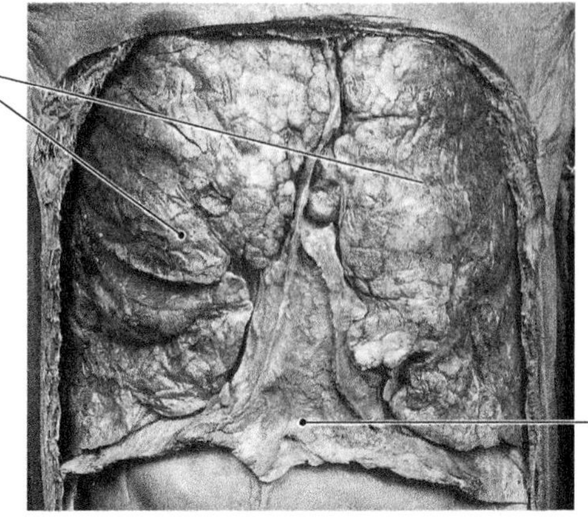

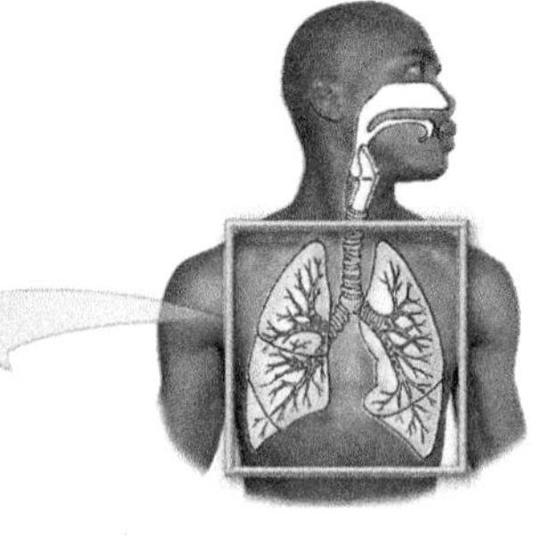

Horizontal fissure

Oblique fissure

2 On each lung, identify the **apex** along its superior margin. The apex of each lung extends into what region of the body?

▶ ______________________________

3 Identify the lobes and the fissures of the right lung.

Lobes

Superior

Middle

Inferior

Fissures

Oblique

Horizontal

4 Identify the lobes and the fissures of the left lung.

Lobes

Superior

Inferior

Fissure

Oblique

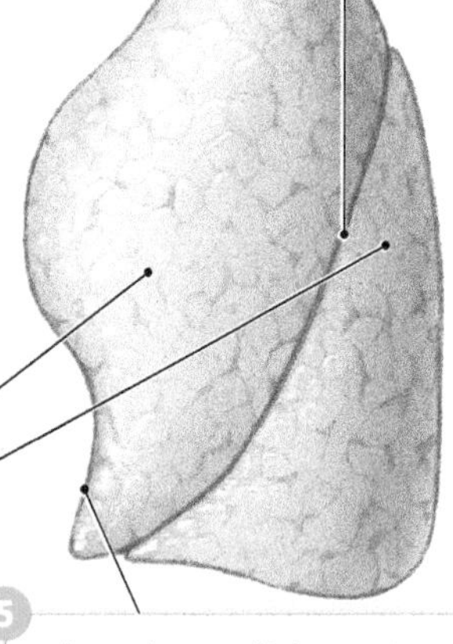

5 Along the medial margin of the left lung, identify the **cardiac notch.**

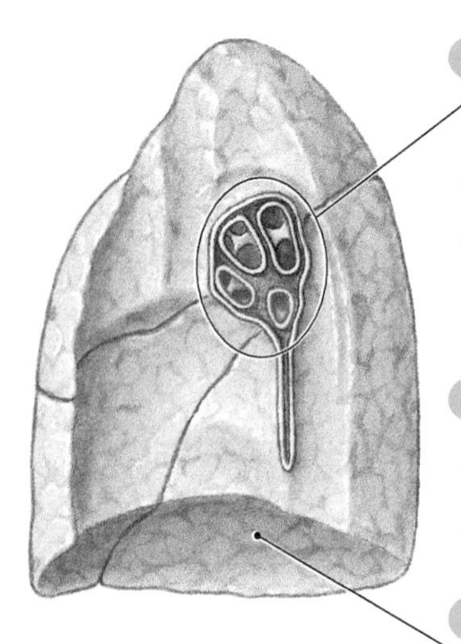

6 On the medial or mediastinal surface of each lung, identify the **hilum.** This region serves as an area of entry or exit for structures that compose the root of the lung. What structures do you think make up the root of the lung?

▶ ______________________________

____

7 On the left lung, identify two unique features:

The **groove for the aorta**

The **cardiac impression** to accommodate the heart

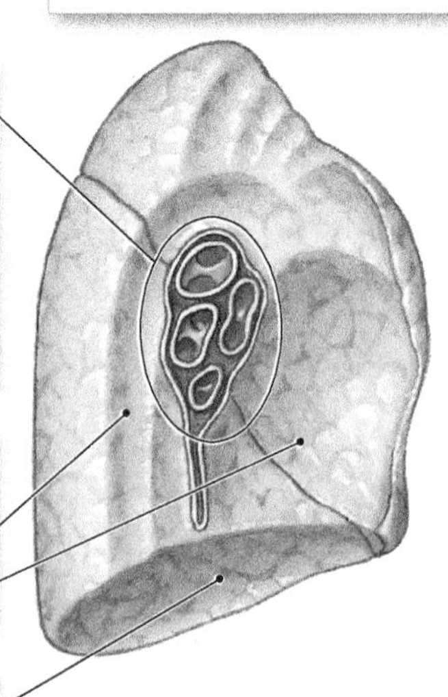

8 On each lung, identify the inferior **base.** It is also called the **diaphragmatic surface** because it rests on the diaphragm.

B The **bronchial tree** refers to the treelike branching of the airways in the lungs. It begins where the trachea divides into the **right** and **left primary bronchi** (singular = **bronchus**) and ends with the **alveoli** (singular = **alveolus**). The bronchial tree contains two functional categories of airways. The **conducting airways** distribute air to a particular region of a lung. They deliver air to the **respiratory airways,** where the exchange of carbon dioxide and oxygen occurs between the lungs and pulmonary capillaries.

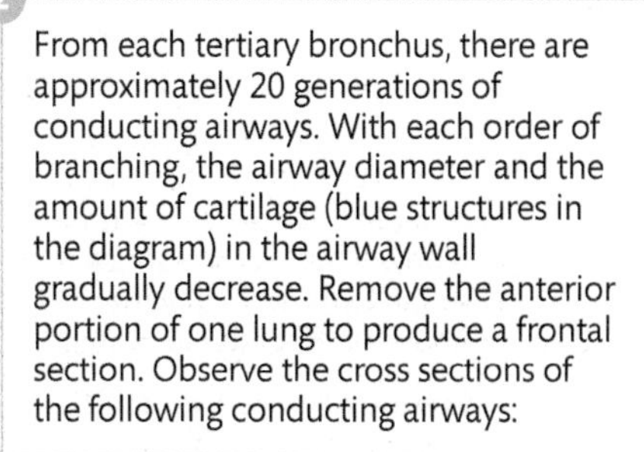

1 On a model of the torso or thoracic cavity, identify the first three generations of conducting airways:

The **right** and **left primary (main) bronchi** deliver air to their respective lungs.

The **secondary (lobar) bronchi** deliver air to the lobes of the lungs.

How many secondary bronchi are found in the right lung?

▶ ______________________________

How many are in the left lung?

▶ ______________________________

The **tertiary (segmental) bronchi** deliver air to bronchopulmonary segments, which are subdivisions within each lobe. There are 10 segments in the right lung and 8 to 10 segments in the left lung.

2 From each tertiary bronchus, there are approximately 20 generations of conducting airways. With each order of branching, the airway diameter and the amount of cartilage (blue structures in the diagram) in the airway wall gradually decrease. Remove the anterior portion of one lung to produce a frontal section. Observe the cross sections of the following conducting airways:

Small bronchi are 1 to 5 mm in diameter and have small cartilage plates embedded in their walls.

Bronchioles have a diameter of 1 mm or less and lack cartilage.

Terminal bronchioles are less than 0.5 mm in diameter and mark the termination of the conducting airways.

3 Notice that the conducting airways make up a relatively small proportion of lung volume and surface area. The vast majority of the bronchial tree consists of the thin-walled respiratory airways, giving the lung a spongelike appearance. Respiratory airways include the following types of airways:

Respiratory bronchiole

Alveolar duct

Alveolar sac

Alveoli in an alveolar sac

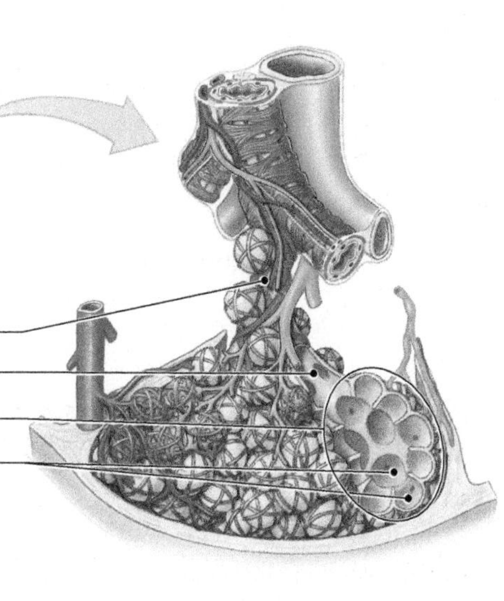

MAKING CONNECTIONS

Why do you think there are more respiratory airways than conducting airways?

▶ ______________________________

____

____

Want more practice? Go to: **MasteringA&P** > Study Area > Menu > Lab Tools > PAL >

- Anatomical Models > Respiratory System
- Human Cadaver > Respiratory System

Activity 25.6

Microscopic Anatomy: Examining the Structure of the Trachea

A The mucous membrane that lines the trachea and many of the respiratory passageways is called the **respiratory mucosa,** which contains the highly specialized **respiratory epithelium.** In the trachea, deep to the respiratory mucosa, is a layer of connective tissue called the submucosa and the adventitia, which contains a thick layer of cartilage.

1. Scan a slide of the trachea under low power. Along the surface that lines the lumen, identify the contents of the respiratory mucosa:

   Pseudostratified, ciliated columnar epithelium

   Lamina propria, a thin layer of connective tissue

2. Deep to the respiratory mucosa, identify the **submucosa.** It contains mucus- and serous-secreting tracheal glands and numerous blood vessels that supply the tracheal wall.

3. Identify an area of **cartilage** that is part of the adventitia. The cartilage that you are observing is a portion of one C-shaped cartilaginous ring.

4. The **perichondrium,** also in the adventitia, is the fibrous connective tissue layer that covers the cartilaginous ring.

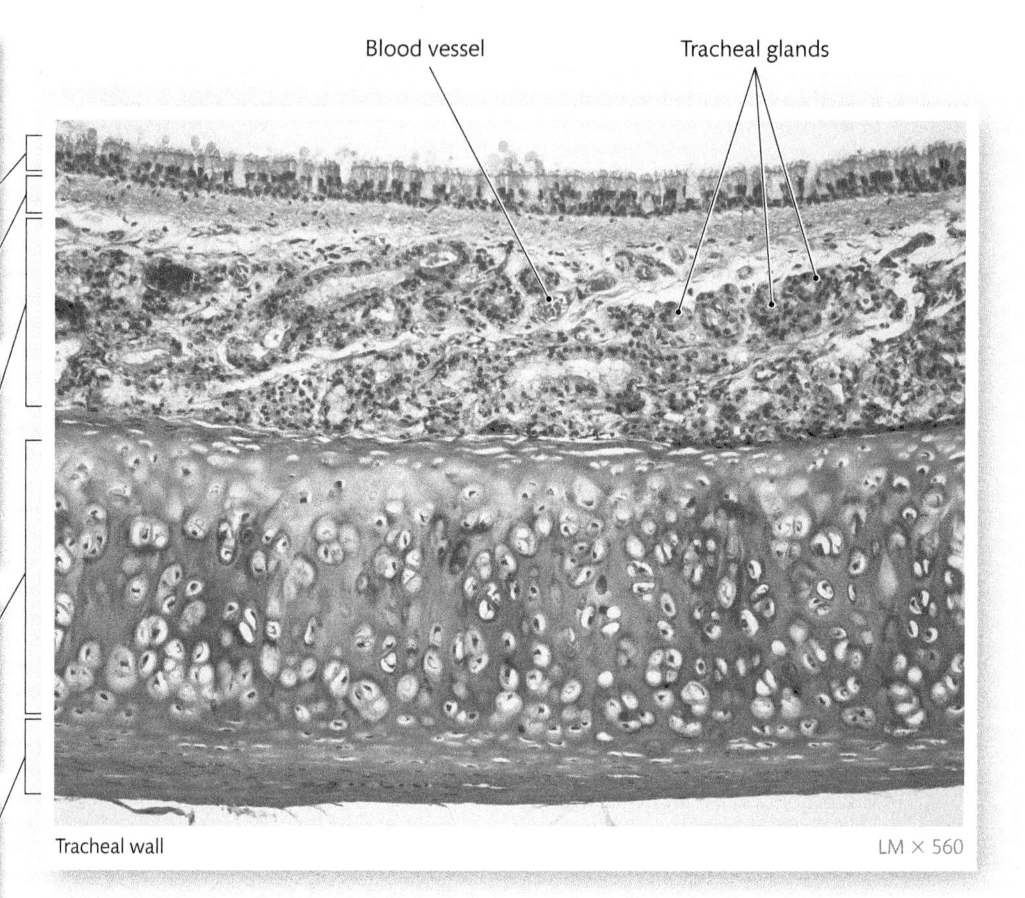

Tracheal wall

IN THE CLINIC

Obstruction of the Trachea

Obstruction of the trachea due to swollen tissues, excessive glandular secretions, or a foreign object can be dangerous. The **Heimlich maneuver** may successfully dislodge foreign objects such as food. In this procedure, the abdomen is compressed to elevate the diaphragm. If done with enough force, the pressure generated may remove the object.

If the obstruction still prevents breathing, a **tracheostomy** must be performed. In this procedure, an incision is made in the anterior wall of the trachea, and a tube is inserted to allow air to enter the lungs.

■ Want more practice? Go to: **MasteringA&P** > Study Area > Menu > Lab Tools > PAL > Histology > Respiratory System > Trachea photos

B The respiratory mucosa and tracheal glands act together to prevent airborne particles from entering the lungs. Mucus secretions trap dust and other particles, and serous secretions contain **lysozyme,** an enzyme that can kill potentially harmful bacteria that enter the respiratory tract. The cilia push the mucus to the pharynx, where it can be swallowed and digested by stomach secretions.

1. Switch to high power and observe the respiratory epithelium:
   - Most of the cells are **ciliated columnar cells.** Observe cilia projecting from their surfaces.
   - Mucus-secreting **goblet cells** are wedged between the cilated cells.
2. In the submucosa, identify the **tracheal glands.** These structures are lined by stratified cuboidal epithelium.

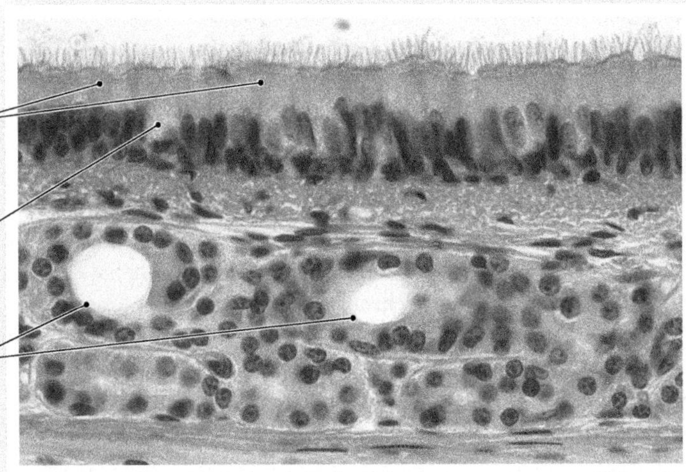

Mucous membrane and submucosa in trachea LM × 570

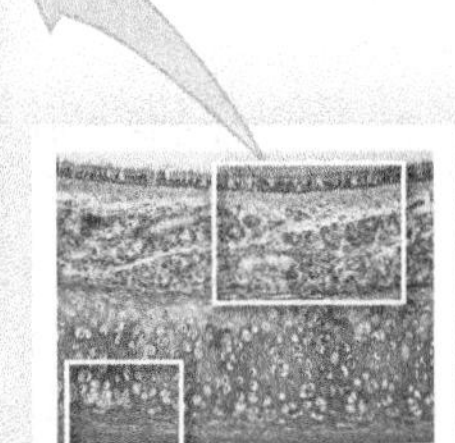

LM × 25

3. Move your slide to observe an area of cartilage.
   - Observe the **lacunae** distributed throughout the purple-staining cartilaginous matrix.
   - Identify a **chondrocyte** within each lacuna.

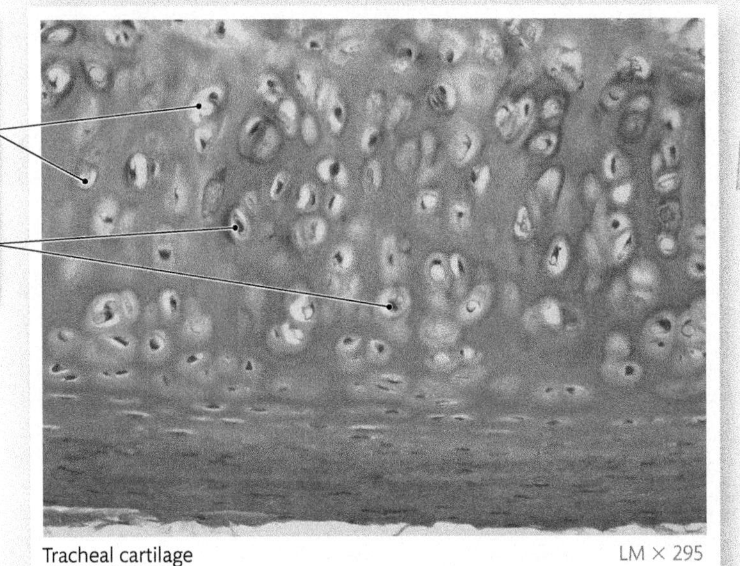

Tracheal cartilage LM × 295

MAKING CONNECTIONS

Why do you think it is essential to have cartilaginous plates in the tracheal wall?

▶ ______________________________

____

____

____

____

Activity **25.7**

Microscopic Anatomy: Examining the Structure of the Lungs

As you observe the conducting and respiratory airways, be aware of the following structural changes that occur as bronchial tubes become smaller and smaller:

- *The amount of cartilage gradually declines.* At the level of the bronchioles, cartilage is completely absent.
- *As the amount of cartilage decreases, smooth muscle becomes more prominent.* In the bronchioles, smooth muscle forms a continuous layer. In the respiratory airways, there is a gradual decline in smooth muscle, and it is absent in the alveoli.
- *The type of epithelium gradually changes.* The epithelium in the trachea and primary bronchi is pseudostratified ciliated columnar. With successive airway generations, the height of the epithelial cells gradually declines from columnar in the small bronchi, to cuboidal in bronchioles, and to squamous in the alveolar passageways. Furthermore, there is a gradual decline in the numbers of ciliated and goblet cells until they are completely absent in the bronchioles and successive generations.

A Most of the lung's volume is occupied by the thin-walled respiratory airways.

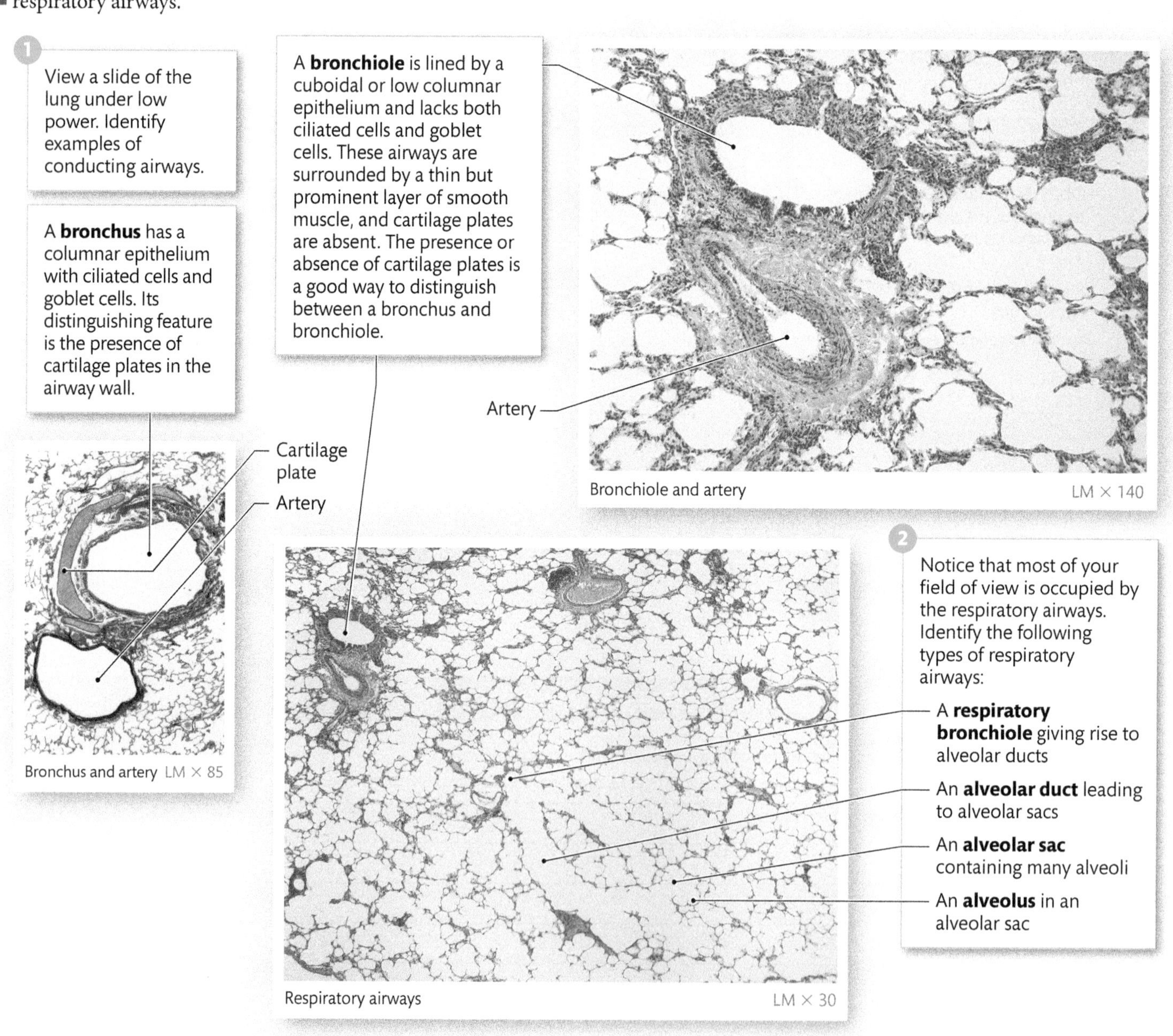

Bronchus and artery LM × 85

Bronchiole and artery LM × 140

Respiratory airways LM × 30

B The type I alveolar cells (type I pneumocytes) in alveoli and endothelial cells in pulmonary capillaries form the respiratory membrane.

1. Use the high-power or oil immersion objective lens to examine the structure of the alveoli more closely. Notice that the epithelium lining the alveoli is a single layer of squamous cells. They are called **type I alveolar cells (type I pneumocytes).**

2. Identify numerous **pulmonary capillaries** as they travel along the walls of the alveoli.

3. Notice the close association between type I alveolar cells and pulmonary capillaries. This relationship forms the **respiratory membrane,** the barrier through which oxygen and carbon dioxide are exchanged between air in an alveolus and blood in the pulmonary circulation. The respiratory membrane contains the following layers:
   - Squamous type I alveolar cells
   - Basal lamina beneath the alveolar cells fused with the basal lamina beneath the capillary endothelial cells
   - Squamous endothelial cells in pulmonary capillaries

4. Carefully scan the epithelia of several alveoli and identify cells that are cuboidal rather than squamous. These **type II alveolar cells (type II pneumocytes)** secrete a phospholipid substance called surfactant, which lowers the surface tension along the surface of alveoli and prevents the air sacs from collapsing.

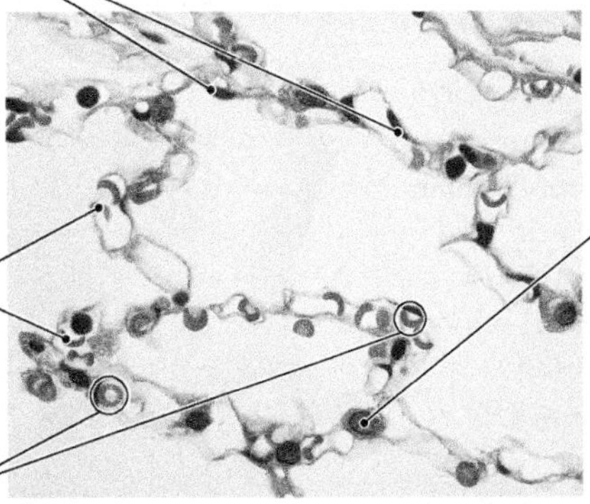

Alveoli LM × 353

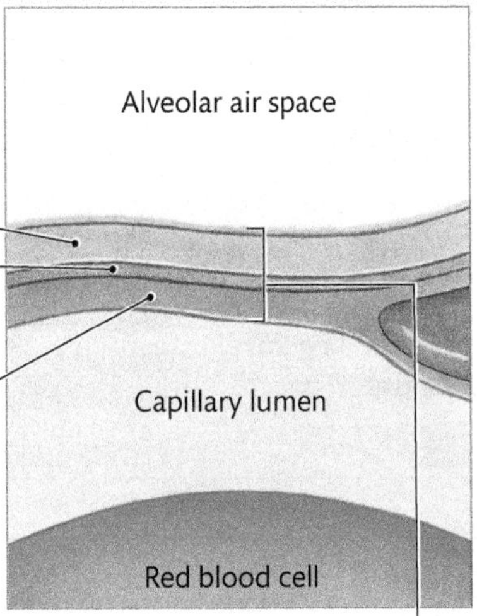

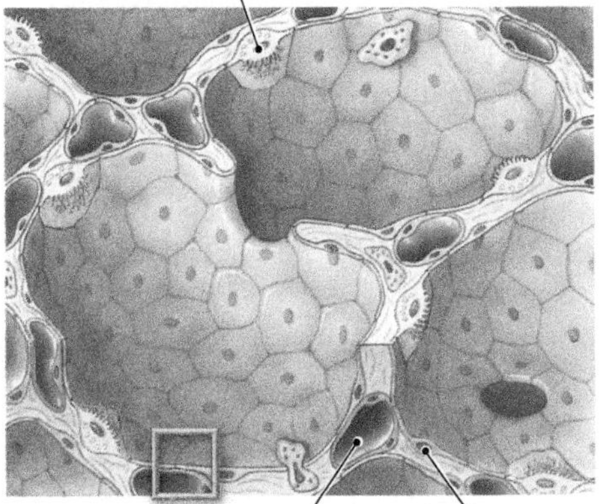

Word Origins

Alveolus comes from the Latin *alveus,* which means "hollow sac" or "cavity." Each alveolus in the lung is a hollow, air-filled sac.

MAKING CONNECTIONS

Emphysema is a condition characterized by damage to alveolar walls. It can develop as a result of long-term exposure to air pollutants or cigarette smoking. Explain how emphysema affects gas exchange between the lungs and pulmonary capillaries.

▶ ______________________________

____

____

____

____

Want more practice? Go to: **MasteringA&P** > Study Area > Menu > Lab Tools > PAL > Histology > Respiratory System > Lung photos

BEFORE YOU MOVE ON . . .

‹‹ LOOKING BACK

There are regional differences in the structure of the mucous membrane that lines the respiratory tract. Similar to what you observed in the trachea, the nasal cavity and nasopharynx have a respiratory mucosa that contains a pseudostratified columnar epithelium with ciliated cells and mucus-secreting goblet cells. The epithelium lining the wall of the oropharynx and laryngopharynx is stratified squamous, however.

In the lungs, the large conducting airways have a respiratory mucosa, with ciliated columnar cells and goblet cells. As the airways diminish in size, the shape of the epithelial cells change from columnar to cuboidal to squamous. In addition, there is a progressive decline in ciliated and goblet cells. Both ciliated and goblet cells are absent in small bronchioles and the respiratory airways.

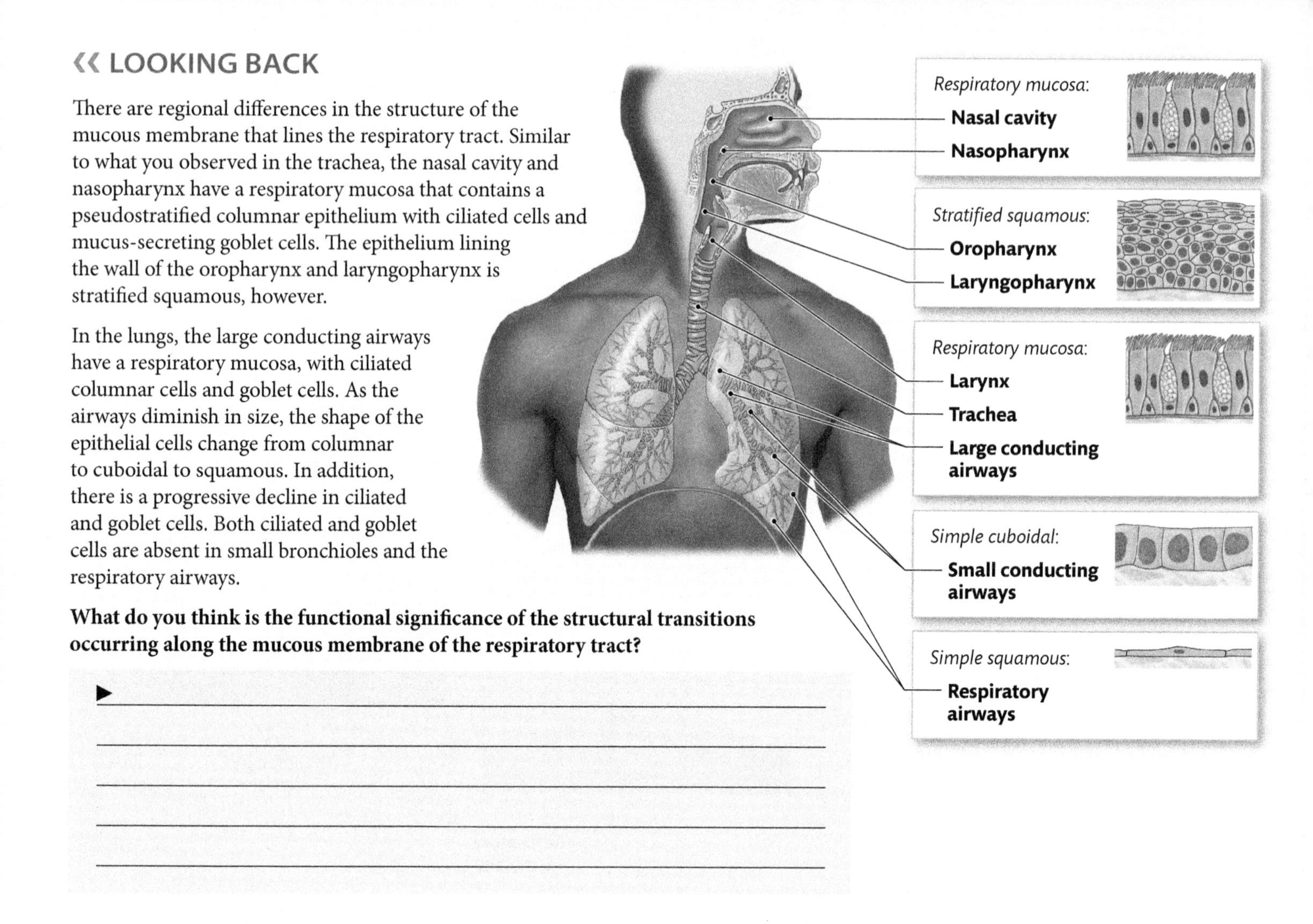

What do you think is the functional significance of the structural transitions occurring along the mucous membrane of the respiratory tract?

▶ ______________________________

____

____

____

____

LOOKING FORWARD ››

In this laboratory exercise, you studied the anatomy of the respiratory system. In the next exercise, you will learn how its unique structure allows the respiratory system to take up oxygen and release carbon dioxide during a process called respiration. You will examine the actions of the respiratory muscles and evaluate pulmonary function by measuring and calculating pulmonary volumes.

Name ______________________

Lab Section ______________________

Date ______________________

EXERCISE 25 REVIEW SHEET

Anatomy of the Respiratory System

1. Ideally, the nasal septum will divide the nasal cavity equally, but it is common for this partition to be slightly off center. A deviated septum, however, is a condition in which the nasal septum is shifted significantly away from the midline. Explain how this abnormality may cause difficulty in breathing.

____

____

____

____

2. The configuration of the conchae and meatuses forms irregular twists and turns along the wall of the nasal cavity. What do you think is the functional advantage of this arrangement?

____

____

____

____

3. Complete the following table.

Types of Airways in the Bronchial Tree

Airway Type	Cartilage (present/absent)	Type of Epithelium (columnar, cuboidal, squamous)	Ciliated Cells (present/absent)	Mucous Cells (present/absent)	Function
a.	b.	c.	Absent	d.	Give rise to alveolar sacs
Primary bronchi	e.	Columnar	f.	g.	h.
i.	Absent	j.	k.	l.	Last generation of conducting airways
Tertiary bronchi	m.	n.	o.	Present	p.
q.	r.	s.	Present	t.	Deliver air to lung lobes
u.	v.	w.	x.	Absent	Groups of airways that form alveolar sacs

4. Which region(s) of the pharynx is not lined by a respiratory mucosa? Explain why.

QUESTIONS 5–15: Identify the labeled structures.

5.

6.

7.

8.

9.

10.

11.

12.

13.

14.

15.

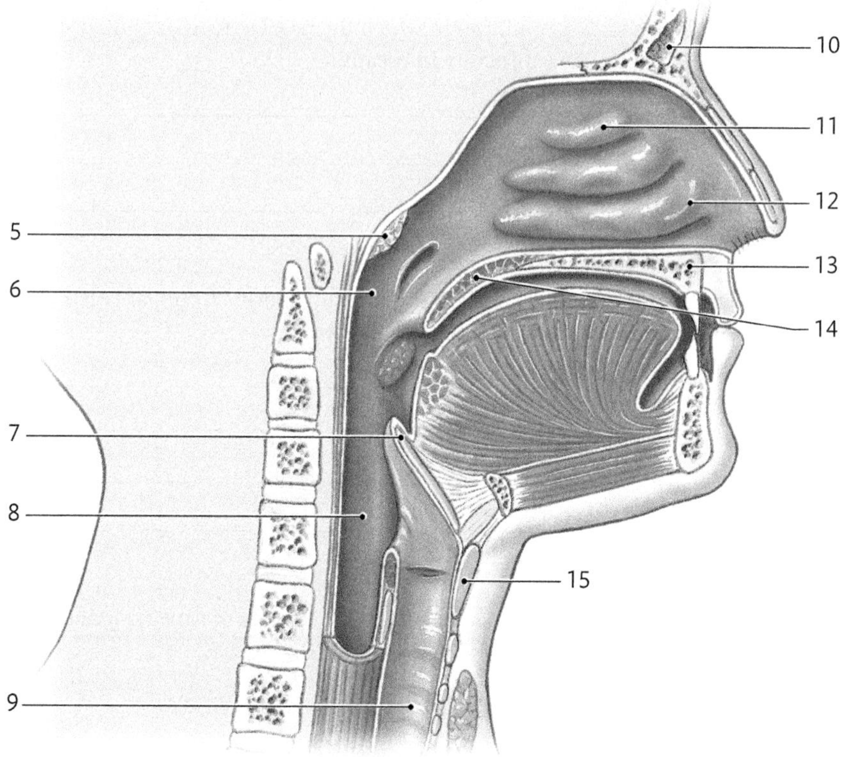

16. Identify the similarities and differences in the gross anatomical structure of the left and right lungs.

Similarities:

Differences:

17. Regarding the structure of the larynx, what functional advantage is there to having several pieces of cartilage connected by soft tissue rather than just a large single ring of cartilage?

____

____

____

____

18. A person who is severely allergic to nuts eats a candy bar without knowing that it contains finely chopped peanuts. Soon after, massive inflammation in her pharynx blocks the passage of air to her trachea and lungs. An emergency tracheostomy must be performed. Explain why knowledge of the surface anatomy of the neck will help you perform this procedure successfully.

____

____

____

____

19. Explain how autonomic nerves that innervate the smooth muscle layer around bronchioles can influence the volume of air reaching a particular region of the lung.

____

____

____

____

20. Is the airway in the photo to the right a bronchus, bronchiole, or alveolus? Discuss specific structural features to explain your answer.

____

____

____

____

____

____

____

____

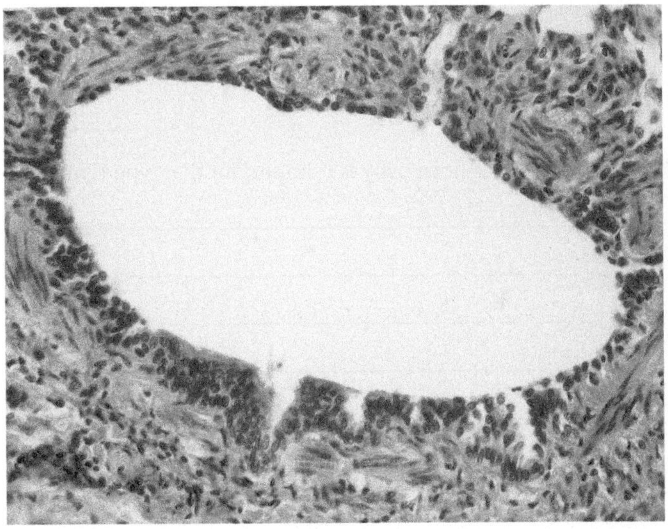

LM × 295

21. In the following diagram, color each structure with the indicated colors.

Trachea = **dark blue**
Primary bronchi = **yellow**
Thyroid cartilage = **dark green**
Right superior lobe = **red**
Secondary bronchi = **purple**
Left inferior lobe = **gray**
Tertiary bronchi = **dark brown**
Right middle lobe = **light blue**
Cricoid cartilage = **orange**
Left superior lobe = **pink**
Hyoid bone = **light brown**
Right inferior lobe = **light green**

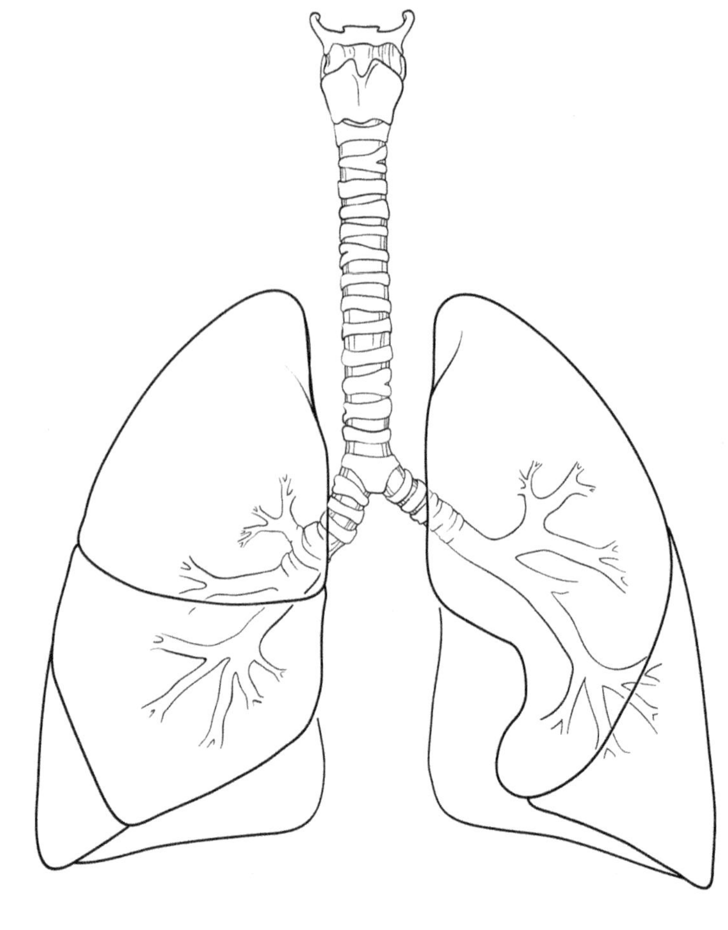

22. When a baby is born, why is it important that type II alveolar cells are functioning normally?

____

____

____

____

____

EXERCISE 26

Respiratory Physiology

LEARNING OUTCOMES

These Learning Outcomes correspond by number to the laboratory activities in this exercise. When you complete the activities, you should be able to:

Activity 26.1 **Explain the significance of Boyle's law as it applies to pulmonary ventilation.**

Activity 26.2 **Identify the muscles involved in pulmonary ventilation and discuss their functions.**

Activity 26.3 **Measure expiratory respiratory volumes and tidal volume.**

Activity 26.4 **Compare predicted and actual vital capacities and explain differences between the two.**

Activity 26.5 **Measure respiratory volumes using the Biopac Student Lab System.**

Activity 26.6 **Calculate respiratory volumes that cannot be directly measured with a handheld spirometer.**

Activity 26.7 **Calculate respiratory minute volume and alveolar ventilation.**

LABORATORY SUPPLIES

- Human skeleton
- Torso model
- Thoracic cavity model
- Microscope slides
- Pipettes
- Water
- Balloons
- Handheld spirometer
- Disposable mouthpieces for handheld spirometers
- Windows or Mac computer system
- Biopac Student Lab System (MP35, MP36, or MP45)
- Biopac airflow transducer (SS11LA or SS11LB)
- Biopac bacteriological filters (AFT1)
- Biopac disposable mouthpieces (AFT2)
- Biopac nose clip (AFT3)
- Biopac calibration syringe: 0.6 liter (AFT6 or AFT6A + AFT11A)

PRE-LAB QUIZ

Before you begin, read all the activities in Exercise 26 and the required reading in your textbook that is assigned by your instructor.

1. During this laboratory exercise, you will study the function of the pleural membranes. What will you use to represent the pleural membranes? ________________
2. True or False: Air flows into the lungs when intrapulmonary pressure is less than atmospheric pressure. __________
3. True or False: The abdominal muscles are involved in forced inhalation. __________
4. During this laboratory exercise, you will use a handheld ________________ to measure respiratory volumes.
5. True or False: In general, vital capacity increases with age and decreases with height. __________
6. The volume of air that enters the alveolar airways each minute is called the __________.
7. True or False: The anatomic dead space refers to the volume of inhaled air that never enters the alveoli. __________
8. During a period of hyperventilation, the carbon dioxide levels __________, and the pH of body fluids __________.
   a. decrease; decrease
   b. decrease; increase
   c. increase; increase
   d. increase; decrease
9. True or False: One respiratory cycle equals one normal inhalation and one normal exhalation. __________
10. During Activity 26.5, the Biopac activity in the laboratory exercise, you will be breathing into an airflow transducer. The transducer converts airflow to air
    a. volume.
    b. mass.
    c. velocity.
    d. acceleration.

Activity **26.1**

Examining Pulmonary Ventilation

Pulmonary ventilation is the movement of air into and out of the lungs. It is governed by pressure gradients that form between the air in the atmosphere **(atmospheric pressure)** and the air in the airways **(intrapulmonary pressure).** Air will flow from an area of high pressure to an area of low pressure.

A According to **Boyle's law,** the pressure of a gas, in a closed container at constant temperature, is inversely proportional to its volume.

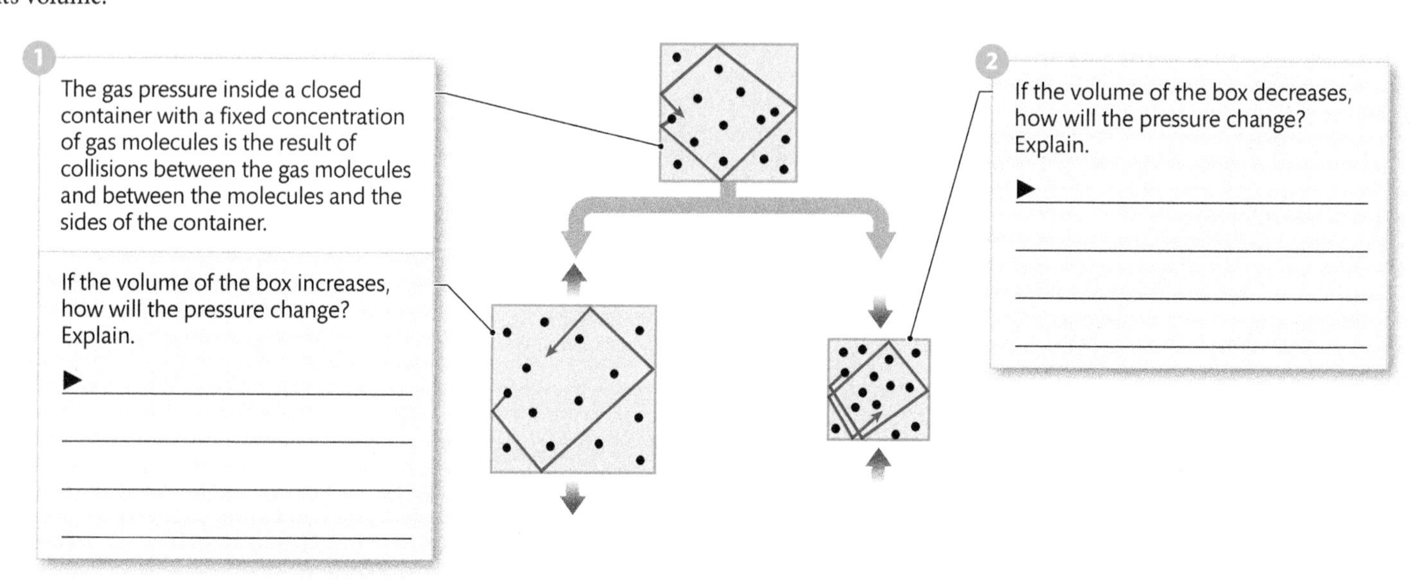

B If intrapulmonary pressure is less than atmospheric pressure, air will flow into the lungs **(inhalation)**. Conversely, if intrapulmonary pressure is greater than atmospheric pressure, air will flow out of the lungs **(exhalation).**

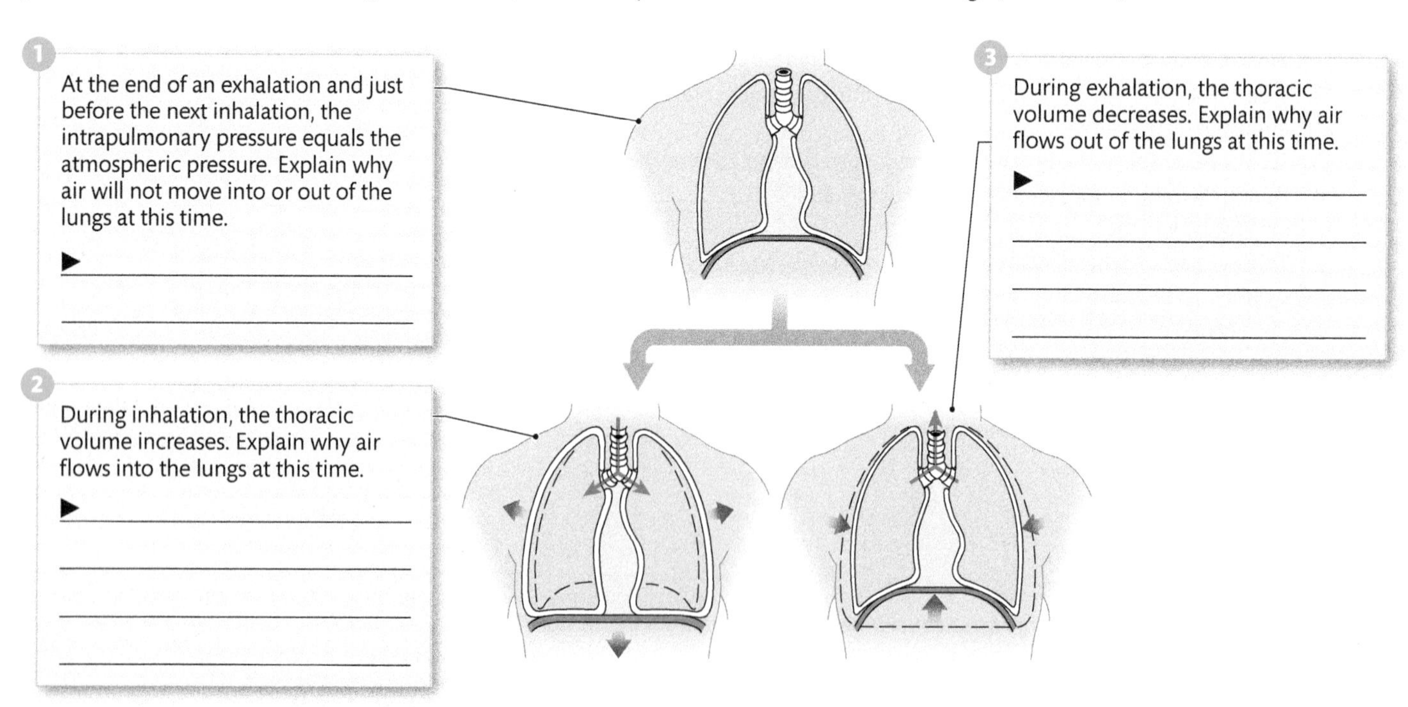

■ Want more practice? Go to: MasteringA&P > Study Area > Menu > Lab Tools > PhysioEx > Exercise 7: Respiratory System Mechanics > Activity 3: Effect of Surfactant and Intrapleural Pressure on Respiration

C Movements of the diaphragm and ribs create changes in thoracic volume.

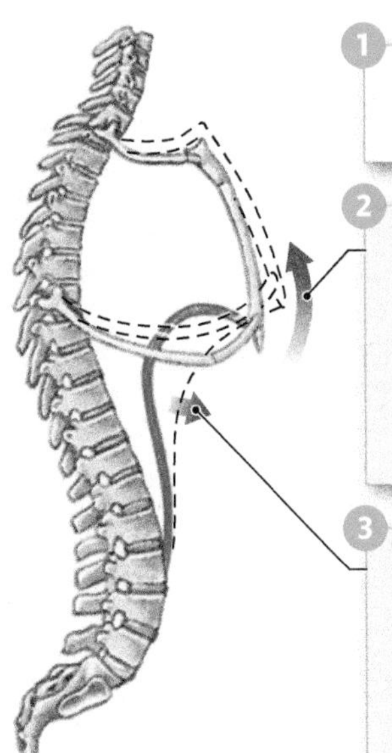

1. With your right hand, palpate two or three ribs on your right side. Place your left hand on your sternum.

2. Inhale and feel your ribs elevating and your sternum moving out. How do these movements change thoracic volume and pressure? ▶

   Thoracic volume: ______________________

   Thoracic pressure: ______________________

3. During inhalation, the diaphragm contracts and moves inferiorly. How does this movement change thoracic volume and pressure? ▶

   Thoracic volume: ______________________

   Thoracic pressure: ______________________

D The serous secretions of the pleural membranes create a bond that prevents the lungs from collapsing.

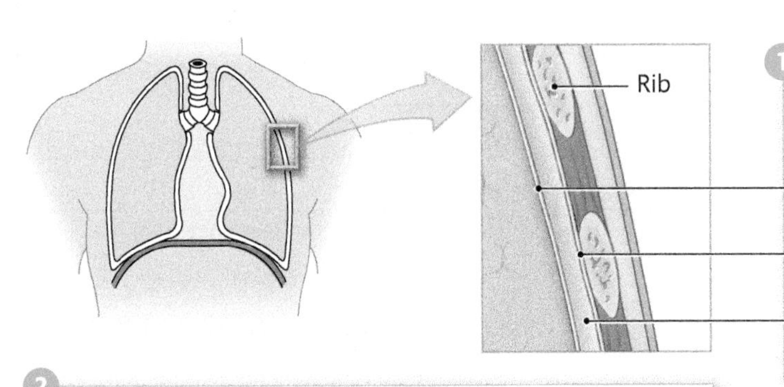

1. On a torso or thoracic cavity model, observe the position of the lungs and the pleural membranes:

   The **visceral pleura** covers the surface of the lungs.

   The **parietal pleura** covers the wall of the thoracic cavity.

   The **pleural cavity** contains a thin layer of serous fluid that keeps the lungs firmly attached to the thoracic wall.

2. Use a medicine dropper or pipette to place a few drops of water on a microscope slide. Mix the drops to form a thin water layer and cover it with a second slide.

3. The microscope slides represent the pleural membranes, and the water layer is the fluid-filled pleural cavity. Without separating the slides, move the glass surfaces back and forth. Notice how easy that is to do. During respiration, when the lungs expand and contract, the fluid layer allows the two pleural membranes to slide smoothly past each other.

4. Try to separate the two slides by pulling them apart. It is difficult to do because the thin film of water forms a strong bond that keeps the two slides together, just as serous fluid in the pleural cavity keeps the pleural membranes together.

MAKING CONNECTIONS

What do you think would happen if air entered the pleural cavity and weakened the bond formed by the serous fluid?

▶ ____

____

____

____

____

____

____

Activity **26.2**

Identifying Respiratory Muscles

A **Inspiratory muscles** increase thoracic volume during inhalation.

1

On a torso model, locate the following muscles involved in **normal inhalation:**

The **diaphragm** is the primary inspiratory muscle in the body. It forms a dome-shaped muscular partition between the thoracic and abdominal cavities. Its convex surface forms the floor of the thoracic cavity.

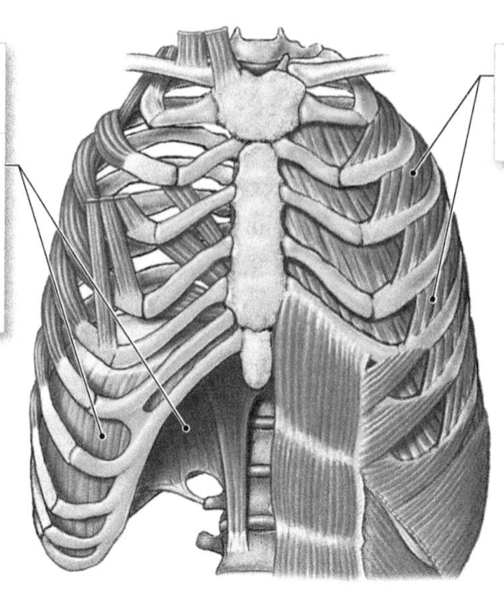

The **external intercostal muscles** are the outermost muscular layers in the intercostal spaces.

2

Describe how the actions of the diaphragm and external intercostal muscles increase the volume of the thoracic cavity. ▶

Diaphragm:

____

____

____

External intercostal muscles:

____

____

____

3

Locate the following muscles involved in **forced inhalation:**

The **sternocleidomastoid** is a superficial neck muscle. From the mastoid process on the temporal bone, it travels inferiorly and medially through the neck and attaches to the clavicle and manubrium.

The **scalenes** are located in the neck. They extend from cervical vertebrae to the first and second ribs.

The **serratus anterior** is attached to the first eight or nine ribs. From there, the muscle passes posteriorly along the rib cage and connects to the medial border of the scapula.

The **pectoralis minor** lies deep to the pectoralis major. From its attachment to ribs 3 through 5, it passes superiorly and laterally to the coracoid process of the scapula.

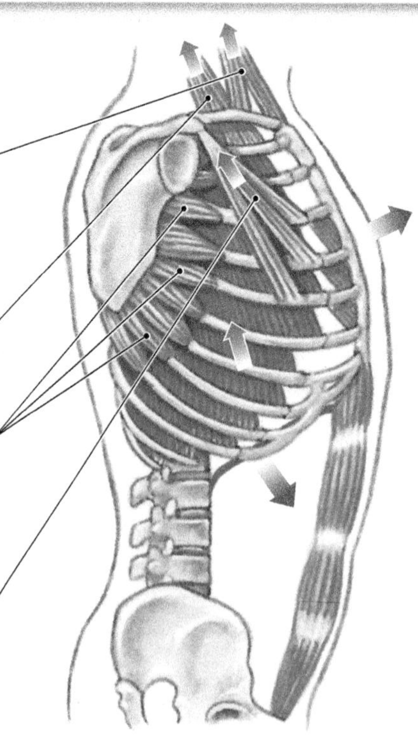

4

What common function do these muscles perform during forced inhalation?

▶ ____

____

____

____

5

Gently palpate the sternocleidomastoid muscle on one side of your neck. Feel the muscle contract as you forcefully inhale. Feel the muscle relax as you exhale.

B Unlike normal inhalation, which occurs when inspiratory muscles are actively contracted, **normal exhalation** is a passive process, resulting from the **elastic recoil** of the lungs. During inhalation, elastic fibers in the walls of bronchial tubes stretch as the lungs expand. When the inspiratory muscles relax, the elastic fibers recoil, and the diaphragm, ribs, and sternum return to their original positions. As a result, thoracic volume decreases, and air moves out of the lungs, following the pressure gradient.

1. Blow up a balloon and compress the opening with your fingers to prevent air from leaking out. The air-filled balloon represents your expanded lungs at the end of an inhalation.
2. Slowly release your fingers from the opening in the balloon. As the balloon recoils to its original position, air will be released, just as the lungs release air when the respiratory muscles relax.

C Expiratory muscles act to decrease thoracic volume during **forced exhalation.**

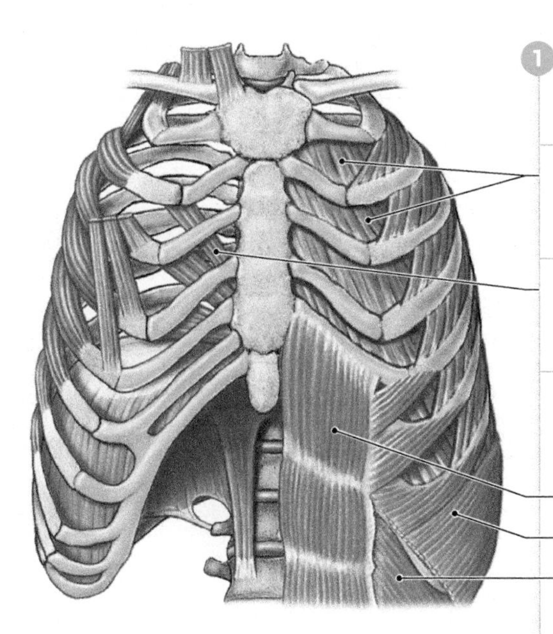

1. On a torso model, locate the following muscles involved in forced exhalation:

The **internal intercostal muscles** are the second layer of muscles in the intercostal spaces. They are deep to the external intercostal muscles.

The **transversus thoracis muscles** are the anterior group of the innermost intercostal muscles.

The abdominal muscles form a strong muscular wall that protects the abdominal viscera. They include the following:

- **Rectus abdominis**
- **External oblique**
- **Internal oblique**
- **Transversus abdominis** (not shown)

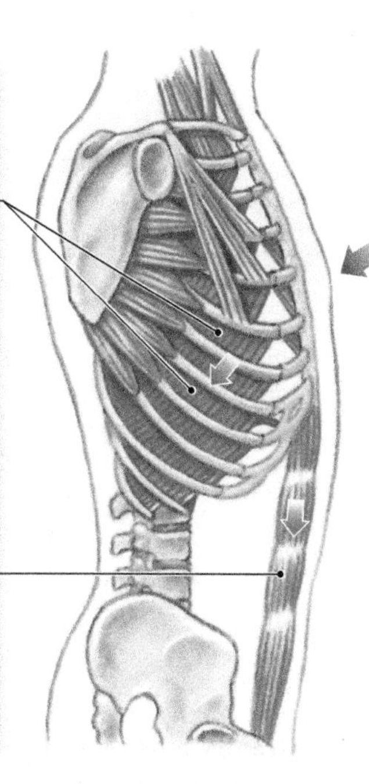

2. What common function do the internal intercostal and transversus thoracis muscles perform during forced exhalation?

▶ ______________________________

____

____

____

3. Palpate your rectus abdominis muscle. Perform a forced inhalation followed by a forced exhalation. Feel the muscle contract during forced exhalation.

MAKING CONNECTIONS

During forced exhalation, contraction of the abdominal muscles increases intraabdominal pressure. Explain how this action helps force air out of the lungs.

▶ ______________________________

____

____

____

____

Want more practice? Go to: **MasteringA&P** > Study Area > Menu > Lab Tools > PAL > Human Cadaver >

- Muscular System > Trunk
- Respiratory System > Thorax photos

Activity 26.3

Measuring Respiratory Volumes

Spirometry is a diagnostic technique used to measure respiratory volumes. The instrument used to measure these volumes is called a **spirometer.** Today, spirometry is usually accomplished using sophisticated computerized airflow transducers. A more traditional device, known as a wet spirometer, consists of a mouthpiece connected with tubing to an air-filled bell inverted in a container of water. When a subject exhales into the mouthpiece, the bell rises in the water. When the subject inhales, the air in the bell returns to the subject's lungs, causing the bell to sink. A pen attached to the bell records the bell's movements as a series of waves, called a **spirogram.** This wavelike recording is a graphic representation of an individual's respiratory volumes.

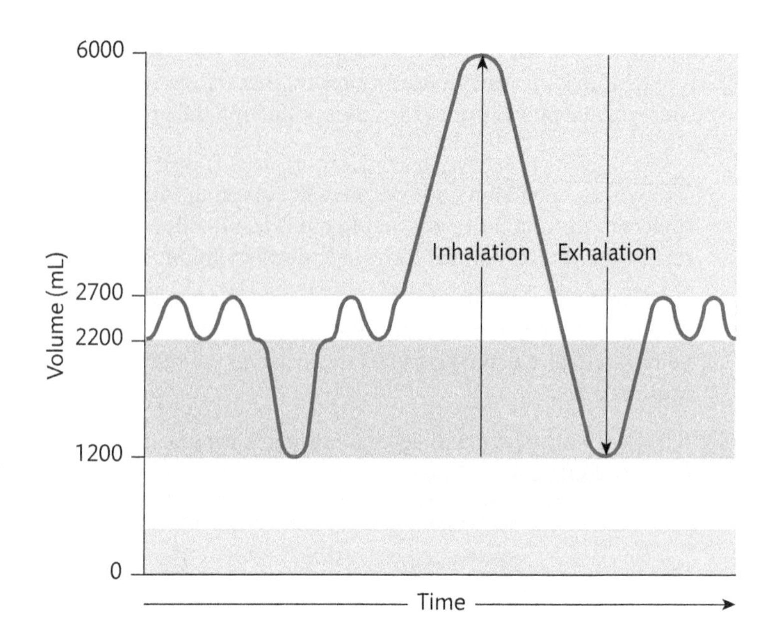

A You will use a **handheld spirometer** to measure or calculate the primary respiratory volumes and capacities. Consult your instructor for specific instructions if other types of spirometers are available in your laboratory.

1 Each person should attach a disposable mouthpiece before using the spirometer.

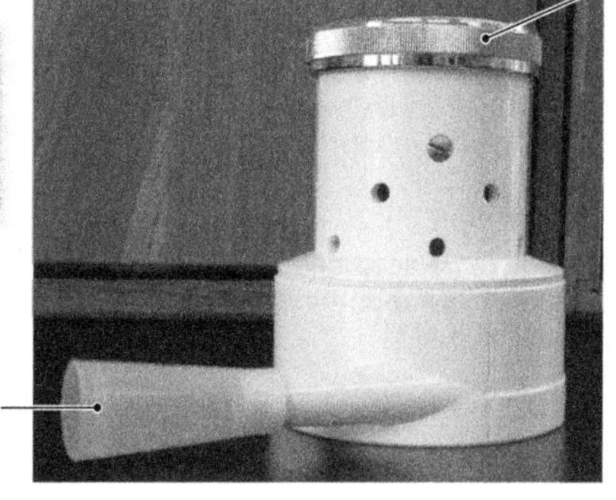

2 Before each volume measurement, set the spirometer dial to the zero mark by turning the calibration knob at the top of the cylinder.

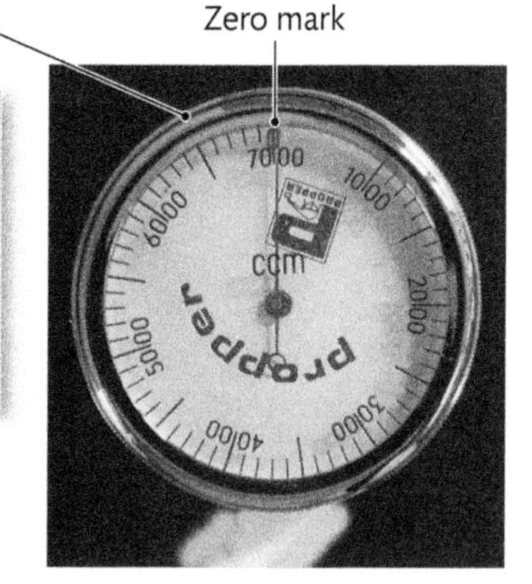

B **Expiratory reserve volume (ERV)** is the maximum volume of air that can be forcefully exhaled after a normal exhalation.

1 Do not begin until you are breathing normally and relaxed. Then, after a normal exhalation, blow into the mouthpiece of the spirometer and forcefully exhale the maximum volume of air that you can release from your lungs. This volume is the expiratory reserve volume.

2 Repeat this procedure two more times and calculate the average value. Record your results in Table 26.1. ▶

TABLE 26.1 Spirometric Measurements of Respiratory Volumes

	Measurements (mL)			
Volume Test	**Trial 1**	**Trial 2**	**Trial 3**	**Average**
Expiratory reserve volume (ERV)				
Expiratory capacity (EC)				
Tidal volume (TV)	X	X	X	

C **Expiratory capacity (EC)** is the maximum volume of air that can be forcefully exhaled after a normal inhalation.

1. Do not begin until you are breathing normally and relaxed. Then, after a normal inhalation, blow into the mouthpiece of the spirometer and forcefully exhale the maximum volume of air that you can release from your lungs. This volume is the expiratory capacity.

2. Repeat this procedure two more times and calculate the average value. Record your results in Table 26.1. ▶

D **Tidal volume (TV)** is the amount of air that moves in (or out of) the lungs during normal inhalation (or exhalation).

1. If you know your expiratory reserve volume and expiratory capacity, how can you calculate your tidal volume?

▶ ______________________________

____

____

____

____

2. Calculate your tidal volume, using your average ERV and your average EC. Record the result in the "Average" column in Table 26.1. ▶

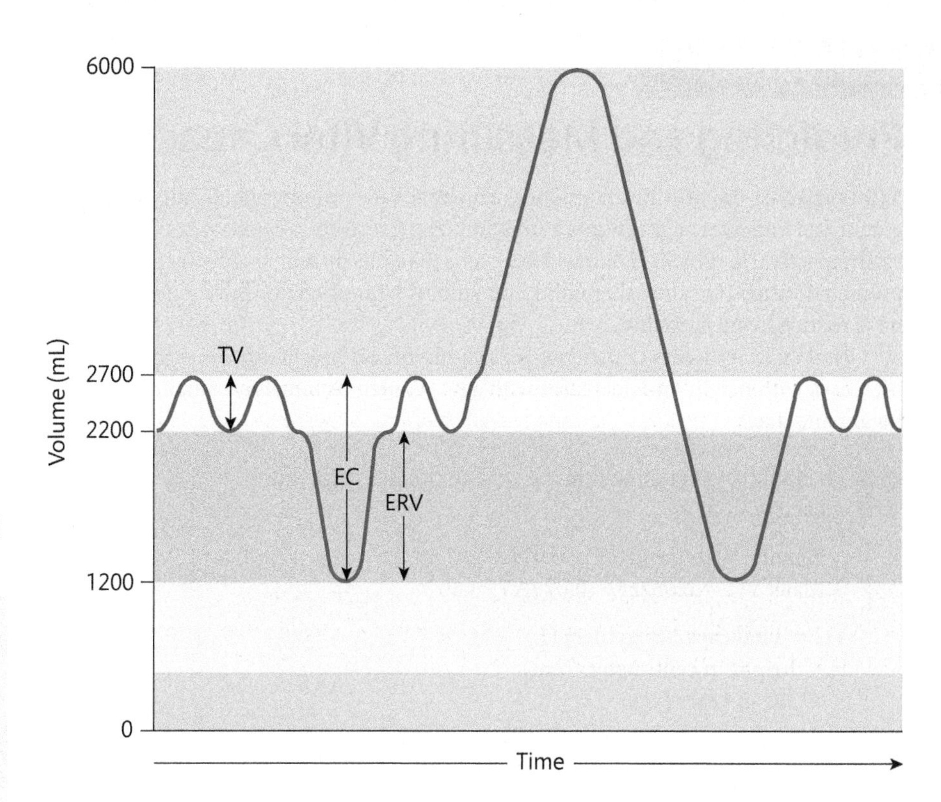

Word Origins

Spirometer comes from the Latin word *spiro* ("to breathe") and the Greek word *metron* ("measure"). The health of a person's respiratory system can be assessed by using a spirometer to measure the volume of air exchanged during inhalation and exhalation.

MAKING CONNECTIONS

A woman has arthritis that affects the articulations of the ribs with the thoracic vertebrae and sternum. How will this condition affect the woman's tidal volume?

▶ ______________________________

____

____

____

____

____

____

____

Want more practice? Go to: **MasteringA&P** > Study Area > Menu > Lab Tools > **PhysioEx** >Exercise 7: Respiratory System Mechanics >

- Activity 1: Measuring Respiratory Volumes and Calculating Capacities
- Activity 2: Comparative Spirometry

Activity **26.4**

Predicting and Measuring Vital Capacity

Vital capacity, the maximum amount of air that a person can exhale after a forced inhalation, is a reliable diagnostic indicator of pulmonary function. A person's vital capacity should be at least 80 percent of the predicted vital capacity. Healthy individuals who exercise regularly will usually have vital capacities that are greater than their predicted values. On the other hand, individuals who smoke or have a pulmonary disease, such as emphysema, have reduced vital capacities.

Vital capacity varies depending on an individual's height, age, and gender. As a general rule, vital capacity increases with height and decreases with age. Women usually have smaller vital capacities than men of comparable height and age.

A The following equations can be used to predict your vital capacity (VC):

- Female: $VC = 0.041H - 0.018A - 2.69$
- Male: $VC = 0.052H - 0.022A - 3.60$

VC = vital capacity in liters (L)
H = height in centimeters (cm)
A = age in years (yr)

1 Apply the appropriate equation, above, to predict your vital capacity. Be sure to convert your height in inches to centimeters (1 inch = 2.54 centimeters). Use the following space to complete your calculations. Convert your answer from liters to milliliters and record this value in the "Predicted Vital Capacity" column in Table 26.2. Record your lab partner's predicted vital capacity in the table as well.

▶

2 **Make a Prediction**

Before you measure your actual vital capacity, make a prediction by placing a check mark after one of the following outcomes: ▶

My actual vital capacity will be less than 80 percent of my predicted vital capacity. ____________

My actual vital capacity will be between 80 percent and 100 percent of my predicted vital capacity. ____________

My actual vital capacity will be greater than 100 percent of my predicted vital capacity. ____________

TABLE 26.2 Comparison of Predicted and Actual Vital Capacities

Subject	Predicted Vital Capacity (mL)	Actual Vital Capacity (mL)	Actual Vital Capacity as a % of Predicted Vital Capacity
Yourself			
Your lab partner			

B Your actual vital capacity (VC) can be measured with a handheld spirometer.

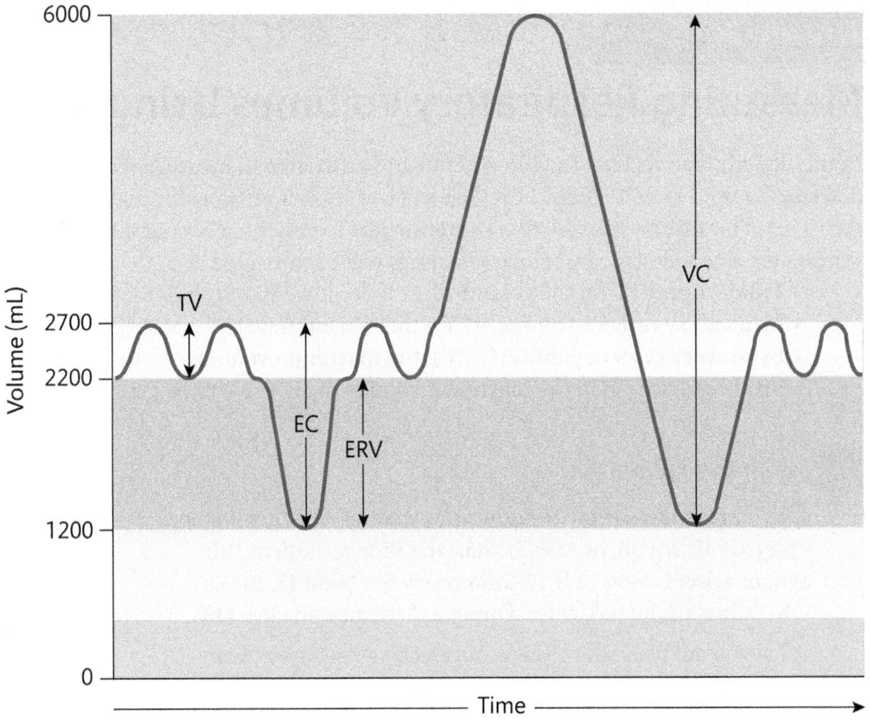

1. Sit in a comfortable position and breathe normally. Then, after a normal exhalation, forcefully inhale so that the maximum volume of air enters the lungs.

2. Immediately after the forced inhalation, blow into the mouthpiece of the spirometer and forcefully exhale the maximum volume of air that you can release from your lungs. Repeat this procedure two more times and record the three values below. ▶

Trial 1: ________ mL

Trial 2: ________ mL

Trial 3: ________ mL

3. Calculate the average value and record the result in the "Actual Vital Capacity" column in Table 26.2. Record your lab partner's actual vital capacity in the table, as well. ▶

4. Calculate your actual vital capacity as a percentage of your predicted vital capacity.

$$\% \text{ predicted VC} = (\text{actual VC}/\text{predicted VC}) \times 100$$

Use the space below to complete your calculations and record the result in the far-right column in Table 26.2. Record your lab partner's result in the table, as well.

▶

5. **Assess the Outcome**

Did your results agree or disagree with your prediction? If your actual values differ from your predicted values, what might be some of the reasons for the variation?

▶ ____

____

____

____

MAKING CONNECTIONS

Consult your textbook or a reliable online source to learn how emphysema and pneumonia affect the lungs. Explain why a person suffering from one of these diseases will most likely have a reduced vital capacity.

▶ ____

____

____

Activity **26.5**

Measuring Respiratory Volumes Using the Biopac Student Lab System

In this activity, you will use the Biopac Student Lab System to measure selected pulmonary volumes. Some of these volumes were described in Activities 26.3 and 26.4. You will use an airflow transducer attached to a calibration syringe/filter assembly (see the figure on the right). The airflow transducer converts airflow to volume. This conversion approximates the volume measurements obtained from a spirometer. The following pulmonary volumes will be measured:

- **Tidal volume (TV):** the volume of air that moves in or out of the lungs during normal inhalation or exhalation
- **Inspiratory reserve volume (IRV):** the maximum volume of air that can be forcefully inhaled after a normal inhalation
- **Expiratory reserve volume (ERV):** the maximum volume of air that can be forcefully exhaled after a normal exhalation
- **Vital capacity (VC):** the maximum volume of air that can be exhaled after a forced inhalation.

A Setup and Calibration

1. Connect the airflow transducer to channel 1 of the Biopac unit (MP35, MP36, or MP45). Start the Biopac Student Lab System, select Lesson 12 (**L12-Pulmonary Function 1),** and click **OK.** When prompted, enter a unique filename and click **OK.**

2. If you are using the SS11A transducer, a bacteriological filter must be placed between the transducer and syringe. The SS11B transducer does not require a filter.

3. **Calibration Stage 1:** Hold the airflow transducer in a vertical or upright position. Make sure it is still and no air is flowing through it. Click **Calibrate.** The calibration will stop automatically after 4 to 8 seconds. The calibration data should show a flat horizontal line traveling across the center of the screen. If the line is not horizontal and flat, click **Redo Calibration.** If the calibration was successful, click **Continue.**

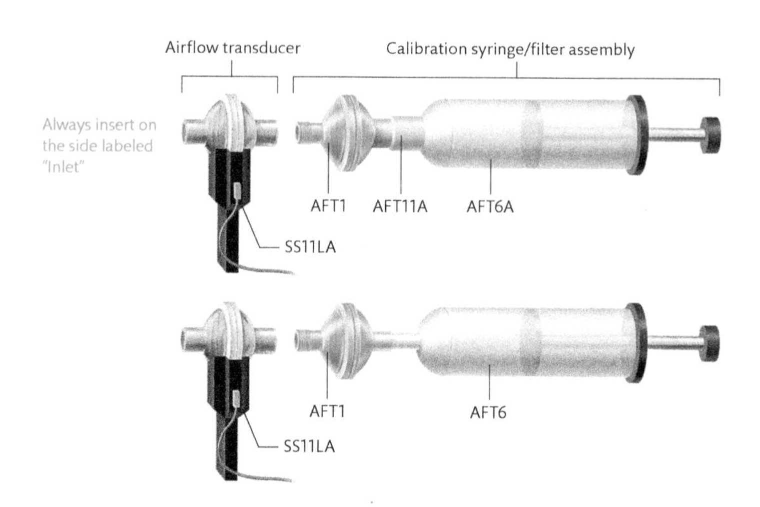

4. Hold the calibration syringe horizontally and attach the airflow transducer as shown in the figure above. Make sure the transducer is attached on the side labeled "**Inlet.**" Pull the calibration syringe plunger all the way out. Hold the syringe in a horizontal position. The airflow transducer should be firmly attached to the syringe, but should not be held. Read the on-screen alert box carefully before beginning calibration stage 2.

5. **Calibration Stage 2:** When you are ready to begin, click **Calibrate** and then click **OK.** Read the on-screen prompts. You will be simulating normal breathing by completing a series of 5 "inhalation–exhalation" cycles. For each cycle, you will push the plunger in all the way, wait 2 seconds, and pull the plunger out all the way. You will wait 2 seconds before beginning the next cycle. When you have completed 5 cycles, click **End Calibration.** If the data look similar to the figure on the left, click **Continue.** If you need to repeat the calibration, click **Redo Calibration.**

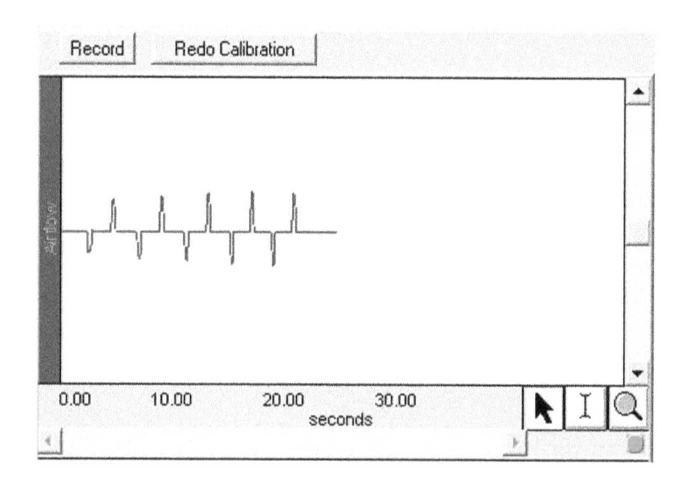

B Recording Data

1. Either you or your lab partner (test subject) will be breathing into the airflow transducer to measure respiratory volumes. Before you begin, insert a filter into the "inlet" side of the transducer and attach a mouthpiece to the same side. Make sure all connections are tight to prevent air leaks. The transducer should be held in a vertical and steady position.

2. The test subject should be seated, relaxed, and looking away from the computer monitor. Place a nose clip on the test subject's nose. Before recording data, the test subject should breathe normally for about 20 seconds. Make sure he or she is breathing through the side labeled "Inlet" and his or her mouth is closed tightly around the mouthpiece. When ready, click **Record** and begin the following breathing sequence:
   - Breathe normally for five cycles (each cycle includes one inhalation and one exhalation).
   - Forcefully inhale as deeply as possible and then forcefully exhale as deeply as possible.
   - Breathe normally for five additional cycles.

   When the test subject finishes the last normal breathing cycle, click **Stop.** Then click **Done.**

C Analyzing the Data

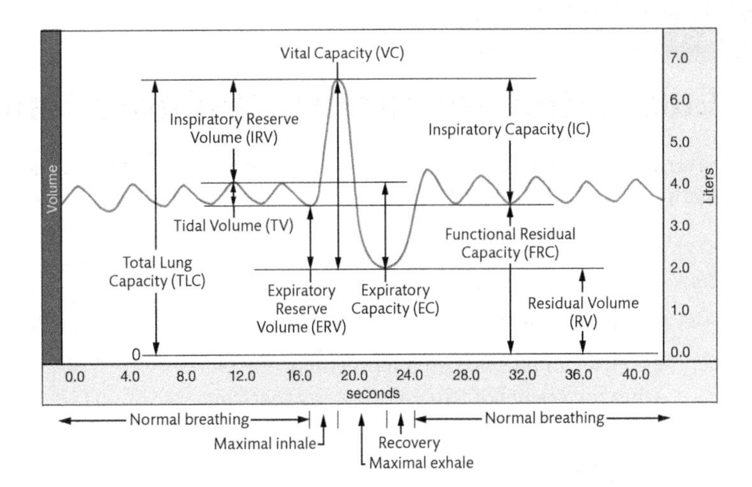

1 Enter the **Review Saved Data** mode and select the appropriate data file for Biopac Lesson 12 (L12). Inspect the recording. The airflow data on channel 1 are hidden. All measurements will be made using the volume measurements on channel 2. At the top of the screen, you will see four volume measurement box settings, all on channel 2:

- **P-P** (peak to peak) is the maximum value minus the minimum value in a selected area.
- **Max** is the maximum value in a selected area.
- **Min** is the minimum value in a selected area.
- **Delta** is the difference in amplitude between the last and first points in a selected area.

The top figure on the right illustrates the pulmonary volumes you will measure by using the I-beam tool to select areas on the computer-generated spirogram.

2 To measure tidal volume (TV), select the area from the valley to the peak of the third breathing cycle during the first period of normal breathing. The P-P measurement is the tidal volume. Record this value in Table 26.3. ▶

3 To measure the inspiratory reserve volume (IRV), select the area from the peak of the normal breathing cycle just before the forced inhalation to the peak of the forced inhalation (see the middle figure on the right). The Delta value is the IRV. Record this value in Table 26.3. ▶

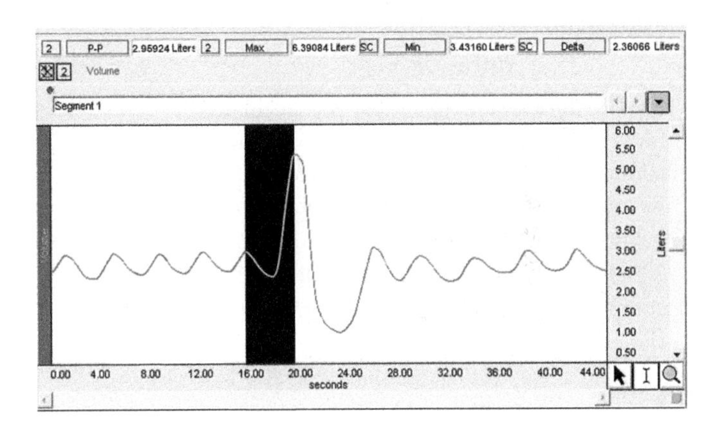

4 To measure the expiratory reserve volume (ERV), select the area from the valley of the normal breathing cycle just before the forced exhalation to the valley of the forced exhalation. The Delta value is the ERV. Record this value in Table 26.3. ▶

5 To measure the vital capacity (VC), select the area from the peak of the forced inhalation to the valley of the forced exhalation (see the bottom figure on the right). The P-P value is the VC. Record this value in Table 26.3. ▶

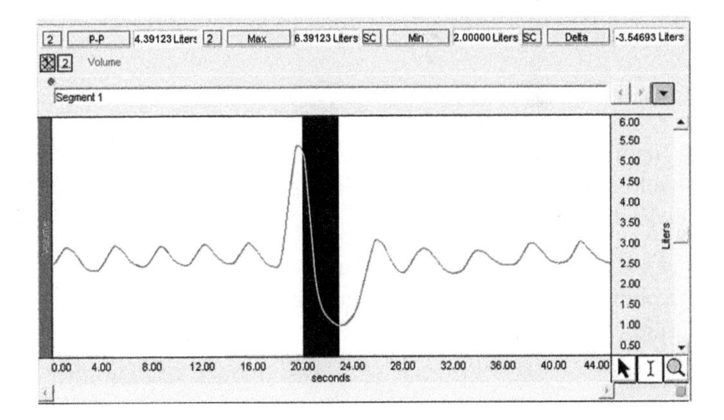

TABLE 26.3 Respiratory Volumes Determined by Biopac Analysis

Subject	TV	IRV	ERV	VC

MAKING CONNECTIONS

Compare your measurements of TV, ERV, and VC with measurements of the same volumes made with the spirometer in Activities 26.3 and 26.4.

▶ ______________________________

____

____

Activity **26.6**

Calculating Respiratory Volumes and Capacities

Several respiratory volumes are not easy to measure with a handheld spirometer, but they can be calculated mathematically.

A **Inspiratory reserve volume (IRV)** is the volume of air that can be forcefully inhaled after a normal inhalation.

1. IRV can be calculated mathematically by using the results of two volume tests that you completed in Activities 26.3 and 26.4. Refer to the spirogram on the right and verify that the inspiratory reserve volume is equal to the vital capacity minus the expiratory capacity. The following equation summarizes this calculation:

$$IRV = VC - EC$$

2. Use the average values that you calculated for vital capacity and expiratory capacity to compute your inspiratory reserve volume. Record your result in the "Measurement" column of Table 26.4. ▶

3. How does your calculation of IRV compare with your Biopac measurement of IRV in Activity 26.5?

▶ ______________________________

____

____

____

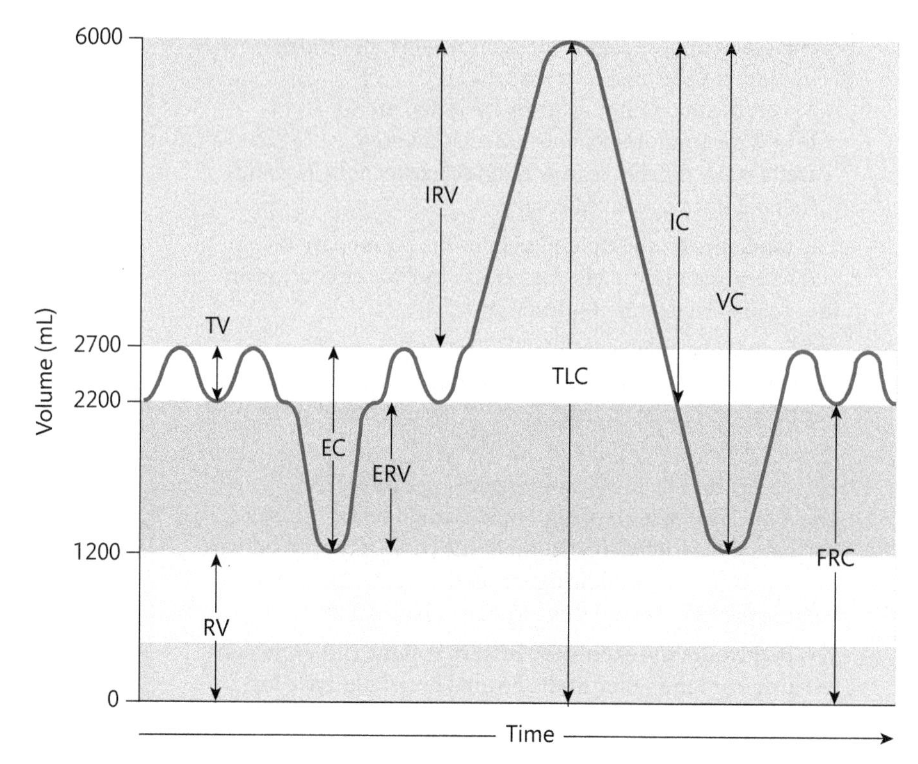

B **Inspiratory capacity (IC)** is the maximum volume of air that can be inhaled after a normal exhalation.

1. Refer to the spirogram. What respiratory volume, calculated in Activity 26.3, can you add to IRV to determine IC?

▶ ______________________________

2. Write the equation for calculating IC in the "Calculation" column of Table 26.4. ▶

3. Calculate IC and record the result in the "Measurement" column of Table 26.4. ▶

TABLE 26.4 Mathematical Calculations of Respiratory Volumes

Volume Test	Calculation	Measurement (mL)
Inspiratory reserve volume (IRV)	IRV = VC − EC	
Inspiratory capacity (IC)		
Functional residual capacity (FRC)		
Total lung capacity (TLC)		

C **Functional residual capacity (FRC)** is the volume of air remaining in the lungs after a normal expiration.

1 To calculate FRC, you must know the value of your **residual volume (RV),** which is the amount of air that remains in the lungs after the most forceful expiration. There is no practical way to measure this volume, but on average, the residual volume is 1200 mL for men and 1100 mL for women. What respiratory volume must be added to RV to calculate FRC?

▶ ______________________________

2 Write the equation for calculating FRC in Table 26.4. ▶

3 Calculate FRC and record the result in Table 26.4. ▶

D **Total lung capacity (TLC)** is the maximum amount of air that the lungs can hold after a maximum forced inspiration.

1 What pulmonary volume can be added to FRC to calculate TLC?

▶ ______________________________

2 Write the equation for calculating TLC in Table 26.4. ▶

3 Calculate TLC and record the result in Table 26.4. ▶

4 The average values of the pulmonary volumes that you calculated in this activity are given in Table 26.5. How do your pulmonary volumes compare with the average values?

▶ ______________________________

____

____

IN THE CLINIC

Respiratory Disorders

Primary respiratory volumes and capacities depend on many factors, including the age, height, and gender of the individual. For this reason, measurements that are within 80 percent of the predicted values are considered normal. Significant reductions in these capacities may indicate a serious pulmonary condition, however. For example, a doctor may diagnose **pulmonary fibrosis** if a patient has a significant decrease in inspiratory and expiratory reserve volumes. Fibrosis is the abnormal buildup of fibrous connective tissue along the airways. It can be caused by chronic inhalation of irritants such as coal dust, silicon, or asbestos.

Not all respiratory diseases are associated with reductions in respiratory volumes. In some cases, the volume of air flowing through the respiratory system is normal, but the air is not flowing quickly enough to meet the body's needs. The restricted airways that are present in a **chronic obstructive pulmonary disease (COPD),** which includes **chronic bronchitis, emphysema,** and **asthma,** reduce the rate at which air can flow in and out of the lungs per minute. Although the vital capacity of a person with COPD may be normal, or near normal, the excessive mucus accumulation and narrow airways slow the time it takes for complete exhalation of that vital capacity. Therefore, a diagnosis of COPD requires measuring the rate of airflow (ventilation) in addition to measuring the respiratory volumes.

TABLE 26.5 Average Respiratory Volumes

Respiratory Volume	Male	Female
Inspiratory capacity (IC)	3800 mL	2400 mL
Functional residual capacity (FRC)	2200 mL	1800 mL
Total lung capacity (TLC)	6000 mL	4200 mL

MAKING CONNECTIONS

A person with chronic bronchitis will have inflamed and swollen airways that produce too much mucus. One symptom of this disease is having skin with a bluish color. Explain why.

▶ ______________________________

____

____

____

____

Activity 26.7

Calculating the Respiratory Minute Volume and Alveolar Ventilation

A normal respiratory rhythm during quiet breathing is called **eupnea.** The average resting respiratory rate is 12 to 14 breaths per minute (bpm). This rate is controlled by respiratory centers in the medulla oblongata. These centers primarily respond to changes in the pH of the blood and cerebrospinal fluid. During a period of **hypoventilation,** a person is breathing too shallowly or too slowly, and the body produces carbon dioxide faster than it can be eliminated through ventilation. Carbon dioxide levels in the blood will rise, causing blood pH to fall. The respiratory control centers respond to the elevated carbon dioxide and low pH by increasing ventilation.

During a period of **hyperventilation,** a person is overventilating and eliminating carbon dioxide faster than it is being produced. Breathing more deeply than necessary is one way a person can hyperventilate. When deep breathing is accompanied by an elevated rate, even more carbon dioxide is eliminated from the body. As carbon dioxide levels decline, the pH of body fluids increases. In response, the respiratory center depresses the respiratory rate. After significant voluntary hyperventilation, there may be a temporary cessation of breathing called **apnea vera.** Over time, the reduced respiration rate will cause carbon dioxide levels to rise, returning the fluid pH, and resulting respiratory rate, to normal.

A **Respiratory minute volume (V_E)** is the volume of air that moves in (or out) of the lungs per minute during normal inhalation (or exhalation).

1. Sit quietly and breathe normally. When you are relaxed, start reading or drawing to distract your attention from your breathing. Have your lab partner count the number of normal respiratory cycles you complete in 1 minute (one cycle = one normal inhalation and one normal exhalation). This value is your **respiratory rate** or **breaths per minute** (f). Record the result here: ▶

   f =________ breaths/minute

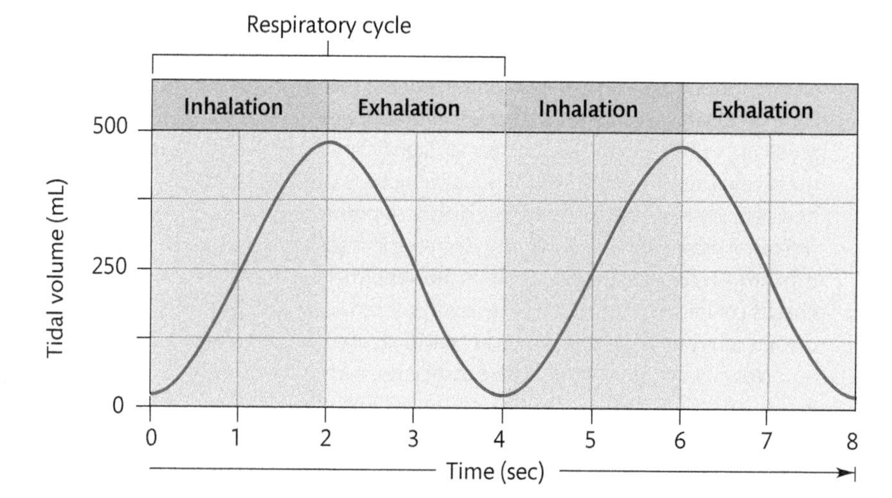

2. In the space provided, calculate your respiratory minute volume by multiplying your respiratory rate by your average tidal volume (calculated in Activity 26.3; see Table 26.1):

$$V_E = f \times TV$$

▶

V_E =________ mL/minute

IN THE CLINIC

Measuring Tidal Volume and Respiratory Rate

Tidal volume can be measured with the spirometer, but the results may not be accurate. It is difficult to breathe normally when you are consciously aware of your breathing. To confirm this fact, try breathing normally while counting your breaths per minute. It is not easy to do that. Therefore, during a physical examination, your respiratory rate is usually taken at the same time your pulse is being taken, when you are not aware of it.

B The volume of air that enters the alveolar airways each minute is called the **alveolar ventilation (V_A)**. Alveolar ventilation is the fraction of the respiratory minute volume involved in gas exchange. Some of the air we inhale remains in the conducting airways and never enters the alveoli. This volume of air is called the **anatomic dead space (V_D)**. On average, the anatomic dead space is 150 mL.

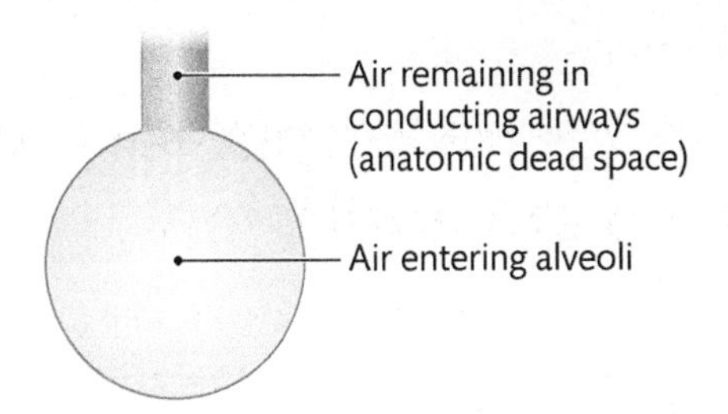

1. To calculate alveolar ventilation, substract the anatomic dead space from the tidal volume and multiply that value by the respiratory rate.

$$V_A = (TV - V_D) \times f$$

In the space provided, calculate your alveolar ventilation. Use your previous results for tidal volume (TV) and respiratory rate (f) and assume that the anatomic dead space is 150 mL.

▶

2. On average, the alveolar ventilation is 70 percent of the respiratory minute volume. Use the space provided to calculate your alveolar ventilation as a percentage of your respiratory minute volume.

$$V_A/V_E \times 100 = \% \text{ of } V_E$$

▶

V_A = ________ % of V_E

How does your V_A compare with the average?

▶ ____

____

3. On average, the anatomic dead space is 30 percent of the tidal volume. Determine the percentage of anatomic dead space in your tidal volume by calculating the V_D /TV ratio.

V_D /TV × 100 = ________ % TV

How does your V_D /TV ratio compare with the average?

▶ ____

____

MAKING CONNECTIONS

If your respiratory rate remains constant but your tidal volume declines, how will your alveolar ventilation change? Will the change improve or worsen pulmonary function? Explain.

▶ ____

____

____

____

____

____

____

____

BEFORE YOU MOVE ON ...

« LOOKING BACK

Respiration involves a series of activities that are the result of an intimate functional relationship between the respiratory and cardiovascular systems. In this laboratory exercise, one of these activities—pulmonary ventilation—was closely examined.

Pulmonary ventilation (breathing) is the movement of air into and out of the lungs along a pressure gradient. Air flows from a region of high pressure to a region of low pressure. During inhalation, thoracic volume increases, causing intrapulmonary pressure to fall below atmospheric pressure. As a result, air flows into the lungs along the pressure gradient. During exhalation, thoracic volume decreases and the intrapulmonary pressure rises above atmospheric pressure, forcing air to move out of the lungs.

In addition to pulmonary ventilation, there are other respiratory activities:

- **External respiration** is the exchange of oxygen and carbon dioxide between the blood in the pulmonary capillaries and the alveoli.
- **Gas transport** is the passage of oxygen and carbon dioxide in the blood between the lungs and all tissues.
- **Internal respiration** is the exchange of oxygen and carbon dioxide between the blood in the systemic capillaries and the body cells.

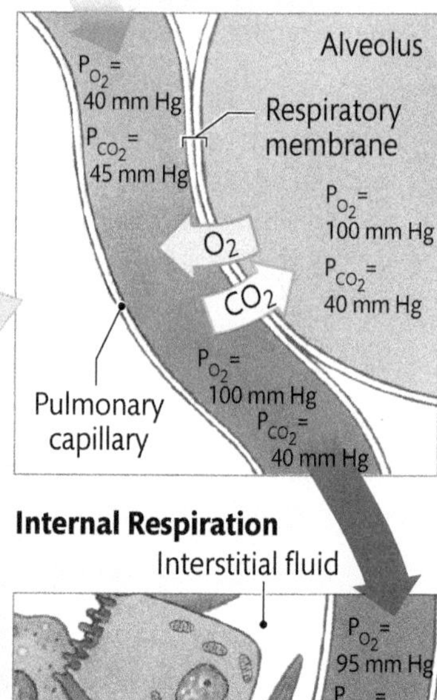

External and internal respiration operate according to the principles of **Dalton's law,** which states that the total pressure of a gas mixture is equal to the partial pressures of all the individual gases in the mixture. During external and internal respiration, oxygen and carbon dioxide diffuse along the partial pressure gradients of each gas.

The partial pressure of atmospheric oxygen will decrease as you go from sea level to higher latitudes. Consider a mountain climber ascending to the summit of a mountain:

- **What do you think will happen to the partial pressure of alveolar oxygen?**
- **How will the change in the partial pressure of oxygen affect each of the respiratory activities described above?**

LOOKING FORWARD »

Our cells rely on an ample supply of oxygen and nutrients to produce ATP, the high-energy fuel molecule that drives all essential metabolic activities. Cells can break down nutrient molecules such as carbohydrates, lipids, and proteins to obtain energy, which can be converted to ATP. Most ATP is produced by the mitochondria in the presence of oxygen. The respiratory system is responsible for taking up oxygen, and the digestive system is responsible for absorbing nutrients. Both oxygen and nutrients are transported by the cardiovascular system to all cells in the body. In the last laboratory exercise, you investigated the way in which the respiratory system delivers oxygen to the body. In the next exercise, you will study the anatomy of the digestive system and understand how nutrients are absorbed and delivered to all cells and tissues.

Name ____________________

Lab Section ____________________

Date ____________________

EXERCISE 26 REVIEW SHEET

Respiratory Physiology

1. Pleurisy is a bacterial, viral, or fungal infection that causes swelling and inflammation of the pleural membranes. One symptom of pleurisy is sharp chest pain while breathing. What do you think causes this pain?

2. Explain how the actions of the inspiratory muscles create a pressure gradient that allows air to enter the lungs.

QUESTIONS 3–9: Identify the labeled respiratory muscles and for each indicate whether it is an inspiratory or expiratory muscle.

Muscle	Inspiratory/Expiratory
3.	
4.	
5.	
6.	
7.	
8.	
9.	

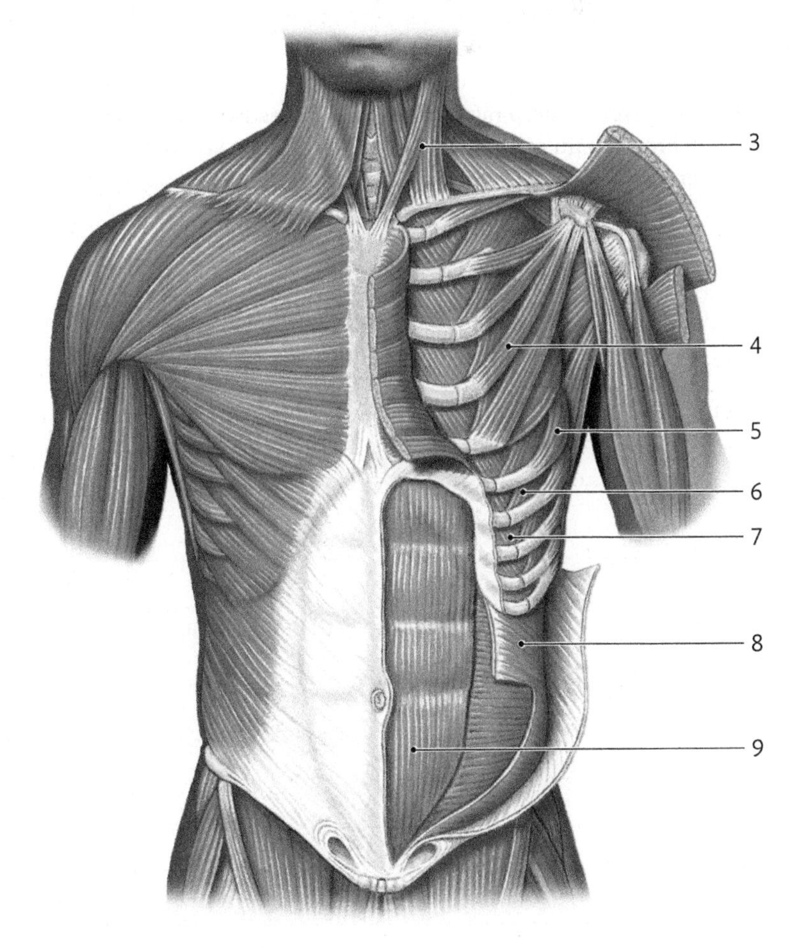

10. During a football game, a running back took a hard hit to the chest and had to leave the game. X-ray and magnetic resonance imaging (MRI) revealed two fractured ribs and adjacent soft tissue trauma. The football player told the examining physician that during normal breathing, he experienced pain while inhaling but not while exhaling. Provide an explanation for this difference.

____

____

____

____

QUESTIONS 11–19: Match the term in column A with the appropriate description in column B.

A	B
11. Residual volume ________	**a.** Amount of air that is normally inspired or expired
12. Expiratory reserve volume ________	**b.** Maximum amount of air that can be inspired after a normal exhalation
13. Vital capacity ________	**c.** Maximum volume of air that can be expired after a normal exhalation
14. Inspiratory capacity ________	**d.** Volume of air that remains in the lungs after the most forceful exhalation
15. Total lung capacity ________	**e.** Volume of air that can be forcefully inspired after a normal inhalation
16. Inspiratory reserve volume ________	**f.** Volume of air remaining in the lungs after a normal exhalation
17. Expiratory capacity ________	**g.** Maximum volume of air that a person can exhale or inhale
18. Tidal volume ________	**h.** Maximum volume of air that can be forcefully exhaled after a normal inhalation
19. Functional residual capacity ________	**i.** Maximum volume of air that the lungs can hold after a maximum forced inhalation

20. Patient A has a respiratory rate (*f*) of 15 breaths/min and a tidal volume (TV) of 600 mL. Patient B has a respiratory rate (*f*) of 18 breaths/min and a tidal volume (TV) of 500 mL. For each patient, calculate the respiratory minute volume (V_E) and alveolar ventilation (V_A). Assume that the anatomic dead space (V_D) is 150 mL.

Which patient has better alveolar ventilation? Explain.

____

____

____

____

EXERCISE

27

Anatomy of the Digestive System

LEARNING OUTCOMES

These Learning Outcomes correspond by number to the laboratory activities in this exercise. When you complete the activities, you should be able to:

Activity 27.1	**Describe the organization of the peritoneum.**
Activities 27.2–27.4	**Identify the organs of the alimentary canal and explain their roles in the digestive process.**
Activities 27.5–27.7	**Identify the accessory digestive organs and explain their functions.**
Activities 27.8–27.10	**Describe the microscopic structure of organs in the alimentary canal.**
Activities 27.11–27.12	**Describe the microscopic structure of accessory digestive organs.**

LABORATORY SUPPLIES

- Torso model
- Skeleton
- Head and neck model, midsagittal section
- Skull
- Tooth model
- Sagittal section of male pelvis
- Sagittal section of female pelvis
- Clear plastic bag
- Tape
- Prepared microscope slides:
  - submandibular gland
  - esophagus
  - stomach
  - three regions—duodenum, jejunum, and ileum—of small intestine
  - liver
  - liver, injected
  - pancreas, islet cells
  - large intestine

PRE-LAB QUIZ

Before you begin, read all the activities in Exercise 27 and the required reading in your textbook that is assigned by your instructor.

1. During this laboratory exercise, you will study the gross anatomy of the alimentary canal, which includes all the following structures *except* the
   a. stomach.
   b. small intestine.
   c. esophagus.
   d. pancreas.
   e. large intestine.
2. The serous membrane associated with the abdominopelvic cavity is called the ______________.
3. True or False: A retroperitoneal organ is suspended to the body wall by a mesentery. __________
4. Chemical digestion of food is completed and nutrients are absorbed in what organ of the alimentary canal? ______________
5. At the ileocecal junction, the ileum of the small intestine joins to the ______________ of the large intestine.
6. The common bile duct and the ______________ duct merge and empty into the duodenum at the same location.
7. True or False: The bare area is the region on the surface of the liver that is not covered by a visceral peritoneum. __________
8. During this laboratory exercise, you will study the microscopic structure of all the following digestive system organs *except* the
   a. stomach.
   b. large intestine.
   c. pharynx.
   d. salivary gland.
   e. liver.
9. You are viewing a slide of a digestive system organ and identify gastric pits and gastric glands in the mucosa. The organ you are viewing is the ______________.
10. You are viewing a slide of a digestive system organ and you identify central veins and portal areas. The organ you are viewing is the ______________.

Activity **27.1**

Examining the Organization of the Peritoneum

A The **peritoneum (peritoneal membrane)** is the serous membrane associated with the abdominopelvic cavity and organs. The **parietal peritoneum** covers the walls of the abdominopelvic cavity and is continuous with the **visceral peritoneum** that covers the outside of most abdominal organs. The narrow potential space between the peritoneal layers is the **peritoneal cavity.**

1. On a torso model, observe the digestive organs in their anatomical position within the **abdominopelvic cavity.** Notice that the organs are in very close contact with one another.

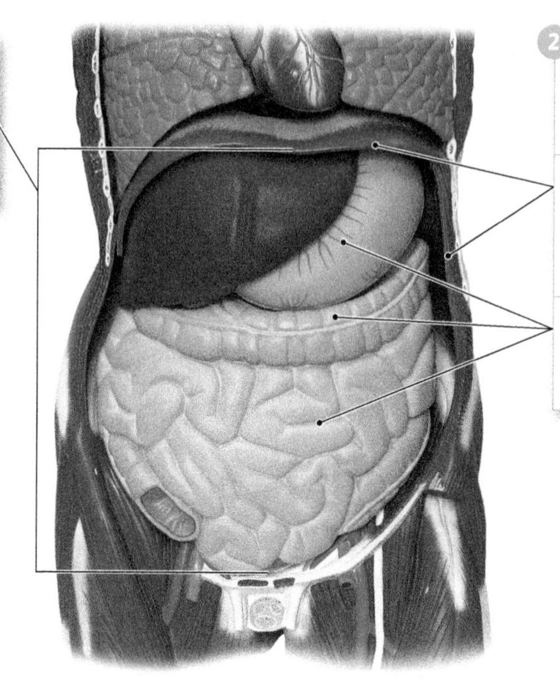

2. Identify the surfaces that are covered by the peritoneum and the orientation of its components.

The **parietal peritoneum** lines the wall of the abdominopelvic cavity, including the inferior surface of the diaphragm.

The **visceral peritoneum** covers the outside surfaces of the digestive organs.

B **Mesenteries** are double layers of parietal peritoneum that attach some digestive organs to the body wall. Mesenteries can be classified as dorsal (posterior) or ventral (anterior), depending on their connections to the body wall during fetal development.

1. On a torso model, identify the **dorsal mesenteries.** If they are not shown on your model, identify the attachment sites.

The **greater omentum** (cut in this figure) is attached to the greater curvature of the stomach. It travels inferiorly to form a membranous covering for the small intestine and other abdominal viscera. In the lower abdominal cavity, the greater omentum loops around, travels superiorly and posteriorly, and attaches to the transverse colon of the large intestine.

The **mesocolons** suspend two sections of the large intestine, the transverse colon and the sigmoid colon, from the posterior body wall.

The **mesentery proper** suspends the second and third portions of the small intestine (jejunum and ileum) to the posterior body wall.

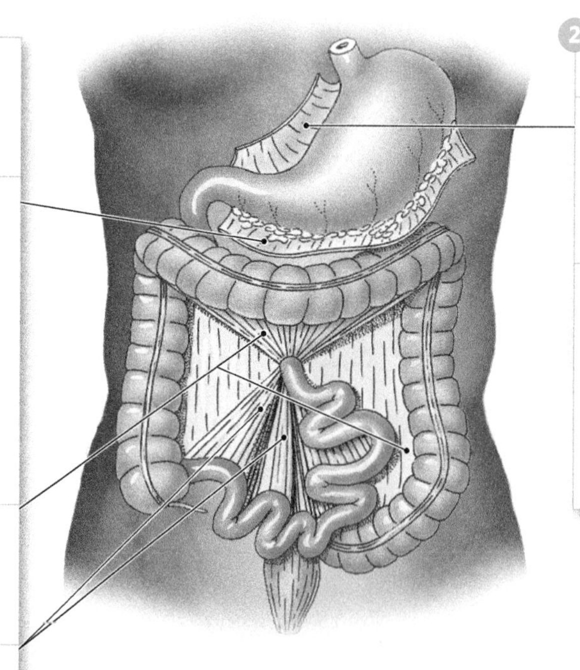

2. Identify the **ventral mesenteries.**

The **lesser omentum** (cut in this figure) is attached to the lesser curvature of the stomach. It connects the stomach to the inferior surface of the liver.

The **falciform ligament** (not shown) is attached to the liver between the organ's left and right lobes. It connects the liver to the diaphragm and anterior body wall. In the adult, the liver is the only organ that is connected to the anterior body wall by a mesentery.

C **Retroperitoneal organs** are not supported by mesenteries. Instead, they are positioned along the posterior body wall, behind the peritoneum.

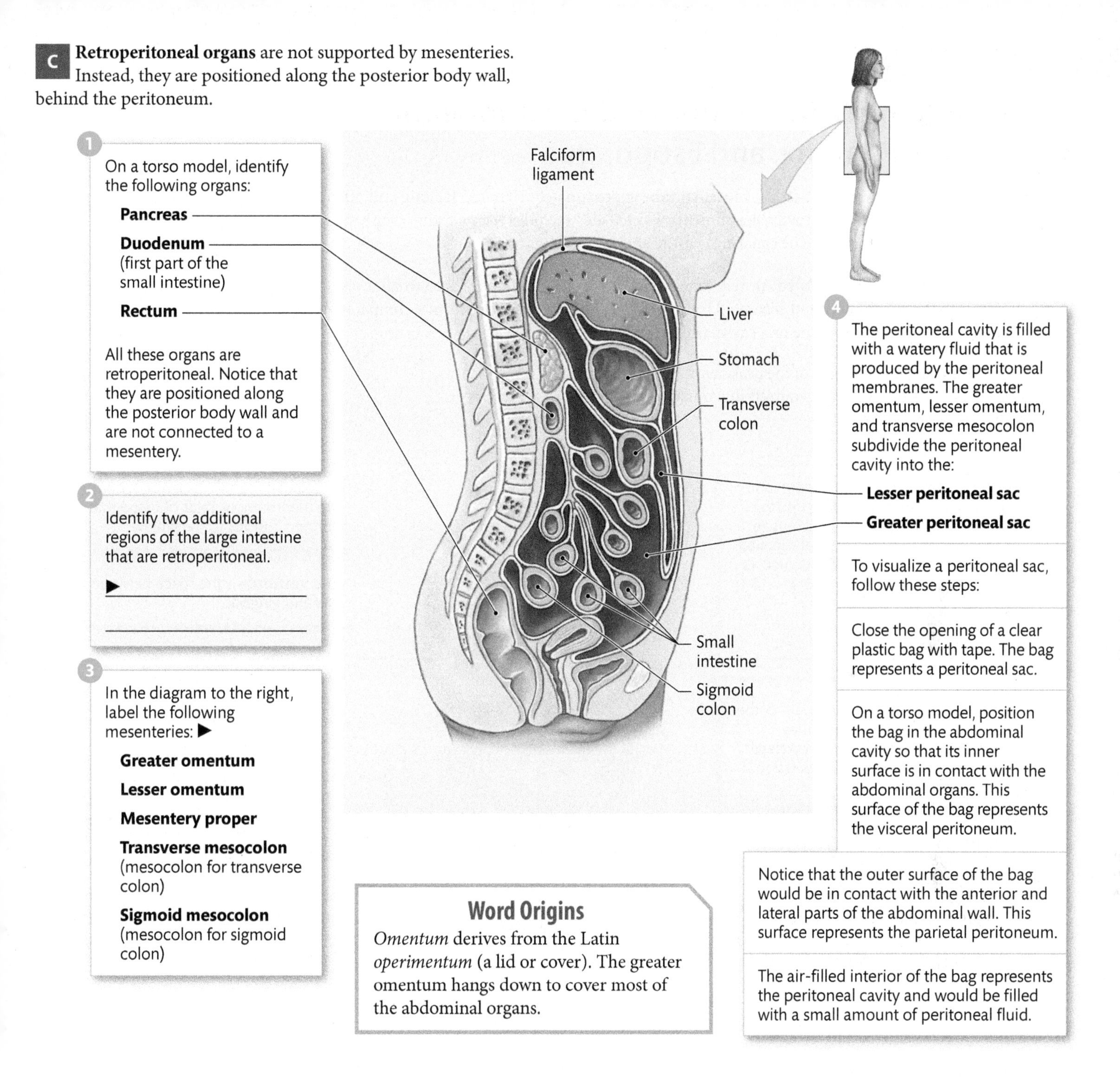

1 On a torso model, identify the following organs:

Pancreas

Duodenum (first part of the small intestine)

Rectum

All these organs are retroperitoneal. Notice that they are positioned along the posterior body wall and are not connected to a mesentery.

2 Identify two additional regions of the large intestine that are retroperitoneal.

▶ ______________________

____

3 In the diagram to the right, label the following mesenteries: ▶

Greater omentum

Lesser omentum

Mesentery proper

Transverse mesocolon (mesocolon for transverse colon)

Sigmoid mesocolon (mesocolon for sigmoid colon)

4 The peritoneal cavity is filled with a watery fluid that is produced by the peritoneal membranes. The greater omentum, lesser omentum, and transverse mesocolon subdivide the peritoneal cavity into the:

Lesser peritoneal sac

Greater peritoneal sac

To visualize a peritoneal sac, follow these steps:

Close the opening of a clear plastic bag with tape. The bag represents a peritoneal sac.

On a torso model, position the bag in the abdominal cavity so that its inner surface is in contact with the abdominal organs. This surface of the bag represents the visceral peritoneum.

Notice that the outer surface of the bag would be in contact with the anterior and lateral parts of the abdominal wall. This surface represents the parietal peritoneum.

The air-filled interior of the bag represents the peritoneal cavity and would be filled with a small amount of peritoneal fluid.

Word Origins

Omentum derives from the Latin *operimentum* (a lid or cover). The greater omentum hangs down to cover most of the abdominal organs.

MAKING CONNECTIONS

The fluid produced by the peritoneum lubricates the membrane surfaces. Why do you think this lubrication is necessary for the normal function of the abdominal organs?

▶ ______________________

____

____

Activity **27.2**

Examining the Gross Anatomy of the Alimentary Canal: Oral Cavity, Pharynx, and Esophagus

The **alimentary canal (digestive tract)** is a muscular tube approximately 10 m (33 ft) long and open to the outside at both ends. Beginning at the oral cavity, the alimentary canal also includes the pharynx, esophagus, stomach, small intestine, and large intestine. The canal ends at the anal opening (anus).

A The **oral cavity** (mouth) is located entirely in the head. It is inferior to the nasal cavity and separated from it by the palate. When food is brought into the oral cavity **(ingestion),** it is chopped into smaller pieces **(chewing)** and manipulated by the tongue to form a compact mass called a **bolus.**

1 Examine an anterior view of the oral cavity by observing your lab partner or looking at yourself in a mirror. Identify the following structures:

The **lips (labia)** surround the anterior opening (oral orifice) leading into the cavity. They consist of skeletal muscle covered with skin and are well supplied with sensory receptors. Why do you think the lips have a pink color?

▶ ______________________________

____

The **cheeks** form the lateral walls of the oral cavity. Deep to the skin, they contain subcutaneous fat and skeletal muscles (muscles of facial expression and mastication).

The **tongue** occupies the floor of the oral cavity.

The **gingiva (gums)** is the mucous membrane that covers the alveolar processes of the mandible and maxillae.

The **labial frenula** (singular, **frenulum**) are membranous folds of tissue that connect the lips to the gingiva.

The **vestibule** is the space between the lips and gingiva.

The **palate,** which forms the roof of the oral cavity, is covered by a mucous membrane.

The **uvula** is the soft tissue extension that hangs down from the posterior tip of the soft palate.

The **lingual frenulum** is a membranous fold that connects the tongue to the floor of the oral cavity.

2 On a model showing a midsagittal section of the head, identify the following structures. Label each one in the figure to the right: ▶

- The bony, anterior portion of the palate, the **hard palate,** is composed of portions of the maxillary and palatine bones.
- The soft tissue posterior portion of the palate, the **soft palate,** is a muscular arch that ends at the uvula.
- The **fauces** is the posterior opening of the oral cavity leading to the pharynx.
- A **palatine tonsil** is located in the posterior region of each cheek at the fauces.
- The **root of the tongue** is anchored to the hyoid bone.
- The **lingual tonsils** are located at the root of the tongue.

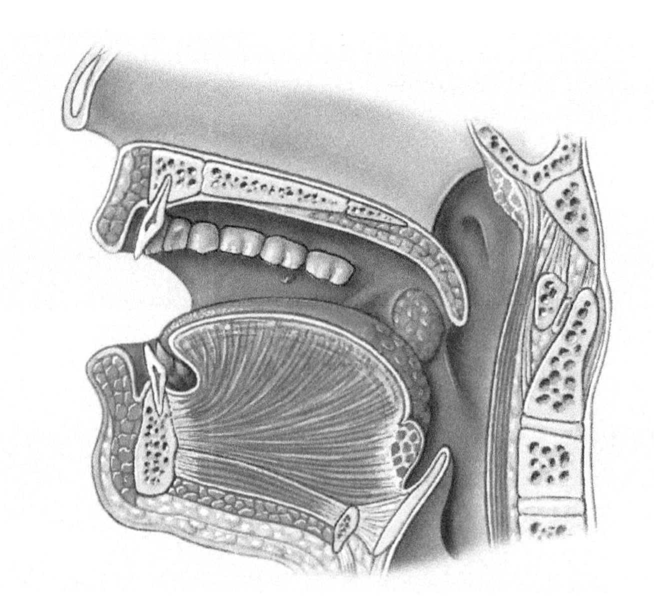

B The **pharynx** is a muscular tube located posterior to the nasal and oral cavities. It begins in the head and ends in the neck.

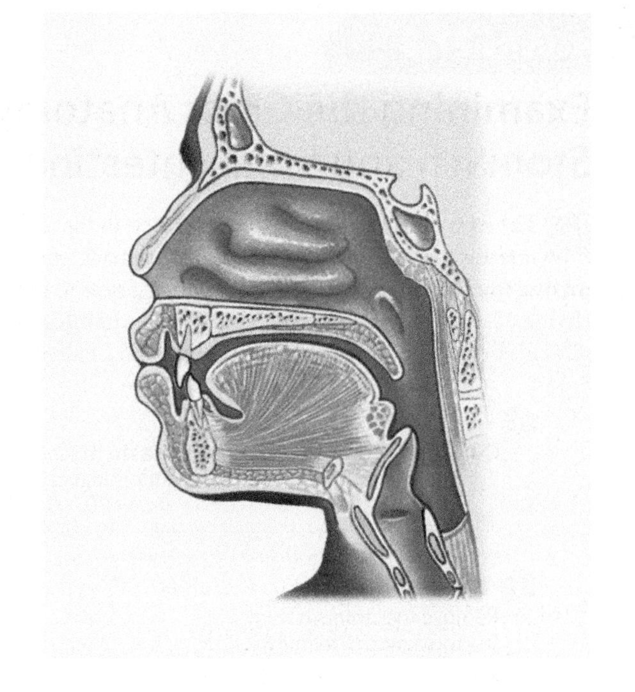

1. On a midsagittal section of the head and neck, identify the three regions of the pharynx:
   - The **nasopharynx** extends from the posterior margin of the nasal cavity to the tip of the **uvula.** Label the nasopharynx and uvula in the figure. ▶
   - The **oropharynx** extends from the tip of the uvula to the superior margin of the **epiglottis.** Label the oropharynx and epiglottis in the figure. ▶
   - The **laryngopharynx** extends from the superior margin of the epiglottis to the bifurcation that gives rise to the **larynx** anteriorly and the **esophagus** posteriorly. Label the laryngopharynx, larynx, and esophagus in the figure. ▶

2. In what region of the pharynx will you find the **pharyngeal tonsil (adenoid)** and the opening to the **auditory (Eustachian) tube?**

▶ ______________________________

Label these structures in the figure. ▶

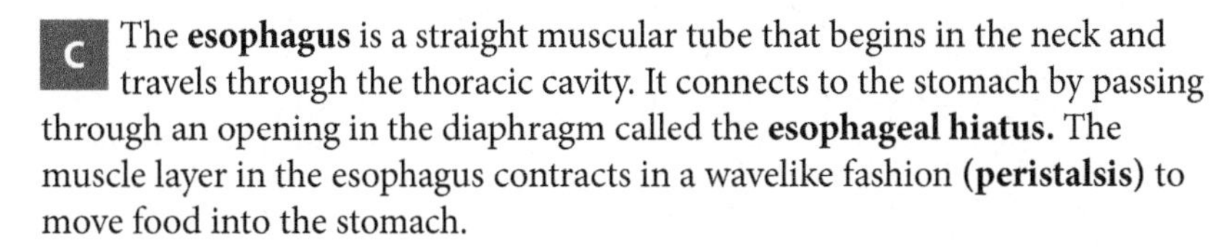

C The **esophagus** is a straight muscular tube that begins in the neck and travels through the thoracic cavity. It connects to the stomach by passing through an opening in the diaphragm called the **esophageal hiatus.** The muscle layer in the esophagus contracts in a wavelike fashion **(peristalsis)** to move food into the stomach.

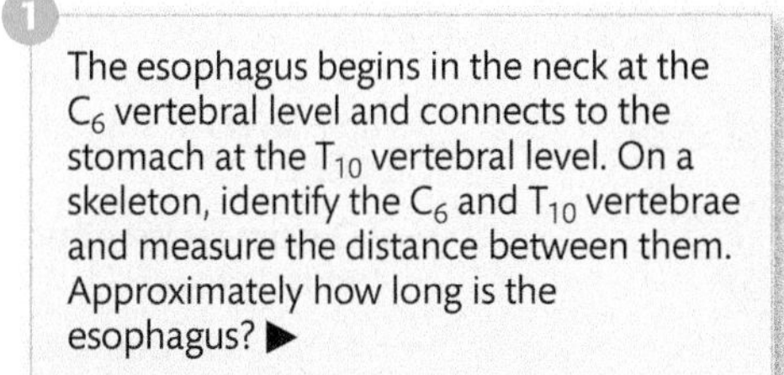

1. The esophagus begins in the neck at the C_6 vertebral level and connects to the stomach at the T_{10} vertebral level. On a skeleton, identify the C_6 and T_{10} vertebrae and measure the distance between them. Approximately how long is the esophagus? ▶

   Length of esophagus: __________ cm

2. On a torso model, identify the **esophagus** and trace its path from the neck into the thoracic cavity. As it travels through the thorax: ▶

   Which structures are anterior to the esophagus?

   ____

   ____

   Which structures are posterior to the esophagus?

   ____

   ____

IN THE CLINIC

Gastroesophageal Reflux

The smooth muscle at the junction between the esophagus and stomach is called the **cardiac** or **lower esophageal sphincter.** While not a true sphincter, it functions like one. When food passes down the esophagus, the sphincter relaxes to allow food into the stomach. If the esophagus is empty, the sphincter remains closed to prevent the backflow of gastric contents. Sometimes, after a heavy meal, regurgitation of acidic gastric material can occur that causes a burning sensation in the chest called **heartburn.** Chronic heartburn is called **gastroesophageal reflux disease,** or **GERD.** Over time, the inflammation and cellular damage to the esophageal wall caused by GERD can lead to esophageal cancer.

MAKING CONNECTIONS

Review the positions of the nasopharynx, oropharynx, and laryngopharynx. Which regions serve a dual function for respiration and digestion? Provide an anatomical explanation for your answer.

▶ ______________________________

____

____

____

■ Want more practice? Go to: **MasteringA&P** > Study Area > Menu > Lab Tools > PAL > Anatomical Models > Digestive System > Alimentary Canal photos

Activity **27.3**

Examining the Gross Anatomy of the Alimentary Canal: Stomach and Small Intestine

A The **stomach** is a J-shaped pouch located in the abdominopelvic cavity just inferior to the diaphragm. Inside the stomach, muscle contractions, called **mixing movements,** blend food with gastric juices to form a pastelike material called **chyme.** Gastric digestive enzymes initiate the **chemical digestion** (breakdown) of proteins before the chyme is transported to the small intestine by peristalsis.

1 On a torso model, identify the **stomach.** It is located almost entirely in which quadrant of the abdominal cavity?

▶ ______________________

2 Remove the stomach from the torso model. Identify the following components of the stomach and label them in the diagram on the right: ▶

The **fundic region (fundus)** is the superior dome-shaped area.

The **cardiac region (cardia)** is a small area near the junction with the esophagus and medial to the fundus.

The **body** is the main portion of the stomach. It is located inferior to the cardiac and fundic regions.

The **pyloric region (pylorus)** is continuous with the body. It narrows to become the pyloric canal, which is continuous with the duodenum.

3 The **pyloric sphincter** is a ring of smooth muscle located at the junction between the pyloric region and the duodenum. What do you think is the function of this sphincter?

▶ ______________________

4 Identify the two curvatures of the stomach and label them in the lower diagram: ▶

The **lesser curvature** is the concave curve along the medial margin of the stomach.

The **greater curvature** is the convex curve along the lateral and inferior margins of the stomach.

5 The stomach wall contains three layers of smooth muscle:

- Outermost **longitudinal muscle layer**
- Middle **circular muscle layer**
- Innermost **oblique muscle layer**

What is the function of these muscle layers?

▶ ______________________

6 Open the stomach and view the numerous tissue folds, known as **rugae,** along its internal surface. As stomach volume increases after a heavy meal is eaten, what do you think happens to the rugae?

▶ ______________________

Want more practice? Go to: **MasteringA&P** > Study Area > Menu > Lab Tools > PAL >
- Anatomical Models > Digestive System > Alimentary Canal photos
- Human Cadaver > Digestive System > Alimentary Canal photos

B The **small intestine** is about 6 m (20 ft) long and 2.5 cm (1 in.) in diameter. The chemical digestion of complex nutrient molecules in food is completed in the small intestine. The end products of chemical digestion are simpler molecules, such as amino acids, monosaccharides, and fatty acids, which can be absorbed into the blood and lymphatic circulations. Undigested material is eventually transported to the large intestine.

1. On a torso model, observe the position of the small intestine. Notice that it is a highly folded tubular structure that occupies much of the abdominal cavity.

2. On your model, identify the regions of the small intestine and label them in the figure on the right: ▶

The **duodenum** is continuous with the pylorus of the stomach and forms a C-shaped tube as it travels around the head of the pancreas.

After the duodenum, the small intestine becomes a highly folded tube. The second region, the **jejunum,** occupies the superior portion of the intestinal folds.

The third region, the **ileum,** occupies the inferior portion of the intestinal folds. At the gross anatomical level, there is no distinct feature that can be used to identify the junction between the jejunum and ileum.

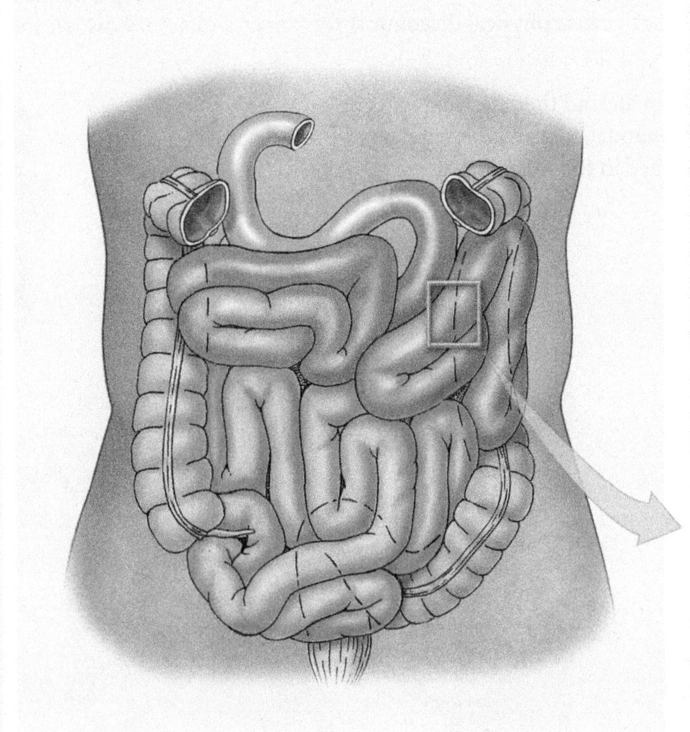

3. If possible on your model, open the small intestine and view the transverse folds along the inner wall. These structures, known as **plicae circulares,** cause a slow, spiraling movement of chyme through the small intestine, allowing more time for digestive enzymes to act and for nutrients to be absorbed. The plicae circulares are most prominent in the jejunum, where most chemical digestion and absorption occur.

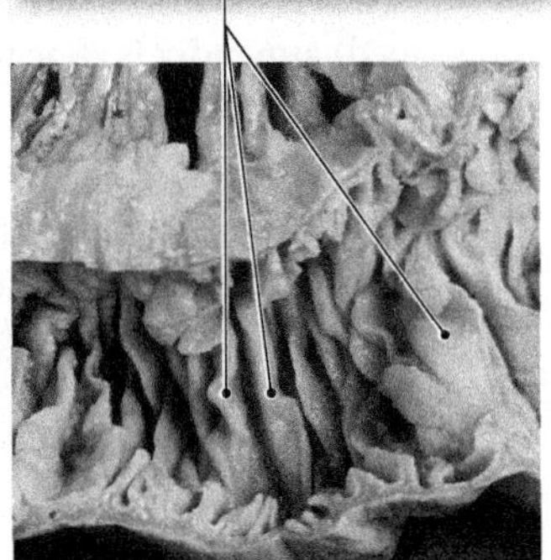

MAKING CONNECTIONS

Explain why the duodenum is attached more firmly in the abdominal cavity than the other two segments of the small intestine, the jejunum and ileum.

▶ ______________________________

____

____

____

____

____

____

Word Origins

Duodenum is derived from the Latin *duodeni,* which means "twelve." The first part of the small intestine is called the duodenum because, on average, its length is about the width of 12 fingers. *Jejunum* comes from the Latin *jejunus,* which means "empty." This portion of the small intestine is usually empty at death. *Ileum* comes from the Greek *eileo,* which means "to roll up or twist." The ileum is the longest segment of the small intestine and has more twists and turns than the duodenum and jejunum.

Activity **27.4**

Examining the Gross Anatomy of the Alimentary Canal: Large Intestine

The **large intestine** is larger in diameter (6.5 cm or 2.5 in.) but shorter in length (1.5 m or 5 ft) than the small intestine. Undigested and unabsorbed material that enters the large intestine is known as **feces.** During **compaction,** many electrolytes and most of the water (about 90 percent) are absorbed from the feces. During **defecation,** the feces is eliminated from the body as it passes through the rectum and anal canal. The large intestine provides a suitable environment for beneficial bacteria that synthesize certain vitamins and break down cellulose (fiber), a material that is not digested by our own enzymes. Bacterial metabolism produces various gases (particularly methane and sulfur-containing compounds) that can cause physical discomfort for you and olfactory discomfort for others who might be nearby!

A The large intestine forms a loop around the intestinal folds of the small intestine. On a torso model, identify the segments of the large intestine that are described in the following steps. Label each segment in the figure. ▶

1. The **cecum** is a pouchlike structure that marks the beginning of the large intestine. It is located in the lower-right quadrant of the abdominal cavity.
2. The **vermiform** (worm-shaped) **appendix** is attached to the posteromedial wall of the cecum.
3. The **ascending colon** is connected to the cecum and travels superiorly on the right side of the abdominal cavity, against the posterior body wall. At its termination, the ascending colon turns sharply to the left (right colic or hepatic flexure) to become the transverse colon.
4. Label the **right colic flexure.** ▶

   The right colic flexure is located just inferior to what organ?

   ▶ ______________________

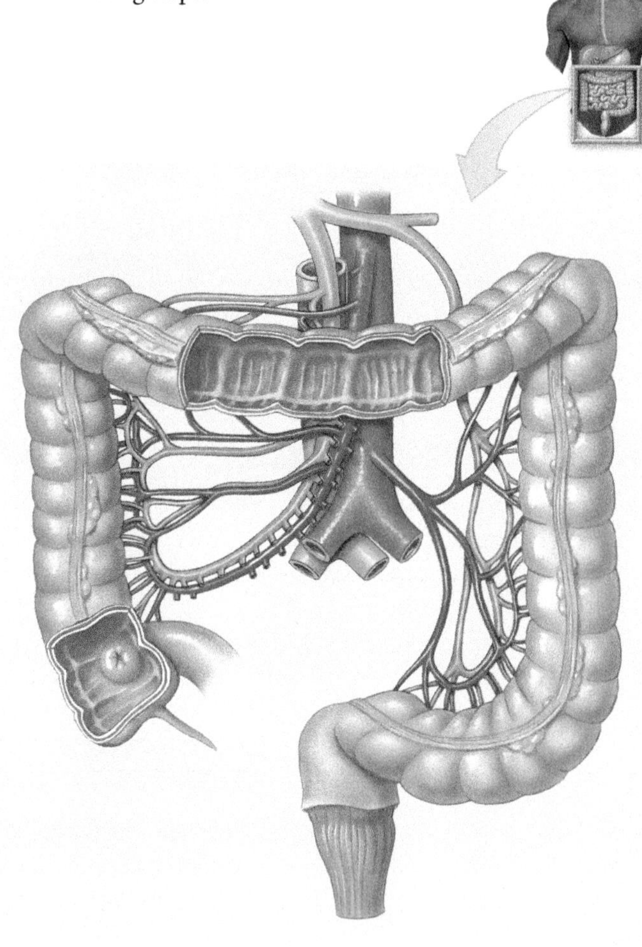

5. The **transverse colon** passes to the left across the abdominal cavity. At its termination, the transverse colon makes a sharp inferior turn (left colic or splenic flexure) to form the descending colon. Along its course, the transverse colon is positioned just inferior to what organs?

   ▶ ______________________

   ____

   ____

6. Label the **left colic flexure.** ▶

   The left colic flexure is located just anterior to what organ?

   ▶ ______________________

7. The **descending colon** travels inferiorly along the left side of the abdominal cavity. As it passes along the left iliac fossa, the descending colon curves to the right at the **sigmoid flexure** and joins the sigmoid colon. Label the sigmoid flexure in the figure. ▶
8. The **sigmoid colon** is S-shaped and passes medially along the left iliac fossa. It joins the rectum at the level of the third sacral vertebra.
9. The **rectum** travels anterior to the sacrum and follows the bone's curvature. Approximately 5 cm (2 in.) inferior to the coccyx, the rectum passes through the muscle on the pelvic floor and gives rise to the **anal canal.**

Want more practice? Go to: **MasteringA&P** > Study Area > Menu > Lab Tools > PAL >

- Anatomical Models > Digestive System > Alimentary Canal photos
- Human Cadaver > Digestive System > Alimentary Canal photos

B The wall of the large intestine has three structural specializations.

1. Identify the **taeniae coli,** three distinct longitudinal bands of smooth muscle. They are found on the wall of the entire large intestine, except the rectum and anal canal.

2. Notice the series of pouches called **haustra.** They are created by the muscle tone of the taeniae coli.

3. Identify the fat-filled tabs, called **fatty appendices,** suspended from the wall.

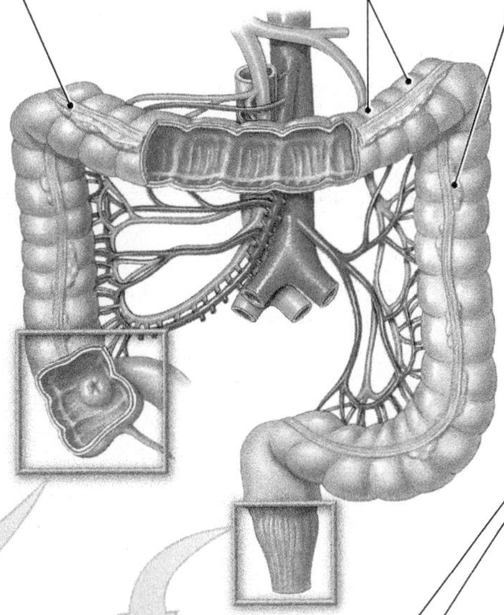

C The small and large intestines are connected at the ileocecal junction.

1. Identify the **ileocecal junction,** where the ileum of the small intestine joins the cecum of the large intestine.

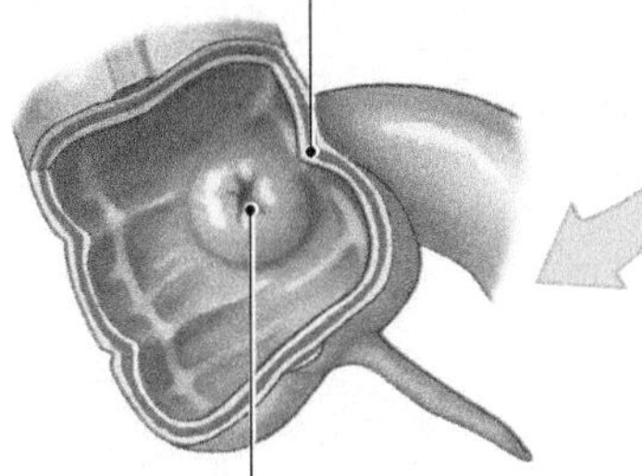

2. View the inside of the cecum and identify the **ileocecal valve,** the smooth muscle sphincter at the junction. Why is it imporant to have a sphincter at this location?

▶ ______________________________

____

____

____

____

____

D The distal end of the rectum is called the **anal canal.**

1. Along the inner wall of the anal canal, identify six to eight longitudinal tissue folds called **anal columns.** These tissue folds contain many small veins. The abrasive forces of bowel movements can damage these veins and cause discomfort and bleeding (hemorrhoids).

2. Two muscular sphinters are located at the anal opening. These sphincters relax to open the anus during defecation and contract to keep it closed at other times. Identify the sphincters on a model.

The **internal anal sphincter** is composed of involuntary smooth muscle and is controlled by the defecation reflex.

The **external anal sphincter** is composed of voluntary skeletal muscle. Thus, if you are not close to a bathroom, you can use your external sphincter to delay defecation until an appropriate time.

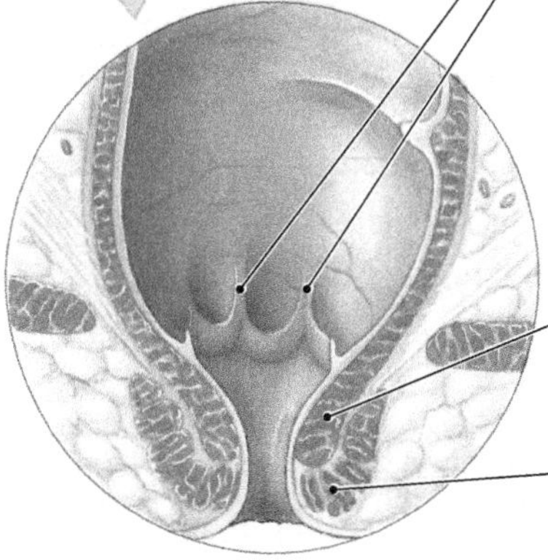

MAKING CONNECTIONS

Compare the functions of the small intestine and large intestine and suggest a reason the small intestine is so much longer than the large intestine.

▶ ______________________________

____

____

____

Activity 27.5

Examining the Gross Anatomy of Accessory Digestive Structures: Tongue and Salivary Glands

A The tongue forms a large part of the floor of the oral cavity. It consists mostly of skeletal muscle and is covered by a mucous membrane.

1 On a model of a midsagittal section of the head and neck, identify the skeletal muscles that form the bulk of the tongue. They are the **intrinsic tongue muscles.** What do you think is the function of these muscles?

▶ ______________________________

____

____

____

____

____

2 On the model and a skeleton, identify the following structures:

Hyoid bone

Soft palate

Styloid process

Mandible

These structures serve as the origins for **extrinsic tongue muscles.** From their origins, these muscles will insert into the connective tissue of the tongue. What do you think is the function of these muscles?

▶ ______________________________

____

____

____

____

____

Want more practice? Go to: **MasteringA&P** > Study Area > Menu > Lab Tools > PAL >

- Anatomical Models > Digestive System > Accessory Organ photos
- Human Cadaver > Digestive System > Accessory Organ photos

B Numerous small salivary glands are scattered throughout the mucosa of the tongue, palate, and cheeks. These glands secrete saliva continuously to keep the lining of the oral cavity moist. In addition, there are three pairs of major salivary glands: parotid glands, submandibular glands, and sublingual glands.

1. On a model of the head and neck, locate the **parotid glands.** They are the largest salivary glands and are located anterior to each ear, between the skin of the cheek and the masseter muscle. They secrete mostly serous fluid with a high concentration of the enzyme salivary amylase. What is the function of salivary amylase?

▶ ____

____

____

2. Locate the **parotid duct** that transports saliva from the parotid gland to the oral cavity. It passes through the buccinator muscle and enters the mouth just opposite the upper second molar on each side of the jaw.

3. The **sublingual glands** are the smallest of the major salivary glands and are located on the floor of the oral cavity near the tongue. They are predominantly mucus-secreting glands and transport saliva to the oral cavity via several small ducts. What important role do the salivary mucous secretions play?

▶ ____

____

____

4. The **submandibular glands** are located on the floor of the oral cavity along the medial surfaces of the mandible. The ducts to the submandibular glands enter the oral cavity inferior to the tongue, near the frenulum. They secrete a mixture of serous and mucous secretions.

5. Saliva also contains antibodies. Explain why.

▶ ____

____

____

____

____

MAKING CONNECTIONS

The disease known as mumps is caused by a virus that usually infects and inflames the parotid salivary glands. Review the location of the parotid glands and explain why a person who has mumps will probably experience pain while eating.

▶ ____

____

____

____

____

____

Activity **27.6**

Examining the Gross Anatomy of Accessory Digestive Structures: Teeth

A The teeth are embedded in the sockets formed by the alveolar processes of the mandible and maxillary bones. They are essential for grinding and chewing food before it is swallowed. The **deciduous** or **primary** teeth appear at intervals between 6 months and 2 years of age. Years later, the roots of the deciduous teeth are reabsorbed, and the teeth are pushed out by the emerging **secondary** or **adult teeth,** which first appear around age 6 but are not completely in place until 17 to 21 years of age.

1. On a skull, observe how the teeth are embedded in the sockets of the maxillary and mandibular **alveolar processes.**

2. In the adult, there are 32 teeth divided into four quadrants. On a skull, identify the eight teeth in a quadrant:

   Central incisor
   Lateral incisor
   Cuspid (canine)
   First premolar
   Second premolar
   First molar
   Second molar
   Third molar (wisdom tooth)

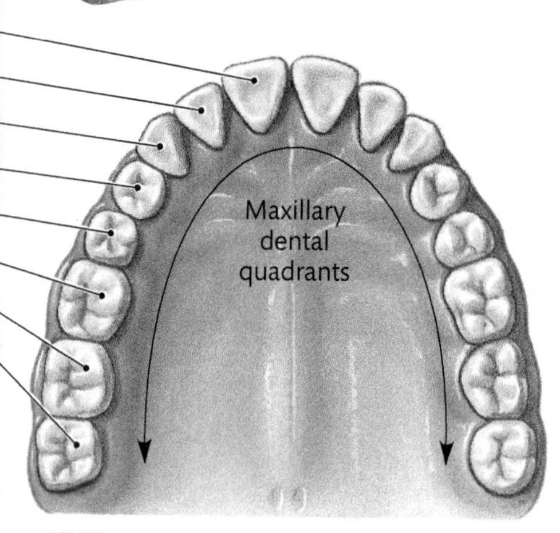

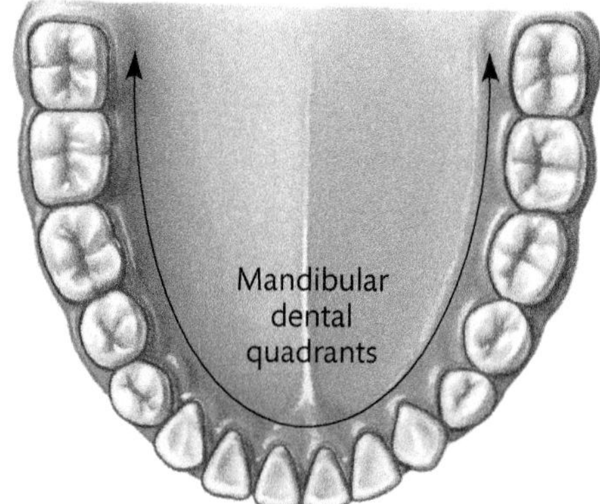

3. On a skull, identify an incisor, a cuspid, and a molar. In the table below, compare their sizes and shapes and predict what special function each type of tooth has while chewing. ▶

Type of Tooth	Size and Shape	Chewing Function
Incisor		
Cuspid		
Molar		

Want more practice? Go to: **MasteringA&P** > Study Area > Menu > Lab Tools > PAL >
- Anatomical Models > Digestive System > Accessory Organ photos
- Human Cadaver > Digestive System > Accessory Organ photos

B The alveolar processes of the mandible and maxillae are covered by a mucous membrane called the gingiva (gums).

Each region of a tooth is identified according to its relationship to the gingival margin (gum line).

1. On a model of a tooth, identify the three regions of a tooth:

   The **crown** is the visible portion of a tooth, above the gum line.

   The **neck** is the region at the gum line where the crown and root meet.

   The **root** is the portion that is below the gum line.

2. The root of a tooth is anchored to the bone by a **periodontal ligament** and is normally not seen. In individuals who have progressive gum disease, the gingival margin recedes, and the roots of some teeth become visible. What type of joint is formed by the articulation of the root of a tooth and its bony socket?

   ▶ ______________________

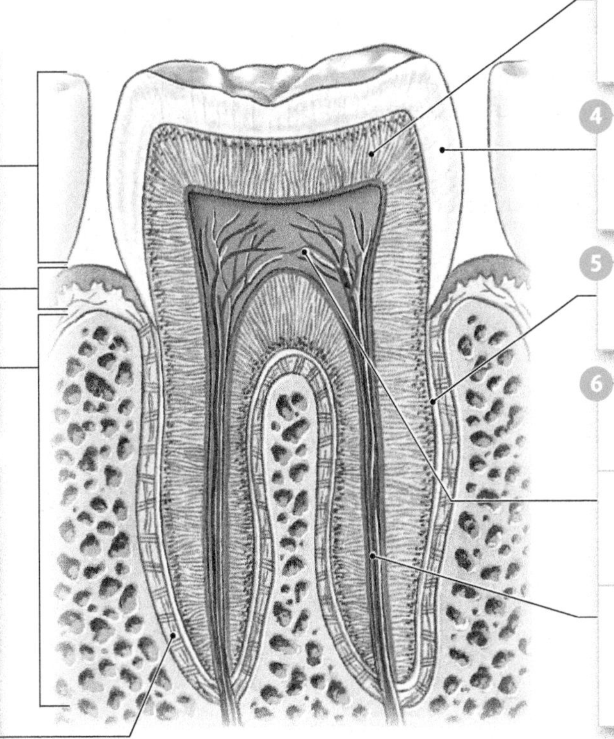

3. Identify the layer of **dentin,** an acellular, bonelike material that makes up the bulk of a tooth.

4. In the crown, identify the **enamel,** a very hard calcified layer that covers the dentin.

5. Identify the **cementum,** which forms a thin covering on the root.

6. Identify the internal cavities of a tooth:

   The **pulp cavity** contains blood vessels, nerves, and connective tissue (pulp).

   Blood vessels and nerves reach the pulp cavity by way of the **root canal,** which extends through the interior of the root.

IN THE CLINIC

Tooth Decay

Enamel is the hardest substance in the body, but it gradually wears away with age and is not replaced when damaged. The progressive decay of enamel and underlying dentin by decalcification forms **dental caries,** or tooth decay. Tooth decay is caused by the metabolic acids released by bacteria, which feed on sugars and other carbohydrates in the food that we eat. If a bacterial infection spreads into the root canal, all the pulp must be removed, and the root canal must be sterilized and filled. This dental procedure is referred to as **root canal therapy** or, simply, a **root canal.**

MAKING CONNECTIONS

Central incisor

Lateral incisor

Cuspid

First molar

Second molar

Compare the arrangement of primary teeth in a child (figure on left) and secondary teeth in an adult. What teeth are present in an adult that are missing in a child?

▶ ______________________

____

____

____

____

____

____

____

Activity **27.7**

Examining the Gross Anatomy of Accessory Digestive Structures: Liver, Gallbladder, and Pancreas

A The **liver** is the largest internal organ in the body. Most of the organ is positioned within the upper-right quadrant of the abdominopelvic cavity and is protected by the rib cage. The liver's primary digestive function is the production of **bile,** which is a derivative of cholesterol and used in the small intestine to emulsify (break apart) fat globules.

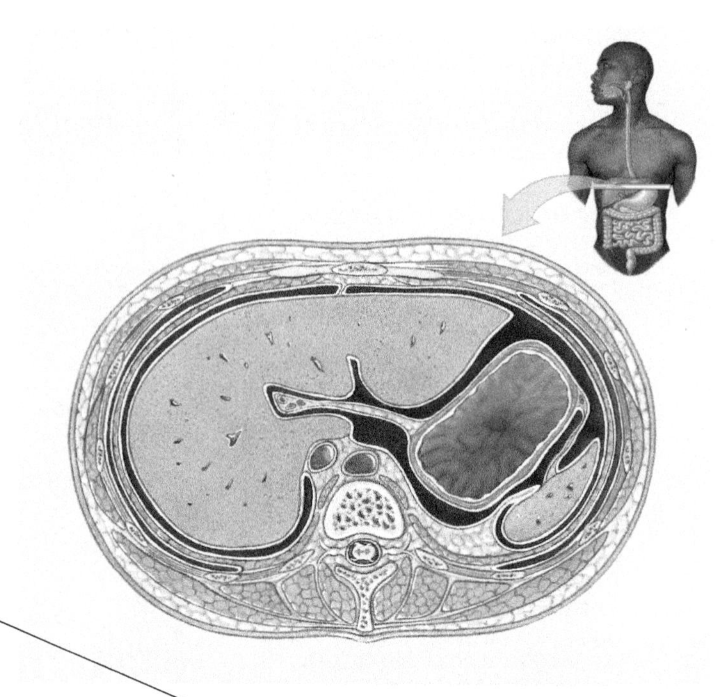

1. On a torso model, identify the liver, stomach, and spleen in the abdominopelvic cavity. Notice that the stomach and spleen are lateral to the left lobe of the liver in the upper left quadrant of the cavity. Label the liver, stomach, and spleen in the cross section on the right. ▶

2. The liver is covered by a visceral peritoneum except on the superior surface, known as the **bare area.** Remove the liver from the torso model and identify the bare area. Identify the structure that is in direct contact with the bare area.

   ▶ ______________________________

3. Observe the extent of the bare area. Around its periphery, the visceral peritoneum folds onto the diaphragm to form the **coronary ligament.**

4. From an anterior view, identify the **right** and **left lobes** and label them in the figure to the right. ▶

5. Locate the border between the right and left lobes. The **falciform ligament** separates these lobes and connects the liver to the anterior body wall.

   Within the posteroinferior free margin of the falciform ligament is the **round ligament (ligamentum teres),** which is a remnant of the umbilical vein in the fetal circulatory system.

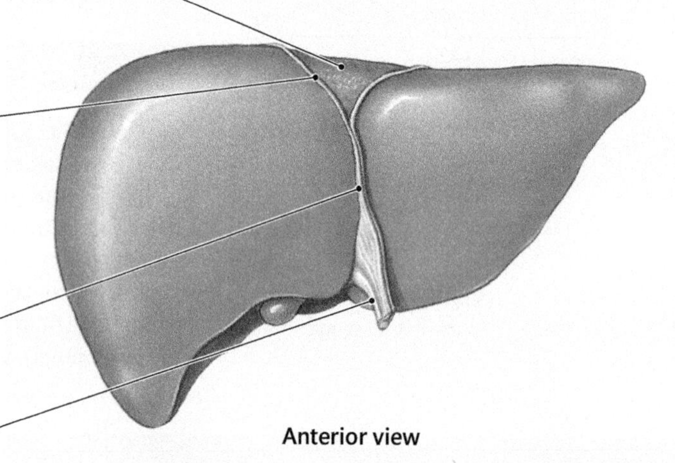

Anterior view

6. Examine the liver from its posteroinferior surface and identify the two smaller lobes:

   The posterior **caudate lobe**

   The anterior **quadrate lobe**

   Label these lobes in the figure to the right. ▶

7. Identify the **porta hepatis.** Notice all four liver lobes meet at this region. The hepatic artery, portal vein, lymphatics, nerves, and hepatic ducts enter or exit the liver at the porta hepatis.

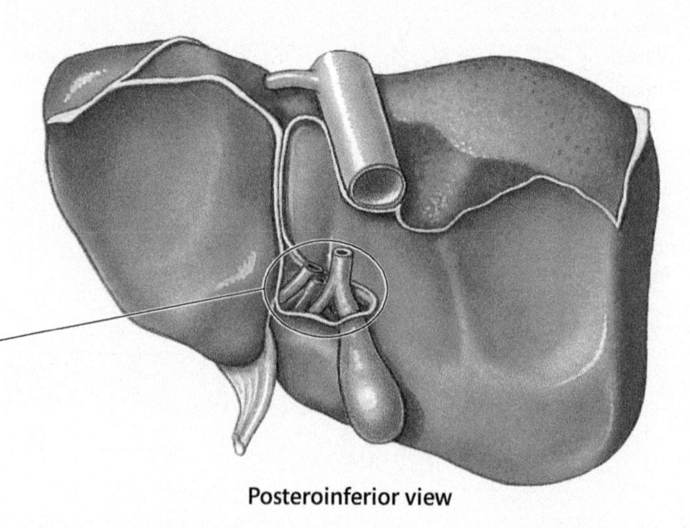

Posteroinferior view

B The **gallbladder** provides a temporary storage site for bile.

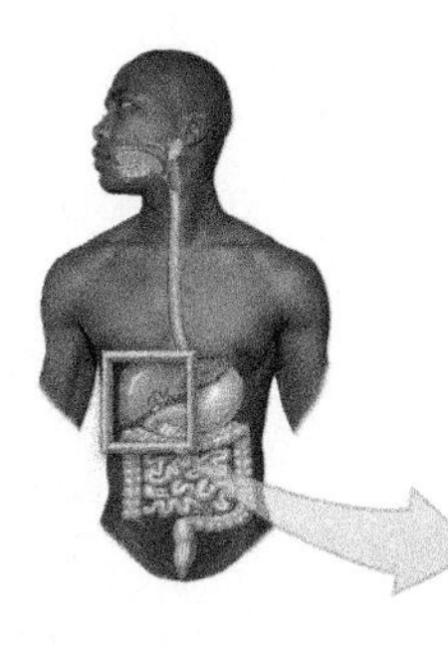

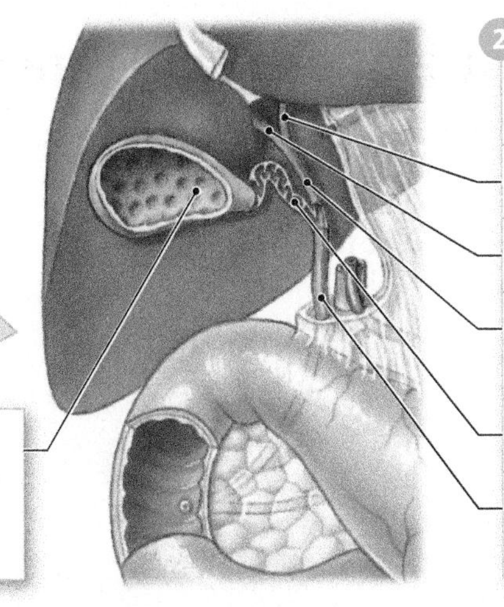

1. Locate the **gallbladder** at the anteroinferior edge of the liver, slightly to the right of the midline.

2. Identify the series of ducts that transport bile from the liver to the gallbladder and small intestine:
   - The **left hepatic duct** originates from the left lobe of the liver.
   - The **right hepatic duct** originates from the right lobe of the liver.
   - The left and right hepatic ducts merge to form the **common hepatic duct.**
   - The **cystic duct** originates from the gallbladder.
   - The common hepatic and cystic ducts merge to form the **common bile duct,** which empties into the duodenum.

C The **pancreas** produces **pancreatic juice,** a watery mixture of digestive enzymes and buffers. In the small intestine, the digestive enzymes chemically digest nutrient molecules and the buffers neutralize stomach acids.

1. On the torso model, remove the stomach to reveal the pancreas positioned horizontally along the posterior abdominal wall. Identify the three regions of the pancreas:

The **head** is the broad right portion. It is surrounded by the C-shaped duodenum.	The **body** is the middle portion.	The elongated **tail** extends to the left side of the abdominal cavity toward the spleen.

2. Identify the **pancreatic duct** traveling within the pancreas along its entire length.

3. Notice that the pancreatic duct merges with the common bile duct to form the **duodenal ampulla.** Identify the ampulla as it empties into the duodenum along its concave surface.

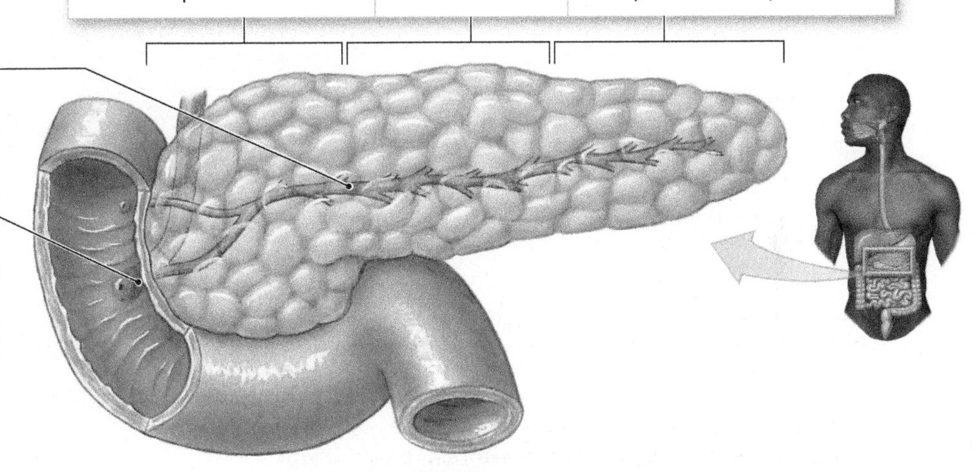

IN THE CLINIC

Gallstones

While stored in the gallbladder, bile becomes concentrated as the gallbladder epithelium absorbs water. If bile is too concentrated or contains too much cholesterol, the cholesterol can crystallize to form **gallstones.** These structures can block the passage of bile in the ducts and cause excessive pain. Gallstones can be dissolved with drugs or pulverized with ultrasound or lasers. If they have to be surgically removed, the gallbladder is sometimes removed as well.

MAKING CONNECTIONS

If the gallbladder is surgically removed, what function(s) will no longer be possible? How will this surgery affect the transport of bile?

▶ ______________________________

____

____

____

____

■ Want more practice? Go to: **MasteringA&P** > Study Area > Menu > Lab Tools > PAL > Anatomical Models > Digestive System > Accessory Organ photos

Activity 27.8

Examining the Microscopic Structure of the Alimentary Canal: Esophagus and Stomach

The wall of the alimentary canal has four tissue layers. From inside to outside, these layers are the **mucosa (mucous membrane), submucosa (connective tissue layer), muscularis externa (muscular layer),** and **serosa (serous layer).** The general arrangement of these layers is similar throughout the alimentary canal, but structural specializations reflect the unique functions of each digestive organ.

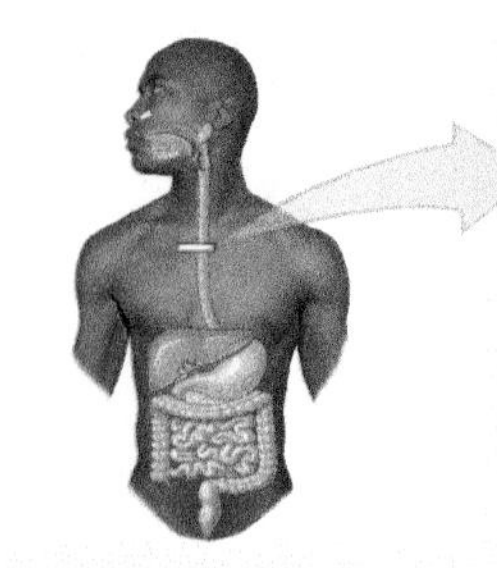

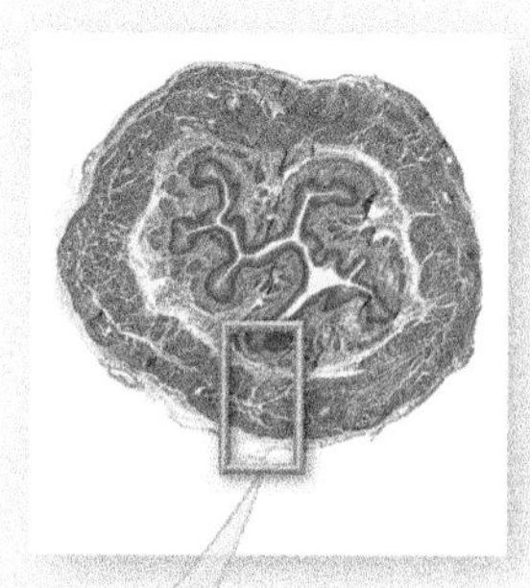

A The esophagus transports food from the pharynx to the stomach.

1 View a slide of the esophagus under low power. Locate the **mucosa** along the top of the section and identify its three components:

A **stratified squamous epithelium** lines the lumen. Why is it important to have a stratified squamous epithelium in the esophagus?

▶ ______________________________

____

The **lamina propria** is a layer of areolar connective tissue.

The **muscularis mucosae** is a thin layer of smooth muscle deep to the lamina propria.

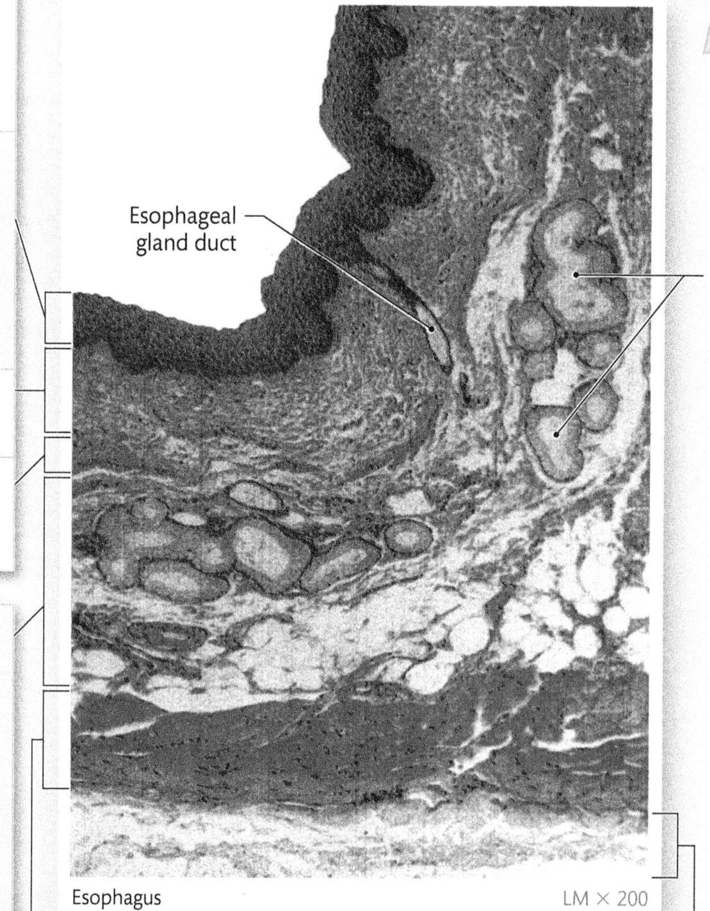

Esophagus LM × 200

2 Identify the **submucosa** deep to the muscularis mucosae. The submucosa is a connective tissue layer that contains blood vessels, nerves, and esophageal glands. The glands secrete mucus onto the luminal surface. What do you think is the function of the mucus secretions?

▶ ______________________________

____

3 Just below the submucosa, locate the **muscularis externa.** It consists of two layers of muscle: an inner circular layer and an outer longitudinal layer. At the superior end, the muscularis externa is entirely skeletal muscle; along the middle portion, it contains a combination of smooth and skeletal muscle; and at the inferior end, only smooth muscle is present. What function does the muscularis externa have in the esophagus?

▶ ______________________________

____

4 Unlike the digestive organs in the abdominal cavity, the esophagus does not have a serous layer. Instead, its outermost layer is the **adventitia.** Identify this layer. It is composed of loose connective tissue that blends in with the connective tissue of adjacent structures.

■ Want more practice? Go to: MasteringA&P > Study Area > Menu > Lab Tools > PAL > Histology > Digestive System > Alimentary Canal photos

B The surface of the stomach contains numerous gastric pits that travel deeply into the mucous membrane and lead to gastric glands.

1 Observe a slide of the fundic or pyloric region of the stomach under low power. Identify the following tissue layers in the stomach wall:

Mucosa

Submucosa

Muscularis externa

Serosa

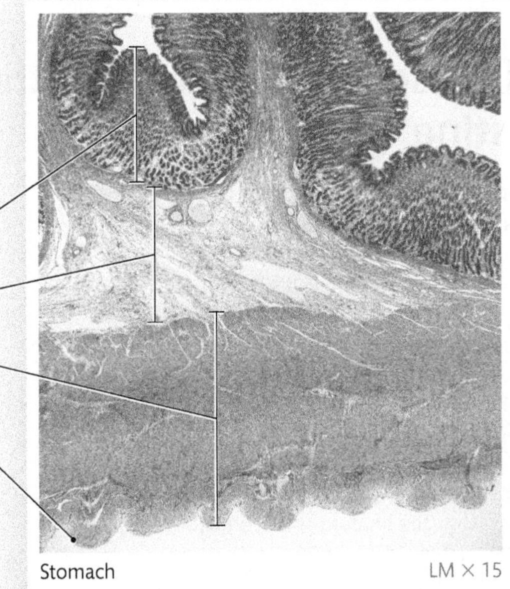

Stomach LM × 15

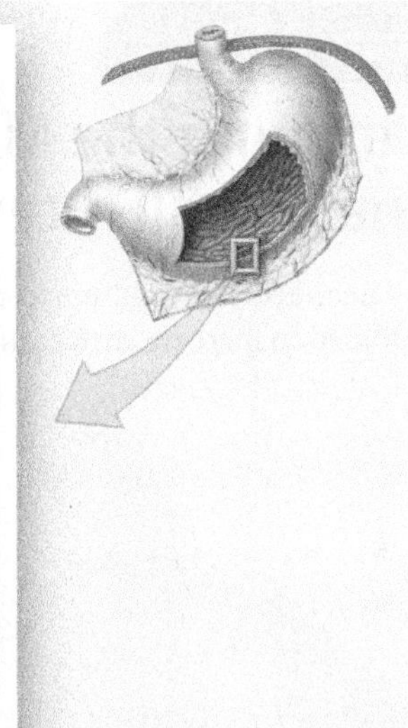

2 Observe the folds of mucosa and submucosa along the surface. These folds are the **rugae** that you identified earlier at the gross anatomical level. Circle and label a ruga in the diagram to the right. ▶

3 Observe the mucosa under high power and identify the following structures:

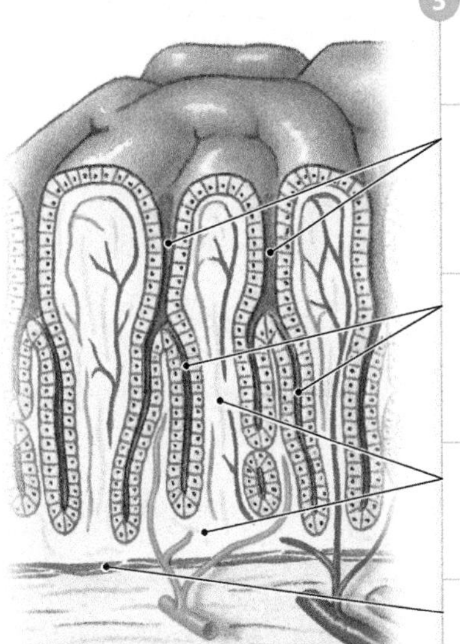

Gastric pits are surface passageways that extend into the mucosa. They are lined by simple columnar epithelial cells. Most of these cells secrete mucus onto the surface.

Each gastric pit gives rise to several **gastric glands.** The gastric glands produce gastric juice: a mixture of mucus, hydrochloric acid, digestive enzymes, and digestive hormones.

The **lamina propria** is the connective tissue between the gastric glands and along the base of the mucosa.

The **muscularis mucosae** is the thin layer of smooth muscle at the base of the mucosa.

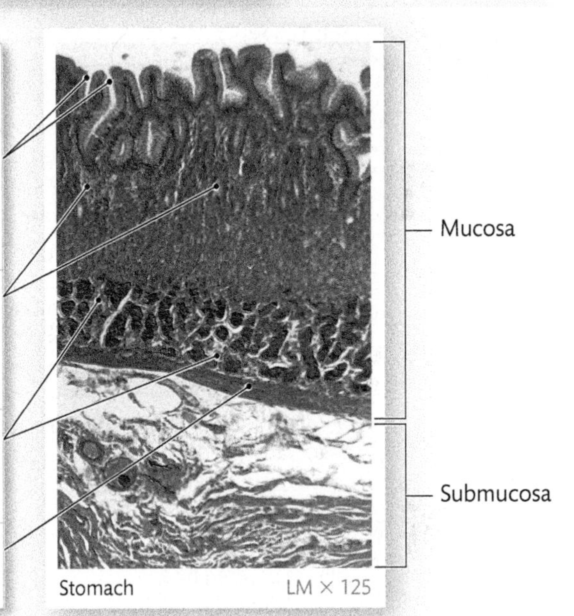

Stomach LM × 125

4 The gastric glands are composed of several cell types. Refer to your textbook and, in the table below, identify the gastric juice component produced by each of the following cell types. ▶

Cell Type	Substance(s) Produced
Mucous cells	
Parietal cells	
Chief cells	
G cells	

MAKING CONNECTIONS

Why do you think it is important that the epithelial cells lining the gastric pits secrete mucus onto the surface of the stomach wall?

▶ ______________________________

____

____

____

Activity 27.9

Examining the Microscopic Structure of the Alimentary Canal: Small Intestine

A The mucosa and submucosa of the small intestine are highly folded to increase surface area for nutrient absorption.

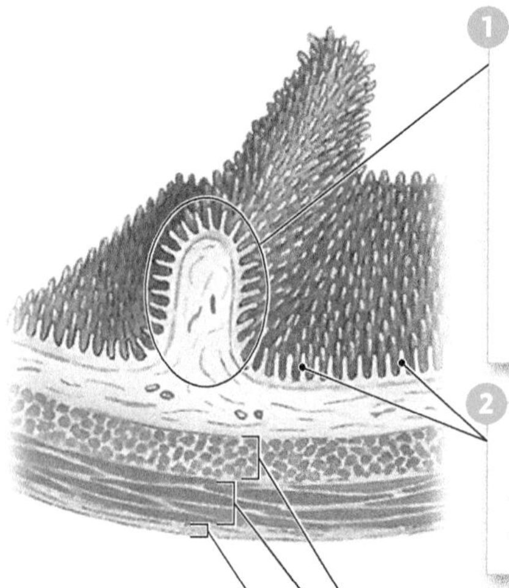

1. Observe a slide of the small intestine under low power. Along the inner surface, observe the folds of mucosa and submucosa. They are the **plicae circulares** that you identified earlier at the gross anatomical level. In the photo on the right, draw a circle around the plica circulare and label the mucosa and submucosa. ▶

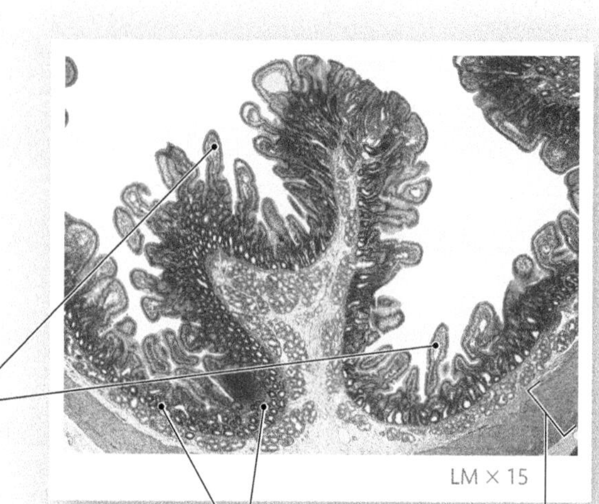

LM × 15

2. Identify the fingerlike projections along the surface of the mucosa. These structures are called **intestinal villi** (singular, **intestinal villus**).

3. At the base of the villi, observe how the epithelium folds into the lamina propria. These deep pockets are **intestinal glands,** which secrete mucus and numerous digestive enzymes.

4. Deep to the submucosa, observe the **muscularis externa.** It contains inner circular and outer longitudinal smooth muscle fibers.

5. The serosa covers the outside wall. It may not be present on your slide.

6. Switch to the high-power objective lens to observe the intestinal villi more closely. Notice that the villi are covered by a simple columnar epithelium that contains the following cell types:

Absorptive cells are the main cell type. They are usually stained pink with dark blue nuclei at the bases. Their primary functions are to complete the final stage of chemical digestion and absorb nutrient molecules.

Mucus-secreting **goblet cells** are prominently interspersed between absorptive cells. These cells contain large oval-shaped regions filled with mucus. They are usually light staining, but on some slides they could be a deep blue or red.

Hormone-producing **enteroendocrine cells** are also found in the epithelium, but unless your slide is specially prepared, you will not be able to identify them.

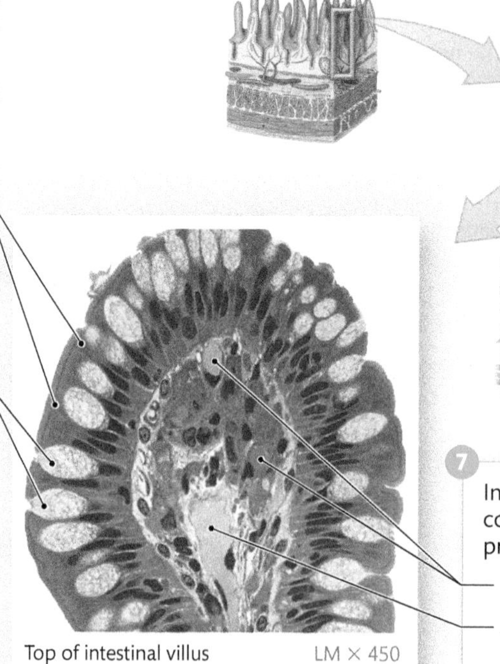

Top of intestinal villus LM × 450

7. In the core of each villus, identify the connective tissue from the lamina propria. In this region, identify:

- **Blood capillaries**
- A lymphatic capillary known as a **lacteal**

■ Want more practice? Go to: **MasteringA&P** > Study Area > Menu > Lab Tools > PAL > Histology > Digestive System > Alimentary Canal photos

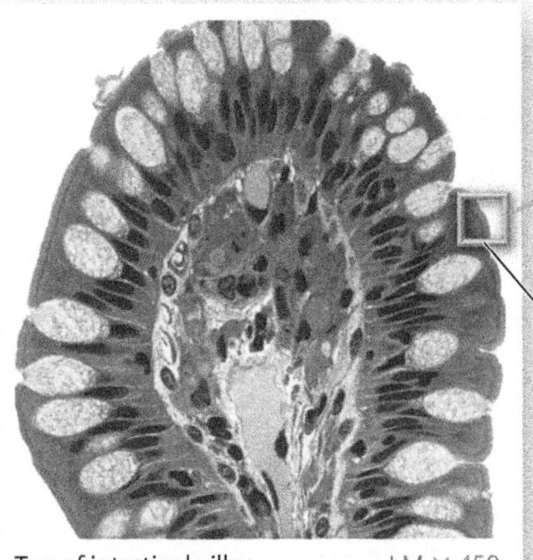
Top of intestinal villus LM × 450

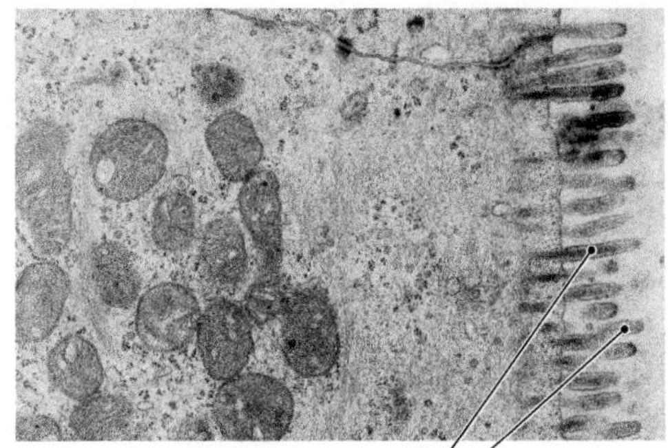
Surface of absorptive cell TEM × 22,170

8 Look carefully along the surface of the absorptive cells and you will notice a dark fringe. That is the **brush border.**

9 The brush border is composed of microscopic projections called **microvilli.** You can see individual microvilli only with an electron microscope. Digestive enzymes along their membranes perform the final steps of chemical digestion, producing nutrient molecules that are small enough to be absorbed. The microvilli make nutrient absorption more efficient. Explain why.

▶ ______________________________

____

B Each region of the small intestine has a unique structural feature that distinguishes it from the other segments.

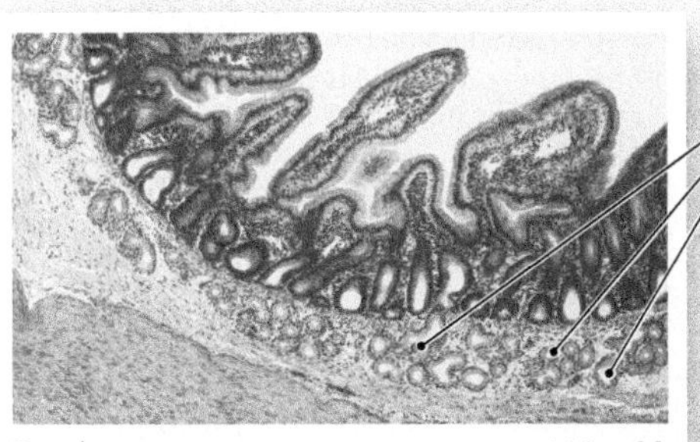
Duodenum LM × 30

1 Observe a slide of the duodenum and identify **duodenal glands** scattered throughout the submucosa. These glands secrete an alkaline mucus onto the surface.

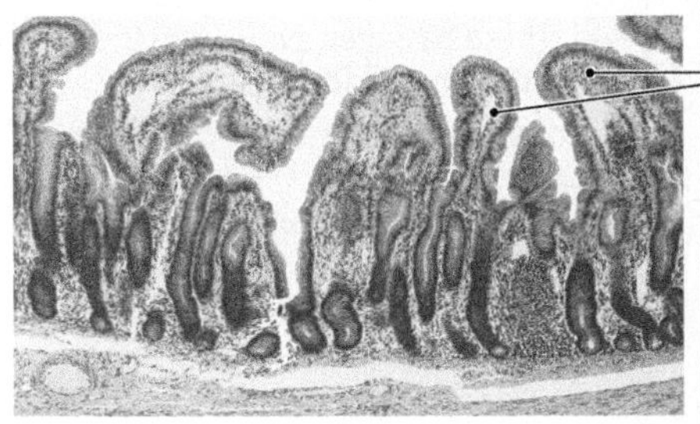
Jejunum LM × 35

2 Compare the size and number of **intestinal villi** in the three segments of the small intestine. They are most prominent in the jejunum.

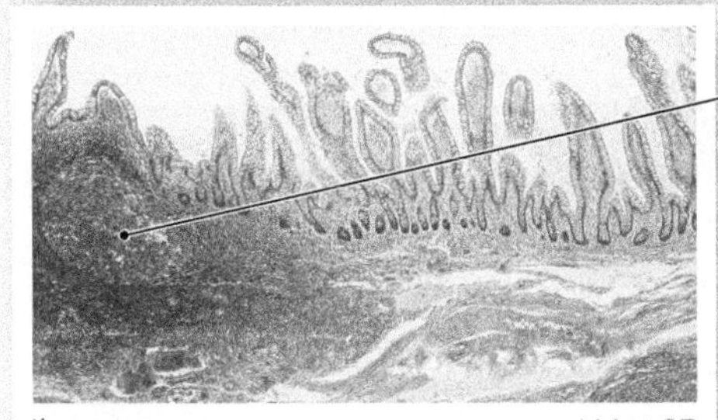
Ileum LM × 37

3 In the ileum, identify the large **aggregated lymphoid nodules (Peyer's patches)** in the submucosa and lamina propria.

MAKING CONNECTIONS

Explain how the following structural features are linked to the primary digestive function of each region of the small intestine. ▶

a. duodenal gland in the duodenum:

____

____

____

b. prominent intestinal villi in the jejunum:

____

____

____

c. aggregated lymphoid nodules in the ileum:

____

____

____

Activity **27.10**

Examining the Microscopic Structure of the Alimentary Canal: Large Intestine

A The inner wall of the large intestine lacks intestinal villi.

1 View a slide of the large intestine under low power. Identify the following tissue layers and label them in the photo below: **mucosa, submucosa, muscularis externa,** and **serosa.** ▶

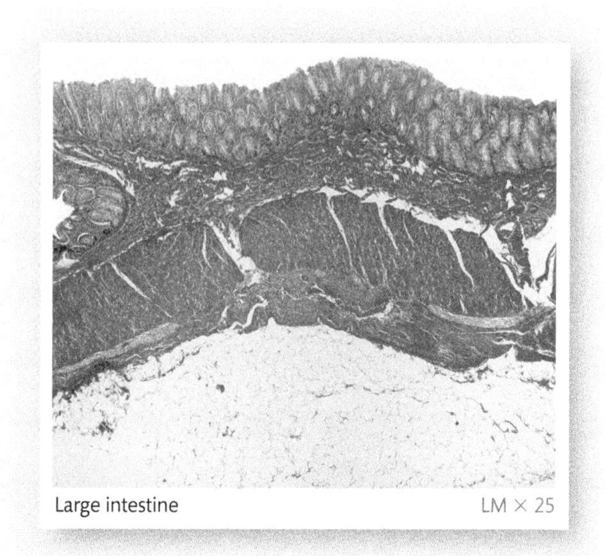
Large intestine LM × 25

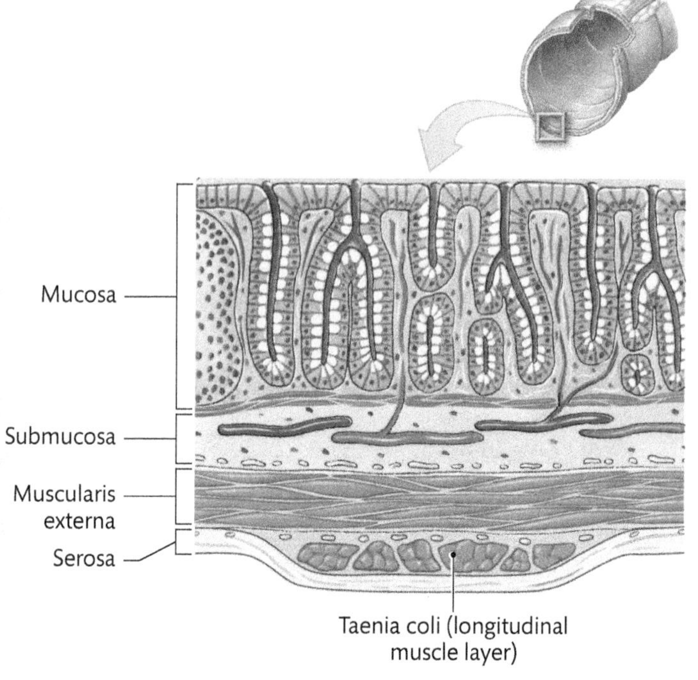

B The mucosa (mucous membrane) of the large intestine contains intestinal glands filled with mucous (goblet) cells.

1 Observe the mucosa under high power and identify the **simple columnar epithelium.**

2 The epithelium is filled with **mucous cells,** which secrete mucus onto the surface. What do you think are the functions of the mucus secretions?

▶ ______________________________

____

____

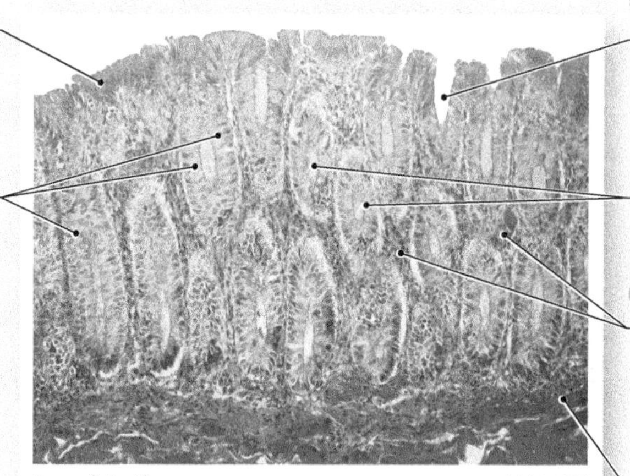
Large intestine LM × 77

3 Along the surface, identify the numerous **invaginations,** which travel deep into the mucosa.

These passageways lead to the **intestinal glands.**

4 Identify the **lamina propria,** the loose connective tissue that fills the spaces between the glands.

The **muscularis mucosae** is a thin pink band of smooth muscle at the base of the mucosa.

MAKING CONNECTIONS

The intestinal villi and the brush border (microvilli), common in the small intestine, are absent in the large intestine. What does this absence tell you about the digestive function of the large intestine?

▶ ______________________________

____

 ■ Want more practice? Go to: **MasteringA&P** > Study Area > Menu > Lab Tools > PAL > Histology > Digestive System > Alimentary Canal photos

Activity **27.11**

Examining the Microscopic Structure of Accessory Digestive Organs: Salivary Glands

A The submandibular gland produces a mixture of mucus and serous secretions.

1. View a slide of the submandibular gland under low power. The glandular cells are arranged in small clusters called **acini** (singular, **acinus**) with a small lumen in the center. The lumen will often be obscured due to the plane of section. Identify the two types of acini:

   Serous cells will have a deep purple-blue or red-staining cytoplasm and centrally located, blue-purple nuclei.

   The **mucous cells** usually have a pale, frothy-looking cytoplasm with blue-purple nuclei pushed to the cell's base.

2. Identify a **salivary gland duct.** A typical duct has a relatively large lumen that is lined by a simple cuboidal epithelium.

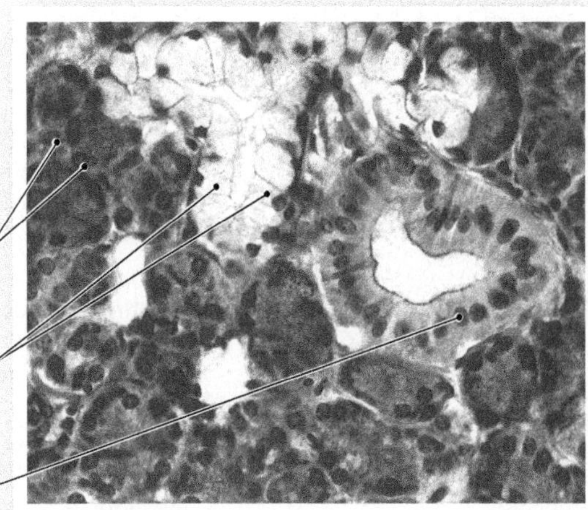

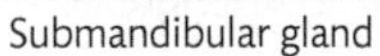

Submandibular gland LM × 435

B The sublingual glands produce a mucus secretion, and the parotid glands produce a mostly serous secretion.

1. In the two photos below, label the following structures: ▶
   - Serous cells
   - Mucous cells
   - Salivary gland duct

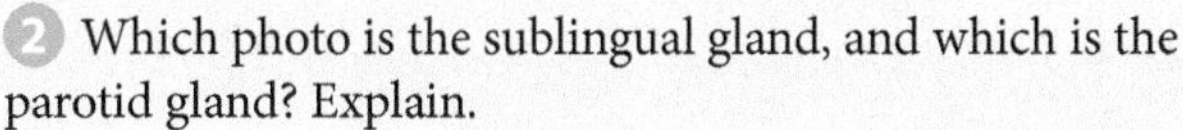

2. Which photo is the sublingual gland, and which is the parotid gland? Explain.

▶ ____

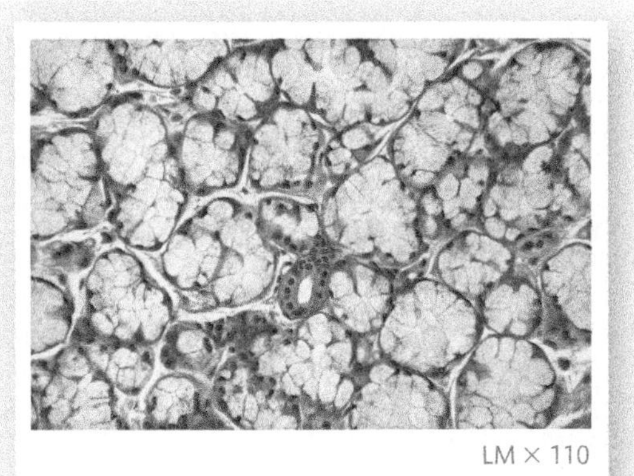

LM × 110

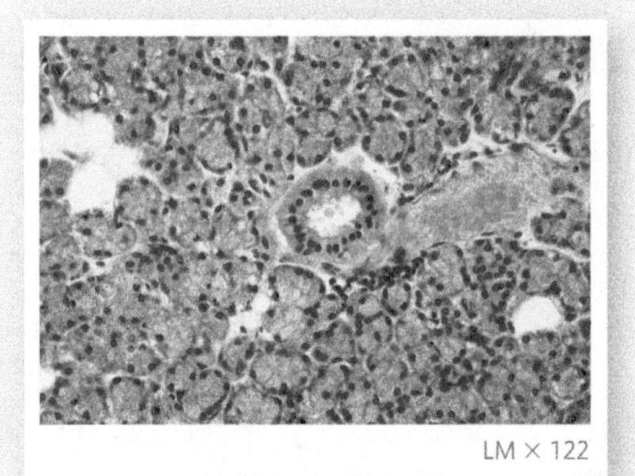

LM × 122

MAKING CONNECTIONS

Infections or the formation of salivary stones (crystallized saliva) can block salivary gland ducts. How would this blockage affect digestion?

▶ ____

Activity 27.12

Examining the Microscopic Structure of Accessory Digestive Organs: Liver and Pancreas

A The functional units of the liver are the **liver lobules.**

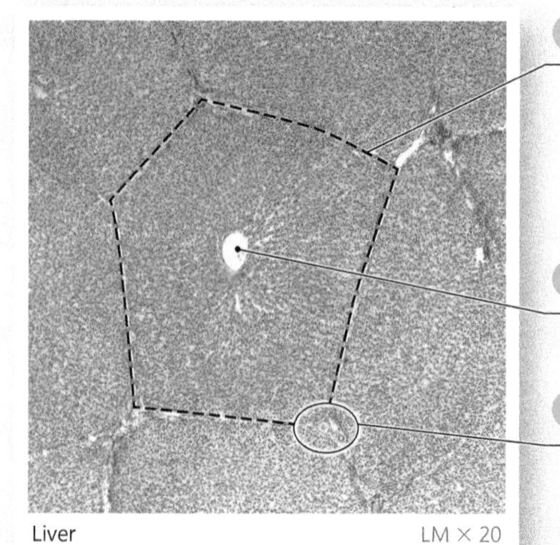

Liver LM × 20

1. View a slide of the liver under low power. Identify a **liver lobule.** Each lobule is a column of tissue, usually with five (pentagon) or six (hexagon) sides. It is separated from adjacent lobules by thin partitions of connective tissue.

2. Identify the **central vein** traveling through the middle of the lobule.

3. Locate the **portal areas (portal triads)** at each corner along the periphery of a liver lobule.

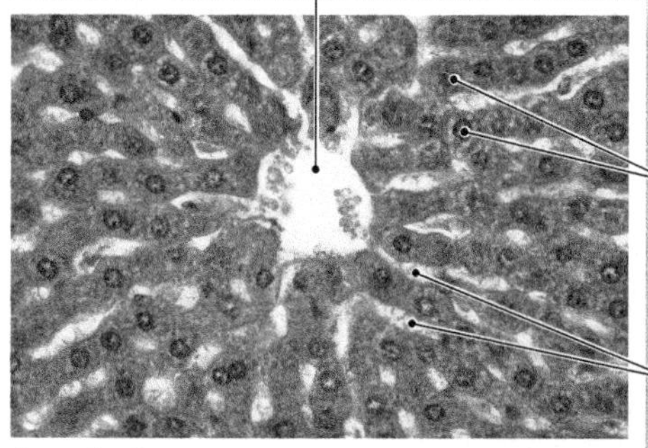

Central vein LM × 260

4. Under high power, observe the central vein and the area adjacent to it.

   Notice how the **hepatocytes (liver cells)** are arranged in rows that radiate from the central vein, like the spokes of a bicycle wheel.

   Observe the **liver sinusoids,** blood capillaries that travel between the rows of hepatocytes.

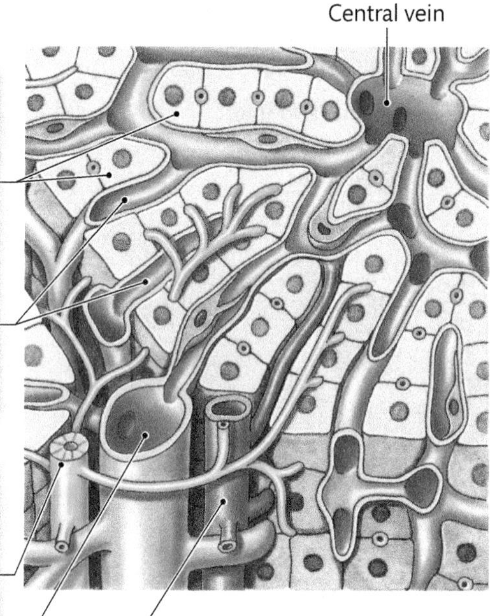

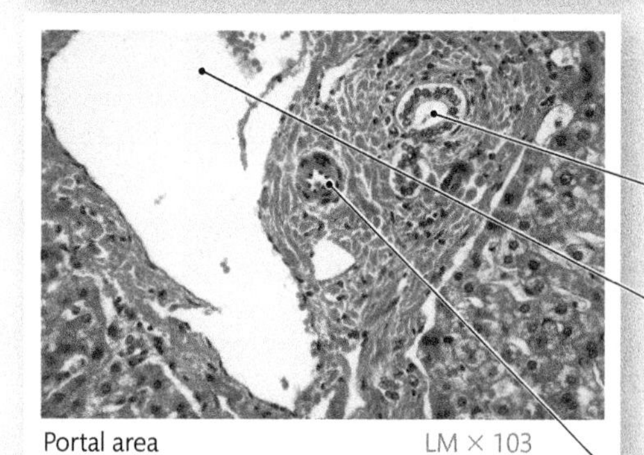

Portal area LM × 103

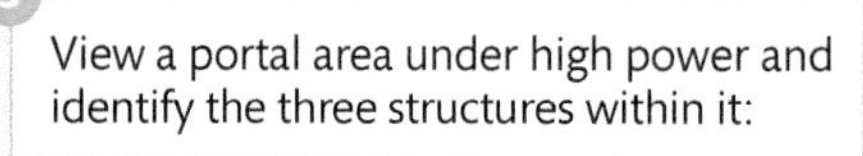

5. View a portal area under high power and identify the three structures within it:

   A **bile duct** transports bile to the duodenum or gallbladder.

   A **branch of the hepatic portal vein** transports venous blood from the gastrointestinal tract to the liver. This blood is rich in nutrients that were absorbed by the small intestine.

   A **branch of the hepatic artery** brings oxygenated arterial blood to the liver.

   Blood from the two blood vessels in the triad percolates through the sinusoids and drains into the central vein.

6

Under low power, examine a slide specially prepared to illustrate the transport of bile through the liver. Identify the tiny ducts called **bile canaliculi** (singular, **canaliculus**). Bile, produced by hepatocytes, is secreted into the canaliculi and flows into the bile ducts in the portal areas. The black-staining pathways (green in the figure) are the canaliculi.

7

Phagocytic cells called **Kupffer cells** are found along the walls of the sinusoids. What do you think is the function of these cells?

▶ ___

___

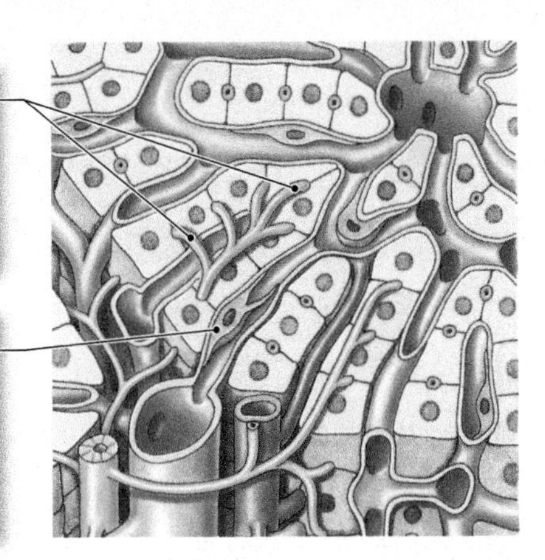

B Most of the pancreas consists of glandular cells that produce a watery mixture of digestive enzymes called **pancreatic juice.**

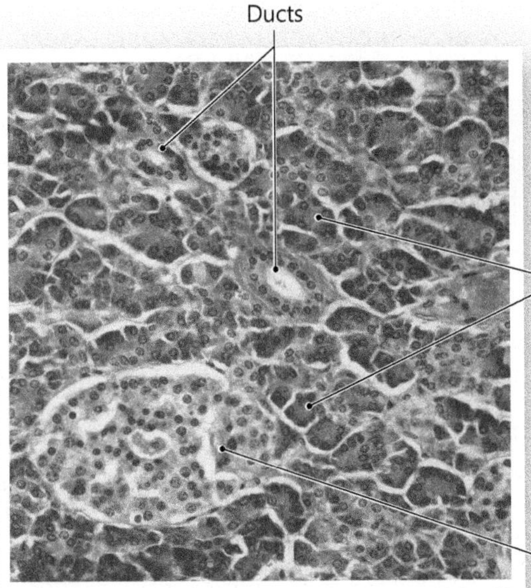

Pancreas LM × 65

1

Under low power, identify the two functional components of the pancreas:

The **pancreatic acini** are clusters of cuboidal cells (pancreatic acinar cells) arranged around a central lumen. The acinar cells produce pancreatic juice that is transported along a network of ducts leading to the pancreatic duct.

The **pancreatic islets (islets of Langerhans)** are the lighter-staining regions of endocrine cells scattered among the pancreatic acini.

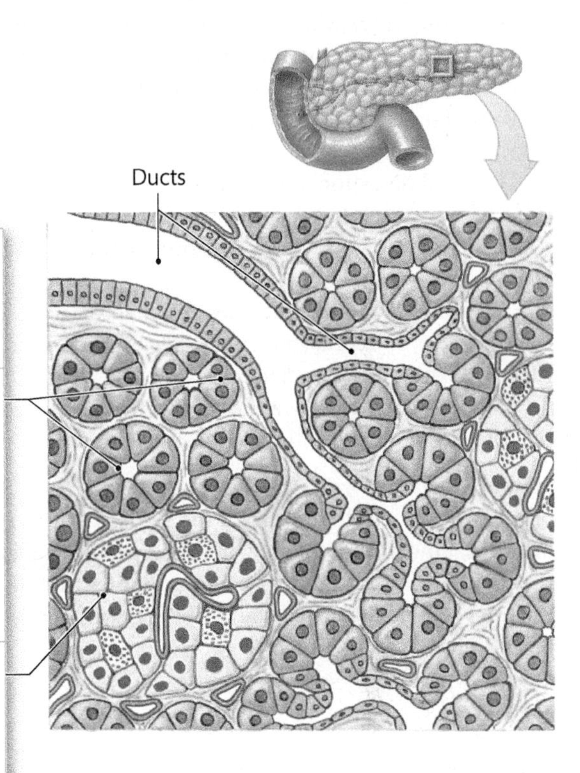

2

Pancreatic juice contains the digestive enzymes listed below. Refer to your textbook and complete the table by identifying the function of each enzyme. ▶

Pancreatic Enzyme	Function
Pancreatic alpha amylase	
Pancreatic lipase	
Nucleases	
Proteolytic enzymes	

MAKING CONNECTIONS

Explain why the liver sinusoids contain a mixture of arterial and venous blood.

▶ ___

___

___

___

___

___

BEFORE YOU MOVE ON . . .

« LOOKING BACK

The process of **digestion** breaks down foods into simpler forms so that nutrients can be delivered to all areas in the body. To perform this function, the digestive system carries out a variety of activities, which occur in various digestive organs:

- **Ingestion** occurs when food is brought into the oral cavity.
- **Mechanical processing** includes three activities. **Chewing** and the **formation of a bolus** take place in the oral cavity. **Muscular actions** (peristalsis and mixing movements) occur throughout the alimentary canal.
- **Chemical digestion** of the nutrient molecules found in food occurs in the oral cavity, stomach, and small intestine.
- **Absorption** of nutrients occurs primarily in the small intestine.
- **Secretion** of digestive enzymes and mucus occurs in the esophagus, stomach, small intestine, large intestine, and accessory digestive organs.
- **Compaction** of feces is a function of the large intestine.
- **Defecation** occurs in the rectum and anal canal.

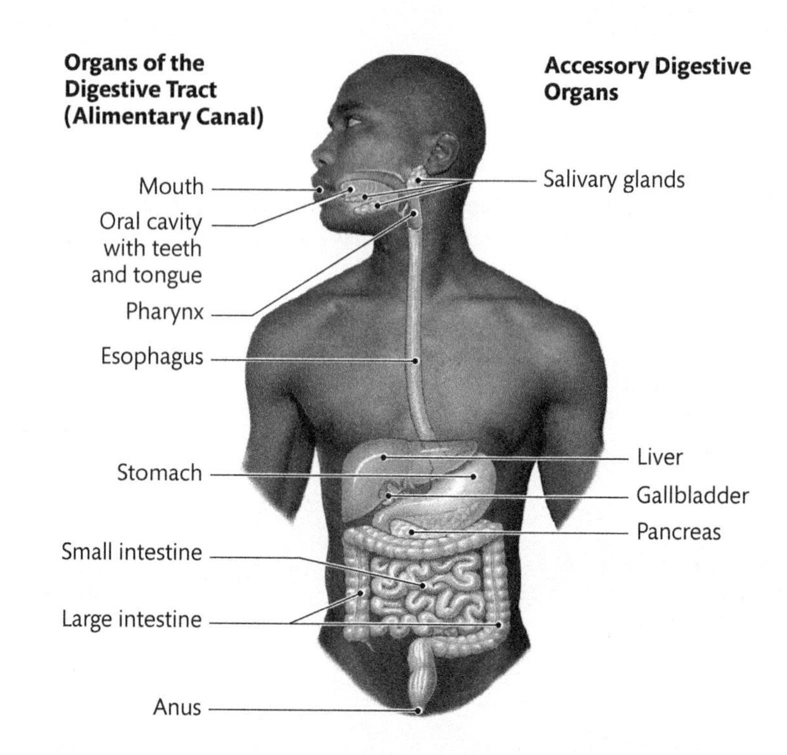

Suppose one or more of the above activities doesn't function properly. How would that affect the digestive process? For instance, how would not chewing food properly affect chemical digestion? Identify other examples that illustrate the interdependence of all the activities involved in digestion.

▶ ____

LOOKING FORWARD »

In this laboratory exercise, you studied the anatomy of the alimentary canal and the accessory organs for digestion. You learned that chemical digestion begins in the oral cavity where amylase, a component of saliva, breaks down complex carbohydrates to disaccharides. In the next laboratory exercise, you will investigate the function of salivary amylase and understand how changes in pH can affect its actions.

Name ______________________

Lab Section ______________________

Date ______________________

EXERCISE **27** **REVIEW SHEET**

Anatomy of the Digestive System

1. For the following structures, identify the region or organ of the digestive system in which it is found and describe its function.

Structure	Location	Function
Lingual tonsils	a.	b.
Root canal	c.	d.
Pancreatic duct	e.	f.
Aggregated lymphoid nodules	g.	h.
Gastric glands	i.	j.
Uvula	k.	l.
Brush border	m.	n.
Taeniae coli	o.	p.
Parietal cells	q.	r.
Duodenal glands	s.	t.
Rugae	u.	v.
Central vein	w.	x.
Cystic duct	y.	z.

2. In the genetic disease cystic fibrosis, ducts in several organs become blocked by an excessive production of thick mucus. Although most cystic fibrosis deaths are due to lung damage, the disease also clogs bile ducts in the liver and pancreatic ducts in the pancreas. Discuss how the disease's effects on the liver and pancreas would affect digestive activity.

____

____

____

____

3. Although many types of bacteria cause serious diseases, some are actually beneficial to humans. In terms of digestion, explain why that statement is true.

____

____

____

____

4. Chronic or severe diarrhea can lead to serious and possibly life-threatening dehydration. Explain why.

____

____

____

____

QUESTIONS 5–16: Match the organ in column A with the appropriate description in column B.

A	B
5. Duodenum ________	**a.** This structure is connected to the large intestine at the cecum.
6. Esophagus ________	**b.** The cells in this organ produce bile.
7. Gallbladder ________	**c.** This structure is posterior to the nasal and oral cavities.
8. Sigmoid colon ________	**d.** This structure is found along the floor of the oral cavity.
9. Pharynx ________	**e.** This structure travels around the head of the pancreas.
10. Pancreas ________	**f.** This structure is the S-shaped portion of the large intestine.
11. Liver ________	**g.** This retroperitoneal structure is posterior to the stomach.
12. Ascending colon ________	**h.** The pyloric sphincter is located at the junction of this structure and the small intestine.
13. Tooth ________	**i.** This structure is a storage site for bile.
14. Ileum ________	**j.** The right colic flexure represents the junction between this structure and the transverse colon.
15. Tongue ________	**k.** In the neck, this structure passes posterior to the trachea.
16. Stomach ________	**l.** Blood vessels and nerves are found in the pulp cavity of this structure.

17. In the following diagram, color the structures with the indicated colors.

Transverse colon = **green**
Liver = **blue**
Jejunum = **yellow**
Cecum = **orange**
Descending colon = **red**
Stomach = **brown**
Ascending colon = **purple**
Sigmoid colon = **light blue**
Gallbladder = **black**
Ileum = **light brown**
Rectum = **light green**
Vermiform appendix = **pink**
Spleen = **gray**

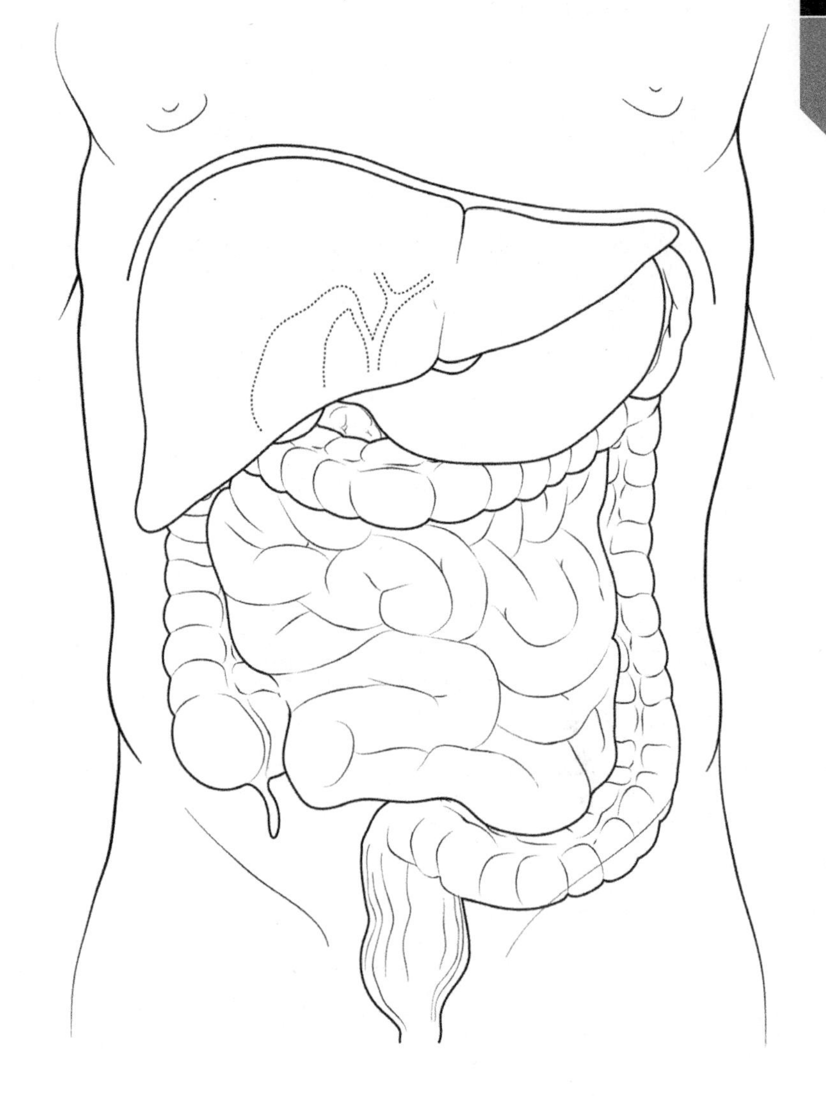

18. In a liver lobule, blood and bile flow in opposite directions. Explain.

____

____

____

19. Conduct the following class activity:
When the enamel and dentin on the crown of a tooth decay, the area of decay must be cleaned and filled with an inert substance. Some individuals are more prone to developing dental caries (cavities) than others. Inspect the oral cavities of other students in the laboratory and note the relative number of fillings in each individual. How much variation exists in your class?

____

____

____

20. Identify the digestive system organs that are illustrated in the following photos. In photo (a), label the mucosa, submucosa, muscularis externa, and serosa.

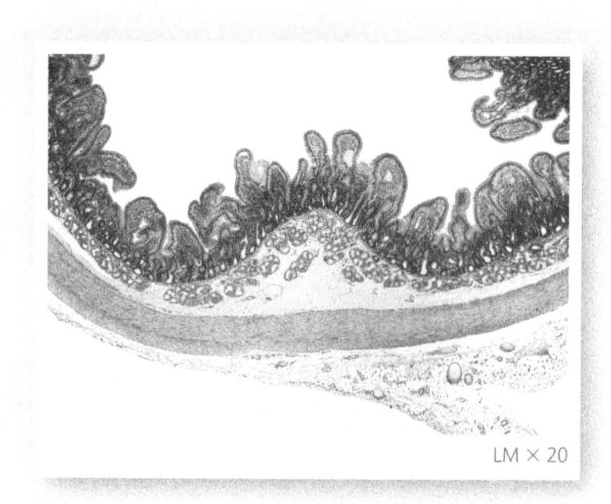

LM × 20

a. ______________________________

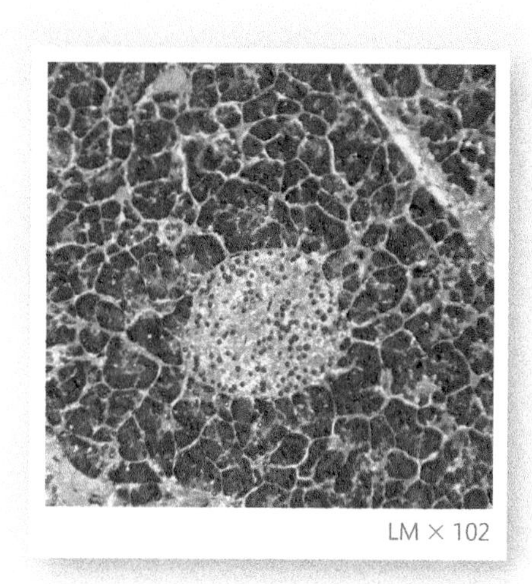

LM × 102

b. ______________________________

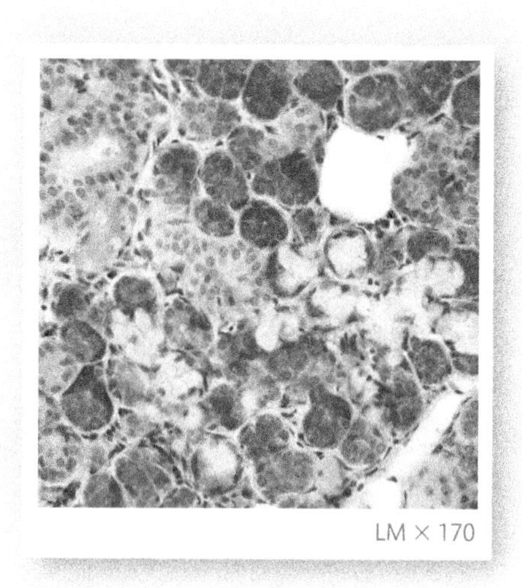

LM × 170

c. ______________________________

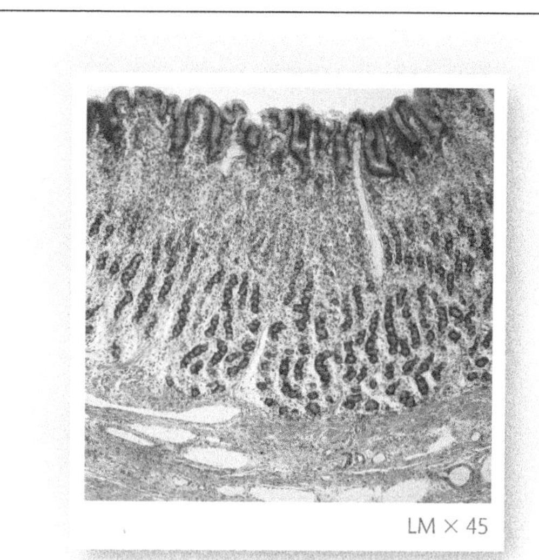

LM × 45

d. ______________________________

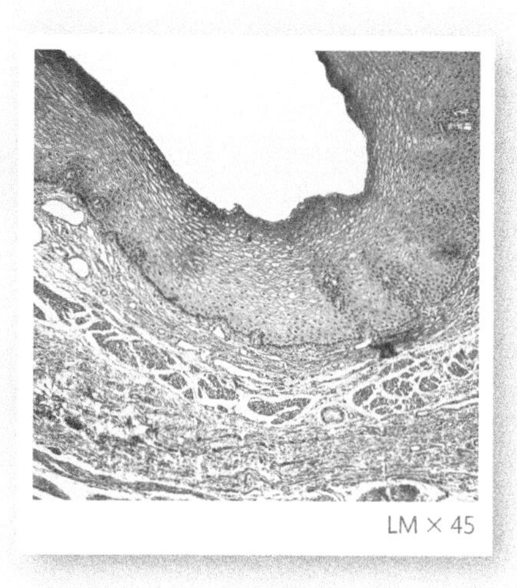

LM × 45

e. ______________________________

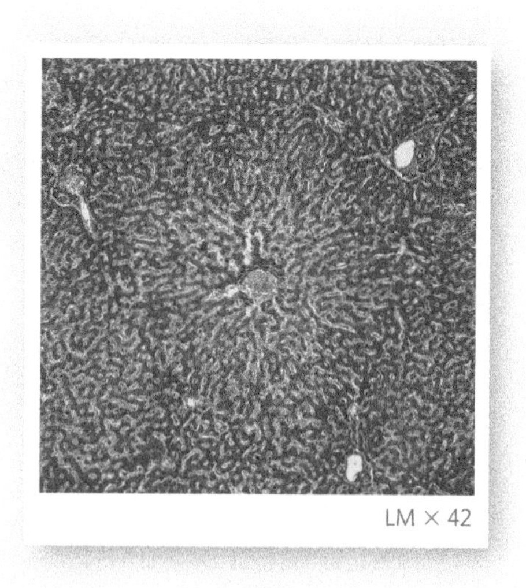

LM × 42

f. ______________________________

EXERCISE

28

Actions of a Digestive Enzyme

LEARNING OUTCOMES

These Learning Outcomes correspond by number to the laboratory activities in this exercise. When you complete the activities, you should be able to:

Activity 28.1 **Conduct positive and negative control tests and explain their significance.**

Activity 28.2 **Perform the IKI and Benedict's tests to detect the presence of starch and disaccharides in various beverages.**

Activity 28.3 **Explain the effect of pH on salivary amylase activity.**

LABORATORY SUPPLIES

- Test tubes
- Test tube rack
- Laboratory marking pencil
- Hot water bath (hot plate and beaker with water)
- Test tube holder/tongs
- 1% solutions of the following:
  - potato starch (pH 4)
  - potato starch (pH 7)
  - potato starch (pH 10)
  - maltose
- Colored pencils
- Gauze for filtering
- Benedict's solution
- 1% iodine, 2% potassium iodide (IKI) solution
- Distilled water
- Cola
- Milk
- Orange juice
- Amylase solution (to be made by the students)
- 10-mL beakers
- Celsius thermometer

PRE-LAB QUIZ

Before you begin, read all the activities in Exercise 28 and the required reading in your textbook that is assigned by your instructor.

1. An enzyme only reacts to a specific reactant substance, known as a ________________.
2. True or False: During this laboratory exercise, you will use a solution that contains iodine and potassium iodine (IKI solution) to test for the presence of disaccharides. ____________
3. True or False: During this laboratory exercise, you will use Benedict's solution to test for the presence of starch. ____________
4. During this laboratory exercise, you will conduct negative control tests for starch and disaccharides. The test solution for both negative control tests will be ____________.
5. True or False: During this laboratory exercise, you will conduct positive control tests for starch and disaccharides. For the positive control test for starch, the test solution must contain starch; for the positive control test for disaccharides, the test solution must contain disaccharides. ____________
6. During this laboratory exercise, you will perform starch and disaccharide tests for all the following beverages *except*
   a. cola.
   b. milk.
   c. coffee.
   d. orange juice.
7. During the chemical digestion of starch with amylase, the starch molecules are first broken down to smaller polysaccharides known as ____________________ and then to the disaccharide ____________________.
   a. dextrin . . . lactose
   b. dextrin . . . maltose
   c. dextrin . . . sucrose
   d. maltose . . . dextrin
   e. lactose . . . dextrin
8. True or False: As amylase digestion of starch proceeds, the concentration of starch decreases, and the concentration of disaccharides increases. ____________
9. True or False: During this laboratory exercise, you will be required to practice universal precautions while analyzing a saliva sample. ____________
10. During this laboratory exercise, you will examine amylase activity in starch solutions of varying pH. The amylase digestion of starch will be conducted at a temperature of ____________________, which approximates body temperature.
    a. 0°C
    b. 37°C
    c. 100°C

Activity 28.1

Conducting Control Tests for Starch and Disaccharides

Enzymes are proteins that act as biological **catalysts:** substances that increase the rate at which chemical reactions occur. In this role, enzymes are essential for regulating all the biochemical reactions in the cells of our bodies. Enzymes are precise in their actions. A particular enzyme will act only on a specific reactant substance, known as the **substrate.**

Amylase, an enzyme found in saliva, initiates chemical digestion of **starch** in the oral cavity. Starch is a **polysaccharide** that contains repeating units of glucose. When starch is digested by amylase, it is broken down by hydrolysis into smaller units called **dextrins.** As hydrolysis continues, the dextrins are broken down to **maltose,** a **disaccharide** consisting of two glucose molecules. During the final stage of chemical digestion, another enzyme in the small intestine, **maltase,** breaks down maltose to glucose molecules.

Before examining the actions of salivary amylase, it is important that you be able to identify the presence of starch and maltose in a solution. To do that, you will prepare control solutions and observe specific color changes that indicate the presence of these substances. These control tests will provide known standards against which comparisons can be made.

A Positive Control Test for Starch

IKI solution is an amber-colored liquid that contains 1% iodine and 2% potassium iodide. When IKI solution is added to starch, the mixture changes color to indicate the presence of starch.

1. Use a marking pencil to label a test tube "Starch (+)." Place 1 mL of 1% starch solution (pH 7) in the test tube.
2. Add 4 drops of IKI solution to the test tube. In the test tube figure on the right, use colored pencils to show the color of the starch solution after the IKI solution has been added. ▶
3. Record the color of the starch-IKI mixture in the "Results" column of Table 28.1. ▶

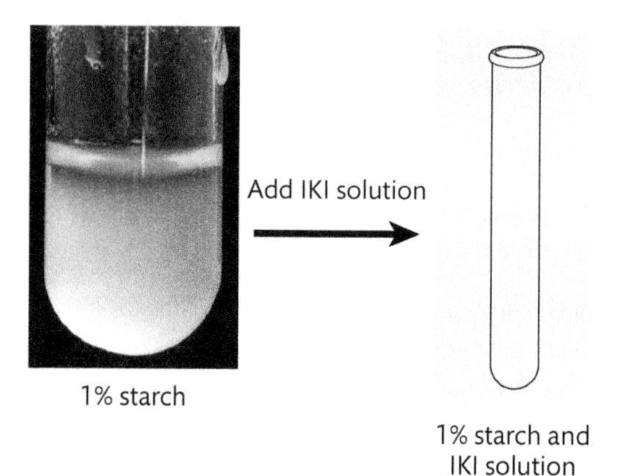

B Negative Control Test for Starch

When IKI solution is added to a liquid that does not contain starch, the color change will not occur.

1. Label a test tube "Starch (−)." Place 1 mL of distilled water in the test tube.
2. Add 4 drops of IKI solution to the test tube. In the test tube figure on the right, use colored pencils to show the color of the distilled water after the IKI solution has been added. ▶
3. Record the color of the water-IKI mixture in the "Results" column of Table 28.1. ▶

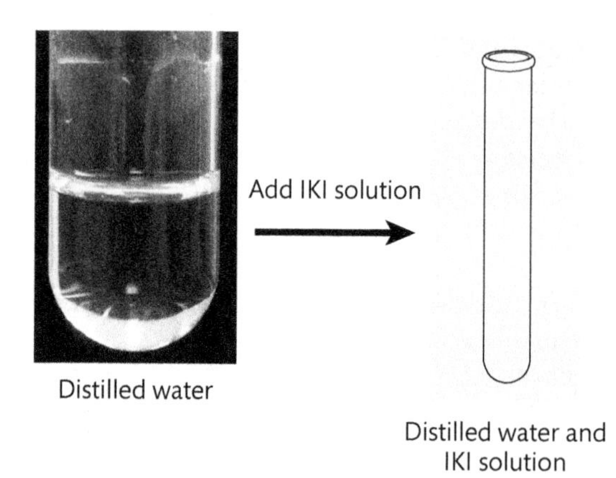

4. Explain how the IKI solution can be used to test for the presence of starch in a solution.

▶ ______________________________

____

____

____

____

Word Origins

Amylase is involved in the chemical breakdown of two forms of starch, called **amylopectin** and **amylose.** Notice that the three words *amylase, amylopectin,* and *amylose* begin with the prefix *amyl-,* which is derived from the Greek word for starch *(amylon).* The *-ase* suffix on a name indicates that the substance is an enzyme. Thus, *amylase* means "starch enzyme."

C Positive Control Test for Disaccharides

Benedict's solution is a clear blue liquid used to test for the presence of disaccharides such as maltose. A colorless fluid that contains disaccharides will turn blue when Benedict's solution is added. After heating in a water bath, the mixture will change color from clear blue to a range of different opaque colors from green to red. The color change indicates the presence of disaccharides.

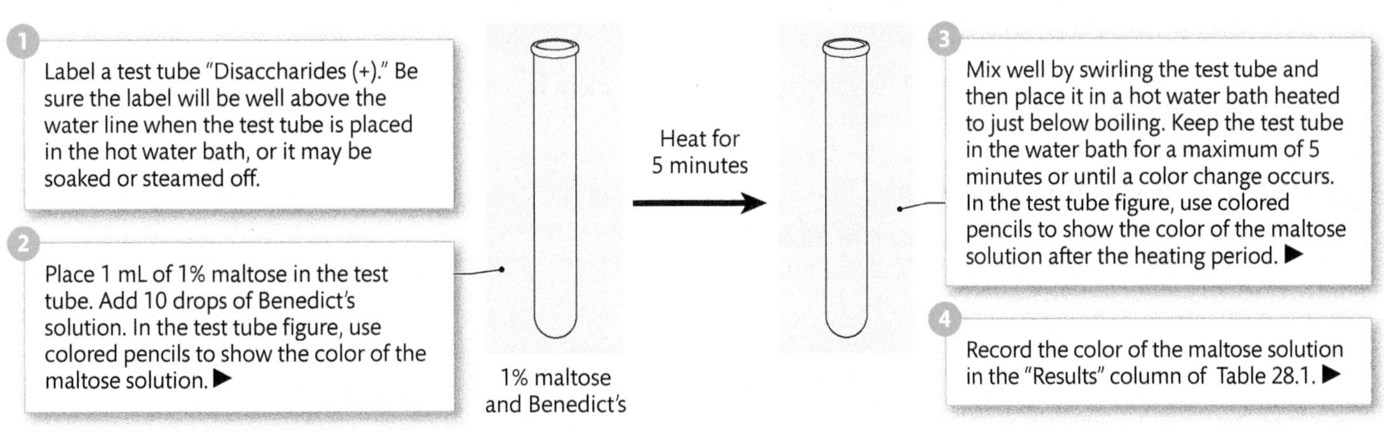

D Negative Control Test for Disaccharides

When Benedict's solution is added to a liquid that does not contain disaccharides, the blue color of the Benedict's solution will remain after heating.

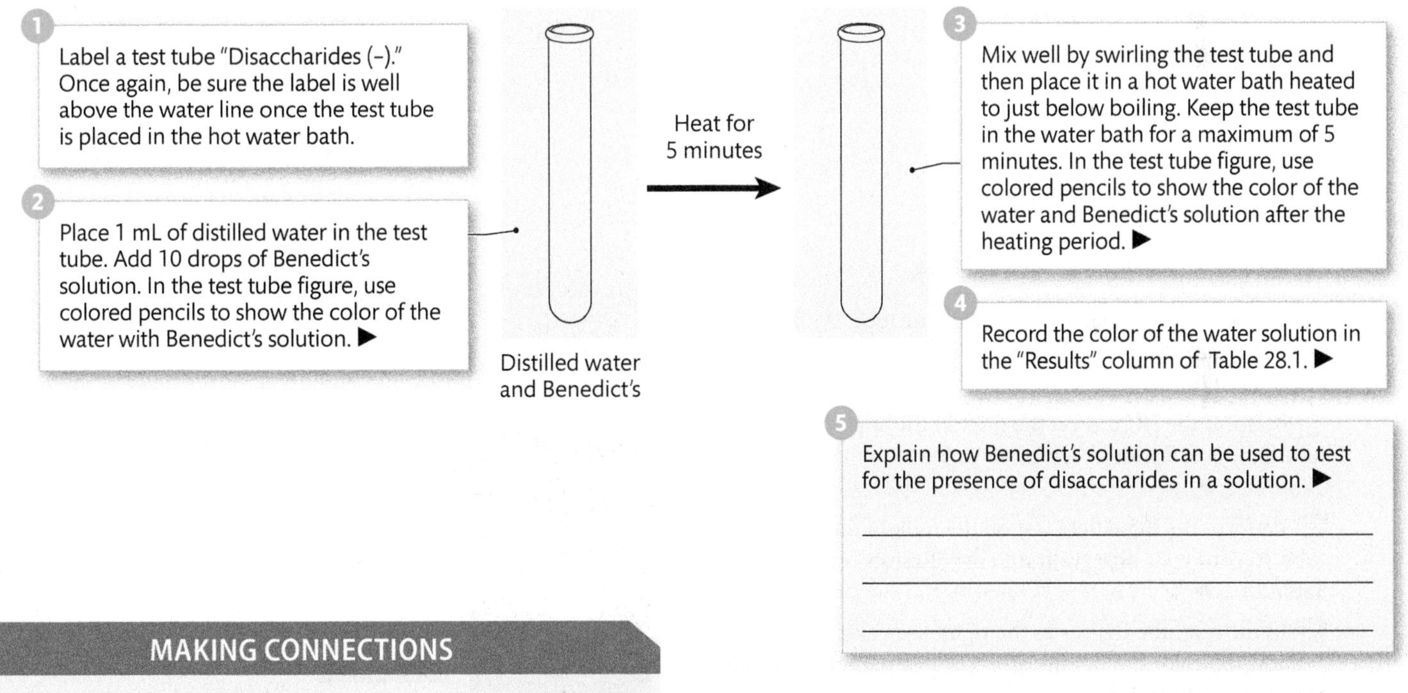

MAKING CONNECTIONS

Why is it important to conduct both a positive control test and a negative control test?

▶ ______________________________

____

____

____

____

____

TABLE 28.1 Control Tests for Starch and Disaccharides

Test	Results
Starch tests	
Positive control	
Negative control	
Disaccharide tests	
Positive control	
Negative control	

Activity 28.2

Testing for the Presence of Starch and Disaccharides in Various Beverages

You can determine the presence or absence of starch or disaccharides in various beverages by observing specific color changes after adding IKI or Benedict's solution.

A Starch and Disaccharide Tests for Cola

1. Obtain two clean test tubes. Label one tube "Cola IKI" and label the other tube "Cola Benedict's." Add 1 mL of cola to each test tube.
2. Perform the IKI test ("starch test") on the cola in the appropriate test tube and record your observation in the "Results" column of Table 28.2. ▶
3. Perform the Benedict's test ("disaccharide test") on the cola in the other test tube, remembering to heat it (to just below boiling) for no more than 5 minutes. Record your observation in the "Results" column of Table 28.2. ▶
4. In the test tube figures to the right, use colored pencils to show the color of the cola after completing the starch and disaccharide tests. ▶

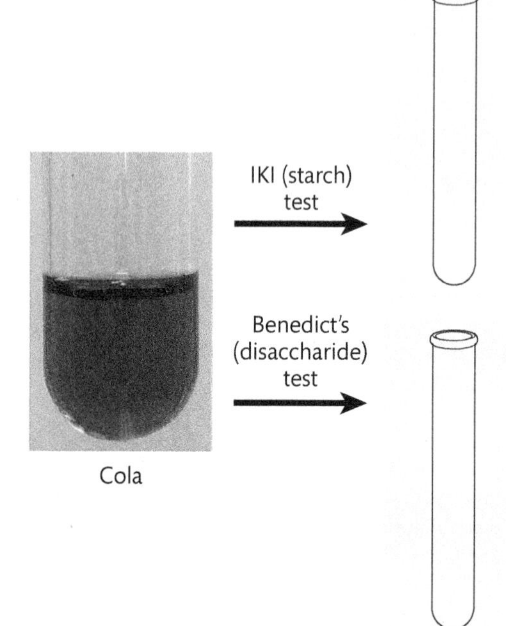

B Starch and Disaccharide Tests for Milk

1. Obtain two clean test tubes. Label one tube "Milk IKI" and label the other tube "Milk Benedict's." Add 1 mL of milk to each test tube.
2. Perform the IKI test on the milk in the appropriate test tube and record your observation in the "Results" column of Table 28.2. ▶
3. Perform the Benedict's test on the milk in the other test tube. Record your observation in the "Results" column of Table 28.2. ▶
4. In the test tube figures to the right, use colored pencils to show the color of the milk after completing the starch and disaccharide tests. ▶

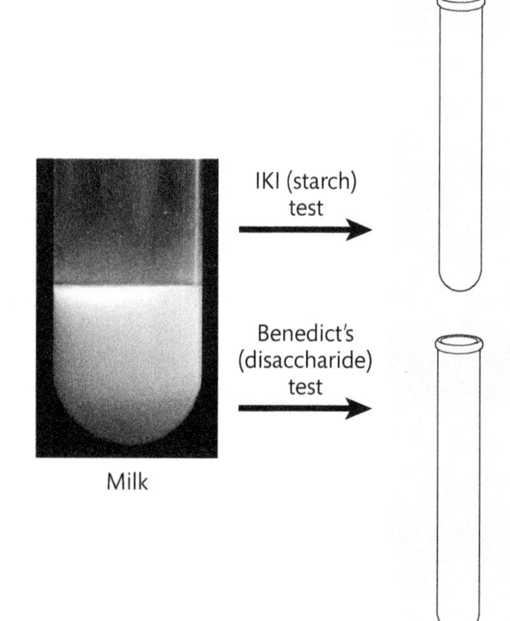

C Starch and Disaccharide Tests for Orange Juice

1. Obtain two clean test tubes. Label one tube "Orange juice IKI" and label the other tube "Orange juice Benedict's." Add 1 mL of orange juice to each test tube.

2. Perform the IKI test on the orange juice in the appropriate test tube and record your observation in the "Results" column of Table 28.2. ▶

3. Perform the Benedict's test on the orange juice in the other test tube. Record your observation in the "Results" column of Table 28.2. ▶

4. In the test tube figures to the right, use colored pencils to show the color of the orange juice after completing the starch and disaccharide tests. ▶

TABLE 28.2 Starch and Disaccharide Tests for Various Beverages

Beverage	Test	Results
Cola	Starch test	
	Disaccharide test	
Milk	Starch test	
	Disaccharide test	
Orange juice	Starch test	
	Disaccharide test	

MAKING CONNECTIONS

Explain how the original colors of the beverages that you tested affected your ability to identify the color changes that indicate the presence of starch or disaccharides.

▶ ______________________________

Activity 28.3

Examining the Effect of pH on Salivary Amylase Activity

You now know that when IKI solution is added to starch, the color of the mixture becomes blue-black or purple. That color change indicates the presence of starch. If amylase is added to a starch-IKI mixture, the chemical digestion of starch will begin. As digestion progresses, starch molecules are broken down to smaller polysaccharides called dextrins. Further digestion leads to the formation of the disaccharide maltose. As more maltose is produced, the dark blue-black color slowly fades and eventually the solution becomes colorless. The reaction occurs in the following sequence:

Chemical digestion of starch with amylase: starch (large polysaccharide) ⟶ dextrins (small polysaccharides) ⟶ maltose (disaccharide)

Color of solution with IKI: blue-black ⟶ colorless

A Amylase Activity at Varying pH

You will examine starch digestion at pH 4, pH 7, and pH 10 by adding salivary amylase to starch-IKI mixtures and observing the color transition from blue-black to colorless. The starch digestion will proceed in a 37°C (body temperature) water bath. The time it takes for the solution to become colorless is an indicator of the enzyme's effectiveness.

Warning: *Saliva is a body fluid that may contain infectious microorganisms. Therefore, universal precautions will be followed while you are analyzing your sample. Before you begin the activities that follow, carefully review universal precautions, described in Appendix A on page A-1 of your lab manual.*

1 Collect 10 mL of your own saliva in a small glass beaker. Add an equal volume of distilled water and mix well.

2 Filter your mixture through a layer of gauze into a second beaker. The resulting **filtrate** is the amylase solution you will use to digest your starch samples.

3 Label three test tubes as follows:

- Starch-IKI, pH 4
- Starch-IKI, pH 7
- Starch-IKI, pH 10

Add 4 drops of IKI solution to each test tube. Then add 0.5 mL of the 1% starch solution, pH 4, to the test tube labeled accordingly. Repeat with the 1% starch solution, pH 7, and the 1% starch solution, pH 10.

4 Add 2 mL of your amylase solution to each tube and mix well by swirling. Place the tubes in a 37°C water bath and observe any color changes.

5 Record the time it takes for the solutions in the tubes to become colorless in Table 28.3. ▶

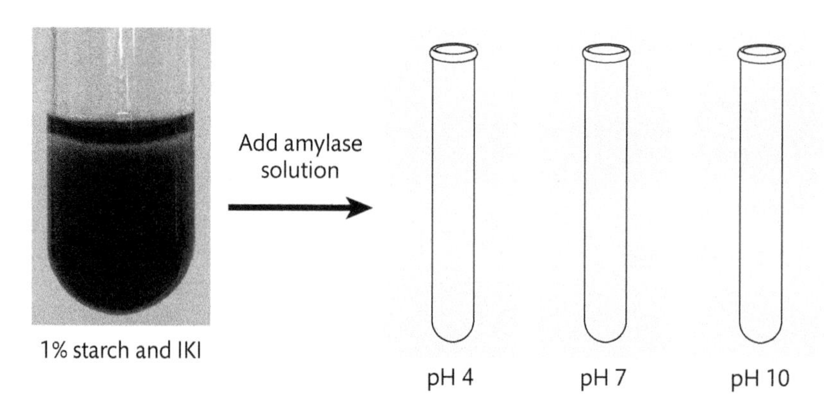

TABLE 28.3 Effect of pH on Amylase Activity

Starch Solution	Time for Solution to Become Colorless
pH 4	
pH 7	
pH 10	

6 After 30 minutes of amylase activity, remove the test tubes from the water bath and observe the three starch solutions. ▶

a. Which solution was first to become colorless?

____

b. Are any of the solutions not completely clear?

____

In the test tube figures above, use colored pencils to show the color of the solutions at each pH after 30 minutes of amylase activity. ▶

Do not discard your solutions at this time.

B Testing for the Presence of Starch and Disaccharides after Amylase Digestion

As amylase digestion proceeds, the concentration of starch decreases, and the concentration of disaccharides (maltose) increases. After a 30-minute period of amylase digestion, you can perform the starch test to determine whether any starch remains in your test solutions and the disaccharide test to detect the presence of maltose.

B1 Testing for the Presence of Starch

1. Label three test tubes as follows:
   - Starch, pH 4
   - Starch, pH 7
   - Starch, pH 10

Transfer 1 mL of each solution from the amylase digestion to the designated test tube.

2. Perform the IKI test: add 4 drops of IKI solution to each test tube. Observe and record the results of this test as positive or negative in Table 28.4. ▶

3. Is there any starch remaining in the test solutions? Explain by comparing the results for each pH.

▶ ______________________________

____

____

____

____

____

____

B2 Testing for the Presence of Maltose

1. Label three test tubes as follows:
   - Disaccharide, pH 4
   - Disaccharide, pH 7
   - Disaccharide, pH 10

Transfer 1 mL of each solution from the amylase digestion to the designated test tube.

2. Perform the Benedict's test: add 10 drops of Benedict's solution to each test tube, mix by swirling, and then heat (to just below boiling) for no more than 5 minutes in a water bath. Observe and record the results of this test as positive or negative in Table 28.4. ▶

3. Are disaccharides (maltose) present in the test solutions? Explain by comparing the results for each pH.

▶ ______________________________

____

____

____

____

____

____

TABLE 28.4 Starch and Disaccharide Tests for Solutions after Amylase Digestion

pH of Test Solution	Starch Test (+/−)	Disaccharide Test (+/−)
pH 4		
pH 7		
pH 10		

MAKING CONNECTIONS

At what pH is the amylase activity most effective? At what pH is it least effective? What do you think accounts for any differences that you observed in amylase activity?

▶ ______________________________

____

____

____

____

____

____

Want more practice? Go to: **MasteringA&P** > Study Area > Menu > Lab Tools > **PhysioEx** > Exercise 8: Chemical and Physical Processes of Digestion >

- Activity 1: Assessing Starch Digestion by Salivary Amylase
- Activity 2: Exploring Amylase Substrate Specificity
- Activity 3: Assessing Pepsin Digestion of Protein
- Activity 4: Assessing Lipase Digestion of Fat

BEFORE YOU MOVE ON . . .

« LOOKING BACK

The salivary glands continuously produce saliva; over a 24-hour period, about 1.0 to 1.5 liters of saliva are secreted into the oral cavity. Saliva contains the digestive enzyme amylase, which breaks down complex carbohydrates (polysaccharides) to disaccharides. In this laboratory exercise, you investigated how amylase activity is affected by pH.

Amylase, like all enzymes, contains a region called the **active site** that binds to a substrate molecule to form the **enzyme–substrate complex**. The active site has a unique three-dimensional shape that is designed for binding only to a specific substrate. Once the enzyme–substrate complex has formed, the enzyme acts to accelerate the formation of a specific **product**. Subsequently, the newly formed product detaches from the active site. The enzyme is not chemically changed during the overall reaction and is recycled for repeated use. The reaction between an enzyme and a substrate can be summarized by the following sequence of events:

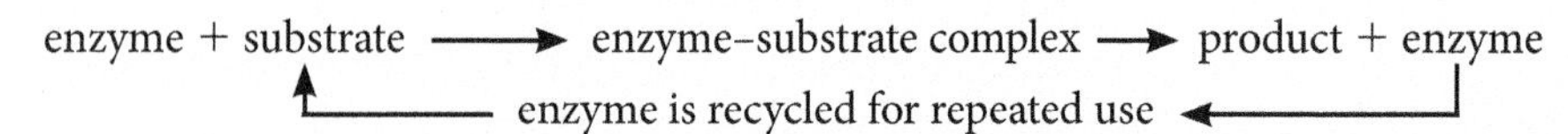

- **Identify the substrate, enzyme–substrate complex, and product for amylase activity.**

  ____

- **How would amylase function be affected if the enzyme were not recycled?**

  ____

- **How would carbohydrate digestion be affected if the active site on amylase were altered?**

  ____

- **There is individual variation in the amount of amylase produced by the salivary glands. Do you think that a person who produces a large amount of salivary amylase can digest carbohydrates more rapidly than a person who produces a small amount of amylase?**

  ____

  ____

  ____

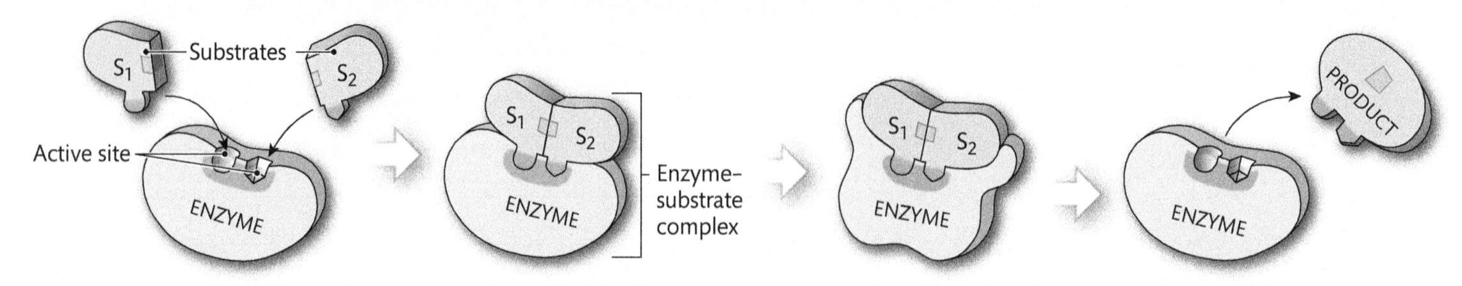

LOOKING FORWARD »

During the process of digestion, one important activity that occurs in the large intestine is the conservation of water and electrolytes through absorption from the feces back to the blood. The undigested substances and other waste products that remain are then released from the body during defecation. In the next laboratory exercise, you will study the anatomy and physiology of the urinary system, which has a similar function. During the process of producing urine, the kidneys conserve water and electrolytes and remove waste products from the blood. The wastes are then eliminated from the body during urination.

Name ______________________

Lab Section ______________________

Date ______________________

EXERCISE **28** **REVIEW SHEET**

Actions of a Digestive Enzyme

QUESTIONS 1–6: Match the term in column A with the appropriate description in column B.

A	B
1. Dextrins ______	**a.** This solution is used to detect the presence of disaccharides in a solution.
2. Amylase ______	**b.** In the small intestine, maltose is broken down to this monosaccharide.
3. Benedict's solution ______	**c.** This solution is used to detect the presence of starch in a solution.
4. Maltose ______	**d.** This disaccharide is the end product of starch digestion in the oral cavity.
5. Glucose ______	**e.** The first step in the digestion of starch in the oral cavity is the formation of these small polysaccharide molecules.
6. IKI solution ______	**f.** This carbohydrate-digesting enzyme is produced by salivary glands.

7. Ninhydrin is a chemical that reacts with amino acids in proteins to produce a purple color. The ninhydrin test was performed on three substances: distilled water, apple cider, and egg whites (which contain albumin protein). Identify each of these substances as one of the following:

a. positive control ______________________

b. negative control ______________________

c. experimental substance ______________________

Provide an explanation for your answers.

____

____

____

____

8. In the stomach, parietal cells produce hydrochloric acid, which is needed to convert the proenzyme pepsinogen to the protein-digesting enzyme pepsin. Hydrochloric acid dramatically lowers the pH of the stomach contents to around 2.5. When saliva is swallowed, will amylase continue to be functional when it enters the stomach? Explain by applying the results you obtained in Activity 28.3.

____

____

____

____

____

9. Like blood and other body fluids, saliva can be collected at a crime scene and used for DNA analysis during a forensic investigation. DNA is found in cells, but saliva is an acellular fluid. What do you think is the origin of the DNA found in saliva?

____

____

____

____

10. Starch and disaccharide tests were completed for three substances: X, Y, and Z. The test results are shown in the following table. Compare the three substances in terms of their starch and disaccharide content.

Starch and Disaccharide Tests for Various Substances

Substance	Test	Results	Conclusions
X	Starch test	Solution turned a dark blue	
	Disaccharide test	No change in color	
Y	Starch test	Solution turned blue, but not as dark as substance X	
	Disaccharide test	Solution changed from blue to yellow and then red	
Z	Starch test	No change in color	
	Disaccharide test	No change in color	

11. Class project:
In this laboratory exercise, you examined the effect of pH on amylase activity. Temperature is another factor that can influence the effectiveness of an enzyme. Design an experiment that tests the effect of temperature on amylase activity. Your experimental design should include the following components:

- An introduction and rationale for the experiment
- A list of materials that will be used
- A description of positive and negative control tests
- A description of methods for testing the effect of temperature on amylase activity
- Tables for recording data

Conduct the experiment that you designed and write a discussion that interprets and critically analyzes the data that you collect.

EXERCISE

29

Anatomy of the Urinary System

LEARNING OUTCOMES

These Learning Outcomes correspond by number to the laboratory activities in this exercise. When you complete the activities, you should be able to:

Activity 29.1 **Describe the structure and anatomical relationships of the organs in the urinary system and explain the function of each organ.**

Activity 29.2 **Compare the structure and function of the two types of nephrons.**

Activity 29.3 **Detail the blood supply of the kidneys.**

Activity 29.4 **Describe the microscopic structure of the organs in the urinary system.**

Activity 29.5 **Dissect a sheep or pig kidney and compare its structure with the structure of a human kidney.**

LABORATORY SUPPLIES

- Torso model
- Dissectible kidney model, coronal section
- Model of a nephron
- Model of the female pelvis
- Model of the male pelvis
- Compound light microscopes
- Prepared microscope slides:
  - human or other mammalian kidney
  - kidney, injected to demonstrate renal blood supply
  - ureter
  - urinary bladder (contracted and distended)
  - urethra
- Injected sheep or pig kidney for dissection
- Dissection tools (scalpel, probe, forceps, dissecting pan)

PRE-LAB QUIZ

Before you begin, read all the activities in Exercise 29 and the required reading in your textbook that is assigned by your instructor.

1. True or False: The kidneys are retroperitoneal organs. ____________
2. In both sexes, the relative position of the urinary bladder to other abdominopelvic organs can best be viewed in what type of section of the female and male pelvic cavities?

   ____
3. True or False: The urethra in the female is longer than the urethra in the male.

   ____
4. The renal sinus includes all the following structures *except*
   a. minor calyces.
   b. renal papillae.
   c. renal pelvis.
   d. major calyces.
5. True or False: Tubular reabsorption refers to substances in the nephron tubules that are returned to the blood. ____________
6. The nephron loops of juxtamedullary nephrons are surrounded by capillaries called
   a. peritubular capillaries.
   b. afferent arterioles.
   c. efferent arterioles.
   d. vasa recta.
7. Renal corpuscles are located in what region of the kidney? ____________
8. What type of epithelium is found in the mucous membrane of the ureter and urinary bladder? ____________
9. During this laboratory exercise, you will dissect a kidney from what animal?

   ____
10. During the kidney dissection, you will be asked to make an incision to give you what type of section? ____________

Activity **29.1**

Examining the Gross Anatomy of the Urinary System

A On a torso model, remove all the organs from the abdominal cavity. Now you can identify the two retroperitoneal kidneys, located on each side of the vertebral column on the posterior abdominal wall. They extend, roughly, from the T_{12} vertebra to the L_3 vertebra.

Why does the right kidney have a more inferior position than the left kidney?

▶ ______________________________

Which organ is directly lateral to the left kidney? ▶ ______________________________

Which organs are anterior to the left kidney? ▶ ______________________________

1. Identify the **adrenal glands** that rest on the superior surface of each kidney.

2. Observe the lateral convex surface and medial concave surface and identify the **hilum** along the center of the concave surface. At the hilum, the renal artery and nerves enter the kidney, and the renal vein and ureter exit the kidney.

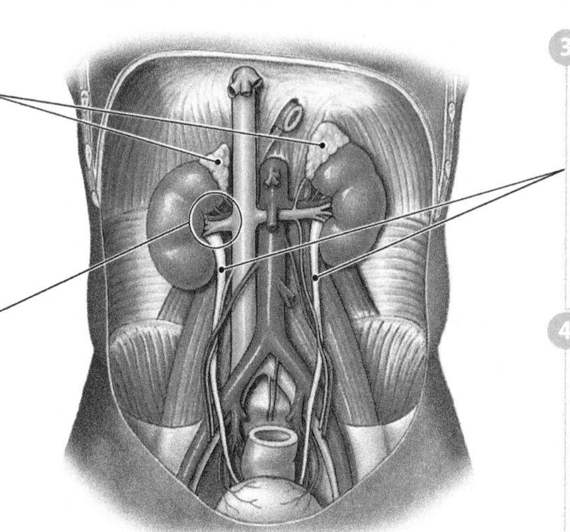

3. Identify the two **ureters** where they exit the kidneys. These retroperitoneal, tubular organs are approximately 25 cm (10 in.) long. Observe that the ureters travel inferiorly along the posterior abdominal wall and enter the pelvic cavity, where they connect to the posterior wall of the urinary bladder.

4. What is the function of the ureters?

▶ ______________________________

____

____

B In a coronal section of the kidney, three regions can be identified. The outer region is the **renal cortex.** The middle region, the **renal medulla,** contains renal pyramids and renal columns. The cortex and medulla contain the various urine-producing tubules that compose the **nephrons.** The inner region of the kidney is a cavity known as the **renal sinus,** which consists of the minor and major **calyces** (singular = **calyx**) and the **renal pelvis.** The renal pelvis collects urine that is produced in the nephrons and transports this fluid to the ureter.

1. Locate the outer **renal cortex.**

2. In the renal medulla, identify the cone-shaped **renal pyramids.**

   Notice that the base of each renal pyramid forms a border with the cortex and that the apex, known as a **renal papilla,** projects inwardly.

3. Neighboring renal pyramids are separated by an inward extension of cortical tissue known as a **renal column.**

4. A **kidney lobe** contains a renal pyramid and the adjacent renal cortex and renal columns. Urine production occurs in the kidney lobes.

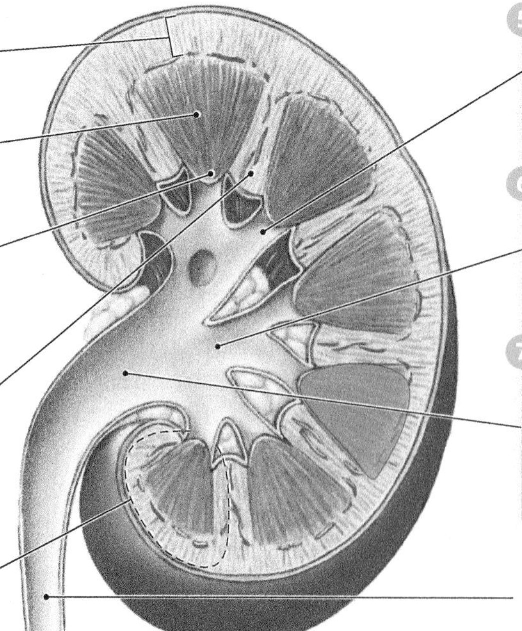

5. Locate a **minor calyx.** Urine formed within a renal pyramid drains through the renal papilla to enter the minor calyx.

6. Follow the minor calyx inward until it joins at least one more minor calyx. The larger area marked by this junction is a **major calyx.**

7. Locate the **renal pelvis,** a funnel-shaped cavity continuous with the ureters. The renal pelvis collects urine from three to six major calyces and transmits it to the ureter.

C In a midsagittal section of the female and male pelvic cavities, the relative position of the urinary bladder to other abdominopelvic organs can be observed. In both sexes: ▶

a. the urinary bladder is anterior to what part of the large intestine (Structure A in the figure below)?

____

b. the urinary bladder is posterior to what cartilaginous joint (Structure B in the figure below)?

____

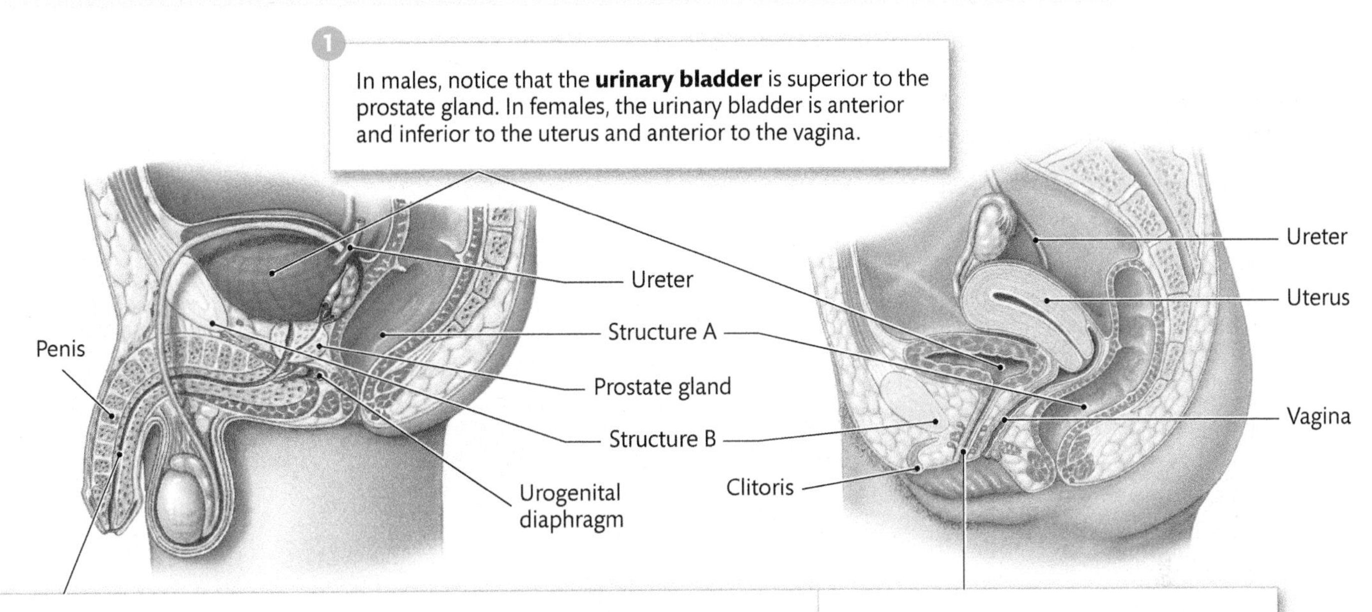

2 The **urethra** is a muscular tube that conveys urine from the bladder to the outside of the body. In males, observe that the urethra is relatively long (approximately 20 cm or 8 in.) and has the following three sections:

- The **prostatic urethra** passes through the prostate gland.
- The **membranous urethra** passes through a band of muscle called the urogenital diaphram.
- The **spongy urethra** travels through the shaft of the penis.

In females, observe that the urethra is relatively short (approximately 3.5 cm or 1.5 in.) and travels anterior to the vagina. The external urethral orifice is located anterior to the vaginal orifice and posterior to the clitoris.

D The mucous membrane of the contracted urinary bladder contains numerous folds called **rugae**. When the bladder fills with urine and the wall stretches, the rugae disappear.

1 What is the function of the urinary bladder?

▶ ____

____

Middle umbilical ligament

Ureter

Lateral umbilical ligament

2 On a dissectible urinary bladder model, identify the rugae along the inner wall.

3 Identify the **trigone,** which is a triangular region marked by the connections of the two ureters and the urethra to the bladder wall. This region lacks rugae and remains smooth at all times. It is noteworthy because bladder infections seem to occur most often in this area. The reason is not known.

MAKING CONNECTIONS

At the junction between the bladder and urethra, a thick band of smooth muscle surrounds the urethra to form the **internal urethral sphincter.** A band of skeletal muscle, the **external urethral sphincter,** is located near the external urethral orifice in females and just inferior to the prostate gland in males. What is the basic function of these sphincters? Because each sphincter is composed of a different muscle type, their functions are slightly different. Explain why.

▶ ____

____

____

____

____

Activity **29.2**

Examining the Nephron

The functional units of the kidney are the nephrons. They consist of various types of tubules that closely interact with capillaries during urine formation. To produce urine, the nephrons perform three activities:

- **Glomerular filtration:** Water, glucose, amino acids, electrolytes, and nitrogen-containing wastes are passed from blood to the nephrons along a pressure gradient.
- **Tubular reabsorption:** Useful substances such as water, glucose, amino acids, and ions are returned to the blood.
- **Tubular secretion:** Unwanted substances such as excess potassium and hydrogen ions, drugs, creatinine, and metabolic acids are transported from the capillaries into the tubules.

A A nephron consists of a glomerular capsule and three types of renal tubules.

1 Each nephron begins with a **glomerular capsule.** On a model, identify one of these cup-shaped structures.

Each capsule surrounds a tuft of capillaries called a **glomerulus** (plural = **glomeruli**).

Together the two structures make up a **renal corpuscle.** In each corpuscle, fluid passes across a filtration membrane from the glomerulus to the glomerular capsule. The fluid entering a glomerular capsule is called the **filtrate.**

2 When the filtrate flows into the **renal tubules,** it is called **tubular fluid.** This fluid is transported by three additional components of a nephron in the following sequence: the **proximal convoluted tubule,** the **nephron loop** (**descending limb** and **ascending limb**), and the **distal convoluted tubule.** Identify these tubules and notice that they form a long, twisting passageway leading to a **collecting duct.** The convoluted tubules and the nephron loop are capable of both reabsorption and secretion. Most reabsorption occurs in the proximal convoluted tubules, however, whereas the distal convoluted tubules have a more important role in secretion.

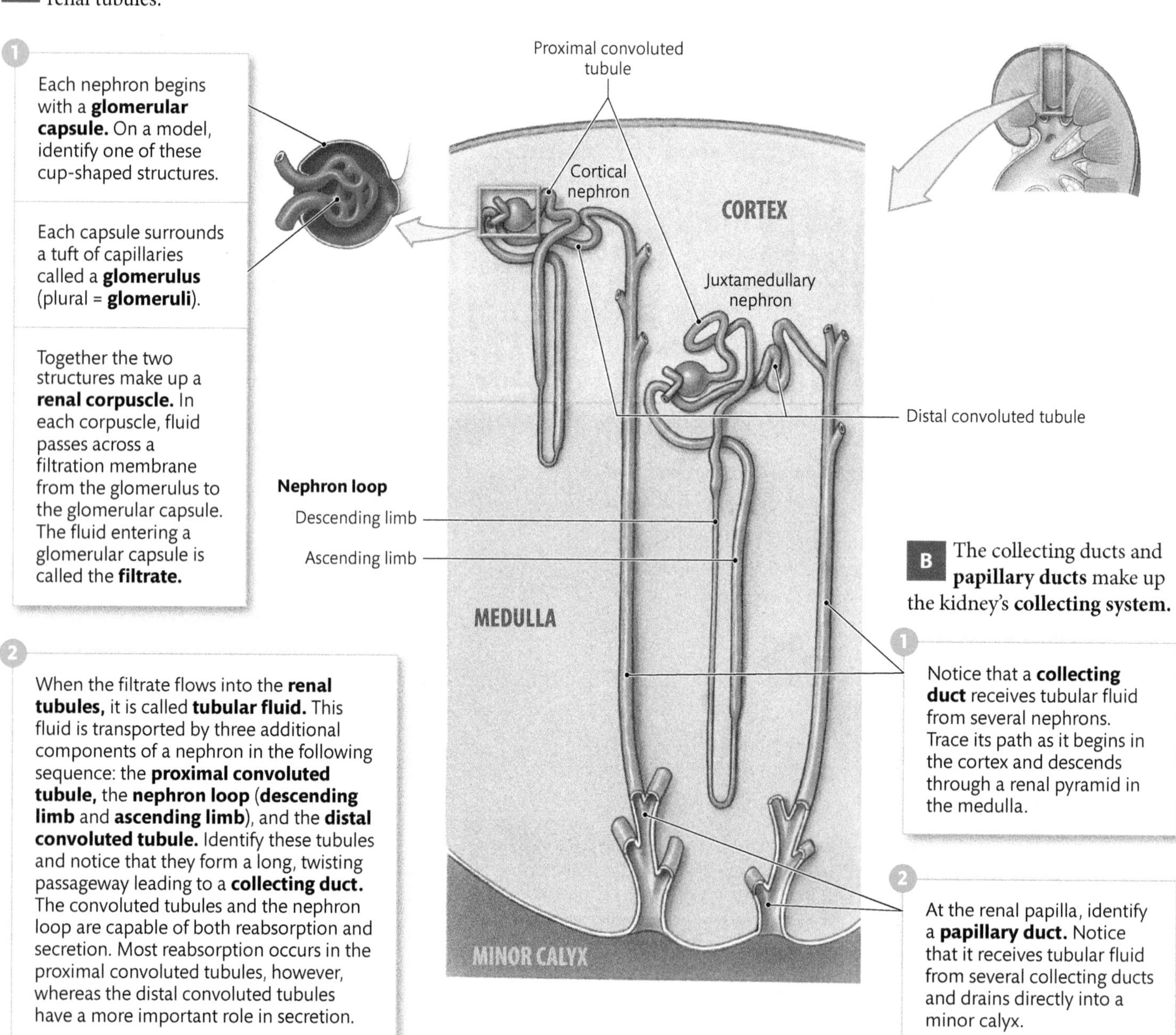

B The collecting ducts and **papillary ducts** make up the kidney's **collecting system.**

1 Notice that a **collecting duct** receives tubular fluid from several nephrons. Trace its path as it begins in the cortex and descends through a renal pyramid in the medulla.

2 At the renal papilla, identify a **papillary duct.** Notice that it receives tubular fluid from several collecting ducts and drains directly into a minor calyx.

■ Want more practice? Go to: **MasteringA&P** > Study Area > Menu > Lab Tools > PAL > Anatomical Models > Urinary System > Renal corpuscle and Nephron photos

C Carefully examine a model of a nephron or the figure on the preceding page. Identify the location of each segment of a nephron and collecting system and complete Table 29.1. ▶

TABLE 29.1 Nephron and Collecting System Structures and Locations

Nephron/Collecting System Structure	Location (cortex, medulla, or both)
Renal corpuscle	
Proximal convoluted tubule	
Nephron loop, descending limb	
Nephron loop, ascending limb	
Distal convoluted tubule	
Collecting duct	
Papillary duct	

D Notice that there are two types of nephrons: **cortical nephrons** and **juxtamedullary nephrons.**

Observe the locations of the two types of nephrons. How do the positions of the renal corpuscles and nephron loops in the cortex and renal pyramid differ in each type?

▶ ______________________________

____

____

____

____

____

____

____

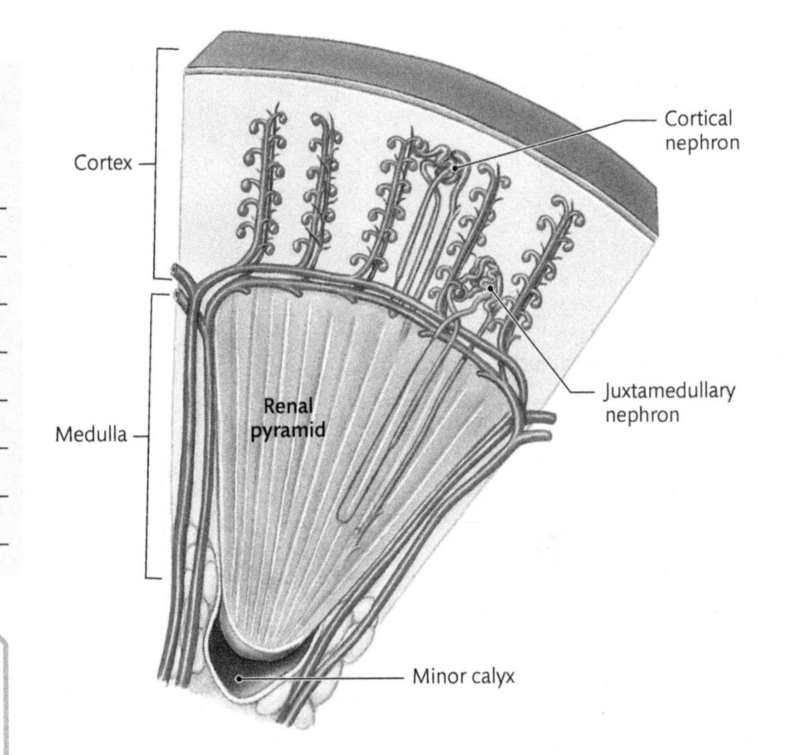

Word Origins

Glomerulus comes from the Latin term for "ball" or "globe." Indeed, the glomerulus is like a ball or tuft of capillaries. This arrangement, rather than a single straight capillary, dramatically increases the amount of capillary surface area through which filtration can occur, allowing the kidneys to filter blood more rapidly.

MAKING CONNECTIONS

Summarize the functional difference between cortical nephrons and juxtamedullary nephrons. Refer to your textbook if necessary.

▶ ______________________________

____

____

____

____

Activity 29.3

Studying the Blood Supply to the Kidney

Under normal conditions, about 25 percent of total cardiac output travels through the kidneys at a rate of about 1.2 liters per minute.

A Each kidney receives its blood supply from a **renal artery.** The renal artery and its branches deliver blood to the glomeruli, which are surrounded by glomerular capsules in the renal cortex and are the sites of blood filtration.

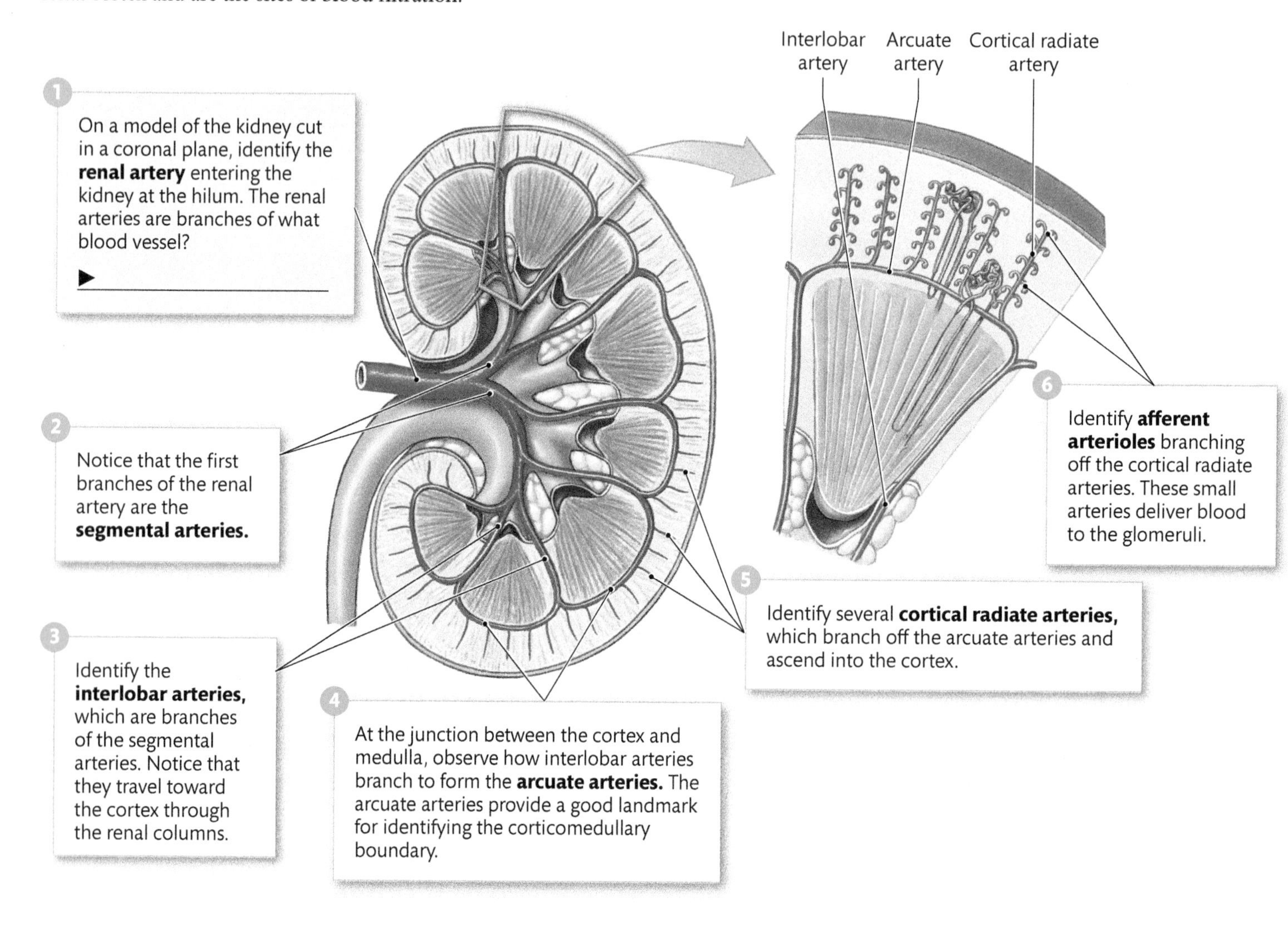

B **Peritubular capillaries** and the **vasa recta** form complex networks of blood vessels around the nephron tubules.

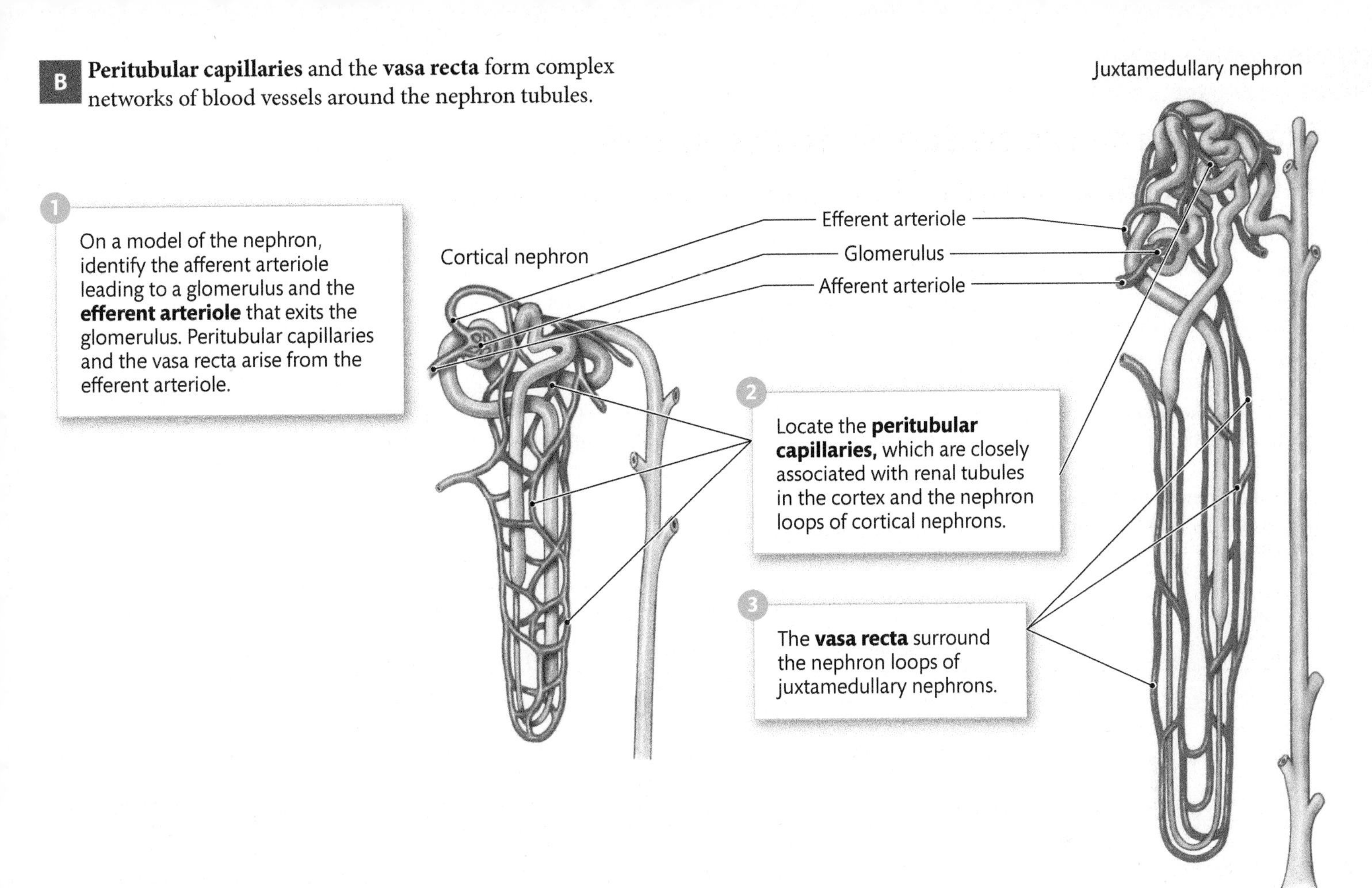

1 On a model of the nephron, identify the afferent arteriole leading to a glomerulus and the **efferent arteriole** that exits the glomerulus. Peritubular capillaries and the vasa recta arise from the efferent arteriole.

2 Locate the **peritubular capillaries,** which are closely associated with renal tubules in the cortex and the nephron loops of cortical nephrons.

3 The **vasa recta** surround the nephron loops of juxtamedullary nephrons.

C The capillaries that surround the nephron tubules empty into the venous system that drains blood from the kidney.

1 The veins in the kidney parallel the arteries and are assigned identical names. Identify these veins on a model.

2 On a torso model, identify the renal vein. Into which vein does the renal vein empty?

▶ ______________________

____

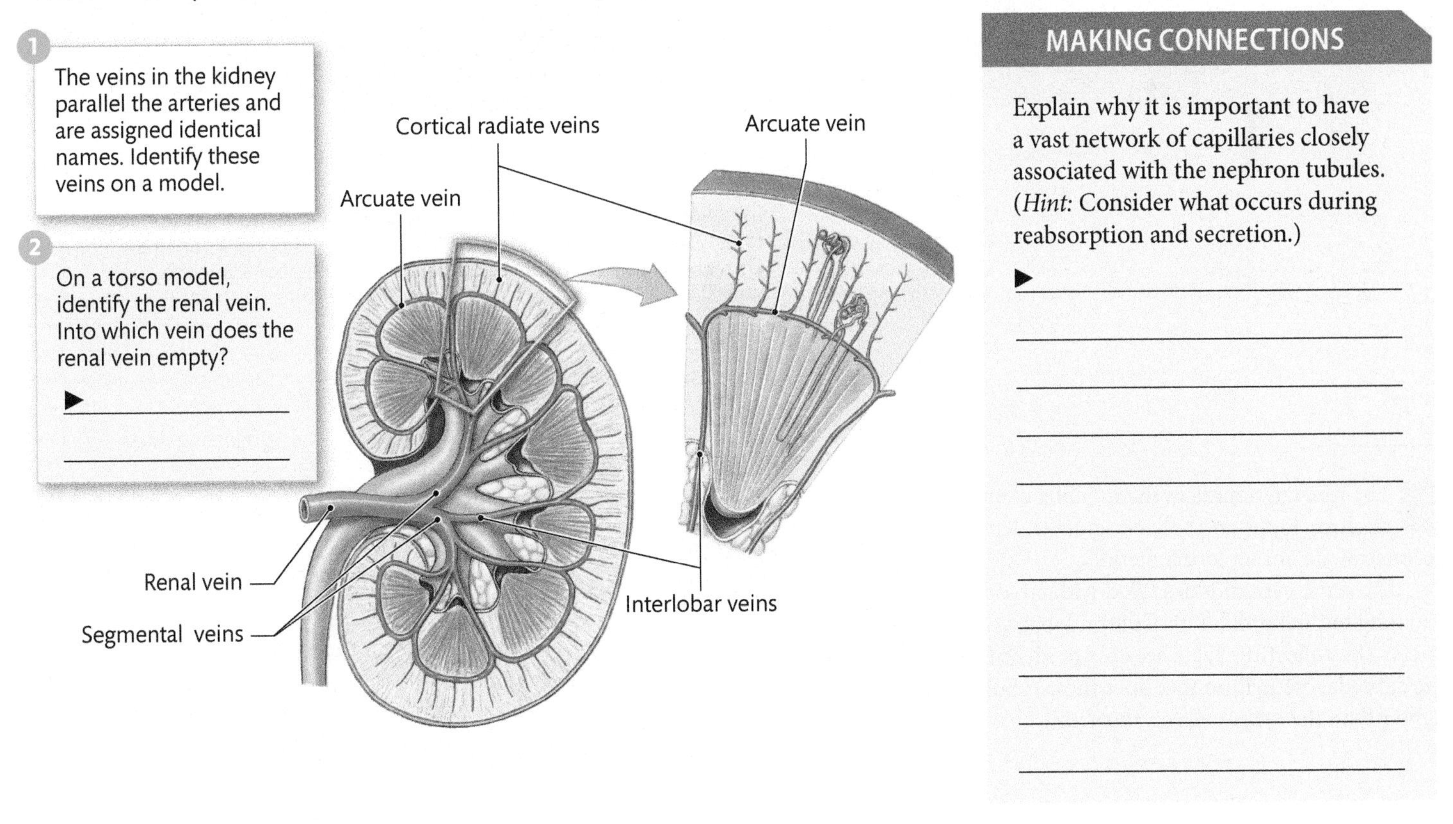

MAKING CONNECTIONS

Explain why it is important to have a vast network of capillaries closely associated with the nephron tubules. (*Hint:* Consider what occurs during reabsorption and secretion.)

▶ ______________________

____

____

____

____

____

____

____

____

____

Examining the Microscopic Structure of the Urinary System

The histology of the kidney is centered on the structure of the nephrons. As described earlier, each renal corpuscle consists of a glomerular capsule and a glomerulus. These structures have a distinctive appearance and are easy to identify throughout the renal cortex. The renal tubules and collecting ducts are found in both the cortex and medulla. They are lined mostly by a simple cuboidal epithelium, although simple squamous and simple columnar epithelia are found in some regions.

The accessory urinary organs—the ureter, urinary bladder, and urethra—are unique because they possess a prominent smooth muscle layer and a mucous membrane with a transitional epithelium. This epithelium can undergo dramatic structural transformations to adjust for the changing physical conditions of the organs where it is located.

A When viewed under a microscope, the renal cortex can be easily identified because it contains numerous renal corpuscles.

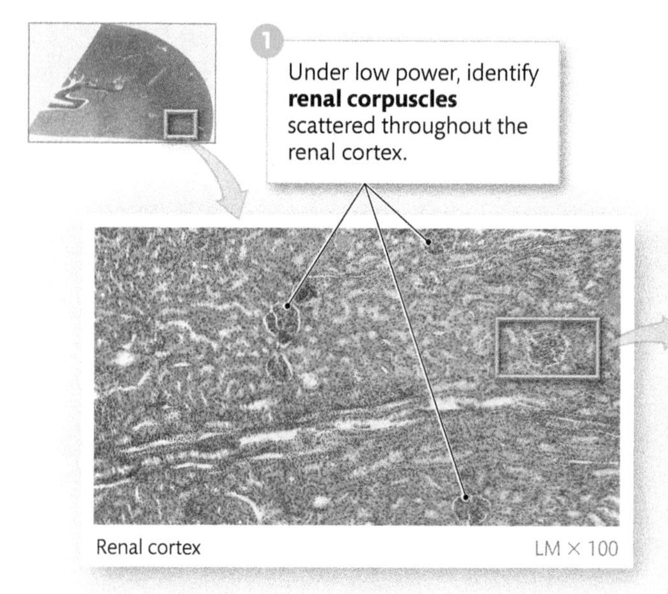

Renal cortex LM × 100

2

Under high power, the **glomerulus** in a renal corpuscle will appear as a spherical mass of pink-to-red-staining tissue surrounded by a clear capsular space.

The **capsular space** is the cavity within the glomerular capsule that receives the filtrate.

Identify the type of **epithelium** that lines the outer wall of the glomerular capsule.

▶ ______________________

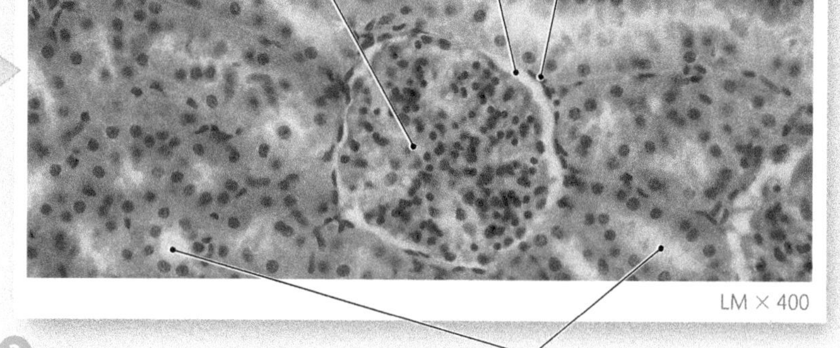

LM × 400

3

In regions surrounding the renal corpuscles, locate cross-sectional and longitudinal profiles of **renal tubules.** Most of them are proximal convoluted tubules or distal convoluted tubules, but some may be collecting ducts. What type of epithelium lines these tubules?

▶ ______________________

B The renal pyramids in the medulla contain nephron loops and collecting ducts. Renal corpuscles are absent in the medulla.

The renal pyramids are filled with cross-sectional and longitudinal profiles of nephron loops and collecting ducts. The collecting ducts are easy to identify because the cuboidal epithelium that lines these tubules stains lighter than the other cells on the slide.

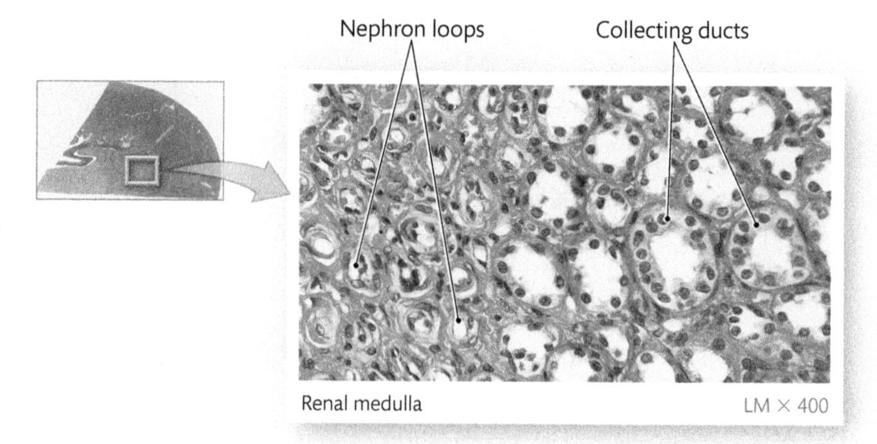

Renal medulla LM × 400

C The accessory urinary structures store urine temporarily and transport it to the outside of the body.

1 Examine a slide of the ureter, first under low power and then high power, and identify the following three tissue layers:

- The **mucous membrane (mucosa)** contains **transitional epithelium** and an underlying **lamina propria** (loose connective tissue).
- The **smooth muscle layer (muscularis)** contains inner longitudinal and outer circular bands.
- The **adventitia** contains fibrous connective tissue.

2 Examine a slide of the urinary bladder under low power. Your slide will have two sections of tissue. Notice that one is much thicker than the other. The thicker section is the contracted wall of an empty bladder; the thinner section is the distended wall of a full bladder.

3 Examine the section of the contracted bladder. Identify the mucous membrane, which consists of transitional epithelium and the lamina propria. Observe that the epithelium is relatively thick, with several layers of cuboidal cells. Also notice that the mucous membrane contains rugae.

Just below the mucous membrane, identify a second layer of connective tissue, called the submucosa, and a region of smooth muscle. The muscle region consists of three layers of intersecting smooth muscle fibers, collectively referred to as the **detrusor muscle.**

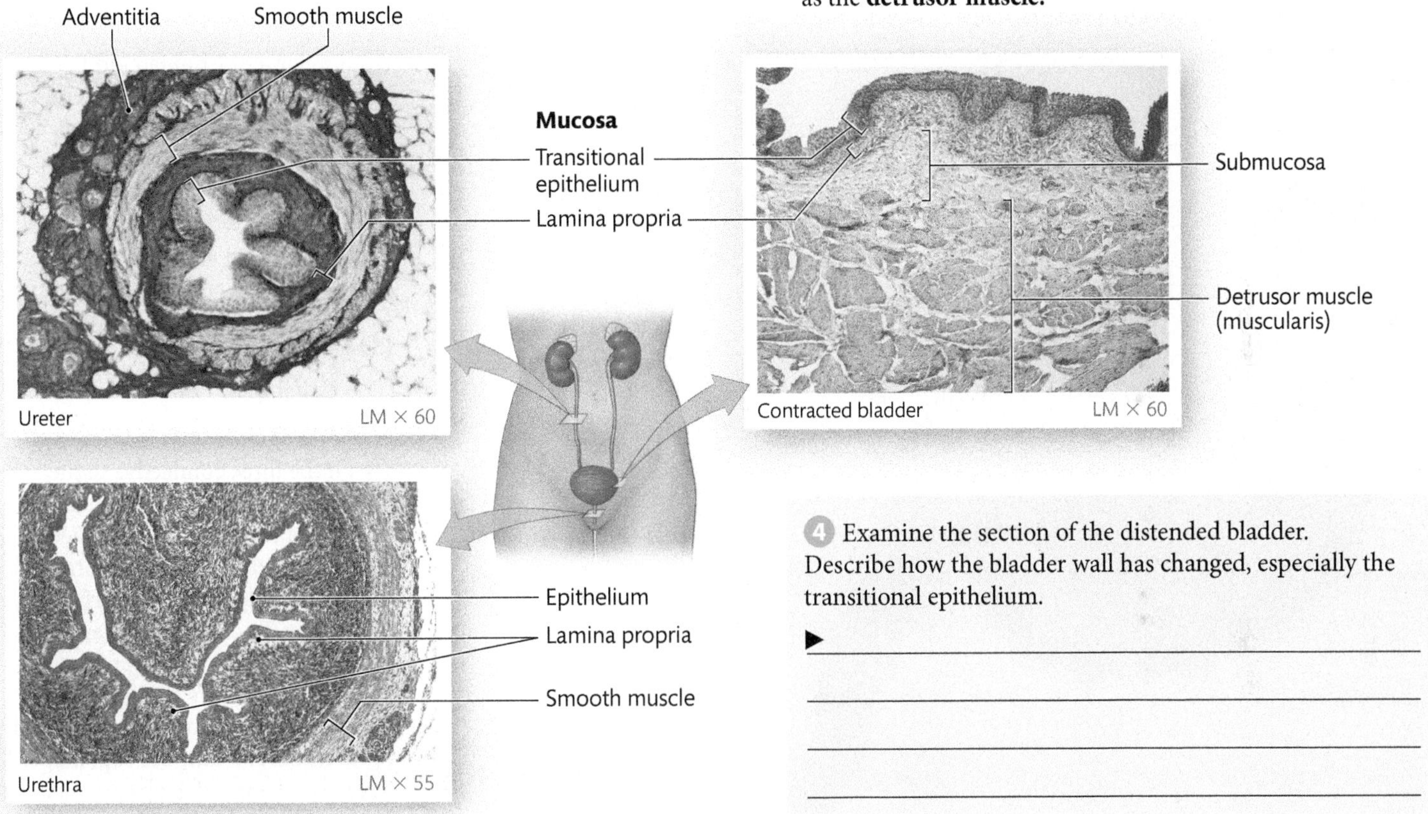

4 Examine the section of the distended bladder. Describe how the bladder wall has changed, especially the transitional epithelium.

▶ ______________________________

____

____

____

5 Examine a slide of the urethra under low power. Identify the epithelium and lamina propria in the mucous membrane. Identify the circular layers of smooth muscle.

6 Switch to high power and examine the mucous membrane more closely. The epithelium lining the mucous membrane will vary depending on the region of the urethra that you are viewing. Near the junction with the urinary bladder, the epithelium is transitional. There is a gradual change to pseudostratified columnar toward the middle section of the urethra. Near the urethral orifice, the epithelium changes to stratified squamous.

Based on the epithelial type, what region of the urethra are you viewing?

▶ ______________________________

MAKING CONNECTIONS

The presence or absence of renal corpuscles is a reliable criterion for determining whether you are viewing the renal cortex or the renal medulla. Explain why.

▶ ______________________________

____

____

____

____

Activity **29.5**

Dissecting a Kidney

In this activity, you will dissect a sheep or pig kidney. Pig kidneys are much larger and flatter than sheep kidneys. Both types are quite similar, anatomically, to the human kidney, however, and thus are excellent models for studying renal structure. As you dissect, have models, illustrations, or photographs of the human kidney available so that you can make structural comparisons. The kidneys you dissect may be double-injected specimens, meaning that the blood vessels have been injected with colored latex for easier identification (red = arteries; blue = veins).

A Before making any incisions, examine the **external anatomy** of the kidney.

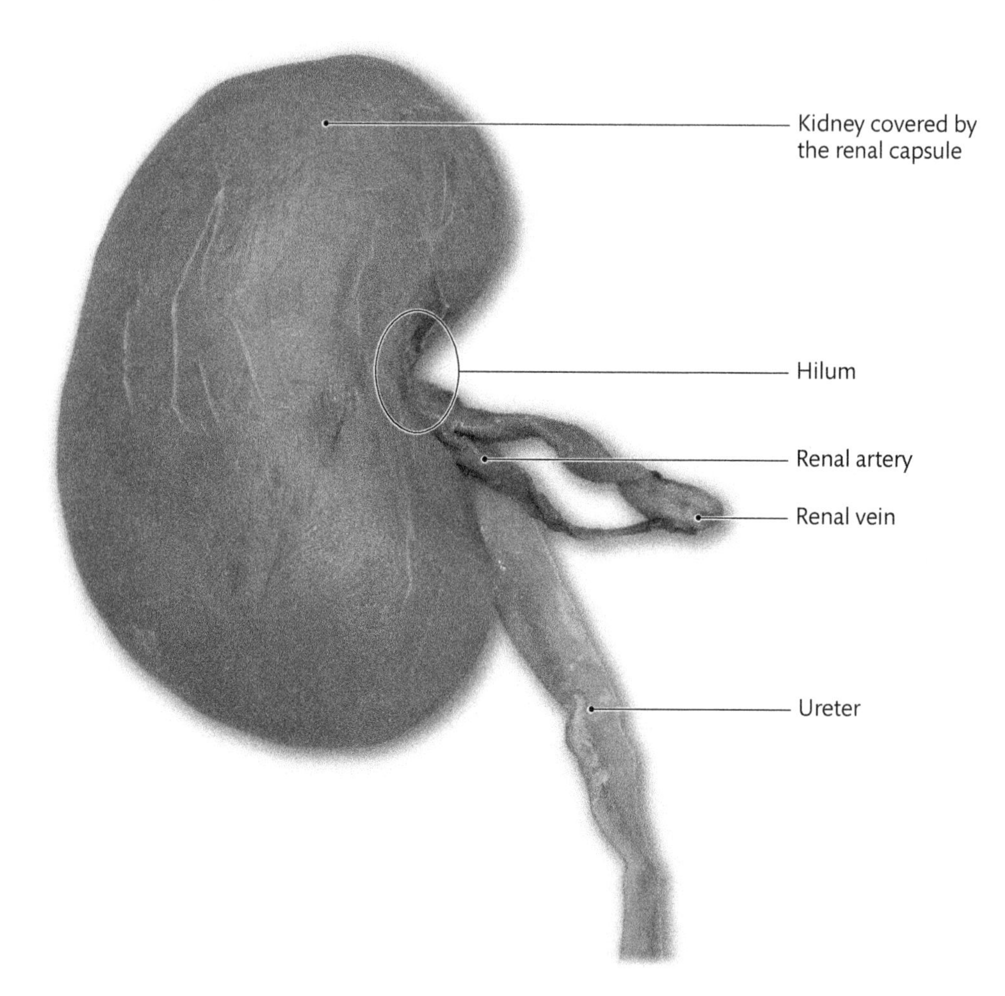

1 Similar to the human kidney, the sheep (or pig) kidney has a lateral convex surface and a medial concave surface. The hilum is located along the medial surface. Identify the ureter, renal artery, and renal vein at the hilum. From superior to inferior, these structures are positioned in the following order:

- Renal artery
- Renal vein
- Ureter

2 Of the three connective tissue layers that cover the kidney, only the innermost **renal capsule** is intact. It appears as a thin, shiny membrane that adheres tightly to the surface of the kidney. To observe this capsule more closely, lift a portion of the membrane from the surface with forceps. The other two coverings, the **renal fascia** and **adipose capsule,** have been removed, but remnants may remain near the hilum.

IN THE CLINIC

Urinary Tract Infections

For two reasons, females are more susceptible to urinary tract infections than males. First, the female urethra is much shorter and inflammatory infections of this structure **(urethritis)** can easily spread to the bladder **(cystitis)** and possibly to the kidneys (**pyelitis** or **pyelonephritis**). Second, the external urethral orifice is very close to the anal opening. Fecal bacteria can be transferred to the urethra, particularly if an individual wipes with toilet paper in a posterior-to-anterior direction after defecation.

■ Want more practice? Go to: MasteringA&P > Study Area > Menu > Lab Tools > PAL > Cat > Urinary System > Sheep Kidney

B To study the **internal anatomy,** use a scalpel to make an incision that will divide the kidney into anterior and posterior halves. Begin at the superior margin and cut along the lateral convex surface, toward the medial concave surface. Try to cut as close to the midline as possible. You can cut completely through the organ or, if you prefer, keep the two halves connected at the hilum. The resulting incision will give you a coronal section of the kidney.

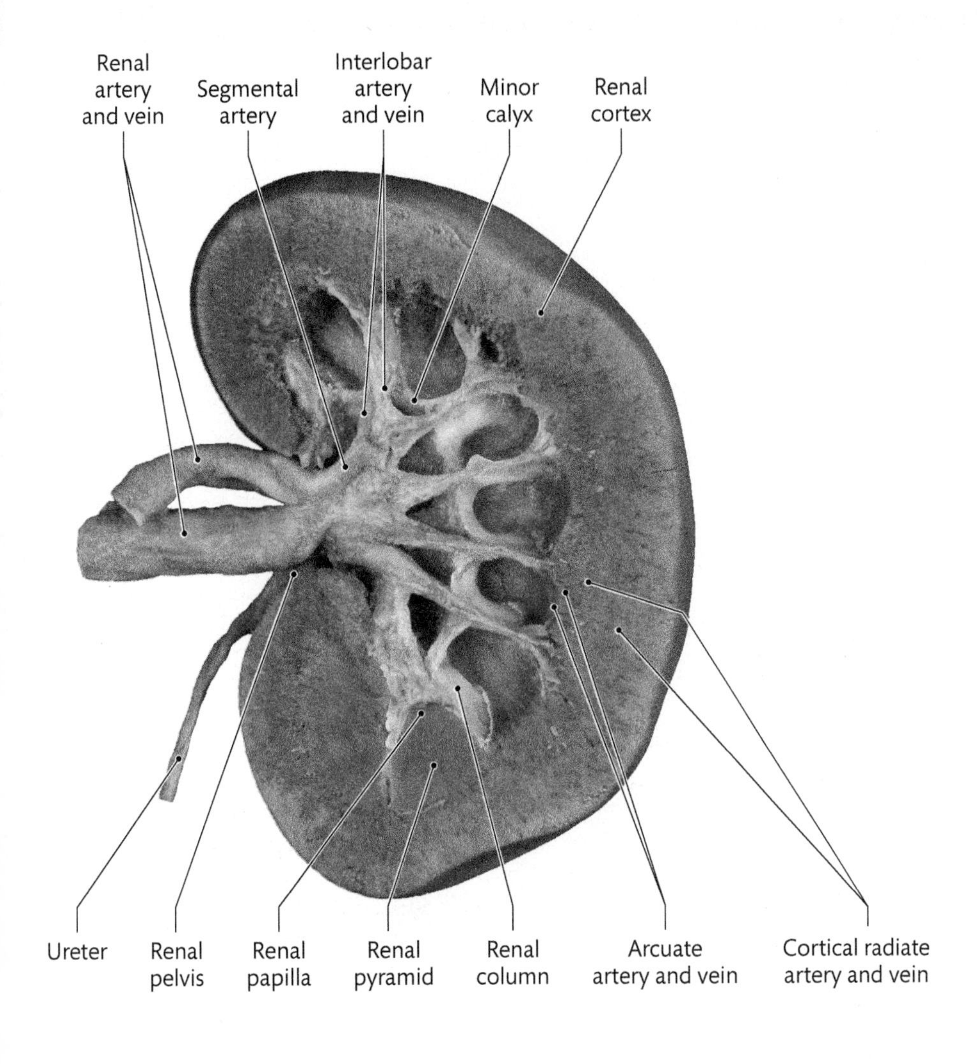

1 Identify the following internal structures:

- Renal cortex, the outer layer with a granular texture and light color.
- Renal medulla, the middle layer, which contains alternating renal pyramids and renal columns. Observe that the pyramids have a darker color than the cortex and have a striped appearance due to the longitudinal arrangement of kidney tubules. The apex of each pyramid (renal papilla) projects inward, and the base borders the cortex. The renal columns, located between the pyramids, are extensions of, and have the same appearance as, the cortex.
- Renal sinus, the funnel-shaped inner region into which urine drains before flowing into the ureter. The sinus consists of minor calyces, major calyces, and the renal pelvis. Identify several minor calyces, located adjacent to the renal papillae. Several minor calyces converge to form a major calyx, and major calyces merge to form the renal pelvis.

2 Identify some of the blood vessels that serve the kidney. Find segmental arteries branching off the renal artery. Carefully dissect within a renal column and locate an interlobar artery and vein. Along the corticomedullary border, identify the arcuate blood vessels. Finally, try to identify the fragile cortical radiate arteries and veins as they travel through the cortex.

MAKING CONNECTIONS

Compare the external anatomy of the human kidney with the sheep or pig kidney by identifying similarities and differences in size, shape, and thickness. Compare the internal anatomy by identifying similarities and differences in the organization of structures as they appear in a coronal section. ▶

	Similarities	Differences
External anatomy		
Internal anatomy		

BEFORE YOU MOVE ON . . .

‹‹ LOOKING BACK

The kidneys are the principal organs of the urinary system. Each kidney contains about one million nephrons that consist of microscopic tubules located in the cortex and renal pyramids of the medulla. The nephrons produce urine by carrying out three activities: glomerular filtration, tubular reabsorption, and tubular secretion. The urine that is produced by the nephrons eventually drains into the renal pelvis, which leads to a ureter. Each ureter transports urine to the urinary bladder where it is stored until its release to the outside, via the urethra, during micturition (urination).

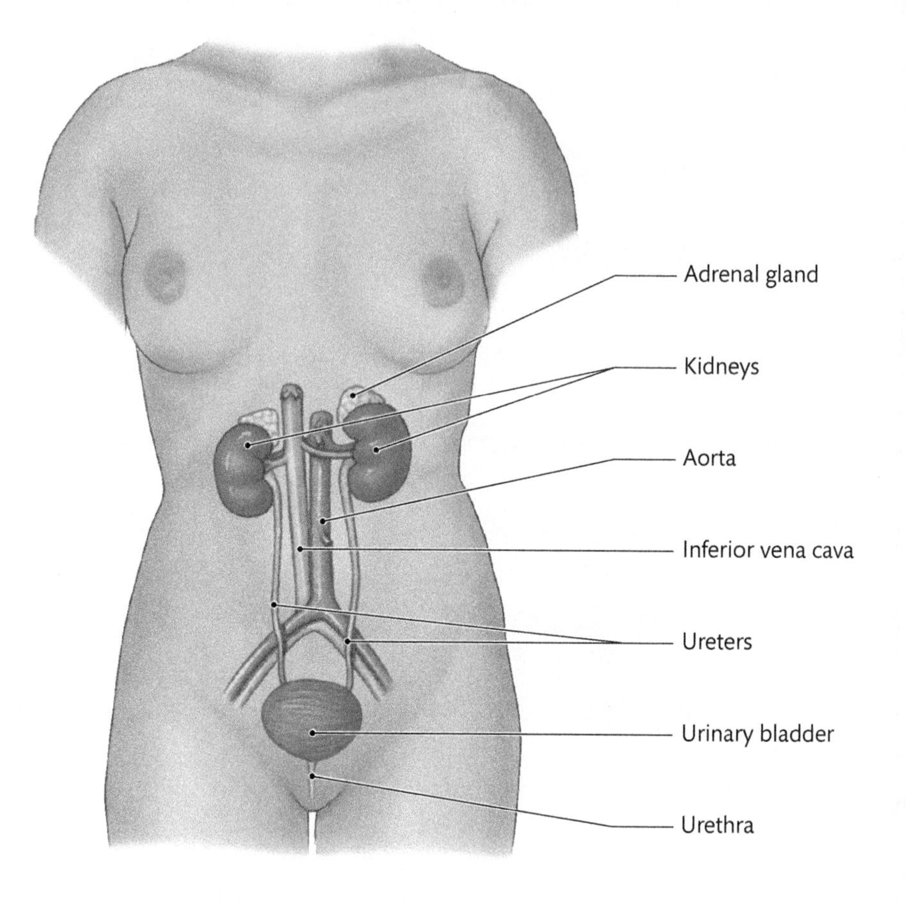

As the kidneys produce urine, they are able to perform the following essential functions:

- removing metabolic wastes and toxins from blood
- conserving glucose, water, and electrolytes
- maintaining osmotic and pH balance in blood

The kidneys also have an endocrine function. They produce and secrete two hormones:

- erythropoietin, which promotes red blood cell production in bone marrow
- renin, which helps maintain blood pressure

Speculate on the connection between the kidneys' urinary and endocrine functions.

▶ ______________________________

____

____

____

____

LOOKING FORWARD ››

In this laboratory exercise, you examined the gross and microscopic anatomy of the organs in the urinary system. In the next laboratory exercise, you will take a closer look at the process of urine production and examine the physical and chemical features of a urine sample.

Name ______________________

Lab Section ______________________

Date ______________________

EXERCISE 29 REVIEW SHEET

Anatomy of the Urinary System

QUESTIONS 1–5: Match the description in column A with the appropriate structure in column B. Answers may be used once or not at all.

A		B
1. In males, the first part of the urethra travels through this structure.	______	**a.** Rectum
2. In both sexes, this structure is posterior to the urinary bladder.	______	**b.** Spleen
3. This structure rests on the superior surface of each kidney.	______	**c.** Liver
4. The right kidney is more inferior than the left kidney due to the position of this structure.	______	**d.** Vagina
5. In females, the urethra is directly anterior to this structure.	______	**e.** Symphysis pubis
		f. Adrenal gland
		g. Prostate gland

QUESTIONS 6–13: Match the labeled structure in the diagram with the appropriate description.

a. The base of this structure borders the renal cortex.

b. Proximal and distal convoluted tubules are located in this region.

c. This structure transports urine to the urinary bladder.

d. Papillary ducts drain into this structure.

e. This structure is an inward extension of the renal cortex.

f. Major calyces drain urine directly into this structure.

g. This structure forms the apex of a renal pyramid.

h. Several minor calyces merge to form this structure.

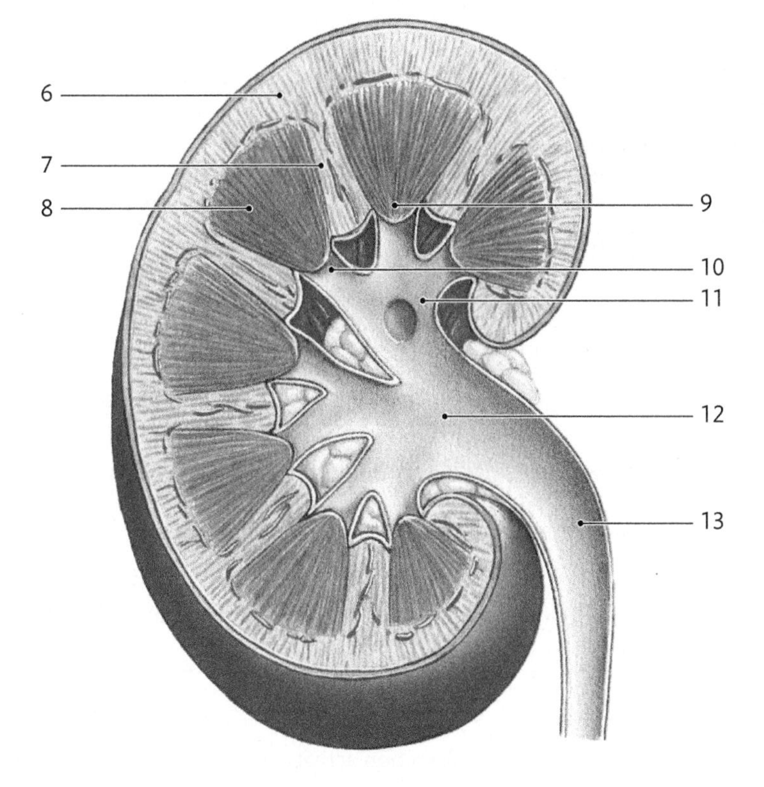

6. ______________

7. ______________

8. ______________

9. ______________

10. ______________

11. ______________

12. ______________

13. ______________

14. In the following diagram, identify each structure by labeling it with the indicated color.

Descending limb of the nephron loop = **green**
Distal convoluted tubule = **blue**
Glomerulus = **red**
Efferent arteriole = **orange**
Proximal convoluted tubule = **brown**
Ascending limb of the nephron loop = **purple**
Afferent arteriole = **black**
Glomerular capsule = **pink**
Cortical radiate artery = **tan**
Collecting duct = **yellow**

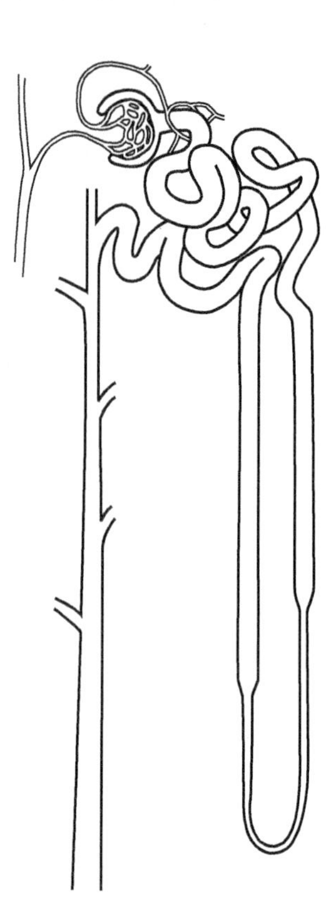

15. Complete the following table.

Artery	Description
a.	Delivers blood to the kidney; a branch of the abdominal aorta
Arcuate arteries	b.
c.	Deliver blood to the glomeruli
Segmental arteries	d.
e.	Arteries that travel through the renal columns
Cortical radiate arteries	f.
g.	Receive blood from the glomeruli

16. a. How is transitional epithelium unique compared with other epithelial types?

____

____

b. Why is it important that this type of epithelium be located in the wall of the urinary bladder?

____

____

17. During your kidney dissection activity, you made an incision to produce a coronal section of the kidney. Why were you instructed to make this type of section rather than a cross or sagittal section?

____

____

____

EXERCISE 30

Urinary Physiology

LEARNING OUTCOMES

These Learning Outcomes correspond by number to the laboratory activities in this exercise. When you complete the activities, you should be able to:

Activity 30.1 **Describe the process of urine production in the kidney.**

Activity 30.2 **Compare the normal and abnormal physical characteristics of urine.**

Activity 30.3 **Analyze the normal and abnormal chemical characteristics of urine.**

Activity 30.4 **Examine and discuss the significance of microscopic sediments in urine.**

LABORATORY SUPPLIES

- Nephron model
- Disposable gloves
- Safety eyewear
- Face masks
- Autoclave bags
- Biohazard waste bags/containers
- Autoclave (if available)
- 70% alcohol solution
- 10% bleach solution
- Student urine samples, collected in 500-mL collection cups
- Individual or combination urinalysis dipsticks
- pH paper (if pH test is not included on the dipsticks)
- Urinometer
- Test tube rack
- Centrifuge and centrifuge tubes
- Pasteur pipettes
- Wax marking pencils
- Compound light microscopes
- Microscope slides and coverslips
- 10% methylene blue solution

PRE-LAB QUIZ

Before you begin, read all the activities in Exercise 30 and the required reading in your textbook that is assigned by your instructor.

1. True or False: The filtrate normally lacks blood cells and large plasma proteins. ______
2. True or False: The blood pressure in the glomerular capillaries is called the capsular hydrostatic pressure. ______
3. The net filtration pressure is
   a. the osmotic pressure that opposes filtration.
   b. the fluid pressure in the capsular space of the glomerular capsule.
   c. the force that moves fluid across the glomerular capillaries into the capsular space.
   d. the osmotic pressure in the capsular space.
4. During this laboratory exercise, you will analyze all the following physical characteristics of urine *except*
   a. color.
   b. volume.
   c. odor.
   d. transparency.
   e. specific gravity.
5. True or False: With regard to the concentration of solutes, the less concentrated a urine sample, the greater is its specific gravity. ______
6. True or False: Renal tubular acidosis refers to a condition in which the pH of urine is less than 4.5. ______
7. A urinary tract infection causes bleeding, and blood appears in the urine. This condition is called
   a. albuminuria.
   b. glucosuria.
   c. ketonuria.
   d. hematuria.
   e. bilirubinuria
8. A normal urine sample will be negative for which substance?
   a. bilirubin
   b. leukocytes
   c. protein
   d. glucose
   e. all the above
9. During this laboratory exercise, what laboratory instrument will you use to separate the solid particles from the liquid portion of your urine sample? ______
10. True or False: It is normal to find a small number of epithelial cells in a urine sample.
    ______

Activity **30.1**

Correlating Nephron Structure with Urine Production

Each kidney contains more than 1 million **nephrons,** the microscopic renal tubules in which urine is produced. Urine production in the nephrons involves three processes: **filtration, reabsorption,** and **secretion.**

A **Filtration** occurs in the **renal corpuscle** (**glomerulus** and **glomerular capsule**).

1 Identify a renal corpuscle on a nephron model.

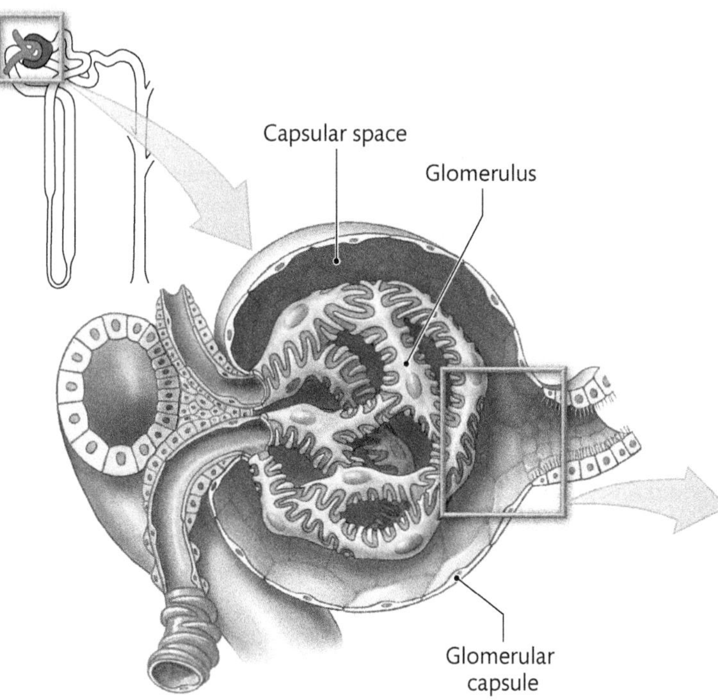

2 During the filtration process, water and small dissolved molecules move from the **glomerular capillaries** to the **capsular space** (lumen of the glomerular capsule). The resulting fluid, called **filtrate,** has the same concentration of small dissolved substances as blood, but normally lacks blood cells and large plasma proteins. It contains mostly water along with excess ions (mostly sodium and potassium), glucose, amino acids, and nitrogenous (nitrogen-containing) metabolic waste products.

Glomerular hydrostatic pressure (GHP)
Filtrate in capsular space
Plasma proteins
Solutes
50
25
15
10 mm Hg
Blood colloid osmotic pressure (BCOP)
Net filtration pressure (NFP)
Capsular hydrostatic pressure (CsHP)

3 Blood pressure in the glomerular capillaries (**glomerular hydrostatic pressure** or **GHP**) forces fluid to move into the capsular space. The **blood colloid osmotic pressure (BCOP)** and the **capsular hydrostatic pressure (CsHP)** force fluid back into the glomerular capillaries and thus oppose the glomerular hydrostatic pressure.

4 The **net filtration pressure (NFP)** is the force that moves fluid across the glomerular capillaries. We can calculate it using the following equation:

$$NFP = GHP - (BCOP + CsHP)$$

Under normal conditions, GHP > BCOP + CsHP; thus, NFP will move fluid into the capsular space to become the filtrate.

5 How will NFP be affected if GHP increases?

▶ ______________________________________

____

____

____

6 How will NFP be affected if plasma protein concentration increases in the glomerular capillary blood?

▶ ______________________________________

____

____

____

B **Reabsorption** occurs throughout the nephron, but mostly in the **proximal convoluted tubule (PCT)** and **nephron loop (loop of Henle).** During reabsorption, 99 percent of the water and other useful substances, such as glucose, amino acids, and various ions, return to the blood. They do so by passing from the renal tubules to the **peritubular capillaries** or **vasa recta.**

1. On a nephron model, identify
   - the PCT
   - the ascending and descending limbs of the nephron loop
   - the peritubular capillaries
   - the vasa recta.
2. Label the figure on the right with the above structures. ▶

C **Secretion** takes place throughout the nephron, but mostly in the **distal convoluted tubule (DCT)** and **collecting duct.** During secretion, unwanted substances such as metabolic wastes, drugs, and excess ions (hydrogen and potassium) are removed from the blood and enter the renal tubules. This process is just the opposite of reabsorption and supplements glomerular filtration.

1. On the nephron model, identify
   - the DCT
   - the collecting duct.
2. Label the figure on the right with the above structures. ▶

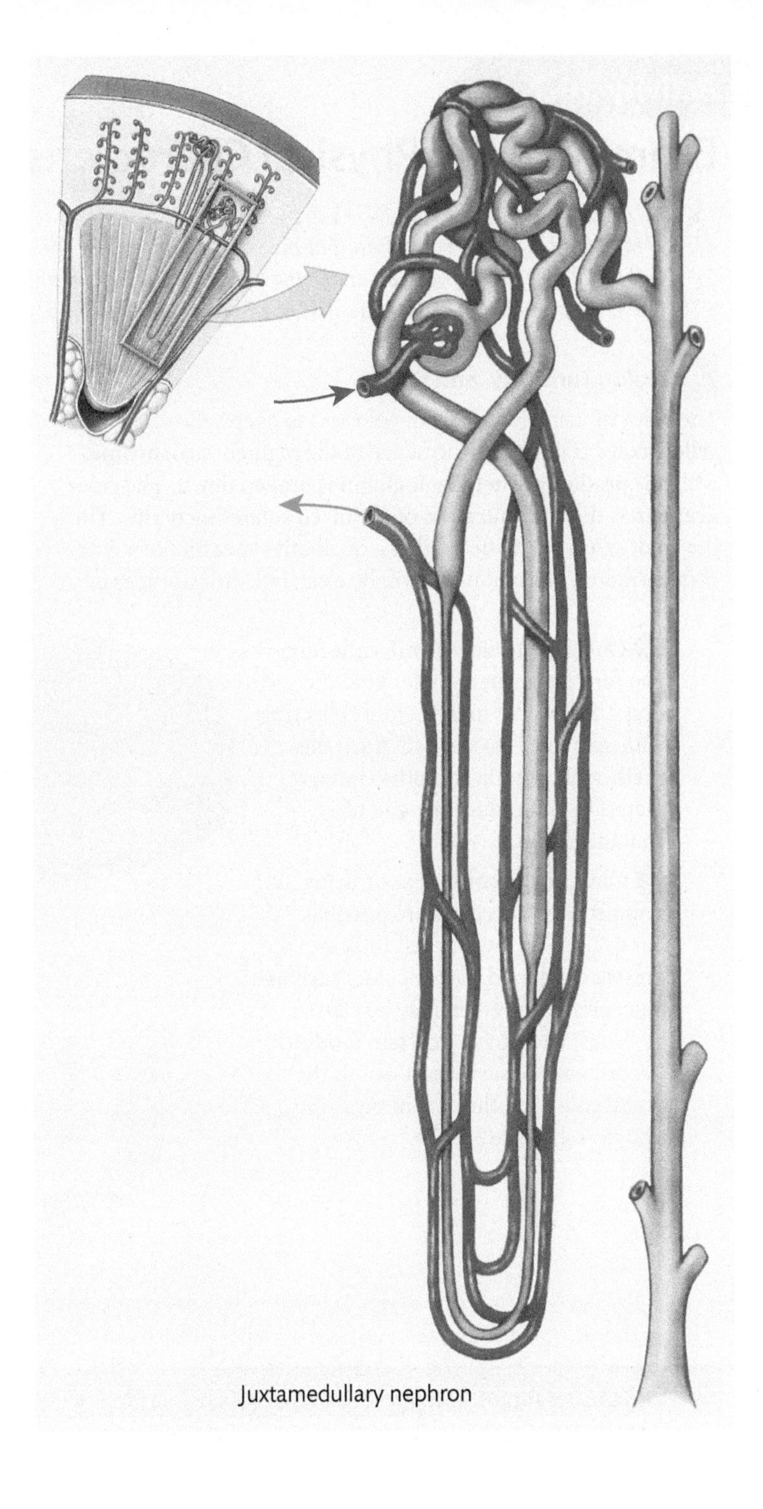

MAKING CONNECTIONS

Urine contains substances that are filtered and secreted by the nephrons, but not substances that are reabsorbed. Explain why.

▶ ____

____

____

____

____

____

____

____

____

____

____

____

Want more practice? Go to: **MasteringA&P** > Study Area > Menu > Lab Tools >
- **PAL** > Anatomical Models > Urinary System > Renal corpuscle and Nephron photos
- **PhysioEx** > Exercise 9: Renal System Physiology > Activities 1–6

Activity 30.2

Examining the Physical Characteristics of Urine

⚠ **Warning:** *Urine is a body fluid that may contain infectious microorganisms. Therefore, universal precautions will be followed while you are analyzing your sample. Before you begin the activities that follow, carefully review universal precautions, which are described in Appendix A on page A-1 of this lab manual.*

A Color, Turbidity, and Odor

The color of urine ranges from colorless to deep yellow. The yellow color is due to the presence of the pigment **urochrome,** which is produced when hemoglobin is broken down. The color deepens as the concentration of dissolved solutes increases. Thus, the color of urine can be used as a qualitative measure of solute concentration. Normal urine will be clear, but various types of suspended particles such as microbes, dead cells, and crystals can cause **turbidity** (cloudiness). Freshly voided urine usually has an aromatic odor. If left to stand, chemical breakdown by bacteria can produce a strong odor of ammonia. Drugs, vitamins, and certain foods may also affect the odor.

1 Obtain a sterile 500-mL collection cup for your urine sample. Void the first 2 to 3 mL of urine before collecting your sample. This step will flush the urethra of bacteria and other foreign materials and reduce the risk of contamination.

2 Observe the **color** of your urine sample. Atypical colors are possible. For example, brown or red indicates the presence of blood. Other colors such as blue, orange, or green may be caused by drugs, vitamins, or certain foods. Record your observations—both the actual color and the darkness of your sample—in Table 30.1. ▶

3 Observe the **transparency** of your sample. Bacteria, mucus, crystals of calcium salts and cholesterol, epithelial cells, and cell casts (hardened cell fragments) are sources of turbidity. Record your results in Table 30.1. ▶

4 Hold your urine sample about 30 cm (12 in.) from your face and wave your hand over the collection cup toward your nose. If you cannot detect an odor, slowly move the cup closer to your nose, but not closer than 20 cm (8 in.). Describe any odor that you can detect in Table 30.1. ▶

IN THE CLINIC

Abnormal Urine Colors

Many types of food can change the color of your urine. For example, asparagus can give a greenish tint, and carrots can add a shade of orange. Beets and blackberries provide reddish tones, and fava beans and rhubarb can turn urine dark brown. Color changes caused by food are temporary and do not have an adverse effect on normal kidney function.

Atypical urine colors can also result from disease such as porphyria. This disease is characterized by the accumulation of porphyrin, a building block for the heme molecule in hemoglobin. Excess porphyrin can give urine a reddish color.

TABLE 30.1 Physical Analysis of a Urine Sample

	Results	
Characteristic	**Normal**	**Your Sample**
Color	Colorless to pale yellow	
Transparency/ turbidity	Clear	
Odor	Slightly aromatic	
Specific gravity	1.003–1.030	

B Specific Gravity

A reliable qualitative method to examine the solute concentration of urine is measuring its **specific gravity,** which is a weight comparison between a given volume of urine and an equal volume of distilled water. Distilled water has a specific gravity of 1.000.

The specific gravity of urine normally ranges from 1.003 to 1.030, depending on the concentration of solutes: the more concentrated a urine sample, the greater its specific gravity.

1 Obtain a **urinometer** to measure the specific gravity of your urine sample. A urinometer consists of two parts:

1. a glass container to hold your urine sample
2. a flotation device called a **hydrometer,** which is used to measure specific gravity.

2 Before you begin, gently swirl your urine sample so that any substances that settled to the bottom are suspended in the fluid. Add a portion of your urine sample to the glass container until the container is at least two-thirds full.

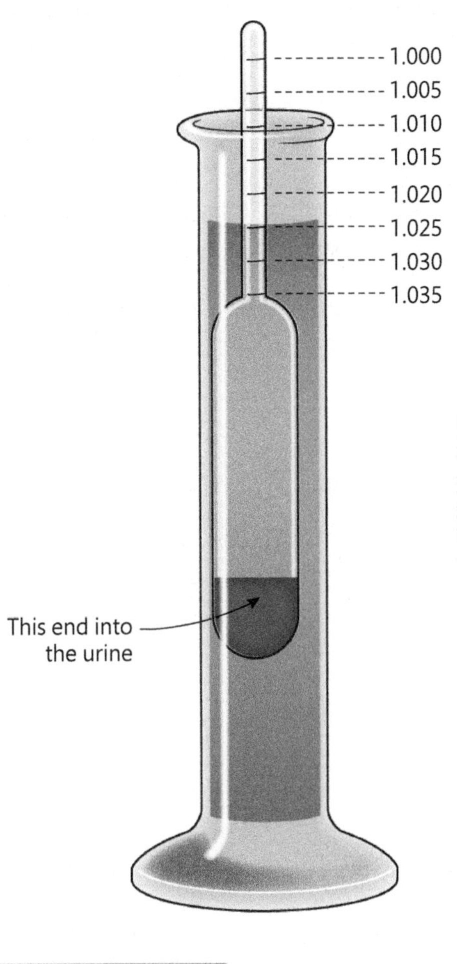

3 Gently place the hydrometer into the urine with the long stem directed upward. The hydrometer must float freely. If it does not, add more urine to the glass container.

4 Determine the specific gravity of your sample by reading the position of the lower margin of the meniscus (curved line on the surface of a fluid) on the hydrometer stem's calibrated scale. Record your results in Table 30.1. ▶

What specific gravity is shown in the figure to the left?

▶ ______________________________

Is this within the normal range?

▶ ______________________________

IN THE CLINIC

Urinalysis

Urinalysis is an examination of the physical, chemical, and microscopic characteristics of urine. It is an important diagnostic tool because abnormalities in urine can reveal various diseases that often go undetected. For example, large blood proteins and red blood cells do not normally appear in the filtrate and are not found in urine. If they are present, it may be due to inflammation of the glomeruli in the renal corpuscles, a condition known as glomerulonephritis.

MAKING CONNECTIONS

Explain why cloudiness (turbidity) in a urine sample could indicate the presence of a urinary tract infection.

▶ ______________________________

____

____

____

Activity 30.3

Examining the Chemical Characteristics of Urine

A As described earlier, urine normally contains a wide range of dissolved ions, small molecules, and nitrogenous wastes. A number of other chemical substances may be present abnormally as a result of disease or infection, however. Table 30.2 summarizes abnormalities in the chemical composition of urine.

TABLE 30.2 Chemical Abnormalities in a Urine Sample

Urine Component	Reason for Abnormality	Clinical Term	Comment
Hydrogen ion concentration (pH)	Low pH (< 4.5): high-protein diet High pH (> 8.0): vegetarian diet or urinary tract infection	Renal tubular acidosis Renal tubular alkalosis	Normal pH range: 4.5–8.0 Average pH: 6.0
Nitrites	Bacterial infection of urinary tract	Nitriuria	Bacteria convert nitrates, a normal urine component, to nitrites. If there is an infection, bacteria convert nitrates, a normal component, to nitrites. Normally nitrite concentration is too low to be detected.
Bilirubin	Blocked bile duct, liver disease (hepatitis, cirrhosis)	Bilirubinuria	Bilirubin is a bile pigment produced in the liver when old red blood cells are recycled. A small amount in the urine is normal.
Urobilinogen	Liver disease (hepatitis, cirrhosis, jaundice)	Urobilinogenuria	In the intestines, bilirubin is converted to urobilinogen, which gives feces its brown color. A small amount of urobilinogen in the urine is normal.
Leukocytes and pus	Inflammation due to urinary tract infection	Pyuria	Pus is a fluid containing leukocytes and the debris of dead cells and pathogens.
Erythrocytes/blood	Inflammation or urinary tract infection that causes bleeding	Hematuria	If enough blood is present, urine will have a reddish or brown color.
Protein (albumin is the most abundant blood protein)	Temporary: physical exercise, high-protein meals, pregnancy Chronic: high blood pressure, kidney infections, poisons	Albuminuria	Blood proteins are usually too large to be filtered and are rarely found in urine.
Glucose	Temporary: high-carbohydrate diet; stress Chronic: diabetes mellitus	Glucosuria	Glucose is found in the filtrate, but normally it is all reabsorbed. Glucose is usually absent in urine or is present in trace amounts.
Ketone bodies	Starvation, low-carbohydrate diet, diabetes mellitus	Ketonuria	Ketone bodies are products of fat metabolism and are normally present in very small amounts in urine.

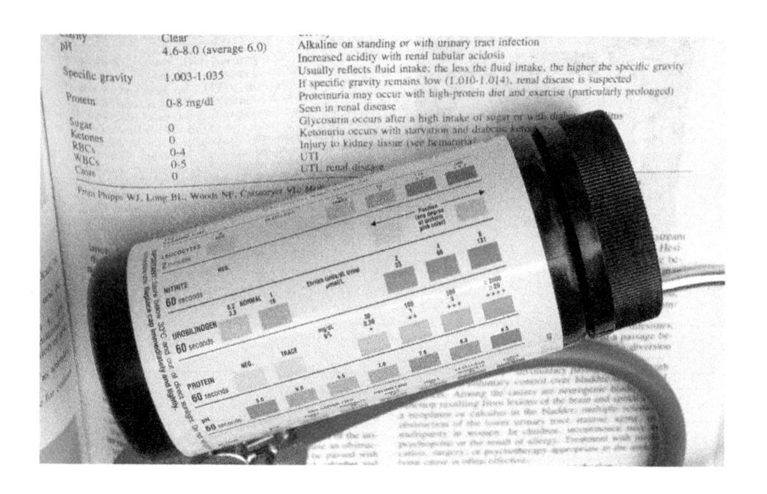

Word Origins

Notice that the name for an abnormally high amount of a substance in urine ends with *-uria*. This term is a derivation of *ouron*, which is the Greek word for urine. Thus, *glucosuria* means "glucose in urine," and *ketonuria* means "ketones in urine."

B In this activity, you will use dipsticks to test for abnormal chemicals in your urine sample. Individual dipsticks have a single test pad to detect one chemical; combination dipsticks have several test pads to identify several chemicals simultaneously. The test pads on the dipstick react with the urine and change color. You will analyze the results by comparing the color of the test pads with a color chart. Depending on the test, results must be taken immediately or after a brief period of time, but never exceeding 2 minutes. Carefully review the times required for each test before proceeding.

1. Gently swirl your sample to suspend any particles that settled to the bottom of the collection cup.
2. Immerse a dipstick into the urine. If you are using a combination dipstick, be sure that all the test pads are completely submerged in the urine.
3. Remove the dipstick and place it on a paper towel with the test pads facing up.
4. After the appropriate time has elapsed, read the results of each test by comparing the color on the test pad with the color chart that is available. Record your results in Table 30.3. ▶

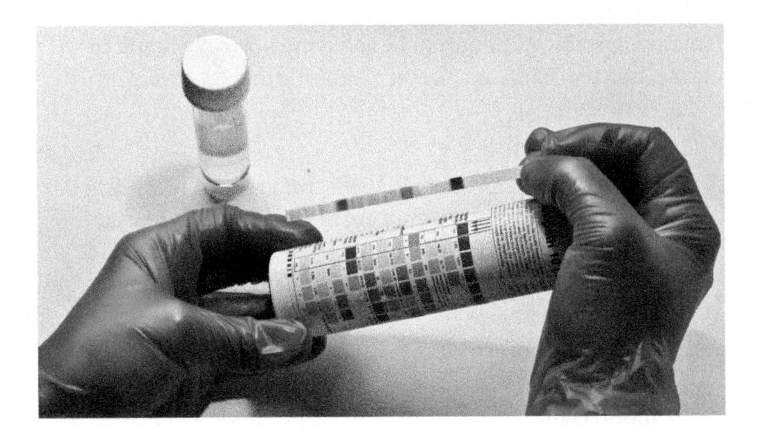

TABLE 30.3 Chemical Analysis of a Urine Sample

	Results	
Characteristic	**Normal**	**Your Sample**
pH	4.5–8.0	
Nitrites	Negative*	
Bilirubin	trace amounts	
Urobilinogen	Positive	
Leukocytes	Negative	
Erythrocytes/ blood	Negative	
Protein	Negative	
Glucose	trace amounts	
Ketone bodies	trace amounts	

*A negative result does not necessarily indicate a complete absence of the substance, but it does indicate that its concentration is too low to be detected.

MAKING CONNECTIONS

Diabetes mellitus is a disease in which the pancreatic islets do not produce enough insulin or the insulin receptors on cell membranes are missing or malfunctioning. In either case, cells are unable to take up glucose from the blood. Explain why glucosuria and ketonuria can be symptoms of diabetes mellitus.

▶ ______________________________

Activity 30.4

Examining a Urine Sample Microscopically

A If a urine sample is centrifuged, the sediment (particles that settle to the bottom of the liquid) can be examined microscopically. Although some solids are normally found in urine, this procedure is a useful diagnostic tool for revealing abnormal sediments linked to various kidney diseases and infections. The sediments can be organized into the following groups:

Epithelial cells
Small numbers of epithelial cells that are shed from various regions of the urinary tract are normally found in urine sediments. They include transitional epithelium from the urinary bladder and ureters, and squamous epithelium from the urethra.

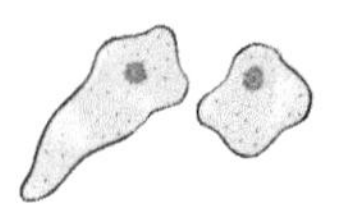
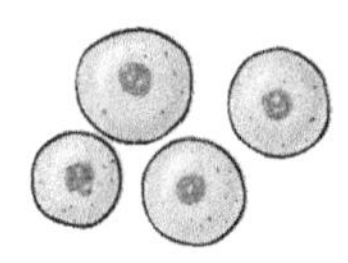
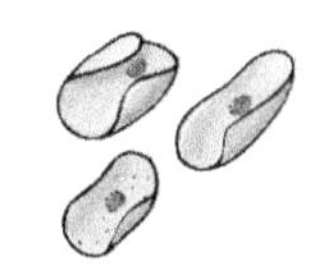

Epithelial cells

Blood cells
Large numbers of **white blood cells** (pus) and any amount of **red blood cells** are abnormal and usually indicate some type of disease or infection.

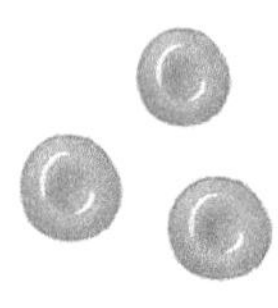
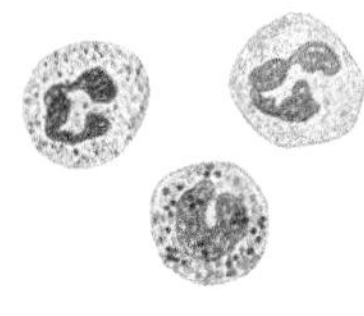
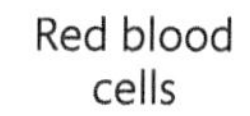

Red blood cells

Leukocytes (pus)

Casts
Casts are hardened organic materials that usually form when cells clump together. They typically form when urine is acidic or contains unusually high levels of proteins or salts. Thus, their presence usually indicates an abnormal or pathological condition. Cellular casts can be formed from epithelial cells, white blood cells, and red blood cells. Granular and waxy casts represent various stages in the breakdown of cellular casts. Hyaline casts form when proteins secreted from nephron tubules clump together.

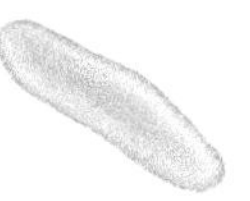
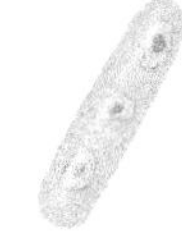
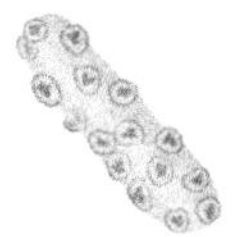

Hyaline cast

Granular cast

Epithelial cast

Pus cast

Red blood cell cast

Crystals
Crystals are small, geometric particles of calcium salts, uric acid, cholesterol, and phosphates. Small amounts of crystals are normal, but various urinary tract infections can increase their presence.

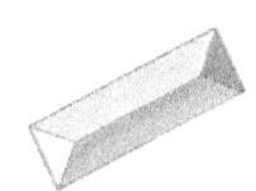

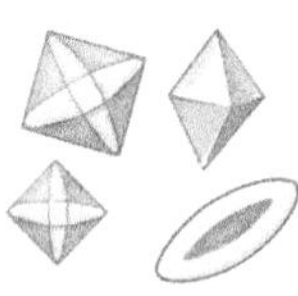
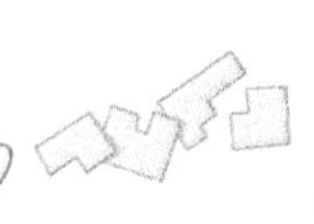
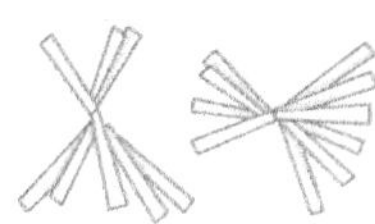

Ammonium-magnesium phosphate crystals

Uric acid crystals

Calcium carbonate crystals

Calcium oxalate crystals

Cholesterol crystals

Calcium phosphate crystals

Microorganisms
A small number of bacteria or other types of microorganisms may also be present if urine is contaminated as it passes along the urethra. A large presence of microbes may indicate an infection.

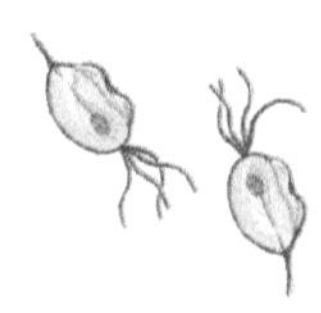

Bacteria

Yeast

Trichomonas

B In this activity, you will view a microscope preparation of sediments from your urine sample.

1. Gently swirl your urine sample to suspend any sediments.
2. Pour a small volume of urine into a centrifuge tube until it is one-half to two-thirds full.
3. Add an equal volume of distilled water to a second centrifuge tube.
4. Place both tubes in centrifuge slots that are opposite each other. Centrifuge the urine sample for about 10 minutes.
5. When centrifugation is complete, carefully pour off the supernatant (liquid portion) in the urine sample and dispose of the fluid as directed by your instructor.
6. Use a Pasteur pipette to obtain a small amount of the sediment.
7. Transfer the sediment to a glass microscope slide, add a drop of 10 percent methylene blue solution, and place a coverslip over the preparation.
8. View your preparation with the low-power objective lens. Reduce the amount of light by decreasing the aperture of the iris diaphragm. Attempt to identify cells, casts, and crystals. To observe crystals more clearly, switch to the high-power objective lens and increase the amount of light.
9. In the space provided, draw some of the sediments that you observe and make comparisons with the examples illustrated in the figures on the previous page.

▶

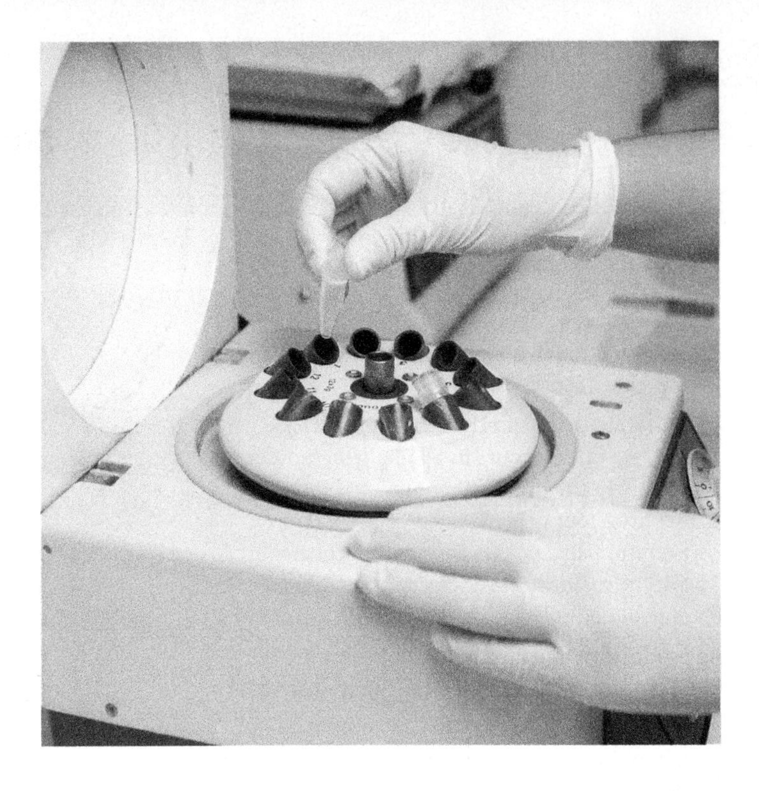

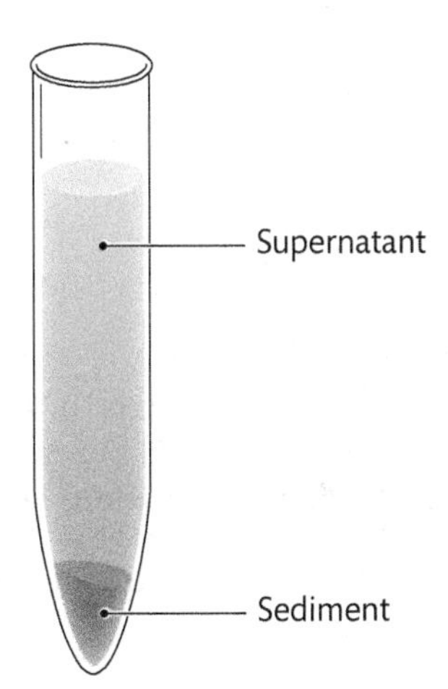

MAKING CONNECTIONS

A kidney stone is composed of calcium and magnesium salts or uric acid. If a person has a kidney stone, you would expect to find an increased amount of what sediment type in the urine sample? Explain.

▶ ______________________________

____

____

____

____

____

BEFORE YOU MOVE ON . . .

❮❮ LOOKING BACK

Urine production, which occurs in the nephrons, includes three activities:

- Filtration occurs in the renal corpuscles.
- Reabsorption occurs in all nephron tubules but mostly in the proximal convoluted tubules and nephron loops.
- Secretion also occurs throughout the nephron, but the majority takes place in the distal convoluted tubules and collecting ducts.

The production and elimination of urine are the bases for several essential kidney functions. These functions

- maintain normal blood pressure, thus keeping fluid concentrations of electrolytes such as sodium, potassium, and calcium within normal ranges
- monitor the solute concentration of body fluids
- keep the pH of body fluids within a narrow range (7.35 to 7.45)
- eliminate metabolic wastes, drugs, and toxic substances from the body.

Why is it important for the kidneys to receive an adequate blood supply? What is the significance of maintaining normal blood concentrations of sodium, potassium, and calcium ions?

▶ ______________________________

____

____

____

____

____

____

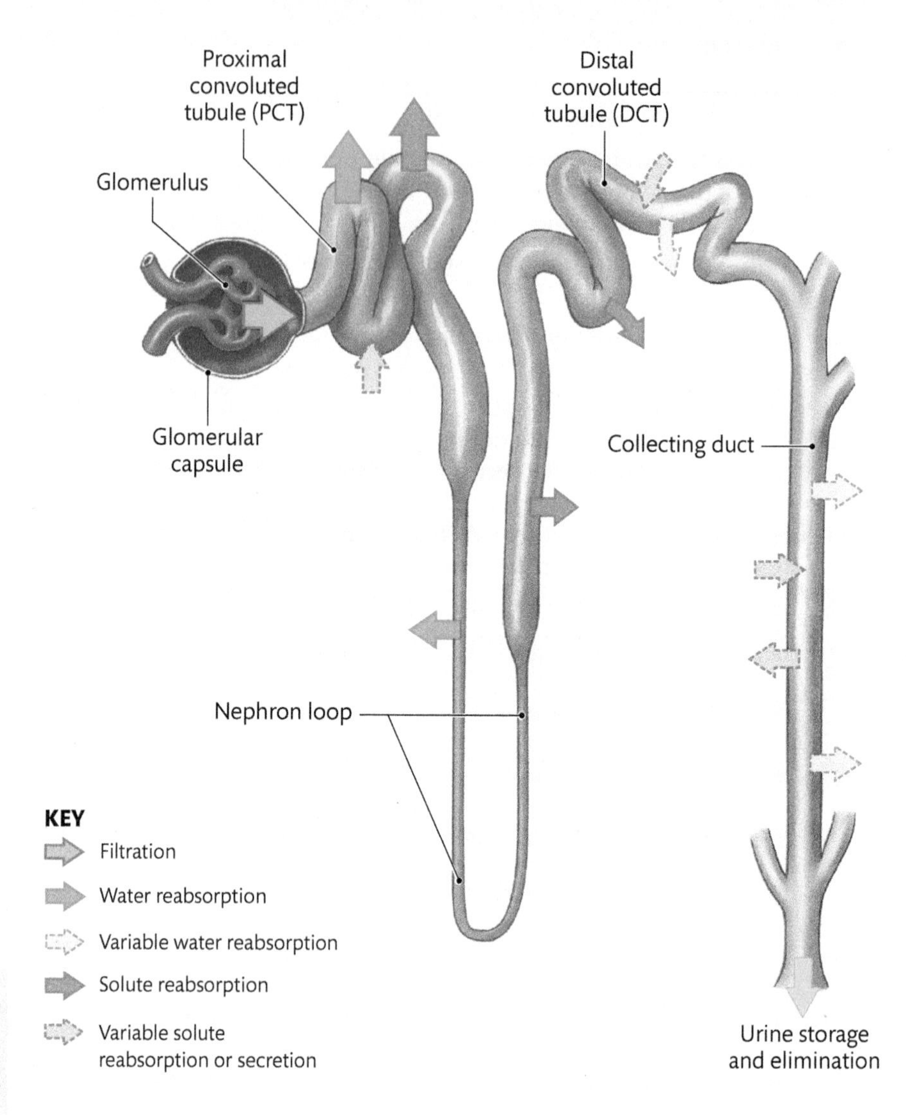

LOOKING FORWARD ❯❯

In the adult, the urinary and reproductive systems are separate and distinct organ systems, each comprising unique structures. During development, urinary and reproductive organs arise from common embryological tissue along the posterior abdominal wall. The term **urogenital** is often used to reflect the common origins of both systems. In the next two laboratory exercises, you will study the anatomy and physiology of the male (Exercise 31) and female (Exercise 32) reproductive systems.

Name ______________________________

Lab Section ______________________________

Date ______________________________

EXERCISE 30 REVIEW SHEET

Urinary Physiology

1. Calculate the net filtration pressure (NFP) in the kidney if the glomerular hydrostatic pressure (GHP) = 45 mm Hg, the blood colloid osmotic pressure (BCOP) = 20 mm Hg, and the capsular hydrostatic pressure (CsHP) = 10 mm Hg.

2. How will the filtration rate in the renal corpuscles be affected if a kidney stone creates a blockage of the urinary tract?

____

____

____

____

____

3. Why is it important to follow strict safety procedures (e.g., universal precautions) when handling a urine sample?

____

____

____

____

____

QUESTIONS 4–9: Match the abnormal urinary condition in column A with the appropriate description in column B.

A	B
4. Bilirubinuria ________________	**a.** This condition indicates the presence of red blood cells in the urine.
5. Pyuria ________________	**b.** This condition could indicate a problem with glomerular filtration.
6. Glucosuria ________________	**c.** This condition could occur temporarily after a high-carbohydrate meal.
7. Hematuria ________________	**d.** This condition could indicate liver disease.
8. Albuminuria ________________	**e.** This condition could occur during a period of starvation.
9. Ketonuria ________________	**f.** This condition indicates the presence of leukocytes and pus in the urine.

10. Why is urinalysis a useful diagnostic tool for identifying certain diseases and infections?

11. Why is it useful to determine the specific gravity of a urine sample?

QUESTIONS 12 AND 13: Identify a possible cause for each of the following abnormal conditions.

12. Renal tubular alkalosis

13. Above-normal amount of crystals in the urine

14. Urinalysis revealed that a patient's urine had a pH of 4.0 and contained numerous casts and albumin. The doctor suggested that the patient reduce the protein in her diet. Explain why.

15. Complete the following table.

Structure	Primary Function during Urine Production
Renal corpuscles	a.
b.	Secretion
Proximal convoluted tubules and nephron loops	c.
d.	During reabsorption, substances pass into these structures.

EXERCISE

31

The Male Reproductive System

LEARNING OUTCOMES

These Learning Outcomes correspond by number to the laboratory activities in this exercise. When you complete the activities, you should be able to:

Activity 31.1 **Describe the structure and function of the testes and accessory ducts in the male reproductive system.**

Activity 31.2 **Describe the structure and function of the accessory glands of the male reproductive system.**

Activity 31.3 **Name the components of the male external genitalia and describe their structure and function.**

Activity 31.4 **Describe the microscopic structure of the testis and epididymis.**

Activity 31.5 **Describe the microscopic structure of the ductus deferens (vas deferens) and the prostate gland.**

Activity 31.6 **Describe the microscopic structure of the penis and spermatozoa.**

LABORATORY SUPPLIES

- Torso model
- Midsagittal section model of the male pelvis
- Anatomical models of male reproductive organs
- Prepared microscope slides:
  - testis
  - epididymis
  - ductus deferens (vas deferens)
  - prostate gland
  - penis
  - sperm smear

PRE-LAB QUIZ

Before you begin, read all the activities in Exercise 31 and the required reading in your textbook that is assigned by your instructor.

1. True or False: Each ejaculatory duct is formed by the merging of the ductus deferens and the duct of the bulbourethral gland. ____________
2. The prostate gland is located inferior to what structure? ____________________
3. True or False: The testes develop from embryonic tissue along the posterior abdominal wall. ____________
4. The spermatic cord contains a portion of what male reproductive structure?

   ____
5. True or False: The two seminal glands are located along the posterior surface of the urinary bladder. ____________
6. True or False: The penile urethra travels through the corpus cavernosum. ____________
7. During this laboratory exercise, you will examine the microscopic structure of all the following structures *except* the
   a. testis.
   b. bulbourethral gland.
   c. epididymis.
   d. ductus deferens.
   e. penis.
8. Spermatogenic cells are found in what male reproductive structure?

   ____
9. You are viewing a slide of a male reproductive structure. You identify a smooth muscle layer that contains three layers: an inner longitudinal layer, a middle circular layer, and an outer longitudinal layer. What structure are you viewing? ____________________
10. You are viewing a slide of a male reproductive structure. You identify three columns of erectile tissue. What structure are you viewing? ____________________

Activity 31.1

Examining the Gross Anatomy of the Male Reproductive System: Testes and Accessory Ducts

The male reproductive system includes the testes, which are the primary sex organs, and accessory ducts, glands, and external genitalia. As a system, these organs function to produce male sex hormones, to produce and store sperm cells, and to transport sperm cells and supporting fluids to the female reproductive tract during sexual intercourse.

A The **testes** are ovoid structures, about 5 cm (2 in.) long and 3 cm (1.2 in.) wide, located within the scrotum (scrotal sac).

1. On a torso model, identify the kidneys on the posterior wall of the abdominal cavity. In the fetus, the **testes** develop from embryonic tissue close to the kidneys.
2. As the fetus grows, the testes slowly move to a more inferior position in the body cavity.
3. During the seventh month of development, the testes make their final descent by traveling through the inguinal canals, passageways that connect the abdominal cavity with the scrotum. This **descent of the testes** is usually complete near the time of birth.
4. On a torso model or a midsagittal section of the male pelvis, identify a **testis,** within the scrotum.
5. As each testis moves into the scrotum, associated structures such as the ductus deferens, blood vessels, nerves, and lymphatics travel with them. Identify the **spermatic cord,** which consists of these associated structures enclosed by connective tissue and muscle.

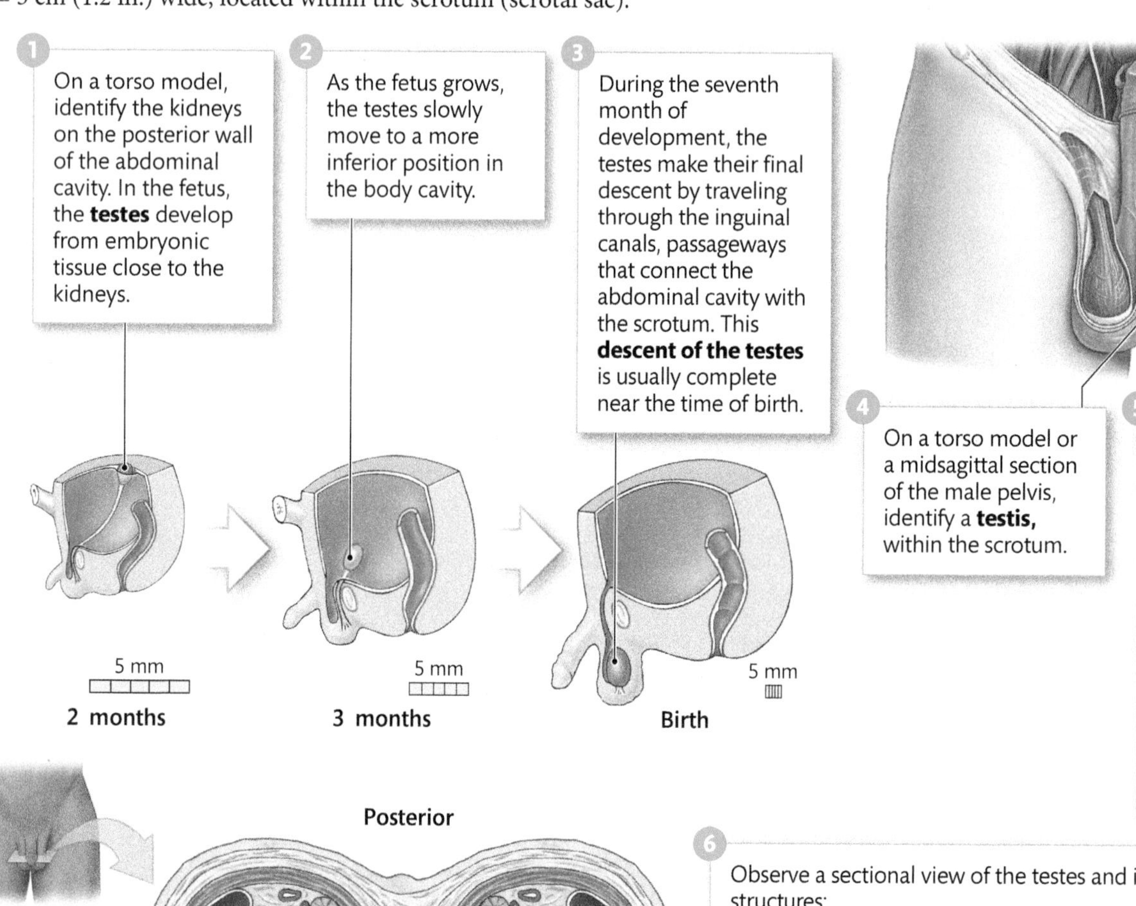

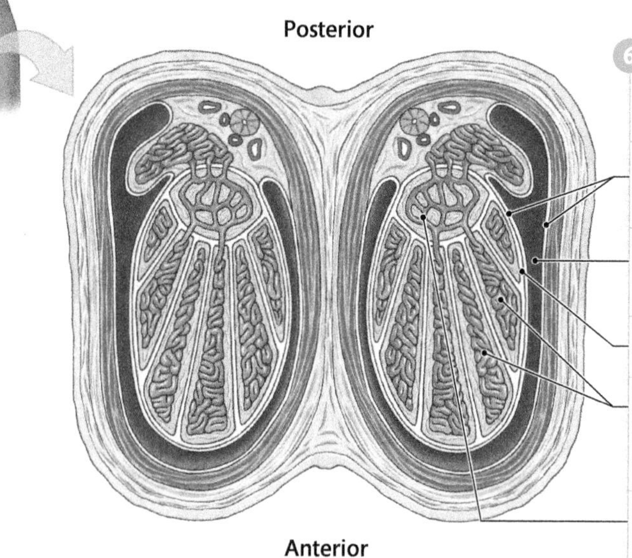

6. Observe a sectional view of the testes and identify the following structures:
   - A **serous membrane,** which is continuous with the peritoneum, lines the wall of each testis and the scrotal sac.
   - A **scrotal cavity** surrounds each testis and is lined by the serous membrane.
   - The **tunica albuginea** is a fibrous capsule that covers each testis.
   - Connective tissue partitions arising from the tunica albuginea divide each testis into about 250 **lobules.** Each lobule contains three or four highly coiled seminiferous tubules.
   - Along the posterior aspect of each testis, the seminiferous tubules converge to form a tubular network called the **rete testis.** The rete testis transports sperm to several short ducts that lead directly into the epididymis.

Want more practice? Go to: **MasteringA&P** > Study Area > Menu > Lab Tools > PAL >
- Human Cadaver > Reproductive System > Male photos
- Anatomical Models > Reproductive System > Male photos

B The **accessory ducts** are muscular tubes that store and transport **semen** (sperm and glandular secretions) to the outside.

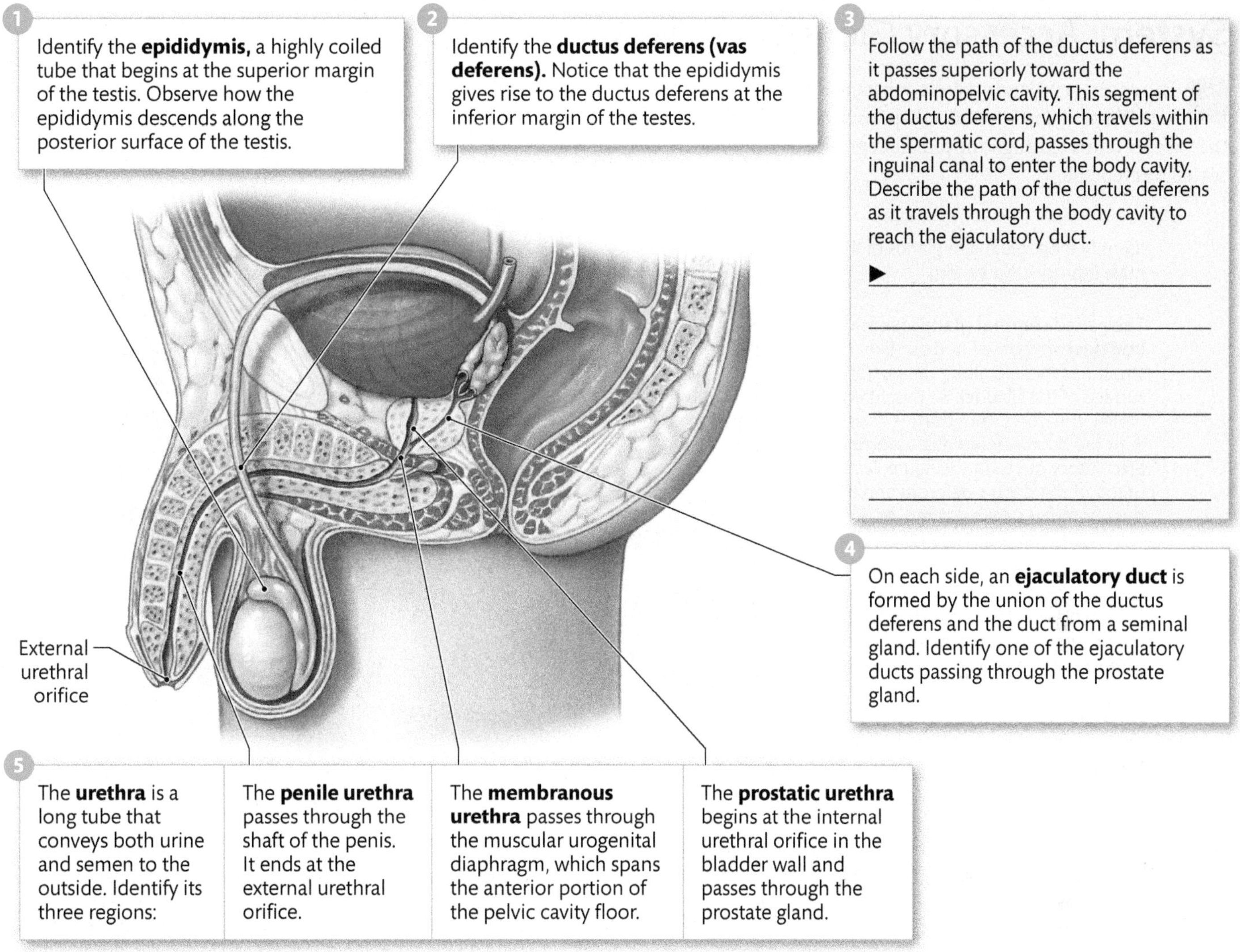

1. Identify the **epididymis,** a highly coiled tube that begins at the superior margin of the testis. Observe how the epididymis descends along the posterior surface of the testis.

2. Identify the **ductus deferens (vas deferens).** Notice that the epididymis gives rise to the ductus deferens at the inferior margin of the testes.

3. Follow the path of the ductus deferens as it passes superiorly toward the abdominopelvic cavity. This segment of the ductus deferens, which travels within the spermatic cord, passes through the inguinal canal to enter the body cavity. Describe the path of the ductus deferens as it travels through the body cavity to reach the ejaculatory duct.

▶ ______________________________

____

____

____

____

____

4. On each side, an **ejaculatory duct** is formed by the union of the ductus deferens and the duct from a seminal gland. Identify one of the ejaculatory ducts passing through the prostate gland.

External urethral orifice

5.

The **urethra** is a long tube that conveys both urine and semen to the outside. Identify its three regions:	The **penile urethra** passes through the shaft of the penis. It ends at the external urethral orifice.	The **membranous urethra** passes through the muscular urogenital diaphragm, which spans the anterior portion of the pelvic cavity floor.	The **prostatic urethra** begins at the internal urethral orifice in the bladder wall and passes through the prostate gland.

MAKING CONNECTIONS

During a vasectomy, incisions are made on each side of the scrotum, and the two ductus deferentia are cut and tied off within the spermatic cord. A man's testes will still function normally, but he will be infertile. Explain why. What other structures could potentially be damaged during a vasectomy?

▶ ______________________________

____

____

____

____

IN THE CLINIC

Inguinal Hernia

In males, the presence of spermatic cords creates areas of weakness along the abdominal wall where the inguinal canals are located. Loops of the small intestine may protrude into the canal and scrotum, resulting in an **inguinal hernia.** Inguinal hernias are less common in females because spermatic cords are absent. In most cases, surgery is required to place the herniated portion of the intestine back in its normal position in the abdominal cavity and repair the weakness in the body wall.

Activity 31.2

Examining the Gross Anatomy of the Male Reproductive System: Accessory Glands

A The accessory glands secrete substances that provide nourishment and support for sperm cells, particularly while they are traveling through the female reproductive tract after sexual intercourse.

1

Identify the three major glands of the male reproductive system:

The paired **seminal glands (seminal vesicles)** are convoluted sac-like structures located along the posterior surface of the bladder. As described earlier, the ducts from these glands unite with the ductus deferentia to form the ejaculatory ducts. In the figure, label the union of the ductus deferens and the duct from the seminal gland. ▶

The **prostate gland** is a chestnut-shaped structure located just inferior to the urinary bladder. The ejaculatory ducts and the prostatic urethra pass through the gland. Label these ducts in the figure. ▶

The **bulbourethral (Cowper's) glands** are pea-size structures located inferior to the prostate gland and within the urogenital diaphragm.

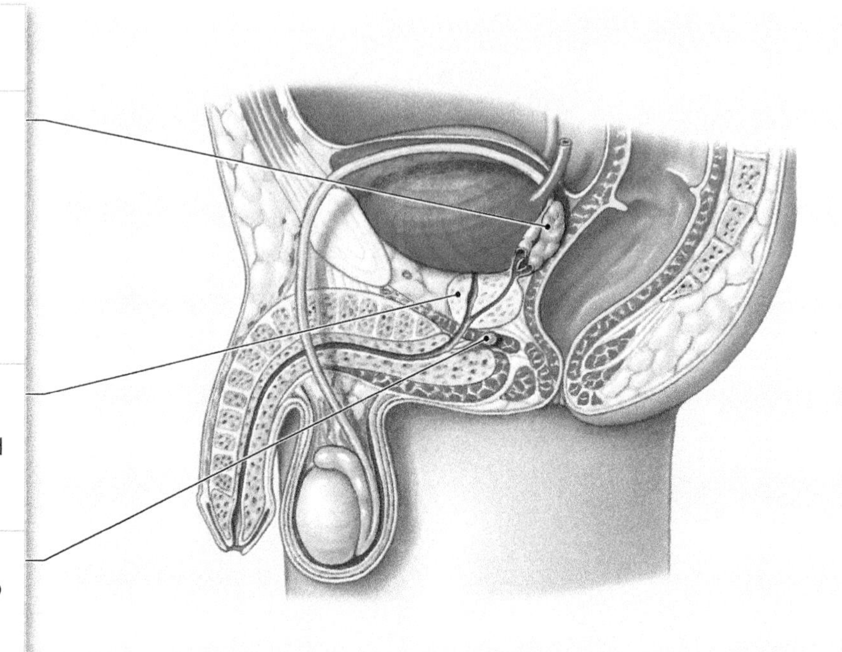

2

Refer to your textbook and describe the functions of the secretions produced by the male reproductive glands in the table below. ▶

Gland	Function of Glandular Secretions
Seminal glands	
Prostate gland	
Bulbourethral glands	

3

A doctor can examine the conditions of some pelvic organs by inserting a finger into the rectum and palpating along the anterior wall. This procedure is known as a **rectal exam.** Label the rectum in the figure above. Explain why a rectal exam would be useful in determining if a man's prostate gland has become enlarged.

▶ ______________________________

____

____

____

____

B The male reproductive glands have a close anatomical relationship with the urinary bladder.

1 From a posterior view, observe the relationship of the urinary bladder with several male reproductive organs. Identify the following structures on a model and label them in the figure to the right. ▶

- Seminal glands
- Prostate gland
- Ductus deferentia
- Bulbourethral glands
- Prostatic urethra
- Membranous urethra
- Ejaculatory ducts

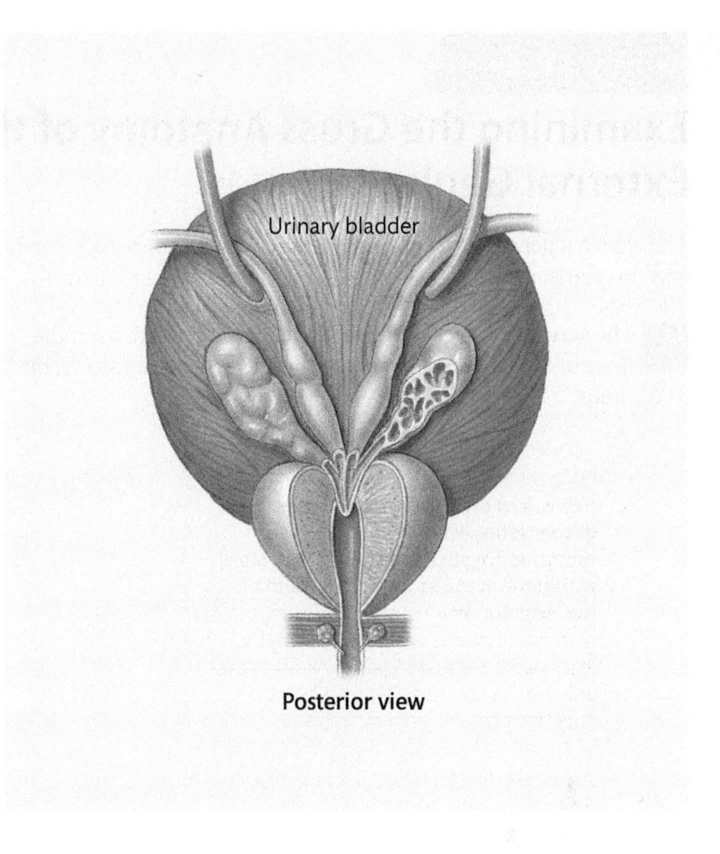

Posterior view

2 The prostate gland is a reproductive organ, but if it becomes enlarged, urinary function will be affected. Explain why.

▶ ______________________________

____

____

____

____

____

Word Origins

Although their name is rather lengthy, the tiny *bulbourethral glands* are aptly named. The term describes their location between the bulb of the penis and the urethra.

MAKING CONNECTIONS

A man is diagnosed with prostate cancer and decides to have his prostate gland surgically removed. The functions of what other structures could be affected by the procedure?

▶ ______________________________

____

____

____

____

Want more practice? Go to: **MasteringA&P** > Study Area > Menu > Lab Tools > PAL >

- Human Cadaver > Reproductive System > Male photos
- Anatomical Models > Reproductive System > Male photos

Activity **31.3**

Examining the Gross Anatomy of the Male External Genitalia

The external genitalia include those structures found outside the body cavity. They include the **scrotum** and the **penis.**

A The **scrotum (scrotal sac)** is suspended inferiorly from the floor of the pelvic cavity, posterior to the penis, and anterior to the anus.

1 On a midsagittal section of the male pelvis, locate the **scrotum.** In addition to the testes, list two other reproductive structures that are located within the scrotum and label them on the diagram. ▶

1. ______________________

2. ______________________

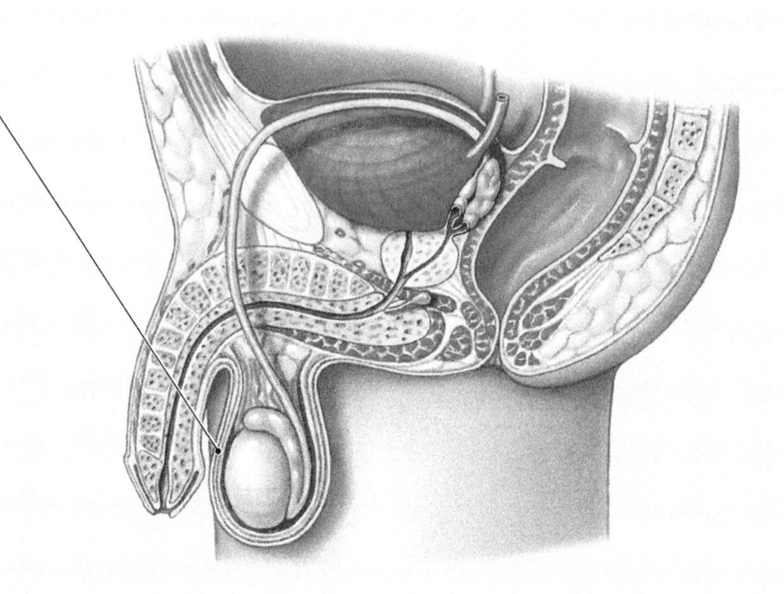

2 In a sectional view, notice that the two **scrotal chambers** are separated by a connective tissue partition called the scrotal septum.

3 Locate the **dartos muscle,** a thin band of smooth muscle deep to the dermis of the skin. The dartos muscles surround the testes by continuing through the scrotal septum. Contractions of the dartos muscle create wrinkling in the skin that covers the scrotum.

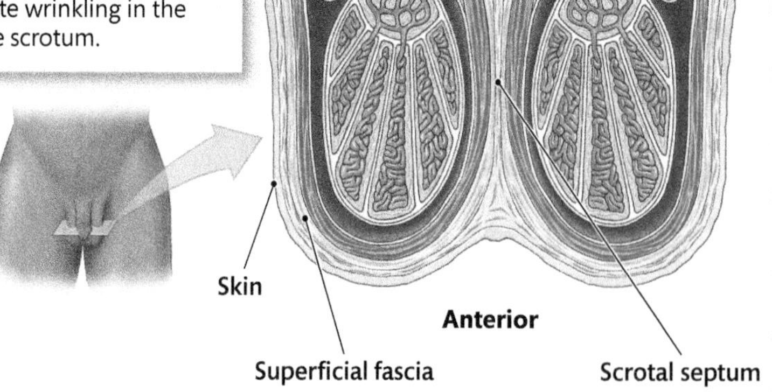

4 Deep to the dermis is a thicker layer of skeletal muscle called the **cremaster muscle.** This muscle, which is an extension of the internal oblique muscle, descends to surround the spermatic cord and testis. If air or body temperature increases, the cremaster muscles relax, and the testes are drawn farther from the body's core. Conversely, a decline in temperature causes the muscles to contract and brings the testes closer to the body.

The actions of the cremaster muscle are critical for normal testicular function. Explain why.

▶ ______________________

____

____

____

____

B The **penis** conveys urine to the outside during micturition (urination) and semen into the female reproductive tract during sexual intercourse.

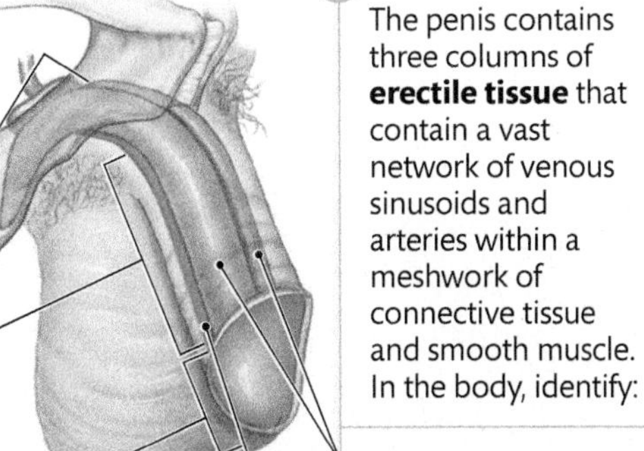

1 On a model, identify the three regions of the **penis:**

The **root** is attached to the pelvic bone and urogenital diaphragm.

The **body** (shaft) of the penis is the elongated cylindrical portion.

The **glans** of the penis is the extended tip of the shaft.

2 The penis contains three columns of **erectile tissue** that contain a vast network of venous sinusoids and arteries within a meshwork of connective tissue and smooth muscle. In the body, identify:

Two dorsal columns known as the **corpora cavernosa** (singular = **corpus cavernosum**), which lie parallel to each other.

A single ventral column called the **corpus spongiosum,** which encloses the penile urethra.

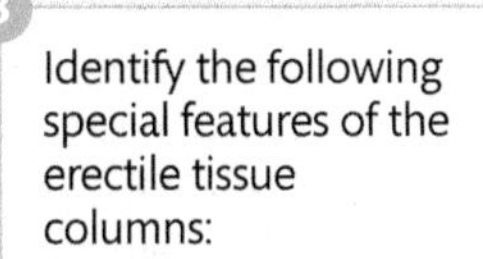

3 Identify the following special features of the erectile tissue columns:

At the root, the corpora cavernosa diverge to form the **crura** (singular = **crus**) **of the penis.**

At the root, the corpus spongiosum enlarges to form the **bulb of the penis.**

Distally, the corpus spongiosum expands to form the **glans** of the penis.

4 Identify the thin fold of skin called the **prepuce (foreskin)** that covers the glans penis. The inner surface of the prepuce contains glands that secrete a waxy material called smegma.

IN THE CLINIC

Circumcision

Smegma, secreted from the prepuce, provides a favorable environment for bacteria. Consequently, bacterial infections are common in this area. To reduce the risk of infections, the prepuce can be surgically removed through a procedure known as **circumcision.** Many circumcisions are also performed for religious or cultural reasons and not necessarily to reduce the infection risk. Removal of the prepuce is commonly performed in newborn males in the United States. The procedure is controversial, however, because opponents believe that it is medically unnecessary and causes undue pain to the baby.

MAKING CONNECTIONS

Although the testes, epididymides, and portions of the ductus deferentia are also external structures, they are not considered to be external genitalia. Can you think of a reason why?

► ______________________________

____

____

____

____

Want more practice? Go to: **MasteringA&P** > Study Area > Menu > Lab Tools > PAL >

- Human Cadaver > Reproductive System > Male photos
- Anatomical Models > Reproductive System > Male photos

Activity 31.4

Examining the Microscopic Anatomy of Male Reproductive Organs: Testis and Epididymis

A The testes contain seminiferous tubules, where the production of sperm occurs.

1. Scan a slide of a human or mammalian testis under low power. Notice that the testis is divided into **lobules** that are separated by connective tissue partitions.

2. Each lobule contains three or four seminiferous tubules. As you scan the slide under low power, observe cross and longitudinal sections of the tubules throughout the field of view.

3. Switch to high power and identify the sectional profiles of **seminiferous tubules.** Notice that the tubules contain several layers of cells surrounding a central lumen. Because of the plane of the section, you might not see the lumen in some tubule profiles.

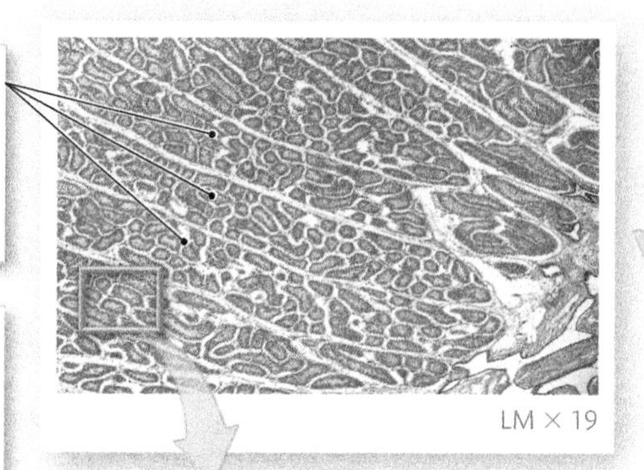

LM × 19

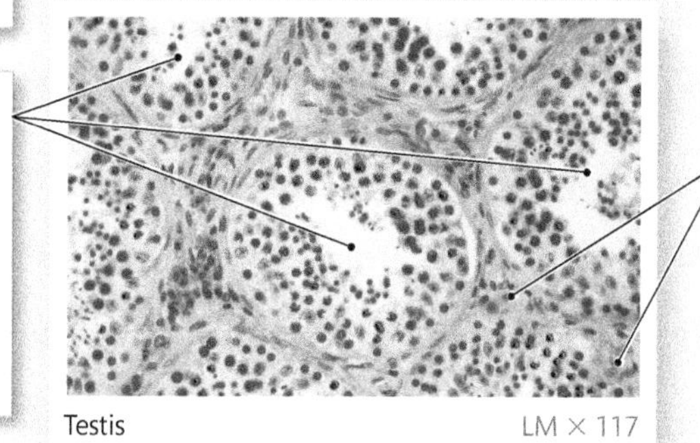

Testis LM × 117

4. Observe areas of connective tissue between seminiferous tubules. In these regions, identify the **interstitial (Leydig) cells.** These cells produce and secrete male sex hormones, or androgens. The primary hormone produced by the interstitial cells is testosterone.

B Most of the cells in the seminiferous tubules are developing sperm cells, also known as **spermatogenic cells.** As the sperm cells develop (**spermatogenesis**), they move from the base to the lumen of the seminiferous tubules.

1. Spermatogenic cells are referred to by several names, according to their stage of development. Examine a seminiferous tubule under high power and identify the following spermatogenic cells:

   - At the base of the tubules are the **spermatogonia.** These undifferentiated cells are continuously dividing by mitosis.
   - Some spermatogonia will differentiate to become **primary spermatocytes** and begin meiosis, a process of sexual cell division that forms cells with half the amount of DNA as the original cell.
   - During meiosis, primary spermatocytes give rise to **secondary spermatocytes.**
   - Secondary spermatocytes form **spermatids.**
   - Spermatids physically mature to become **spermatozoa (sperm cells).**

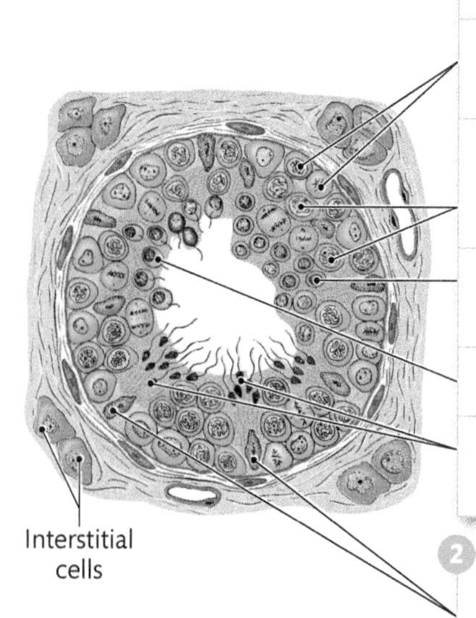

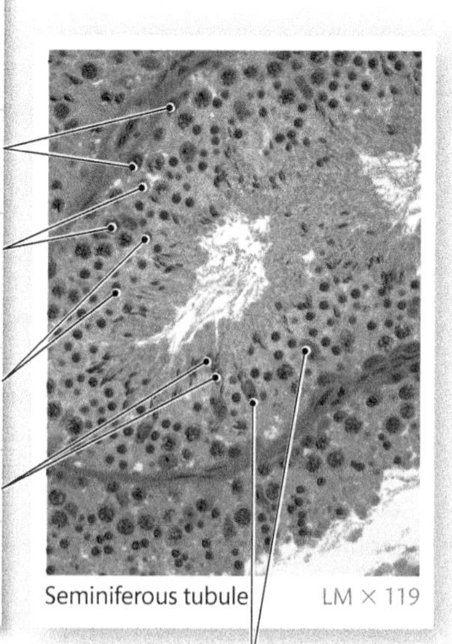

Seminiferous tubule LM × 119

2. In addition to the spermatogenic cells, identify **nurse (sustentacular) cells,** which extend from the base to the lumen of the tubules and completely surround the spermatogenic cells. Nurse cells nourish the spermatogenic cells and regulate spermatogenesis.

C It takes about 2 weeks for sperm cells to travel the entire length of the epididymis. During this time, sperm cells become functionally mature.

1 Scan a slide of the epididymis under low power and observe numerous sectional profiles of **tubules in the epididymis.** Understand that you are actually viewing one continuous tubule that is highly convoluted.

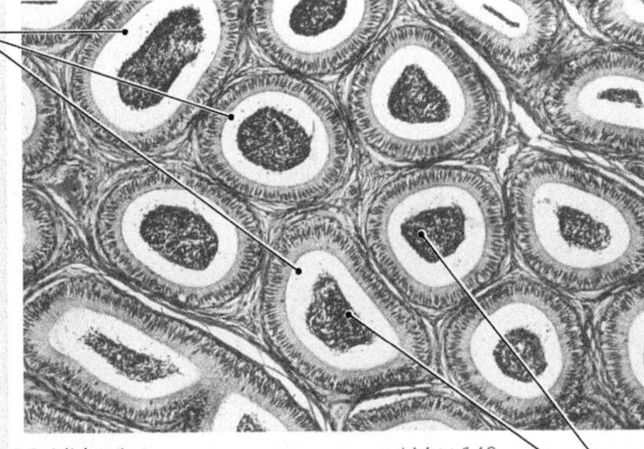

Epididymis LM × 140

2 Switch to the high-power objective lens and identify the epithelium lining the lumen. What type of epithelium do you observe?

▶ ______________________________

3 Identify the elongated microvilli, known as **stereocilia,** extending into the lumen from the surface of the epithelial cells. The stereocilia absorb testicular fluid and provide nutrients to the sperm cells. (Despite their name, these structures are not true cilia.)

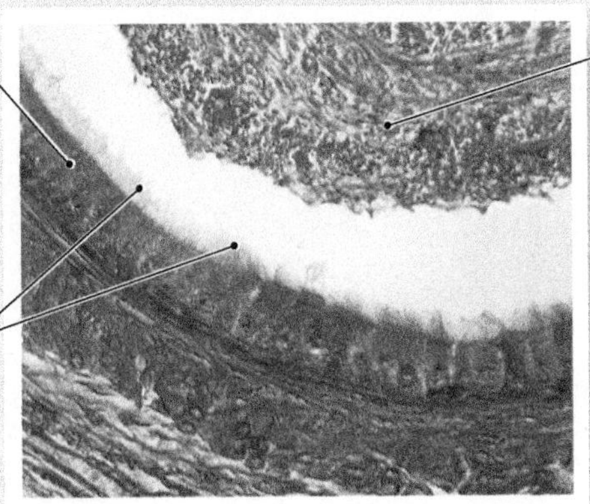

Epididymis LM × 410

4 In the lumina of some tubule profiles, you will see masses of **sperm cells.** The epididymis is about 7 m (23 ft) long. The sperm that enter it from the testes are not fully mature and are nonmotile. If the sperm cannot actively move, how do you think they are able to traverse the entire length of the epididymis?

▶ ______________________________

____

____

MAKING CONNECTIONS

In the testes, nurse cells are joined together by tight junctions that divide the seminiferous tubules into basal and luminal compartments (see the figure on the right). This division forms the **blood–testis barrier,** which avoids any contact between developing sperm cells and immune cells. Why is this division important?

▶ ______________________________

____

____

____

____

____

____

Luminal compartment

Basal compartment

Blood–testis barrier

Nucleus of nurse cell

Activity 31.5

Examining the Microscopic Anatomy of Male Reproductive Organs: Ductus Deferens and Prostate Gland

A The wall of the ductus deferens is composed mostly of smooth muscle.

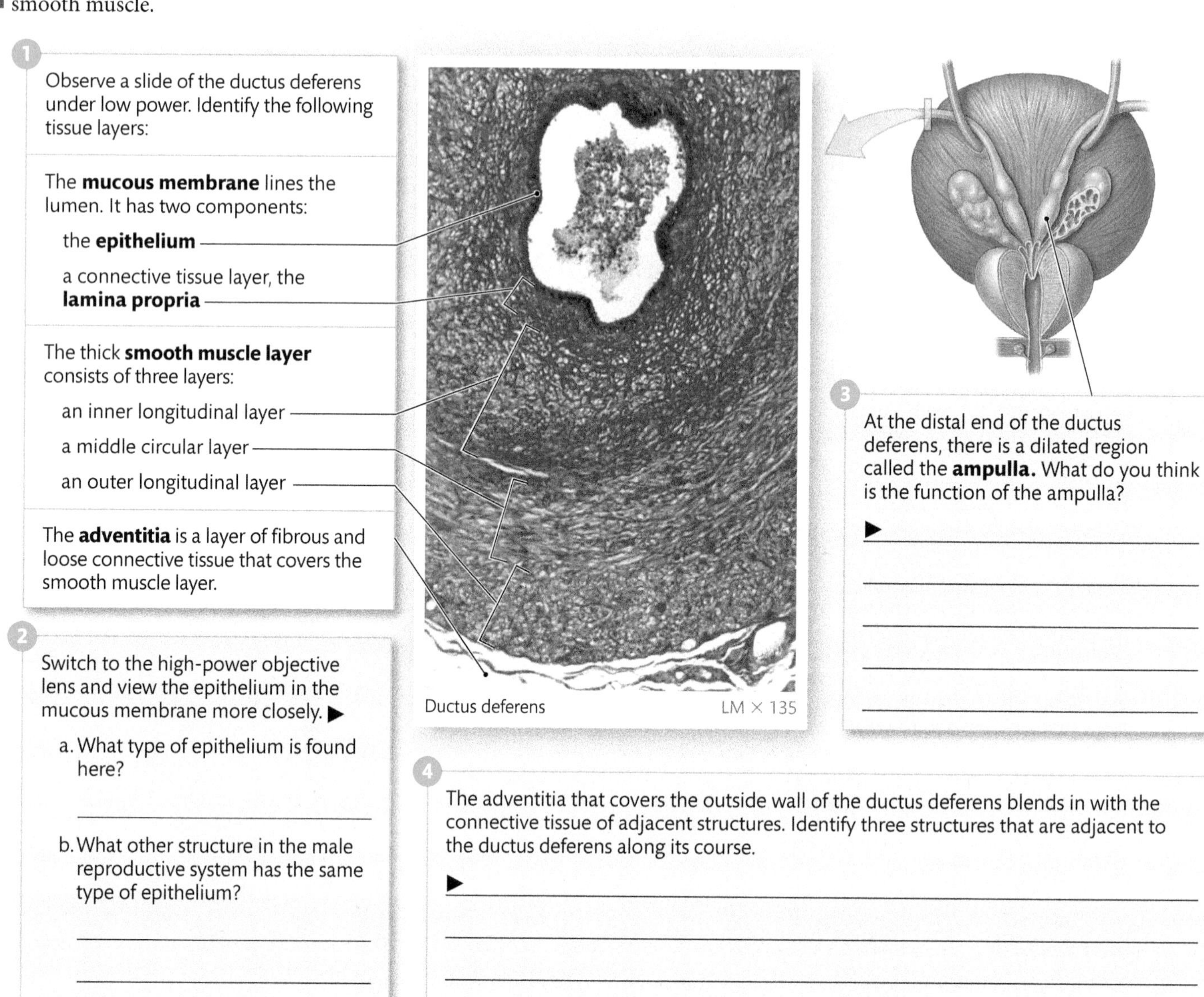

Ductus deferens LM × 135

1 Observe a slide of the ductus deferens under low power. Identify the following tissue layers:

The **mucous membrane** lines the lumen. It has two components:

- the **epithelium**
- a connective tissue layer, the **lamina propria**

The thick **smooth muscle layer** consists of three layers:

- an inner longitudinal layer
- a middle circular layer
- an outer longitudinal layer

The **adventitia** is a layer of fibrous and loose connective tissue that covers the smooth muscle layer.

2 Switch to the high-power objective lens and view the epithelium in the mucous membrane more closely. ▶

a. What type of epithelium is found here?

____

b. What other structure in the male reproductive system has the same type of epithelium?

____

____

3 At the distal end of the ductus deferens, there is a dilated region called the **ampulla.** What do you think is the function of the ampulla?

▶ _______________

____

____

____

____

4 The adventitia that covers the outside wall of the ductus deferens blends in with the connective tissue of adjacent structures. Identify three structures that are adjacent to the ductus deferens along its course.

▶ _______________

____

____

B The glandular regions in the prostate gland are separated by thick bands of smooth muscle.

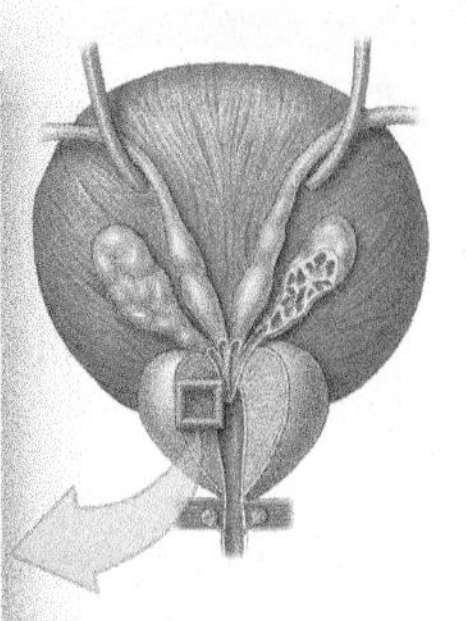

1 Examine a slide of the prostate gland under low power. Notice the following features:

The prostate gland contains numerous irregularly shaped **glands.**

The glands are separated by regions of **smooth muscle** and **connective tissue.**

2 View a glandular region with the high-power objective lens. What type of epithelium lines the gland?

▶ ______________________________

Switch back to the low-power objective lens and locate the **prostatic urethra** (this structure may be absent on your slide). Notice that the epithelium lining the urethra is transitional. Secretions from the prostate gland drain into the prostatic urethra. In addition to the prostate gland, what other structures empty into the prostatic urethra?

▶ ______________________________

____

Prostate gland LM × 120

IN THE CLINIC

Prostate Gland Diseases

Diseases of the prostate gland are common, particularly among men over age 50. **Benign prostatic hyperplasia (BPH)** is a noncancerous tumor that causes enlargement of the prostate and constriction of the urethra. Individuals with this condition can experience difficulty with micturition (urination), retention of urine in the bladder, urinary tract infections, and development of kidney stones. Treatment for BPH ranges from drug therapies that reduce the tumor to surgical removal of the prostate.

Prostate cancer also causes enlargement of the prostate and constriction of the urethra. During its early stages, there are usually no symptoms, but early detection may be possible by performing a blood test for elevated levels of a prostate enzyme known as **prostate-specific antigen (PSA).** PSA test evaluation must be done carefully because false positive and false negative results are possible. Treatment for prostate cancer includes surgical removal of the gland or radiation therapy, in which small radioactive pellets are placed into the gland and specifically attack the cancerous tumor. The prognosis is very encouraging if treatment occurs before a secondary tumor forms. If a metastasis is detected, treatments are not as effective. Most types of prostate cancer grow so slowly, however, that if it occurs after the age of 70, men usually die of natural causes or other diseases related to aging rather than the cancer itself. For this reason, sometimes the best treatment for prostate cancer may be no treatment at all.

MAKING CONNECTIONS

Why do you think the ductus deferens has a thick, well-developed muscle layer?

▶ ______________________________

Activity 31.6

Examining the Microscopic Anatomy of Male Reproductive Organs: Penis and Spermatozoa

A Erectile tissue is the dominant structural feature of the penis.

1 Scan a slide of the penis with the low-power objective lens and observe its general organization. Identify the three columns of erectile tissue:

the paired dorsal **corpora cavernosa**

the single ventral **corpus spongiosum**

2 Notice that the **penile urethra** passes through the corpus spongiosum.

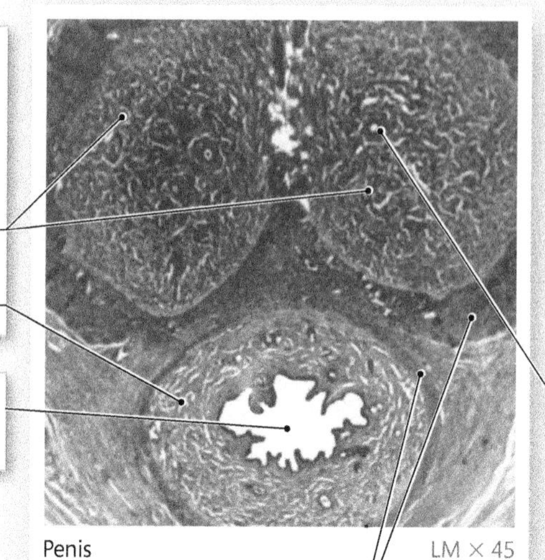

Penis LM × 45

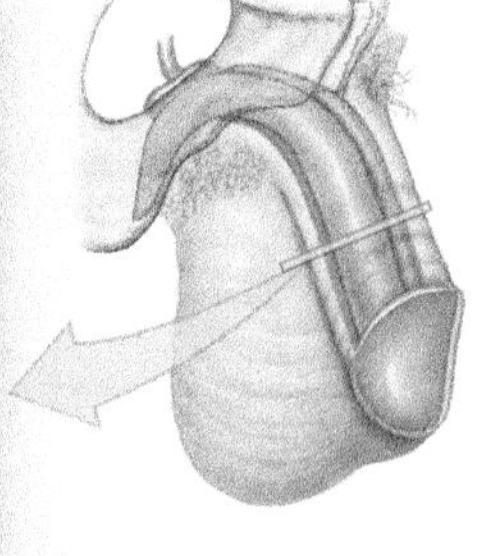

3 Locate a **deep (central) artery** traveling through each corpus cavernosum. These arteries provide the main blood supply to the venous sinusoids in the erectile tissue.

4 Identify the **tunica albuginea,** a tough, fibrous connective tissue sheath that surrounds the corpora cavernosa and the corpus spongiosum.

5 Upon sexual arousal, parasympathetic nerves relax the smooth muscle surrounding the erectile tissue and the arteries supplying the penis. As the arteries dilate, blood flow to the penis increases, the venous sinusoids become engorged with blood, and the penis swells, elongates, and becomes erect.

In the figure on the left, label the following structures: ▶

- Erectile tissue in the corpora cavernosa
- Deep arteries of the penis
- Tunica albuginea
- Erectile tissue in the corpus spongiosum
- Penile urethra

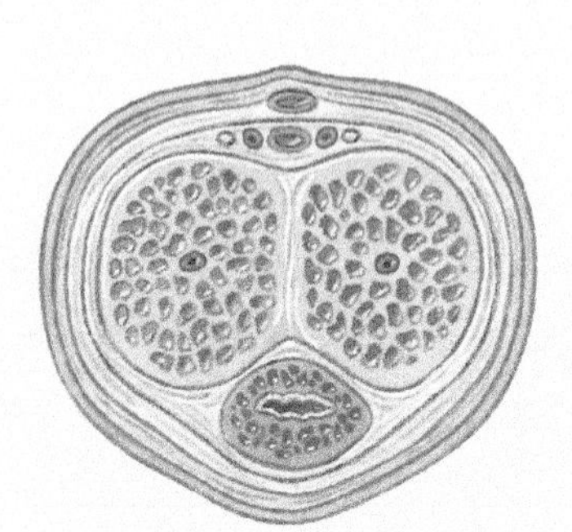

6 In the penis, the tunica albuginea surrounding the corpus spongiosum is less dense and more elastic compared with the same structure surrounding the corpora cavernosa. As a result, the corpus spongiosum is less turgid (swollen) upon erection than the corpora cavernosa. What do you think is the functional significance of this difference?

▶ ____________________

____

____

____

B Sperm cells contain half the amount of DNA found in other cells of the body.

1 Focus a slide of a **human** (or **mammalian**) **sperm smear** under low magnification. When the spermatozoa (sperm cells) can be identified, switch to high magnification.

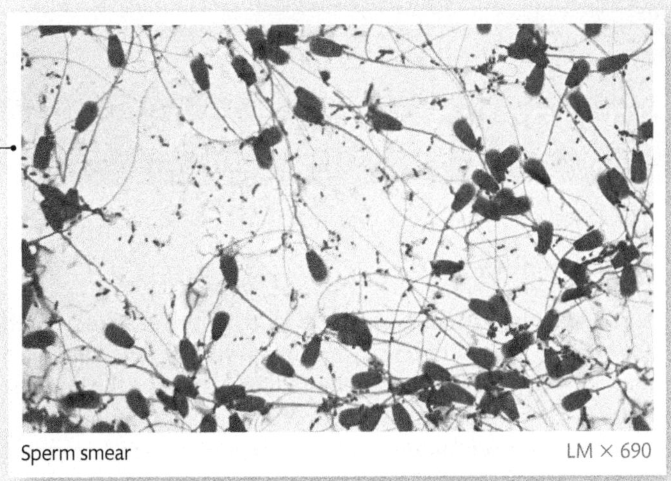
Sperm smear LM × 690

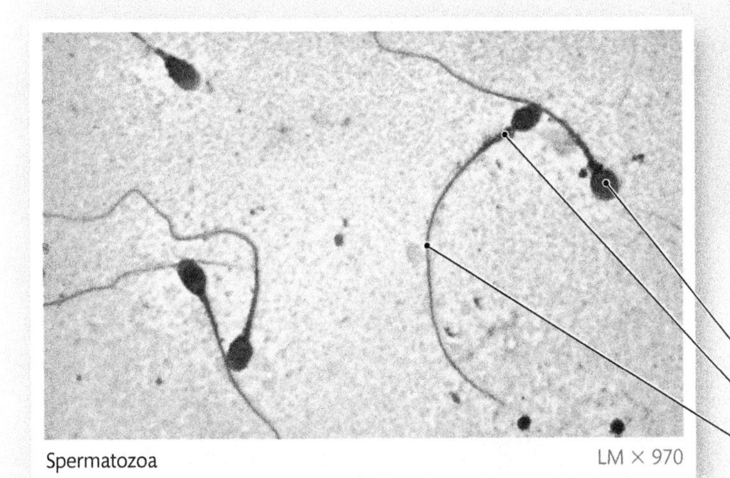
Spermatozoa LM × 970

2 Under high magnification, identify the parts of a sperm cell:

- **head** and **acrosome**
- **middle piece** and **neck**
- **tail**

3 Identify the acrosome, head, neck, middle piece, and tail on a model of a sperm cell and label the parts in the diagram on the right. ▶

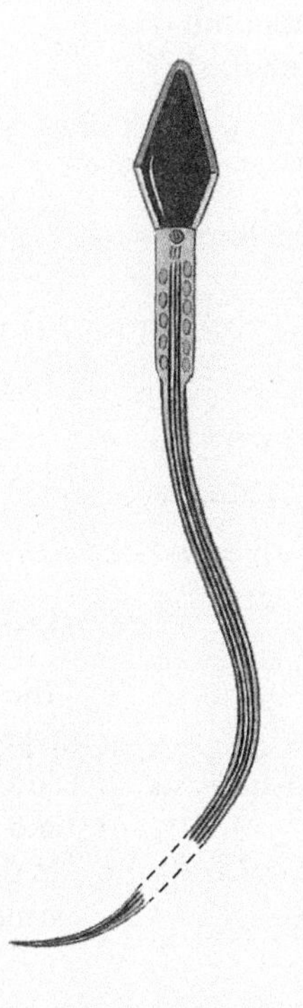

4 Refer to your textbook and complete the following table. ▶

Sperm Cell Structure	Function
Acrosome	
Head	
Neck	
Middle piece	
Tail	

MAKING CONNECTIONS

Prescription drugs such as Viagra and Cialis are available for men who cannot achieve or sustain an erection (erectile dysfunction). How do you think these drugs work?

▶ ______________________________

BEFORE YOU MOVE ON . . .

‹‹ LOOKING BACK

The reproductive system is unique to all other organ systems in the body because it is not necessary for the survival of the individual, but its activities are absolutely required for sustaining the human species. Unlike other organs that are functional throughout life, the reproductive organs are inactive until puberty, which normally occurs between 11 and 15 years of age. At the onset of puberty, the reproductive organs respond to increased levels of **sex hormones** by growing rapidly and becoming functionally mature structures. In males, the sex hormones are called **androgens.** The primary androgen is **testosterone.** In addition to its role in the development and maturation of male sex organs, testosterone also stimulates the growth of bones and skeletal muscle and affects the development of the central nervous system.

Recently, androgens have been the subject of intense media coverage because of their use as anabolic steroids by professional (and amateur) athletes. The rationale for using these drugs is that they increase muscle mass, strength, and the ability to train longer and harder. Some people claim that anabolic steroids are safe to use because they are compounds that are produced naturally by the body. Do you think this rationale is scientifically sound? Explain your answer.

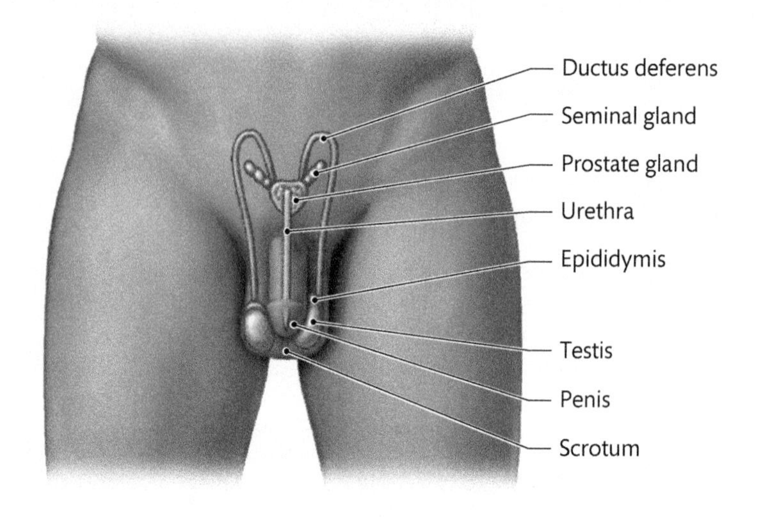

LOOKING FORWARD ››

The adult reproductive system is unique because male and female reproductive organs have distinct differences in both structure and function. In this laboratory exercise, you studied the anatomy and physiology of the organs in the male reproductive system. In the next exercise, you will complete a similar examination of structures in the female reproductive system.

Name ______________________

Lab Section ______________________

Date ______________________

EXERCISE 31 REVIEW SHEET

The Male Reproductive System

1. Explain why the urethra has both a reproductive function and a urinary function in males.

____

____

____

2. Complete the following table.

Structure	Location	Function
Seminiferous tubules	a.	b.
Erectile tissue	c.	d.
Cremaster muscle	e.	f.
Ejaculatory duct	g.	h.

3. If one or both testes do not complete the migration into the scrotum and remain in the abdominal cavity, they must be surgically moved into the scrotum. If the testes remain in the abdominal cavity, the individual will become sterile. Explain why.

____

____

____

QUESTIONS 4–11: Match the structure in column A with the appropriate description in column B.

A

4. Testis ______

5. Ductus deferens ______

6. Scrotum ______

7. Urethra ______

8. Ejaculatory duct ______

9. Penis ______

10. Epididymis ______

11. Seminal gland ______

B

a. In addition to the urethra, this muscular duct travels through the prostate gland.

b. Spermatozoa become functionally mature while in this structure.

c. This structure has a urinary and reproductive function.

d. The corpus spongiosum is found in this structure.

e. Interstitial (Leydig) cells, found in this structure, produce testosterone.

f. This structure transports sperm cells from the epididymis to the ejaculatory duct.

g. This two-chambered structure contains the testes.

h. Secretions from this structure are a component of semen.

12. Identify the structure that is shown in each of the following photos.

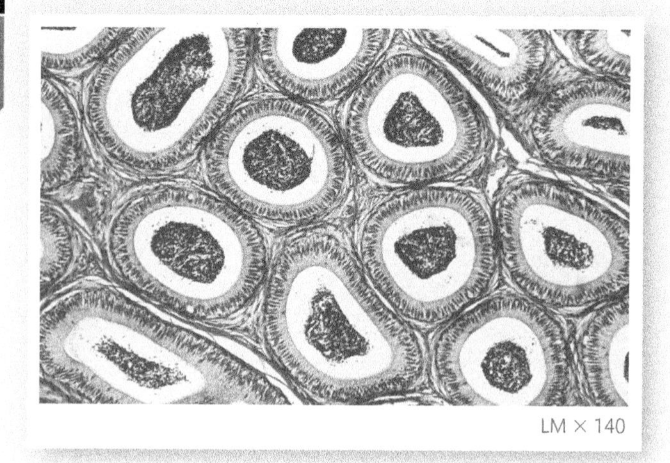

LM × 140

a. ______________________________

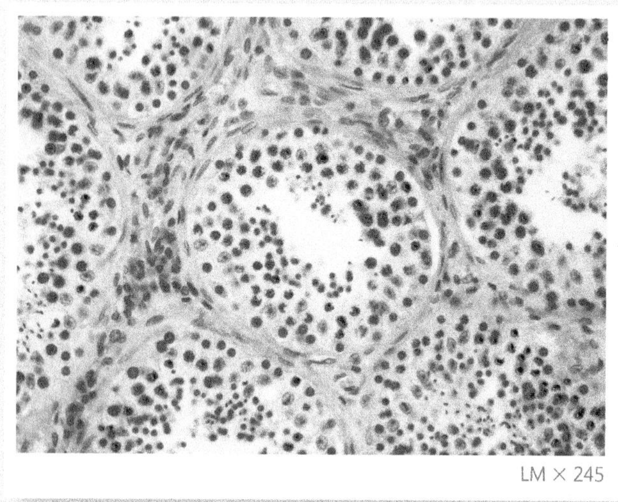

LM × 245

b. ______________________________

LM × 175

c. ______________________________

13. In the diagram on the right, color the structures with the indicated colors.

Prostate = **dark green**
Epididymis = **dark blue**
Corpus cavernosum = **red**
Ejaculatory duct = **brown**
Testis = **yellow**
Seminal gland = **orange**
Corpus spongiosum = **light blue**
Bulbourethral gland = **tan**
Ductus deferens = **light green**
Scrotum = **gray**
Urinary bladder = **purple**
Urethra = **black**

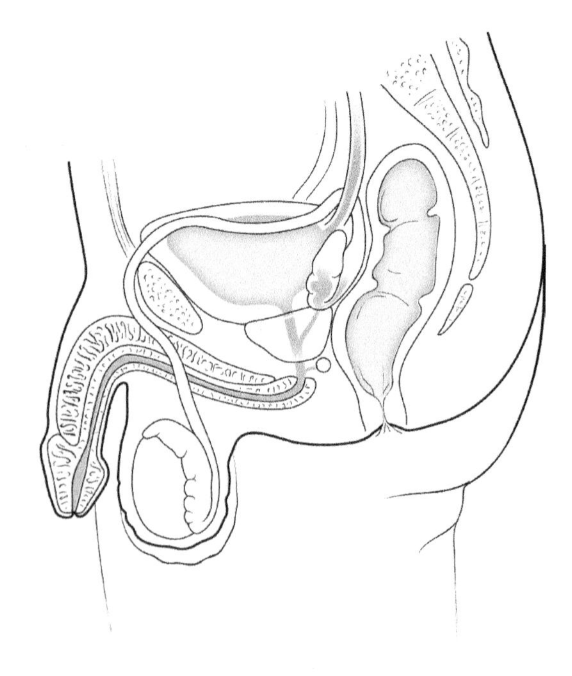

EXERCISE 32

The Female Reproductive System

LEARNING OUTCOMES

These Learning Outcomes correspond by number to the laboratory activities in this exercise. When you complete the activities, you should be able to:

Activity 32.1 **Describe the structure and function of the ovaries and accessory organs in the female reproductive system.**

Activity 32.2 **Describe the structure and function of the external genitalia in the female reproductive system.**

Activity 32.3 **Describe the structure and function of the mammary glands in the female reproductive system.**

Activity 32.4 **Describe the microscopic structure of the ovary.**

Activity 32.5 **Describe the microscopic structure of the uterus.**

Activity 32.6 **Describe the microscopic structure of the uterine tubes and vagina.**

LABORATORY SUPPLIES

- Anatomical models:
  - female torso
  - female pelvis, midsagittal section
  - female external genitalia
  - ovary, showing follicular development
- Whole skeleton or model of the bony pelvis
- Three pieces of string, approximately 30 cm (12 in.) in length
- Prepared microscope slides:
  - ovary
  - uterus, proliferative stage
  - uterus, secretory stage
  - uterus, menstrual stage
  - uterine tube, cross section
  - vagina, cross section

PRE-LAB QUIZ

Before you begin, read all the activities in Exercise 32 and the required reading in your textbook that is assigned by your instructor.

1. The ____________ ligament attaches the ovary to the uterus.
2. The body of the uterus narrows and gives rise to the ____________, which protrudes into the superior end of the vagina.
3. True or False: Each mammary gland covers the pectoralis major muscle on the anterior thoracic wall. ____________
4. During this laboratory exercise, you will examine the microscopic structure of all the following structures *except* the
   a. ovary.
   b. vagina.
   c. uterus.
   d. mammary gland.
   e. uterine tube.
5. True or False: Along the floor to the vestibule, the vaginal orifice is anterior to the urethral orifice. ____________

Questions 6–10: Match the structure in column A with the organ in which it is found in column B.

A	B
6. Erectile tissue ____________	a. Mammary gland
7. Lactiferous duct ____________	b. Vagina
8. Primary follicle ____________	c. Uterus
9. Endometrium ____________	d. Clitoris
10. Stratified squamous epithelium ____________	e. Ovary

Activity 32.1

Examining the Gross Anatomy of the Female Reproductive System: Ovaries and Accessory Organs

The female reproductive system produces and maintains egg cells, transports egg cells to the site of fertilization, provides a favorable environment for the developing fetus, facilitates the birth process, and produces female sex hormones.

A The ovaries are solid, ovoid structures, about 2 cm (0.8 in.) long and 1 cm (0.4 in.) wide. In the adult, they are positioned along the lateral walls of the pelvic cavity.

1. On a model of a female torso or pelvic cavity, identify the **ovaries.** Notice that each ovary is located on the lateral wall of the pelvis in the region between the internal and external iliac arteries.

2. Like the testes, the ovaries develop from embryonic tissue along the posterior abdominal wall, near the kidneys, and descend to their adult position. How does the descent of the ovaries differ from the descent of the testes?

▶ ______________________________

____

B The accessory female reproductive organs include the uterine tubes, uterus, and vagina.

1. On a model, locate the **uterine tubes (oviducts, fallopian tubes).** Notice that each uterine tube is connected to the uterine wall and extends laterally toward the ovary.

2. Identify the **uterus.** In a nonpregnant adult female, it is shaped like an inverted pear that is bent anteriorly. Describe the position of the uterus relative to the urinary bladder.

▶ ______________________________

____

3. Identify the **vagina,** which is a muscular tube between 7.5 and 9 cm (3 and 3.5 in.) long extending from the uterus to the vaginal orifice.

Describe the position of the vagina relative to the urinary bladder and urethra.

▶ ______________________________

____

____

____

Describe its position relative to the rectum.

▶ ______________________________

____

____

____

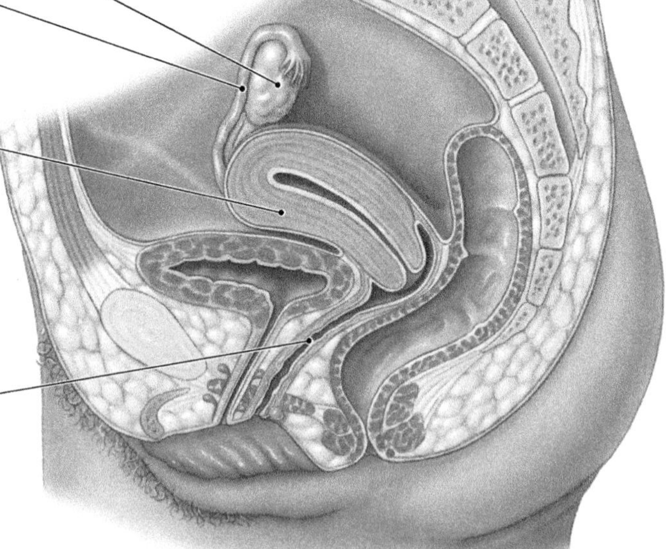

4. Refer to your textbook and describe the functions of the female reproductive organs in the table below. ▶

Structure	Function
Ovaries	
Uterine tubes	
Uterus	
Vagina	

C The uterine tubes and the uterus each have three well-defined regions.

D Various supporting mesenteries and ligaments connect the female reproductive organs to the body wall.

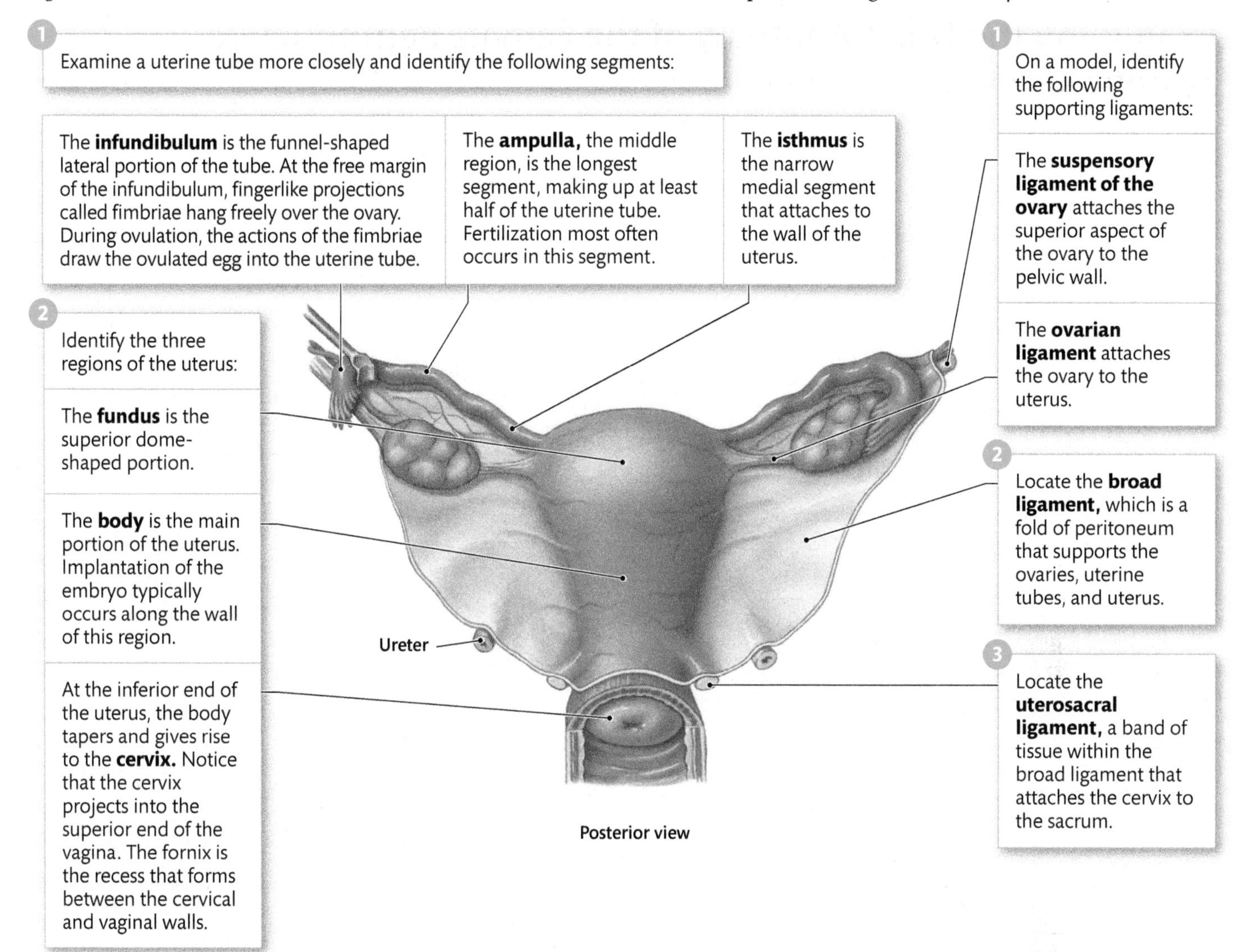

Word Origins

In Latin, the word *vagina* means "sheath." The vagina serves as a sheath or covering that surrounds the erect penis during sexual intercourse. It also forms a sheath around the full-term fetus as it passes from the uterus to the outside during childbirth.

IN THE CLINIC

Position of the Uterus

The anterior bend, known as anteflexion, is the normal position of the uterus. In some women, however, the uterus bends posteriorly toward the sacrum. This position is referred to as retroflexion of the uterus. A posterior bend of the uterus has no negative clinical consequences. In fact, a retroflexed uterus typically becomes anteflexed during pregnancy.

MAKING CONNECTIONS

Explain how a bacterial infection that develops in the vagina can spread to the abdominopelvic cavity. How can such an infection (e.g., pelvic inflammatory disease) cause infertility?

▶ ______________________________

____

____

____

____

____

____

Want more practice? Go to: **MasteringA&P** > Study Area > Menu > Lab Tools > PAL >

- Human Cadaver > Reproductive System > Female photos
- Anatomical Models > Reproductive System > Female photos

Activity 32.2

Examining the Gross Anatomy of the Female Reproductive System: External Genitalia

A The vaginal and urethral openings are enclosed by the external genitalia.

1. On a model of the female external genitalia, identify the **mons pubis,** a mound of fat and skin that covers the symphysis pubis. After puberty, the skin is covered with pubic hair.

2. Locate the **labia majora,** two fatty skin folds that are covered with pubic hair. They project posteriorly from the mons pubis.

3. Locate the **labia minora,** which are two smaller skin folds located within the borders of the labia majora. These structures lack hair and have a pinkish color due to their rich vascular supply.

4. The **vestibule** is the space enclosed by the labia minora.

5. Identify the **clitoris.** It is located anterior to the vestibule, adjacent to the area where the labia minora meet. Similar to the penis, the clitoris is a cylindrical structure composed mostly of erectile tissue. The glans clitoris, located at the distal end of the clitoris, is covered by a fold of skin, the prepuce.

6. Along the floor of the vestibule, identify two openings:

   The **urethral orifice** is the anterior opening.

   The **vaginal orifice** is the posterior opening.

7. The **vestibular bulbs** (not shown) are deep elongated masses of erectile tissue located on either side of the vestibule and extending from the vaginal opening to the clitoris. In the figure, label the positions of the two vestibular bulbs. ▶

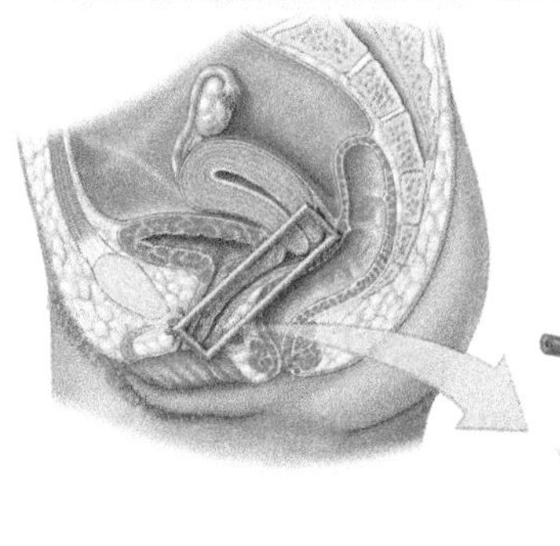

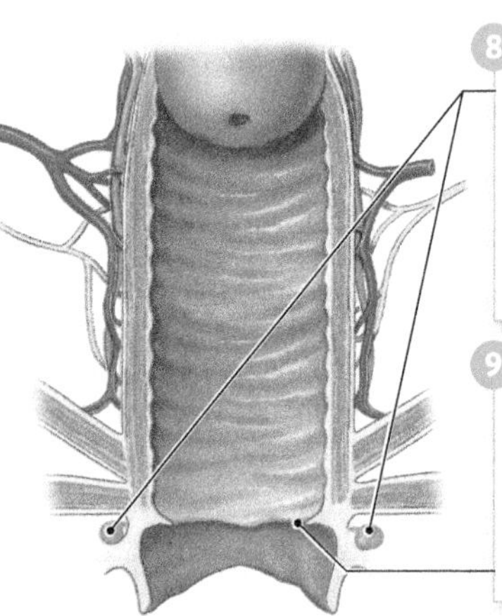

8. On a model of the female pelvic cavity, locate the distal end of the vagina. A pair of **greater vestibular glands** is located deep to the posterior portion of the labia minora on each side of the vaginal orifice. The secretions from these glands moisten the area around the vaginal orifice to facilitate the insertion of the penis during sexual intercourse.

9. The vaginal orifice is partially closed by the **hymen,** a thin connective tissue membrane. The hymen usually ruptures and causes bleeding after the first sexual intercourse or other activity that exerts excess pressure on the membrane, such as inserting a tampon or physical exercise.

■ Want more practice? Go to: MasteringA&P > Study Area > Menu > Lab Tools > PAL > Human Cadaver > Reproductive System > Female photos

B The external genitalia are located in the **urogenital triangle**, which is the anterior region of the perineum.

1. On a skeleton or model of the bony pelvis, connect the two **ischial tuberosities** with a piece of string.

2. With two additional pieces of string, connect each ischial tuberosity to the **symphysis pubis.**

3. The area enclosed by the string is the **urogenital triangle.** The female external genitalia are located within this region.

Anal triangle

4. What male reproductive structures are located in the male urogenital triangle?

▶ ____

____

____

Word Origins

Labium (plural = *labia*) means "lip" in Latin. Anatomists use the word to describe any lip-shaped structure. The two pairs of skin folds that are parts of the female external genitalia are good examples. Thus, the larger pair is called the *labia majora* and the smaller pair the *labia minora.*

MAKING CONNECTIONS

During sexual arousal, the clitoris, like the penis, becomes erect. Explain why.

▶ ____

____

____

____

____

____

Activity **32.3**

Examining the Gross Anatomy of the Female Reproductive System: Mammary Glands

A The mammary glands are located in the subcutaneous tissue on the anterior thoracic wall.

1 On a torso model, identify the position of the mammary glands on the anterior thoracic wall. Each gland overlies the **pectoralis major** muscle.

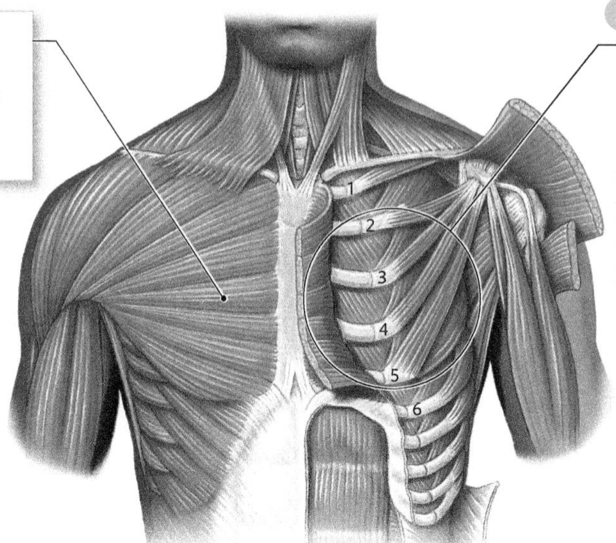

2 On a skeleton or torso model, identify the following structures on the anterior thoracic wall that define the position of the mammary gland's circular base. The base extends from

- the second to fifth or sixth ribs
- the sternum to the axilla.

IN THE CLINIC

Breast Cancer

Breast cancer affects about one in eight women in the United States. The most common type occurs in the epithelium that lines the ducts in the lobules. Potentially, it can spread quickly if cancerous cells dislodge from the primary tumor and enter lymphatic vessels to the axillary and parasternal (next to the sternum) lymph nodes. Free communication exists between these lymph nodes and other body regions. Consequently, metastases can spread to a variety of areas, the most common being the vertebrae, brain, liver, and lungs.

Risk factors for breast cancer include family history and various factors that tend to increase one's exposure to estrogens, such as early puberty, late menopause, having a first pregnancy after age 35, or receiving estrogen replacement therapy. Recent evidence has indicated that exposure to environmental toxins that mimic estrogen may also increase one's risk.

Early detection is the primary weapon against breast cancer because the survival rate declines dramatically if the cancer has spread to the axillary lymph nodes. The American Cancer Society recommends that women, age 45 through 54, should receive an annual mammogram. Younger women, age 40 through 44, should be able to begin annual mammograms if they choose. Older women, age 55 and over, should get mammograms every two years. It was once believed that a breast self-exam (BSE) was vital for early breast cancer detection and thus could save lives. There is no evidence to support this claim, however, so now BSE is considered optional. The focus has shifted to *breast awareness*, which means that a woman should be familiar with how her breasts normally look and feel and report any changes to her doctor as soon as possible.

3 Read the "In the Clinic" box on the left and answer the following question:

Do you think it is possible for a man to get breast cancer? Explain your answer.

▶ ______________________________

____

____

____

____

____

____

____

B The mammary glands are modified sweat glands. They produce milk that provides nourishment for a baby after birth.

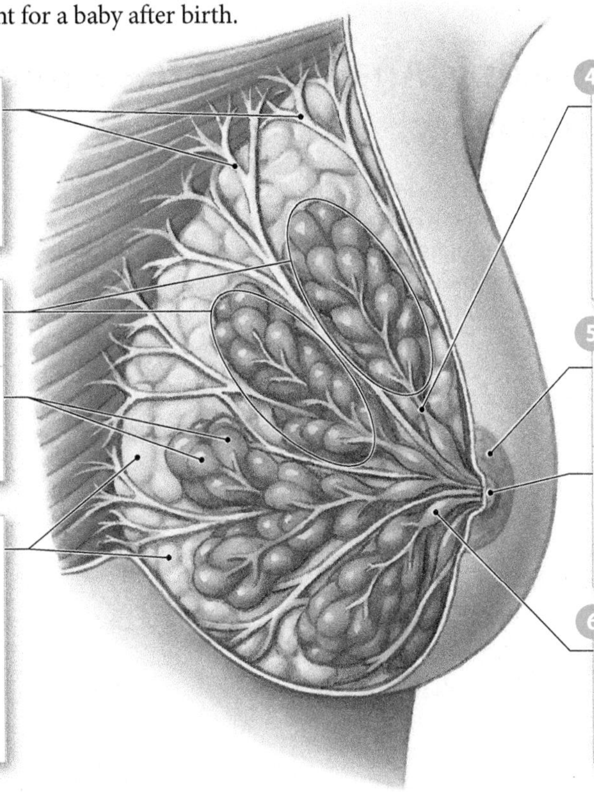

1 On a model of the mammary gland, identify the **suspensory ligaments** that support the gland. They travel from the skin to the deep fascia.

2 Notice that the mammary gland is divided into 15 to 20 **lobes.**

Each lobe contains numerous **lobules** with clusters of glandular cells that produce milk.

3 Notice that the lobes of glandular tissue are divided by connective tissue and fat. The amount of glandular material is fairly constant among nonpregnant females. The amount of fat deposition differs greatly, however, and is the basis for the variation in breast size.

4 Milk is secreted into ducts, which progressively increase in diameter and eventually drain into a **lactiferous duct.** Identify a lactiferous duct in each lobe. Notice how these ducts converge at the nipple.

5 At the apex of each breast, identify the **areola,** a darkly pigmented circular area of skin.

The **nipple,** projecting from the center of the areola, is the structure from which the baby sucks milk.

6 At the nipple, observe how each lactiferous duct forms an expanded **lactiferous sinus,** where milk is stored prior to breastfeeding.

IN THE CLINIC

Polymastia

Embryologically, the mammary glands can develop anywhere along the **milk** or **mammary line,** a thickened ridge of embryonic tissue that extends from the axilla to the groin. Usually, only a small segment of the milk line along the midthoracic region persists. It is from this tissue that the mammary glands form. They develop in both males and females, but are only functional in females during periods of breastfeeding. Occasionally, other portions of the milk line will remain active and give rise to accessory nipples **(polythelia)** or complete mammary glands **(polymastia).**

MAKING CONNECTIONS

During pregnancy, the glandular cells and ducts grow and develop in preparation for breastfeeding. The first secretion produced by the mammary glands at birth is a protein-rich liquid called **colostrum.** After 2–3 days of breastfeeding, the glands begin to produce **breast milk.** Both colostrum and breast milk contain antibodies and other antibiotic substances. Why is it important for colostrum and breast milk to contain antibodies?

▶ ______________________________

____

____

____

____

____

■ Want more practice? Go to: **MasteringA&P** > Study Area > Menu > Lab Tools > PAL > Anatomical Models > Reproductive System > Mammary gland photo

Activity 32.4

Examining the Microscopic Anatomy of the Female Reproductive System: Ovary

A The **ovarian cycle** refers to the cyclic monthly changes that occur in the ovary during a woman's reproductive life. During the first half of the cycle, known as the **follicular phase,** about 6 to 12 primordial follicles are induced by **follicle-stimulating hormone (FSH)** from the anterior pituitary to grow and mature.

1 Scan a slide of a human or mammalian ovary under low power. Identify the two regions of the ovary:

The outer **ovarian cortex** contains the developing egg cells (oocytes) that are enclosed by multicellular capsules called follicles.

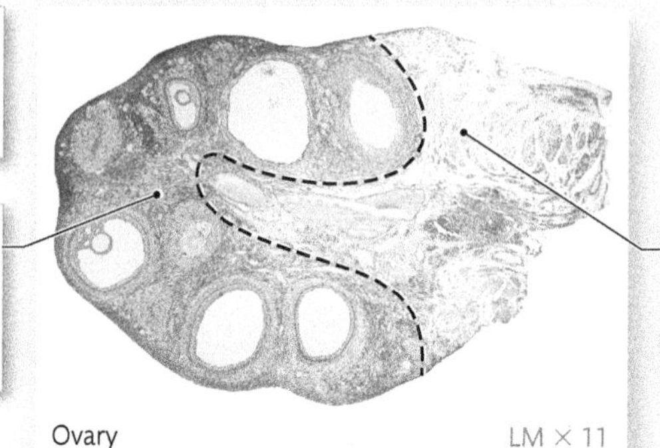

Ovary LM × 11

The inner **ovarian medulla** is a core of loose connective tissue that contains blood vessels, nerves, and lymphatics that supply the ovary.

2 In the cortex, identify the various stages of follicular development. It is unlikely that all the stages will appear on one slide, so viewing several slides will be necessary.

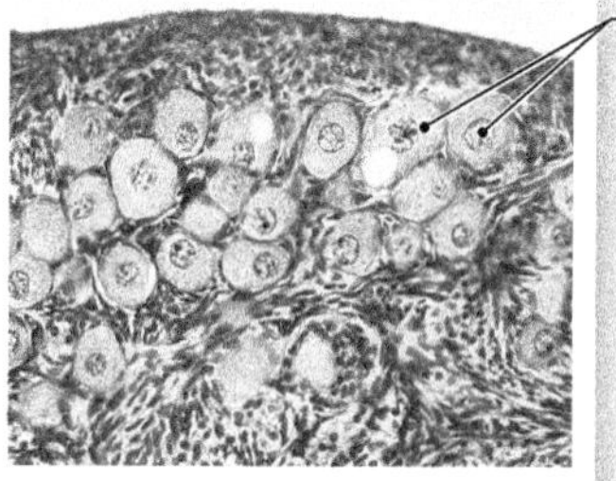

Primordial follicles LM × 145

Primordial follicles are produced during fetal development and located in the outer portion of the cortex. Each primordial follicle contains an oocyte (egg cell) surrounded by a single layer of squamous follicle (follicular) cells.

In the transition to a **primary follicle,** the follicular cells divide to form multiple layers and become cuboidal or columnar. These cells are now referred to as granulosa cells. The granulosa cells and the thecal cells, which form a layer around the follicle, begin to produce female sex hormones known as estrogens. A protective layer of glycoprotein, called the zona pellucida, forms around the egg, separating it from the granulosa cells.

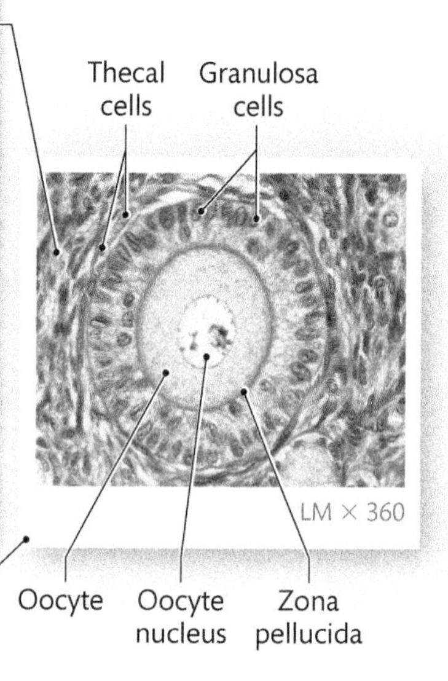

LM × 360

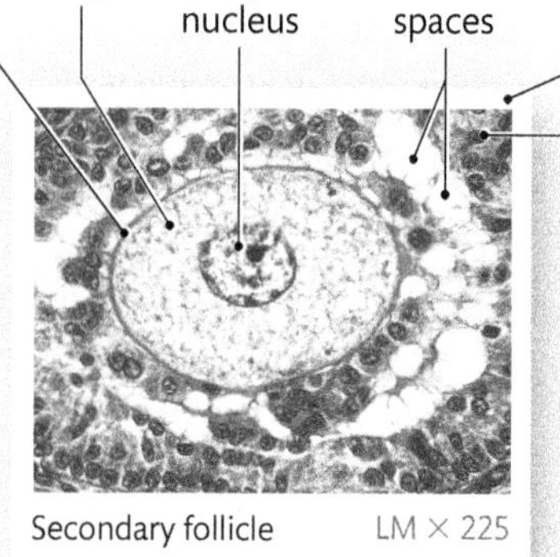

Secondary follicle LM × 225

A **secondary follicle** forms when fluid-filled spaces appear between the granulosa cells. These spaces eventually merge to form a single fluid-filled chamber known as the antrum.

In a **tertiary follicle,** granulosa cells surround the large antrum. The egg is located at one end of the antrum. It is surrounded by a ring of granulosa cells called the corona radiata and is connected to the wall of the follicle by an additional stalk of granulosa cells.

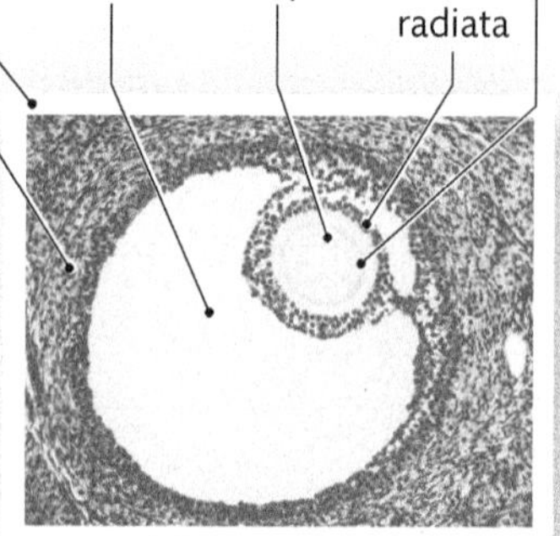

Tertiary follicle LM × 50

B The release of the egg from the ovary, known as **ovulation,** is influenced by **luteinizing hormone (LH)** from the anterior pituitary. Ovulation typically occurs at about the midpoint of the ovarian cycle. After ovulation, the remnants of the follicle are transformed into a structure called the **corpus luteum** to begin the **luteal phase.**

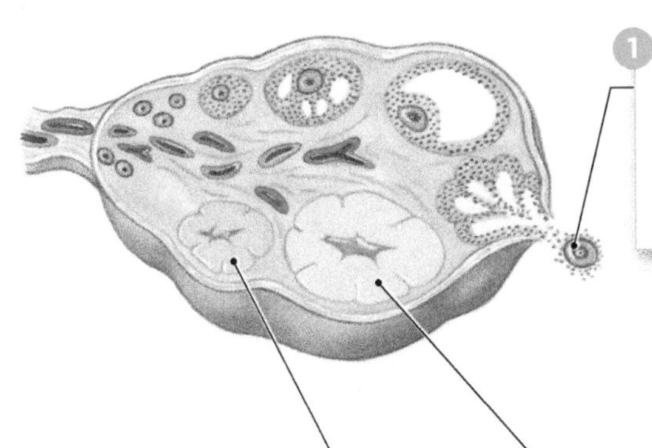

1. On a slide of the ovary, identify the egg and its surrounding corona radiata in a tertiary follicle. At ovulation, the follicle releases the **egg** and **corona radiata** into the pelvic cavity.

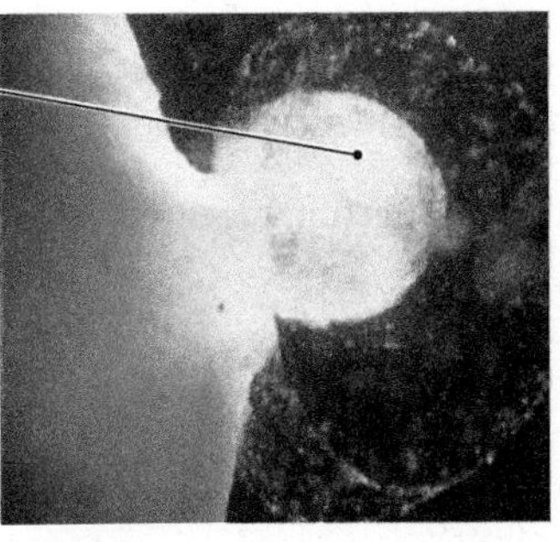

Ovulation

2. Identify the **corpus luteum.** This structure forms from the remnants of the follicle after ovulation. If a pregnancy occurs, the corpus luteum produces progesterone and estrogen to maintain the wall of the uterus during the early period of development.

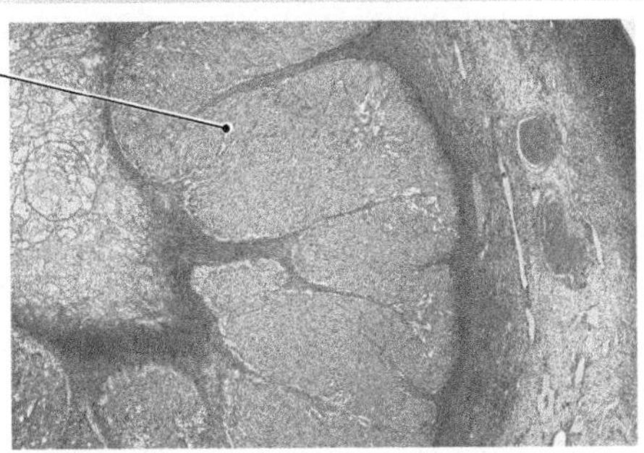

Corpus luteum LM × 15

3. If fertilization does not occur, the corpus luteum will begin to break down about 2 weeks after ovulation. Degeneration occurs when fibroblasts enter the corpus luteum, and a clump of scar tissue called the **corpus albicans** is formed. Identify the corpus albicans on a slide.

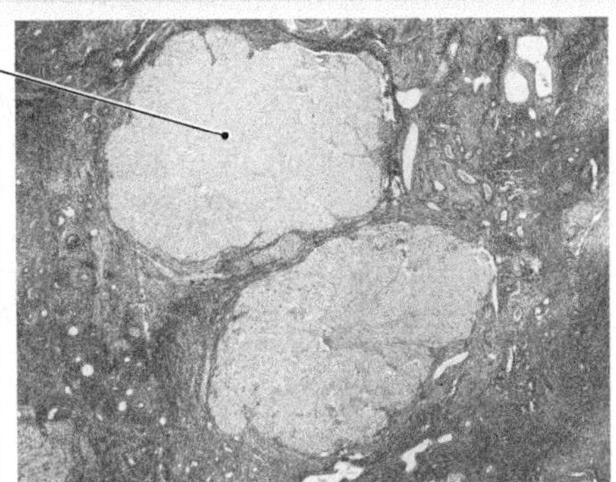

Corpus albicans LM × 15

MAKING CONNECTIONS

The surfaces of the ovaries in a prepubescent girl are smooth. Several years after puberty, however, they become rough and uneven, with numerous regions of scar tissue. Provide an explanation for this transformation.

▶ ____

____

____

■ Want more practice? Go to: **MasteringA&P** > Study Area > Menu > Lab Tools > PAL > Histology > Reproductive System > Ovary photos

Activity **32.5**

Examining the Microscopic Anatomy of the Female Reproductive System: Uterus

The primary functions of the uterus are to receive the embryo that results from fertilization, nourish and support the embryo/fetus during its development, and facilitate the birth of the full-term fetus.

A The uterine wall contains three tissue layers.

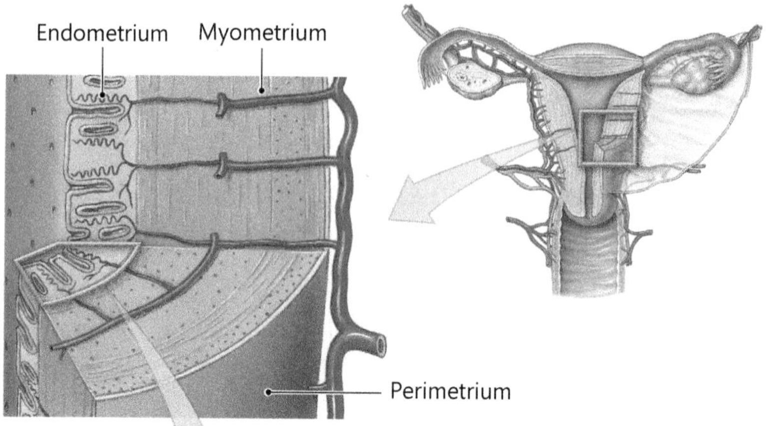

Endometrium
Myometrium
Perimetrium

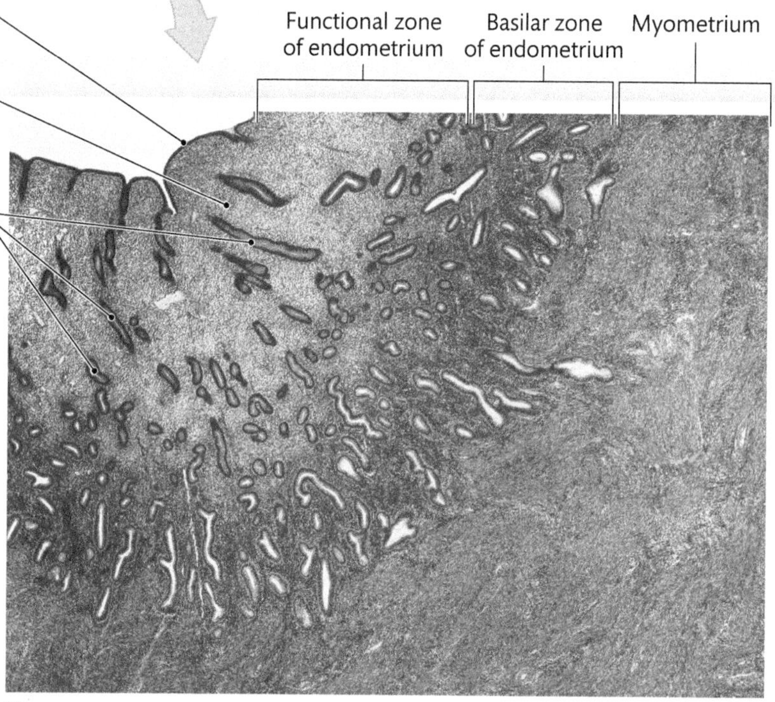

Functional zone of endometrium
Basilar zone of endometrium
Myometrium

Uterus
LM × 40

1 Observe a slide of the uterus under low power. Locate the **endometrium,** which is the inner layer of the uterine wall. Identify the following parts of the endometrium:

The simple columnar **epithelium** lines the lumen of the uterus.

The **lamina propria** is a thick layer of connective tissue deep to the epithelium.

Numerous **uterine glands** are scattered throughout the lamina propria.

2 In the endometrium, differentiate between the superficial **functional zone,** which contains most of the glands, and the deeper **basilar zone,** which borders the smooth muscle layer (myometrium). What roles do these two zones have during the menstrual cycle?

▶ ____________________

____

____

____

____

3 Move the slide to the next layer in the uterine wall, the **myometrium,** and identify its interlacing fibers of smooth muscle. What function does the myometrium have during the birth process?

▶ ____________________

____

____

4 The outermost layer, the **perimetrium** (see top-left figure), is a serous membrane and is a part of the visceral peritoneum. This layer may not be present on your slides.

B The **uterine** or **menstrual cycle** refers to the cyclic changes that occur in the endometrium over a period of approximately 28 days. The events that occur are influenced by hormones and correspond to specific stages of the ovarian cycle.

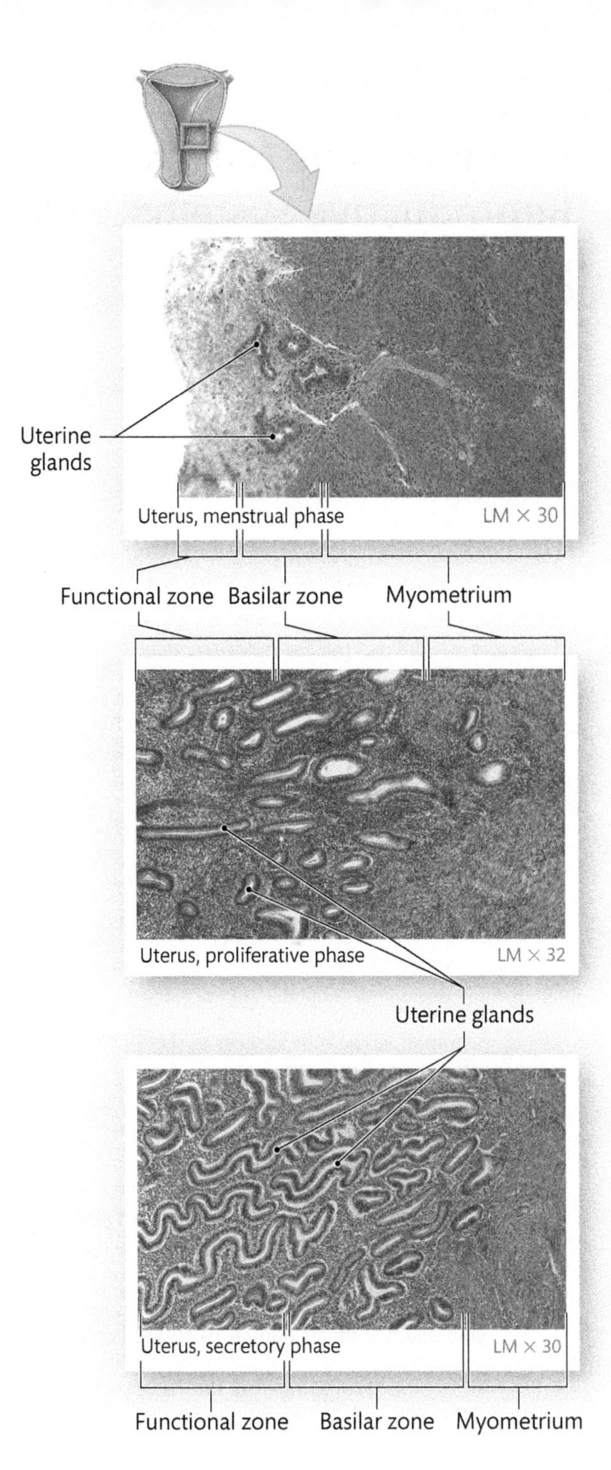

1. Observe a slide of the uterus in the **menstrual phase.** Notice the degeneration of the endometrial glands and blood vessels in the functional zone. During the menstrual phase (approximately days 1–5), the functional zone breaks down, and endometrial tissue and blood pass out of the vagina as the menstrual flow or menses.

2. Observe a slide of the uterus in the **proliferative phase.** Identify the uterine glands throughout the functional zone. Notice that the glands are simple in structure with little or no branching. During the proliferative phase (approximately days 6–14), the basilar zone gives rise to a new functional zone. During this process, new glands form, the endometrium thickens, and blood vessels infiltrate the new tissue. Estrogens that are produced by cells in the ovarian follicles induce these rebuilding activities.

3. Observe a slide of the uterus in the **secretory phase.** Notice that the uterine glands are now enlarged and have numerous branches. The secretory phase is the final stage of the uterine cycle (approximately days 15–28). The events during this period are designed to prepare the endometrium for implantation. Uterine glands are larger and more complex and begin to secrete glycoproteins and other nutrients. The arteries provide a rich resource of maternal blood to support the early embryo. These activities are supported by progesterone that is produced by the corpus luteum.

MAKING CONNECTIONS

Sometimes regions of endometrial tissue can appear in areas outside the uterus, such as on the ovaries or along the peritoneum that covers the pelvic cavity. This condition, called endometriosis, can cause severe pain during menstruation. Explain why.

▶ ______________________________

____

____

Activity **32.6**

Examining the Microscopic Anatomy of the Female Reproductive System: Uterine Tubes and Vagina

A The **uterine tubes** are approximately 13 cm (5 in.) long. They are muscular tubes that are lined by an epithelium with numerous ciliated cells.

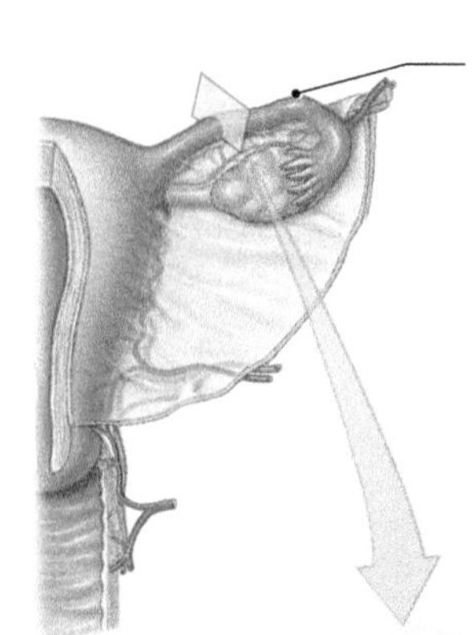

1. Scan a slide of the uterine tube in cross section with the low-power objective lens. Notice that the **lumen** has an irregular shape.

2. The **mucous membrane** lines the lumen. Identify its two components:
   - the **simple columnar epithelium**
   - the loose connective tissue in the deeper **lamina propria**

3. Identify the **smooth muscle layer** surrounding the mucous membrane. The muscle layer thickens as the uterine tube approaches the uterus.

4. The **serosa** (see top figure) is a serous membrane that covers the outside wall of the uterine tube. It consists of a thin layer of connective tissue and simple squamous epithelium.

5. Switch to the high-power objective lens and notice that the epithelium contains numerous **ciliated cells.** The bottom figure shows the epithelial surface of the uterine tube, with cilia (light blue and yellow) projecting from the surface of ciliated cells.

6. Why are the muscle layer and ciliated cells essential for normal functioning of the uterine tubes?

▶ ______________________________

____

____

____

____

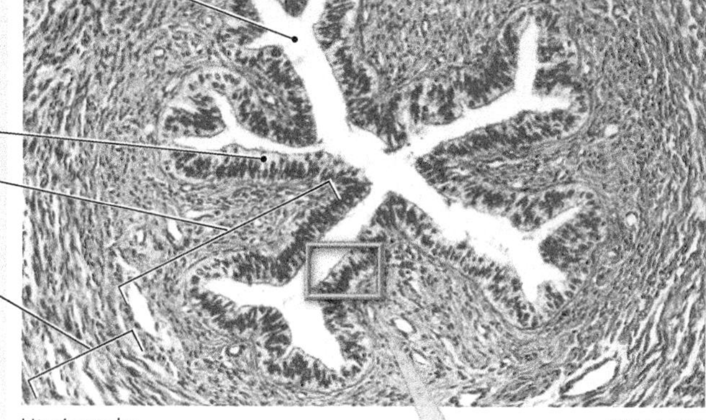

Uterine tube LM × 325

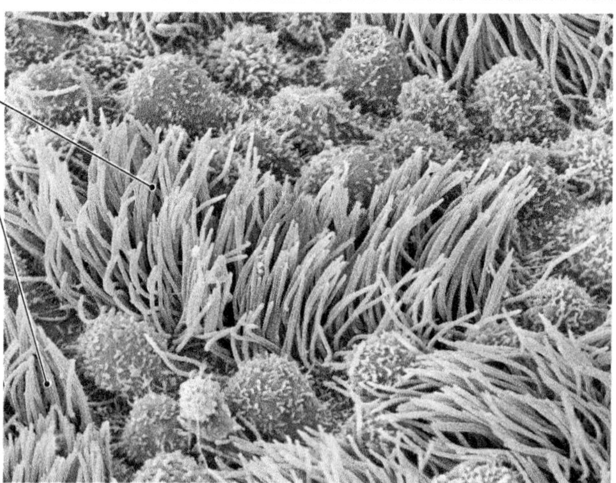

Uterine tube, epithelial surface SEM × 15,000

B The wall of the **vagina** has a thick smooth muscle layer and is lined by a stratified squamous epithelium.

1 Scan a slide of the vagina with the low-power objective lens and locate the two tissue layers of the mucous membrane:

A **stratified squamous epithelium** lines the lumen.

The underlying **lamina propria** is a layer of loose connective tissue.

2 A thick **smooth muscle layer** is deep to the lamina propria. Notice that bundles of smooth muscle are arranged in both circular and longitudinal fashion. Fibrous connective tissue is interspersed between the muscle bundles.

3 The **adventitia** (not shown) covers the outside wall of the vagina. It consists of fibrous connective tissue that blends in with connective tissue from neighboring structures.

Vagina LM × 20

4 Switch to high power and examine the mucous membrane more carefully. Notice the thickness of the epithelium. Why do you think the vagina is lined by a stratified squamous epithelium?

▶ ______________________________

____

____

Vagina LM × 40

5 Notice that the lamina propria contains numerous blood vessels.

MAKING CONNECTIONS

Identify similarities and differences in the microscopic structure of the uterine tube and vagina. ▶

Similarities:

____

____

____

Differences:

____

____

____

■ Want more practice? Go to: **MasteringA&P** > Study Area > Menu > Lab Tools > PAL > Histology > Reproductive System > Vagina photos

BEFORE YOU MOVE ON . . .

« LOOKING BACK

Both the male and female reproductive systems develop from similar embryonic tissue. In fact, during the first few weeks of development, male embryos are indistinguishable from female embryos. As development proceeds, however, two distinct and special organ systems form. The male reproductive system contains a pair of testes, accessory ducts and glands, and external genitalia, which include the scrotum and penis. The female reproductive system contains a pair of ovaries and uterine tubes, a uterus, a vagina, and external genitalia, which include the labia majora, labia minora, vestibule, and clitoris.

Despite the obvious structural differences between the sexes, can you identify any functional similarities that the male reproductive system shares with the female reproductive system?

▶ ____

____

____

____

What are the functional differences between the two reproductive systems?

▶ ____

____

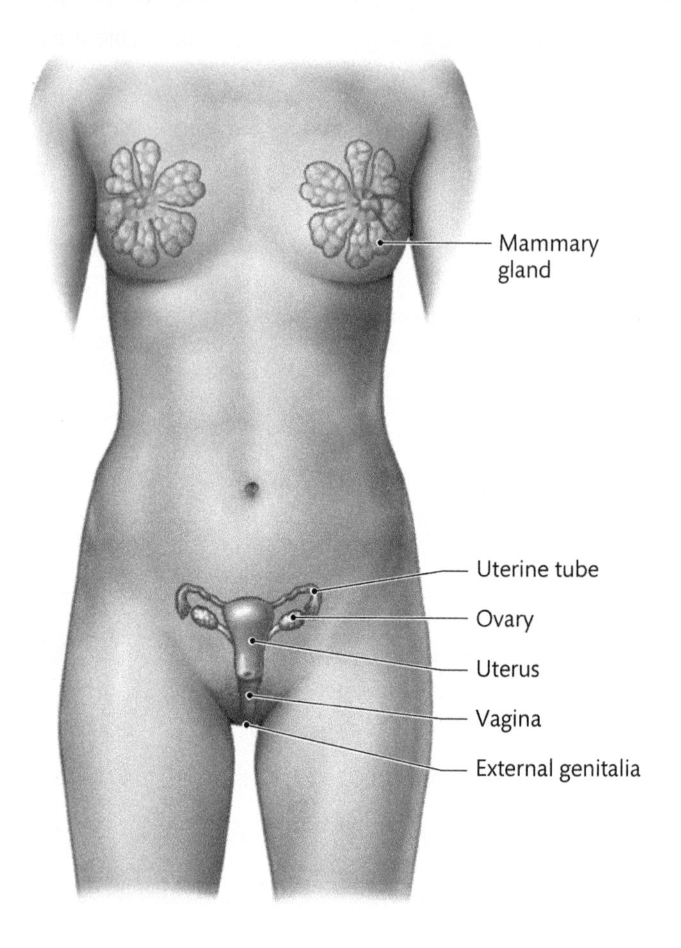

LOOKING FORWARD »

In the first laboratory exercise, you were introduced to the idea that the human body can be studied from six **levels of organization.** This fundamental biological principle for studying anatomy and physiology was emphasized and highlighted throughout this laboratory manual. Exercises 2 through 5 focused on the first three organizational levels: the **chemical level,** the **cellular level,** and the **tissue level.** You learned that molecules are organized in a unique way to form organelles and that organelles are the structural and functional components of cells. At the tissue level, the first multicellular level of organization that you studied, collections of cells are grouped together to carry out a common function.

Beginning with Exercise 6, and for all the remaining laboratory exercises, you used your knowledge of cells and tissues to study the fourth and fifth organizational levels: the **organ level** and the **organ system level.** You discovered that an organ contains at least two, and often all four, types of tissues that are uniquely organized to carry out a specific function. An organ system is a collection of organs that work as a unit by performing a variety of functions that collectively carry out a common activity.

The sixth organizational level, the **organism level,** includes the anatomy and physiology of all the organ systems in the body. Although you studied each organ system separately, the functional connections between organ systems were discussed on several occasions. It is critical to your understanding of human biology to realize that the function of each organ system is closely integrated with and dependent on the functions of other systems. For example, the skeletal system works with the muscular system to produce movement, provides protection for the brain and spinal cord in the nervous system, and contains red bone marrow where blood cells are produced for the cardiovascular and lymphatic systems. The digestive system absorbs calcium and phosphate ions that are used to produce new bone tissue in the skeletal system, absorbs fluids to maintain normal blood volume in the cardiovascular system, and absorbs nutrients that are transported by the cardiovascular system and distributed to cells and tissues in all the organ systems. As you move forward in your professional career, recognize how the activities of each organ system influence and regulate the activities of the other organ systems. This close integration of function is the key to understanding and appreciating how the body works.

Name ______________________

Lab Section ______________________

Date ______________________

EXERCISE **32** **REVIEW SHEET**

The Female Reproductive System

1. An ectopic pregnancy occurs when the early embryo implants at a location outside the uterus. A common site for an ectopic pregnancy is in a uterine tube. Explain why an ectopic pregnancy that occurs in a uterine tube will not be successful.

____

____

____

____

2. Predict what might happen if the broad ligament and other supporting ligaments were damaged by infection or injury.

____

____

____

____

QUESTIONS 3–14: Match the structure in column A with the correct description in column B.

A

3. Uterine tube ______________

4. Clitoris ______________

5. Vestibule ______________

6. Ovary ______________

7. Vagina ______________

8. Labia minora ______________

9. Suspensory ligaments ______________

10. Urogenital triangle ______________

11. Cervix ______________

12. Broad ligament ______________

13. Greater vestibular glands ______________

14. Lactiferous ducts ______________

B

a. A Pap smear is a test for detecting cancer of this structure.

b. This structure passes posterior to the urinary bladder and urethra and anterior to the rectum.

c. Milk produced by the mammary glands drains into these structures.

d. The external genitalia are located in this region.

e. Fertilization usually occurs in the ampulla of this structure.

f. This structure supports the internal female reproductive organs in the pelvic cavity.

g. Secretions from these structures moisten the area near the vaginal orifice during sexual intercourse.

h. Like the penis, this structure has corpora cavernosa.

i. This structure encloses the vestibule.

j. During fetal development, primordial follicles are produced by this structure.

k. The vaginal and urethral openings are located in this region.

l. These structures provide support for the mammary glands.

15. What would happen if fertilization occurred normally but the corpus luteum did not produce progesterone?

____

____

____

____

16. Identify the female reproductive organ that is shown in each of the following photos.

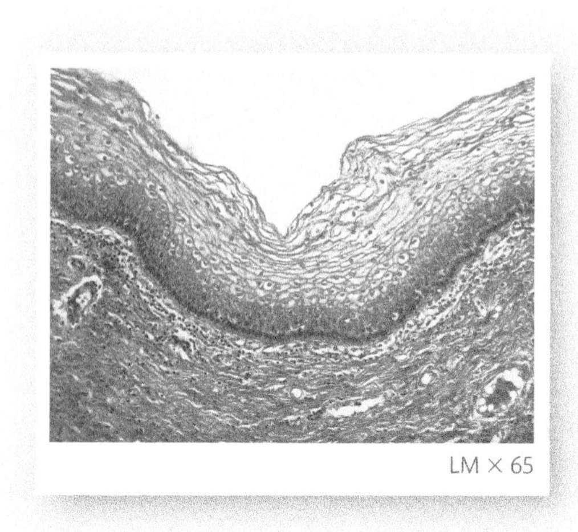

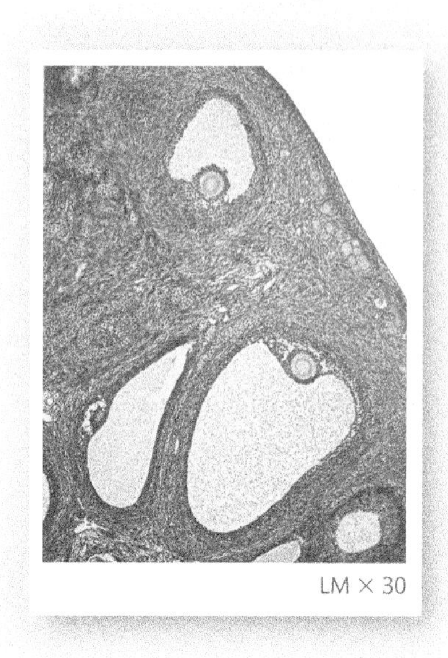

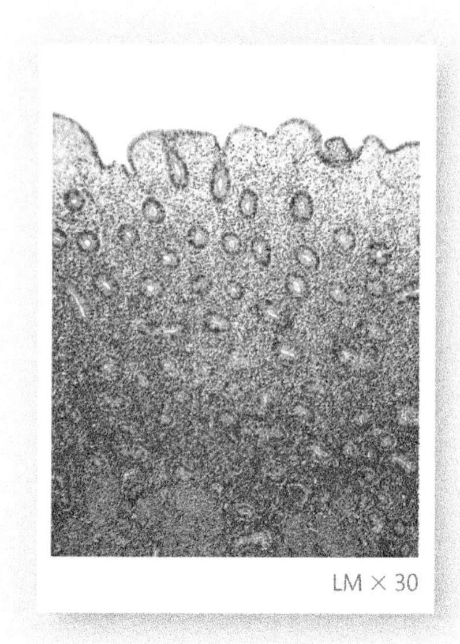

a. ______________________ b. ______________________ c. ______________________

QUESTIONS 17–27: In the following diagram, identify the labeled structures.

17. ______________________

18. ______________________

19. ______________________

20. ______________________

21. ______________________

22. ______________________

23. ______________________

24. ______________________

25. ______________________

26. ______________________

27. ______________________

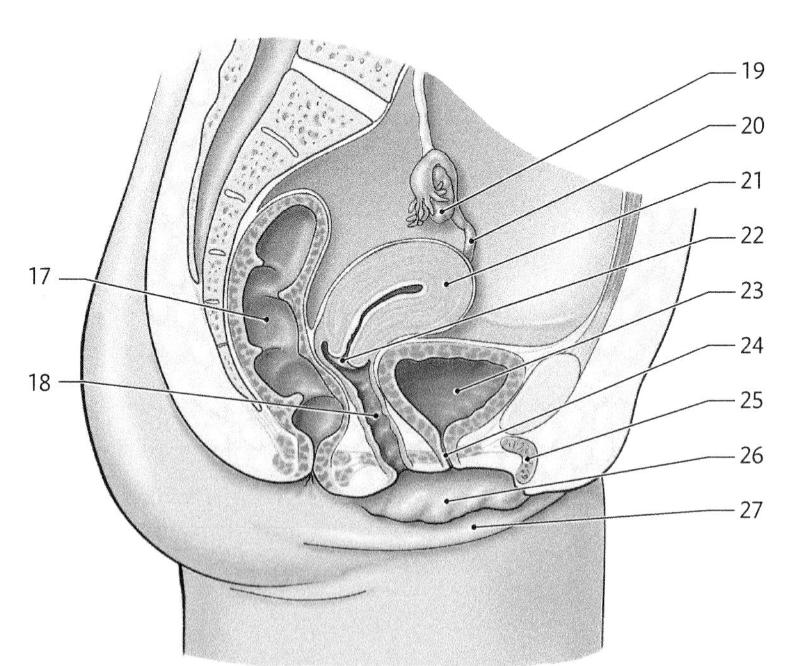

PRACTICAL SAFETY GUIDELINES FOR DISSECTION

Dissection can be a rewarding learning experience for any student who studies anatomy and physiology. The following guidelines will help you to successfully complete your dissection of the cat and to get the most out of your work:

- Always wear dissecting gloves to protect your hands from the preservative fluids. You may also use additional protective gear such as a laboratory mask and safety eyewear. You should always wear safety eyewear when cutting animal bones.
- Wear a laboratory coat to protect your clothing. If laboratory coats are not available, wear a work shirt or sweatshirt, rather than your best clothes. Avoid wearing any article of clothing made from wool, which tends to absorb the chemicals from animal preservatives.
- If you have long hair, make sure it is tied back and away from your face.
- If you are working with a lab partner, establish a team-oriented approach. You might want to assign specific tasks so that both of you will have an active, hands-on role in the dissection activities. For example, during the first half of the session, one person can read the instructions while the other person dissects. During the second half of the class, you could switch roles.
- Dissection does not exclusively refer to cutting and making incisions. Use your sharp instruments—scalpel and scissors—only when necessary. Make every effort to avoid damaging important structures that you will be studying later. If you are ever in doubt about making a cut, always check with your instructor first.
- Probes and dissecting needles are useful dissecting tools. Use these instruments to trace the paths of blood vessels and nerves or to separate muscles or muscle groups.
- At the conclusion of each dissection session, wrap the cat in a moist cloth and place in the storage bag. Seal the bag as tightly as possible with an elastic band or a piece of string, and store in a refrigerator until your next dissection. Place any loose animal tissues in a separate plastic bag or other container designated for this purpose. Never throw animal tissues into a general trash container or down the drain.
- Before leaving the laboratory, clean your work areas with a 10% bleach solution or a commercially prepared disinfectant.

DISSECTION EXERCISE

C1

Introduction to the Cat and Removal of the Skin

LEARNING OUTCOMES

These Learning Outcomes correspond by number to the laboratory activities in this exercise. When you complete the activities, you should be able to:

Activity C1.1 **Compare the terms of direction for a bipedal organism and a quadrupedal organism.**

Activity C1.2 **Name the bones of the cat skeleton and compare the cat and human skeletal systems.**

Activity C1.3 **Identify important external structures on the cat and determine its sex.**

Activity C1.4 **Remove the skin of the cat to expose the underlying skeletal muscles.**

LABORATORY SUPPLIES

- Cat skeleton
- Preserved cats
- Dissecting equipment: trays, tools, gloves
- Protective eyewear
- Face mask

PRE-LAB QUIZ

Before you begin, read all the activities in Exercise C1 and the required reading in your textbook that is assigned by your instructor.

Questions 1–4: Match the directional terms in column A with the appropriate meaning for a quadrupedal animal, such as the cat, in column B.

A	B
1. Dorsal ______	a. Toward the belly
2. Caudal ______	b. Toward the back
3. Ventral ______	c. Toward the head
4. Cranial ______	d. Toward the tail

5. True or False: The cat has fewer vertebrae than the human. ______
6. True or False: In the cat, the clavicles do not articulate with any other bone in the skeleton. ______
7. List the three regions of the cat's trunk. ______
8. True or False: The mammary papillae are found in female cats only. ______
9. During this dissection exercise, you will remove the skin from the cat. Your first skin incision will be a
   a. lateral incision along the medial surface of a forelimb.
   b. circular incision around the perineum.
   c. midline incision along the ventral surface.
   d. lateral incision along the medial surface of a hindlimb.
   e. lateral incision along the inferior margin of the mandible.
10. True or False: The loose connective tissue that separates the skin from the underlying muscle is called the deep fascia. ______

Comparing Directional Terms for Bipedal and Quadrupedal Organisms

The cat that you dissect will be packed in a clear plastic bag. When you receive your animal, obtain a tag and write your name and any other important information that will help identify your specimen, such as your lab partner's name, instructor's name, and section number. Attach the tag to the plastic bag with string or tape.

When you remove your cat from its bag, a small amount of preservative fluid may remain. You may be instructed to keep this fluid in the bag to keep your animal moist during storage. Alternatively, your instructor may ask you to drain this fluid into a storage container for safe disposal at a later date.

Before you begin dissecting your cat, carefully review the dissection guidelines that appear on page C-1. These guidelines offer useful suggestions for maintaining your cat over the long term, dissecting with care and safety in mind, and enhancing your learning experience.

A Directional terms are used differently for a quadrupedal organism, such as a cat, compared with a bipedal organism, such as a human. Remove your cat from its storage bag and place it on a dissecting tray. Study the following directional terms.

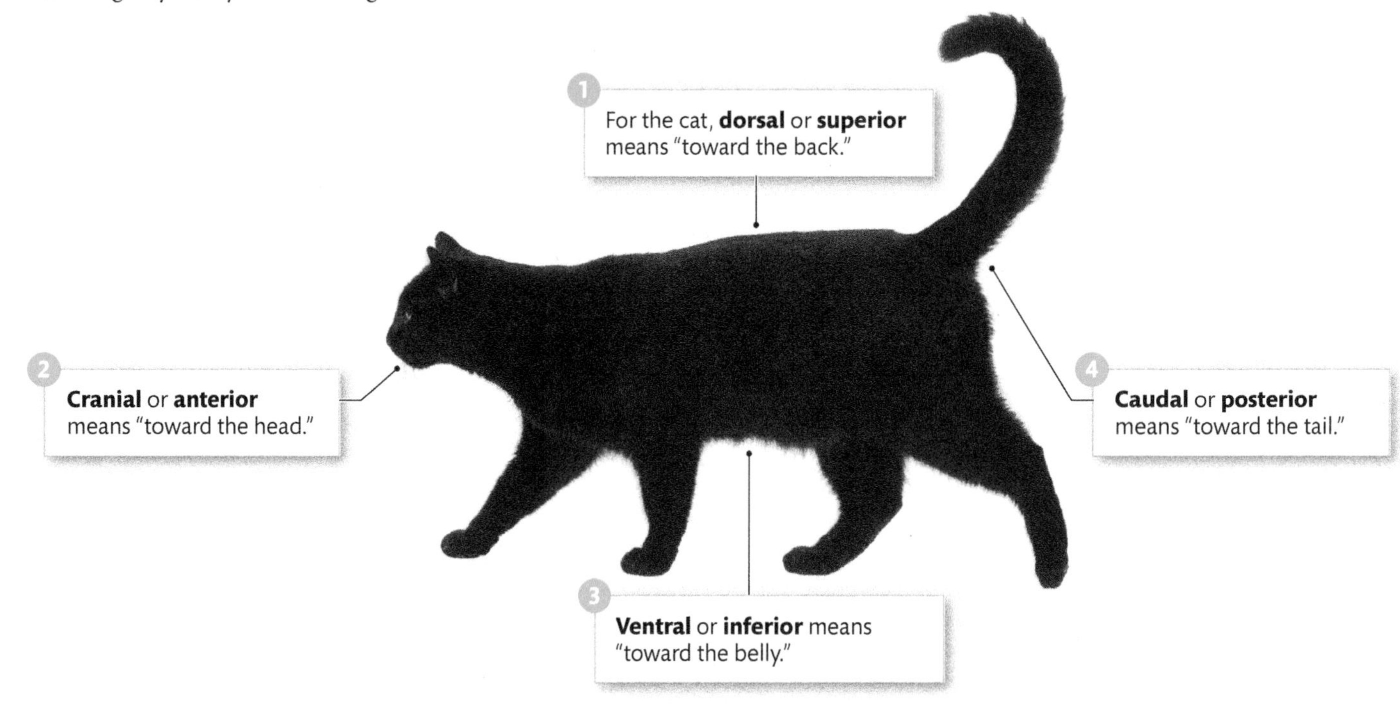

B In the human, superior/inferior and anterior/posterior have different meanings compared with the cat.

1 Consider how directional terms are used for a human and correctly pair superior, inferior, anterior, and posterior with the directional terms listed below. ▶

Dorsal	
Ventral	
Cranial	
Caudal	

2 Label the four pairs of directional terms in the figure to the right. ▶

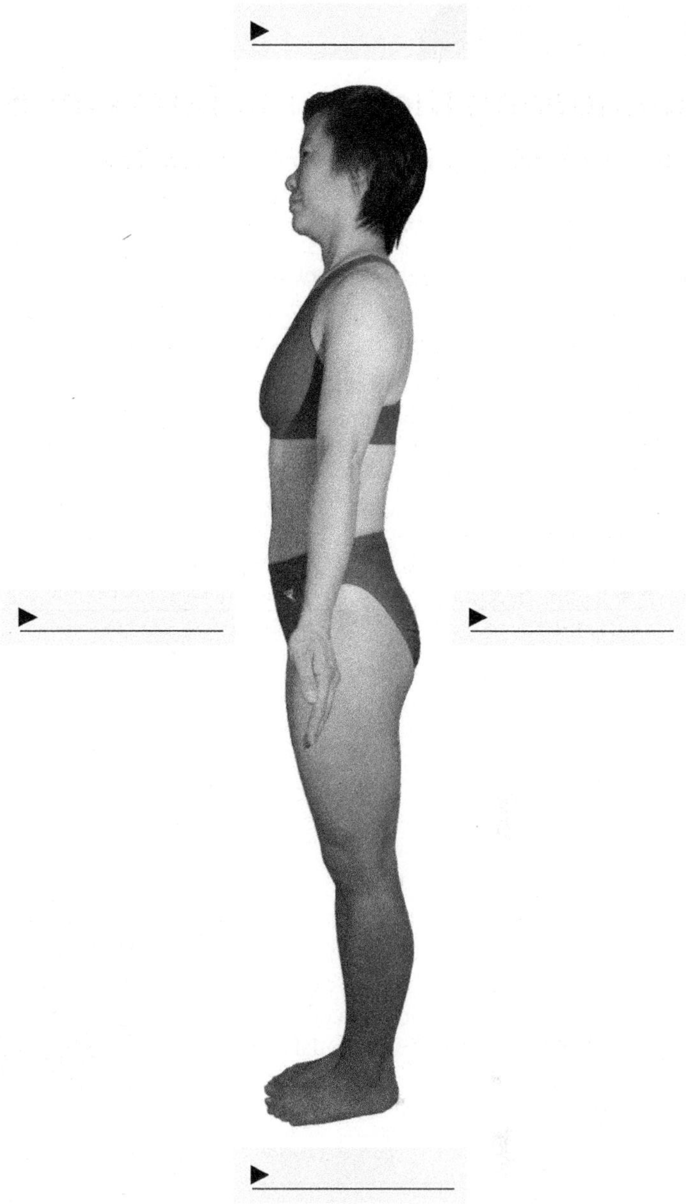

▶ ____________

▶ ____________

▶ ____________

▶ ____________

MAKING CONNECTIONS

In the human, the esophagus is posterior to the trachea. How would you describe the position of the esophagus in relation to the trachea in the cat? Provide a rationale for your answer.

▶ ____

Activity **C1.2**

Identifying the Bones of the Cat Skeleton

The bones of the cat skeleton have names that are identical to those of the human skeleton. Disparities in the number and arrangement of these bones are due to differences in how the body is supported for bipedal (human) and quadrupedal (cat) locomotion.

A The axial skeleton of the cat contains more bones than that of the human.

1 Observe a complete skeleton of the cat. The **skull** contains 11 **cranial bones** and 15 **facial bones.** How many cranial bones and facial bones are in the human skull?

Cranial bones

▶ ______________________________

Facial bones

▶ ______________________________

2 Locate the **vertebral column.** It contains between 51 and 55 vertebrae and is divided into cervical, thoracic, lumbar, sacral, and caudal regions. Compare the cat and human vertebral columns by completing the following table. ▶

Comparison of the Vertebral Column in the Cat and Human Skeletons

	Number of Vertebrae	
Vertebral Region	**Cat**	**Human**
Cervical	7	
Thoracic	13	
Lumbar	7	
Sacral	3	
Caudal	21–25	
Total	51–55	

3 In the human skeleton, the sacral and caudal (coccygeal) vertebrae are fused. Is that true for the cat as well? Explain what you observe as you inspect the sacral and caudal vertebrae in the cat.

▶ ______________________________

____

____

____

____

____

____

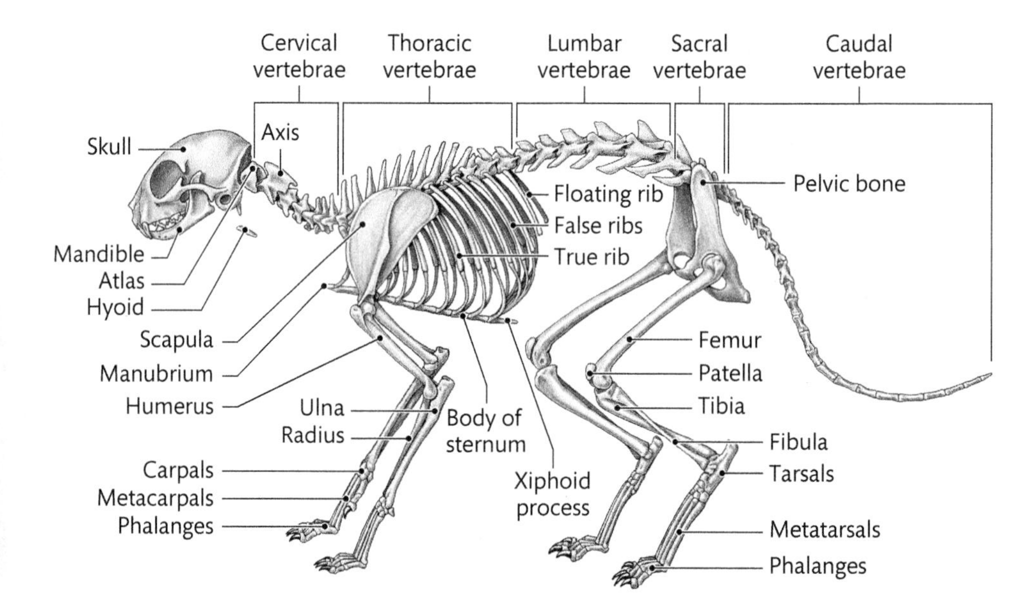

B The bones of the cat forelimbs and hindlimbs are, for the most part, analogous to the bones of the human upper and lower limbs, but there are some differences.

1. Each **pectoral girdle** in the human skeleton contains a scapula and a clavicle. In the cat skeleton, the slender clavicles are embedded in muscle. Notice that they do not articulate with any other bone in the skeleton.

2. The cat has seven **carpal bones** in the wrist. How many carpal bones are at each wrist in the human skeleton?

▶ ______________________________

3. The **pelvic girdle** in the cat is narrower and longer than in the human. Why do you think there is a difference?

▶ ______________________________

____

4. On the human skeleton, the hands and feet each have five digits. Notice that the cat skeleton has five digits on each forelimb. How many digits are on each hindlimb?

▶ ______________________________

5. When humans walk, the tarsal bones, metatarsal bones, and phalanges are in contact with the ground. Observe the cat skeleton. Identify the bones that contact the ground when the cat walks.

▶ ______________________________

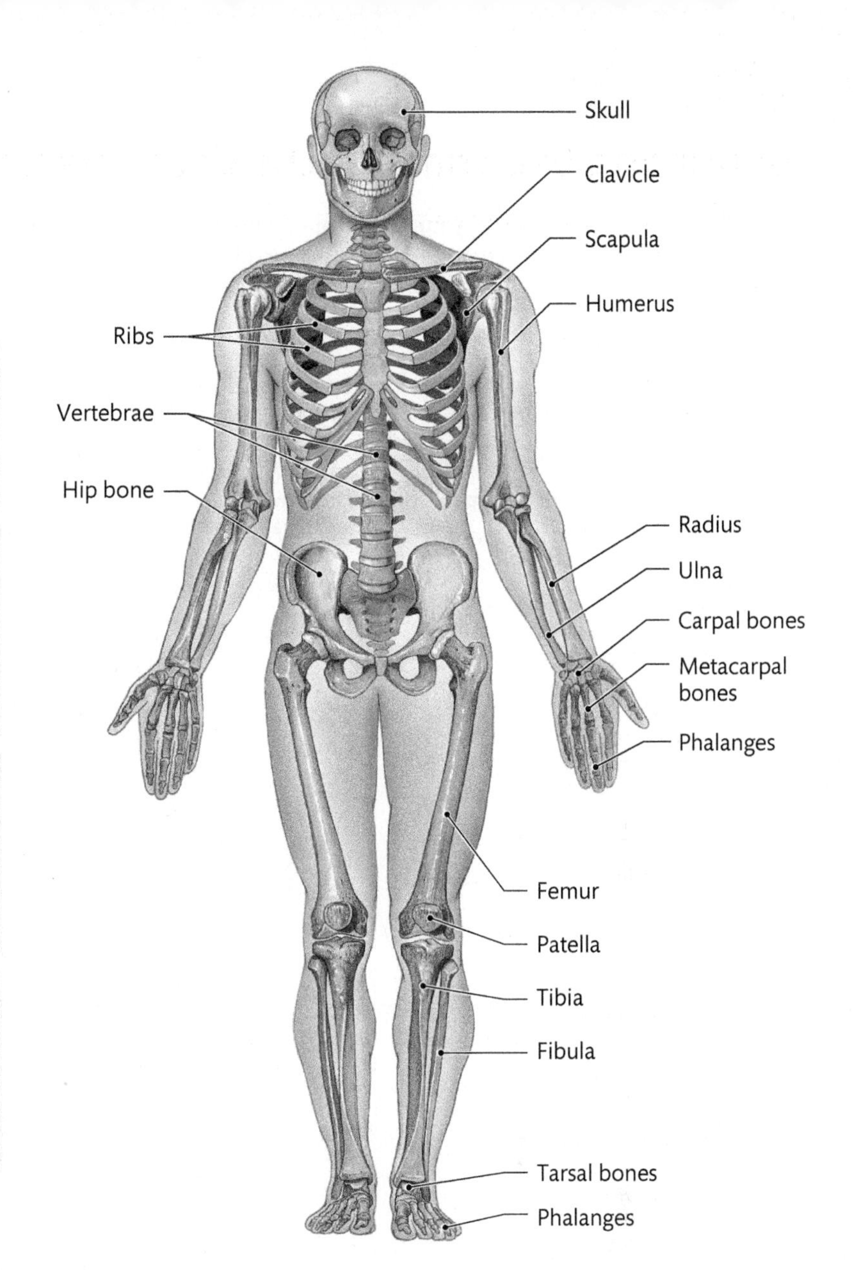

MAKING CONNECTIONS

The cat has fewer metatarsals and phalanges on the hindlimbs than on the feet of human lower limbs. Can you think of a reason why there is a difference?

▶ ______________________________

____

____

____

Activity **C1.3**

Identifying External Structures on the Cat

A An external examination of your cat will allow you to identify many structures that should be obvious, but others may be unfamiliar to you.

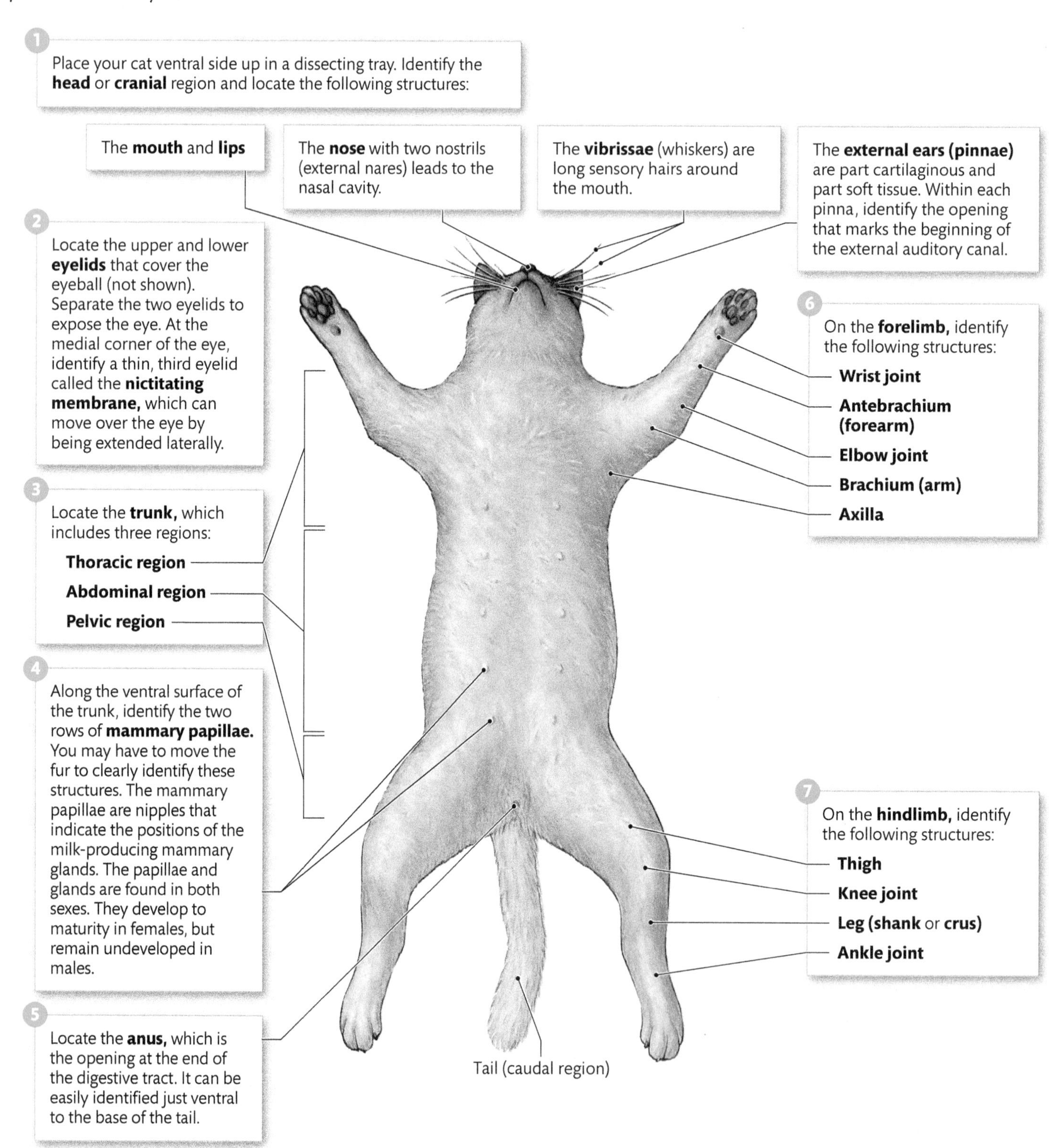

B Certain external structural differences will allow you to determine the sex of your cat. Because both sexes have mammary papillae, this characteristic cannot be used to discriminate between the sexes.

1. Locate the anus and the region that surrounds it.

2. In males, the **scrotum** is an external pouch that contains the **testes**. It is located just anterior to the anus. The **penis** is usually retracted but the **prepuce** (fold of skin covering the tip of the penis) can be identified as a small elevation anterior to the scrotum.

3. In females, the **urogenital aperture** is just ventral to the anus and is much larger than the male opening at the tip of the penis.

4. After determining the sex of your cat, examine another cat of the opposite sex and notice the differences.

MAKING CONNECTIONS

Count the number of pairs of mammary papillae on your cat and compare that number with those of the others in your class. How much variation, if any, is there in the number of mammary papillae on the cats in your class? Is there any relationship between the number of papillae and the sex of the cat? (Specifically, is there any evidence that suggests that one sex has more mammary papillae than the other?) Explain your results.

▶ ____

Activity C1.4

Removing the Skin from the Cat

To dissect the muscles of the cat, the skin must be removed first. The skin of the cat is thin and fragile in some regions of the body, such as the abdominal wall, so be particularly careful when dissecting in these areas.

A Making the Skin Incisions

1. Place your cat ventral side up in a dissecting tray. Because most cats are preserved with the limbs extended, you may not have to secure your specimen to the dissecting tray with string. As you proceed with the steps that follow, ensure that you cut the skin only and avoid damaging the underlying muscle tissue.
2. Make an initial incision along the midline of the ventral surface from the jaw to the external genitalia.
3. Cut around the **perineum** (area between the anus and urogenital opening) and continue the incision along the proximal one-third of the tail.
4. If you have a female cat, make circular incisions around each of the mammary papillae.
5. Make the following lateral incisions on each side of the body:
   - Along the inferior margin of the mandible, across the neck to the shoulder
   - Along the medial surface of the forelimb to the wrist
   - Along the medial surface of the hindlimb to the ankle

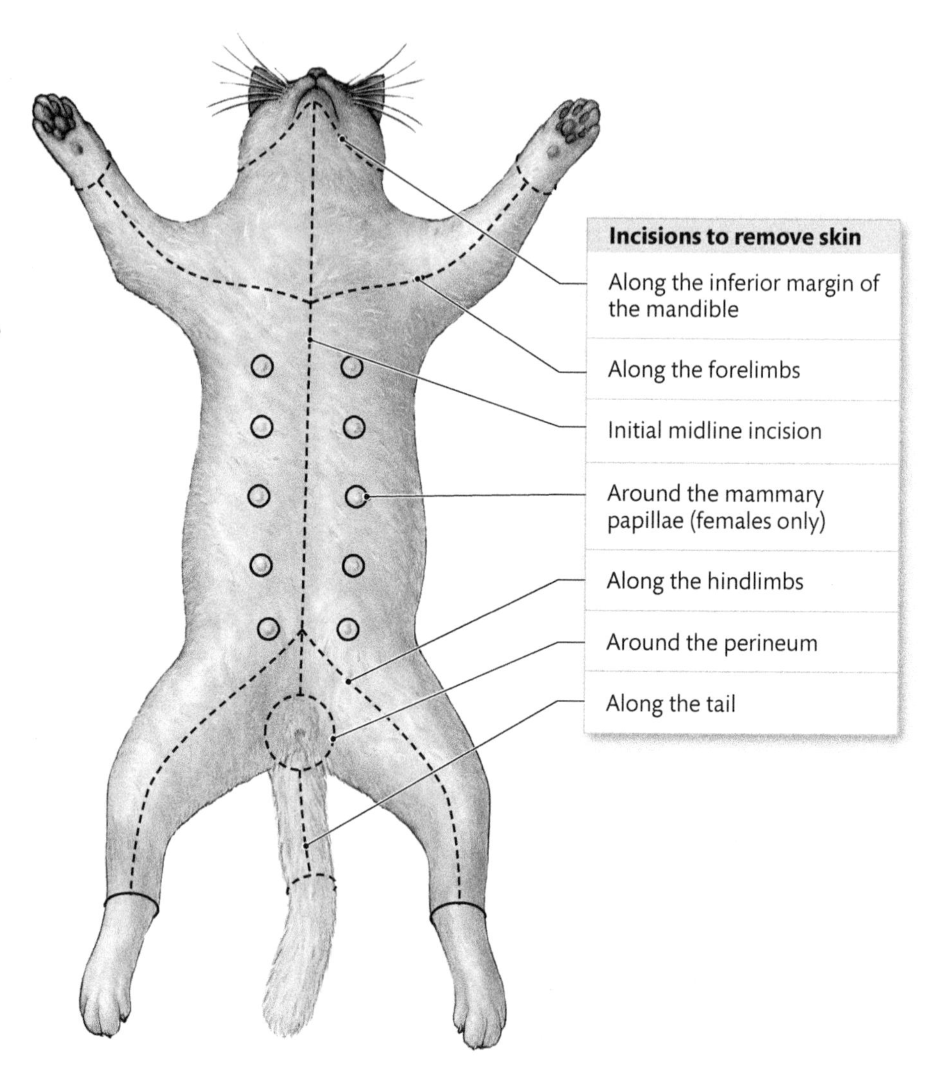

B Removing the Skin

1 After making the incisions just described, lift the skin with forceps while carefully cutting the skin away from the underlying muscle with a scalpel. You should be cutting through a whitish, loose connective tissue layer that separates skin from muscle. This layer is called the **superficial fascia.**

2 As the skinning progresses, you might find it easier to use a blunt probe and your gloved fingers in place of a scalpel and forceps. Your goal is to remove skin from the trunk, forelimbs, hindlimbs, and the proximal one-third of the tail. Leave intact the skin on the skull, feet, perineum, distal end of the tail, and, in females, around the mammary papillae. As you approach the dorsal surface, turn your cat over to complete the process.

3 Do not discard the skin when it is removed. You will use it, along with wet towels, to wrap the cat after each dissection session.

4 Inspect the internal surface of the skin and attempt to identify numerous delicate, brown fibers. These fibers belong to cutaneous muscles that act to wrinkle and move the skin. The cat has two cutaneous muscles:

- The **cutaneous maximus** is found along the lateral aspect of the thoracic and abdominal body wall. This muscle is not found in humans.
- The **platysma,** which is also found in humans, is located in the skin that covers the ventral surface of the neck and mandible.

5 Inspect your skinned cat and notice that various amounts of superficial fascia and fat cover the muscles. When you dissect the muscles, the first step will always be to remove as much of this tissue as possible. If time permits, begin removing the superficial fascia and fat. Proceed with care and patience to avoid damaging important structures.

6 After the superficial fascia has been removed, you will notice a thin, translucent connective tissue layer covering the muscles. This layer is called the **deep fascia.** The deep fascia not only surrounds individual muscles, it also forms partitions around muscle groups (muscle compartments) that may have a common function, blood supply, or innervation.

7 If your dissection session is ending here, wrap your cat with the skin that you just removed. Cover the animal with moist towels and return it to the storage bag.

MAKING CONNECTIONS

Why is the superficial fascia significant when removing the skin from a cat?

▶ ____

____

____

____

____

____

BEFORE YOU MOVE ON . . .

« LOOKING BACK

There is no better way to learn and understand human anatomy than to dissect, palpate (touch and manipulate), and observe actual human tissue. Unfortunately, most undergraduate human anatomy and physiology laboratories do not have the facilities and resources to perform dissections on a human body (cadaver). As an alternative, other mammalian bodies with similar anatomical features can be studied and compared with human form and function. The cat is an excellent (and popular) choice for animal dissection because its skeleton, musculature, and internal organ systems correspond closely to the human.

One significant difference between the cat and the human is the way each moves: The cat is a quadrupedal organism, and the human is a bipedal organism. What anatomical differences would you expect to find as a result of these different modes of locomotion?

LOOKING FORWARD »

In this cat dissection exercise, you removed the skin to reveal the cat's musculature. In the next dissection exercise, you will examine the cat muscular system. Because the muscles of the cat are similar in structure and function to the muscles of the human, the cat is an excellent animal model for helping you understand the human muscular system.

Name ________________

Lab Section ________________

Date ________________

EXERCISE C1 REVIEW SHEET

Introduction to the Cat and Removal of the Skin

1. Explain how the definitions of superior/inferior and anterior/posterior differ in a bipedal organism, such as a human, and a quadrupedal organism, such as a cat.

2. Compare the skeletons of the cat and human by noting similarities and differences in structure.

a. Similarities:

b. Differences:

3. Explain how to determine the sex of a cat.

4. Discuss the difference between the superficial fascia and deep fascia.

QUESTIONS 5–10: Match the terms in column A with the appropriate description in column B.

A	B
5. Vibrissae ______	**a.** Thin, third eyelid that is found in cats, but not humans
6. Pinnae ______	**b.** Region between the anus and urogenital opening
7. Nictitating membrane ______	**c.** Long sensory hairs found around the mouth of the cat
8. Mammary papillae ______	**d.** Nipples that identify the locations of the mammary glands
9. Urogenital aperture ______	**e.** External ears
10. Perineum ______	**f.** Common opening for the urinary and reproductive systems

QUESTIONS 11–38: Identify the labeled bones in the diagram below.

11. ______

12. ______

13. ______

14. ______

15. ______

16. ______

17. ______

18. ______

19. ______

20. ______

21. ______

22. ______

23. ______

24. ______

25. ______

26. ______

27. ______

28. ______

29. ______

30. ______

31. ______

32. ______

33. ______

34. ______

35. ______

36. ______

37. ______

38. ______

DISSECTION EXERCISE

C2

Dissection of the Cat Muscular System

LEARNING OUTCOMES

These Learning Outcomes correspond by number to the laboratory activities in this exercise. When you complete the activities, you should be able to:

Activities **C2.1–C2.2**	**Identify the skeletal muscles on the thorax and abdomen of the cat, describe their functions, and make comparisons with human musculature.**
Activity **C2.3**	**Identify the skeletal muscles on the head and neck of the cat, describe their functions, and make comparisons with human musculature.**
Activity **C2.4**	**Identify the skeletal muscles on the forelimb of the cat, describe their functions, and make comparisons with human musculature.**
Activity **C2.5**	**Identify the skeletal muscles on the hindlimb of the cat, describe their functions, and make comparisons with human musculature.**

LABORATORY SUPPLIES

- Preserved cats
- Dissecting equipment: trays, tools, gloves
- Protective eyewear
- Face mask
- Anatomical models of the human muscular system

PRE-LAB QUIZ

Before you begin, read all the activities in Exercise C2 and the required reading in your textbook that is assigned by your instructor.

Questions 1–8: Match the muscle in column A with the cat body region in which it is found in column B.

A	B
1. Sternothyroid ______	a. Lower forelimb
2. Pectoantebrachialis ______	b. Dorsal thorax
3. Coracobrachialis ______	c. Thigh
4. Acromiotrapezius ______	d. Head
5. Temporalis ______	e. Neck
6. Sartorius ______	f. Upper forelimb
7. Soleus ______	g. Ventral thorax
8. Brachioradialis ______	h. Leg

9. In humans, there is one trapezius muscle. In the cat, how many muscles are in the trapezius group? ______
10. All the following muscles that you will dissect in the cat are also found in the human *except* the
    a. semimembranosus.
    b. internal oblique.
    c. supraspinatus.
    d. pectoralis major.
    e. epitrochlearis.

Activity **C2.1**

Identifying Muscles of the Ventral Thorax and Abdomen

As you advance through this dissection, use your probe to separate individual muscles and notice the direction or grain of the muscle fibers. Many muscles can be distinguished by noticing changes in the grain of the fibers. When a muscle has been successfully isolated, attempt to locate its origin and insertion. Gently grip the muscle with forceps and pull it to see what bones or body parts move. This activity will help you visualize specific muscle actions.

To identify deep muscles, superficial muscles must first be cut (transected) with a scalpel or scissors. With forceps, grip the midsection of the muscle belly and gently lift up a free margin. Carefully slip a probe deep to the free margin, along the midline of the muscle belly. Use a scalpel to carefully transect the muscle along the path of the underlying probe. Folding back (reflecting) the cut ends of the muscle will expose the deeper muscles that you wish to study. When transecting a muscle, never cut the origin or insertion unless you are specifically instructed to do so.

Try to avoid damaging blood vessels and nerves that you will be asked to identify in subsequent dissections. To lower the risk of damaging an important structure, you could dissect muscles on one side of the body only. The other side could then be used to locate blood vessels and peripheral nerves and later for reviewing the musculature.

A Superficial muscles of the ventral thorax include the **pectoral group,** which adduct the forelimb.

1. Place your cat ventral side up in a dissecting tray. Carefully remove as much fat and superficial fascia as possible from the ventral surface of the thorax and abdomen.

2. Along the thorax, identify the **pectoral group** of muscles:

The **pectoantebrachialis** is the most anterior muscle of this group. It is a narrow band of muscle that travels across the thorax to the forelimb. It originates on the manubrium and inserts on fascia covering the ulna near the elbow. Humans do not have this muscle.

The **pectoralis major** travels deep to the pectoantebrachialis. It originates on the sternum and inserts on the humerus. Transect and reflect the pectoantebrachialis to reveal the pectoralis major in its entirety.

The **pectoralis minor** is the largest pectoral muscle in the cat. It is positioned just posterior to the pectoralis major and has similar attachments.

The **xiphihumeralis** is the most posterior muscle in the pectoral group. It often fuses with the pectoralis minor. It extends from the xiphoid process to the humerus.

B The abdominal wall muscles flex the trunk and compress the abdominal wall to support internal organs.

1. Along the ventral and lateral surfaces of the abdomen, identify the **external oblique,** a thin, broad sheet that forms the outermost of three abdominal muscular layers.

2. Notice that the external oblique gives rise to a broad fascial membrane (an **aponeurosis**).

3. Identify the **linea alba** along the midventral surface. This structure is a band of connective tissue formed by the intersecting fibers of the aponeuroses from both sides.

4. Lift up the area of the external oblique with forceps. Make a small incision and reflect the muscle to expose the second abdominal muscle layer, the **internal oblique.** Notice that the fibers of the internal oblique travel at a right angle to the fibers of the external oblique.

5. Lift the internal oblique with forceps. Make a small incision and reflect the muscle to expose the third and deepest abdominal muscle layer, the **transversus abdominis.** Notice that its fibers run in a different direction than the first two layers.

6. The two **rectus abdominis** muscles travel parallel to each other on either side of the linea alba. Identify the muscle on one side of the body as it travels from the pubic bone to the costal cartilages of the ribs and the sternum. The muscle is positioned between the aponeuroses of the internal oblique and transversus abdominis for much of its length. Carefully remove the aponeurosis in one area to get a better view of the muscle.

C Muscles that move the scapula and ribs and flex the neck are located deep to the pectoral and abdominal muscles.

1 Transect and reflect all the pectoral muscles. The following deep muscles can now be identified:

The **scalenus** is a group of three muscles—the scalenus anterior, the scalenus medius, and the scalenus posterior—that travel vertically from the ribs to the cervical vertebrae. These muscles flex the neck.

The **serratus ventralis** is a large, fan-shaped muscle that can be seen along the lateral side of the rib cage. Notice the sawlike or "serrated" appearance of this muscle at its origin on the ribs. The serratus ventralis, which is equivalent to the serratus anterior found in humans, inserts on the vertebral border of the scapula. It moves the scapula ventrally and anteriorly.

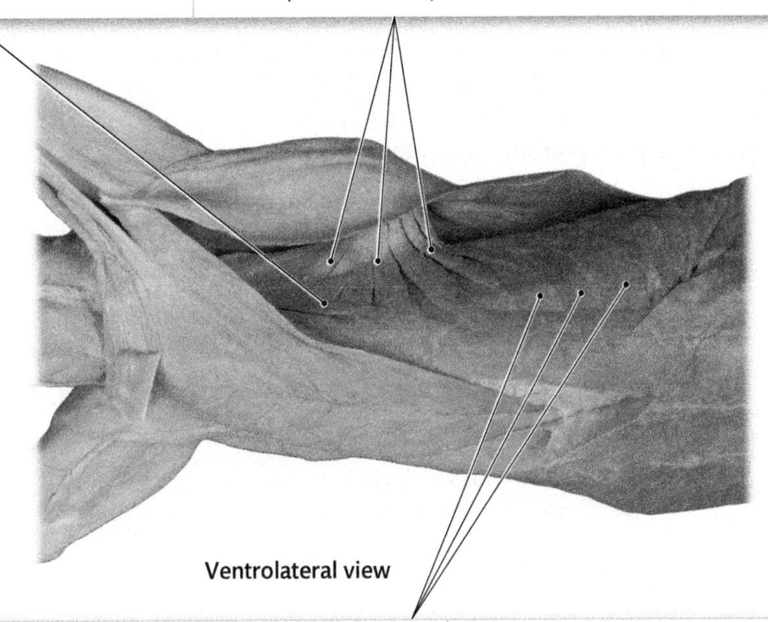
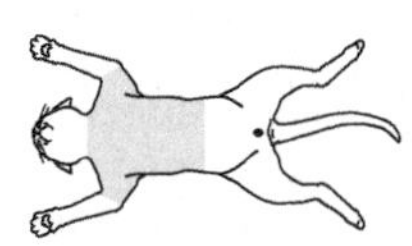

Ventrolateral view

2 Transect and reflect the external oblique and rectus abdominis near their attachments to the ribs. Identify the muscles located in the **intercostal spaces** between the ribs:

The **external intercostal muscles** form the outer layer that is visible within each space. These muscles elevate the ribs during inspiration.

In one intercostal space, cut the external intercostal muscle near its attachment to one of the ribs. Reflect the muscle to reveal the **internal intercostal muscle** (not shown). Notice the difference in the direction of the fibers between the two muscle layers. These muscles lower the ribs during expiration.

Word Origins

When you identify the muscles pectoralis major and pectoralis minor, you would logically assume that the pectoralis major is the larger muscle. In the cat, however, pectoralis major is the smaller of the two. The terms *major* and *minor* apply to the human, where pectoralis major is a much larger muscle than pectoralis minor.

MAKING CONNECTIONS

Compare and contrast the anatomy of the pectoralis muscles in the cat and the human.

▶ ______________________________

____

____

____

____

Identifying Muscles of the Dorsal Thorax and Abdomen

A Superficial muscles on the dorsal surface of the thorax and abdomen act on the scapula and forelimb.

1. Place your cat dorsal side up in a dissecting tray. Carefully remove any excess fat and superficial fascia on the dorsal thorax and abdomen.

2. The muscles that compose the **trapezius** group have a broad origin that extends from the posterior surface of the skull to the thoracic vertebrae. They insert on the clavicle **(clavotrapezius)** and scapula (**acromiotrapezius** and **spinotrapezius**). In humans, these muscles are fused to form one large muscle, known simply as the trapezius. These muscles move the scapula medially, anteriorly, and posteriorly. From anterior to posterior, identify the three trapezius muscles:

Clavotrapezius **Acromiotrapezius** **Spinotrapezius**

3. The **levator scapulae ventralis** travels from the occipital bone on the skull to the vertebral (medial) border of the scapula. Identify a portion of this muscle as it passes between the clavotrapezius and acromiotrapezius. This muscle moves the scapula anteriorly.

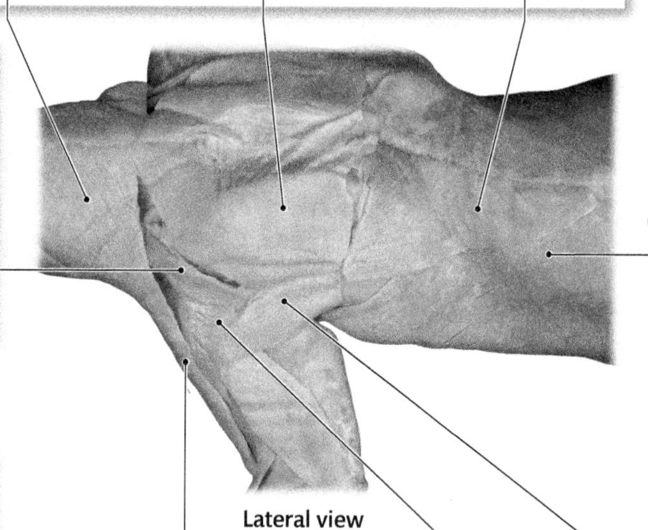

Lateral view

4. Identify the three **deltoid** muscles. These muscles originate from the pectoral girdle (clavicle or scapula) and drape over the shoulder joint to insert on bones in the forelimb (humerus or ulna). These muscles elevate and rotate the humerus.

The anterior muscle is the **clavodeltoid.** Notice that the fibers of this muscle are continuous with those of the clavotrapezius.

The **acromiodeltoid** is a short, bulky muscle, just posterior to the clavodeltoid.

The most posterior muscle in this group is the **spinodeltoid.** Notice that the fibers of this muscle travel at roughly a right angle to the fibers of the other two deltoid muscles.

5. Identify the **lumbodorsal fascia** (not shown), an extensive region of deep fascia on the dorsal abdominal surface.

6. Arising from the lumbodorsal fascia, posterior to the spinotrapezius, is the **latissimus dorsi.** Notice that this muscle covers much of the dorsal surface, but tapers as it travels laterally and ventrally toward the humerus. The latissimus dorsi moves the forelimb posteriorly.

B The **rhomboideus group** and **rotator cuff** muscles are deep to the trapezius and deltoid muscle groups.

1. Transect and reflect the muscles of the trapezius group and deltoid group. That will expose the rhomboids and muscles on the dorsal surface of the scapula.

2. Identify the three muscles belonging to the rhomboideus group. These muscles connect the skull and cervical and thoracic vertebrae to the scapula. As a group, they move the scapula anteriorly and medially. From anterior to posterior, these muscles are:
Rhomboideus capitis **Rhomboideus minor** **Rhomboideus major**

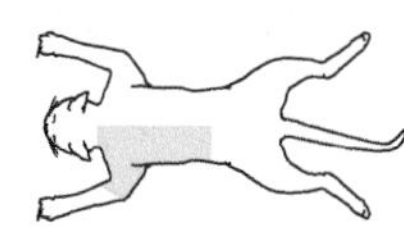

3. Transect and reflect the rhomboideus capitis to expose the **splenius,** the cranial extension of the deep back muscles (erector spinae). It extends and laterally flexes the head. Identify its origin on the thoracic vertebral column and its insertion on the occipital bone.

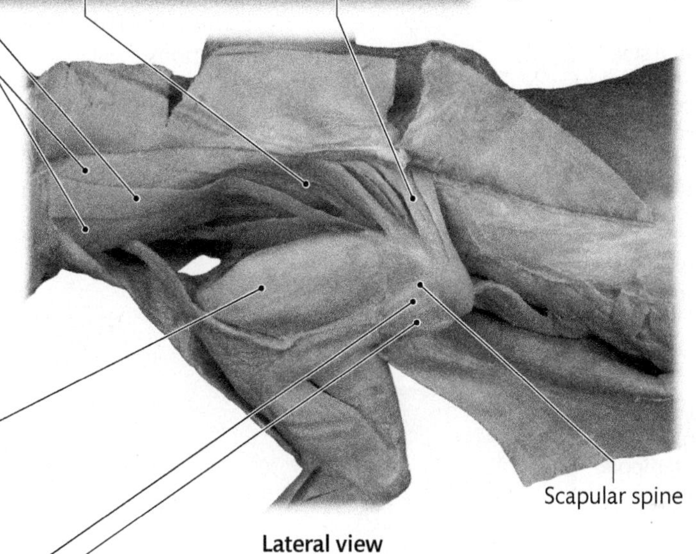

Lateral view

4. By palpating, locate the **scapular spine.**

The muscle anterior to the spine, the **supraspinatus,** occupies the supraspinous fossa. Follow the muscle to its attachment on the humerus.

The muscle posterior to the spine, the **infraspinatus,** occupies the infraspinous fossa and extends to the humerus.

These two muscles laterally rotate the humerus.

5. The **teres major** is attached to the posterior border of the scapula and travels to the humerus. Identify this muscle, which is located lateral and posterior to the infraspinatus. The teres major rotates the humerus and moves the forelimb posteriorly.

6. The **teres minor** (not shown) is a much smaller muscle than the teres major. Identify this muscle where it originates on the lateral border of the scapula, anterior to the teres major. Attempt to follow the muscle to its insertion on the humerus.

MAKING CONNECTIONS

Review the muscles of the dorsal thorax and abdomen in the cat and identify similarities and differences with comparable human muscles. ▶

Similarities:

____

____

____

Differences:

____

____

____

Activity **C2.3**

Identifying Muscles of the Head and Neck

For most cats injected with latex dyes, the site of injection is in the neck. If that is the case with your cat, it may be difficult to identify some muscles in this region. Select the side with the least damage and proceed with your dissection.

A Two important muscles of mastication (chewing) are located on the skull.

1. Place your cat ventral side up in a dissecting tray. Remove the skin from the cheek and temporal regions of the head. Identify the **masseter,** which is the large chewing muscle located in the cheek. It originates on the zygomatic arch and inserts on the mandible.

2. Locate the **temporalis** muscle (not shown) traveling along the temporal bone, dorsal to the masseter. This muscle also inserts on the mandible.

B Palpate the **mandible** and move your fingers posteriorly until you can feel the **hyoid bone.** From the hyoid bone, move your fingers posteriorly along the neck until you can feel the **thyroid cartilage.** These structures are important anatomical landmarks for identifying many of the neck muscles and understanding their actions.

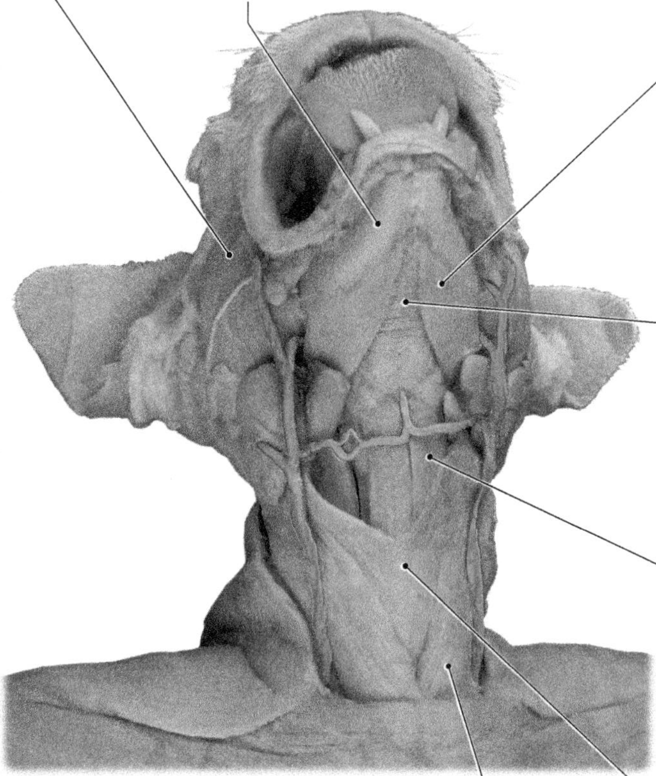

1. The **digastric** is a V-shaped muscle that travels from the temporal and occipital bones to the mandible. Identify this muscle as it travels just inferior to the mandible. Its main action is depression of the mandible.

2. Identify the **mylohyoid.** Its fibers run transversely between the two digastric muscles. It is attached to the mandible and hyoid bone and forms the muscular floor of the mouth.

3. Locate the **sternohyoid,** a superficial muscle that runs along the midventral line of the neck from the sternum to the hyoid bone. This muscle moves the larynx and hyoid bone posteriorly.

4. Identify the **sternomastoid.** This superficial muscle travels anteriorly and dorsally across the neck from the sternum to the mastoid process.

5. The **cleidomastoid** travels laterally and parallel to the sternomastoid. It extends from the clavicle to the mastoid process. The cleidomastoid and sternomastoid flex the head and bend the head to one side (lateral flexion).

Ventral view

C Two important neck muscles lie deep to the sternohyoid.

1. Transect and reflect the sternohyoid muscle and identify the **thyroid cartilage** of the larynx.

2. Identify the following two muscles that travel deep to the sternohyoid. Both muscles are attached to the thyroid cartilage.

   The **thyrohyoid** extends from the thyroid cartilage to the hyoid bone. It moves the hyoid posteriorly.

   The **sternothyroid** extends from the sternum to the thyroid cartilage. It moves the larynx posteriorly.

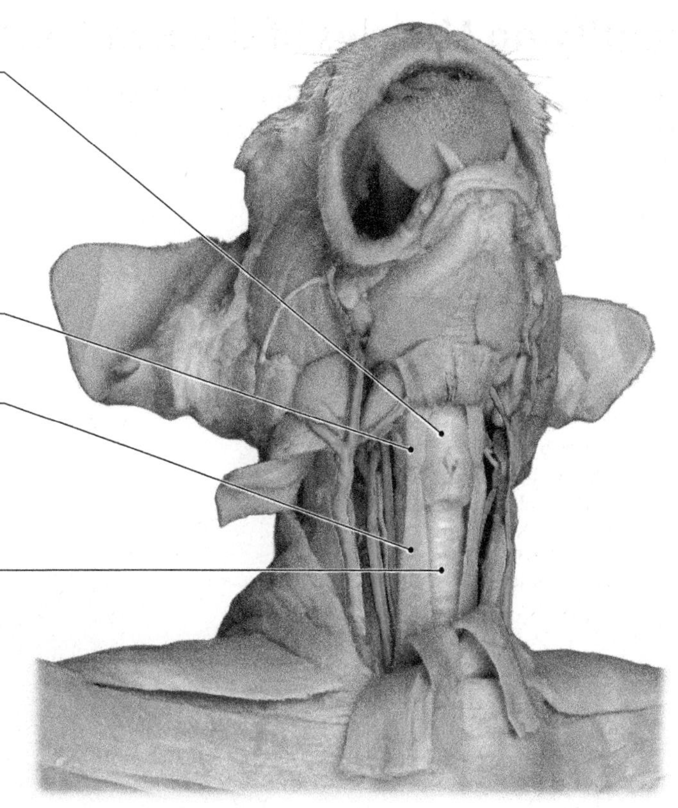

Ventral view

Word Origins

The names *sternomastoid* and *cleidomastoid* provide clues for the origins and insertions of these muscles. The sternomastoid originates on the manubrium of the sternum and inserts on the mastoid process of the temporal bone. The cleidomastoid originates on the clavicle (cleid means "clavicle") and also inserts on the mastoid process. In humans, the sternomastoid and cleidomastoid are fused to form one muscle. As you might predict, the human muscle is called the *sternocleidomastoid*. As you continue to dissect, look for similar clues in the names of other muscles.

MAKING CONNECTIONS

Consider the actions of the head and neck muscles and explain why most of these muscles are important for digestion.

▶ ____

____

____

____

____

____

____

____

Activity **C2.4**

Identifying Muscles of the Forelimb

A The muscles found on the medial aspect of the upper forelimb flex and pronate the lower forelimb and move the forelimb anteriorly.

1. Place your cat ventral side up in a dissecting tray. If you have not already done so, transect and reflect the clavodeltoid, the pectoantebrachialis, and pectoralis major near their insertions on the humerus. That will expose the deeper muscles on the upper forelimb.

2. Identify the **biceps brachii** along the anteromedial surface of the humerus. It originates on the scapula and inserts on the proximal end of the radius.

3. Locate the **brachialis** (see the figure on the next page), which passes laterally to the biceps brachii. It originates on the humerus and inserts on the ulna.

4. Find the slender **coracobrachialis** that passes medial to the biceps brachii. It originates on the scapula and inserts on the humerus.

5. Identify the **epitrochlearis,** which is located posterior to the biceps brachii. This muscle acts with the triceps brachii to extend the lower forelimb. Humans do not have this muscle.

B The muscles found on the medial aspect of the lower forelimb flex the foot and digits.

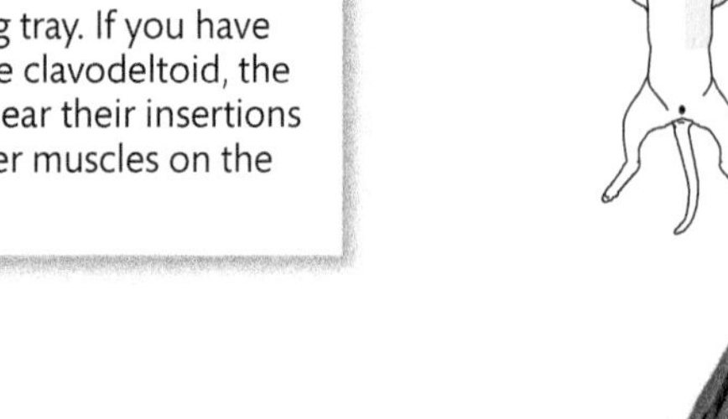

1. The four superficial flexor muscles originate on the distal end of the humerus. The brachioradialis inserts on the distal end of the radius. The other three muscles insert on carpal and metacarpal bones. From anterior to posterior, identify the following muscles:
   - **Brachioradialis**
   - **Flexor carpi radialis**
   - **Palmaris longus**
   - **Flexor carpi ulnaris**

2. Locate the **pronator teres** near the medial aspect of the elbow. From its origin at the medial epicondyle of the humerus, it travels distally to the radius. This muscle pronates the lower forelimb.

3. The **flexor digitorum profundus** is a deep muscle that originates on the medial epicondyle and inserts on the phalanges. It can be seen when the flexor carpi radialis, palmaris longus, and flexor carpi ulnaris are transected and reflected.

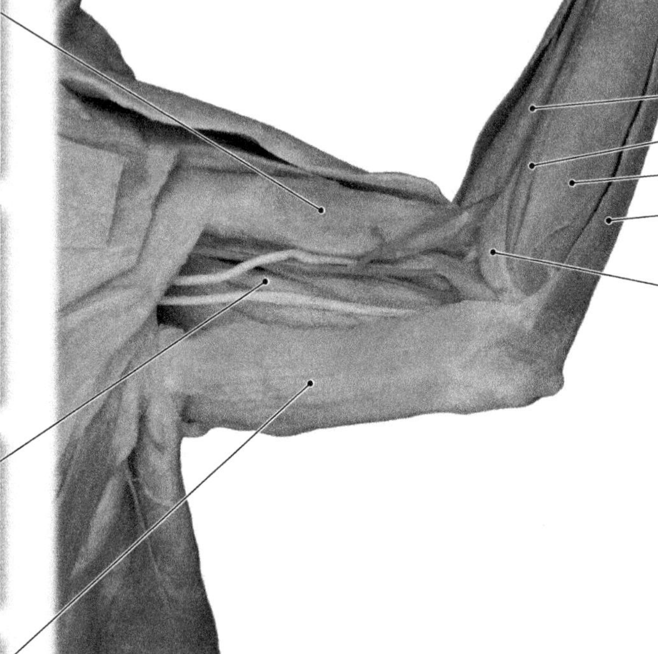

Medial view

Word Origins

Biceps and *triceps* are derived from Latin and mean "two heads" and "three heads," respectively. In the human, the biceps brachii has two heads of origin, and the triceps brachii has three heads, as the names indicate. In the cat, the triceps brachii has three heads of origin. Once again, the name correctly describes the anatomy of the muscle. The biceps brachii in the cat does not have two heads, however. It has only one head that originates from the glenoid fossa of the scapula. Because the biceps brachii in the cat is homologous to its counterpart in the human, the names have remained the same. It is nevertheless a rare example of an anatomical name that does not correctly describe the nature of the structure.

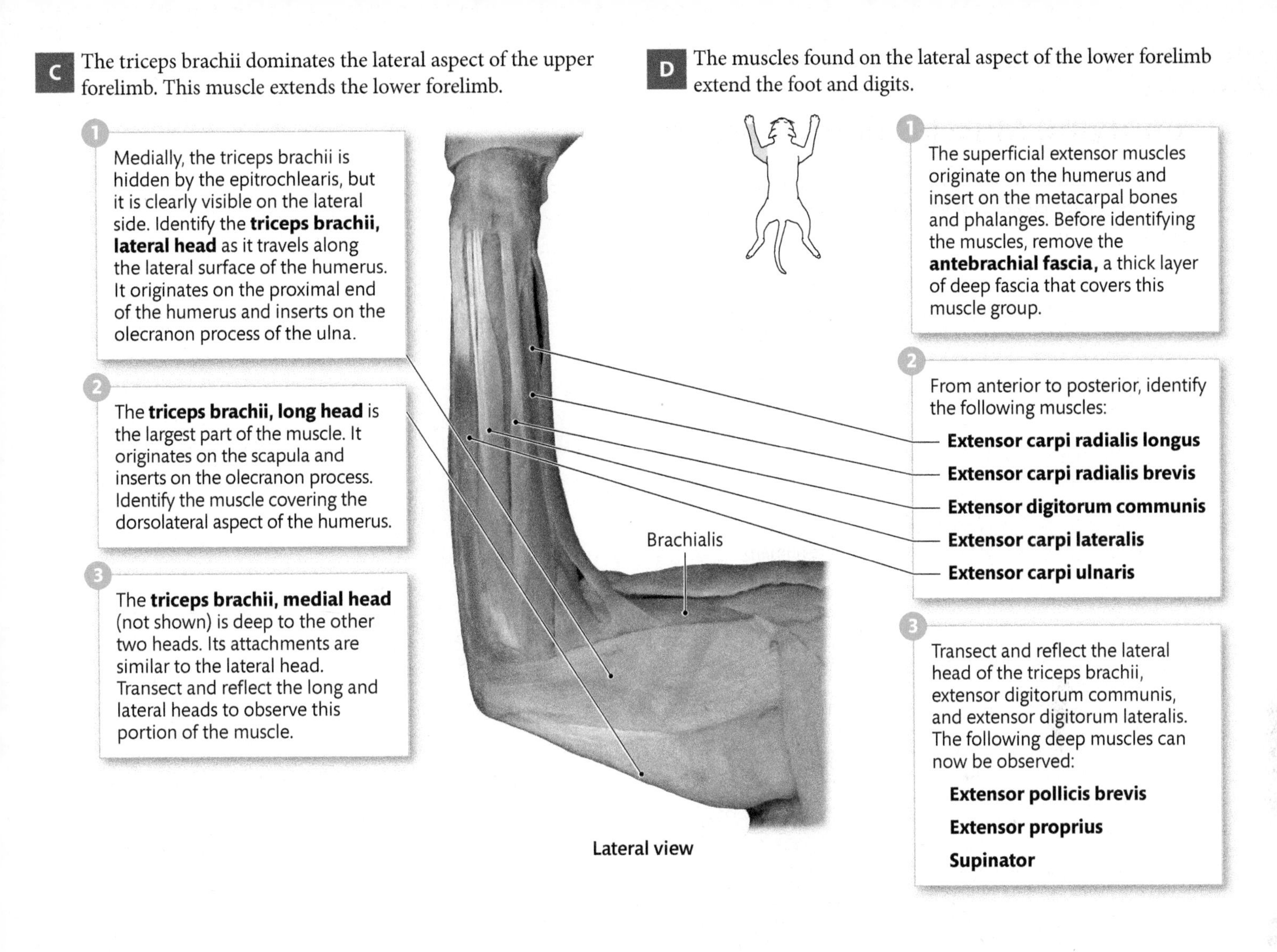

Lateral view

C The triceps brachii dominates the lateral aspect of the upper forelimb. This muscle extends the lower forelimb.

1. Medially, the triceps brachii is hidden by the epitrochlearis, but it is clearly visible on the lateral side. Identify the **triceps brachii, lateral head** as it travels along the lateral surface of the humerus. It originates on the proximal end of the humerus and inserts on the olecranon process of the ulna.

2. The **triceps brachii, long head** is the largest part of the muscle. It originates on the scapula and inserts on the olecranon process. Identify the muscle covering the dorsolateral aspect of the humerus.

3. The **triceps brachii, medial head** (not shown) is deep to the other two heads. Its attachments are similar to the lateral head. Transect and reflect the long and lateral heads to observe this portion of the muscle.

D The muscles found on the lateral aspect of the lower forelimb extend the foot and digits.

1. The superficial extensor muscles originate on the humerus and insert on the metacarpal bones and phalanges. Before identifying the muscles, remove the **antebrachial fascia,** a thick layer of deep fascia that covers this muscle group.

2. From anterior to posterior, identify the following muscles:

   Extensor carpi radialis longus
   Extensor carpi radialis brevis
   Extensor digitorum communis
   Extensor carpi lateralis
   Extensor carpi ulnaris

3. Transect and reflect the lateral head of the triceps brachii, extensor digitorum communis, and extensor digitorum lateralis. The following deep muscles can now be observed:

   Extensor pollicis brevis
   Extensor proprius
   Supinator

MAKING CONNECTIONS

Review the muscles of the forelimb in the cat and identify similarities and differences with comparable human muscles in the arm and forearm. ▶

Similarities:

____

____

____

Differences:

____

____

____

Activity **C2.5**

Identifying Muscles of the Hindlimb

A Adductor muscles, hamstrings, and the quadriceps femoris can be identified along the medial and ventral aspects of the thigh.

1 Position your cat so that you can view the medial and ventral aspects of the hindlimb. Two broad, flat muscular bands occupy the superficial surface of the medial thigh. Identify these muscles:

The **sartorius** is the anterior muscle. It extends from the iliac crest to the tibia.

The **gracilis** is the posterior muscle. It travels from the pubic bone to the tibia.

Both muscles adduct the thigh and extend the leg. Notice that they are much larger in the cat than in the human.

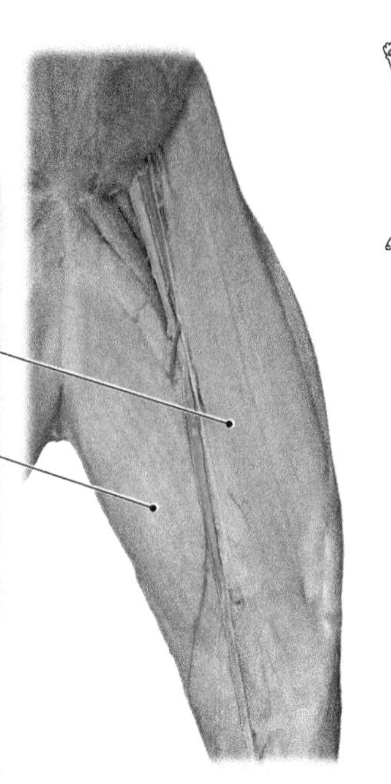

Medial view

2 Transect and reflect the sartorius and gracilis. Identify the following deep muscles on the medial aspect of the thigh. A small portion of the **iliopsoas** can be seen, just medial to sartorius. The iliopsoas is a powerful hip flexor. It is composed of two separate muscles, the **iliacus** and **psoas major.** Observe the muscle as it travels to its insertion on the lesser trochanter of the femur.

3 From anterior to posterior, identify the following three muscles:

Pectineus

Adductor longus

Adductor femoris

Notice that the adductor femoris is much larger than the other two muscles. All these muscles originate on the pubis and insert on the femur. They are all adductors of the thigh.

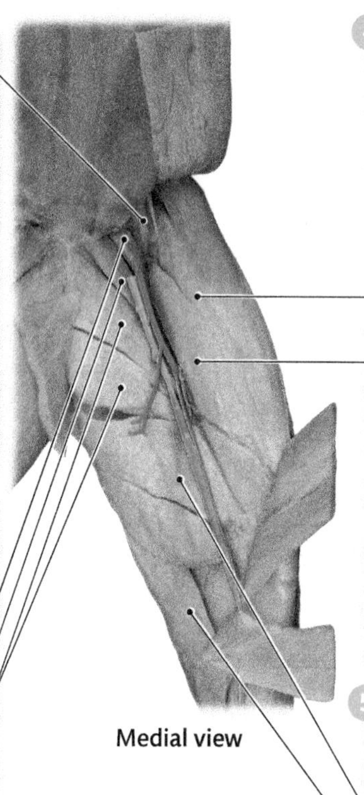

Medial view

4 The **quadriceps femoris** is a group of four anterior thigh muscles that lie deep to the sartorius, medially, and the biceps femoris and tensor fasciae latae, laterally (see part B of this activity on the next page). These muscles originate on the ilium and femur and insert on the tibia via the patellar ligament. Their main action is extension of the leg. From the medial aspect of the thigh, identify the following muscles of the quadriceps femoris group:

The **rectus femoris** lies on the anteromedial aspect of the thigh.

The **vastus medialis** is medial to the rectus femoris.

The **vastus lateralis** (see middle figure on the next page) is lateral to the rectus femoris. This muscle can best be viewed from the lateral aspect of the thigh, after the biceps femoris and the tensor fasciae latae are reflected.

The **vastus intermedius** (not shown) is deep to the rectus femoris. To observe this muscle, the rectus femoris must be transected and reflected.

5 Posterior to the adductor muscles, identify two of the hamstring muscles:

Semimembranosus

Semitendinosus

Observe the relative sizes of these muscles. The much larger semimembranosus occupies nearly half of the posterior thigh.

■ Want more practice? Go to: **MasteringA&P** > Study Area > Menu > Lab Tools > PAL > Cat > Muscular System > Lower limb (hindlimb) photos

B The tensor fasciae latae, biceps femoris, and gluteal muscles can be identified along the lateral and posterior aspects of the thigh.

1. Postition your cat so that you can view the lateral and posterior aspects of the hindlimb. Separate and identify three gluteal muscles:

   The **gluteus medius** is the anterior muscle.

   The **gluteus maximus** is a small muscle posterior to the gluteus medius. Unlike the human muscle, the gluteus maximus in the cat is much smaller than the gluteus medius.

   The **caudofemoralis** is posterior to the gluteus maximus. In humans, this muscle is fused to and becomes part of the gluteus maximus.

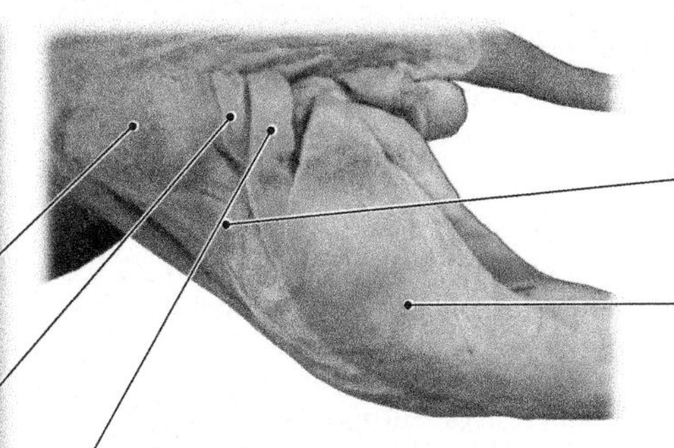

Lateral view

2. Identify two broad muscles, which occupy most of the superficial surface.

   The **tensor fasciae latae** is the anterior muscle.

   The **biceps femoris,** one of three hamstring muscles, is posterior to the tensor fasciae latae.

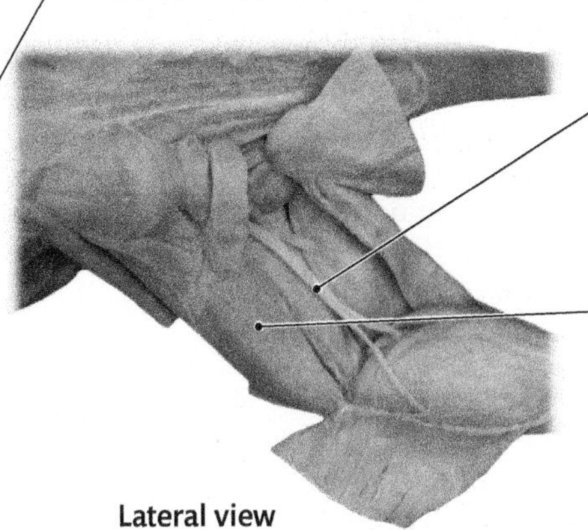

Lateral view

3. Transect and reflect the tensor fasciae latae and the biceps femoris. The **sciatic nerve** can be clearly seen traveling diagonally through the thigh.

4. The **vastus lateralis,** a muscle that you dissected earlier, can be better seen from this view than from the medial view.

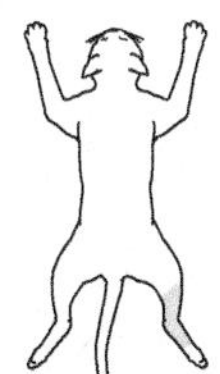

C Muscles in the leg extend and flex the foot and digits.

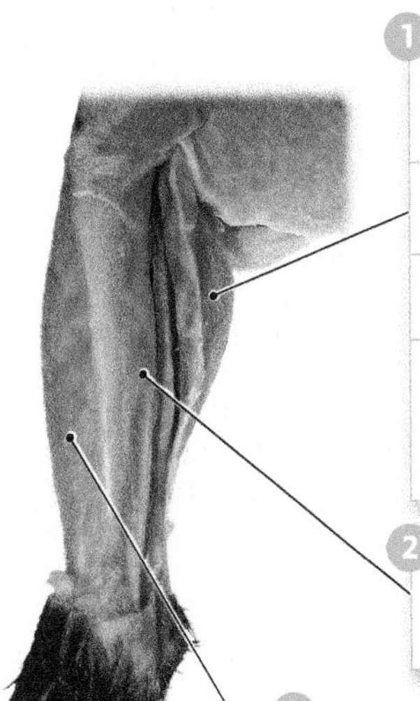

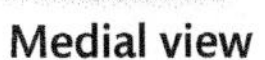

Medial view

1. Identify the two large muscles along the posterior aspect of the leg:

   The **gastrocnemius** is superficial.

   The **soleus** is deep.

   Notice that both insert on the calcaneus by way of the **calcaneal (Achilles) tendon.**

2. Deep to the soleus, identify the **flexor digitorum longus.**

3. Along the ventral surface of the leg, identify:

   Tibialis anterior

   Extensor digitorum longus

4. If your dissection session is ending here, wrap your cat with the skin. Cover the animal with moist towels, and return it to the storage bag.

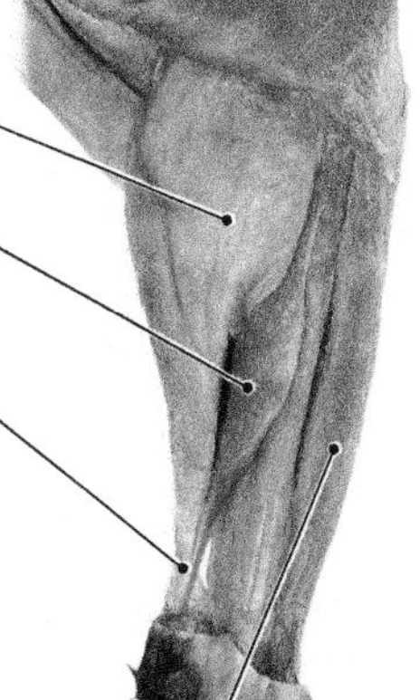

Lateral view

MAKING CONNECTIONS

The cat hindlimb, particularly the thigh, is bulkier than the human lower extremity. Using models or figures of the human musculature for comparison, identify the muscles that reveal significant species differences in size. Speculate on the functional advantage that the cat might have for having a massive or bulky hindlimb.

▶ ____________________

____

____

____

____

____

____

____

____

____

BEFORE YOU MOVE ON . . .

« LOOKING BACK

Most muscles in the cat are comparable in structure and function to those in the human. For example, the abdominal wall and intercostal muscles are identical, and most muscles in the head and neck and the limbs are the same.

Despite these similarities, there are some notable differences. Humans have fewer individual muscles than cats. Consider the following examples:

- Both cats and humans have pectoralis major and pectoralis minor muscles, but cats have two additional muscles in the pectoral group: the pectoantebrachialis and the xiphihumeralis.
- Humans have only one deltoid muscle, but cats have a deltoid group consisting of three muscles: the clavodeltoid, the acromiodeltoid, and the spinodeltoid.
- Similarly, humans have one trapezius muscle, whereas cats have a trapezius group: the clavotrapezius, the acromiotrapezius, and the spinotrapezius.

There are also differences in the size of some muscles, particularly in the hindlimb. Humans have a massive gluteus maximus muscle, but in the cat, the gluteus maximus is smaller than the gluteus medius. In cats, the sartorius and gracilis muscles are large and powerful, but the same muscles in the human are slender, straplike bands and are relatively weak.

What do you think is the reason for the differences between human musculature and cat musculature?

▶ ______________________________

____

____

____

____

____

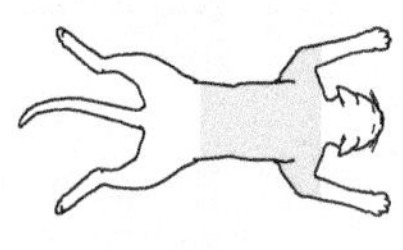

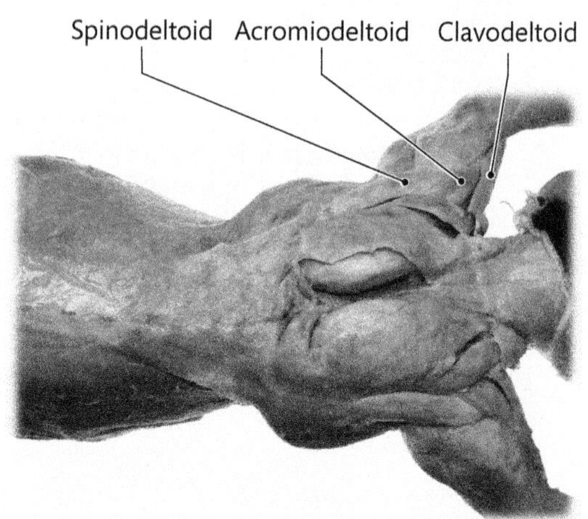

Dorsal view

LOOKING FORWARD »

The muscles that you dissected in this cat dissection exercise are innervated by peripheral nerves originating from the brain (**cranial nerves**) and spinal cord (**spinal nerves**). In the next cat dissection exercise, you will dissect and identify the major peripheral nerves derived from spinal nerves in the cat nervous system.

Name ______________________

Lab Section ______________________

Date ______________________

EXERCISE **C2** **REVIEW SHEET**

Dissection of the Cat Muscular System

1. The names of some muscles provide clues about their location. Identify the body region where the following muscles are located.

a. Sternomastoid ______________________

b. Rectus abdominis ______________________

c. Supraspinatus ______________________

d. Tibialis anterior ______________________

e. Brachialis ______________________

f. Temporalis ______________________

g. Pectoralis major ______________________

h. Biceps femoris ______________________

i. Flexor carpi radialis ______________________

j. Gluteus maximus ______________________

2. Discuss the following structures in terms of their significance to the muscular system.

a. Lumbodorsal fascia ______________________

____

b. Linea alba ______________________

____

c. Hyoid bone ______________________

____

d. Scapula ______________________

____

e. Calcaneal (Achilles) tendon ______________________

____

f. Mandible ______________________

____

QUESTIONS 3–11: Identify the labeled muscles in the diagram. Select your answers from the list below.

a. Xiphihumeralis
b. Pectoantebrachialis
c. Internal oblique
d. Transversus abdominis
e. Latissimus dorsi
f. Pectoralis major
g. External oblique
h. Rectus abdominis
i. Pectoralis minor

3. ______________________________
4. ______________________________
5. ______________________________
6. ______________________________
7. ______________________________
8. ______________________________
9. ______________________________
10. ______________________________
11. ______________________________

Ventral view

QUESTIONS 12–18: Identify the labeled muscles in the diagram. Select your answers from the list below.

a. Clavodeltoid
b. Latissimus dorsi
c. Acromiotrapezius
d. Clavotrapezius
e. Spinodeltoid
f. Spinotrapezius
g. Acromiodeltoid

12. ______________________________
13. ______________________________
14. ______________________________
15. ______________________________
16. ______________________________
17. ______________________________
18. ______________________________

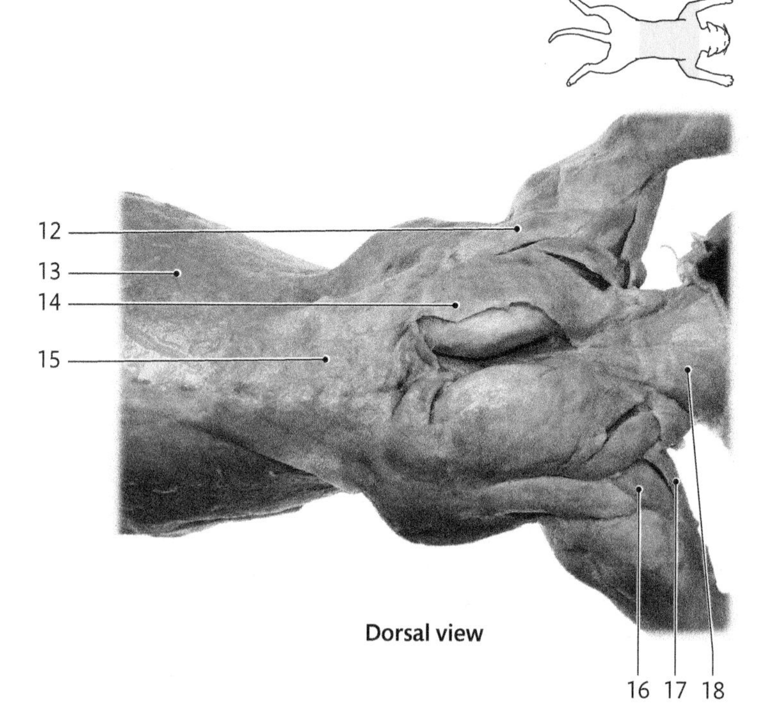

Dorsal view

DISSECTION EXERCISE

C3

Dissection of the Cat Peripheral Nervous System

LEARNING OUTCOMES

These Learning Outcomes correspond by number to the laboratory activities in this exercise. When you complete the activities, you should be able to:

Activity C3.1 **Identify the brachial plexus and locate the major peripheral nerves that are derived from it.**

Activity C3.2 **Identify the lumbar plexus and locate the major peripheral nerves that are derived from it.**

Activity C3.3 **Identify the sacral plexus and locate the major peripheral nerves that are derived from it. Identify the termination of the spinal cord and the cauda equina.**

LABORATORY SUPPLIES

- Preserved cats
- Dissecting equipment: trays, tools, gloves
- Bone cutters
- Protective eyewear
- Face mask

PRE-LAB QUIZZ

Before you begin, read all the activities in Exercise C3 and the required reading in your textbook that is assigned by your instructor.

Questions 1–6: Match the peripheral nerve in column A with the spinal nerve plexus in which it is found in column B. Answers in column B may be used more than once.

A	B
1. Obturator nerve ______	a. Brachial plexus
2. Ulnar nerve ______	b. Lumbar plexus
3. Femoral nerve ______	c. Sacral plexus
4. Sciatic nerve ______	
5. Median nerve ______	
6. Radial nerve ______	

7. True or False: The obturator nerve travels through the femoral triangle. ______
8. The tibial and common fibular nerves are the terminal branches of what nerve?
   a. femoral nerve
   b. sciatic nerve
   c. radial nerve
   d. obturator nerve
   e. median nerve
9. During this cat dissection exercise, you will remove the vertebral arches from lumbar vertebrae and the sacrum to expose the cauda equina. What tool will you use to remove the vertebral arches? ______
10. The cauda equina is formed by the nerve roots of all the following spinal nerves *except* the
    a. caudal spinal nerves.
    b. lumbar spinal nerves.
    c. thoracic spinal nerves.
    d. sacral spinal nerves.

Activity **C3.1**

Identifying Major Peripheral Nerves of the Brachial Plexus

The brachial plexus in the cat is formed by the ventral rami of the fifth to eighth cervical spinal nerves (C_5–C_8) and the first thoracic spinal nerve (T_1). The plexus is located medial to the shoulder and anterior to the first rib.

A Several branches of the brachial plexus innervate muscles that act at the shoulder and move the scapula.

1. Place your cat ventral side up in a dissecting tray. If not already done, transect and reflect the pectoralis group of muscles to reveal the **brachial plexus** in the axillary region. (C_5 is not shown in the figure.)

2. Identify the **spinal nerve roots** of the brachial plexus, close to where they emerge from the vertebral column.

3. The **long thoracic nerve** supplies the serratus ventralis. Identify the nerve as it passes along this muscle.

4. Locate the **dorsal thoracic nerve** as it travels to the latissimus dorsi.

5. Identify the **subscapular nerve.** You can see it as it travels lateral to the beginning of the external jugular vein. This nerve innervates the subscapularis and teres major muscles.

6. Identify the **ventral thoracic nerve.** It gives off several branches that supply the pectoralis muscles.

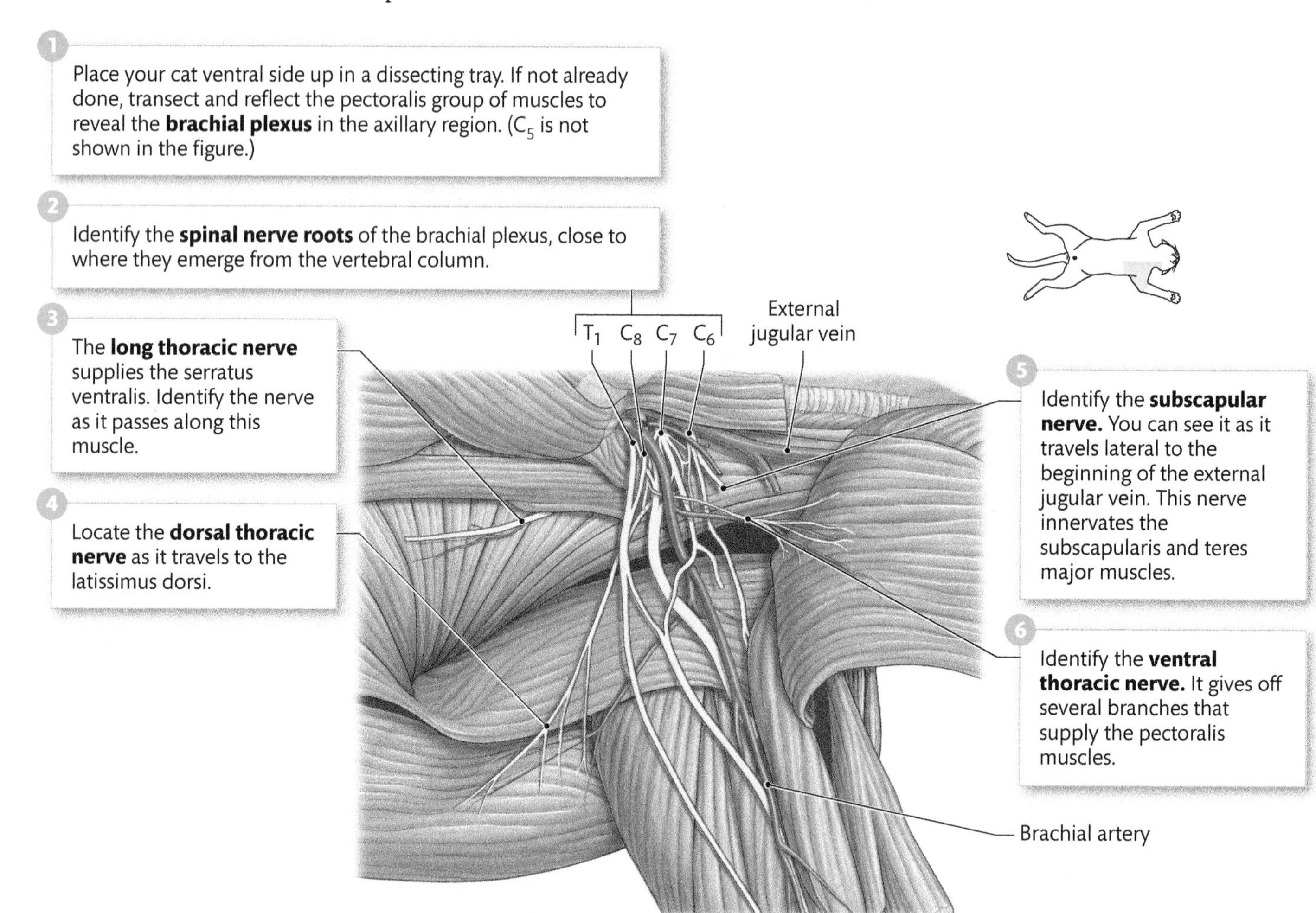

Ventrolateral view

Word Origins

Plexus is the Latin word for "a braid." A spinal nerve plexus is formed when the ventral rami of spinal nerves interweave with one another in a way that is similar to forming a braid with hair.

B The brachial plexus has five terminal branches that innervate muscles in the upper and lower forelimb.

1. The **ulnar nerve** is the most posterior terminal branch. Identify the nerve as it travels toward the elbow. It supplies the flexor carpi ulnaris, a portion of the flexor digitorum profundus, and some small muscles in the paw (in humans, the hand).

2. The **median nerve** can be identified in the forelimb as it travels with the brachial artery. It innervates most of the ventral muscles of the forelimb and some of the small muscles in the paw.

3. The **musculocutaneous nerve** can be seen traveling to the biceps brachii and coracobrachialis muscles.

4. The **axillary nerve** travels lateral to the musculocutaneous nerve. It innervates the spinodeltoid muscle.

5. The **radial nerve** is the largest peripheral nerve of the brachial plexus. You can see it traveling to the dorsal aspect of the forelimb, inferior to the musculocutaneous nerve. The radial nerve innervates all the dorsal muscles of the forelimb.

Long thoracic nerve

Spinal nerve roots

External jugular vein

Ventral thoracic nerve

Dorsal thoracic nerve

Brachial artery

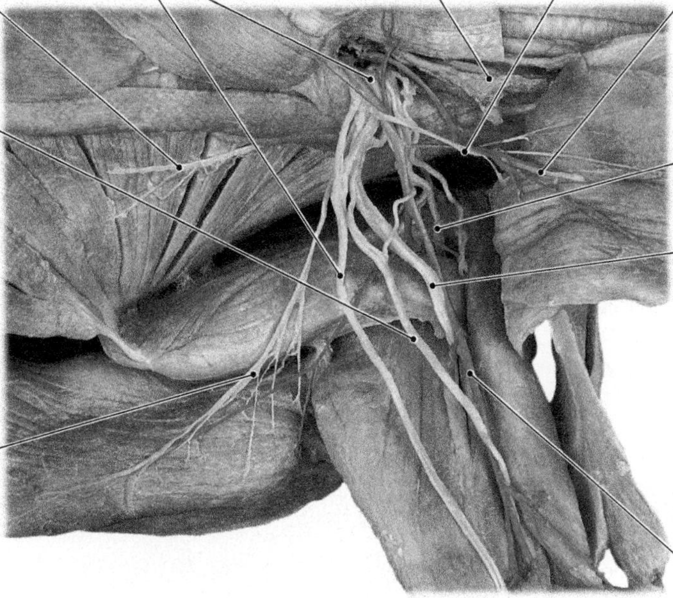

Ventrolateral view

IN THE CLINIC

Ulnar Nerve Injury

The ulnar nerve travels posterior to the medial epicondyle of the humerus before entering the forearm. The nerve is vulnerable to injury at this point because of its subcutaneous position. Mild trauma to the medial epicondyle (e.g., banging your elbow on a table) can cause ulnar nerve compression. The phrase "hitting your funny bone" refers to the numbness and tingling sensation along the medial part of the palm that occur as a result of this type of injury. If the medial epicondyle is fractured, causing more serious damage to the ulnar nerve, long-term or permanent motor and sensory loss to the hand could occur.

MAKING CONNECTIONS

Compare the functional loss in the cat and the human if the musculocutaneous nerve is damaged.

▶ ______________________________

Activity C3.2

Identifying Major Peripheral Nerves of the Lumbar Plexus

The **lumbar plexus** in the cat is formed by the ventral rami of the fourth to seventh lumbar spinal nerves (L_4–L_7) and all three sacral spinal nerves (S_1–S_3). Components of the lumbar plexus can be identified in the anterior thigh.

A The femoral nerve is the largest branch of the lumbar plexus. It travels through the femoral triangle accompanied by the femoral blood vessels.

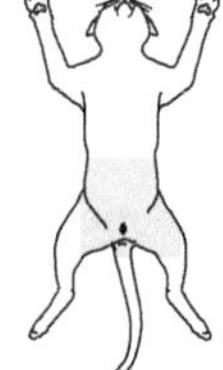

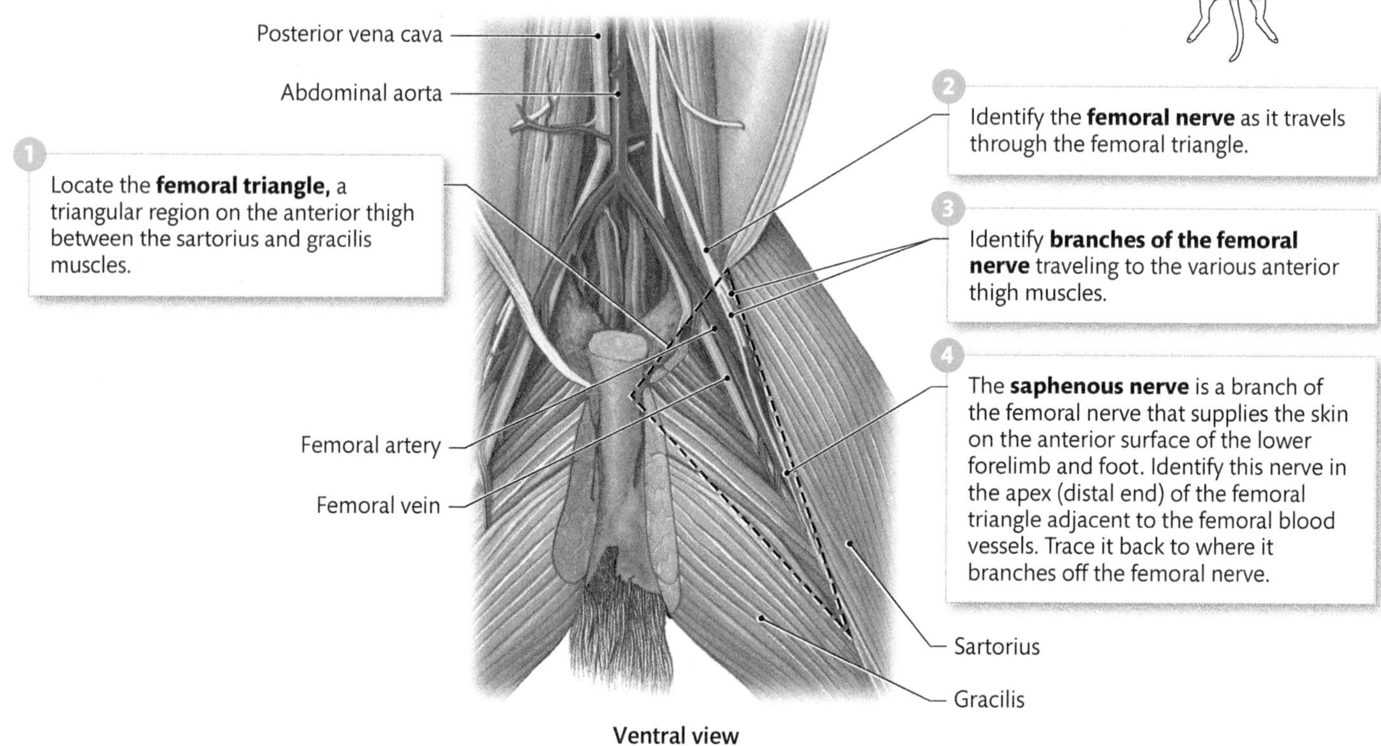

Ventral view

B The obturator nerve is a second major branch of the lumbar plexus. It passes through the obturator foramen to reach the medial thigh muscles.

1. If not already done, transect and reflect the iliopsoas muscle at approximately the level where it enters the anterior thigh. Deep to the iliopsoas, identify the **obturator nerve.**

2. The obturator nerve travels posteriorly into the pelvic cavity. It enters the thigh by passing through the **obturator foramen** and gives off branches to the adductors, pectineus, and gracilis. On a cat skeleton, identify the obturator foramen, a large opening in each pelvic bone.

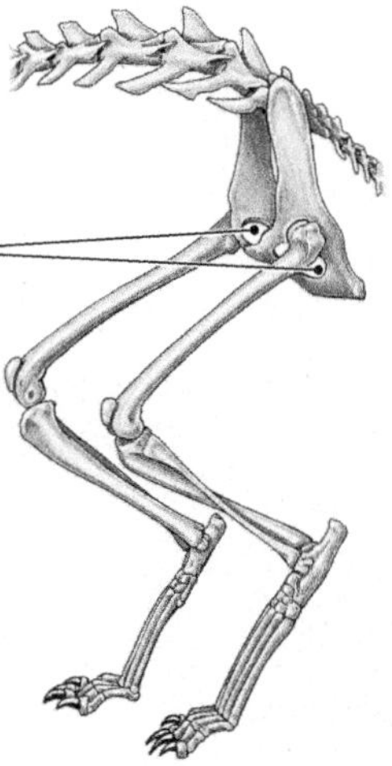

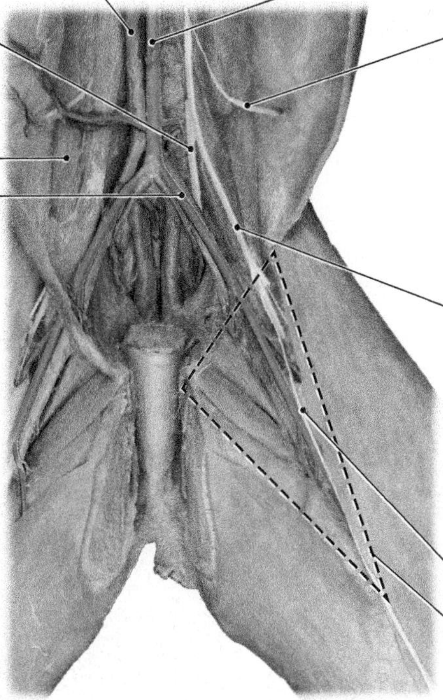

3. From this dissection view, find the **lateral cutaneous nerve,** which supplies the skin on the lateral surface of the thigh.

4. You can also see a more proximal portion of the **femoral nerve** running parallel to the external iliac artery before entering the thigh.

Posterior vena cava
Abdominal aorta
Iliopsoas muscle
External iliac artery
Saphenous nerve
Femoral triangle

Ventral view

MAKING CONNECTIONS

Your pet cat is infected with a mysterious virus that attacks nervous tissue. The infection has been localized to the L_6 spinal cord level and the L_6 spinal nerves. The infection caused a partial loss of function of both the femoral and obturator nerves. Based on this information, what can you say about the origin and structure of each nerve?

▶ ______________________________

____

____

____

____

Activity C3.3

Identifying Major Peripheral Nerves of the Sacral Plexus

The sacral plexus in the cat is formed by the ventral rami of the sixth and seventh lumbar spinal nerves (L_6–L_7) and all three sacral spinal nerves (S_1–S_3). Components of the sacral plexus can be identified in the posterior thigh.

A The sciatic nerve is the largest branch of the sacral plexus and is the largest peripheral nerve in the cat (and the human).

1. Place your cat dorsal side up in a dissecting tray. If not already done, transect and reflect the tensor fasciae latae and the biceps femoris muscles.

2. Identify the **sciatic nerve.** It can be clearly seen traveling diagonally through the posterior thigh, between the vastus lateralis and the semimembranosus.

3. Trace the sciatic nerve distally until it diverges to form its two terminal branches.

   The medial branch is the **tibial nerve.** It supplies muscles in the posterior aspect of the lower hindlimb and most muscles in the foot.

   The lateral branch is the **common fibular nerve.** It supplies muscles in the anterior and lateral aspects of the lower hindlimb.

4. In the thigh, identify **branches of the sciatic nerve** that innervate the hamstring muscles.

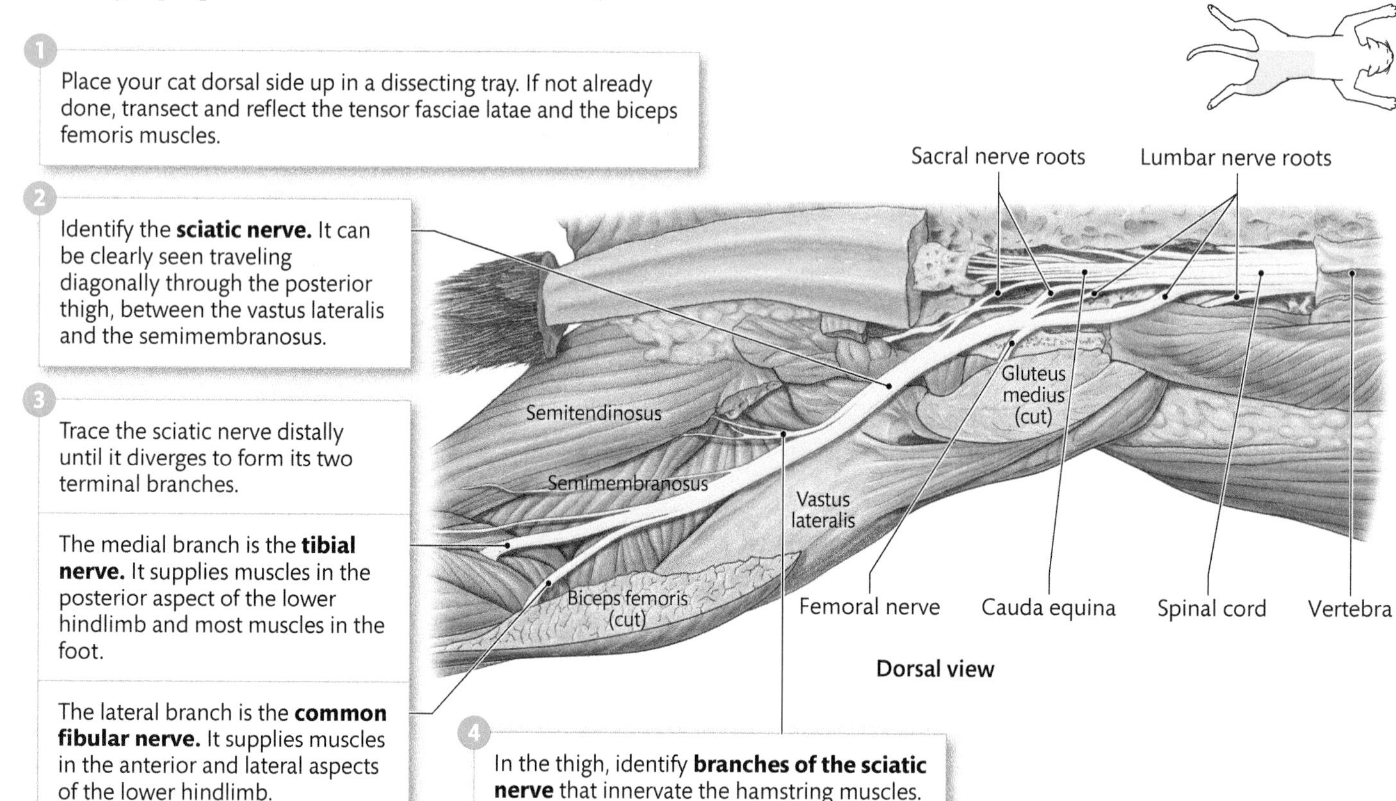

Dorsal view

 ■ Want more practice? Go to: **MasteringA&P** > Study Area > Menu > Lab Tools > PAL > Cat > Nervous System > Nerves photos

B Lumbar, sacral, and caudal nerve roots form the cauda equina.

1. Remove the skin and muscles that cover the distal end of the lumbar vertebrae and the three fused vertebrae that form the sacrum. Make sure the vertebral arches (pedicles and laminae) are fully exposed.

2. With bone cutters, cut the bony pedicles on each side of the vertebrae to remove the vertebral arches. In the sacrum, the vertebral arches will be fused. Work on one vertebra at a time. Be especially careful to avoid damaging the underlying spinal cord.

3. The thick, opaque membrane that covers the spinal cord is the **dura mater,** the outermost meningeal layer. Notice that the dura mater also covers the nerve roots of spinal nerves that extend from the spinal cord.

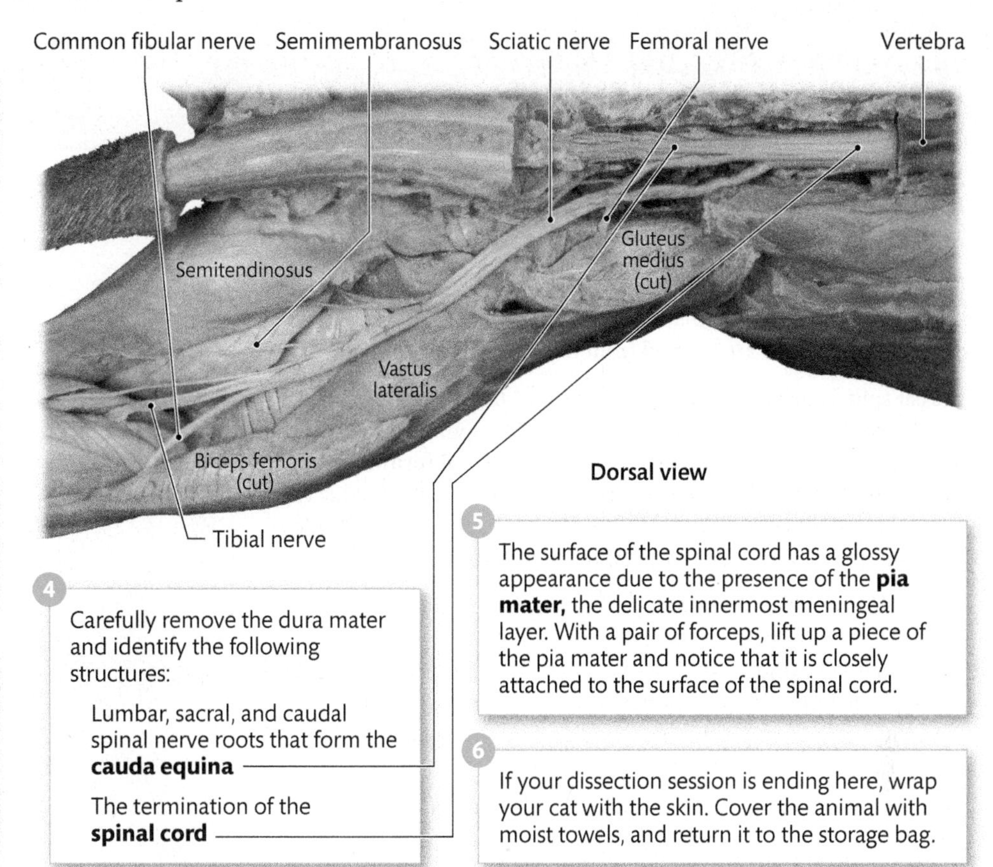

4. Carefully remove the dura mater and identify the following structures:

   Lumbar, sacral, and caudal spinal nerve roots that form the **cauda equina**

   The termination of the **spinal cord**

5. The surface of the spinal cord has a glossy appearance due to the presence of the **pia mater,** the delicate innermost meningeal layer. With a pair of forceps, lift up a piece of the pia mater and notice that it is closely attached to the surface of the spinal cord.

6. If your dissection session is ending here, wrap your cat with the skin. Cover the animal with moist towels, and return it to the storage bag.

Dissection note: *The middle meningeal layer, the **arachnoid mater,** does not preserve well during commercial preparation and cannot be identified.*

MAKING CONNECTIONS

A veterinarian uses an anesthetic on a cat to block the ventral rami of L_6 and L_7. Explain why the anesthetic will have an effect on nerves from both the lumbar and sacral plexuses.

▶ ______________________________

____

____

____

____

BEFORE YOU MOVE ON . . .

« LOOKING BACK

The **peripheral nervous system (PNS)** of the cat includes 12 pairs of cranial nerves, which originate from the brain, and 38 or 39 pairs of spinal nerves, which are derived from the spinal cord.

The cat spinal cord gives rise to 8 cervical, 13 thoracic, 7 lumbar, 3 sacral, and 7 or 8 caudal spinal nerve pairs. The spinal nerves exit the vertebral column by passing through **intervertebral foramina,** which are openings on each side of the vertebral column between adjacent vertebrae. After exiting the vertebral column, a spinal nerve divides into two main branches, or **rami** (singular = **ramus**):

- The **dorsal ramus** is the smaller branch. It supplies the deep back muscles and skin on the dorsal body wall.
- The **ventral ramus** is the larger branch. It supplies all thoracic and abdominal muscles (except the deep back muscles), all structures in the forelimb and hindlimb, and the skin on the lateral and ventral aspects of the body wall.

Similar to the cat, there are 12 pairs of cranial nerves in the human PNS. The human spinal cord, however, gives rise to only 31 spinal nerves pairs: 8 cervical, 12 thoracic, 5 lumbar, 5 sacral, and 1 coccygeal.

Although the spinal nerves arise from the spinal cord in a segmental fashion, the ventral rami of most form complex networks or **plexuses** (singular = **plexus**). In the cat (and the human), there are four spinal nerve plexuses: **cervical, brachial, lumbar,** and **sacral.** The cervical plexus gives rise to peripheral nerves that innervate structures in the neck. Peripheral nerves from the brachial plexus supply structures in the forelimbs, and nerves from the lumbar and sacral plexuses innervate the hindlimbs.

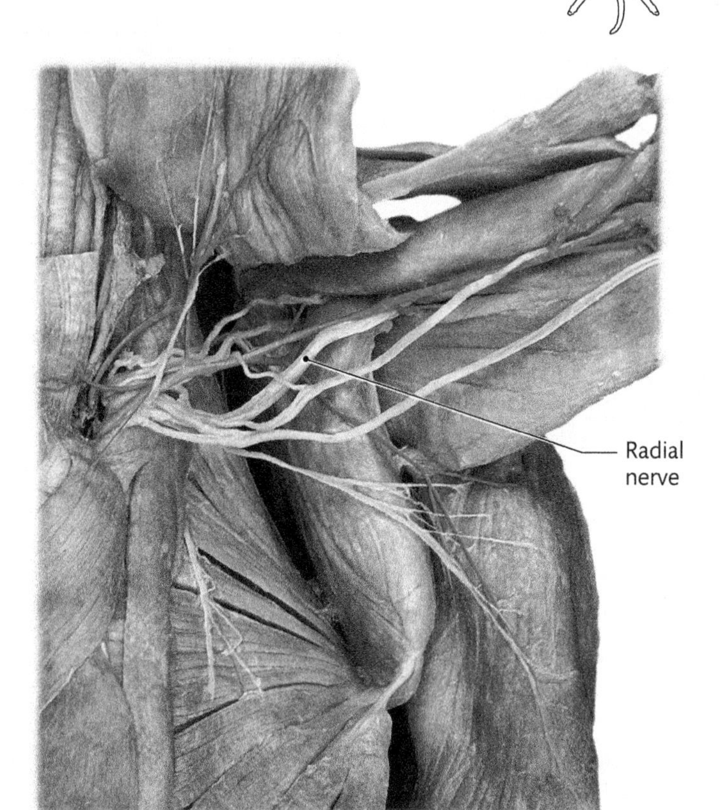

Many of the peripheral nerves derived from a spinal nerve plexus contain components from several spinal nerves. The radial nerve, for example, receives nerve fibers from every spinal nerve in the brachial plexus. What functional advantage would this structural pattern have over another type of peripheral nerve that has nerve fibers from only one spinal nerve?

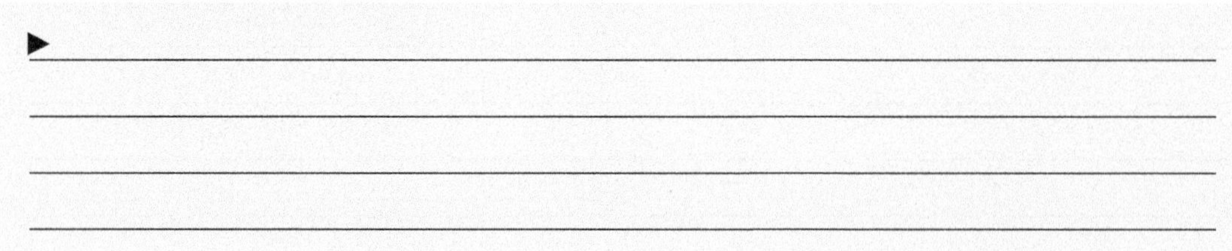

LOOKING FORWARD »

In this cat dissection exercise, you identified peripheral nerves that supply skeletal muscles. Peripheral nerves also supply the internal organs in the thoracic and abdominopelvic body cavities. In the next cat dissection exercise, you will open the body cavities to expose the internal organs and identify the organs of the cat endocrine system.

Name ______________________

Lab Section ______________________

Date ______________________

EXERCISE **C3** **REVIEW SHEET**

Dissection of the Cat Peripheral Nervous System

1. Complete the table below by identifying the affected muscles and the resulting functional loss if the following peripheral nerves were damaged.

Damaged Nerve	Affected Muscles	Functional Loss
Axillary nerve	a.	b.
Median nerve	c.	d.
Femoral nerve	e.	f.
Obturator nerve	g.	h.
Tibial nerve	i.	j.
Common fibular nerve	k.	l.

2. What is the femoral triangle? What important structures pass through this region?

____

____

____

QUESTIONS 3–9: Match the peripheral nerve in column A with the cat muscles it innervates in column B.

A

3. Musculocutaneous nerve ______________________

4. Radial nerve ______________________

5. Median nerve ______________________

6. Ulnar nerve ______________________

7. Femoral nerve ______________________

8. Obturator nerve ______________________

9. Sciatic nerve ______________________

B

a. Quadriceps femoris muscles

b. Semimembranosus and semitendinosus

c. Biceps brachii and coracobrachialis

d. Most ventral forelimb muscles

e. Pectineus and gracilis

f. Triceps brachii

g. Flexor carpi ulnaris

QUESTIONS 10–18: Identify the labeled peripheral nerves in the three photos. Select your answer from the list below.

a. Saphenous nerve
b. Sciatic nerve
c. Obturator nerve
d. Tibial nerve
e. Median nerve
f. Ulnar nerve
g. Common fibular nerve
h. Radial nerve
i. Femoral nerve

10. ______________
11. ______________
12. ______________
13. ______________
14. ______________
15. ______________
16. ______________
17. ______________
18. ______________

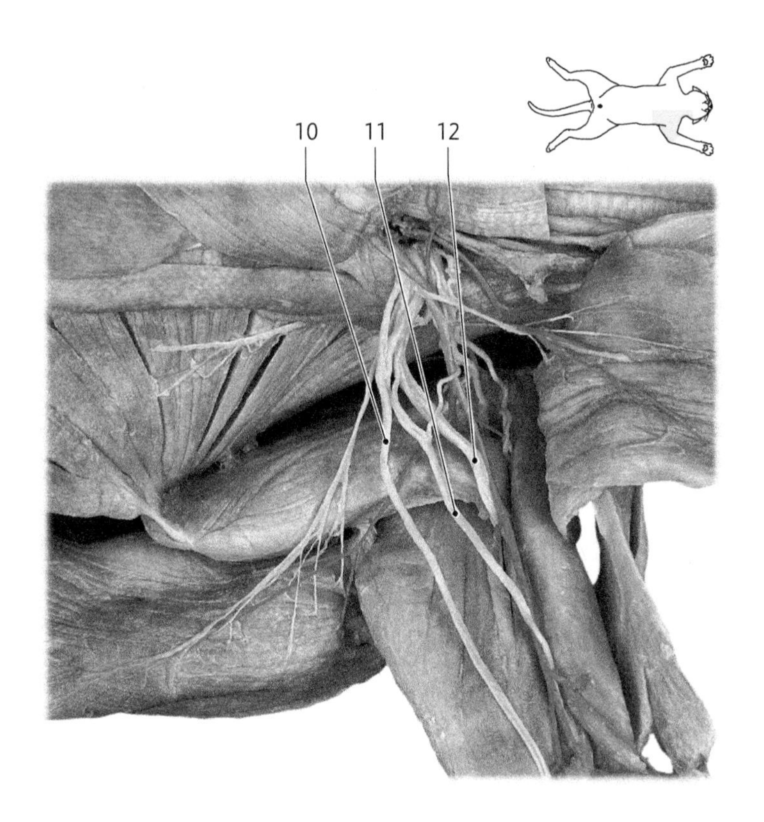

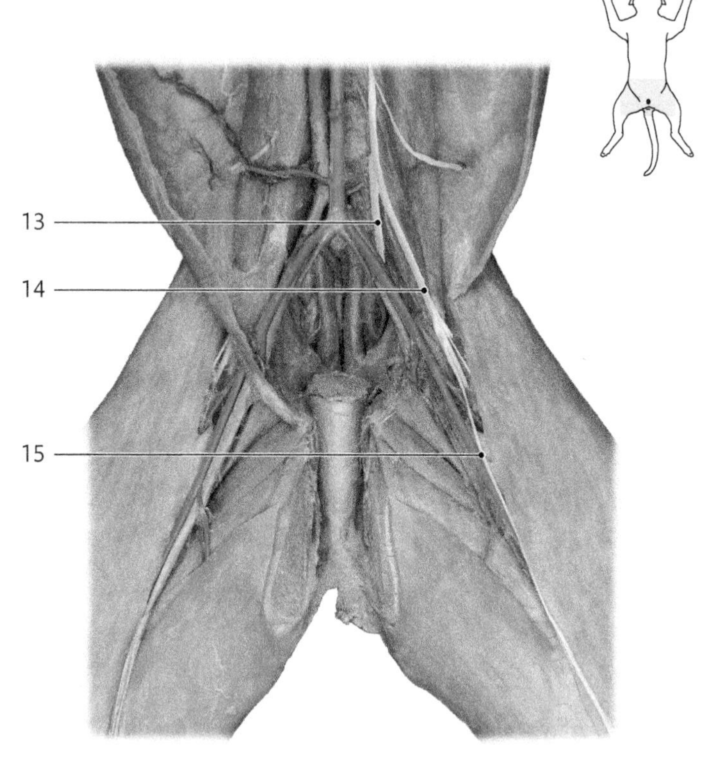

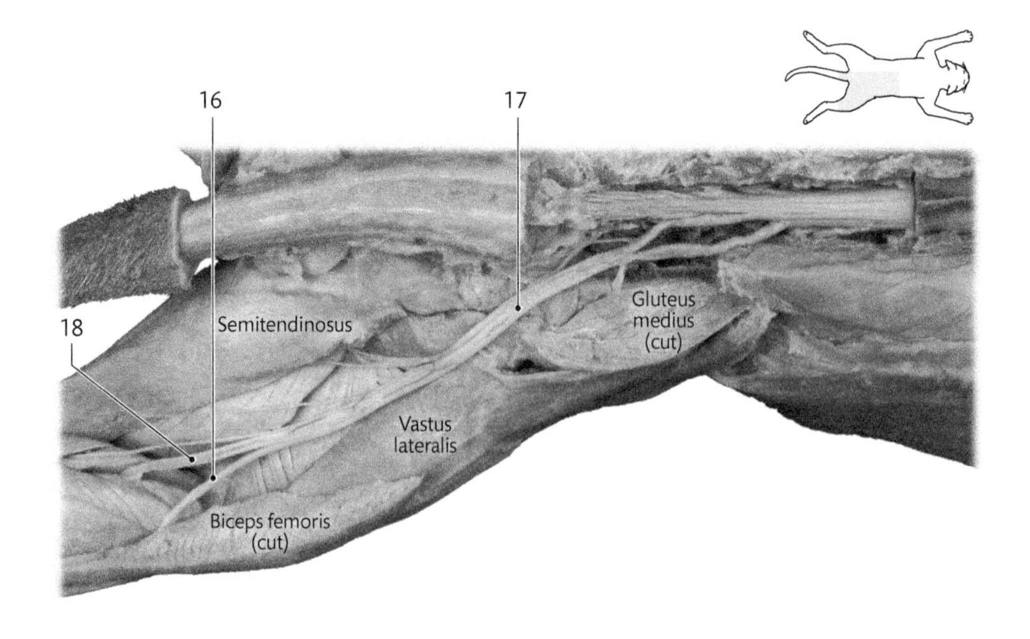

DISSECTION EXERCISE

Dissection of the Cat Ventral Body Cavities and Endocrine System

C4

LEARNING OUTCOMES

These Learning Outcomes correspond by number to the laboratory activities in this exercise. When you complete the activities, you should be able to:

Activity C4.1 **Identify the major organs in the cat thoracic and abdominopelvic cavities.**

Activity C4.2 **Identify the organs of the cat endocrine system.**

LABORATORY SUPPLIES

- Preserved cats
- Dissecting equipment: trays, tools, gloves
- Protective eyewear
- Face mask
- Paper towels
- Anatomical models of the brain
- Dissecting microscopes

PRE-LAB QUIZ

Before you begin, read all the activities in Exercise C4 and the required reading in your textbook that is assigned by your instructor.

1. During this cat dissection exercise, you will make an incision along the ventral thoracic wall to the right or left of the midline to avoid cutting through what structure?

   ____

2. True or False: The heart is located in the central region of the thorax, known as the mediastinum. ____________
3. During this cat dissection exercise, you will identify two organs that are divided into lobes. Name these two organs. ______________________
4. True or False: The visceral peritoneum lines the body wall of the abdominopelvic cavity, and the parietal peritoneum covers the surfaces of the organs in the cavity.

   ____

5. True or False: The greater omentum is the peritoneal membrane that covers most of the abdominopelvic organs. ____________
6. True or False: Similar to humans, the cat adrenal glands rest on the superior margins of the kidneys. ____________

Questions 7–10: Match the endocrine organ in column A with the correct description in column B.

A	B
7. Pituitary gland ________	a. This structure partially covers the heart and extends into the neck to cover the posterior end of the trachea.
8. Thymus ________	b. This structure is located in the neck just posterior to the larynx.
9. Pancreas ________	c. During your dissection, you will identify this structure when you lift the stomach from its anatomical position.
10. Thyroid ________	d. The infundibulum connects this structure to the hypothalamus.

Activity **C4.1**

Exploring the Contents of the Thoracic and Abdominopelvic Cavities

A When making your incisions to open the ventral body cavities, it is easy for the untrained dissector to damage the internal organs by cutting too deeply. This mistake is more likely to occur if you have a small cat with thin and fragile abdominal muscles. To reduce the likelihood of an error, proceed with caution by using the scissors with the blunt end on the inside of the body. When the body cavities are opened, your cat may contain excess preservative fluids or coagulated blood. If so, use paper towels to absorb the excess fluids and to remove the coagulated blood. Fluids and solid material that are removed from the body cavity should be stored in a secure container designated by your instructor until the materials can be safely discarded.

1. Place your cat ventral side up in a dissecting tray. With your fingers, locate the superior margins of the sternum and rib cage. Free any muscle attachments to this area with a scalpel.
2. With a pair of scissors, cut posteriorly (toward the tail) along the ventral thoracic wall. Make your incision slightly to the right or left of the midline so that you cut through the softer costal cartilages rather than the sternum. Return to the midline when you reach the abdomen and continue the incision along the midventral surface until you reach the pubic bone.
3. Just posterior to the forelimbs and anterior to the hindlimbs, make lateral incisions to the left and right sides.

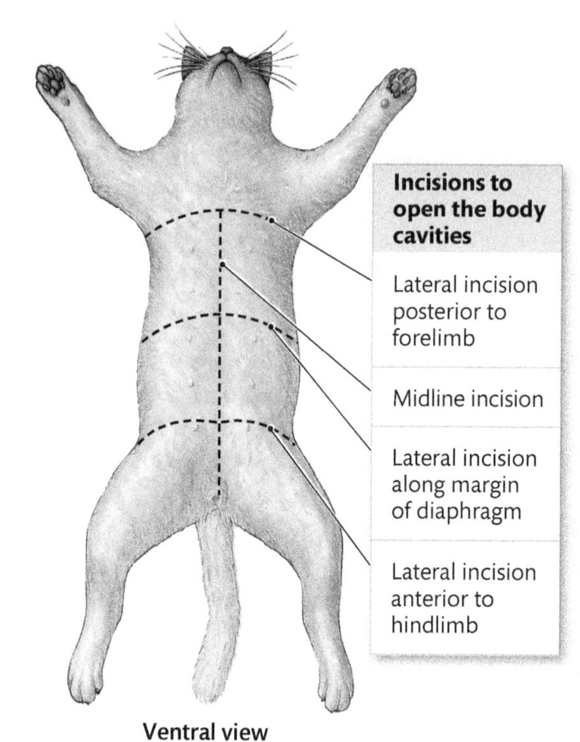

Ventral view

4. Use your fingers to feel and identify the inferior margin of the rib cage. This margin marks the line of attachment for the diaphragm. Carefully cut laterally along the margin to detach the diaphragm from the body wall.
5. Your incisions have made two pairs of lateral flaps that you can open to view the internal organs in the thoracic and abdominopelvic cavities. You can close the flaps at the end of your dissection session to protect the body cavities.
6. When you open the flaps for the thoracic cavity for the first time, you might find that they will not remain open. If that is the case, you may have to break the ribs by gently pressing down with the palm of your hand as you spread the flaps open.

B Now you can make general observations of the ventral body cavities to identify the major thoracic and abdominopelvic organs. More detailed inspections of these structures will be completed in dissection exercises to follow. Unless otherwise instructed, all internal organs that you observe should be left intact.

1. Open the lateral flaps that you produced earlier and fully expose the thoracic and abdominopelvic regions.
2. Identify the following structures in the thoracic cavity:

   The **heart** is located in the central region of the thorax, known as the mediastinum.

   The two **lungs** are located on either side of the heart. Each is divided into lobes.

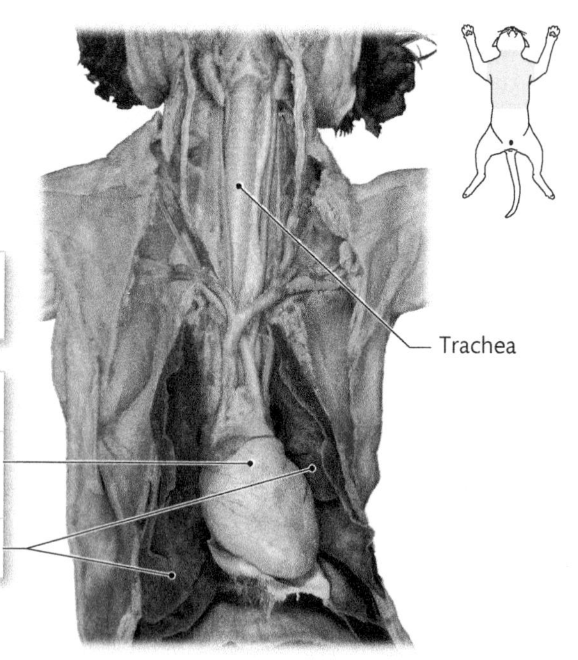

Ventral view

3

In the abdominopelvic cavity, notice that the surfaces of the organs and body wall are lined by a shiny, translucent membrane, which is known as the **peritoneum (peritoneal membrane).** The portion of the membrane that lines the body wall is the **parietal peritoneum;** the portion covering the surfaces of organs is the **visceral peritoneum.**

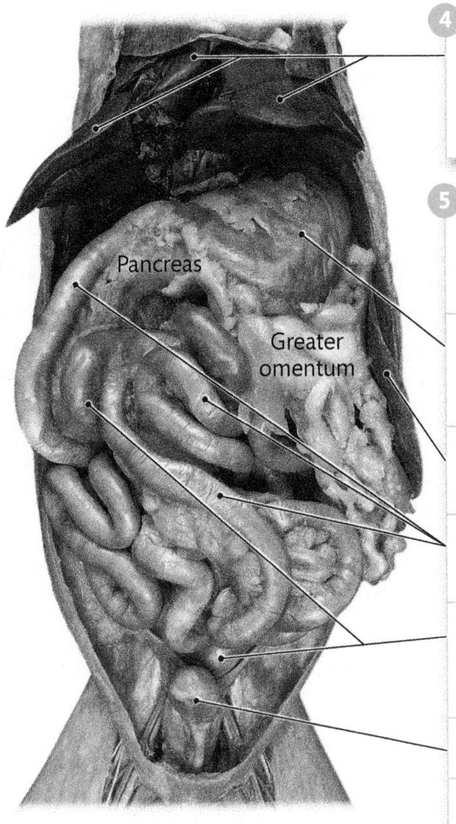

Ventral view

4

Identify the **liver,** which takes up much of the anterior region of the abdominopelvic cavity. Its anterior surface is covered by the diaphragm. Similar to the lungs, the liver is divided into several lobes.

5

The **greater omentum** is a fat-laden peritoneal membrane that covers most of the abdominopelvic organs. Carefully lift this membrane to identify the following structures:

The **stomach** is located in the anterior region of the abdominopelovic cavity on the left side. It is partially covered by the liver.

The **spleen** is a large, flat organ that is adjacent to the lateral aspect of the stomach.

The **small intestine** is a highly convoluted tubular organ that fills the central region of the abdominopelvic cavity.

The **large intestine** forms an inverted U by traveling around the small intestine. It ends as a straight tubular organ, known as the rectum.

The **urinary bladder** is ventral to the rectum.

A bean-shaped **kidney** (not shown) can be found along the dorsal body wall by carefully moving the intestines to one side. Shifting the intestines to the other side will reveal the second kidney.

MAKING CONNECTIONS

For each of the following organs, describe its position in relation to neighboring organs: ▶

Heart ______________________________

Lungs ______________________________

Liver ______________________________

Stomach ______________________________

Spleen ______________________________

Small intestine ______________________________

Large intestine ______________________________

Urinary bladder ______________________________

Activity C4.2

Identifying Organs of the Cat Endocrine System

You will identify only the organs that are traditionally included in the endocrine system. Keep in mind that many other organs, such as the heart, lungs, stomach, small intestine, and kidneys, also possess hormone-producing endocrine cells.

A Endocrine Organs in the Brain

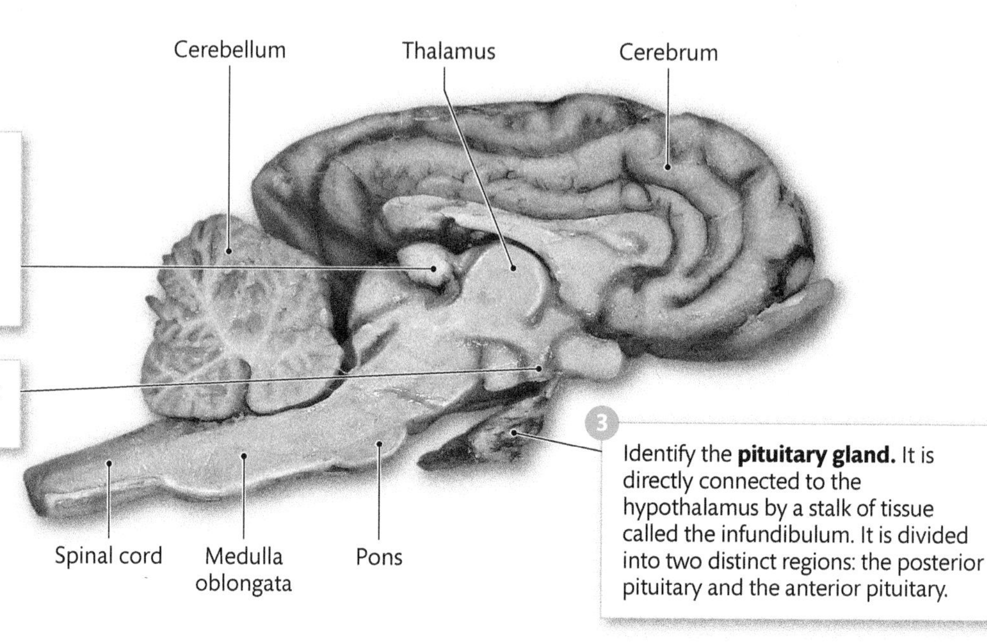

1. On a midsagittal section of a human brain model or dissected mammalian (e.g., cat or sheep) brain, identify the **pineal gland** located along the roof of the third ventricle.

2. The **hypothalamus** is located just inferior to the thalamus.

3. Identify the **pituitary gland.** It is directly connected to the hypothalamus by a stalk of tissue called the infundibulum. It is divided into two distinct regions: the posterior pituitary and the anterior pituitary.

B Endocrine Organs in the Neck and Thoracic Cavity

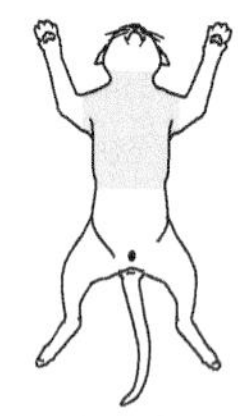

1. Return to your cat dissection. If you have not already done so, transect and reflect the muscles in the neck to provide a clear view of the deeper structures in this area. Locate the **thyroid gland.** It consists of two elongated lobes that are located on each side of the trachea, just posterior to the larynx. An isthmus travels across the ventral surface of the trachea and connects the two lobes of the gland. The isthmus is significantly reduced in older cats and may not be present in your animal.

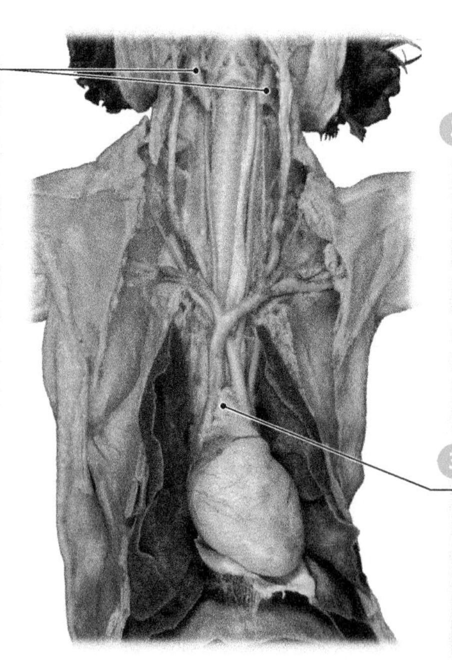

Ventral view

2. The two pairs of **parathyroid glands** (not shown) are embedded along the dorsal surface of each lobe of the thyroid gland. They are quite small and difficult to identify. Carefully remove one of the thyroid gland lobes and observe its dorsal surface with a dissecting microscope. Locate the parathyroid glands, which will appear lighter than the surrounding thyroid tissue.

3. Open the thoracic wall body flaps that you produced earlier to fully expose the thoracic organs. Identify the **thymus gland,** a whitish glandular structure that partially covers the heart and the posterior end of the trachea.

C Endocrine Organs in the Abdominal Cavity

1. Open the abdominal wall body flaps that you produced earlier to fully expose the abdominal organs. Lift the stomach from its anatomical position and identify the **pancreas** along the dorsal body wall between the duodenum (first part of the small intestine) and spleen. The pancreas is a glandular structure that appears lighter in color than the neighboring small intestine.

2. Identify the **adrenal glands,** which are narrow bands of glandular tissue, located anterior and medial to the kidneys. Be careful not to mistake them for adipose (fat) tissue, which appears whiter and has a softer texture than the glands. In the human, the adrenal glands rest on the superior margins of the kidneys. In the cat, however, the adrenal glands and kidneys are separated from each other.

3. If your dissection session is ending here, close the body flaps and wrap your cat with the skin. Cover the animal with moist towels, and return it to the storage bag.

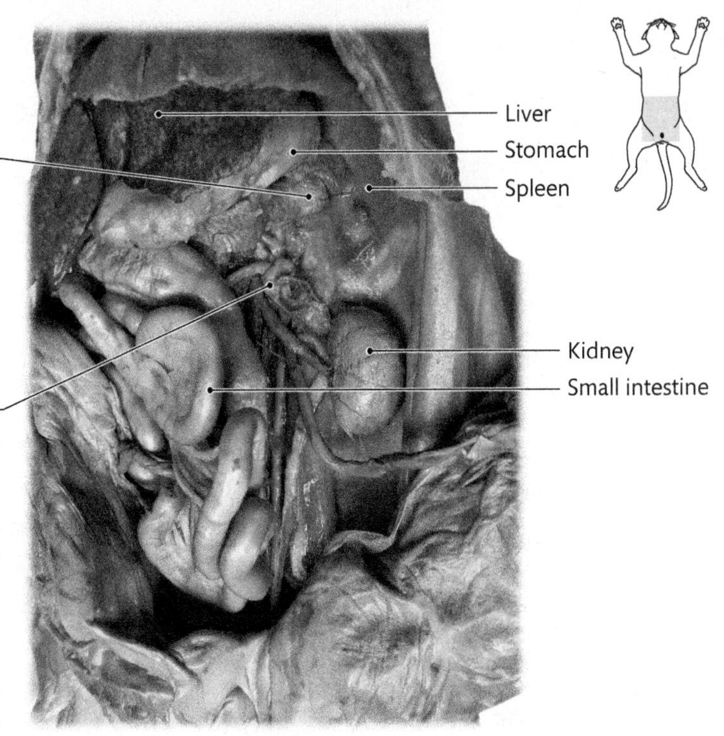

Ventral view

MAKING CONNECTIONS

Refer to your textbook to review the functions of the endocrine organs that you dissected in this activity. Complete the following table. ▶

Endocrine Organ	Hormones Produced	Function
Pineal gland		
Hypothalamus		
Anterior pituitary		
Thyroid gland		
Parathyroid gland		
Thymus gland		
Pancreas		
Adrenal cortex		
Adrenal medulla		

BEFORE YOU MOVE ON . . .

« LOOKING BACK

In this cat dissection exercise, you made general observations of the thoracic and abdominopelvic organs in the cat. The organs that you identified are parts of several organ systems, including the cardiovascular, respiratory, digestive, urinary, and lymphatic systems.

You also completed a more detailed examination of the cat endocrine system, which is similar in structure and function to the human endocrine system. One notable difference is the relative position of the adrenal glands and kidneys. In humans, each adrenal gland sits directly on the superior surface of a kidney. In cats, the adrenal glands and kidneys are separated from each other.

During the procedure to open the thoracic and abdominopelvic cavities (the ventral body cavities), you were instructed on two occasions to use your fingers to feel important anatomical landmarks before making an incision. This type of manual examination is called palpation. Anatomists consider palpation to be a practical and essential dissection tool. Can you explain why?

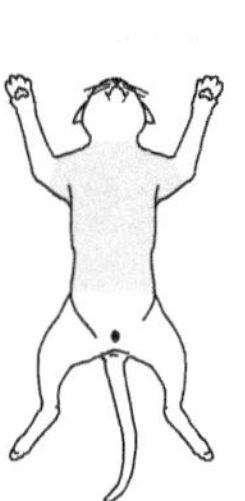

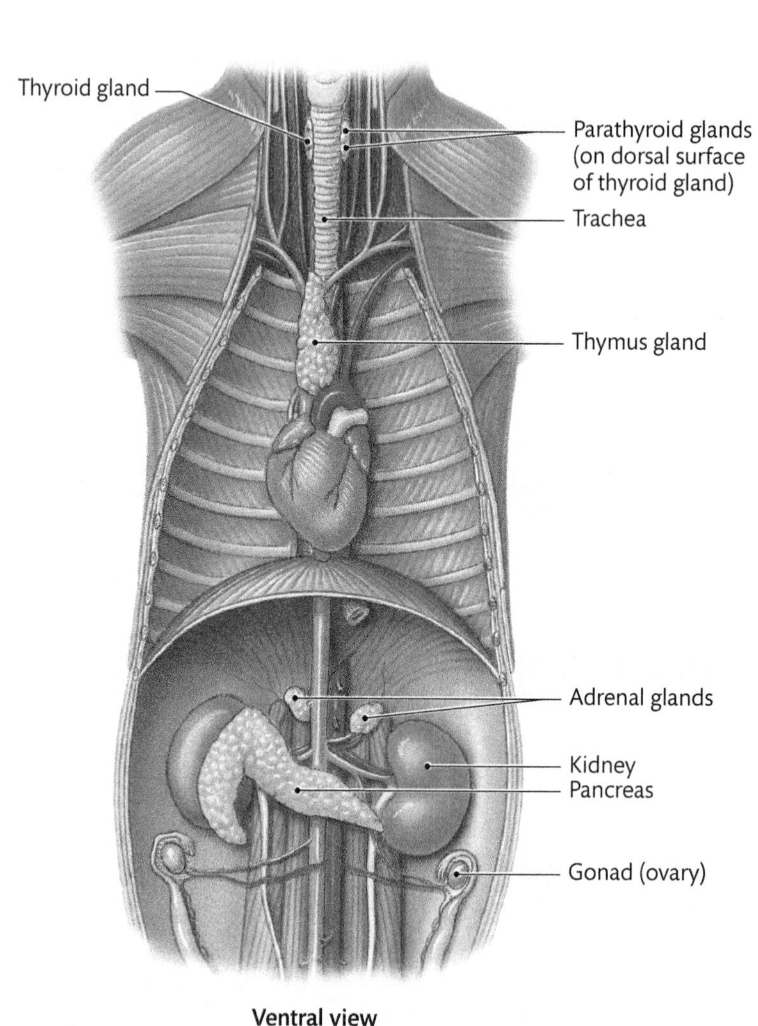

Ventral view

LOOKING FORWARD »

The endocrine system includes all the cells and tissues in the body that produce and secrete hormones. Hormones are chemical messengers that exert specific effects on target cells some distance from their origin. They are able to reach their target destinations via transport in the bloodstream. In the next cat dissection exercise, you will identify the major blood vessels in the cat cardiovascular system.

Name ______________________

Lab Section ______________________

Date ______________________

EXERCISE **C4** **REVIEW SHEET**

Dissection of the Cat Ventral Body Cavities and Endocrine System

QUESTIONS 1–8: Match the cat endocrine organ in column A with the body region where it is located in column B. Answers may be used more than once.

A	B
1. Pancreas ____________	**a.** Thoracic cavity
2. Pituitary gland ____________	**b.** Cranial cavity
3. Thymus gland ____________	**c.** Abdominopelvic cavity
4. Pineal gland ____________	**d.** Neck
5. Thyroid gland ____________	
6. Adrenal glands ____________	
7. Parathyroid glands ____________	
8. Hypothalamus ____________	

QUESTIONS 9–14: Identify the cat endocrine organ that is located close or adjacent to the following structures.

9. Stomach

____

10. Kidneys

____

11. Heart and posterior end of the trachea

____

12. Anterior end of the trachea and larynx

____

13. Third brain ventricle

____

14. Thalamus

____

15. Review the anatomy of the human endocrine system and make comparisons with the cat endocrine system by identifying similarities and differences.

Similarities:

____

____

Differences:

____

____

QUESTIONS 16–33: Identify the labeled structures in the photos. Select your answers from the list below. Some answers may be used more than once.

a. Spleen
b. Urinary bladder
c. Thyroid gland
d. Stomach
e. Liver, right lobes
f. Right lung
g. Large intestine
h. Liver, left lobes
i. Left lung
j. Small intestine
k. Pancreas
l. Thymus gland
m. Greater omentum
n. Heart

16. ____________________
17. ____________________
18. ____________________
19. ____________________
20. ____________________
21. ____________________
22. ____________________
23. ____________________
24. ____________________
25. ____________________
26. ____________________
27. ____________________
28. ____________________
29. ____________________
30. ____________________
31. ____________________
32. ____________________
33. ____________________

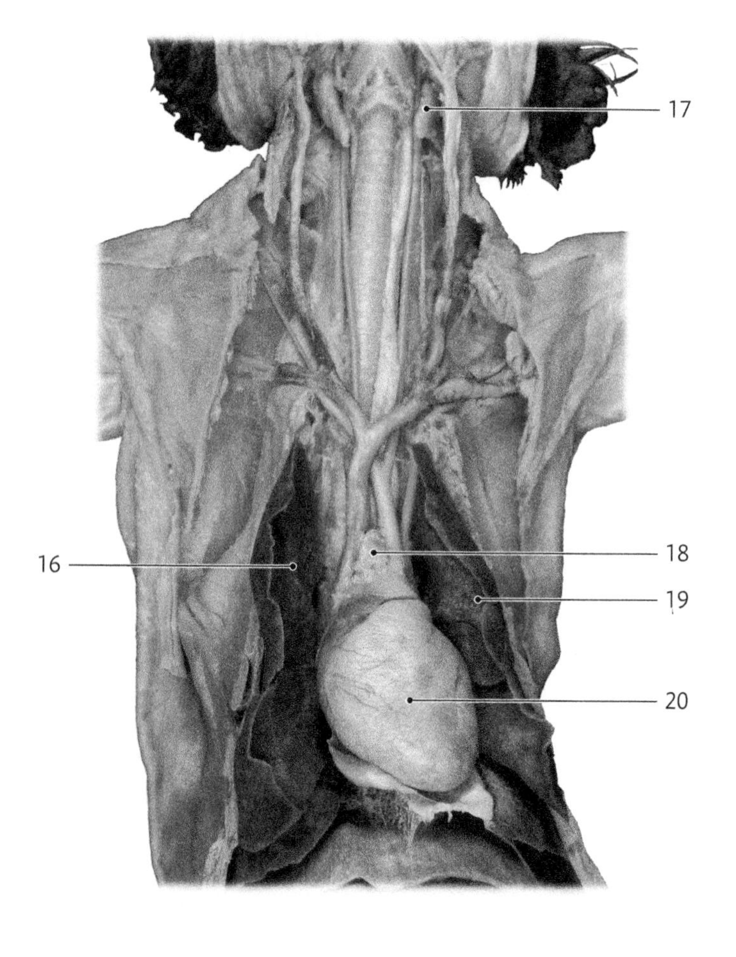

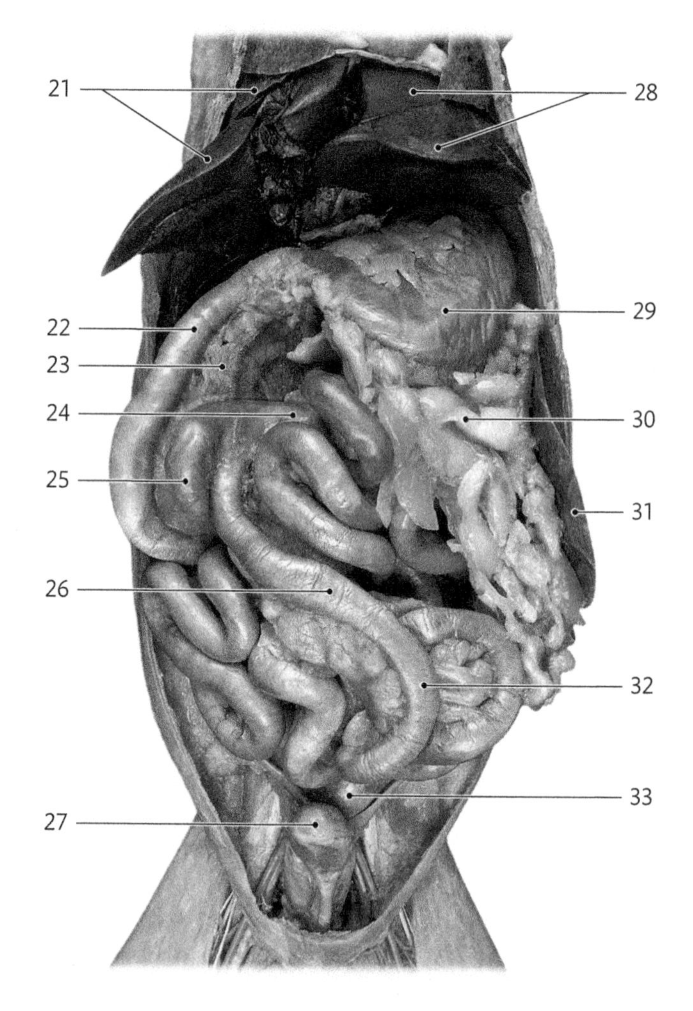

DISSECTION EXERCISE

C5

Dissection of the Cat Cardiovascular System

LEARNING OUTCOMES

These Learning Outcomes correspond by number to the laboratory activities in this exercise. When you complete the activities, you should be able to:

Activity C5.1 **Describe the anatomy of the cat heart.**

Identify major blood vessels in the cat and compare the blood circulatory patterns in the cat and the human.

Activity C5.2 **Identify the arteries and veins in the cat head, neck, and thorax.**

Activity C5.3 **Identify the arteries and veins in the cat forelimb.**

Activity C5.4 **Identify the arteries and veins in the cat abdomen.**

Activity C5.5 **Identify the arteries and veins in the cat pelvis and hindlimb.**

LABORATORY SUPPLIES

- Preserved cats
- Dissecting equipment: trays, tools, gloves
- Protective eyewear
- Face mask
- Anatomical models of the human heart
- Anatomical models of the major blood vessels in the human

PRE-LAB QUIZ

Before you begin, read all the activities in Exercise C5 and the required reading in your textbook that is assigned by your instructor.

1. True or False: The coronary sulcus is a groove that separates the atria from the ventricles. ____________
2. True or False: The tricuspid valve is located at the opening between the left atrium and left ventricle. ____________
3. During this cat dissection exercise, you will be identifying major arteries and veins in the cat cardiovascular system. To lower the risk of damaging important blood vessels, you should avoid using which dissecting tool(s)?
   a. scalpel
   b. forceps
   c. scissors
   d. probe
   e. both (a) and (c)
4. In the cat, which pair of arteries are both direct branches of the aortic arch?
   a. brachiocephalic and left subclavian arteries
   b. left and right common carotid arteries
   c. brachiocephalic and right subclavian arteries
   d. brachiocephalic and right common carotid arteries
   e. left and right subclavian arteries
5. True or False: The anterior vena cava is formed by the union of the left and right brachiocephalic veins. ____________
6. Identify the two terminal branches of the brachial artery. ____________
7. The suprarenal arteries supply blood to what structures? ____________
8. True or False: Veins that drain blood from most abdominal organs empty into the hepatic portal vein. ____________
9. When the external iliac artery enters the thigh, it becomes the ____________ artery.
10. True or False: The anterior and posterior tibial veins unite to form the femoral vein. ____________

Activity **C5.1**

Examining the Structure of the Cat Heart

The cat heart is remarkably similar, anatomically, to the human and other mammalian hearts. The typical mammalian heart consists of four chambers: two atria that receive blood returning to the heart and two ventricles that pump blood away from the heart. The heart pumps blood through two interconnected circulatory pathways:

- The **pulmonary circuit** transports blood to and from the heart and lungs.
- The **systemic circuit** transports blood to and from the heart and all the tissues in the body.

A If you have already dissected a mammalian heart (e.g., sheep or pig) or have human heart models available, use these resources for comparison as you study the external features of the cat heart.

Dissection notes:

- *If you have not already opened the ventral body cavities and made preliminary observations of their contents, these activities must first be completed. Complete Activity C4.1 before you proceed with this dissection.*
- *In this exercise, you will dissect and identify the external structures of the heart and the major arteries and veins of the cat. If your animal has been injected with latex dyes, the arteries will be red, and the veins will be blue.*

1. Place your cat ventral side up in a dissecting tray. Separate the body wall flaps to expose the contents of the thoracic cavity. Secure the flaps by pinning them to the dissecting pan.

2. Remove the pericardium to expose the heart sitting in the mediastinum. Remove any remnants of the thymus gland from the anterior part of the thorax and in the neck by pulling it out with your forceps.

3. Locate the walls of the four heart chambers. The two smaller, anterior chambers are the atria (auricles), which receive blood from the great veins.
   - **Right atrium**
   - **Left atrium**

   The two posterior muscular chambers are the ventricles, which pump blood into the great arteries.
   - **Right ventricle**
   - **Left ventricle**

4. Identify the **interventricular sulcus,** a groove that travels diagonally along the ventral surface of the heart. This groove represents the position of the interventricular septum, a muscular wall that separates the left and right ventricles. Observe a coronary artery and vein traveling along the groove.

5. Identify the **apex of the heart,** the posterior tip of the left ventricle.

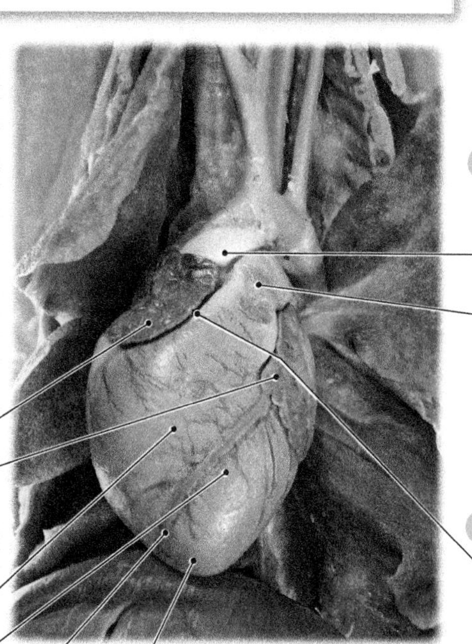

Ventral view

6. Locate the two great arteries:
   - The **aorta** arises from the left ventricle.
   - The **pulmonary trunk** is ventral to the aorta and arises from the right ventricle.
   - The **ligamentum arteriosum** (not shown) is a short solid cord that connects the two great arteries.

7. Identify the **coronary (atrioventricular) sulcus,** a groove that separates the atria from the ventricles.

8. Move the heart slightly to the left and observe the two venae cavae (not shown) as they enter the right atrium:
   - The **anterior vena cava** (**precava;** superior vena cava in humans) returns blood from body regions that are anterior to the diaphragm.
   - The **posterior vena cava** (**postcava;** inferior vena cava in humans) returns blood from body regions that are posterior to the diaphragm.

B Dissecting the internal structures of the cat heart is challenging because of its small size. Dissecting a larger sheep heart will allow you to study the internal anatomy of the heart in greater detail. Your instructor may want you to attempt the internal dissection of the cat heart so that you can compare its structure with that of the human and sheep hearts. If so, have anatomical models of the human heart and dissected sheep hearts available for comparison and proceed with the following dissection.

1. Place your cat ventral side up in a dissecting tray. Within the thoracic cavity, identify the heart's ventral surface.

2. With a scalpel, make a longitudinal incision of the heart, beginning at the apex and cutting anteriorly toward the great vessels. Your plane of section should run parallel to the heart's ventral surface. Stop your incision just before the great vessels. Reflect the ventral wall of the heart to expose the internal structures.

3. Identify the four heart chambers:
   - **Left atrium**
   - **Right atrium**
   - **Right ventricle**
   - **Left ventricle**

4. Locate the **interatrial septum** (not shown) that separates the two atria. Identify the **fossa ovalis,** an oval-shaped depression inside the right atrium, along the interatrial septum.

5. Identify the **interventricular septum,** which separates the two ventricles.

6. Notice the relative thickness of the chamber walls: The atrial walls are much thinner than the ventricular walls, and the left ventricular wall is thicker than the right ventricular wall.

7. Identify the heart valves located at the junctions of the ventricles and the great arteries. They are called **semilunar valves** because each contains three crescent-shaped cusps.

   The **pulmonary semilunar valve** is positioned between the right ventricle and pulmonary trunk.

   The **aortic semilunar valve** is located between the left ventricle and aorta.

8. Locate the **atrioventricular (AV) valves** guarding the openings between atria and ventricles on each side of the heart.

   The **bicuspid (mitral) valve** contains two cusps. It is located at the opening between the left atrium and left ventricle.

   The **tricuspid valve** contains three cusps. It is located at the opening connecting the right atrium and right ventricle.

9. Identify the **chordae tendineae,** slender cords that attach to the free margins of the AV valve cusps.

10. Follow the chordae tendineae to the ventricular wall where they are connected to the **papillary muscles.**

    The arrangement of the chordae tendineae and papillary muscles prevents the cusps from everting into the atria during ventricular contraction, thus preventing the backflow of blood from the ventricles to the atria.

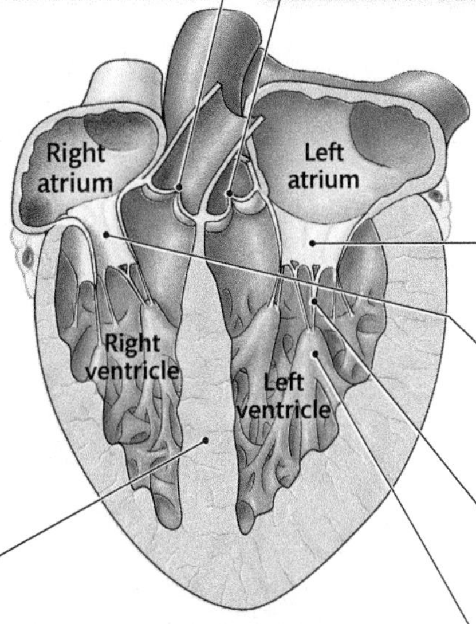

Ventral view

MAKING CONNECTIONS

In the fetal circulation, the ligamentum arteriosum and the fossa ovalis are passageways through which blood flow bypasses the pulmonary circuit. Suggest a reason why it is important for both of these circulatory bypasses to close after birth.

▶ ______________________________

____

____

Activity **C5.2**

Identifying Arteries and Veins in the Cat Head, Neck, and Thorax

A The blood supply to the head, neck, and thorax originates from branches of the aortic arch and thoracic division of the descending aorta.

Dissection note: *During the remaining activities in this exercise, you will be instructed to dissect and identify various arteries and veins in the circulatory system. Unless specifically instructed, do not use scissors or a scalpel for these dissections. You will be less likely to destroy important structures if you use forceps and probes rather than sharp cutting instruments. As you dissect the arteries and veins in the cat, have anatomical models of the major blood vessels in the human available for comparison.*

1. Identify the curved portion of the aorta, known as the **aortic arch.** Coming off the aortic arch are two large arteries that extend cranially.

   The larger first branch is the **brachiocephalic (innominate) artery.**

   The second branch is the **left subclavian artery.**

2. Identify the major branches of the brachiocephalic artery:

   The **left and right common carotid arteries** travel cranially and supply blood to structures in the neck and head.

   The **right subclavian artery** travels to the right forelimb. In some cats, this blood vessel branches off the right common carotid artery.

3. Follow the subclavian artery toward the forelimb. When the blood vessel travels past the first rib, it becomes the **axillary artery.** Identify branches of the axillary artery traveling to the pectoral and posterior shoulder muscles and latissimus dorsi.

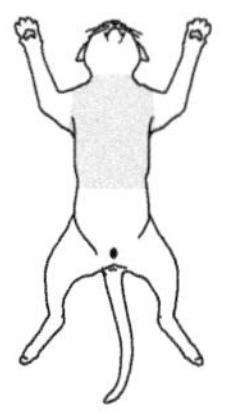

Left subclavian artery

Brachiocephalic artery

Aortic arch

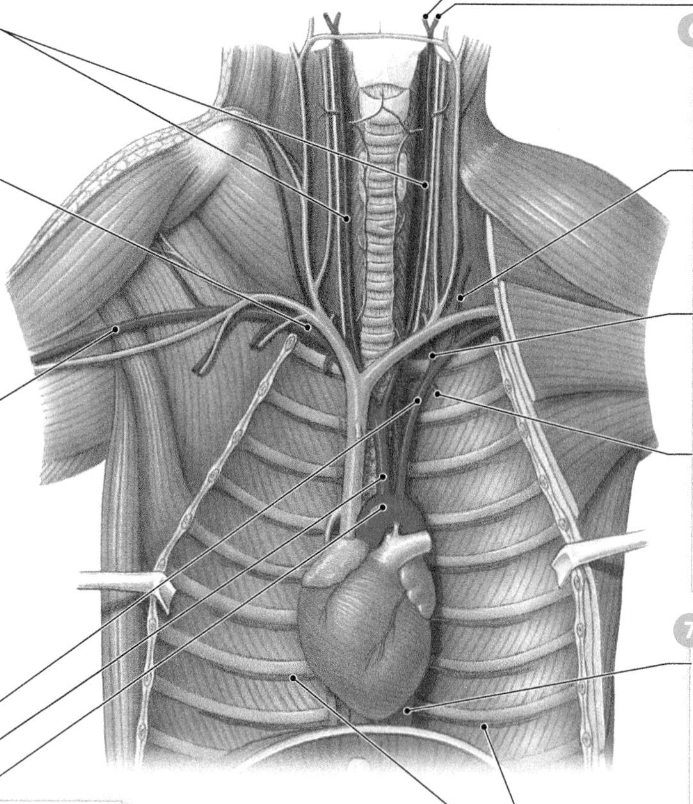

4. Trace the paths of the two common carotid arteries as they travel along either side of the trachea. Dissect one common carotid artery and identify branches that supply the trachea, larynx, muscles in the neck, and occipital region of the scalp.

5. Dissect the common carotid artery cranially to the thyroid cartilage and identify its two terminal branches:

   The **internal carotid artery** supplies blood to the brain.

   The **external carotid artery** gives off branches to the neck, face, and scalp.

6. Return to one of the subclavian arteries and identify the following branches:

   The **thyrocervical (costocervical) trunk** travels cranially to supply the neck and shoulders.

   The **vertebral artery** travels through the transverse foramina of the cervical vertebrae to supply the brain.

   The **internal mammary (internal thoracic) artery** travels just lateral to the sternum to supply structures on the ventral body wall.

7. Return to the aortic arch. Pull the thoracic organs to the right side and follow the aortic arch as it descends into the thoracic cavity. This portion of the aorta is known as the **thoracic division of the descending aorta.**

Ventral view

8. Identify the paired **intercostal arteries** that branch off the descending aorta and travel within the intercostal spaces.

9. Identify **bronchial arteries** that supply the lungs, **esophageal arteries** that supply the esophagus, and **phrenic arteries** that supply the diaphragm (not shown). These branches also originate from the descending aorta in the thorax.

B The veins that drain the head, neck, and thorax empty into the anterior vena cava.

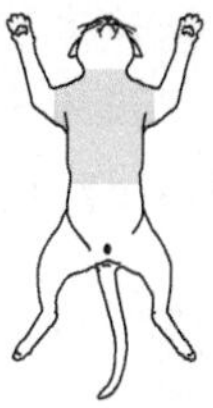

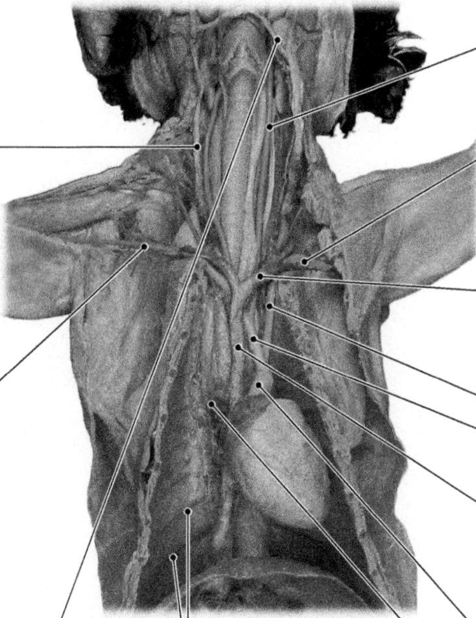

1. In the neck, on each side of the trachea, identify two large veins traveling parallel to each other.

   The more lateral vessel is the **external jugular vein.** It drains superficial structures in the head and neck.

   The medial vessel is the **internal jugular vein** (not shown), which drains the brain.

2. On one side of the body, follow the external and internal jugular veins caudally. Notice that the internal jugular vein empties into the external jugular vein.

3. At the level of the larynx, identify the **transverse jugular vein,** which crosses the ventral surface of the neck and connects the two external jugular veins.

4. Identify the **subclavian vein** passing over the first rib and notice that it joins the external jugular vein.

5. Locate the **brachiocephalic vein.** It is formed by the union of the subclavian and external jugular veins.

6. Identify the **anterior vena cava (precava)** and notice that it is formed by the union of the left and right brachiocephalic veins.

7. Reflect the right lung and heart away from each other to expose the dorsal thoracic wall. Traveling along this wall to the right of the midline is the **azygos vein.** Follow this vein to where it joins the anterior vena cava. In humans, a hemiazygos vein is present on the left side of the thoracic cavity in addition to the azygos vein. The cat does not have a hemiazygos vein.

8. Identify several pairs of **intercostal veins** traveling in the intercostal spaces. Notice how the intercostal veins drain into the azygos vein.

MAKING CONNECTIONS

Explain how the arterial branches emerging from the aortic arch differ in the cat and the human. Is there any significant difference between the two in the regions that are supplied by these arterial branches?

▶ ____

____

____

____

____

Activity **C5.3**

Identifying Arteries and Veins in the Cat Forelimb

A The blood supply to the forelimbs originates from the subclavian arteries.

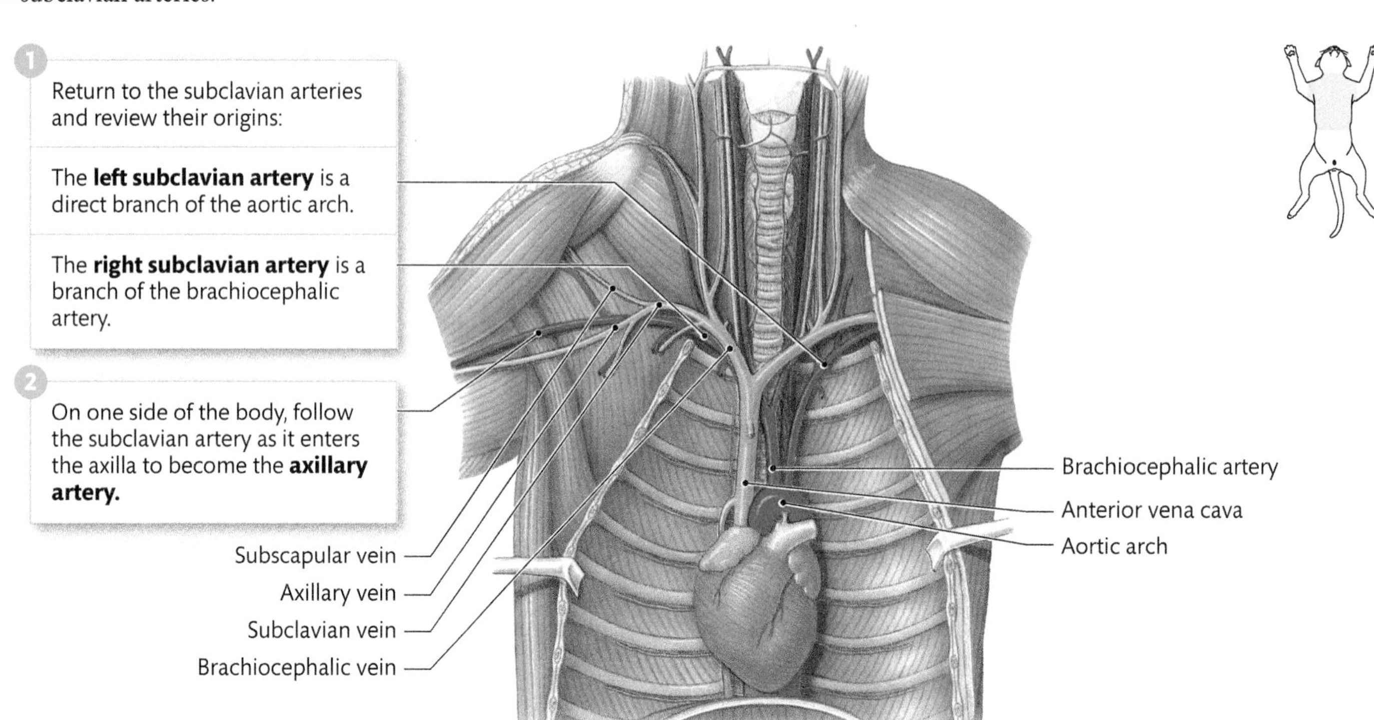

Ventral view

3 Follow the axillary artery into the upper forelimb (the brachium) where it becomes the **brachial artery.**

4 Identify the **deep brachial artery** (not shown), a branch of the brachial artery that passes deep to the biceps brachii muscle.

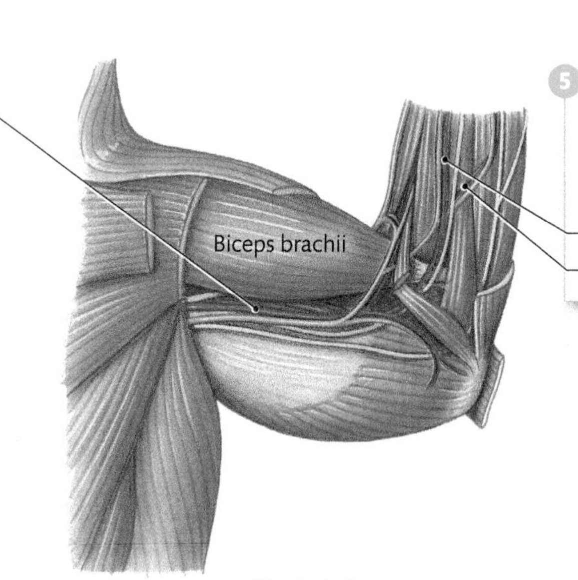

Ventral view

5 Trace the brachial artery to the elbow and into the lower forelimb. Identify the two terminal branches of the brachial artery:

- The lateral **radial artery**
- The medial **ulnar artery**

B The veins that drain the forelimb empty into the anterior vena cava by way of the subclavian veins.

1 Return to where the subclavian and external jugular veins come together. Follow the subclavian vein laterally as it travels with the subclavian artery. At the first rib, identify the two veins that converge to form the subclavian vein: the **subscapular vein** and the **axillary vein** (see the top figure on the previous page). The subscapular vein receives branches that drain the muscles at the shoulder. The axillary vein continues laterally toward the forelimb.

2 When the axillary vein enters the upper forelimb (the brachium), it becomes the **brachial vein.** Find the brachial vein traveling with the brachial artery.

3 Locate the **deep brachial vein** (not shown) traveling with the deep brachial artery, deep to the biceps brachii muscle.

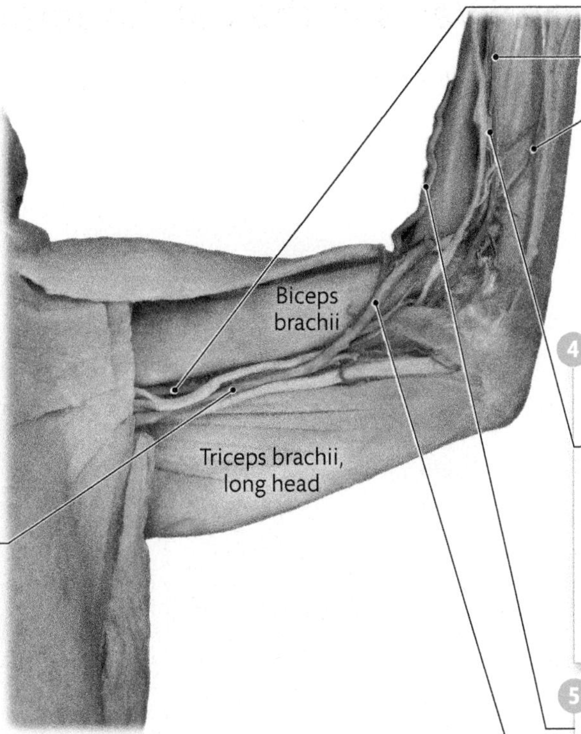

Brachial artery
Radial artery
Ulnar artery
Biceps brachii
Triceps brachii, long head

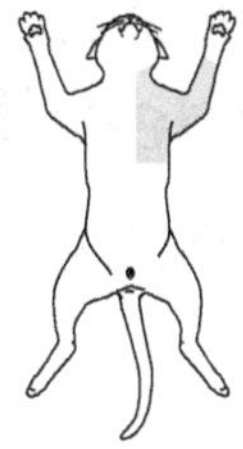

4 Identify two veins that drain blood from the paw and lower forelimb:

The lateral **radial vein**

The medial **ulnar vein** (not shown)

These veins travel with their respective arteries. Follow them proximally to the elbow where they converge to form the brachial vein.

5 Identify the **cephalic vein,** a large superficial vein traveling along the lateral surface of the forelimb.

6 Locate the **median cubital vein,** which connects the cephalic and brachial veins.

Ventral view

MAKING CONNECTIONS

List variations, if any, that you found in the branching patterns of arteries and veins in the cat forelimb. Share your results with other students in your laboratory. Do any variations appear more than once?

▶ ____

____

____

____

____

____

Activity C5.4

Identifying Arteries and Veins in the Cat Abdomen

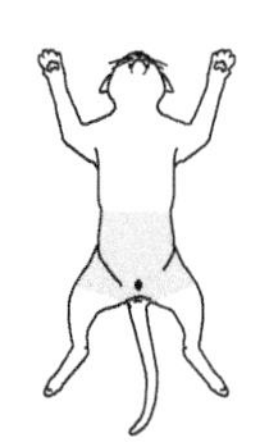

A Reflect the abdominal organs to the right side to get a clear view of the descending aorta as it travels through the abdominal cavity. This portion of the aorta, known as the **abdominal division of the descending aorta (abdominal aorta),** gives off branches that supply abdominal organs.

1. Locate the **celiac trunk (artery).** It is a single midline branch located just caudal to the diaphragm. Branches of the celiac trunk supply the liver, gallbladder, stomach, duodenum, pancreas, and spleen.

2. Find the **superior (anterior) mesenteric artery.** It is another single midline branch that arises from the aorta just caudal to the celiac trunk. It supplies the small intestine and portions of the large intestine.

3. Identify the paired **suprarenal (adrenolumbar) arteries,** which supply the adrenal glands. Branches from these arteries supply the posterior body wall, and occasionally a branch (inferior phrenic artery) travels to the diaphragm.

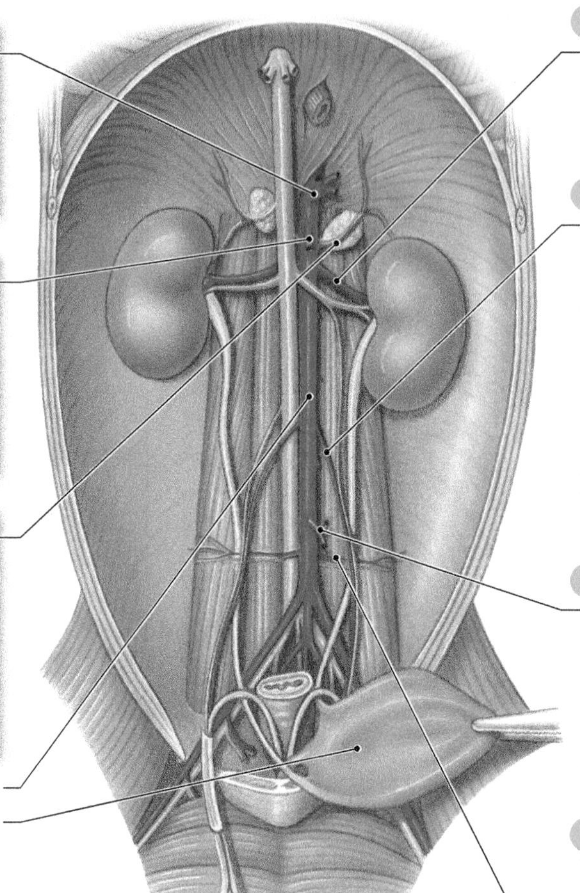

Ventral view

4. Identify the paired **renal arteries,** which supply the kidneys. In some cats, these arteries send branches to the adrenal glands.

5. Locate the paired **gonadal arteries.** If your cat is a male, these blood vessels are called **spermatic (testicular) arteries;** if your cat is a female, they are called **ovarian arteries.** The spermatic arteries travel through the inguinal canal, within the spermatic cord, to supply the scrotum and testes; the ovarian arteries travel laterally along the dorsal abdominal wall to supply the ovaries and uterus.

6. Find a single midline branch, the **inferior (posterior) mesenteric artery** as it arises from the aorta slightly caudal to the gonadal arteries. This artery supplies the last part of the large intestine, including the rectum.

7. Along both sides of the abdominal aorta, locate several pairs of small **lumbar arteries** (not shown), which supply the dorsal muscular wall.

   Identify the paired **iliolumbar arteries,** which also supply the dorsal body wall. These arteries are larger than the lumbar arteries.

Word Origins

Mesenteric refers to the mesentery, a double-layered extension of the peritoneum that supports the intestines. Branches of the superior and inferior mesenteric arteries travel through the mesenteries to reach the organs they supply: the small and large intestines. Identify a mesentery and observe the numerous arterial branches traveling through it.

B Reflect the abdominal organs to the left side of the abdominal cavity and observe the **posterior vena cava (postcava)** traveling along the middorsal body wall. Veins that drain abdominal organs empty into the posterior vena cava.

1. Locate the paired **hepatic veins,** which drain blood from the hepatic portal system in the liver (see below). Gently scrape away some of the liver to expose these veins and observe their connections to the posterior vena cava just caudal to the diaphragm.

2. Identify the paired **suprarenal (adrenolumbar) veins,** which drain the adrenal glands. There is considerable variation in the drainage pattern of these veins. Typically, the right vein drains directly into the vena cava, and the left vein empties into the renal vein.

3. From the medial side of each kidney, identify the **renal vein** traveling alongside the renal artery. Follow the vein to the vena cava. In some cases, two renal veins drain each kidney. Be aware of this variation.

4. The paired gonadal veins (**ovarian** or **spermatic veins**) drain the ovaries or testes. The best way to locate them is to start at the ovaries or testes and trace them toward the vena cava. Typically, the left gonadal vein drains into the renal vein, and the right gonadal vein drains directly into the vena cava.

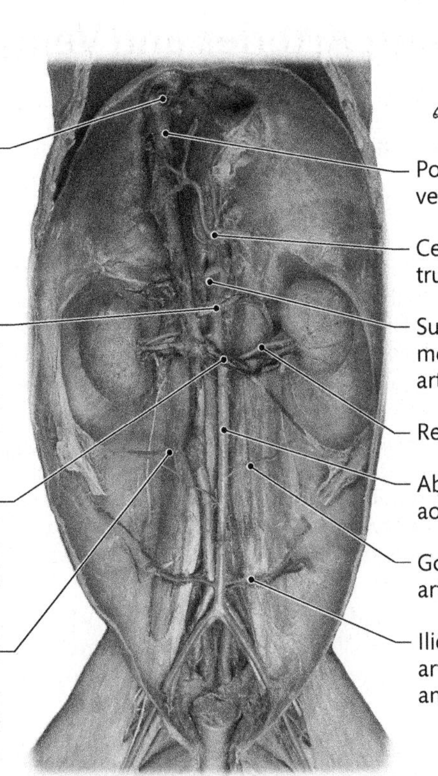

Ventral view

C Veins that drain most of the abdominal organs empty into the **hepatic portal vein,** the main vein of the **hepatic portal system.** The hepatic portal vein delivers blood to liver sinusoids (capillaries). After passing through the sinusoids, the blood drains into the hepatic veins, which lead to the posterior vena cava.

1. Identify the **hepatic portal vein** where it enters the liver.

2. Identify the two main veins that empty into the hepatic portal vein:

   The **gastrosplenic vein** drains most of the stomach and spleen.

   The **anterior (superior) mesenteric vein** drains the small intestine, large intestine, and pancreas.

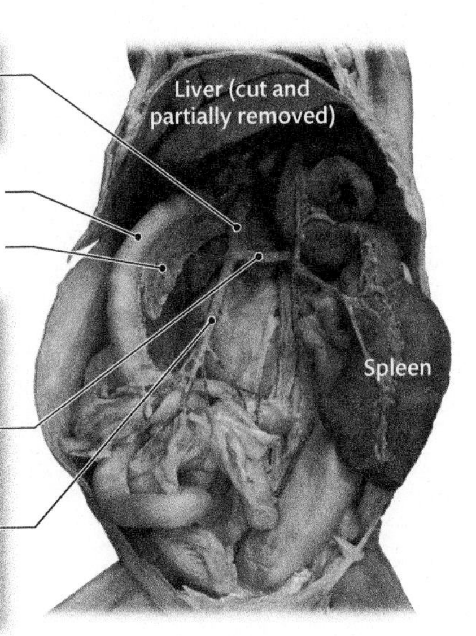

Ventral view

MAKING CONNECTIONS

List variations, if any, that you found in the branching patterns of arteries and veins in the abdomen of the cat. Share your results with other students in your laboratory. Do any variations appear more than once?

▶ ____________________

____

____

____

____

____

____

____

____

____

____

Activity **C5.5**

Identifying Arteries and Veins in the Cat Pelvis and Hindlimb

A The **external iliac** and **internal iliac arteries** arise from the caudal end of the abdominal aorta. They supply blood to the pelvis and hindlimb.

1. Caudal to the inferior mesenteric artery, identify the two relatively large **external iliac arteries.** These blood vessels deliver blood to the hindlimb.

2. Find the paired **internal iliac arteries.** They arise just caudal to the external iliac arteries. The internal iliac arteries supply blood to the gluteal muscles and organs in the pelvic cavity.

3. Locate the **caudal (median sacral) artery,** which arises from the end of the abdominal aorta and travels into the tail.

4. Follow the external iliac artery from its origin at the abdominal aorta. Before entering the thigh, this artery gives off the **deep femoral artery,** which supplies the urinary bladder, ventral abdominal wall, and medial thigh muscles.

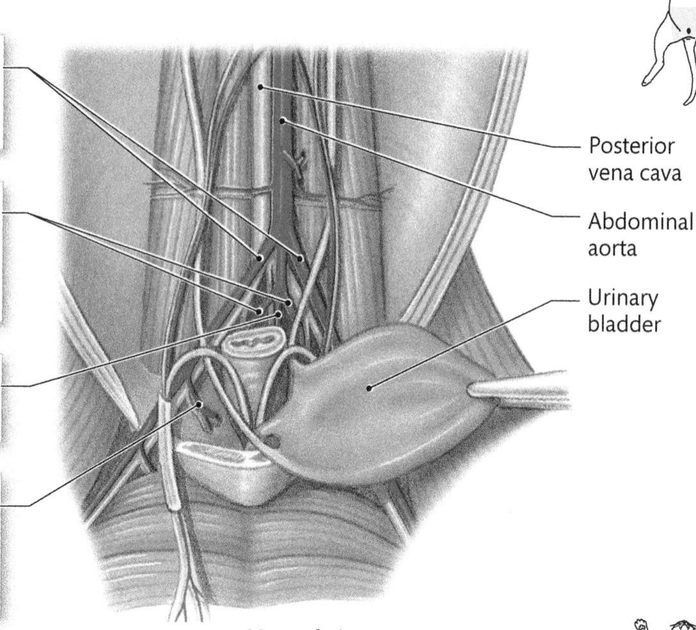

Ventral view

5. Follow the external iliac artery into the thigh, where it becomes the **femoral artery.** Locate the femoral artery and attempt to identify two of its branches:

   The **saphenous artery** arises from the femoral artery just proximal to the knee. Follow it as it travels to the medial aspect of the lower hindlimb.

   The **popliteal artery** is a continuation of the femoral artery as it travels into the popliteal fossa, posterior to the knee.

6. Trace the path of the popliteal artery until it divides to form the

   Anterior tibial artery

   Posterior tibial artery

   The two tibial arteries supply blood to the lower hindlimb and the foot.

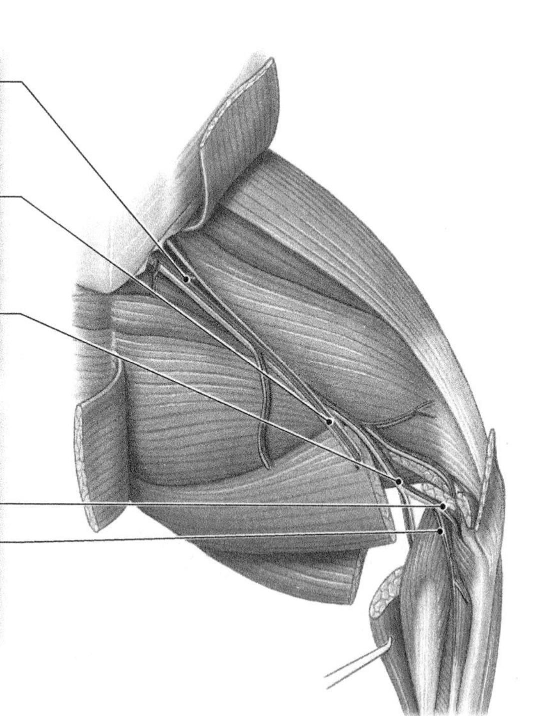

Ventral view

B Veins that drain the pelvis and hindlimb empty into the common iliac vein.

1 Identify the external and internal iliac veins.

The **external iliac vein** is the larger, more lateral vein that drains blood from the hindlimb.

The smaller **internal iliac vein** (not shown) drains structures in the pelvis and the gluteal region. Identify this vessel traveling posteriorly into the pelvic cavity.

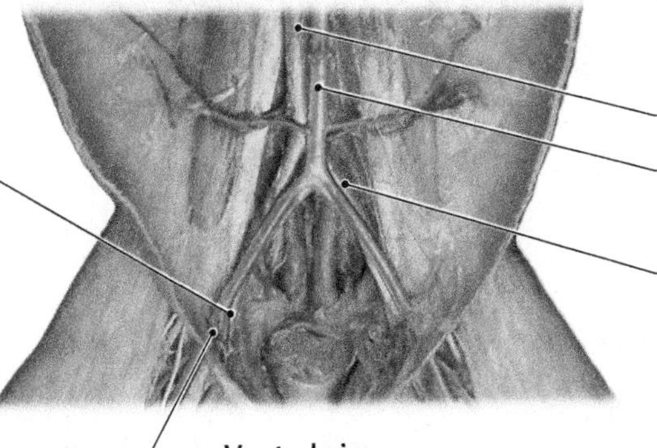

Ventral view

2 On one side, trace the internal and external iliac veins cranially until they merge to form the **common iliac vein.**

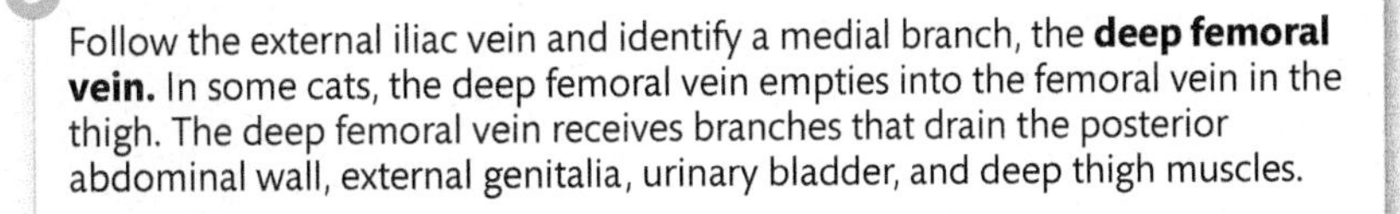

3 Follow the external iliac vein and identify a medial branch, the **deep femoral vein.** In some cats, the deep femoral vein empties into the femoral vein in the thigh. The deep femoral vein receives branches that drain the posterior abdominal wall, external genitalia, urinary bladder, and deep thigh muscles.

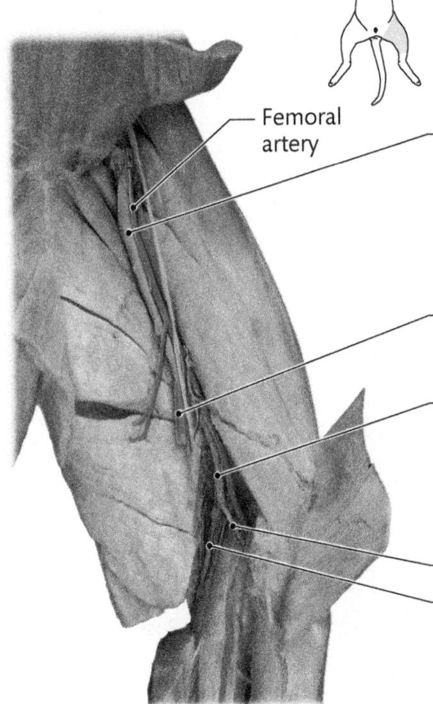

Ventral view

4 When the external iliac vein enters the thigh, it becomes the **femoral vein.** Identify this vein traveling with the femoral artery.

5 Locate the two branches that converge to form the femoral vein:

The **saphenous vein** is a superficial vein that travels along the medial aspect of the thigh and lower hindlimb. It accompanies the saphenous artery and nerve.

The **popliteal vein** travels posterior to the knee, in the popliteal fossa. Locate this vein traveling with its corresponding artery.

6 In the lower hindlimb, identify two veins that converge to form the popliteal vein:

Anterior tibial vein

Posterior tibial vein

The two tibial veins drain blood from the lower hindlimb and foot.

7 If your dissection session is ending here, close the body flaps and wrap your cat with the skin. Cover the animal with moist towels, and return it to the storage bag.

MAKING CONNECTIONS

Explain how the branching pattern of the iliac arteries differs between the cat and the human. Is there any significant difference between the cat and the human in the regions that are supplied by these arterial branches?

▶ ____

____

____

____

BEFORE YOU MOVE ON ...

« LOOKING BACK

In this cat dissection exercise, you examined the structure of the heart and identified major blood vessels in the cat cardiovascular system. The structure of the cat heart is similar to the human and other mammalian hearts. Each side of the heart has two chambers: an atrium that receives blood entering the heart and a ventricle that pumps blood leaving the heart. The right ventricle pumps blood into the pulmonary circuit, and the left ventricle pumps blood into the systemic circuit.

The typical branching patterns of blood vessels in the cat are similar to the human. There are, however, differences between the cat and human in the arterial branches coming off the aortic arch and the branching pattern of the iliac arteries. You observed these variations and were asked to describe the changes.

It is very likely that during this dissection you observed modifications in the typical circulatory patterns of the major blood vessels. For example, you might have noticed variations in blood drainage from abdominal organs or a disparity in the characteristic branching of arteries and veins in the forelimb.

Would you expect to find more variation in the distribution pattern of arteries or the distribution pattern of veins? Explain.

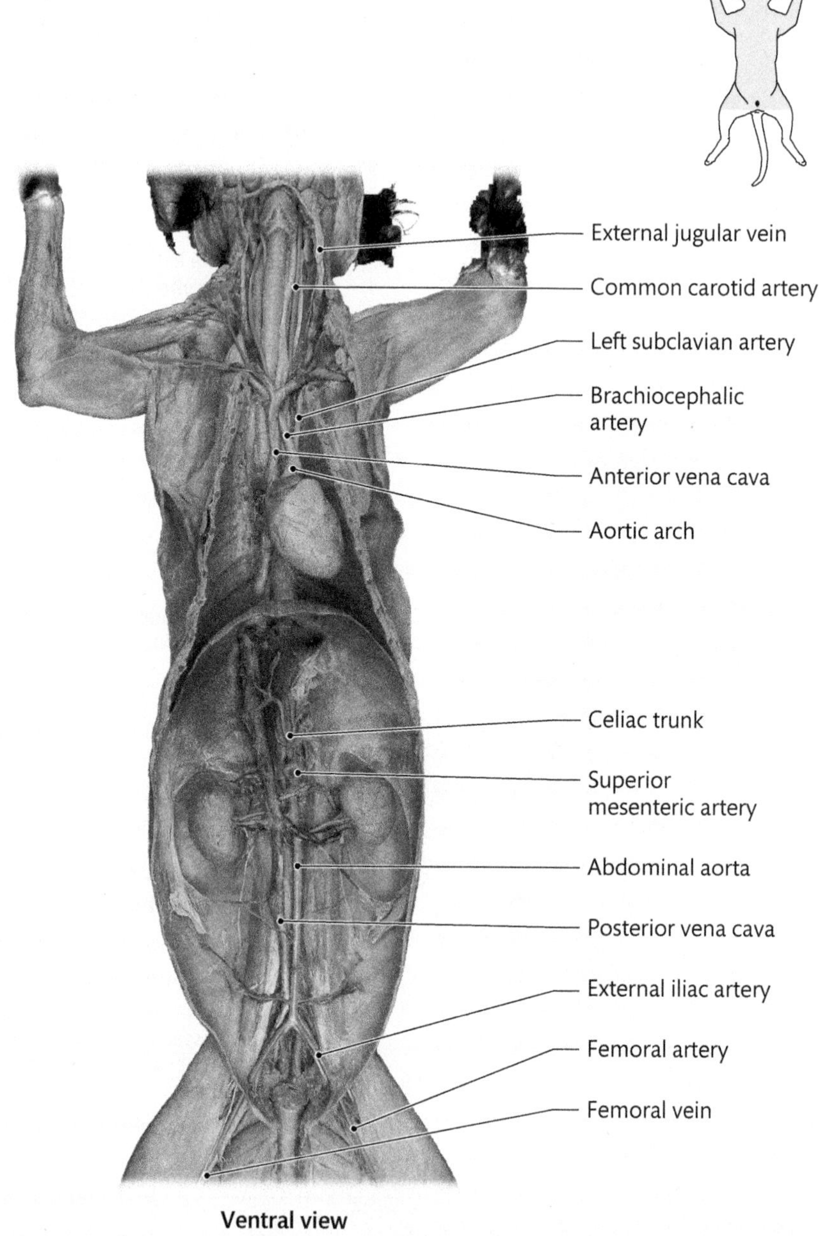

Ventral view

LOOKING FORWARD »

Similar to humans and other mammals, the cat has two circulatory systems: one for blood and one for lymph. In the next cat dissection exercise, you will identify the major structures of the lymphatic system, including the **thoracic duct,** a major lymphatic vessel that transports lymph.

Name ______________________

Lab Section ______________________

Date ______________________

EXERCISE **C5** **REVIEW SHEET**

Dissection of the Cat Cardiovascular System

QUESTIONS 1–3: Identify the heart chamber(s) where the following structures are located.

1. Papillary muscles

____

2. Fossa ovalis

____

3. Apex of the heart

____

QUESTIONS 4–10: Match the cat arteries in column A with the correct branch arteries in column B.

A	B
4. Arch of the aorta ______	**a.** Superior mesenteric artery
5. Descending aorta in abdomen ______	**b.** Radial artery
6. Common carotid artery ______	**c.** Brachiocephalic artery
7. Subclavian artery ______	**d.** Internal mammary artery
8. External iliac artery ______	**e.** Saphenous artery
9. Brachial artery ______	**f.** Deep femoral artery
10. Femoral artery ______	**g.** External carotid artery

QUESTIONS 11–18: Identify the body region or structure that is supplied by the following arteries.

11. Internal carotid arteries

____

12. Brachial arteries

____

13. Femoral arteries

____

14. Internal iliac arteries

____

15. Intercostal arteries

____

16. Celiac trunk

____

17. Superior mesenteric artery

____

18. Gonadal arteries

____

QUESTIONS 19–24: Identify the vein in the cat that is formed when the following pairs of veins converge.

19. External jugular vein and subclavian vein

____

20. Left and right brachiocephalic (innominate) veins

____

21. Radial vein and ulnar vein

____

22. Left and right common iliac veins

____

23. Gastrosplenic vein and anterior mesenteric vein

____

24. Popliteal vein and saphenous vein

____

QUESTIONS 25–30: Identify the labeled blood vessels in the photo. Select your answers from the list below.

a. Median cubital vein
b. Brachial artery
c. Radial artery
d. Brachial vein
e. Ulnar artery
f. Cephalic vein

25. ______________

26. ______________

27. ______________

28. ______________

29. ______________

30. ______________

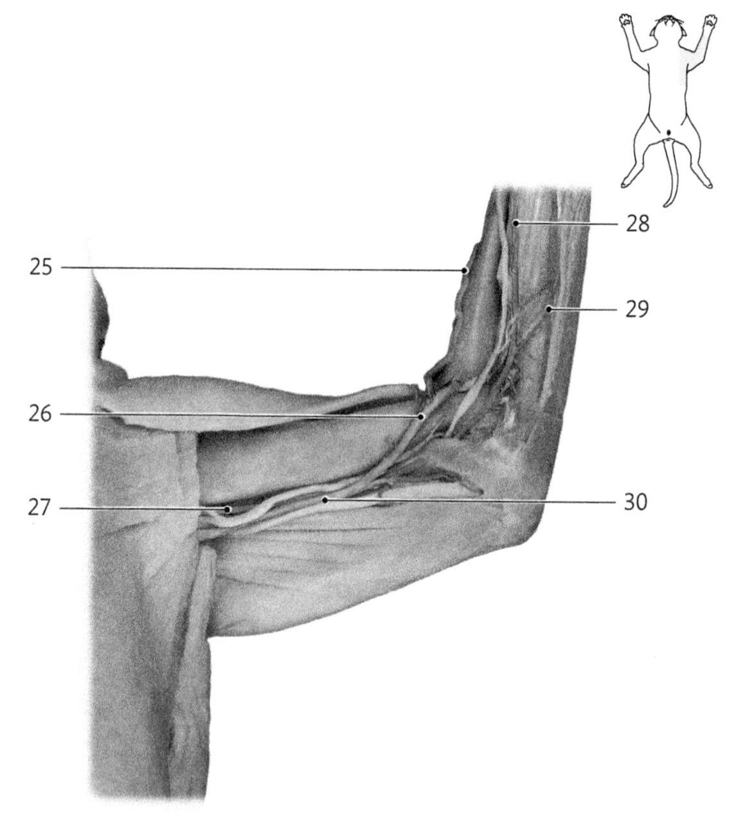

Ventral view

QUESTIONS 31–40: Identify the labeled blood vessels in the photo. Select your answers from the list below.

a. Axillary artery
b. Transverse jugular vein
c. Common carotid artery
d. Brachiocephalic artery
e. Posterior (inferior) vena cava
f. Aortic arch
g. External jugular vein
h. Left subclavian artery
i. Brachiocephalic vein
j. Azygos vein

31. ______________________

32. ______________________

33. ______________________

34. ______________________

35. ______________________

36. ______________________

37. ______________________

38. ______________________

39. ______________________

40. ______________________

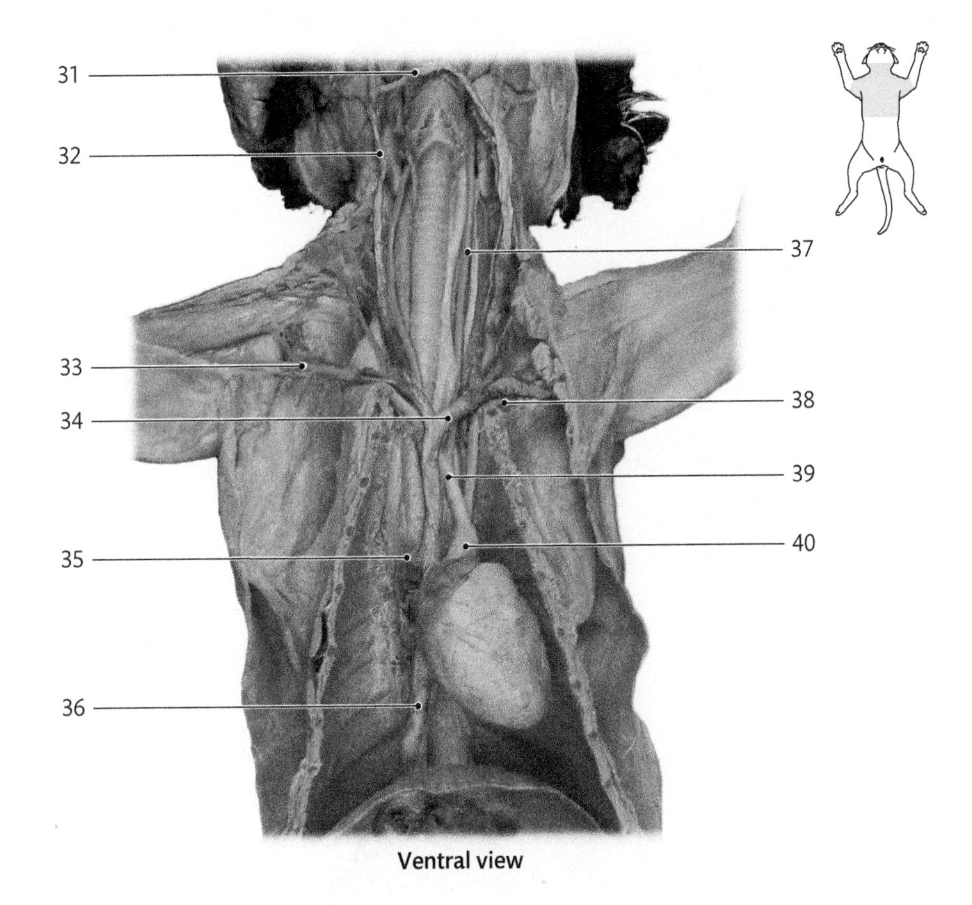

Ventral view

QUESTIONS 41–50: Identify the labeled blood vessels in the photo. Select your answers from the list below.

a. Superior mesenteric artery
b. Gonadal artery and vein
c. External iliac artery
d. Common iliac vein
e. Posterior vena cava
f. Renal artery and vein
g. Iliolumbar artery and vein
h. Femoral artery
i. Abdominal aorta
j. Celiac trunk

41. ______________________
42. ______________________
43. ______________________
44. ______________________
45. ______________________
46. ______________________
47. ______________________
48. ______________________
49. ______________________
50. ______________________

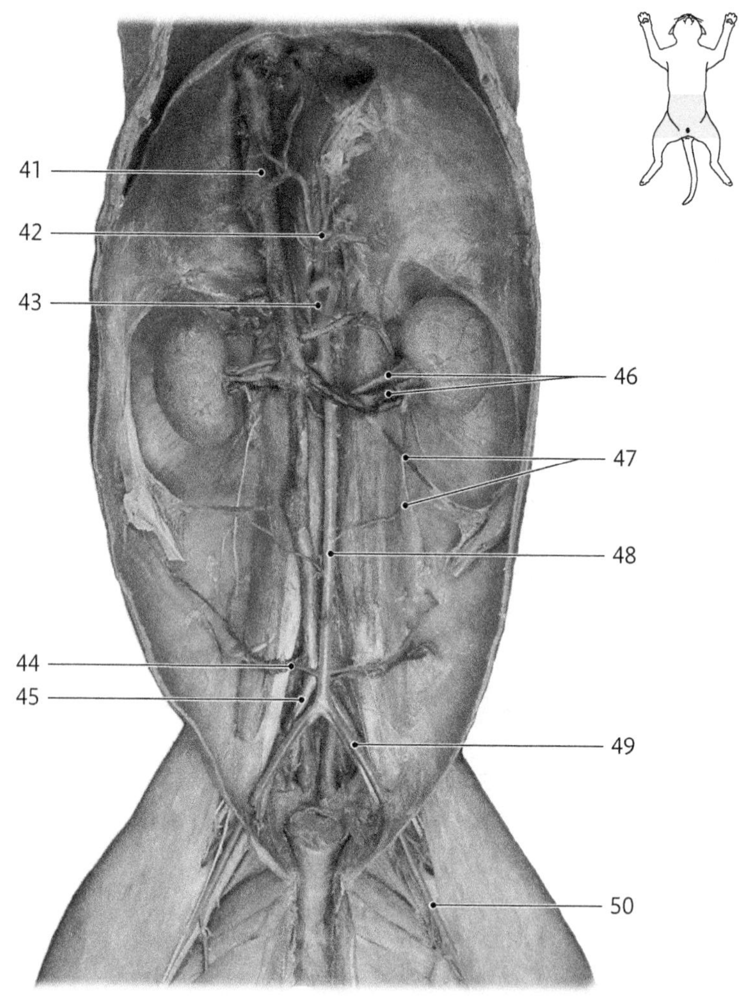

Ventral view

DISSECTION EXERCISE

C6

Dissection of the Cat Lymphatic System

LEARNING OUTCOMES

These Learning Outcomes correspond by number to the laboratory activities in this exercise. When you complete the activities, you should be able to:

Activity C6.1 **Identify the cat thymus gland and spleen and describe the functions of each.**

Activity C6.2 **Identify the cat mandibular lymph nodes and thoracic duct and describe the functions of each.**

LABORATORY SUPPLIES

- Preserved cats
- Dissecting equipment: trays, tools, gloves
- Protective eyewear
- Face mask

PRE-LAB QUIZ

Before you begin, read all the activities in Exercise C6 and the required reading in your textbook that is assigned by your instructor.

1. During this cat dissection exercise, you will identify all the following structures *except* the
   a. mandibular lymph nodes.
   b. spleen.
   c. tonsils.
   d. thymus.
   e. thoracic duct.
2. True or False: The gastrosplenic ligament connects the spleen to the stomach.
   ____
3. True or False: During your cat dissection, the thymus is identified, but it is very small and consists mostly of fat tissue. Based on this observation, you conclude that your cat is young. ______________
4. True or False: Unlike humans, cats do not have an appendix. ______________
5. True or False: Proportionally, the spleen is much larger in the human than in the cat.
   ____
6. The mandibular lymph node can be found along the posterior margin of what muscle?
   ____

Questions 7–10: Match the structure in column A with the correct description in column B.

A

7. Cisterna chyli ________
8. Aorta ________
9. Spleen ________
10. Thymus ________

B

a. This structure is located partly in the neck and partly in the thorax.
b. This structure marks the beginning of the thoracic duct.
c. During the dissection, you will identify this structure on the left side of the abdominal cavity.
d. The thoracic duct travels parallel to this structure.

Activity **C6.1**

Examining the Cat Thymus and Spleen

In both cats and humans, the major organs of the lymphatic system include the **thymus, spleen,** and **lymph nodes.** Other lymphatic structures include the **tonsils,** located in the pharynx, and lymphoid tissue in the mucous membrane of the digestive tract, collectively known as **mucosa-associated lymphoid tissue** (MALT). Specific examples of MALT include **Peyer's patches** in the small intestine and the **appendix** in the large intestine. The appendix, however, is not found in the cat.

A The thymus gland is located in the thoracic cavity.

Dissection note: *If you have not already opened the ventral body cavities and made preliminary observations of their contents, these activities must first be completed. Complete Activity C4.1 before you proceed with this dissection.*

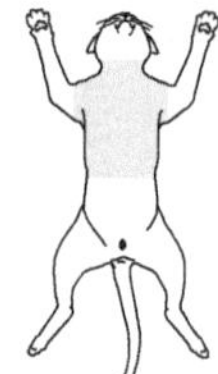

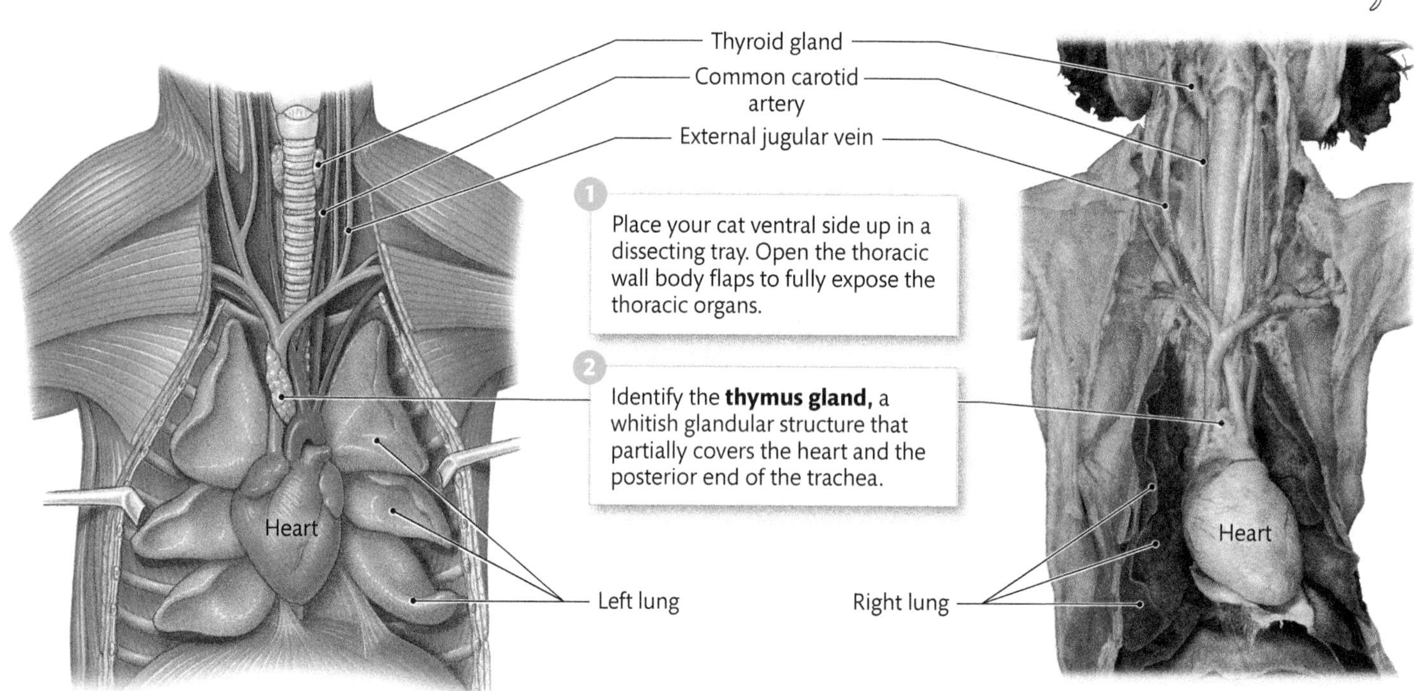

1. Place your cat ventral side up in a dissecting tray. Open the thoracic wall body flaps to fully expose the thoracic organs.

2. Identify the **thymus gland,** a whitish glandular structure that partially covers the heart and the posterior end of the trachea.

3. Examine the size and texture of the thymus gland. In young animals, it is relatively large, but as the cat matures and ages, the gland is reduced in size and replaced by fat. Compare the appearance of the thymus in your cat with others in the laboratory.

IN THE CLINIC

Thymus Gland Function

In humans and cats, the thymus gland has both endocrine and lymphatic functions. As an endocrine gland, the thymus produces and secretes thymic hormones. As a lymphatic organ, the thymus uses thymic hormones to induce normal development of T lymphocytes, which play a critical role in fighting off disease and infection. The thymus gland is therefore considered a component of two organ systems: the endocrine system and the lymphatic system.

B The spleen is located in the abdominal cavity. Proportionally, the spleen is much larger in the cat than in the human.

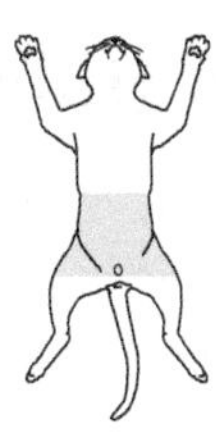

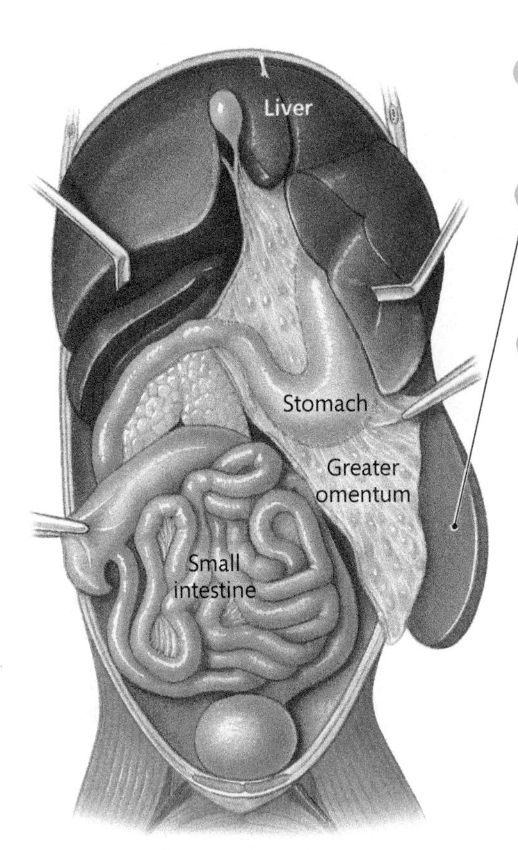

1. Open the abdominal wall body flaps to fully expose the abdominal organs.

2. Locate the **spleen,** a reddish-brown elongated organ found on the left side of the abdominal cavity.

3. Identify the greater omentum and notice its attachment to the greater curvature of the stomach. Retract the spleen and the left lobes of the liver to the left side to see the **gastrosplenic ligament** (not shown). It is a continuation of the greater omentum that extends from the medial side of the spleen to the greater curvature of the stomach.

4. Discuss with other members of your dissection group how the spleen provides protection from pathogens found in the blood.

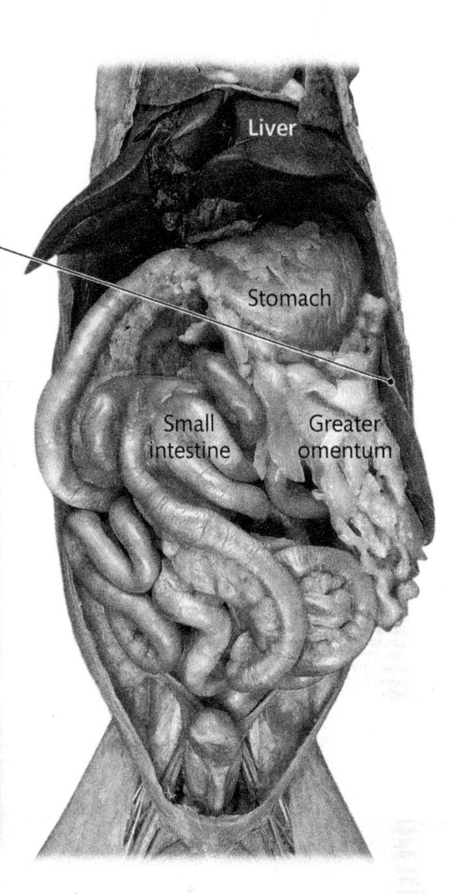

MAKING CONNECTIONS

In addition to its lymphatic function, the spleen serves as a blood reservoir; it can store a volume of blood in reserve. Suggest a practical significance for this function.

▶ ____

____

____

____

____

____

Examining Cat Lymph Nodes and the Cat Thoracic Duct

A **Lymph nodes** are located throughout the cat's body along the paths of lymphatic collecting vessels. They are generally either difficult to identify or are inadvertently destroyed during other dissections. Relatively large lymph nodes, found in the head and neck region of the cat, can be identified with careful dissection.

1. Turn your cat on its side. If not done previously, carefully cut the skin along the side of the head from the ear to the corner of the mouth. The skin is very thin in this area. Exercise caution to avoid cutting too deeply and damaging underlying structures.

2. In the cheek, identify the large **masseter muscle.** If you dissected the cat muscular system earlier, this muscle was identified at that time.

3. Locate the **mandibular lymph node** along the posterior margin of the masseter muscle. It appears as a dark, bean-shaped structure. Do not confuse it with the salivary glands, which are lighter in color.

4. Discuss with other members of your dissection group how lymph nodes provide protection from pathogens found in lymph.

5. Similar to humans, **tonsils** are located along the wall of the cat's pharynx. The tonsils are designed to fight off infections originating in the upper respiratory and digestive tracts. Do not attempt to locate the tonsils at this time.

Salivary glands

Ventrolateral view

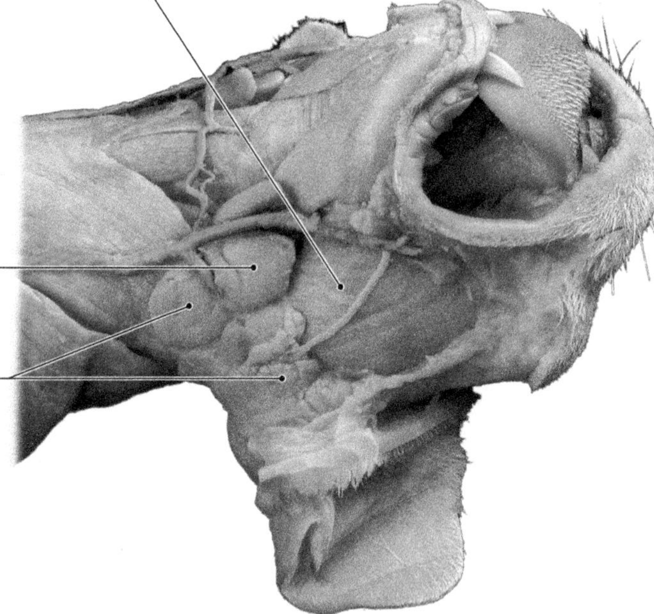

Ventrolateral view

B Small **lymphatic capillaries** provide drainage for the accumulating fluids (extracellular fluids) that escape from blood capillaries. Lymphatic capillaries transport lymph to superficial and deep **lymphatic collecting vessels,** which usually travel alongside arteries and veins. **Lymphatic trunks** are large lymphatic vessels positioned at strategic locations throughout the body. Each lymphatic trunk is formed by the union of several collecting vessels and drains the lymph from a specific region of the body. On each side of the body, the lymphatic trunks unite to form a **lymphatic duct.** The **right lymphatic duct** drains lymph from the right side of the head and neck, the right forelimb, and the right half of the thorax. It empties into the venous circulation by joining with the right subclavian vein. The **thoracic duct** drains lymph from all other body regions. It empties into the venous circulation at the junction of the left external jugular and left subclavian veins.

Most lymphatic vessels are too small to identify with confidence. The **thoracic duct,** however, is large enough to identify if you dissect with care and patience.

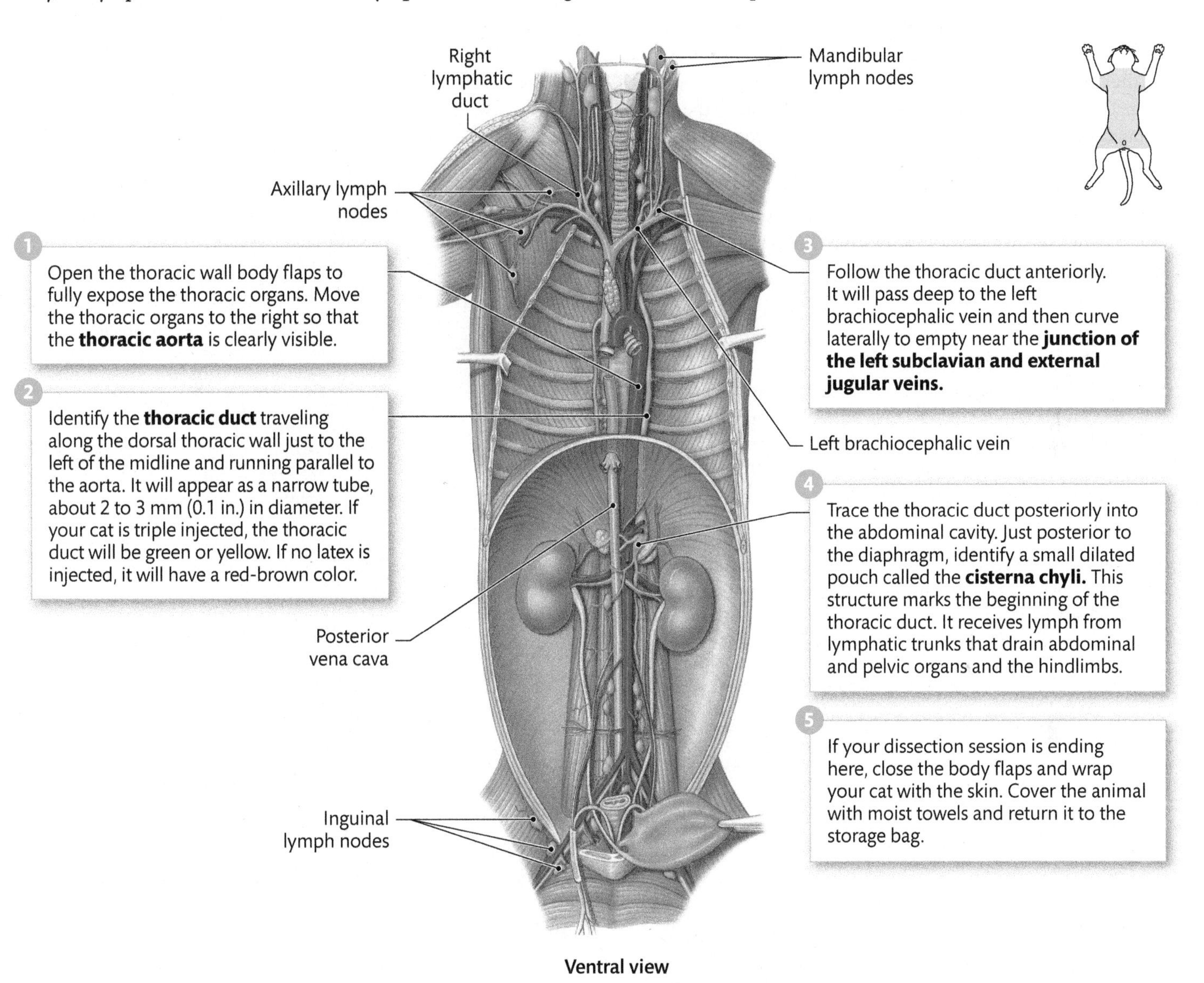

Ventral view

MAKING CONNECTIONS

The mandibular lymph nodes are much larger than the other lymph nodes in the cat. Why do you think they are so large?

▶ ____

____

____

____

____

Word Origins

Cisterna chyli is derived from both Latin and Greek. *Cisterna* is the Latin word for "an underground space for water." *Chyli* is derived from the Greek word *chylos,* which means "juice." The cisterna chyli is a deep space within the abdominal cavity that receives chyle, a fat-rich juice that drains from the small intestine.

BEFORE YOU MOVE ON . . .

« LOOKING BACK

The structure and relative positions of the major lymphatic organs are similar in cats and humans. The thymus partially covers the heart in the thoracic cavity and extends into the neck, covering part of the trachea. The spleen is located on the left side of the abdominal cavity. In cats, it is a very large organ in proportion to other organs in the abdominal cavity. Lymph nodes are scattered throughout the body. In the dissection, you were able to identify the large mandibular lymph nodes located on each side of the head near the masseter muscles. Mandibular lymph nodes are not present in humans.

As in humans and other mammals, the lymphatic circulation in the cat drains extracellular fluids that escape from blood capillaries and recycles the fluid, called lymph, back into the blood. In both cats and humans, two **lymphatic ducts** return lymph to the subclavian veins. On the right side, the **right lymphatic duct** drains lymph from the upper right quadrant of the body and empties into the right subclavian vein. The **thoracic duct** is much larger and drains lymph from all other body regions. It empties into left subclavian vein.

Speculate on the significance of the lymphatic circulation. How would the circulation of body fluids be affected if the lymphatic circulation were blocked?

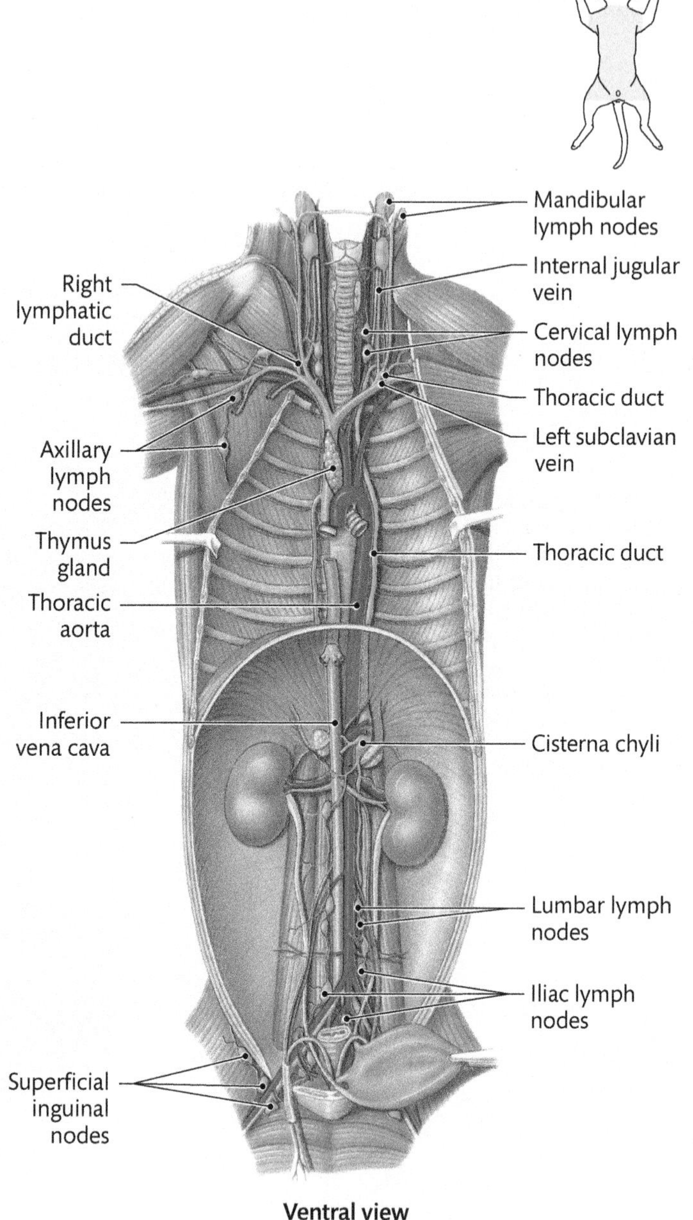

Ventral view

LOOKING FORWARD »

Tonsils are masses of lymphatic tissue that are located in the pharynx of humans and cats. They are strategically located to protect the respiratory system from disease-causing microorganisms. In the next cat dissection exercise, you will review the anatomy of the cat respiratory system and make comparisons with the human.

Name ______________________

Lab Section ______________________

Date ______________________

EXERCISE **C6** **REVIEW SHEET**

Dissection of the Cat Lymphatic System

1. How does the function of the spleen compare with that of the lymph nodes?

2. If the thymus did not function normally, what would be the effect on the body's ability to ward off diseases and infections?

3. What is the function of the thoracic duct? How does it compare with the function of the right lymphatic duct?

4. Explain why it is difficult to identify lymphatic vessels in the cat.

5. What is the function of the cisterna chyli?

QUESTIONS 6–10: Match the term in column A with the appropriate description in column B.

A		B
6. Lymphatic capillaries	______	**a.** Lymph nodes are located along the paths of these structures
7. Lymphatic trunks	______	**b.** Drain excess extracellular fluids
8. Thoracic duct	______	**c.** Drains lymph from the upper-right quadrant of the body
9. Lymphatic collecting vessels	______	**d.** Drain lymph from a specific body region
10. Right lymphatic duct	______	**e.** The cisterna chyli is found at the beginning of this structure

QUESTIONS 11–17: Identify the labeled structures in the photos. Select your answers from the list below.

a. Mandibular lymph node	**11.** ______
b. Thymus gland	**12.** ______
c. Liver	**13.** ______
d. Masseter muscle	**14.** ______
e. Thoracic duct	**15.** ______
f. Stomach	**16.** ______
g. Spleen	**17.** ______

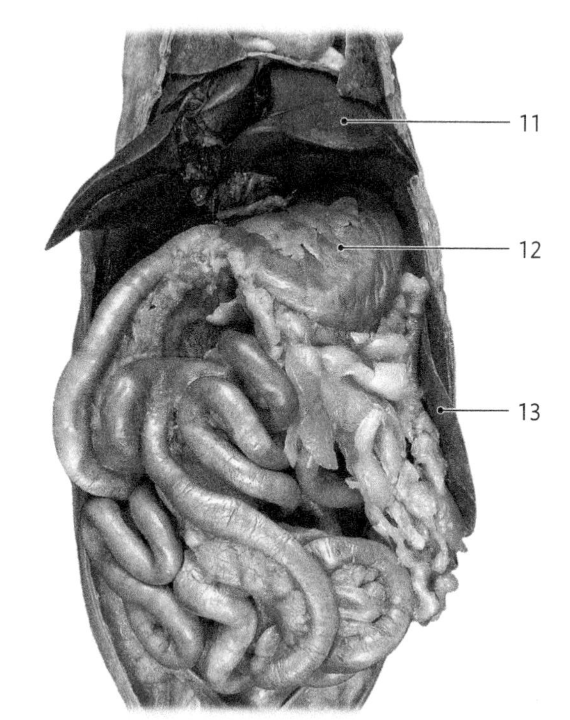

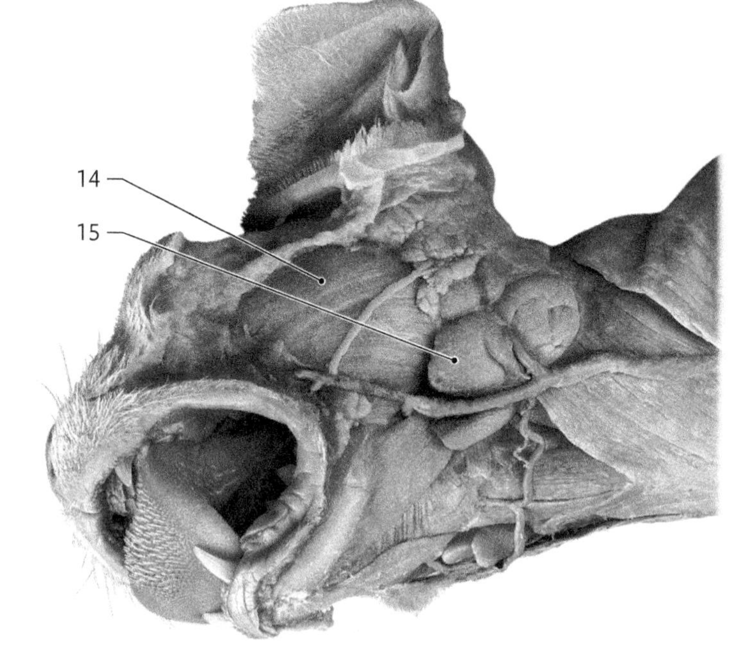

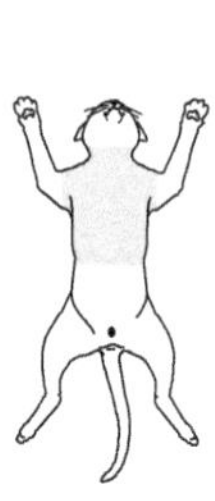

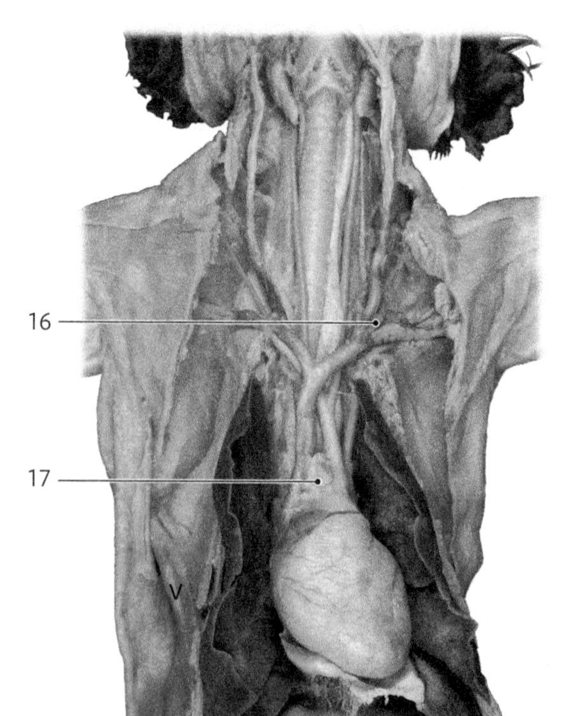

DISSECTION EXERCISE

Dissection of the Cat Respiratory System

C7

LEARNING OUTCOMES

These Learning Outcomes correspond by number to the laboratory activities in this exercise. When you complete the activities, you should be able to:

Activity C7.1 **Describe the structure and function of the cat respiratory organs in the head and neck and make comparisons with similar organs in the human.**

Activity C7.2 **Describe the structure and function of the cat lungs and make comparisons with the human lungs.**

LABORATORY SUPPLIES

- Preserved cats
- Dissecting equipment: trays, tools, gloves
- Protective eyewear
- Face mask
- Hand lens or dissecting microscope
- Anatomical models of the human respiratory system

PRE-LAB QUIZ

Before you begin, read all the activities in Exercise C7 and the required reading in your textbook that is assigned by your instructor.

1. During this cat dissection exercise, you will examine all the following structures *except* the
   a. external nares.
   b. diaphragm.
   c. thyroid cartilage.
   d. oral cavity.
   e. true vocal cords.
2. True or False: The tracheal wall contains a series of C-shaped cartilaginous rings, but the esophageal wall does not. ____________
3. True or False: In the cat, the right lung has three lobes, and the left lung has four lobes.
   ____

Questions 4–10: Match the structure in column A with the correct description in column B.

A	B
4. Pulmonary arteries ________	a. The glottis is the space between these structures.
5. Pulmonary veins ________	b. This structure travels dorsal to the trachea.
6. Diaphragm ________	c. This structure covers the wall of the thoracic cavity.
7. Cricoid cartilage ________	d. This blood vessel transports oxygen rich blood.
8. True vocal cords ________	e. This structure forms a portion of the wall of the larynx.
9. Esophagus ________	f. This blood vessel transports oxygen-poor blood.
10. Parietal pleura ________	g. This structure separates the thoracic and abdominal cavities.

Activity **C7.1**

Examining Cat Respiratory Organs in the Head and Neck

A The respiratory structures in the head and neck include the nose, the nasal cavity, the pharynx, the larynx, and most of the trachea.

Dissection note: *If you have not already opened the ventral body cavities and made preliminary observations of their contents, these activities must first be completed. Complete Activity C4.1 before you proceed with this dissection.*

1. Place your cat ventral side up in a dissecting tray. Identify the **external nares** or **nostrils** on the nose. These openings allow air to enter the nasal cavity. From there, air passes into the pharynx. The nasal cavity and pharynx will be examined in conjunction with the oral cavity during the dissection of the digestive system.

2. If you have not already done so, transect and reflect the muscles in the neck to provide a clear view of the deeper structures in this area.

3. The **larynx** is at the anterior end of the neck, between the pharynx (throat) and trachea. The wall of the larynx is formed by two plates of cartilage that are joined by connective tissue.

   Locate the **thyroid cartilage,** which is the largest cartilaginous plate.

   Identify the **cricoid cartilage,** just posterior to the thyroid cartilage.

Ventral view

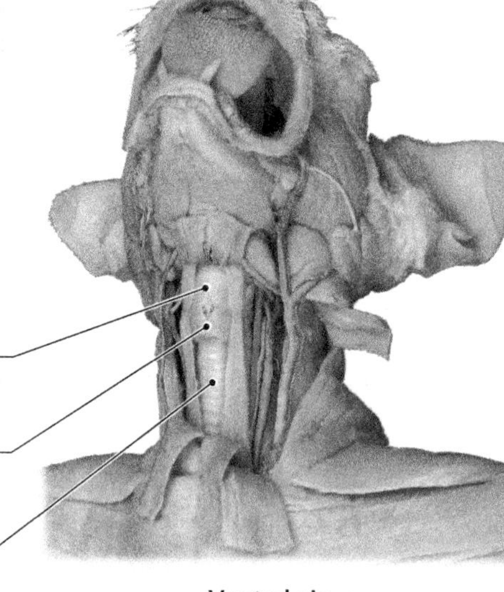

Ventral view

4. Identify the **trachea.** It begins at the posterior border of the cricoid cartilage. The tracheal wall contains a series of C-shaped cartilaginous rings that are connected by soft tissue. Run your finger along the surface of the trachea. You should be able to distinguish between the alternating hard surfaces of the cartilaginous rings and the soft tissue surfaces between them.

Word Origins

Trachea is derived from the Greek term *tracheia arteria,* which means "rough artery." If you run your finger along the surface of the cat trachea, it has a rough feel due to the presence of the incomplete cartilaginous rings in its wall.

B The vocal cords are located along the wall of the larynx.

Dissection note: *The thyroid gland was studied earlier when you dissected the cat endocrine system. If you did not dissect the endocrine system, examine the thyroid gland at this time.*

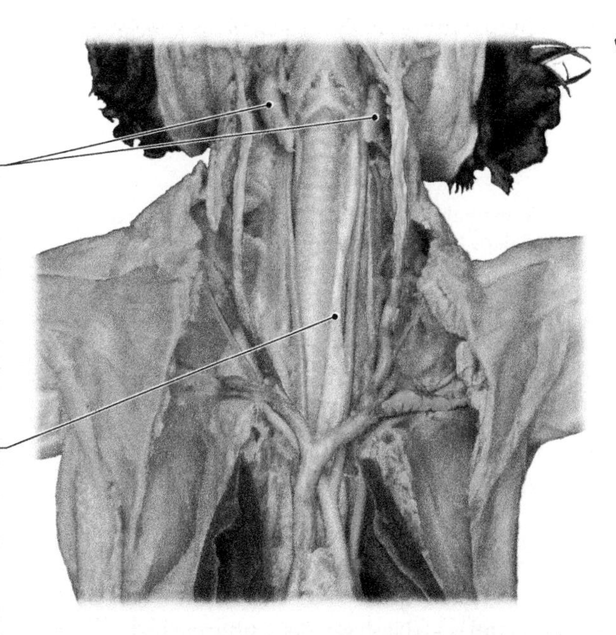

1. Locate the **thyroid gland** in the neck. It consists of two elongated lobes located on each side of the trachea, just posterior to the thyroid cartilage. An isthmus crosses the ventral surface of the trachea and connects the two lobes. (This structure may not be present in older cats.) The thyroid gland produces hormones that regulate cell metabolism, general growth and development, and the maturation of the nervous system.

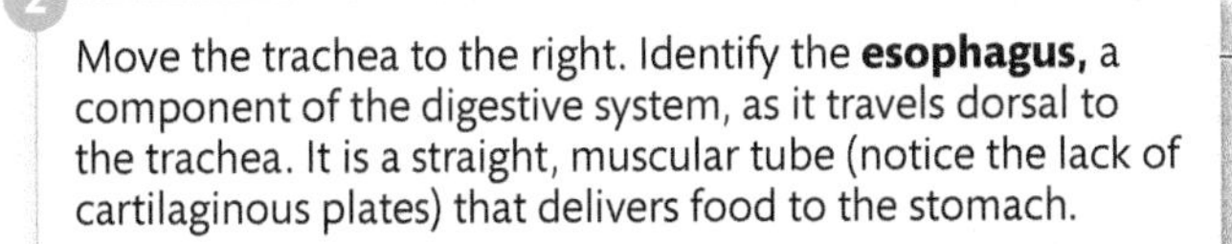

2. Move the trachea to the right. Identify the **esophagus,** a component of the digestive system, as it travels dorsal to the trachea. It is a straight, muscular tube (notice the lack of cartilaginous plates) that delivers food to the stomach.

3. Make an incision along the midventral line of the cricoid and thyroid cartilages. Spread the two sides apart to get a clear view of the mucous membranes inside the larynx.

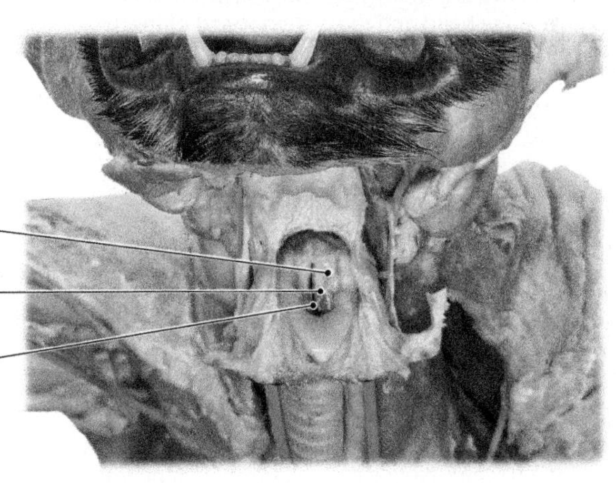

4. Identify two pairs of membranous folds along the inner laryngeal wall.

   The anterior folds are the **false vocal cords.**

   The posterior folds are the **true vocal cords.**

   The space between the true vocal cords is the **glottis.** Sound is produced when air passes through the glottis, causing the true vocal cords to vibrate. The false vocal cords have no role in sound production.

MAKING CONNECTIONS

Suggest a reason why the trachea has cartilaginous plates in its wall, but the esophagus does not.

▶ ______________________________

____

____

____

____

Activity **C7.2**

Examining Cat Respiratory Organs in the Thoracic Cavity

A The trachea ends in the thoracic cavity, where it divides into the right and left primary bronchi.

Dissection notes:

- *Before you begin to dissect the thoracic cavity, remove any excess preservative fluids or coagulated blood that might have accumulated in this area. This step can be done by gently rinsing your cat in the sink and using paper towels to absorb excess rinse water. Do not allow any solid material to flow into the sink's drain pipes. Any solid material should be disposed of in a secure container designated by your instructor until it can be finally disposed of correctly.*
- *The thymus gland was studied earlier if you dissected the endocrine and lymphatic systems. If you did not dissect these systems, you will examine the thymus gland at this time.*

1 Place your cat ventral side up in a dissecting tray. Separate the body wall flaps to expose the contents of the thoracic cavity. Notice the thin, shiny membrane that covers the inner surface of the thoracic wall. It is a serous membrane called the **parietal pleura.** A similar membrane that covers the surface of the lungs is called the **visceral pleura.** The two pleural membranes are continuous with each other where blood vessels and airways enter and exit the lung. The space between the visceral and parietal pleura is not a true space. It is a potential space known as the **pleural cavity.** The two membranes secrete a watery fluid, which fills the cavity and allows the two pleural membranes to slide along each other during breathing movements.

2 The **thymus gland** is a whitish glandular structure that partially covers the heart and the posterior end of the trachea. The thymus is important for the maturation of lymphocytes, a type of white blood cell that is vital for fighting infections. The thymus will be relatively large in young animals. As the cat matures, the size of the gland is reduced.

3 Notice the position of the mediastinum, which is the central space between the lungs, where the **heart** is located.

4 If the heart has not yet been removed, shift it slightly to the left. With forceps or a blunt probe, scrape away the connective tissue from the posterior thoracic wall to expose the end of the trachea. Notice that the trachea branches to form the **right** and **left primary bronchi** (not shown), which supply air to the right and left lungs, respectively.

5 The **diaphragm** is the musculotendinous partition that divides the thoracic and abdominal cavities. Notice the dome-shaped surface of the diaphragm, with the concave side facing the abdominal cavity.

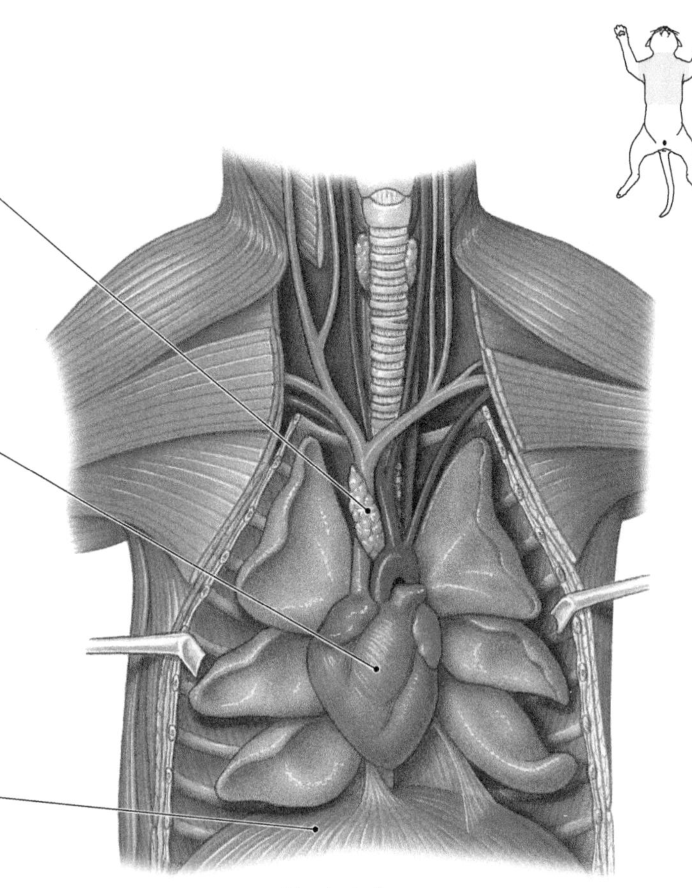

Ventral view

B In the cat, the right lung has four lobes, and the left lung has three lobes.

1. The **lungs** are located on each side of the thorax. Examine each lung and identify the individual lobes.

2. Cut out a small, thin section of lung and examine it with a hand lens or dissecting microscope. If your cat is double injected with latex dyes, you will be able to observe three types of vessels:
   - Blood vessels transporting blood that is low in oxygen will contain blue dye. These vessels could be **pulmonary arteries** (pulmonary circulation) or **bronchial veins** (systemic circulation).
   - Blood vessels transporting blood that is high in oxygen will contain red dye. These vessels could be **pulmonary veins** (pulmonary circulation) or **bronchial arteries** (systemic circulation).
   - Small airways, called **bronchioles,** distribute air throughout the lungs. These vessels will be hollow.

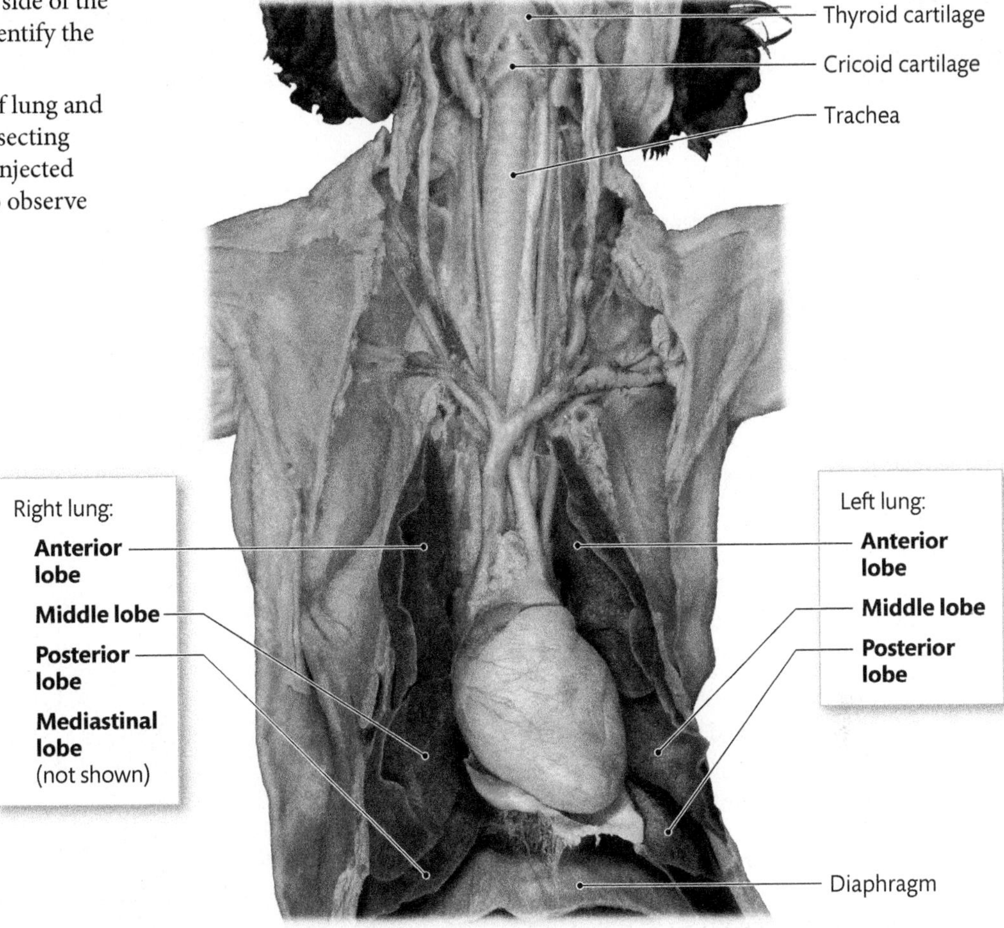

Ventral view

3. When you have completed the thoracic cavity dissection, close the body wall flaps. If your dissection session is ending here, wrap your cat with the skin. Cover the animal with moist towels, and return it to the storage bag.

MAKING CONNECTIONS

Compare the thoracic anatomy of the human and the cat and identify similarities and differences between the two. ▶

Similarities:

____

____

____

____

Differences:

____

____

____

____

BEFORE YOU MOVE ON ...

« LOOKING BACK

The cat respiratory system is organized in much the same way as the human respiratory system. For example, the respiratory organs in both are grouped into two subdivisions:

- The **upper respiratory system** includes the nose, nasal cavity, paranasal sinuses, and pharynx.
- The **lower respiratory system** includes the larynx, trachea, and lungs.

If you made structural comparisons with human respiratory structures during your dissection, you saw that the anatomy of the cat larynx, trachea, and lungs parallels the structure of these organs in the human.

Now that you have completed this dissection, consider how the respiratory system is functionally integrated with other organ systems you have dissected thus far. For example, as a result of gas exchange between alveoli and pulmonary capillaries, the respiratory system, in conjunction with the cardiovascular system, supports skeletal, muscular, and nervous system function by providing oxygen to bone, muscle, and nerve cells and releasing carbon dioxide, a metabolic waste. In addition, the respiratory system supports cardiovascular function by contributing bicarbonate ions, which are important buffering agents that regulate pH in the blood.

The other organ systems play key roles to support normal respiratory function. Identify one function of each organ system mentioned above—skeletal, muscular, nervous, and cardiovascular—that will promote normal respiratory function.

▶ ______________________________

____

____

____

____

____

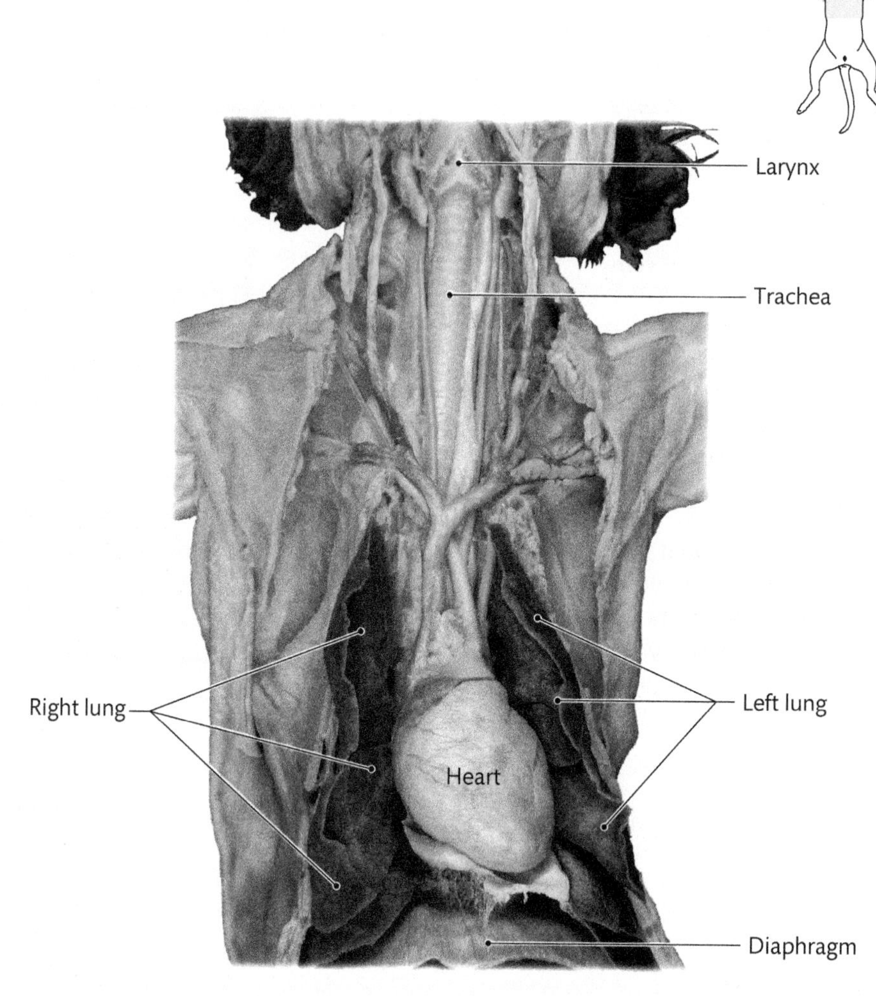

Ventral view

LOOKING FORWARD »

All the respiratory structures that you studied in this dissection exercise are located in the head, neck, and thoracic cavity. In the next cat dissection exercise, you will examine the digestive system. Most of your dissecting activities will shift to the abdominopelvic cavity, where most of the digestive organs are found.

Name ____________________

Lab Section ____________________

Date ____________________

EXERCISE **C7** **REVIEW SHEET**

Dissection of the Cat Respiratory System

1. Describe the anatomical relationship between the trachea and esophagus in the neck of the cat.

____

____

2. Identify similarities and differences between the true and false vocal cords.

a. Similarities

____

____

b. Differences

____

____

3. Describe the functional difference between the following pairs of blood vessels:

a. Bronchial arteries and pulmonary arteries

____

____

b. Bronchial veins and pulmonary veins

____

____

QUESTIONS 4–10: Match the term in column A with the the appropriate description in column B.

A	B
4. External nares ________	**a.** Small conducting airways in the lungs
5. Bronchioles ________	**b.** The space between the two true vocal cords
6. Pleural membrane ________	**c.** Large airways that branch off the trachea
7. Trachea ________	**d.** The serous membrane that covers the surface of the lungs
8. Thyroid cartilage ________	**e.** Openings on the nose that lead into the nasal cavity
9. Glottis ________	**f.** Airway that contains a series of C-shaped cartilaginous plates in its wall
10. Primary bronchi ________	**g.** Large cartilaginous plate that forms a portion of the laryngeal wall

QUESTIONS 11–23: Identify the labeled structures in the photo. Select your answers from the list below.

a. Thyroid gland
b. Cricoid cartilage
c. Right lung, anterior lobe
d. Left lung, middle lobe
e. Right lung, posterior lobe
f. Thymus
g. Diaphragm
h. Thyroid cartilage
i. Right lung, middle lobe
j. Esophagus
k. Left lung, anterior lobe
l. Trachea
m. Left lung, posterior lobe

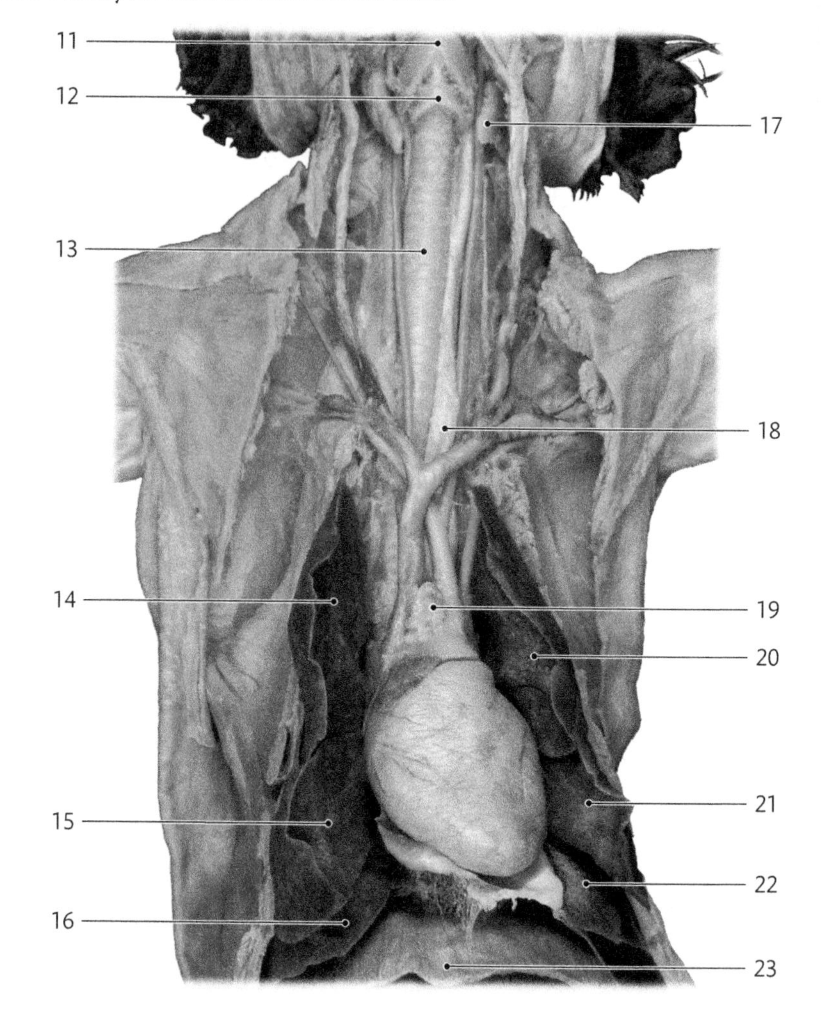

11. ______________________

12. ______________________

13. ______________________

14. ______________________

15. ______________________

16. ______________________

17. ______________________

18. ______________________

19. ______________________

20. ______________________

21. ______________________

22. ______________________

23. ______________________

DISSECTION EXERCISE

C8

Dissection of the Cat Digestive System

LEARNING OUTCOMES

These Learning Outcomes correspond by number to the laboratory activities in this exercise. When you complete the activities, you should be able to:

Activity C8.1	**Identify and describe the function of cat digestive organs in the head and make comparisons with similar organs in the human.**
Activity C8.2	**Identify the esophagus in the cat and describe its relationship with adjacent structures.**
Activity C8.3	**Describe the organization of the cat peritoneum in the abdominal cavity.**
Activities C8.4–C8.5	**Describe the structure and function of cat digestive organs in the abdominal cavity and make comparisons with similar organs in the human.**

LABORATORY SUPPLIES

- Preserved cats
- Dissecting equipment: trays, tools, gloves
- Bone cutters
- Protective eyewear
- Face mask
- Hand lens or dissecting microscope
- Anatomical models of the human digestive system

PRE-LAB QUIZ

Before you begin, read all the activities in Exercise C8 and the required reading in your textbook that is assigned by your instructor.

1. During this cat dissection exercise, you will examine all the following structures *except* the
   a. stomach.
   b. large intestine.
   c. urinary bladder.
   d. tongue.
   e. esophagus.
2. During this cat dissection exercise, you will expose the interior of the oral cavity by using bone cutters to cut through what structure? ____________
3. True or False: During swallowing, the epiglottis closes off the entrance to the larynx. ____________
4. In both humans and cats, the largest salivary gland is the
   a. submandibular gland.
   b. parotid gland.
   c. sublingual gland.
   d. molar gland.
   e. infraorbital gland.
5. The peritoneal membrane structure that covers most abdominal organs is called the ____________.
6. True or False: The pyloric sphincter is a ring of smooth muscle that controls the passage of material from the ileum to the cecum. ____________
7. The teeth are embedded in bony sockets of what bones? ____________
8. True or False: The ileum is the second portion of the small intestine. ____________
9. The primary function of the ____________ is to store bile.
10. All the following structures are segments of the large intestine *except* the
    a. cecum.
    b. transverse colon.
    c. descending colon.
    d. rectum.
    e. duodenum.

Activity **C8.1**

Examining Digestive Structures in the Head

A Most salivary glands in the cat are located close to the masseter muscle.

Dissection notes:

- *If you have not already opened the ventral body cavities and made preliminary observations of their contents, these activities must first be completed. Complete Activity C4.1 before you proceed with this dissection.*
- *As you dissect the cat digestive system, have anatomical models of the human digestive system available for comparison.*

1 Place your cat on its side in a dissecting tray. If not done previously, carefully cut the skin along the side of the head from the ear to the corner of the mouth. Because the skin is very thin in this area, exercise caution to avoid cutting too deeply and damaging underlying structures.

2 In the cheek, identify the large **masseter muscle.** If you dissected the cat muscular system, this muscle was identified at that time.

3 Use the masseter as an anatomical landmark to help you identify the following salivary glands:

The **parotid gland** is the largest salivary gland. It appears as a large glandular structure, posterior to the masseter and ventral to the external ear.

The **parotid duct** emerges from the anterior margin of the parotid gland and crosses the masseter near the mandible. The parotid duct transports saliva, produced in the parotid gland, to the oral cavity.

The **submandibular gland** is a smaller salivary gland that lies just ventral to the parotid gland, posterior to the angle of the mouth.

The **sublingual gland** (not shown) is located just anterior to the submandibular gland. The mandibular lymph node may partially cover the sublingual gland and obscure its presence. If so, remove the lymph node to obtain a better view.

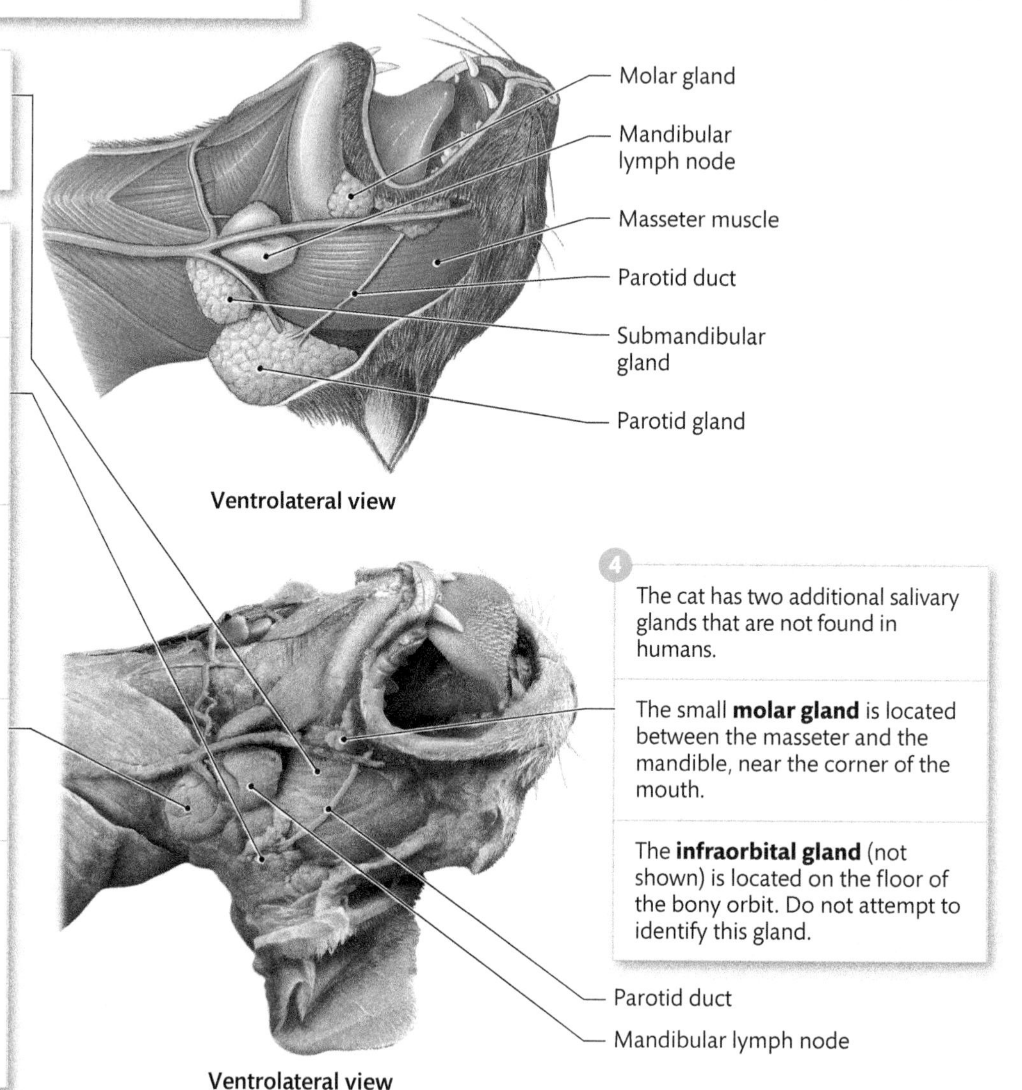

4 The cat has two additional salivary glands that are not found in humans.

The small **molar gland** is located between the masseter and the mandible, near the corner of the mouth.

The **infraorbital gland** (not shown) is located on the floor of the bony orbit. Do not attempt to identify this gland.

■ Want more practice? Go to: **MasteringA&P** > Study Area > Menu > Lab Tools > PAL > Cat > Digestive System > Head and neck photos; salivary glands photos

B The head contains several important structures associated with the digestive and respiratory systems. Some structures have overlapping functions.

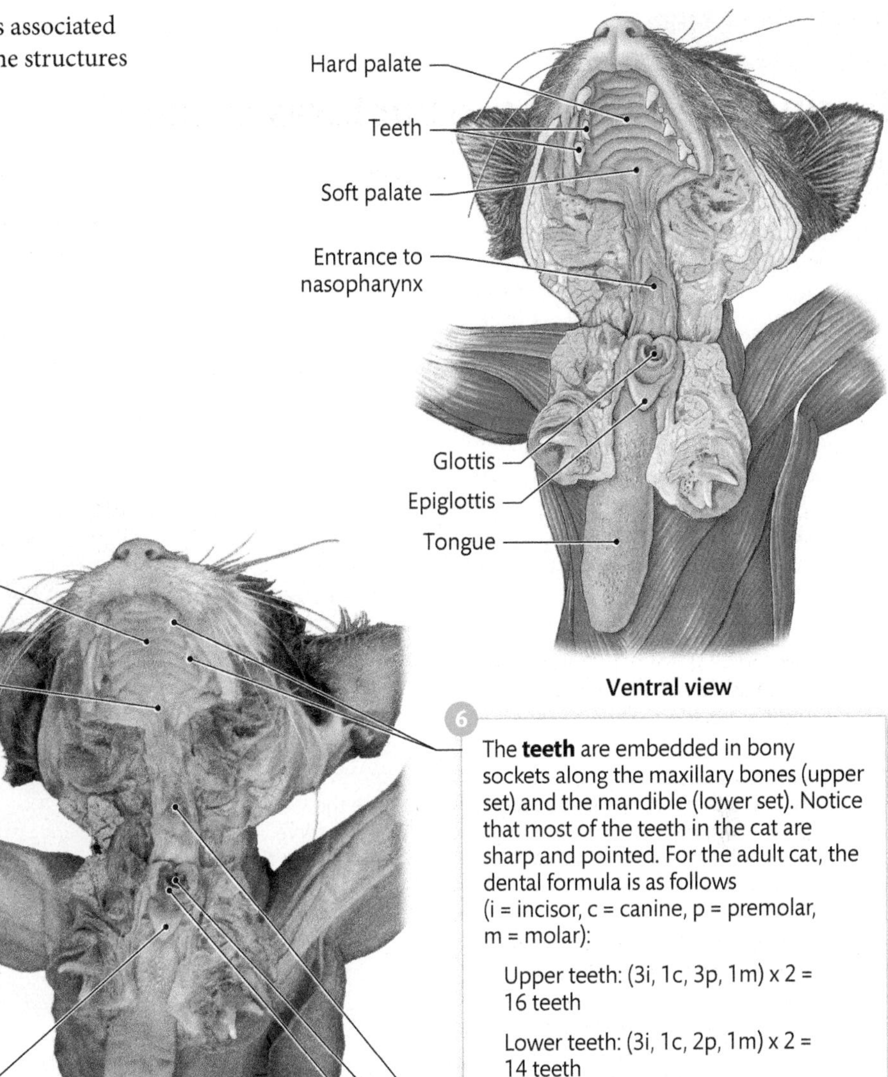

1 With a pair of scissors, cut posteriorly on both sides of the head from the corners of the mouth to the angles of the mandible. Use bone cutters to cut through the mandible. Pull the mandible inferiorly to expose the interior of the **oral cavity.**

2 Identify the **palate,** which forms the roof of the mouth.

The **hard palate** is the anterior bony portion, characterized by conspicuous transverse ridges.

The **soft palate** is the posterior soft tissue portion. The soft palate separates the oropharynx ventrally from the nasopharynx dorsally. The uvula, the posterior tip of the soft palate in humans, is not present in cats.

3 Make a small incision through the soft palate to expose the nasopharynx. Observe the openings (internal nares) that lead to the **nasal cavity** (not shown).

4 Locate the **epiglottis.** It is the flap of cartilage located at the anterior end of the larynx, near the base of the tongue. During swallowing, the epiglottis closes off the entrance to the larynx.

5 Identify the **tongue.** It is a muscular structure attached to the floor of the mouth by a fold of soft tissue known as the lingual frenulum. Posteriorly, it is attached to the hyoid bone. Run your finger across the tongue's surface and feel its rough, uneven surface. The texture of the surface is due to the presence of sensory papillae where the taste buds are located.

6 The **teeth** are embedded in bony sockets along the maxillary bones (upper set) and the mandible (lower set). Notice that most of the teeth in the cat are sharp and pointed. For the adult cat, the dental formula is as follows (i = incisor, c = canine, p = premolar, m = molar):

Upper teeth: (3i, 1c, 3p, 1m) x 2 = 16 teeth

Lower teeth: (3i, 1c, 2p, 1m) x 2 = 14 teeth

MAKING CONNECTIONS

Compare the dentition between cats and humans. How do cat teeth differ from human teeth? What is the functional significance of this difference?

▶ ______________________________

____

____

____

____

Activity **C8.2**

Tracing the Path of the Cat Esophagus

A The esophagus is a straight, muscular tube that travels through the neck and thoracic cavity and enters the abdominal cavity where it joins the stomach. As you follow its path, pay close attention to its anatomical relationships with other structures.

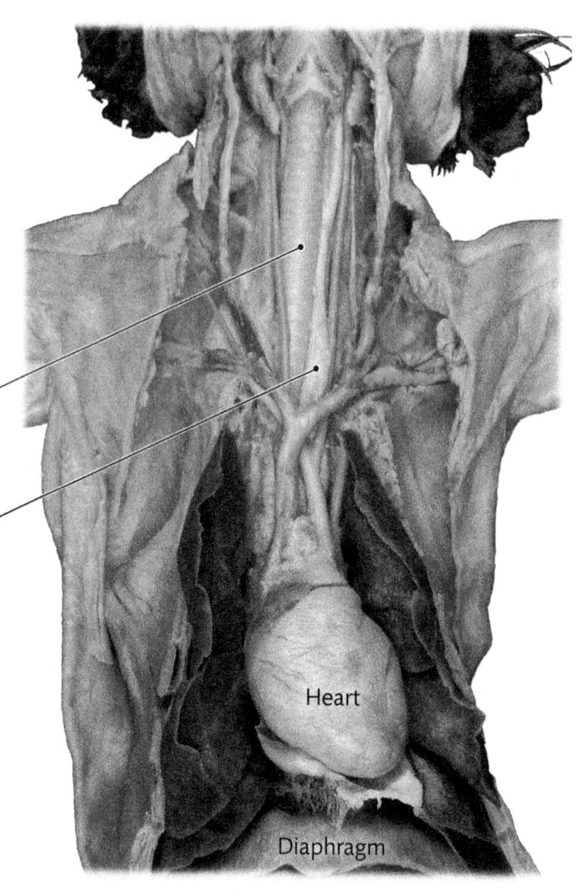

Ventral view

1. Place your cat ventral side up in a dissecting tray. If you have not already done so, transect and reflect the muscles in the neck to provide a clear view of the deeper structures in this area.

2. Identify the **trachea** traveling along the midventral line of the neck, between the larynx and the thoracic cavity.

3. Move the trachea to one side to expose the **esophagus.** The esophagus is dorsal to the trachea for its entire course through the neck.

4. Open the thoracic wall body flaps to fully expose the thoracic organs. Move the left lung medially and observe the esophagus traveling along the dorsal thoracic wall to the left of the midline (not shown). The thoracic division of the descending aorta can also be seen traveling along the dorsal body wall alongside the esophagus.

5. Trace the path of the esophagus through the thoracic cavity. Observe how it passes through an opening in the diaphragm **(esophageal hiatus)** to enter the abdominal cavity where it connects to the stomach.

Diaphragm
Kidneys
Abdominal aorta
Posterior vena cava

Ventral view

6. Notice that the esophagus does not appear as a round, tubelike structure, but instead is flattened. In life, the esophagus is a flat tube, which means that the lumen (internal space of a tubelike structure) is closed. The lumen will open only when the esophagus is transporting food and fluids to the stomach.

MAKING CONNECTIONS

As you have seen in this dissection activity, the esophagus travels dorsal to the trachea. Speculate on a reason why the absence of cartilage in the dorsal wall of the trachea is necessary for the esophagus to function normally.

▶ ______________________________

Activity **C8.3**

Examining the Organization of the Cat Peritoneum

A The **peritoneum (peritoneal membrane)** is a serous membrane that covers the wall of the abdominal cavity and the outer walls of abdominal organs.

1. Place your cat ventral side up in a dissecting tray. Separate the body wall flaps to expose the contents of the abdominal cavity.
2. Notice the thin, shiny membrane that covers the inner surface of the abdominal wall. This membrane is the **parietal peritoneum.**
3. Identify a similar membrane that covers the surfaces of the abdominal organs. This membrane is called the **visceral peritoneum.**

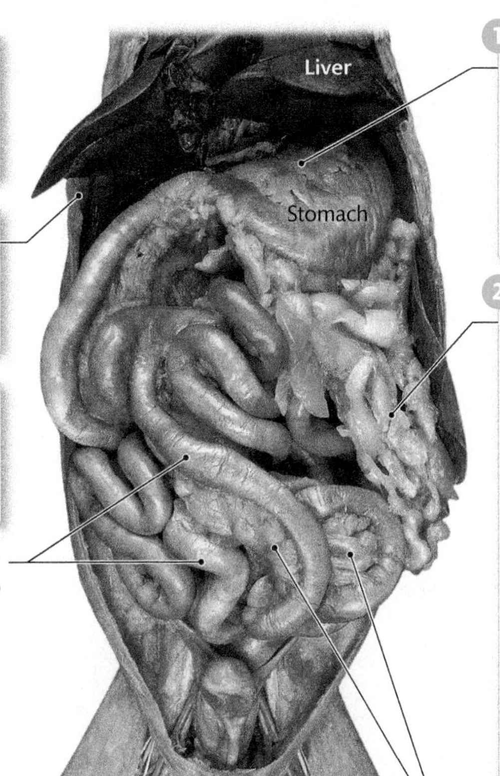

Ventral view

B Special peritoneal structures—the **lesser omentum, greater omentum,** and **mesenteries**—provide protection and support for the abdominal organs. Each structure is composed of a double layer of peritoneum.

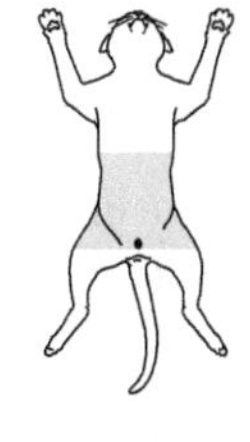

1. Identify the **lesser omentum,** which connects the stomach and duodenum (first part of the small intestine) to the liver. Keep this membrane intact.
2. Identify the **greater omentum** (intact in right figure; retracted in left figure). When you first look into the abdominal cavity, it is one of the first structures that you will notice because it covers most of the abdominal organs. The greater omentum might have been removed during an earlier dissection. If not, cut the greater omentum at its attachment to the greater curvature of the stomach and fold it posteriorly to expose the internal organs.

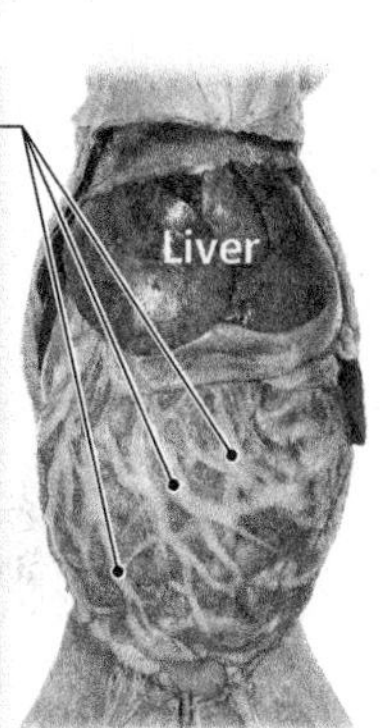

Ventral view

3. Locate the **mesenteries.** They are peritoneal membranes that connect portions of the small and large intestines to the body wall. Lift up the small intestine and identify the mesentery connected to it. Notice the arteries and veins that travel through the mesentery to supply the intestine.

MAKING CONNECTIONS

As you observed in this dissection activity, the greater omentum is a large, fat-laden membrane that covers many of the abdominal organs. Speculate on a possible function for this structure.

▶ ______________________________

____

____

____

Examining Cat Digestive Organs in the Anterior Abdominal Cavity

The digestive organs in the cat abdominal cavity include components of the alimentary canal as well as accessory digestive organs. Unlike the thoracic organs, which are surrounded and protected by the ribs and sternum, the abdominal organs are supported by a body wall that consists entirely of soft tissue. The three layers of flat muscles and the rectus abdominis provide the bulk of this support. The posterior ribs, however, protect portions of some abdominal organs, such as the liver, which are adjacent to the diaphragm.

A The liver, gallbladder, and pancreas are accessory digestive organs.

1. Identify the **liver,** which is the largest internal organ in the body. In the cat, it is divided into five lobes (the human liver has four lobes).

2. The **falciform ligament** is a peritoneal membrane that connects the liver to the ventral body wall. The ligament's attachment to the body wall has probably been cut. Identify its attachment to the liver within a fissure that divides the organ into right and left sides.

3. Lift the two right lobes of the liver to locate the **gallbladder.** It is a sac-like structure that lies in a depression on the posterior surface of the liver. Its primary function is to store bile that is produced in the liver.

4. Identify the following ducts that transport bile:

   The **cystic duct** extends from the gallbladder.

   The **hepatic duct** from the liver merges with the cystic duct.

   The union of the cystic and hepatic ducts forms the **common bile duct,** which transports bile to the duodenum.

5. The **spleen** is a reddish-brown, elongated organ found on the left side of the abdominal cavity. Notice that the spleen is connected to the stomach by a peritoneal membrane called the gastrosplenic ligament. The spleen is not a digestive organ. Instead, it plays an important role in fighting disease by destroying pathogenic viruses and bacteria found in the blood. If you dissected the cat lymphatic system, you examined this organ at that time.

6. The **pancreas** is a glandular structure that appears lighter in color than the neighboring small intestine. If you dissected the cat endocrine system, you examined this structure at that time. Lift the stomach from its anatomical position and identify the pancreas along the dorsal body wall between the duodenum and spleen.

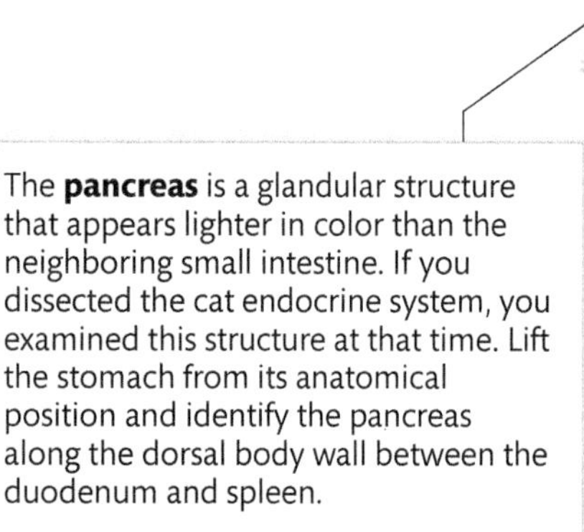

Ventral view

B The stomach is a large muscular pouch on the left side of the abdominal cavity. It is partially hidden by the liver.

Pancreas Gallbladder Liver

Ventral view

1 Identify the following regions of the **stomach:**

The **cardiac region** is the area where the esophagus joins the stomach (hidden by the left lobes of the liver).

The anterior dome-shaped region is the **fundus** (partially hidden by the left lobes of the liver).

The **body** is the main portion of the stomach.

From the body, the stomach narrows down to form a tube, the **pyloric region,** which joins the duodenum. The pyloric sphincter (see below) is located at the stomach–duodenum junction.

2 With scissors, open the stomach by cutting along the posterior curved margin (the greater curvature). Wash out any debris that is inside.

3 Return to the pyloric region, where the stomach joins the duodenum. Identify the **pyloric sphincter,** a ring of smooth muscle that controls the passage of partially digested food (chyme) into the duodenum.

4 Notice the complex pattern of folds, or **rugae,** along the inner surface of the stomach.

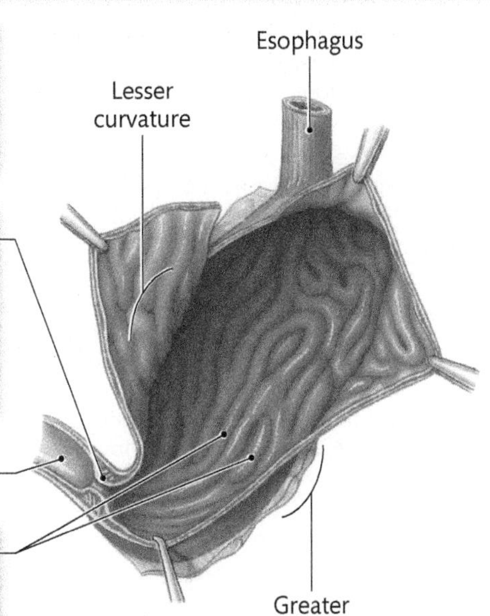

MAKING CONNECTIONS

What is the function of the stomach? What role do you think the rugae have during normal stomach activity?

▶ ______________________________

____

____

____

Activity **C8.5**

Examining Cat Digestive Organs in the Posterior Abdominal Cavity

The small and large intestines occupy the posterior abdominal cavity. Along with the stomach, the small and large intestines are the abdominal organs of the alimentary canal and are often referred to as the **gastrointestinal tract.**

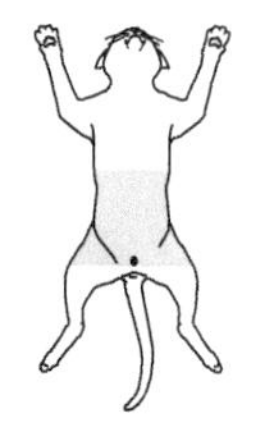

A The small intestine is a highly coiled tubular structure that takes up much of the central region in the abdominal cavity.

1 The **small intestine** extends from the stomach to the large intestine. Identify its three parts:

The first segment, the **duodenum,** is a short C-shaped section connected to the pyloric region of the stomach.

The second portion, the **jejunum,** is the largest segment of the cat small intestine. It comprises about 50 percent of the small intestine's total length.

The last section, the **ileum,** connects to the cecum of the large intestine. (In humans, the ileum is the largest segment of the small intestine.) Notice, as you did earlier, how the small intestine is held in place by the mesentery.

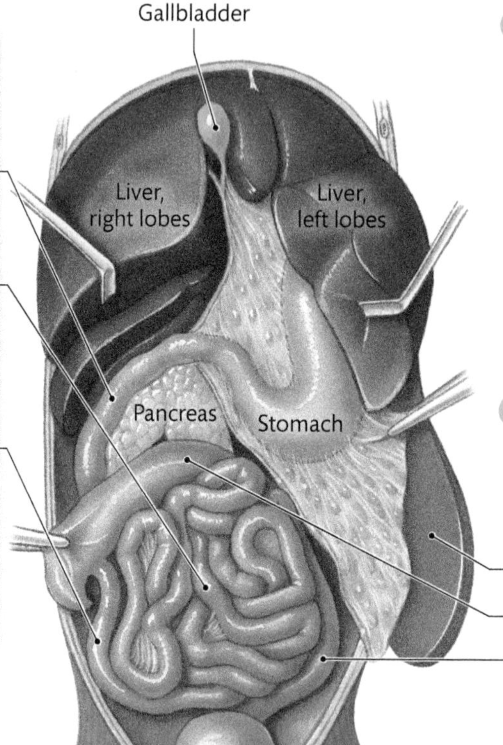

Ventral view

2 With a pair of scissors, carefully cut along the wall of the jejunum or ileum. Your incision should be about 2.5 cm (1 in.) in length. Expose the inside of the small intestine and wash out the contents. Move your gloved finger over the inner surface and feel its soft, velvety quality. This delicate texture is due to the presence of numerous fingerlike projections, called **villi** (singular = **villus**), along the surface.

3 Cut away a small piece of the intestine and observe the villi more closely with a hand lens or dissecting microscope.

Word Origins

Alimentary has its roots in the Latin word *alimentarius,* which means "nourishment." The alimentary canal is responsible for ingesting, digesting, and absorbing the nutrients (nourishment) in the food you eat.

Want more practice? Go to: **MasteringA&P** > Study Area > Menu > Lab Tools > PAL > Cat > Digestive System > Abdominal cavity photos

B The large intestine is fairly easy to distinguish from the small intestine because it has less coiling and a larger diameter.

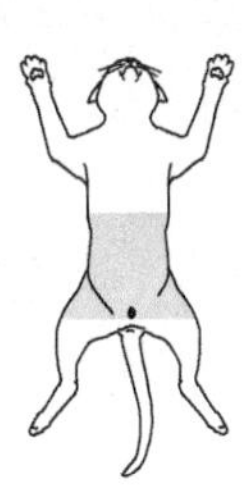

1 Identify the **cecum,** a blind sac that marks the beginning of the large intestine.

2 Locate the **ileocecal junction,** the connection between the ileum of the small intestine and the cecum of the large intestine.

Ileum Ascending colon Spleen

Ventral view

3 Cut open the cecum and wash out its contents. Locate the connection with the ileum and identify the **ileocecal valve.** This structure is a smooth, muscular sphincter, which regulates the passage of material (feces) into the large intestine.

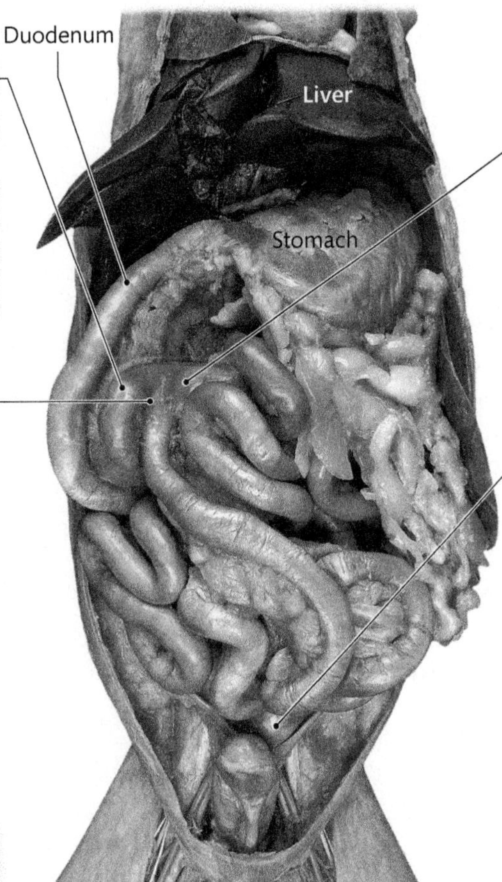

Ventral view

4 Identify the other segments of the large intestine:

From the cecum, the short **ascending colon** travels cranially along the right side of the abdominal cavity.

When it reaches the liver, the ascending colon bends to the left and gives rise to the **transverse colon,** which passes across the abdominal cavity toward the spleen (see the figure on the previous page).

Near the spleen, the transverse colon bends caudally and becomes the **descending colon,** which travels along the left side of the abdominal cavity.

Near its termination, the descending colon makes a slight medial bend as it empties into the **rectum** (not shown). The rectum descends along the dorsal midline of the abdominal and pelvic cavities and ends at the anus (not shown).

5 If your dissection session is ending here, close the body flaps and wrap your cat with the skin. Cover the animal with moist towels, and return it to the storage bag.

MAKING CONNECTIONS

Describe similarities and differences between the abdominal anatomy of the cat and the human. ▶

Similarities:

____

____

____

Differences:

____

____

____

BEFORE YOU MOVE ON ...

‹‹ LOOKING BACK

The digestive systems of cats and humans share many anatomical similarities. Both include the organs of the alimentary canal and various accessory digestive organs. The alimentary canal is a muscular tube that begins at the mouth and ends at the anus. It travels through the head, neck, and thoracic and abdominopelvic cavities. The alimentary canal includes the oral cavity, pharynx, esophagus, stomach, small intestine, and large intestine. The accessory digestive organs are the teeth, tongue, salivary glands, liver, gallbladder, and pancreas.

Many digestive organs have overlapping functions with other organ systems. For example, salivary glands have digestive and lymphatic functions. Salivary gland secretions contain the enzyme amylase, which chemically digests carbohydrates. Saliva also contains lysozyme, an antibacterial enzyme that provides protection from infections. The oropharynx and laryngopharynx are passageways for both food and air. Thus, they have digestive and respiratory functions.

Identify digestive organs in the abdominal cavity that have overlapping functions with other organ systems. How are these additional functions significant for normal digestive activity?

▶ ____

____

____

____

____

____

____

____

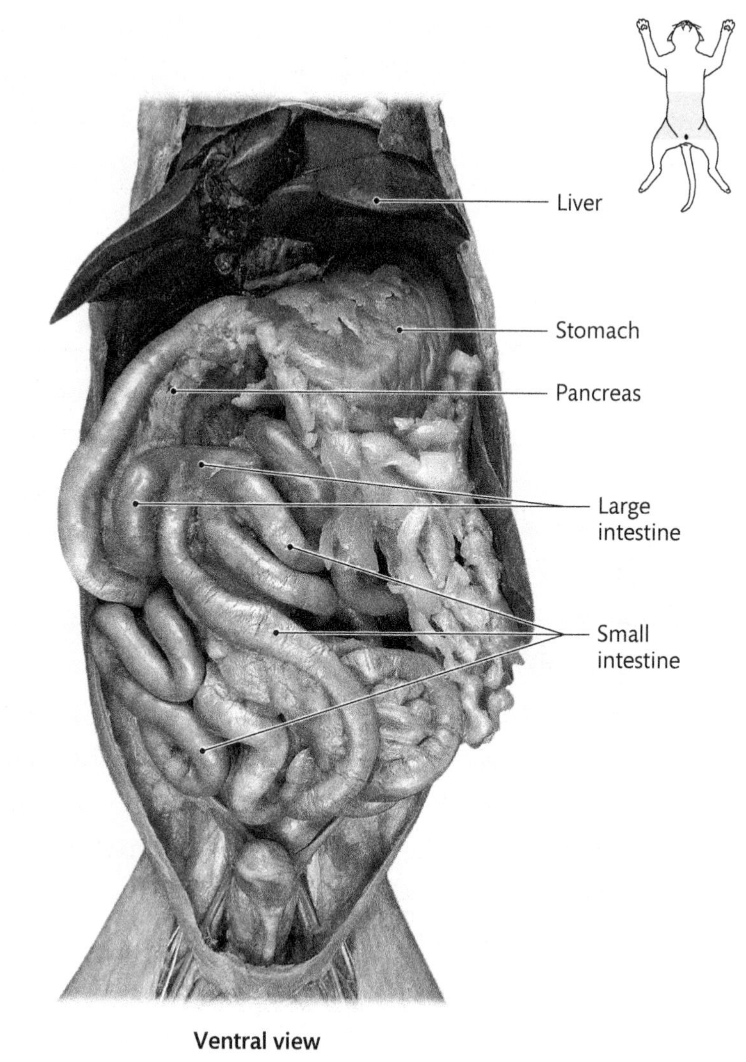

Ventral view

LOOKING FORWARD ››

One function of the digestive system is to release undigested substances and other wastes to the outside during defecation. In the next cat dissection exercise, you will dissect the cat urinary system, which eliminates waste products from the blood during urination.

Name ___________________________

Lab Section ___________________________

Date ___________________________

EXERCISE **C8** **REVIEW SHEET**

Dissection of the Cat Digestive System

1. What do you think would happen if the epiglottis did not function properly while swallowing food?

____

____

2. Describe the position of the esophagus in relation to the following organs:

a. Trachea

____

b. Heart

____

3. **a.** Describe the primary digestive functions of the small intestine.

____

____

b. What significance do the villi have in the digestive activities of the small intestine?

____

____

4. Define the following terms:

a. Alimentary canal

____

____

b. Peritoneum

____

____

QUESTIONS 5–11: Match the structure in column A with the organ or region in column B in which it is found. Answers in column B may be used more than once.

A	B
5. Duodenum ______	**a.** Stomach
6. Cecum ______	**b.** Oral cavity
7. Pyloric region ______	**c.** Head
8. Molar gland ______	**d.** Large intestine
9. Hard palate ______	**e.** Small intestine
10. Ileum ______	
11. Cardiac region ______	

QUESTIONS 12–22: Identify the labeled structures in the photo. Select your answers from the list below.

a. Ascending colon
b. Ileum
c. Spleen
d. Descending colon
e. Duodenum
f. Liver
g. Jejunum
h. Gallbladder
i. Stomach
j. Pancreas
k. Cecum

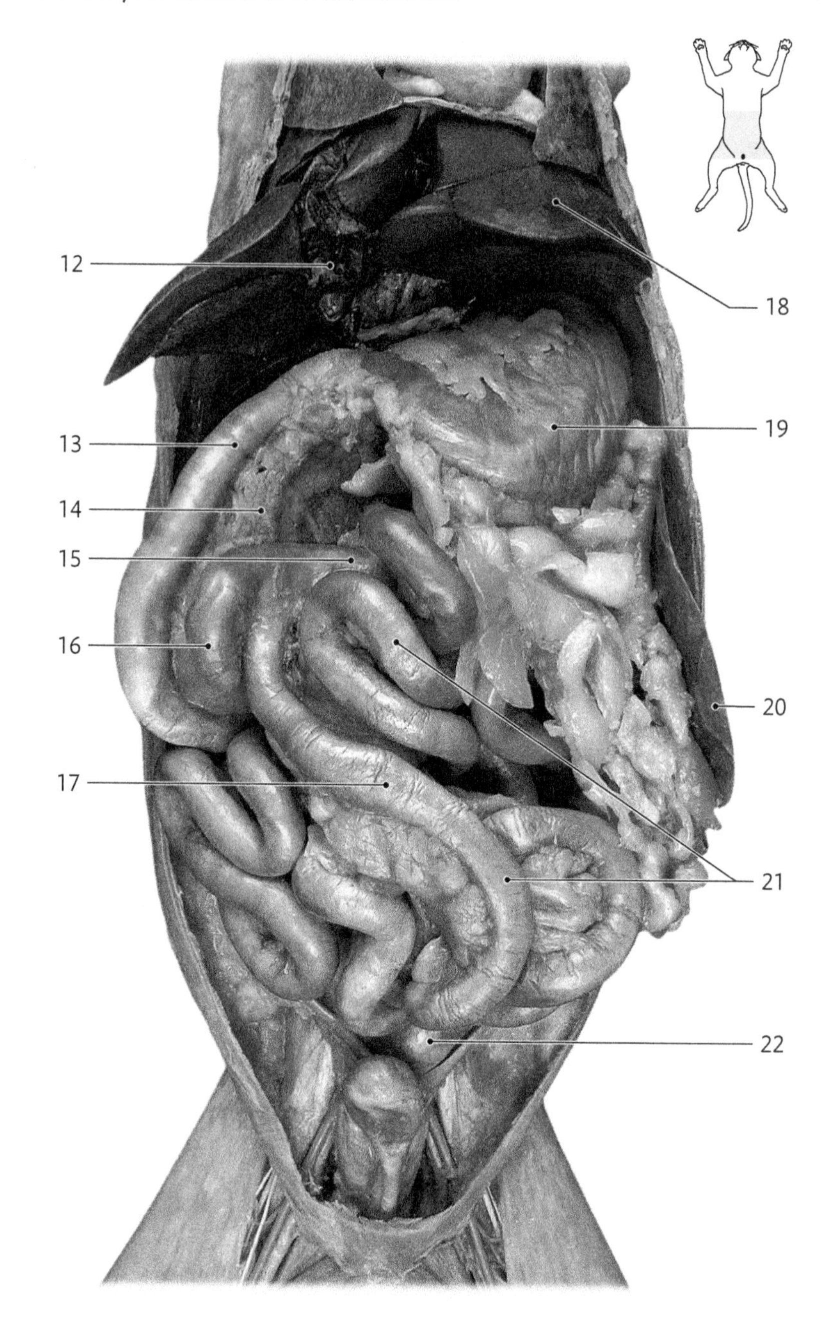

12. ______________________

13. ______________________

14. ______________________

15. ______________________

16. ______________________

17. ______________________

18. ______________________

19. ______________________

20. ______________________

21. ______________________

22. ______________________

DISSECTION EXERCISE

Dissection of the Cat Urinary System

LEARNING OUTCOMES

These Learning Outcomes correspond by number to the laboratory activities in this exercise. When you complete the activities, you should be able to:

Activity C9.1 **Describe the structure and function of the cat urinary system and make comparisons with similar organs in the human.**

Activity C9.2 **Describe the internal anatomy of the cat kidney and make comparisons with the human kidney.**

LABORATORY SUPPLIES

- Preserved cats
- Dissecting equipment: trays, tools, gloves
- Protective eyewear
- Face mask
- Anatomical models of the human urinary system

PRE-LAB QUIZ

Before you begin, read all the activities in Exercise C9 and the required reading in your textbook that is assigned by your instructor.

1. During this dissection exercise, you will identify all the following structures *except* the
   a. renal artery.
   b. duodenum.
   c. urethra.
   d. renal sinus.
   e. urinary bladder.
2. True or False: While dissecting the urinary system in the male cat, care should be taken to avoid damaging the ductus deferens (vas deferens) as it loops over the ureter.

   ____
3. True or False: While dissecting the urinary system in the female cat, care should be taken to not confuse the ureter with the vagina because both have a similar structure and are located in the same area. ______________
4. True or False: The hilum of the kidney is located on the lateral convex surface.

   ____
5. During this dissection exercise, you will examine the internal anatomy of the cat kidney by first making an incision along what plane of section? ______________

Questions 6–10: Match the structure in column A with the correct description in column B.

A

6. Urinary bladder ________
7. Renal vein ________
8. Prostate gland ________
9. Urethra ________
10. Renal papilla ________

B

a. In the male cat, a portion of the urethra travels through this structure.
b. In the kidney, this structure is the apex of the renal pyramid.
c. This structure transports blood from the kidney to the posterior vena cava.
d. The ureters transport urine to this structure.
e. In females, the vagina and this structure empty into the urogenital sinus.

Activity **C9.1**

Examining the Cat Urinary System

The cat urinary system shares a common embryological origin with the reproductive system. Accordingly, the term **urogenital system** is often used when referring to both urinary and reproductive structures. This exercise will highlight urinary system anatomy, but you will also see reproductive structures that are in proximity to the area of dissection.

Dissection notes:

- *If you have not already opened the ventral body cavities and made preliminary observations of their contents, these activities must first be completed. Complete Activity C4.1 before you proceed with this dissection.*
- *As you dissect the cat urinary system, have anatomical models of the human urinary system available for comparison.*

A The Urinary System in the Female Cat

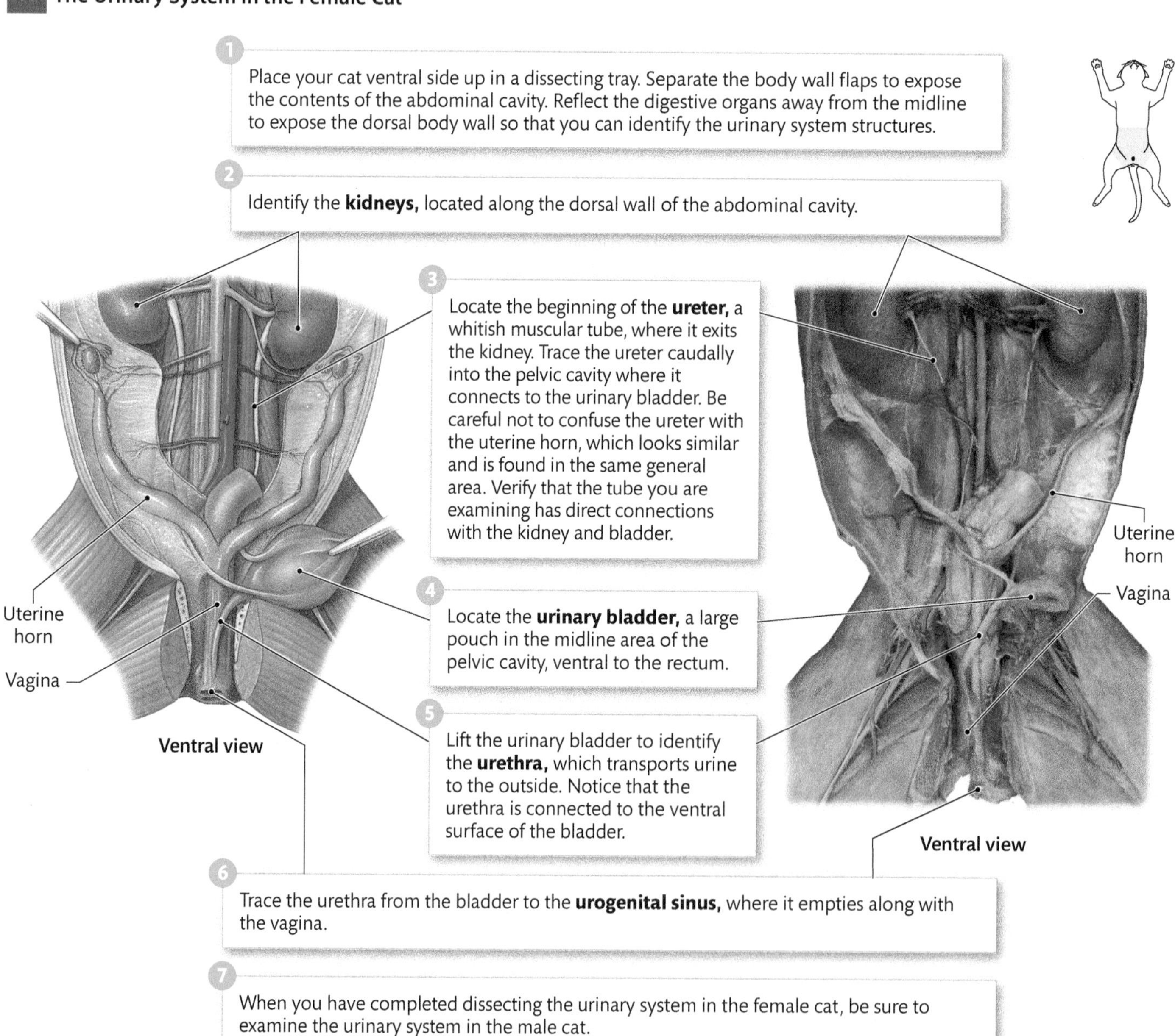

1. Place your cat ventral side up in a dissecting tray. Separate the body wall flaps to expose the contents of the abdominal cavity. Reflect the digestive organs away from the midline to expose the dorsal body wall so that you can identify the urinary system structures.

2. Identify the **kidneys,** located along the dorsal wall of the abdominal cavity.

3. Locate the beginning of the **ureter,** a whitish muscular tube, where it exits the kidney. Trace the ureter caudally into the pelvic cavity where it connects to the urinary bladder. Be careful not to confuse the ureter with the uterine horn, which looks similar and is found in the same general area. Verify that the tube you are examining has direct connections with the kidney and bladder.

4. Locate the **urinary bladder,** a large pouch in the midline area of the pelvic cavity, ventral to the rectum.

5. Lift the urinary bladder to identify the **urethra,** which transports urine to the outside. Notice that the urethra is connected to the ventral surface of the bladder.

6. Trace the urethra from the bladder to the **urogenital sinus,** where it empties along with the vagina.

7. When you have completed dissecting the urinary system in the female cat, be sure to examine the urinary system in the male cat.

B The Urinary System in the Male Cat

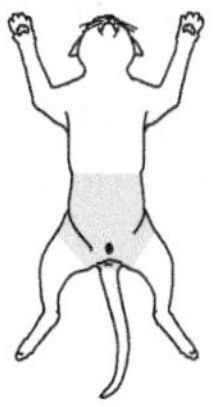

1. Place your cat ventral side up in a dissecting tray. Separate the body wall flaps to expose the contents of the abdominal cavity. Reflect the digestive organs away from the midline to expose the dorsal body wall so that you can identify the urinary system structures.

2. Identify the **kidneys,** located along the dorsal wall of the abdominal cavity.

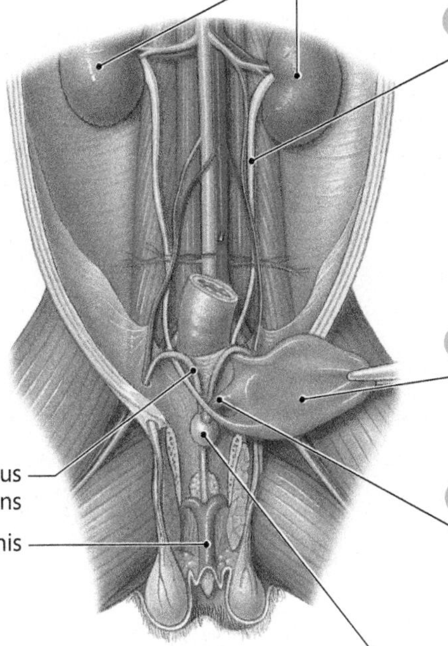

3. Locate the beginning of the **ureter,** a whitish muscular tube, where it exits the kidney. Trace the ureter caudally into the pelvic cavity where it connects to the urinary bladder. Be careful not to damage the ductus deferens (vas deferens), which is embedded in fat as it loops over the ureter dorsal to the bladder.

4. Locate the **urinary bladder,** a large pouch found in the midline area of the pelvic cavity, ventral to the rectum.

5. Lift the urinary bladder to identify the **urethra,** which transports urine to the outside. Notice that the urethra is connected to the ventral surface of the bladder.

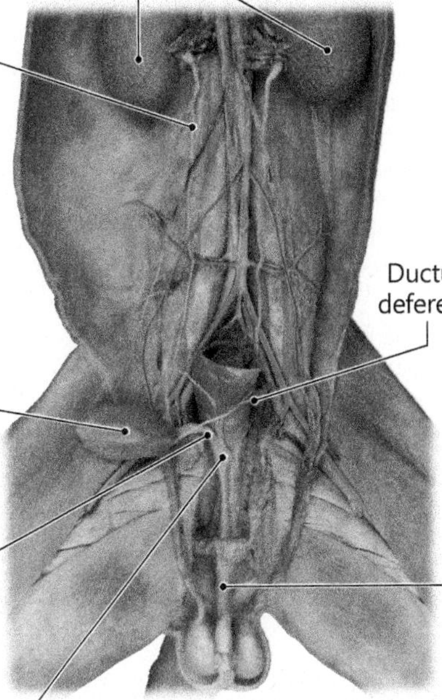

Ductus deferens

Penis

Ventral view

Ductus deferens

Penis

Ventral view

6. Trace the urethra from the bladder to the **prostate gland,** through which it travels. From the prostate gland, the urethra continues toward and enters the penis. Do not dissect this portion of the urethra at this time.

7. When you have completed dissecting the urinary system in the male cat, be sure to examine the urinary system in the female cat.

MAKING CONNECTIONS

Crystals in your pet cat's urine are partially blocking his urethra. The veterinarian recommends feeding him moist rather than dry foods and providing him with plenty of water. Why do you think the veterinarian made this recommendation?

▶ ____

____

____

____

Activity **C9.2**

Examining the Structure of the Cat Kidney

A The cat kidneys are covered by the peritoneal membrane and thus are retroperitoneal organs.

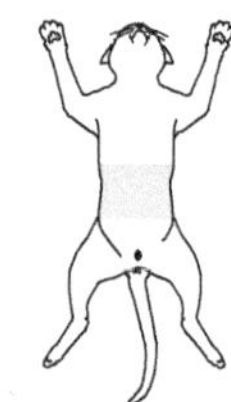

1. Carefully cut away the peritoneum that covers one kidney. Notice that the kidney has a lateral convex surface and a medial concave surface. The indentation on the medial surface is called the **hilum** of the kidney.

2. The hilum is the region where blood vessels, nerves, and the ureter enter or exit the kidney. Identify the following structures as they enter or exit the kidney:
   - **Renal artery**
   - **Renal vein**
   - **Ureter**

3. Cut the renal artery, renal vein, and ureter near the hilum and remove the kidney from the body cavity.

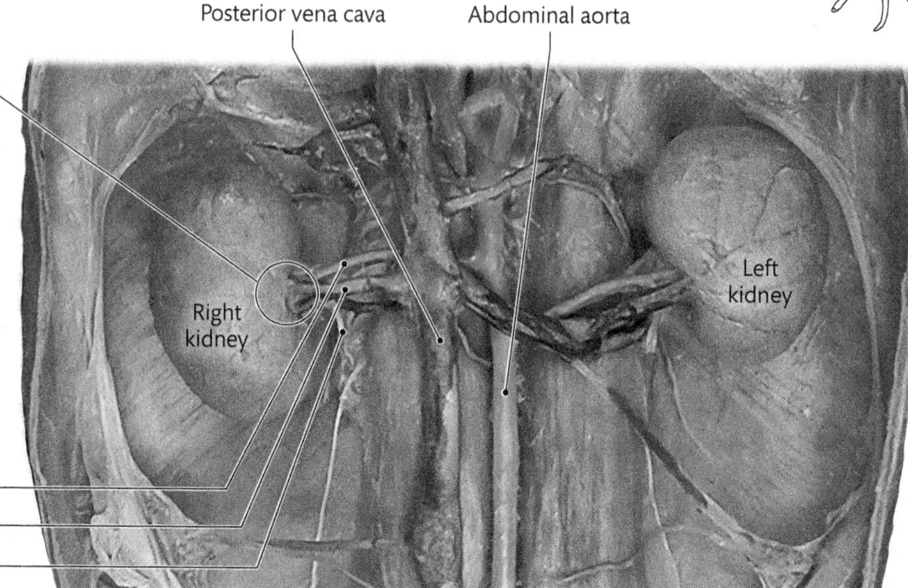

Ventral view

IN THE CLINIC

The Adipose Capsule of the Kidney

In both cats and humans, a pad of fat known as the **adipose capsule** surrounds the kidneys. Human kidneys are usually secured within their capsule, but in the cat, the kidneys can move more freely. This increase in mobility does not cause any harm to the cat. A mobile human kidney could result in serious urinary problems, however. For example, excessive movement could create an abnormal bend in the ureter (like a kink in a garden hose), which would obstruct the flow of urine to the urinary bladder.

B Internally, the cat kidney contains three regions: the renal cortex, the renal medulla, and the renal sinus.

1. Using a scalpel, make a longitudinal incision through the kidney, beginning at the lateral border and cutting toward the hilum. The plane of the section (frontal or coronal) should run parallel to the anterior surface of the kidney.

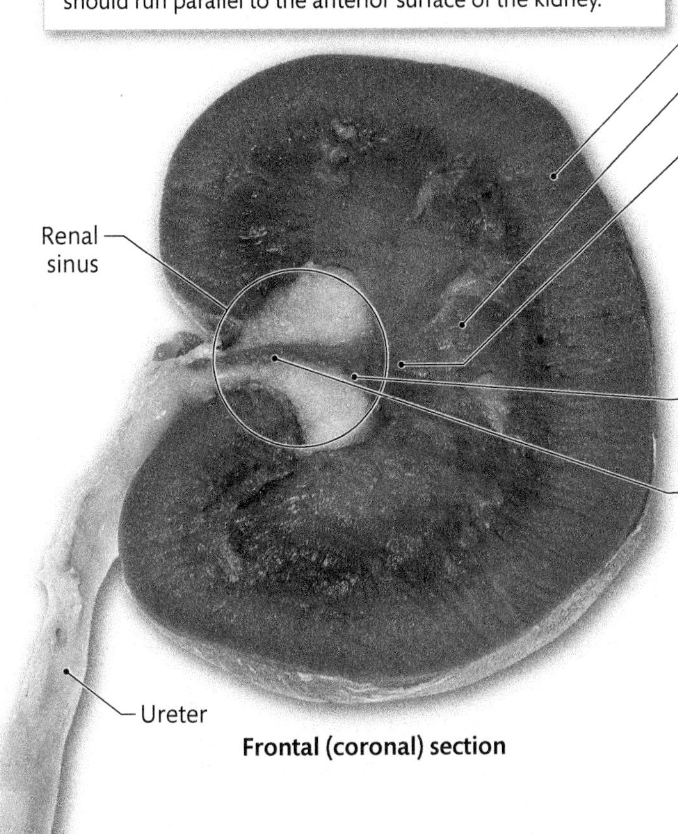

Frontal (coronal) section

2. Identify the **renal cortex,** the outer region of kidney tissue.

3. Identify the renal medulla, the darker region deep to the cortex. It consists of a single **renal pyramid.**

   The apex of the renal pyramid, the **renal papilla,** is directed inward, toward the hilum.

4. The **renal sinus** is the central cavity where urine is drained before entering the ureter. Remove the connective tissue that covers the sinus by lifting the tissue with tweezers and carefully cutting it away with a scalpel. When the connective tissue is removed, identify the two regions of the renal sinus:

   The **calyx** is the region that surrounds the renal papilla.

   The **renal pelvis** is the funnel-shaped region that leads directly into the ureter.

5. If your dissection session is ending here, close the body flaps and wrap your cat with the skin. Cover the animal with moist towels, and return it to the storage bag.

Word Origins

Ureter has its origin in the Greek word *oureter,* which means "urinary canal." The ureter is a muscular tube or canal that transports urine from the kidney to the urinary bladder.

MAKING CONNECTIONS

Examine the internal anatomy of the human kidney on models that are available in the laboratory and compare it with the internal anatomy of the cat kidney. Identify similarities and differences between the two species. ▶

Similarities:

____

____

____

Differences:

____

____

____

■ Want more practice? Go to: **MasteringA&P** > Study Area > Menu > Lab Tools > PAL > Cat > Urinary System > Sheep Kidney

BEFORE YOU MOVE ON ...

« LOOKING BACK

The principal organs of the cat urinary system are the **kidneys.** During the process of producing urine, the kidneys remove metabolic wastes and toxins from the blood but conserve water, electrolytes, and glucose. The kidneys also have an endocrine function: They produce the hormones erythropoietin, renin, and calcitriol.

The **ureters, urinary bladder,** and **urethra** are accessory organs of the urinary system. Collectively, these structures transport, store, and excrete urine.

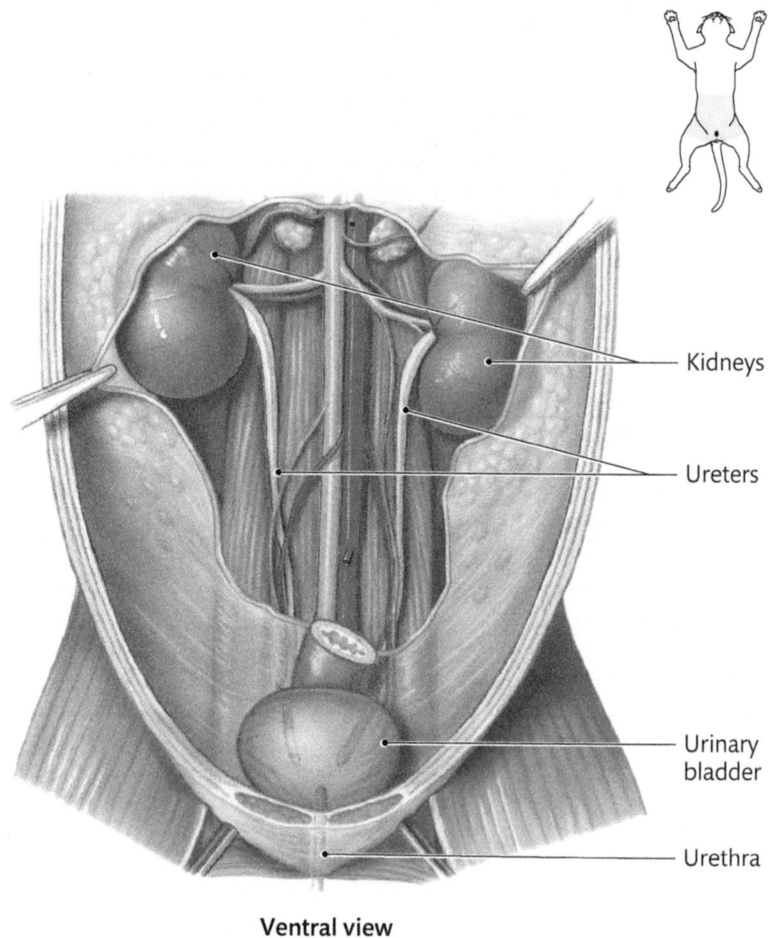

Ventral view

- **Urine production is important for maintaining osmotic and pH balance in blood plasma. Can you think of a reason why?**

▶ ______________________________

____

____

____

____

____

- **How is the kidney's endocrine function connected to cardiovascular function?**

▶ ______________________________

____

____

____

____

____

LOOKING FORWARD »

The urinary and reproductive systems arise from similar embryological tissues, and many of the adult structures are found in the pelvic cavity. In this cat dissection exercise, your focus was on the anatomy of the urinary system. In the next cat dissection exercise, you will turn your attention to the reproductive system.

Name ________________________________

Lab Section ________________________________

Date ________________________________

EXERCISE C9 REVIEW SHEET

Dissection of the Cat Urinary System

1. The kidneys and the ureters are retroperitoneal. Is the urinary bladder also a retroperitoneal organ, or is it connected to the body wall by a mesentery? Explain your answer.

____

____

____

2. Most structures in the body receive blood from collateral arteries, which establish alternate blood pathways and maintain normal blood flow if the main pathway is blocked. Is that true for the blood supply to the kidneys? Explain your answer.

____

____

____

3. In your dissection, you observed that the ureters have close anatomical relationships with structures in the reproductive system. For each sex, identify one nearby reproductive system structure and its relation to the ureter.

a. Female cat

____

____

b. Male cat

____

____

4. In the female cat, the urethra and the vagina empty into a common space, the urogenital sinus. How does this arrangement compare with the arrangement in human females?

____

____

____

5. Compare the pathway of the male urethra in the cat with the pathway of the male urethra in the human.

____

____

____

QUESTIONS 6–14: Match the term in column A with the appropriate description in column B.

A	B
6. Renal sinus ____________	**a.** In the cat kidney, only one of these structures is found in the medulla
7. Ureter ____________	**b.** In the cat kidney, the portion of the renal sinus that surrounds the renal papilla
8. Renal pyramid ____________	**c.** Region along the medial surface of the kidney where blood vessels and the ureter enter or exit
9. Renal pelvis ____________	**d.** Central cavity in the kidney where urine is drained
10. Renal cortex ____________	**e.** The outer tissue layer in the kidney
11. Urethra ____________	**f.** In the female cat, the region where both the urethra and the vagina empty
12. Hilum of the kidney ____________	**g.** Funnel-shaped portion of the renal sinus that leads to the ureter
13. Urogenital sinus ____________	**h.** In the male cat, a portion of this duct travels through the prostate gland
14. Calyx ____________	**i.** A muscular tube that transports urine from a kidney to the urinary bladder

QUESTIONS 15–24: Identify the labeled structures in the photo. Select your answers from the list below.

a. Urethra
b. Posterior vena cava
c. Right kidney
d. Prostate gland
e. Left ureter
f. Right ureter
g. Abdominal aorta
h. Urinary bladder
i. Left kidney
j. Renal vein

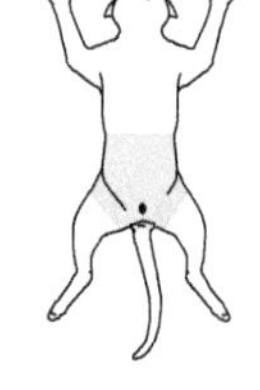

15. ____________________

16. ____________________

17. ____________________

18. ____________________

19. ____________________

20. ____________________

21. ____________________

22. ____________________

23. ____________________

24. ____________________

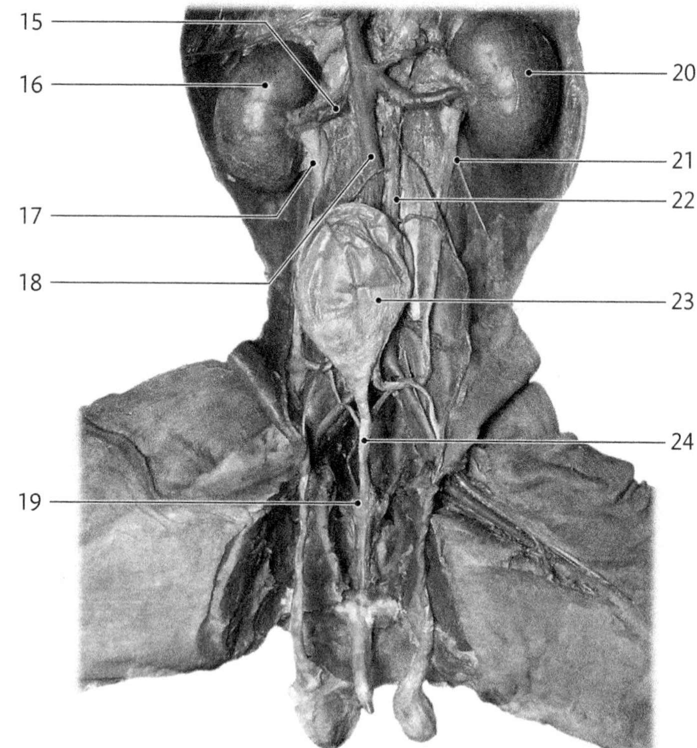

Ventral view

C10

Dissection of the Cat Reproductive System

LEARNING OUTCOMES

These Learning Outcomes correspond by number to the laboratory activities in this exercise. When you complete the activities, you should be able to:

Activity C10.1 **Describe the structure and function of the male reproductive system in the cat and make comparisons with similar organs in the human.**

Activity C10.2 **Describe the structure and function of the female reproductive system in the cat and make comparisons with similar organs in the human.**

LABORATORY SUPPLIES

- Preserved cats
- Dissecting equipment: trays, tools, gloves
- Bone cutters
- Protective eyewear
- Face mask
- Anatomical models of the male and female human reproductive systems

PRE-LAB QUIZ

Before you begin, read all the activities in Exercise C10 and the required reading in your textbook that is assigned by your instructor.

1. During this cat dissection exercise, you will identify all the following structures in the male *except* the
   a. prostate.
   b. testes.
   c. urethra.
   d. urogenital sinus.
   e. shaft of the penis.
2. During this cat dissection exercise, you will identify all the following structures in the female *except* the
   a. urethra.
   b. uterine body.
   c. bulbourethral glands.
   d. uterine tubes.
   e. vagina.
3. During this dissection exercise, you will get a clear view of the reproductive systems by using bone cutters to cut through what structure? ______________________
4. True or False: In both cat and human female reproductive systems, the vagina and the urethra unite and have a common opening. ____________
5. True or False: In males of both cats and humans, the urethra has a reproductive function and a urinary function. ____________
6. All the following structures are found in the female reproductive system of both the cat and the human *except* the
   a. uterine horns.
   b. uterine tubes.
   c. uterine body.
   d. cervix.
   e. vagina.
7. True or False: In males of both cats and humans, the testes originate in the abdominal cavity and travel into the scrotum during fetal development. ____________
8. The funnel-shaped opening at the lateral end of the uterine tube is called the cervical opening. ______________________
9. In female cats, the folds of skin on either side of the urogenital orifice are called the ______________________.
10. In the male cat, a portion of the ductus deferens, along with testicular blood vessels and nerves, forms the ______________________.

Activity **C10.1**

Examining Structures in the Reproductive System of the Male Cat

Both the male and female reproductive systems in the cat develop from similar embryonic tissue. During the first few weeks of development, male embryos are indistinguishable from female embryos. As development proceeds, the reproductive structures differentiate into two distinct organ systems.

Dissection notes:

- *If you have not already opened the ventral body cavities and made preliminary observations of their contents, these activities must first be completed. Complete Activity C4.1 before you proceed with this dissection.*
- *As you dissect the cat reproductive systems, have anatomical models of the human reproductive systems available for comparison.*

A The testes in the cat (and the human) form in the abdominal cavity and move into the scrotum during fetal development.

1. Place your cat ventral side up in a dissecting tray. Use a pair of bone cutters to cut through the **symphysis pubis,** the cartilaginous joint between the two pubic bones, and spread the bones apart. This procedure will provide a clear view of the entire male reproductive system.

Ureter

Entrances to inguinal canal

Pubic bone (cut)

2. The **scrotum** will appear as a swollen sac ventral to the base of the tail and anus. Carefully cut the wall of the scrotum. The membrane that lines its inner wall is an extension of the peritoneum.

3. Notice that the scrotum is divided into two compartments, each one containing a **testis** (plural = **testes**).

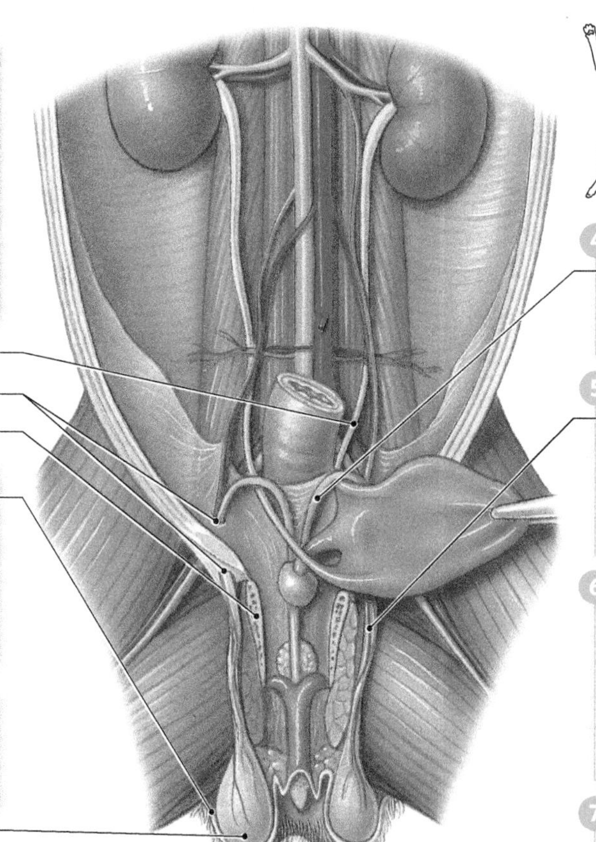

Ventral view

4. Locate the **ductus (vas) deferens** as it crosses over the ureter close to the base of the urinary bladder.

5. Trace the pathway of the ductus deferens laterally. Notice that it is joined by testicular blood vessels and nerves to form the **spermatic cord.**

6. Carefully dissect and trace the pathway of the spermatic cord to the testes. Each spermatic cord will leave the abdominal cavity and enter the scrotum by traveling through the **inguinal canal.**

7. Locate the **epididymis** (not shown). It is a highly coiled tube that arises from several small ducts in the testis and rests on its dorsal surface. Follow the convoluted pathway of the epididymis and notice its connection to the ductus deferens.

B Similar to the human, the cat penis becomes erect when the erectile tissue in the corpora cavernosa and corpus spongiosum becomes engorged with blood.

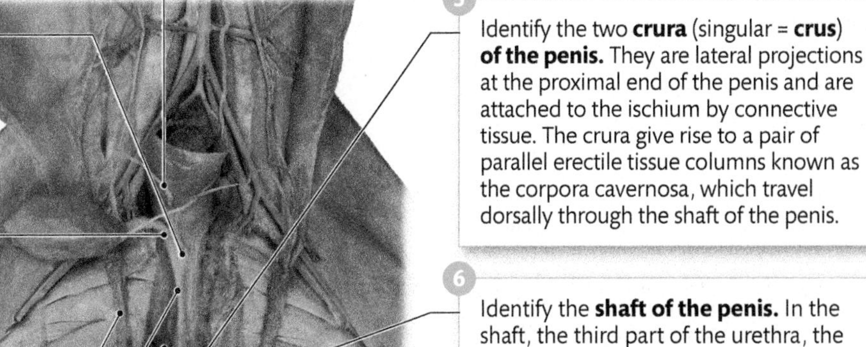

1. Identify the **prostate gland.** It is a round glandular structure that can be found in the midline, dorsal to the symphysis pubis, and near the proximal end of the penis.

2. Identify the **urethra,** where it originates from the bladder. Notice that it travels with the ductus deferens for a short distance before entering the prostate gland, where the two ducts converge. The portion of the urethra that passes through the prostate gland is called the prostatic urethra. Inside the prostate, the urethra becomes a urogenital tube that transports sperm and seminal fluids from the ductus deferens and prostate as well as urine from the urinary bladder.

3. Identify the urethra as it exits the prostate gland and passes caudally toward the penis. This segment of the urethra is known as the **membranous urethra.**

4. Two small **bulbourethral glands** are located on each side of the membranous urethra as it approaches the penis. Identify one of these glands. Their secretions enter the membranous urethra and contribute to the seminal fluids.

5. Identify the two **crura** (singular = **crus**) **of the penis.** They are lateral projections at the proximal end of the penis and are attached to the ischium by connective tissue. The crura give rise to a pair of parallel erectile tissue columns known as the corpora cavernosa, which travel dorsally through the shaft of the penis.

6. Identify the **shaft of the penis.** In the shaft, the third part of the urethra, the penile urethra, travels through a ventral column of erectile tissue called the corpus spongiosum.

7. Identify the **glans penis,** which is the cone-shaped structure at the distal end of the shaft. At the tip of the glans penis, the penile urethra opens to the outside at the urogenital orifice. The shaft of the penis is usually retracted in a fold of skin known as the prepuce, but it emerges when the cat copulates.

8. When you have completed dissecting the reproductive organs in your cat, be sure to examine the reproductive organs in a female cat.

9. If your dissection session is ending here, close the body flaps and wrap your cat with the skin. Cover the animal with moist towels and return it to the storage bag.

MAKING CONNECTIONS

Because of our upright, bipedal position, humans (but not cats and other quadrupeds) are susceptible to an inguinal hernia. This condition is characterized by a loop of intestine being displaced into the inguinal canal. Speculate on a reason why males are more vulnerable than females to this disorder.

▶ ______________________________

____

____

Activity **C10.2**

Examining Structures in the Reproductive System of the Female Cat

A In cats and other quadruped mammals, two uterine horns extend from the body of the uterus. In humans, the uterine horns are not present.

1. Place your cat ventral side up in a dissecting tray. Use a pair of bone cutters to cut through the **symphysis pubis,** the cartilaginous joint between the two pubic bones, and spread the bones apart. This procedure will provide a clear view of the entire female reproductive system.

2. The uterus is located along the dorsal body wall. Identify the following regions of the uterus:

The **uterine horns** extend from the uterine body in a wavy course to the left and right along the dorsal body wall.

The **uterine body** can be found along the midline, just dorsal to the urinary bladder.

The **cervix** is the tapered caudal end of the uterus, which projects into the vagina (not shown).

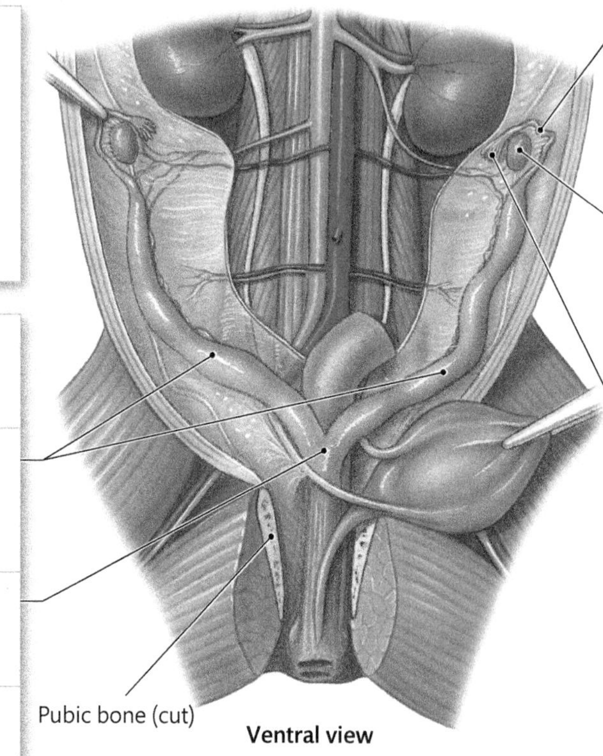

3. The narrow distal ends of the uterine horns are the **uterine tubes (oviducts).** Locate the uterine tube on one side of the body.

4. The **ovaries** are oval-shaped structures located along the lateral walls of the body cavity. Locate an ovary on one side of the body.

5. At the lateral end of the uterine tube, identify a funnel-shaped opening called the **infundibulum.** At the free margin of the infundibulum, fingerlike projections called fimbriae drape over the ovary. During ovulation, maturing egg cells are released from the ovary and are drawn into the oviduct by the wavelike actions of the fimbriae. Fertilization and early development of the embryo occur in the oviduct prior to implantation in the uterine horn.

IN THE CLINIC

Multiple Births

In cats and many other mammals, multiple births are more common than in humans. Two factors could account for this difference. First, in humans, usually only one mature egg is ovulated during each ovarian cycle, but in other mammalian species, several eggs can be released from the ovary. Second, in humans, the lack of uterine horns restricts fetal development to the body of the uterus. In other mammals, the presence of uterine horns provides more surface area for the implantation of embryos into the uterine wall and more space for the development of several fetuses at once.

 Want more practice? Go to: MasteringA&P > Study Area > Menu > Lab Tools > PAL > Cat > Reproductive System > Female

B In the cat, the vagina and the urethra have a common opening to the outside.

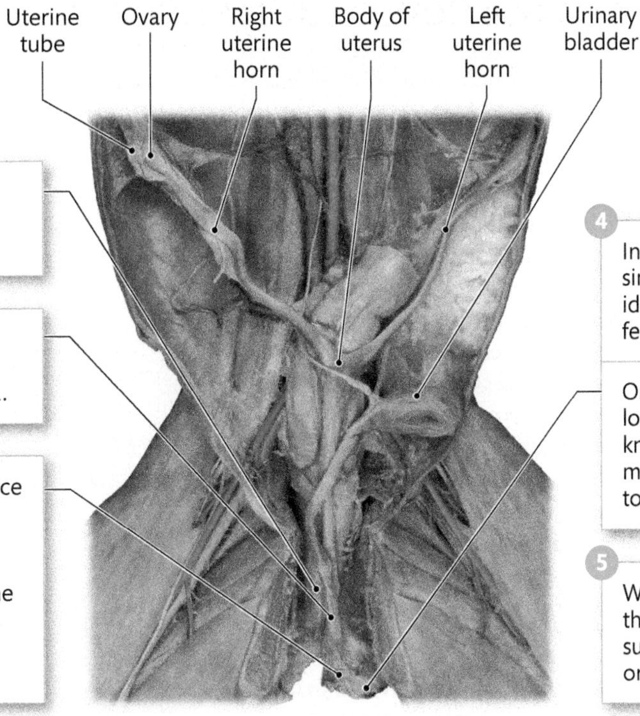

Ventral view

1. Locate the **vagina,** which is a straight, muscular tube that passes caudally from the cervix.

2. Identify the **urethra.** From the urinary bladder, it travels along the ventral surface of the vagina.

3. Find the **urogenital sinus.** Notice that it is a common passageway for both the vagina and the urethra. Its termination is at the urogenital orifice. In humans, the vagina and urethra do not unite and therefore have separate openings in the vestibule.

4. In the ventral wall of the urogenital sinus, near the urogenital orifice, identify the **clitoris** (not shown), the female homologue of the penis.

   On either side of the urogenital orifice, locate the longitudinal folds of skin known as the **labia majora.** The labia minora are also present but are probably too small for a clear identification.

5. When you have completed dissecting the reproductive organs in your cat, be sure to examine the reproductive organs in a male cat.

6. If your dissection session is ending here, close the body flaps and wrap your cat with the skin. Cover the animal with moist towels, and return it to the storage bag.

MAKING CONNECTIONS

Compare the female reproductive systems in cats and humans by identifying similarities and differences between the two species. ▶

Similarities:

____

____

____

Differences:

____

____

____

BEFORE YOU MOVE ON ...

« LOOKING BACK

Despite the obvious structural differences between the sexes that you observed in this dissection, the adult reproductive systems share some general similarities. For example, both male and female reproductive systems consist of primary sex organs, or gonads, and accessory sex organs. The gonads include the testes in the male and the ovaries in the female. In both sexes, the gonads produce gametes and sex hormones. The accessory sex organs include various internal glands and ducts as well as the external genitalia that are found outside the pelvic cavity. In both males and females, these structures nourish, support, and transport the gametes. The accessory organs in the female reproductive system also have the unique function of supporting the development of the fetus during pregnancy and facilitating the birth process.

The sex hormones produced by the testes and ovaries (androgens and estrogens, respectively) are vital for the development, maturation, and maintenance of the reproductive organs. In addition, sex hormones have a wide range of effects on other organ systems in the body. How do sex hormones influence the structure and function of the following organ systems?

- **Integumentary system**
- **Skeletal system**
- **Muscular system**
- **Nervous system**
- **Endocrine system**

▶ ____

____

____

____

____

____

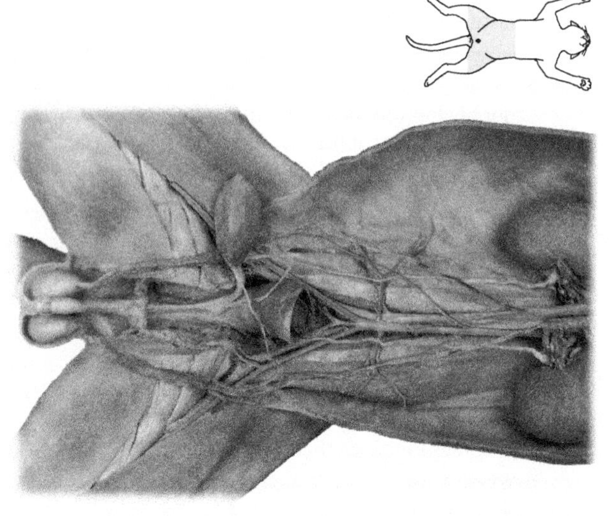

Male

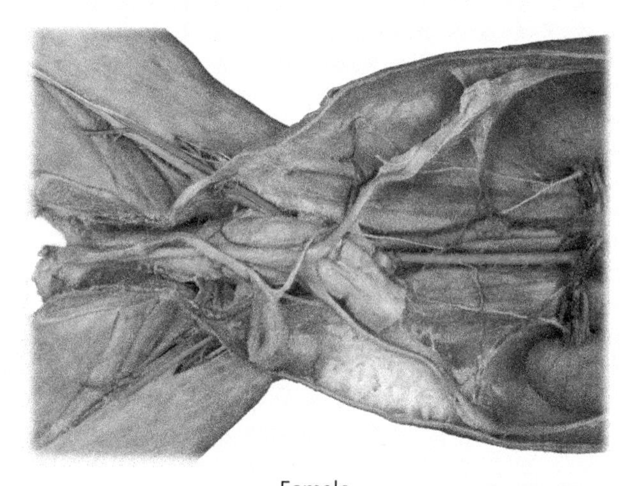

Female

Ventral views

LOOKING FORWARD »

Dissecting a whole body, whether it is a human cadaver or a cat, provides students a unique opportunity to study the three-dimensional relationships of organs and organ systems. Now that you have completed a whole body dissection, you likely have an appreciation for how much can be learned with a little patience, dissecting skill, and keen observation. Also, the new knowledge you have gained from this dissection will help you better understand the intimate connection between anatomy and physiology at the highest level of organization: the organism. One key concept that was underscored throughout this cat dissection was the integration of function among all the organ systems. Rather than operating as independent entities, the normal operation of each organ system is dependent on the normal operation of the others, and they all must work together to sustain the life of the entire organism.

Name ______________________

Lab Section ______________________

Date ______________________

EXERCISE **C10** **REVIEW SHEET**

Dissection of the Cat Reproductive System

1. In both the male and female cats, identify structures that have a urogenital function.

a. Male cat

____

____

____

b. Female cat

____

____

____

QUESTIONS 2–11: Match the cat reproductive structure in column A with the appropriate description in column B.

A

2. Prostate gland ______________________

3. Testes ______________________

4. Crura of the penis ______________________

5. Bulbourethral glands ______________________

6. Body of the uterus ______________________

7. Uterine horns ______________________

8. Ovaries ______________________

9. Vagina ______________________

10. Urogenital orifice ______________________

11. Uterine tubes ______________________

B

a. The fimbriae of these structures drape over the ovaries

b. In the male, secretions from these structures enter the membranous urethra

c. Lateral extensions of the uterus

d. In the female, the urogenital sinus is a common passageway for the urethra and this structure

e. In the male, this structure is located along the midline dorsal to the symphysis pubis

f. In the female, this structure is located along the midline, dorsal to the urinary bladder

g. Lateral projections at the proximal end of the penis

h. Maturing egg cells are released from these structures

i. These structures develop in the abdominal cavity but migrate to the scrotum

j. Common opening to the outside for the urinary and reproductive systems

QUESTIONS 12–19: Identify the labeled structures in the photo. Select your answers from the list below.

a. Testis
b. Glans penis
c. Prostate gland
d. Spermatic cord
e. Ductus deferens
f. Membranous urethra
g. Bulbourethral gland
h. Shaft of the penis

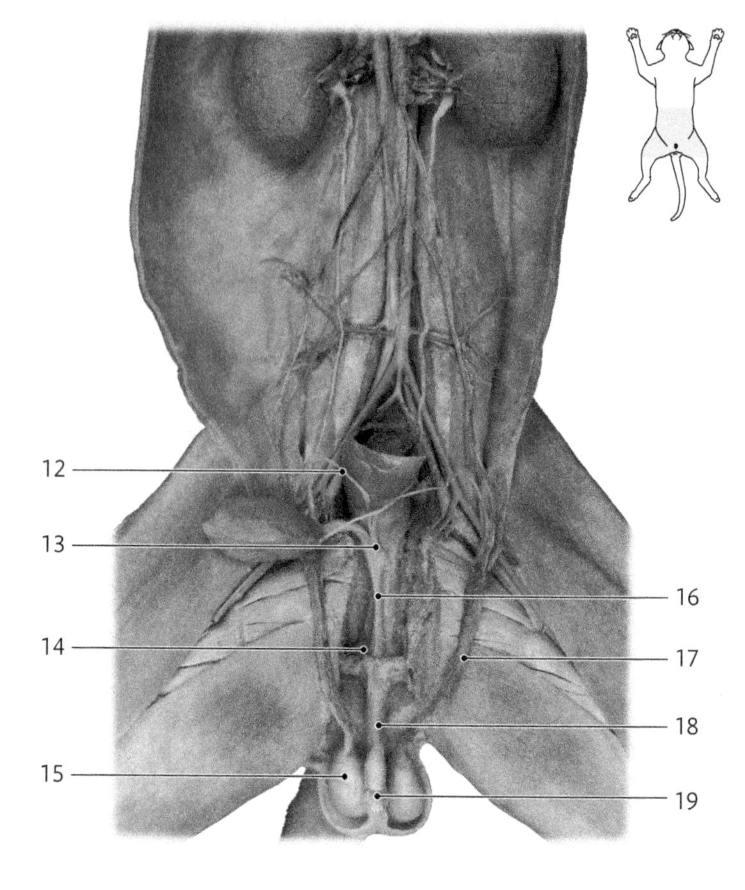

12. ______________________________
13. ______________________________
14. ______________________________
15. ______________________________
16. ______________________________
17. ______________________________
18. ______________________________
19. ______________________________

QUESTIONS 20–25: Identify the labeled structures in the photo. Select your answers from the list below.

a. Vagina
b. Left uterine horn
c. Uterine body
d. Ovary
e. Uterine tube (oviduct)
f. Right uterine horn

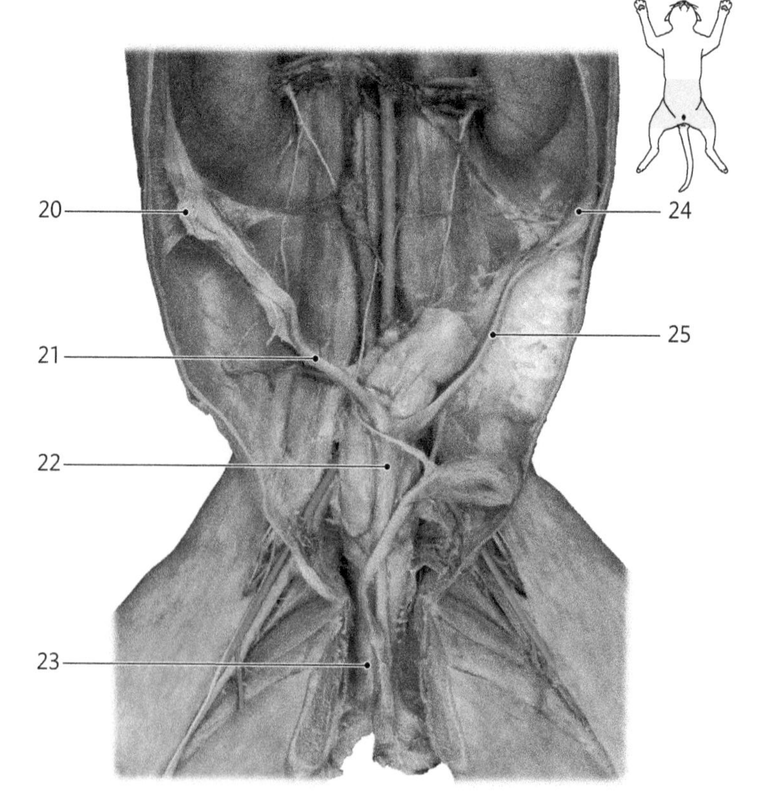

20. ______________________________
21. ______________________________
22. ______________________________
23. ______________________________
24. ______________________________
25. ______________________________

Appendix A

Understanding Universal Precautions for Handling Biospecimens

Universal precautions should be used when you are handling or if you could potentially be exposed to any human body fluids. Review the following universal precautions guidelines and be sure that you understand them before proceeding with any activities in this laboratory manual that involve collecting and handling human body fluids.

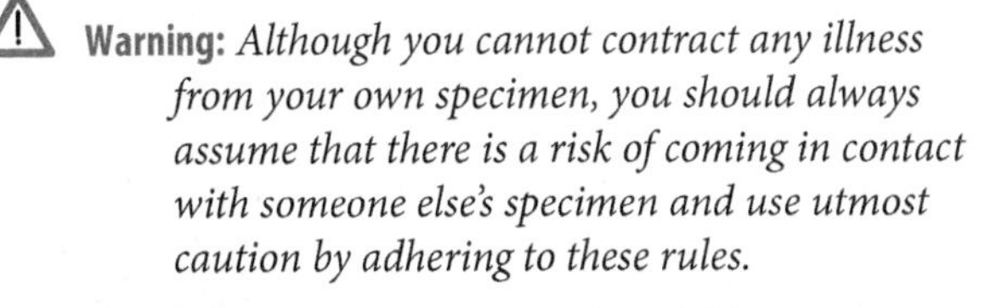

Warning: *Although you cannot contract any illness from your own specimen, you should always assume that there is a risk of coming in contact with someone else's specimen and use utmost caution by adhering to these rules.*

A Work with extreme caution when using human body fluids in the laboratory.

1. Always assume that the body fluid with which you are working is infected with a disease-causing organism.
2. Work with your own body fluids only. Under no circumstances should you collect or conduct experiments with the body fluids of another individual.
3. Always wear **gloves, safety eyewear,** and a **mask** when working with body fluids. Never allow body fluids to come in contact with unprotected skin. Protective gear should be worn throughout the activities and during cleanup.

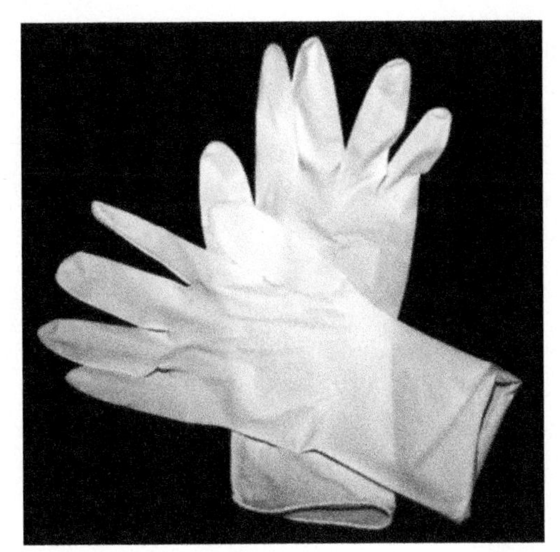

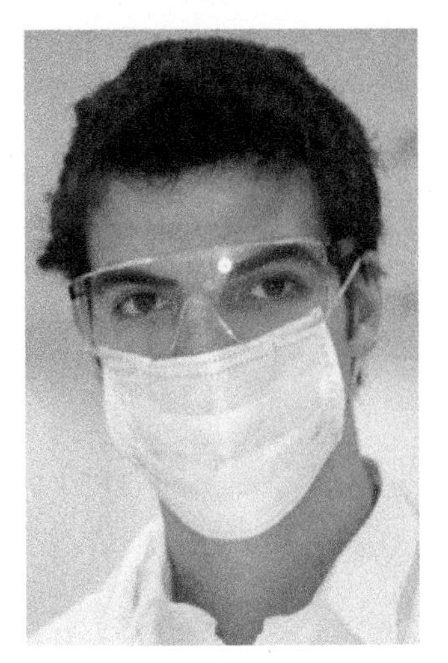

B Protect yourself from exposure to body fluids.

1. If any part of your skin is accidentally contaminated with a body fluid, disinfect the area immediately with a **70% alcohol solution** for at least 30 seconds followed by a 1-minute soap scrub and then rinsing.
2. Any body fluid that spills onto your work area should immediately be disinfected with a **10% bleach solution** or a commercially prepared disinfectant. The contaminated area should remain covered with the bleach/disinfectant for at least 30 minutes before wiping it off.

C When your work is completed, thoroughly clean your equipment and work area.

1. Disinfect all reusable glassware and other instruments in a 10% bleach solution for 30 minutes and then wash in hot, soapy water. If available, autoclaving this equipment before washing, rather than disinfecting with bleach, is recommended.
2. Clean work areas with a 10% bleach solution or a commercially prepared disinfectant.
3. Deposit gloves, masks, and any toweling used for cleanup in an appropriate biohazard container. Never place this material in the regular garbage.

D If you prefer not to work with body fluids, alternatives are available.

1. Simulated blood and urine are commercially available.
2. Simulated body fluid samples can be used to complete the activities in this laboratory manual without the risk of being exposed to human body fluids.

Appendix B

Weights and Measures

TABLE 1 The U.S. System of Measurement

Physical Property	Unit	Relationship to Other U.S. Units	Relationship to Household Units
Length	inch (in.)	1 in. = 0.083 ft	
	foot (ft)	1 ft = 12 in.	
		= 0.33 yd	
	yard (yd)	1 yd = 36 in.	
		= 3 ft	
	mile (mi)	1 mi = 5280 ft	
		= 1760 yd	
Volume	fluidram (fl dr)	1 fl dr = 0.125 fl oz	
	fluid ounce (fl oz)	1 fl oz = 8 fl dr	= 6 teaspoons (tsp)
		= 0.0625 pt	= 2 tablespoons (tbsp)
	pint (pt)	1 pt = 128 fl dr	= 32 tbsp
		= 16 fl oz	= 2 cups (c)
		= 0.5 qt	
	quart (qt)	1 qt = 256 fl dr	= 4 c
		= 32 fl oz	
		= 2 pt	
		= 0.25 pt	
	gallon (gal)	1 gal = 128 fl oz	
		= 8 pt	
		= 4 qt	
Mass	grain (gr)	1 gr = 0.002 oz	
	dram (dr)	1 dr = 27.3 gr	
		= 0.063 oz	
	ounce (oz)	1 oz = 437.5 gr	
		= 16 dr	
	pound (lb)	1 lb = 7000 gr	
		= 256 dr	
		= 16 oz	
	ton (t)	1 t = 2000 lb	

TABLE 2 The Metric System of Measurement

Physical Property	Unit	Relationship to Standard Metric Units	Conversion to U.S. Units	
Length	nanometer (nm)	1 nm = 0.000000001 m (10^{-9})	= 3.94×10^{-8} in.	25,400,000 nm = 1 in.
	micrometer (µm)	1 µm = 0.000001 m (10^{-6})	= 3.94×10^{-5} in.	25,400 µm = 1 in.
	millimeter (mm)	1 mm = 0.001 m (10^{-3})	= 0.0394 in.	25.4 mm = 1 in.
	centimeter (cm)	1 cm = 0.01 m (10^{-2})	= 0.394 in.	2.54 cm = 1 in.
	decimeter (dm)	1 dm = 0.1 m (10^{-1})	= 3.94 in.	0.25 dm = 1 in.
	meter (m)	standard unit of length	= 39.4 in.	0.0254 m = 1 in.
			= 3.28 ft	0.3048 m = 1 ft
			= 1.093 yd	0.914 m = 1 yd
	kilometer (km)	1 km = 1000 m	= 3280 ft	1.609 km = 1 mi
			= 1093 yd	
			= 0.62 mi	
Volume	microliter (µL)	1 µL = 0.000001 L (10^{-6})		
		= 1 cubic millimeter (mm^3)		
	milliliter (mL)	1 mL = 0.001 L (10^{-3})	= 0.0338 fl oz	5 mL = 1 tsp
		= 1 cubic centimeter (cm^3 or cc)		15 mL = 1 tbsp
				30 mL = 1 fl oz
	centiliter (cL)	1 cL = 0.01 L (10^{-2})	= 0.338 fl oz	2.95 cL = 1 fl oz
	deciliter (dL)	1 dL = 0.1 L (10^{-1})	= 3.38 fl oz	0.295 dL = 1 fl oz
	liter (L)	standard unit of volume	= 33.8 fl oz	0.0295 L = 1 fl oz
			= 2.11 pt	0.473 L = 1 pt
			= 1.06 qt	0.946 L = 1 qt
Mass	picogram (pg)	1 pg = 0.000000000001 g (10^{-12})		
	nanogram (ng)	1 ng = 0.000000001 g (10^{-9})	= 0.000000015 gr	66,666,666 ng = 1 gr
	microgram (µg)	1 µg = 0.000001 g (10^{-6})	= 0.000015 gr	66,666 µg = 1 gr
	milligram (mg)	1 mg = 0.001 g (10^{-3})	= 0.015 gr	66.7 mg = 1 gr
	centigram (cg)	1 cg = 0.01 g (10^{-2})	= 0.15 gr	6.67 cg = 1 gr
	decigram (dg)	1 dg = 0.1 g (10^{-1})	= 1.5 gr	0.667 dg = 1 gr
	gram (g)	standard unit of mass	= 0.035 oz	28.4 g = 1 oz
			= 0.0022 lb	454 g = 1 lb
	dekagram (dag)	1 dag = 10 g		
	hectogram (hg)	1 hg = 100 g		
	kilogram (kg)	1 kg = 1000 g	= 2.2 lb	0.454 kg = 1 lb
	metric ton (mt)	1 mt = 1000 kg	= 1.1 t	0.907 mt = 1 t
			= 2205 lb	

Temperature	Celsius	Fahrenheit
Freezing point of pure water	0°	32°
Normal body temperature	36.8°	98.6°
Boiling point of pure water	100°	212°
Conversion	°C → °F: °F = (1.8 × °C) + 32	°F → °C: °C = (°F − 32) × 0.56

Photo Credits

Note: Unless otherwise stated, all illustrations are © Pearson Education.

Front Matter Author Photo Stephen N. Sarikas

Exercise 1: 01-04-02t Sfio Cracho/Shutterstock; **01-04-02m** Sfio Cracho/Shutterstock; **01-04-02b** Sfio Cracho/Shutterstock; **01-05-02** Anita Impagliazzo/Pearson Education, Inc.; **01-BYMO-01** William C. Ober

Exercise 2: 02-01-01 Peter Skinner/Science Source; **02-02-01** Mikhail Melnikov/Shutterstock; **02-02-02** Mikhail Melnikov/Shutterstock; **02-02-03** Scenics & Science/Alamy Stock Photo; **02-02-04** Lisa Lee/Pearson Education, Inc.; **02-02-05** Roman Milert/Alamy Stock Photo; **02-04-01** Brent Selinger/Pearson Education, Inc.; **02-BYMO-01** Juice Images/AGE Fotostock; **02-RS-01** Charles D. Winters/Science Source

Exercise 3: 03-02-02 Biophoto Associates/Science Source; **03-02-03** Ed Reschke/Photolibrary/Getty Images; **03-03-01** Steve Downing/Pearson Education, Inc.; **03-03-02** Steve Downing/Pearson Education, Inc.; **03-03-03** Stephen N. Sarikas; **03-03-04** Stephen N. Sarikas; **03-03-05** Stephen N. Sarikas; **03-04-02-01** Ed Reschke/Getty Images; **03-04-02-02** Ed Reschke/Getty Images; **03-04-02-03** Ed Reschke/Photolibrary/Getty Images; **03-04-03-01** Ed Reschke/Getty Images; **03-04-03-02** Ed Reschke/Getty Images; **03-04-03-03** Ed Reschke/Getty Images; **03-04-03-04** Ed Reschke/Getty Images; **03-RS-02** Robert B. Tallitsch; **03-RS-03** Don W. Fawcett/Science Source

Exercise 4: 04-RS-03a Steve Gschmeissner/Science Source; **04-RS-03b** David M. Phillips/Science Source; **04-RS-03c** Steve Gschmeissner/Science Source

Exercise 5: 05-01-01-01 Lisa Lee/Pearson Education, Inc.; **05-01-01-02** Lisa Lee/Pearson Education, Inc.; **05-01-02** Robert B. Tallitsch; **05-01-03** Robert B. Tallitsch; **05-02-01** Steve Downing/Pearson Education, Inc.; **05-02-02** Courtesy of Dr. Gregory N. Fuller Chief Neuropathologist U T MD Anderson Cancer Center; **05-02-03** Robert B. Tallitsch; **05-02-04** Robert B. Tallitsch; **05-03-01-01** Steve Downing/Pearson Education, Inc.; **05-03-01-02** Anita Impagliazzo/Pearson Education, Inc.; **05-03-02-01** Steve Downing/Pearson Education, Inc.; **05-03-02-02** Anita Impagliazzo/Pearson Education, Inc.; **05-04-01** Steve Downing/Pearson Education, Inc.; **05-04-02** Lisa Lee/Pearson Education, Inc.; **05-04-03-01** Frederic H. Martini; **05-04-03-02** Anita Impagliazzo/Pearson Education, Inc.; **05-05-01** Steve Downing/Pearson Education, Inc.; **05-05-02** Lisa Lee/Pearson Education, Inc.; **05-05-03** Steve Downing/Pearson Education, Inc.; **05-06-01-01** Lisa Lee/Pearson Education, Inc.; **05-06-01-02** Steve Downing/Pearson Education, Inc.; **05-06-01-03** Anita Impagliazzo/Pearson Education, Inc.; **05-06-02-01** Lisa Lee/Pearson Education, Inc.; **05-06-02-02** William C. Ober; **05-06-03** Lisa Lee/Pearson Education, Inc.; **05-07-01** Steve Downing/Pearson Education, Inc.; **05-07-02** Steve Downing/Pearson Education, Inc.; **05-08-01** Lisa Lee/Pearson Education, Inc.; **05-09-01** Nina Zanetti/Pearson Education, Inc.; **05-09-02** Steve Downing/Pearson Education, Inc.; **05-09-03** Lisa Lee/Pearson Education, Inc.; **05-10-01-01** Lisa Lee/Pearson Education, Inc.; **05-10-01-02** Lisa Lee/Pearson Education, Inc.; **05-RS-06** Steve Downing/Pearson Education, Inc.; **05-RS-07** Steve Downing/Pearson Education, Inc.; **05-RS-08** Nina Zanetti/Pearson Education, Inc.; **05-RS-09** Steve Downing/Pearson Education, Inc.

Exercise 6: 06-01-01 Lisa Lee/Pearson Education, Inc.; **06-01-02** Lisa Lee/Pearson Education, Inc.; **06-01-03** Lisa Lee/Pearson Education, Inc.; **06-01-04** Robert B. Tallitsch; **06-02-01-01** Lisa Lee/Pearson Education, Inc.; **06-02-01-02** Lisa Lee/Pearson Education, Inc.; **06-02-01-03** Lisa Lee/Pearson Education, Inc.; **06-02-02** Steve Downing/Pearson Education, Inc.; **06-02-03** David M. Phillips/Science Source; **06-03-02** James Stevenson/Science Source; **06-04-01** Jeremy Burgess/Science Source; **06-BYMO-01-01** Lisa Lee/Pearson Education, Inc.; **06-BYMO-01-02** Lisa Lee/Pearson Education, Inc.

Exercise 7: 07-03-01 Robert B. Tallitsch; **07-04-01** Ralph T. Hutchings; **07-07-01** Ralph T. Hutchings; **07-07-02** Ralph T. Hutchings; **07-07-03** Ralph T. Hutchings; **07-11-01** Anita Impagliazzo/Pearson Education, Inc.; **07-11-02a** Ian Hooton/Science Source; **07-11-02b** Kristin Piljay/Pearson Education, Inc.; **07-11-02c** Princess Margaret Rose Orthopaedic Hospital/Science Source; **07-12-01** Ralph T. Hutchings; **07-12-02** Ralph T. Hutchings; **07-12-04** Ralph T. Hutchings; **07-12-05** Ralph T. Hutchings; **07-12-06** Ralph T. Hutchings; **07-13-01** Ralph T. Hutchings; **07-13-02l** Ralph T. Hutchings; **07-13-02r** Ralph T. Hutchings; **07-13-03** Ralph T. Hutchings; **07-13-04** Ralph T. Hutchings; **07-13-05** Bryon Spencer/Pearson Education, Inc.; **07-14-01** Ralph T. Hutchings; **07-14-02** Ralph T. Hutchings; **07-RS-01** Larry DeLay/Pearson Education, Inc.; **07-RS-04** Michael J. Timmons

Exercise 8: 08-01-02t Ralph T. Hutchings; **08-01-02b** Ralph T. Hutchings; **08-01-03l** Ralph T. Hutchings; **08-01-03m** Ralph T. Hutchings; **08-01-03r** Ralph T. Hutchings; **08-02-01l** Ralph T. Hutchings; **08-02-01r** Ralph T. Hutchings; **08-02-02l** Ralph T. Hutchings; **08-02-02r** Ralph T. Hutchings; **08-03-01l** Ralph T. Hutchings; **08-03-01r** Ralph T. Hutchings; **08-03-02** Ralph T. Hutchings; **08-04-01l** Ralph T. Hutchings; **08-04-01r** Ralph T. Hutchings; **08-04-02** Ralph T. Hutchings; **08-05-01l** Ralph T. Hutchings; **08-05-01r** Ralph T. Hutchings; **08-05-02l** Ralph T. Hutchings; **08-05-02r** Ralph T. Hutchings; **08-05-03l** Ralph T. Hutchings; **08-05-03r** Ralph T. Hutchings; **08-06-01** Ralph T. Hutchings; **08-06-02t** Ralph T. Hutchings; **08-06-02b** Ralph T. Hutchings

Exercise 9: 09-01-03l Ralph T. Hutchings; **09-01-03r** Ralph T. Hutchings; **09-02-01** Ralph T. Hutchings; **09-03-02** Anita Impagliazzo/Pearson Education, Inc.; **09-03-03** Anita Impagliazzo/Pearson Education, Inc.; **09-03-04** Ralph T. Hutchings; **09-04-01** Anita Impagliazzo/Pearson Education, Inc.; **09-04-03** Ralph T. Hutchings; **09-04-04** Ralph T. Hutchings; **09-04-05** Ralph T. Hutchings; **09-04-06** Ralph T. Hutchings; **09-04-07** Ralph T. Hutchings; **09-05-01** Ralph T. Hutchings; **09-05-02t** Ralph T. Hutchings; **09-05-02b** Ralph T. Hutchings; **09-06-01** Ralph T. Hutchings; **09-06-02** Ralph T. Hutchings; **09-07-01t** Ralph T. Hutchings; **09-07-01b1** Stephen N. Sarikas; **09-07-01b2** Stephen N. Sarikas; **09-07-01b3** Stephen N. Sarikas; **09-07-01b4** Stephen N. Sarikas; **09-07-01b5** Stephen N. Sarikas; **09-08-03** Ralph T. Hutchings; **09-09-06** Ralph T. Hutchings; **09-RS-04** Ralph T. Hutchings

Exercise 10: 10-01-01 Steve Downing/Pearson Education, Inc.; **10-02-02** Kent Wood/Science Source; **10-03-01** Biophoto Associates/Science Source

Exercise 11: 11-01-04 Anita Impagliazzo/Pearson Education, Inc.; **11-02-03** Ralph T. Hutchings; **11-03-04** William C. Ober; **11-04-01** William C. Ober; **11-04-03** William C. Ober; **11-06-01** Anita Impagliazzo/Pearson Education, Inc.; **11-07-02** Karen Krabbenhoft/Pearson Education, Inc.; **11-RS-01** Anita Impagliazzo/Pearson Education, Inc.; **11-RS-03** William C. Ober; **11-RS-04** William C. Ober

Exercise 12: 12-01-01 Anita Impagliazzo/Pearson Education, Inc.; **12-01-02** Anita Impagliazzo/Pearson Education, Inc.; **12-02-01** Anita Impagliazzo/Pearson Education, Inc.; **12-02-02** Anita Impagliazzo/Pearson Education, Inc.; **12-02-03l** Anita Impagliazzo/Pearson Education, Inc.; **12-02-03r** Anita Impagliazzo/Pearson Education, Inc.; **12-02-04l** Anita Impagliazzo/Pearson Education, Inc.; **12-02-04r** Anita Impagliazzo/Pearson Education, Inc.; **12-03-01** Anita Impagliazzo/Pearson Education, Inc.; **12-03-02** Anita Impagliazzo/Pearson Education, Inc.; **12-04-01** Anita Impagliazzo/Pearson Education, Inc.; **12-04-02** Anita Impagliazzo/Pearson Education, Inc.; **12-04-03** Anita Impagliazzo/Pearson Education, Inc.; **12-05-01** Anita Impagliazzo/Pearson Education, Inc.; **12-05-02** Anita Impagliazzo/Pearson Education, Inc.; **12-10-02** Artur Bogacki/Shutterstock; **12-11-02** Anita Impagliazzo/Pearson Education, Inc.; **12-12-01l** Anita Impagliazzo/Pearson Education, Inc.; **12-12-01r** Anita Impagliazzo/Pearson Education, Inc.; **12-12-02** Anita Impagliazzo/Pearson Education, Inc.; **12-14-01** Hanna Monika/Shutterstock; **12-14-02** Pearson Education, Inc.; **12-14-03** Pearson Education, Inc.; **12-14-04** Pearson Education, Inc.; **12-14-05** Kristin Piljay/Pearson Education, Inc.; **12-14-06** Pearson Education, Inc.; **12-14-07** Kristin Piljay/Pearson Education, Inc.; **12-15-01** Pearson Education, Inc.; **12-15-02** JRP Studio/Shutterstock; **12-15-03** Pearson Education, Inc.; **12-15-04** Pearson Education, Inc.; **12-BYMO-01** Anita Impagliazzo/Pearson Education, Inc.; **12-RS-01** Anita Impagliazzo/Pearson Education, Inc.; **12-RS-02l** Anita Impagliazzo/Pearson Education, Inc.; **12-RS-02r** Anita Impagliazzo/Pearson Education, Inc.; **12-RS-04l** Anita Impagliazzo/Pearson Education, Inc.; **12-RS-04r** Anita Impagliazzo/Pearson Education, Inc.

Exercise 13: 13-01-01 Kristin Piljay/Pearson Education, Inc.; **13-01-02** Kristin Piljay/Pearson Education, Inc.; **13-02-01** Anita Impagliazzo/Pearson Education, Inc.; **13-04-02** Stephen N. Sarikas; **13-04-03** Kristin Piljay/Pearson Education, Inc.; **13-05-01** Kristin Piljay/Pearson Education, Inc.; **13-05-02** Kristin Piljay/Pearson Education, Inc.; **13-06-01** Kristin Piljay/Pearson Education, Inc.; **13-RS-02** Anita Impagliazzo/Pearson Education, Inc.

Exercise 14: 14-01-02 Lisa Lee/Pearson Education, Inc.; **14-01-03** Steve Downing/Pearson Education, Inc.; **14-02-01t** Lisa Lee/Pearson Education, Inc.; **14-02-01l** Michael Abbey/Science Source; **14-02-01r** Biophoto Associates/Science Source; **14-02-02** CRNI/Science Source; **14-03-01** Lisa Lee/Pearson Education, Inc.; **14-03-02t** Lisa Lee/ Pearson Education, Inc.; **14-03-02b** Lisa Lee/ Pearson Education, Inc.; **14-04-01t** Lisa Lee/ Pearson Education, Inc.; **14-04-01b** Steve Downing/ Pearson Education, Inc.; **14-04-02t** Lisa Lee/Pearson Education, Inc.; **14-04-02b** Robert Tallitsch; **14-05-01t** Michael J. Timmons; **14-05-01b** Nina Zanetti/ Pearson Education, Inc.; **14-06-01** Lisa Lee/Pearson Education, Inc.; **14-BYMO-01** William C. Ober

Exercise 15: 15-01-01t Yvonne Baptiste-Syzmanski/ Pearson Education, Inc.; **15-01-01m** William C. Ober; **15-01-01b** Yvonne Baptiste-Syzmanski/ Pearson Education, Inc.; **15-02-01** Yvonne Baptiste-Syzmanski/Pearson Education, Inc.; **15-02-03** Yvonne Baptiste-Syzmanski/Pearson Education, Inc.; **15-02-04** Yvonne Baptiste-Syzmanski/Pearson Education, Inc.; **15-02-05** Winston Charles Poulton/ Pearson Education, Inc.; **15-03-01** Yvonne Baptiste-Syzmanski/Pearson Education, Inc.; **15-03-02** Yvonne Baptiste-Syzmanski/Pearson Education, Inc.; **15-03-03** Ralph T. Hutchings; **15-04-01** Stephen N. Sarikas; **15-04-02** Winston Charles Poulton/Pearson Education, Inc.; **15-07-01** Winston Charles Poulton/ Pearson Education, Inc.; **15-07-02** Winston Charles Poulton/Pearson Education, Inc.; **15-09-01** William C. Ober; **15-09-02** William C. Ober; **15-10-01** William C. Ober; **15-10-02** William C. Ober; **15-10-03** William C. Ober; **15-RS-03** Ralph T. Hutchings

Exercise 16: 16-02-04 Living Art Enterprises LLC/ Science Source; **16-04-02** Anita Impagliazzo/Pearson Education, Inc.; **16-BYMO-01** Anita Impagliazzo/ Pearson Education, Inc.

Exercise 17: 17-01-02 Kristin Piljay/Pearson Education, Inc.; **17-01-03** Kristin Piljay/Pearson Education, Inc.; **17-02-01** Kristin Piljay/Pearson Education, Inc.; **17-02-02** Kristin Piljay/Pearson Education, Inc.; **17-03-01** Kristin Piljay/Pearson Education, Inc.; 17-**03-02** Kristin Piljay/Pearson Education, Inc.

Exercise 18: 18-01-01 Winston Charles Poulton/ Pearson Education, Inc.; **18-01-02** Biophoto Associates/Science Source; **18-02-01** Stephen N. Sarikas; **18-02-02** Kristin Piljay/Pearson Education, Inc.; **18-03-01** Pearson Education, Inc.; **18-03-02l** Robert B. Tallitsch; **18-03-02r** Biophoto Associates/ Science Source; **18-04-01** Kristin Piljay/Pearson Education, Inc.; **18-04-02** Kristin Piljay/Pearson Education, Inc.; **18-04-03** Kristin Piljay/Pearson Education, Inc.; **18-04-04** Kristin Piljay/Pearson Education, Inc.; **18-05-01** Ralph T. Hutchings; **18-05-03** William C. Ober; **18-06-02** Memorisz/ Shutterstock; **18-07-01t** Yvonne Baptiste-Syzmanski/ Pearson Education, Inc.; **18-07-01b** Elena Dorfman/ Pearson Education, Inc.; **18-07-02l** Minerva Studio/ Shutterstock; **18-07-02r** Sue Ford/Science Source; **18-07-03t** Yvonne Baptiste-Syzmanski/Pearson Education, Inc.; **18-07-03b** Elena Dorfman/Pearson Education, Inc. **18-08-01** Lisa Lee/Pearson Education, Inc.; **18-09-02** Kristin Piljay/Pearson Education, Inc.; **18-09-03** Kristin Piljay/Pearson Education, Inc.; **18-11-01** Photomak/Shutterstock; **18-12-02l** Diane Hirsch/Fundamental Photographs, NYC; **18-12-02r** Diane Hirsch/Fundamental Photographs, NYC; **18-12-03** Stephen N. Sarikas; **18-13-01** Creative Digital Visions LLC/Pearson Education, Inc.; **18-14-01** Michael J. Timmons; **18-14-02** Lisa Lee/Pearson Education, Inc.; **18-15-01** Kristin Piljay/Pearson Education, Inc.; **18-15-02** Kristin Piljay/Pearson Education, Inc.; **18-15-03** Kristin Piljay/Pearson Education, Inc.; **18-16-01** Kristin Piljay/Pearson Education, Inc.; **18-16-03** Kristin Piljay/Pearson Education, Inc.; **18-16-04** Kristin Piljay/Pearson Education, Inc.

Exercise 19: 19-01-02 Creative Digital Visions LLC/Pearson Education, Inc.; **19-02-01t** Pearson Education, Inc.; **19-02-01b** Pearson Education, Inc.; **19-02-02** Anita Impagliazzo/Pearson Education, Inc.; **19-03-01t** Nikolay Pozdeev/iStock/Getty Images; **19-03-01b** Pearson Education, Inc.; **19-03-02** Anita Impagliazzo/Pearson Education, Inc.; **19-03-03t** Pearson Education, Inc.; **19-03-03b** Nikolay Pozdeev/iStock/Getty Images; **19-04-01** Lisa Lee/ Pearson Education, Inc.; **19-04-02** Lisa Lee/Pearson Education, Inc.; **19-05-01t** Steve Downing/Pearson Education, Inc.; **19-05-01b** Steve Downing/Pearson Education, Inc.; **19-05-02t** Anita Impagliazzo/ Pearson Education, Inc.; **19-05-02m** Frederic H. Martini; **19-05-02b** Frederic H. Martini; **19-06-01t** Steve Downing/Pearson Education, Inc.; **19-06-01b** Steve Downing/Pearson Education, Inc.; **19-06-02l** Lisa Lee/Pearson Education, Inc.; **19-06-02r** Ed Reschke/Oxford Scientific/Getty Images; **19-RS-02a** Lisa Lee/Pearson Education, Inc.; **19-RS-02b** Lisa Lee/Pearson Education, Inc.; **19-RS-02c** Lisa Lee/ Pearson Education, Inc.; **19-RS-02d** Robert B. Tallitsch

Exercise 20: 20-01-01 Lisa Lee/Pearson Education, Inc.; **20-01-02t** Steve Downing/Pearson Education, Inc.; **20-01-02b** Susumu Nishinaga/Science Source; **20-02-01** Steve Downing/Pearson Education, Inc.; **20-02-02** Robert B. Tallitsch; **20-02-03** Steve Downing/Pearson Education, Inc.; **20-02-04** Steve Downing/Pearson Education, Inc.; **20-02-05** Steve Downing/Pearson Education, Inc.; **20-03-01** Robert B. Tallitsch; **20-03-02** Robert B. Tallitsch; **20-04-03-01** Jack Scanlan/Pearson Education, Inc.; **20-04-03-02** Jack Scanlan/Pearson Education, Inc.; **20-04-03-03** Jack Scanlan/Pearson Education, Inc.; **20-04-03-04** Jack Scanlan/Pearson Education, Inc.; **20-BYMO-01** Steve Downing/Pearson Education, Inc.; **20-RS-07-01** Karen E. Petersen; **20-RS-07-02** Karen E. Petersen; **20-RS-07-03** Karen E. Petersen

Exercise 21: 21-02-01 Anita Impagliazzo/Pearson Education, Inc.; **21-08-01** William C. Ober; **21-09-01** Shawn Miller & Mark Nielsen/Pearson Education, Inc.; **21-09-02** Shawn Miller & Mark Nielsen/Pearson Education, Inc.; **21-10-01** Shawn Miller & Mark Nielsen/Pearson Education, Inc.; **21-10-02** Shawn Miller & Mark Nielsen/Pearson Education, Inc.; **21-BYMO-01** Ralph T. Hutchings; **21-RS-02** Yvonne Baptiste-Syzmanski/Pearson Education, Inc.; **21-RS-04** Ralph T. Hutchings; **21-RS-05** Karen Krabbenhoft/ Winston Charles Poulton/Pearson Education, Inc.

Exercise 22: 22-01-01 Steve Downing/Pearson Education, Inc.; **22-01-02-01** Lisa Lee/Pearson Education, Inc.; **22-01-02-02** Lisa Lee/Pearson Education, Inc.; **22-01-03** Steve Downing/Pearson Education, Inc.; **22-02-01-01** Lisa Lee/Pearson Education, Inc.; **22-02-01-02** Lisa Lee/Pearson Education, Inc.; **22-02-01-03** Steve Downing/Pearson Education, Inc.; **22-02-03-01** Steve Downing/Pearson Education, Inc.; **22-02-03-02** Robert B. Tallitsch; **22-03-02** Anita Impagliazzo/Pearson Education, Inc.; **22-03-03** Anita Impagliazzo/Pearson Education, Inc.; **22-04-01** Anita Impagliazzo/Pearson Education, Inc.; **22-04-02** William C. Ober; **22-04-03** Anita Impagliazzo/Pearson Education, Inc.; **22-04-04** William C. Ober; **22-06-01** Anita Impagliazzo/ Pearson Education, Inc.; **22-06-02** Anita Impagliazzo/ Pearson Education, Inc.; **22-07-01-01** Anita Impagliazzo/Pearson Education, Inc.; **22-07-01-02** Anita Impagliazzo/Pearson Education, Inc.; **22-07-02** Anita Impagliazzo/Pearson Education, Inc.; **22-07-03** Anita Impagliazzo/Pearson Education, Inc.; **22-08-01** William C. Ober; **22-08-02** Anita Impagliazzo/ Pearson Education, Inc.; **22-08-03** William C. Ober; **22-09-01** Anita Impagliazzo/Pearson Education, Inc.; **22-10-01** Kristin Piljay/Pearson Education, Inc.; **22-10-02** Kristin Piljay/Pearson Education, Inc.; **22-10-03** Kristin Piljay/Pearson Education, Inc.; **22-10-04** Stephen N. Sarikas

Exercise 23: 23-01-02 William C. Ober; **23-02-01** Privilege/Fotolia; **23-03-03** William C. Ober; **23-04-01** Wavebreakmedia/Shutterstock

Exercise 24: 24-01-01 Pearson Education, Inc.; **24-01-02** William C. Ober; **24-02-02-01** Anita Impagliazzo/Pearson Education, Inc.; **24-02-02-02** Anita Impagliazzo/Pearson Education, Inc.; **24-03-02** Ralph T. Hutchings; **24-03-03** William C. Ober; **24-04-01-01** Lisa Lee/Pearson Education, Inc.; **24-04-01-02** Robert B. Tallitsch; **24-04-02-01** Lisa Lee/Pearson Education, Inc.; **24-04-02-02** Lisa Lee/Pearson Education, Inc.; **24-05-01-01** Lisa Lee/ Pearson Education, Inc.; **24-05-01-02** Lisa Lee/ Pearson Education, Inc.; **24-05-02-01** Lisa Lee/ Pearson Education, Inc.; **24-05-02-02** Lisa Lee/ Pearson Education, Inc.; **24-05-03** Lisa Lee/ Pearson Education, Inc.; **24-RS-01** Lisa Lee/Pearson Education, Inc.; **24-RS-02** Robert B. Tallitsch; **24-RS-03** Lisa Lee/Pearson Education, Inc.; **24-RS-04** Lisa Lee/Pearson Education, Inc.; **24-RS-05** Lisa Lee/ Pearson Education, Inc.; **24-RS-06** Winston Charles Poulton/Pearson Education, Inc.

Exercise 25: 25-01-01 William C. Ober; **25-01-02** William C. Ober; **25-02-01** William C. Ober; **25-03-01-01** Anita Impagliazzo/Pearson Education, Inc.; **25-03-01-02** Lester V. Bergman/Corbis Documentary/Getty Images; **25-04-01** Blend Images/Alamy Stock Photo; **25-05-01-01** Ralph T. Hutchings; **25-05-01-02** Anita Impagliazzo/Pearson Education, Inc.; **25-05-02** Anita Impagliazzo/Pearson Education, Inc.; **25-06-01** Lisa Lee/Pearson Education, Inc.; **25-06-02-01** Nina Zanetti/Pearson Education, Inc.; **25-06-02-02** Lisa Lee/Pearson Education, Inc.; **25-06-02-03** Lisa Lee/ Pearson Education, Inc.; **25-07-01-01** Lisa Lee/ Pearson Education, Inc.; **25-07-01-02** Biophoto Associates/Science Source; **25-07-01-03** Lisa Lee/ Pearson Education, Inc.; **25-07-02** Steve Downing/ Pearson Education, Inc.; **25-BYMO-01** Anita Impagliazzo/Pearson Education, Inc.; **25-RS-02** Lisa Lee/Pearson Education, Inc.

Exercise 26: 26-02-02 Jessica Peterson/Tetra Images/ Brand X Pictures/Getty Images; **26-03-02-01** Stephen N. Sarikas; **26-03-02-02** Stephen N. Sarikas; **26-BYMO-01** Anita Impagliazzo/Pearson Education, Inc.; **26-RS-01** Anita Impagliazzo/Pearson Education, Inc.

Exercise 27: 27-01-01 Leif Saul/Pearson Education, Inc.; **27-01-02** Anita Impagliazzo/Pearson Education, Inc.; **27-02-03** Anita Impagliazzo/Pearson Education, Inc.; **27-02-04** Anita Impagliazzo/Pearson Education, Inc.; **27-03-01** Anita Impagliazzo/Pearson Education, Inc.; **27-03-02-01** Anita Impagliazzo/Pearson Education, Inc.; **27-03-02-02** Ralph T. Hutchings; **27-04-01** Anita Impagliazzo/Pearson Education, Inc.;

27-05-01-01 Anita Impagliazzo/Pearson Education, Inc.; **27-05-01-02** Anita Impagliazzo/Pearson Education, Inc.; **27-05-02** Anita Impagliazzo/Pearson Education, Inc.; **27-07-01** Anita Impagliazzo/Pearson Education, Inc.; **27-07-02** Anita Impagliazzo/Pearson Education, Inc.; **27-07-03** Anita Impagliazzo/Pearson Education, Inc.; **27-08-01-01** Anita Impagliazzo/ Pearson Education, Inc.; **27-08-01-02** Alfred Pasieka/ Photolibrary/Getty Images; **27-08-01-03** Robert B. Tallitsch; **27-08-02-01** Nina Zanetti/Pearson Education, Inc.; **27-08-02-02** Robert B. Tallitsch; **27-09-01-01** Lisa Lee/Pearson Education, Inc.; **27-09-01-02** M.I. Walker/Science Source; **27-09-02-01** M.I. Walker/Science Source; **27-09-02-02** Stephen N. Sarikas; **27-09-03-01** Lisa Lee/Pearson Education, Inc.; **27-09-03-02** Lisa Lee/Pearson Education, Inc.; **27-09-03-03** Lisa Lee/Pearson Education, Inc.; **27-10-01** Lisa Lee/Pearson Education, Inc.; **27-10-02** Lisa Lee/Pearson Education, Inc.; **27-11-01-01** Anita Impagliazzo/Pearson Education, Inc.; **27-11-01-02** Robert B. Tallitsch; **27-11-02** Steve Downing/ Pearson Education, Inc.; **27-11-03** Lisa Lee/Pearson Education, Inc.; **27-12-01-01** Nina Zanetti/Pearson Education, Inc.; **27-12-01-02** Steve Downing/Pearson Education, Inc.; **27-12-01-03** Steve Downing/ Pearson Education, Inc.; **27-12-03** Lisa Lee/Pearson Education, Inc.; **27-BYMO-01** Anita Impagliazzo/ Pearson Education, Inc.; **27-RS-02** Lisa Lee/Pearson Education, Inc.; **27-RS-03** Steve Downing/Pearson Education, Inc.; **27-RS-04** Lisa Lee/Pearson Education, Inc.; **27-RS-05** Lisa Lee/Pearson Education, Inc.; **27-RS-06** Lisa Lee/Pearson Education, Inc.; **27-RS-07** Lisa Lee/Pearson Education, Inc.

Exercise 28: 28-01-01 Stephen N. Sarikas; **28-01-02** Stephen N. Sarikas; **28-02-01** Stephen N. Sarikas; **28-02-02** Stephen N. Sarikas; **28-02-03** Stephen N. Sarikas; **28-03-01** Stephen N. Sarikas

Exercise 29: 29-01-01 Anita Impagliazzo/Pearson Education, Inc.; **29-04-01-01** Lisa Lee/Pearson Education, Inc.; **29-04-01-02** Lisa Lee/Pearson Education, Inc.; **29-04-01-03** Lisa Lee/Pearson Education, Inc.; **29-04-02-01** Lisa Lee/Pearson Education, Inc.; **29-04-02-02** Steve Downing/Pearson Education, Inc.; **29-04-03-01** Biophoto Associates/ Science Source; **29-04-03-02** Lisa Lee/Pearson Education, Inc.; **29-04-03-03** Robert B. Tallitsch; **29-05-01** Yvonne Baptiste-Syzmanski/Pearson Education, Inc.; **29-05-02** Shawn Miller & Mark Nielsen/Pearson Education, Inc.

Exercise 30: 30-02-01 Alexey Filatov/Shutterstock; **30-03-01** Keith Frith/Fotolia; **30-03-02** Martin William Allen/Alamy Stock Photo; **30-04-06** Thelefty/Shutterstock

Exercise 31: 31-04-01-01 Frederic H. Martini; **31-04-01-02** Lisa Lee/Pearson Education, Inc.; **31-04-02** Nina Zanetti/Pearson Education, Inc.; **31-04-03-01** Biophoto Associates/Science Source; **31-04-03-02** Frederic H. Martini; **31-05-01** Robert B. Tallitsch; **31-05-02** Michael Abbey/Science Source; **31-06-01** Biophoto Associates/Science Source; **31-06-02-01** Lisa Lee/Pearson Education, Inc.; **31-06-02-02** Steve Downing/Pearson Education, Inc.; **31-RS-01** Biophoto Associates/Science Source; **31-RS-02** Lisa Lee/Pearson Education, Inc.; **31-RS-03** Michael Abbey/Science Source

Exercise 32: 32-01-01 Anita Impagliazzo/Pearson Education, Inc.; **32-03-01** Pearson Education, Inc.; **32-04-01-01** Lisa Lee/Pearson Education, Inc.; **32-04-01-02** Steve Downing/Pearson Education, Inc.; **32-04-01-03** Steve Downing/Pearson Education, Inc.; **32-04-01-04** Frederic H. Martini; **32-04-01-05** Steve Downing/Pearson Education, Inc.; **32-04-02-01** C. Edelmann/Science Source; **32-04-02-02** Lisa Lee/Pearson Education, Inc.; **32-04-02-03** Lisa Lee/Pearson Education, Inc.; **32-05-01** Lisa Lee/ Pearson Education, Inc.; **32-05-02-01** Frederic H. Martini; **32-05-02-02** Lisa Lee/Pearson Education, Inc.; **32-05-02-03** Lisa Lee/Pearson Education, Inc.; **32-06-01-01** Frederic H. Martini; **32-06-01-02** Steve Gschmeissner/Science Source; **32-06-02-01** Lisa Lee/Pearson Education, Inc.; **32-06-02-02** Lisa Lee/ Pearson Education, Inc.; **32-RS-01** Steve Downing/ Pearson Education, Inc.; **32-RS-02** Lisa Lee/Pearson Education, Inc.; **32-RS-03** Frederic H. Martini

Cat 1: C01-01-01 Eric Isselee/Shutterstock; **C01-01-02** Shawn Miller & Mark Nielsen/Pearson Education, Inc.; **C01-BYMO-01** Eric Isselee/Shutterstock

Cat 2: C02-01-01 Shawn Miller & Mark Nielsen/ Pearson Education, Inc.; **C02-01-02** Shawn Miller & Mark Nielsen/Pearson Education, Inc.; **C02-02-01** Shawn Miller & Mark Nielsen/Pearson Education, Inc.; **C02-02-02** Shawn Miller & Mark Nielsen/ Pearson Education, Inc.; **C02-03-01** Shawn Miller & Mark Nielsen/Pearson Education, Inc.; **C02-03-02** Shawn Miller & Mark Nielsen/Pearson Education, Inc.; **C02-04-01** Shawn Miller & Mark Nielsen/ Pearson Education, Inc.; **C02-04-02** Shawn Miller & Mark Nielsen/Pearson Education, Inc.; **C02-05-01-01** Shawn Miller & Mark Nielsen/Pearson Education, Inc.; **C02-05-01-02** Shawn Miller & Mark Nielsen/ Pearson Education, Inc.; **C02-05-02-01** Shawn Miller & Mark Nielsen/Pearson Education, Inc.; **C02-05-02-02** Shawn Miller & Mark Nielsen/Pearson Education, Inc.; **C02-05-03-01** Shawn Miller & Mark Nielsen/ Pearson Education, Inc.; **C02-05-03-02** Shawn Miller & Mark Nielsen/Pearson Education, Inc.; **C02-BYMO-01** Shawn Miller & Mark Nielsen/Pearson Education, Inc.; **C02-RS-01** Shawn Miller & Mark Nielsen/Pearson Education, Inc.; **C02-RS-02** Shawn Miller & Mark Nielsen/Pearson Education, Inc.

Cat 3: C03-01-02 Shawn Miller & Mark Nielsen/ Pearson Education, Inc.; **C03-02-02** Shawn Miller & Mark Nielsen/Pearson Education, Inc.; **C03-03-02** Shawn Miller & Mark Nielsen/Pearson Education, Inc.; **C03-BYMO-01** Shawn Miller & Mark Nielsen/ Pearson Education, Inc.; **C03-RS-01** Shawn Miller & Mark Nielsen/Pearson Education, Inc.; **C03-RS-02** Shawn Miller & Mark Nielsen/Pearson Education, Inc.; **C03-RS-03** Shawn Miller & Mark Nielsen/ Pearson Education, Inc.

Cat 4: C04-01-02 Shawn Miller & Mark Nielsen/ Pearson Education, Inc.; **C04-01-03** Shawn Miller & Mark Nielsen/Pearson Education, Inc.; **C04-02-01** Shawn Miller & Mark Nielsen/Pearson Education, Inc.; **C04-02-02** Shawn Miller & Mark Nielsen/ Pearson Education, Inc.; **C04-02-03** Shawn Miller & Mark Nielsen/Pearson Education, Inc.; **C04-RS-01** Shawn Miller & Mark Nielsen/Pearson Education, Inc.; **C04-RS-02** Shawn Miller & Mark Nielsen/ Pearson Education, Inc.

Cat 5: C05-01-01 Shawn Miller & Mark Nielsen/ Pearson Education, Inc.; **C05-02-02** Shawn Miller & Mark Nielsen/Pearson Education, Inc.; **C05-03-02** Shawn Miller & Mark Nielsen/Pearson Education, Inc.; **C05-04-02** Shawn Miller & Mark Nielsen/ Pearson Education, Inc.; **C05-04-03** Shawn Miller & Mark Nielsen/Pearson Education, Inc.; **C05-05-02-01** Shawn Miller & Mark Nielsen/Pearson Education, Inc.; **C05-05-02-02** Shawn Miller & Mark Nielsen/ Pearson Education, Inc.; **C05-BYMO-01** Shawn Miller & Mark Nielsen/Pearson Education, Inc.; **C05-RS-01** Shawn Miller & Mark Nielsen/Pearson Education, Inc.; **C05-RS-02** Shawn Miller & Mark Nielsen/Pearson Education, Inc.; **C05-RS-03** Shawn Miller & Mark Nielsen/Pearson Education, Inc.

Cat 6: C06-01-01 Shawn Miller & Mark Nielsen, Pearson Education, Inc.; **C06-01-02** Shawn Miller & Mark Nielsen/Pearson Education, Inc.; **C06-02-01** Shawn Miller & Mark Nielsen/Pearson Education, Inc.; **C06-RS-01** Shawn Miller & Mark Nielsen/ Pearson Education, Inc.; **C06-RS-02** Shawn Miller & Mark Nielsen/Pearson Education, Inc.; **C06-RS-03** Shawn Miller & Mark Nielsen/Pearson Education, Inc.

Cat 7: C07-01-01 Shawn Miller & Mark Nielsen/ Pearson Education, Inc.; **C07-01-02-01** Shawn Miller & Mark Nielsen/Pearson Education, Inc.; **C07-01-02-02** Shawn Miller & Mark Nielsen/Pearson Education, Inc.; **C07-02-02** Shawn Miller & Mark Nielsen/ Pearson Education, Inc.; **C07-BYMO-01** Shawn Miller & Mark Nielsen/Pearson Education, Inc.; **C07-RS-01** Shawn Miller & Mark Nielsen/Pearson Education, Inc.

Cat 8: C08-01-01 Shawn Miller & Mark Nielsen/ Pearson Education, Inc.; **C08-01-02** Shawn Miller & Mark Nielsen/Pearson Education, Inc.; **C08-02-01** Shawn Miller & Mark Nielsen/Pearson Education, Inc.; **C08-03-01-01** Shawn Miller & Mark Nielsen/ Pearson Education, Inc.; **C08-03-01-02** Shawn Miller & Mark Nielsen/Pearson Education, Inc.; **C08-04-02** Shawn Miller & Mark Nielsen/Pearson Education, Inc.; **C08-05-02-01** Shawn Miller & Mark Nielsen/ Pearson Education, Inc.; **C08-05-02-02** Shawn Miller & Mark Nielsen/Pearson Education, Inc.; **C08-BYMO-01** Shawn Miller & Mark Nielsen/Pearson Education, Inc.; **C08-RS-01** Shawn Miller & Mark Nielsen/Pearson Education, Inc.

Cat 9: C09-01-01 Shawn Miller & Mark Nielsen/ Pearson Education, Inc.; **C09-01-02** Shawn Miller & Mark Nielsen/Pearson Education, Inc.; **C09-02-01** Shawn Miller & Mark Nielsen/Pearson Education, Inc.; **C09-02-02** Shawn Miller & Mark Nielsen/ Pearson Education, Inc.; **C09-RS-01** Shawn Miller & Mark Nielsen/Pearson Education, Inc.

Cat 10: C10-01-02 Shawn Miller & Mark Nielsen/ Pearson Education, Inc.; **C10-02-02** Shawn Miller & Mark Nielsen/Pearson Education, Inc.; **C10-BYMO-01-01** Shawn Miller & Mark Nielsen/Pearson Education, Inc.; **C10-BYMO-01-02** Shawn Miller & Mark Nielsen/Pearson Education, Inc.; **C10-RS-01** Shawn Miller & Mark Nielsen/Pearson Education, Inc.; **C10-RS-02** Shawn Miller & Mark Nielsen/ Pearson Education, Inc.

Appendix: AppA.01 Martin Lee/Mediablitzimages/ Alamy Stock Photo; **AppA.02** John Powell Photographer/Alamy; **AppA.03** Kristin Piljay/ Pearson Education; **AppA.04** design56/Shutterstock

Index

Note: An italicized "t" following a page number indicates that the item is referenced in a table on that page. A page number containing "C-" indicates that the item is referenced in the Cat exercises.

A

C

D

J

K

N

Q

R

T

X

Y

Z